MATHEMATICS:

ANALYSIS AND APPROACHES

ENHANCED ONLINE

STANDARD LEVEL
COURSE COMPANION

Natasha Awada
Paul Belcher
Laurie Buchanan
Jennifer Chang Wathall
Phil Duxbury
Jane Forrest
Josip Harcet
Rose Harrison
Lorraine Heinrichs
Ed Kemp
Paul La Rondie
Palmira Mariz Seiler
Jill Stevens
Ellen Thompson
Marlene Torres-Skoumal

OXFORD
UNIVERSITY PRESS

OXFORD
UNIVERSITY PRESS

Great Clarendon Street, Oxford, OX2 6DP, United Kingdom

Oxford University Press is a department of the University of Oxford. It furthers the University's objective of excellence in research, scholarship, and education by publishing worldwide. Oxford is a registered trade mark of Oxford University Press in the UK and in certain other countries

First published in 2019

British Library Cataloguing in Publication Data
Data available

978-0-19-842711-7

10 9

Paper used in the production of this book is a natural, recyclable product made from wood grown in sustainable forests. The manufacturing process conforms to the environmental regulations of the country of origin.

Printed in India by Multivista Global Pvt. Ltd

Acknowledgements

The publisher would like to thank the following authors for contributions to digital resources:

Alexander Aits
Natasha Awada
Laurie Buchanan
Eliza Casapopol
Tom Edinburgh
Jim Fensom
Neil Hendry
Georgios Ioannadis
Alissa Kamilova
Ed Kemp
Paul La Rondie
Martin Noon
Jill Stevens
Ellen Thompson
Felix Weitkamper
Daniel Wilson-Nunn

Cover: TTStock/iStockphoto. All other photos © Shutterstock, except: **p67:** The Picture Art Collection/Alamy Stock Photo; **p83:** Dave Porter/Alamy Stock Photo; **p86(l):** Peter Polk/123RF; **p213(r):** mgkaya/iStockphoto; **p214(tr):** NASA/SCIENCE PHOTO LIBRARY; **p214(bl):** Chictype/iStockphoto; **p230:** Blackfox Images /Alamy Stock Photo; **p582(l):** dpa picture alliance/ Alamy Stock Photo; **p582(r):** Xinhua/Alamy Stock Photo; **p639(l):** Oxford University Press ANZ/Brent Parker Jones Food Photographer/Sebastian Sedlak Food Stylist.

Every effort has been made to contact copyright holders of material reproduced in this book. Any omissions will be rectified in subsequent printings if notice is given to the publisher.

Course Companion definition

The IB Diploma Programme Course Companions are designed to support students throughout their two-year Diploma Programme. They will help students gain an understanding of what is expected from their subject studies while presenting content in a way that illustrates the purpose and aims of the IB. They reflect the philosophy and approach of the IB and encourage a deep understanding of each subject by making connections to wider issues and providing opportunities for critical thinking.

The books mirror the IB philosophy of viewing the curriculum in terms of a whole-course approach and include support for international mindedness, the IB learner profile and the IB Diploma Programme core requirements, theory of knowledge, the extended essay and creativity, activity, service (CAS).

IB mission statement

The International Baccalaureate aims to develop inquiring, knowledgable and caring young people who help to create a better and more peaceful world through intercultural understanding and respect.

To this end the IB works with schools, governments and international organisations to develop challenging programmes of international education and rigorous assessment.

These programmes encourage students across the world to become active, compassionate, and lifelong learners who understand that other people, with their differences, can also be right.

The IB learner profile

The aim of all IB programmes is to develop internationally minded people who, recognising their common humanity and shared guardianship of the planet, help to create a better and more peaceful world. IB learners strive to be:

Inquirers They develop their natural curiosity. They acquire the skills necessary to conduct inquiry and research and show independence in learning. They actively enjoy learning and this love of learning will be sustained throughout their lives.

Knowledgeable They explore concepts, ideas, and issues that have local and global significance. In so doing, they acquire in-depth knowledge and develop understanding across a broad and balanced range of disciplines.

Thinkers They exercise initiative in applying thinking skills critically and creatively to recognise and approach complex problems, and make reasoned, ethical decisions.

Communicators They understand and express ideas and information confidently and creatively in more than one language and in a variety of modes of communication. They work effectively and willingly in collaboration with others.

Principled They act with integrity and honesty, with a strong sense of fairness, justice, and respect for the dignity of the individual, groups, and communities. They take responsibility for their own actions and the consequences that accompany them.

Open-minded They understand and appreciate their own cultures and personal histories, and are open to the perspectives, values, and traditions of other individuals and communities. They are accustomed to seeking and evaluating a range of points of view, and are willing to grow from the experience.

Caring They show empathy, compassion, and respect towards the needs and feelings of others. They have a personal commitment to service, and act to make a positive difference to the lives of others and to the environment.

Risk-takers They approach unfamiliar situations and uncertainty with courage and forethought, and have the independence of spirit to explore new roles, ideas, and strategies. They are brave and articulate in defending their beliefs.

Balanced They understand the importance of intellectual, physical, and emotional balance to achieve personal well-being for themselves and others.

Reflective They give thoughtful consideration to their own learning and experience. They are able to assess and understand their strengths and limitations in order to support their learning and professional development.

Contents

Number and algebra
Functions
Geometry and trigonometry
Statistics and probability
Calculus
Exploration

Digital contents

Digital content overview
Click on this icon here to see a list of all the digital resources in your enhanced online course book. To learn more about the different digital resource types included in each of the chapters and how to get the most out of your enhanced online course book, go to page ix.

Syllabus coverage
This book covers all the content of the Mathematics: analysis and approaches SL course. Click on this icon here for a document showing you the syllabus statements covered in each chapter.

Practice exam papers
Click on this icon here for an additional set of practice exam papers.

Worked solutions
Click on this icon here for worked solutions for all the questions in the book.

Introduction

The new IB diploma mathematics courses have been designed to support the evolution in mathematics pedagogy and encourage teachers to develop students' conceptual understanding using the content and skills of mathematics, in order to promote deep learning. The new syllabus provides suggestions of conceptual understandings for teachers to use when designing unit plans and overall, the goal is to foster more depth, as opposed to breadth, of understanding of mathematics.

What is teaching for conceptual understanding in mathematics?

Traditional mathematics learning has often focused on rote memorization of facts and algorithms, with little attention paid to understanding the underlying concepts in mathematics. As a consequence, many learners have not been exposed to the beauty and creativity of mathematics which, inherently, is a network of interconnected conceptual relationships.

Teaching for conceptual understanding is a framework for learning mathematics that frames the factual content and skills; lower order thinking, with disciplinary and non-disciplinary concepts and statements of conceptual understanding promoting higher order thinking. Concepts represent powerful, organizing ideas that are not locked in a particular place, time or situation. In this model, the development of intellect is achieved by creating a synergy between the factual, lower levels of thinking and the conceptual higher levels of thinking. Facts and skills are used as a foundation to build deep conceptual understanding through inquiry.

The IB Approaches to Teaching and Learning (ATLs) include teaching focused on conceptual understanding and using inquiry-based approaches. These books provide a structured inquiry-based approach in which learners can develop an understanding of the purpose of what they are learning by asking the questions: why or how? Due to this sense of purpose, which is always situated within a context, research shows that learners are more motivated and supported to construct their own conceptual understandings and develop higher levels of thinking as they relate facts, skills and topics.

The DP mathematics courses identify twelve possible fundamental concepts which relate to the five mathematical topic areas, and that teachers can use to develop connections across the mathematics and wider curriculum:

Approximation	Modelling	Representation
Change	Patterns	Space
Equivalence	Quantity	Systems
Generalization	Relationships	Validity

Each chapter explores two of these concepts, which are reflected in the chapter titles and also listed at the start of the chapter.

The DP syllabus states the essential understandings for each topic, and suggests some content-specific conceptual understandings relevant to the topic content. For this series of books, we have identified important topical understandings that link to these and underpin the syllabus, and created investigations that enable students to develop this understanding. These investigations, which are a key element of every chapter, include factual and conceptual questions to prompt students to develop and articulate these topical conceptual understandings for themselves.

A tenet of teaching for conceptual understanding in mathematics is that the teacher does not **tell** the student what the topical understandings are at any stage of the learning process, but provides investigations that guide students to discover these for themselves. The teacher notes on the ebook provide additional support for teachers new to this approach.

A concept-based mathematics framework gives students opportunities to think more deeply and critically, and develop skills necessary for the 21st century and future success.

Jennifer Chang Wathall

Investigation 3

1 On your GDC, plot the graphs of each of these functions:

a $f(x)=\frac{1}{x}$ b $g(x)=\frac{2}{x}$ c $h(x)=\frac{3}{x}$

d $j(x)=\frac{-1}{x}$ e $k(x)=\frac{-2}{x}$ f $m(x)=\frac{-3}{x}$

2 **Factual** What features of each graph in question **1** are:

a the same for each graph? b different for each of the three graphs?

3 **Conceptual** What effect does changing the magnitude of the parameter k in the function $f(x)=\frac{k}{x}$ have on the graph of $y=f(x)$?

In every chapter, investigations provide inquiry activities and factual and conceptual questions that enable students to construct and communicate their own conceptual understanding in their own words. The key to concept-based teaching and learning, the investigations allow students to develop a deep conceptual understanding. Each investigation has full supporting teacher notes on the enhanced online course book.

Gives students the opportunity to reflect on what they have learned and deepen their understanding.

Reflect Why does zero not have a reciprocal?

Developing inquiry skills

Does mathematics always reflect reality? Are fractals such as the Koch snowflake invented or discovered?

Think about the questions in this opening problem and answer any you can. As you work through the chapter, you will gain mathematical knowledge and skills that will help you to answer them all.

Every chapter starts with a question that students can begin to think about from the start, and answer more fully as the chapter progresses. The developing inquiry skills boxes prompt them to think of their own inquiry topics and use the mathematics they are learning to investigate them further.

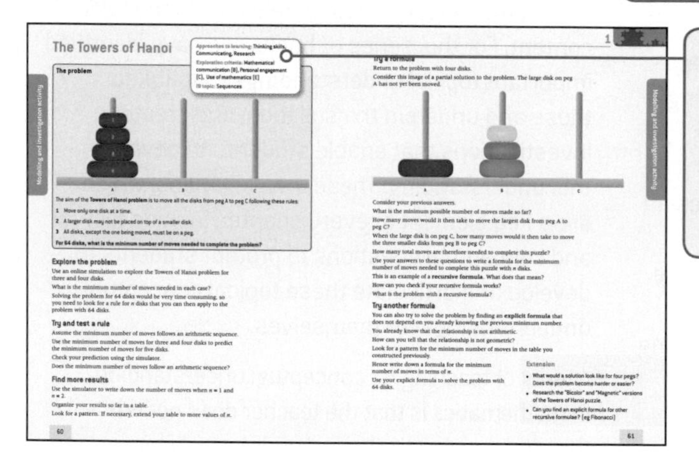

The Towers of Hanoi

Approaches to learning: Thinking skills, Communicating, Research
Exploration criteria: Mathematical communication (B), Personal engagement (C), Use of mathematics (E)
IB topic: Sequences

Modelling and investigation activity

The problem

The aim of the **Towers of Hanoi problem** is to move all the disks from peg A to peg C following these rules:

1 Move only one disk at a time.
2 A larger disk may not be placed on top of a smaller disk.
3 All disks, except the one being moved, must be on a peg.

For 64 disks, what is the minimum number of moves needed to complete the problem?

Explore the problem

Use an online simulation to explore the Towers of Hanoi problem for three and four disks.

What is the minimum number of moves needed in each case?

Solving the problem for 64 disks would be very time consuming, so you need to look for a rule for n disks that you can then apply to the problem with 64 disks.

Try and test a rule

Assume the minimum number of moves follows an arithmetic sequence.

Use the minimum number of moves for three and four disks to predict the minimum number of moves for five disks.

Check your prediction using the simulator.

Does the minimum number of moves follow an arithmetic sequence?

Find more results

Use the simulator to write down the number of moves when $n = 1$ and $n = 2$.

Organize your results so far in a table.

Look for a pattern. If necessary, extend your table to more values of n.

60

1

Try a formula

Return to the problem with four disks.

Consider this image of a partial solution to the problem. The large disk on peg A has not yet been moved.

Consider your previous answers.

What is the minimum possible number of moves made so far?

How many moves would it then take to move the largest disk from peg A to peg C?

When the large disk is on peg C, how many moves would it then take to move the three smaller disks from peg B to peg C?

How many total moves are therefore needed to complete this puzzle?

Use your answers to these questions to write a formula for the minimum number of moves needed to complete this puzzle with n disks.

This is an example of a **recursive formula**. What does that mean?

How can you check if your recursive formula works?

What is the problem with a recursive formula?

Try another formula

You can also try to solve the problem by finding an **explicit formula** that does not depend on you already knowing the previous minimum number.

You already know that the relationship is not arithmetic.

How can you tell that the relationship is not geometric?

Look for a pattern for the minimum number of moves in the table you constructed previously.

Hence write down a formula for the minimum number of moves in terms of n.

Use your explicit formula to solve the problem with 64 disks.

Extension

- What would a solution look like for four pegs? Does the problem become harder or easier?
- Research the "Bicolor" and "Magnetic" versions of the Towers of Hanoi puzzle.
- Can you find an explicit formula for other recursive formulae? (eg Fibonacci)

Modelling and investigation activity

61

The modelling and investigation activities are open-ended activities that use mathematics in a range of engaging contexts and to develop students' mathematical toolkit and build the skills they need for the IA. They appear at the end of each chapter.

International-mindedness

How do you use the Babylonian method of multiplication?

Try 36 × 14

TOK

Is it possible to know things about which we can have no experience, such as infinity?

TOK and International-mindedness are integrated into all the chapters.

The chapters in this book have been written to provide logical progression through the content, but you may prefer to use them in a different order, to match your own scheme of work. The Mathematics: analysis and approaches Standard and Higher Level books follow a similar chapter order, to make teaching easier when you have SL and HL students in the same class. Moreover, where possible, SL and HL chapters start with the same inquiry questions, contain similar investigations and share some questions in the chapter reviews and mixed reviews – just as the HL exams will include some of the same questions as the SL paper.

How to use your enhanced online course book

Throughout the book you will find the following icons. By clicking on these in your enhanced online course book you can access the associated activity or document.

Prior learning

Clicking on the icon next to the "Before you start" section in each chapter takes you to one or more worksheets containing short explanations, examples and practice exercises on topics that you should know before starting, or links to other chapters in the book to revise the prior learning you need.

Additional exercises

The icon by the last exercise at the end of each section of a chapter takes you to additional exercises for more practice, with questions at the same difficulty levels as those in the book.

Animated worked examples

This icon leads you to an animated worked example, explaining how the solution is derived step-by-step, while also pointing out common errors and how to avoid them.

Click on the icon on the page to launch the animation. The animated worked example will appear in a second screen.

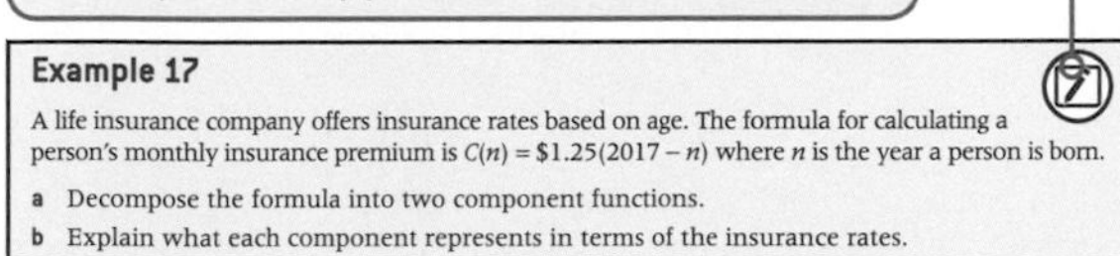

Example 17

A life insurance company offers insurance rates based on age. The formula for calculating a person's monthly insurance premium is $C(n) = \$1.25(2017 - n)$ where n is the year a person is born.

a Decompose the formula into two component functions.

b Explain what each component represents in terms of the insurance rates.

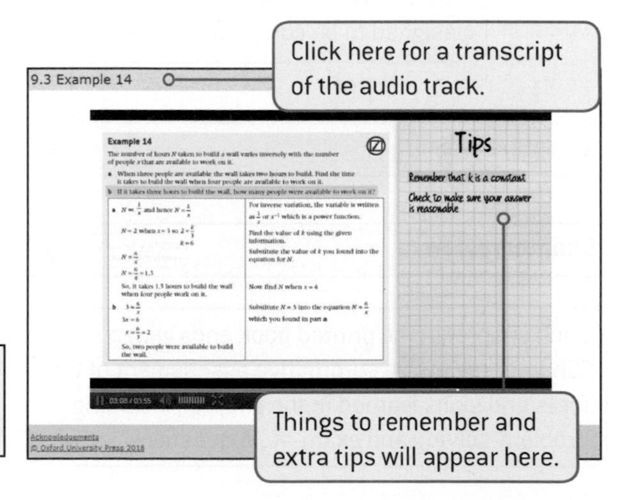

Graphical display calculator support

Supporting you to make the most of your TI-Nspire CX, TI-84+ C Silver Edition or Casio fx-CG50 graphical display calculator (GDC), this icon takes you to step-by-step instructions for using tecÜology to solve specific examples in the book.

Click on the icon for the menu and then select your GDC model.

Example 9

Find the equation of the following graph.

There are two methods you can use to do this on your GDC.

Method 1: Using **Quadratic regression**

TI-Nspire CX

TI-84+

Casio fx-CG50

Teacher notes

This icon appears at the beginning of each chapter and opens a set of comprehensive teaching notes for the investigations, reflection questions, TOK items, and the modelling and investigation activities in the chapter.

Assessment opportunities

This Mathematics: analysis and approaches enhanced online course book is designed to prepare you for your assessments by giving you a wide range of practice. In addition to the activities you will find in this book, further practice and support are available on the enhanced online course book.

End of chapter tests and mixed review exercises

This icon appears twice in each chapter: first, next to the "Chapter summary" section and then next to the "Chapter review" heading.

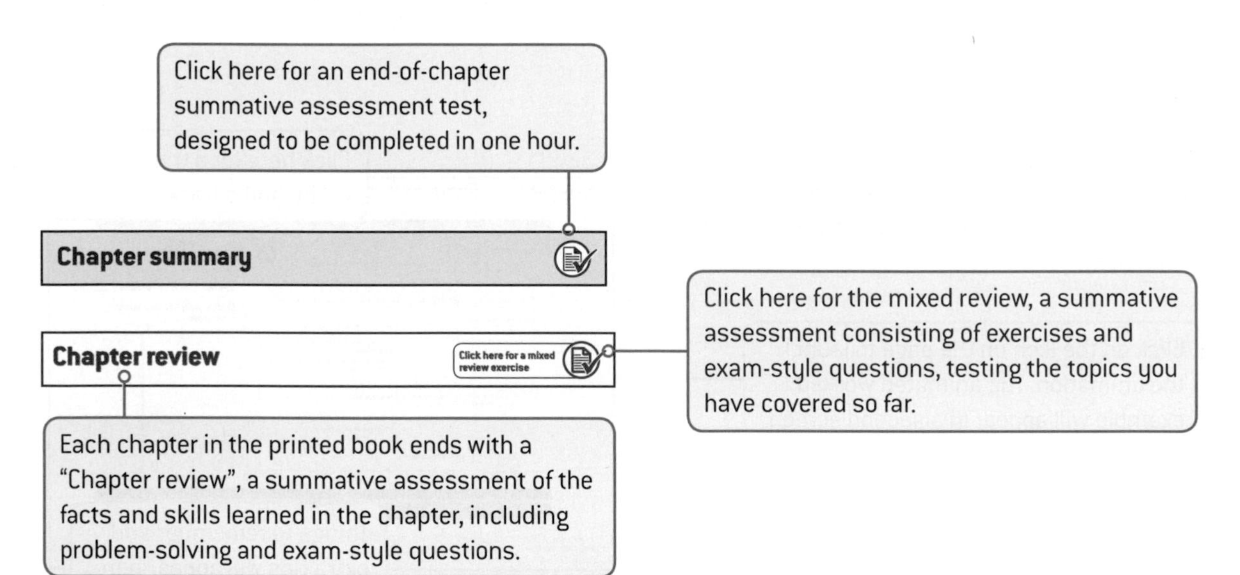

Exam-style questions

Exam-style questions

15 **P1:** Find an expression for derivative of each of the following functions.

a $f(x) = x^4 - 2x^3 - x^2 + 3x - 4$ (1 mark)

Plenty of exam practice questions, in Paper 1 (P1) or Paper 2 (P2) style. Each question in this section has a mark scheme in the worked solutions document found on the enhanced online course book, which will help you see how marks are awarded.

The number of darker bars shows the difficulty of the question (one dark bar = easy; three dark bars = difficult).

Click here for further exam practice

Exam practice exercises provide exam style questions for Papers 1 and 2 on topics from all the preceding chapters. Click on the icon for the exam practice found at the end of chapters 5, 10 and 14 in this book.

Answers and worked solutions

Answers to the book questions

Concise answer to all the questions in this book can be found on page 648.

Worked solutions

Worked solutions for **all** questions in the book can be accessed by clicking the icon found on the Contents page or the first page of the Answers section.

Answers and worked solutions for the digital resources

Answers, worked solutions and mark schemes (where applicable) for the additional exercises, end-of-chapter tests and mixed reviews are included with the questions themselves.

1 From patterns to generalizations: sequences and series

Concepts

- Patterns
- Generalization

Microconcepts

- Arithmetic and geometric sequences
- Arithmetic and geometric series
- Common difference
- Sigma notation
- Common ratio
- Sum of sequences
- Binomial theorem
- Proof
- Sum to infinity

You do not have to look far and wide to find visual patterns—they are everywhere!

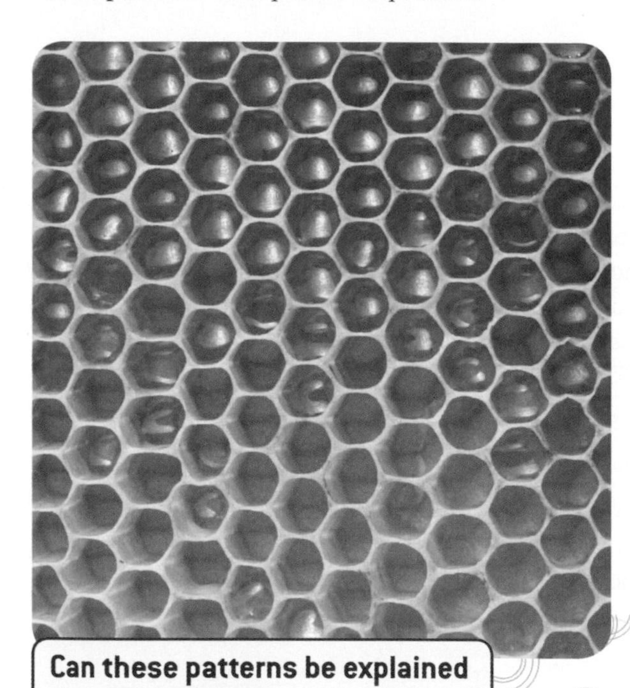

Can these patterns be explained mathematically?

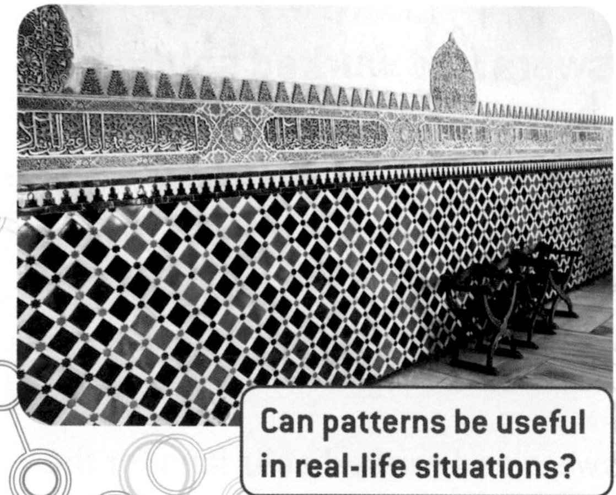

Can patterns be useful in real-life situations?

If you take out a loan to buy a car how can you determine the actual amount it will cost?

What information would you require in order to choose the best loan offer? What other scenarios could this be applied to?

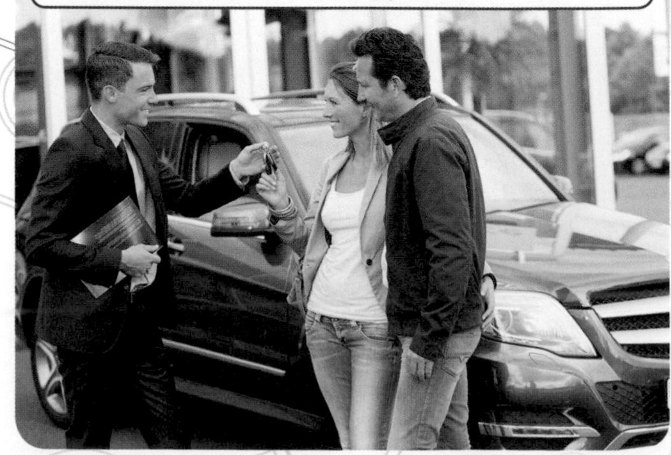

The diagrams shown here are the first four iterations of a fractal called the Koch snowflake.

What do you notice about:

- how each pattern is created from the previous one?
- the perimeter as you move from the first iteration through the fourth iteration? How is it changing?
- the area enclosed as you move from the first iteration to the fourth iteration? How is it changing?

What changes would you expect in the fifth iteration?

How would you measure the perimeter at the fifth iteration if the original triangle had sides of 1 m in length?

If this process continues forever, how can an infinite perimeter enclose a finite area?

Developing inquiry skills

Does mathematics always reflect reality? Are fractals such as the Koch snowflake invented or discovered?

Think about the questions in this opening problem and answer any you can. As you work through the chapter, you will gain mathematical knowledge and skills that will help you to answer them all.

Before you start

Click here for help with this skills check

You should know how to:	Skills check
1 Solve linear equations: eg $2(x-2)-3(3x+7)=4$ $2x-4-9x-21=4$ $-7x-25=4$ $-7x=29$ $x=\frac{-29}{7}$	**1** Solve each equation: **a** $2x+10=-4x-8$ **b** $3a-2(2a+5)=-12$ **c** $(x+2)(x-1)=(x+5)(x-2)$
2 Perform operations ($+ - \times \div$) with fractions and simplify using order of operations: eg $\frac{1}{2}+\frac{3}{4}\times\frac{2}{5}$ $=\frac{1}{2}+\frac{6}{20}=\frac{1}{2}+\frac{3}{10}=\frac{5}{10}+\frac{3}{10}=\frac{8}{10}=\frac{4}{5}$	**2** Simplify: **a** $\frac{3}{4}+\frac{1}{2}-\frac{5}{6}$ **b** $\frac{1}{2}\times\frac{3}{4}+\frac{5}{6}\times\frac{7}{8}$ **c** $\frac{-3}{4}\div\frac{5}{8}$
3 Substitute into a formula and simplify using order of operations: eg $P=-2(x+4)^2+3x$, $x=-1$ $P=-2(-1+4)^2+3(-1)$ $P=-2(3)^2+3(-1)$ $P=-2(9)+3(-1)$ $P=-18-3$ $P=-21$	**3** Substitute the given value(s) of the variable(s) and simplify the expression using order of operations: **a** $A=2(x-3)^2$, $x=-5$ **b** $S=2^x+(x-1)(x+3)$, $x=2$ **c** $P=3d-2(n+4)^2$, $d=-3$, $n=2$

1.1 Number patterns and sigma notation

TOK

Where did numbers come from?

Sequences

Investigation 1

Luis wants to join a gym. He finds one that charges a $100 membership fee to join and $5 every time he uses the facilities. What formula will help Luis to calculate how much he will pay in a given number of visits? If he goes to the gym 12 times in the first month, how much will he pay that month?

Luis's friend Elijah decides to join a different gym. His has no membership fee but charges $15 per use. How many times does each of them need to use their gym before they have paid the same total amount?

1 Complete the table below of Luis's and Elijah's gym fees:

Number of visits to the gym	Total fees paid by Luis	Total fees paid by Elijah
0	$100	$0
1	$105	$15
2	$110	$30
3		
4		
5		
6		
7		
8		

2 What patterns do you see emerging?

3 Which pattern is increasing faster?

4 When will the fees paid by Luis and Elijah be the same?

5 Are there any limitations to these patterns? Will they eventually reach a maximum value?

The fees paid in the investigation above create a sequence of numbers.

A **sequence** is a list of numbers that is written in a defined order, ascending or descending, following a specific rule.

Look at the following sequences and see if you can determine the next few terms in the pattern:

i 2, 4, 6, 8, ...　　**ii** 14, 11, 8, 5, ...

iii 1, 4, 9, 16, ...　　**iv** 1, 3, 9, 27, ...

A sequence can be either finite or infinite.

A **finite sequence** has a fixed number of terms.

An **infinite sequence** has an infinite number of terms.

For example, 7, 5, 3, 1, −1, −3 is finite because it ends after the sixth term.

For example, 7, 5, 3, −1, ... is infinite because the three dots at the end (called an ellipsis) indicate that the sequence is never ending or continues indefinitely.

A sequence can also be described in words. For example, the positive multiples of 3 less than 40 is a finite sequence because the terms are 3, 6, 9, ... 39.

A term in a sequence is named using the notation u_n, where n is the position of the term in the sequence. The first term is called u_1, the second term is u_2, etc.

A formula or expression that mathematically describes the pattern of the sequence can be found for the general term, u_n.

Let's return to the opening investigation and determine an expression for the general term for each sequence.

Investigation 1 continued

1 Write down the first few terms for Luis.

2 Describe the pattern.

3 Justify the general expression (u_n) for Luis at the bottom of the table below.

4 Write down the first few terms for Elijah.

5 Describe the pattern.

6 Complete the table below:

Term number	Total fees paid by Luis	Pattern for Luis	Total fees paid by Elijah	Pattern for Elijah
u_1	\$100	$100+(5\times0)$	\$0	15×0
u_2	\$105	$100+(5\times1)$	\$15	
u_3	\$110	$100+(5\times2)$	\$30	
u_4	\$115	$100+(5\times3)$	\$45	
u_5	\$120	$100+(5\times4)$	\$60	
u_n	—	$100+5(n-1)$	—	

7 The general term for Luis can be simplified from $u_n = 100+5(n-1)$ to $u_n = 5n+95$. Justify this simplification.

Continued on next page

8 What is the simplified form of Elijah's general term?

9 How did the pattern help you to write the general term for Elijah's sequence?

A sequence is called **arithmetic** when the same value is added to each term to get the next term.

For example, 100, 105, 110, 115, 120, … (+5, +5, +5, +5)

Another example of an arithmetic sequence is 1, 4, 7, 10, … because 3 is added to each term. The value added could also be negative, such as −5. For example, 20, 15, 10, 5, …

TOK

Do the names that we give things impact how we understand them?

Investigation 2

Let's look at a new type of sequence.

1 Complete this table of values for the next few terms of each sequence:

Sequence 1	Sequence 2	Sequence 3
5	2	36
10	−4	12
20	8	4
40	−16	$\frac{4}{3}$

2 Describe the pattern. For any of these sequences, how do you calculate a term using the previous term?

Look at the pattern for the first sequence as shown in the table below:

Term number	Sequence 1	Pattern	Sequence 2	Pattern	Sequence 3	Pattern
u_1	5	$5\times1=5\times2^0$	2		36	
u_2	10	$5\times2=5\times2^1$	−4		12	
u_3	20	$5\times4=5\times2^2$	8		4	
u_4	40	$5\times8=5\times2^3$	−16		$\frac{4}{3}$	
u_5	80	$5\times16=5\times2^4$	32		$\frac{4}{9}$	
u_n	—	$5\times2^{n-1}$	—		—	

3 Describe the relationship between the term number and the power of 2 for that term.

4 Justify the expression for the general term for the first pattern.

5 Complete the table for the other two patterns.

6 **Factual** How do you find the general term of **this type** of sequence?

7 **Conceptual** How does looking at the pattern help you find the general term of any sequence?

A sequence is called **geometric** when each term is multiplied by the same value to get the next term.

For example, 5, 10, 20, 40, … (×2 ×2 ×2) or 200, 100, 50, 25, … ($\times\frac{1}{2}$ $\times\frac{1}{2}$ $\times\frac{1}{2}$)

Example 1

Find an expression for the general term for each of the following sequences and state whether they are arithmetic, geometric or neither.

a 3, 6, 9, 12, … **b** 3, −12, 48, −192, … **c** 2, 10, 50, 250, … **d** $\frac{1}{3}, \frac{2}{5}, \frac{3}{7}, \frac{4}{9}, \ldots$

a The sequence can be written as $3 \times 1, 3 \times 2, 3 \times 3, \ldots$ The general term is $u_n = 3n$. This is an arithmetic sequence.	This sequence is the positive multiples of 3. This is an arithmetic sequence because 3 is being added to each term.
b The sequence can be written as $3 \times 1, 3 \times -4, 3 \times 16, 3 \times -64$ which can be expressed as $3 \times (-4)^0, 3 \times (-4)^1, 3 \times (-4)^2, 3 \times (-4)^3, \ldots$ The general term is $u_n = 3(-4)^{n-1}$ This is a geometric sequence.	This is a geometric sequence because each term is multiplied by −4.
c The sequence can be written as $2 \times 1, 2 \times 5, 2 \times 25, 2 \times 125, \ldots$ or $2 \times 5^0, 2 \times 5^1, 2 \times 5^2, 2 \times 5^3, \ldots$ The general term is $u_n = 2 \times 5^{n-1}$ This is a geometric sequence.	This is a geometric sequence because each term is multiplied by 5.

Continued on next page

d The sequence can be written as $\frac{1}{3}, \frac{2}{2+3}, \frac{3}{4+3}, \frac{4}{6+3}, \ldots$ or	Consider the numerators and denominators as two separate sequences.
$\frac{1}{3+(2\times 0)}, \frac{2}{3+(2\times 1)}, \frac{3}{3+(2\times 2)}, \frac{4}{3+(2\times 3)}, \ldots$	The numerators are the positive integers.
The general term is $u_n = \frac{n}{3+2(n-1)} = \frac{n}{3+2n-2} = \frac{n}{2n+1}$	The denominators, 3, 5, 7, 9, … , are the odd positive integers with $u_1 = 3$ and 2 being added to each term.
The sequence is neither arithmetic nor geometric.	Each sequence has a first term, a second term, a third term, etc, so n will always be a positive integer.

Exercise 1A

1 Write down the next three terms in each sequence:

a −8, −11, −14, −17, … **b** 9, 16, 25, 36, …

c 6, 12, 18, 24, … **d** 1000, 500, 250, 125, …

e $\frac{1}{2}, \frac{2}{3}, \frac{3}{4}, \frac{4}{5}, \ldots$ **f** $\frac{1}{3}, \frac{2}{9}, \frac{3}{27}, \frac{4}{81}, \ldots$

2 For each of the following sequences, find an expression for the general term and state whether the sequence is arithmetic, geometric or neither:

a 10, 50, 250, 1250 **b** 41, 35, 29, 23, …

c $\frac{1}{3}, -\frac{1}{9}, \frac{1}{27}, -\frac{1}{81}, \ldots$ **d** 1, 1, 2, 3, 5, 8, …

e $\frac{1}{2}, \frac{3}{4}, \frac{5}{6}, \frac{7}{8}, \ldots$ **f** −12, −36, −108, −324, …

3 For each of the following real-life situations, give the general term and state whether the sequence is arithmetic, geometric or neither.

a Ahmad deposits $100 in a savings account every month. After the first month, the balance is $100; after the second month, the balance is $200; and so on.

b Luciana is trying to lose weight. The first month, she loses 6 kg, and she continues to lose half as much each subsequent month.

c The temperature of the water in the swimming pool in your backyard is only 70°F. It is too cold, so you decide to turn up the temperature by 10% every hour until it is warm enough.

Consider the sequence −4, −3, −1, 3, … Notice that it is neither arithmetic or geometric. This is a new type of sequence called recursive.

A **recursive sequence** uses the previous term or terms to find the next term. The general term will include the notation u_{n-1}, which means "the previous term".

For example, given the general term $u_n = 2u_{n-1} + 5$ and $u_1 = -4$, the second term can be found by replacing u_{n-1} in the general term with u_1:

$$u_2 = 2u_1 + 5 = 2(-4) + 5 = -3$$

The third term can then be found by repeating the process with u_2:

$$u_3 = 2u_2 + 5 = 2(-3) + 5 = -1$$

Here is another example:

Given $u_n = (u_{n-1})^2 - 2$ and $u_1 = 5$, find the next three terms in the sequence.

$$u_2 = (u_1)^2 - 2 = 5^2 - 2 = 23$$

$$u_3 = (u_2)^2 - 2 = 23^2 - 2 = 527$$

$$u_4 = (u_3)^2 - 2 = 527^2 - 2 = 277727$$

TOK

Who would you call the founder of algebra?

Example 2

Find the first five terms for each of the following recursive sequences:

a $u_n = \frac{u_{n-1}}{3} + 3$ and $u_1 = 9$ **b** $u_n = 2u_{n-1} - 3$ and $u_1 = 1$

a $u_2 = \frac{u_1}{3} + 3 = \frac{9}{3} + 3 = 3 + 3 = 6$	To find the second term, replace u_{n-1} in the general term with the value of u_1 given in the question.
$u_3 = \frac{u_2}{3} + 3 = \frac{6}{3} + 3 = 2 + 3 = 5$	To find the third term, now replace u_{n-1} with the answer you found for u_2.
$u_4 = \frac{u_3}{3} + 3 = \frac{5}{3} + 3 = \frac{5}{3} + \frac{9}{3} = \frac{14}{3}$	Repeat this process until you have the first five terms.
$u_5 = \frac{u_4}{3} + 3 = \frac{\frac{14}{3}}{3} + 3 = \frac{14}{9} + \frac{27}{9} = \frac{41}{9}$	
$9, 6, 5, \frac{14}{3}, \frac{41}{9}$	
b $u_2 = 2u_1 - 3 = 2(1) - 3 = -1$ $u_3 = 2u_2 - 3 = 2(-1) - 3 = -5$ $u_4 = 2u_3 - 3 = 2(-5) - 3 = -13$ $u_5 = 2u_4 - 3 = 2(-13) - 3 = -29$ $1, -1, -5, -13, -29$	

Example 3

For each of the recursive sequences below, find a recursive formula for the general term:

a 12, 8, 4, 0, ... **b** −0.32, 3.2, −32, 320, ... **c** 500, 100, 20, 4, ...

a To calculate the next term, add −4. $u_n = u_{n-1} - 4$, where $u_1 = 12$	First, consider whether the sequence is arithmetic or geometric. This can be determined by checking whether a value is being added or multiplied to get the consecutive terms. Since −4 is being added each time, the recursive general term is $u_n = u_{n-1} - 4$, where $u_1 = 12$. It is important to state the first term so that you know the first value to substitute into the general term.
b To calculate the next term, multiply by −10. $u_n = -10u_{n-1}$, where $u_1 = -0.32$	
c To calculate the next term, multiply by $\frac{1}{5}$. $u_n = \frac{1}{5}u_{n-1}$, where $u_1 = 500$	

Exercise 1B

1 Find the first five terms for each of the following recursive sequences.

a $u_n = -4u_{n-1}$, $u_1 = 1$

b $u_n = \frac{-2}{u_{n-1}}$, $u_1 = 3$

c $u_n = 2(u_{n-1})^2$, $u_1 = -1$

d $u_n = 3u_{n-1} + 5$, $u_1 = m$

2 For each of the recursive sequences below, find a recursive formula for the general term.

a −2, −4, −6, −8, ...

b 1, 4, 16, 64, ...

c 52, 5.2, 0.52, 0.052, ...

d 14, 19, 24, 29, ...

e 2, 3, 6, 18, 108, 1944, ...

f 1, 2, 6, 24, ...

Series

A **series** is created when the terms of a sequence are added together.

A sequence or series can be either finite or infinite.

A **finite series** has a fixed number of terms.

For example, $7 + 5 + 3 + 1 + -1 + -3$ is finite because it ends after the sixth term.

An **infinite series** continues indefinitely.

For example, the series $10 + 8 + 6 + 4 + \ldots$ is infinite because the ellipsis indicates that the series continues indefinitely.

A series can be written in a form called **sigma notation**.

Sigma is the 18th letter of the Greek alphabet, and the capital letter, Σ, is used to represent a sum. Here is an example of a finite series written in sigma notation:

"5" is the upper limit of this series.

"$3n - 2$" is the general term of this series.

$$\sum_{n=1}^{5} 3n-2$$

"n" is called the index and represents a variable. The values of n will be consecutive integers.

"1" is called the lower limit and is the first n value that is substituted into the general term.

Consecutive n values are substituted, until the "5" or upper limit is reached. It will be the last value substituted.

$$\sum_{n=1}^{5} 3n-2 = (3(1)-2)+(3(2)-2)+(3(3)-2)+(3(4)-2)+(3(5)-2)$$

$$\sum_{n=1}^{5} 3n-2 = 1+4+7+10+13 = 35$$

For an infinite series, the upper limit is ∞. An example of an infinite geometric series is $\sum_{n=1}^{\infty} 3 \times 2^n$.

TOK

Is mathematics a language?

Example 4

For each of the following finite series in sigma notation, find the terms and calculate the sum:

a $\sum_{n=1}^{4} (-1)^n n^2$ **b** $\sum_{n=3}^{7} (-2)^n$ **c** $\sum_{n=1}^{3} 2n - n^2$

a $(-1)^1(1)^2 + (-1)^2(2)^2 + (-1)^3(3)^2 + (-1)^4(4)^2$ $= -1 + 4 - 9 + 16$ $= 10$	Substitute $n = 1$ to find the first term, $n = 2$ for the second term and so on. The last term is found when you substitute the upper limit of $n = 4$. Remember to add up all the terms once you have found them.
b $(-2)^3 + (-2)^4 + (-2)^5 + (-2)^6 + (-2)^7$ $= (-8) + 16 + (-32) + 64 + (-128)$ $= -88$	Substitute $n = 3$ to find the first term, $n = 4$ for the second term and so.
c $(2(1) - 1^2) + (2(2) - 2^2) + (2(3) - 3^2)$ $= 1 + 0 + (-3)$ $= -2$	

Example 5

Write each of the following series in sigma notation:

a $-3+5+13+21+29$ **b** $-3+6-12+24$ **c** $8+12+16+20+\ldots$

a This finite arithmetic series can be rewritten as:

$$-3+(-3+8)+(-3+16)+(-3+24)+(-3+32)$$
$$=(-3+0\times 8)+(-3+1\times 8)+(-3+2\times 8)+(-3+3\times 8)+(-3+4\times 8)$$

The general term is $-3+8n$.

Therefore, the sigma notation is $\sum_{n=0}^{4} -3+8n$.

b This finite geometric series can be rewritten as $-3\times 1+3\times 2+-3\times 4+3\times 8$
$=-3\times 2^0+3\times 2^1+-3\times 2^2+3\times 2^3$

The general term must include 3×2^n.

Since the first term is negative and the signs of following terms alternate, you need to multiply by $(-1)^{n+1}$.

If the first term was positive, or if you were starting with $n=1$, you would multiply by $(-1)^n$ to make the signs alternate.

Therefore, the sigma notation is:

$$\sum_{n=0}^{3} 3(-1)^{n+1}\,2^n$$

The answers to parts **a** and **b** are just one way to write each sequence using sigma notation. For example, another correct expression for part **a** would be

$$\sum_{n=1}^{5} -3+8(n-1)$$

c $\sum_{n=2}^{\infty} 4n$

or

$\sum_{n=1}^{\infty} 4(n+1)$

This infinite arithmetic series is the multiples of 4, starting with the second multiple.

To make the lower limit 1, you can also write the notation like this.

Note that because this is an infinite series, the upper limit is ∞.

Exercise 1C

1 For each of the following series in sigma notation, find the terms and calculate the sum:

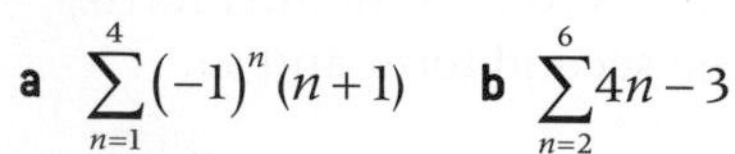

a $\sum_{n=1}^{4}(-1)^n(n+1)$ **b** $\sum_{n=2}^{6} 4n-3$

c $\sum_{n=1}^{3} n(n+1)$ **d** $\sum_{n=3}^{5}\frac{(-1)^{n+1}}{n-2}$

2 Write each of the following series in sigma notation:

a $4+16+64+256+\ldots$

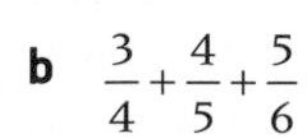

b $\frac{3}{4}+\frac{4}{5}+\frac{5}{6}$ **c** $-1+\frac{1}{2}-\frac{1}{3}+\frac{1}{4}-\cdots+\frac{1}{100}$

d $-2-2-2-2-2-2-2-2$

e $5+10+17+26+\ldots$

f $49m^6+64m^7+81m^8+100m^9+121m^{10}$

Developing inquiry skills

Let's return to the chapter opening problem about the Koch snowflake.

- **i** How many sides does the initial triangle have?
- **ii** How many sides does the second iteration have? What about the third iteration?
- **iii** What kind of sequence do these numbers form?
- **iv** Write the general term for the number of sides in any iteration.
- **v** You can use your general term to find the number of sides for a given iteration by substituting the value for n. Find the number of sides the 12th iteration will have.

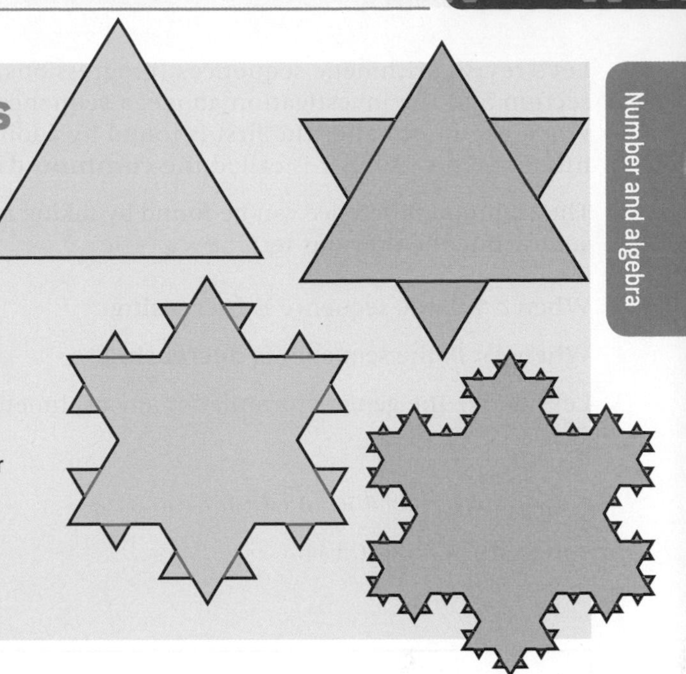

1.2 Arithmetic and geometric sequences

Investigation 3

A restaurant is discussing how many people they can fit at their tables. One table seats four people. If they push two tables together, they can seat six people. If three tables are joined, they can seat eight people.

1 Copy and complete the table as shown below:

Number of tables	1	2	3	4	5	6	7	8
Number of seats	4	6	8					

2 For every new table added, how many extra seats are created?

3 Write a general formula to calculate the number of seats for any number of joined tables.

4 You can use your general term to find the number of seats for a given number of tables by substituting the value for n in your general formula. How many seats will there be with twenty tables?

Let's revisit arithmetic sequences (progressions). From the last section and the investigation above, a sequence is arithmetic when each term after the first is found by adding a fixed non-zero number. This number is called the **common difference (d)**.

The common difference can be found by taking any term and subtracting the previous term: $d = u_n - u_{n-1}$.

When $d > 0$, the sequence is **increasing**.

When $d < 0$, the sequence is **decreasing**.

Let's derive the general formula for any arithmetic sequence:

$u_1, u_2, u_3, \ldots$

$= u_1, u_1+d, u_1+d+d, u_1+d+d+d, \ldots$

$= u_1, u_1+d, u_1+2d, u_1+3d, \ldots$

TOK

Is all knowledge concerned with identification and use of patterns?

Looking at the pattern between the term number and the coefficient of d, the formula for any term in the sequence is

$$u_n = u_1 + (n-1)d$$

Example 6

For each arithmetic progression, write down the value of d and find the term indicated.

a 14, 11, 8, ... Find u_{12} **b** 2, 7, 12, ... Find u_9 **c** −7, −5, −3, ... Find u_{21}

a $d = 8 - 11$ or $11 - 14 = -3$ $u_n = \text{u1} + (n-1)d$ $u_{12} = 14 + (12-1)(-3)$ $u_{12} = 14 + (11)(-3)$ $u_{12} = 14 - 33$ $u_{12} = -19$	To find d, take any term and subtract the previous term. Substitute u_1, n and d into the general term formula and simplify.
b $d = 12 - 7$ or $7 - 2 = 5$ $u_n = u_1 + (n-1)d$ $u_9 = 2 + (9-1)(5)$ $u_9 = 2 + (8)(5)$ $u_9 = 2 + 40$ $u_9 = 42$	

c $d = -3 - (-5)$ or $-5 - (-7) = 2$ $u_n = u_1 + (n-1)d$ $u_{21} = -7 + (21-1)(2)$ $u_{21} = -7 + (20)(2)$ $u_{21} = -7 + 40$ $u_{21} = 33$	

Exercise 1D

For each sequence below, use the general formula to find the term indicated.

1 5, 13, 21, ... u_9

2 40, 32, 24, ... u_{11}

3 5.05, 5.37, 5.69, ... u_7

4 $\frac{1}{2}, \frac{5}{6}, \frac{7}{6}, \ldots u_6$

5 $x+2, x+5, x+8, \ldots u_9$

6 $3a, 6a, 9a, \ldots u_{12}$

The general formula can be used to find more than just a given term. It can also be used to solve for u_1, n or d.

Example 7

Given an arithmetic sequence in which $u_1 = 14$ and $d = -3$, find the value of n such that $u_n = 2$

$u_n = u_1 + (n-1)d$ $2 = 14 + (n-1)(-3)$ $2 = 14 - 3n + 3$ $-3n = -15$ $n = 5$	Substitute u_1, u_n and d into the general term formula and solve for n. When solving for n, remember that your answer needs to be a natural number!

Example 8

Find the number of terms in the following finite arithmetic sequence.
−8, −4, 0, ... 36

$u_n = u_1 + (n-1)d$ $36 = -8 + (n-1)(4)$ $44 = 4n - 4$ $48 = 4n$ $n = 12$	Substitute u_1, u_n and d into the general term formula and solve for n.

Example 9

Two terms in an arithmetic sequence are $u_6 = 4$ and $u_{11} = 34$. Find u_{15}.

$u_6 + 5d = u_{11}$	The issue here is that you do not have u_1 or d. However, you do know that u_{11} is five terms higher in the sequence than u_6. Therefore, the difference between these terms is $5d$.
$4 + 5d = 34$ $5d = 30$ $d = 6$	Substitute values for the terms and rearrange to find d.
$u_{15} = u_{11} + 4d$ $u_{15} = 34 + 4(6)$ $u_{15} = 58$	Similarly, u_{15} is four terms higher in the sequence than u_{11}.

Exercise 1E

1. Given an arithmetic progression with $u_{21} = 65$ and $d = -2$, find the value of the first term.
2. Given that two terms of an arithmetic sequence are $u_5 = -3.7$ and $u_{15} = -52.3$, find the value of the 19th term.
3. Given an arithmetic sequence in which $u_1 = 11$ and $d = -3$, find the term that has a value of 2.
4. There exists an arithmetic sequence with $t_3 = 4$ and $t_6 = 184$. Find the 14th term.
5. Find the number of terms in the finite arithmetic sequence 6, −1, −8, ..., −36.
6. A movie theatre with 12 rows of seating has 30 seats in the first row. Each row behind it has two additional seats. How many seats are in the last row?
7. The Winter Olympics are held every four years and were held in Vancouver, Canada in 2010. When will the Winter Olympics be held for the first time after 2050?

8. Yinting wants to start adding weight lifting to her exercise routine. Her trainer suggests that she starts with 40 repetitions for the first week and then increases by six repetitions each week after. Find the number of weeks *before* Yinting will be doing 82 repetitions.

A sequence is **geometric** when each term after the first is found by multiplying by a non-zero number, called the **common ratio (r)**.

The common ratio can be found by taking any term and dividing by the previous term: $r = \frac{u_n}{u_{n-1}}$.

When $r < -1$ or $r > 1$, the sequence is **diverging**.

When $-1 < r < -1$, $r \neq 0$, the sequence is **converging**.

The general formula for any geometric sequence can be derived similarly to arithmetic sequences on page 14:

$u_1, u_2, u_3, u_4, u_5, \ldots$

$= u_1, r \times u_1, r \times r \times u_1, r \times r \times r \times u_1, r \times r \times r \times r \times u_1, \ldots$

$= u_1, u_1 r, u_1 r^2, u_1 r^3, u_1 r^4, \ldots$

EXAM HINT

If $r = 1$, then you will have a constant sequence, not a progression, as all the terms will be the same!

Looking at the pattern between the term number and the exponent of r, we can see that the formula for any term in the sequence is

$$u_n = u_1 r^{n-1}.$$

Investigation 4

For each of the given conditions below, generate the first five terms of the geometric sequence using a table:

1 $u_1 = 3, r = 2$

2 $u_1 = -4, r = -2$

3 $u_1 = 100, r = \frac{1}{2}$

4 $u_1 = -7299, r = -\frac{1}{3}$

5 $u_1 = 10\,000, r = \frac{1}{10}$

6 $u_1 = 1, r = 3$

A sequence is called **diverging** if the absolute value of each subsequent term is getting larger, tending away from 0.

A sequence is called **converging** if the absolute value of each subsequent term is getting smaller, approaching 0.

7 **Factual** Sort the sequences into two sets: convergent or divergent.

8 **Conceptual** What do you notice about the value of r in each of the sets?

9 **Conceptual** How can you tell whether a sequence is converging or diverging?

Example 10

For each geometric progression, write down the value of r and find the term indicated.

a $40, 20, 10, \dots u_{12}$ **b** $\frac{1}{2}, \frac{-1}{4}, \frac{1}{8}, \dots u_9$ **c** $87, 8.7, 0.87, \dots u_6$ **d** $7x, 14x^2, 21x^3, \dots u_8$

a $r = \frac{u_n}{u_{n-1}} = \frac{20}{40} = \frac{1}{2}$	First calculate r by dividing any term by the previous term.
$u_n = u_1 r^{n-1}$ $u_{12} = 40\left(\frac{1}{2}\right)^{12-1}$ $u_{12} = 40\left(\frac{1}{2}\right)^{11}$ $u_{12} = 40\left(\frac{1}{2048}\right)$ $u_{12} = \frac{5}{256}$	Substitute u_1, n and r into the general term formula and simplify.
b $r = \frac{-\frac{1}{4}}{\frac{1}{2}} = \frac{-1}{2}$ $u_n = u_1 r^{n-1}$ $u_9 = \frac{1}{2}\left(-\frac{1}{2}\right)^{9-1}$ $u_9 = \frac{1}{2}\left(\frac{-1}{2}\right)^{8}$ $u_9 = \frac{1}{512}$	
c $r = \frac{u_n}{u_{n-1}} = \frac{8.7}{87} = 0.10$ $u_n = u_1 r^{n-1}$ $u_6 = 87(0.10)^{6-1}$ $u_6 = 87(0.10)^5$ $u_6 = 87(0.00001)$ $u_6 = 0.00087$	
d $r = \frac{u_n}{u_{n-1}} = \frac{14x^2}{7x} = 2x$ $u_n = u_1 r^{n-1}$ $u_8 = 7x(2x)^{8-1}$ $u_8 = 7x(2x)^7$ $u_6 = 7x(128x^7)$ $u_6 = 896x^8$	

Exercise 1F

1 For the following sequences, determine whether they are geometric and if so, find the term indicated.

a $9, 27, 81, \ldots u_6$

b $2, -12, 18, \ldots u_9$

c $6, 4.5, 3.375, \ldots u_7$

d $-4, 6, -9, \ldots u_8$

e $500, 100, 20, \ldots u_{13}$

f $4, 9, 16, \ldots u_7$

g $3, 3m, 3m^2, \ldots u_{12}$

2 Suppose you find one cent on the first day of September and two cents on the second day, four cents on the third day, and so on. How much money will you find on the last day of September?

3 Write the first five terms of a geometric sequence in which the sixth term is 64.

Example 11

Given $u_1 = 81$ and $u_6 = \frac{1}{729}$ find the values of u_2, u_3, u_4 and u_5 such that the sequence is geometric.

$u_n = u_1 r^{n-1}$ $u_6 = u_1 r^{6-1}$ $\frac{1}{729} = 81r^5$ $r^5 = \frac{1}{59\,049}$ $r = \sqrt[5]{\frac{1}{59\,049}} = \frac{1}{9}$	This question is asking you to complete the blanks: 81, ___, ___, ___, ___, $\frac{1}{729}$ Note: The term between two existing terms is sometimes called "geometric means". This makes $u_1 = 81$ and $u_6 = \frac{1}{729}$. In order to find the missing terms, you first need to find r.
$u_2 = r \times u_1 = 81\left(\frac{1}{9}\right) = 9$ $u_3 = r \times u_2 = 9\left(\frac{1}{9}\right) = 1$ $u_4 = r \times u_3 = 1\left(\frac{1}{9}\right) = \frac{1}{9}$ $u_5 = r \times u_4 = \left(\frac{1}{9}\right)\left(\frac{1}{9}\right) = \frac{1}{81}$ Therefore, the sequence is $81, 9, 1, \frac{1}{9}, \frac{1}{81}, \frac{1}{729}$.	Now you can find the four missing terms by multiplying each term by r, starting with u_2.

Example 12

In a geometric sequence, $u_1 = 5$ and $u_5 = 1\,280$. The last term of the sequence is 20 480. Given that r is positive, find the number of terms in the sequence.

$u_n = u_1 r^{n-1}$ $u_5 = u_1 r^{5-1}$	You first need to find r:

Continued on next page

$1280 = 5r^4$ $r^4 = 256$ $r = \sqrt[4]{256}$ $r = 4$ $u_n = u_1 r^{n-1}$ $20\,480 = 5 \times 4^{n-1}$ $4096 = 4^{n-1}$ $4^6 = 4^{n-1}$ $6 = n - 1$ $n = 7$	Now you can use $u_1 = 5$, $r = 4$ and $u_n = 20\,480$ to find n.

Example 13

The value of a car depreciates at a rate of 19% each year. If you buy a new car today for \$33 560, how much will it be worth after four years?

$r = 100\% - 19\% = 81\% = 0.81$ $u_n = u_1 r^{n-1}$ $u_5 = 33\,560(0.81)^{5-1}$ $u_5 = 33\,560(0.81)^4$ $u_5 = 14\,446.47956...$ The car is worth $\approx$ \$14 446.48	Since you want to find the value of the car, you need to subtract the depreciation from 100% to find r. Since u_1 is when time = 0, after four years would be the fifth term. Note: Be sure to use approximation signs for any rounded values. Unless otherwise stated, monetary values should be rounded to two decimal places.

Exercise 1G

1 If a geometric sequence has $u_5 = 40$ and $u_{10} = 303.75$, find the value of u_{15}.

2 For a geometric sequence that has $r = -\frac{4}{5}$ and $u_6 = -1\,280$, find the 20th term.

3 If 16, $x + 2$, 1 are the first three terms of a geometric sequence, find all possible values of x.

4 Find the number of terms in the geometric sequence

6, 12, 24, ..., 1 536

5 Find the possible values of the common ratio of a geometric sequence whose first term is 2 and whose fifth term is 32.

6 In 2017, the number of students enrolled in a high school was 232. It is estimated that the student population will increase by 3% every year. Estimate the number of students that will be enrolled in 2027.

7 An old legend states that a peasant won a reward from a king. The peasant asked to be paid in rice; one grain on the first square of a chessboard, two grains on the second, four on the third square, and so on.

a How many grains of rice would be on the 30th square?

b Which square would contain exactly 512 grains of rice?

8 Given $u_1 = 8$ and $u_5 = 128$, find the values of u_2, u_3 and u_4 such that the sequence is geometric.

Investigation 5

Consider each of the sequences below:

a 20, 16, 12, ... **b** 20, 10, 5, ...

c −4, −12, −36, ... **d** −16, −18, −20, ...

e 1, 2, 6, ... **f** $\frac{1}{2}, \frac{1}{4}, \frac{1}{8}, \ldots$

g 2, 4, 6, ... **h** 3, 6, 7, ...

1. Decide whether each sequence is geometric, arithmetic or neither.
2. Find the general term for each sequence, if possible.
3. **Factual** What is a common difference? What is a common ratio?
4. **Conceptual** How can you determine whether a sequence is geometric, arithmetic or neither?
5. **Conceptual** How do the common difference or common ratio help you to describe and generalize a sequence?

Exercise 1H

For each question below, first decide whether the situation is arithmetic or geometric, then solve accordingly.

1. A frog fell into a 1 m well and wanted to go back up to the top of the well. Every day it moved up half the distance to the top. After 10 days, how much did the frog have left to climb?
2. Your grandparents deposit $2000 into a bank account to start a college fund for you. They will continue to deposit a fixed amount each month if you deposit $5 a month as well. In 36 months, you would like to have $6500 in the account. How much will they have to contribute each month?
3. The Chinese zodiac associates years with animals, based on a 12-year cycle. Samu was born in 1962, the Year of the Tiger. He lives in Finland, which celebrated its centennial in 2017. Did Finland gain its independence in the Year of the Tiger?
4. A scientist puts six bacteria, which multiply at a constant rate, in a Petri dish. She records the number of bacteria each minute thereafter. If she counts 324 bacteria 20 minutes later, at what rate are the bacteria reproducing?

1.3 Arithmetic and geometric series

Arithmetic series

Investigation 6

Let's make a triangle out of coins. Place one coin on the table.

Underneath it, create a row of two coins.

Place three in the next row.

Continue adding rows, each with one more coin than the row before.

1. Is the sequence between the row number and number of coins in that row arithmetic or geometric?
2. Write the general formula for the number of coins in any row.
3. Complete the table showing the **total number** of coins it will take to make a triangle with the given number of rows:

Number of rows	Total number of coins
1	1
2	3
3	
4	
5	

4. Hypothesize how many coins are needed to create a triangle with 10 rows. How many coins are needed for 20 rows? How many coins are needed for n rows?

The sum of the terms of an arithmetic sequence is called an **arithmetic series**.

An example of an arithmetic series is $3 + 6 + 9 + 12 + \ldots 30$.

Carl Gauss, a German mathematician, was the first to determine the formula to calculate the sum of an arithmetic sequence. When he tried to add up the natural numbers from 1 to 100, he noticed he could match them into pairs with equal sum:

TOK

How is intuition used in mathematics?

$$1 + 2 + 3 + \cdots 98 + 99 + 100$$

$1 + 100 = 101$ $\quad$ $2 + 99 = 101$ $\quad$ $3 + 98 = 101$

Since there would be 50 pairs whose sum is 101, the total sum must be $50 \times 101 = 5050$.

Consider the same process with the general arithmetic series:

$$u_1 + u_2 + u_3 + \dots u_n$$

We can match the pairs into equal sums:

$$u_1 + u_n \qquad u_2 + u_{(n-1)} \qquad u_3 + u_{(n-2)}$$

Since we started with n terms, there would be $\frac{n}{2}$ sums.

The sum, S_n, would be expressed as $S_n = \frac{n}{2}(u_1 + u_n)$.

Since we already have the formula $u_n = u_1 + (n - 1)d$, we can substitute it into the sum formula:

$$S_n = \frac{n}{2}(u_1 + u_n) = \frac{n}{2}(u_1 + u_1 + (n-1)d) = \frac{n}{2}(2u_1 + (n-1)d)$$

This gives us a second formula for the sum:

$$S_n = \frac{n}{2}[2u_1 + (n-1)d]$$

Geometric series

For geometric series, $u_1 + u_1r + u_1r^2 + \cdots u_1r^{n-2} + u_1r^{n-1}$, the formula for the sum is:

$$S_n = \frac{u_1(r^n - 1)}{r - 1} = \frac{u_1(1 - r^n)}{1 - r}, r \neq 1$$

Example 14

For each infinite series below, decide whether it is arithmetic or geometric, then find the sum to the given term:

a $1 + 5 + 9 + \dots u_8$

b $6 + 12 + 18 + \dots u_7$

c $-2 - 4 - 8 - \dots u_9$

d $100 + 50 + 25 + \dots u_6$

e $132 + 124 + 116 + \dots u_7$

f $\frac{1}{8} - \frac{1}{4} + \frac{1}{2} - \dots u_{12}$

Continued on next page

a Arithmetic	Since the difference between each term is +4, this series is arithmetic.
$d = u_n - u_{n-1}$ $d = 5 - 1 = 4$	Find the value of d.
$S_n = \frac{n}{2}\left[2u_1 + (n-1)d\right]$ $S_8 = \frac{8}{2}\left[2(1) + (8-1)(4)\right]$ $S_8 = 4[2 + (7)(4)]$ $S_8 = 4[2 + 28]$ $S_8 = 4(30)$ $S_8 = 120$	Substitute n, u_1 and d into the sum formula and simplify.
b Arithmetic	Since the difference between each term is +6, this series is arithmetic.
$d = u_n - u_{n-1}$ $d = 12 - 6 = 6$	Find the value of d.
$S_n = \frac{n}{2}\left[2u_1 + (n-1)d\right]$ $S_7 = \frac{7}{2}\left[2(6) + (7-1)(6)\right]$ $S_7 = 3.5[12 + (6)(6)]$ $S_7 = 3.5[12 + 36]$ $S_7 = 3.5(48)$ $S_7 = 168$	Substitute n, u_1 and d into the sum formula and simplify.
c Geometric	Since each term is found by multiplying the previous term by 2, this series is geometric.
$r = \frac{u_n}{u_{n-1}} = \frac{-4}{-2} = 2$	Find the value of r.
$S_n = u_1\left(\frac{1-r^n}{1-r}\right)$ $S_9 = -2\left(\frac{1-2^9}{1-2}\right)$ $S_9 = -2\left(\frac{1-512}{-1}\right)$ $S_9 = -2(511)$ $S_9 = -1022$	Substitute n, u_1 and r into the sum formula and simplify.

d Geometric

Since each term is found by multiplying the previous term by $\frac{1}{2}$, this series is geometric.

$r = \frac{50}{100} = \frac{1}{2}$

Find the value of r.

$S_n = u_1\left(\frac{1-r^n}{1-r}\right)$

Substitute n, u_1 and r into the sum formula and simplify.

$S_6 = 100\left(\frac{1-\left(\frac{1}{2}\right)^6}{1-\frac{1}{2}}\right)$

$S_6 = 100\left(\frac{1-\frac{1}{64}}{\frac{1}{2}}\right)$

$S_6 = 100\left(\frac{63}{64}\times\frac{2}{1}\right)$

$S_6 = 100\left(\frac{63}{32}\right)$

$S_6 = \frac{6300}{32} = \frac{1575}{8}$

e Arithmetic

Since the difference between each term is −8, this series is arithmetic.

$d = u_n - u_{n-1}$

Find the value of d.

$d = 124 - 132 = -8$

$S_n = \frac{n}{2}\left[2u_1 + (n-1)d\right]$

Substitute n, u_1 and d into the sum formula and simplify.

$S_7 = \frac{7}{2}\left[2(132) + (7-1)(-8)\right]$

$S_7 = 3.5[264 + (6)(-8)]$

$S_7 = 3.5(264 - 48)$

$S_7 = 3.5(216)$

$S_7 = 756$

Continued on next page

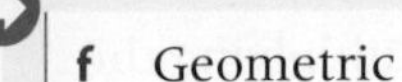

f Geometric	Since each term is found by multiplying the previous term by −2, this series is geometric.
$r = \frac{-\frac{1}{4}}{\frac{1}{8}} = -2$	Find the value of r.
$S_n = u_1\left(\frac{1-r^n}{1-r}\right)$ $S_{12} = \frac{1}{8}\left(\frac{1-(-2)^{12}}{1-(-2)}\right)$ $S_{12} = \frac{1}{8}\left(\frac{1-4096}{3}\right)$ $S_{12} = \frac{1}{8}(-1365)$ $S_{12} = \frac{-1365}{8}$	Substitute n, u_1 and r into the sum formula and simplify.

You can also use these formulas to find the sums of finite series.

Example 15

For each finite series below, decide whether it is arithmetic or geometric, then find the sum:

a $5+7+9+\cdots 39$ **b** $\sum_{n=1}^{15} -2(3)^n$ **c** $\frac{5}{4}+\frac{5}{8}+\frac{5}{16}+\cdots \frac{5}{256}$

a Arithmetic $d = u_n - u_{n-1}$ $d = 7 - 5 = 2$	Since the difference between each term is 2, this series is arithmetic.
$u_n = u_1 + (n-1)d$ $39 = 5 + (n-1)(2)$ $34 = 2(n-1)$ $17 = n - 1$ $n = 18$	Since you do not have n, you need to find this first using the general term formula.
$S_{18} = \frac{18}{2}(5+39)$ $S_{18} = 9(44)$ $S_{18} = 396$	Substitute n, u_1 and u_n into the sum formula and simplify.

b Geometric	Since each term is found by multiplying the previous term by 3, this series is geometric.
$r = \frac{-2(3)^2}{-2(3)} = 3$	Find the value of r.
$S_{15} = -6\left(\frac{1-3^{15}}{1-3}\right)$ $S_{15} = -6\left(\frac{1-14\,348\,907}{-2}\right)$ $S_{15} = 3(-14\,348\,906)$ $S_{15} = -\,43\,046\,718$	Substitute n, u_1 and r into the sum formula and simplify.
c Geometric	Since each term is found by multiplying the previous term by $\frac{1}{2}$, this series is geometric.
$r = \frac{u_n}{u_{n-1}} = \frac{\frac{5}{8}}{\frac{5}{4}} = \frac{1}{2}$	Find the value of r.
$u_n = u_1 r^{n-1}$ $\frac{5}{256} = \frac{5}{4}\left(\frac{1}{2}\right)^{n-1}$ $\frac{1}{64} = \left(\frac{1}{2}\right)^{n-1}$ $\left(\frac{1}{2}\right)^6 = \left(\frac{1}{2}\right)^{n-1}$ $n-1=6$ $n=7$	Since you do not have n, you need to find this first using the general term formula. Substitute n, u_1 and r into the sum formula and simplify.
$S_7 = \frac{5}{4}\left(\frac{1-\left(\frac{1}{2}\right)^7}{1-\frac{1}{2}}\right)$ $S_7 = \frac{5}{4}\left(\frac{1-\frac{1}{128}}{\frac{1}{2}}\right)$ $S_7 = \frac{5}{4}\left(\frac{127}{128} \times \frac{2}{1}\right)$ $S_7 = \frac{5}{4}\left(\frac{127}{64}\right)$ $S_7 = \frac{635}{256}$	

Exercise 1I

1 For each series below, decide whether it is arithmetic or geometric, then find the sum.

a $\frac{1}{5}+\frac{8}{15}+\frac{13}{15}+\cdots \quad S_7$

b $\sum_{n=1}^{8}\frac{1}{2}(-3)^n$

c $0.1 + 0.05 + 0.025 + \ldots \quad S_8$

d $6 + 12 + 18 + \ldots + 288$

e All multiples of 4 between 1 and 999

f $\sum_{n=1}^{6}(-1)^{n-1}(2)^n$

2 A theatre has 30 rows of seats. There are 22 seats in the first row, 26 in the second row, 30 in the third row, etc. How many people will the theatre hold?

3 Every human has two biological parents, four biological grandparents, eight biological great-grandparents and so on. How many biological ancestors are in your family for the last six generations if you are the first generation?

4 Each hour, a grandfather clock chimes the number of times that corresponds to the time of day. For example, at 5 am, it will chime five times. How many times will the clock chime in 24 hours?

5 Consider the following sequence of figures made up of line segments.

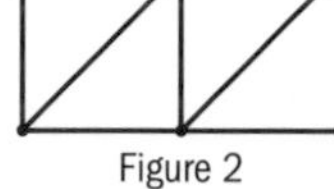
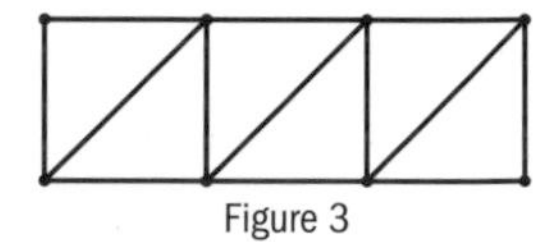

Figure 1 Figure 2 Figure 3

Find the total number of line segments in the first 48 figures.

Let's return to the coin triangle investigation from the beginning of this section.

Can you now find the total number of coins needed to form a triangle with 10 rows? What about for 20 rows?

Investigation 7

Let's return to infinite geometric series.

Recall from section 1.2 that a geometric sequence is diverging if $|r|>1$ and is converging if $|r|<1$. This holds true for series as well.

Part 1:

1 Copy and complete the chart below for a geometric series with $u_1 = 1$ and $r = 0.5$:

Term number	Term	Sum
1	1	1
2	0.5	1.5
3	0.25	1.75
4		
5		
6		
7		
8		
9		
10		
—	—	—

2 As the term number increases, what do you notice about the value of the term?

3 Looking at the second column, is the sequence converging or diverging?

4 How does this affect the sum?

5 What value is the sum approaching?

6 Using your GDC, plot the first sequence with the term number on the x-axis and the value of the term on the y-axis.

7 How does a graph show that the sum of a sequence is converging? What value is the sum of the first sequence converging to?

Part 2:

1 Repeat steps above for the following sequences:

a $u_1 = 10,\ r = \frac{1}{10}$ **b** $u_1 = -3,\ r = 2$

c $u_1 = 248,\ r = -\frac{1}{2}$ **d** $u_1 = 1,\ r = 4$

2 **Conceptual** When does an infinite geometric sequence have a finite sum?

3 Find the first four terms of the arithmetic sequence below:

a $u_1 = 10,\ d = 5$ **b** $u_1 = 10,\ d = -2$

c $u_1 = 10,\ d = -0.5$ **d** $u_1 = 10,\ d = 0.4$

4 **Conceptual** Which of the arithmetic sequences are decreasing? Which are increasing?

5 **Conceptual** Does an infinite arithmetic sequence converge to a finite sum?

6 **Conceptual** When does an infinite sequence have a finite sum?

Let's look at the sum of converging geometric series.

Consider the formula for the sum of a finite geometric series:

$$S_n = \frac{u_1(1-r^n)}{1-r}$$

Distributing the numerator:

$$S_n = \frac{u_1 - u_1 r^n}{1-r}$$

Splitting into two fractions:

$$S_n = \frac{u_1}{1-r} - \frac{u_1 r^n}{1-r}$$

As you saw in the previous investigation, with a converging geometric sequence, as the term number increases, the value of the term becomes increasingly smaller, which has almost no impact on the sum. Therefore, as the number of terms approaches infinity, the limit of $\frac{u_1 r^n}{1-r}$ will approach zero.

> The sum of a converging infinite geometric series is $S_\infty = \frac{u_1}{1-r}, |r| < 1$

Example 16

Decide whether the infinite geometric series below are converging, and if so, find the sum:

a $20 + 10 + 5 + \cdots$ **b** $\frac{1}{3} + \frac{2}{3} + \frac{4}{3} + \cdots$ **c** $\frac{1}{4} - \frac{1}{8} + \frac{1}{16} + \cdots$ **d** $\sum_{k=1}^{\infty} -2\left(\frac{-1}{3}\right)^n$

a $r = \frac{u_n}{u_{n-1}} = \frac{10}{20} = \frac{1}{2}$ Converging $S_\infty = \frac{u_1}{1-r}$ $S_\infty = \frac{20}{1-\frac{1}{2}}$ $S_\infty = \frac{20}{\frac{1}{2}}$ $S_\infty = 20 \times \frac{2}{1}$ $S_\infty = 40$	This series is converging since $\lvert r \rvert < 1$.
b $r = \frac{u_n}{u_{n-1}} = \frac{\frac{2}{3}}{\frac{1}{3}} = \frac{2}{3} \times \frac{3}{1} = 2$ Not converging	This series is not converging since $r > 1$.
c $r = \frac{u_n}{u_{n-1}} = \frac{-\frac{1}{8}}{\frac{1}{4}} = -\frac{1}{8} \times \frac{4}{1} = -\frac{1}{2}$	This series is converging since $\lvert r \rvert < 1$.

Converging $S_\infty = \frac{u_1}{1-r}$ $S_\infty = \frac{\frac{1}{4}}{1 - -\frac{1}{2}}$ $S_\infty = \frac{\frac{1}{4}}{1+\frac{1}{2}}$ $S_\infty = \frac{\frac{1}{4}}{\frac{3}{2}}$ $S_\infty = \frac{1}{4} \times \frac{2}{3}$ $S_\infty = \frac{1}{6}$	
d $u_1 = -2\left(\frac{-1}{3}\right)^1 = \frac{2}{3}$ $u_2 = -2\left(\frac{-1}{3}\right)^2 = \frac{-2}{9}$ $r = \frac{u_n}{u_{n-1}} = \frac{-\frac{2}{9}}{\frac{2}{3}} = -\frac{2}{9} \times \frac{3}{2} = -\frac{1}{3}$	
Converging $S_\infty = \frac{\frac{2}{3}}{1-\left(-\frac{1}{3}\right)}$ $S_\infty = \frac{\frac{2}{3}}{1+\frac{1}{3}}$ $S_\infty = \frac{\frac{2}{3}}{\frac{4}{3}}$ $S_\infty = \frac{2}{3} \times \frac{3}{4}$ $S_\infty = \frac{1}{2}$	This series is converging since $\lvert r \rvert < 1$.

Example 17

A bouncy ball is dropped straight down from a height of 1.5 metres. With each bounce, it rebounds 75% of its height. What is the total distance it travels until it comes to rest?

$u_1 = 1.5$ $r = 0.75$ $S_\infty = ?$ $S_\infty = \frac{u_1}{1-r} = \frac{1.5}{1-0.75}$ $S_\infty = \frac{1.5}{0.25}$ $S_\infty = 6$	
Distance travelled $= 2 \times 6 = 12$	However, this only represents the distance travelled on the down bounces. Because the ball also travels up, you must double the sum.
Distance travelled $= 12 - 1.5 =$ 10.5 metres	But since the ball was dropped from above, you must subtract the original 1.5 metres.

Example 18

Avi does chores around his house to earn money to buy his family presents for Hanukkah. On the first day of December, he earns \$0.50. On the second, he earns \$1.75 and on the third, he earns \$3. If this pattern continues for the first two weeks of December, how much money will he have in total on the first night of Hanukkah on 14 December?

The sequence is 0.50, 1.75, 3, ... $d = u_n - u_{n-1}$ $d = 1.75 - 0.50 = 1.25$ $S_n = \frac{n}{2}\left[2u_1 + (n-1)d\right]$ $S_{14} = \frac{14}{2}\left[2(0.50) + (14-1)(1.25)\right]$ $S_{14} = 7[1 + (13)(1.25)]$ $S_{14} = \$120.75$	

TOK

Is it possible to know things about which we can have no experience, such as infinity?

Exercise 1J

1 Decide whether the infinite geometric series below are converging, and if so, find the sum:

a $0.25 + 0.375 + 0.5625 + \ldots$

b $-\frac{3}{8} + \frac{9}{32} - \frac{27}{128} + \ldots$

c $\sum_{n=1}^{\infty} -5\left(\frac{1}{2}\right)^n$

d $\sum_{n=1}^{\infty} (-1)^{n+1}(2)^n$

e $(27x - 27) + (9x - 9) + (3x - 3) + \ldots$

f $\frac{1}{\sqrt{2}} + \frac{2}{\sqrt{2}} + \frac{4}{\sqrt{2}} + \cdots$

g $\frac{1}{\sqrt{2}} + \frac{1}{2\sqrt{2}} + \frac{1}{4\sqrt{2}} + \cdots$

2 Give an example of an infinite geometric series with a finite sum.

3 Give an example of an infinite geometric series where it is impossible to find the sum.

4 A ball is dropped from a height of 12 ft. Each time it hits the ground, it rebounds to $\frac{3}{5}$ of its previous height. Find how far the ball will travel before coming to a stop.

5 An oil well in Texas, USA produces 426 barrels in its first day of production. Its production decreases by 0.1% each day. If there are 42 gallons in one barrel of oil, find how many gallons this well will produce before it runs dry.

Example 19

For an arithmetic series, $S_n = -126$, $u_1 = -1$ and $u_n = -20$. Find the value of n.

$S_n = \frac{n}{2}[u_1 + u_n]$ $-126 = \frac{n}{2}[-1 - 20]$ $-252 = -21n$ $n = 12$	Remember that there are two formulas for the sum of an arithmetic series. Since you are given u_n, use the formula $S_n = \frac{n}{2}(u_1 + u_n)$

Example 20

The second term of an arithmetic sequence is 7. The sum of the first four terms is 12. Find the first term and the common difference.

$u_2 = 7$ $S_4 = 12$ $7 = u_1 + (2 - 1)d$ $7 = u_1 + d$ $u_1 = 7 - d$	

Continued on next page

$12=\frac{4}{2}\left[2u_1+(4-1)d\right]$ $12 = 2[2u_1 + 3d]$ $12 = 2[2(7 - d) + 3d]$ $12 = 2[14 - 2d + 3d]$ $12 = 2[14 + d]$ $6 = d + 14$ $d = -8$ $u_1 = 7 - d = 7 - (-8) = 15$	

Example 21

For the series 3 + 9 + 27 + …, what is the minimum number of terms needed for the sum to exceed 1000?

$\frac{3(3^n-1)}{3-1}>1000$ Using the GDC: $n = 5.92...$ The minimum number of terms is 6.	When using a graph or solver on the GDC, be sure to give the decimal answer first, then give the final answer.
or $\frac{3(3^n-1)}{3-1}>1000$ Using a table on the GDC: when $n=5$, $S_n=363$ when $n=6$, $S_n=1092$ The minimum number of terms is 6.	When using a table on the GDC, be sure to give both crossover values, then give the final answer.

Exercise 1K

1 In an arithmetic sequence, the first term is −8 and the sum of the first 20 terms is 790.

a Find the common difference.

b **i** Find u_{28}.

ii Hence, find S_{28}.

c Find how many terms it takes for the sum to exceed 2000.

2 In an arithmetic series, $S_{40} = 1900$ and $u_{40} = 106$. Find the value of the first term and the common difference.

3 The sum of an infinite geometric series is 20, and the common ratio is 0.2. Find the first term of this series.

4 The sum of an infinite geometric series is three times the first term. Find the common ratio of this series.

5 Create two different infinite geometric series that each have a sum of 8.

6 In a geometric sequence, the fourth term is 8 times the first term. The sum of the first 10 terms is 2557.5. Find the 10th term of this sequence.

7 A large company created a phone tree to contact all employees in case of an emergency. Each of the five vice presidents calls five employees, who in turn each call five other employees, and so on. How many rounds of phone calls are needed to reach all 2375 employees?

8 A geometric sequence has all positive terms. The sum of the first two terms is 15 and the sum to infinity is 27.

a Find the value of the common ratio.

b Hence, find the first term.

9 The first three terms of an infinite geometric sequence are $m - 1$, 6, $m + 8$.

a Write down two expressions for r.

b **i** Find two possible values of m.

ii Hence, find two possible values of r.

c **i** Only one of these r values forms a geometric sequence where an infinite sum can be found. Justify your choice for r.

ii Hence, calculate the sum to infinity.

TOK

How do mathematicians reconcile the fact that some conclusions conflict with intuition?

Developing inquiry skills

Returning to the chapter opening investigation about the Koch snowflake, the enclosed area can be found using the sum of an infinite series.

In the second iteration, since the sides of the new triangles are $\frac{1}{3}$ the length of the sides of the original triangle, their areas must be $\left(\frac{1}{3}\right)^2 = \frac{1}{9}$ of its area.

If the area of the original triangle is 1 square unit, then the total area of the three new triangles is $3\left(\frac{1}{9}\right)$.

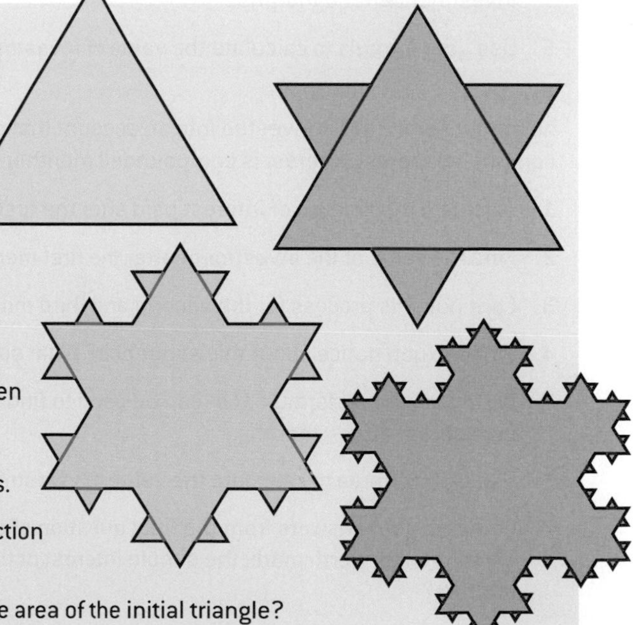

i Find the total area for the third and fourth iterations.

ii How can you use what you have learned in this section to find the total area of the Koch snowflake?

iii How does the area of a Koch snowflake relate to the area of the initial triangle?

1.4 Applications of arithmetic and geometric patterns

As you have seen in previous sections, arithmetic and geometric sequences and series can be applied to many real-life situations. In this section, you will explore two of these applications: interest and population growth.

Interest is the charge for borrowing money. Interest can also be used to describe money earned on an investment account. You will focus on two types of interest, simple and compound.

Simple interest is interest paid on the initial amount borrowed, saved or invested (called the **principal**) only and not on past interest.

Compound interest is paid on the principal and the accumulated interest. Interest paid on interest!

Investigation 8

Part 1:

An amount of $1000 is invested into an account that pays 5% per annum simple interest on a monthly basis.

1 Calculate the amount of interest paid per month.

2 Find the value of this investment after the first three months.

3 What do you notice about this sequence? What sort of sequence is it?

4 Write the general formula that can be used to find the amount of the investment after n months.

5 Use your formula to calculate the value of investment after two years.

Part 2:

An amount of $1000 is invested into an account that pays 5% per annum compound interest. Interest is compounded monthly.

1 Calculate the amount of interest paid after the first month.

2 Find the value of the investment after the first month.

3 Continue this process for the second and third months.

4 What do you notice about this sequence? What sort of sequence is it?

5 Write the general formula that can be used to find the amount of the investment after n months.

6 Use your formula to calculate the value of investment after two years.

7 Compare your answers from the final question of each part. Which investment is worth more; the simple interest or the compound interest?

8 **Conceptual** Which type of series models simple interest? Which type models compound interest?

9 **Conceptual** How are outcomes and growth patterns related?

From the previous investigation, you should have concluded that simple interest can be modelled by an arithmetic series, and compound interest can be modelled by a geometric series.

For simple interest, assuming A represents the accumulated amount, P is the principal, r is the annual rate and n is the time in years:

Initial amount:
$A = P$

After the first year:
$A = P + Pr = P(1 + r)$

After the second year:
$A = P(1 + r) + Pr = P(1 + r + r) = P(1 + 2r)$

After the third year:
$A = P(1 + 2r) + Pr = P(1 + 2r + r) = P(1 + 3r)$

After the nth year:
$\mathbf{A = P(1 + nr)}$

For compound interest, making the same assumptions as above:

Initial amount:
$A = P$

After the first year:
$A = P + Pr = P(1 + r)$

After the second year:
$A = P(1 + r) + P(1 + r)r = P(1 + r)(1 + r) = P(1 + r)^2$

After the third year:
$A = P(1 + r)^2 + P(1 + r)^2 r = P(1 + r)^2(1 + r) = P(1 + r)^3$

After the nth year:
$\mathbf{A = P(1 + r)^n}$

Interest can be paid over any time period—for example, yearly, monthly, quarterly, and so on.

To calculate compound interest over a non-year period, we use this formula:

$A = P\left(1 + \frac{r}{n}\right)^{nt}$, where A is the final amount (principal + interest), P is the principal, r is the annual interest rate expressed as a decimal, n is the number of compoundings in a year, and t is the total number of years.

Example 22

Sebastian took out a loan for a new car that cost \$35 000. The bank offered him 2.5% per annum simple interest for five years. Calculate the total value Sebastian has to repay the bank.

$A = P(1 + nr)$

$A = 35\,000(1 + 5(0.025))$

$A = 35\,000(1.125)$

$A = \$39\,375$

Example 23

Habib put \$5000 into a savings account that pays 4% interest per annum, compounded monthly. How much will be in the account after four years if Habib does not deposit or withdraw any money?

Since the interest is paid monthly,

$r = \frac{0.04}{12}$ and

$nt = 4 \times 12 = 48$

$A = 5000\left(1 + \frac{0.04}{12}\right)^{48}$

$A \approx \$5865.99$

Example 24

Leslie got a student loan of \$12 090 CAD to pay her tuition at the University of Toronto this year. If Leslie has to repay \$16 000 CAD in two years, what is the annual interest rate if the interest on the loan is compounded monthly?

$16\,000 = 12\,090\left(1 + \frac{r}{12}\right)^{24}$

$1.3234\ldots = \left(1 + \frac{r}{12}\right)^{24}$

$1 + \frac{r}{12} = \sqrt[24]{1.3234\ldots}$

$1 + \frac{r}{12} = 1.0117438\ldots$

$\frac{r}{12} = 0.0117438\ldots$

$r = 0.14092\ldots$

$r \approx 14.1\%$

Example 25

Oliver's savings account now has a balance of \$900. If the interest rate was 1.5% per annum, compounded quarterly, how much did Oliver originally deposit six years ago if he has not withdrawn or deposited any money since?

If the interest is compounded quarterly,

$r = \frac{0.015}{4}$

and $nt = 6 \times 4 = 24$

$900 = P\left(1 + \frac{0.015}{4}\right)^{24}$

$900 = P(1.00375)^{24}$

$900 = P(1.09399...)$

$P = \$822.6766...$

$P \approx \$822.68$

Exercise 1L

1 For each amount given below, calculate the amount of interest paid/earned:

a 1500 USD, simple interest of 6%, paid annually for ten years.

b 32 000 GBP, simple interest of 1.25%, paid for eight years.

c 14 168 000 Yen, compound interest of 2%, compounded monthly for three years.

d 300 000 Mexican Peso, compound interest of 4%, compounded daily for two years.

e 250 000 Swiss Francs, compound interest of 2.25%, compounded monthly for 25 years.

2 Fernando wants to purchase a new laptop for 2 323 000 Columbian Pesos. The electronics store has a financing offer. Fernando pays the electronics store in weekly installments over two years.

i What is the annual simple interest rate the electronics store is charging if he ends up paying it 2 480 000 Pesos in total?

ii How much is his weekly payment to the electronics store? Round your answer to the nearest Peso.

3 A teacher makes a one-time investment of \$90 000 in a retirement account for five years. The annual interest rate is 2.25%, compounded monthly. What will the final balance of the account be?

4 After five years of making quarterly payments in her car loan, Riley has paid off her loan. She has paid a total of \$32 546 CAD. If the per annum compound rate was 4.2%, how much was the loan for?

5 If you invest 5000 Swiss Francs today in an investment that pays 3.25% per annum, compounded

TOK

Do all societies view investment and interest in the same way?

What is your stance?

for the investment to double in value, assuming no additional withdrawals or deposits are made?

6 Isabella's parents are starting a college fund for her today, on her fifth birthday. Assuming they make no additional deposits or withdrawals, if they want to give her 50 000 Brazilian Real on her eighteenth birthday, how much do they need to deposit now if the bank is offering an interest rate of 5.5%, compounded monthly. Round your answer to the nearest whole number.

7 Oliver and Harry were each given 400 GBP for their birthday. Oliver puts his in a savings account that pays 1.25% interest, compounded monthly. Harry chooses to invest in a mutual fund paying 1.75% interest, compounded annually. If each brother does not touch the money for five years, who will have earned more?

8 Ahmed has 40 000 Egyptian Pounds to invest. He wants to split his money equally between a savings account with simple interest rate of 1.2% paid annually and a two-year guaranteed investment certificates (GIC) with an interest rate of 3.5%, compounded monthly. How long will he have to leave the money in the savings account so that it earns the same amount as the GIC?

Population growth

Growth can follow different patterns and we can use these patterns to predict and compare outcomes, in order to make decisions.

Population growth is defined as the increase of the number of individuals in a population over time. It is important to study population growth and predict future trends so that decisions can be made about allocation of resources.

Population growth is based on the number of births and deaths over a certain amount of time for a given population.

Investigation 9

The population of Canada can be found in the table below:

Year	2013	2014	2015	2016	2017
Population (millions)	35.152	35.535	35.832	36.264	36.708

Source: Statistics Canada

Part 1:

1 Calculate the percentage growth rate between each year in the table.

2 Find the average percentage growth rate.

3 Use this average rate to create a general formula for the population of Canada in any given year.

4 Use your general formula to estimate the population of Canada in 2025. Is your value realistic? What factors might affect this value?

Part 2:

1 Graph the data above with time on the x-axis and population on the y-axis.

2 On the same axes, graph your general formula from question **3** above.

3 Comment on how closely your formula models the actual data.

4 **Conceptual** How can you fit a model to real-life growth data?

5 **Conceptual** How can using sequences to model growth help us make predictions and comparisons?

Students could find data for a different country or region and model in the same way as in this investigation. This could be turned into a mini-exploration.

Example 26

In 2015, the population of the Greater Tokyo Area, Japan, was 13.491 million people. In 2016, 405 000 people moved into the city while 331 000 moved out.

a Find the net increase in the population of the city.

b Calculate the annual population growth rate.

c Assuming the growth rate stays the same, find a general formula for the population.

d Estimate the population of Toyko in 2025. Is this value reasonable? Explain your answer.

Continued on next page

a $405\,000 - 331\,000 = 74\,000$	The net increase is the number of people who moved in minus the number who moved out.
b $\frac{74\,000}{13\,491\,000} = 0.005485... \approx 0.00549$	The growth rate is the net increase divided by the population.
c $u_1 = 13\,491\,000$ $r = 1.005485... \approx 1.00549$ This makes the general formula: $P = 13\,491\,000\,(1.00549)^t$, where P is the population and t is number of years after 2015.	Other variables can be used for the general formula.
d $P = 13\,491\,000(1.00549)^{10}$ $P = 14\,249\,167.99967...$ $P \approx 14\,249\,168$ people This is not reasonable as it does not account for any births or deaths.	

Example 27

A biologist is growing a culture of bacteria in a lab. She counts the number of bacteria in a dish every six hours. Below are her results.

Time	0	6	12	18
Number of bacteria	125	350	980	2744

a Create a general formula to model the number of bacteria at any given time.

b Use your general formula to predict the number of bacteria at the end of the second day.

c Determine when the number of bacteria will exceed 250 000.

d Will the bacteria continue to multiply indefinitely? Explain your answer.

a $u_1 = 125$ $r = \frac{350}{125} = 2.8$ This makes the general formula: $N = 125(2.8)^t$, where N is number of bacteria and t is the number of six-hour periods that have elapsed.	Other variables can be used for the general formula. Be sure to define the variables you choose.

b The end of the second day would be 48 hours. Since we have defined t as the number of six-hour periods that have elapsed: $t = \frac{48}{6} = 8.$ $N = 125(2.8)^8$ $N = 472252.4979$ $N \approx 472\ 252$ bacteria	Round down to the nearest whole number here, as it is not possible to have part of a bacteria.
c $125(2.8)^t > 250\ 000$ By GDC: $t > 7.3822...$ Since t is every six hours: $t > 7.3822... \times 6 = 44.3$ hours or $t > 1.85$ days	
d No. The bacteria will eventually start to die because of factors such as overcrowding.	

Exercise 1M

1 A high white blood cell count can indicate that the patient is fighting an infection. A doctor is monitoring the number of white blood cells in one of her patients after receiving antibiotics. The lab returns the following data.

Hour	0	12	24	36
White blood cells (cells mcL^{-1})	12 500	11 000	9680	8518.4

a Create a general formula to model the patient's white blood cell count at any given time.

b Use your general formula to calculate the number of white blood cells this patient will have after three days.

c Discuss the limitations of your general formula.

2 The US unemployment rate for the first three months of a year is shown in the table below.

Month	January	February	March
Rate	7.9%	7.7%	7.5%

a What sort of sequence is this?

b Write a formula to calculate the unemployment rate for any month.

c Use your general formula to estimate the unemployment rate in December of the same year.

d Is it realistic to expect the unemployment rate to continue to decrease forever? Explain your answer.

3 Half-life is the time required for a substance to decay to half of its original amount.

a A radioactive isotope has a half-life of 1.23 years. Explain what this means.

b Write a general formula to calculate the amount remaining of the substance.

c Use your GDC to sketch a graph of this situation.

d If you start with a 52-gram sample of the isotope, how much will remain in 7.2 years?

1.5 The binomial theorem

Investigation 10

A taxi cab is situated at point A and must travel to point B. It may only move right or down from any given spot, without moving back on itself. You want to figure out how many shortest possible routes there are from A to B.

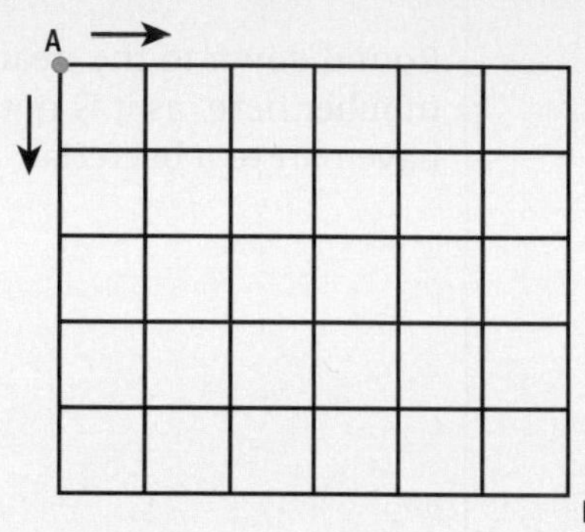

Start with the green dot. There are two possible routes because the taxi can move right and down or down and right.

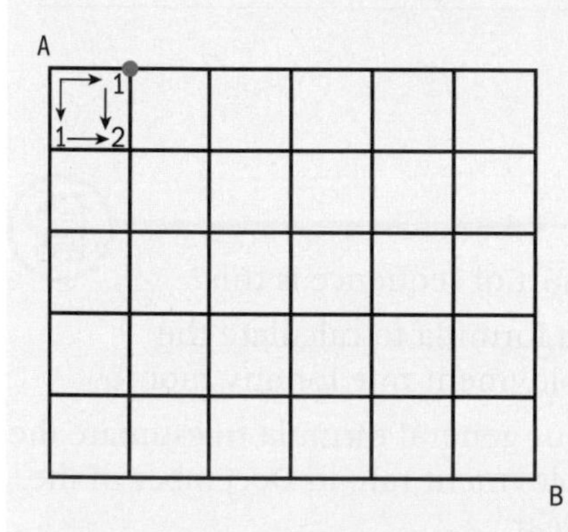

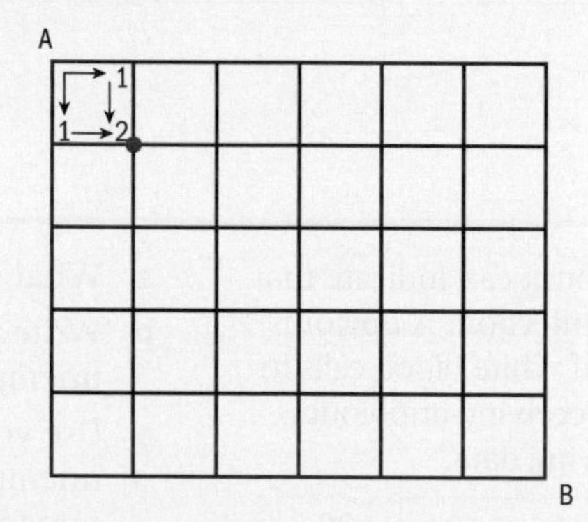

1 Can you figure out the number of shortest possible routes from A to the red dot? Mark each intersection between A and the red dot to help you.

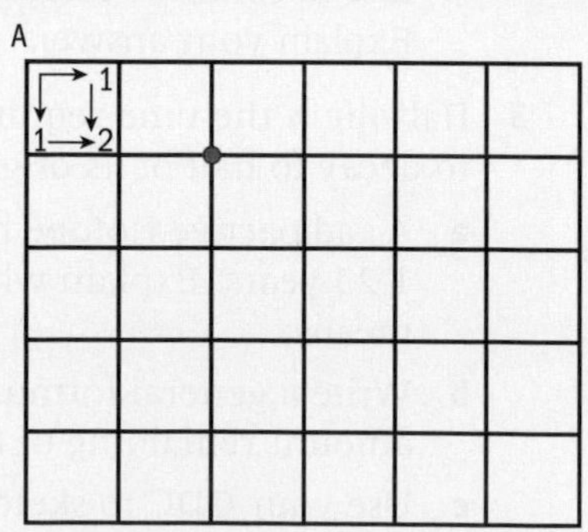

2 Use the same notation as in the first box to fill in the shaded boxes in the grid below.

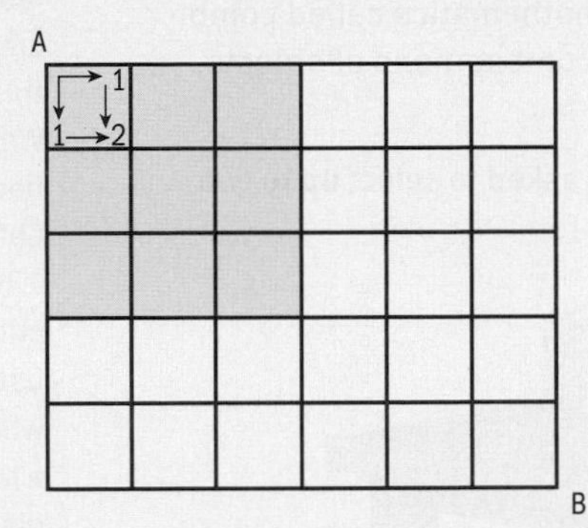

3 Turn the grid 45° so that the diagram is now a diamond with point A at the top. Describe the patterns you see in the numbers you have written so far.

4 Describe how these patterns can help you to complete the rest of the grid.

5 Complete the grid and state how many possible routes there are from A to B.

Investigation 11

The pattern you saw in the last investigation is called Pascal's triangle.

Part 1:

Blaise Pascal (1623–1662), a French mathematician, physicist and inventor, is credited with the triangular pattern below:

1
1 1
1 2 1
1 3 3 1
1 4 6 4 1

1 Describe the pattern used to create each successive row of the triangle.

2 Copy the triangle above and add five more rows.

3 What do you notice about the terms on the left end and on the right end of each row?

Part 2:

1 Find the sum of each of the rows in the triangle above.

2 What do you notice about these values?

3 State the general formula to find the sum of any row of Pascal's triangle.

4 Find the sum of the 16th row.

TOK

Why do we call this Pascal's triangle when it was in use before Pascal was born?

Are mathematical theories merely the collective opinions of different mathematicians, or do such theories give us genuine knowledge of the real world?

Investigation 12

Pascal's triangle is related to another branch of mathematics called combinatorics, which deals with the arrangement and combinations of objects.

Part 1:

Imagine there are two boxes on a table and you are asked to select **up to** two boxes.

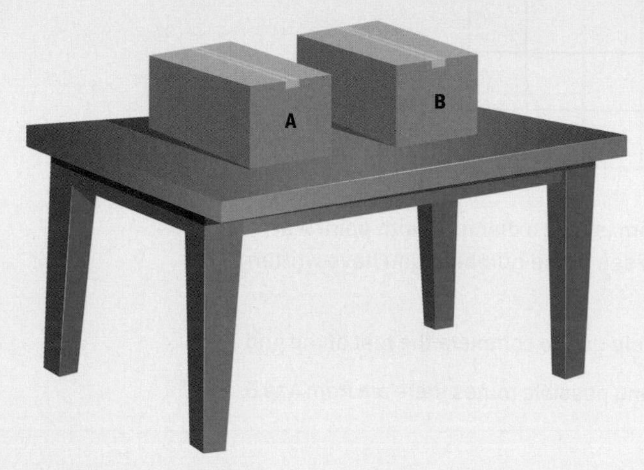

You could choose to take no boxes, one box or two boxes.

If we examine the number of ways to choose the above number of boxes, we see that there is one way to choose no boxes and one way to choose both boxes. However, if we choose one box, there are two possibilities, box A or box B.

So, the number of possible combinations is:

1 2 1

1 Imagine there are three boxes, labelled A, B and C. Find the different combinations for:

 a choosing no boxes b choosing one box
 c choosing two boxes d choosing three boxes.

2 Total the number of combinations for each number of boxes in question **1**.

3 Repeat the process for four boxes.

4 Look back at the last investigation on Pascal's triangle.

5 What do you notice about your combinations in question **2** and **3**?

6 How can Pascal's triangle be used to calculate the number of combinations for choosing 0, 1, 2, 3, … items from a given number of items?

7 Rewrite Pascal's triangle using ${}_nC_r$ notation.

The notation for a combination can be $\binom{n}{r}$, ${}^nC_r$ or ${}_nC_r$, where n is the number of objects available and r is the number of objects we want to choose.

8 **Conceptual** How would you use Pascal's triangle to explain why ${}_nC_r = {}_nC_{n-r}$?

TOK

Although Pascal is credited with bringing this triangle to the Western world, it had been known in China as early as the 13th century. What criteria should be used to determine who "invented" a mathematical discovery?

TOK

How many different tickets are possible in a lottery? What does this tell us about the ethics of selling lottery tickets to those who do not understand the implications of these large numbers?

Example 28

Calculate the following combinations using Pascal's triangle:

a ${}_4C_2$ **b** ${}^6C_5$ **c** $\binom{3}{3}$ **d** $\binom{5}{3}$

a Since $n = 4$, you need the fifth row.

1 4 6 4 1

Since $r = 2$, you need the third term.

${}_4C_2 = 6$

b ${}^6C_5$ means seventh row, sixth term.

${}_6C_5 = 6$

c $\binom{3}{3}$ means fourth row, fourth term, or you could use the fact ${}_nC_n = 1$

d $\binom{5}{3}$ means sixth row, fourth term.

$\binom{5}{3} = 10$

Another way to evaluate a combination is to use the formula:

$$^nC_r = \frac{n!}{r!(n-r)!}$$

Where $n!$, called "n factorial", represents the multiplication of each preceding integer starting at n down to 1.

$$n! = n(n-1)(n-2)(n-3)\ldots 3 \times 2 \times 1$$

For example, $7! = 7 \times 6 \times 5 \times 4 \times 3 \times 2 \times 1$.

Example 29

a Calculate the following combinations using the formula without using your GDC:

i $\binom{7}{3}$ **ii** ${}_8C_5$ **iii** $\binom{4}{1}$ **iv** ${}^6C_4$

b Find the following combinations by using your GDC:

i $\binom{32}{9}$ **ii** ${}_{25}C_{12}$

Continued on next page

a **i** $\binom{7}{3}=\frac{7!}{3!(7-3)!}$	Expand each factorial.
$\binom{7}{3}=\frac{7!}{3!4!}$	Cancel factors as appropriate.
$\binom{7}{3}=\frac{7\times6\times5\times4\times3\times2\times1}{(3\times2\times1)(4\times3\times2\times1)}$	Multiply the remaining factors.
$\binom{7}{3}=\frac{7\times6\times5\times\cancel{4\times3\times2\times1}}{(3\times2\times1)(\cancel{4\times3\times2\times1})}$	
$\binom{7}{3}=\frac{7\times\cancel{6}\times5}{\cancel{(3\times2\times1)}}$	
$\binom{7}{3}=7\times5$	
$\binom{7}{3}=35$	
ii ${}_8C_5=\frac{8!}{5!(8-5)!}$	
${}_8C_5=\frac{8!}{5!3!}$	
${}_8C_5=\frac{8\times7\times6\times5\times4\times\cancel{3\times2\times1}}{(5\times4\times3\times2\times1)(\cancel{3\times2\times1})}$	
${}_8C_5=\frac{8\times7\times\cancel{6\times5\times4}}{\cancel{5\times4\times3\times2}\times1}$	
${}_8C_5=8\times7$	
${}_8C_5=56$	
iii $\binom{4}{1}=\frac{4!}{1!(4-1)!}$	
$\binom{4}{1}=\frac{4!}{1!3!}$	
$\binom{4}{1}=\frac{4\times\cancel{3\times2\times1}}{\cancel{3\times2\times1}}$	
$\binom{4}{1}=4$	

iv $^6C_4 = \frac{6!}{4!(6-4)!}$

$^6C_4 = \frac{6!}{4!2!}$

$^6C_4 = \frac{6\times5\times4\times3\times2\times1}{(4\times3\times2\times1)(2\times1)}$

$^6C_4 = \frac{6\times5\times\cancel{4\times3\times2\times1}}{\cancel{(4\times3\times2\times1)}(2\times1)}$

$^6C_4 = \frac{30}{2}$

$^6C_4 = 15$

b **i** 28 048 800

ii 5 200 300

Investigation 13

Recall that a **binomial** is a simplified algebraic expression of the sum or difference of two terms. For example,

$$x+3,\quad -2a+4b,\quad x^2+6,\quad x+y$$

A **binomial expansion** is the result when we expand a binomial raised to a power.

Look at the binomial expansions of $(x+y)^n$ as shown in the table below:

$(x+y)^0$	1
$(x+y)^1$	$x+y$
$(x+y)^2$	$x^2+2xy+y^2$
$(x+y)^3$	$x^3+3x^2y+3xy^2+y^3$
$(x+y)^4$	$x^4+4x^3y+6x^2y^2+4xy^3+y^4$
$(x+y)^5$	$x^5+5x^4y+10x^3y^2+10x^2y^3+5xy^4+y^5$

1 What do you notice about:

- **a** the relationship between the value of n and the number of terms in the expansion
- **b** the exponents of x in each expansion
- **c** the exponents of y in each expansion
- **d** the sum of the exponents of x and y in each term of each expansion
- **e** the coefficients of the terms in each expansion?

2 Hypothesize the expansion of $(x+y)^6$.

Using these patterns is a generalization called the **binomial theorem** which provides an efficient method for expanding binomial expressions.

3 **Conceptual** How can you explain the coefficients of terms in the binomial expansion using Pascal's triangle?

4 **Conceptual** How does the binomial theorem use combinations?

Example 30

Use the binomial theorem to expand the following:

a $(x+2)^5$ **b** $(3-a)^6$ **c** $(2x+y)^4$ **d** $(3x-2y)^6$

a Setting up each term using the patterns discussed in the previous investigation:

Binomial coefficient	x	2
${}_5C_0$	x^5	2^0
${}_5C_1$	x^4	2^1
${}_5C_2$	x^3	2^2
${}_5C_3$	x^2	2^3
${}_5C_4$	x^1	2^4
${}_5C_5$	x^0	2^5

Binomial coefficient	x	2
1	x^5	1
5	x^4	2
10	x^3	4
10	x^2	8
5	x	16
1	1	32

Multiplying each term:

$(1)x^5(1) = x^5$

$(5)x^4(2) = 10x^4$

$(10)x^3(4) = 40x^3$

$(10)x^2(8) = 80x^2$

$(5)x(16) = 80x$

$(1)(1)(32) = 32$

$\therefore (x+2)^5 = x^5 + 10x^4 + 40x^3 + 80x^2 + 80x + 32$

For the binomial coefficients, use the formula ${}_nC_r$, where n is the exponent in the expansion and r starts at 0 and increases by 1 each term.

Be careful to distinguish between the "coefficient" and the "binomial coefficient". If, for example, you were asked to find the coefficient in x^4, the answer here would be 10.

b

Binomial coefficient	3	$-a$
${}_6C_0$	3^6	$(-a)^0$
${}_6C_1$	3^5	$(-a)^1$
${}_6C_2$	3^4	$(-a)^2$
${}_6C_3$	3^3	$(-a)^3$
${}_6C_4$	3^2	$(-a)^4$
${}_6C_5$	3^1	$(-a)^5$
${}_6C_6$	3^0	$(-a)^6$

Binomial coefficient	3	$-a$
1	729	1
6	243	$-a$
15	81	a^2
20	27	$-a^3$
15	9	a^4
6	3	$-a^5$
1	1	a^6

Multiplying each term:

$$(1)(729)(1) = 729$$
$$(6)(243)(-a) = -1458a$$
$$(15)(81)(a^2) = 1215a^2$$
$$(20)(27)(-a^3) = -540a^3$$
$$(15)(9)(a^4) = 135a^4$$
$$(6)(3)(-a^5) = -18a^5$$
$$(1)(1)(a^6) = a^6$$

$$\therefore (3 - a)^6 = 729 - 1458a + 1215a^2 - 540a^3 + 135a^4 - 18a^5 + a^6$$

Note: When one of the terms in the binomial is negative, remember to use parentheses when applying the exponent.

c

$$\binom{4}{0}(2x)^4 y^0 + \binom{4}{1}(2x)^3 y^1 + \binom{4}{2}(2x)^2 y^2 + \binom{4}{3}(2x)^1 y^3 + \binom{4}{4}(2x)^0 y^4$$

$$= 16x^4 + (4)(8x^3)y + 6(4x^2)y^2 + 4(2x)y^3 + y^4$$

$$= 16x^4 + 32x^3y + 24x^2y^2 + 8xy^3 + y^4$$

Continued on next page

d $\binom{6}{0}(3x)^6(-2y)^0 = 729x^6$

$\binom{6}{1}(3x)^5(-2y)^1 = 6(243x^5)(-2y)$
$= -2916x^5y$

$\binom{6}{2}(3x)^4(-2y)^2 = 15(81x^4)(4y^2)$
$= 4860x^4y^2$

$\binom{6}{3}(3x)^3(-2y)^3 = 20(27x^3)(-8y^3)$
$= -4320x^3y^3$

$\binom{6}{4}(3x)^2(-2y)^4 = 15(9x^2)(16y^4)$
$= 2160x^2y^4$

$\binom{6}{5}(3x)^1(-2y)^5 = 6(3x)(-32y^5)$
$= -576xy^5$

$\binom{6}{6}(3x)^0(-2y)^6 = 64y^6$

$\therefore 729x^6 - 2916x^5y + 4860x^4y^2 - 4320x^3y^3 + 2160x^2y^4 - 576xy^5 + 64y^6$

Exercise 1N

Expand the following using the binomial theorem:

1 $(x+5)^4$

2 $(2-b)^5$

3 $(2x-1)^6$

4 $(4x+y)^4$

5 $(x-3y)^3$

6 $(3x+4y)^5$

When finding just a single term in an expansion, it is easier to use the binomial theorem patterns, rather than calculate the entire expansion. For any term in the expansion of $(a+b)^n$:

$${}_nC_r(a)^{n-r}(b)^r$$

where n is the binomial power and r is one less than the term number, or r = term number −1.

Example 31

Find the seventh term in the expansion of $(3x - 4y)^{10}$.

For the seventh term,

$n = 10$

$r = 7 - 1 = 6$

${}_{10}C_6(3x)^4(-4y)^6$

$210(81x^4)(4096y^6)$

$69\,672\,960x^4y^6$

Example 32

The fifth term in the expavnsion of $(ax + by)^n$ is $560x^3y^4$. Find the values of a and b, given that a and b are integers.

Since it is the fifth term, r must be $5 - 1 = 4$, and n must be $3 + 4 = 7$.

$\therefore$ The binomial coefficient is ${}_7C_4 = 35$

$\frac{560}{35} = 16$

Since 16 is the fourth power of an integer, it must be the coefficient of y.

$\therefore 35(x)^3\left(\sqrt[4]{16}y\right)^4$

$35(x)^3(2y)^4$

This makes the original binomial $(x + 2y)^7$.

$a = 1$ and $b = 2$

Example 33

In the expansion of $\left(x+\frac{1}{x}\right)^9$, find the term number that will contain x^3.

In order for the final simplified term to contain x^3, the only possibility is: $(x)^6\left(\frac{1}{x}\right)^3 = x^3$

Since the second exponent is 3, the term number must be $3 + 1 = 4$.

Example 34

Use the binomial theorem to estimate $(1.87)^4$ to five decimal places.

$(1.87)^4 = (1 + 0.87)^4$ $(1+0.87)^4 = \binom{4}{0}1^4 0.87^0 + \binom{4}{1}1^3 0.87^1 + \binom{4}{2}1^2 0.87^2 + \binom{4}{3}1^1 0.87^3 + \binom{4}{4}1^0 0.87^4$ $(1 + 0.87)^4 = 1 + 4(0.87)^1 + 6(0.87)^2 + 4(0.87)^3 + 0.87^4$	In order to use the binomial theorem, we need to create a binomial from 1.87. The easiest way is to split 1.87 into 1 + 0.87. You can use any two numbers whose sum is 1.87.
$(1 + 0.87)^4 = 1 + 4(0.87) + 6(0.7569) + 4(0.658503) + 0.57289761$	
$(1 + 0.87)^4 = 1 + 3.48 + 4.5414 + 2.634012 + 0.57289761$	
$(1 + 0.87)^4 = 12.22830961$	
$(1 + 0.87)^4 \approx 12.22831$	

Exercise 10

1 Find the indicated term for the expansions below:

  a The fifth term of $(3x - 5)^{11}$

  b The ninth term of $(x + 6y)^{10}$

  c The middle term of $(2 - 3y)^6$

  d The constant term of $(x^4 - 3)^9$

  e The last term of $\left(-2x^2 - \frac{3}{x}\right)^7$

2 The values in the fourth row of Pascal's triangle are 1, 3, 3, 1.

  a Write down the values in the next row of the triangle.

  b Hence, find the term in x in $(3x - 2)^4$.

3 a Find the term in x^3 in the expansion of $(x - 3)^8$.

  b Find the term in x^5 in the expansion of $-2x(x - 3)^8$.

4 Find the term in x^9 in the expansion of $\left(x^3 - \frac{3}{x}\right)^{11}$.

5 In the expansion of $x^7\left(\frac{x^4}{2} + \frac{k}{x^3}\right)^7$, the constant term is 168. Find the value of k.

6 The fourth term in the expansion of $(2x - k)^8$ is $-387\,072x^5$. Find the value of k.

7 In the expansion of $(a - b^2)^6$, what is the coefficient of the term in a^3b^6?

8 Use the binomial theorem to estimate $(2.52)^3$ to three decimal places.

9 a Expand $(x - 5)^5$.

  b Hence find the term in x^4 in $(2x - 1)(x - 5)^5$.

10 a Show that the expansions of $(3 - 2x)^4$ and $(-2x + 3)^4$ are the same.

  b Will $(b - ax)^n$ and $(-ax + b)^n$ always have the same expansion? Explain your answer.

11 Find the binomial power, in the form $(a + b)^n$, with expansion $27x^3 - 108x^2y + 144xy^2 - 64y^3$.

12 Show that ${}_nC_0 + {}_nC_1 + {}_nC_2 + \dots {}_nC_{n-1} + {}_nC_n = 2^n$.

Developing your toolkit

Now do the Modelling and investigation activity on page 60.

1.6 Proofs

Investigation 14

How do we know that $3(x-2)-4(3x+5)+2x-2=-7(x+4)$ is true for all values of x?

1 Using substitution, show the statement holds true for $x=1$.

2 Using substitution, show the statement holds true for $x=-2$.

3 **Factual** If the proof holds true for two values of x, does this mean it will hold true for **all** values of x?

4 Draw a vertical line down the middle of your page, to create two columns. At the top, label the left-hand side column with "LHS" and the right-hand side column with "RHS". Write the LHS side of the statement above under the first column and the RHS under the second column.

Use algebra to simplify the LHS. When it is fully simplified, think of a way to make it look exactly like the RHS.

5 **Conceptual** Explain why using algebra is a better method than using substitution.

TOK

"Mathematics may be defined as the economy of counting. There is no problem in the whole of mathematics which cannot be solved by direct counting."

—E. Mach

To what extent do you agree with this quote?

A mathematical **proof** is a series of logical steps that show one side of a mathematical statement is equivalent to the other side for all values of the variable.

We need proofs in mathematics to show that the mathematics we use every day is correct, logical and sound.

There are many different types of proofs (deductive, inductive, by contradiction), but we will use **algebraic proofs** in this section.

The goal of an algebraic proof is to transform one side of the mathematical statement until it looks exactly like the other side.

One rule is that you **cannot** move terms from one side to the other. Imagine that there is an imaginary fence between the two sides of the statement and terms cannot jump the fence!

At the end of a proof, we write a concluding statement, such as LHS $\equiv$ RHS or QED.

EXAM HINT

QED is an abbreviation for the Latin phrase "quod erat demonstrandum" which means "what was to be demonstrated" or "what was to be shown". The sign with three bars ($\equiv$) is the identity symbol, which means that the two sides are equal by definition.

Example 35

Prove that $\frac{6}{8} = \frac{1}{4} + \frac{1}{2}$.

LHS	RHS	
$\frac{6}{8}$	$\frac{1}{4} + \frac{1}{2}$	First, simplify the RHS using a common denominator.
	$\frac{1}{4} + \frac{2}{4}$	
	$\frac{3}{4}$	
	$\frac{3}{4} \times \frac{2}{2}$	Once simplified completely, in order for the RHS to match the LHS, multiply by $\frac{2}{2}$ (which is equivalent to 1).
	$\frac{6}{8}$	
	QED	Don't forget a concluding statement at the end.

Exercise 1P

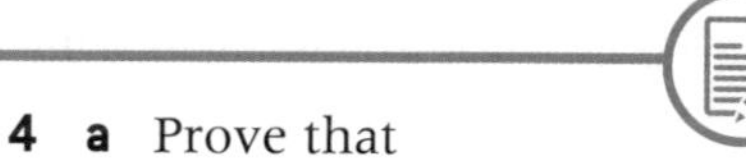

1 Prove that $-2(a-4) + 3(2a+6) - 6(a-5) = -2(a-28)$.

2 Prove that $(x-3)^2 + 5 = x^2 - 6x + 14$.

3 Prove that $\frac{1}{m} = \frac{1}{m+1} + \frac{1}{m^2+m}$.

4 a Prove that $\frac{x-2}{x} \div \frac{3x-6}{x^2+x} = \frac{x+1}{3}$.

b For what values of x does this mathematical statement **not** hold true?

TOK

What is the role of the mathematical community in determining the validity of a mathematical proof?

Chapter summary

- A **sequence** (also called a **progression**) is a list of numbers written in a particular order. Each number in a sequence is called a **term**.
- A **finite sequence** has a fixed number of terms.
  An **infinite sequence** has an infinite number of terms.
- A formula or expression that mathematically describes the pattern of the sequence can be found for the general term, u_n.
- A sequence is called **arithmetic** when the same value is added to each term to get the next term.
- A sequence is called **geometric** when each term is multiplied by the same value to get the next term.
- A **recursive sequence** uses the previous term or terms to find the next term. The general term will include the notation u_{n-1}, which means "the previous term".
- A **series** is created when the terms of a sequence are added together.
  A sequence or series can be either finite or infinite.
  A **finite series** has a fixed number of terms.
  For example, $7 + 5 + 3 + 1 + -1 + -3$ is finite because it ends after the sixth term.
  An **infinite series** continues indefinitely.
  For example, the series $10 + 8 + 6 + 4 + \ldots$ is infinite because the ellipsis indicates that the series continues indefinitely.

- The formula for any term in an arithmetic sequence is: $u_n = u_1 + (n-1)d$
- The formula for any term in a geometric sequence is: $u_n = u_1 r^{n-1}$.
- The sum, S_n, of an arithmetic series can be expressed as: $S_n = \frac{n}{2}(u_1 + u_n)$.
- The sum, S_n, of an arithmetic series can also be expressed as: $S_n = \frac{n}{2}\left[2u_1 + (n-1)d\right]$
- For geometric series, the formula for the sum is:

$$S_n = \frac{u_1(r^n - 1)}{r-1} = \frac{u_1(1-r^n)}{1-r},\ r \neq 1$$

- The sum of a converging infinite geometric series is: $S_\infty = \frac{u_1}{1-r}, |r| < 1$
- **Simple interest** is interest paid on the initial amount borrowed, saved or invested (called the **principal**) only and not on past interest.

  Compound interest is paid on the principal and the accumulated interest. Interest paid on interest!
- We can calculate combinations using the formula ${}^nC_r = \frac{n!}{r!(n-r)!}$
- A mathematical **proof** is a series of logical steps that show one side of a mathematical statement is equivalent to the other side for all values of the variable.
- We need proofs in mathematics to show that the mathematics we use every day is correct, logical and sound.

  The goal of an algebraic proof is to transform one side of the mathematical statement until it looks exactly like the other side.

  One rule is that you **cannot** move terms from one side to the other.

  At the end of a proof, we write a concluding statement, such as LHS ≡ RHS or QED.

Developing inquiry skills

Return to the opening problem. How has your understanding of the Koch snowflake changed as you have worked through this chapter? What features of, for example, the ninth iteration can you now work out from what you have learned?

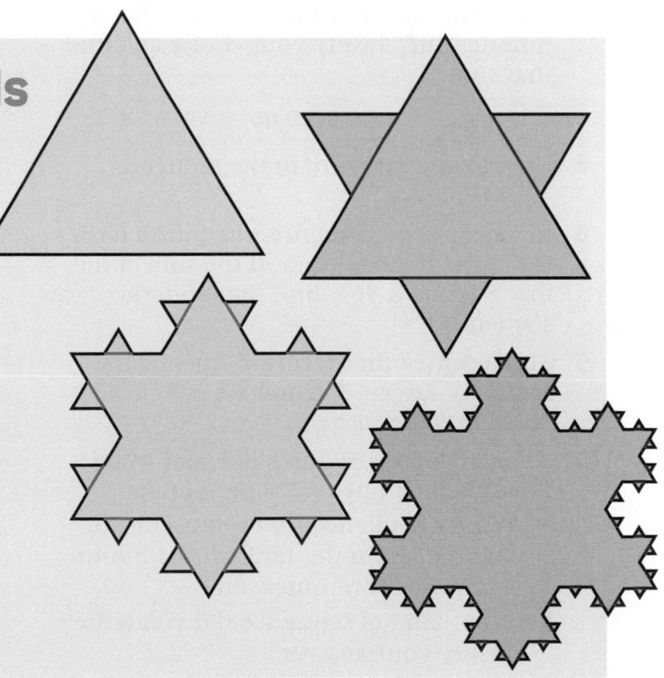

Chapter review

1 For each of the following sequences,

a Identify whether the sequence is arithmetic, geometric or neither.

b If it is arithmetic or geometric, find an expression for u_n.

c If it is arithmetic or geometric, find the indicated term.

d If it is arithmetic or geometric, find the indicated sum.

i 3, 6, 18, ... u_8, S_{12}

ii −16, −14, −12, ... u_{10}, S_8 ,$u_{10} = 2$

iii 2000, 1000, 500, ... u_9, S_7

iv $\sum_{n=1}^{\infty} 3 \times 2^{n-1}$, u_5, S_{10}

v The consecutive multiples of 5 greater than 104, u_7, S_9

2 In an arithmetic sequence, $u_6 = -5$ and $u_9 = -20$. Find S_{20}.

3 Write down the first five terms for the recursive sequence $u_n = -2u_{n-1} + 3$ with $u_1 = -4$.

4 For the geometric series $0.5 - 0.1 + 0.02 \ldots$ $S_n = 0.416$. Find the number of terms in the series.

5 For a geometric progression, $u_3 = 4.5$ and $u_7 = 22.78125$. Find the value of the common ratio and the first term.

6 Which of the following sequences has an infinite sum? Justify your choice and find that sum.

A $\frac{1}{4}, -\frac{1}{8}, \frac{1}{16}, \ldots$ **B** 0.06, 0.12, 0.24, ...

7 How many terms are in the sequence 4, 7, 10, ..., 61?

8 In a geometric sequence, the fourth term is 8 times the first term. If the sum of the first 8 terms is 765, find the 9th term of the sequence.

9 Three consecutive terms of a geometric sequence are $x - 3$, 6 and $x + 2$. Find all possible values of x.

10 A tank contains 55 litres of water. Water flows out at a rate of 7% per minute.

a Write a sequence that represents the volume left in the tank after 1 minute, 2 minutes, 3 minutes, etc.

b What kind of sequence did you write? Justify your answer.

c How much water is left in the tank after 10 minutes?

d How much water has been drained after 15 minutes?

e How long will it take to drain the tank?

11 Marie-Jeanne is experimenting with weights of 100 g attached to a spring.

She records the weight and length of the spring after attaching the weight.

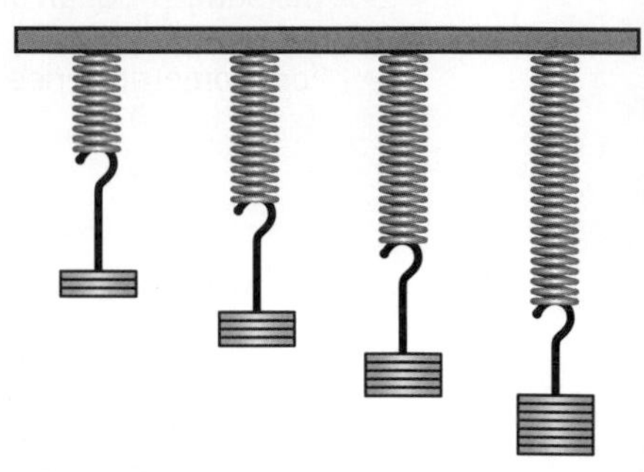

Number of 100 g weights attached to the spring	1	2	3	4
Length of spring (cm)	45	49	53	57

a Write a general formula that represents the length of the spring if n weights are attached.

b Calculate the supposed length of a spring if a 1 kg weight is attached.

c Explain any limitations of Marie-Jeanne's experiment.

d If the spring stretches to 101 cm, what is the total weight that was attached to the spring?

12 Kostas gets a four-year bank loan to buy a new car that is priced at €20 987. After the four years, Kostas will have paid the bank a total of €22 960. What annual interest rate did the bank give him if the interest was compounded monthly?

13 The first seven numbers in row 14 of Pascal's triangle are 1, 13, 78, 286, 715, 1287, 1716.

a Complete the row and explain your strategy.

b Explain how you can use your answer in part **a** to find the 15th row. State the terms in the 15th row.

14 Using the binomial theorem, expand $(3x - y)^6$.

15 Find the coefficient of the term in x^2 in the expansion of $\left(\frac{3}{x^2} - 4x^4\right)^8$.

16 In the binomial expansion of $\left(\frac{2}{x^2} - 5x^5\right)^n$, the sixth term contains x^{25}. Solve for n.

17 **a** Find the term in x^5 in the expansion of $(x-3)^9$.

b Hence, find the term in x^6 in the expansion of $-2x(x-3)^9$.

18 In the expansion of $\left(\frac{x^3}{3}+\frac{k}{x}\right)^{12}$, the constant is $\frac{112\,640}{27}$. Find the value of k.

19 Prove that:

$(2x-1)(x-3)-3(x-4)^2=-x^2+17x-45$.

20 **a** Prove that $\frac{x^2-x-6}{x+4}\cdot\frac{x^2-16}{x^2+2}=\frac{x^2-7x+12}{x}$.

b For what values of x does this mathematical statement **not** hold true?

Exam-style questions

21 **P1: a** Find the binomial expansion of $\left(1-\frac{x}{4}\right)^5$ in ascending powers of x. (3 marks)

b Using the first three terms from the above expansion, find an approximation for 0.975^5. (3 marks)

22 **P1:** The 15th term of an arithmetic series is 143 and the 31st term is 183.

a Find the first term and the common difference. (5 marks)

b Find the 100th term of the series. (2 marks)

23 **P2:** Angelina deposits \$3000 in a savings account on 1 January 2019, earning compound interest of 1.5% per year.

a Calculate how much interest (to the nearest dollar) Angelina would earn after 10 years if she leaves the money alone. (3 marks)

b In addition to the \$3000 deposited on January 1st 2019, Angelina deposits a further amount of \$1200 into the same account on an annual basis, beginning on 1st January 2020. Calculate the total amount of money in her account at the start of January 2030 (before she has deposited her money for that year). (4 marks)

24 **P2:** Brad deposits \$5500 in a savings account which earns 2.75% compound interest per year.

a Determine how much Brad's investment will be worth after 4 years. (3 marks)

b Calculate, to the nearest year, how long Brad must wait for the value of the investment to reach \$12 000. (5 marks)

25 **P2:** Find the coefficient of the term in x^5 in the binomial expansion of $(3+x)(4+2x)^8$. (4 marks)

26 **P1:** The coefficient of x^2 in the binomial expansion of $(1+3x)^n$ is 495.

Determine the value of n. (6 marks)

27 **P2:** Find the constant term in the expansion of $\left(x^3-\frac{2}{x}\right)^8$. (4 marks)

28 **P2: a** Find the binomial expansion of $\left(\frac{1}{2x}-x\right)^4$ in ascending powers of x. (3 marks)

b Hence, or otherwise, find the term independent of x in the binomial expansion of $(3-x)^3\left(\frac{1}{2x}-x\right)^4$. (4 marks)

29 **P2:** A convergent geometric series has sum to infinity of 120.

Find the 6th term in the series, given that the common ratio is 0.2. (5 marks)

30 **P1:** The second term in a geometric series is 180 and the sixth term is $\frac{20}{9}$.

Find the sum to infinity of the series. (7 marks)

31 **P2:** Find the value of $\sum_{n=0}^{n=15}\left(1.6^n-12n+1\right)$, giving your answer correct to 1 decimal place. (6 marks)

32 **P1:** A ball is dropped from a vertical height of 20 m.

Following each bounce, it rebounds to a vertical height of $\frac{5}{6}$ its previous height.

Assuming that the ball continues to bounce indefinitely, show that the maximum distance it can travel is 220 m. (5 marks)

33 **P1:** Prove the binomial coefficient identity $\binom{n}{k}=\binom{n-1}{k}+\binom{n-1}{k-1}$. (6 marks)

34 **P2:** Find the sum of all integers between 500 and 1400 (inclusive) that are not divisible by 7. (7 marks)

The Towers of Hanoi

Approaches to learning: Thinking skills, Communicating, Research
Exploration criteria: Mathematical communication (B), Personal engagement (C), Use of mathematics (E)
IB topic: Sequences

The problem

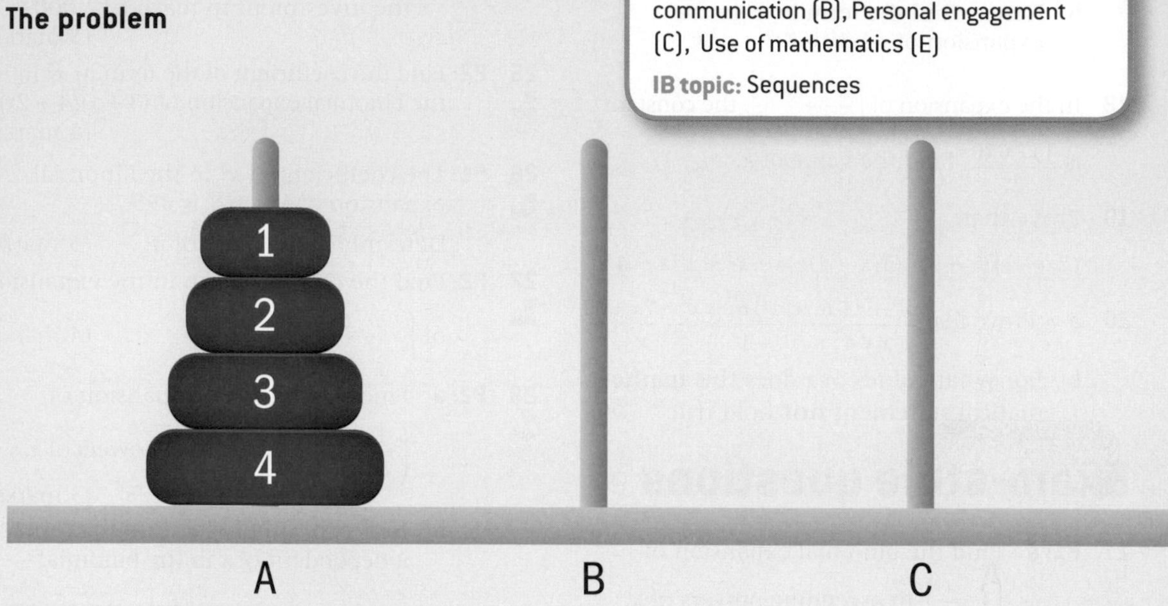

The aim of the **Towers of Hanoi problem** is to move all the disks from peg A to peg C following these rules:

1. Move only one disk at a time.
2. A larger disk may not be placed on top of a smaller disk.
3. All disks, except the one being moved, must be on a peg.

For 64 disks, what is the minimum number of moves needed to complete the problem?

Explore the problem

Use an online simulation to explore the Towers of Hanoi problem for three and four disks.

What is the minimum number of moves needed in each case?

Solving the problem for 64 disks would be very time consuming, so you need to look for a rule for n disks that you can then apply to the problem with 64 disks.

Try and test a rule

Assume the minimum number of moves follows an arithmetic sequence.

Use the minimum number of moves for three and four disks to predict the minimum number of moves for five disks.

Check your prediction using the simulator.

Does the minimum number of moves follow an arithmetic sequence?

Find more results

Use the simulator to write down the number of moves when $n = 1$ and $n = 2$.

Organize your results so far in a table.

Look for a pattern. If necessary, extend your table to more values of n.

Try a formula

Return to the problem with four disks.

Consider this image of a partial solution to the problem. The large disk on peg A has not yet been moved.

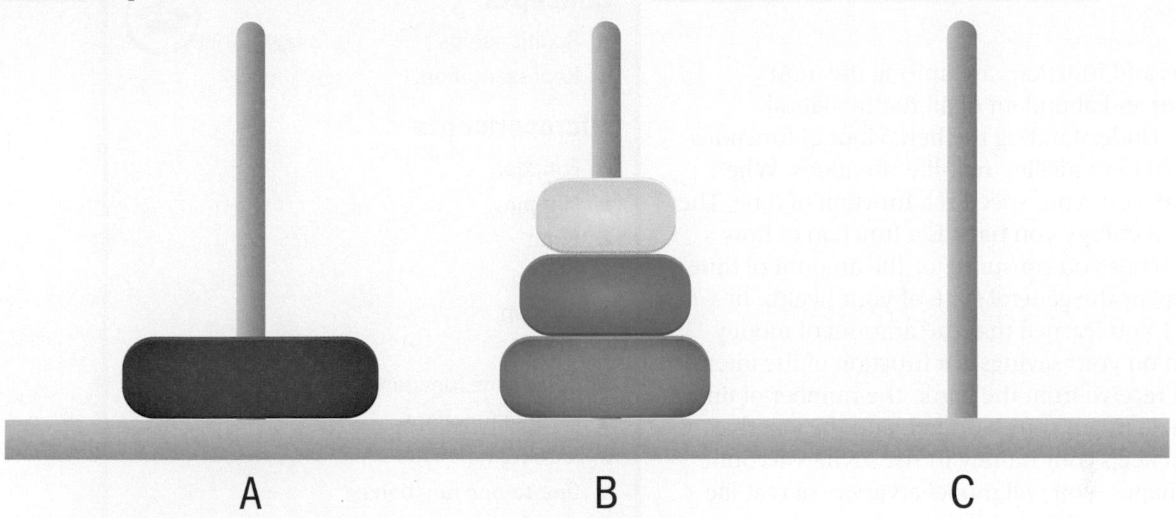

Consider your previous answers.

What is the minimum possible number of moves made so far?

How many moves would it then take to move the largest disk from peg A to peg C?

When the large disk is on peg C, how many moves would it then take to move the three smaller disks from peg B to peg C?

How many total moves are therefore needed to complete this puzzle?

Use your answers to these questions to write a formula for the minimum number of moves needed to complete this puzzle with n disks.

This is an example of a **recursive formula**. What does that mean?

How can you check if your recursive formula works?

What is the problem with a recursive formula?

Try another formula

You can also try to solve the problem by finding an **explicit formula** that does not depend on you already knowing the previous minimum number.

You already know that the relationship is not arithmetic.

How can you tell that the relationship is not geometric?

Look for a pattern for the minimum number of moves
in the table you constructed previously.

Hence write down a formula for the minimum number of moves in terms of n.

Use your explicit formula to solve the problem with 64 disks.

Extension

- What would a solution look like for four pegs? Does the problem become harder or easier?
- Research the "Bicolor" and "Magnetic" versions of the Towers of Hanoi puzzle.
- Can you find an explicit formula for other recursive formulae? (eg Fibonacci)

2 Representing relationships: introducing functions

Relations and functions are among the most important and abundant of all mathematical patterns. Understanding the behaviour of functions is essential to modelling real-life situations. When you drive a car, your speed is a function of time. The amount of energy you have is a function of how many calories you consume, or the amount of time you sleep, or the general state of your health. In chapter 1 you learned that the amount of money you earn on your savings is a function of the interest rate you receive from the bank, the number of times the interest rate is compounded, and the length of time you keep your money in the savings account. In this chapter you will model a variety of real-life problems using different forms of functions.

Concepts

- Relationships
- Representation

Microconcepts

- Function
- Mapping
- Input
- Output
- Domain
- Range
- Composite functions
- Inverse functions
- Identity functions
- One-to-one functions
- Self-inverses
- Variables

How far will a car drive on a fuel tank of fuel?

What kind of relationships exist between two quantities or variables?

Can shadows be modelled using functions?

One of the most important concepts in economics is supply and demand. Supply is the amount of goods available for people to purchase. Demand is the actual amount of goods people will buy at a given price. An example of a supply and demand relationship is shown in the graph.

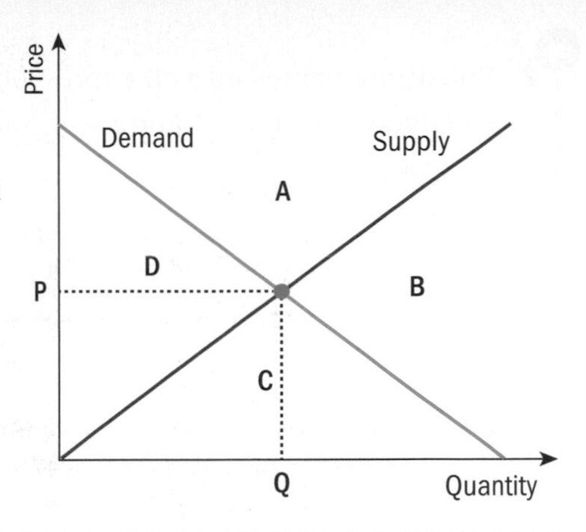

- How does the graph help you explain the relationship between the two variables "quantity" and "price"?
- Which variable would you label as "independent" and which one as "dependent"?
- How do you interpret the point where the two lines meet?
- How can you interpret the regions A, B, C and D in terms of the given variables?
- How can you express the relationship shown in the graph between quantity and price in other ways, for example, numerically or algebraically?

Developing inquiry skills

Imagine you work in a sports store and you need to decide on the price of a box of tennis balls. What kind of inquiry questions would you ask? For example, how many tennis balls are in each box? How can the value of one tennis ball be determined? Might the value of the box change over time? What further information do you need?

Think about the questions in this opening problem and answer any you can. As you work through the chapter, you will gain mathematical knowledge and skills that will help you to answer them all.

Before you start

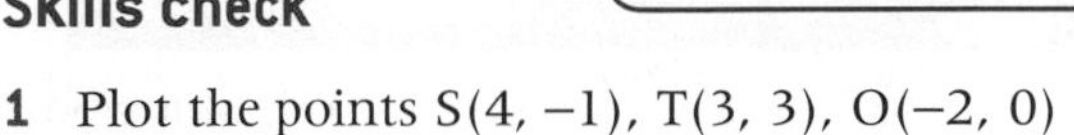

You should know how to:

1 Plot coordinates.
For example, plot the points A(1, 4), B(−2, 4), C(0, −3) and D(2, −3) on a coordinate plane.

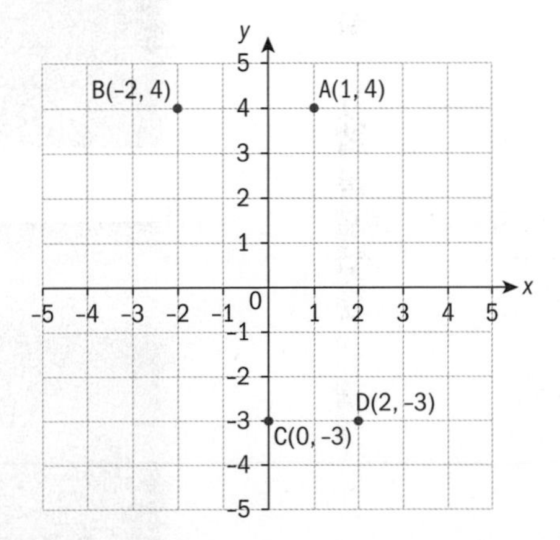

Skills check

1 Plot the points S(4, −1), T(3, 3), O(−2, 0) and P(−1, −2) on a coordinate plane.

2 List the coordinates of each point on the grid below:

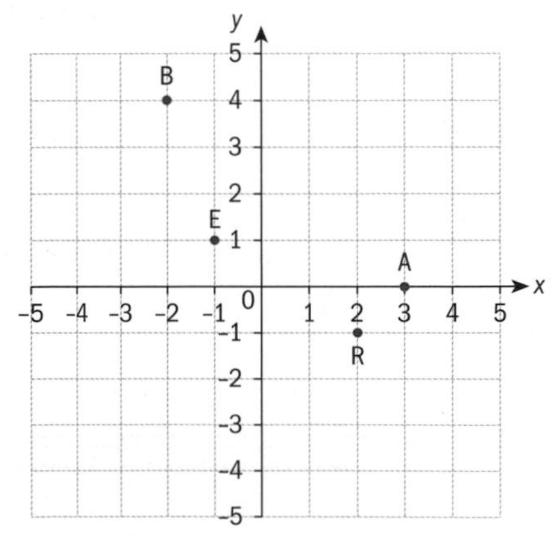

Continued on next page

2 Substitute values into an expression.

eg Given $x = -1, y = 3$ and $z = \frac{1}{2}$, find the values of:

a $2x^2 - 5y$

$$\begin{aligned} 2x^2 - 5y &= 2(-1)^2 - 5(3) \\ &= 2(1) - 5(3) \\ &= 2 - 15 \\ &= -13 \end{aligned}$$

b $xy - y$

$$\begin{aligned} xy - y &= (-1)(3) - 3 \\ &= -3 - 3 \\ &= -6 \end{aligned}$$

c $8z^2$

$$8z^2 = 8\left(\frac{1}{2}\right)^2 = 8\left(\frac{1}{4}\right) = \frac{8}{4} = 2$$

3 Solve linear equations.

eg Solve $2(x - 3) = 4x + 10$

$$2(x - 3) = 4x + 10$$

$$2x - 6 = 4x + 10$$

$$-2x = 16$$

$$x = -8$$

4 Use your GDC to graph an equation.

eg Graph $y = -3x - 4$

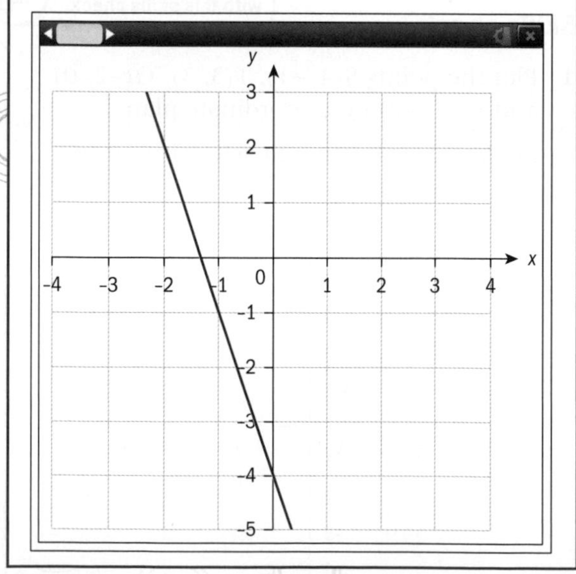

3 Given $x = 2, y = -3$ and $z = -\frac{1}{2}$, find the values of:

a $4x - 3y$

b $x^2 - y^2$

c $x + y + z$

d $-6z^2$

4 Solve the following equations:

a $-2x + 1 = 6x - 15$

b $-2(x + 4) = 2(x - 5)$

c $-3(x + 2) + 4(x - 1) = 3$

d $(x - 2)^2 = (x + 4)^2$

5 Use your GDC to graph the following equations:

a $y = 2x + 2$

b $y = x^2 + 1$

c $y = 2\sqrt{x} - 1$

d $y = -x^2 + 5x - 6$

2.1 What is a function?

A mathematical **relation** is a relationship between any set of ordered pairs and can be expressed as a mapping and graph.

Investigation 1

Various relations are sorted into two columns below. Compare the relations in each column.

	Column 1	Column 2
a	$\{(3,2), (3,5), (4,1), (4,2)\}$	$\{(3,6), (2,5), (4,1), (7,2)\}$
b	$\{(1,2), (3,2), (1,5), (3,5)\}$	$\{(1,2), (3,5), (2,2), (4,5)\}$
c	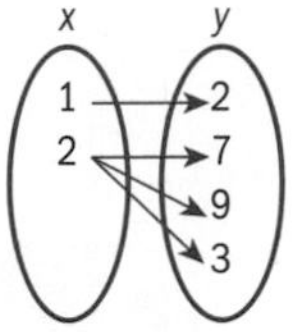 Mapping $x \to y$: 1 → 2; 2 → 7, 9, 3	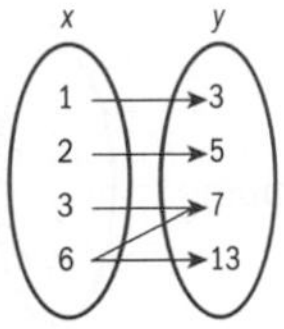 Mapping $x \to y$: 1 → 3; 2 → 5; 3 → 7; 6 → 13
d	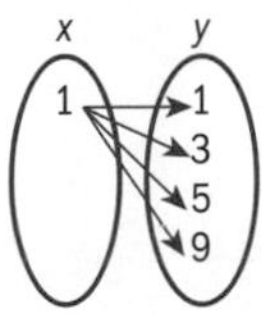 Mapping $x \to y$: 1 → 1, 3, 5, 9	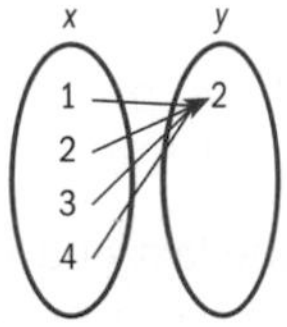 Mapping $x \to y$: 1, 2, 3, 4 → 2
e	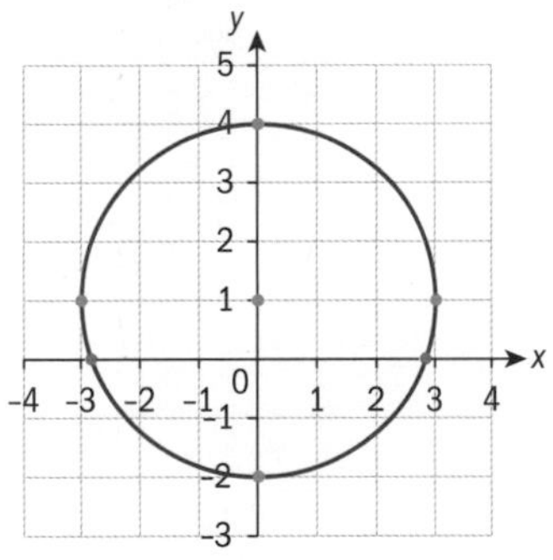 Graph of a circle through (0, 4), (−3, 1), (3, 1), (0, −2)	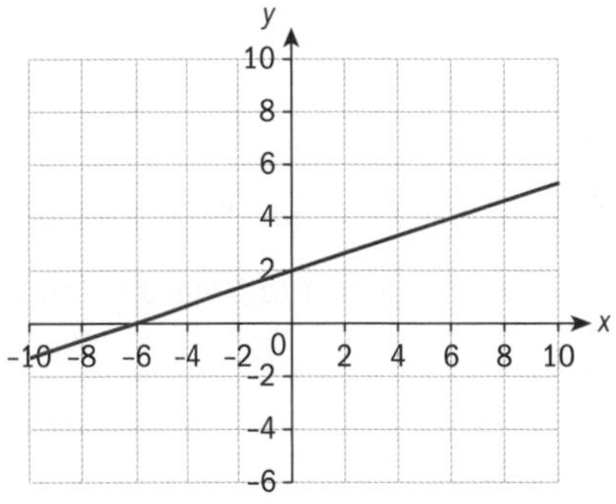 Graph of a straight line crossing the y-axis at 2
f	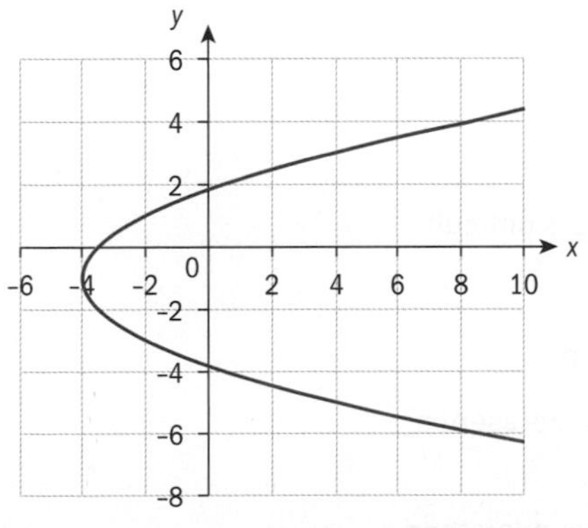 Graph of a sideways parabola with vertex at (−4, 0)	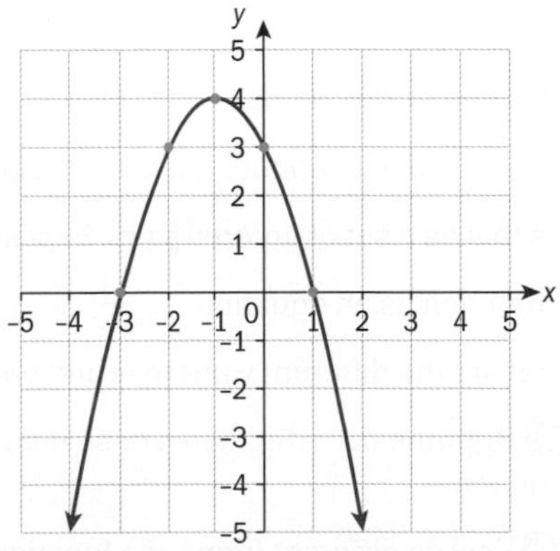 Graph of a downward parabola with vertex at (−1, 4), through (−3, 0), (1, 0) and (0, 3)

Continued on next page

Functions

1 Choose some input values in each column and state their corresponding outputs.

2 What difference do you notice between the two columns?

The relations in the second column are called functions.

3 **Conceptual** What is a function?

4 **Conceptual** How would you describe a function using mappings?

5 Give three new examples (various form) of functions.

6 **Factual** What are functions used for?

A **function** is a special type of relation. A function can be a one-to-one mapping

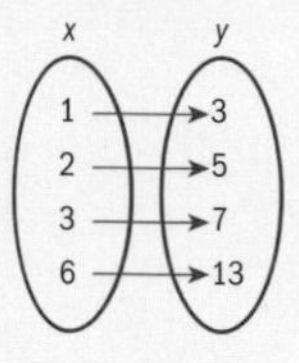

or a many-to-one mapping

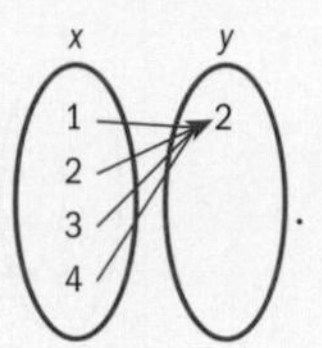

.

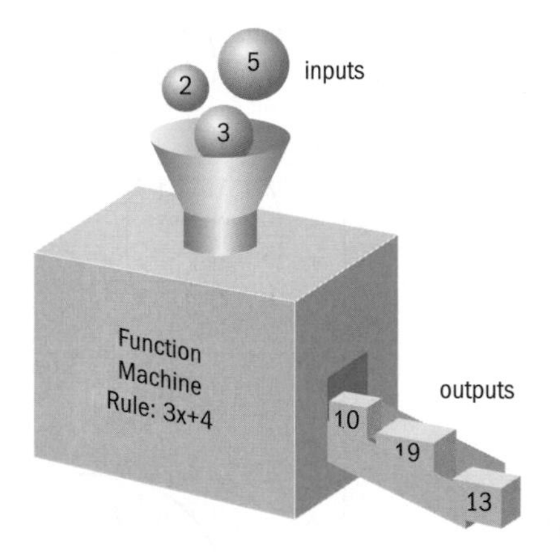

Think of a function like a machine. If we put in one value for x, we want one value for y to come out each time.

Think back to compound interest in the last chapter. What would happen if two investments of the same amount, with the same rate and same time frame gave two different returns?

Investigation 2

Functions can be expressed in words. For example:

Every number from 1 to 6 maps to three times itself.

1 Express this function as:

  i a set of ordered pairs ii a table of values

  iii a mapping diagram

2 For each of the representations above, describe how you can tell whether they are a function or not.

Here is another function in words:

Every real number maps to its square.

3 Try to express this as a set of ordered pairs. Explain why this is difficult.

4 Express this function as an equation.

5 **Factual** What are the different ways to represent a function?

6 **Conceptual** Hypothesize why there are so many ways to represent a relation or function.

7 **Conceptual** How do different forms of a function help you understand a real-life problem?

Example 1

Determine whether each relation below is a function or not.

a $\{(6,12), (8,16), (10,20), (12,24)\}$

b $\{(5,20), (5,25), (10,40), (10,100)\}$

c

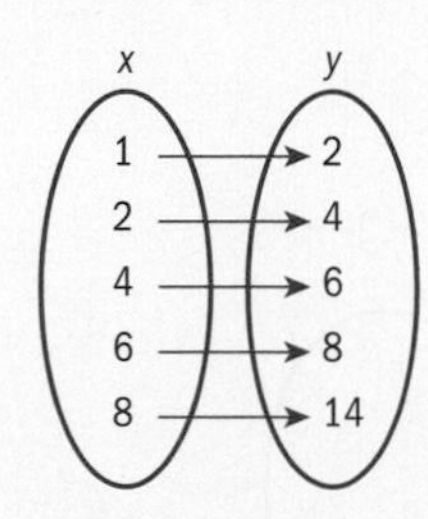

d

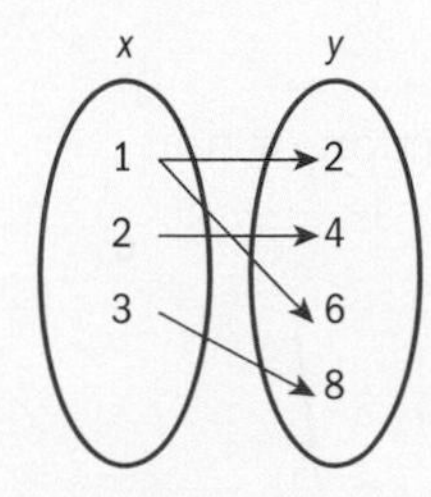

e $y = -3x + 7$

f $y = \sqrt{x}$

a Function	When a relation is given as a set of ordered pairs check to see whether the relation is one-to-one or many-to-one.
b Not a function	This is not a function as this is a one-to-many relation.
c Function	The relation is one-to-one so this is a function.
d Not a function	This is a one-to-many so this is not a function.
e Function	This is a function because every value of x maps to a different value of y.
f Function	This is a function because every value of x maps to a different value of y.

International-mindedness

One of the first mathematicians to study the concept of functions was Frenchman Nicole Oresme in the 14th century. Below is a page from Oresme's *Livre du ciel et du monde*, 1377.

Investigation 3

Look at the graphs below:

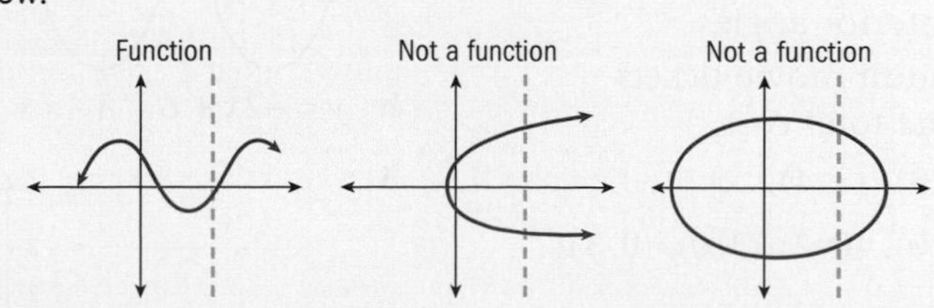

Continued on next page

Functions

1 Hypothesize why the second and third graphs are not functions.

2 We use the **vertical line test** to determine whether a graph is of a function. Using the diagrams above and what you know about the definition of a function, explain the vertical line test.

3 **Conceptual** How does the vertical line test help to identify functions?

Example 2

Determine whether each relation below is a function or not:

a

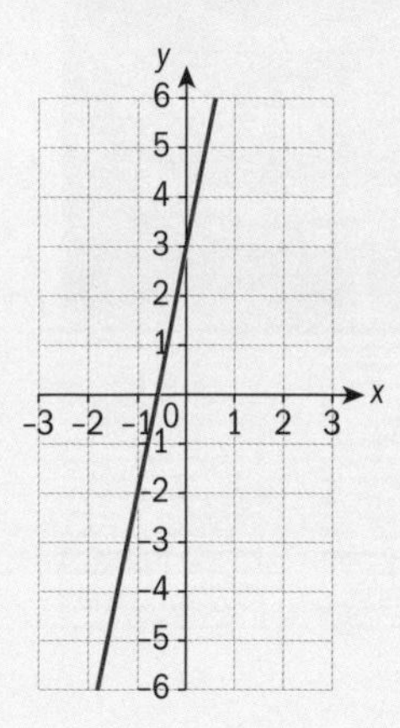

b

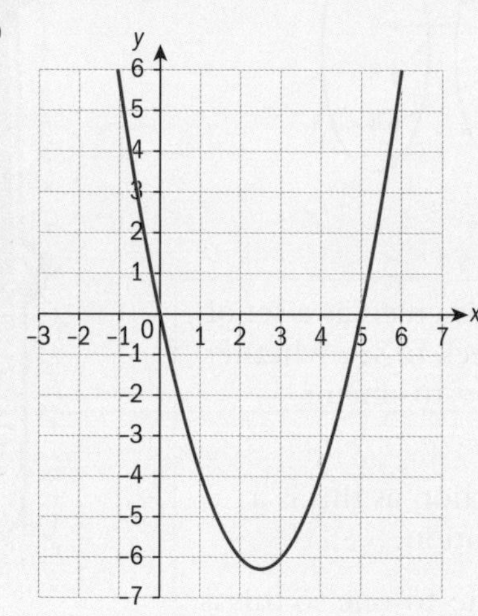

c

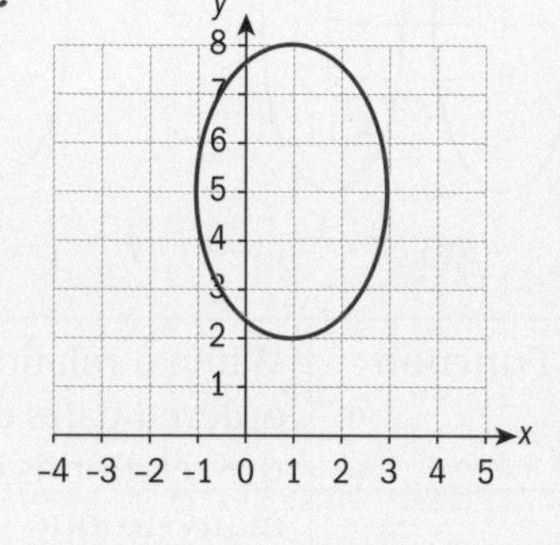

a Function	It passes the vertical line test.
b Function	It passes the vertical line test.
c Not a function	It fails the vertical line test.

Exercise 2A

1 Determine whether each relation below is a function or not. If it is not a function, state the reason why.

a The number of identical marbles and the total mass of the marbles.

b The number of sides in a polygon and the sum of the interior angles.

c The number of adult movie tickets purchased and the total cost.

d $\{(0,1), (1,2), (2,3), (3,4), \ldots\}$

e $\{(1,0), (0,1), (2,0), (0,2), (3,0), (0,3)\}$

f

x	2	3	4	6	7	9
y	1	1	1	1	1	1

g

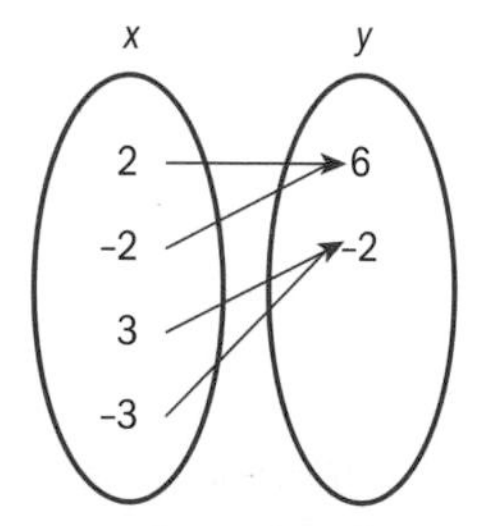

h $y = -2x + 6$ **i** $x = 3$ **j** $y = x^2$

k

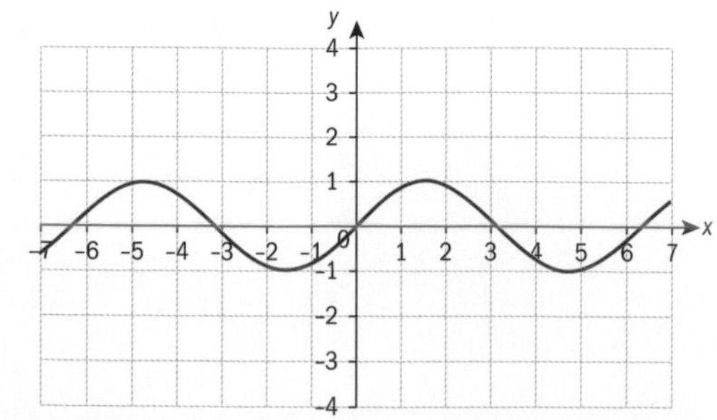

l

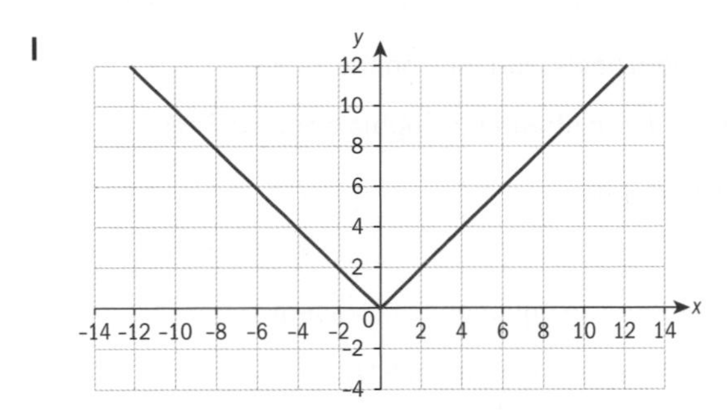

m

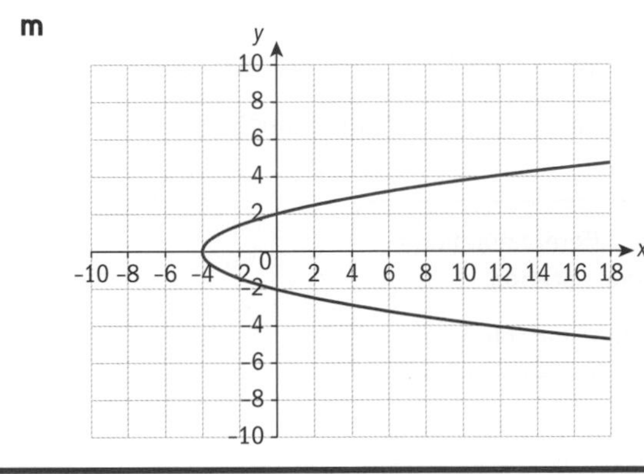

2 For each of the relation types below, give an example that is **not** a function:

a table of values

b mapping diagram

c equation

d graph

3 Which of the following statements is true? Justify your choice.

All functions are relations, but not all relations are functions.

All relations are functions, but not all functions are relations.

Developing inquiry skills

Let's return to the chapter opener about economics and supply and demand.

Look at the graph on the right. Is this relation a function? Explain how you know. What does this mean in terms of this real-life situation?

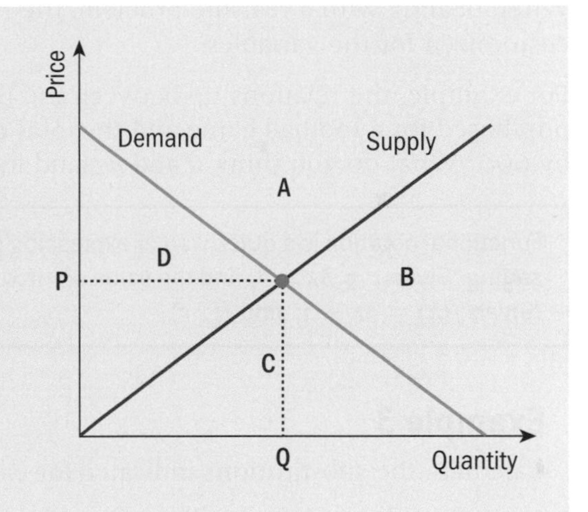

2.2 Functional notation

Investigation 4

Two neighbours, Jacob and Sophia, are each draining their backyard swimming pool in preparation for winter. The table below shows the water level (depth) at various times.

Time (hours)	0	5	10	14
Depth of water in Jacob's pool (m)	1.4	0.9	0.4	0
Depth of water in Sophia's pool (m)	2	1	0	0

Continued on next page

1 On the same set of axes, graph the depths of water in both pools with time on the x-axis.

2 Label the depth of water in Jacob's pool as $J(t)$ and the depth of water in Sophia's pool as $S(t)$.

3 Make as many observations as possible about the two graphs.

Are the graphs functions? Explain your answer.

4 Jacob drained his pool at a rate of 0.1 metres per hour. Write an equation to represent $J(t)$.

Sophia drained her pool at a rate of $\frac{1}{5}$ metres per hour. Write an equation for $S(t)$.

5 Explain what the following notations mean:

i $J(t) = S(t)$ **ii** $J(2)$ **iii** $S(8)$ **iv** $J(t) = 1$ **v** $S(t) > J(t)$

6 **Conceptual** What is the purpose of having different notation for different functions?

When writing an expression for a function, there are several different notations. For example, $f(x)$, $g(x)$, $f: \rightarrow$. These are all called **functional notation**.

$f(x)$, said as "f of x", represents a function f with independent variable x.

We use functional notation instead of just "$y =$" to be able to distinguish between different functions, in the same way people have different names.

When dealing with a real-life problem, the functional notation can be customized for the variables.

For example, the relationship between the number of tickets purchased for a football game and the total cost, could be represented by $C(n)$. What do you think C and n stand for?

Functional notation is a quick way of expressing substitution. Rather than saying "Given $y = 3x - 5$, find the value of y if $x = 2$", we can simply state: "Given $f(x) = 3x - 5$, find $f(2)$".

Example 3

Calculate the substitutions indicated for each given function.

a $g(x) = \{(2, 3), (4, 5), (6, 7), (8, 9), (10, 11)\}$, $g(8)$

b $f(x) = -2x - 1$, $f(-3)$

c $f{:}x \rightarrow 3x + 7$, $f(1)$

d $h(x) = 3$, $h(-1)$

e $f(4)$ for the graph.

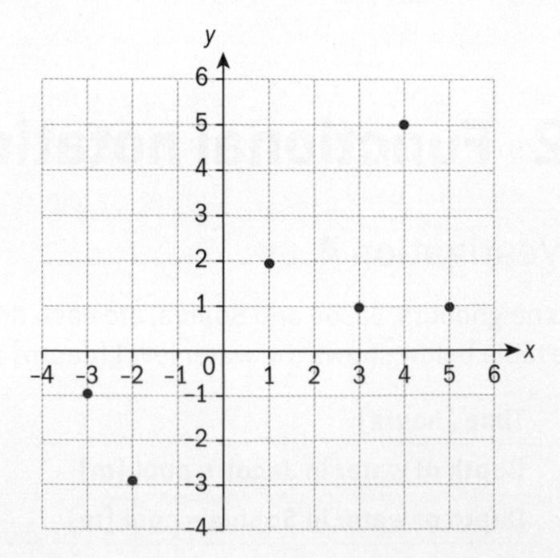

a 9	When the function is given as a set of ordered pairs, find the point with the given x-value and state its corresponding y-value. $g(8)$ means "the y-coordinate of the point that has an x-coordinate of 8". In the example, the point is (8, 9), so $g(8) = 9$.
b $f(x) = -2x - 1, f(-3)$ $f(-3) = -2(-3) - 1$ $f(-3) = 6 - 1$ $f(-3) = 5$ This represents the point $(-3, 5)$ for this function.	Don't forget to use parentheses when substituting!
c $f: x \rightarrow 3x + 7, f(1)$ $f: 1 \rightarrow 3(1) + 7$ $f: 1 \rightarrow 3 + 7$ $f: 1 \rightarrow 10$	
d 3	Since any function in the form of $y = n$ is a constant function or a horizontal line, the answer will also be n for any x-value. Therefore $h(-1) = 3$.
e 5	Similar to a set of ordered pairs, find the given x-value and write down its corresponding y-value. For $f(4)$, the point on the graph is (4, 5), so $f(4) = 5$.

Example 4

If $f(x) = -3x^2 - 1$, find:

a $f(-1)$ **b** $f(0)$ **c** $f(100)$ **d** $f(a)$ **e** $f(x + 1)$

a $f(-1) = -3(-1)^2 - 1$ $f(-1) = -3(1) - 1$ $f(-1) = -3 - 1$ $f(-1) = -4$	Be sure to follow the order of operations!
b $f(0) = -3(0)^2 - 1$ $f(0) = -3(0) - 1$ $f(0) = 0 - 1$ $f(0) = -1$	
c $f(100) = -3(100)^2 - 1$ $f(100) = -3(10\,000) - 1$ $f(100) = -30\,000 - 1$ $f(100) = -30\,001$	

Continued on next page

d $f(a) = -3(a)^2 - 1$ $f(a) = -3a^2 - 1$	
e $f(x+1) = -3(x+1)^2 - 1$ $f(x+1) = -3(x^2 + 2x + 1) - 1$ $f(x+1) = -3x^2 - 6x - 3 - 1$ $f(x+1) = -3x^2 - 6x - 4$	Remember: $(x+a)^2 \neq x^2 + a^2$ $(x+a)^2 = (x+a)(x+a)$ $(x+a)^2 = x^2 + 2ax + a^2$

When using functional notation, there are two types of questions that can be asked.

The first, as seen in the previous examples, is $f(n) = ?$. This means "to substitute n for x and simplify to find $f(n)$ or y".

The second, $f(?) = n$, is the opposite. This means to "find the x-value for the given y-value of n".

TOK

What is the relationship between real-life problems and mathematical models?

Example 5

The formula to calculate the volume of gas left in a car's tank in litres, after travelling d kilometres, is $V(d) = -0.115d + 60$.

a Explain what $V(250)$ means in the context of the question and calculate its value.

b Explain what $V(d) = 10$ means in the context of the question and calculate its value.

c What values of d do not make sense for this situation?

a $V(250)$ means "the volume of gas in the tank after travelling 250 km". $V(250) = -0.115(250) + 60$ $V(250) = 31.25$	
b $V(d) = 10$ means "how far a car has travelled when there are 10 litres left in the tank". $V(d) = -0.115d + 60 = 10$ $-0.115d = -50$ $d = 434.7826086...$ $d \approx 435$ km	Note: Remember to round to three significant figures unless the question otherwise specifies and to use an approximation sign.
c d cannot be negative as you cannot drive a negative distance. Also, d cannot be greater than the distance a car can travel on a full tank of gas. $0 < d \leq n$, where n is the distance a vehicle can drive on a full tank of gas.	

Example 6

Nikita is planning a sport banquet. He must pay \$320 to rent the room and \$20 per person for the dinner.

a Express the total cost, C, as a function of the number of people, n, who attend the banquet.

b What notation could be used to find the cost of the banquet if 125 people attend? Calculate this value.

c What notation could be used to find the number of people who can attend the banquet if Nikita's budget for the banquet is \$1250? Calculate this value.

d What values of n do not make sense for this situation?

a $C(n) = 20n + 320$	
b $C(125) = 20(125) + 320$ $C(125) = 2820$ \$2820	
c $C(n) = 1250$ $20n + 320 = 1250$ $20n = 930$ $n = 46.5$ 46 people	Since rounding up will put the total over \$1250, we round down to 46 people.
d $0 \le n \le a$ where a is the capacity of the room	n cannot be negative and cannot be larger than the number of people the room can hold or the number of people who can be covered by Nikita's budget.

Exercise 2B

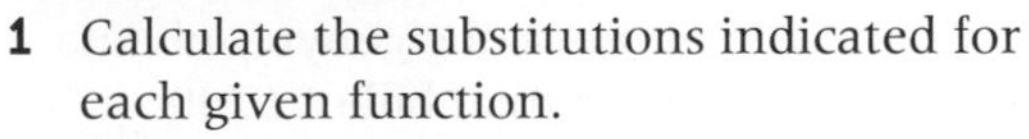

1 Calculate the substitutions indicated for each given function.

a $g(x) = -x^2 + 2,\ g(-4)$

b $f: x \to 5x - 1,\ f(-9)$

c $C(n) = 20n + 250,\ C(100)$

d $h(x) = -4,\ h(5)$

e $f(2)$ for the mapping diagram below

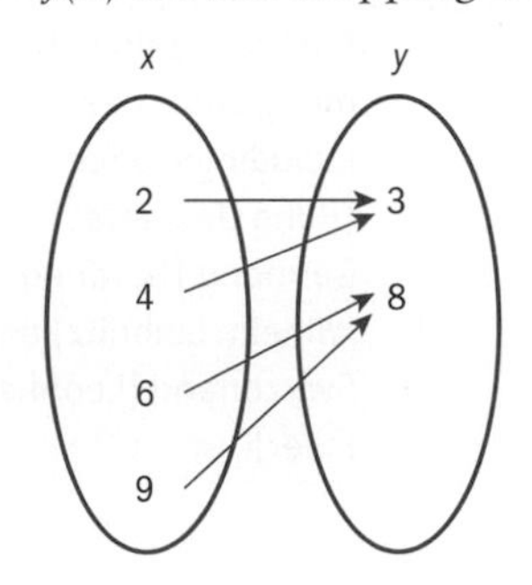

f $f(-3)$ for the graph below

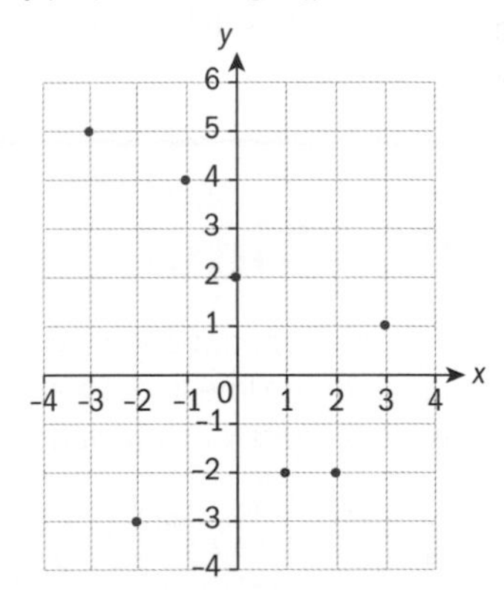

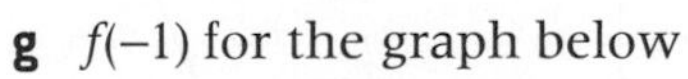

g $f(-1)$ for the graph below

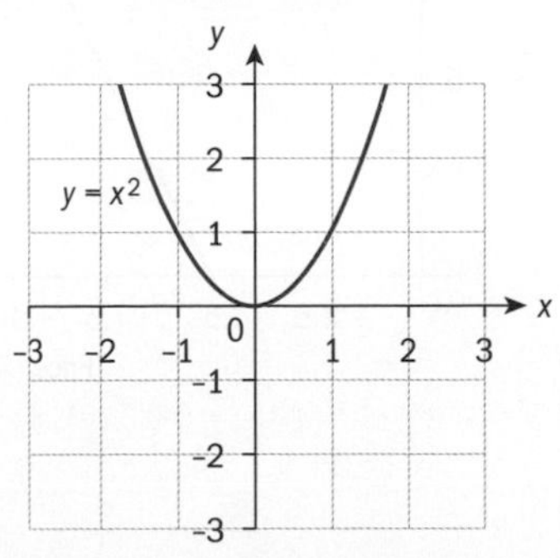

2 If $f(x) = -3x^2 - 1$, $g(x) = -4x + 7$ and $h(x) = 6$, find:

a $f(-3)$ b $g(15)$
c $f(1) + g(-1)$ d $h(0)$
e $f(x - 2)$ f $g(n)$
g $f(1) \times h(1)$ h $f(x + 1) \times g(x - 2)$

3 The journey of a car travelling towards Perth, Australia, can be expressed by the relation $d = -75t + 275$, where d is the distance from Perth in km and t is driving time in hours.

a Is this relation a function? Justify your answer.

b Rewrite the equation in functional notation.

c At the start of the journey, how far was the car from Perth?

d What values of t do not make sense for this situation?

4 To convert a temperature from Celsius to Fahrenheit, we use the formula:

$$F(C) = \frac{9}{5}C + 32 .$$

a Is this formula a function? Justify your answer.

b Describe what $F(17)$ is asking and calculate the value.

c Describe what $F(C) = 100$ is asking and calculate the value of C.

d Use the formula to calculate the temperature in degrees Fahrenheit at which water freezes.

e Use the formula to calculate the temperature in degrees Fahrenheit at which water boils.

f The average body temperature of a dog is 38.75°C. Convert this to degrees Fahrenheit.

g Cookies are typically baked at 350°F. Convert this to degrees Celsius.

5 A British telecommunications company offers the following roaming data package to its customers: a flat fee of £25 plus £10 per gigabyte of data

a Express the total cost, C, as a function of the number of gigabytes of data (g).

b What values of g do not make sense in this context?

c State the notation that could be used to find the roaming cost for a trip where 14 gb of data is used? Calculate this value.

d State the notation that could be used to find the number of gigabytes of data one can use for a total bill of £100. Calculate this value.

Developing inquiry skills

Let's return to the chapter opener about economics and supply and demand. The equation of the line in the graph on the right is $y = -10 + 2x$. Rewrite this in functional notation and define the variables you choose.

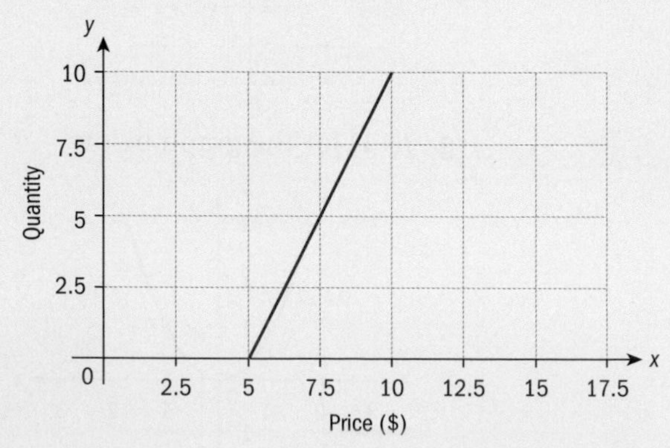

International-mindedness

The development of functions bridged many countries including France (Rene Descartes), Germany (Gottfried Wilhelm Leibnitz) and Switzerland (Leonhard Euler).

2.3 Drawing graphs of functions

TOK

Does a graph without labels have meaning?

"Draw" and "sketch" are two of the IB command terms you need to know.

Draw: Represent by means of a labelled, accurate diagram or graph, using a pencil. A ruler (straight edge) should be used for straight lines. Diagrams should be drawn to scale. Graphs should have points correctly plotted (if appropriate) and joined in a straight line or smooth curve.

Sketch: Represent by means of a diagram or graph (labelled as appropriate). The sketch should give a general idea of the required shape or relationship, and should include relevant features.

Example 7

Using the information below, draw the graph for each line:

a x-intercept at $(-3, 0)$ and y-intercept at $(0, 5)$

b y-intercept at $(0, -1)$ and slope of $\frac{1}{2}$

c $y = -x + 1$

d $C(n) = 40 + 5n$

a

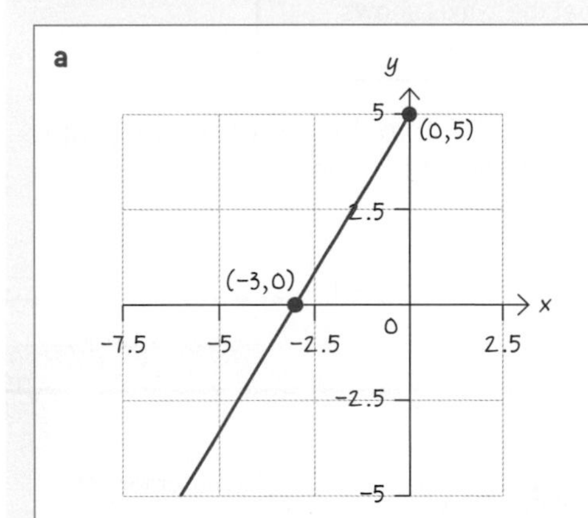

c

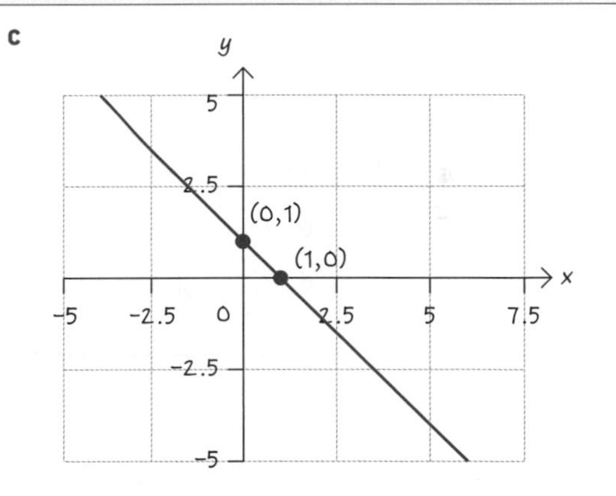

b

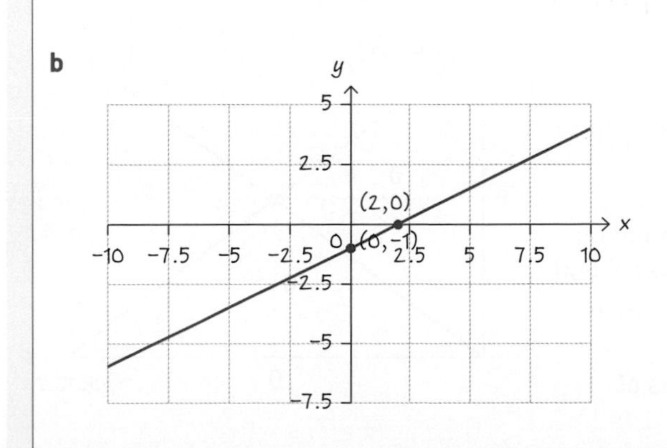

d

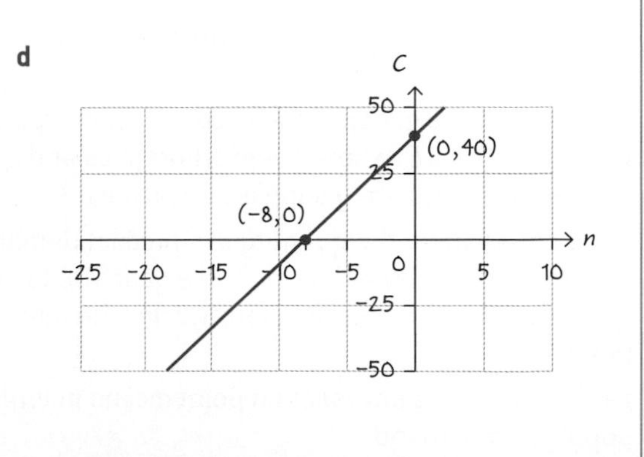

Example 8

Using your GDC, sketch the following graphs:

a $f(x) = -x^2 + 4$ **b** $y = \sqrt{x-3}$

a

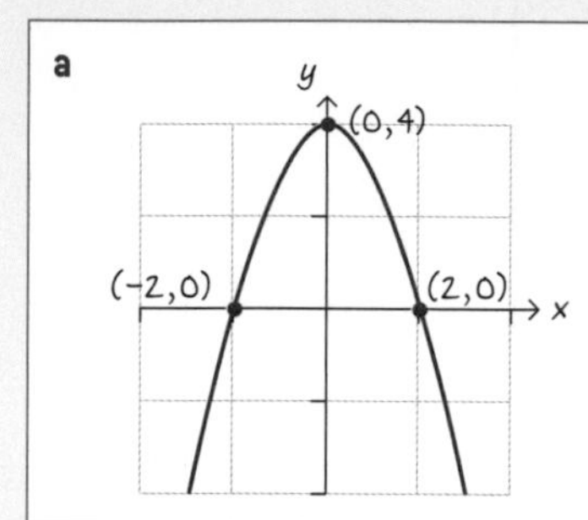

b

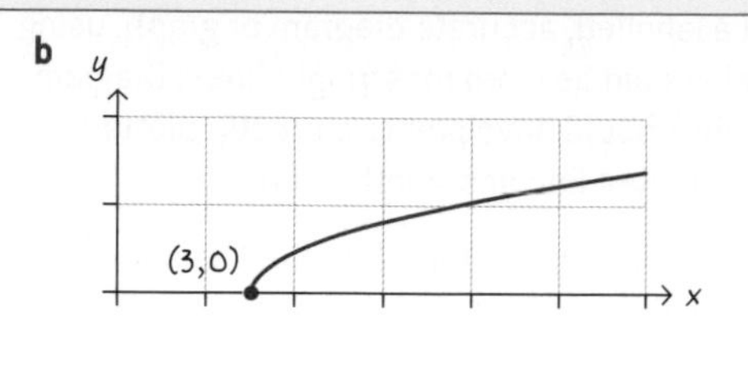

When sketching the graph of a function, be sure to include the relevant features of the graph, including the x- and y-intercepts.

Exercise 2C

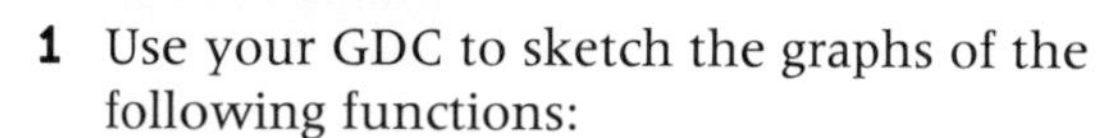

1 Use your GDC to sketch the graphs of the following functions:

a $f(x) = -0.01x^2 + 0.5x + 2.56$

b $g : x \to 0.44\sqrt{3.2x - 4.1}$

c $y = 2\sin(x + 3) + 2$

HINT

For part **1c**, be sure your calculator is in radians, and set the window of your GDC so that the x-axis shows from -2π to 2π.

2 Given $f(x) = x^2 + 1$, $g(x) = -3x + 2$ and $h(x) = \sqrt{x}$, use your GDC to sketch the graphs of the following:

a $f(x) + g(x)$ **b** $f(x) - h(x)$

c $f(x) \times g(x)$ **d** $f(x) + g(x) + h(x)$

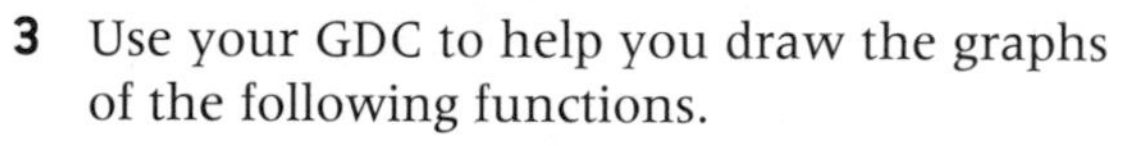

3 Use your GDC to help you draw the graphs of the following functions.

a $y = 0.2x^2 + 1.3x - 4.5$

b $f(x) = 3 \times 2^{0.5x+3} - 1$

Developing inquiry skills

Let's return to our opening problem about supply and demand.

The equation we generated in section 2.2 for supply was $Q(P) = -10 + 2P$, where P represents price and Q represents the quantity supplies by producers.

If the equation for the quantity of a product demanded by consumers is $D(P) = 30 - 3P$, use your GDC to sketch both graphs on the same set of axes and find the intersection point.

Explain what this intersection point means in terms of supply and demand.

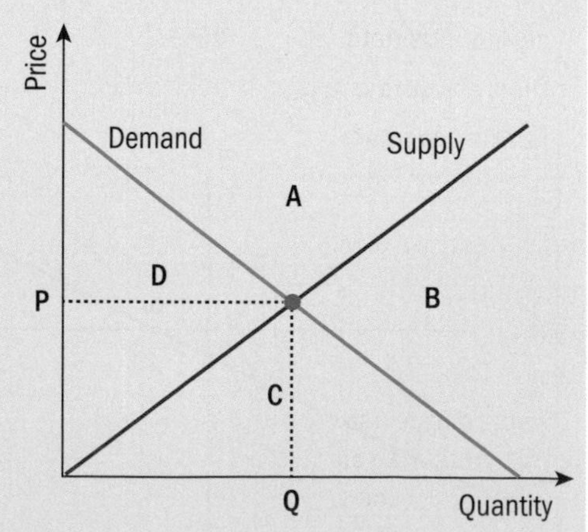

2.4 The domain and range of a function

In section 2.2, we looked at what values of the variables do not make sense for a given situation. We will now examine this idea in more depth.

Investigation 5

Over 100 years ago, crickets were used to determine the temperature outside. Farmers would count the number of chirps in 25 seconds. They would then divide the number by 3 and add 4. This would give the temperature outside in degrees Celsius.

1 Identify the independent and dependent variables in this situation.

2 Write an equation using functional notation to represent this situation and define the variables.

3 Copy and complete the table below:

Number of chirps							
Temperature (°C)	−10°C	0°C	3°C	4°C	10°C	20°C	30°C

4 Which values from the table above do not make sense?

5 Is there a limit to the °C that can be calculated using this formula?

6 Hypothesize the maximum number of times a cricket could chirp in 25 seconds.

7 Is there a limit to the number of chirps that can work for this function?

In this situation, the number of chirps is the "domain" and the temperature is the "range".

8 **Factual** What do "domain" and "range" mean?

9 **Conceptual** How do the domain and range vary depending on the problem being modelled?

The **domain** of a function is all possible values of the independent variable

The **range** of a function is all possible values of the dependent variable

There are two different types of data we may encounter: discrete data and continuous data.

Discrete data is graphed as separate, distinct points.

Continuous data is graphed as lines or curves.

For continuous data, there are different ways to express the domain and range of a function: as an inequality or member of a set, and interval notation.

Inequalities uses the symbols $>$, $\geq$, $<$, $\leq$.

Numbers may also be expressed as part of a **set**, such the set of real numbers or the set of positive integers.

Interval notation uses brackets and prentheses:

][or ( ) mean the values are not included.

[ ] means the values are included.

TOK

Around the world you will often encounter different words for the same object, like "trapezium" and "trapezoid" or "root" and "surd".

Sometimes more than one type of symbol might have the same meaning, such as "interval" and "set notation".

To what extent does the language we use shape the way we think?

Example	Inequality of number set	Interval notation
x is less than 3	$x < 3$	$]-\infty, 3[$ or $(-\infty, 3)$
x is less than or equal to 3	$x \leq 3$	$]-\infty, 3]$ or $(-\infty, 3]$
x is greater than −4	$x > -4$	$]-4, \infty[$ or $(4, \infty)$
x is greater than or equal to −4	$x \geq -4$	$[-4, \infty[$ or $[-4, \infty)$
x is between −1 and 5	$-1 < x < 5$	$]-1, 5[$ or $(-1, 5)$
x is between 6 and 9, inclusive	$6 \leq x \leq 9$	$[6, 9]$
x is all real numbers	$x \in \mathbb{R}$	$]-\infty, \infty[$ or $(-\infty, \infty)$

For interval notation, we usually prefer to use brackets [ ] over parentheses ( ), because a domain or range in this form (a, b) could be mistaken for a point.

Example 9

State the following descriptions in both inequality (or set) and interval notation:

a x is from −3 and 8, inclusive
b a is less than negative six
c y is greater than or equal to 9
d x is between −2 and 14
e n is all real values
f x is all integer values

a $-3 \leq x \leq 8$
$[-3, 8]$

b $a < -6$
$]-\infty, -6[$ or $(-\infty, -6)$

c $y \geq 9$
$[9, \infty[$ or $[9, \infty)$

d $-2 < x < 14$
$]-2, 14[$ or $(-2, 14)$

e $n \in \mathbb{R}$
$]-\infty, \infty[$or $(-\infty, \infty)$

f $x \in \mathbb{Z}$
$]\ldots -2, -1, 0, 1, 2, \ldots[$ or $(\ldots -2, -1, 0, 1, 2, \ldots)$

This is an example of discrete data.

Let's look at how to state the domain and range of functions in various forms.

Type of function	Example	Domain	Range
a Tables of values	t: 2, 5, 7, 9; s: 3, 1, 5, 5	Since this is discrete data, list the x-values in order from smallest to largest for either set or interval notation. $\{2, 5, 7, 9\}$	There is no need to write a repeating value more than once. $\{1, 3, 5\}$
b Mapping diagram	x: -3, -2, 5, 7; y: 4, 6, 8	$\{-3, -2, 5, 7\}$	$\{4, 6, 8\}$

c	Set of ordered pairs	$\{(-3, 5), (-1, 9), (3, 5)\}$	$\{-3, -1, 3\}$	$\{5, 9\}$
d	Equation	$y = 3x - 6$	Since there is no restriction on the value of x or y that we can substitute into this equation, all real values are possible. $x \in \mathbb{R}$ $]-\infty, \infty[$ or $(-\infty, \infty)$	$y \in \mathbb{R}$ $]-\infty, \infty[$ or $(-\infty, \infty)$
e	Discrete graph	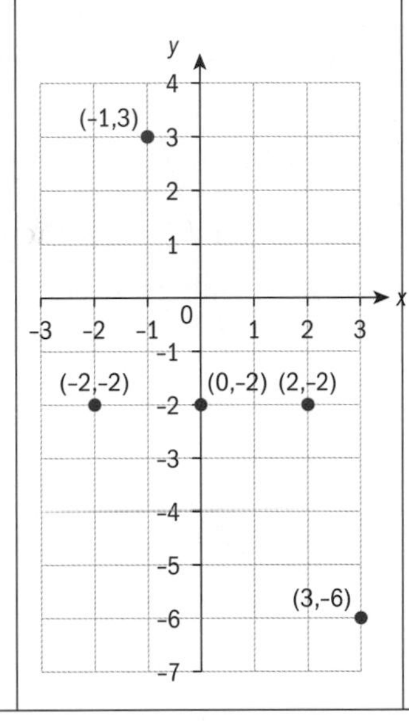	$\{-2, -1, 0, 2, 3\}$	$\{-6, -2, 3\}$

Investigation 6

Let's look at the graphs of functions of continuous data.

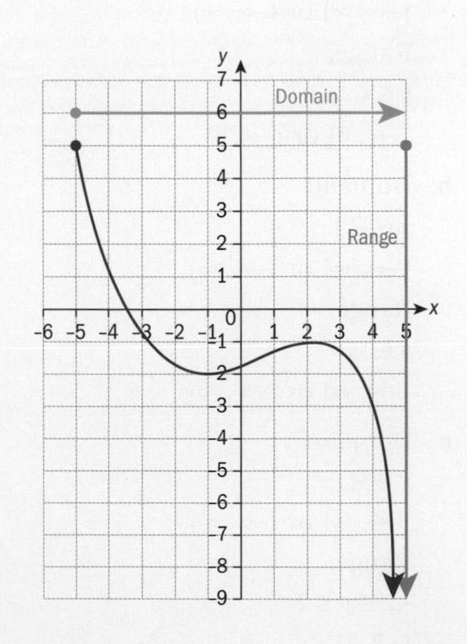

1. Look at the left-hand side of the graph. What is the smallest x-value?
2. Look at the right-hand side of the graph. What is the largest x-value?

These values form the domain.

3. Write the domain of this function as both an inequality and in interval notation.
4. Look at bottom of the graph. What is the smallest y-value?
5. Look at the top of the graph. What is the largest y-value?

These values form the range.

6. Write the range of this function as both an inequality and in interval notation.
7. How does a graph help you find the domain and range?

Functions

Example 10

State the domain and range for the following functions:

a

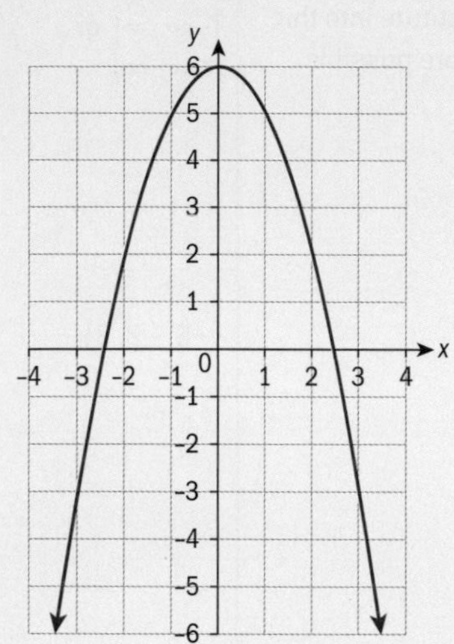

b

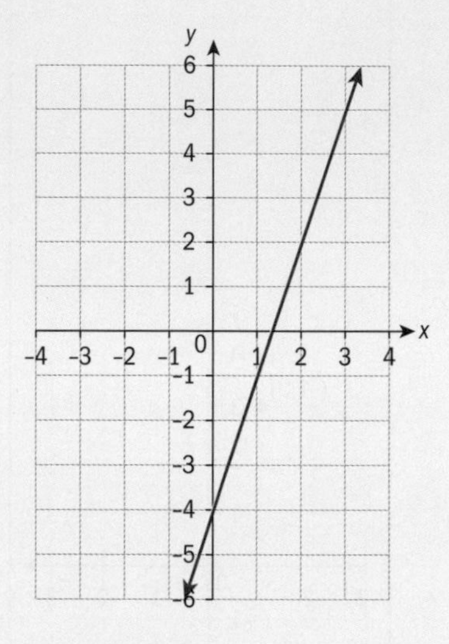

c

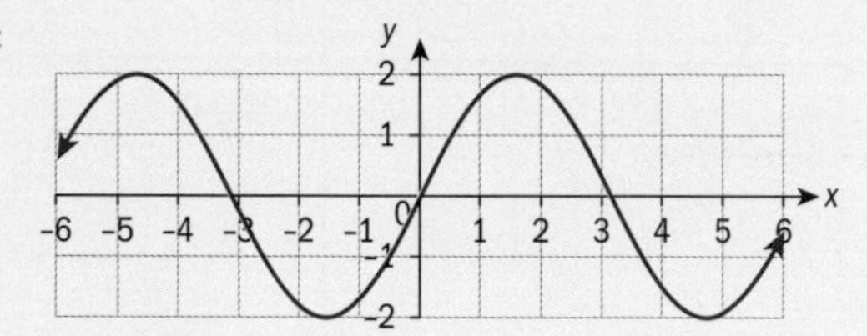

d

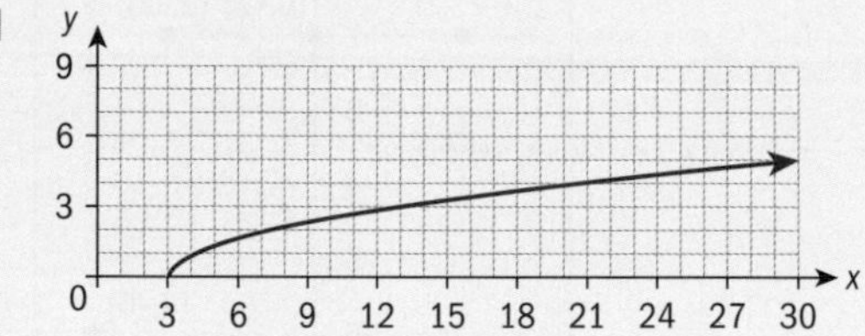

a Domain:

$x \in \mathbb{R}$

$]-\infty, \infty[$ or $(-\infty, \infty)$

Range:

$y \leq 6$

$]-\infty, 6]$ or $(-\infty, 6)$

b Domain:

$x \in \mathbb{R}$

$]-\infty, \infty[$ or $(-\infty, \infty)$

Range:

$y \in \mathbb{R}$

$]-\infty, \infty[$ or $(-\infty, \infty)$

c Domain:

$x \in \mathbb{R}$

$]-\infty, \infty[$ or $(-\infty, \infty)$

Range:

$-2 \leq y \leq 2$

$[-2, 2]$

d Domain: $x \geq 3$ $[3, \infty[$ or $[3, \infty)$ Range: $y \geq 0$ $[0, \infty[$ or $[0, \infty)$	

A **piecewise function** is a function that has two or more equations for different intervals of the domain of the function.

Example 11

Consider the piecewise function: $f(x) = \begin{cases} 2x, & 0 \leq x \leq 2 \\ 4, & 2 < x \leq 4 \\ \frac{1}{2}x + 2, & 4 < x \leq 8 \end{cases}$

a Find $f(2)$, $f(3)$ and $f(7)$. **b** Sketch the graph of f.

a $f(2) = 2 \times 2$
$= 4$

Use the equation $f(x) = 2x$, since 2 is in the interval $0 \leq x \leq 2$.

$f(3) = 4$

Use the equation $f(x) = 4$, since 3 is in the interval $2 < x \leq 4$.

$f(7) = \frac{1}{2}(7) + 2$
$= \frac{11}{2}$

Use the equation $f(x) = \frac{1}{2}x + 2$, since 7 is in the interval $4 < x \leq 8$.

b

Graph each equation for the restricted interval.

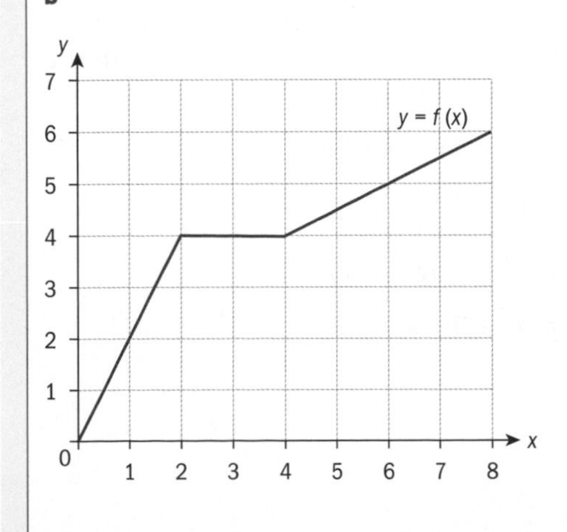

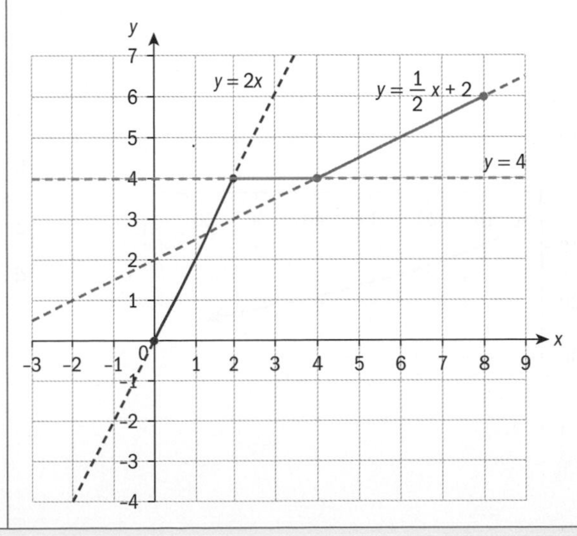

Exercise 2D

1 For each relation below, state whether it is a function or not. If yes, state the domain and range.

a

x	6	8	8	12
y	1	6	4	−1

b $\{(-5, 6), (-2, 4), (3, 14)\}$

c $\{(-12, 7), (-8, -8), (-5, -8)\}$

d

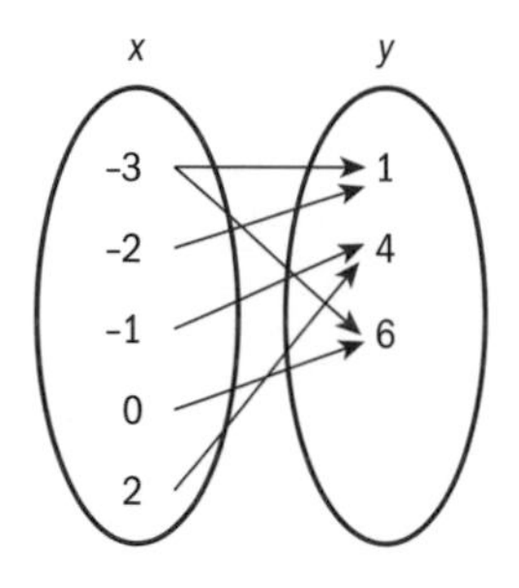

e $f(x) = -2x$

f $g(x) = -4$

g $x = 3$

h

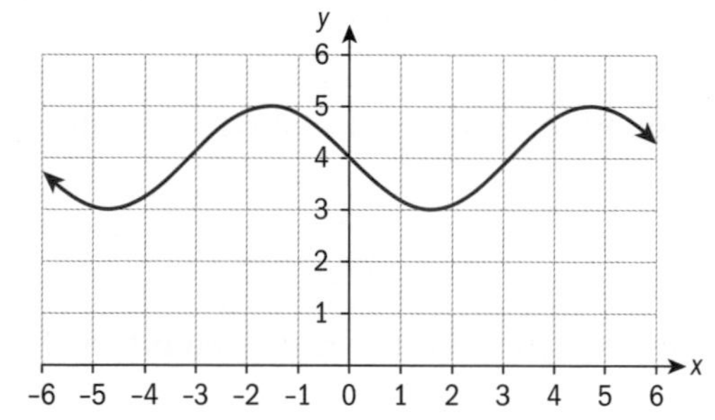

i

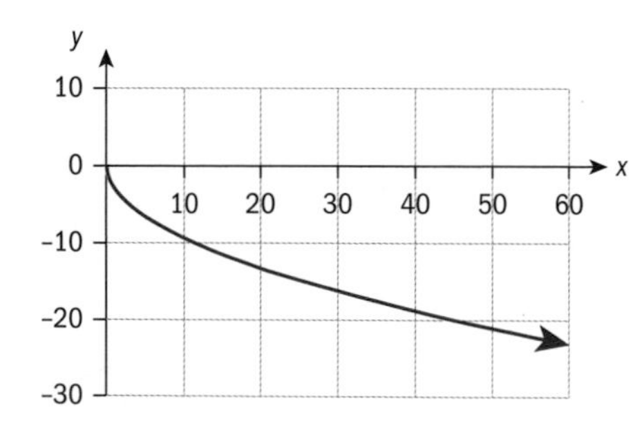

j

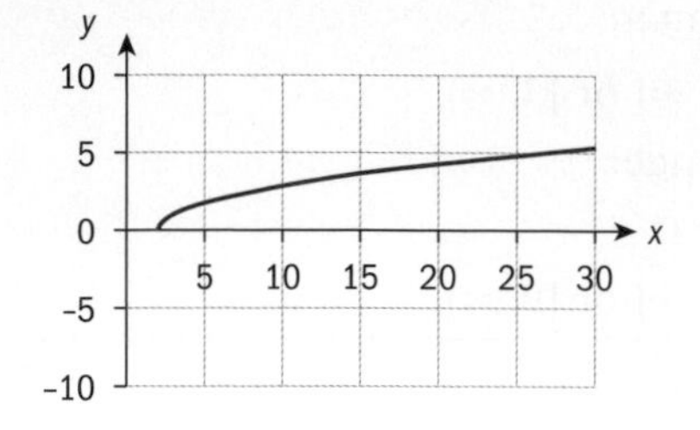

k

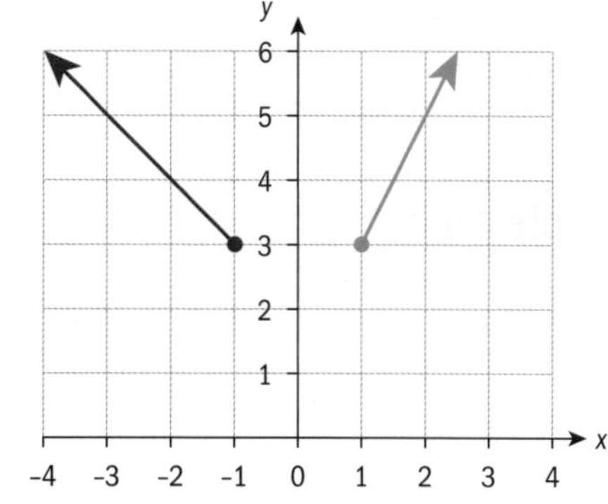

l

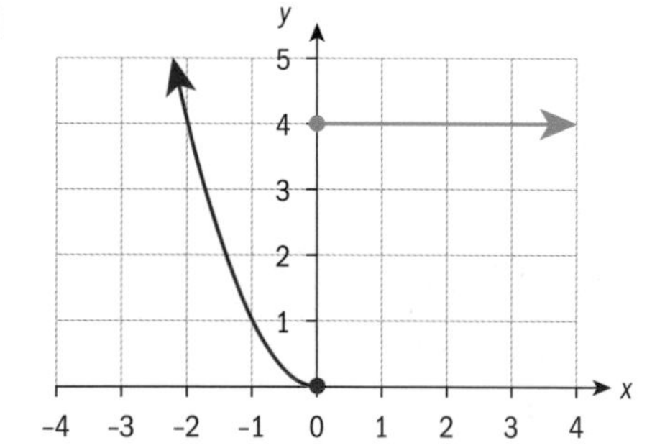

2 Use your GDC to graph the following functions. Sketch the graph and state the domain and range.

a $f(x) = -(x - 4)(x + 2)(x - 1)$

b $g(x) = \dfrac{x+2}{x-1}$

c $f: x \to -\log(x - 1)$

d $g: x \to 3\cos x + 2$

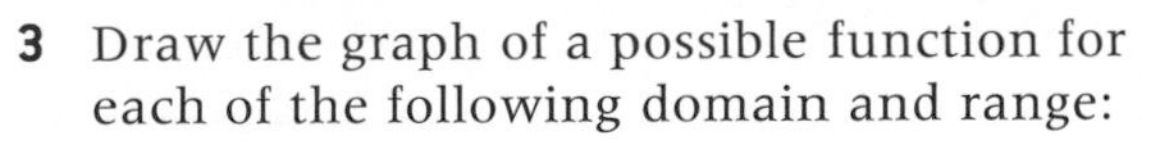

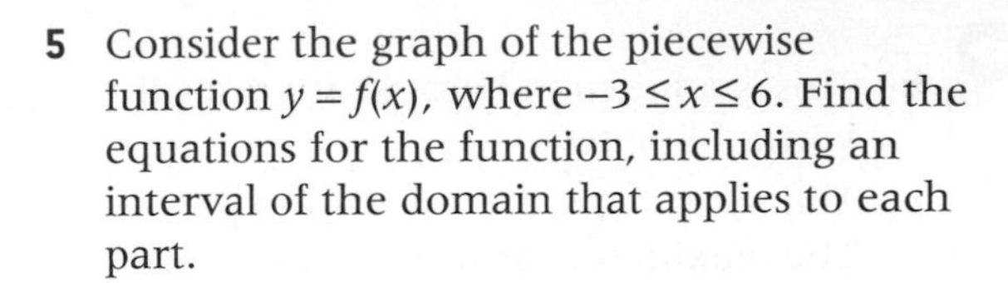

3 Draw the graph of a possible function for each of the following domain and range:

a domain: $x \in \mathbb{R}$

range: $y \in \mathbb{R}$

b domain: $]-\infty, \infty[$

range: $[3, \infty[$

c domain: $0 \le x < 6$

range: $-3 \le y < 2$

d domain: $x > 0$

range: $y > 0$

4 Consider the piecewise function

$$f(x) = \begin{cases} \frac{1}{3}x + 2, & 0 \le x \le 6 \\ -x + 10, & 6 < x \le 10 \end{cases}$$

a Find each value: **i** $f(6)$ **ii** $f(8)$

b Sketch the graph of f.

c Write down the domain and range of f.

5 Consider the graph of the piecewise function $y = f(x)$, where $-3 \le x \le 6$. Find the equations for the function, including an interval of the domain that applies to each part.

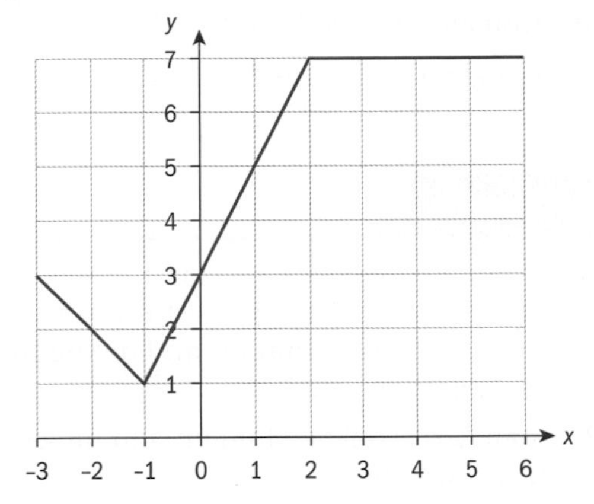

Functions

Example 12

The path of a golf ball after being hit by a club can be modelled by the function $h(t) = 60t - 16.1t^2$, where t is the time in seconds and h is the height of the ball in feet.

a Use your GDC to sketch a graph of the ball's path. Label the axes.

b Explain why this graph is a function.

c Use your GDC to find the t-intercepts and maximum value.

d Write down the domain and range for this situation.

a

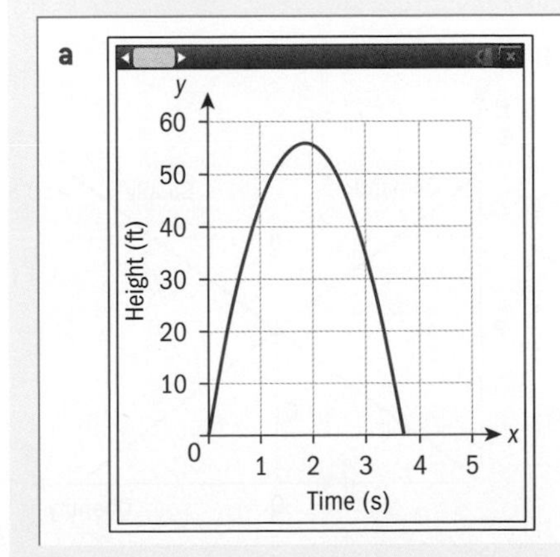

Continued on next page

b This is a function because it passes the vertical line test.	
c The t-intercepts are $x = 0$ and $x \approx 3.73$. The maximum value is $y \approx 55.9$.	
d Domain: $0 \le x \le 3.73$, $[0, 3.73]$ Range: $0 \le x \le 55.9$, $[0, 55.9]$	

Exercise 2E

1 **a** A plumber charges \$40 for a service call plus \$21 an hour. Write a function that represents this situation and define the variables you used.

b State a reasonable domain and range for this situation.

c How much will have you have to pay the plumber if he is at your house for four hours?

2 Forensic scientists can determine the height of a person based on the length of their femur. The equation is $h(f) = 2.47f + 54.10$, where f is the length of the femur (cm) and h is the person's height (cm).

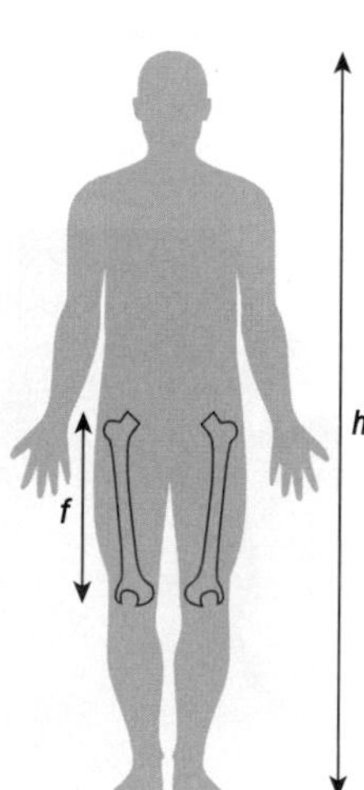

a Use your GDC to sketch a graph of this function. Label the axes.

b State a reasonable domain and range for this function.

c Determine the height of a person with femur length of 51 cm.

d If someone is 161 cm tall, what is the length of their femur?

3 Javier invests €10 000 in an investment fund. The annual interest rate is 2.5%, compounded monthly.

a Write an equation to represent this situation.

b Explain why this equation is a function.

c State the domain and range of this function.

d How long will it take Javier to double his money, if he does not make any additional deposits or withdrawals?

Developing inquiry skills

Let's return to the chapter opener about supply and demand.

While $D(P)$ is a linear function, the domain is not.

$x \in \mathbb{R}$ because it is a real-life situation. What would the lower and upper limit be for the domain? What about the range?

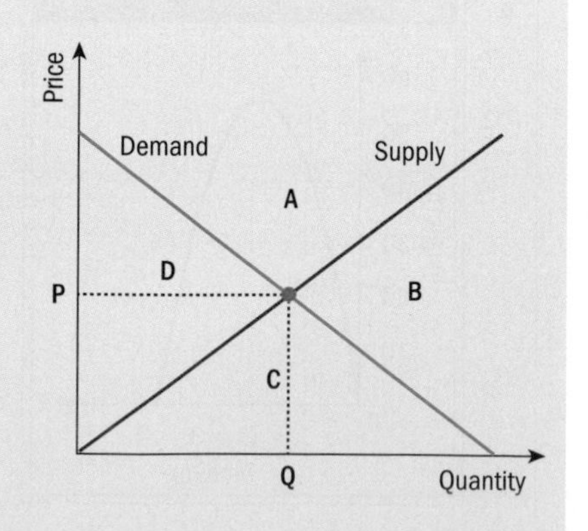

2.5 Composite functions

Investigation 7

Harper works for an insurance company. She works on commission and earns 5% on all sales above \$1000.

1 a Calculate the amount of money on which Harper will earn in commission if her sales last month were \$13 500.

b Calculate her commission.

2 Calculate the commission earned for monthly sales of \$24 300.

3 Following the pattern in questions **1** and **2**:

a Write a function, $f(x)$, for x amount of sales on which Harper earns commission.

b Using your function from part **a,** write a second function, $g(x)$, that can be used to calculate how much Harper earns in commission.

4 Instead of calculating her monthly commission in two steps as above, combine your functions from question **3** into one function, $h(x)$.

When we combine functions like this, it is called a **composition of functions** or **composite function.**

Input
x−3
function f:
x - 1000
Output
f(x)=9
Input
function g:
x × 0.05
Output
g(f(x))=10

5 **Factual** How do you write a composite function given two (or more) functions?

6 **Conceptual** How can composition of functions help to represent the stages or order of a real-life problem?

7 **Conceptual** Does the order you perform the functions matter? Explain your answer.

8 Use your composite function from question **4** to calculate Harper's commission for sales of:

a \$17 000 b \$9550 c \$950

9 **Factual** How do you calculate a value from a given composite function?

TOK

"The object of mathematical rigour is to sanction and legitimize the conquests of intuition" – Jacques Hadamard.

Do you think that studying the graph of a function contains the same level of mathematical rigour as studying the function algebraically?

As we have seen previously in this chapter, a function is like a machine or a set of instructions for the operations performed on a variable.

A composite function denoted as $f(g(x))$ or $f \circ g(x)$, can be thought of as a series of operations or machines.

We read both of the notations as "f of g of x".

For example, to make fries, you first have to put the potatoes through a slicer, then take the slices and put them in a fryer.

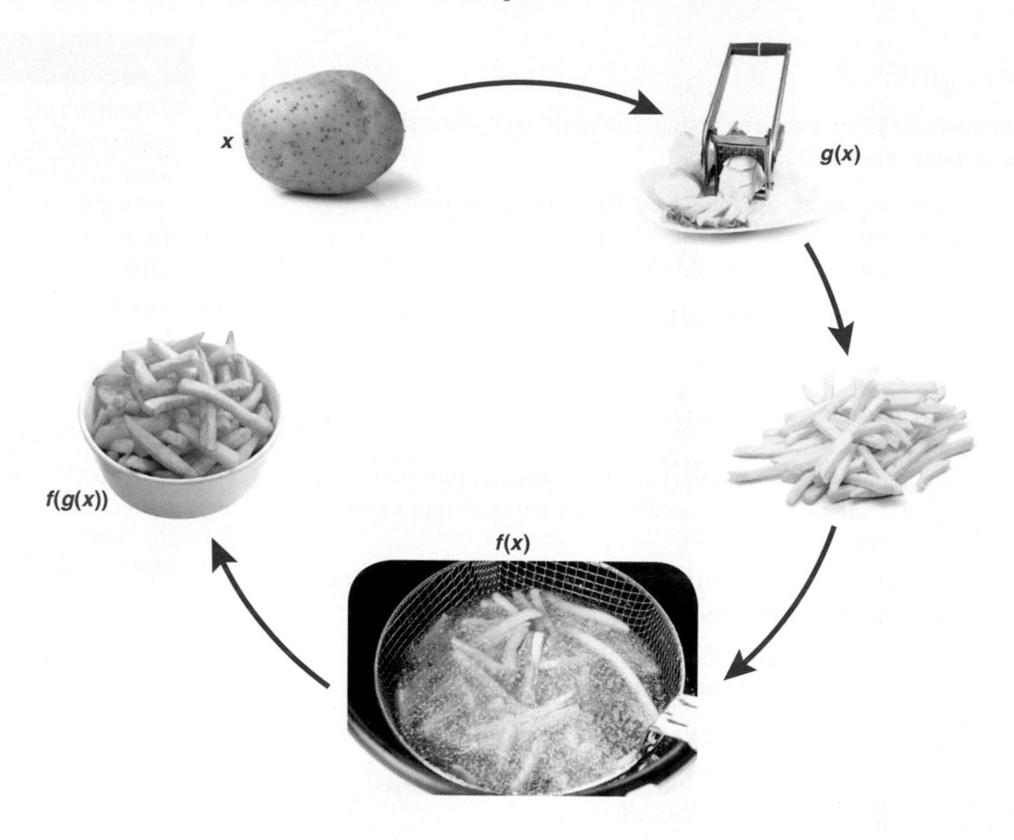

In this example, x is the potatoes, $g(x)$ is the slicer and $f(x)$ is the fryer. The notation $f(g(x))$ represents first slicing the potatoes and then putting them in the fryer.

When performing a composition of functions, the order matters. Imagine if we put the potatoes in the fryer, then tried to slice them!

Example 13

Given the functions:

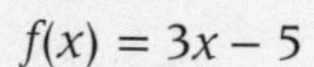 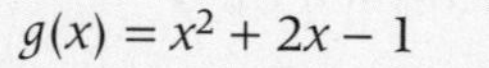 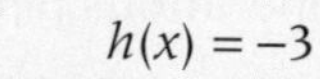

$f(x) = 3x - 5$ $\quad$ $g(x) = x^2 + 2x - 1$ $\quad$ $h(x) = -3$

Find:

a $f(g(x))$ **b** $(g \circ f)(x)$ **c** $f(h(x))$

d $f(g(-1))$ **e** $(f \circ f \circ g)(x)$ **f** $(f \circ g \circ h)(1)$

a $f(g(x)) = 3(x^2 + 2x - 1) - 5$ $f(g(x)) = 3x^2 + 6x - 3 - 5$ $f(g(x)) = 3x^2 + 6x - 8$	The notation $f(g(x))$ means that we put the g function inside the f function. In other words, replace all the variables in the outer function with the inner function and simplify.
b $(g \circ f)(x) = (3x - 5)^2 + 2(3x - 5) - 1$ $(g \circ f)(x) = 9x^2 - 30x + 25 + 6x - 10 - 1$ $(g \circ f)(x) = 9x^2 - 24x + 14$	
c $f(h(x)) = 3(-3) - 5 = -9 - 5 = -14$	
d $g(-1) = (-1)^2 + 2(-1) - 1$ $g(-1) = 1 - 2 - 1 = -2$ $f(g(-1)) = f(-2) = 3(-2) - 5$ $f(g(-1)) = -6 - 5 = -11$ Or $f(g(x)) = 3x^2 + 6x - 8$ (from part **a**) $f(g(-1)) = 3(-1)^2 + 6(-1) - 8$ $f(g(-1)) = 3 - 6 - 8$ $f(g(-1)) = -11$	When there is a value given for x, there are two ways to do this question: You can substitute -1 for x into $g(x)$ first and then substitute $(g -1)$ into $f(x)$. Or You can put $g(x)$ into $f(x)$, and then substitute -1 for x.
e $(f \circ g)(x) = 3x^2 + 6x - 8$ (from part **a**) $(f \circ f \circ g)(x) = 3(3x^2 + 6x - 8) - 5$ $(f \circ f \circ g)(x) = 9x^2 + 18x - 24 - 5$ $(f \circ f \circ g)(x) = 9x^2 + 18x - 29$	With more than two functions, start with the innermost function, or start from the right and work left. For the example, this means put g into f, then put the simplified answer back into f again.
f $(f \circ g \circ h)(1)$ Working from the inside out: $h(1) = -3$ $(g \circ h)(1) = g(-3)$ $= (-3)^2 + 2(-3) - 1$ $= 9 - 6 - 1$ $= 2$ $(f \circ g \circ h)(1) = f(2)$ $= 3(2) - 5$ $= 1$	

When working with a composite function such as $f(g(x))$, start with the innermost function and work outwards.

Example 14

Given $f(x) = \sqrt{x}$ and $g(x) = x^2 + 3x$:

a find a simplified expression for $(g \circ f)(x)$ **b** state the domain of $(g \circ f)(x)$

c find a simplified expression for $(f \circ g)(x)$ **d** state the domain of $(f \circ g)(x)$

e explain why the domains of $(g \circ f)(x)$ and $(f \circ g)(x)$ are not the same.

a $(g \circ f)(x) = \left(\sqrt{x}\right)^2 + 3\sqrt{x}$

$(g \circ f)(x) = x + 3\sqrt{x}$

b $x \geq 0$ or $[0, \infty[$ or $[0, \infty)$

For the domain, we need to look at the inside function $f(x) = \sqrt{x}$, which has a domain of $x \geq 0$.

This forces the composition to have this restriction as well.

c $(f \circ g)(x) = \sqrt{x^2 + 3x}$

d $x^2 + 3x \geq 0$

$x(x + 3) \geq 0$

$x = -3, 0$

Using a test point:

If $x = -10$, $x^2 + 3x > 0$

If $x = -1$, $x^2 + 3x < 0$

If $x = 10$, $x^2 + 3x > 0$

So, $x^2 + 3x \geq 0$ when $x \leq -3$ or $x \geq 0$

Therefore the domain is

$x \leq -3 \cup x \geq 0$

or

$]-\infty, -3] \cup [0, \infty[$

or

$(-\infty, -3] \cup [0, \infty)$

The domain of the inside function is $x \in \mathbb{R}$, which has no restriction. However, when we look at the composition, due to the square root, we need to solve $x^2 + 3x \geq 0$.

e The domains are not the same because the domains of the inside functions are not the same, and because the composite function $f(g(x))$ restricts the values from the range of g which "fit" into the domain of f.

The domain and range of a composite function are based on the domain and range of the inside function, as well as of the composite function itself.

Exercise 2F

1 **a** Given the functions

$f(x) = -x^2 + 5x$

$g(x) = 4x - 2$

$h(x) = \sqrt{x} + 1$,

find:

i $f(g(x))$ **ii** $f(f(x))$

iii $f(h(x))$ **iv** $(g \circ h)(x)$

v $(f \circ f \circ f)(-1)$ **vi** $g(h(9))$

vii $(g \circ f)(2) + (f \circ g)(2)$

b State the domain of parts **ai–iv.**

2 Create two different functions, $f(x)$ and $g(x)$, such that

a $f(g(x)) = g(f(x))$

b $f(g(x)) \neq g(f(x))$

c $f(2) = g(2)$

3 Given $f(x) = -2x + 5$ and $g(x) = 4x - 1$:

a find an expression for $f(g(x))$

b solve $f(g(x)) = 12$.

4 Given $f(x) = 3x^2 - 6$ and $g(x) = -x + 4$:

a find $f(g(x))$

b using your GDC, sketch the graph of $f(g(x))$

c state the domain and range of $f(g(x))$

5 You purchase a new refrigerator. You have no way to get it home, so you pay the delivery fee of \$25. The sales tax on the delivery fee is 6%.

a Write a function $f(x)$ that represents the cost of the refrigerator and the delivery fee.

b Write a function $g(x)$ that represents the cost of the refrigerator and the tax.

c Find and interpret both $f(g(x))$ and $g(f(x))$.

d If taxes cannot be charged on delivery, which composite function from part **c** should be used to calculate the total amount you must pay to the store?

A composite function can be formed with more than just equations. Let's look at other forms for functions, such as ordered pairs, diagrams and graphs.

Example 15

If $f(x) = \{(-4,-2), (3,4), (9,1), (10,-2)\}$ and

$g(x) = \{(-4,5), (-2,3), (0,6), (7,-4), (11,-4)\}$, find $f(g(-2))$.

$g(-2) = 3$ $f(g(-2)) = f(3) = 4$	Start with the inside function, $g(-2)$. Find the point in the function g that has an x-coordinate of -2. The answer to $g(-2)$ will be the y-coordinate of this point. Now go to the function f and find the point with an x-coordinate of 3. The y-coordinate is the final answer. This method can also be used with a table of values or mapping diagram.

Example 16

Using the graph of two functions $f(x)$ and $g(x)$ on the right, find:

a $f(g(2))$ **b** $(g \circ f)(-1)$ **c** $g(g(0))$

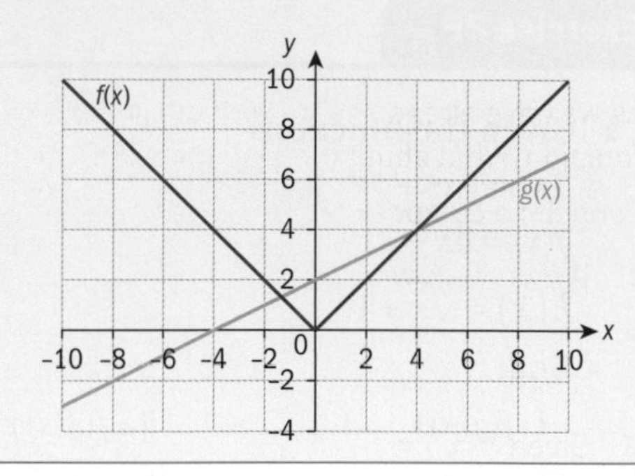

a $g(2) = 3$	Start with the inside $g(2)$. Look at the graph of the g function and find where $x = 2$. The value will be the y-coordinate of this point.
$f(g(2)) = f(3)$ $f(3) = 3$	Now look at the graph of the f function and find where $x = 2$. The final answer will be the y-coordinate of this point.
b $f(-1) = 1$ $(g \circ f)(-1) = g(1)$ $g(1) = 2.5$	
c $g(0) = 2$ $g(g(0)) = g(2)$ $g(2) = 3$	

Exercise 2G

1 Given the functions

$f(x) = \{(2, 1), (3, 2), (5, 6), (10, -4)\}$

$g(x) = \{(-2, 3), (3, 11), (6, -3), (11, 0)\}$

a find:

i $f(g(-2))$ **ii** $g(f(5))$

iii $g(g(3))$

b state the domain and range of both functions.

2 Using the table of values below, find:

x	-1	3	7	9
$f(x)$	9	6	-2	-1
$g(x)$	-5	7	0	-4

a $(f \circ g)(3)$ **b** $(g \circ f)(-1)$ **c** $(f \circ f)(9)$

3 Using the graph below, find:

a $g(f(0))$ **b** $g(f(1))$

c $f(g(-2))$ **d** $f(g(0))$

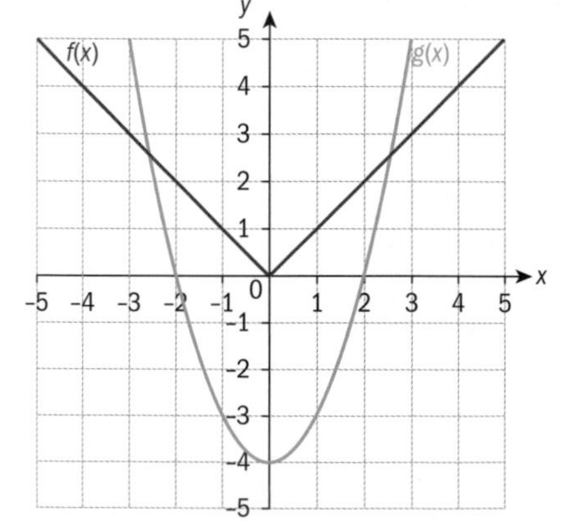

4 Create two functions, $f(x)$ and $g(x)$, each a set of ordered pairs, such that $f(g(-1)) = 2$.

Investigation 8

As we have already seen, we can combine two functions into a composite function. What about the decomposition of functions?

Consider a composite function such that: $h(x) = f(g(x)) = (x-1)^2$

1 If $f(x) = x^2$, write down the function for $g(x)$.

2 **Factual** Explain how you can decompose a composite function into different, smaller component functions.

3 Given $h(x) = f(g(x))$, find two possible component functions, $f(x)$ and $g(x)$, for the following functions:

a $h(x) = \sqrt{x-5}$ **b** $h(a) = (a+2)^2 - 3(a+2)$

c $h: x \to \dfrac{1}{2x+2}$ **d** $h(x) = 4(x-4) - 5$

4 How many different functions could make up the composite function of $f(x) = \sqrt{(x-2)^4}$?

5 Can all composite functions be decomposed? Explain your answer.

6 **Conceptual** Hypothesize why one might need to decompose a composite function.

TOK

Is mathematics independent of culture?

Example 17

A life insurance company offers insurance rates based on age. The formula for calculating a person's monthly insurance premium is $C(n) = \$1.25(2017 - n)$ where n is the year a person is born.

a Decompose the formula into two component functions.

b Explain what each component represents in terms of the insurance rates.

c **i** How old is someone in 2017 if they were born in 1968?

ii How much will their monthly insurance premium be?

d Manuel pays \$41.25 a month for life insurance. In what year was he born?

a $f(n) = 2017 - n$

$g(n) = 1.25n$

b $f(n)$ calculates a person's age in 2017.

$g(n)$ calculates the monthly premium of \$1.25 for each year the person has been alive (their age in 2017).

c **i** $2017 - 1968 = 49$

ii $1.25 \times 49 = \$61.25$

d $41.25 = \$1.25(2017 - n)$

$33 = 2017 - n$

$n = 2017 - 33$

$n = 1984$

Exercise 2H

1 You work for a credit card company, and you receive a base salary plus commission based on how many people you sign up for a credit card each month. You are paid in GBP. Your boss calculates your monthly salary using the formula $S(n) = 2.20(n - 100)$, where n is the number of people who sign up for a credit card.

a Decompose the formula into two component functions.

b Explain what each component represents in relation to your monthly salary.

c i If you get 224 people to sign up for a credit card, how many people will you be paid for?

ii How much will you be paid?

d i How much will you earn if you get 276 people to sign up for credit cards this month?

ii Your colleague was paid £114.40 last month. How many people did she get to sign up for a credit card?

2 The amount, $P, paid for a new TV is given by the formula $P(x) = 1.10(x - 25)$, for $x > 25$, where x is the original price of the TV.

a State the component functions and explain what they might represent.

b i How much did you pay for the TV before tax if the price tag was $699.99?

ii How much did the TV cost after tax?

c i The store offers insurance on the TV for $49.99. If they cannot charge tax on insurance, rewrite $P(x)$ to include this fee.

ii Use the new $P(x)$ to calculate how much you paid with the insurance if the price tag of the TV was $525.99.

3 a Create a word problem that can be modelled using the function $f(x) = 0.75(x - 250)$. Define the variables and state a reasonable domain for your function.

b State the component functions and explain what they represent in the context of your word problem.

2.6 Inverse functions

Investigation 9

Consider the following two groups of relations:

Group 1	Group 2
$y = -4x$	$y = -\frac{1}{4}x$
$y = \frac{1}{2}x$	$y = 2x$
$y = 2x - 1$	$y = \frac{x+1}{2}$
$y = \frac{1}{3}x - 2$	$y = 3x + 6$
$y = x^2$, for $x \geq 0$	$y = \sqrt{x}$
$y = x^2 - 4$, for $x \geq 0$	$y = \sqrt{x+4}$

Part 1:

1 Compare each function in Group 1 with its corresponding function in Group 2. Hypothesize the relationship.

Each function in Group 1 and its corresponding function in Group 2 are called **inverses**.

2 **Factual** What is an inverse function? What does an inverse function map? What does the inverse map? Think of input (independent) and output (dependent).

3 Using the patterns in the table, explain how to use the equation of a function to find the equation of its inverse.

Part 2:

1 Using your GDC, sketch the graph of each function from Group 1 and its corresponding inverse on the same set of axes.

2 Explain the graphical relationship between a function and its inverse.

3 **Conceptual** How does a graph help you to determine whether two functions are inverses?

4 For the last two functions and their inverses in the table, state their domains and ranges.

5 **Conceptual** How are the domains and ranges of two inverse functions related?

6 Recall from section 2.1 that the vertical line test will tell you whether a graph of a relation is a function. What do you think the **horizontal line test** tells you? Think of inputs mapping to outputs.

The **inverse** of a function $f(x)$ is denoted $f^{-1}(x)$. It reverses the action of that function, so $(f \circ f^{-1})(x) = x$.

In order for a function to have an inverse function, it must be a **one-to-one function**. This means it will have one value of y for every value of x, and one value of x for every value of y. A one-to-one function will pass both the vertical line test and the horizontal line test.

Two functions, $f(x)$ and $g(x)$ are inverses if and only if:

$$(f \circ g)(x) = (g \circ f)(x) = x$$

Example 18

a Using composition of functions, show that $f(x) = 2x + 4$ and $g(x) = \frac{x-4}{2}$ are inverses.

b Is $f(x)$ a one-to-one function? Explain your answer.

a $(f \circ g)(x) = 2\left(\frac{x-4}{2}\right) + 4$

$(f \circ g)(x) = x - 4 + 4$

$(f \circ g)(x) = x$

$(g \circ f)(x) = \frac{(2x+4)-4}{2}$

$(g \circ f)(x) = \frac{2x}{2}$

$(g \circ f)(x) = x$

Since $(f \circ g)(x) = (g \circ f)(x) = x$, these are inverses.

Show that $(f \circ g)(x) = x$ and $(g \circ f)(x) = x$. If both of these compositions equal x, then the two functions are inverses.

b

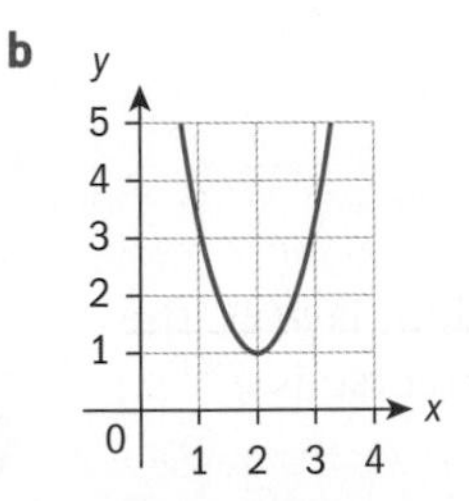

Since the graph of f passes both the vertical and horizontal line tests, it is a one-to-one function.

Note: We could also say that if f and its inverse both pass the vertical line test, then f is a one-to-one function.

Example 19

i Determine if each function below is a one-to-one function.

ii If it is a one-to-one function, find its inverse.

a $f(x) = 4x + 2$ **b** $f: x \to x^2 + 1$

c $g(x) = \sqrt{x+2} - 3$ **d** $h: x \to (x-1)^3 - 2$

a **i** Yes, f is a one-to-one function because it passes both the vertical and horizontal line test.	Remember from section 2.2 that $f(x)$ or other functional notation is a replacement for y in a function.
ii $f(x) = 4x + 2$ $y = 4x + 2$ $x = 4y + 2$ $4y = x - 2$ $y = \frac{x-2}{4}$ $f^{-1}(x) = x - \frac{2}{4}$	Simplify this line into gradient-intercept form $y = mx + b$. In the last step, be sure to change y for the inverse notation $f^{-1}(x)$. Otherwise you are saying that two different expression equal y.
b **i** No, f is not a one-to-one function because it does not pass the horizontal line test. **ii** Because f is not a one-to-one function, it does not have an inverse function.	Graph the function on your GDC, and use the horizontal line test to determine whether it is a one-to-one function.
c **i** Yes, $g(x)$ is a one-to-one function, because it passes both the vertical and horizontal line tests.	
ii $g(x) = \sqrt{x+2} - 3$ $y = \sqrt{x+2} - 3$ $x = \sqrt{y+2} - 3$ $\sqrt{y+2} = x + 3$ $\left(\sqrt{(y+2)}\right)^2 = (x+3)^2$ $y + 2 = (x+3)^2$ $y = (x+3)^2 - 2$ $g^{-1}(x) = (x+3)^2 - 2$, for $x \geq -3$	Because the original is $g(x)$, the notation for its inverse is $g^{-1}(x)$.

d **i** Yes, h is a one-to-one function, because it passes both the vertical and horizontal line tests.

ii $h: x \to (x-1)^3 - 2$

$$y = (x-1)^3 - 2$$
$$x = (y-1)^3 - 2$$
$$(y-1)^3 = x + 2$$
$$\sqrt[3]{(y-1)^3} = \sqrt[3]{x+2}$$
$$y - 1 = \sqrt[3]{x+2}$$
$$y = \sqrt[3]{x+2} + 1$$
$$h^{-1}(x) = \sqrt[3]{x+2} + 1$$

To find the equation of an inverse function algebraically, switch x and y, and then isolate y.

Example 20

i For each function below, sketch its graph.

ii On the same set of axes, sketch the graph of its inverse.

a $f(x) = -2x + 2$ **b** $f: x \to -x^2 + 4$, for $x \geq 0$

a **i** This is the graph of $f(x) = -2x + 2$.

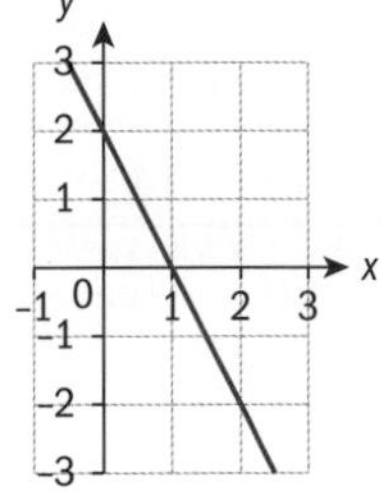

Notice that the x-intercept is $(1, 0)$ and the y-intercept is $(0, 2)$.

ii For the inverse, the points will be $(0, 1)$ and $(2, 0)$.

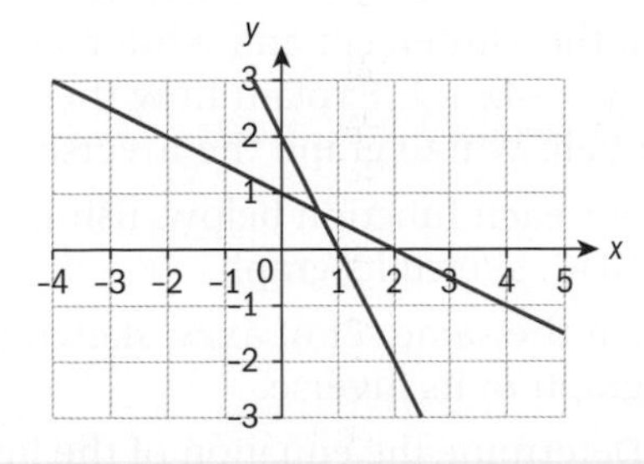

Continued on next page

b i This is the graph of $f: x \rightarrow -x^2 + 4$, for $x \geq 0$.

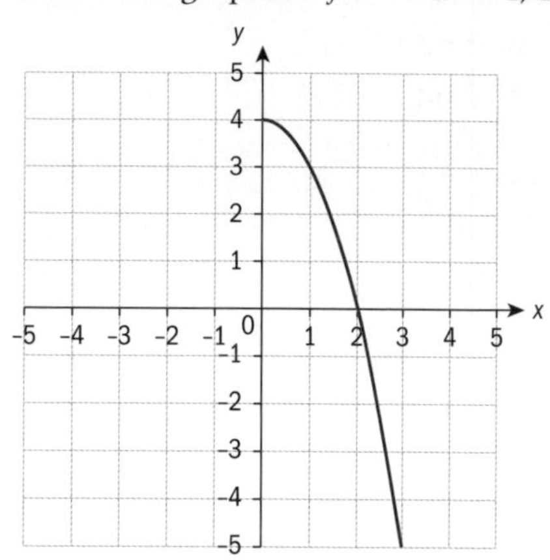

Notice that the x-intercept is $(2, 0)$ and the y-intercept is $(0, 4)$.

ii For the inverse, the axes intercepts will be $(0, 2)$ and $(4, 0)$.

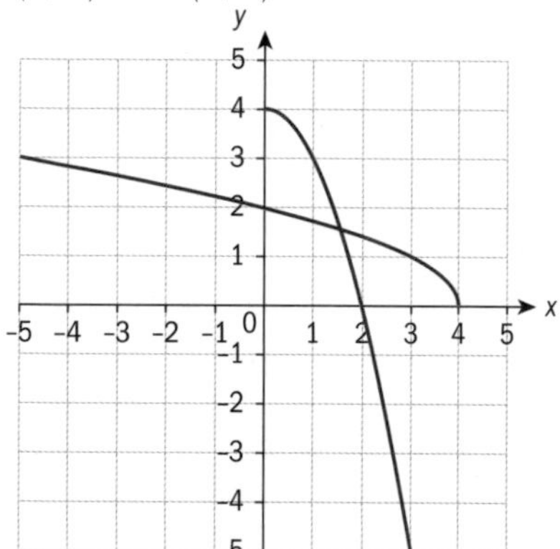

To sketch the graph of an inverse, graph the original function and pick some points (like the x-intercepts, y-intercepts, etc) and switch the coordinates. This will result in a reflection over the line $y = x$.

$$(x, y) \rightarrow (y, x)$$

Exercise 2I

1 Determine algebraically whether the following pairs of functions are inverses.

a $f(x) = -2x + 2$ and $g(x) = \frac{1}{2}x + 2$

b $f: x \rightarrow 2x^3 + 4$ and $g: x \rightarrow \sqrt[3]{\frac{x-3}{2}}$

c $f(x) = \sqrt{3x - 2}$ and $g(x) = \frac{x^2}{3} + \frac{2}{3}$

d $g(x) = -\frac{3}{4}x + 5$ and $h(x) = -\frac{4x - 20}{3}$

2 Find the x-intercept and y-intercepts of the line $y = -4x + 2$. Explain how these points can help you to graph the inverse.

3 i For each function below, using your GDC, sketch its graph.

ii On the same set of axes, sketch the graph of its inverse.

iii Determine the equation of the inverse.

a $f(x) = -4x^2 + 4$, for $x \geq 0$

b $g(x) = -2\sqrt{x} + 5$ **c** $g: x \rightarrow \frac{1}{2}x + 6$

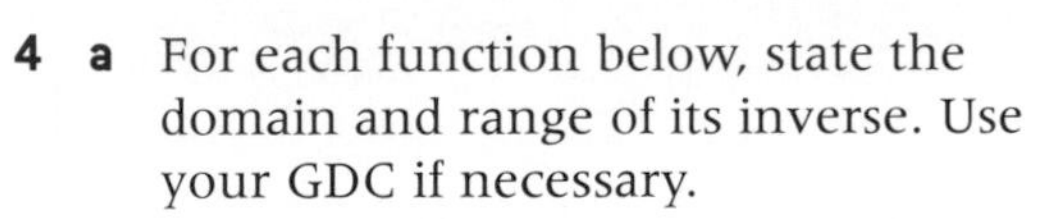

4 **a** For each function below, state the domain and range of its inverse. Use your GDC if necessary.

  i $f(x) = 2x^2 - 5x + 6$ for $x \geq 1.25$

  ii $f: x \to -3x + 1$ **iii** $g: x \to 2^x + 3$

  iv $g(x) = -\sqrt{-x+2} + 1$

 b Explain how you can use the domain and range of a function to find the domain and range of its inverse.

5 Create a one-to-one function. State the equation of the function and of its inverse and explain why the inverse is also a one-to-one function.

6 If $f(x) = 2x - 5$:

 a solve $f(x) = 11$

 b find $f^{-1}(x)$

 c find $f^{-1}(11)$.

 d What do you notice about your answers to parts **a** and **c**?

 e Create a general rule to explain your answer for part **d**.

7 Given $f(x) = -2x - 1$ and $g(x) = -3x^2$, show that $g \circ f^{-1}(-1) = 0$.

Functions

Investigation 10

1 Draw the graph of the following functions:

 a $f(x) = x$ **b** $f: x \to \frac{1}{x}$ **c** $g(x) = 3 - x$

2 For each function above, draw its inverse. What do you notice?

3 All of these functions are called **self-inverse functions.**

 a Write a definition of a self-inverse function.

 b Write an algebraic definition similar to the algebraic definition for regular inverses: $f(g(x)) = g(f(x)) = x$.

4 **Conceptual** How does a graph help you identify a self-inverse function?

5 The self inverse function $f(x) = x$ is also called the **identity function**. Write a definition of the identity function.

6 What does the identity function map? Think of inputs and outputs.

7 **Factual** What do you think the identity line is?

TOK

Do you think that mathematics is just the manipulation of symbols under a set of rules?

Example 21

Show that the function $f(x) = \frac{x}{x-1}$ is a self-inverse function.

$f(f(x)) = x$	We must show that $f(f(x)) = x$.
$\frac{\left(\frac{x}{x-1}\right)}{\left(\frac{x}{x-1}\right) - 1} = x$	
$\frac{\left(\frac{x}{x-1}\right)}{\left(\frac{x}{x-1}\right) - \left(\frac{x-1}{x-1}\right)} = x$	

Continued on next page

$\dfrac{\left(\dfrac{x}{x-1}\right)}{\left(\dfrac{x-x+1}{x-1}\right)}=x$ $\dfrac{\left(\dfrac{x}{x-1}\right)}{\left(\dfrac{1}{x-1}\right)}=x$ $\left(\dfrac{x}{x-1}\right)\left(\dfrac{x-1}{1}\right)=x$ $x=x$ Therefore $f(x)=\dfrac{x}{x-1}$ is a self-inverse function.	Because $f(f(x))=x$, it is a self-inverse function.

Example 22

Find the value of k such that $f: x \rightarrow \dfrac{3x-5}{x+k}$ is a self-inverse function.

To be a self-inverse function, $f(f(x))=x$, $\dfrac{3\left(\dfrac{3x-5}{x+k}\right)-5}{\left(\dfrac{3x-5}{x+k}\right)+k}=x$ $\dfrac{\dfrac{(9x-15)}{(x+k)}-5}{\dfrac{(3x-5)}{(x+k)}+k}=x$ $\dfrac{\dfrac{(9x-15)}{(x+k)}-\dfrac{5(x+k)}{(x+k)}}{\dfrac{(3x-5)}{(x+k)}+\dfrac{k(x+k)}{(x+k)}}=x$ $\dfrac{\dfrac{9x-15-5x-5k}{x+k}}{\dfrac{3x-5+kx+k^2}{x+k}}=x$ $\dfrac{4x-5k-15}{3x-5+kx+k^2}=x$ $4x-5k-15=3x^2-5x+kx^2+k^2x$ $4x-5k-15=(3+k)x^2+(k^2-5)x$ Since there is no x^2 on the left-hand side: $3+k=0$ $k=-3$	Note that if you equate the coefficients of x, you will have $k^2=9$.

Exercise 2J

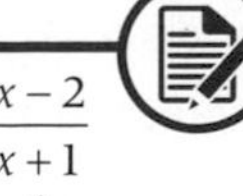

1 Show graphically that $f(x) = -x$ is a self-inverse function.

2 a Show that $y = 3 - x$ is a self-inverse function.

b Show that $y = -2 - x$ is a self-inverse function.

c Show that $y = \frac{1}{2} - x$ is a self-inverse function.

d Write a generalization from your answers in parts **a–c**.

3 Show that $f(x) = \frac{-x-2}{5x+1}$ is a self-inverse function.

4 Find the value of m such that $g(x) = \frac{2x-4}{x+m}$ is a self-inverse function.

International-mindedness

Which do you think is superior; the Bourbaki group analytical approach or the Mandelbrot visual approach to mathematics?

Functions

Developing your toolkit

Now do the Modelling and investigation activity on page 106.

Chapter summary

- A mathematical **relation** is a relationship between any set of ordered pairs.
- A **function** is a special type of relation. A relation is a function if each input relates to only one output. In other words, one value of the independent variable does not give more than one value of the dependent variable. A function cannot have a repeating x-value with different y-values.
- When writing an expression for a function, there are several different notations. For example, $f(x)$, $g(x)$, $f: \rightarrow$. These are all called **functional notation**.

  $f(x)$, said as "f of x", represents a function f with independent variable x.
- Functional notation is a quick way of expressing substitution. Rather than saying "Given $y = 3x - 5$, find the value of y if $x = 2$", we can simply state: "Given $f(x) = 3x - 5$, find $f(2)$".
- **Draw** and **sketch** are two of the IB command terms you need to know.

  Draw: Represent by means of a labelled, accurate diagram or graph, using a pencil. A ruler (straight edge) should be used for straight lines. Diagrams should be drawn to scale. Graphs should have points correctly plotted (if appropriate) and joined in a straight line or smooth curve.

  Sketch: Represent by means of a diagram or graph (labelled as appropriate). The sketch should give a general idea of the required shape or relationship, and should include relevant features.
- The **domain** of a function is all possible values of the independent variable, or x.

  The **range** of a function is all possible values of the dependent variable, or y.
- There are two different types of data we may encounter: discrete data and continuous data.

  Discrete data is graphed as separate, distinct points.

  Continuous data is graphed as lines or curves.

Continued on next page

- **Inequalities** use the symbols $>$, $\geq$, $<$, $\leq$.

  Numbers may also be expressed as part of a **set**, such the set of real numbers or the set of positive integers.

  Interval notation uses brackets and parentheses:

  ][or ( ) mean the values are not included.

  [ ] means the values are included.
- A **piecewise function** is a function that has two or more equations for different intervals of the domain of the function.
- A composite function denoted as $f(g(x))$ or $(f \circ g)(x)$, can be thought of as a series of operations or machines. We read both of the notations as "f of g of x".
- When working with a composite function such as $f(g(x))$, start with the innermost function and work outwards.
- The domain and range of a composite function are based on the domain and range of the inside function, as well as of the composite function itself.
- The **inverse** of a function $f(x)$ is denoted $f^{-1}(x)$. It reverses the action of that function, so $(f \circ f^{-1})(x) = x$.

  In order for a function to have an inverse function, it must be a **one-to-one function**. This means it will have one value of y for every value of x, and one value of x for every value of y. A one-to-one function will pass both the vertical line test and the horizontal line test.
- Two functions, $f(x)$ and $g(x)$ are inverses if and only if: $(f \circ g)(x) = (g \circ f)(x) = x$
- To find the equation of an inverse function algebraically, switch x and y, and then isolate y.
- To sketch the graph of an inverse, graph the original function and pick some points (like the x-intercepts, y-intercepts, etc) and switch the coordinates. This will result in a reflection over the line $y = x$.

  $$(x, y) \rightarrow (y, x)$$

Developing inquiry skills

Return to the opening problem. How is what you have learned in this chapter useful in real-life situations such as this? Why is understanding the behaviour of functions essential to successfully modelling real-life situations?

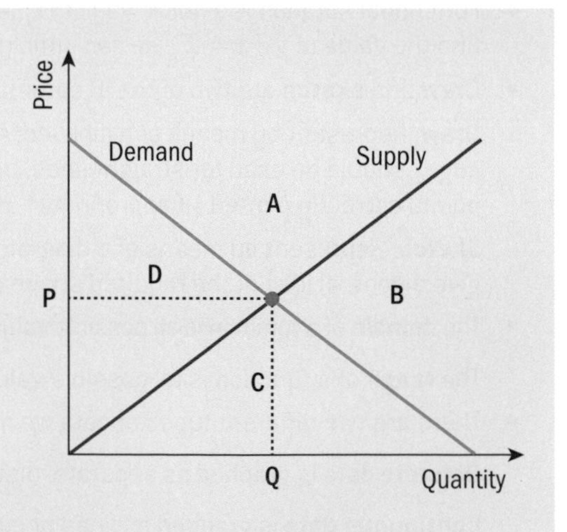

Chapter review

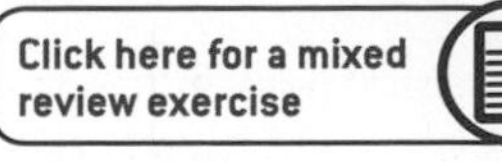

1 Decide whether the following relations are functions or not.

a

x	−9	−5	0	1	3	5
y	−8	−6	−9	−1	0	5

b

x	−11	−1	−1	11
y	5	7	0	8

c

x: -6, -4, 2, 3, 5

y: 2, 5, 7, 8

d

x: -1

y: 5, 6, 7, 8

e

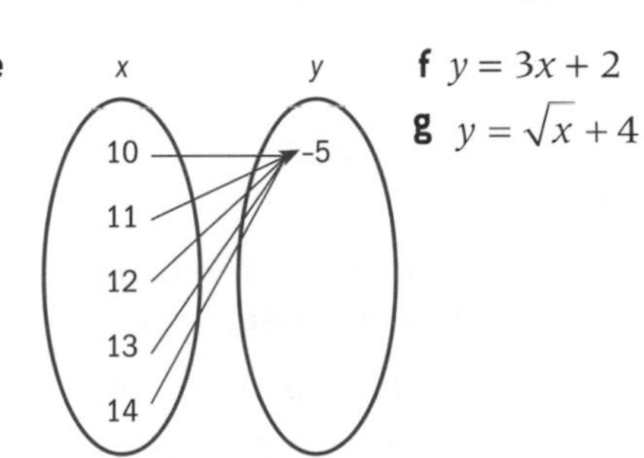

f $y = 3x + 2$

g $y = \sqrt{x} + 4$

h

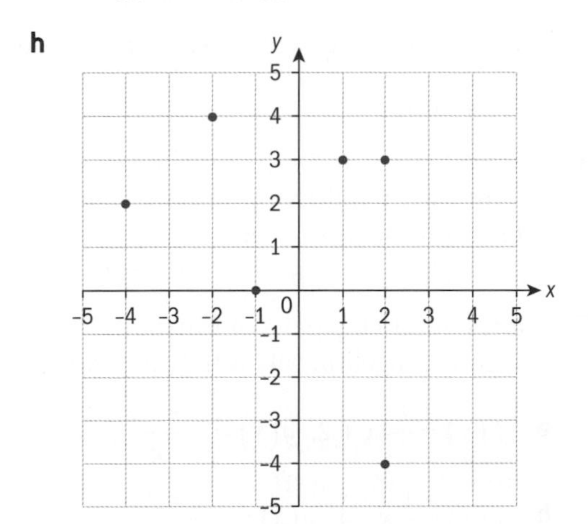

i

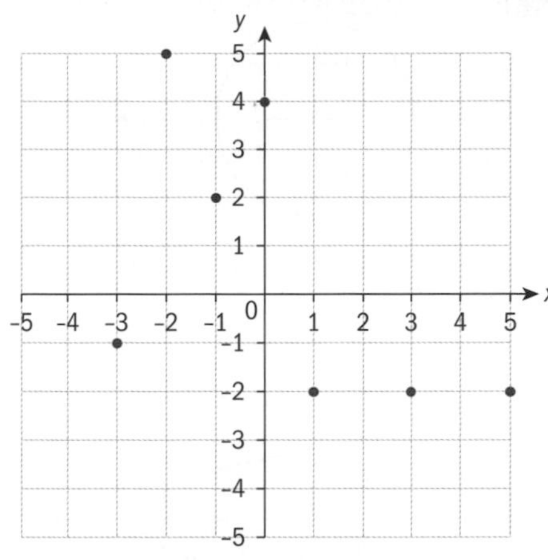

j

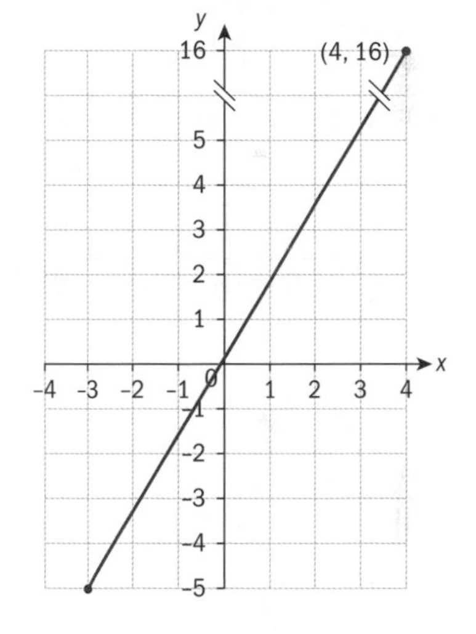

k

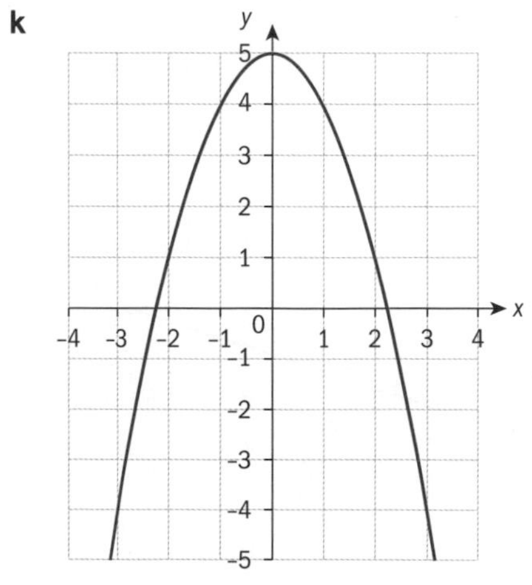

l

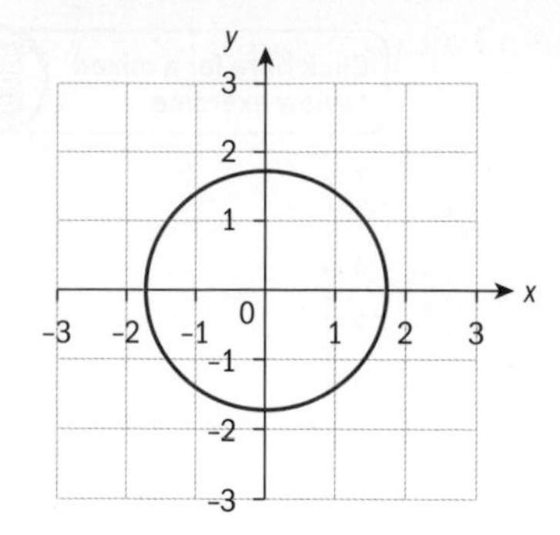

m

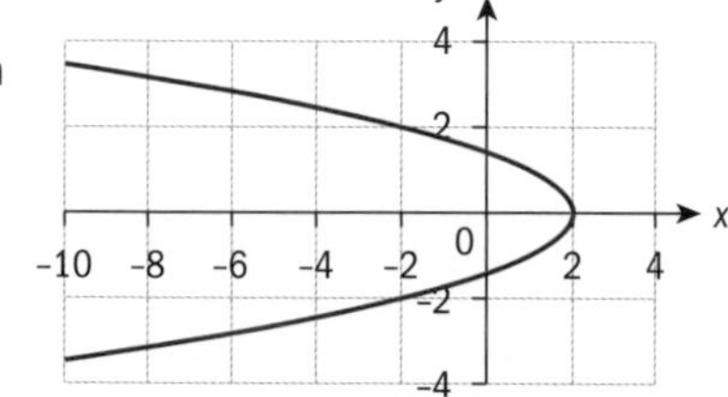

f

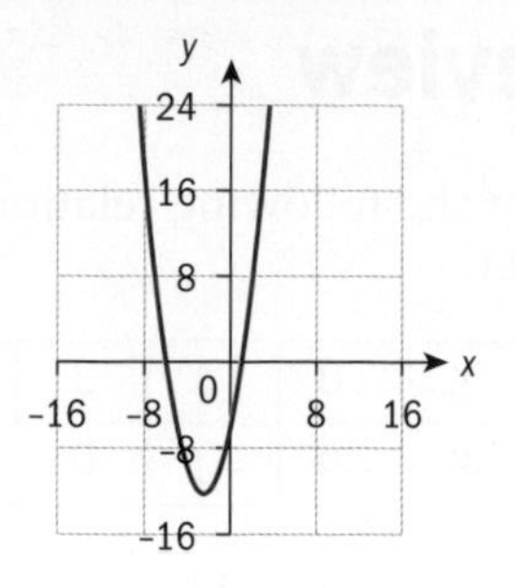

g

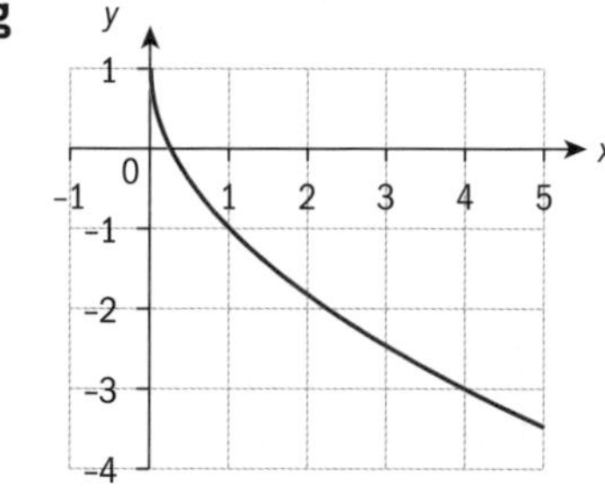

h

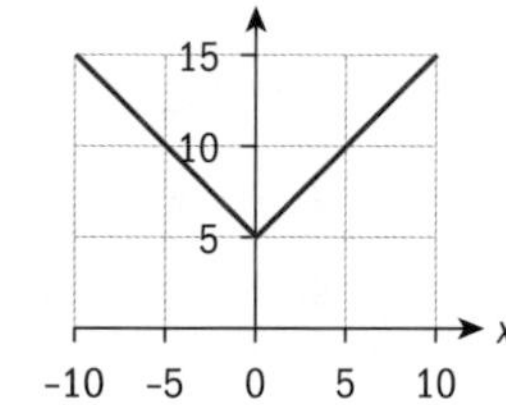

2 State the domain and range for the functions below:

a

x	-5	-1	0	1	4	9
y	-8	0	6	-1	9	1

b

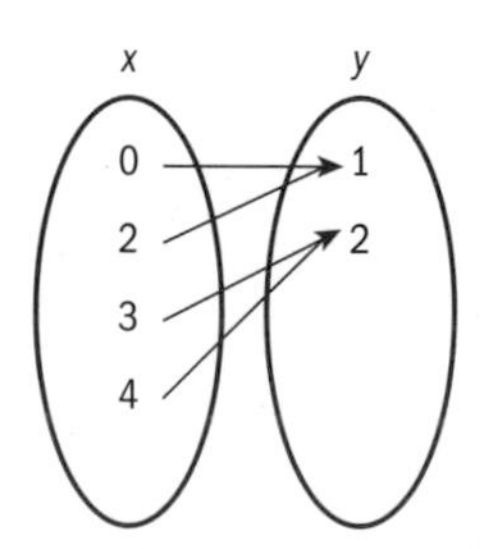

c $\{(-8, 2), (-5, 3), (0, 2), (1, -2)\}$

d $y = -\frac{1}{2}x + 1$

e

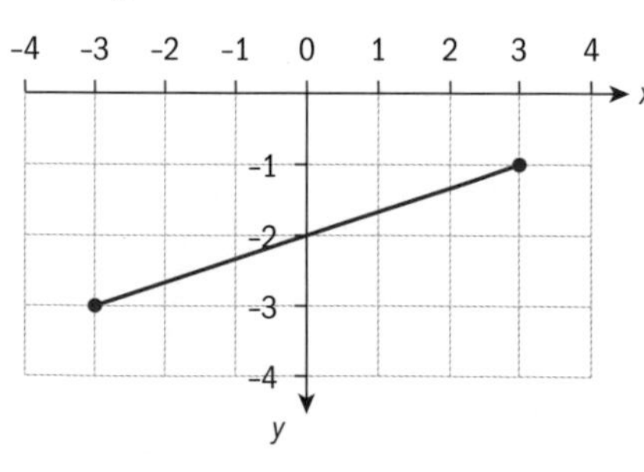

3 Use the functions $f(x) = x^2 - 6$, $g(x) = -2x$ and $h(x) = -4$ to evaluate:

a $f(3)$ **b** $f(-2)$ **c** $g(-6)$

d $f(1) + h(2)$ **e** $2f(0) - 2g(-1)$

f $h(0) \times f(-1)$ **g** $g^{-1}(-3)$ **h** $f(g(x))$

i $f \circ g^{-1}(x)$

4 Use your GDC to help you sketch the graphs of the following functions and state their domain and range:

a $y = |x^3| - 2$ **b** $y = 2x^4 - 5x^3 + x - 2$

5 For the following pairs of functions, determine algebraically if they are inverses:

a $f(x) = -4x + 2$, $g(x) = -\frac{x-2}{4}$

b $f(x) = \frac{1}{2}x - 4$, $g(x) = -\frac{x-2}{4}$

c $f(x) = \frac{1}{2}x^2 + 4$, $g(x) = 2x + \frac{1}{4}$

d $f(x) = \frac{2x+3}{3x-1}$, $g(x) = \frac{3+x}{3x-2}$

6 The graph below shows $y = f(x)$ for $-3 \le x \le 2$.

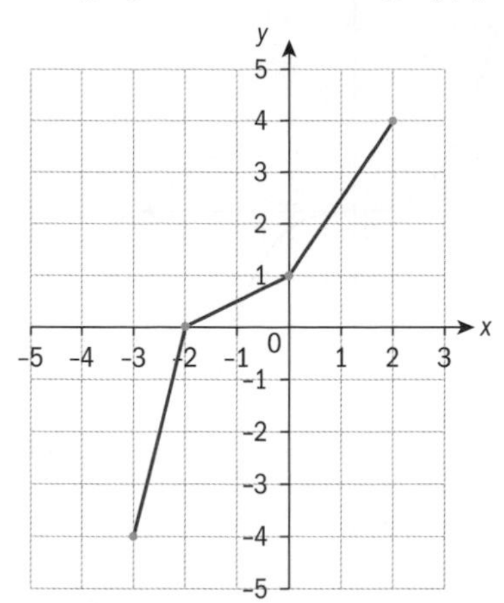

a i Write down the value of $f(-3)$.
 ii Write down the value of $f(2)$.

b Find the domain of f^{-1}.

c Sketch the graph of f^{-1}.

7 Let $f(x) = (x + 2)^3$. Let g be a function such that $(f \circ g)(x) = -8x^6$.

8 Let $f(x) = 2\sqrt{x} + x^2$. Let h be a function such that $h(16) = -2$. Find $(f \circ h^{-1})(-2)$.

9 Show that $f(x) = -\frac{3}{x}$ is a self-inverse function.

Exam-style questions

10 **P1:** Find the range of the following functions.

a $f(x) = 5x + 1$, domain $\{x \in \mathbb{R}, -5 \le x \le 5\}$ (2 marks)

b $f(x) = 4 - 2x$, domain $\{x = -1, 0, 1, 2, 3, 4\}$ (2 marks)

c $f(x) = x^2$, domain $\{x \in \mathbb{R}, 0 \le x \le 10\}$ (2 marks)

d $f(x) = 250 - 12.5x$, domain $\{x \in \mathbb{R}, 0 \le x \le 10\}$ (2 marks)

11 **P2:** $f(x) = 4x - 2$, $x \in \mathbb{R}$ and $g(x) = x^2 - 8x + 15$, $x \in \mathbb{R}$.

a Find $f(-2)$. (2 marks)

b Find $g(-2)$. (2 marks)

c Find an expression for $f^{-1}(x)$. (2 marks)

d Solve the equation $g(x) = 35$. (4 marks)

12 **P1:** A function is given by $f(x) = 128x - 15$, $-3 < x < 15$.

a Determine the value of $f\left(\frac{3}{2}\right)$. (2 marks)

b Determine the range of the function f. (4 marks)

c Determine the value of a such that $f(a) = 1162.6$ (2 marks)

13 **P1:** State **i** the domain, and **ii** the range for each of the following functions.

a

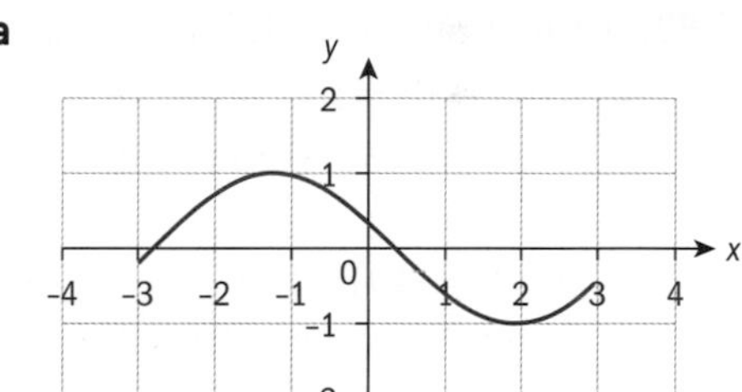

(2 marks)

b

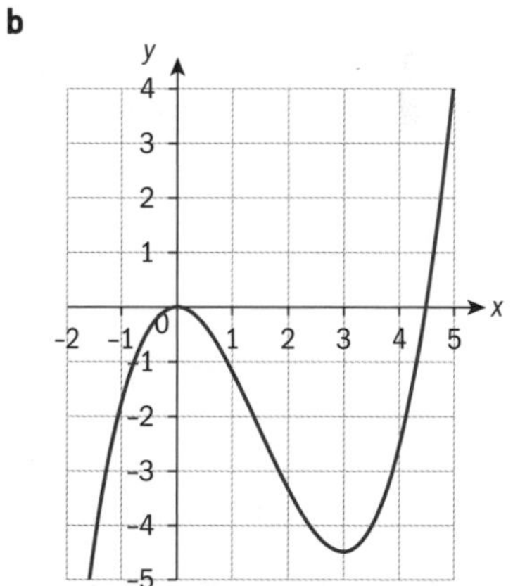

(2 marks)

c

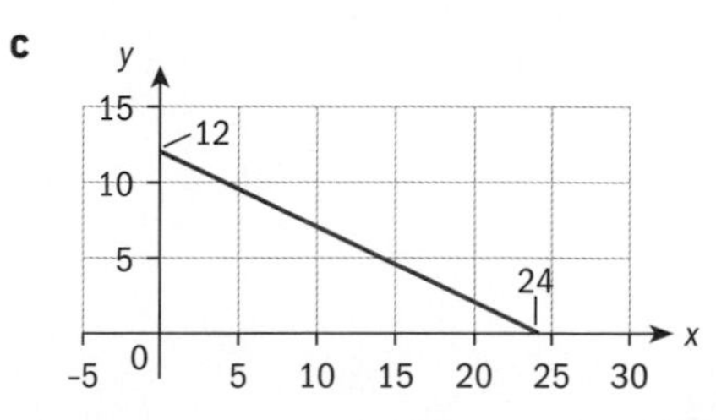

(2 marks)

Functions

d

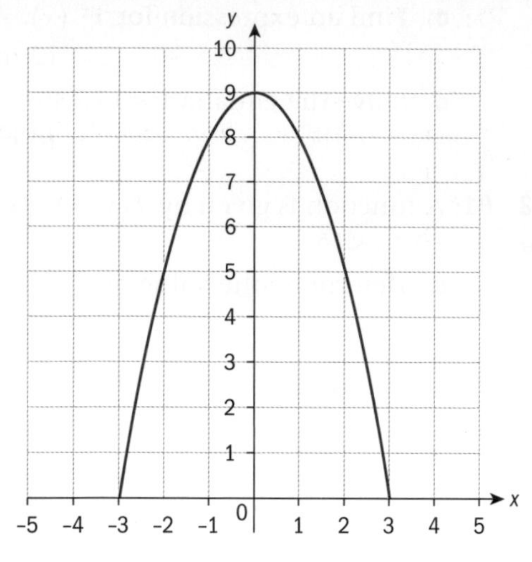

(2 marks)

14 P2: A function $f(x)$ is defined as $f(x) = 3x - 10$

a Given that the range of $f(x)$ is $5 < f(x) < 50$, find the domain of $f(x)$. (4 marks)

b Find $ff(10)$. (3 marks)

c Find the inverse function, $f^{-1}(x)$. (2 marks)

d State the range of the inverse function. (2 marks)

15 P1: State which of the following graphs represent functions, giving reasons for your answers.

a

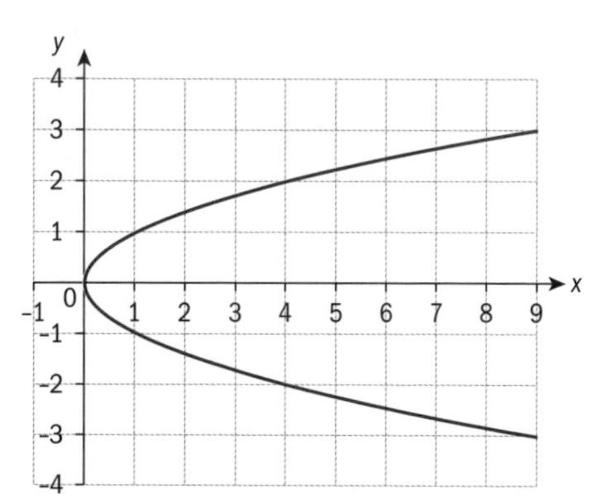

(2 marks)

b

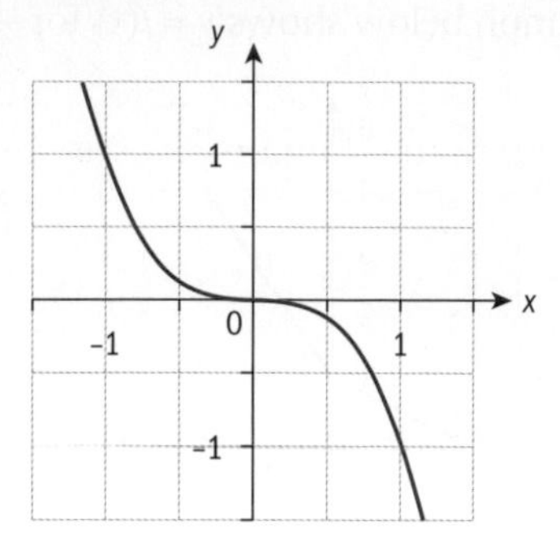

(2 marks)

c

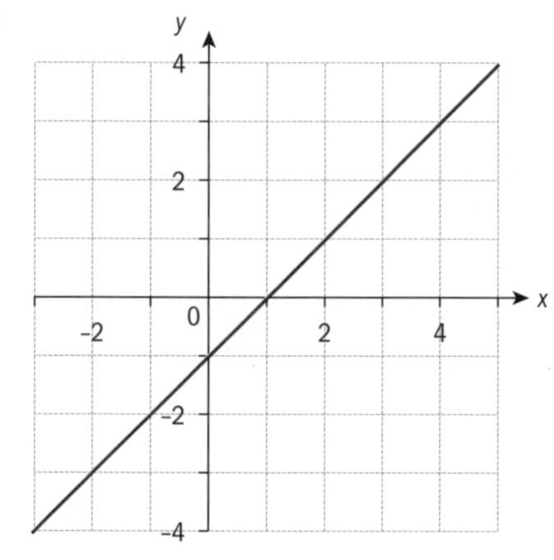

(2 marks)

d

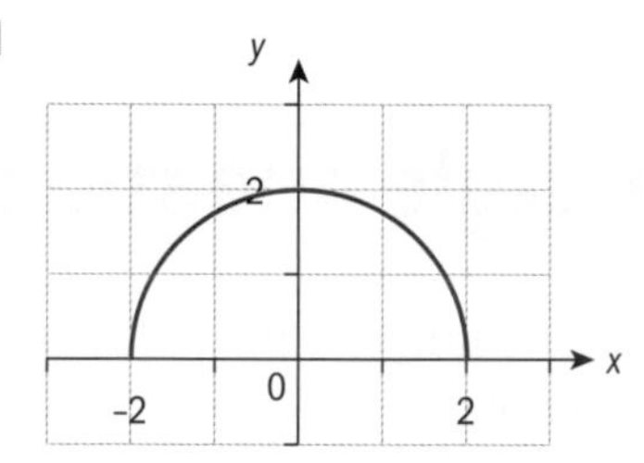

(2 marks)

16 P1: Consider the function $f(x) = \dfrac{k}{x-1} + 1$, $x > 1$, $x \in \mathbb{R}$, $k \in \mathbb{R}$.

a Show that $f(x)$ is a self-inverse function. (3 marks)

b State the range of f. (2 marks)

c Sketch the graph of $y = f(x)$. (2 marks)

17 P1: Consider the function $f(x) = \dfrac{1-2x}{3x+6}$, $x \neq -2$, $x \in \mathbb{R}$.

a State the range of f. (1 mark)

b Find the inverse function $f^{-1}(x)$. (3 marks)

c State the domain and the range of $f^{-1}(x)$. (2 marks)

18 P1: Consider the functions $f(x) = x^2$, $x \in \mathbb{R}$; and $g(x) = 2x - 1$, $x \in \mathbb{R}$.

a Solve the equation $f(x) = g(x)$. (3 marks)

b Solve the equation $fg(x) = gf(x)$. (5 marks)

19 P2: Katie organises a party for her work colleagues. She has a maximum budget of \$1000.

The cost to rent a local hall is \$430 for the evening.

She also has to budget for food, which will cost approximately \$14.50 per person.

a Write down a formula connecting the total cost of the party (\$$C$) with the number of people attending the party(p). (2 marks)

b Explain why $C = f(p)$ is a function. (1 mark)

c Derive an expression for p in terms of C. (2 marks)

d Hence, calculate the greatest number of people Katie is able to invite. (2 marks)

e Given that only 16 people attend the party, calculate how much each guest should be charged so that Katie covers her costs. (3 marks)

20 P1: The function $h(x)$ is defined as $h(x) = \frac{x}{3} + 2$, $x \geq 0$, $x \in \mathbb{R}$.

a State the range of $h(x)$. (1 mark)

b Derive an expression for the inverse function, $h^{-1}(x)$. (3 marks)

c Find an expression for $hh(x)$ in the form $hh(x) = ax + b$, where a and b are constants. (3 marks)

d Solve the equation $h(x) = h^{-1}(x)$. (2 marks)

e Explain why the equation $h(x) = h^{-1}(x)$ has the same solutions as the equation $h(x) = x$. (1 mark)

21 P1: The function $p(x)$ is defined by $p(x) = x^2 + 4x - 11$, $x \in \mathbb{R}$.

Given that $p(x) = fgh(x)$ and $f(x) \neq x$, $g(x) \neq x$, $h(x) \neq x$, find possible functions for $f(x)$, $g(x)$, and $h(x)$. (5 marks)

22 P1: Consider the functions $f(x) = x^2 - 4$, $g(x) = \frac{1}{x+1}$, $h(x) = 2^x$, $x \in \mathbb{R}$.

a Find the range of $f(x)$. (1 mark)

b Find the range of $g(x)$. (1 mark)

c Find the range of $h(x)$. (1 mark)

d Find an expression for $gf(x)$. (2 marks)

e Solve the equation $gf(x) = 9$. (2 marks)

f Solve the inequality $gh(x) > \frac{1}{17}$. (5 marks)

23 P1: Consider the function $p(x) = x^3$, $-2 \leq x \leq 2$, $x \in \mathbb{R}$.

a Find the range of $p(x)$. (2 marks)

b Find an expression for the inverse function $p^{-1}(x)$. (2 marks)

c Find all the solutions to the equation $p(x) = p^{-1}(x)$. (2 marks)

d Sketch the graphs of $y = p(x)$ and $y = p^{-1}(x)$ on the same axes. (2 marks)

24 P1: a Show that $r(x) = \frac{3x+5}{4x-3}$ $x \in \mathbb{R}$, $x \neq \frac{3}{4}$ is a self-inverse function. (3 marks)

b Hence determine the value of $rr(5)$. (2 marks)

25 P1: The function $f(x)$ is one-to-one and defined such that $f(x) = x^2 - 6x + 13$, $x \geq k$, $x \in \mathbb{R}$, $k \in \mathbb{R}$.

a Find the least possible value for k. (3 marks)

b Find an expression for the inverse function $f^{-1}(x)$. (3 marks)

c State the domain and the range of $f^{-1}(x)$. (2 marks)

26 P1: Given that $f(x) = x - 3$ and $gf(x) = 2x^2 + 18$, derive an expression for the function $g(x)$. (4 marks)

Graphs of functions: describing the "what" and researching the "why"

Approaches to learning: Thinking skills, Communicating, Research
Exploration criteria: Presentation (A), Mathematical communication (B), Personal engagement (C)
IB topic: Graphs, Functions, Domain

Bulgaria population data

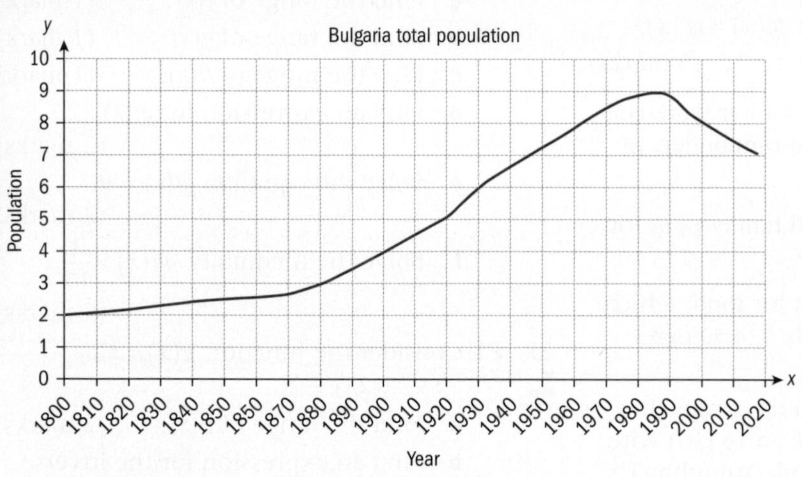

This graph includes two essential elements:

- a title
- x- and y-axes labels with units.

Without any research, write a paragraph to describe this graph.

Do not just describe the graph, but also explain why it might have this shape. Include any interesting points and regions on the graph "where things happen".

Using sources

You can use your general knowledge, printed sources and internet sources to research data. Different sources can often give different explanations, and not all sources are valid, useful or accurate.

How do you know if a source is reliable?

Use Internet research to find out more precisely what happened at the key dates shown on the Bulgaria population graph.

Keep a record of any sources that you use.

Could you use the graph to predict what might happen to the population of Bulgaria in the future? Explain your answer.

Worldwide Wii console sales

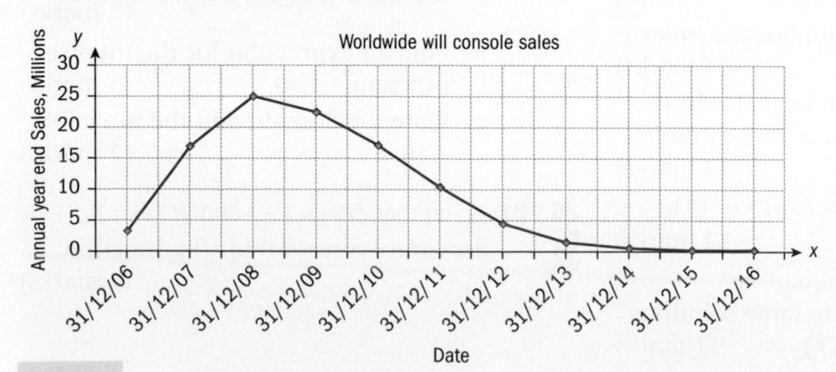

Initially **without research** write a paragraph about this graph, then research the reasons for the shape of this graph.

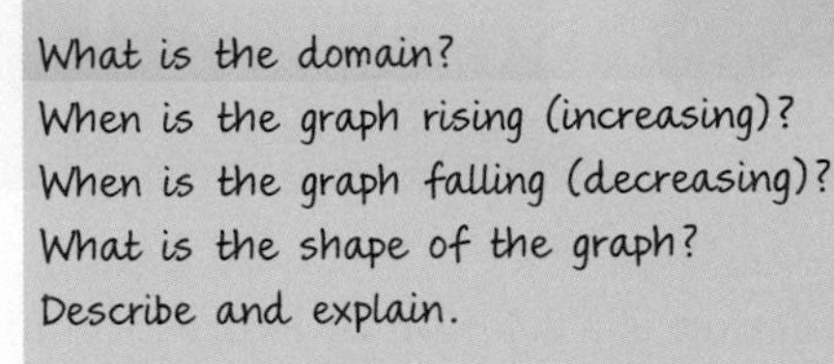

Global mean temperature anomaly

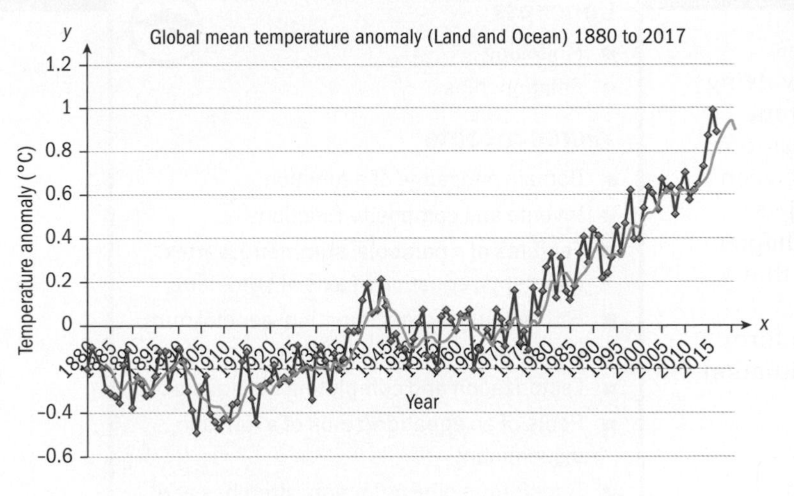

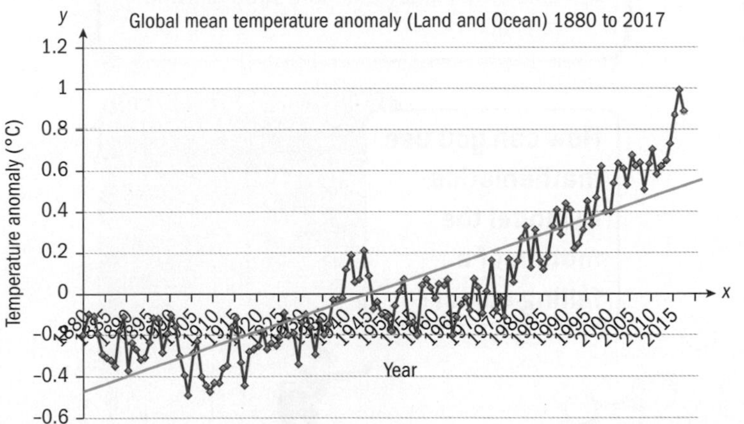

Research what is meant by "global mean temperature anomaly".

This data is based on deviations from the base average for 1951 to 1980.

On the graph in red the 5-year moving average trend line is included. What is a 5-year moving average?

On the second graph is a linear trend line. What is a linear trend line?

What are the advantages and disadvantages of each of the representations of the data shown?

Describe the data, note any interesting points or trends and to try to explain and investigate why the trends may be as they are.

TOK

This is a potentially controversial topic with many opinions and theories. How can you protect against your own biases?

Extension

Find and research a graph from the news or an academic journal from one of your subjects or another source.

Describe and explain the trends and the reason for the shape of the graph.

Now display or print out the graph.

Write a series of questions for other students to answer that encourage them to describe and explain the graph.

3 Modelling relationships: linear and quadratic functions

A function is a relation or expression involving one or more variables. Linear and quadratic functions are used to model real-life situations such as exchange rates, relationships between temperature scales, motion of falling objects and projected income or profit. In this chapter you will study three forms of equations that represent linear functions and three forms that represent quadratic functions. Each form provides useful information about the situation the function models.

Concepts

- Modelling
- Relationships

Microconcepts

- Domain and range of a function
- Inverse and composite functions
- Features of a parabola: symmetry, vertex, intercepts, equation of axis of symmetry
- Forms of a quadratic function: general form, intercept form and vertex form
- Factorization and completing the square
- Roots of an equation/zeros of a function, discriminant
- Transformations: reflection, stretches and translation

How can you represent the relationship between two currencies?

How can you use mathematics to model the motion of a falling object?

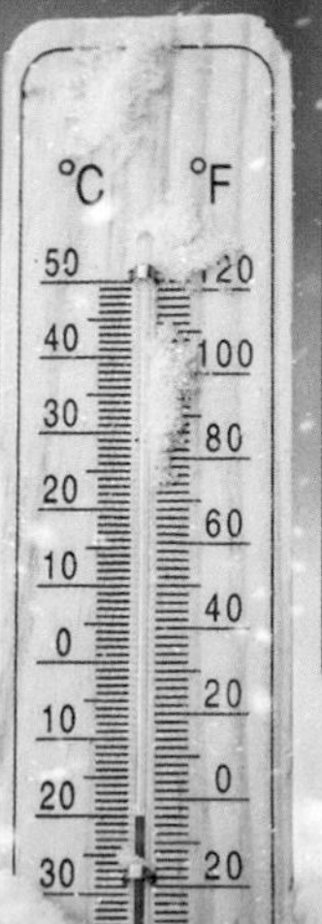

How much does the temperature change in the Celsius scale for each one-degree increase in the Fahrenheit scale?

How does raising the ticket price for an event change the projected income from the event?

Crates of emergency supplies are sometimes dropped from cargo planes to humanitarian aid workers.

Each crate has its own parachute. Pilots must determine the height and time of the drop that will allow the containers to reach the ground intact. The crate is said to be in free fall during the time before the parachute opens. The functions h and g give the crate's height above the ground, measured in metres, t seconds after the crate leaves the plane.

$$\text{During free fall: } h(t) = -4.9t^2 + 720$$

$$\text{With parachute open: } g(t) = -5t + 670$$

Why are two models needed to represent the height of the crate?

For the model representing motion during free fall, what does the constant value 720 represent?

For the model representing motion with the parachute open, what does -5, the coefficient of t, represent?

How long after the crate leaves the plane does the parachute open?

What is the domain of each function in this context?

Think about the questions in this opening problem and answer any you can. As you work through the chapter, you will gain mathematical knowledge and skills that will help you to answer them all.

Developing inquiry skills

Write down any similar inquiry questions you might ask to model the path of something in sport, for example: determining where an archer's arrow would land, deciding whether a tennis ball would land within the baseline, or considering whether a high-jumper would pass over the bar successfully.

Before you start

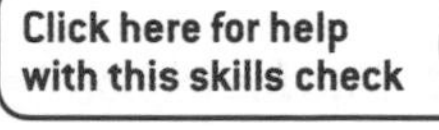

You should know how to:	Skills check
1 Solve simple equations for a given variable. **a** Solve for x: $4x - 5 = 0$ $4x = 5$ $x = \frac{5}{4}$ **b** Solve for n: $n^2 + 6 = 11$ $n^2 = 5$ $n = \pm\sqrt{5}$	**1** Solve each equation for the given variable: **a** $4x + 6 = x - 3$ **b** $3t^2 - 20 = 1$ **c** $2(a - 3) = 6(a + 2)$
2 Factorize mathematical expressions: eg Factorize $4x^2 - 12x$ $4x^2 - 12x = 4x(x - 3)$ eg Factorize $x^2 - 4x - 12$ $x^2 - 4x - 12 = (x + 2)(x - 6)$ eg Factorize $9a^2 - 16$ This is an example of the difference of two perfect squares. $9a^2 - 16 = (3a - 4)(3a + 4)$	**2** Factorize each expression: **a** $3m^2 - 15m$ **b** $x^2 - 36$ **c** $n^2 + 8n + 7$ **d** $4x^2 + x - 3$ **e** $9x^2 + 18x$ **f** $2a^2 - 3a - 5$ **g** $12x^2 + 5x - 2$ **h** $16a^2 - 49b^2$

3.1 Gradient of a linear function

Gradient

Investigation 1

The three tables below each show two linked quantities. The quantity in the first column is called the independent variable. As the independent variable changes, the quantity in the second column (the dependent variable) changes as a result.

For example, in table 1 the number of days that you rent the car is the independent variable—you choose how long you rent the car. The cost of the car rental is the dependent variable—it increases with the number of days you rent the car.

Cost of renting a car

Number of days	Cost in euros
1	30.25
2	60.50
3	90.75
4	121.00
5	151.25

Table 1

Equivalent temperatures

Degrees Celsius	Degrees Fahrenheit
0	32
15	59
20	68
30	86
35	95

Table 2

Predicted model rocket height

Time after launch (s)	Height (m)
1	35
3	75
4	80
5	75
8	0

Table 3

The **rate of change** tells you how quickly the dependent variable changes as a result of the independent variable changing.

In table 1, the rate of change of the rental cost between days 1 and 2 is given by:

$$\text{Rate of change} = \frac{\text{the change in the dependent variable}}{\text{the change in the independent variable}} = \frac{60.50-30.25}{2-1} = \frac{30.25}{1} = 30.25.$$

1 Find the rate of change in the cost of car rental between days 2 and 3, days 3 and 4, and days 4 and 5. Is this rate of change constant between all points?

2 Explain what information the rate of change you found in table 1 gives you.

3 Find the rate of change between each consecutive pair of values for tables 2 and 3. Are these rates of change constant between all points?

4 Explain what information the rates of change you found in tables 2 and 3 gives you.

5 Use your GDC to plot a scatter graph of the data in tables 1, 2 and 3.

6 What do you notice about the graphs of quantities where the rate of change is constant between all points in the table? What do you notice about the shape of graphs where the rate of change is *not* constant between all the points in the table?

7 **Factual** If you are given a graph which shows the relationship between two variables, how can you tell from the shape of the graph whether the rate of change between the variables is constant?

8 **Factual** If you are given a table of values which shows the relationship between two variables, how can you tell whether the rate of change between the variables is constant?

9 **Conceptual** Why does a linear graph represent a constant rate of change?

10 In table 1, the number of days and cost in euros have a linear relationship. Explain what this means.

11 **Conceptual** What is the relationship between real-life variables that have a constant rate of change, and how can this relationship be represented?

There is a constant rate of change between any two points on the graph of a line. This constant rate of change is called the **gradient** or **slope** of the line.

Reflect What are some other real-life examples that show how the gradient of a line can be interpreted as a rate of change?

Functions

The gradient of a straight line is a measure of how steep the line is. The steeper the graph is, the quicker the dependent variable changes with respect to the change in the independent variable.

The gradient is constant along the entire length of any line. You can find the gradient by plotting the graph of the line, choosing any two points on the line, and finding the horizontal change and vertical change between these two points. The gradient is then given by:

$$\text{gradient} = \frac{\text{vertical change}}{\text{horizontal change}}$$

Example 1

Find the gradient of each line.

a

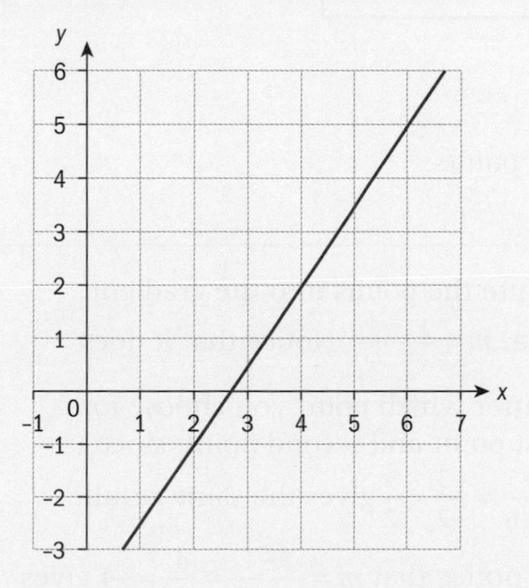

b

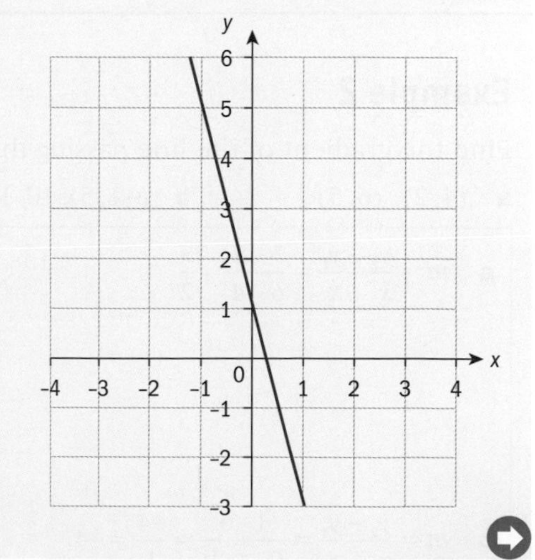

Continued on next page

a gradient $= \dfrac{\text{vertical change}}{\text{horizontal change}} = \dfrac{3}{2}$	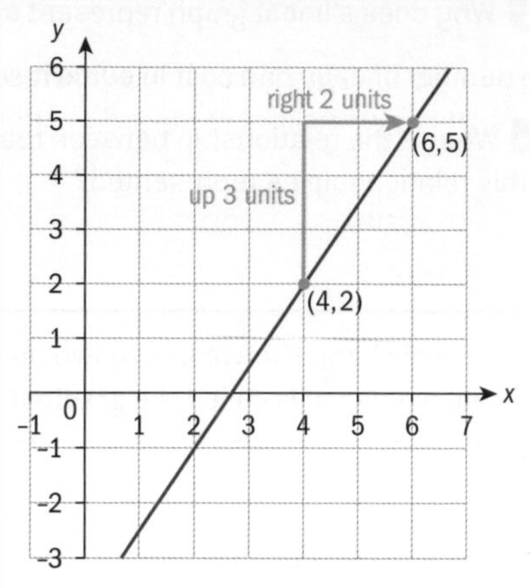
b gradient $= \dfrac{\text{vertical change}}{\text{horizontal change}} = \dfrac{-4}{1} = -4$	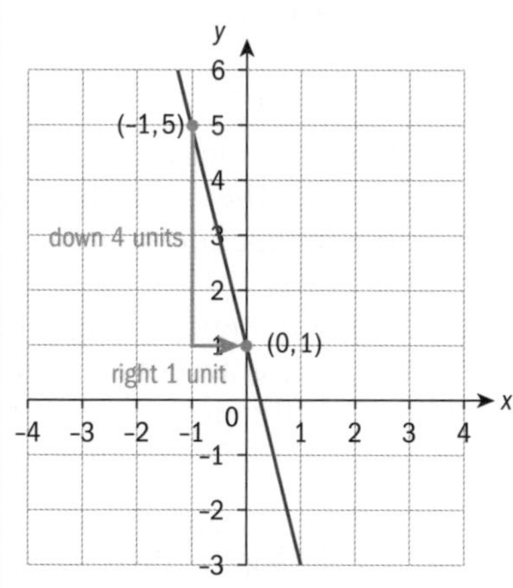

The formula for the gradient (m) of a line is $m = \dfrac{y_2 - y_1}{x_2 - x_1}$, where (x_1, y_1) and (x_2, y_2) are any two points on the line.

Example 2

Find the gradient of the line passing through the given points:

a $(4, 2), (6, 5)$ **b** $(-1, 5), (0, 1)$

a $m = \dfrac{y_2 - y_1}{x_2 - x_1} = \dfrac{5-2}{6-4} = \dfrac{3}{2}$	Substitute the points into the gradient formula, $m = \dfrac{y_2 - y_1}{x_2 - x_1}$. Notice that it does not matter which point you choose for the first point and second point, since $m = \dfrac{2-5}{4-6} = \dfrac{-3}{-2} = \dfrac{3}{2}$ gives the same result.
b $m = \dfrac{y_2 - y_1}{x_2 - x_1} = \dfrac{1-5}{0-(-1)} = \dfrac{-4}{1} = -4$	Again, notice that $m = \dfrac{5-1}{-1-0} = \dfrac{4}{-1} = -4$ gives the same result.

Example 3

On a certain day in 2017, four people exchanged US dollars (USD) to Canadian dollars (CAD). The amounts they exchanged and received are shown in the table.

USD	CAD
5	6.25
10	12.50
20	25.00
50	62.50

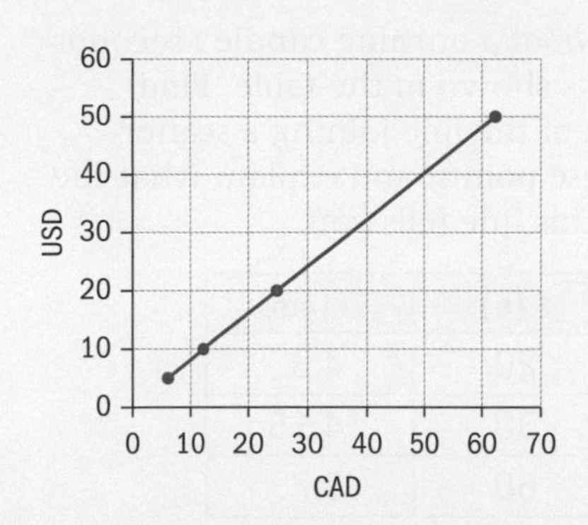

This data is linear. A scatter plot is graphed to show these values of USD against CAD. Find the gradient of the line joining these points, and explain what the gradient of this line tells you.

$\text{Gradient} = \frac{62.50 - 12.50}{50 - 10} = \frac{50}{40} = \frac{1.25}{1}$	Since the data is linear, the gradient is constant. So, you can choose any two points in the table to substitute into the gradient formula. The points (50, 62.50) and (10, 12.5) are used here.
This gradient tells you that one USD is worth 1.25 CAD.	This value is called the "exchange rate". You might also say that for each increase of one USD, the equivalent amount in CAD increases by 1.25.

Exercise 3A

1 Find the gradient of each line:

a

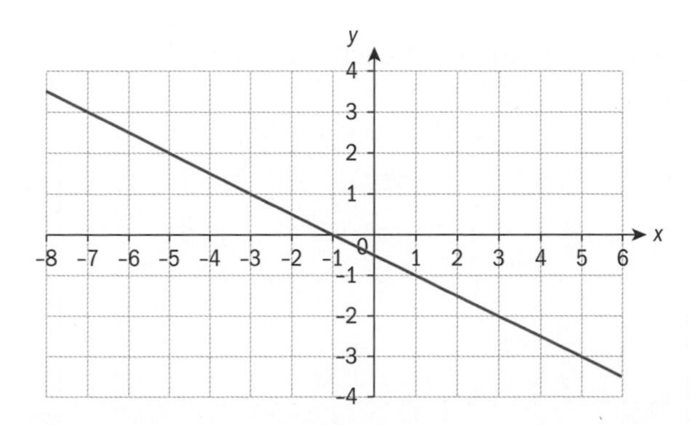

b

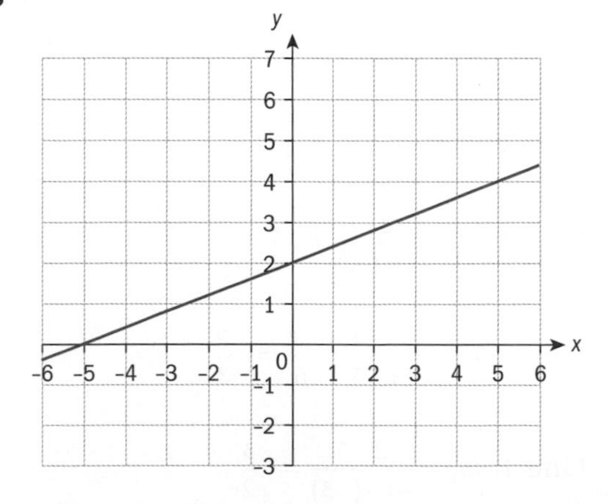

2 Find the gradient of the line passing through the given points:

a $(4, 8), (8, 11)$ **b** $(-2, 2), (4, -4)$

c $(-7, 1), (7, 8)$

3 The height, h, of a burning candle t seconds after it is lit is shown in the table. Find the gradient of the line joining a scatter graph of these points, and explain what the gradient of this line tells you.

t (s)	h (cm)
20	4.3
30	4.15
60	3.7

4 Baldwin Street in Dunedin, New Zealand, is known as the steepest street in the world.

Baldwin Street is a short residential street about 350 metres long. From its base at approximately 30 metres above sea level, it rises to a height of about 100 m above sea level. The segment shown in the following diagram models Baldwin Street.

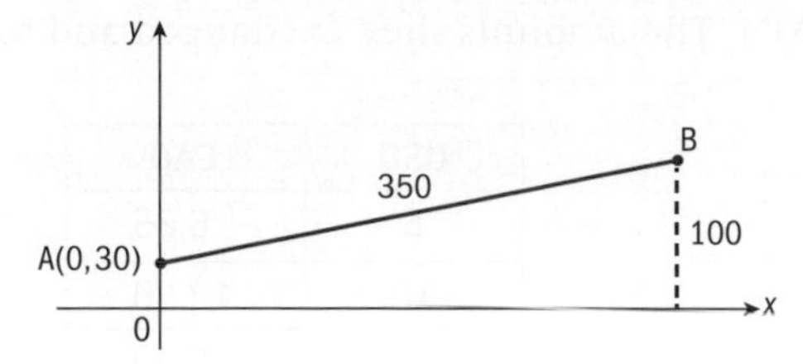

a Find the coordinates of B, to the nearest hundredth.

b Find the gradient of segment AB, to the nearest hundredth.

The grade of a road is given as a percent.

It is calculated by the formula:

$\frac{\text{rise}}{\text{run}} \times 100$ (or $\frac{\text{vertical change}}{\text{horizontal change}} \times 100$).

c Find the grade of Baldwin Street.

For two lines L_1 and L_2 with gradients m_1 and m_2 respectively:

- L_1 and L_2 are parallel if $m_1 = m_2$.
- L_1 and L_2 are perpendicular if $m_1 \times m_2 = -1$.

Example 4

Find gradient of the lines passing through the given points.

Identify any lines that are parallel or perpendicular, and justify your answer.

Line 1: $(-4, 3)$ and $(2, 7)$ Line 2: $(-2, 0)$ and $(1, -2)$

Line 3: $(5, 1)$ and $(11, -3)$ Line 4: $(-3, 5)$ and $(-1, 2)$

Line 1: $m_1 = \frac{7-3}{2-(-4)} = \frac{4}{6} = \frac{2}{3}$ Line 2: $m_2 = \frac{-2-0}{1-(-2)} = -\frac{2}{3}$ Line 3: $m_3 = \frac{-3-1}{11-5} = \frac{-4}{6} = -\frac{2}{3}$ Line 4: $m_4 = \frac{2-5}{-1-(-3)} = -\frac{3}{2}$	Find the gradient of each line using the formula $m = \frac{y_2 - y_1}{x_2 - x_1}$.

Lines 1 and 4 are perpendicular, since $m_1 \times m_4 = \left(\frac{2}{3}\right) \times \left(-\frac{3}{2}\right) = -1$.	Since two lines L_1 and L_2 are perpendicular if $m_1 \times m_2 = -1$
Lines 2 and 3 are parallel, since $m_2 = m_3 = -\frac{2}{3}$.	Since two lines L_1 and L_2 are parallel if $m_1 = m_2$

Exercise 3B

1 Determine whether lines 1 and 2 which pass through the given points are parallel, perpendicular or neither.

a Line 1: $(3, 6)$ and $(6, 11)$
Line 2: $(4, -1)$ and $(9, 2)$

b Line 1: $(5, -1)$ and $(3, 7)$
Line 2: $(-1, 4)$ and $(0, 0)$

2 A line passes through the points $(3, 2)$ and $(x, 5)$ and is perpendicular to a line with gradient $\frac{4}{3}$. Find the value of x.

3 Liam works up to 60 hours each week. His weekly pay, in dollars, depends on the number of hours he works, as shown in the graph.

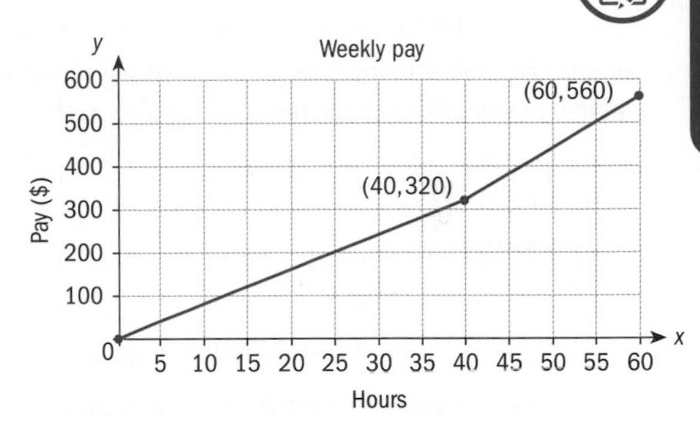

a Find the gradient for each line segment in the graph.

b Explain the meaning of each gradient in the context of Liam's work.

3.2 Linear functions

Why do we have more than one scale for measuring temperature?

What represents a greater increase in temperature: an increase of one-degree Celsius or an increase of one-degree Fahrenheit?

At what temperature is degrees Celsius equal to degrees Fahrenheit?

In section 3.1 you looked at the rate of change, or gradient, between two connected variables.

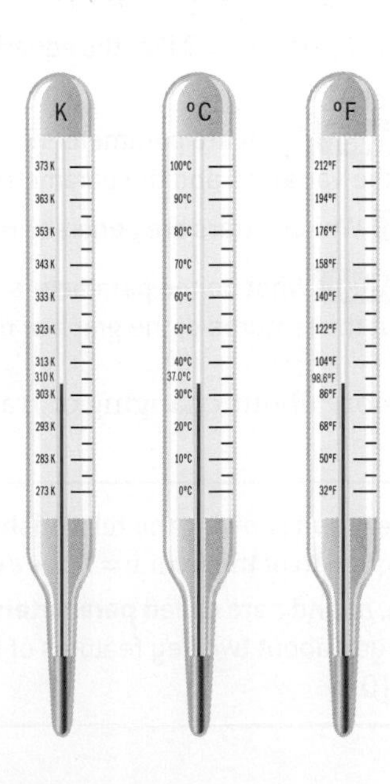

This section will take that idea further by finding how you can represent the relationship between two connected variables using an equation. If x represents the independent variable and y represents the dependent variable, you will learn how you can express y in terms of x.

Parameters of a linear graph

Investigation 2

1 Plot the graph of $y = x$ on your GDC (Note: $y = x$ is called the "parent graph" for linear functions. For every point on the line $y = x$, the dependent variable y is always equal to the independent variable x.)

a Write down the gradient of this line.

b Write down the coordinates of the point at which this graph crosses the y-axis (This is called the y-intercept of the line).

Consider equations of the form $y = x + c$ for a given constant c. Here, the dependent variable y is equal to the independent variable, x, plus a constant value c.

2 On your GDC, plot the graphs of $y = x$, $y = x + 2$, $y = x - 2$ and $y = x + 4$ on the same axes.

3 **Factual** Compare and contrast the graph of $y = x$ with the graphs of each of the other equations.

4 **Conceptual** What feature of the graph is defined by the constant c?

5 Consider equations of the form $y = mx$. Here, the dependent variable y is equal to the independent variable, x, multiplied by a constant value m.

a Sketch the graphs of $y = x$, $y = 5x$, $y = 2x$, $y = \frac{1}{3}x$ and $y = -2x$ on the same axes.

b Compare and contrast the graph of $y = x$ with the graphs of each of the other equations.

c **Factual** What feature of the graph is defined by the constant m?

6 Consider equations of the form $y = mx + c$.

a Without using your GDC, can you state the gradient and y-intercept of the graph of $y = \frac{1}{2}x - 3$? Plot the equation on your GDC to see if you are correct.

b **Conceptual** What are parameters?
Identify the variables and the parameters in the equation $y = mx + c$.

c **Factual** Which letters are generally used to represent parameters?

d **Conceptual** What do the parameters in the equation $y = mx + c$ tell you about the features of the graph of the line?

You will learn more about changing or transforming parent graphs in section 3.3.

Any straight-line graph showing the relationship between two variables x and y will have an equation of the form $y = mx + c$ where m and c are constants. In this equation, m and c are called **parameters** of the function. These parameters tell you about two key features of the graph: the gradient m and the y-intercept $(0, c)$.

Example 5

Find the equation of the following lines:

a

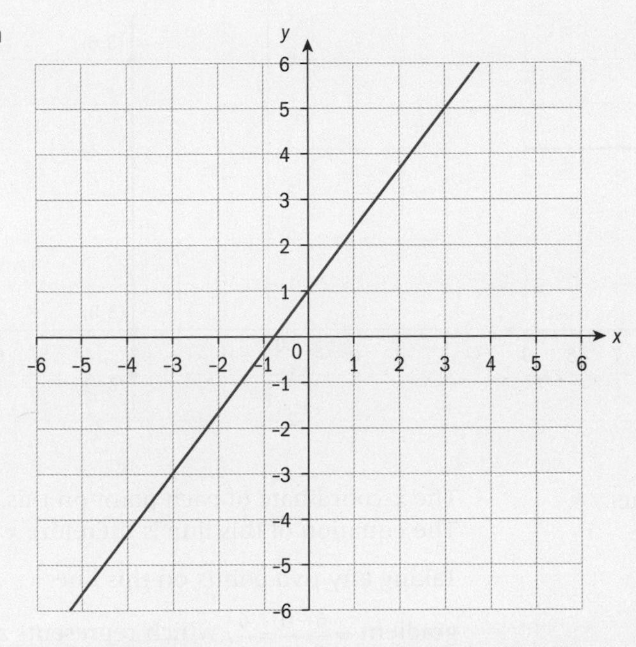

b The line with y-intercept $(0,-4)$ that is perpendicular to the line with equation $y = -\frac{7}{2}x + 6$

a The line crosses the y-axis at $(0,1)$, so $c = 1$	Find the value of c from the y-intercept.
Gradient $= \frac{5-1}{3-0} = \frac{4}{3}$	Find another point which lies on the line, such as $(3, 5)$, and use this along with $(0, 1)$ to find the gradient using the equation $m = \frac{y_2 - y_1}{x_2 - x_1}$.
$y = \frac{4}{3}x + 1$	Substitute m and c into the equation $y = mx + c$.
b $c = -4$	Since the y-intercept is $(0, -4)$, $c = -4$.
$m \times \left(-\frac{7}{2}\right) = -1$ $m = \frac{2}{7}$	Since two lines l_1 and l_2 are perpendicular if $m_1 \times m_2 = -1$
$y = \frac{2}{7}x - 4$	Substitute m and c into the equation $y = mx + c$.

Horizontal and vertical lines

Look at these two graphs.

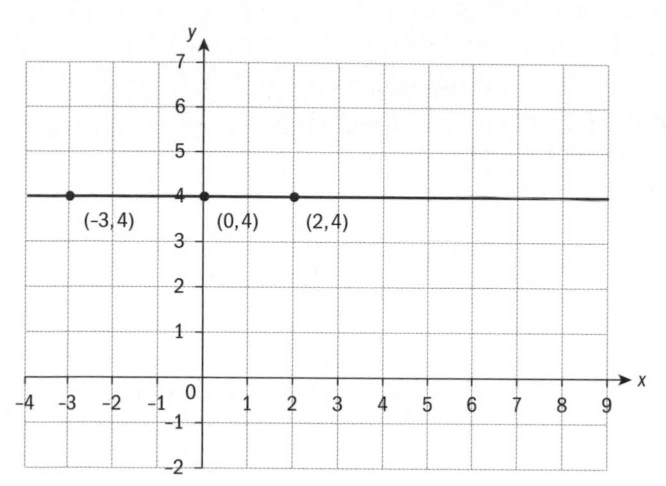

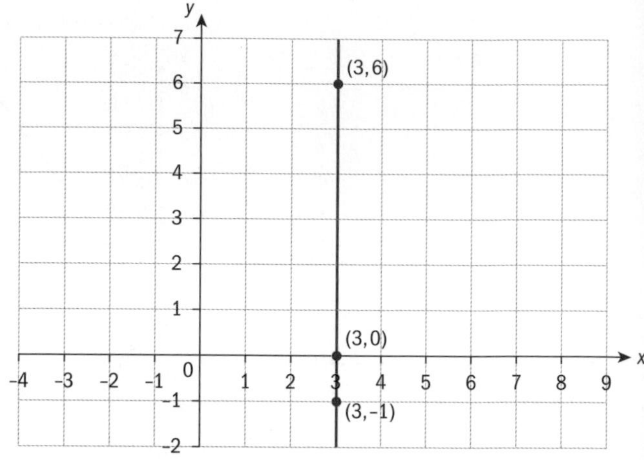

The y-coordinate of each point on this line is 4. The equation of this line is therefore $y = 4$.

Taking any two points on this line,

gradient $= \frac{4-4}{2-0} = \frac{0}{2} = 0$.

The x-coordinate of each point on this line is 3. The equation of this line is therefore $x = 3$.

Taking any two points on this line,

gradient $= \frac{6-0}{3-3} = \frac{6}{0}$, which represents an infinite value and so is undefined.

> The equation of a horizontal line with y-intercept $(0, c)$ is $y = c$.
> The equation of a vertical line with x-intercept $(b, 0)$ is $x = b$.
> The gradient of a horizontal line is 0.
> The gradient of a vertical line is undefined.

Example 6

Write down the following:

a the equation of the horizontal line that passes through $(0, -2)$

b the equation of the vertical line that passes through the point $(3, 6)$

c the equation of the line with gradient 0 which passes through the point $(-7, 4)$

a $y = -2$	The equation of a horizontal line with y-intercept $(0, -2)$ is $y = -2$.
b $x = 3$	For a vertical line, the x-coordinate remains constant so the equation of a vertical line passing through $(3, 6)$ is $x = 3$.
c $y = 4$	A line with gradient of 0 is horizontal and so the y-coordinate remains constant.

Exercise 3C

1 Write down the gradient and y-intercept for the following lines:

a line 1: $y = 3x - 7$

b line 2: $y = -\frac{2}{3}x + 4$

c line 3: $y = -2$

2 Write down the equation of the line with gradient $\frac{1}{5}$ passing through $(0, 1)$.

3 Find the equation, in gradient-intercept form, of the following lines:

a the line that passes through the point $(0, -1)$ and is parallel to the line $y = 4x - 3$

b the line that passes through the points $(-3, -2)$ and $(1, 10)$

4 Write down the following:

a the equation of the vertical line that passes through $(8, -1)$

b the equation of the horizontal line that passes through the point $(-3, -10)$

c the equation of the line perpendicular to $y = 1$ that passes through $(9, 5)$

d the point of intersection of the lines $x = -2$ and $y = 7$

Different forms for the equation of a straight line

The equation of a straight line written in the form $y = mx + c$ is said to be written in **gradient-intercept form**.

Reflect Can you explain why this form of the equation of a straight line is called the gradient-intercept form?

TOK

Descartes showed that geometric problems could be solved algebraically and vice versa.

What does this tell us about mathematical representation and mathematical knowledge?

There are, however, other ways in which you can write the equation of a straight line.

Consider a line that passes through the point (x_1, y_1) and let (x, y) be any other point on the line. Then the gradient of the line is given by $m = \frac{y - y_1}{x - x_1}$. This can be rewritten as $y - y_1 = m(x - x_1)$.

The **point-gradient form** of the equation of a line is $y - y_1 = m(x - x_1)$. In this equation the parameters tell you the gradient m and a point (x_1, y_1) that lies on the line.

Example 7

Find the equation, in point-gradient form, of the following lines:

a the line that passes through $(-4, 5)$ and is parallel to the line with equation $y = -\frac{1}{2}x - 3$

b the line that passes through the points $(-1, 2)$ and $(3, -4)$.

Continued on next page

a $m = -\frac{1}{2}$	Parallel lines have the same gradient.
$y - 5 = -\frac{1}{2}(x + 4)$	Substitute the point $(-4, 5)$ and $m = -\frac{1}{2}$ into the equation $y - y_1 = m(x - x_1)$.
b $m = \frac{-4 - 2}{3 - (-1)} = -\frac{3}{2}$	Determine the gradient of the line using the equation $m = \frac{y_2 - y_1}{x_2 - x_1}$.
$y - 2 = -\frac{3}{2}(x + 1)$ or $y + 4 = -\frac{3}{2}(x - 3)$	Substitute the gradient and either of the given points into the equation $y - y_1 = m(x - x_1)$.

Reflect How could you verify that both answers shown for part **b** are equations of the same line?

Example 8

Draw the graph of the line with the given equation:

a $y = 4x - 3$ **b** $y = -\frac{2}{3}x$ **c** $y + 4 = -2(x - 3)$ **d** $y = 3$

a $y = 4x - 3$ 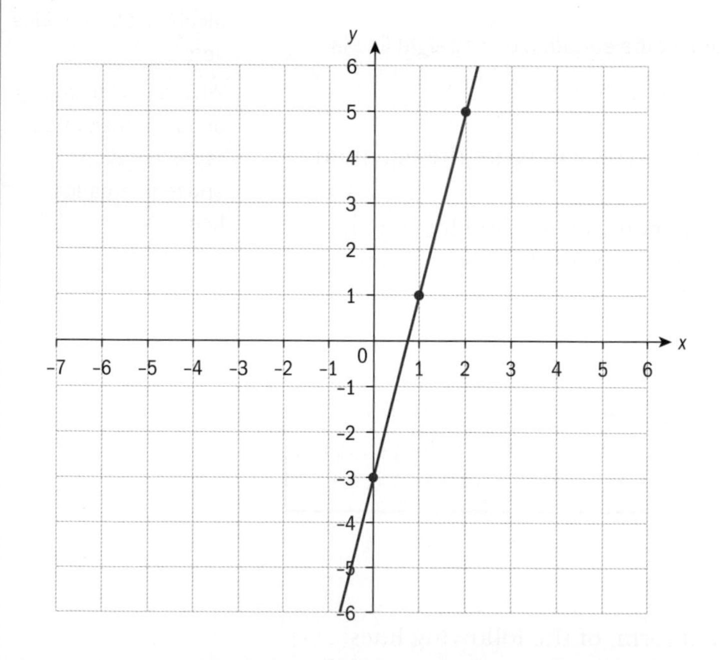	From the parameters in the equation, you know that the y-intercept is $(0, -3)$ and the gradient is 4 or $\frac{4}{1}$. Plot the point $(0, -3)$. From this point, move up 4 and right 1 to plot a second point. Since $\frac{-4}{-1} = 4$, you can also move down 4 and left 1 to get from one point to another point on the line.

b $y = -\frac{2}{3}x$

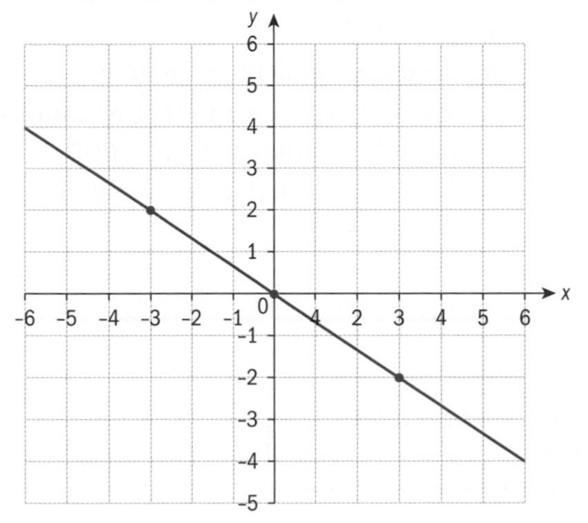

The equation $y = -\frac{2}{3}x$ can be written as $y = -\frac{2}{3}x + 0$. Therefore, the y-intercept is $(0, 0)$ and the gradient is $-\frac{2}{3}$.

Plot the point $(0, 0)$.

From this point, move down 2 and right 3 since $-\frac{2}{3} = \frac{-2}{3}$. Alternatively, you can move up 2 and left 3, since $-\frac{2}{3} = \frac{2}{-3}$.

c $y + 4 = -2(x - 3)$

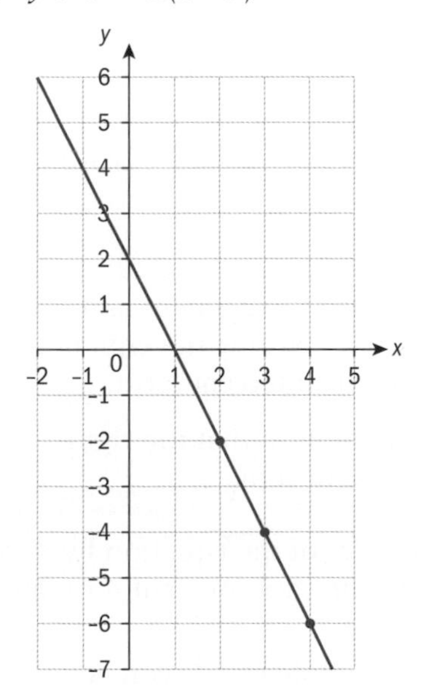

From the point-gradient form of the equation of the line you can determine that the line passes through $(3, -4)$ and has gradient $-2 = \frac{-2}{1}$ or $\frac{2}{-1}$.

Plot the point $(3, -4)$. From this point, move down 2 and right 1 to plot a second point. Alternatively, you could move up 2 and left 1.

d $y = 3$

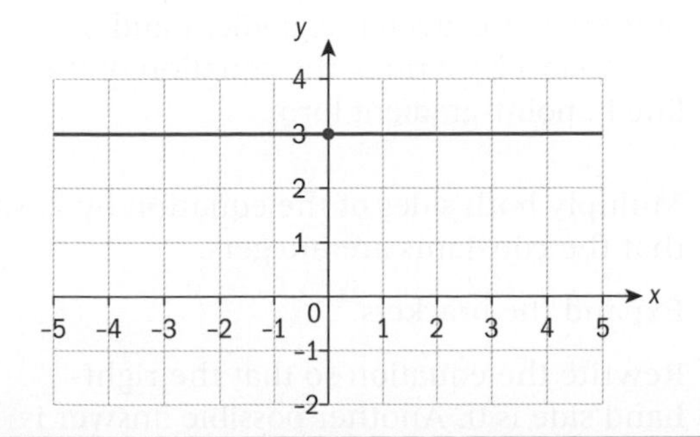

When y is constant, the graph is a horizontal line. The line passes though the point $(0, 3)$.

Notice, you could write $y = 3$ as $y = 0x + 3$. When the gradient is 0, the graph of the line is horizontal.

Exercise 3D

1 Draw the graph of the line with the given equation:

a $y = -3x$

b $y - 2 = \frac{1}{3}(x + 4)$

c $y = \frac{1}{2}$

d $y = -\frac{3}{4}x + 5$

2 Find the equation, in point-gradient form, of the line with gradient -3 that passes through $(2, 6)$.

3 Consider the line passing though the points $(-3, -4)$ and $(-5, 2)$.

a Find the gradient of the line.

b Write down two different equations for the line in point-gradient form.

c Verify that the two equations represent the same line.

A third form of the equation of a line is the **general form**, $ax + by + d = 0$. In this form you may be told to use integers for a, b and d.

Example 9

Find an equation for each of the following lines. Give your answers in the form $ax + by + d = 0$ where a, b and d are integers.

a the line with gradient $-\frac{3}{4}$ and y-intercept $(0, 2)$

b the line with gradient $-\frac{1}{2}$ that passes through $(3, 6)$

a $y = -\frac{3}{4}x + 2$	Since you are given the gradient and y-intercept, start by writing the equation of the line in gradient-intercept form.
$\frac{3}{4}x + y - 2 = 0$	Rewrite the equation so that the right-hand side is equal to zero.
$4\left(\frac{3}{4}x + y - 2\right) = 4(0)$	Multiply both sides of the equation by 4, so that the coefficients and constant term all become integers.
$3x + 4y - 8 = 0$	If instead you multiply by -4, you get an equivalent answer of $-3x - 4y + 8 = 0$.
b $y - 6 = -\frac{1}{2}(x - 3)$	Since you are given the gradient and a point, start by writing the equation of the line in point-gradient form.
$2(y - 6) = 2 \times \left(-\frac{1}{2}\right)(x - 3)$	Multiply both sides of the equation by 2, so that the constants are integers.
$2y - 12 = -x + 3$	Expand the brackets.
$x + 2y - 15 = 0$	Rewrite the equation so that the right-hand side is 0. Another possible answer is $-x - 2y + 15 = 0$.

Example 10

a Draw the graph of $3x + 4y - 8 = 0$. **b** Sketch the graph of $3x - 2y + 9 = 0$.

a

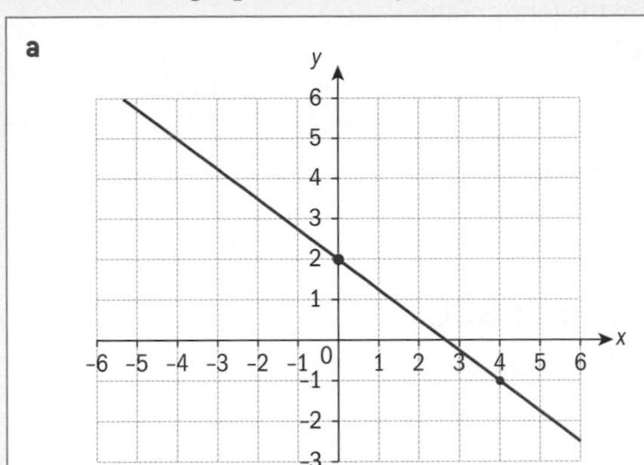

One method of graphing a line given in general form is to rewrite the equation in gradient-intercept form.

$$3x + 4y - 8 = 0$$
$$4y = -3x + 8$$
$$y = -\frac{3}{4}x + 2$$

You can now see that the y-intercept is $(0, 2)$ and gradient is $-\frac{3}{4}$. Use this information to draw the graph of the line.

b

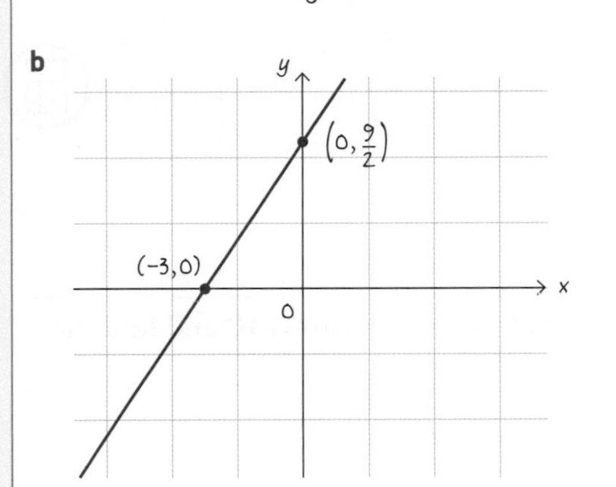

Another method you can use to graph a line given in general form is to find the intercepts.

To find the x-intercept, substitute 0 for y.

$$3x - 2(0) + 9 = 0$$
$$3x = -9$$
$$x = -3$$

To find the y-intercept, substitute 0 for x.

$$3(0) - 2y + 9 = 0$$
$$-2y = -9$$
$$y = \frac{9}{2} = 4\frac{1}{2}$$

Notice that you were directed to **sketch,** rather than **draw** the graph. When sketching, you should show the key features of the graph, such as the axial intercepts of the line and the correct general shape of the graph.

Exercise 3E

1 Write the equation of each of these lines in the general form $ax + by + d = 0$ where a, b and d are integers.

a $y = \frac{1}{6}x - 3$

b The line with gradient $-\frac{2}{3}$ and y-intercept $(0, 4)$.

c The line with gradient -1 that passes through $(-3, 2)$.

2 Change the general form of the equation of each line to gradient-intercept form, $y = mx + c$.

a $3x + y - 5 = 0$

b $2x - 4y + 8 = 0$

c $5x + 2y + 7 = 0$

3 Sketch the graph of the line and label the coordinates of the axial intercepts.

a $x + 2y + 6 = 0$

b $2x - 6y + 8 = 0$

Reflect What are the key features of the graph of a line?
What are three different forms for the equation of a line?
What is the relationship between the parameters in each of the forms of the equation of a line and key features of the graph of the line?

Reflect Which form(s) of the equation of a line have parameters giving a point on the line and the gradient?

Finding the intersection of two lines using your GDC

If two lines intersect, you can use your GDC to find the point of intersection:

Example 11

Use your GDC to find the point of intersection for each pair of lines:

a $y = -3x + 1$ and $y = -5x + 3$

b $y = 2x + 5$ and $3x + 2y - 12 = 0$

a 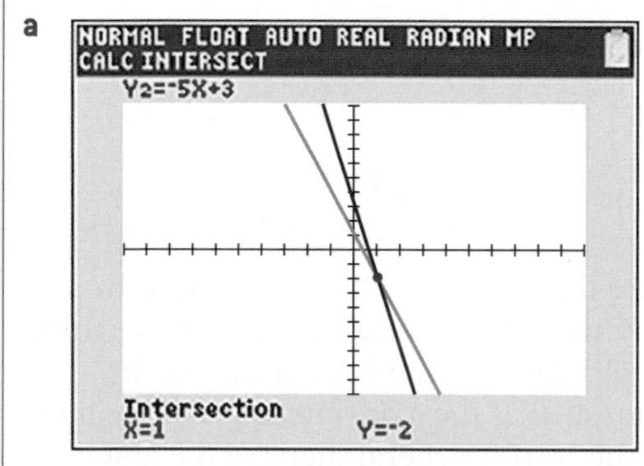 Point of intersection: $(1,-2)$	Graph both lines on your GDC and find the point of intersection.
b 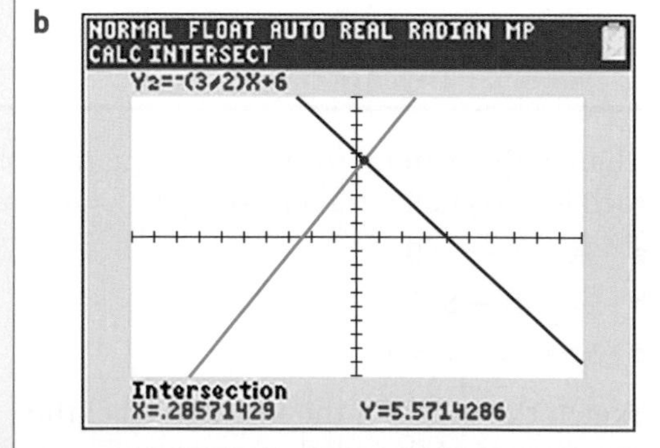 Point of intersection: $(0.286, 5.57)$	Change $3x + 2y - 12 = 0$ to gradient-intercept form: $y = -\frac{3}{2}x + 6$. Graph both lines on your GDC and find the point of intersection. The answer is shown correct to 3 s.f. The exact answer is $\left(\frac{2}{7}, \frac{39}{7}\right)$.

Example 12

Solve $1.26x - 15.3 = -1.37x + 26.5$ by using your GDC to find a point of intersection.

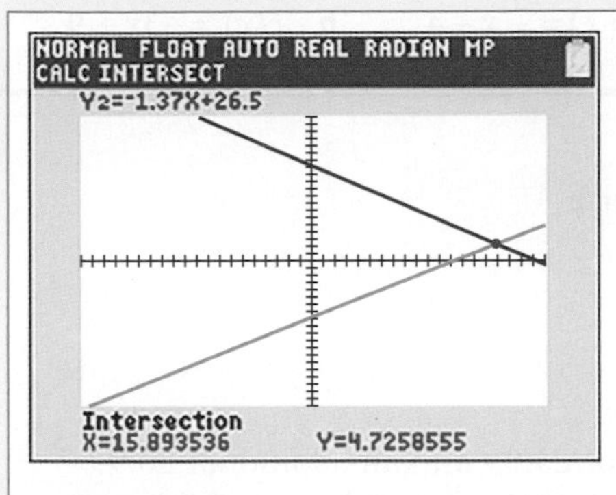

$x = 15.9$

Graph the left-hand side and the right-hand side of the equation: $y = 1.26x - 15.3$ and $y = -1.37x + 26.5$.

You may need to adjust your viewing window to see the point of intersection.

Find the point of intersection. The x-coordinate of the point of intersection is the solution to the equation.

The answer is shown correct to 3 s.f.

Exercise 3F

1 Use your GDC to find the point of intersection for each pair of lines:

a $y = 2x - 1$ and $y = 3x + 1$

b $y = 2x + 1$ and $4x + 2y = 8$

c $y = 4.3x + 7.2$ and $y = 0.5x - 6.4$

d $2x - 3y = -1$ and $y = -\frac{3}{4}x + 2$

2 Solve each equation by using your GDC to find a point of intersection:

a $3x - 4 = 0.5x - 1.75$

b $6.28x + 15.3 = 2.29x - 4.85$

3 The following equations give the weekly salary an employee can earn at two different sales jobs, where x is the amount of sales in euros and y is the weekly salary in euros. Find the amount of sales for which the weekly salaries would be equal.

$$y = 0.16x + 200$$
$$y = 0.10x + 300$$

Linear functions

A function whose graph is a line is called a **linear function**. You can write a linear function as an equation, such as $y = 2x + 3$, or using function notation, $f(x) = 2x + 3$. You can think of $f(x)$ as another representation for y.

Exercise 3G

1 Consider the functions $f(x) = -x + 5$, $g(x) = 2x + 3$ and $h(x) = \frac{1}{3}x - 4$. Find the following:

a $f(3)$ **b** $g(0)$

c $h(6) - g(1)$ **d** $f(2) + g(-1)$

e $(f \circ g)(4)$ **f** $(h \circ f)(-7)$

g $(f \circ g)(x)$ **h** $(h \circ f)(x)$

2 Give the domain and range of each function:

a $f(x) = 3x + 8$ **b** $h(x) = x - 6$

3 Sketch a graph of the following:

a a linear function with range {6}

b a line that is not a function.

4 Find $f^{-1}(x)$ for each of the following linear functions. Give your answers in the form $f^{-1}(x) = mx + c$.

a $f(x) = \frac{1}{2}x + 4$ **b** $f(x) = -3x + 9$

Reflect Why are the gradients of a linear function and its inverse reciprocals?

Example 13

A video streaming service charges a monthly service fee of £6.99 and an additional £0.99 for each 24-hour download rental of a premium movie. The total monthly cost of this service, in GBP, is $f(x) = 0.99x + 6.99$, where x is the number of premium movie rentals that month.

a Find the total monthly cost for a month in which there are five premium movie rentals.

b Find $f^{-1}(x)$ and tell what x and $f^{-1}(x)$ represents in this function.

c Find the number of premium move rentals downloaded in a month where the total monthly cost was £18.87.

a $f(5) = 0.99(5) + 6.99 = 11.94$ The total monthly cost is £11.94.	Substitute 5 for x and evaluate.
b $f(x) = 0.99x + 6.99$	
$y = 0.99x + 6.99$	Write the function in terms of x and y.
$x = 0.99y + 6.99$	Interchange x and y.
$x - 6.99 = 0.99y$ $y = \frac{x - 6.99}{0.99}$	Solve for y.
$f^{-1}(x) = \frac{x - 6.99}{0.99}$, where £$x$ is the total monthly cost and $f^{-1}(x)$ is the number of premium move rentals downloaded during the month	Express with the notation $f^{-1}(x)$.
c $f^{-1}(18.87) = \frac{18.87 - 6.99}{0.99}$ $= 12$ premium movie rentals	Find $f^{-1}(18.87)$.

Exercise 3H

1 Find $f^{-1}(x)$ for each of the following linear functions. Give your answers in the form $f^{-1}(x) = mx + c$.

a $f(x) = 4x - 5$ **b** $f(x) = -\frac{1}{6}x + 3$

c $f(x) = 0.25x + 1.75$

2 The graph of a linear function $y = f(x)$ and the line $y = x$ is shown below. Copy the graphs and then add a sketch of the graph of $y = f^{-1}(x)$.

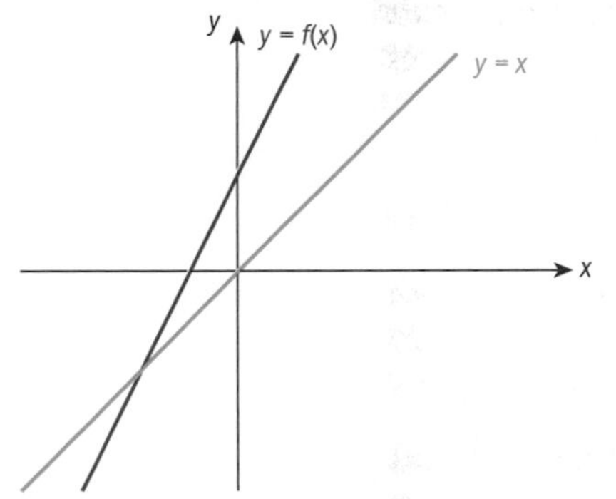

3 A t-shirt company imprints logos on t-shirts. The company charges a one-time set-up fee of $65 and $10 per shirt. The total cost of x shirts, in CAD, is given by $f(x) = 10x + 65$.

a Find the total cost for 55 t-shirts.

b Find $f^{-1}(x)$ and tell what x and $f^{-1}(x)$ represent in this function.

c Find the number of t-shirts in an order with a total cost of $5065.

Linear models

A linear function which describes the relationship between two variables which are connected in real-life is called a **linear model**.

A linear model is used to analyse and predict how the dependent variable will change in response to the independent variable changing.

International mindedness

The term "function" was introduced by the German mathematician Gottfried Wilhelm Leibniz in the 17th century and the notation was coined by Swiss Leonard Euler in the 18th century.

Example 14

The table shows the relationship between the number of days a car is rented and the cost of the rental.

Number of days (x)	Cost in euros ($f(x)$)
1	30.25
2	60.50
3	90.75
4	121.00
5	151.25

a Without plotting a graph, show algebraically that the relationship between the number of days and the cost is a linear relationship.

b Find a linear model for this relationship. Give your answer as a function in gradient-intercept form.

c Find the cost of renting a car for seven days.

a $m = \dfrac{60.50 - 30.25}{2 - 1} = 30.25$ $m = \dfrac{90.75 - 60.50}{3 - 2} = 30.25$	You need to find the gradient of each line segment joining consecutive points in the table.

Continued on next page

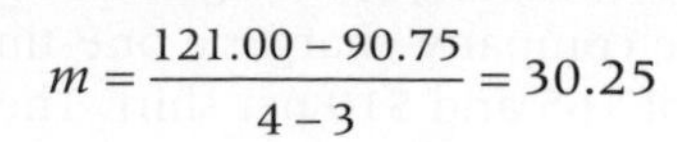

$m = \dfrac{121.00 - 90.75}{4 - 3} = 30.25$ $m = \dfrac{151.25 - 121.00}{5 - 4} = 30.25$ The gradient of each line segment between consecutive data points is the same, so the relationship between number of days and cost is a linear relationship.	
b $m = 30.25$ $f(x) - 30.25 = 30.25(x - 1)$	You found the gradient in part **a**. This is the point-gradient form of the equation of the line, using the point (1,30.25). Alternatively, you could use the gradient-intercept form $30.25 = 30.25(1) + c$, and solve for c.
$f(x) - 30.25 = 30.25x - 30.25$	Expand the brackets.
$f(x) = 30.25x$	Solve for $f(x)$.
c $f(x) = 30.25(7) = 211.75$ The cost of renting the car for seven days is €211.75.	Substitute 7 for x in the model and evaluate.

Example 15

There is a linear relationship between degrees Celsius and degrees Fahrenheit. Some equivalent values are shown in the table below.

Degrees Celsius	Degrees Fahrenheit
0	32
15	59
20	68
30	86
35	95

a Find a linear model that relates the temperature, F, in degrees Fahrenheit, to the temperature, C, in degrees Celsius. Give your answer in the form $F(C) = mC + b$, where m and b are constants to be determined.

b Use your model to find the temperature in degrees Fahrenheit when the temperature in degrees Celsius is 27°C.

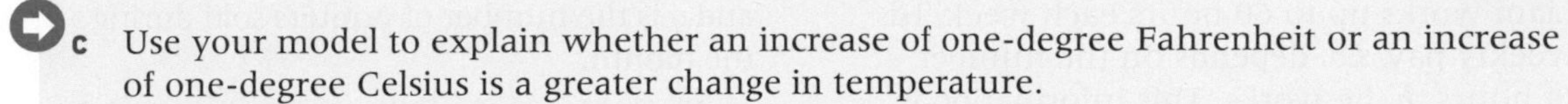

c Use your model to explain whether an increase of one-degree Fahrenheit or an increase of one-degree Celsius is a greater change in temperature.

d Find the value at which the temperature in degrees Celsius is equal to the temperature in degrees Fahrenheit.

a $m = \frac{59-32}{15-0} = \frac{9}{5} = 1.8$	Find the gradient. You could use any two values given in the table.
$F(C) = 1.8C + 32$	The y-intercept is 32, so $b = 32$.
b $F(27) = 1.8(27) + 32 = 80.6$ The equivalent measure is 80.6°F.	Substitute 27 for C.
c For every increase of 1 degree Celsius, there is an increase of 1.8 degrees Fahrenheit. This indicates that an increase of one-degree Celsius is a greater change in temperature than an increase of one-degree Fahrenheit.	
d $x = 1.8x + 32$	Let C and F both equal x.
$-0.8x = 32$	Solve for x.
$x = -40$ −40°C is the same temperature as −40°F.	

Exercise 3I

1 The force applied to a spring and the extension of the spring are connected by a linear relationship.

When a spring holds no mass, its extension is zero. When a force of 160 newtons (N) is applied, the extension of the spring is 5 cm.

a Find a linear model for the extension of the spring in terms of the force applied. Make sure you state clearly the variables you use to represent each quantity.

b Find the extension of the spring when a force of 370 N is applied.

2 Frank is a salesman. He is paid a basic weekly salary, and he also earns a percentage of commission on every sale he makes.

In a certain week, Frank makes sales totalling £1500. His total pay for that week is £600.

In another week, Frank makes sales totalling £2000. His total pay for that week is £680.

a Find, in gradient-intercept form, an equation that relates Frank's total weekly pay, £y, with his total sales revenue, £x.

b Explain the meaning of both the gradient and y-intercept in your model.

c Find Frank's total weekly pay when his weekly sales total £900.

3 A new fitness gym is offering two membership plans.

Plan A: A one-off enrollment fee of \$79.99, and a further monthly fee of \$9.99 per month

Plan B: no enrollment fee, and monthly fees of \$20.00 per month

a Find a linear model for each plan, where total cost is a function of number of months. Identify the variables you use.

After a certain number of months, Plan A becomes more cost-effective than Plan B.

b Use the models from part **a** to determine how many months a person needs to be a member before Plan A becomes more cost-effective than Plan B.

4 Liam works up to 60 hours each week. His weekly pay, £p, depends on the number of hours, h, he works. This information is presented in the graph shown.

a Find the equations for the piecewise function.

b Find the amount Liam is paid in a week that he works:

i 22 hours

ii 47 hours

5 Office Resource has 3000 printers available to sell in a certain month. On average, sales drop by 6.5 printers for each €1 increase in price. This is modelled by the demand function, $q = -6.5p + 3000$, where €p is the sales price and q is the number of printers sold during the month.

a Find the number of printers the model predicts Office Resource sells in a month, if the selling price is €200.

b Explain how raising the sales price by €20 affects the sales.

The manufacturer supplying the printers to Office Resource controls supply according to the function $q = 48p - 1600$, where q is the number of printers supplied during the month and €p is the price Office Resource must pay the manufacturer for each printer they supply. This function is known as the supply function.

c Find the price per printer Office Resource must pay to be supplied with 2000 printers a month.

d Graph the supply and demand functions on your GDC and then sketch the graphs on your paper.

e When the quantity supplied equals the quantity demanded the market is said to be in equilibrium. Find the equilibrium price and the equilibrium demand.

Reflect What type of real-life situations can be modelled by linear functions?

How can you use linear models in real-life situations?

Developing inquiry skills

In the opening scenario for this chapter you looked at how crates of emergency supplies were dropped from a plane.

The function $g(t) = -5t + 670$ gives the height of the crate when the parachute is open as a function of the number of seconds after the crate leaves the plane.

You were asked to explain what -5, the coefficient of t, represents. Do you agree with the answer you gave? If not, what is your answer now?

3.3 Transformations of functions

The function $f(x) = ax^2 + bx + c$; $a \neq 0$, is called a quadratic function. Notice that the highest power of x in a quadratic function is 2.

In section 3.1 you saw that the data in the following table is not linear. The data gives the predicted height, h, of a model rocket t seconds after it is launched. Variables h and t are connected by a quadratic function, and the graph of h against t is a curved shape called a **parabola**.

Time after launch (s)	Height (m)
1	35
3	75
4	80
5	75
8	0

Predicted model rocket height

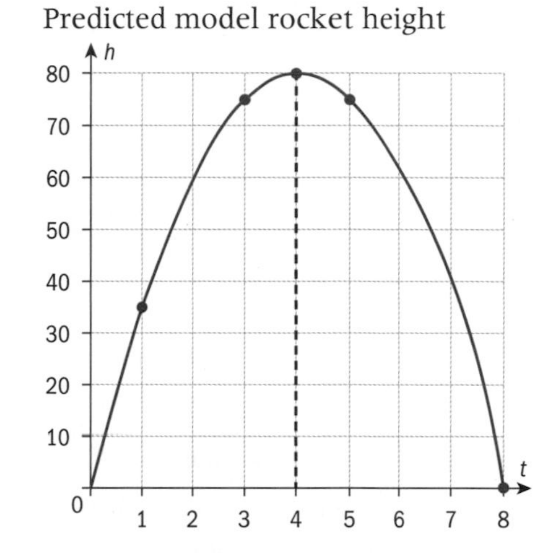

International-mindedness

The Shulba Sutras in ancient India and the Bakhshali manuscript contained an algebraic formula for solving quadratic equations.

This parabola is **concave down** and has a **maximum point**. A parabola may also be **concave up** and have a **minimum point**.

The maximum or minimum point on the graph of a quadratic function is called the **vertex** of the parabola.

The dashed vertical line passing through the vertex is called the **axis of symmetry**.

The axis of symmetry of a parabola is a line of symmetry which passes through the **vertex**.

The simplest quadratic function, or "parent quadratic", is the function $f(x) = x^2$.

The graph of $y = x^2$ is shown.

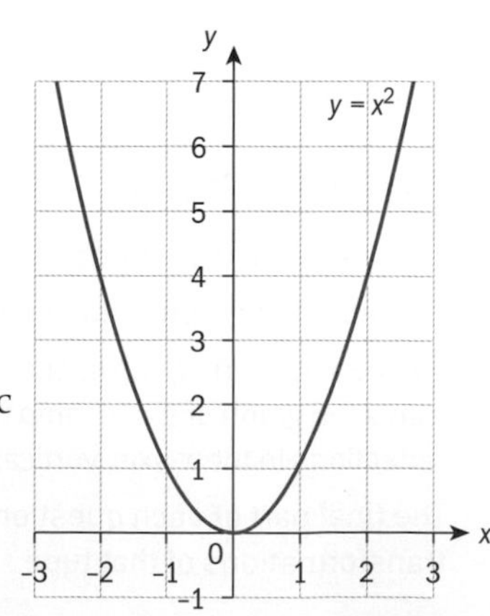

The vertex is (0, 0) and it is a minimum point.

The axis of symmetry is the y-axis. The equation of the axis of symmetry is $x = 0$.

In this section you are going to obtain the equations of other quadratic graphs by transforming the graph of $y = x^2$.

You can transform the parent graph $y = x^2$ to the graph of another parabola $y = g(x)$ by applying one or more of the following three transformations.

Reflections	Stretches
The graph of $y = g(x)$ is a **reflection in the x-axis** of the graph of $y = x^2$. 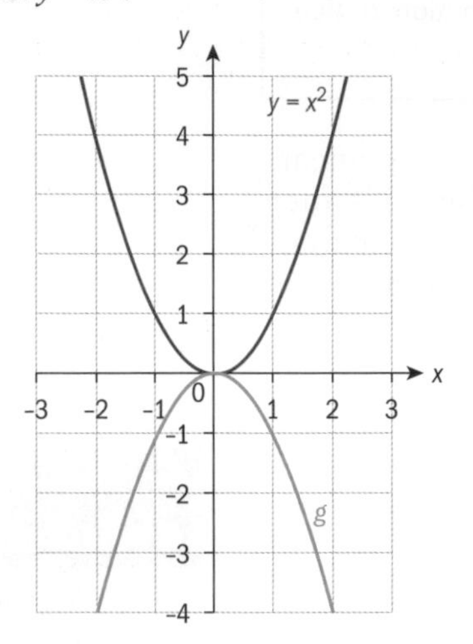	The graph of $y = g(x)$ is a **vertical stretch** of the graph of $y = x^2$, with scale factor 2. Each point on the graph of g is twice the distance from the x-axis as the points on the graph of $y = x^2$. 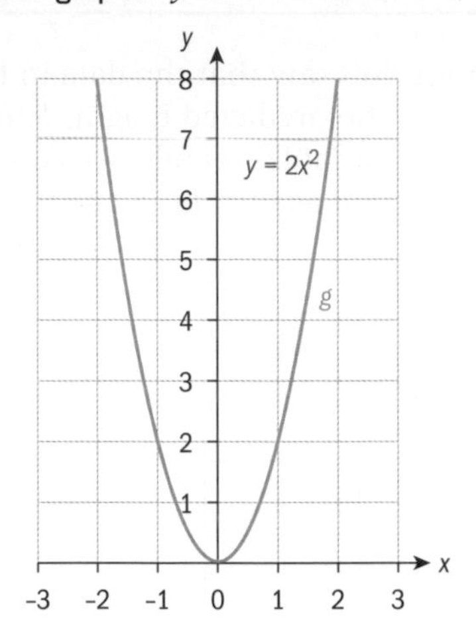 When the scale factor is less than 1, the graph of $y = g(x)$ is a **vertical compression**.

Translations	
The graph of $y = g(x)$ is a **horizontal translation** of the graph of $y = x^2$, shifted 3 units to the right. 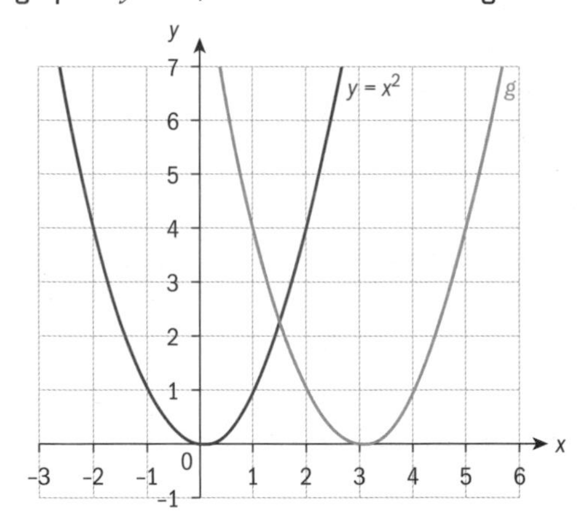	The graph of $y = g(x)$ is a **vertical translation** of the graph of $y = x^2$, shifted 2 units down. 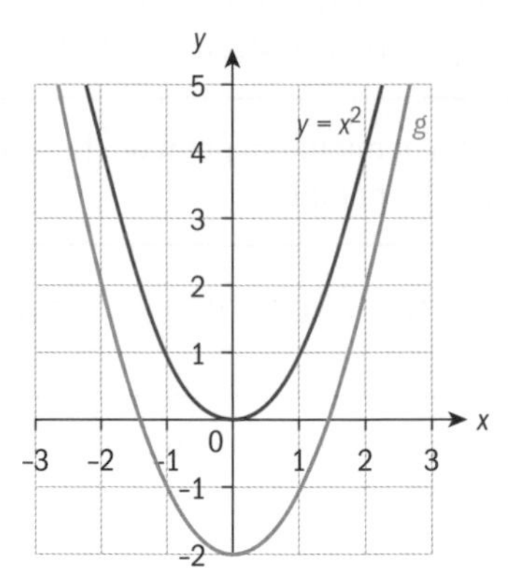

Investigation 3

Use your GDC to plot graphs of the equations in questions **1** – **4**.

Compare the graph of each equation with the graph of $y = x^2$.

For the equations given in questions **1** – **4**, describe the transformation which maps the graph of $y = x^2$ onto the given graph in terms of: horizontal translations, reflections in the x-axis, vertical stretches/compressions and vertical translations.

The final part of each question allows you to make a generalization about transformations of that type.

1 a $y = (x - 3)^2$ b $y = (x + 4)^2$ c $y = (x - h)^2$

2 a $y = x^2 + 1$ b $y = x^2 - 2$ c $y = x^2 + k$

3 a $y = -x^2$ b $y = 2x^2$ c $y = \frac{1}{3}x^2$
d $y = -3x^2$ e $y = ax^2$

4 a $y = (x + 4)^2 - 2$ b $y = 2(x - 3)^2$ c $y = -\frac{1}{2}x^2 + 4$
d $y = 3(x - 2)^2 - 4$ e $y = a(x - h)^2 + k$

5 **Factual** What effect do a, h and k have in transforming the graph of $y = x^2$ to the graph of $y = a(x - h)^2 + k$?

6 Without using your GDC, sketch a graph of $y = x^2$. Then apply the transformations in the order given.
a Vertical stretch with scale factor 2; horizontal translation right 3; vertical translation down 4. Write down the equation of the final graph.
b Horizontal translation right 3; vertical translation down 4; vertical stretch with scale factor 2. Write down the equation of the final graph.

7 **Conceptual** Does the order in which the transformations of the graph of $y = x^2$ to the graph of $f(x) = a(x - h)^2 + k$ are completed matter? Explain why (or why not) this is the case.

8 **Conceptual** Do you think that the graph of any quadratic function can be obtained through transformations of the graph of $y = x^2$?

The graph of *any* quadratic function can be obtained by applying a combination of the following transformations to the graph of the parent quadratic, $y = x^2$.

Transformations of $y = x^2$	
Reflection in the x-axis $y = -x^2$	**Vertical translation** $y = x^2 + k$ Up k units for $k > 0$ Down $\lvert k\rvert$ units for $k < 0$
Vertical stretch with scale factor $\lvert a\rvert$ $y = ax^2$ Vertical stretch for $\lvert a\rvert > 1$ Vertical compression for $0 < \lvert a\rvert < 1$	**Horizontal translation** $y = (x - h)^2$ Right h units for $h > 0$ Left $\lvert h\rvert$ units for $h < 0$

Example 16

Sketch the parent quadratic, $y = x^2$, and the graph of $y = g(x)$ on the same axes. Then write down the coordinates of the vertex and the equation of the axis of symmetry for the graph of g.

a $g(x) = \frac{1}{2}x^2$ b $g(x) = (x + 3)^2 - 2$ c $g(x) = 2x^2 + 3$ d $g(x) = -3x^2$

Continued on next page

a

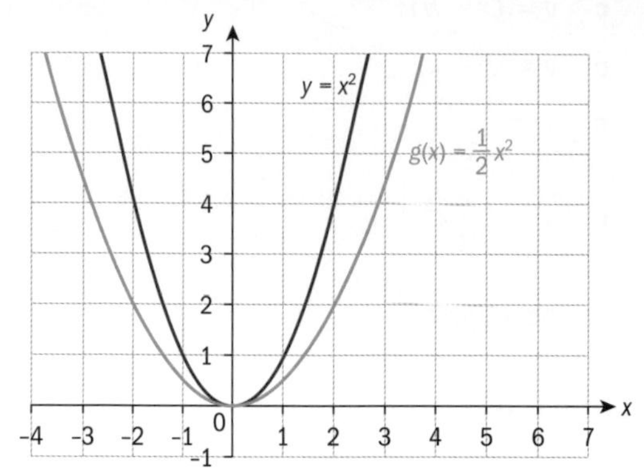

axis of symmetry: $x = 0$ vertex: $(0, 0)$

The vertex is the minimum point on the graph of g.

The axis of symmetry is the vertical line passing through the vertex.

$g(x) = \frac{1}{2}x^2$

$a = \frac{1}{2} \Rightarrow$ vertical compression with scale factor $\frac{1}{2}$.

Each point on the graph of g is half the distance to the x-axis as the corresponding point on the graph of $y = x^2$.

b

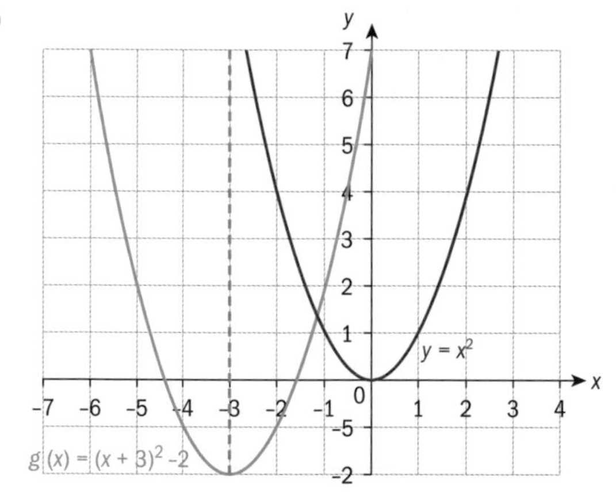

axis of symmetry: $x = -3$ vertex: $(-3, -2)$

$g(x) = (x + 3)^2 - 2$

$h = -3 \Rightarrow$ horizontal translation to the left 3 units.

$k = -2 \Rightarrow$ vertical translation down 2 units.

c

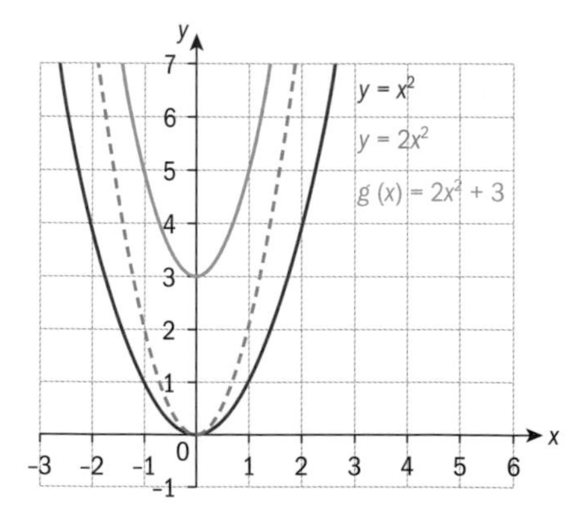

axis of symmetry: $x = 0$ vertex: $(0, 3)$

$g(x) = 2x^2 + 3$

$|a| = 2 \Rightarrow$ vertical stretch with scale factor 2.

Each point on the graph of $y = 2x^2$ is two times the distance from the x-axis as the corresponding point on the graph of $y = x^2$.

$k = 3 \Rightarrow$ horizontal translation up 3 units.

d

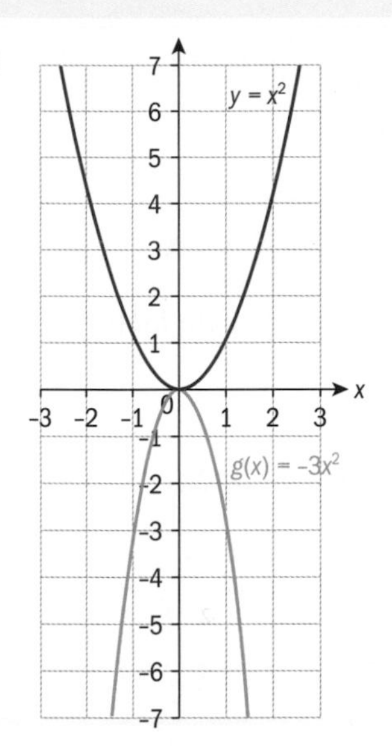

axis of symmetry: $x = 0$ vertex: $(0, 0)$

$g(x) = -3x^2 = -1 \times 3x^2$

$a = -3 \Rightarrow$ reflection in the x-axis.

$|a| = 3 \Rightarrow$ vertical stretch with scale factor 3.

Each point on the graph of $y = g(x)$ is three times the distance from the x-axis as the corresponding point on the graph of $y = x^2$.

HINT

When a question asks for the equation of the axis of symmetry, the equation will be given in the form $x = \ldots$. Students often lose marks on exams for writing things like "axis of symmetry = 2", or just writing a number rather than an equation.

HINT

Translations can be represented by vectors in the form $\begin{pmatrix} h \\ k \end{pmatrix}$. For example, $\begin{pmatrix} 0 \\ 2 \end{pmatrix}$ represent a vertical translation of 2 units up, and $\begin{pmatrix} -5 \\ -3 \end{pmatrix}$ represent a horizontal translation of 5 units to the left and a vertical translation of 3 units down.

Example 17

Let $f(x) = x^2$.

Describe a series of transformations which map the graph of $y = f(x)$ onto the graph of $y = g(x)$. Then write an expression for the function $g(x)$.

a

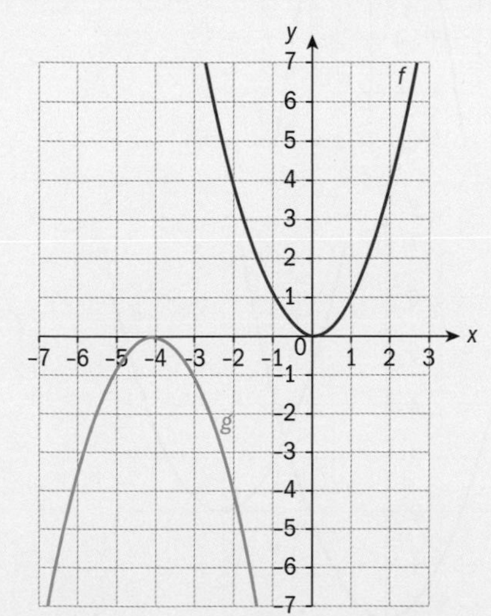

b

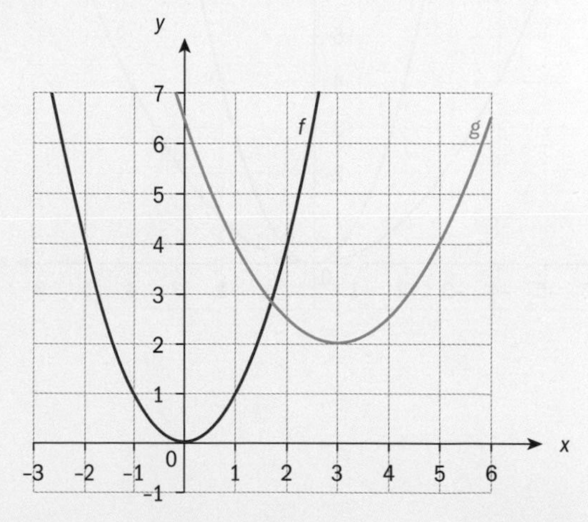

Continued on next page

a The graph of $y = g(x)$ can be obtained from the graph of $y = f(x)$ by translating left 4 units and a reflection in the x-axis. $g(x) = -(x+4)^2$	Translation 4 units left $\Rightarrow h = -4$. Reflection in x-axis $\Rightarrow a = -1$.
b The graph of $y = g(x)$ can be obtained from the graph of $y = f(x)$ by translating right 3 units, a vertical compression with scale factor $\frac{1}{2}$, and translating up 2 units. $g(x) = \frac{1}{2}(x-3)^2 + 2$	Translation right 3 units $\Rightarrow h = 3$. Vertical compression with scale factor $\frac{1}{2}$ $\Rightarrow a = \frac{1}{2}$. Translation up 2 units $\Rightarrow h = 2$.

Exercise 3J

1 Sketch the parent quadratic, $y = x^2$, and the graph of $y = g(x)$ on the same axes. Then write down the coordinates of the vertex and the equation of the axis of symmetry for the graph of g.

a $g(x) = (x+3)^2$ **b** $g(x) = -x^2 + 4$

c $g(x) = \frac{1}{4}x^2$ **d** $g(x) = 2(x-4)^2 - 3$

2 Describe the transformations of the graph of $f(x) = x^2$ that lead to the graph of g. Then write an equation for $g(x)$.

a

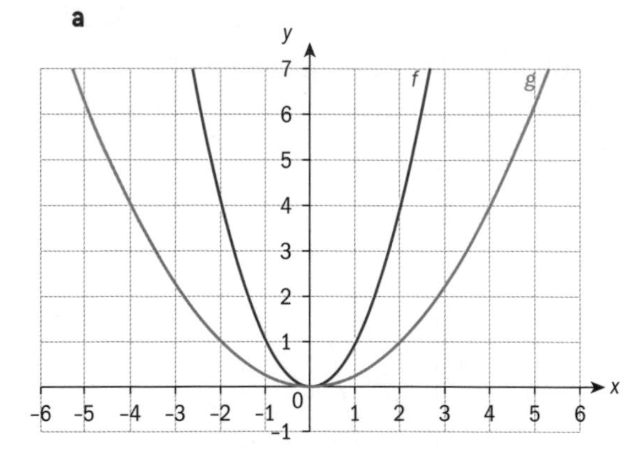

b

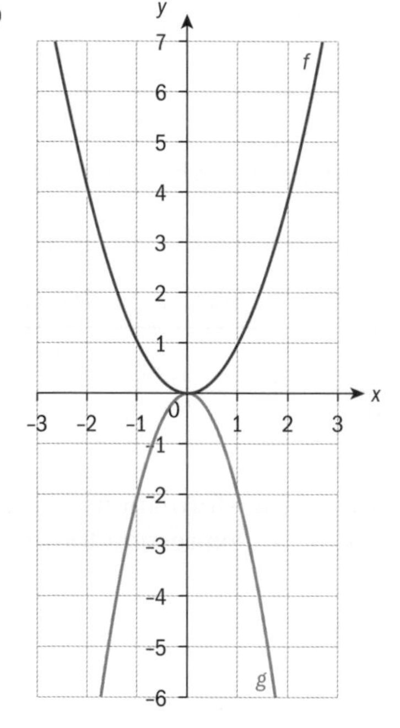

c

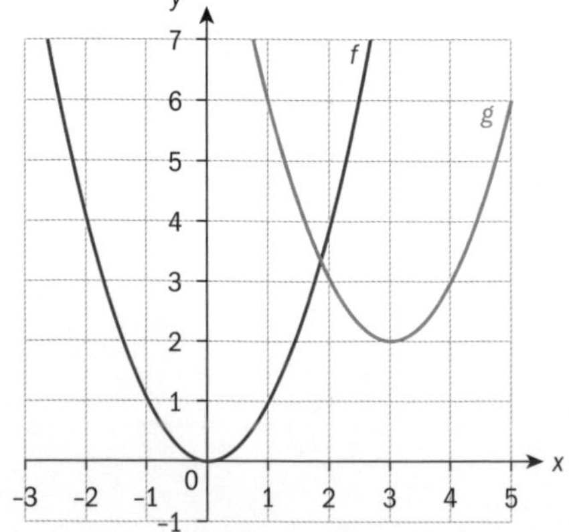

d

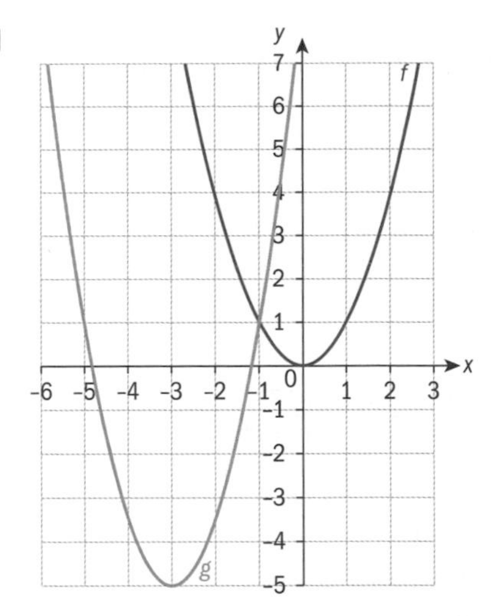

You have seen that vertical stretches and compressions change the distance between each point on a graph of a function and the x-axis. Horizontal stretches and compression change the distance between each point on a graph of a function and the y-axis.

Investigation 4

1 Use your GDC to graph $f(x) = (x)^2$ and $g(x) = (-x)^2$. Explain why it appears that there is only one graph.

2 For each pair of the functions given below, the graph of g can be obtained through a **vertical** stretch or compression of the graph of f. Graph each pair of functions on your GDC and write down the scale factor for the **vertical** stretch or compression.

a $f(x) = x^2$ and $g(x) = \frac{1}{4}x^2$ b $f(x) = x^2$ and $g(x) = 4x^2$

3 You could rewrite the functions g in question **2** as $g(x) = \left(\frac{1}{2}x\right)^2$ and $g(x) = (2x)^2$. The graph of g can be obtained through a **horizontal** stretch or compression of the graph of f. Graph each pair of functions on your GDC and write down the scale factor for the **horizontal** stretch or compression.

a $f(x) = x^2$ and $g(x) = \left(\frac{1}{2}x\right)^2$ b $f(x) = x^2$ and $g(x) = (2x)^2$

4 **Conceptual** Why it is unnecessary to use reflections in the y-axis or horizontal stretches and compressions when obtaining graphs of quadratics through transformations of the graph of $f(x) = x^2$?

5 Graph each pair of functions on your GDC and describe the transformation that changes the graph of f to the graph of g.

a $f(x) = \sqrt{x}$ and $g(x) = \sqrt{-x}$ b $f(x) = 2^x$ and $g(x) = 2^{-x}$

6 For each pair of functions in question **5**, $g(x) = f(-x)$. What transformation changes the graph of $y = f(x)$ to the graph of $y = f(-x)$?

Continued on next page

7 The diagram below shows part of the graphs of $f(x) = \sin(x)$ and $g(x) = \sin(2x)$.

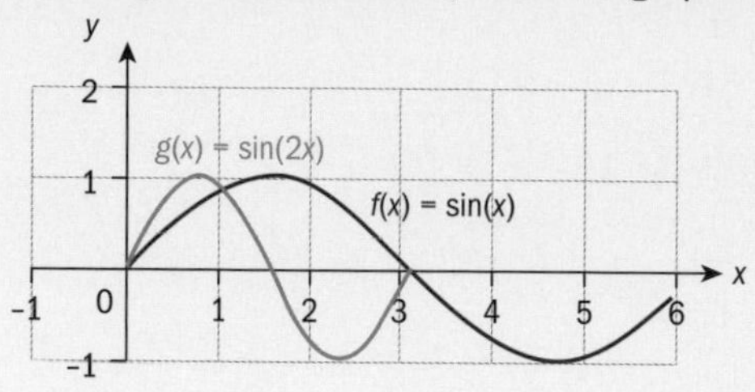

a Can the graph of g be obtained from a vertical compression of the graph of f?
b Can the graph of g be obtained from a horizontal compression of the graph of f?

In investigation 4, you saw that when transforming the graph of $f(x) = x^2$, it is not necessary to use reflections in the y-axis or horizontal stretches and compressions. You also saw some functions where reflections in the y-axis or horizontal stretches and compressions led to graphs that could not be obtained by other types of transformations.

A summary of all transformations you will encounter in this course is shown in the table.

Transformations of $y = f(x)$	
Reflection in the x-axis $y = -f(x)$	**Reflection in the y-axis** $y = f(-x)$
Vertical dilations with scale factor $\|a\|$ $y = af(x)$ Vertical stretch for $\|a\| > 1$ Vertical compression for $0 < \|a\| < 1$	**Horizontal dilations with scale factor $\frac{1}{\|q\|}$** $y = f(qx)$ Horizontal stretch for $0 < \|q\| < 1$ Horizontal compression for $\|q\| > 1$
Vertical translations $y = f(x) + k$ Up k units for $k > 0$ Down $\|k\|$ units for $k < 0$	**Horizontal translations** $y = f(x - h)$ Right h units for $h > 0$ Left $\|h\|$ units for $h < 0$

Transforming piecewise linear functions

Example 18

Consider the graph of the function $y = f(x)$, where $-2 \le x \le 6$.

Sketch the graph of:

a $y = 2f(x)$ **b** $y = f(-x)$
c $y = f(2x)$ **d** $y = f(x - 1) - 3$

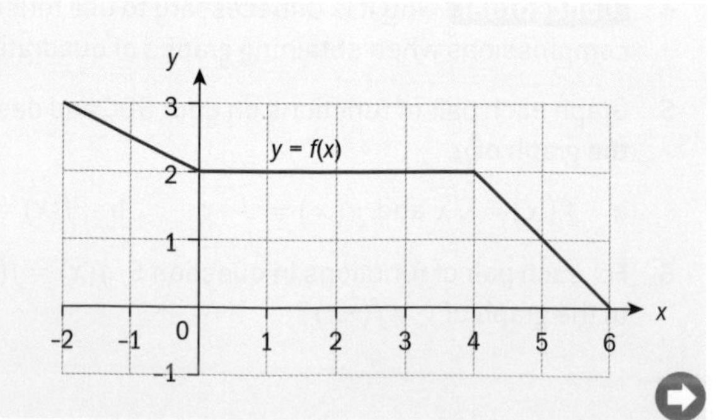

a

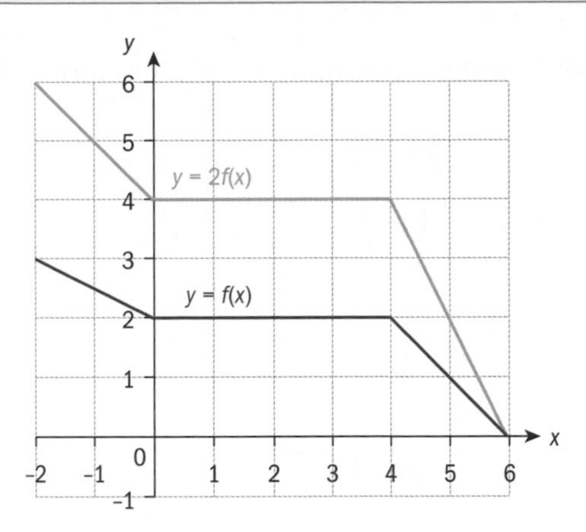

$y = 2f(x)$

$a = 2 \Rightarrow$ vertical stretch with scale factor 2.

Each point on the graph of $y = 2f(x)$ is 2 times the distance to the x-axis as the corresponding point on the graph of $y = f(x)$.

b

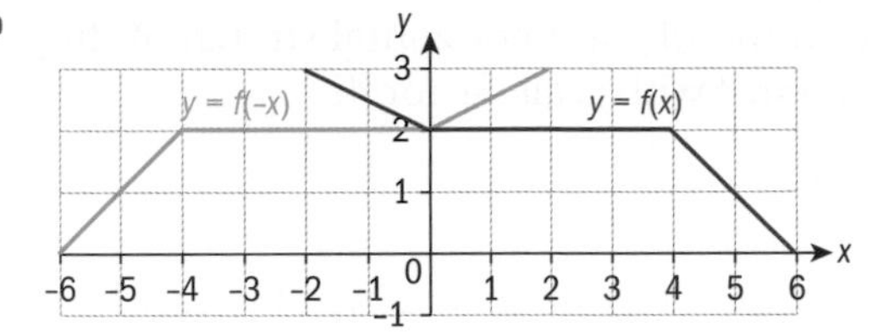

$y = f(-x)$

Reflection of $y = f(x)$ in the y-axis

c

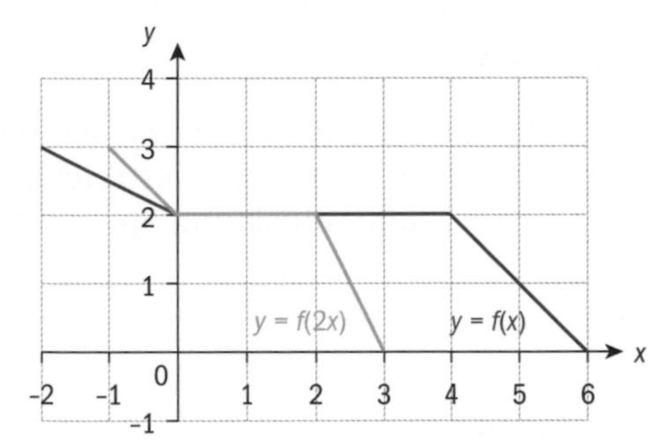

$y = f(2x)$

$q = 2 \Rightarrow$ horizontal compression with scale factor $\frac{1}{2}$.

Each point on the graph of $y = f(2x)$ is half the distance to the y-axis as the corresponding point on the graph of $y = f(x)$.

d

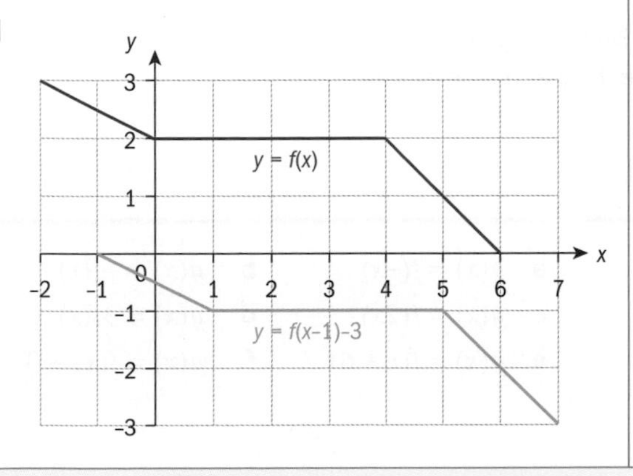

$y = f(x - 1) - 3$

$h = 1 \Rightarrow$ horizontal translation right 1 unit.

$k = -3 \Rightarrow$ and vertical translation down 3 units.

Example 19

Functions g, r and s are transformation of the graph of f. Find the functions g, r and s in terms of f.

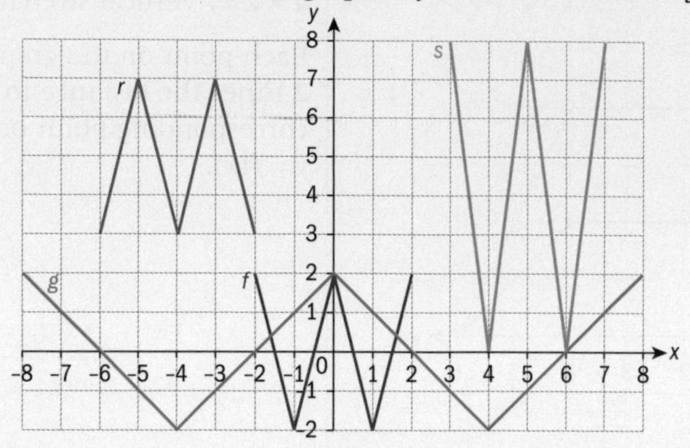

$g(x) = f\left(\frac{1}{4}x\right)$	The graph of g is a horizontal stretch of the graph of f with scale factor 4. $\frac{1}{q} = 4 \Rightarrow q = \frac{1}{4}$
$r(x) = -f(x + 4) + 5$	The graph of r can be obtained by translating the graph of f left 4 units, reflecting in the x-axis, and then translating up 5 units. $h = -4$ and $k = 5$
$s(x) = 2f(x - 5) + 4$	The graph of g can be obtained by translating the graph of f right 5 units, vertical stretching with scale factor 2, and translating up 4 units. $a = 2 \quad h = 5 \quad k = 4$

Reflect What is the relationship between the graphs of $y = f(x)$ and $y = f(-x)$?

The graph of $y = f(qx)$ is a horizontal stretch or compression of the graph of $y = f(x)$. For which values of q is the transformation a stretch rather than a compression?

Exercise 3K

1 The graph of $y = f(x)$, where $-3 \le x \le 6$, is shown. Copy the graph of f and draw these functions on the same axes.

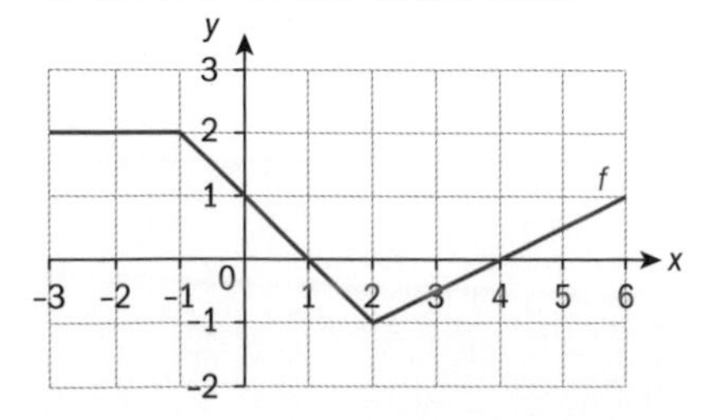

a $g(x) = f(-x)$ **b** $g(x) = -f(x)$

c $g(x) = f(2x)$ **d** $g(x) = 3f(x)$

e $g(x) = f(x + 6)$ **f** $g(x) = f(x) - 3$

2 The graphs of functions r and s are transformation of the graph of f. Find the functions r and s in terms of f.

a

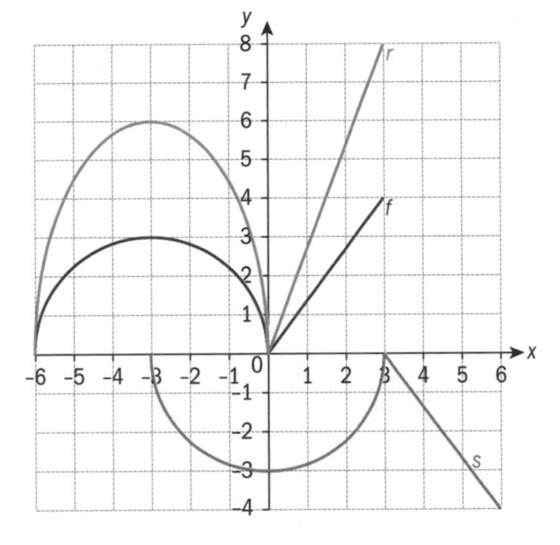

b

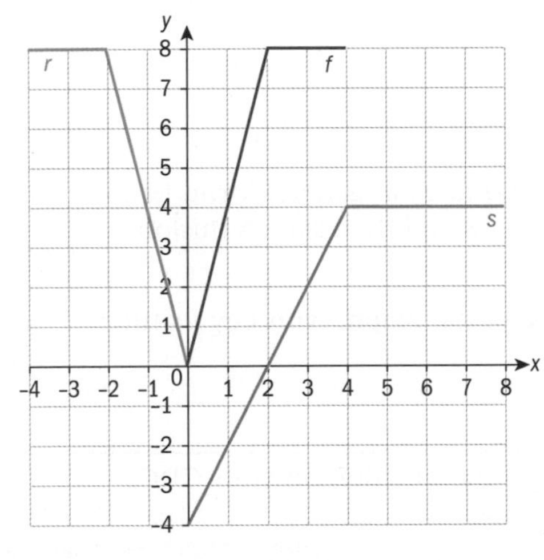

3 The diagram shows the graph of $y = f(x)$, for $2 \le x \le 8$.

a Write down the range of f.

Let $g(x) = f(-x)$.

b Sketch the graph of g.

c Write down the domain of g.

The graph of h can be obtained by a vertical translation of the graph of g. The range of h is $-4 \le y \le 2$.

d Find the equation for h in terms of g.

e Find the equation for h in terms of f.

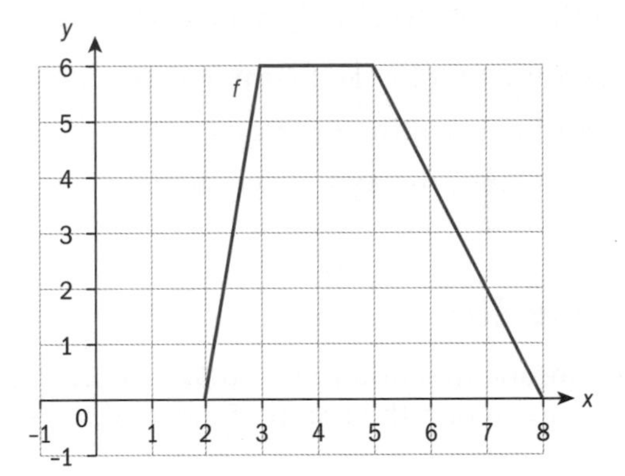

Developing inquiry skills

In the opening scenario for this chapter you looked at how crates of emergency supplies were dropped from a plane.

The function $h(t) = -4.9t^2 + 720$ gives the height of the crate during free fall.

How could you transform the parent graph $h(t) = t^2$ to give the function $h(t) = -4.9t^2 + 720$? What do these transformations tell you about the motion of the crate in this context?

3.4 Graphing quadratic functions

In section 3.3, you studied transformations of quadratic graphs. In this section, you will further study the graphs of quadratic functions and different forms of the equations for quadratics. First you will graph with a GDC and then without.

Transferring a graph from GDC to paper

You may be asked to sketch, on paper, the graph of a quadratic function. You could do this by plotting the graph on your GDC and then carefully re-creating a sketch of the graph on paper.

When you "**sketch**" a parabola, your sketch should show the general shape of the graph accurately, and label the key features including:

- the coordinates of x- and y-intercepts
- the coordinates of the vertex
- the equation of the axis of symmetry

Example 20

The quadratic function $f(x) = -0.5x^2 + 7.5x - 18$ is said to be in general form. Use technology to plot the graph of $f(x) = -0.5x^2 + 7.5x - 18$ and then sketch this on paper.

Your sketch should show the general correct shape of the graph, with key features labelled.

Also state the domain and range of this function.

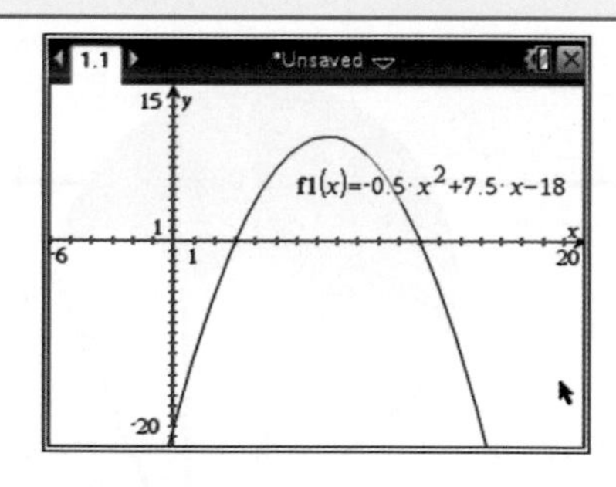	Graph the function on your GDC.
y-intercept is $(0, -18)$	Read the y-intercept from your GDC.
The zeros are at $x = 3$ and $x = 12$. So, $(3, 0)$ and $(12, 0)$ are also points on the graph.	Use your GDC to find the zeros of the function.
Vertex is $(7.5, 10.125)$	Use your GDC to find the coordinates of the vertex.

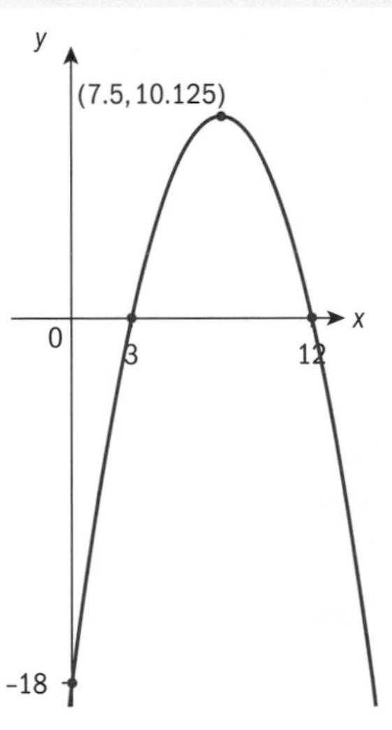

Domain of f is $\{x \in \mathbb{R}\}$.	You can evaluate the function f at any real value of x, so the domain is the set of all real numbers.
Range of f is $\{y \in \mathbb{R} \mid y \leq 10.125\}$.	The vertex of the graph is (7.5, 10.125), so the maximum value of the function f is 10.125

Exercise 3L

Use your GDC to plot the graph of the quadratic function. Then use the graph to find the coordinates of the x-intercept, the y-intercept and the vertex of the graph.

1 $f(x) = 3x^2 + 7x - 4$

2 $f(x) = -4.2x^2 + 6.1x - 3$

Use your GDC to plot the graph of the quadratic function. Sketch the graph on your paper, labelling the coordinates of key features of the graph. Then write down the domain and range of the function.

3 $f(x) = -2x^2 - 7x + 3$

4 $f(x) = 1.25x^2 - 12.4x$

Use your GDC to plot the graph of the quadratic function over the given domain. Sketch the graph on your paper, labelling the coordinates of the key features of the graph. Then write down the range of the function.

5 $f(x) = -3.6x^2 + 8.1$, for $-1.5 \leq x \leq 1.5$

6 $f(x) = x^2 - 2x - 5$, for $-2 \leq x \leq 4$

EXAM HINT

In examination questions where the domain is restricted for a given function, you must not sketch points outside of that domain.

When sketching the graph of a quadratic without the use of a GDC, different forms of the equation will help you to identify the key features of the graph.

Vertex form

In the previous section, you graphed functions of the form $f(x) = a(x - h)^2 + k$ by transforming the graph of $f(x) = x^2$. Complete the following investigation to learn what the parameters of the equation tell you about the graph of the function.

Investigation 5

1 Graph each function by transforming the graph of $y = x^2$.

a $f(x) = 2(x+1)^2 - 4$

b $f(x) = -\frac{1}{2}(x-2)^2 + 5$

c $f(x) = -3(x-1)^2 - 6$

2 Copy the table and record the parameters a, h and k. Use your graphs from question **1** to determine whether the graph is concave up or down, the equation of the axis of symmetry and the coordinates of the vertex,

$f(x) = a(x-h)^2 + k$	a	h	k	Up/ down	Axis of symmetry	Vertex
a $f(x) = 2(x+1)^2 - 4$						
b $f(x) = -\frac{1}{2}(x-2)^2 + 5$						
c $f(x) = -3(x-1)^2 - 6$						

Look for patterns in your table and answer the following questions.

3 **Conceptual** Which parameter in the function $f(x) = a(x-h)^2 + k$ determines whether the graph of the function is concave up or concave down? Explain how you determine whether it is up or down.

4 **Conceptual** What is the equation of the axis of symmetry and what are the coordinates of the vertex of the graph of $f(x) = a(x-h)^2 + k$?

5 Consider the function $f(x) = 4(x-3)^2 + 2$. **Without** using your GDC, say whether the graph is concave up or concave down. Write down the equation of the axis of symmetry and the coordinates of the vertex of the graph. Then graph the equation on your GDC to verify your answers.

A quadratic function written in the form $f(x) = a(x-h)^2 + k$, $a \neq 0$, is said to be in **vertex form**. The coordinates of the vertex are (h,k) and the equation of the axis of symmetry is $x = h$.

Example 21

Write down the coordinates of the vertex and the equation of the axis of symmetry for each function:

a $f(x) = -3(x-4)^2 + 1$ b $f(x) = 2(x+3)^2 - 6$

a $f(x) = -3(x-4)^2 + 1$	
vertex: (4, 1)	The vertex is (h, k), where $h = 4$ and $k = 1$.
axis of symmetry: $x = 4$	The axis of symmetry is $x = h$.
b $f(x) = 2(x+3)^2 - 6$	$f(x) = 2(x-(-3))^2 + (-6)$, so $h = -3$ and
vertex: $(-3, -6)$	$k = -6$.
axis of symmetry: $x = -3$	

General form

A quadratic function in the form $f(x) = ax^2 + bx + c$, $a \neq 0$, is said to be in general form. You will learn about the parameters of this form in the following investigation.

Investigation 6

1 Expand each of the following equations to write them in the general form, $f(x) = ax^2 + bx + c$.

a $f(x) = 2(x+1)^2 - 4$

b $f(x) = -\frac{1}{2}(x-2)^2 + 5$

c $f(x) = -3(x-1)^2 - 6$

2 Complete a table like the one below. Record your answers from question 1 in the first column. Find the values of $-\frac{b}{2a}$ and $f\left(-\frac{b}{2a}\right)$. Recall that you found the equation of the axis of symmetry and coordinates of the vertex of each graph in investigation 5.

$f(x) = ax^2 + bx + c$	$-\frac{b}{2a}$	$f\left(-\frac{b}{2a}\right)$	Axis of symmetry	Vertex
a				
b				
c				

3 **Conceptual** What is the equation of the axis of symmetry of the graph $f(x) = ax^2 + bx + c$?

4 **Conceptual** What are the coordinates of the vertex of the graph of $f(x) = ax^2 + bx + c$?

5 **Conceptual** What is the x-coordinate of the y-intercept of the graph of any function? Hence, find the y-intercept of the graph of $f(x) = ax^2 + bx + c$.

International-mindedness

How do you use the Babylonian method of multiplication?

Try 36×14

$f(x) = ax^2 + bx + c, a \neq 0$, is called the **general form** of the equation for a quadratic function.

The equation of the axis of symmetry is $x = -\frac{b}{2a}$. The coordinates of the vertex are $\left(-\frac{b}{2a}, f\left(-\frac{b}{2a}\right)\right)$ and the y-intercept is $(0, c)$.

Example 22

Find the equation of the axis of symmetry, the coordinates of the vertex, and the y-intercept. Use these features of the graph of the quadratic and a point of symmetry to the y-intercept to sketch a graph.

a $f(x) = -x^2 + 6x - 4$ **b** $f(x) = 2x^2 + 4x - 1$

a

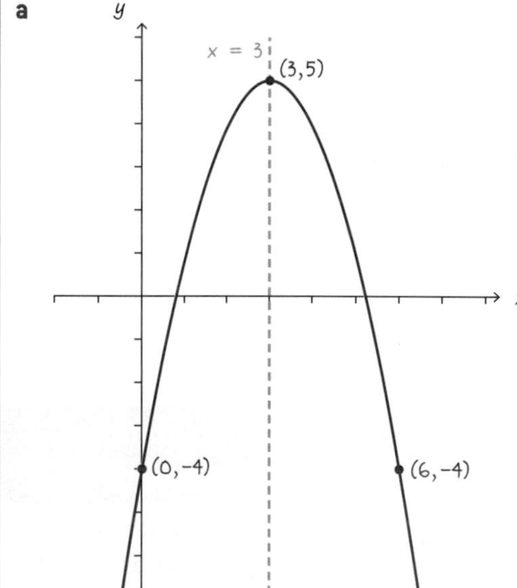

$f(x) = -x^2 + 6x - 4$

$\frac{-b}{2a} = \frac{-6}{2(-1)} = 3 \Rightarrow$ the axis of symmetry is $x = 3$.

$f\left(\frac{-b}{2a}\right) = f(3) = -(3)^2 + 6(3) - 4 = 5 \Rightarrow$ the vertex is $(3, 5)$.

$c = -4 \Rightarrow$ the y-intercept is $(0, -4)$.

The point symmetric to the y-intercept is $(6, -4)$.

$a = -1 < 0 \Rightarrow$ the parabola is concave down and the vertex is a maximum point.

b

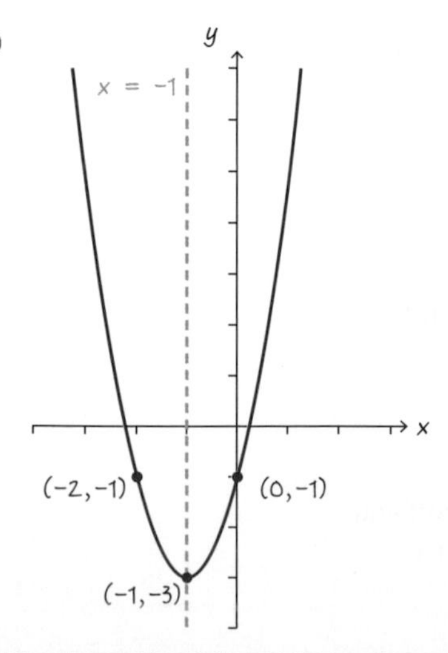

$f(x) = 2x^2 + 4x - 1$

$\frac{-b}{2a} = \frac{-4}{2(2)} = -1 \Rightarrow$ the axis of symmetry is $x = -1$.

$f\left(\frac{-b}{2a}\right) = f(-1) = 2(-1)^2 + 4(-1) - 1 = -3 \Rightarrow$ the vertex is $(-1, -3)$.

$c = -1 \Rightarrow$ the y-intercept is $(0, -1)$.

The point symmetric to the y-intercept is $(-2, -1)$.

$a = 2 > 0 \Rightarrow$ the parabola is concave up and the vertex is a minimum point.

Factorized form

For the graph of a quadratic function $y = f(x)$, the x-intercepts are the x-values where $y = f(x) = 0$.

Writing a quadratic function in factorized form allows us to easily find the values of x for which $f(x) = 0$.

A quadratic function which is written in the form $f(x) = a(x - p)(x - q)$, $a \neq 0$, is said to be written in **factorized form** (or **intercept form**). The coordinates of the x-intercepts of the graph are $(p, 0)$ and $(q, 0)$.

Here, $f(x)$ is written as the product of three linear factors: a, $(x - p)$ and $(x - q)$.

So $f(x) = 0$ only when one or more of the three factors is itself equal to zero.

Since $a \neq 0$, it follows that either:

$(x - p) = 0$ and hence $x = p$, or

$(x - q) = 0$ and hence $x = q$

$(p, 0)$ and $(q, 0)$ are the x-intercepts of the graph $y = f(x)$.

You can also find the equation of the axis of symmetry from the factorized form. Due to the symmetry of a parabola, the axis of symmetry will pass through the midpoint between the x-intercepts.

If the x-coordinates of the x-intercepts are p and q then the equation of the axis of symmetry is $x = \frac{p+q}{2}$,
and the vertex of the graph has coordinates $\left(\frac{p+q}{2}, f\left(\frac{p+q}{2}\right)\right)$.

Example 23

Find the equation of the axis of symmetry and the coordinates of the x-intercepts, the y-intercept and the vertex. Use these features to sketch the graph of the parabola.

a $f(x) = (x - 4)(x - 2)$ **b** $f(x) = -2(x + 3)(x + 1)$

a

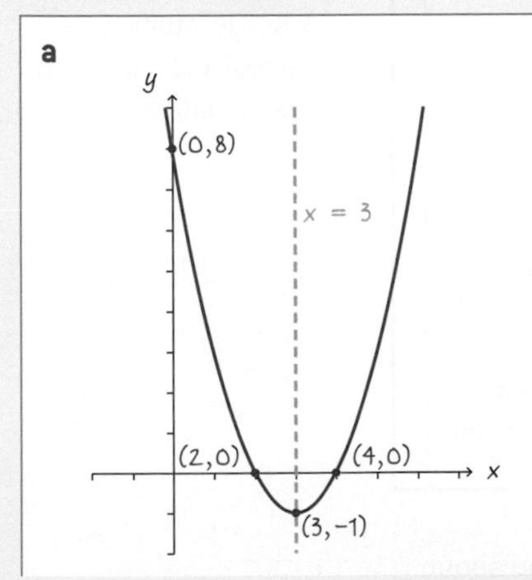

$f(x) = (x - 4)(x - 2)$

$a = 1 > 0 \Rightarrow$ the parabola is concave up and the vertex is a minimum point.

$p = 4$ and $q = 2 \Rightarrow$ the x-intercepts are $(4, 0)$ and $(2, 0)$.

$\frac{p+q}{2} = \frac{4+2}{2} = 3 \Rightarrow$ the axis of symmetry is $x = 3$.

$f\left(\frac{p+q}{2}\right) = f(3) = (3-4)(3-2) = -1 \Rightarrow$ the vertex is $(3, -1)$.

Continued on next page

Note that you find the y-intercept of the graph of any function by finding $f(0)$.

$f(0) = (0-4)(0-2) = 8 \Rightarrow$ the y-intercept is $(0, 8)$.

b

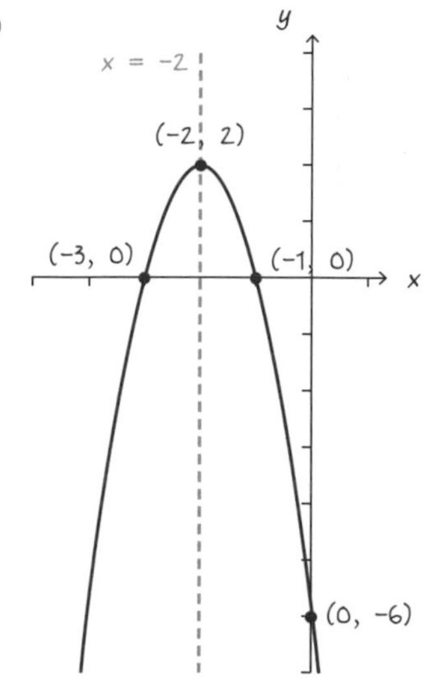

$f(x) = -2(x+3)(x+1)$

$a = -2 < 0 \Rightarrow$ the parabola is concave down and the vertex is a maximum point.

$p = -3$ and $q = -1 \Rightarrow$ the x-intercepts are $(-3, 0)$ and $(-1, 0)$.

$\frac{p+q}{2} = \frac{-3+-1}{2} = -2 \Rightarrow$ the axis of symmetry is $x = -2$.

$f\left(\frac{p+q}{2}\right) = f(-2) = -2(-2+3)(-2+1) = 2 \Rightarrow$ the vertex is $(-2, 2)$.

$f(0) = -2(0+3)(0+1) = -6 \Rightarrow$ the y-intercept is $(0, -6)$.

Reflect What is the largest possible domain of a quadratic function? What is the range of the function $y = a(x-h)^2 + k$?

A summary of the relationship between the parameters of a quadratic function and key features of the graph of the function is shown in the table.

Quadratic functions and their graphs		
General form $f(x) = ax^2 + bx + c$, $a \neq 0$ y-intercept: $(0, c)$ axis of symmetry: $x = -\frac{b}{2a}$ vertex: $\left(-\frac{b}{2a}, f\left(-\frac{b}{2a}\right)\right)$	**Vertex form** $f(x) = a(x-h)^2 + k$, $a \neq 0$ axis of symmetry: $x = h$ vertex: (h, k)	**Intercept form** $f(x) = a(x-p)(x-q)$, $a \neq 0$ x-intercepts: $(p, 0)$, $(q, 0)$ axis of symmetry: $x = \frac{p+q}{2}$ vertex: $\left(\frac{p+q}{2}, f\left(\frac{p+q}{2}\right)\right)$
$a > 0 \Rightarrow$ the parabola is concave up and the vertex is a minimum point. $a < 0 \Rightarrow$ the parabola is concave down and the vertex is a maximum point.		

TOK

How would you choose which formula to use?

When is intuition helpful and harmful in mathematics?

Reflect What do the parameters in each of the different forms tell you about the graph?

Reflect Why is it useful to have more than one form for representing a quadratic function?

Exercise 3M

1 Write down the equation of the axis of symmetry and the coordinates of vertex for the graph of each function:

a $f(x) = 2(x - 3)^2 + 4$

b $f(x) = (x - 1)^2 - 5$

c $f(x) = -4(x + 3)^2 + 2$

d $f(x) = 3(x + 6)^2 - 5$

2 Find the coordinates of the y-intercept, the equation of the axis of symmetry and the coordinates of the vertex for the graph of each function:

a $f(x) = x^2 - 8x + 5$

b $f(x) = 3x^2 - 6x + 2$

c $f(x) = -2x^2 - 8x - 11$

d $f(x) = 2x^2 + 6x + 3$

3 Find the coordinates of the x-intercepts, the equation of the axis of symmetry and the coordinates of the vertex for the graph of each function:

a $f(x) = (x - 2)(x - 4)$

b $f(x) = 4(x + 3)(x - 1)$

c $f(x) = -(x + 5)(x - 3)$

d $f(x) = 2(x + 3)(x + 2)$

4 Sketch the graph of each function and label the key features of the graph:

a $f(x) = 2(x + 3)(x - 1)$

b $f(x) = -2x^2 - 8x - 11$

c $f(x) = -3(x - 2)^2 + 5$

d $f(x) = 3x^2 + 12x + 8$

Changing to general form or factorized form

You now know the three forms of a quadratic function. At times you may want to change one form to another to determine certain key features of the graph of the functon.

If a quadratic expression will factorize, then you can factorize it in order to write in factorized form.

Example 24

Each of the following functions can be written in the form $f(x) = a(x - p)(x - q)$.

Find the values of a, p and q and then write down the coordinates of the x- and y-intercepts of the graph of $y = f(x)$.

a $f(x) = x^2 + 6x - 16$ b $f(x) = -4x^2 + 2x$

a $f(x) = x^2 + 6x - 16$	Factorize.
$f(x) = (x + 8)(x - 2)$	
$a = 1, p = -8, q = 2$	Note that $a = 1, p = 2, q = -8$ is also correct.
x-intercepts: $(-8, 0)$ and $(2, 0)$	
y-intercept: $(0, -16)$	$f(x) = x^2 + 6x - 16 \Rightarrow c = -16$

Continued on next page

Functions

b $f(x) = -4x^2 + 2x$	Factorize.
$f(x) = -2x(2x - 1)$	
$f(x) = -4x\left(x - \frac{1}{2}\right)$	Factor out the coefficient of x in the first factor.
$a = -4,\ p = 0,\ q = \frac{1}{2}$	$-4x\left(x - \frac{1}{2}\right) = -4(x-0)\left(x - \frac{1}{2}\right)$
x-intercepts: $(0, 0)$ and $\left(\frac{1}{2}, 0\right)$	
y-intercept: $(0, 0)$	$-4x^2 + 2x = -4x^2 + 2x + 0 \Rightarrow c = 0$

A quadratic function given in factorized form can be expanded to change it to general form.

Example 25

a The function $f(x) = 3(x-1)(x+2)$ can be written in the form $f(x) = ax^2 + bx + c$. Find the values of a, b and c and then write down the coordinates of the x- and y-intercepts of the graph $y = f(x)$.

b The function $f(x) = -\frac{1}{2}(x-4)^2 - 2$ can be written in the form $f(x) = ax^2 + bx + c$. Find the values of a, b and c and the coordinates of the vertex and the y-intercept of the graph $y = f(x)$.

a $f(x) = 3(x-1)(x+2)$	
$f(x) = 3(x^2 + x - 2)$	Expand the brackets.
$f(x) = 3x^2 + 3x - 6$	Multiply out the bracket.
$a = 3,\ b = 3,\ c = -6$	
x-intercepts: $(1,0)$ and $(-2,0)$	
y-intercept: $(0,-6)$	$f(x) = 3(x-1)(x+2) \Rightarrow p = 1$ and $q = -2$
	$f(x) = 3x^2 + 3x - 6 \Rightarrow c = -6$
b $f(x) = -\frac{1}{2}(x-4)^2 - 2$	Expand the brackets.
$f(x) = -\frac{1}{2}\left(x^2 - 8x + 16\right) - 2$	
$f(x) = -\frac{1}{2}x^2 + 4x - 10$	Multiply out and simplify.
$a = -\frac{1}{2},\ b = 4,\ c = -10$	
vertex: $(4, -2)$	$f(x) = -\frac{1}{2}(x-4)^2 - 2 \Rightarrow h = 4$ and $k = -2$
y-intercept: $(0, -10)$	$f(x) = -\frac{1}{2}x^2 + 4x - 10 \Rightarrow c = -10$

Exercise 3N

1 Each function can be written in the form $f(x) = a(x-p)(x-q)$, where $p > q$.

Find the values of a, p and q and then write down the coordinates of the x- and y-intercepts of the graph of $y = f(x)$.

a $f(x) = x^2 + 7x - 18$

b $f(x) = 3x^2 - 11x + 10$

c $f(x) = 0.5x^2 + 3x + 4$

d $f(x) = -4x^2 + 18x - 8$

2 Each function can be written in the form $f(x) = ax^2 + bx + c$. Find the values of a, b and c and then write down the coordinates of the x- and y-intercepts of the graph of $y = f(x)$.

a $f(x) = 4(x - 1)(x + 5)$

b $f(x) = -2(x + 7)(x + 1)$

3 Each function can be written in the form $f(x) = ax^2 + bx + c$. Find the values of a, b and c. Find the coordinates of the vertex and of the y-intercept of the graph of $y = f(x)$.

a $f(x) = -3(x + 1)^2 - 6$

b $f(x) = \frac{1}{2}(x - 4)^2 + 3$

4 The function $f(x) = x^2 - 2x - 8$ can be written in the form $f(x) = a(x - p)(x - q)$, where $p > q$.

a Find the values of:

i a

ii p

iii q

b Write down the coordinates of the:

i x-intercepts

ii y-intercept

c Find the coordinates of the vertex of the graph of $y = f(x)$.

d Sketch the graph of $y = f(x)$.

5 Let $f(x) = (x - 3)^2 - 2$. Part of the graph of $y = f(x)$ is shown.

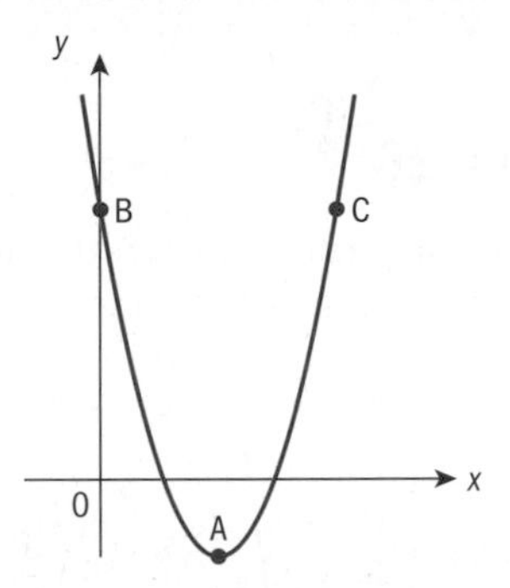

a The vertex of the graph of $y = f(x)$ is A.

i Write down the coordinates of A.

ii Write down the equation of the axis of symmetry for the graph of $y = f(x)$.

b Find the equation for the function f in the form $f(x) = ax^2 + bx + c$.

c The coordinates of B are $(0,q)$. Write down the value of q.

e The coordinates of C are (p,q). Find the value of p.

6 Let $f(x) = x^2 - 2x - 3$ and $g(x) = x - 2$.

a Let $h(x) = (f \circ g)(x)$. Show that $h(x) = x^2 - 6x + 5$.

b Find the equation of the axis of symmetry for the graph of h.

c Find the coordinates of the vertex of the graph of h.

d Find an equation for h in the form $h(x) = (x - p)(x - q)$, where p and q are integers.

e Sketch a graph of $y = -h(x)$, for $1 \le x \le 5$.

Fitting a quadratic function to a graph

Now you will find the equation of a quadratic function given some information about its graph.

Reflect What are three different forms of an equation of a quadratic function?

For each form of an equation of a quadratic function, which features of the graph of the function help you determine the parameters in the equation?

Example 26

Use the information shown in the graph to find an equation for the quadratic function. Write your final answer in general form, $f(x) = ax^2 + bx + c$.

a

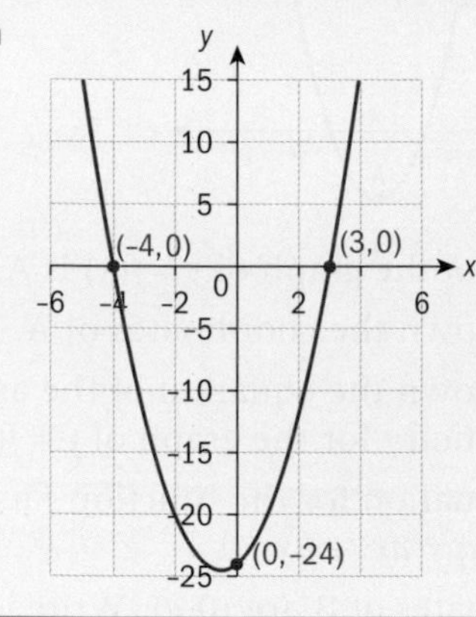

b

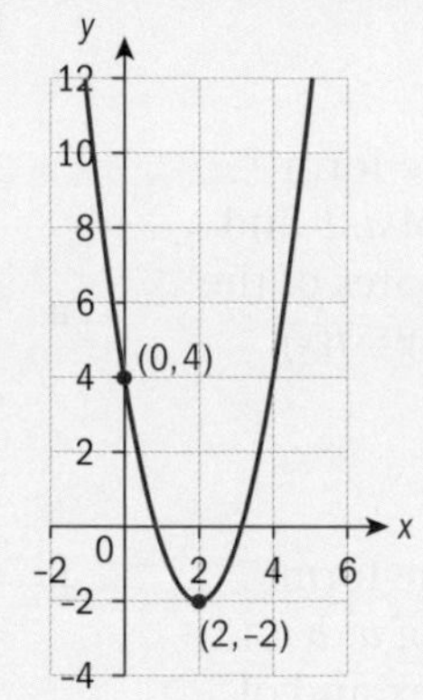

a $f(x) = a(x-p)(x-q)$ $f(x) = a(x+4)(x-3)$	Since both x-intercepts are given, you can start with intercept form and substitute -4 for p and 2 for q.
$-24 = a(0+4)(0-3)$ $-24 = -12a$ $a = 2$	From the point $(0,-24)$ you know that $y = -24$ when $x = 0$. Substitute and solve for a.
$f(x) = 2(x+4)(x-3)$ $f(x) = 2(x^2 + x - 12)$ $f(x) = 2x^2 + 2x - 24$	You now have the function in intercept form. Expand to write f in general form. You can check your answer by graphing this answer on your GDC and verifying the graph has the correct intercepts.
b $f(x) = a(x-h)^2 + k$ $f(x) = a(x-2)^2 - 2$	Since the vertex is given, you can start with vertex form and substitute 2 for h and -2 for k.
$4 = a(0-2)^2 - 2$ $4 = 4a - 2$ $a = 1.5$	From the point $(0, 4)$ you know that $y = 4$ when $x = 0$. Substitute and solve for a.
$f(x) = 1.5(x-2)^2 - 2$ $f(x) = 1.5(x^2 - 4x + 4) - 2$ $f(x) = 1.5x^2 - 6x + 4$	You now have the function in vertex form. Expand to write f in general form. You can check your answer by graphing this answer on your GDC and verifying the graph has the correct vertex and y-intercept.

Exercise 30

1 Use the information shown in the graph to find an expression for the quadratic function. Write your final answer in the form $f(x) = ax^2 + bx + c$.

a

(0, −12)

(2, −16)

b

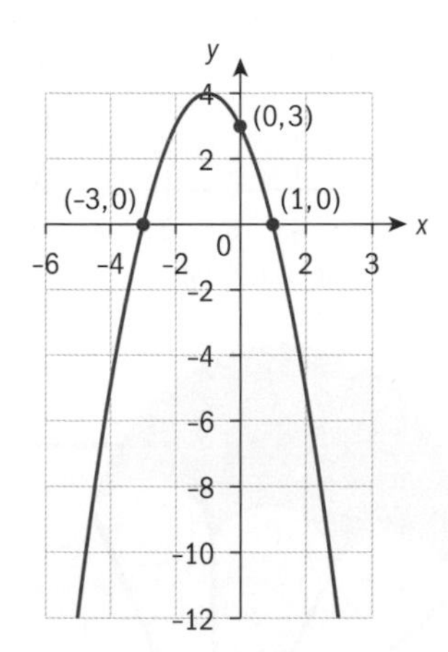

c

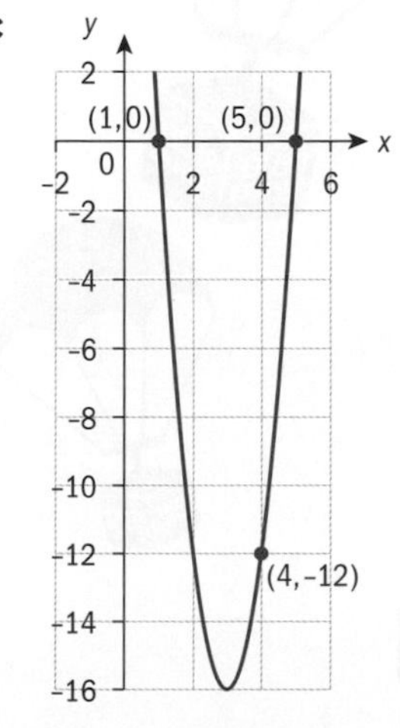

d

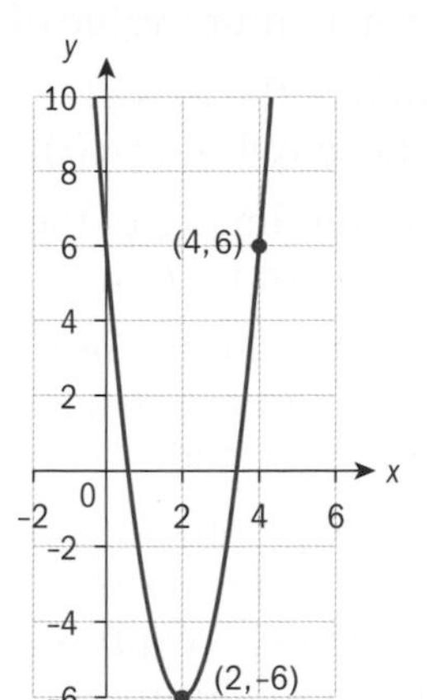

e

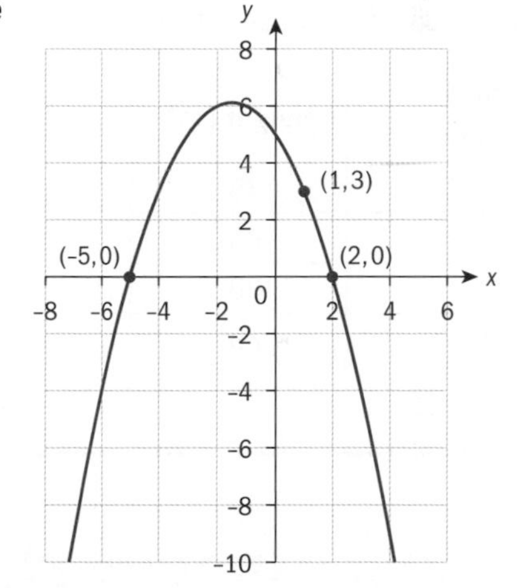

f

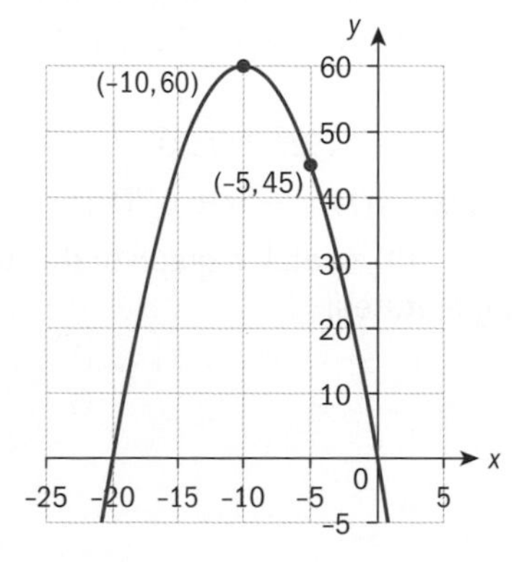

2 The graph of the quadratic function $y = f(x)$ has x-intercepts $(-1, 0)$ and $(3, 0)$. The function has a maximum value of 4.

a Find the equation of the axis of symmetry for the graph of $y = f(x)$.

b Write down the coordinates of the vertex for the graph of $y = f(x)$.

c Find an equation for f in the form $f(x) = a(x-h)^2 + k$, where a, h and k are constants to be determined.

d A translation of the graph of $y = f(x)$ right 4 units and down 5 units results in the graph of $y = g(x)$. Find an expression for the function $g(x)$ in the form $f(x) = ax^2 + bx + c$.

3 The table and graph are representations of the predicted height, h m, of a model rocket, t seconds after it is launched.

Time after launch (s)	Height (m)
1	35
3	75
4	80
5	75
8	0

a Write down the coordinates of the vertex of the graph and then explain its meaning in terms of the context of the graph.

b Find an equation for $h(t)$ and give the domain.

c Find the predicted height of the model rocket 2.4 seconds after launch.

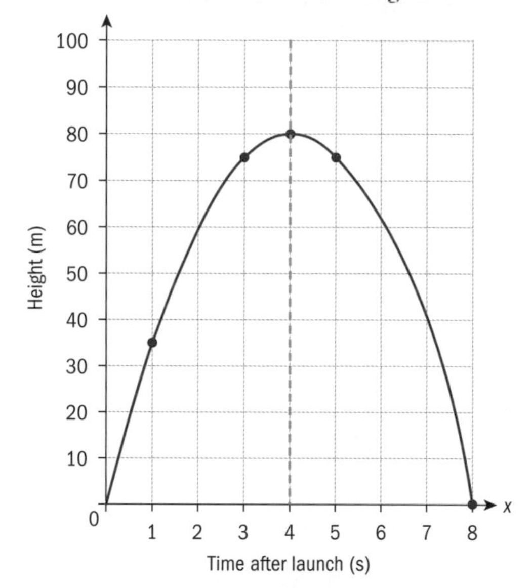

Developing inquiry skills

In the opening scenario for this chapter you looked at how crates of emergency supplies were dropped from a plane. The functions give the height of the crate.

During free fall: $h(t) = -4.9t^2 + 720$

With parachute open: $g(t) = -5t + 670$

Without using your GDC, sketch a graph of this piecewise function and label the key features.

TOK

How can you deal with the ethical dilemma of using mathematics to plot the course of a missile?

3.5 Solving quadratic equations by factorization and completing the square

Break-even point, profit, revenue and costs are four related business concepts.

What does each of these terms mean?

Why would business owners want to find break-even points?

The publisher of a newsletter uses the following models to estimate monthly revenue, R, and monthly cost, C. Cost and revenue are in thousands of euros and x is the number of subscribers in thousands.

$$R(x) = 35x - 0.25x^2$$
$$C(x) = 300 + 15x$$

How would you find the number of subscribers that determines the break-even point?

In sections 3.3 and 3.4 you saw that any equation that can be written in the form $ax^2 + bx + c = 0$, where $a \neq 0$ is called a **quadratic equation**. In this section you will study two methods of solving quadratic equations.

Solving by factorization

If the left-hand side of $ax^2 + bx + c = 0$ can be factorized to the form $a(x - p)(x - q) = 0$, then the equation can be solved using the **zero-product property**. You have already met this concept in section 3.4 when you learned to write a quadratic equation in factorized form.

$a(x - p)(x - q)$ is the product of three linear factors: a, $(x - p)$ and $(x - q)$.

So $a(x - p)(x - q) = 0$ only when one or more of the three factors is itself equal to zero.

Since $a \neq 0$, it follows that either:

$(x - p) = 0$ and hence $x = p$, or
$(x - q) = 0$ and hence $x = q$

International-mindedness

Ancient Babylonians and Egyptians studied quadratic equations like these thousands of years ago to find, for example, solutions to problems concerning the area of a rectangle.

Example 27

Solve each equation using the factorization method:

a $x^2 + 5x + 6 = 0$ **b** $2x^2 - 7x - 4 = 0$ **c** $9x^2 + 6x + 1 = 0$

a $x^2 + 5x + 6 = 0$	
$(x + 2)(x + 3) = 0$	Factorize the left-hand side of the equation.
$x + 2 = 0$ or $x + 3 = 0$	Use the zero-product property to set each factor equal to zero and solve.
$x = -2$ or $x = -3$	
$x = -2, -3$	-2 and -3 are called the **solutions** or **roots** of the equation.

Continued on next page

b $2x^2 - 7x - 4 = 0$ $(2x+1)(x-4) = 0$ $2x + 1 = 0$ or $x - 4 = 0$ $x = -\frac{1}{2}$ or $x = 4$ $x = -\frac{1}{2}, 4$	Factorize the left-hand side of the equation. Use the zero-product property to set each factor equal to zero and solve. **HINT** In parts **b** and **c**, the expression has not been factored into the form $a(x-p)(x-q)$, though it could be. Sometimes (like in these examples) it is better not to factor out the "a".
c $9x^2 + 6x + 1 = 0$ $(3x+1)(3x+1) = 0$ $3x + 1 = 0$ $x = -\frac{1}{3}$	Factorize the left-hand side of the equation. This can be written as $(3x+1)^2 = 0$ and so $9x^2 + 6x + 1$ is called a perfect square trinomial. When a quadratic is a perfect square you can say the equation has **two equal roots**.

Exercise 3P

1 Solve by factorization:

a $x^2 - 4x + 3 = 0$ **b** $x^2 - x - 20 = 0$
c $x^2 - 8x + 12 = 0$ **d** $x^2 - 121 = 0$
e $x^2 + x - 42 = 0$ **f** $x^2 - 8x + 16 = 0$

2 Solve by factorization:

a $2x^2 + x - 3 = 0$ **b** $3x^2 + 5x - 12 = 0$
c $4x^2 + 11x + 6 = 0$ **d** $9x^2 - 49 = 0$
e $4x^2 - 16x + 7 = 0$ **f** $12x^2 + 11x - 5 = 0$

Quadratic equations not given in the form $ax^2 + bx + c = 0$ must re-written before solving by factorization.

Example 28

Find the roots of these equations using the factorization method:

a $3x^2 - 4 = 16x + 8$ **b** $n(n+8) = 5(n+2)$

a $3x^2 - 4 = 16x + 8$	
$3x^2 - 16x - 12 = 0$	Collect like terms on one side of the equation and write the quadratic in general form.
$(3x+2)(x-6) = 0$ $3x + 2 = 0$ or $x - 6 = 0$	Factorize and solve for x.
$x = -\frac{2}{3}, 6$	Remember that **roots** is another term for the solutions of an equation.

b $n(n+8) = 5(n+2)$	Quadratic equations do not need to be written in terms of x. The variable in this equation is n.
$n^2 + 8n = 5n + 10$ $n^2 + 3n - 10 = 0$	Expand the brackets and collect like terms.
$(n+5)(n-2) = 0$ $n = -5, 2$	Factorize and solve for n.

Exercise 3Q

1 Solve the following quadratic equations by factorization:

a $x^2 - x - 20 = 2x + 8$

b $2x^2 - 3x - 8 = -x^2 + 2x$

c $4(x^2 + 5) = 3x^2 + 10x - 4$

d $3x(x+5) = -(x+5)$

e $3(x+2)(x-2) = 5x$

f $x + 8 = \frac{-15}{x}$

2 Let $f(x) = x^2 - 2$, $g(x) = 2x + 1$ and $h(x) = x^2 + 5x + 3$.

a Show that $(f \circ g)(x) = 4x^2 + 4x - 1$.

b Find the values of x for which $(f \circ g)(x) = h(x)$.

Quadratic equations involving perfect squares

Consider the equation $x^2 = 4$. You could rewrite the equation as $x^2 - 4 = 0$ and factorize, $(x+2)(x-2) = 0$. The solutions are $x = -2$, $x = 2$. Alternatively, you could take the square roots of both sides of $x^2 = 4$. You must take both the positive and negative square root of 4 to obtain both solutions, $\sqrt{x^2} = \pm\sqrt{4}$. The solutions are $x = \pm 2$. The plus or minus symbol allows you to show both roots in a condensed form.

The quadratic expression $x^2 + 10x + 25$ is called a **perfect square** because $x^2 + 10x + 25 = (x+5)^2$.

Quadratic equations which involve a perfect square can be solved by taking square roots.

Example 29

Find the solutions to each equation:

a $x^2 + 10x + 25 = 11$ **b** $x^2 - 4x + 4 = 8$

a $x^2 + 10x + 25 = 11$	Notice that if you collect like terms, $x^2 + 10x + 14 = 0$ does not factorize.
$(x+5)^2 = 11$	Factorize the perfect square trinomial on the left-hand side of the equation.
$x + 5 = \pm\sqrt{11}$	Take the square roots of each side of the equation.
$x = -5 \pm \sqrt{11}$	Solve for x. The answer represents the two solutions, $-5 + \sqrt{11}$ and $-5 - \sqrt{11}$.
b $x^2 - 4x + 4 = 8$ $(x-2)^2 = 8$	Factorize the perfect square trinomial on the left-hand side of the equation.

Continued on next page

$x-2=\pm\sqrt{8}$ $x=2\pm\sqrt{8}$	Take the square roots of each side of the equation and solve for x.
$x=2\pm2\sqrt{2}$	The solution may be rewritten as $2\pm2\sqrt{2}$, since $\sqrt{8}=\sqrt{4\times2}=2\sqrt{2}$.

Points of intersection

You can use your GDC to find points of intersection of curves and lines. You can also solve equations by finding points of intersection.

Example 30

a Find the points of intersection of the graphs of $f(x) = 4x^2 - 2x - 5$ and $g(x) = 3x + 2$.

b Solve $x^2 + 10x + 25 = 11$ by using your GDC to find points of intersection.

a 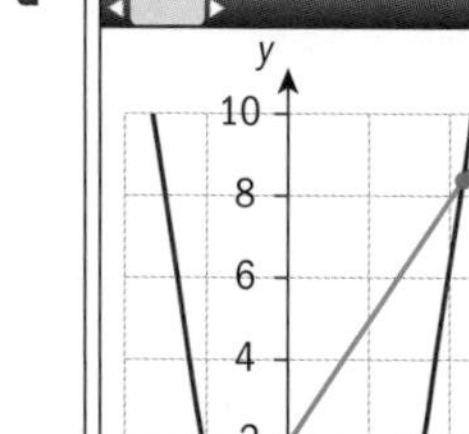 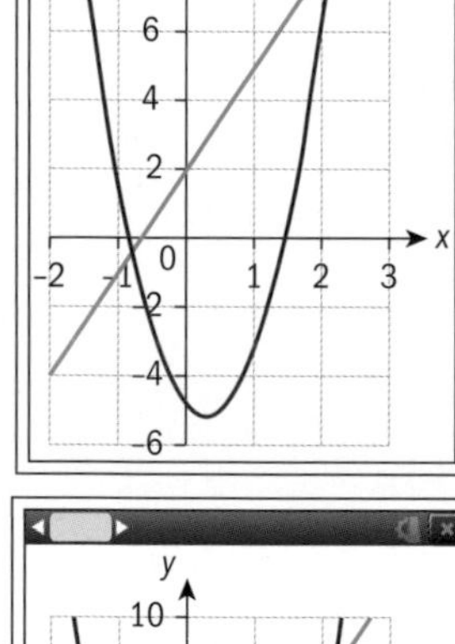	Graph $f(x) = 4x^2 - 2x - 5$ and $g(x) = 3x + 2$. There are two points of intersection. Find the coordinates of both points.
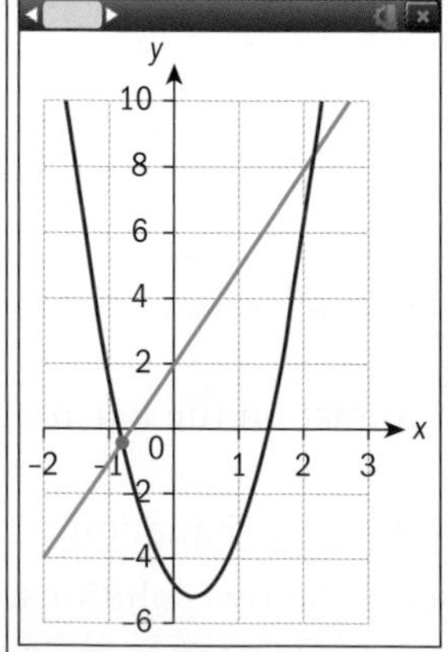 The points of intersection are (2.09, 8.26) and (−0.838, −0.514).	The answers are shown correct to 3 s.f.

b

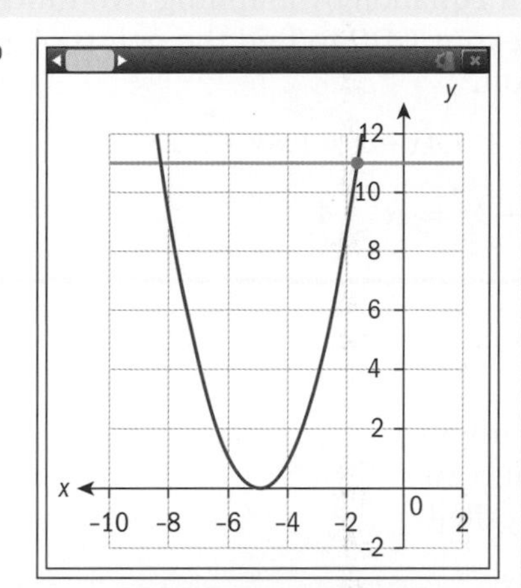

$x^2 + 10x + 25 = 11$

Graph the left- and right-hand sides of the equation: $y = x^2 + 10x + 25$ and $y = 11$.

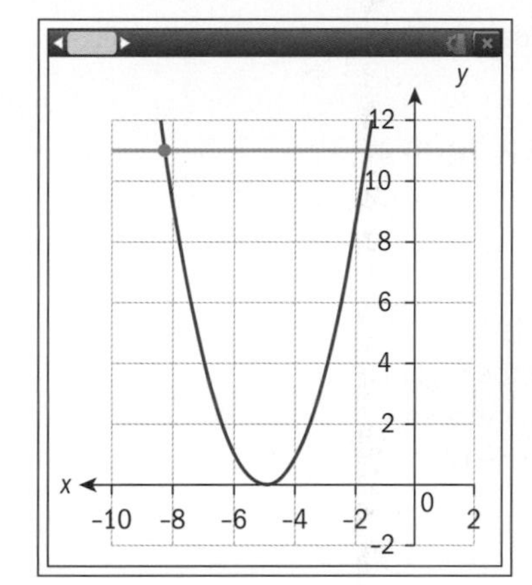

$x = -1.68$, $x = -8.32$

The x-coordinates the points of intersection are the solutions to the equation. You must find both solutions.

The answers are shown correct to 3 s.f.

Notice that you found the exact values to the solution of the equation $x^2 + 10x + 25 = 11$ in example 29 part **a**.

You can check to see that your exact answers have the approximate values you found here.

$x = -5 + \sqrt{11} \approx -1.68$

$x = -5 - \sqrt{11} \approx -8.32$

Alternatively, you could rewrite $x^2 + 10x + 25 = 11$ as $x^2 + 10x + 14 = 0$ and solve by finding the x-intercepts of the graph of $f(x) = x^2 + 10x + 14$.

Exercise 3R

Find the exact value of the solutions to each equation, without using your GDC. Then use your GDC to approximate and graphically check your solutions.

1 $x^2 - 8x + 16 = 10$ **2** $x^2 + 20x + 100 = 15$

3 $x^2 + 12x + 36 = 12$ **4** $x^2 - 10x + 25 = 27$

Find the points of intersection for the following pairs of functions.

5 $f(x) = -x + 5$ and $g(x) = 4x^2 + 3x + 2$

6 $f(x) = 4.25x^2 + 5.35x - 4.81$ and $g(x) = 2x + 5$

7 $f(x) = -2x^2 + 3x + 4$ and $g(x) = -x + 6$

8 $f(x) = 2x^2 - 3x + 1$ and $g(x) = -x^2 + 7x - 4$

Solve each equation by graphing two functions and using your GDC to find the points of intersection.

9 $-3.6x^2 + 5.4x - 2 = 1.8x - 7.2$

10 $0.5x^2 + 3x = -x^2 + 4$

Solving equations by completing the square

Many quadratic equations that you will come across do not involve a perfect square. However, they can easily be transformed into an expression which *does* involve a perfect square by a process called "Completing the square".

You can then use the technique you have just studied in order to solve them.

First you will find the value of c that can be added to $x^2 + bx$ to form a perfect square trinomial.

The following area models will show you how to do this.

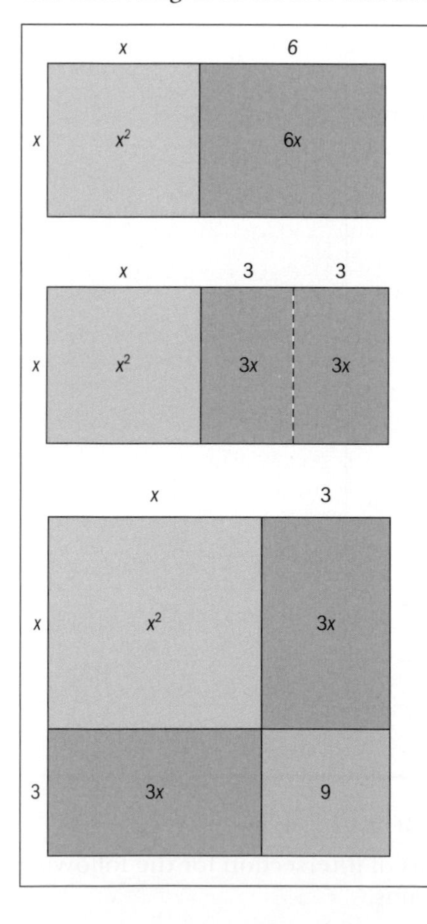

The area of the region shaded in green is x^2 and the area of the region shaded in pink is $6x$. Thus, the area of the whole rectangle is $x^2 + 6x$.

Divide the pink region into two equal parts.

Rearrange the parts.

The large square formed has area $(x + 3)^2$.

So the expression $x^2 + 6x$ becomes a perfect square when you add 9.

Investigation 7

For each of the following diagrams, find the value of c which makes the expression into a perfect square, and then write the expression as a perfect square.

1 $x^2 + 6x + c$

x 3
x x^2 3x
3 3x

2 $x^2 + 8x + c$

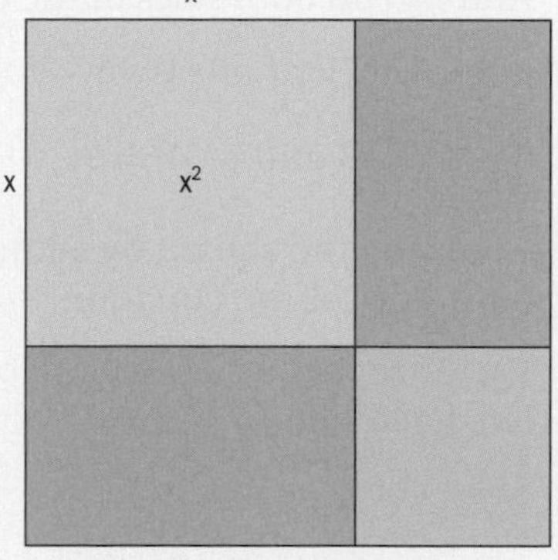

3 $x^2 + 10x + c$

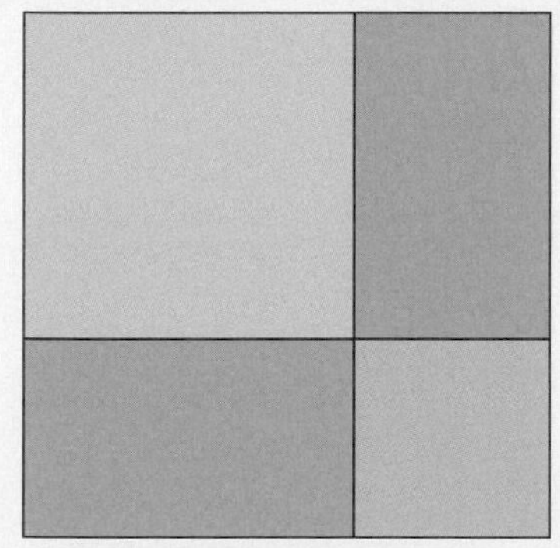

4 $x^2 + bx + c$

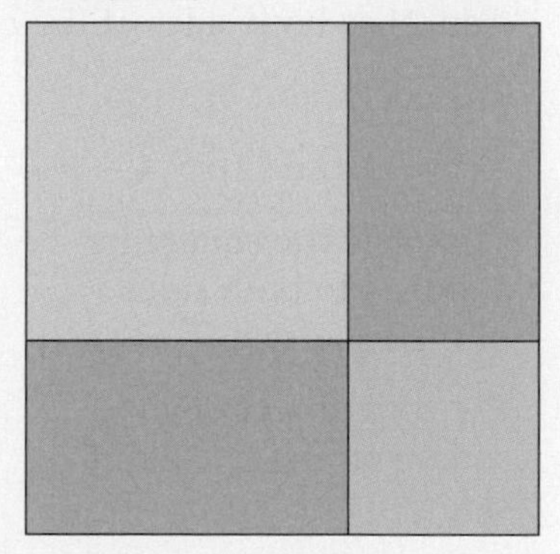

5 Copy and fill in the blank: To complete the square for $x^2 + bx$, add ______.

6 **Conceptual question** How does the area model help you understand the process of completing the square and why is the process called completing the square?

To solve a quadratic equation in the form $x^2 + bx + c = 0$ by completing the square:

1. Write the equation in the form $x^2 + bx = c$.
2. Add $\left(\frac{b}{2}\right)^2$ to both sides and write the equation as:

$$x^2 + bx + \left(\frac{b}{2}\right)^2 = c + \left(\frac{b}{2}\right)^2$$

3. Factorize the perfect square on the left-hand side: $\left(x + \frac{b}{2}\right)^2 = c + \left(\frac{b}{2}\right)^2$.
4. Take the square root of both sides and solve for x.

Example 31

Solve these quadratic equations by completing the square:

a $x^2 - 6x = 10$ **b** $x^2 + 4x - 21 = 0$ **c** $x^2 - 5x + 3 = 0$

a $x^2 - 6x = 10$	Add $\left(\frac{b}{2}\right)^2$ to both sides of the equation. The coefficient of x is −6. $\frac{1}{2}(-6) = -3$ and $(-3)^2 = 9$.
$x^2 - 6x + 9 = 10 + 9$	Complete the square by adding 9 to both sides of the equation.
$(x - 3)^2 = 19$	Factorize the perfect square on the left-hand side.
$x - 3 = \pm\sqrt{19}$ $x = 3 \pm \sqrt{19}$	Solve for x.
b $x^2 + 4x - 21 = 0$	Add 21 to both sides of the equation.
$x^2 + 4x = 21$	The coefficient of x is 4. $\frac{1}{2}(4) = 2$ and $(2)^2 = 4$.
$x^2 + 4x + 4 = 21 + 4$	Complete the square by adding 4 to both sides of the equation.
$(x + 2)^2 = 25$	Factorize the perfect square on the left-hand side.
$x + 2 = \pm\sqrt{25}$ $x = -2 \pm 5$ $x = -7, 3$	Solve for x.
c $x^2 - 5x + 3 = 0$	Subtract 3 from both sides of the equation.
$x^2 - 5x = -3$	The coefficient of x is −5. $\frac{1}{2}(-5) = -\frac{5}{2}$ and $\left(-\frac{5}{2}\right)^2 = \frac{25}{4}$.
$x^2 - 5x + \frac{25}{4} = -3 + \frac{25}{4}$	Complete the square by adding $\frac{25}{4}$ to both sides of the equation.
$\left(x - \frac{5}{2}\right)^2 = \frac{13}{4}$	Factorize the perfect square on the left-hand side.
$x - \frac{5}{2} = \pm\frac{\sqrt{13}}{2}$ $x = \frac{5 \pm \sqrt{13}}{2}$	Solve for x.

HINT

When the solutions are rational numbers you could also solve the equation by factorization.

$x^2 + 4x - 21 = 0$
$\Rightarrow (x + 7)(x - 3)$
$= 0 \Rightarrow x = -7, 3$

Exercise 3S

Solve these quadratic equations by completing the square:

1 $x^2 + 12x = 2$

2 $x^2 - 3x = 2$

3 $x^2 - 6x + 4 = 0$

4 $x^2 - 12x + 4 = 0$

5 $x^2 + 5x - 4 = 0$

6 $x^2 + x - 11 = 0$

International-mindedness

Over one thousand years ago, Arab and Hindu mathematicians were developing methods similar to completing the square to solve quadratic equations. They were finding solutions to mathematical problems such as "What must be the square which, when increased by 10 of its own roots, amounts to 39?" This is written as $x^2 + 10x = 39$

To solve $ax^2 + bx + c = 0$ by completing the square, when the coefficient of x^2 is not 1, you must first divide both sides of the equation by the coefficient of x^2.

Example 32

Solve by completing the square:

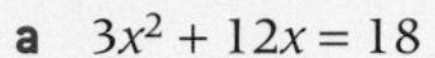

a $3x^2 + 12x = 18$

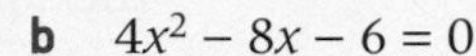

b $4x^2 - 8x - 6 = 0$

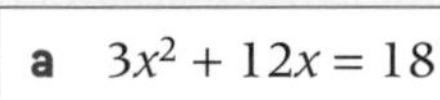

a $3x^2 + 12x = 18$	The coefficient of x^2 is 3.
$x^2 + 4x = 6$	Divide both sides of the equation by 3.
	Now the coefficient of x is 4.
	$\frac{1}{2}(4) = 2$ and $(2)^2 = 4$.
$x^2 + 4x + 4 = 6 + 4$	Complete the square by adding 4 to both sides of the equation.
$(x + 2)^2 = 10$	Factorize the perfect square on the left-hand side.
$x + 2 = \pm\sqrt{10}$	
$x = -2 \pm \sqrt{10}$	Solve for x.
b $4x^2 - 8x - 6 = 0$	
$x^2 - 2x = \frac{3}{2}$	Divide both sides of the equation by 4 and add the constant term to both sides.
	Now the coefficient of x is -2.
	$\frac{1}{2}(-2) = -1$ and $(-1)^2 = 1$.
$x^2 - 2x + 1 = \frac{3}{2} + 1$	Complete the square by adding 1 to both sides of the equation.
$(x-1)^2 = \frac{5}{2}$	Factorize the perfect square on the left-hand side.
$x - 1 = \pm\sqrt{\frac{5}{2}}$	Solve for x.
$x = 1 \pm \sqrt{\frac{5}{2}}$	The answer could also be written as $x = \frac{2 \pm \sqrt{10}}{2}$.

Functions

Reflect What can be said about the roots of a quadratic equation that can be solved by the factorization method?

What can be said about the roots of a quadratic equation that can be solved by completing the square?

Can you use factorization to find the roots of any quadratic equation? Can you use completing the square to find the roots of any equation?

Exercise 3T

Solve each quadratic equations by completing the square:

1 $2x^2 + 16x = 10$

2 $5x^2 - 30x = 10$

3 $6x^2 - 12x - 3 = 0$

4 $6x(x + 8) = 12$

5 $2x^2 + x - 6 = 0$

6 $2x(x + 8) + 12 = 0$

7 The publisher of a newsletter uses the following models to estimate monthly sales revenue (R) and the monthly cost of production (C), in terms of the number of people who subscribe to the newspaper (x).

$$R(x) = 35x - 0.25x^2$$
$$C(x) = 300 + 15x$$

C and R are both given in thousands of euros. The values of x for which revenue is equal to cost are called "break-even" points.

a Write down an equation, in terms of x, to find the break-even points.

b Solve your equation in **a** by completing the square.

c Write down the number of subscribers at which newspaper sales break even.

Profit is equal to revenue minus cost.

d Find the number of subscribers that yields maximum profit.

e Find the maximum profit.

Developing inquiry skills

In the opening scenario for this chapter you looked at how crates of emergency supplies were dropped from a plane. The functions $h(t)$ and $g(t)$ give the height of the crate:

During free fall: $h(t) = -4.9t^2 + 720$

With parachute open: $g(t) = -5t + 670$

How long after the crate leaves the plane does the parachute open? Write a quadratic equation you could solve to answer this question.

How could you solve this equation?

3.6 The quadratic formula and the discriminant

Investigation 8

Copy the following table and fill in the missing steps or explanations to complete the square for the general quadratic equation $ax^2 + bx + c = 0$.

$ax^2 + bx + c = 0$	Solve by completing the square.
1	1 Divide both sides of the equation by a.
2	2 Subtract $\frac{c}{a}$ from both sides of the equation.
3 $x^2 + \frac{b}{a}x + \left(\frac{1}{2} \times \frac{b}{a}\right)^2 = -\frac{c}{a} + \left(\frac{1}{2} \times \frac{b}{a}\right)^2$	3
4 $x^2 + \frac{b}{a}x + \frac{b^2}{4a^2} = \frac{b^2}{4a^2} - \frac{c}{a}$	4
5 $x^2 + \frac{b}{a}x + \frac{b^2}{4a^2} = \frac{b^2 - 4ac}{4a^2}$	5
6	6 Factorize the left-hand side of the equation.
7 $x + \frac{b}{2a} = \pm\frac{\sqrt{b^2 - 4ac}}{2a}$	7
8	8 Solve for x.

9 **Conceptual** What is the result of completing the square for the general equation $ax^2 + bx + c = 0$?

10 **Conceptual** Could you use this formula to find the roots of any quadratic equation?

11 **Conceptual** How does completing the square for the general form allow you to find the roots of any quadratic equation?

The solutions to any quadratic equation given in the form $ax^2 + bx + c = 0$, where $a \neq 0$, are given by the **quadratic formula** $x = \frac{-b \pm \sqrt{b^2 - 4ac}}{2a}$.

Example 33

Find the exact solutions to each equation by using the quadratic formula:

a $x^2 - 6x + 4 = 0$ **b** $3x^2 + 6x = -2$ **c** $6x^2 = 7x + 5$ **d** $x^2 - 4x + 5 = 0$

a $x^2 - 6x + 4 = 0$ $x = \frac{6 \pm \sqrt{(-6)^2 - 4(1)(4)}}{2(1)}$	Use the quadratic formula with $a = 1$, $b = -6$ and $c = 4$.
$x = \frac{6 \pm \sqrt{20}}{2} = \frac{6 \pm 2\sqrt{5}}{2} = 3 \pm \sqrt{5}$	Give your solution in exact form and simplified as far as is possible.

Continued on next page

b $3x^2 + 6x = -2$	
$3x^2 + 6x + 2 = 0$	Write the equation in general form.
$x = \dfrac{-6 \pm \sqrt{(6)^2 - 4(3)(2)}}{2(3)}$	Use the quadratic formula with $a = 3$, $b = 6$ and $c = 2$.
$x = \dfrac{-6 \pm \sqrt{12}}{6}$	
$x = \dfrac{-6 \pm 2\sqrt{3}}{6} = \dfrac{-3 \pm \sqrt{3}}{3}$	Give your solution in exact form and simplified as far as is possible.
c $6x^2 = 7x + 5$	
$6x^2 - 7x - 5 = 0$	Write the equation in general form.
$x = \dfrac{7 \pm \sqrt{(-7)^2 - 4(6)(-5)}}{2(6)}$	Use the quadratic formula with $a = 6$, $b = -7$ and $c = -5$.
$x = \dfrac{7 \pm \sqrt{169}}{12} = \dfrac{7 \pm 13}{12}$	
$x = \dfrac{5}{3}, -\dfrac{1}{2}$	The solutions are not surds, so this equation could also have been solved by factorization.
d $x^2 - 4x + 5 = 0$	
$x = \dfrac{4 \pm \sqrt{(-4)^2 - 4(1)(5)}}{2(1)}$	Use the quadratic formula with $a = 1$, $b = -4$ and $c = 5$.
$x = \dfrac{4 \pm \sqrt{-4}}{2}$	There is no real number with a square of -4, so this equation has no real solutions.
no real solutions	

The solutions of a quadratic equation $ax^2 + bx + c = 0$ are also known as the **roots** of the equation. These values are also called the **zeros** of the quadratic function $y = ax^2 + bx + c$.
If p and q are the real zeros of $y = ax^2 + bx + c$, then the points $(p, 0)$ and $(q, 0)$ are x-intercepts of the graph of $y = ax^2 + bx + c$.

Exercise 3U

1 Use the quadratic formula to find the roots of each equation:

a $x^2 + 4x - 2 = 0$

b $3x^2 - 8x + 5 = 0$

c $2x^2 - 5x - 2 = 0$

2 Solve each equation using the quadratic formula:

a $x^2 + 3x = 9$

b $3x^2 = 4x + 2$

c $2 + 2x - x^2 = 0$

d $3x^2 + 4x = -10$

e $-2(x - 3)^2 = 2x - 9$

f $9 - \dfrac{9}{x} = 2x$

g $\dfrac{x+3}{x-1} = \dfrac{2x}{x+1}$

3 Use the quadratic formula to find the zeros of each function:

a $y = 6x^2 + 5x - 6$

b $y = 2x^2 - 4x + 1$

c $y = -x^2 + 2x + 4$

4 Let $f(x) = 2x^2 - 4x + c$. The y-intercept of the graph $y = f(x)$ is $(0, -2)$.

a Write down the value of c.

b Find the vertex of the graph $y = f(x)$.

c The x-intercepts of the graph $y = f(x)$ are $\left(r+\sqrt{s},0\right)$ and $\left(r-\sqrt{s},0\right)$. Find the value of r and of s.

Investigation 9

A quadratic equation may have two distinct real roots, two equal real roots or no real roots. Complete the investigation to explore the nature of the roots a quadratic equation.

1 For each equation write down the value of $b^2 - 4ac$, use the quadratic formula to find the real roots, and then describe the nature of the roots. Use your GDC to make a rough sketch of the graph of the related function. Record your answers in a table like the one below. The first row of the table has been completed for you

$ax^2 + bx + c = 0$	$b^2 - 4ac$	Roots	Nature of roots	Sketch of $y = ax^2 + bx + c$
a $-5x^2 + 6x + 2 = 0$	76	$\frac{-6\pm\sqrt{76}}{-10}=\frac{3\pm\sqrt{19}}{5}$	two distinct real roots	y, x
b $3x^2 + 2x - 1 = 0$				
c $x^2 - 4x - 2 = 0$				
d $-x^2 - 6x - 9 = 0$				
e $4x^2 + 12x + 9 = 0$				
f $-2x^2 - 4x - 5 = 0$			no real roots	
g $4x^2 - 2x + 3 = 0$				

2 The quadratic equations in parts **f** and **g** have no real roots. Use their graphs to explain why.

3 **Conceptual** What are the three types of roots that a quadratic equation can have?

4 **Factual** What is the relationship between the nature of the roots of the equation $ax^2 + bx + c = 0$ and the value of $b^2 - 4ac$?

5 **Factual** What is the relationship between the value of $b^2 - 4ac$ and the number of x-intercepts of the graph of the function. $y = ax^2 + bx + c$?

A quadratic equation $ax^2+bx+c=0$ has **discriminant** $\Delta = b^2 - 4ac$.

The symbol Δ is used to represent the discriminant.

HINT

A quadratic equation may have complex roots, which include a real and imaginary part, but complex numbers are beyond the scope of the SL syllabus. So, we say "no real roots", because "no roots" may not be accurate.

The discriminant can be used to determine (or *discriminate*) the number and nature of the roots of the equation $ax^2+bx+c=0$, or the number of x-intercepts of the graph of the equation $y = ax^2+bx+c$.

For...	The equation $ax^2+bx+c=0$ has ...	The graph of $y=ax^2+bx+c$ has ...
$b^2-4ac>0$	two distinct real roots	two x-intercepts
$b^2-4ac=0$	two equal real roots (one repeated root)	one x-intercept
$b^2-4ac<0$	no real roots	no x-intercepts

Reflect Why does a quadratic equation with a discriminant of 0 have two equal real roots?

Why does a quadratic equation with a discriminant greater than 0 have two distinct real roots?

Why does a quadratic equation with a discriminant less than 0 have no real roots?

How can you use the discriminant to determine when the roots of a quadratic equation will be rational?

Why does the discriminant determine the nature of the roots?

Example 34

Find the value of the discriminant of the following quadratic equations, and then state the number and nature of the roots of each equation.

a $2x^2 + 5x - 1 = 0$ **b** $x^2 + 10x = -25$

a $2x^2 + 5x - 1 = 0$	This equation is already in general form, with $a = 2$, $b = 5$ and $c = -1$.
$\Delta = 5^2 - 4(2)(-1) = 25+ 8 = 33$	Calculate $\Delta = b^2 - 4ac$.
$\Delta = 33 > 0 \Rightarrow$ two distinct real roots	You must state the value of the discriminant, and whether it is greater than, equal to or less than zero.
b $x^2 + 10x = -25$	
$x^2 + 10x + 25 = 0$	Write the equation in general form, with $a = 1$, $b = 10$ and $c = 25$.
$\Delta = 10^2 - 4(1)(25) = 100 - 100 = 0$	Calculate $\Delta = b^2 - 4ac$.
$\Delta = 0 \Rightarrow$ two equal real roots	

Example 35

Find the value(s) of k for which the graph of $y = x^2 + 9x + k$ has two distinct real roots.

a $y = x^2 + 9x + k$	Find the discriminant of $x^2 + 9x + k = 0$.
$\Delta = 9^2 - 4(1)(k) = 81 - 4k$	$\Delta = b^2 - 4ac$ where $a = 1$, $b = 9$ and $c = k$.
$81 - 4k > 0$ $-4k > -81$	The graph has two distinct real roots when $\Delta > 0$.
$k < \frac{81}{4}$	Remember to reverse the inequality symbol when you divide by a negative value.

Exercise 3V

1 Use your GDC to sketch the following graphs. Find the value of the discriminant, and then state the nature of the roots of each equation.

a $x^2 -5x + 9 = 0$ **b** $6x^2 + 7x - 3 = 0$

c $x^2 - 4x + 15 = 0$ **d** $3x^2 + 4x = 8$

e $x^2 - 4x + 4 = 0$ **f** $5x^2 = x - 10$

2 For each equation, find the value(s) of k such that the equation has two distinct real roots.

a $x^2 + 3x + k = 0$ **b** $kx^2 + 20x + 5 = 0$

3 For each equation, find the value(s) of p such that the equation has two equal real roots.

a $x^2 + 5x + p = 0$ **b** $3x^2 - 12x + p = 0$

c $2x^2 - 2px + 4 = 0$ **d** $x^2 - 3px - 2p = 0$

4 For each equation, find the value(s) of m such that the equation has no real roots.

a $x^2 - 2x + m = 0$ **b** $3mx^2 - 6x + 1 = 0$

c $x^2 + 5x + m - 2 = 0$

Solving quadratic inequalities

You have now solved quadratic equations by factorizing, completing the square, using a GDC, and using the quadratic formula. You can also use any of these methods to help solve a quadratic inequality.

Investigation 10

1 For each of these quadratic equations:

$x^2 + 10x + 25 = 0$ $x^2 + 5x + 6 = 0$ $2x^2 - 3x + 17 = 0$
$x^2 - 4x + 11 = 0$ $5x^2 - 6x + 9 = 0$

a Choose a method to solve the equation.
b Was your method quick or efficient?
c How else could you have solved it?

2 Draw a flow diagram that helps you decide when to use each method, depending on the type of quadratic equation you are faced with.

3 **Conceptual** How do you choose which method to use to solve a quadratic equation?

4 **Conceptual** Which methods of solving a quadratic equation work for any quadratic equation?

5 **Conceptual** How is technology beneficial to help solve quadratic equations?

International mindedness

Over 2000 years ago, Babylonians and Egyptians used quadratics to work with land area.

Example 36

Solve each inequality:

a $x^2 - 6x + 8 < 0$

b $2x^2 - 5 \geq x - 3$

a $x^2 - 6x + 8 < 0$	To solve this inequality, you can sketch the graph of $y = x^2 - 6x + 8$.
$(x-2)(x-4) < 0$ When $(x-2)(x-4) = 0$, then $x = 2$ or $x = 4$	First, find the x-intercepts of the graph by factorizing $x^2 - 6x + 8$.
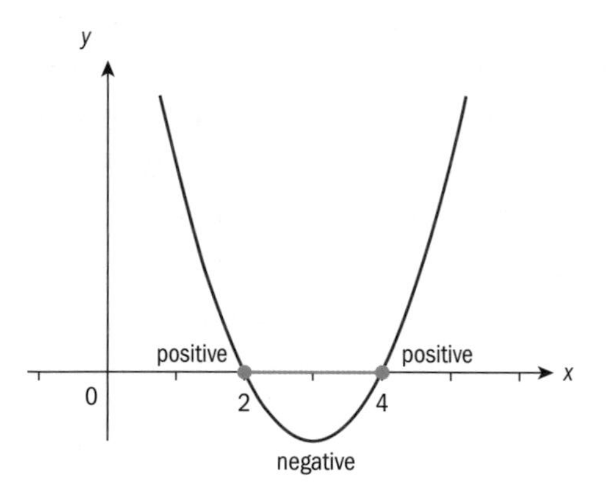 	Make a rough sketch of $y = (x-2)(x-4)$ by plotting the x-intercepts and using the fact that the parabola is concave up. Since you are looking for values where y is less than zero, the solution to the inequality consists of the x-values for which the graph lies below the x-axis.
The solution is $2 < x < 4$.	
b $2x^2 - 5 \geq x - 3$ $2x^2 - x - 2 \geq 0$	Write the quadratic inequality in general form and consider the graph of $y = 2x^2 - x - 2$.
When $2x^2 - x - 2 = 0$, then $x = \frac{1 \pm \sqrt{17}}{4}$	To find the x-intercepts of the graph, use either the quadratic formula or completing the square.
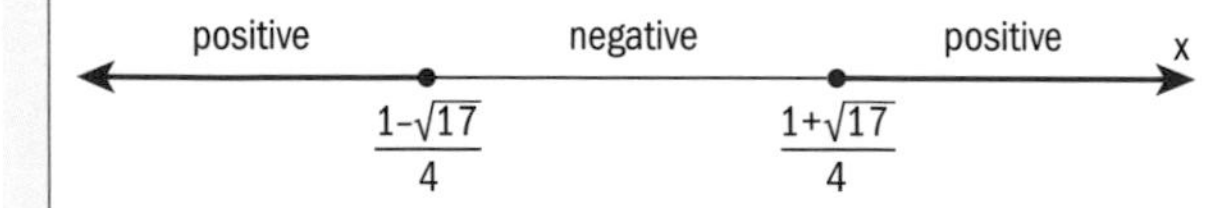 	You can use a number line to analyse the inequality, by plotting the x-intercepts and considering the concavity of the graph or by testing a value in each interval.
The solution is $x \leq \frac{1-\sqrt{17}}{4}$ or $x \geq \frac{1+\sqrt{17}}{4}$.	Since you are looking for values where y is greater than or equal to zero, the solution to the inequality consists of the x-values for which the graph lies above the x-axis.

Example 37

Solve each inequality graphically using your GDC. Give your answers correct to 3 significant figures.

a $2x^2 \geq 7x - 4$

b $3x^2 - 7x < 2x - 2$

a $2x^2 \geq 7x - 4$

$2x^2 - 7x + 4 \geq 0$

Write the quadratic inequality in general form and plot the graph of $y = 2x^2 - 7x + 4$ on your GDC.

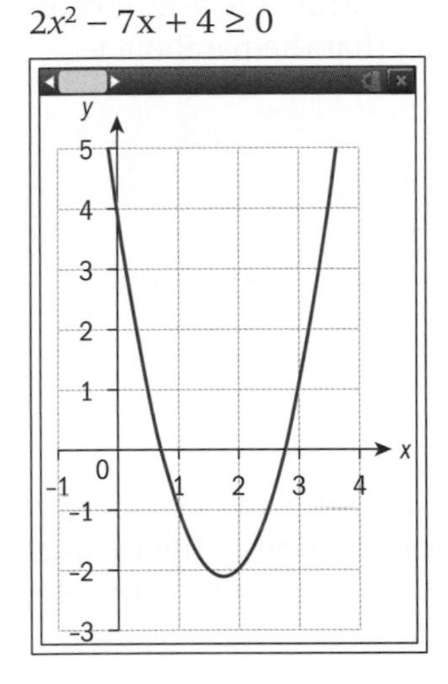

Use your GDC to find the x-intercepts correct to 3 s.f.

Since you are looking for values where y is greater than or equal to zero, the solution to the inequality consists of the x-values for which the graph lies above the x-axis.

The solution is $x \leq 0.719$ or $x \geq 2.78$.

b $3x^2 - 7x < 2x - 2$

Rather than writing the inequality in general form, you can graph the expressions on each side of the inequality. Use your GDC to graph $y = 3x^2 - 7x$ and $y = 2x - 2$. Find the x-coordinates of the points of intersection.

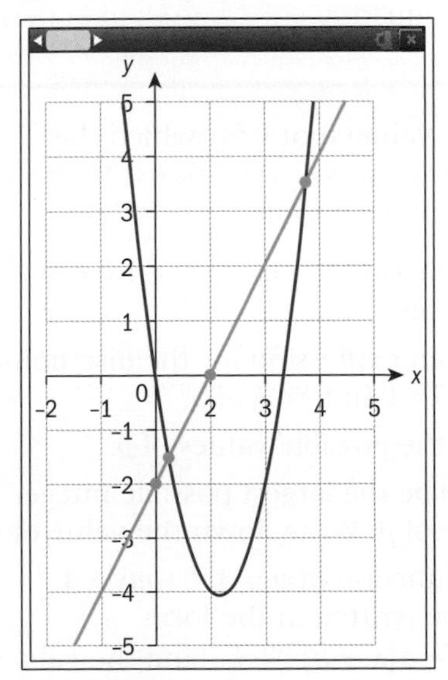

The solution consists of the values of x for which the graph of $y = 3x^2 - 7x$ lies below (that is, is less than) the graph of $y = 2x - 2$.

The solution is $0.242 < x < 2.76$.

Example 38

Find the value(s) of k for which the graph of $y = x^2 + kx + 9$ has no x-intercepts.

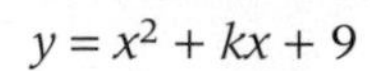

$y = x^2 + kx + 9$

Find the discriminant of $x^2 + kx + 9 = 0$.

$\Delta = k^2 - 4(1)(9) = k^2 - 36$

$\Delta = b^2 - 4ac$ where $a = 1$, $b = k$ and $c = 9$.

$k^2 - 36 < 0$

The graph has no x-intercepts when $\Delta < 0$.

Make a rough sketch of $y = k^2 - 36 = (k + 6)(k - 6)$, by plotting the x-intercepts and using the fact that the parabola is concave up.

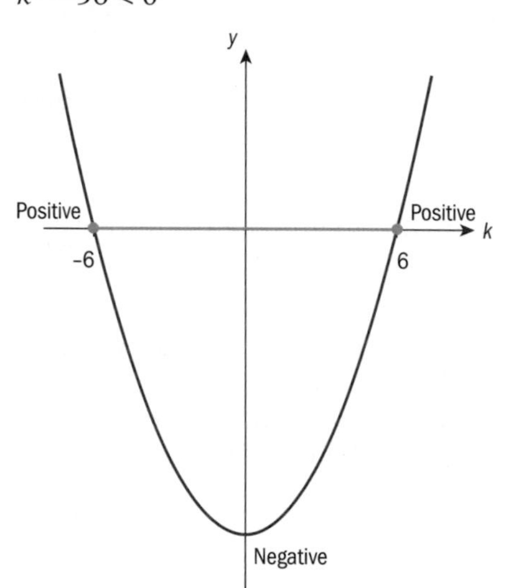

Since you are looking for values where y is less than zero, the solution to the inequality consists of the k-values for which the graph lies below the k-axis.

Therefore, $-6 < k < 6$.

Exercise 3W

1 Solve each inequality:

a $3x^2 + 5x - 2 \geq 0$ **b** $x^2 - 5 \leq 0$
c $2x^2 + 6x - 6 < x^2 + 2x$

2 Solve each inequality graphically using your GDC.

a $x^2 - 12x \geq 3$ **b** $3x^2 < 7x + 6$
c $8x^2 - 3x < 9$

3 For each equation, find the value(s) of k such that the equation has two distinct real roots.

a $-x^2 + kx - 4 = 0$ **b** $x^2 + 2kx + 3 = 0$

4 Find the value(s) of m such that the equation $x^2 + 6mx + m = 0$ has no real roots.

5 Find the value(s) of k for which the quadratic equation $kx^2 - 6kx + 2 + k = 0$ has two distinct real roots.

6 The graph of $f(x) = 3x^2 + px + 4$ has no x-intercepts.

a Find an expression for the discriminant of $f(x) = 0$ in terms of p.

b Find the possible values of p.

c Let m be the largest possible integer value of p. Write down the value of m.

d The function $h(x) = 3x^2 + mx + 4$ can be written in the form $h(x) = a(x - h)^2 + k$. Find the values of a, h and k.

Developing inquiry skills

The opening problem for this chapter considered the height of a crate dropped from a cargo plane. The functions h and g give the crate's height above the ground, measured in metres, t seconds after the crate leaves the plane.

During free fall: $h(t) = -4.9t^2 + 720$

With parachute open: $g(t) = -5t + 670$

Now that you have learned several methods for solving a quadratic equation you can answer the following questions.

How long after the crate leaves the plane does it reach the ground?

What is the domain of h and g in the context of the problem?

3.7 Applications of quadratics

In a suspension bridge, a supporting cable is attached to two towers. The road deck hangs on vertical cables attached to the supporting cable. The supporting cable is in the shape of a parabola.

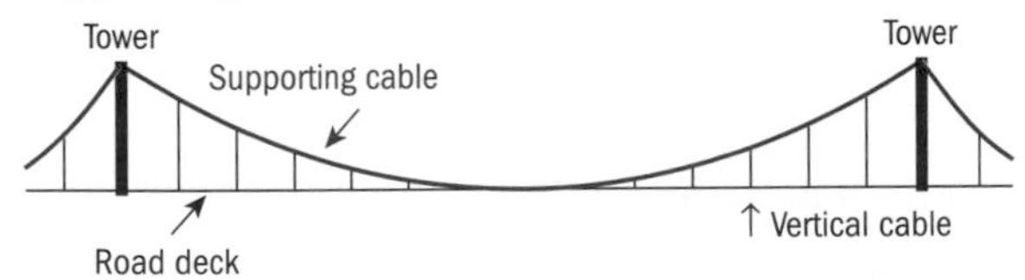

What information is needed to write a function that models the shape of the supporting cable?

Investigation 11

The main span of the Clifton Suspension Bridge in Bristol, UK is about 194 m long. The height of the towers is 26.2 m. The supporting cable for the main span of the bridge touches the road deck at the centre of the span.

1 The parabola shown in the graph models the supporting cable of the main span of the Clifton Suspension Bridge and the green line segments represent the towers.

Write down the values of r, s and t.

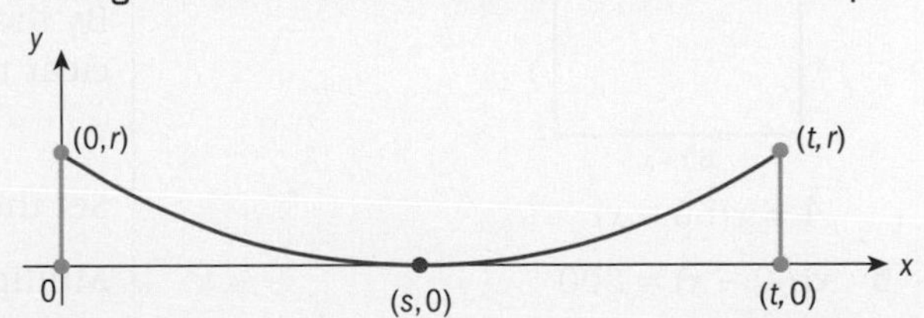

2 Consider the three forms of the equation of a quadratic function you have studied. For each of these forms the values of some of the parameters are equal to r, s or t.

$y = a(x-h)^2 + k$ $\quad$ $y = ax^2 + bx + c$ $\quad$ $y = a(x-p)(x-q)$

Replace the appropriate parameters in each equation with the values of r, s and t.

3 Choose one of the equations from question **2** and find any other missing parameters to complete a quadratic function that models the supporting cable.

Continued on next page

4 **Factual** Which form of the quadratic function did you use and why?

5 Define the variables in your function.

6 **Conceptual question** How can the quadratic function help you answer questions about the supporting cable of the Clifton Suspension Bridge?

7 Find the height of the supporting cable from the road deck at a point 24 m from a tower.

8 Suppose there are vertical cables with length 2.75 m, attaching the supporting cable to the road deck. Find the distance of these cables from the left-hand tower.

Reflect How can quadratic relationships be represented? What type of real-life relationships can be modelled by quadratic equations?
How do you decide which form of quadratic function to use for your model?

You have learned about different forms of quadratic functions and different methods for solving quadratic equations. Now you will use this knowledge to solve a variety of real-life problems. While one solution method is shown in each example, other methods may be possible.

Example 39

A rancher plans to use 120 m of fencing to build a rectangular pen.

a Let x represent the width of the pen. Find the length and area of the pen in terms of x.

b Find the dimensions of the pen if the area is 800 m^2.

c Find the maximum possible area of the pen.

a Let l represent the length of the pen and A represent the area. $2l + 2x = 120$ $l + x = 60$ $l = 60 - x$	Use the fact that the perimeter of the pen is 120 m to write an equation for the perimeter in terms of l and x.
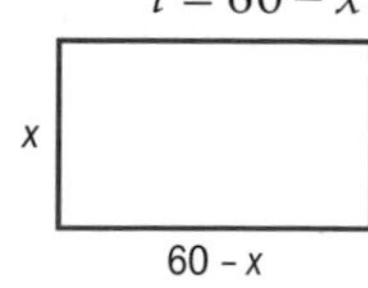 $A = x(60 - x)$	By drawing a simple diagram of the pen, it is clear that the area is length × width. Set the $A = 800$ in your expression from part **a**.
b $x(60 - x) = 800$ $x(60 - x) = 800$ $60x - x^2 = 800$ $x^2 - 60x + 800 = 0$ $(x - 20)(x - 40) = 0$ $x = 20, 40$ The pen is 20 m by 40 m.	Multiply out the bracket and write the quadratic equation in general form. Solve by factorization. If the width is 20, the length is 40. If the width is 40, the length is 20.

c $A = x(60 - x) = -x^2 + 60x$	The maximum occurs at the vertex of the graph of the area function.
$\frac{-b}{2a} = \frac{-60}{-2} = 30$	The maximum area occurs when the width of the pen is 30.
$A = 30(60 - 30) = 30^2 = 900$	
The maximum possible area is 900 m².	Substitute 30 for the width in the area equation.

Example 40

A bakery sells apple pies for \$12 each, and sells an average of 40 apple pies per day. The owner estimates that for each \$0.50 increase in price, the average sales will decrease by one pie a day.

Suppose there have been x increases of \$0.50 in the price of an apple pie above the initial price of \$12.

a Write down an expression for the price, \$$P$, of an apple pie after x increases of \$0.50 above the initial price of \$12.

b Write down the number of apple pies sold per day in terms of x.

c Find a function that gives the estimated revenue, R, from apple pie sales in terms of x.

d Use the revenue function to find the sale price the bakery owner should charge to maximize the revenue from apple pie sales.

a $P = 12 + 0.50x$	Think about the pattern. One increase = $12 + 0.50(1)$ Two increases = $12 + 0.50(2)$ x increases = $12 + 0.50x$
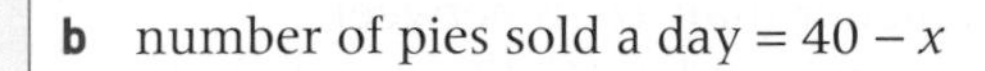 **b** number of pies sold a day = $40 - x$	After x increases of \$0.50, the owner will sell $40 - x$ apple pies per day.
c $R(x) = (12 + 0.50x)(40 - x)$	The revenue equals the selling price times the number of pies sold.
d 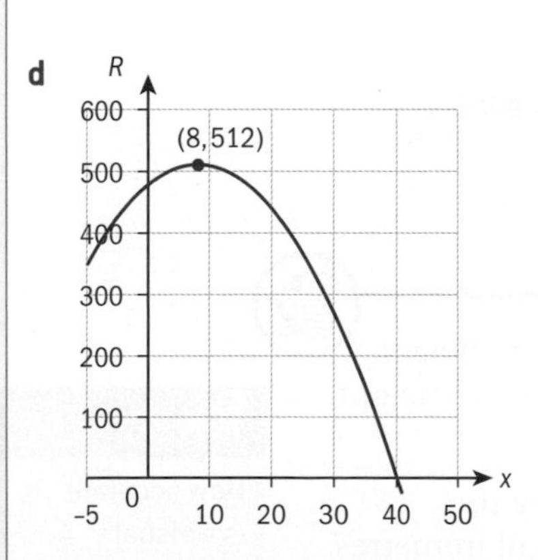	You can find the maximum as you did in example 39 or you can use your GDC to graph the revenue function and find the maximum.
The sale price that maximizes revenue is \$16.	The maximum revenue of \$512 occurs when there are 8 increases of \$0.50. $12 + 0.50(8) = 16$.

Example 41

The height, h, of a ball t seconds after it is thrown is modelled by the function $h(t) = 1.5 + 23t - 4.9t^2$, where h is the height of the ball in metres.

a Write down the height of the ball at the time it is released from the thrower's hand.

b Find the number of seconds it takes the ball to reach the ground.

c Find the length of time the ball is higher than 16 m.

a The height of the ball is 1.5 m.

When $t = 0$, $h = 1.5$.

b $1.5 + 23t - 4.9t^2 = 0$

When the ball reaches the ground, the height is 0.

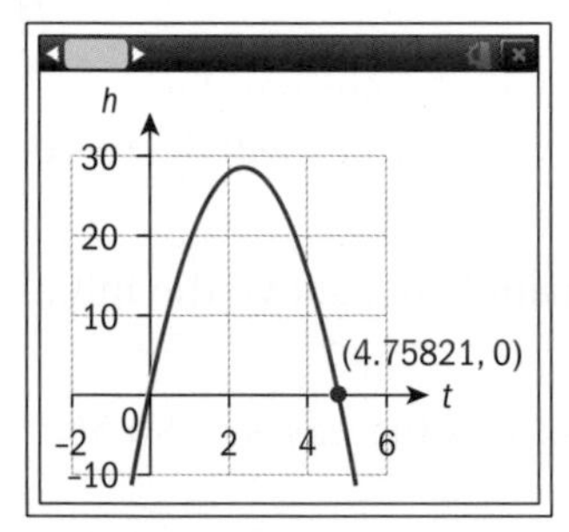

Use your GDC to graph the function $h(t) = 1.5 + 23t - 4.9t^2$ and find the positive x-intercept.

The ball reaches the ground 4.76 seconds after it is thrown.

c $1.5 + 23t - 4.9t^2 > 16$

Find the values of t where h is greater than 16.

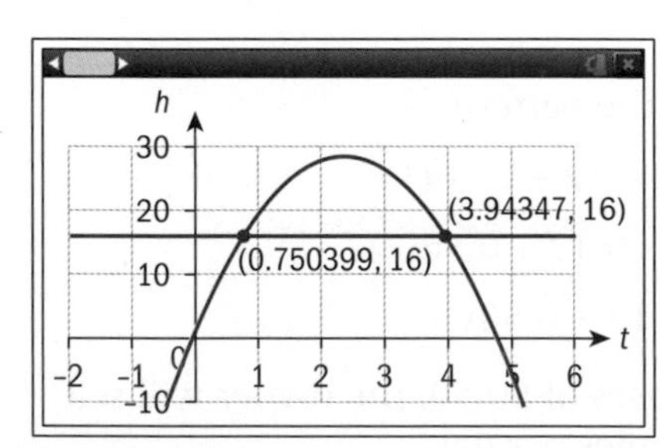

Use your GDC to graph $h(t) = 1.5 + 23t - 4.9t^2$ and $h(t) = 16$.

Find the points of intersection.

The ball is higher than 16 m between 0.750399 seconds and 3.94347 seconds and $3.94347 - 0.750399 = 3.193071$.

The ball is higher than 16 m for 3.19 seconds.

Reflect When solving a real-life problem modelled by a quadratic equation, how do you determine which method of solution to use?

Exercise 3X

1 The length of the base of a triangle is four more than twice the height. The area of the triangle is 24 m^2. Find the lengths of the base and height of the triangle.

2 The height of a ball t seconds after it is thrown is modelled by the function $h(t) = 2 + 20t - 4.9t^2$, where h is the height of the ball in metres.

a Find the height of the ball 3 seconds after it is thrown.

b Find the times at which the ball has a height of 6 m.

c Find the maximum height of the ball.

TOK

How accurate is a visual representation of a mathematical concept?

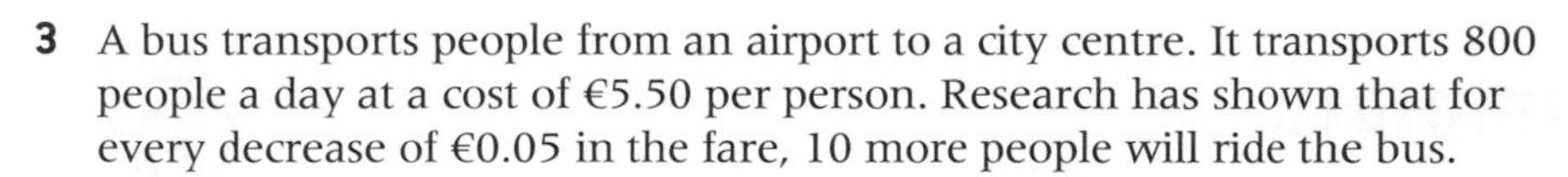

3 A bus transports people from an airport to a city centre. It transports 800 people a day at a cost of €5.50 per person. Research has shown that for every decrease of €0.05 in the fare, 10 more people will ride the bus.

Suppose there have been x decreases of €0.05 below the initial cost of €5.50.

a Find an expression, in terms of x, for the bus fare.

b Find an expression, in terms of x, for the number of people who ride the bus in a day.

c Find an expression, in terms of x, for the daily revenue generated by people riding the bus.

d Find the number of fare decreases that result in a revenue of €4500.

e Find an appropriate domain for this context.

4 The shape of an archway is modelled by the graph of a quadratic function. The maximum height of the archway is 4 m and the maximum width is 4 m. The graph shows a model of the archway.

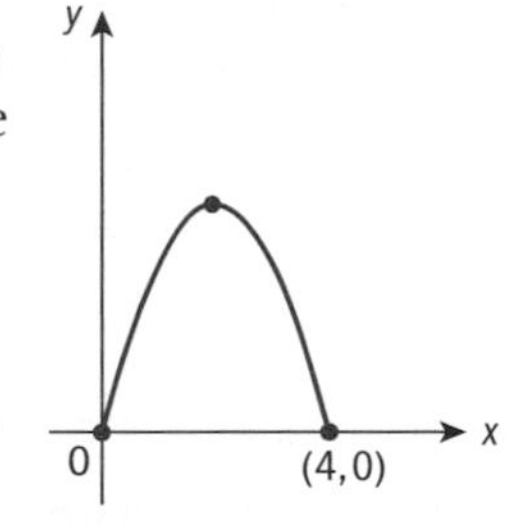

a Find a function to model the archway.

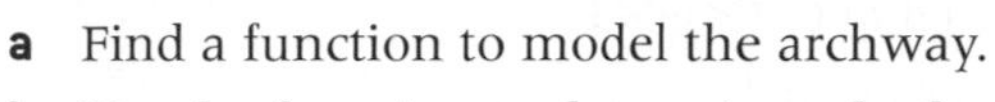

b Use the function to determine whether an object 3 m wide and 1.6 m tall will fit through the archway.

5 A rectangle has length x m and perimeter 310 m.

a Find a function for the area of the rectangle in terms of x.

b Use the function to find the dimensions of the rectangle with the maximum possible area.

FIFA (Fédération Internationale de Football Association), the governing body for international football (soccer), sets restrictions on the dimensions of the rectangular playing field.

FIFA's rules give the following limitations on the dimensions of a regulation football field.

The length of the touch line must be greater than the length of the goal line.

Touch line: minimum 90 m maximum 120 m

Goal line: minimum 45 m maximum 90 m

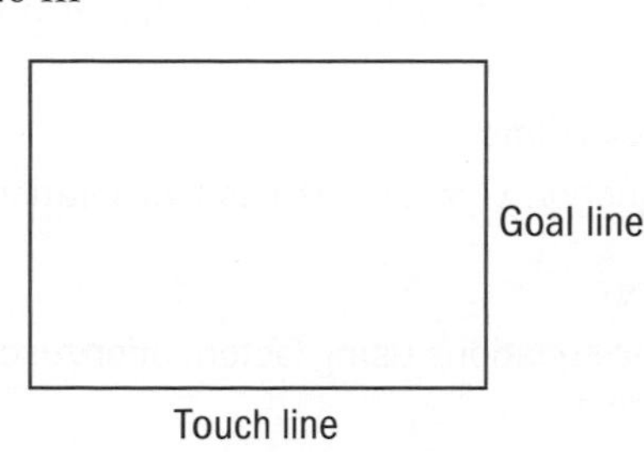

c Suppose a football field has perimeter 310 m. Determine whether the dimensions you found in part **b** meet the requirements for a FIFA regulation field width. Explain your reasoning.

d Suppose the function you found in part **a** is used to model the area of a regulation FIFA football field with perimeter 310 m and a touch line of length x. Write down the appropriate domain for this function.

e Find the maximum area of a football field with perimeter 310 m that meets FIFA's regulations.

TOK

We have seen the involvement of several nationalities in the development of quadratics in the chapter.

To what extent do you believe that mathematics is a product of human social collaboration?

Developing your toolkit

Now do the Modelling and investigation activity on page 182.

Chapter summary

Graphs of linear functions

- The gradient of a line passing through the points (x_1, y_1) and (x_2, y_2) is given by $m = \frac{y_2 - y_1}{x_2 - x_1}$.
- The gradients of parallel lines are equal.
- The product of the gradients of two perpendicular lines is -1.
- For a graph of a linear function in the form $y = mx + c$, the gradient is m and the y-intercept is $(0, c)$.
- For a graph of a linear function in the form $y - y_1 = m(x - x_1)$, the line passes through the point (x_1, y_1) and has gradient m.
- The general form of the equation of a line is $ax + by + d = 0$.

Transformations of the graph of $y = f(x)$

- The graph of $y = -f(x)$ is a reflection in the x-axis.
- The graph of $y = f(-x)$ is a reflection in the y-axis.
- The graph of $y = af(x)$ is a vertical dilation with scale factor $|a|$.
  - vertical stretch when $|a|>1$ and vertical compression when $0<|a|<1$
- The graph of $y = f(qx)$ is a horizontal dilation with scale factor $\frac{1}{|q|}$.
  - horizontal stretch for $0 < |q| < 1$ and horizontal compression for $|q| > 1$
- The graph of $y = f(x) + k$ is a vertical translation.
  - up k units for $k > 0$ and down $|k|$ units for $k < 0$
- The graph of $y = f(x - h)$ is a horizontal translation.
  - right h units for $h > 0$ and left $|h|$ units for $h < 0$

Graphs of quadratic functions

- For the graph of a quadratic function in the form $f(x) = a(x-h)^2 + k$, the vertex is (h, k) and the equation of the axis of symmetry is $x = h$. Transformations of the graph of $y = x^2$ can also be used to graph this form of the equation.
- For the graph of a quadratic function in the form $f(x) = ax^2 + bx + c$, the equation of the axis of symmetry is $x = -\frac{b}{2a}$ and the y-intercept is $(0, c)$.
- For the graph of a quadratic function in the form $f(x) = a(x-p)(x-q)$, the x-intercepts are $(p, 0)$ and $(q, 0)$ and the equation of the axis of symmetry is $x = \frac{p+q}{2}$.

Solving quadratic equations

- The quadratic equation $x^2 = c$, $c > 0$, has two solutions, $x = \sqrt{c}$ and $x = -\sqrt{c}$. The solutions can be written as $x = \pm\sqrt{c}$.
- To solve quadratic equations using factorization, use the zero-product property to set each factor equal to 0 and solve.
- To solve $ax^2 + bx + c = 0$ by completing the square: (1) divide both sides of the equation by a; (2) subtract the constant term from both sides of the equation; (3) add $\left(\frac{1}{2}\text{ the coefficient of } x\right)^2$ to both sides of the equation; (4) factorize the left-hand side of the equation; and (5) solve for x.

- The solutions of $ax^2 + bx + c = 0$, given by the quadratic formula are $x = \frac{-b \pm \sqrt{b^2 - 4ac}}{2a}$.
  - If $b^2 - 4ac > 0$, the equation will have two distinct real roots.
  - If $b^2 - 4ac = 0$, the equation will have two equal real roots.
  - If $b^2 - 4ac < 0$, the equation will have no real roots.

Developing inquiry skills

In the opening scenario for this chapter you looked at how crates of emergency supplies were dropped from a plane. The functions $h(t)$ and $g(t)$ give the height of the crate:

During free fall: $h(t) = -4.9t^2 + 720$

With parachute open: $g(t) = -5t + 670$

How realistic is this model in representing this real-life situation? State at least two advantages of the model, and give at least one criticism.

Suppose a heavier crate was dropped, but its parachute was the same. How would the model be different for this heavier crate?

Suggest suitable functions to model the path of a javelin through the air, or the path of a basketball as it leaves a player's hands and passes through a hoop. What do you need to consider in finding a model for each? How could each model help to make predictions in a real-life scenario?

Chapter review

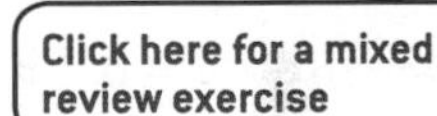

1 Sketch each line, marking on your sketch the axial intercepts:

a $y = 2x + 4$

b $y - 3 = -4(x + 2)$

c $2x + 3y - 6 = 0$

2 Find an equation for each line:

a the line which passes through the points $(-4, 2)$ and $(8, -1)$

b the line which is parallel to the line $y = \frac{1}{2}x + 3$ and has y-intercept $(0, -5)$

c the line which is perpendicular to $y = -\frac{2}{3}x + 7$ and passing through the point $(2, 4)$

d the line which is passes through $(-3, -4)$ with gradient 0

3 Consider the function $f(x) = \begin{cases} 2x + 1, & -2 \le x < 1 \\ 3, & 1 \le x \le 3 \end{cases}$

a Find $f(1)$ and $f(2)$.

b Graph $y = f(x)$.

4 Describe the series of transformations of the graph of $y = f(x)$ that lead to the graph of the given functions.

a $y = 2f(x - 3)$

b $y = \frac{1}{2}f(x) + 5$

c $y = -f(x + 2) - 1$

d $y = f(3x)$

e $y = f(-x) + 6$

5 In each case, find the indicated features of the graph of $y = f(x)$.

a $f(x) = 2(x - 3)(x + 7)$; x-intercepts, equation of the axis of symmetry

b $f(x) = -3(x - 4)^2 + 2$; equation of the axis of symmetry, vertex

c $f(x) = -x^2 - 4x + 6$; equation of the axis of symmetry, y-intercept

6 The function $f(x) = 3x^2 + 18x + 20$ can be written in the form $f(x) = a(x - h)^2 + k$.

a Find the values of:

i a **ii** h **iii** k

b Write down the vertex of the graph of $y = f(x)$.

c The graph of $y = g(x)$ is a translation of the graph of $y = f(x)$ right 5 units and down 3 units. Find the vertex of the graph of $y = g(x)$.

7 Solve each equation:

a $(x - 3)^2 = 64$ **b** $(x + 2)^2 = 7$

c $x^2 + 14x + 49 = 0$ **d** $x^2 + x - 12 = 0$

e $3x^2 + 4x - 7 = 0$

8 The equation $-x^2 + 3kx - 4 = 0$ has two equal real roots. Find the possible values of k.

9 The y-intercept of the graph of a quadratic function is $(0, -16)$ and the x-intercepts are $(-4,0)$ and $(2,0)$. Find the equation of the function in the form $f(x) = ax^2 + bx + c$ where a, b and c are constants.

10 Solve each equation. Give your answers correct to 3 significant figures.

a $2x^2 - 6x - 5 = 0$ **b** $-x^2 - 3x = 0.5x - 7$

11 The height, h metres above the water, of a stone thrown from a bridge is modelled by the function $h(t) = 18 + 13t - 4.9t^2$, where t is the time in seconds after the stone is thrown.

a Find the initial height from which the stone is thrown.

b Find the maximum height reached by the stone.

c Find the amount of time it takes for the stone to hit the water below the bridge.

d Write down the domain of the function h in the context of this real-life scenario.

e Find the length of time for which the height of the stone is greater than 23 m.

12 A rectangle is inscribed in isosceles triangle ABC as shown in the diagram.

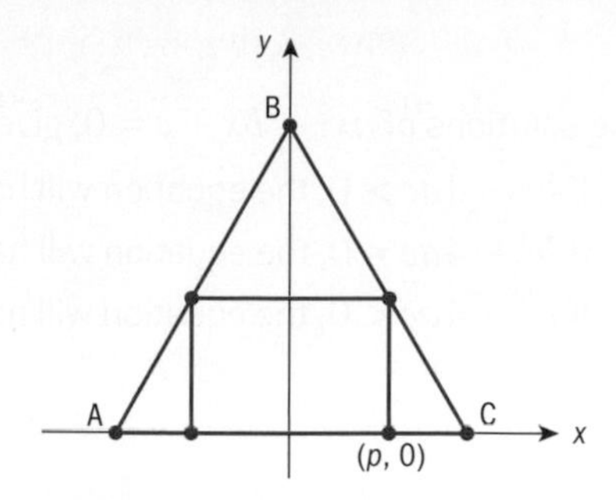

The altitude of triangle ABC from B to side AC is 7 cm and AC = 8 cm. The coordinates of one of the vertices of the inscribed rectangle are $(p, 0)$.

a Write down the coordinates of points A, B and C.

b Find the equation of the line passing through points B and C.

c Find the dimensions of the rectangle inscribed in the triangle, in terms of p.

d Write down an expression for the area of the inscribed rectangle in terms of p.

e Find the dimensions of the rectangle with maximum possible area.

f Find the maximum possible area of the inscribed rectangle.

Exam-style questions

13 **P2:** A line has equation $-7x - 12y + 168 = 0$

a Write down the equation of the line in the form $y = mx + c$. (2 marks)

b Given that the line intersects the x-axis at point A and the y-axis at point B, find the coordinates of A and B. (2 marks)

c Calculate the area of triangle OAB. (2 marks)

14 **P2: a** Using your GDC, sketch the curve of $y = -2.9x^2 + 4.1x + 5.9$ for $-1 \le x \le 2$. (2 marks)

b Write down the coordinates of the points where the curve intersects the x- or y-axis. (2 marks)

c Write down the range of y. (2 marks)

15 P2: a Show that the solutions to the equation $x^2 - 6x - 43 = 0$ may be written in the form $x = p \pm q\sqrt{13}$, where p and q are positive integers. (4 marks)

b Hence, or otherwise, solve the inequality $x^2 - 6x - 43 \leq 0$ (2 marks)

16 P1: A function f is defined by $f(x) = 3(x-1)^2 - 18,\ x \in \mathbb{R}$.

a Write $f(x)$ in the form $ax^2 + bx + c$, where a, b and c are constants. (2 marks)

b Find the coordinates of the vertex of the graph of f. (1 mark)

c Find the equation of the axis of symmetry of the graph of f. (1 mark)

d State the range of f. (2 marks)

e The graph of f is translated through the vector $\begin{pmatrix} 2 \\ -1 \end{pmatrix}$ to form a curve representing a new function $g(x)$.

Find $g(x)$ in the form $px^2 + qx + r$, where p, q and r are constants. (3 marks)

17 P1: a Solve the equation $8x^2 + 6x - 5 = 0$ by factorization. (4 marks)

b Determine the range of values of k for which $8x^2 + 6x - 5 = k$ has no real solutions. (3 marks)

18 P1: Consider the function $f(x) = x^2 - 10x + 27,\ x \in \mathbb{R}$.

a Show that the function f can be expressed in the form $f(x) = a(x-h)^2 + k$, where a, h and k are constants. (3 marks)

b Hence write down the coordinates of the vertex of the graph of $y = f(x)$. (1 mark)

c Hence write down the equation of the line of symmetry of the graph of $y = f(x)$. (1 mark)

19 P1: The quadratic curve $y = x^2 + bx + c$ intersects the x-axis at $(10, 0)$ and has equation of line of symmetry $x = \frac{5}{2}$.

a Find the values of b and c. (4 marks)

b Hence, or otherwise, find the other two coordinates where the curve intersects the coordinate axes. (2 marks)

20 P1: Consider the function $f(x) = 2x^2 - 4x - 8,\ x \in \mathbb{R}$.

a Show that the function f can be expressed in the form $f(x) = a(x-h)^2 + k$, where a, h and k are constants. (3 marks)

b The function $f(x)$ may be obtained through a sequence of transformations of $g(x) = x^2$. Describe each transformation in turn. (3 marks)

21 P1: Consider the equation $f(x) = 2kx^2 + 6x + k,\ x \in \mathbb{R}$.

a In the case that the equation $f(x) = 0$ has two equal real roots, find the possible values of k. (4 marks)

b In the case that the equation of the line of symmetry of the curve $y = f(x)$ is $x + 1 = 0$, find the value of k.

c Solve the equation $f(x) = 0$ when $k = 2$. (3 marks)

22 P1: A curve $y = f(x)$ passes through the points with coordinates $A(-12, 10)$, $B(0, -16)$, $C(2, 9)$, and $D(14, -10)$

a Write down the coordinates of each point after the curve has been transformed by $f(x) \mapsto f(2x)$. (4 marks)

b Write down the coordinates of each point after the curve has been transformed by $f(x) \mapsto f(-x) + 3$. (4 marks)

Hanging around

Approaches to learning: Thinking skills: Create, Generating, Planning, Producing
Exploration criteria: Presentation (A), Personal engagement (C), Reflection (D)
IB topic: Quadratic modelling, Using technology

Investigate

Hang a piece of rope or chain by its two ends. It must be free hanging under its own weight. It doesn't matter how long it is or how far apart the ends are.

1 What possible shaped curve does the hanging chain resemble?

2 What form might its equation take?

3 How could you test this?

Import the curve into a graphing package

A graphing package can fit an equation of a curve to a photograph.

Take a photograph of your hanging rope/chain.

What do you need to consider when taking this photo?

Import the image into a graphing package.

Carefully follow the instructions for the graphing package you are using.

The image should appear in the graphing screen.

Fit an equation to three points on the curve

Select three points that lie on the curve.

Does it matter what three points you select?

Would two points be enough?

In your graphing package, enter your three points as x- and y-coordinates.

Now use the graphing package to find the best fit quadratic model to your three chosen points.

Carefully follow the instructions for the graphing package you are using.

Test the fit of your curve

Did you find a curve which fits the shape of your image exactly?

What reasons are there that may mean that you did not get a perfect fit?

The shape that a free-hanging chain or rope makes is actually a catenary and not a parabola at all. This is why you did not get a perfect fit.

Research the difference between the shape of a catenary and a parabola.

International-mindedness

The word "catenary" comes from the Latin word for "chain".

Extension

Explore one or more of the following – are they quadratic?

The cross section of a football field.

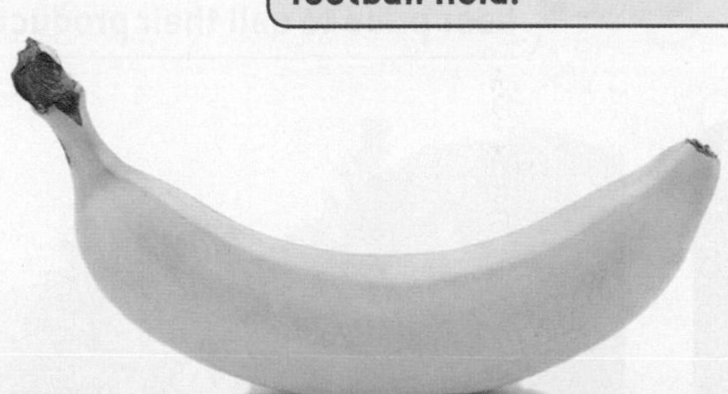

The curve of a banana.

The path of a football when kicked in the air. Here you would need to be able to use available software to trace the path of the ball that is moving.

A well-known landmark – perhaps the Sydney Harbour Bridge or the arches at the bottom of the Eiffel Tower.

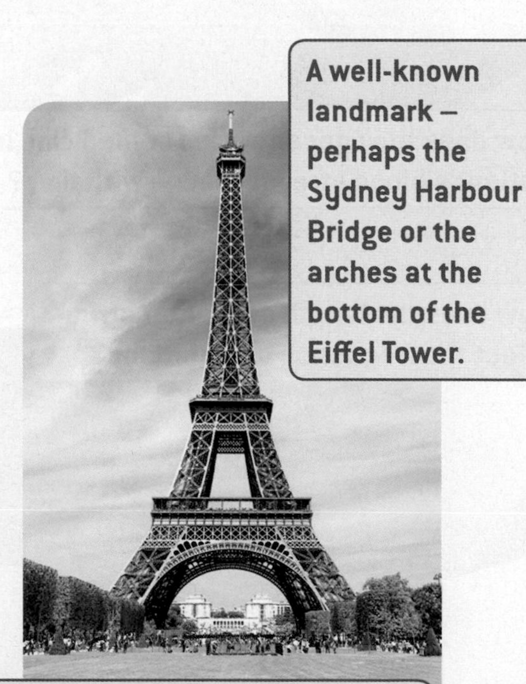

Other objects that look like a parabola – for example, the arch of a rainbow, water coming from a fountain, the arc of a Satellite dish.

4 Equivalent representations: rational functions

From business to music, astronomy to architecture, rational functions are used to represent relationships between two real-life variables and hence help make predictions and influence decisions. In this chapter you will learn about reciprocal and rational functions, their graphs, and where they appear in the world around you.

Concepts

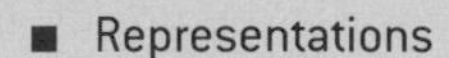

- Representations
- Equivalence

Microconcepts

- Domain and range of a rational function
- Features of reciprocal and rational functions: symmetry, intercepts, horizontal and vertical asymptotes
- Modelling with reciprocal and rational functions

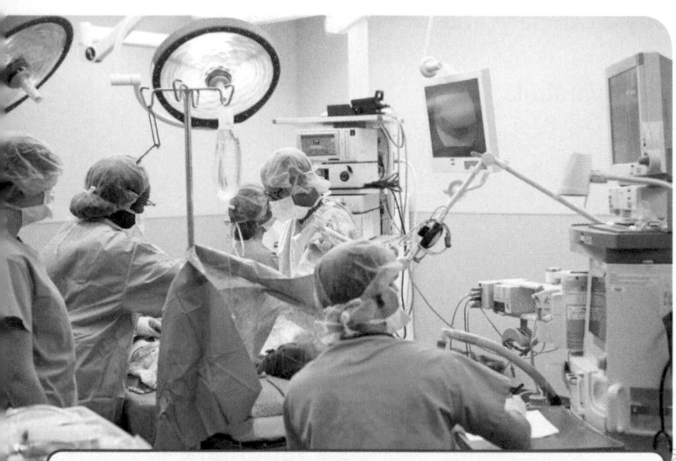

How does the concentration of medicine in a patient's blood stream change with time?

How can a business determine the best price to sell their products?

How does the length of a guitar string affect the frequency of vibrations?

What is the optimum file size to save photos to your phone in order to get the best compromise between photo quality and available space?

A water park is designing a new water slide.

The slide will be very steep to begin with, and then will level out so that a person is travelling almost horizontally when they hit the water at the bottom.

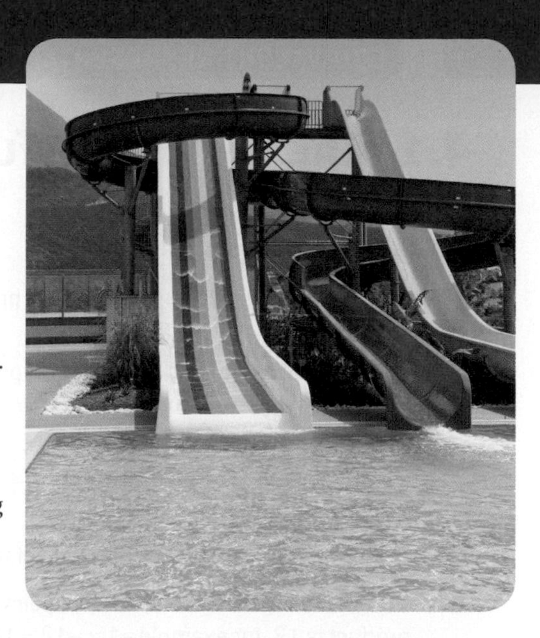

- How could you describe the shape of this slide? Sketch a path that a person on the slide would travel, from setting off at the top to reaching the water at the bottom.
- How gradual should the curve in the slide be to ensure a person's safety, but to make sure they travel very fast to begin with?
- How could you model the shape of the waterslide using mathematics?
- How might a suitable model help the slide's designer to choose the best overall shape for the slide?

Developing inquiry skills

Think of other physical objects that might have a similar shape to this water slide – for example, the scoop on a bulldozer.

Think of any abstract quantities that are related in such a way that the *graph* showing one quantity plotted against the other has the same shape as the water slide.

Think about the questions in this opening problem and answer any you can. As you work through the chapter, you will gain mathematical knowledge and skills that will help you to answer them all.

Before you start

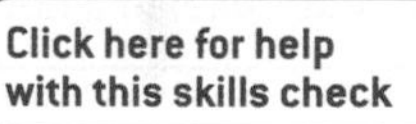

You should know how to:

1 Solve simple equations

eg $x - 3 = 0$

$x = 3$

eg $2x + 1 = 0$

$2x = -1$

$x = -\frac{1}{2}$

2 Sketch horizontal and vertical lines

eg sketch the lines $x = 2$, $x = -1$, $y = 3$ and $y = -2$ on the same axes.

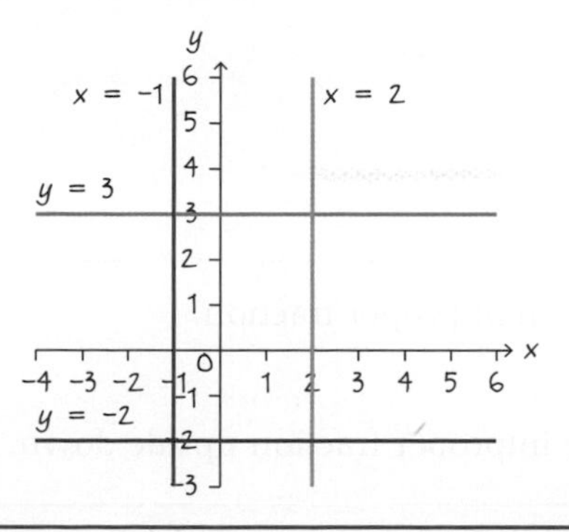

Skills check

1 Solve these equations:

a $x + 5 = 0$

b $6 - x = 0$

c $2x - 5 = 0$

2 Sketch these lines on the same set of axes: $x = 3$, $x = -2$, $y = -3$, $y = 4$.

4.1 The reciprocal function

Investigation 1

Think of pairs of positive numbers whose product is 12, for example
1×12, 2×6, 3×4, 4×3.

1 Copy this table and add more pairs of positive numbers (integers or fractions) whose product is 12.

x	1	2	3	4					
y	12	6	4	3					

2 Use your GDC to graph a scatter plot of the coordinate pairs in the table.

3 Now make a similar table, but find pairs of negative numbers whose product is 12, for example $-1\times -12 = 12$. Use your GDC to graph these points on the same set of axes.

4 Explain what you notice about
   a the value of x as y gets bigger
   b the value of y as x gets bigger.

The **end behaviour** of a graph is the appearance of a graph as it is followed further and further in either direction.

5 What do you notice about the end behaviour of your graph? Can you explain why this is the case?

The **reciprocal** of a number is 1 divided by that number.

For example, the reciprocal of 4 is $\frac{1}{4}$.

Taking the reciprocal of a fraction "turns it upside down".

For example, the reciprocal of $\frac{2}{3}$ is $1\div\frac{2}{3}=1\times\frac{3}{2}=\frac{3}{2}$.

The product of any number and its reciprocal is equal to 1.

Reflect Why does zero not have a reciprocal?

TOK

Is zero the same as nothing?

Example 1

Find the reciprocal of $4\frac{1}{2}$.

$4\frac{1}{2}=\frac{9}{2}$	Write as an improper fraction.
The reciprocal of $\frac{9}{2}$ is $\frac{2}{9}$.	Turn the improper fraction upside down.

Exercise 4A

1 Find the reciprocal of each number:

a 3 b 5 c -2 d -1

e $\frac{3}{5}$ f $\frac{22}{7}$ g $-\frac{8}{9}$ h $2\frac{3}{4}$

2 Find the reciprocal of each expression:

a 1.5 b x c $2x$ d $4y$

e $\frac{3x}{4}$ f $\frac{d}{t}$ g $\frac{3}{4d}$ h $\frac{x+2}{x-3}$

3 Show that the product of each number and its reciprocal is equal to 1:

a 4 b $\frac{7}{11}$ c $\frac{2}{x}$ d $\frac{x-1}{x-2}$

EXAM HINT
Remember that the IB command term **show that** requires you to show all the steps leading to the answer. You may not simply write down the final answer.

Domain and range of the reciprocal function

The **reciprocal function** is $f(x)=\frac{k}{x}$, where k is a non-zero constant.

Reflect Are there any values of x for which the reciprocal function is not defined?

Recall that the domain of a function is all the values of x for which the function $f(x)$ is defined.

In your studies, you have already found that you cannot divide a number by zero. If you try to calculate $1 \div 0$, you get an undefined answer. Therefore, 0 is not in the domain of $f(x)=\frac{1}{x}$.

However, when x is any real number other than 0, then $\frac{1}{x}$ exists. So, the domain of $f(x)=\frac{1}{x}$ is all real numbers other than zero.

The **domain** of the reciprocal function $f(x)=\frac{1}{x}$ is $x \in \mathbb{R} \setminus 0$.

This means "x can be any value within the set of real numbers, other than zero".

Investigation 2

Look again at investigation 1, where you found pairs of numbers whose product was 12. This function can be written as $xy = 12$, or as a reciprocal function $y = \frac{12}{x}$.

In this investigation you will consider the range of the function $y = \frac{12}{x}$.

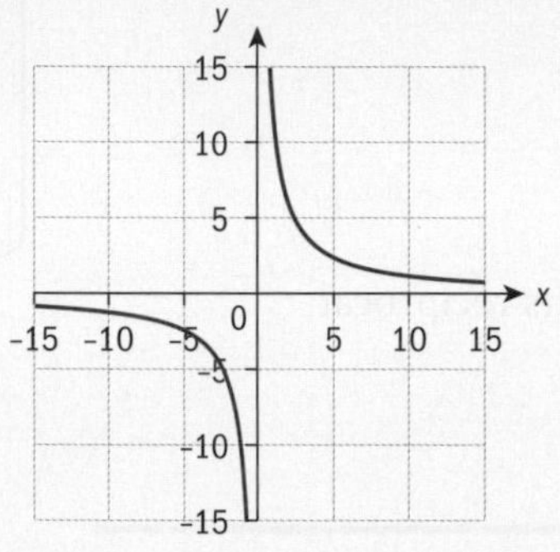

First, you will consider what happens to the value of y as the absolute value of x becomes very large.

1 a Find y when x is **i** 24 **ii** 240 **iii** 2400 **iv** −24 **v** −240 **vi** −2400

b What happens to the value of y as x gets larger?

c **Factual** Is there a value of x which would make the value of y zero?

d **Conceptual** What can you say about the value of y when the absolute value of x becomes infinitely large?

Now consider what happens to the y value as the x value gets closer to zero (remember, x cannot equal zero, but it can get close to it!).

2 a Find the value of y when **i** $x = 0.1$ **ii** $x = 0.001$ **iii** $x = 0.00001$ **iv** -0.1 **v** -0.001 **vi** 0.00001

b **Conceptual** Explain what is happening to y as the absolute value of x approaches zero.

c **Factual** Given any large value of y, is it possible to find a value of x for which $y = \frac{1}{x}$? What does this tell you about the range of $f(x)$?

The **range** of the reciprocal function $f(x) = \frac{1}{x}$ is $y \in \mathbb{R} \setminus 0$.

This means "$f(x)$ can be any value y within the set of real numbers, other than zero".

Limits and asymptotes

Look back at investigation 2. It is clear that $f(x)$ is approaching zero as the absolute value of x gets very large. A **limit** is the value, y, which a function $f(x)$ is *tending towards*, but does not reach, as x tends towards a certain value.

The first limit of the function $f(x)=\frac{1}{x}$ is $\lim_{x\to\pm\infty} f(x)=0$

This means that as x tends towards either positive or negative infinity, $f(x)$ tends towards zero.

The second limit of the function $f(x)=\frac{1}{x}$ are $\lim_{x\to 0^+} f(x)=\infty$ and $\lim_{x\to 0^-} f(x)=-\infty$

This means that as x tends towards zero from the positive side, $f(x)$ tends towards infinity; and as x tends towards zero from the negative side, $f(x)$ tends towards negative infinity.

Note: We say "tends towards" as it never actually reaches this value.

When sketching the graph of $y=f(x)$, we represent the limit $\lim_{x\to\pm\infty} f(x)=0$ by drawing a dashed horizontal line (called an asymptote) along $y=0$. This indicates that $f(x)$ tends towards, but never reaches, zero as x gets very large.

We also represent the limits $\lim_{x\to 0^+} f(x)=\infty$ and $\lim_{x\to 0^-} f(x)=-\infty$ by drawing a vertical asymptote along the line $x=0$. This indicates that $f(x)$ tends towards, but never reaches, infinity as x gets close to zero.

International-mindedness

The word *asymptote* is derived from the Greek *asymptotos*, which means "not falling together".

Functions

Sketching a graph of $y=\frac{k}{x}$ using technology

Using your GDC, plot the graph of $y=\frac{12}{x}$, which you studied in investigations 1 and 2.

You are now going to sketch this graph on paper. Your sketch should give a general idea of the required shape of the graph, and should show how the graph behaves as it approaches each asymptote.

The **graph of a reciprocal function** $f(x)=\frac{1}{x}$ has the following features:

The x-axis is a horizontal asymptote.

The y-axis is a vertical asymptote.

Both the range and domain are all real numbers except zero.

Continued on next page

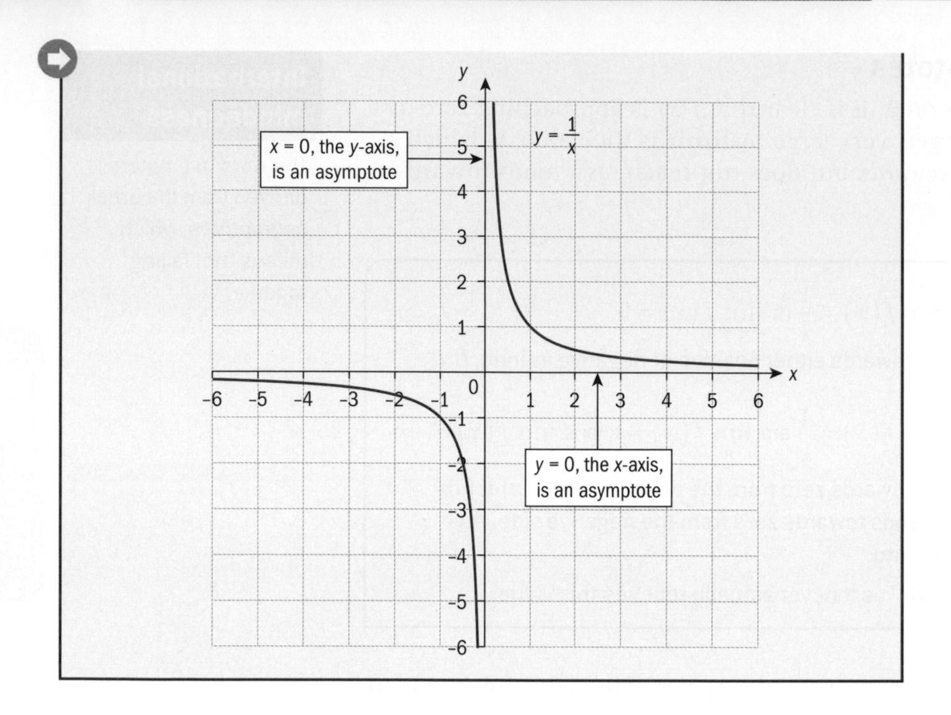

First draw and label the x- and y-axes on paper.

Now sketch the graph of $y = \frac{12}{x}$ on your axes, making sure that the curve gets closer and closer to, but does not touch, the axes.

Many reciprocal functions have the same shape. Because they do not cross the coordinate axes, they might look the same. To tell the difference between them, it is helpful to label one point on each branch of the curve, near to the center of the curve of that branch (such as $(4, 3)$ and $(-4, -3)$ in the graph of $y = \frac{12}{x}$).

Your resulting sketch should look like the graph shown on the right.

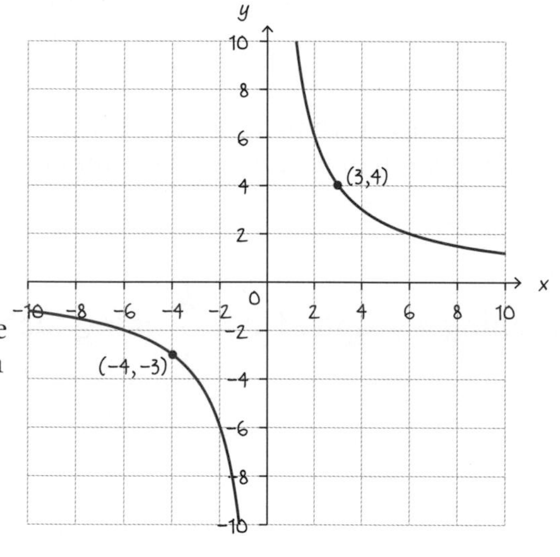

The parameter of the reciprocal function

For the reciprocal function $f(x) = \frac{k}{x}$, k is called the **parameter** of the function.

Investigation 3

1 On your GDC, plot the graphs of each of these functions:

a $f(x)=\frac{1}{x}$ b $g(x)=\frac{2}{x}$ c $h(x)=\frac{3}{x}$

d $j(x)=\frac{-1}{x}$ e $k(x)=\frac{-2}{x}$ f $m(x)=\frac{-3}{x}$

2 **Factual** What features of each graph in question **1** are:

a the same for each graph? b different for each of the three graphs?

3 **Conceptual** What effect does changing the magnitude of the parameter k in the function $f(x)=\frac{k}{x}$ have on the graph of $y=f(x)$?

4 On your GDC, plot the graph of $p(x)=\frac{-1}{x}$. Using what you know about transformations of graphs from chapters 2 and 3, describe a transformation that maps the graph of $f(x)=\frac{1}{x}$ onto the graph of $p(x)=\frac{-1}{x}$.

5 In a similar way, describe a transformation that maps the graph of $f(x)=\frac{1}{x}$ onto the graph of $q(x)=\frac{1}{(-x)}$.

6 **Factual** State what you notice about the graphs of $p(x)$ and $q(x)$. Can you explain why this is the case?

7 **Conceptual** What effect does the sign of the parameter have on the graph of a reciprocal function?

8 **Reflect** What are the equations of the asymptotes of $y=\frac{1}{x}$?

9 Using the idea that an asymptote is a graphical representation of a limit, explain whether the graph of $y=\frac{1}{x}$ will intercept either the x-axis or the y-axis at any point.

10 **Conceptual** How does the graph of $y=\frac{k}{x}$ differ from the graph of $y=\frac{1}{x}$?

TOK

"Film is one of the three universal languages, the other two: mathematics and music", is a quote from movie director, Frank Capra. To what extent do you agree?

Exercise 4B

1 Sketch the graph of each function:

a $y=\frac{8}{x}$

b $y=\frac{12}{x}$

c $xy=16$

2 On the same set of axes, sketch the graphs of $y=\frac{10}{x}$ and $y=-\frac{10}{x}$.

3 Sketch the graph of each function:

a $y=-\frac{8}{x}$ b $y=-\frac{12}{x}$ c $xy=-16$

d Compare your answers to question **1**. What do you notice?

4 Sketch the curve $y=\frac{5}{x}$. Write down the equations of its asymptotes, and state the domain and range.

Functions

Inverse of a reciprocal function

Investigation 4

1 Copy and complete this table for $f(x) = \frac{4}{x}$.

x	0.25	0.4	0.5	1	2	4	8	10	16
$f(x)$									

2 **Factual** What do you notice about the top and bottom rows of the table?

3 Use your GDC to plot the graph of the function $f(x) = \frac{4}{x}$.

4 Plot the line $y = x$ on the same graph.

5 Reflect $f(x) = \frac{4}{x}$ in the line $y = x$. What do you notice?

6 **Factual** What does this tell you about the inverse function $f^{-1}(x)$?

You will now consider an algebraic way of showing the same result.

7 If $f(x) = \frac{k}{x}$ find $(f \circ f)(x)$.

Remember that $(f \circ f^{-1})(x) = x$ for any function x. However, you have just shown that $(f \circ f)(x) = x$ for the reciprocal function $f(x) = \frac{k}{x}$.

8 **Conceptual** What does this tell you about the inverse of the reciprocal function?

The reciprocal function $f(x) = \frac{k}{x}$ is a self-inverse function because $f(x) = f^{-1}(x)$ for all values of x in the domain of f.

Modelling with the reciprocal function

There are many pairs of real-life quantities which vary with one another according to the reciprocal function.

Reflect For the skate ramp on the right, state which two quantities vary with one another in a reciprocal relationship.

The shape of this skate ramp could be modelled by the function $f(x) = \frac{1}{x}$.

Investigation 5

Your student's council is planning a new car park for your school but the local government laws state that the total area of the car park must be no more than 800 m^2. The students decide that it will be rectangular, and now have to decide on the length and width of the car park.

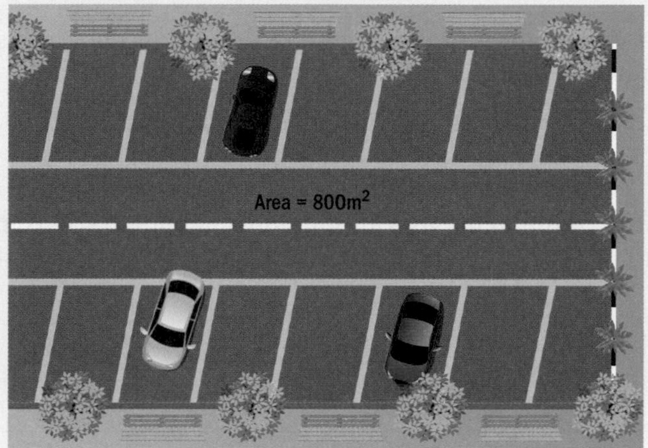

1 Work with a classmate to write, in a copy of the table below, some suggestions for the length and width of the car park which would give an area of 800 m^2.

Length (m)									
Width (m)									

2 **Factual** What type of relationship is this? How can you be sure? Can you write an equation linking the width (x metres) and the length (y metres) of the car park?

3 Plot a graph of this function on your GDC.

For the car park to fit alongside the back entrance to school, the car park's length has to be twice its width.

4 Write this criterion as an equation in x and y. Plot the graph of this equation on the same axes as your original function.

5 Hence determine what the length and width of the car park will need to be.

6 **Factual** Are the values you found in question **5** the only possible values for the length and width, given the restrictions placed on the students? How can you be sure?

7 **Conceptual** How can you tell from a table of values that a real-life situation could be represented by a reciprocal function?

Example 2

Adam's mobile phone has a limited amount of free space which he can use to store photos. Adam's phone has 240 MB (megabytes) of storage space.

The phone settings allow Adam to choose the default photo size (in MB). The photo size is the amount of storage space that each photo occupies on the phone.

Continued on next page

a Copy and complete this table, which shows the default photo size and the number of photos that Adam can store on his phone.

Photo size (MB)	2	3	4	5	6	8	10	12	30
Number of photos	120				40				

b Write down a function for *y*, the number of photos Adam can store in his phone, in terms of the individual photo size, *x*.

c Use your GDC to plot a graph of the function you found in part **b**.

Adam wants to have space to store 50 photos on his phone.

d Use a graphical approach to find the maximum individual photo size (hint: plot an appropriate straight line on your GDC graph in part **c** to help you solve this problem graphically).

e Use an algebraic approach to find the maximum individual photo size.

a

Size (MB)	2	3	4	5	6	8	10	12	30
Number of photos	120	80	60	48	40	30	24	20	8

b $y = \dfrac{240}{x}$

From the table of values, you can see that $xy = 240$. The question asks you to give y in terms of x, so you should make y the subject of the equation.

c

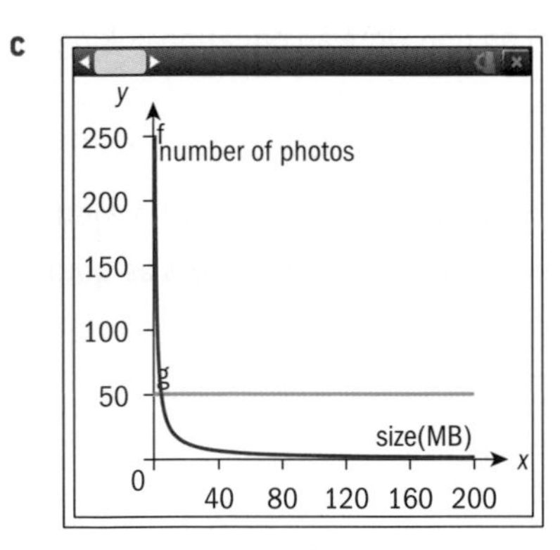

You only need to consider positive values of *x*, as negative values are not possible in this real-life context.

d Maximum file size is 4.8 MB

Plot the line $y = 50$ on your graph. The intersection of $y = \dfrac{240}{x}$ with $y = 50$ will give you the size of photo to choose.

e $\dfrac{240}{x} = 50$

$x = \dfrac{240}{50}$

$x = 4.8$ MB

Solve the equations $y = \dfrac{240}{x}$ and $y = 50$ algebraically by equating them and solving for *x*.

Exercise 4C

1 Ben is a dentist. He believes there is a relationship between the average number of cavities developed in each of his patients' mouths per year (y) and the number of minutes they brush their teeth each morning and evening (x). Ben believes the function $y = \frac{2}{x}$ represents this relationship.

Alita and Chamse are Ben's patients.

a Determine how many cavities Alita is likely to develop in one year if she brushes her teeth for two minutes each morning and evening.

b Chamse develops four cavities in a year. Find the number of minutes she likely spent brushing her teeth each morning and evening.

2 On a string instrument, the length of a string (l cm) and the frequency of its vibrations (v Hz) are modelled by the equation $v = \frac{1000}{l}$.

a Use technology to sketch the function for $0 < l \leq 20$.

b Find the frequency of vibrations for a string which is 10 cm long.

c Draw the line $l = 5$ on your sketch. Find the intersection point and explain what it means in the context of this problem.

3 Tak's phone can store 64 videos which are each one minute long; or 32 videos of length two minutes; or 16 videos of length four minutes; etc.

a Find the number of videos of length 16 minutes Tak could store on his phone.

b Find a function to represent the relationship between the number of videos on Tak's phone (x) and the length of each video (y minutes).

c Use your GDC to plot a graph of the function you found in part **b**.

d Tak wants to store 48 videos on his phone. Plot an appropriate straight line on the same axes to help you find the maximum length of each video.

TOK

How can a mathematical model give us knowledge even if it does not yield accurate predictions?

Developing inquiry skills

Look back at the opening problem for this chapter, in which a water park was designing a water slide.

Using what you have learned about reciprocal functions, what type of function could you use to model the shape of the waterslide?

How would the parameter of the function affect the shape of the waterslide? What parameter might you choose to ensure the slide is both exciting, and safe? What other things might you need to take into consideration when modelling the shape of the slide?

4.2 Transforming the reciprocal function

In the previous exercise you started with the parent function for reciprocal graphs, $y = \frac{1}{x}$ and then worked with functions of the form $y = \frac{k}{x}$.

You will now look at other functions that can be obtained by transforming the reciprocal function $y = \frac{1}{x}$ in different ways.

Investigation 6

1 Use your GDC to plot the graphs of $y = \frac{1}{x}$, $y = \frac{1}{x} + 2$ and $y = \frac{1}{x} - 3$.

2 Use your plotted graphs to copy and complete this table.

Rational function	Vertical asymptote	Horizontal asymptote	Domain	Range
$y = \frac{1}{x}$				
$y = \frac{1}{x} + 2$				
$y = \frac{1}{x} - 3$				

3 **Factual** When you transform the graph of $y = \frac{1}{x}$ onto

a $y = \frac{1}{x} + 2$ and **b** $y = \frac{1}{x} - 3$:

i Describe the transformation that maps the graph of $y = \frac{1}{x}$ onto this function.

ii What happens to the horizontal asymptote? Can you justify why this is the case by considering what y tends towards as x becomes very large?

iii What happens to the vertical asymptote?

iv Which axis does this graph now cross? What are the coordinates of the point where the graph crosses this axis? Can you use algebra to explain why the graph crosses the axis at this point?

4 **Conceptual** What are the asymptotes of a rational function of the form $y = \frac{p}{x} + d$?

5 Use your GDC to plot the graphs of $y = \frac{1}{x}$, $y = \frac{1}{x+2}$, $y = \frac{1}{x-3}$ and $y = \frac{4}{x-3}$, and $y = \frac{4}{(2x-3)}$, and then copy and complete this table.

Rational function	Vertical asymptote	Horizontal asymptote	Domain	Range
$y = \frac{1}{x}$				
$y = \frac{1}{x+2}$				
$y = \frac{1}{x-3}$				
$y = \frac{4}{x-3}$				
$y = \frac{4}{2x-3}$				

6 **Factual** When you transform the graph of $y = \frac{1}{x}$ onto $y = \frac{p}{cx+d}$:

a Describe the series of transformations that map the first graph to the second. Be careful about the order in which you perform transformations!

b What happens to the horizontal asymptote?

c What happens to the vertical asymptote? Can you use algebra to justify why this is the case?

d Which axis does this graph now cross? What are the coordinates of the point where the graph crosses this axis. Can you use algebra to explain why the graph crosses the axis at this point?

7 **Conceptual** Using what you have learned about transformations of the graph $y = \frac{1}{x}$, can you find the equations of the asymptotes of the function $f(x) = \frac{p}{cx+d} + q$, and hence state the domain and range?

8 Describe the transformation that maps the graph of $y = \frac{1}{x}$ onto the graph of $y = \frac{k}{x}$ when:

i $k > 0$ ii $k < 0$

Reciprocal functions of the form $y = \frac{k}{x+d} + c$, where k, c and d are constants, will have a vertical asymptote with equation $x = -d$ and a horizontal asymptote with equation $y = c$.

TOK

When students see a familiar equation with a transformation, they will often get a "gut feeling" about what the function looks like.

Respond to this question.

Is intuition helpful or harmful in mathematics?

Sketching the graph of a function of the form $f(x) = \dfrac{p}{cx+d} + q$

In order to sketch the graph of a reciprocal function, you need to know:

- the equations of the horizontal and vertical asymptotes
- the coordinates of any points where the graph crosses the axes.

Example 3

a Sketch the function $y = \dfrac{1}{x-2}$. Show any asymptotes, not on the axes, as dotted lines.

b Write down the equations of the horizontal and vertical asymptotes.

c State the domain and range.

a 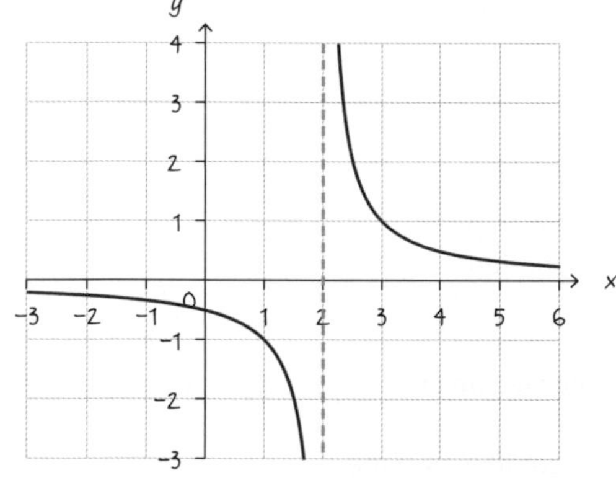	Draw the axes and asymptotes. Sketch the general shape of the graph and label any axes intercepts.
b **Method 1: by transforming graphs** $y = \dfrac{1}{x-2}$ is a horizontal translation of $y = \dfrac{1}{x}$ by 2 units in the x-direction, so the vertical asymptote is $x = 2$, and the horizontal asymptote remains unchanged. **Method 2: by using algebra** The vertical asymptote occurs when the function is not defined, ie: $x - 2 = 0 \Rightarrow x = 2$ The horizontal asymptote occurs at the limit of y as $x \to \pm\infty$, ie $y = 0$	Either transform graphs, or use algebra, to find the equations of the asymptotes.
c Domain: $x \in \mathbb{R} \setminus 2$ Range: $y \in \mathbb{R} \setminus 0$	Use the equations of the horizontal and vertical asymptotes to determine the domain and range.

Exercise 4D

You should not use your GDC for questions **1**, **2** and **3**.

1 Identify the horizontal and vertical asymptotes of these functions, and state the range and domain.

a $y = \frac{1}{x+1}$ **b** $y = \frac{1}{x-5}$

c $y = \frac{-1}{x-4}$ **d** $y = \frac{5}{x+5}$

e $y = \frac{12}{x+1} + 2$ **f** $y = \frac{12}{x+1} - 2$

g $y = \frac{4}{x-3} + 2$ **h** $y = \frac{-4}{x-4} - 4$

2 Sketch the graph of each function. Show the asymptotes as dotted lines and state the domain and range.

a $y = \frac{1}{x+4}$ **b** $y = \frac{-1}{x+4}$

c $y = \frac{1}{x+4} + 1$ **d** $y = \frac{2}{x+5} - 1$

e $y = \frac{1}{2x+1}$ **f** $y = \frac{3}{4x+8}$

g $y = \frac{5}{3x-6} + 2$ **h** $y = \frac{1}{2-x} + 1$

3 Each function is a transformation $y = \frac{1}{x}$.

Match the function with its graph. Explain your reasoning and then describe the transformation.

a $y = \frac{1}{x-2}$ **b** $y = \frac{-1}{x-2}$

c $y = \frac{1}{x-2} + 2$ **d** $y = \frac{1}{x-2} - 2$

e $y = \frac{3}{x-2}$

i

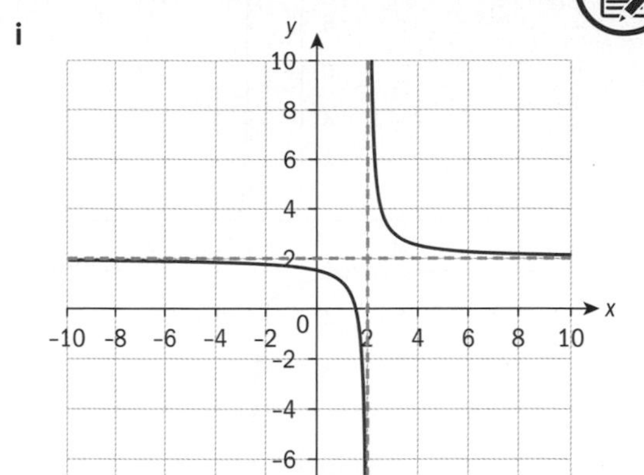

ii

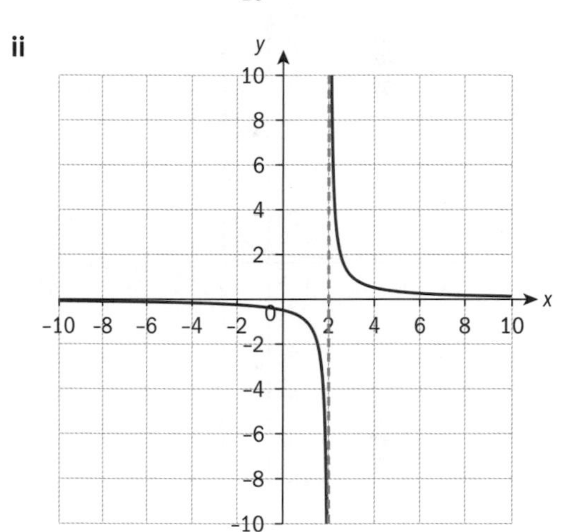

iii

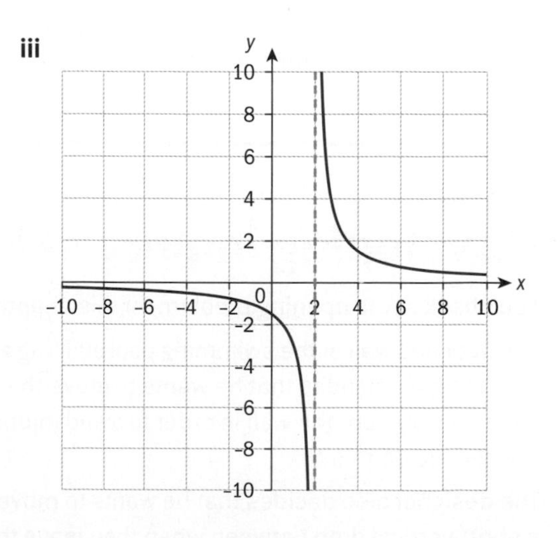

Functions

iv

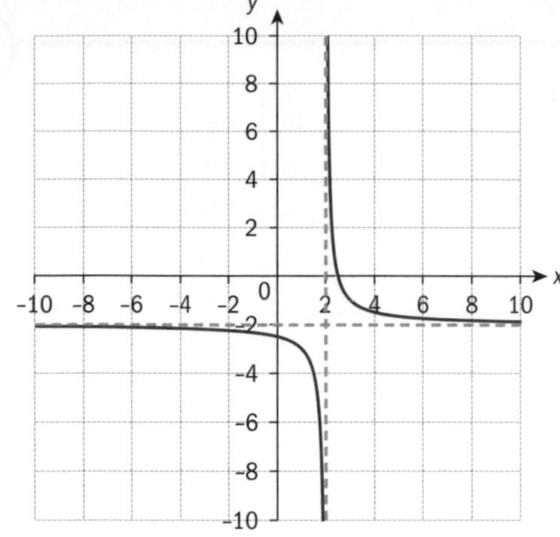

v

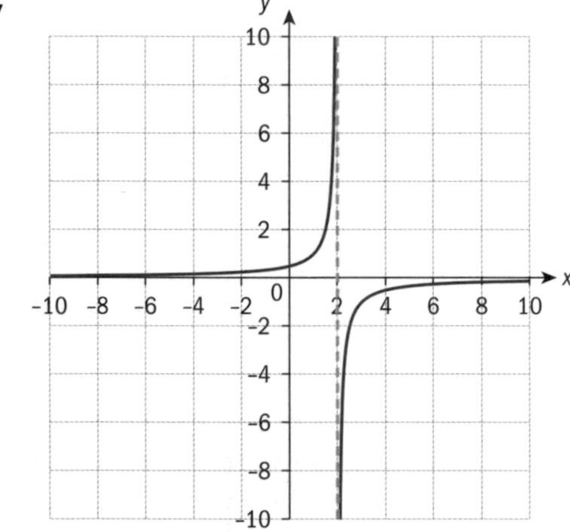

4 When lightning strikes, you see the flash almost instantly, but the sound of the thunder reaches you later, as sound travels at approximately 330 m s^{-1}. The speed of sound is affected by temperature of the surrounding air. The time sound takes to travel one kilometre is modelled by $t = \dfrac{1000}{0.6c + 330}$, where t is the time in seconds and c is the air temperature in Celsius.

a Sketch the graph of t for $-20 \le t \le 40$, where t is the temperature in °C.

b Draw the line $t = 3$ to find the temperature of the surrounding air when you hear the sound of thunder 3 seconds after the flash of lighting from one kilometre away.

c What is the air temperature when you hear the sound after 6 seconds?

5 When you join a fitness centre for a trial, you get the first five sessions for free. After the first five sessions, the cost to join is 200 euros. The average cost (c) per session (s) is modelled by $c = \dfrac{200}{(s-5)}$.

a Sketch the function and identify the horizontal and vertical asymptotes.

b Draw a line on your sketch to find when the cost per session averages 20 euros.

6 On the same set of axes sketch $y = x + 3$ and $y = \dfrac{1}{x}$ and on another set of axes sketch $y = x - 2$ and $y = \dfrac{1}{x-2}$.

Discuss what you notice about symmetry and intercepts.

Developing inquiry skills

Look back at the opening problem for this chapter, in which a water park was designing a water slide.

The floor and wall of the swimming pool building are to be modelled by the x- and y-axes respectively. The designer decides that he wants to move the point where a person sets off—at the top of the slide—away from the wall in order to avoid injuries. How could he develop his mathematical model of the slide to do this?

The designer also decides that he wants to move the end of the slide up a little, so that a person has a short vertical drop between when they leave the end of the slide and when they hit the water. How could he develop his mathematical model of the slide to do this?

4.3 Rational functions of the form $f(x)=\frac{ax+b}{cx+d}$

A **rational function** is a function of the form $f(x)=\frac{g(x)}{h(x)}$, where g and h are polynomials.

In this course, $g(x)$ and $h(x)$ will be restricted to linear functions and $h(x) \neq 0$.

So, we have $f(x)=\frac{ax+b}{cx+d}$ where a, b, c and d are constants.

Investigation 7

1 **Factual** Look again at the function $f(x)=\frac{p}{cx+d}+q$ which you studied at the end of section 4.2. Is this a rational function?

2 Describe, using words, how you could write the expression $\frac{p}{cx+d}+q$ as a single fraction over a common denominator.

3 Express $\frac{p}{cx+d}+q$ as a single fraction of the form $\frac{ax+b}{cx+d}$. State the values of a and b.

4 When studying the function $f(x)=\frac{p}{cx+d}+q$, you found that the horizontal asymptote was $y=q$, and the vertical asymptote was $x=\frac{-d}{c}$.

Can you use this to predict the equations of the horizontal and vertical asymptotes of $g(x)=\frac{ax+b}{cx+d}$ in terms of the parameters a, b, c and d?

5 Use your GDC to plot the graphs of these rational functions, and then use your GDC to copy and complete the table below.

$y=\frac{x}{x+2}$ $y=\frac{x+1}{x+2}$ $y=\frac{2x}{x+2}$ $y=\frac{2x-1}{x+2}$

Rational function	Vertical asymptote	Horizontal asymptote	Domain	Range
$y=\frac{x}{x+2}$				
$y=\frac{x+1}{x+2}$				
$y=\frac{2x}{x+2}$				
$y=\frac{2x-1}{x+2}$				

Continued on next page

International-mindedness

The first Chinese abacus was invented around 500 BC.

The Moroccan scholar Ibn al-Bannā' al-Marrākush used a lattice multiplication in the 14th century that John Napier, a Scottish mathematician, adapted in the 17th century. Edmund Gunter of Oxford then produced the first slide rule before Blaise Pascal started to develop a mechanical calculator in France in the 17th century which led to our present day GDC.

6 Check that the equation you found in question **4** for the horizontal asymptote gives the same equation as those you recorded in the table above.

7 How can you find the vertical asymptote of a rational function?

8 How can you find the equations of the asymptotes of a rational function algebraically?

There is a way to determine the equation of the horizontal asymptote algebraically from the equation $g(x) = \frac{ax+b}{cx+d}$.

Remember that the *vertical* asymptote is found when the denominator is equal to zero. However, if you reflect the function in the line $y = x$, the *horizontal* asymptote will become a *vertical* asymptote, so you need only find the *vertical* asymptote of the new expression. This is a neat little trick!

$$\begin{aligned} y &= \frac{ax+b}{cx+d} \\ y(cx+d) &= ax+b \\ cxy+dy &= ax+b \\ cxy-ax &= b-dy \\ x(cy-a) &= b-dy \\ x &= \frac{b-dy}{cy-a} \end{aligned}$$

Rearranging the equation to make x the subject.

The horizontal asymptote occurs when the denominator is zero, hence:

$$\begin{aligned} cy-a &= 0 \\ y &= \frac{a}{c}. \end{aligned}$$

For a rational function $y = \frac{ax+b}{cx+d}$ the vertical asymptote is the line $x = -\frac{d}{c}$, and the horizontal asymptote is $y = \frac{a}{c}$.

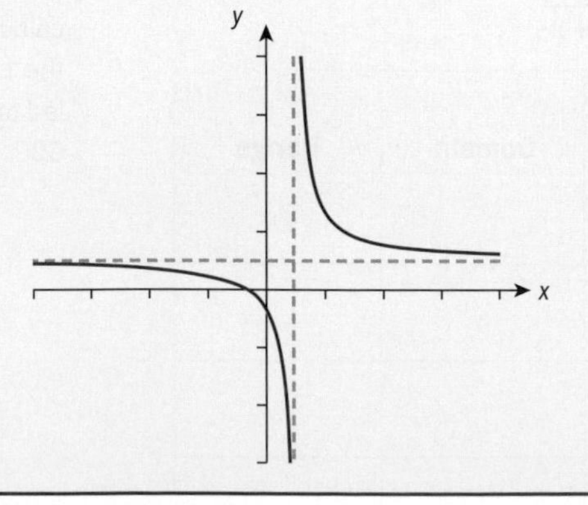

Reflect What values could you choose for the parameters a, b, c and d to make the rational function $y = \frac{ax+b}{cx+d}$ equal to the reciprocal function $y = \frac{1}{x}$?

What is the relationship between a reciprocal function and a rational function?

Sketching rational functions from the equation without the use of technology

The general equation of a rational function is $y = \dfrac{ax+b}{cx+d}$. To sketch this function without the use of your GDC, you should:

1 Find the vertical asymptote, $x = -\dfrac{d}{c}$.

2 Find the horizontal asymptote, $y = \dfrac{a}{c}$.

3 Find the y-intercept by letting $x = 0$ and solving for y.

4 Find the x-intercept by letting $y = 0$ and solving for x.

5 Test values on either side of the vertical asymptote to determine whether the curve tends towards positive or negative infinity on that side.

6 Draw the axes and asymptotes, then sketch the curve.

TOK

Look at the development of calculating technology on page 201, and then respond to this question

What are the ethical considerations when sharing mathematical knowledge?

Example 4

a Find the equations of the horizontal and vertical asymptotes for $y = \dfrac{4x+1}{2x-6}$.

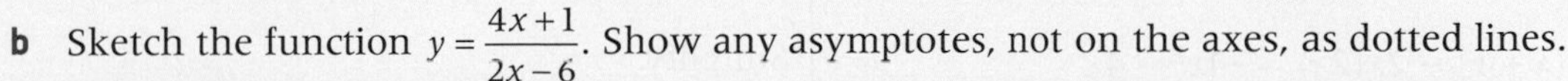

b Sketch the function $y = \dfrac{4x+1}{2x-6}$. Show any asymptotes, not on the axes, as dotted lines.

c State the domain and range.

a $y = \dfrac{4x+1}{2x-6}$ $a = 4,\ c = 2,\ d = -6.$	Determine the parameters of this equation given the general equation of a rational function $y = \dfrac{ax+b}{cx+d}$
Horizontal asymptote $y = \dfrac{a}{c} = \dfrac{4}{2} = 2$ Vertical asymptote $x = -\dfrac{d}{c} = -\dfrac{-6}{2} = 3$	State the equations of the asymptotes.
b $y = \dfrac{4x+1}{2x-6} = 0$	Find the x-intercept by letting $y = 0$.
$\Rightarrow 4x+1 = 0 \Rightarrow x = -\dfrac{1}{4}$	Find the y-intercept by letting $x = 0$.
$x = 0 \Rightarrow y = -\dfrac{1}{6}$ 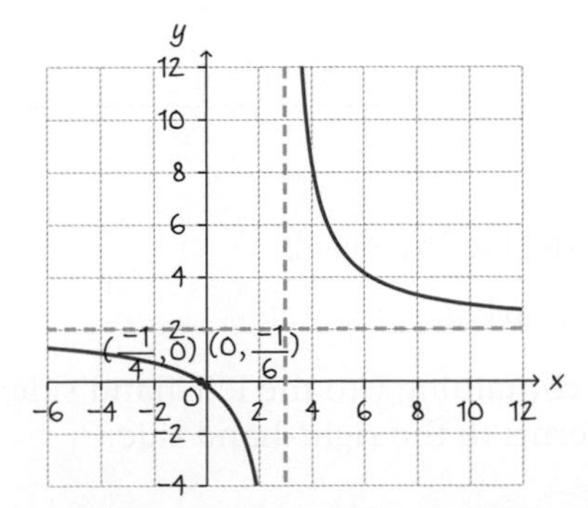	Sketch the axes, asymptotes and curve. Label the axial intercepts.

Continued on next page

c Domain $x \in \mathbb{R}, x \neq 3$ Range $y \in \mathbb{R}, y \neq 2$	State the domain and range, using the asymptotes to help you.

Solving rational equations

You can solve rational equations by multiplying both sides of the equation by a common denominator.

Example 5

Using the least common denominator of all the fractions in the equation, solve $\frac{3}{2x} - \frac{2x}{x+1} = -2$

$(2x)(x+1)\left(\frac{3}{2x} - \frac{2x}{x+1}\right) = (2x)(x+1)(-2)$	Note that $x \neq -1$ and $x \neq 0$. The LCD of the fractions is $2x(x+1)$. Multiply both sides by the LCD.
$3(x+1) - 2x(2x) = -4x^2 - 4x$ $3x + 3 - 4x^2 = -4x^2 - 4x$ $7x = -3$ $x = -\frac{3}{7}$	Simplify.

The inverse of a rational function

You find the inverse of a rational function in the same way that you find the inverse of a linear function. You learned how to do this in Chapter 2.

Example 6

If $f(x) = \frac{x-4}{3x-2}$, find $f^{-1}(x)$.

$x = \frac{y-4}{3y-2}$	Swap x and y.
$x(3y-2) = y-4$	Multiply both sides by $3y-2$.
$3xy - 2x = y - 4$	Expand the bracket.
$3xy - y = 2x - 4$	Take terms containing y to the left-hand side; take any other terms to the right-hand side.

$y(3x-1) = 2x-4$	Take y as a common factor on the left-hand side.
$y = \dfrac{2x-4}{3x-1}$	Divide both sides by $3x-1$.
$f^{-1}(x) = \dfrac{2x-4}{3x-1}$	

Example 7

Chas and Dev are planning to print their sports team logo on baseball caps, and sell the caps. They have to hire a printing machine, which costs $500, and it will cost them an additional $5 for every cap to be bought and printed.

a Develop a model which links the number of caps they produce with average cost, per cap, they incur.

b Use your GDC to determine how many caps must be printed and sold if they are to sell the caps at $7 per cap and just break even.

a Let the number of caps produced be x, and the average cost incurred, per cap, be y.	Define the variables.
$y = \dfrac{500+5x}{x}$	Form an equation for the average cost: $y = \dfrac{\text{(hiring the machine)} + \text{(cost per cap)(number produced)}}{\text{number of caps produced}}$
b 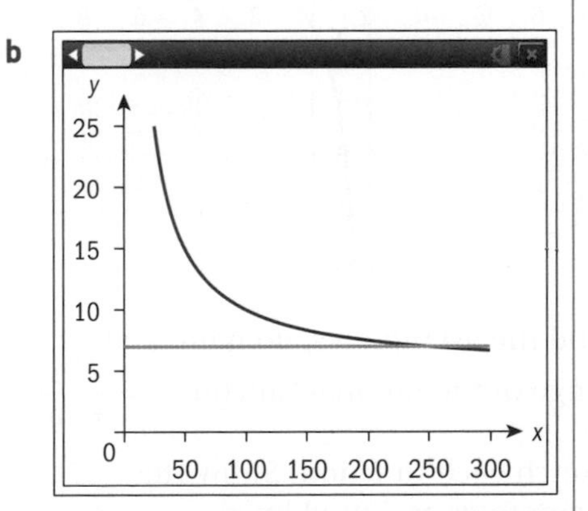	Use your GDC to plot a graph of y against x. This is a rational function, but you only need to consider the part of the function where x is positive as negative x is meaningless in this context. Plot the line $y = 7$ on the same axes and find the intersection point.
Chas and Dev must produce 250 caps to break even.	State your answer in the context of the problem.

Functions

Exercise 4E

1 For each function, find the equations of the horizontal and vertical asymptotes, then write down the domain and range.

a $y=\dfrac{x+1}{x-1}$ **b** $y=\dfrac{2x+3}{x+1}$ **c** $y=\dfrac{6x-1}{2x+4}$

d $y=\dfrac{2-3x}{5-4x}$ **e** $y=\dfrac{9x-2}{6-3x}$

2 Match these equations to their graphs and give reasons for your answer.

a $y=\dfrac{4}{x}$ **b** $y=\dfrac{x-3}{x+2}$ **c** $y=\dfrac{2x-3}{x+2}$

d $y=\dfrac{3-2x}{x+2}$

i

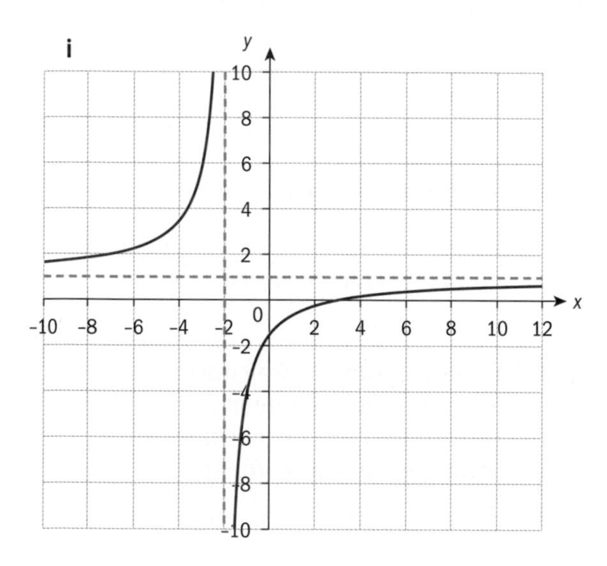

ii

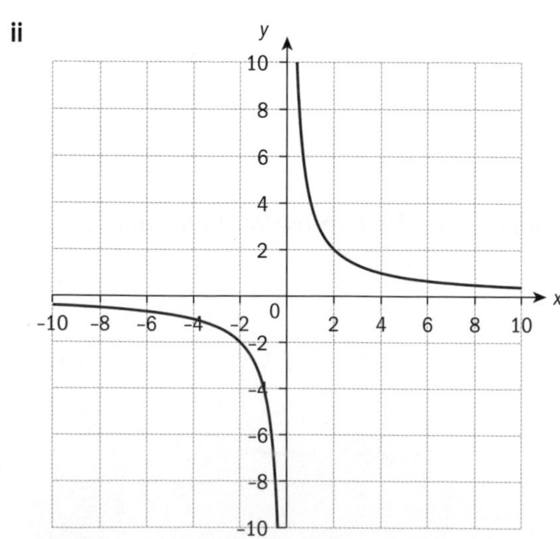

iii

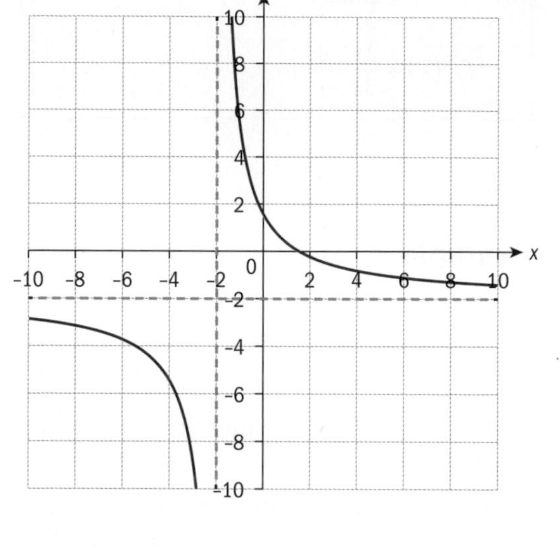

iv

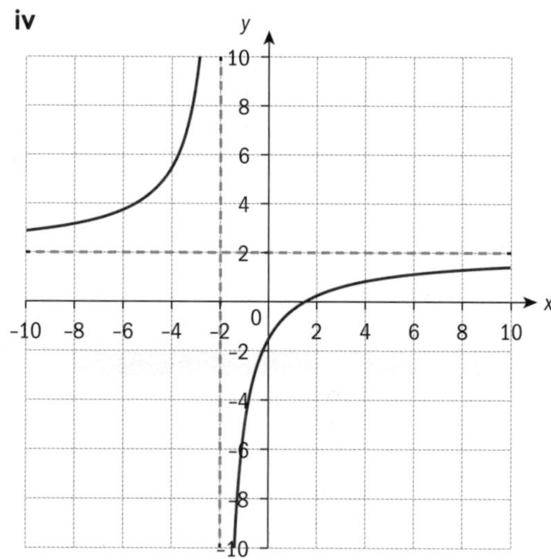

3 Find the asymptotes, domain, and range of the rational function $y=\dfrac{x-p}{x-q}$.

4 Sketch each function. Show the asymptotes as dotted lines.

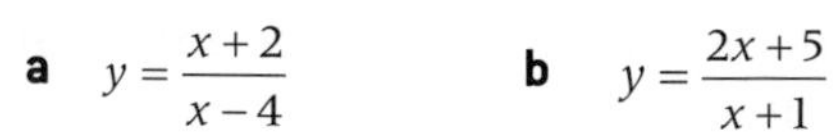

a $y=\dfrac{x+2}{x-4}$ **b** $y=\dfrac{2x+5}{x+1}$

c $y=\dfrac{x-7}{2x-1}$ **d** $y=\dfrac{1-x}{1+x}$

5 Solve:

a $\dfrac{5}{2x}+\dfrac{x+7}{x+4}=2$ **b** $\dfrac{2x-3}{x+1}=\dfrac{x+6}{x-2}$

c $7-\dfrac{5}{x-2}=\dfrac{10}{x+2}$ **d** $\dfrac{x+5}{x+8}=1+\dfrac{6}{x+1}$

6 Will solved $\frac{2}{x-3} = \frac{x}{x-3}$ to get $x = 2$ and $x = 3$. One of these solutions is called *extraneous*. Extraneous solutions are solutions that do not satisfy the original form of the equation because they make the denominator equal to 0.

What is the solution to Will's equation?

7 Find the inverse of each function.

a $f(x) = \frac{x+3}{x-2}$ **b** $f(x) = \frac{7-2x}{x}$

c $f(x) = \frac{1+7x}{9-x}$ **d** $f(x) = \frac{5-11x}{x+6}$

8 Emily is setting up a company to make football shirts. It costs \$500 to purchase the equipment, and costs a further \$10 per shirt to purchase material. The average production cost (\$*M*) to produce *s* shirts can be modelled by the equation $M(s) = \frac{10s+500}{s}$.

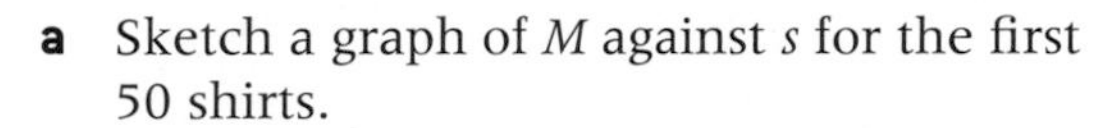

a Sketch a graph of *M* against *s* for the first 50 shirts.

b Draw the line $M(s) = 35$ to find the number of shirts that Emily needs to make so that the average production cost per shirt is \$35.

c Find the number of shirts Emily needs to make so that the average production cost per shirt is \$20?

9 An Internet security service charges a 20 AUD set-up fee and then a monthly charge of 10 AUD.

a Construct a model to show the average monthly cost.

b Sketch a graph of the model.

c Use your model to find how long you have to use the service to average 15 AUD per month.

d Tim uses the service for many years at the same price. What will his average cost per month get closer and closer to?

10 Let $f(x) = m + \frac{6}{x-n}$. The line $x = 5$ is an asymptote to the graph of *f*.

a Write down the value of *n*.

The graph passes through the point (7, 7).

b Find the value of *m*.

c Write down the equation of the horizontal asymptote.

11 Consider the function $y = \frac{4}{x-2} + 3$.

a Write down the equation of the horizontal asymptote.

b Find the vertical asymptote.

c Find the coordinates of the axial intercepts.

d Hence, sketch the function. Label the asymptotes and axes intercepts.

12 Let $f(x) = \frac{2x+1}{x-1}$.

a Sketch the graph of *f* for $-4 \le x \le 5$.

b Write down the equations of the asymptotes.

c Find the *x*-intercept of *f*.

13 $f(x) = \frac{x+2}{x+3}$ and $g(x) = \frac{1}{x}$:

a Find $(g \circ f)(x)$.

b Plot $(g \circ f)(x)$ and $f(x)$ on your GDC, and use this to solve $f(x) = (g \circ f)(x)$.

Developing your toolkit

Now do the Modelling and investigation activity on page 212.

Chapter summary

- The **reciprocal** of a number is 1 divided by that number.
- The **reciprocal function** is $f(x)=\frac{k}{x}$, where k is a non-zero constant.
- The **domain** of the reciprocal function $f(x)=\frac{1}{x}$ is $x\in\mathbb{R}, x\neq 0$.
- The **range** of the reciprocal function $f(x)=\frac{1}{x}$ is $y\in\mathbb{R}, y\neq 0$.
- Limits of the function $f(x)=\frac{1}{x}$ are
  - $\lim_{x\to\pm\infty} f(x)=0$; this corresponds to the horizontal asymptote $y=0$.
  - $\lim_{x\to 0^+} f(x)=\infty$ and $\lim_{x\to 0^-} f(x)=-\infty$, which correspond to the vertical asymptote $x=0$.
- The reciprocal function $f(x)=\frac{k}{x}$ is self-inverse function because $f(x)=f^{-1}(x)$ for all values of x in the domain of f. You can also see this graphically because the graph is a reflection of itself in the line $y=x$.
- A **rational function** is a function of the form $f(x)=\frac{g(x)}{h(x)}$, where g and h are polynomials.
- You can transform the graph of the rational function $y=\frac{1}{x}$ to generate rational functions of the form $y=\frac{p}{cx+d}+q$ or $y=\frac{ax+b}{cx+d}$.
- For a rational function $y=\frac{ax+b}{cx+d}$, the vertical asymptote is $x=-\frac{d}{c}$ and the horizontal asymptote is $y=\frac{a}{c}$.
- To sketch the function $y=\frac{ax+b}{cx+d}$ without the use of your GDC, you should:
  - Find the vertical asymptote, $x=-\frac{d}{c}$.
  - Find the horizontal asymptote, $y=\frac{a}{c}$.
  - Find the y-intercept by letting $x=0$ and solving for y.
  - Find the x-intercept by letting $y=0$ and solving for x.
  - Test values on either side of the vertical asymptote to determine whether the curve tends towards positive or negative infinity on that side.
  - Draw the axes and asymptotes, then sketch the curve.
- Reciprocal and rational functions can be used to model many real-life concepts, particularly in the fields of business, science and medicine.

TOK

What is the biggest number that you know? Million? Billion? What about the prefixes used in storage like giga and tera? What are they?

What is a googol? What is larger?

In this chapter we have seen the symbol used for infinity.

Research Hilbert's paradox of the Grand Hotel.

What do you think is meant by infinity?

Developing inquiry skills

Look back at the designer's progress with the waterslide at the end of section 4.2. How could you model the shape of this waterslide as a rational function of the form $f(x)=\dfrac{ax+b}{cx+d}$?

If the swimming pool is in a building 20 metres high and there is 15 metres of space between the wall and the edge of the pool, suggest suitable values for the parameters a, b, c and d and suggest a suitable domain for the model.

Chapter review

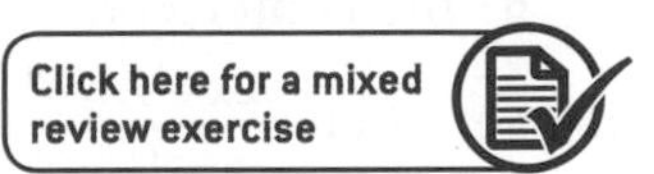

1 For each function, find the equations of the horizontal and vertical asymptotes, and write down the domain and range.

a $y=\dfrac{2}{x}$ **b** $y=\dfrac{1}{x+8}$

c $y=\dfrac{x}{2x-10}$ **d** $y=\dfrac{3}{x-2}+3$

e $y=\dfrac{2x}{x-9}$ **f** $y=\dfrac{8x-5}{2x+4}$

g $y=\dfrac{1-x}{x+4}$ **h** $y=\dfrac{2x-1}{2x+6}-4$

2 Sketch each function, showing the asymptotes as dotted lines.

a $y=\dfrac{3}{x}$ **b** $y=-\dfrac{3}{x}$

c $y=-\dfrac{3}{x-2}$ **d** $y=\dfrac{x+2}{x-2}+3$

3 Consider the function $f(x)=\dfrac{1}{x-1}+2$.

a Write down the equations of the asymptotes.

b Find the axial intercepts.

c Sketch the graph of $f(x)$. Show the asymptotes as dotted lines.

4 Use your GDC to plot the graphs of $f(x)=\dfrac{2x+1}{2x-1}$ and $g(x)=x+2$ on the same set of axes.

By considering the points of intersection, solve the equation $\dfrac{2x+1}{2x-1}=x+2$.

5 Use your GCD to solve each equation graphically.

a $\dfrac{x+3}{2x-1}=4-x$ **b** $x^2-2x-3=\dfrac{x+4}{x-1}-5$

c $\dfrac{10x+3}{2x+4}=4-x^2$

6 Let $f(x)=\dfrac{2x-8}{1-x}$.

a Find the x-intercept.

b Write down the equation of the vertical asymptote.

c Find the equation of the horizontal asymptote.

7 a Given that $f(x)=\dfrac{ax+b}{x-d}$, find the values of a and d for which $f(x)$ has horizontal and vertical asymptotes which intersect at (3, 2).

b Find the value of b for which $f(x)$ passes through the point (1,–4).

8 Let $f(x)=\frac{5}{x-m}+n$. The line $x = 4$ is a vertical asymptote of $f(x)$.

a Write down the value of m.

The graph of $y = f(x)$ has a y-intercept at $(0,7)$.

b Find the value of n.

c Write down the equation of the horizontal asymptote of $y = f(x)$.

9 The function $f(x)$ is defined as $f(x)=4+\frac{1}{2x-5}$.

a Sketch the curve of f for $-2 \le x \le 5$.

Label on your sketch:

b the equation of each asymptote

c the x-intercept

d the y-intercept.

10 Let $f(x)=\frac{2x+1}{x-1}$.

a Find $f^{-1}(x)$.

b Sketch the graph of f for $-3 \le x \le 5$.

c Write down the equations of the asymptotes.

d Write down the x-intercept.

e Solve $f(x) = f^{-1}(x)$.

11 **a** Find the inverse function, f^{-1}, of $f(x)=\frac{1}{x-2}$, $x > 2$.

b On the same set of axes, sketch $f(x)$ and $f^{-1}(x)$.

c Solve the equation $f(x) = f^{-1}(x)$ in the given domain.

12 Let $f(x)=\frac{1}{x-1}$.

a Sketch the graph of f.

The graph of f is transformed to the graph of g by a translation of $\begin{pmatrix}2\\3\end{pmatrix}$.

b Find an expression for $g(x)$.

c Find the intercepts of $g(x)$.

d Write down the equations of the asymptotes of $g(x)$.

e Sketch the graph of $g(x)$.

13 Let $f(x) = 2x + 3$ and $g(x)=\frac{5}{4x}$.

a Find $f^{-1}(x)$.

b Show that $(g \circ f^{-1})(x)=\frac{5}{2x-6}$.

Let $h(x)=\frac{5}{2x-6}$ for $x \ge 0$.

c Find the y-intercept of h.

d Sketch the graph of h.

For the graph of h^{-1}.

e Write down the x-intercept.

f Write down the equation of the vertical asymptote.

14 Consider the function $f(x)=2+\frac{10}{x-4}$.

a Write down the equations of the horizontal and vertical asymptotes.

b State the domain and range of f.

c Find the x- and y-intercepts.

d Sketch the graph of f, showing horizontal and vertical asymptotes and the coordinates of the intercepts.

The function g is defined as $g(x)=\frac{10}{x}$.

e Write down two transformations that will transform the graph of g onto the graph of f.

Exam-style questions

15 **P1:** Consider the function given by

$$f(x)=\frac{3x-20}{4+2x}.$$

a State the largest possible domain for the function. (1 mark)

b State the largest possible range for the function. (1 mark)

c Determine the coordinates of any points where the curve $y = f(x)$ intersects the x and y-axes. (2 marks)

16 **P1:** Write down the greatest possible domain and range of each of the following functions.

a $f(x)=\frac{-4}{3x+6}$ (2 marks)

b $g(x)=\frac{-4+12x}{3x+6}$ (2 marks)

c $h(x)=\frac{-4+12x}{3x}$ (2 marks)

d $p(x)=-\frac{4}{3x}$ (2 marks)

17 **P1:** The function f is defined by

$$f(x)=\frac{3x+8}{x-1}, x\in\mathbb{R}, x\neq 1.$$

a Write down the equation of the vertical asymptote. (1 mark)

b Write down the equation of the horizontal asymptote. (1 mark)

c Hence, sketch the function $y=f(x)$. (3 marks)

18 **P1:** The function f is defined by

$$f(x)=10+\frac{3}{2-x}, x\in\mathbb{R}. x\neq 2.$$

a Write down the equation of the horizontal asymptote. (1 mark)

b Write down the equation of the vertical asymptote. (1 mark)

c Express $f(x)$ in the form $f(x)=\frac{ax+b}{cx+d}$ where a, b, c, d are constants. (3 marks)

19 **P2:** The function f is defined by

$$f(x)=\frac{a+bx}{c+8x}.$$

a Given that f has a vertical asymptote at $x=-\frac{3}{4}$, determine the value of c. (2 marks)

b Given that the curve $y=f(x)$ passes through the points $\left(\frac{1}{2},\frac{2}{5}\right)$ and $\left(4,-\frac{3}{38}\right)$, determine the values of a and b. (5 marks)

20 **P2:** The number of butterflies kept in a private conservatory may be modelled by the formula $P=\frac{18(1+0.82t)}{3+0.034t}$, where p is the population of butterflies after time t months.

a Write down the initial butterfly population. (1 mark)

b Use the formula to estimate the number of butterflies in the conservatory after one year. (2 marks)

c Calculate the number of months that will have passed when the butterfly population reaches 100. (2 marks)

d Show that, when governed by this model, the number of butterflies in the conservatory cannot exceed 435. (3 marks)

21 **P1:** The function f is defined by

$$f(x)=\frac{17-10x}{2x-1}, x\in\mathbb{R}, x\neq\frac{1}{2}.$$

a Show that f can be written in the form $f(x)=\frac{12}{2x-1}-5$. (4 marks)

b Hence state the equation of the vertical asymptote. (1 mark)

c Hence state the equation of the horizontal asymptote. (1 mark)

d Sketch the graph of $y=f(x)$. (3 marks)

22 **P1:** The function f is defined by

$$f(x)=\frac{1+4x}{3+2x}, x\in\mathbb{R}, x\neq-\frac{3}{2}.$$

Sketch the graph of $y=f(x)$, stating clearly the equations of any asymptotes and the coordinates of the points where the graph intersects the coordinate axes. (6 marks)

To infinity and ... !

Approaches to learning: Thinking skills, Communication, Research, Collaboration

Exploration criteria: Mathematical communication (B), Personal engagement (C), Use of mathematics (E)

IB topic: Linking different areas

What is infinity?

Discuss these questions with a partner and then share your ideas with the whole class:

1 What is your current understanding of "Infinity" in mathematics?

2 Do you think an understanding of infinity is important? Interesting? Necessary?

3 Explain your reasoning.

Processes which go on forever are quite common in mathematics.

Where have you already met the concept of infinity in this course? Write down any other examples of where you have met the concept of infinity in mathematics, in your other academic studies or outside of your academic studies.

The three games on the following page are based on the idea of infinity. However, each game looks at infinity in a slightly different way.

What is the difference in the meaning of infinity in these three games?

To consider this question, discuss these three questions:

- In game 1, if you counted forever would you miss any numbers out?
- In game 2, is it possible to count all of the rational numbers?
- In game 3, what is the difficulty in counting the set of numbers?

Let's think further about infinity

Game 1

The winner is the person who names the biggest positive natural number.

The first person names the biggest natural number (positive whole number) they can.

The next person names a natural number that is bigger than the previous number if they can.

This continues around the class.

Who will win the game?

Game 2

The winner is the person who names the closest rational number to 0.

Follow the same steps as in game 1.

Game 3

The winner is the person who names the closest real number to 1.

Follow the same steps as in game 1.

Conclusion

With your reasoning from these games in mind, answer the following questions:

1 Are two sets that contain an infinite number of numbers necessarily the same size?

2 How can one infinity be larger than another?

Extension

Research one of these concepts, historical developments, applications or paradoxes that result from the existence of infinity. They are all conceptually difficult.

You can use both online sources or printed resources.

Present your ideas to your class

The representation of infinity in art

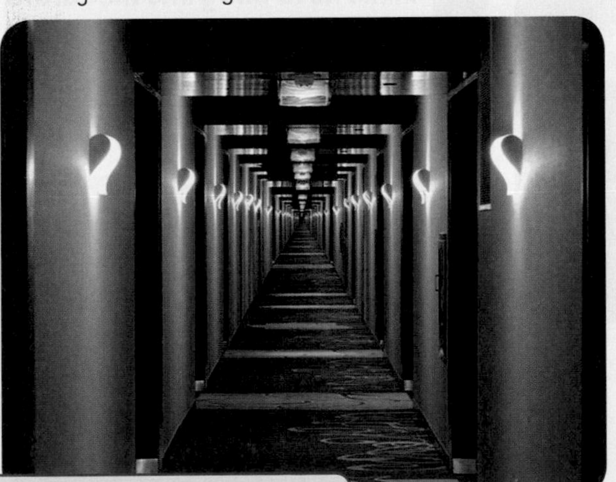

Hilbert's paradox of the grand hotel

Cantor's diagonal proof

Arithmetic properties of Infinity

Cantor's orders of infinity

Division by 0

$0.999999\ldots = 1$

The Infinitude of primes

Fractals

5 Measuring change: differentiation

The motion of people and objects, a company's profits and losses, and how rates of change of two phenomena are related may change from moment to moment. This chapter is about differential calculus: a way of measuring instantaneous change. Differential calculus provides a framework for you to model, interpret and make predictions about real-life problems. Calculus was first formulated in the 19th century and today it is perhaps the most important area of mathematics which aids the development of modern science.

Concepts

- Change
- Relationships

Microconcepts

- Limit of a function at a point
- Derivative function as the gradient of a curve, and as a rate of change
- The power rule for differentiating polynomial functions
- Tangents and normals to curves
- Differentiation rules: chain, product and quotient rules
- Maxima, minima and points of inflexion
- Kinematics problems
- Optimization

How is a runner's speed measured with respect to time?

At which instant must a rocket make its re-entry into the Earth's atmosphere, or risk being caught in the Earth's orbit?

How can a company maximize its profits and minimize its losses?

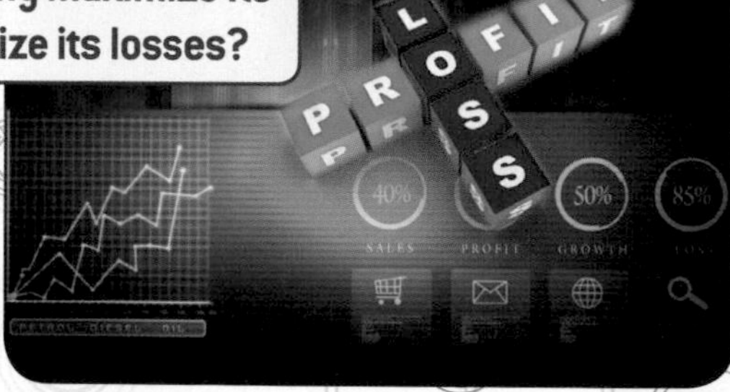

What are the dimensions of a cylindrical object for maximum volume using minimum material?

How does the rate at which water flows out of a reservoir affect the rate of change of the water depth in the reservoir?

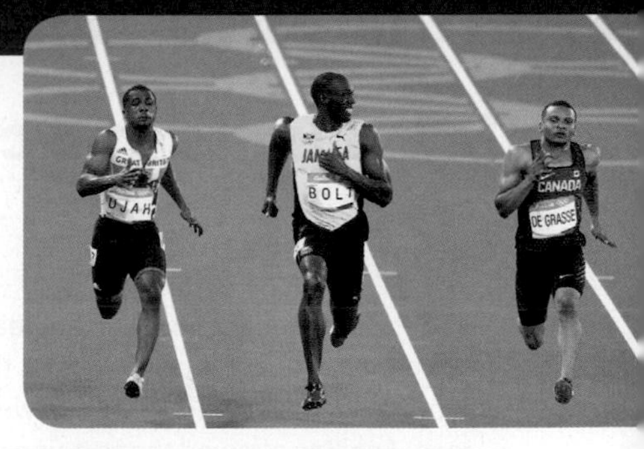

At the 2009 World Athletics Championships in Berlin, Germany, the Jamaican sprinter Usain Bolt posted a time of 9.58 s in the men's 100m sprint. After this record-breaking time, Bolt was hailed as the fastest man in the world. But what does this mean?

- Was Bolt running at the same speed throughout the whole race?
- How could you determine Bolt's speed at any particular moment in the race?
- How could you determine Bolt's fastest speed in the race?

To answer the above questions, think about the following:

- How is "fastest speed" measured?
- What is the relationship between "fastest speed" and "average speed"?

Developing inquiry skills

List some sports in which the *direction* of something changes rapidly—for example, a tennis ball after it is hit. What different inquiry questions might you ask in this situation? For example, how can you determine the speed of the tennis ball before and after it changes direction? How can you determine the time at which the ball changes direction?

Before you start

Click here for help with this skills check

You should know how to:	Skills check
1 Find the gradient of a line passing through points A(1, 2) and B(0, −3). eg gradient of (AB) is $\frac{-3-2}{0-1} = 5$	**1** Find the gradient of a line passing through each pair of points: **a** A (0, 0) and B (−4, −3) **b** $C\left(\frac{-3}{4}, 2\right)$ and D (4, −1)
2 Use rational exponents to rewrite expressions in the form cx^n. eg $\frac{2}{x^3} = 2x^{-3}$; $\sqrt[7]{x} = x^{\frac{1}{7}}$	**2** Use rational exponents to rewrite expressions in the form cx^n. **a** $7\sqrt{x}$ **b** $\frac{1}{x^2}$ **c** $\frac{8}{5\sqrt{x^3}}$
3 Sketch graphs of functions, clearly labelling any intercepts and asymptotes eg $y = \frac{1}{x-2}$ 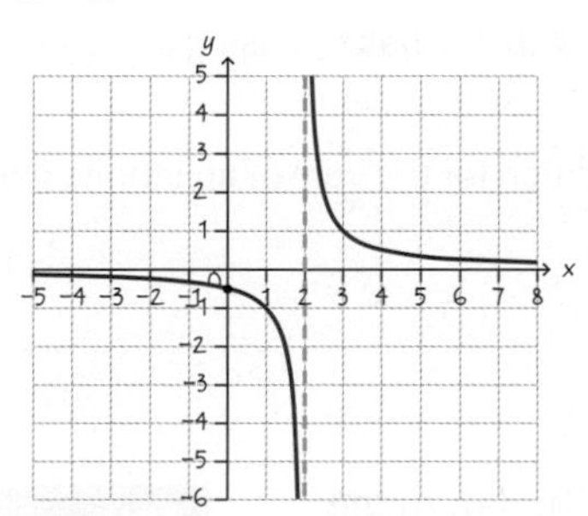	**3** Sketch the graph of $f(x) = \frac{2}{x+3}$.
4 Find the sum of an infinite geometric series. eg $\sum_{n=0}^{\infty}\left(\frac{1}{3}\right)^n$ Since $\lvert r\rvert < 1$, $S_\infty = \frac{1}{1-\frac{1}{3}} = \frac{3}{2}$.	**4** Find $\sum_{n=0}^{\infty} 5\left(\frac{1}{2}\right)^n$.

5.1 Limits and convergence

Investigation 1

About 2500 years ago, Zeno of Elea—a philosopher and logician—posed the following problem: Achilles and a tortoise were running a race. Achilles allowed the tortoise a head start of 10 metres. Both started running at a constant speed: Achilles at 10 ms^{-1}, and the tortoise at 1 ms^{-1}. After how many metres would Achilles overtake the tortoise?

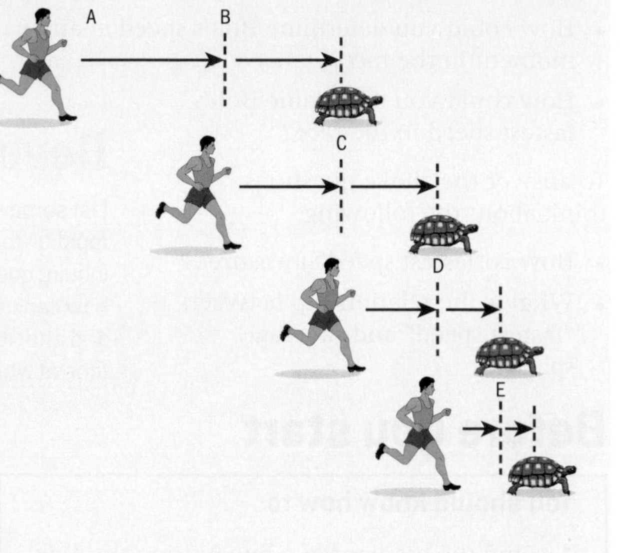

Zeno analysed the problem this way:

- Achilles starts at point A and runs to the tortoise's starting point, B, which is 10 m from A.
- When Achilles reaches B, because the tortoise runs at $\frac{1}{10}$ th the speed of Achilles, the tortoise is now at C, which is 1 m from B.
- When Achilles reaches C, the tortoise has run to D, which is 0.1 m from C.

1 List the distance that Achilles covers as he runs from:

a A to B **b** B to C **c** C to D.

Since this pattern continues, whenever Achilles reaches any point where the tortoise had been, he still has to run further to catch up, because the tortoise has advanced to the next point. Zeno's conclusion therefore was that Achilles would never overtake the tortoise!

Since Zeno knew intuitively that Achilles *would* overtake the tortoise, he concluded that the problem needed to be analysed in a different way. Here, you will use a geometric series to analyse the problem.

Copy questions **2** and **3** and fill in the blanks,

2 When Achilles reaches point D, the tortoise has run to point E, which is ______m from D.

3 When Achilles reaches point E, the tortoise has run to point F, which is _____m from E.

4 Using the previous information, represent the distances that Achilles has run using a geometric series. What are the first term and the common ratio?

5 Using what you learned in chapter 1, state a necessary condition for the sum of an infinite geometric series to exist.

6 Can you find the sum to infinity of this series?

7 Does Achilles overtake the tortoise?

It took several millennia for mathematicians to arrive at the language and concepts needed to satisfactorily solve this paradox. In this section you will learn some of the mathematics developed by 17th and 18th century mathematicians in an attempt to deal with the concepts of time and infinity.

TOK

Should paradoxes change the way that mathematics is viewed as an area of knowledge?

Limit of a function

Limits describe the output of a function as the input approaches a certain value.

If $f(x)$ approaches a real number L as x gets closer to (but not equal to) a real value a (for both $x < a$ and $x > a$), then we say that the limit of $f(x)$ as x approaches a is L.

We write this as $\lim_{x \to a} f(x) = L$.

Investigation 2

1. Use your GDC to plot the graph of the function $f(x) = \dfrac{x^2 - 1}{x - 1}$.
2. Copy this table and, beginning to the left of $x = 1$, write down the value of $f(x)$ at each x-value in the table for $0.6 \leq x < 1$.

   Then, beginning to the right of $x = 1$, write down the value of $f(x)$ at each x-value in the table for $1 < x \leq 1.4$.

x	0.6	0.7	0.8	0.9	1.0	1.1	1.2	1.3	1.4
$f(x) = \dfrac{x^2 - 1}{x - 1}$									

3. Write down the value that $f(x)$ is approaching as you move along your graph from the left and the right of $x = 1$.
4. **Conceptual** What is meant by the limit of a function at a particular point?
5. Summarize your observations about the values of the function as it approaches $x = 1$ from the left and from the right, and state the value of the function at $x = 1$.
6. **Conceptual** What condition is placed on the left and right limits of f at $x = a$ for the limit of f to exist at a?
7. Plot the graph of $g(x) = x + 1$ on your GDC. Evaluate $g(x)$ at each of the points where you evaluated $f(x)$ (in question **2**).
8. **Conceptual** Does a function need to be defined at $x = c$ in order to have a limit at $x = c$?
9. **Factual** Write down graphical similarities and differences between f and g, and explain. Can you see the graphical difference between f and g on your GDC or graphing software? If not, explain why not.

In investigation 2, f is *not* defined at $x = 1$ and g *is* defined at $x = 1$. However, as x approaches 1 from the left and the right, both f and g approach 2. This means that the limit of the functions as x approaches 1 *both* from the left and from the right is 2.

We can write this result using the following notation:

$$\lim_{x \to 1} \frac{x^2 - 1}{x - 1} = 2$$

For the limit of a function to exist as x approaches a certain value a, it is necessary that the value of the limit of the function is the same as x approaches a from the left and the right-hand sides.

Sometimes, a function f can approach one value M as x tends to a from the left, and another value N as x tends to a from the right. If this is the case, we write

$$\lim_{x\to a^-} f(x) = M \quad \text{and} \quad \lim_{x\to a^+} f(x) = N.$$

$x \to a^-$ means x approaches a from values of x which are less than a; we say "x tends to a from the left".
$x \to a^+$ means x approaches a from values of x which are greater than a; we say "x tends to a from the right".

The limit, L, of a function f exists as x approaches a real value a if and only if $\lim_{x\to a^+} f(x) = L$ **and** $\lim_{x\to a^-} f(x) = L$.

We then write $\lim_{x\to a} f(x) = L$.

Example 1

Use your GDC to help you answer parts **a** and **b**.

a Sketch the graph of $y = \dfrac{x^2 - 9}{x - 3}$, $x \neq 3$. **b** Find $\lim_{x\to 3} \dfrac{x^2 - 9}{x - 3}$ numerically.

a

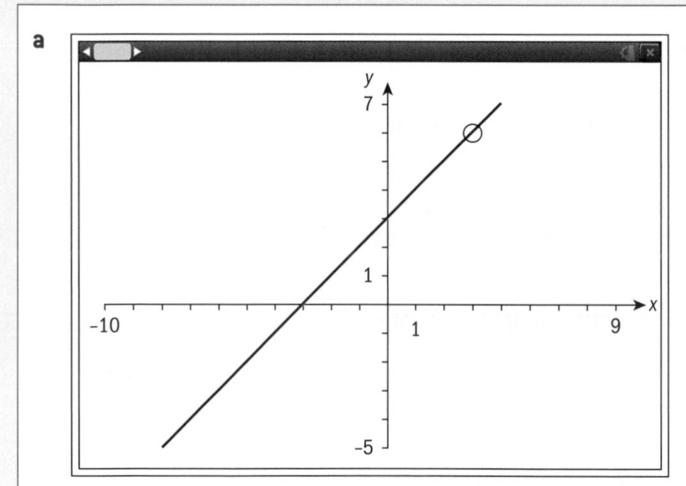

a There are open circles on the graph of the function at $x = 3$, since the rational function is not defined at $x = 3$.

b From the table of values, since the limits from both the left and right are the same to the nearest whole number, $\lim_{x\to 3} \dfrac{x^2 - 9}{x - 3} = 6$.

Example 2

Using your GDC:

a Sketch the graph of $y = \dfrac{3^x - 1}{x}$, $x \neq 0$.

b Find $\lim_{x\to 0^-} \dfrac{3^x - 1}{x}$ and $\lim_{x\to 0^+} \dfrac{3^x - 1}{x}$ numerically, giving your answer to 1 decimal place.

a

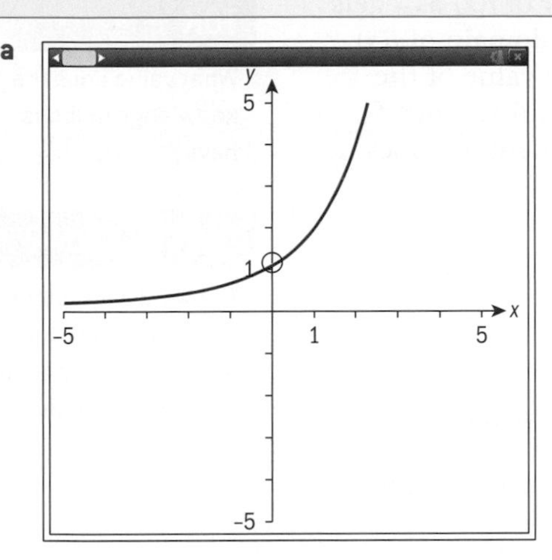

b From the table of values, since the limits from both the left and right are the same to 1d.p.,

$$\lim_{x \to 0} \frac{3^x - 1}{x} \approx 1.1$$

Exercise 5A

Use technology to examine each function numerically and graphically. Find the limit as x tends toward the given value, if it exists.

1 $\lim_{x \to 3} (x^2 + 1)$

2 $\lim_{x \to 1} (5 - 2x)$

3 $\lim_{x \to 0} \frac{2x^2 - x}{x}$

4 $\lim_{x \to 1} \frac{x^2 - x}{x - 1}$

Limits at infinity

The concept of limits helps us to understand the behaviour of functions for extreme values of x.

In chapter 4 you studied rational functions, such as $f(x) = \frac{1}{x^2}, x \neq 0$, and its asymptotes.

The **horizontal asymptote** $y = k$ tells you the behaviour of the function for very large values of x, ie the value of y that $f(x)$ approaches as $x \to \infty$ or $x \to -\infty$.

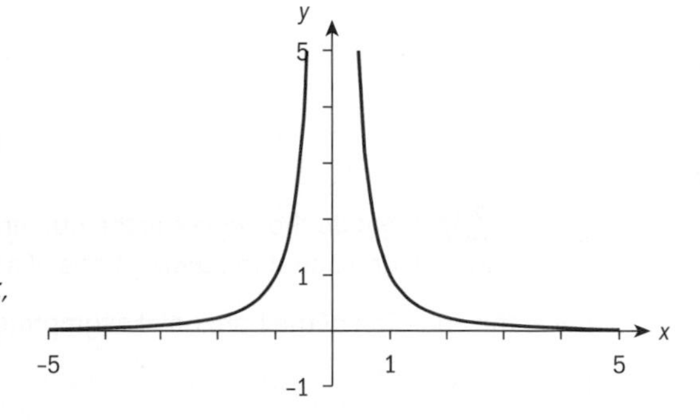

Calculus

The **vertical asymptote** $x = c$ describes the behaviour of $f(x)$ as x gets infinitesimally close to the value c which is not in the domain of $f(x)$. As x approaches c from the left and from the right, the value of the function increases without bound, and approaches infinity. Since f increases without bound as x approaches k, the limit therefore does not exist, since the limit is not a real number.

TOK

What value does the knowledge of limits have?

Did you know?

The word asymptote comes from the Greek *asymptotes*, meaning "not falling together".

Investigation 3

1 For the rational functions $f(x) = \frac{p(x)}{q(x)}$, $q(x) \neq 0$, find algebraically the horizontal asymptotes, if they exist, by finding the limits of the functions as $x \to \infty$, and as $x \to -\infty$.

a $y = \frac{-2x+3}{x+1}$ **b** $y = \frac{x+2}{-3x^2+3x-1}$ **c** $y = \frac{x^3-25}{x^2+25}$

d $y = \frac{3x+2}{x^2-1}$ **e** $y = \frac{x^3+x-4}{4x^3+3x^2+x}$ **f** $y = \frac{x^2+1}{x-1}$

g $y = \frac{\sqrt{4x^2+1}}{3x-1}$ **h** $y = \frac{x^2+3x}{x^5-1}$ **i** $y = \frac{x^5-2x}{x^3-x+1}$

2 Summarize your findings for the functions in question **1** by copying and completing the following table. The first one has been done for you.

	Degree ($f(x)$)	Leading coefficient	Degree ($g(x)$)	Leading coefficient	Horizontal asymptote
a	1	−2	1	1	$y=-2$
b					
c					
d					
e					
f					
g					
h					
i					

3 **Conceptual** What do the degree of the numerator and the degree of the denominator tell you about the asymptotes of a function?

4 What is the equation of the horizontal asymptote of the function $f(x) = \frac{1}{x^2}, x \neq 0$?

5 What is the equation of the vertical asymptote of the function $f(x) = \frac{1}{x^2}, x \neq 0$?

Example 3

Consider the function $f(x)=\frac{2x}{x-1}, x\neq 1$.

a Sketch the function. **b** State the equations of any asymptotes.

a

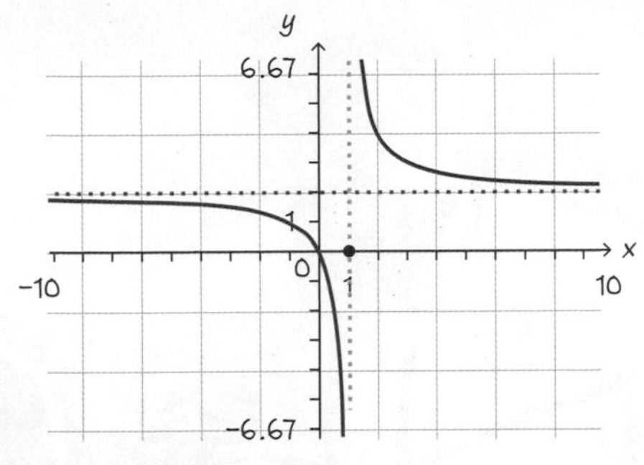

b From the graph of $f(x)$, you can see that the function has a vertical asymptote at $x = 1$, and a horizontal asymptote at $y = 2$.

As $x \to \infty$, $f(x) \to 2$

As $x \to -\infty$, $f(x) \to 2$

That is, $\lim_{x\to\pm\infty} f(x) = 2$. Therefore, $y = 2$ is a horizontal asymptote.

As $x \to 1^+$, $f(x) \to \infty$

As $x \to 1^-$, $f(x) \to -\infty$

Therefore, $x = 1$ is a vertical asymptote.

Exercise 5B

State the equations of any vertical and horizontal asymptotes for each of the following functions.

1 $f(x)=\frac{3x}{6x-1}$ **2** $g(x)=\frac{x^2+3}{3-x^2}$ **3** $h(x)=\frac{1-x-x^3}{x^3-1}$ **4** $k(x)=-\frac{5x}{x^2-2}$

Limits of sequences

In chapter 1 you learned that if a geometric series has a finite sum, then it converges to its sum. Recall the formula for finding the sum of a finite geometric series: $S_n=\frac{u_1(1-r^n)}{1-r}$.

When $|r|<1$, then $\lim_{n\to\infty}\left(\frac{u_1(1-r^n)}{1-r}\right)=\frac{u_1}{1-r}$.

> For an infinite geometric series, when $|r|<1$ the series converges to its sum, $S_\infty=\frac{u_1}{1-r}$. If $|r|>1$, then the series diverges.

5.2 The derivative function

Investigation 4

The Jamaican sprinter Usain Bolt was the first person, since speed times were fully automated, to hold both the 100 m and 200 m world records at the same time.

The data shows the time, in seconds, which had elapsed when Usain Bolt passed every 10 m mark in the 100 m final at the World Championships in Berlin (2009).

The graph shows Bolt's time (plotted on the x-axis) against the distance he had run (plotted on the y-axis).

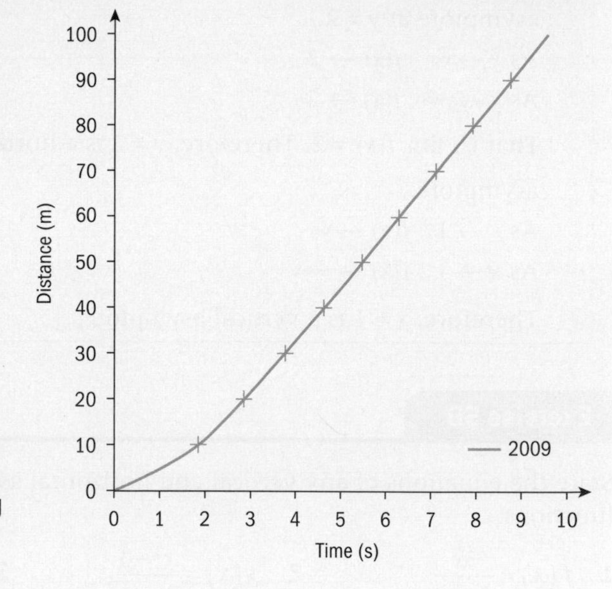

1 Write down what you notice about the time Bolt took to cover each 10 m distance in 2009.

2 Estimate Bolt's average running speed over the 100 m.

3 Estimate Bolt's fastest average speed over any 10 m interval in the race, and state the 10 m interval over which he achieved this.

4 **Factual** Compare Bolt's fastest speed with his average speed in that race. If you were to examine Bolt's speed at *any* point in that race, which of the two speeds you found would likely be closest to the speed that he was running at that instant?

5 **Factual** From the given information, is it possible to find his fastest speed at any particular moment in time, for example, at 3.25 s?

6 **Conceptual** What is an average rate of change between output and input values of a function?

In investigation 4, the distance Usain Bolt had covered was a function of time. You found his average speed over any particular 10 m interval by dividing the change in the distance he had covered—that is, 10 m—by the corresponding change in time.

For any two points (t_1, d_1) and (t_2, d_2) on the graph, Bolt's average speed between those two points is given by $\frac{d_2 - d_1}{t_2 - t_1}$.

You may have noticed that this is the same as the formula for the gradient of a straight line joining the points (t_1, d_1) and (t_2, d_2).

Therefore, the average rate of change of a function f between the points x_1 and x_2 is the gradient of the line segment joining $(x_1, f(x_1))$ and $(x_2, f(x_2))$.

> The average rate of change of a function f between two values x_1 and x_2 is given by $\frac{\Delta y}{\Delta x} = \frac{f(x_2) - f(x_1)}{x_2 - x_1}$.
>
> (Note that $\frac{\Delta y}{\Delta x}$ means "the change in y divided by the change in x", where Δ is the Greek letter "delta".)

If a function is linear, the gradient between any two points is the same, and hence the rate of change of the function is always constant.

This, however, is not the case for non-linear functions. In order to explore the gradient or rate of change of a non-linear function, you first need to learn some essential definitions.

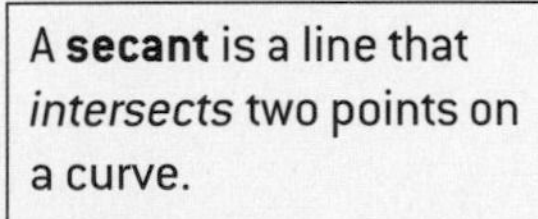 A **secant** is a line that *intersects* two points on a curve. A **tangent** to a curve at a specific point is a line that *touches* the curve at that point.	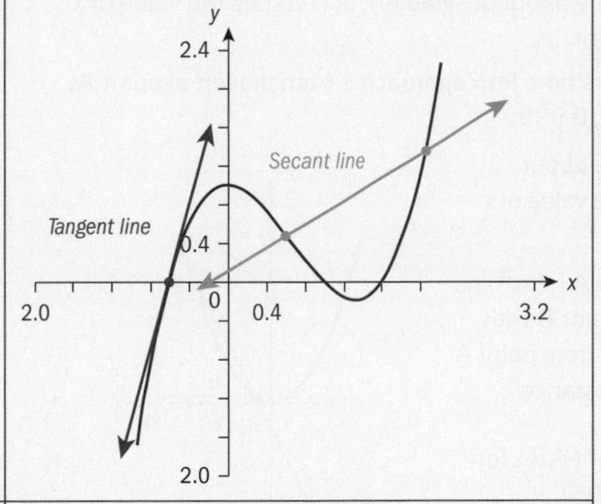
The green line is a tangent to the curve at the pink marked point. It also intersects the curve at another point.	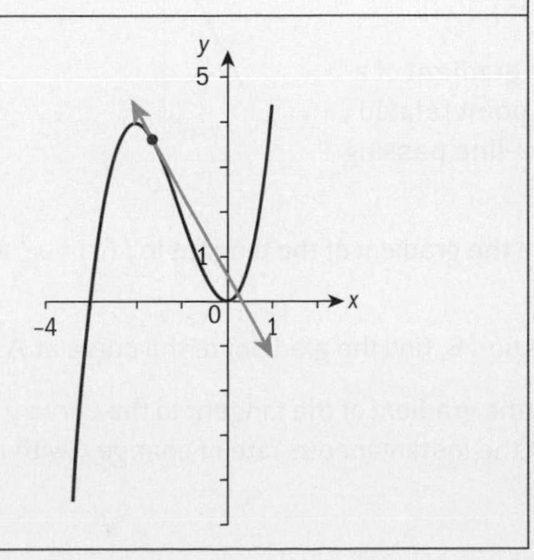

Calculus

Investigation 5

Let the function f be defined by $f(x) = x^2$.

$A(1, 1)$ is a point which lies on the graph of $y = f(x)$, and B is any other arbitrary point (x, x^2) on the curve.

Let (AB) represent a secant line joining points A and B.

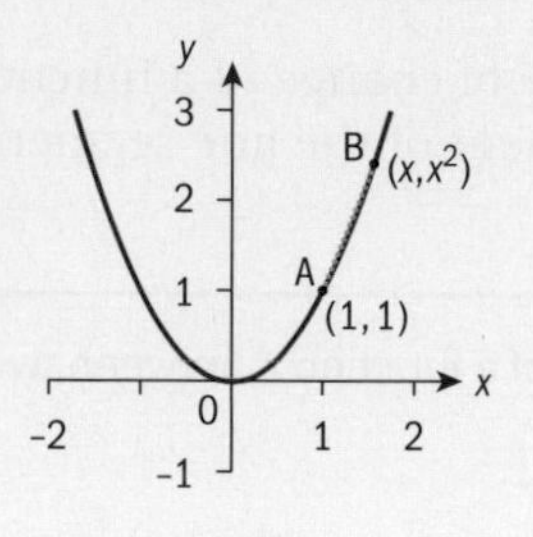

1 Calculate the gradient of (AB) for each of the different values of x given in the table. The first one has been done for you.

x	$B(x, f(x))$	Gradient of (AB)
2	$(2, 4)$	$\frac{4-1}{2-1} = 3$
1.5		
1.1		
1.01		
1.001		

HINT

The formula for finding the gradient, m, of the line joining $(a, f(a))$ and $(b, f(b))$ is $m = \frac{f(b) - f(a)}{b - a}$.

In the table, the x-values of B approach 1 from the right (that is for values of x greater than 1); which is the x-coordinate of A. Hence, the point $B(x, f(x))$ gets closer and closer to point A.

2 Write down what you notice about the gradient of (AB) as the value of x approaches 1 from the right.

3 Repeat question **1**, but this time let x approach 1 from the left of point A, using x values: 0, 0.8, 0.9, 0.999.

Write down what you notice about the gradient of (AB) as the value of x approaches 1 from the left.

Now consider a fixed point $A(x, x^2)$ on the curve, and a second point B with x-coordinate h units away from point A (where h is a very small distance).

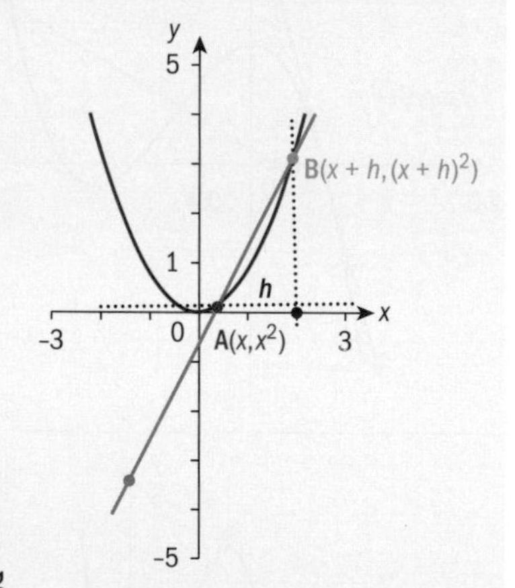

4 Write down the gradient of (AB) for $A(x, x^2)$ and $B(x, (x + h)^2)$.

5 **Conceptual** How is the gradient of a tangent to the curve at a point related to the gradient of a secant line passing through the same point?

6 Letting h tend to 0, deduce the gradient of the tangent to $f(x) = x^2$ at any point x.

7 Using your answer to question **6**, find the gradient of the curve at $A(1, 1)$.

8 **Conceptual** What does the gradient of the tangent to the curve $y = f(x)$ at any point tell you about the instantaneous rate of change y with respect to x at that point?

The gradient of the secant line (AB) is $\frac{f(x+h)-f(x)}{h}$

This gradient is a measure of the **average rate of change** of $f(x)$ between x and $x + h$.

As B approaches A, the gradient of (AB) approaches the gradient of the tangent line to the curve at A.

So, the gradient of the tangent to the curve at A is the value of the limit of $\frac{f(x+h)-f(x)}{h}$ as h approaches 0.

The gradient of the tangent to the curve at A is the instantaneous rate of change of $f(x)$ at A. This instantaneous rate of change is called the **derivative** of f with respect to x.

We sometimes use different notation for the derivative function:

- $f'(x)$ (this is called "f dash of x" or "f prime of x"); or
- $\frac{dy}{dx}$ (meaning "the derivative of y with respect to x")

In the scenario where you studied Usain Bolt, finding the speed over shorter intervals of time and distance should give a more accurate approximation of the sprinter's speed at a given time in the race.

The smaller the time interval you take, the closer you are to finding the speed at a particular instant.

Example 4

A particle moves in a straight line so that its position from its starting point at any time t in seconds is given by $s = 3t^2$, where s is in metres. The particle passes though a certain point when $t = a$ and then sometime later though another point when $t = a + h$.

a Find the average speed of the particle from $t = a$ to $t = a + h$.

b Explain how you could find the speed of the particle at the instant $t = a$ and find the speed at this point.

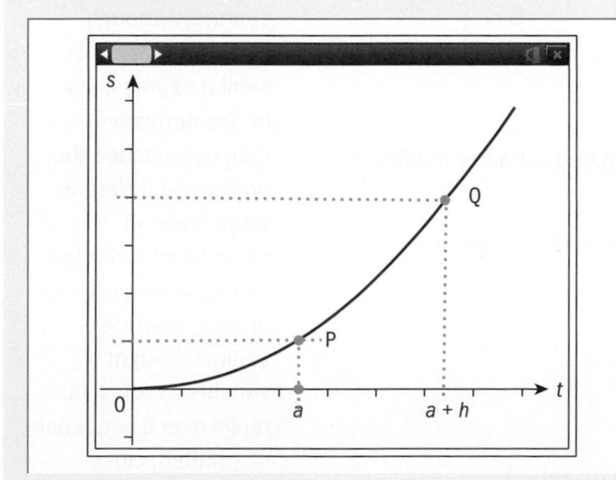

Use your GDC to plot a graph of the function.

The y-coordinates of points P and Q on the graph represent the position of the particle at times $t = a$ and $t = a + h$, respectively.

Continued on next page

Calculus

a $P(a, 3a^2)$ and $Q(a+h, 3(a+h)^2)$	Write the coordinates of P and Q.
Average speed $= \frac{3(a+h)^2 - 3a^2}{(a+h)-a}$	Average speed $= \frac{\text{total distance travelled}}{\text{total time}}$
$= \frac{3(a^2+2ah+h^2)-3a^2}{h}$	
$= \frac{3h^2+6ah}{h}$	Simplify.
$= (3h+6a)\ \text{ms}^{-1}$	
b The speed of the particle when $t = a$ is equal to the gradient of the tangent to the curve at P. This tangent is found when Q and P are the same point, ie when $h = 0$. Letting $h = 0$, speed when $t = a$ is $6a\ \text{ms}^{-1}$	Explain your reasoning carefully.

The power rule for differentiating polynomials

The process of finding the derivative of a function is called **differentiation**.

Investigation 6

1 Use your GDC to plot a graph of the function $y = x^2$ and find the derivative (gradient) of the graph at each of the x-values given in the table.

Copy and complete this table to record your results.

x	−2	−1	0	1	2	3
$\frac{dy}{dx}$	−4	−2	0	2	4	6

2 Plot a graph of x against $\frac{dy}{dx}$.

3 What is the gradient function for $y = x^2$? Hint: Find the equation of the line you plotted in question **2**.

4 Repeat steps **1–3** for the functions $y = x^3$ and $y = x^4$ and record your results in a table like this:

Function	**Gradient function**
$y = x^2$	$2x$
$y = x^3$	
$y = x^4$	

International-mindedness

Sir Isaac Newton and Gottfried Leibniz independently developed the idea of the derivative function—sometimes called the **differential calculus**—in the 17th century. Although they both arrived at similar expressions for the derivative, they approached the problem in different ways. Newton considered variables changing with respect to time; whereas Leibniz thought of variables x and y that range over a sequence of infinitely close values.

5 Using the table you completed in question **4**, can you suggest a rule for derivative of $f(x) = x^n$?

6 **Conceptual** How is the power rule useful in differentiating a polynomial function?

In investigation 6, you considered only positive integer values for n, but the power rule is true for any real number n.

The power rule for differentiating x^n:
If n is any real number and $f(x) = x^n$, then $f'(x) = nx^{n-1}$.

Example 5

Find the derivative of each function.

a $f(x) = x^{11}$ **b** $f(x) = \frac{1}{x^4}$ **c** $f(x) = \sqrt{x}$

a $f(x) = x^{11}$	
$f'(x) = 11x^{11-1}$	Use the power rule.
$f'(x) = 11x^{10}$	Simplify.
b $f(x) = \frac{1}{x^4}$	
$f(x) = x^{-4}$	Rewrite using rational exponents.
$f(x) = -4x^{-4-1}$	Use the power rule.
$f(x) = -4x^{-5}$	Simplify.
c $f(x) = \sqrt{x}$	
$f(x) = x^{\frac{1}{2}}$	Rewrite using rational exponents.
$f'(x) = \frac{1}{2}x^{\frac{1}{2}-1}$	Use the power rule.
$f'(x) = \frac{1}{2}x^{-\frac{1}{2}} = \frac{1}{2\sqrt{x}}$	Simplify.

Exercise 5C

Find the derivative of each function:

1 $f(x) = x^7$ **2** $f(x) = x^{18}$ **3** $f(x) = x^{\frac{-1}{2}}$

4 $f(x) = \sqrt[5]{x}$ **5** $f(x) = \frac{1}{\sqrt{x}}$ **6** $f(x) = \sqrt[4]{x^3}$

Derivatives of constant and linear functions

Investigation 7

Note: Where necessary, in this investigation you should use the power rule for differentiating polynomials.

1 **Factual** What is the *gradient* of a straight line parallel to the x-axis?

2 Write down the derivative of:

a $f(x) = -7$ b $f(x) = 3$
c $f(x) = 0$ d $f(x) = \pi$

3 **Conceptual** What is the *derivative* of a constant function?

4 **Factual** For a straight line with equation $y = mx + c$, which parameter tells you the gradient of the line?

5 Find the derivative of:

a $y = -x + 2$ b $4x - 2y + 1 = 0$ c $f(x) = \frac{3-2x}{4}$

6 **Conceptual** How is the derivative of a linear function related to the gradient of its graph?

If $f(x) = c$ where $c \in \mathbb{R}$, then $f'(x) = 0$.
If $f(x) = mx + c$ where $m, c \in \mathbb{R}$, then $f'(x) = m$.

Derivative of a constant multiple of a function

For $c \in \mathbb{R}$, $(cf)'(x) = c(f'(x))$, provided $f'(x)$ exists.

Derivative of the sum of difference of functions

If $f(x) = u(x) \pm v(x)$, then $f'(x) = u'(x) \pm v'(x)$.

International-mindedness

Maria Agnesi, an 18th century, Italian mathematician, published a text on calculus and also studied curves of the form

$$y = \frac{a^2}{x^2} + a^2$$

Example 6

Find the derivative of each function.

a $f(x) = \frac{3}{x^{12}}$ b $y = 5 + 3x^2$

c $s = (t+1)(t-2)$ d $f(x) = \frac{2x^4 - 3x^3 + 1}{x^2}, x \neq 0$

a $f(x) = \frac{3}{x^{12}}$	
$f(x) = 3x^{-12}$	Rewrite using rational exponents.
$f'(x) = 3(-12x^{-12-1})$	Use the power rule.
$f'(x) = -36x^{-13}$	Simplify.

b $y = 5 + 3x^2$	Differentiate each term. Note that the derivative of the constant term is 0.
$\frac{dy}{dx} = 0 + 3\left(2x^{2-1}\right)$	Simplify.
$\frac{dy}{dx} = 6x$	
c $s = (t + 1)(t - 2)$	Expand so that the function is the sum or difference of terms in the form at^n.
$s = t^2 - t - 2$	
$\frac{ds}{dt} = 2t - 1$	Differentiate each term.
d $f(x) = \frac{2x^4 - 3x^3 + 1}{x^2} = \frac{2x^4}{x^2} - \frac{3x^3}{x^2} + \frac{1}{x^2}$ $= 2x^2 - 3x + x^{-2}$	Rewrite so that the function is the sum or difference of terms in the form ax^n.
$f'(x) = 4x - 3 - 2x^{-3}$	Differentiate each term.

Exercise 5D

1 Find the gradient function for each of the following functions.

a $y = x^4 - \frac{1}{2}x^2$

b $f(x) = 5x(x^2 - 1)$

c $f(x) = 6x^4 - 3x^2 - 10$

d $s = 2t^2 + 3t$

e $v = -9.8t + 4.9$

f $c = 24x + 10$

2 Find $f'(x)$ for each function.

a $f(x) = 6\sqrt{x}$

b $f(x) = 5\sqrt[5]{x^3}$

c $f(x) = \frac{2}{x} - 3\sqrt{x}$

3 Differentiate with respect to x.

a $f(x) = \frac{3}{2x^2}$

b $f(x) = \frac{3}{(2x)^2}$

c $f(x) = 4\pi x^3 + 3\pi$

d $f(x) = (x + 1)^2$

e $f(x) = \frac{x^3 + x - 3}{x}$

f $f(x) = (2x-1)(x^2+3)$

4 Find $\frac{dy}{dx}$ for each function.

a $y = 1 + x\sqrt{x}$

b $y = \frac{7}{x^2} - \frac{1}{\sqrt{x}}$

c $y = \sqrt[3]{x} + \sqrt[4]{x}$

TOK

Mathematics and the real world

The seemingly abstract concept of calculus allows us to create mathematical models that permit human feats, such as getting a man on the Moon. What does this tell us about the links between mathematical models and physical reality?

Tangents and normals

How much can the cyclist lean inwards, in order to negotiate a bend, without falling off?

In order for a cyclist to move in a circular path, a certain force is required to keep him from shooting away from the circle at a tangent. This force is **perpendicular** to the tangent to the curve of the circular path.

The **normal line** at a point on a curve is the line perpendicular to the curve's tangent at that point.

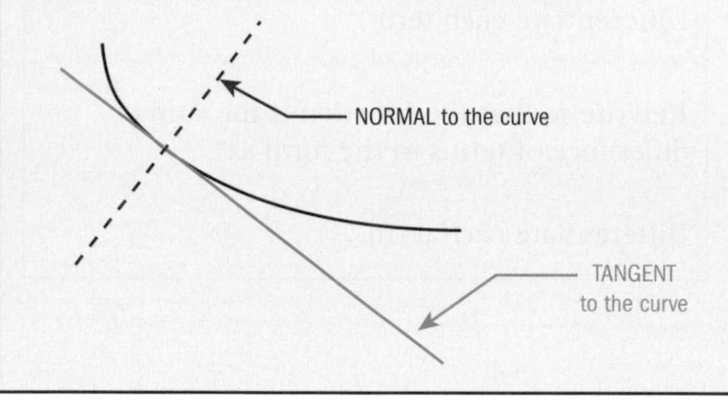

Can you find other examples as to how tangent and normal lines relate to real-life problems?

$f(x)$ is a non-linear function and (x_1, y_1) is a point which lies on the curve $y = f(x)$. If $f'(x_1) = m$, where m is a constant, then m is the gradient of the tangent to $y = f(x)$ at (x_1, y_1), and the equation of the tangent line is $(y - y_1) = m(x - x_1)$.

You can also find the equation of the normal to the function at a particular point. As you saw in the example above where the cyclist travels around a bend, the normal to the curve at any point is perpendicular to the tangent at that point.

HINT

Recall the different forms of the equation of a straight line:
$y = mx + c$ (gradient-intercept form)
$ax + by + d = 0$ (general form)
$y = mx + c$ (point-gradient form).

Example 7

Find the gradient function of $f(x) = x^2 - \frac{4}{\sqrt{x}}$.

Hence find the gradient of the tangent to $f(x)$ at the point where $x = 4$.

$f(x) = x^2 - 4x^{\frac{-1}{2}}$	Rewrite using rational exponents.
$f'(x) = 2x - 4\left(\frac{-1}{2}x^{\frac{-1}{2}-1}\right)$	Find the derivative of each term.
$f'(x) = 2x + 2x^{\frac{-3}{2}}$	
$f'(4) = 2(4) + 2(4)^{\frac{-3}{2}}$	Simplify.
$f'(4) = \frac{33}{4}$	Substitute $x = 4$ into the expression for $f'(x)$ to find the gradient of the tangent to $f(x)$ at $x = 4$.

Example 8

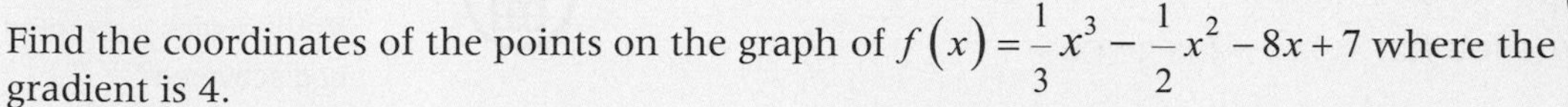

Find the coordinates of the points on the graph of $f(x) = \frac{1}{3}x^3 - \frac{1}{2}x^2 - 8x + 7$ where the gradient is 4.

$f(x) = \frac{1}{3}x^3 - \frac{1}{2}x^2 - 8x + 7$	Find the derivative of each term.
$f'(x) = x^2 - x - 8$ $4 = x^2 - x - 8$	Substitute $f'(x) = 4$. Simplify to a three-term quadratic.
$x^2 - x - 12 = 0$ $(x - 4)(x + 3) = 0$	Factorize to find the roots of the quadratic.
$x = 4$ or $x = -3$	$f'(x) = 4$ at these two values of x.
$f(4) = \frac{1}{3}4^3 - \frac{1}{2}4^2 - 8(4) + 7$ $f(4) = \frac{-35}{3}$ $f(-3) = \frac{1}{3}(-3)^3 - \frac{1}{2}(-3)^2 - 8(-3) + 7$ $f(-3) = \frac{35}{2}$ The coordinates are $(4, \frac{-35}{2})$ and $(-3, \frac{35}{2})$.	Substitute the two values of x into the expression for $f(x)$ in order to find the y-coordinates at these points.

Exercise 5E

1 Find the gradient of the tangent to each function at the given value of x.

a $y = x^2 - 4x + 2$ at $x = -1$

b $y = \dfrac{2x^2 - 5}{x}$ at the point $(1, -3)$

c $f(x) = \sqrt[4]{x} + \dfrac{8}{\sqrt{x}}$ at $x = 1$

2 Find the coordinates of the points on the graph of $f(x) = \frac{2}{3}x^3 - \frac{9}{2}x^2 - 3x + 8$ where the gradient is 2.

Recall that for perpendicular lines with gradients m_1 and m_2, then $m_1 m_2 = -1$.

If a tangent line has gradient m, the gradient of the normal line is $\dfrac{-1}{m}$.

Example 9

The function $f(x) = 3x^2 - 2$ is given.

Find:

a the derivative of the function at $x = 1$

b the equation of the tangent to the curve at $x = 1$

c the equation of the normal to the curve at $x = 1$

a $f(x) = 3x^2 - 2$ $f'(x) = 6x$ $f'(1) = 6$	Find the derivative using the power rule.
b $m_{\text{tangent}} = f'(1)$	Evaluate the derivative at $x = 1$. The gradient of the tangent at $x = 1$ is the derivative of the curve at $x = 1$.
Equation of the tangent is: $y - 1 = 6(x - 1)$ or $y = 6x - 5$	Substitute $x = 1$ into the original formula to find the value for y. Then use $y - y_1 = m(x - x_1)$, where $(1, 1)$ is a point on the tangent line.
c $m_{\text{normal}} = \frac{-1}{m_{\text{tangent}}}$ $m_{\text{normal}} = \frac{-1}{6}$	Since normal and tangent lines are perpendicular, the gradient of the normal is the negative reciprocal of the gradient of the tangent line.
Equation of the normal is: $y - 1 = \frac{-1}{6}(x - 1)$ or $y = \frac{-x}{6} + \frac{7}{6}$	Using $y - y_1 = m(x - x_1)$, where $(1, 1)$ is a point on the normal line.

TOK

Mathematics – invented or discovered?

If mathematics is created by people, why do we sometimes feel that mathematical truths are objective facts about the world rather than something constructed by human beings?

Example 10

Find the equation(s) of any horizontal tangents to the curve $y = 2x^3 + 3x^2 - 12x + 1$.

$y = 2x^3 + 3x^2 - 12x + 1$	
$y' = 6x^2 + 6x - 12$	Find the derivative using the power rule.
$y' = 6(x^2 + x - 2)$	Simplify and factorize.

$y' = 6(x-1)(x+2)$	
$y' = 0$	Horizontal tangent lines have gradient 0
$6(x-1)(x+2) = 0$	Solve the equation.
$x = 1$ or $x = -2$ $(1, -6), (-2, 21)$	Find the y-coordinate by substituting the x-values in the equation of the curve.
Equations of the tangents are:	
$y = -6$ and $y = 21$	

Exercise 5F

1 Find the equation of the normal to $y = \frac{1-2x}{x^2}$ at the point $(2, \frac{-3}{4})$.

2 Find the equation(s) of any tangents to $y = -x^3 + 2x^2 + 1$ which are parallel to the tangent at $x = -1$.

3 Find the equation(s) of any normals to the curve $y = x + \frac{1}{x}$ at points where the tangent is parallel to the line $y = -3x$.

4 The gradient of the tangent line to the graph of $f(x) = 3x^2 - 2kx - 9$ at $x = 1$ is 10. Find the value of k.

5 Given that $f(x) = x^3 - x^2 - 2x + 1$, find the coordinates of all the points where $f(x)$ has a horizontal tangent.

6 Given that $g(x) = \frac{1}{x^n}$, show that $xg'(x) + ng(x) = 0$.

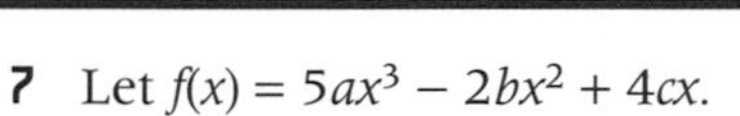

7 Let $f(x) = 5ax^3 - 2bx^2 + 4cx$.

a Find $f'(x)$.

b Given that $f'(x) \geq 0$ and $a > 0$, show that $b^2 \leq 15ac$.

8 Let $f(x) = \frac{-20}{x} + 1$ for $x > 0$.

Let $g(x) = 5x + 3$ for $x \in \mathbb{R}$.

Find the values of x for which the graphs of f and g have the same gradient.

Developing inquiry skills

Is it possible to find Usain Bolt's speed at any particular moment during a race? How could you use what you have learned in this section to suggest a method to find his instantaneous speed?

Calculus

5.3 Differentiation rules

The chain rule

The function $y = (3 - x)^3$ can be written as a composite of two functions g and f such that $y = f(g(x))$ or $(f \circ g)(x)$ where $g(x) = (3 - x)$ and $f(x) = x^3$.

HINT

Refer to composite functions in Chapter 1.

Investigation 8

The chain rule

1. Expand $y = (1 + x)^2$ and hence find $\frac{dy}{dx}$. Give your answer in factorized form.
2. Write $y = (1 + x)^2$ as the composition of two functions f and g, such that $y = f(g(x))$.
3. Let $u = g(x)$. Find $\frac{du}{dx}$.
4. Write $y = f(g(x))$ in terms of the variable u, that is, $y = f(u)$.
   Hence, find $\frac{dy}{du}$.
5. Use your answers to questions **3** and **4** to find $\frac{dy}{du} \times \frac{du}{dx}$. Give your answer in terms of x.

$$\begin{aligned}\frac{dy}{du} \times \frac{du}{dx} &= 2u \cdot 1 \\ &= 2(1+x) \cdot 1 \\ &= 2(1+x)\end{aligned}$$

6. **Factual** Compare the expression you obtained for $\frac{dy}{du} \cdot \frac{du}{dx}$ to the expression you found for $\frac{dy}{dx}$ in question **1**.
7. Repeat questions **1** – **6** for each of these functions. Record your results in a table like this one.

Question	$y = (1 - 2x)^2$	$y = (3x - 1)^2$	$y = (1 + ax)^2$
1			
2			
3			
4			
5			
6			

8. **Conceptual** How can you find the derivative of a composite function $y = f(g(x))$?

TOK

What is the difference between inductive and deductive reasoning?

Did you know?

In the relationship $\frac{dy}{dx} = \frac{dy}{du} \cdot \frac{du}{dx}$, $\frac{dy}{du}$ and $\frac{du}{dx}$ are not fractions. Therefore, $\frac{dy}{dx}$ is not arrived at by cancelling du top and bottom. However, since these are rates of change, you can intuitively see that if, for example, y changes twice as fast as u and u changes twice as fast as x, then y would change four times as fast as x.

The chain rule states that if $y = f(u)$ and $u = g(x)$, then $\frac{dy}{dx} = \frac{dy}{du} \times \frac{du}{dx}$.

Another definition for the chain rule is $(f \circ g)'(x) = f'(g(x)) \cdot g'(x)$.

Example 11

Differentiate each function.

a $y = (x^2 - 4x)^3$ **b** $y = \sqrt{5x^2 - 2}$ **c** $y = \dfrac{2}{\sqrt{1-3x}}$

a $y = (x^2 - 4x)^3$	
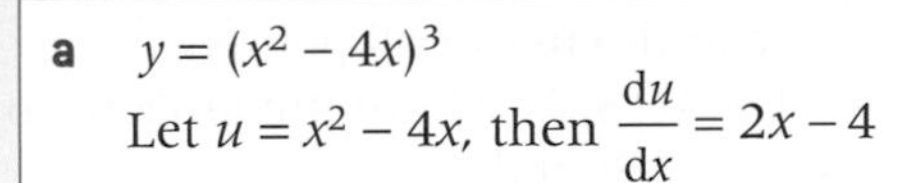 Let $u = x^2 - 4x$, then $\dfrac{du}{dx} = 2x - 4$	Define u and find $\dfrac{du}{dx}$.
Hence, $y = u^3$ and $\dfrac{dy}{du} = 3u^2$	Write y in terms of u and find $\dfrac{dy}{du}$.
$\dfrac{dy}{dx} = \dfrac{dy}{du} \cdot \dfrac{du}{dx} = 3u^2 \cdot (2x - 4)$	Use the chain rule.
$\dfrac{dy}{dx} = 3(x^2 - 4x)^2 (2x - 4)$	Substitute for u and simplify.
b $y = \sqrt{5x^2 - 2}$	
Let $u = 5x^2 - 2$, then $\dfrac{du}{dx} = 10x$	Define u and find $\dfrac{du}{dx}$.
Hence, $y = \sqrt{u}$ and $\dfrac{dy}{du} = \dfrac{1}{2} u^{\frac{-1}{2}}$	Write y in terms of u and find $\dfrac{dy}{du}$.
$\dfrac{dy}{dx} = \dfrac{dy}{du} \times \dfrac{du}{dx} = \dfrac{1}{2} u^{\frac{-1}{2}} (10x)$	Use the chain rule.
$\dfrac{dy}{dx} = \dfrac{1}{2}(5x^2 - 2)^{\frac{-1}{2}} (10x)$	Substitute for u and simplify.
$\dfrac{dy}{dx} = \dfrac{5x}{\sqrt{5x^2 - 2}}$	
c $y = \dfrac{2}{\sqrt{1-3x}}$	
Let $u = 1 - 3x$, then $\dfrac{du}{dx} = -3$	Define u and find $\dfrac{du}{dx}$.
Hence, $y = 2u^{\frac{-1}{2}}$ and $\dfrac{dy}{dx} = -u^{\frac{-3}{2}}$	Write y in terms of u and find $\dfrac{dy}{du}$.
$\dfrac{dy}{dx} = \dfrac{dy}{du} \times \dfrac{du}{dx} = -u^{\frac{-3}{2}} (-3)$	Use the chain rule.
$\dfrac{dy}{dx} = -(1 - 3x)^{\frac{-3}{2}} (-3)$	Substitute for u and simplify.
$\dfrac{dy}{dx} = \dfrac{3}{\sqrt{(1-3x)^3}}$	

Exercise 5G

1 Use the chain rule to find $\frac{dy}{dx}$ for each function.

a $y = (2x + 3)^5$

b $y = \sqrt{1 - 2x}$

c $y = \frac{-3}{\sqrt{2x^2 - 1}}$

d $y = 2\left(x^2 - \frac{2}{x}\right)^3$

2 Find the equation of the tangent to the curve $y = 6\sqrt[3]{1 - 2x}$ at $x = 0$.

3 Consider $y = \frac{a}{\sqrt{1 + bx}}$ where $a, b \in \mathbb{R}$ and $a > 0$.

Find a and b such that $\frac{dy}{dx} = \frac{-3}{8}$ at point $(1, 1)$.

4 Find the equation of the normal to the curve $y = \frac{4}{(3 - x)^3}$ at the point where $x = 1$.

5 Find the x-coordinates of all points on the curve $y = 1 - 3x^3 + 2x$ where the tangent to the curve is horizontal.

Investigation 9

1 Use the differentiation rules from section 5.2 to copy and complete the table below.

$f(x)$	Expand	$f'(x)$
x^3		
$x(x + 1)$		
$(x - 1)(x + 1)$		
$(2 - x)(3 - x^2)$		

The table below shows each function $f(x)$ from question **1** as the product of two functions $u(x)$ and $v(x)$, that is $f(x) = u(x) \times v(x)$.

2 Copy and complete the table below.

$f(x)$	$u(x)$	$u'(x)$	$v(x)$	$v'(x)$	$u(x)\,v'(x) + v(x)\,u'(x)$
x^3	x		x^2		
$x(x + 1)$	x		$(x + 1)$		
$(x - 1)(x + 1)$	$(x - 1)$		$(x + 1)$		
$(2 - x)(3 - x^2)$	$(2 - x)$		$(3 - x^2)$		

3 **Factual** For each function $f(x)$, compare the expression you found for $f'(x)$ in question **1** with the expression you found for $u(x)v'(x) + v(x)u'(x)$ in question **2**.

The result you discovered in question **3** is called the **product rule**.

4 **Conceptual** How is the product rule useful to help you differentiate the product of two polynomial functions?

The **product rule** states that if $f(x) = u(x)v(x)$, then $f'(x) = u(x)v'(x) + v(x)u'(x)$.

Another way of writing this is:

if $y = uv \Rightarrow \frac{dy}{dx} = u\frac{dv}{dx} + v\frac{du}{dx}$

Example 12

Use the product rule to find the derivative of these functions:

a $f(x) = (x^3 + 3x^2 + 6)(2x - 1)$ **b** $y = x^2\sqrt{x^2+1}$

a $f(x) = (x^3 + 3x^2 + 6)(2x - 1)$	
$u(x) = x^3 + 3x^2 + 6$ $v(x) = 2x-1$	Define $u(x)$ and $v(x)$.
$u'(x) = 3x^2 + 6x$ $v'(x) = 2$	Differentiate each separately.
$f'(x) = u(x)v'(x) + v(x)u'(x)$	Use the product rule.
$f'(x) = (x^3 + 3x^2 + 6)(2) + (2x - 1)(3x^2 + 6x)$ $= 2x^3 + 6x^2 + 12 + 6x^3 + 12x^2 - 3x^2 - 6x$ $= 8x^3 + 15x^2 - 6x + 12$	Simplify.
b $y = x^2\sqrt{x^2+1}$	
$u(x) = x^2$ $v(x) = \sqrt{x^2+1}$	Define $u(x)$ and $v(x)$.
$= (x^2+1)^{\frac{1}{2}}$	Write each term as x raised to a power so you can differentiate.
$u'(x) = 2x$ $v'(x) = \frac{1}{2}(x^2+1)^{\frac{-1}{2}}(2x)$ $v'(x) = \frac{x}{\sqrt{x^2+1}}$	Differentiate each separately. You need to use the chain rule for $v(x)$.
$y = uv \Rightarrow y' = uv' + vu'$	Use the product rule.
$y'(x) = (x^2) \times \frac{x}{\sqrt{x^2+1}} + \sqrt{x^2+1} \times (2x)$ $= \frac{x^3}{\sqrt{x^2+1}} + 2x\sqrt{x^2+1}$	Simplify.

Example 13

Find the equation of the normal to the curve

$y = \frac{\sqrt{x+1}}{(2x+1)^3}$, $x \neq \frac{-1}{2}$ at $(0, 1)$.

$y = \frac{\sqrt{x+1}}{(2x+1)^3} = \sqrt{x+1}(2x+1)^{-3}$	Change quotient to product.
$u(x) = \sqrt{x+1}$ $\quad v(x) = (2x+1)^{-3}$	Define $u(x)$ and $v(x)$.
$u'(x) = \frac{1}{2\sqrt{x+1}}$ $\quad v'(x) = -3(2)(2x+1)^{-4}$	Differentiate each term (using chain rule for v).
$y' = \frac{1}{2\sqrt{(x+1)}(2x+1)^3} - \frac{6\sqrt{(x+1)}}{(2x+1)^4}$	Use product rule.
at $x = 0 \Rightarrow y' = \frac{-11}{2}$	
Gradient of normal is $\frac{2}{11}$	Since normal and tangent are perpendicular.
Equation of normal: $y = \frac{2}{11}x + 1$	Using $y - y_1 = m(x - x_1)$ with $(0, 1)$

Exercise 5H

1 Differentiate each function with respect to x.

a $y = x^2(2x - 1)$

b $y = (2x - 3)(x + 3)^3$

c $y = x\sqrt{2-3x}$

d $y = (2x + 1)(x^2 - x + 1)^2$

e $y = \sqrt{(x+2)}(2-3x)$

2 $y = \sqrt{x+1}(3-x)^2$

a Show that $y' = \frac{(x-3)(5x+1)}{2\sqrt{x+1}}$.

b Find the x-coordinates of all points on the graph of y where the tangent to the curve is parallel to the x-axis.

3 Find the equation of the normal to the curve $y = \frac{x}{1-2x}$, $x \neq \frac{1}{2}$ at $(0, 0)$.

Investigation 10

As you saw in example 13, you can use the product rule to differentiate rational functions. However, there are times when it's easier not to change a quotient to a product, and so you also need to know the **quotient rule**, which allows you to differentiate a quotient directly.

In this investigation you will derive an expression for the quotient rule.

A function, $Q(x)$, can be written in the form $Q(x) = \frac{u(x)}{v(x)}$ where $v(x) \neq 0$.

TOK

Who do you think should be considered the discoverer of calculus?

1 Make $u(x)$ the subject of $Q(x) = \frac{u(x)}{v(x)}$.
2 Differentiate your expression for $u(x)$ using the product rule.
3 Make $Q'(x)$ the subject of your equation in question **2**.
4 Substitute $\frac{u(x)}{v(x)}$ for $Q(x)$.

Now you should have an expression for the quotient rule.

5 **Conceptual** Which method (the product rule or the quotient rule) would be more efficient to use for differentiating rational functions?

TOK

How can causal relationships be established in mathematics?

Quotient rule

If $f(x) = \frac{u(x)}{v(x)}$, $v(x) \neq 0$ then $f'(x) = \frac{v(x) \cdot u'(x) - u(x) \cdot v'(x)}{[v(x)]^2}$

Example 14

Use the quotient rule to differentiate $y = \frac{x^2 - 3}{3x - x^2}$.

$y = \frac{x^2 - 3}{3x - x^2}$	
$u(x) = x^2 - 3$ $\quad v(x) = 3x - x^2$	Define $u(x)$ and $v(x)$.
$u'(x) = 2x$ $\quad v'(x) = 3 - 2x$	Differentiate each term.
$y' = \frac{v(x)u'(x) - u(x)v'(x)}{v(x)^2}$	Use the quotient rule.
$y' = \frac{(3x - x^2)(2x) - (x^2 - 3)(3 - 2x)}{(3x - x^2)^2}$	
$y' = \frac{3x^2 - 6x + 9}{(3x - x^2)^2}$	Simplify. Although you may need to combine like terms in the numerator, you may leave the denominator in factored form.

Exercise 5I

1 Differentiate each expression using the quotient rule.

a $y = \frac{1 + 3x}{5 - x}$ **b** $y = \frac{\sqrt{x}}{2 - x}$

c $y = \frac{1 + 2x}{\sqrt{1 - x^2}}$ **d** $y = \frac{1 + 3x}{x^2 + 1}$

2 Find the equation of the normal to the curve $f(x) = \frac{3x - 2}{x^2 + 1}$ at the point where $x = 0$.

Calculus

3 Consider the "trident of Newton" curve below:

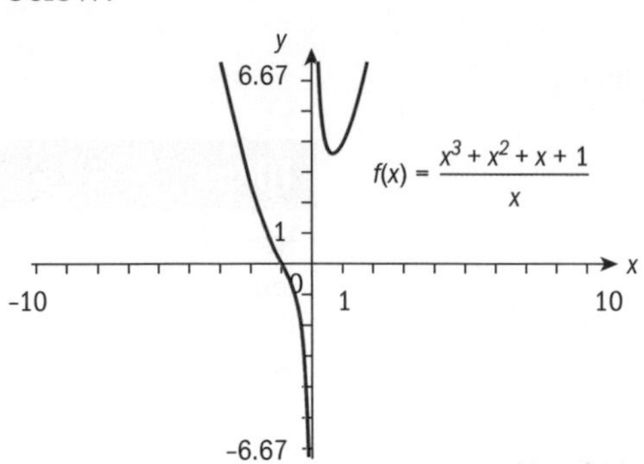

Find the value of x where the tangent to the curve is parallel to the line $y = x$.

Reflect If you are asked to differentiate an expression in an examination, how could you determine which of the three differentiation rules (chain, product and quotient rules) to use?

Exercise 5J

1 Differentiate each function with respect to x.

a $y = (x-1)(x+3)^2$

b $y = (x+1)\sqrt{1-2x}$

c $y = \dfrac{x+1}{x-1}$

d $y = \dfrac{2}{x^4 - 2x + 1}$

2 Find the equations of the tangent and the normal to the curve $f(x) = \dfrac{1+\sqrt{x}}{x-1}, x \neq 1$ at the point $(9, \frac{1}{2})$.

Did you know?

The "trident of Newton", which you encountered in exercise 5I question **3**, was the 66th curve in Newton's classification of cubics. Can you identify why the curve has this name? Research how this, and other famous curves, have been used to model real-life situations.

5.4 Graphical interpretation of first and second derivatives

Increasing and decreasing functions

A function is **increasing** on an interval if as the x-value *increases* the y-value *increases*. A function is **decreasing** on as the x-value *increases* the y-value *decreases*.

Investigation 11

The diagram shows the graph of $f(x) = x^3 - 6x^2 + 2$.

1 Differentiate f and find the value of $f'(x)$ at each of the x-values given in the table below. Copy and complete the table to show your answers.

x	−2	−1	0	1	3	4	5
$f'(x)$							

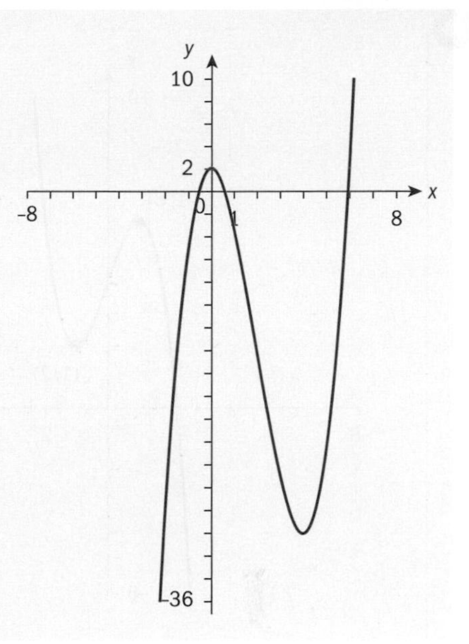

2 What do you notice about the sign of the gradient of $f(x)$ when $x < 0$? Is $f(x)$ increasing or decreasing for $x < 0$?

3 What do you notice about the sign of the gradient of $f(x)$ when $0 < x < 4$? Is $f(x)$ increasing or decreasing for $0 < x < 4$?

4 What do you notice about the sign of the gradient of $f(x)$ when $x > 4$? Is $f(x)$ increasing or decreasing for $x > 4$?

5 **Conceptual** What do the signs of the gradient of $f(x)$ tell you about whether $f(x)$ is increasing or decreasing on a given interval?

If $f'(x) > 0$ for all $x \in]a, b[$ then f is increasing on $]a, b[$.
If $f'(x) < 0$ for all $x \in]a, b[$ then f is decreasing on $]a, b[$.

Sign diagrams for the gradient function can be used to determine the intervals where a function is increasing or decreasing. You will learn how to use these in the following examples.

Example 15

Find the intervals where each function is increasing or decreasing:

a $f(x) = x^3 - 3x + 4$ **b** $y = x^2 - 1$

a $f(x) = x^3 - 3x + 4$

$f'(x) = 3x^2 - 3$ — Find $f'(x)$.

$f'(x) = 3(x - 1)(x + 1)$

$f'(x) = 0 \Rightarrow x = \pm 1$ — Finding the points where $f'(x) = 0$ allows you to find the boundaries of the intervals where f is increasing or decreasing.

x	$x < -1$	$-1 < x < 1$	$x > 1$
Sign of $f'(x)$	$f'(-2) > 0$ +	$f'(0) < 0$ −	$f'(2) > 0$ +
$f(x)$	increasing	decreasing	increasing

Test the values to the left and right of the roots of $f'(x)$.

The function is increasing on $]-\infty, -1[\cup]1, \infty[$.

The function is decreasing on $]-1, 1[$.

Identify the intervals where f is increasing or decreasing.

We could also say that the function is increasing for $x < -1$ and $x > 1$, and it is decreasing for $-1 < x < 1$.

Continued on next page

Calculus

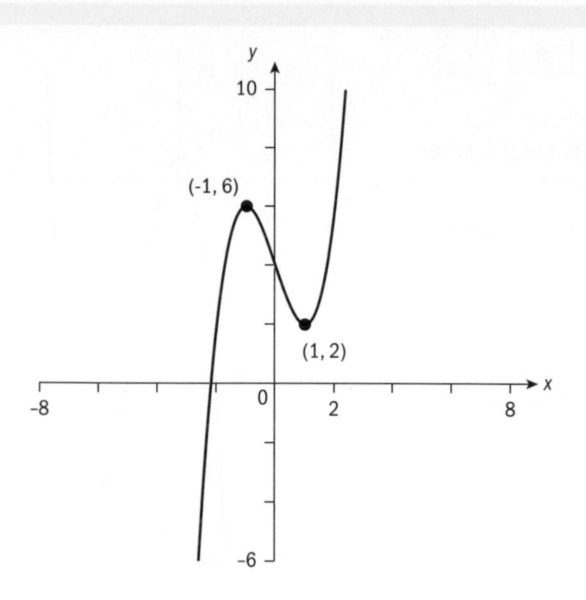

Check your answer graphically.

b $y = x^2 - 1$

$y' = 2x$

$y' = 0 \Rightarrow x = 0$

Find y'.

Find the root of $y' = 0$.

Test the values to the left and right of the root of y'.

x	$x < 0$	$x > 0$
Sign of $f'(x)$	$f'(-1) < 0$	$f'(1) > 0$
$f(x)$	decreasing	increasing

The function is increasing on $]0, \infty[$.

The function is decreasing on $]-\infty, 0[$.

Identify the intervals where f is increasing or decreasing.

Check your answer graphically.

The answer could also be given as the function is increasing for $x < 0$, and increasing for $x > 0$.

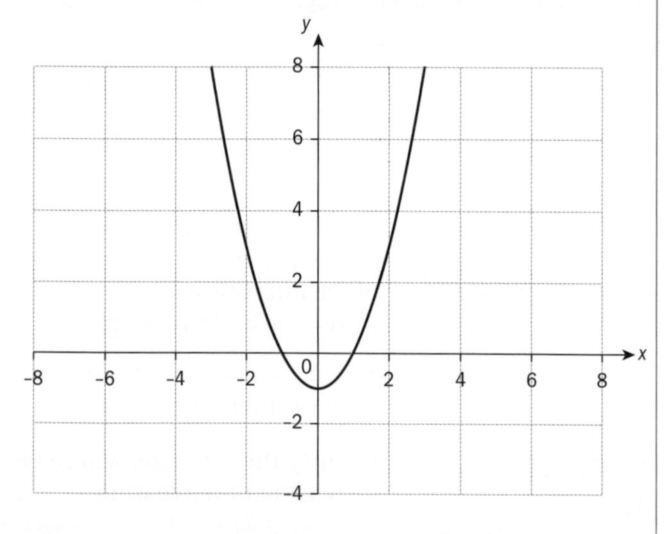

Exercise 5K

1 For each function **a–d**, write down the intervals where the functions are:

i increasing **ii** decreasing.

a

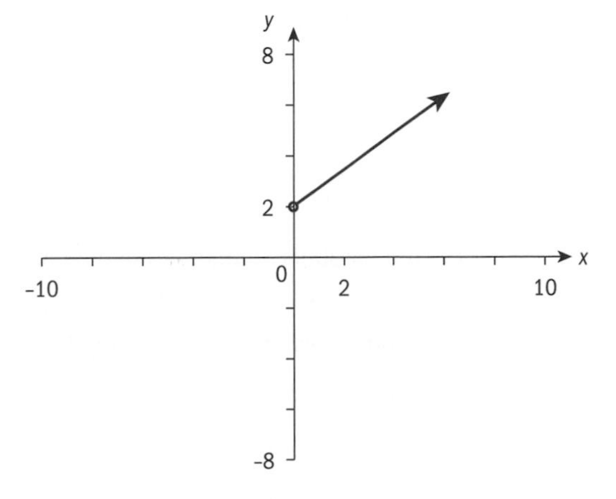

b

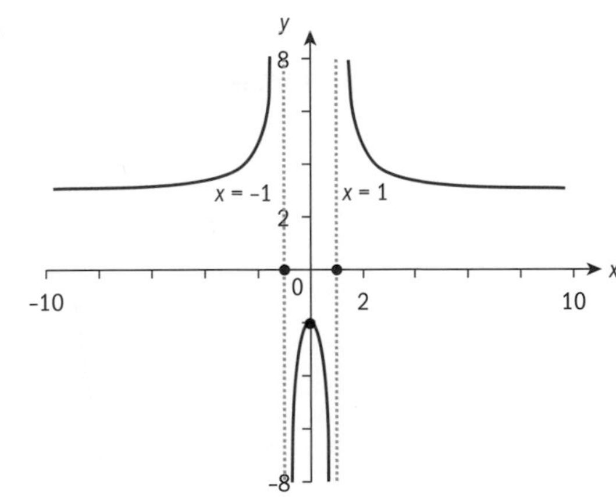

c

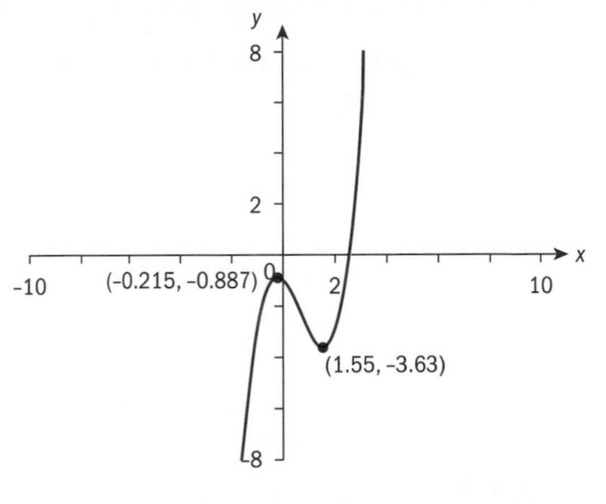

d

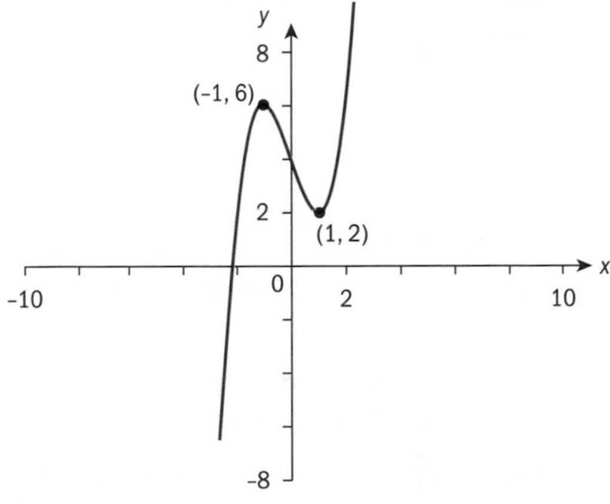

2 Find the intervals where $f(x)$ is increasing or decreasing.

a $f(x) = -x^3$

b $f(x) = 2x^2 - 8$

c $f(x) = \frac{1}{\sqrt{x-1}}$

d $f(x) = \sqrt{x} - 2x$

Investigation 12

Consider the graph of a quadratic function whose leading coefficient is positive, that is, $a > 0$.

For $a > 0$, the vertex is a minimum point.

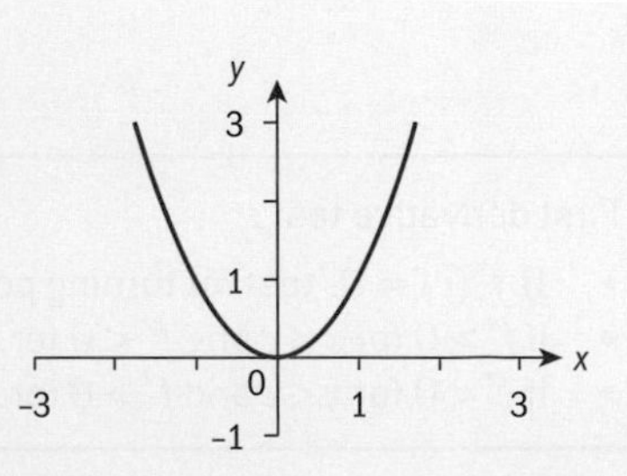

1 What is the gradient of the parabola at the minimum point?

2 What sign do the gradients left of the minimum point on the parabola have?

3 What sign do the gradients right of the minimum point on the parabola have?

4 State the subset of the domain where the function is **i** increasing and **ii** decreasing.

Continued on next page

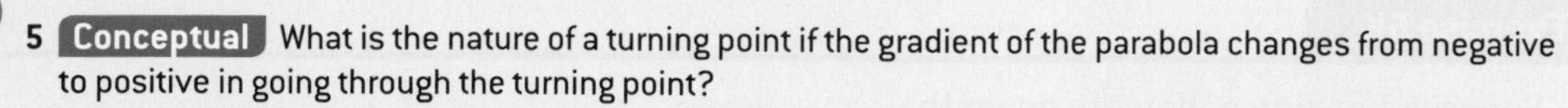

5 **Conceptual** What is the nature of a turning point if the gradient of the parabola changes from negative to positive in going through the turning point?

Consider the graph of a quadratic function whose leading coefficient is negative, that is, $a < 0$.

For $a < 0$, the vertex is a maximum point.

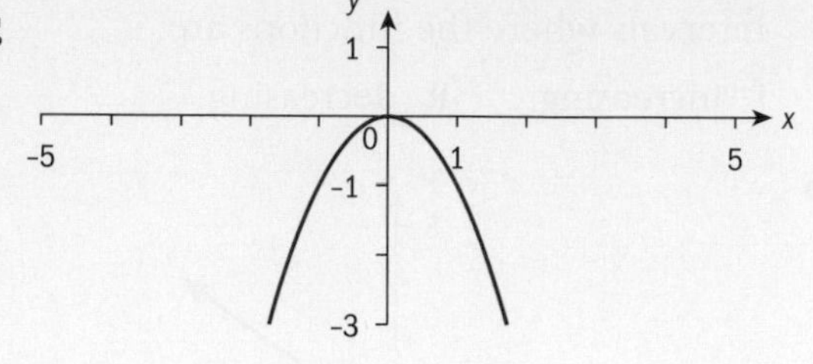

6 What is the gradient of the parabola at the maximum point?

7 What sign do the gradients left of the maximum point on the parabola have?

8 What sign do the gradients right of the maximum point on the parabola have?

9 **Conceptual** What is the nature of a turning point if the gradient of the parabola changes from positive to negative in going through the turning point?

10 **Factual** In the interval where the gradients are negative, is the function increasing or decreasing?

11 **Factual** In the interval where the gradients are positive, is the function increasing or decreasing?

12 State the subset of the domain where the function is **i** increasing and **ii** decreasing.

13 **Conceptual** How can you identify the intervals on which a function is increasing/decreasing using the first derivative test?

14 **Conceptual** How does the first derivative test help in classifying local extrema and identifying intervals where a function is increasing/decreasing?

A is a **global minimum:** at A, the value of y is the least it attains on the entire domain.

B is a **local maximum:** it is a turning point where the gradient function has a positive sign to the left of B and a negative sign to the right of B.

C is a **local minimum:** it is a turning point where the gradient function has a negative sign to the left of C and a positive sign to the right of C.

D is a **global maximum:** at D, the value of y is the greatest it attains on the entire domain.

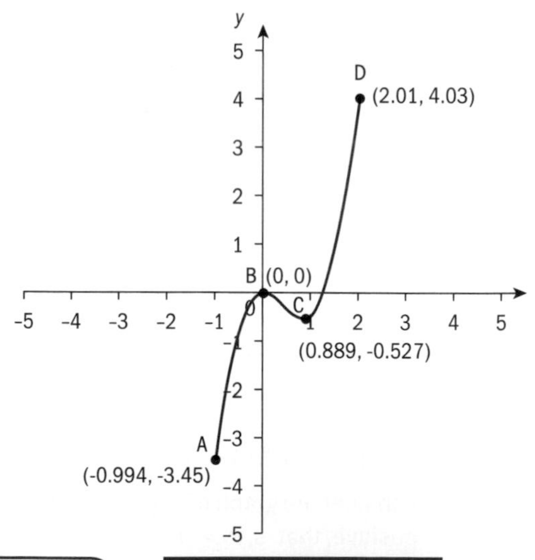

First derivative test :

- If $f'(c) = 0$, test for turning points.
- If $f' > 0$ for $x < c$ and $f' < 0$ for $x > c$, then f has a local maximum at $x = c$.
- If $f' < 0$ for $x < c$ and $f' > 0$ for $x > c$, then f has a local minimum at $x = c$.

TOK

Does the fact that Leibniz and Newton came across the Calculus at similar times support the argument of Platonists over Constructivists?

Example 16

Consider the function $f(x) = 4 - 3x^2 + x^3$ for $-2 \leq x \leq 3$.

a Find and classify the nature of any turning points.

b State the intervals for which the function is increasing or decreasing.

c State if the points found in part **a** are local or global extrema.

a $f(x) = 4 - 3x^2 + x^3$

$f'(x) = -6x + 3x^2$

$f'(x) = -3x(2 - x)$

$f'(x) = 0$ when $x = 0$ and $x = 2$

Find $f'(x)$.

x	$x < 0$	$0 < x < 2$	$x > 2$
Sign of $f'(x)$	$f'(-1) > 0$	$f'(1) < 0$	$f'(3) > 0$
$f(x)$	increasing	decreasing	increasing

Use the first derivative test to identify the nature of the turning points.

$f(0) = 4 \Rightarrow (0, 4)$ and

$f(2) = 0 \Rightarrow (2, 0)$

(0, 4) is a local maximum point

(2, 0) is a local minimum point.

Solve $f'(x) = 0$ to find the possible turning points.

Find the y-coordinates of the possible turning points.

Test the values of $f'(x)$ to the left and right of the possible turning points.

b f is increasing for $x \in [-2, 0[\cup]2, 3]$.

f is decreasing for $x \in]0, 2[$.

$0 < x \leq -2$ or $2 < x \leq 3$.
When $f' > 0 \Rightarrow f$ is increasing.

$0 < x < 2$. When $f' < 0 \Rightarrow f$ is decreasing.

c $f(-2) = 4 - 3(-2)^2 + (-2)^3$

$f(-2) = -16$

$f(-2) < f(2) \Rightarrow (-2, -16)$ is the global minimum.

$f(3) = 4 - 3(3)^2 + (3)^3$

$f(3) = 4$

$f(0) = f(3)$ and hence is both a local *and* global maximum $(2, 0)$ is a local minimum point.

Check the y-values for the end points of the domain to see if they are greater or less than the y-values at the turning points.

Example 17

Use the derivative to show that the graph of $y = \frac{x-5}{x}$ has no turning points and state the interval(s) where y is increasing/decreasing. Then, sketch the graph of the function.

$\frac{dy}{dx} = \frac{5}{x^2}$.

Write $y = 1 - \frac{5}{x}$ and differentiate using the power rule, or use the quotient rule.

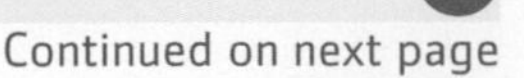

Continued on next page

Calculus

Since $\frac{5}{x^2} \neq 0$ for all x, y has no turning points.	$\frac{dy}{dx} \neq 0$
Since $\frac{5}{x^2} > 0$ for all x, y is increasing throughout its domain.	Identify the intervals where f is increasing or decreasing.
$y = \frac{x-5}{x}$ has a vertical asymptote at $x = 0$ and a horizontal asymptote at $y = 1$. 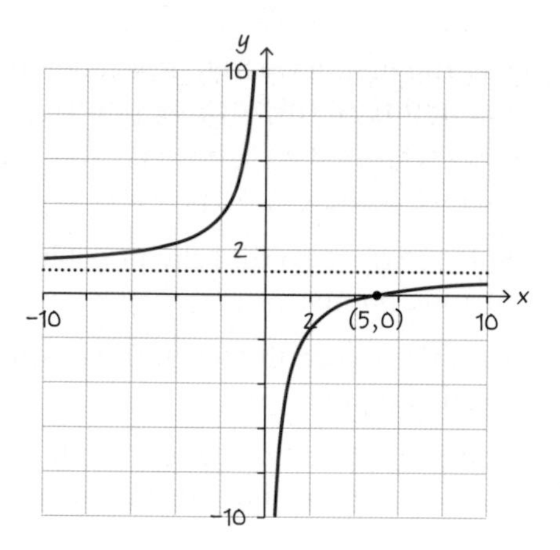	A vertical asymptote occurs at a value of x for which the function is not defined – here where the denominator is zero. A horizontal asymptote occurs at the limit of the function as $x \to \pm\infty$.

Exercise 5L

1 For each of the following functions, find and classify any turning points. Confirm your answers graphically.

a $f(x) = x^2 - 2$

b $f(x) = x - 2\sqrt{x}$

c $f(x) = x^3 - 6x^2 + 2$

2 The function $f(x) = ax^3 + 2x^2 - 6$ has a local maximum at $x = 1$. Find the value of a.

3 The cubic function $p(x) = ax^3 + bx^2 + cx + d$ has a tangent with equation $y = 3x + 1$ at the point $(0, 1)$ and has a turning point at $(-1, -3)$. Find the values of a, b, c and d.

4 The graph of the function $y = x^3 + ax^2 + b$ has a local minimum point at $(4, -11)$. Find the coordinates of the local maximum.

Second derivative

Given a function $f(x)$, the derivative $f'(x)$ is known as the **first derivative** of $f(x)$. The gradient of the first derivative, $f''(x)$, is called the **second derivative** of $f(x)$.

You can also use $\frac{dy}{dx}$ notation where $f'(x) = \frac{dy}{dx}$, and $f''(x) = \frac{d^2y}{dx^2}$.

Investigation 13

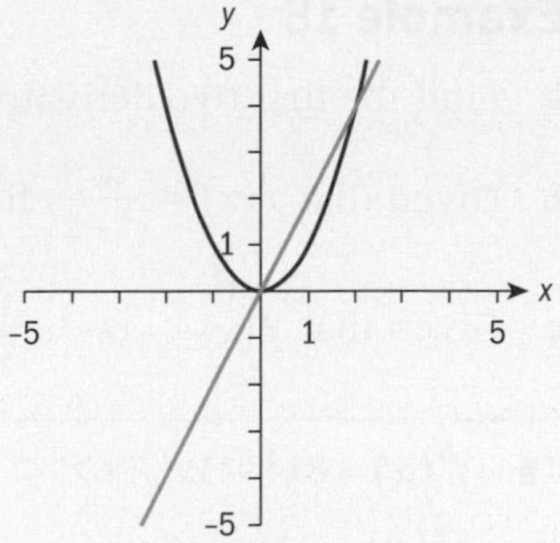

Use the graph of $f(x) = x^2$ and the graph of its derivative to answer the following questions.

1 State the change in the signs of the gradients of f as you move from the points left of the minimum point, through the minimum, then right of the minimum point. Explain how your answer is demonstrated in the graph of $f'(x)$.

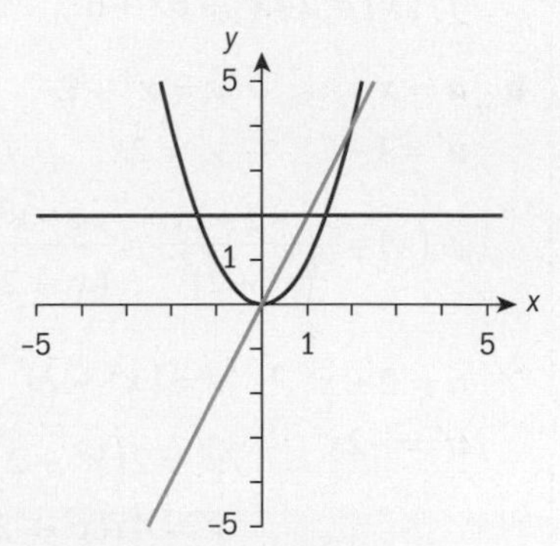

2 The pink horizontal line in the second graph shows $f''(x)$, the derivative of $f'(x)$.

What is the sign of the gradient of $f'(x)$ at any value of x? In particular, what is the sign of the second derivative at the value of x where $f(x)$ has its minimum point?

3 When the second derivative of a function is positive, is the *gradient* of $f(x)$ increasing or decreasing as you go from left to right?

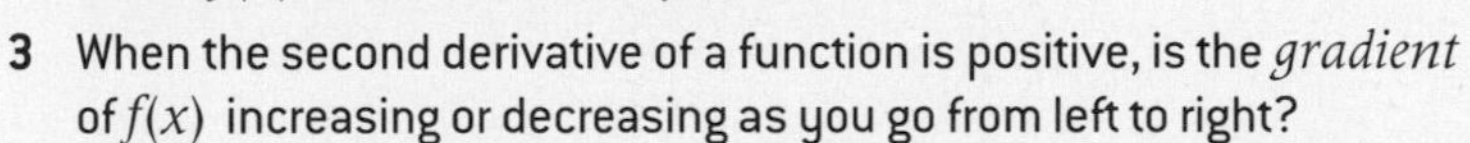

4 Sketch the graph of $f(x) = -x^2$ together with its first two derivative functions. Answer questions **1** and **2** for this function.

5 When the second derivative of a function is negative, is the *gradient* of $f(x)$ increasing or decreasing as you go from left to right?

6 Summarize your findings by stating what you think the sign of the second derivative of a function will be at a minimum point, and at a maximum point.

7 Analyse the graph of $f(x) = 3 + x + \frac{1}{x}$ and its first two derivative functions. Do your findings from question **6** still hold true?

In this investigation, you have discovered the second derivative test for classifying local extrema.

8 **Conceptual** How is the second derivative useful in classifying local extrema?

9 Consider now the graph of $f(x) = x^4$.

Graph its first two derivatives. Does your conjecture work for this function? Why, or why not?

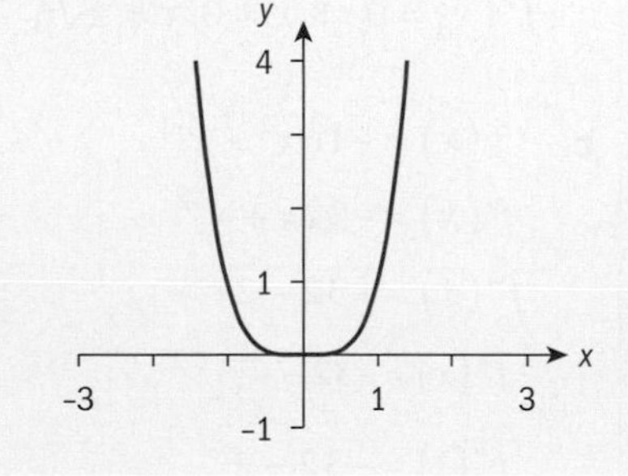

10 **Conceptual** How can you determine the nature of a turning point when the second derivative test is inconclusive?

Second derivative test for maxima and minima

If $f'(c) = 0$ and $f''(c) < 0$ then $f(x)$ has a local maximum at $x = c$.

If $f'(c) = 0$ and $f''(c) > 0$ then $f(x)$ has a local minimum at $x = c$.

If $f'(c) = 0$ and $f''(c) = 0$ then the second derivative test is inconclusive. If the second derivative test is inconclusive, revert to the first derivative test for local extrema.

Example 18

a Find the first two derivatives of $f(x) = 2x^4 - x^3 + 3x^2 + 1$.

b Given that $f(x) = \frac{x}{x^2+2}$, find the value of x for which $f''(x) = 0$.

c Given that $f(x) = -16x^2 - \frac{1}{x}$, find $f''(1)$.

a $f'(x) = 8x^3 - 3x^2 + 6x$	Differentiate f to find f'.
$f''(x) = 24x^2 - 6x + 6$	Differentiate f' to find f''.
b $u = x \quad v = x^2 + 2$ $u' = 1 \quad v' = 2x$	Find the first derivative using the quotient rule.
$f'(x) = \frac{x^2+2-2x^2}{(x^2+2)^2} = \frac{2-x^2}{(x^2+2)^2}$	
$u = 2 - x^2 \quad v = (x^2+2)^2$ $u' = -2x \quad v' = 2(x^2+2)(2x)$ $v' = 4x(x^2 + 2)$	Find the second derivative by using the quotient rule again.
$f''(x) = \frac{-2x(x^2+2)^2 - 4x(x^2+2)(2-x^2)}{(x^2+2)^4}$	You also need to use the chain rule to find v'.
$f''(x) = \frac{-2x(x^2+2)(x^2+2+4-2x^2)}{(x^2+2)^4}$	
$= \frac{-2x(x^2+2)(-x^2+6)}{(x^2+2)^4}$	
$f''(x) = 0 \Rightarrow x = 0, x = \pm\sqrt{6}$	Solve $f''(x) = 0$. Note that there are no real solutions to $x^2 + 2 = 0$.
c $f(x) = -16x^2 - x^{-1}$	Rewrite the function so you can use the power rule.
$f'(x) = -32x + x^{-2}$	Find the first derivative.
$f''(x) = -32 - 2x^{-3}$ $f''(x) = -32 - \frac{2}{x^3}$	Find the second derivative.
$f''(1) = -32 - 2$ $f''(1) = -34$	Evaluate the second derivative when $x = 1$.

In example 16, you found the turning points on the graph of $f(x) = 4 - 3x^2 + x^3$, and used the first derivative test to justify whether they were minima or maxima.

Here, you will use the second derivative test to justify the nature of each turning point.

Example 19

Use the second derivative test to determine the nature of the turning points on the graph of $f(x) = 4 - 3x^2 + x^3$.

In example 16, you found that the turning points were (0, 4) and (2, 0). $f'(x) = -6x + 3x^2$ $f''(x) = -6 + 6x$	Find the second derivative.
The turning points are: $f''(0) = -6 + 6(0) = -6 < 0 \Rightarrow (0, 4)$ is a maximum point $f''(2) = -6 + 6(2) = 6 > 0 \Rightarrow (2, 0)$ is a minimum point	Use the second derivative test to determine the nature of the points.

Exercise 5M

1 Find the second derivative of $f(x) = 2x^{\frac{5}{2}}$.

2 Given $f(x) = \frac{x^4}{12} - 2x^2 + 5x + 1$, find the values of x for which $f''(x) = 0$.

3 Find the gradient function $f(x) = \sqrt{5 - 4x}$.

4 Given that $y = (1 - ax)^2$ and $\frac{d^2y}{dx^2} = 8$, find the value(s) of a.

Investigation 14

Sketch the function $f(x) = x^3 - 3x + 1$.

1 Find the first and second derivatives of f and sketch their graphs.

There is a point on the graph of f where the first derivative has a value of 0.

2 Write down the x-coordinate of this point and state the gradient of f' at this point.

3 Does the sign of the gradient of f change as you move from left to right through this point?

4 State the value of f'' at this point.

5 Describe the concavity of f to the left and to the right of this point.

6 State the signs of f'' to the left and to the right of this point on f.

7 Repeat steps **1–8** for $y = -f(x)$.

A point where the graph of a function is continuous and where the concavity changes is a point of inflexion.

Continued on next page

HINT

Points where the first derivative equals zero are sometimes called *stationary points*.

8 **Conceptual** Why does the sign of a function's second derivative at any point indicate the concavity of the function at that point?

9 Conjecture the value of the second derivative of a function at a point of inflexion.

10 Does the converse hold? In other words, if the second derivative of a function a point c is the value you obtained in question **9**, is c a point of inflexion of the function? To test this, use the function $f(x) = x^4$.

11 **Conceptual** Analyse the concavity of $f(x) = x^3 - 3x + 1$ on both sides of the point of inflexion, and the concavity of $f(x) = x^4$ on both sides of its turning point. Can you determine the additional condition necessary for f to have a point of inflexion at $x = c$.

A point of inflexion corresponds to a change in concavity of the graph of f.

If f has a point of inflexion at $x = c$, then $f''(c) = 0$.

If $f''(c) = 0$ **and** f changes concavity at $x = c$, then f has a point of inflexion at $x = c$.

f is concave down in an interval if, for all x in the interval, $f''(c) < 0$.

f is concave up in an interval if, for all x in the interval, $f''(c) > 0$.

HINT

Points of inflexion can also be called **inflection points.**

Example 20

Consider the function $f(x) = 2x^4 - 4x^2 + 1$.

a Find all turning points, and determine their nature (you should justify your answers).

b Find the intervals where f is **i** increasing and **ii** decreasing.

c Find the intervals where the graph of f is **i** concave up and **ii** concave down.

d Sketch the graph of f, indicating any maxima, minima and points of inflexion.

a $f'(x) = 8x^3 - 8x$. $8x^3 - 8x = 8x(x^2 - 1) = 0 \Rightarrow x = 0, \pm 1$	Set $f'(x) = 0$ to find any possible turning points.
$f''(x) = 24x^2 - 8$ $f''(0) = -8 \Rightarrow f$ has a local maximum at $x = 0$. $f''(-1) = 16 \Rightarrow f$ has a local minimum at $x = -1$. $f''(1) = 16 \Rightarrow f$ has a local minimum at $x = 1$.	Use the second derivative test to find the nature of the turning points.
Turning points are $(0, 1)$, $(-1, -1)$ and $(1, -1)$.	You can find the y-coordinates of each point by substituting the x-values into the original function.

b Sign diagram

x	$x<-1$	$-1<x<0$	$0<x<1$	$x>1$
Sign of f'	−	+	−	+
Behaviour of f	decreasing	increasing	decreasing	increasing

Hence, **i** f is decreasing in the intervals $]-\infty,-1[\cup]0,1[$ and **ii** f is increasing in the intervals $]-1,0[\cup]1,\infty[$.

We could also say, f is decreasing when $x<-1$ and when $0<x<1$. f is increasing when $-1<x<0$ and when $x>1$.

c $24x^2-8=0 \Rightarrow x=\pm\frac{1}{\sqrt{3}}$

Set $f''=0$ to find any possible points of inflexion.

x	$x<-\frac{1}{\sqrt{3}}$	$-\frac{1}{\sqrt{3}}<x<\frac{1}{\sqrt{3}}$	$x>\frac{1}{\sqrt{3}}$
Sign of f''	+	−	+
Concavity of f	concave up	concave down	concave up

f is concave up on the interval

$$\left]-\infty,-\frac{1}{\sqrt{3}}\right[\cup\left]\frac{1}{\sqrt{3}},\infty\right[.$$

f is concave down on the interval $\left]-\frac{1}{\sqrt{3}},\frac{1}{\sqrt{3}}\right[$.

d

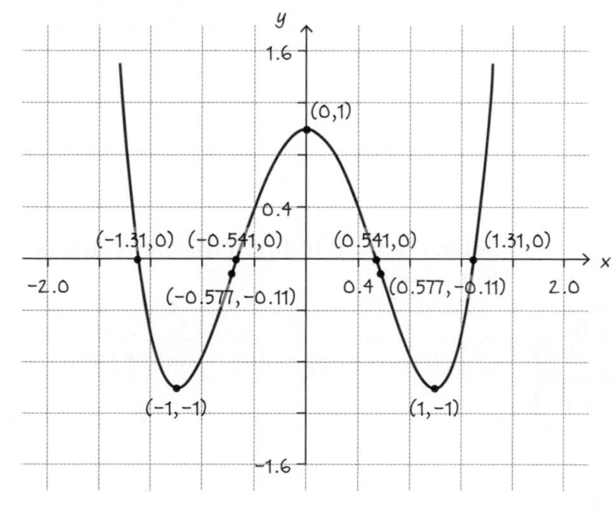

Although the question does not ask for zeros in the sketch, you should indicate on the graph where they are, and if possible, indicate their value.

Calculus

Investigation 15

Consider the function $y = x^3$.

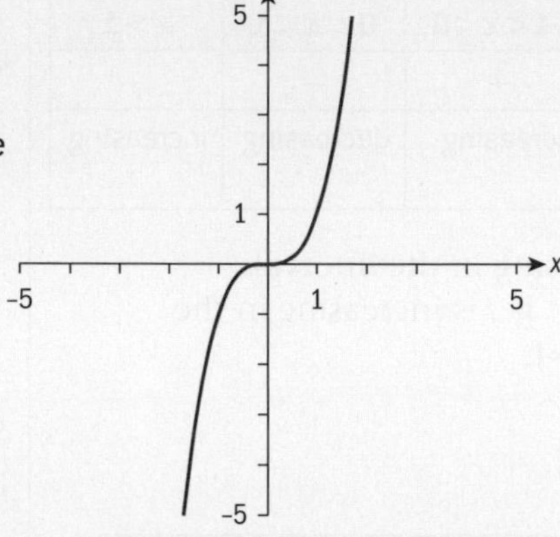

1. Explain why the graph has a point of inflexion at (0, 0).
2. What is the value of the first derivative of the function at (0, 0)?
3. What are the signs of $f'(x)$ to the left and right of (0, 0)?
4. If (0, 0) were a turning point, what would happen to the signs of $f'(x)$ as you move along the graph from left to right through (0, 0)?
5. What is the nature of a point where the $f'(x) = 0$ if the sign of $f'(x)$ does not change in going through the point whose second derivative is 0?

This kind of point of inflexion is called a horizontal point of inflexion.

6. **Conceptual** What three conditions are necessary for a function $f(x)$ to have a horizontal point of inflexion at $x = c$?

If f has a point of inflexion at $x = c$ and $f'(c) = 0$, the point is called a **horizontal point of inflexion** since its gradient is parallel to the x-axis.

HINT

A **horizontal point of inflexion** can also be called a **stationary point of inflexion** or a **stationary inflection point.**

Example 21

Find and classify all points of inflexion of $f(x) = 3x^4 + 4x^3 - 2$.

$f(x) = 3x^4 + 4x^3 - 2$

$f'(x) = 12x^3 + 12x^2$

$f''(x) = 36x^2 + 24x$

$f''(x) = 12x(3x + 2)$

$f''(x) = 0 \Rightarrow x = 0$ or $x = \frac{-2}{3}$

Solve $f''(x) = 0$ to find possible points of inflexion.

x	$x < \frac{-2}{3}$	$\frac{-2}{3} < x < 0$	$x > 0$
$f''(x)$	$f''(x) > 0$	$f''(x) < 0$	$f''(x) > 0$

Check the change of the sign of $f''(x)$.

Since $f''(x)$ changes sign at $x = 0$ and $x = \frac{-2}{3}$,

f has a point of inflexion at these two points.

Find the y-values at the points of inflexion.

$f(0) = -2$ and $f\left(\frac{-2}{3}\right) = \frac{-70}{27}$

Hence, $(0, -2)$ and $\left(\frac{-2}{3}, \frac{-70}{27}\right)$ are points of inflexion. $f'(0) = 0$, therefore $(0, -2)$ is a horizontal point of inflexion. $f'\left(\frac{-2}{3}\right) \neq 0$, therefore $\left(\frac{-2}{3}, \frac{-70}{27}\right)$ is a non-horizontal point of inflexion.	Find the value of $f'(x)$ at each point of inflexion in order to determine whether each is a horizontal or non-horizontal point of inflexion.

Example 22

Find and classify any turning points and points of inflexion of $y = (x+1)(x-3)^3$ and justify your answers. Confirm your answers graphically.

$\frac{dy}{dx} = 4x(x-3)^2$ $\frac{dy}{dx} = 0 \Rightarrow x = 0, x = 3$	Differentiate using the product and chain rules (or by multiplying out), set $\frac{dy}{dx} = 0$, and solve.
$y(0) = -27$, $y(3) = 0$	Find y.
$\frac{d^2y}{dx^2} = 12(x-3)(x-1)$	Again, use the chain and product rules.
$\frac{d^2y}{dx^2} = 36 > 0$ at $x = 0$, hence f has a local local minimum at $(0, -27)$. $\frac{d^2y}{dx^2} = 0$ at $x = 3$, hence the second derivative test is inconclusive.	Apply the second derivative test.
$\frac{d^2y}{dx^2} = -2.28 < 0$ at $x = 2.9$ $\frac{d^2y}{dx^2} = 2.52 > 0$ at $x = 3.1$ Since f changes concavity at $x = 3$ and $f'(3) = f''(3) = 0$, $(3, 0)$ is a horizontal point of inflexion.	Test for a change in concavity of f at $x = 3$. $\frac{d^2y}{dx^2}$ is negative for small values to the left of $x = 3$ and positive for small values to the right of $x = 3$.

Example 23

$f(x) = 2x^3 + x^4$

a Find all turning points and points of inflexion; determine their nature and justify your answers.

Continued on next page

Calculus

b Find the intervals where the function is **i** concave up and **ii** concave down.

c Sketch the function, indicating any maxima, minima and points of inflexion.

a $f'(x) = 6x^2 + 4x^3$

$6x^2 + 4x^3 = 0 \Rightarrow x = 0; x = -\frac{3}{2}$

$f''(x) = 12x + 12x^2$

Set the first derivative equal to zero and solve for x.

$f''\left(-\frac{3}{2}\right) = 9 > 0 \Rightarrow f$ has a local min at $x = -\frac{3}{2}$.

$f''(0) = 0 \Rightarrow$ at $x = 0$, test is inconclusive

Evaluate the second derivative at both points.

x	$x < 0$	$x > 0$
$f'(x)$	+	+

Use the first derivative test to determine that $x = 0$ is not an extrema.

Since the first derivative does not change sign at $x = 0$, $(0, 0)$ is a possible point of inflexion.

To be sure it is a point of inflexion, we need to check that the concavity of the graph of f changes at $x = 0$.

b $f''(x) = 12x + 12x^2$

$12x + 12x^2 = 0 \Rightarrow x = 0, -1$

x	$x < -1$	$-1 < x < 0$	$x > 0$
Sign of f''	+	−	+
Concavity of f	up	down	up

Consider the sign of $f''(x)$ in order to determine concavity change.

f is concave up on $]-\infty, -1[$ and $]0, \infty[$; and concave down on $]-1, 0[$.

Since $f''(0) = 0$, and concavity changes at $(0, 0)$, hence $(0, 0)$ is a horizontal point of inflexion.

c

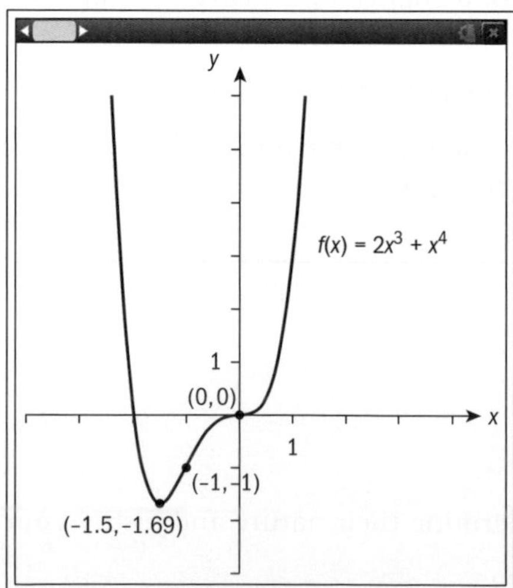

International-mindedness

The Greeks' mistrust of zero meant that Archimedes' work did not lead to calculus.

Exercise 5N

For each function in questions **1–8**, find:

a any points of inflexion

b the intervals where the function is concave up

c the intervals where the function is concave down

Justify your answers.

1 $y = x^3 - x$ **2** $y = x^4 - 3x + 2$

3 $y = x^3 - 6x^2 - 12x + 2$ **4** $y = x^3 + x^2 - 1$

5 $y = 4x^3 - x^4$ **6** $y = x^3 - 3x^2 + 3x - 1$

7 $y = 2x^4 + x^3 + 1$ **8** $f(x) = x^4 - 4x^3 + 16x - 16$

9 For each of the following functions, find and classify all points of inflexion.

a $f(x) = x^3 + 2x^2 + 1$ **b** $f(x) = (x - 1)^3$

c $f(x) = -3x^4 - 8x^3 + 2$ **d** $f(x) = 3 - \frac{1}{\sqrt{x}}$

10 For each of the following functions:

i Find and classify all points where the first derivative is equal to zero.

ii Find and classify all points of inflexion.

iii Find intervals where the function is increasing or decreasing.

iv Find intervals where the function is concave up or concave down.

a $f(x) = x^3 - 3x^2 - 6x + 1$

b $f(x) = (x - 1)^2$

c $f(x) = 3x^4 + 4x^3 - 2$

Curve sketching

Investigation 16

1 Copy the graph of $y = f(x)$, shown on the right.

2 On the same set of axes, mark points which show the value of the $f'(x)$ for points A, B, C and D, labelling them A′, B′, C′ and D′.

3 What are intervals in which f is **i** increasing and **ii** decreasing? What does this tell you about the sign of f' in each of these intervals?

4 Use your answer to question **3** to help you complete a sketch of the graph of f'.

5 What key points of f do the maximum and minimum points of f' correspond to? Label these points as E, F and G on your sketch of the graph of f'.

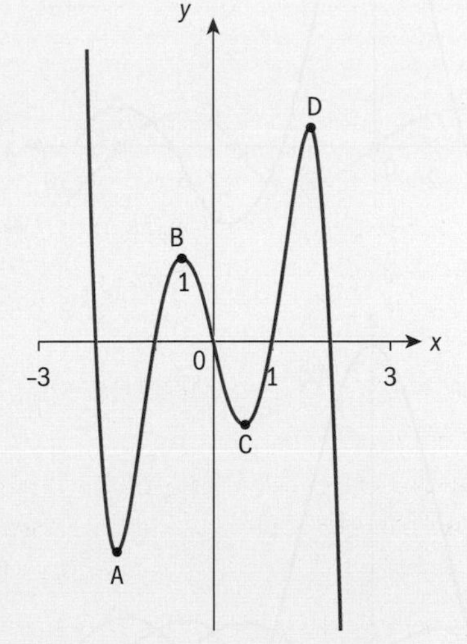

6 What are the intervals for which f is **i** concave up and **ii** concave down? What does this tell you about the values of f'' in these intervals?

7 On the same axes, mark points which show the value of the $f''(x)$ at points E, F and G. Label these points E′, F′ and G′. Complete a sketch of the graph of f''.

8 **Conceptual** Describe how you can sketch the graphs of $y = f'(x)$ and $y = f''(x)$ from the graph of $y = f(x)$.

When asked to make a sketch of $y = f'(x)$ or $y = f''(x)$ from the graph of $y = f(x)$, you do not need to draw exact graphs. However, you should label your graph with the following information, where possible.

- axes
- x-intercepts
- turning points and points of inflexion
- asymptotes.

Example 24

The following diagram shows the graph of $y = f(x)$. Copy the graph, and sketch the graphs of the first and second derivatives of f on the same set of axes.

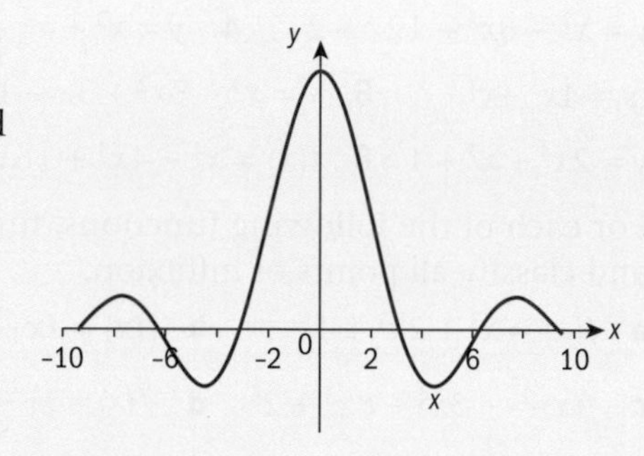

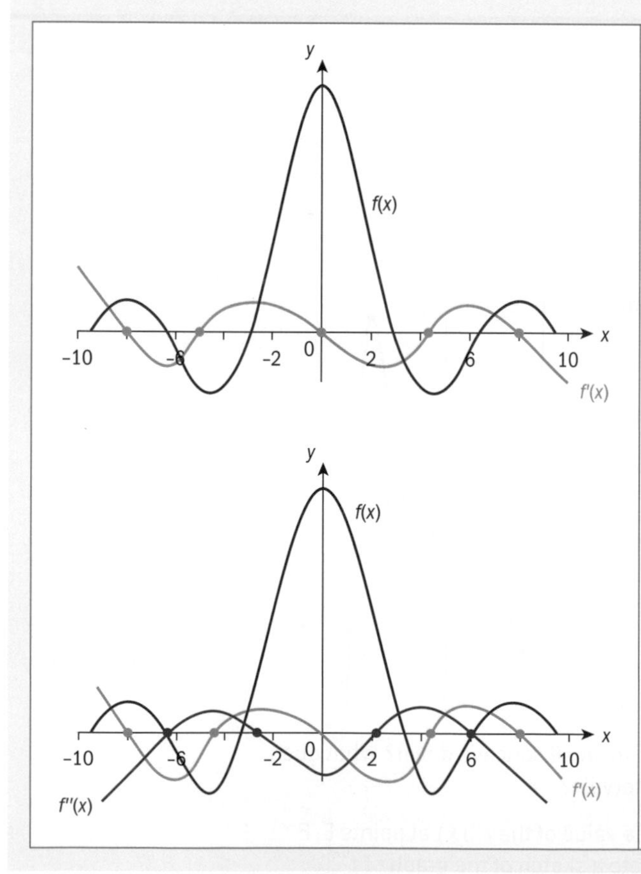

- The turning points of f are the zeros of f'.
- The intervals in which f is increasing tell you where the value of f' is positive, and the intervals where f is decreasing tell you where the value of f' is negative.
- The points of inflexion of f correspond to the turning points of f'.

- The turning points of f' are the zeros of f''.
- Where f is concave up, f'' is positive; where f is concave down, f'' is negative.

Exercise 50

1 Copy the graph of each function given. On the same set of axes, sketch the graphs of the first and second derivatives of the function.

a

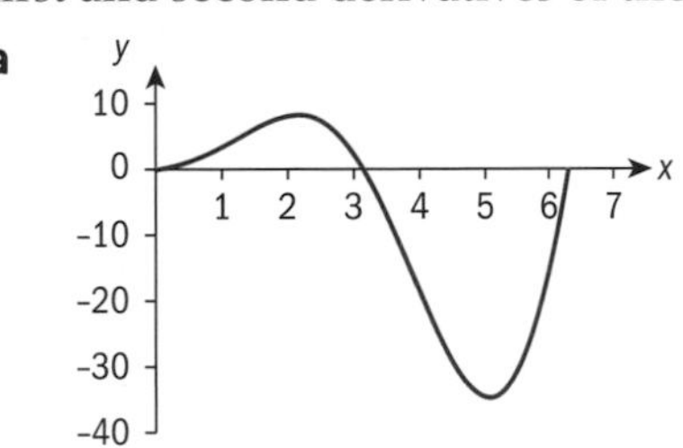

b

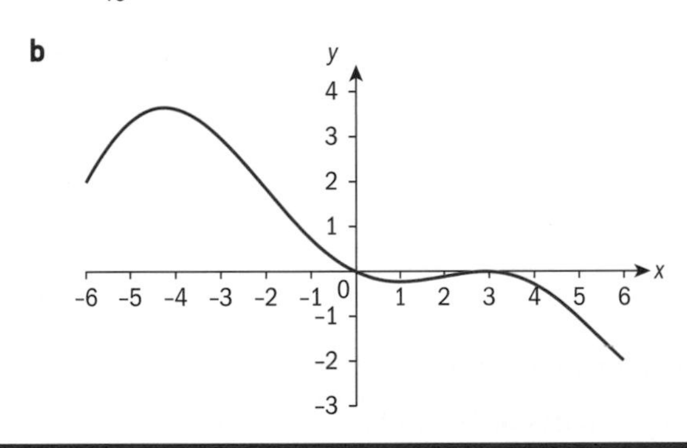

c

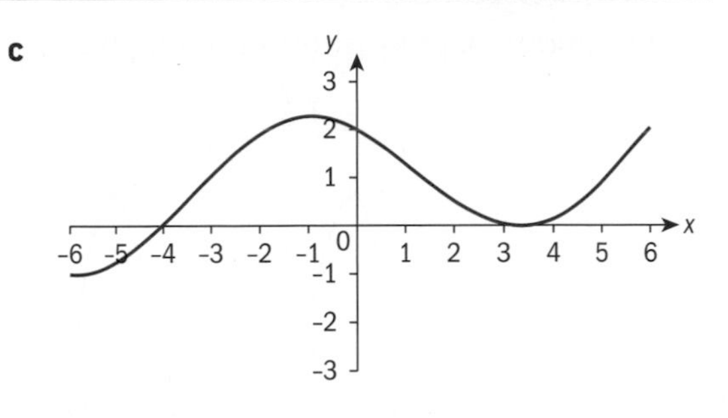

d

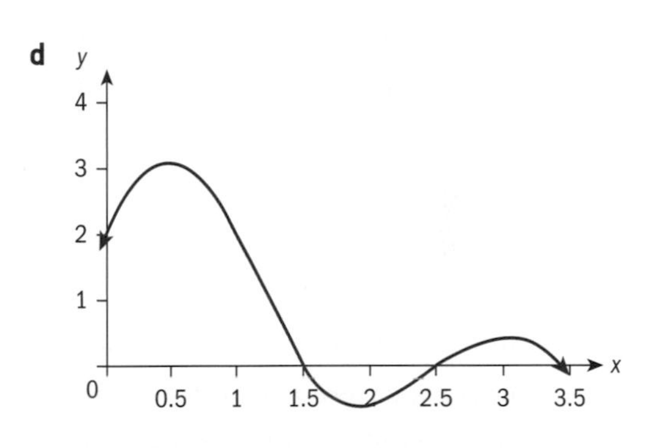

You have worked on sketching the first two derivatives of a function from the graph of the function. How do you sketch the function from its derivatives?

Investigation 17

1 Sketch the graph of $f'(x) = 4x^3 - 12x^2$.

Using the graph of $y = f'(x)$:

2 Describe how can you find the x values of any turning points or horizontal points of inflexion of f, and find these points.

3 Describe how can you find the intervals on which f is **i** increasing and **ii** decreasing, and find these intervals.

4 Describe how can you find the intervals on which f is **i** concave up and **ii** concave down.

5 Sketch a possible graph for $f(x)$ on the same graph.

6 On the same graph, sketch two other functions for f whose derivative is f'.

7 How are the different possible graphs of f related to one another?

8 **Conceptual** Given the graph $y = f'(x)$, describe how can you sketch a possible graph of $y = f(x)$.

TOK

Mathematics and Knowledge Claims: Euler was able to make important advances in mathematical analysis before Calculus had been put on a solid theoretical foundation by Cauchy and others. However, some work was not possible until after Cauchy's work.

What does this suggest regarding intuition and imagination in Mathematics?

Example 25

The follow diagram shows the graph of $y = f'(x)$ for a function f.

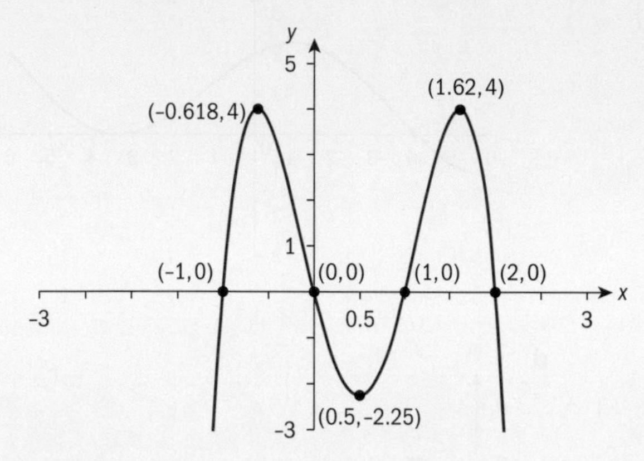

From the graph, indicate:

a the x-coordinate of any points where f has turning points and determine the nature of these points

b the intervals where f is **i** increasing and **ii** decreasing

c the intervals where f is **i** concave up and **ii** concave down.

d Sketch a possible graph of f using your answers to parts **a**, **b** and **c**.

a $x = -1, 0, 1, 2$. At $x = -1$, f has a local minimum, at $x = 0$ f has a local maximum, and at $x = 1$ f has a local minimum, at $x = 2$ f has a local maximum.	Identify the points where $f'(x) = 0$. At $x = -1$ and $x = 1$, f' goes from negative to positive, hence these points are local minima. At $x = 0$ and $x = 2$, f' goes from negative to positive, hence these points are local maxima.
b i $-1 < x < 0$; $1 < x < 2$ **ii** $x < -1$; $0 < x < 1$; $x > 2$	f is increasing where $f'(x) > 0$ and decreasing where $f'(x) < 0$.
c i $x < -0.618$; $0.5 < x < 1.62$ **ii** $-0.618 < x < 0.5$; $x > 1.62$	f is concave up when $f''(x) > 0$. f is concave down when $f''(x) < 0$.
d 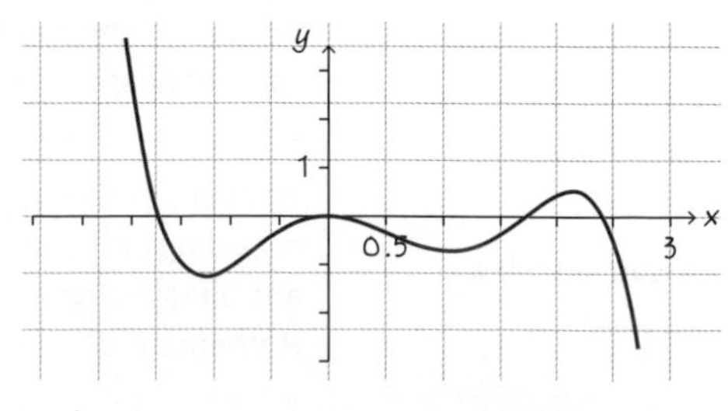	

Investigation 18

The graph of f'' is given.

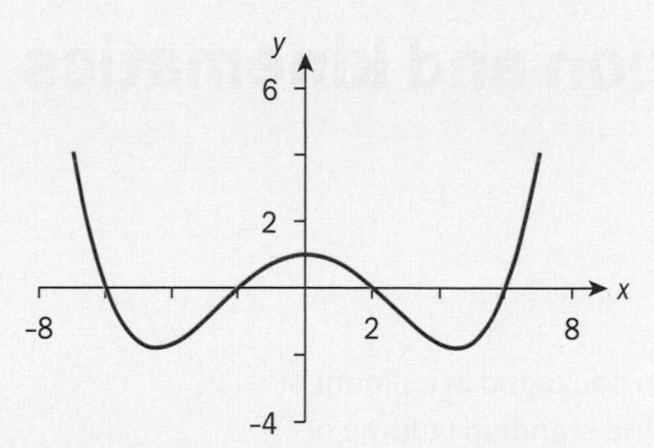

1 Using the graph, how do you identify
   a where f has a point(s) of inflexion?
   b the intervals where f is
      i concave up
      ii concave down?
2 **Conceptual** What information can you deduce about $f'(x)$ from the graph of $f''(x)$?
3 Copy the graph of f'' and sketch a possible graph of f' on the same graph.
4 Sketch a possible graph of f using all your answers.
5 Compare the graphs you made to the graphs of one of your classmates. What is the same about these graphs? What is different? Can different graphs both be correct?

Exercise 5P

1 For each graph of $y = f'(x)$ given below, sketch a possible graph of $y = f(x)$.

a

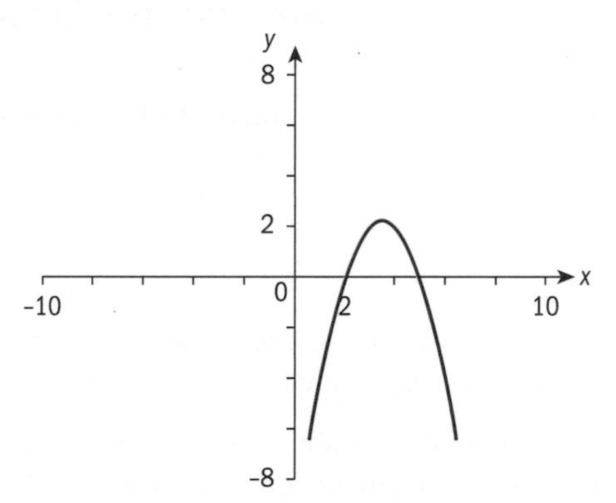

b

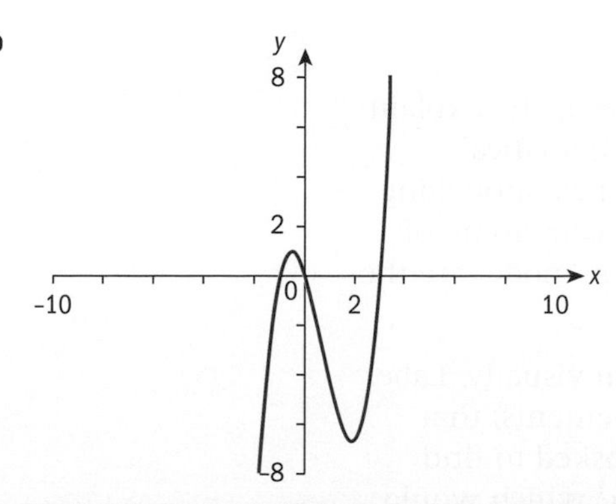

2 Copy the given graph of f', and on the same set of axes, sketch the possible graphs of f and f''.

a

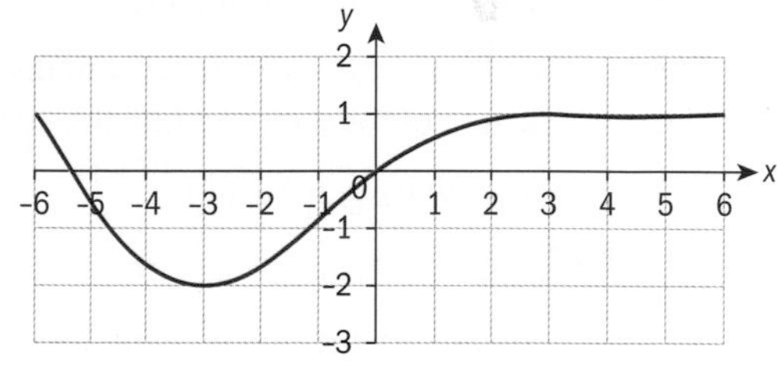

b

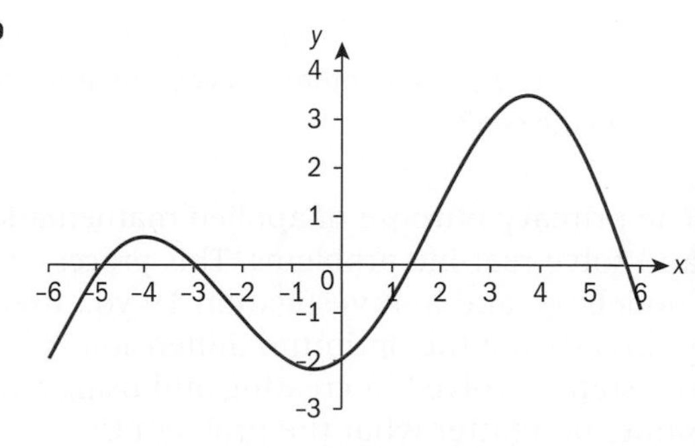

Calculus

5.5 Application of differential calculus: optimization and kinematics

Optimization

Investigation 19

Soft drink beverages are often packaged in cylindrical cans made from aluminium. The standard volume of such a can is 330 ml (this may vary among countries). Companies are always interested in minimizing production costs in order to maximize profits. Are companies using the least amount of aluminium in order to produce the 330 ml cans?

Your task is to find the height and diameter of a 330 ml can that will minimize the can's surface area.

1 Write down an expression for the surface area A of a cylinder. This expression should contain two variables. Which variables are these?

2 Write down an expression, containing the same two variables as used in question **1**, for the volume V of a cylinder. Equate your expression to 330 cm^3, since the volume of a can must be 330 cm^3.

3 Use the expression for V (from question **2**) to rewrite the expression for A (from question **1**) in terms of r, and state a reasonable domain for r.

4 Differentiate your expression for A with respect to r. Hence, find the values of r and h which minimize the surface area of the can, and find the surface area in this case.

5 Check your answers graphically.

6 **Conceptual** What is optimization in calculus?

7 Research actual dimensions and surface areas of your favorite beverages, and deduce whether or not companies are using the least amount of material in order to have the desired volume of 330 ml.

8 If companies are using more than the necessary amount of aluminium, give some reasons as to why this might be so. What other considerations might they need to make in the production of the size and shape of the can?

The primary purpose of applied mathematics is to investigate, explain and solve real-life problems. This process is called mathematical modelling, and in investigation 19 you used mathematical modelling to investigate the optimum dimensions of a soft drink can. Some of the steps involved in creating and using a mathematical model are the same, no matter what the problem is:

- If possible, draw a diagram to represent the problem visually. Label all the known elements, as well as the unknown elements, that you need to find (for the soft drink can, you were asked to find the height and radius of a can with volume 330 cm^3 which would minimize the surface area).

- Identify the independent variable(s) and their constraints (for the soft drink can, this was the radius of the can, which had domain $r > 0$).
- Identify all other constraints on the problem (for the soft drink can, you were told that the volume must be 330 cm^3).
- Translate the real-life problem into a mathematical function(s) (you used expressions for the area and volume of a cylinder).
- Carry out the mathematics necessary in order to solve the problem (you needed to differentiate the expression for A with respect to r, and find the value of r which minimized A).
- Reflect on the reasonableness of your results (using your GDC to check your results can be helpful. You then compared your theoretical results against the actual dimensions of a soft drink can).
- Apply your methods to other similar problems (what other problems could the mathematical model for the surface area of a soft drink can be applied to?).

In particular, optimization problems deal with finding the most efficient and effective solutions to real-life problems.

Investigation 20

You have 20 m of fencing to enclose a part of your yard as an outdoor, rectangular-shaped enclosure for your rabbits. You want to enclose the largest possible area with this amount of fencing.

1. Draw a suitable diagram to represent this scenario. Letting x represent the width of the enclosure, express the sides of the rectangle in terms of x.
2. Write down an expression for the area of the enclosure in terms of x. State the domain of x and explain why it has this domain.
3. Find the dimensions of the enclosure which maximizes its area, and find the maximum area.
4. Comment on the geometric significance of the dimensions of the largest enclosure.
5. Show that for any rectangular enclosure with fixed perimeter P, the maximum area will always be the same geometric shape as that which you found in question **3**.
6. Repeat questions **1–3** to find the maximum area that can be enclosed by an isosceles triangle with a perimeter of 20 m. What can you say about the three sides of the triangle?
7. **Conceptual** How does finding a mathematical model for a general case help you apply solutions to particular cases?

Example 26

A cardboard box manufacturer makes open boxes by cutting equal squares of side length x cm from the corners of a rectangular piece of cardboard measuring 15 cm by 24 cm. The sides are then folded up, as shown in the diagram. Find x so that the volume of the box is maximized, and find the maximum volume of the box. Check your answers graphically.

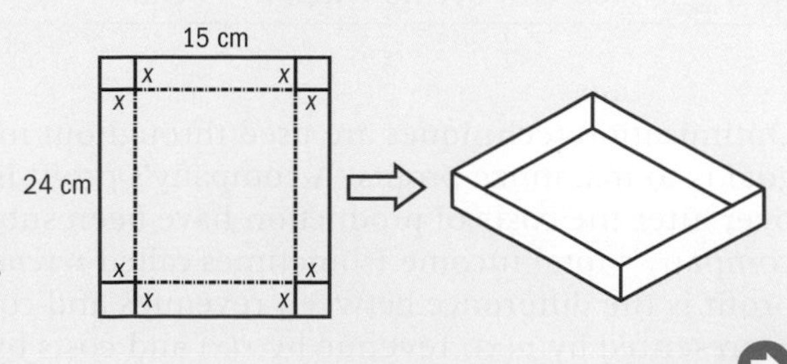

Continued on next page

Calculus

length: $24 - 2x$ width: $15 - 2x$	The length and width of the base of the box are the original dimensions of the piece of cardboard, minus the box sides.
$x > 0$ $24 - 2x > 0 \Rightarrow x < 12$ $15 - 2x > 0 \Rightarrow x < 7.5$ So, $0\text{ cm} < x < 7.5\text{ cm}$	In this context, all side lengths must be positive. You can use this to find the domain of x.
Volume $V = \text{length} \times \text{width} \times \text{height}$ $= (24 - 2x) \times (15 - 2x) \times x$ $= (360 - 48x - 30x + 4x^2)x$ $= (360x - 78x^2 + 4x^3)\text{cm}^3$	Find an expression for the volume of the box in terms of x.
$\frac{dV}{dx} = 360 - 156x + 12x^2$ $\frac{dV}{dx} = 12(x-10)(x-3)$ $\quad \frac{dV}{dx} = 0$ when $x = 3$ or $x = 10$ $x = 10$ is not within the domain of possible x.	To maximize the volume, we need to find the maximum point on the graph of V. Hence, find $\frac{dV}{dx}$.
$\frac{d^2V}{dx^2} = -156 + 24x.$	Set $\frac{dV}{dx} = 0$ to find the turning points.
When $x = 3$ $\quad \frac{d^2V}{dx^2} = -84 < 0 \Rightarrow V$ has a local maximum at $x = 3$	Recall that $0\text{ cm} < x < 7.5\text{ cm}$.
Hence, $V_{max} = 360(3) - 78(3)^2 + 4(3)^3$ $V_{max} = 486\text{ cm}^3$	Find $\frac{d^2V}{dx^2}$ and use it to determine the nature of the turning point at $x = 3$. Substitute $x = 3$ into the equation for V and calculate the maximum volume.
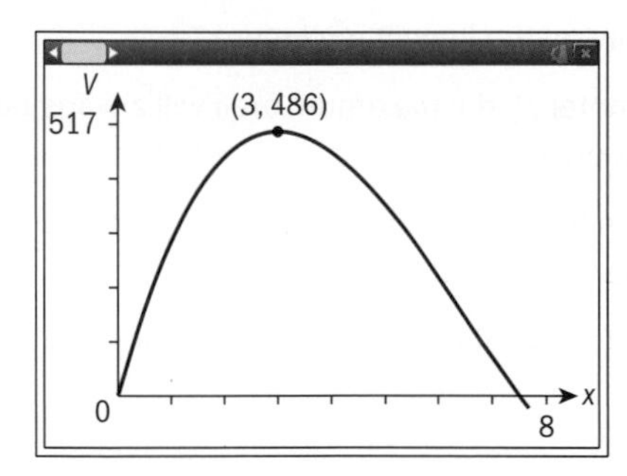 The graph shows that the maximum volume $V_{max} = 486\text{ cm}^3$ occurs when $x = 3$ cm.	Use your GDC to plot a graph of V against x for $0\text{ cm} < x < 7.5\text{ cm}$. Check that V has a maximum at (3, 486).

Optimization techniques are used throughout industry, since the goal is to maximize profits. A company's profit is the amount left over after the costs of production have been subtracted from the company's total income (sometimes called *revenue*). In other words, profit is the difference between revenues and costs. If profit is represented by $p(x)$, revenue by $r(x)$ and costs by $c(x)$, then $p(x) = r(x) - c(x)$.

TOK

How can you justify the raise in tax for plastic containers eg plastic bags, plastic bottles etc. using optimization?

Example 27

A small company manufactures and sells fishing poles. The cost of manufacturing fishing poles can be modelled by the function $c(x) = 7x + 3$, where x is the number of batches (each containing 1000 poles) manufactured. Revenue is modelled by $r(x) = x^3 - 10x^2 + 20x$. The company has enough workers to produce a maximum of 1200 fishing poles.

a State the domain of both the cost and revenue functions.

b Find the number of fishing poles that the company should manufacture in order to maximize its profits.

The company introduces a new process which allows them to produce 6000 poles.

c Find the number of poles that would cause the company to minimize profits (or maximizes losses).

d By graphing the cost and revenue functions, find the number of poles which should be manufactured if the company is to just break even (that is, the production level at which the costs and revenues are equal).

e It would not maximize the company's profits to produce as many poles as workers are capable of producing. Use your graph to explain why.

a For both functions, $0 \le x \le 1.2$	The company can produce a maximum of 1200 poles, which is 1.2 thousand units. $p(x) = r(x) - c(x)$
b $p(x) = (x^3 - 10x^2 + 20x) - (7x + 3)$ $= x^3 - 10x^2 + 13x - 3$	Find the x-values where $p'(x) = 0$.
$p'(x) = 3x^2 - 20x + 13$ $3x^2 - 20x + 13 = 0 \Rightarrow x_1 = 0.730;$ $x_2 = 5.94$	You can use the polynomial root finder on your GDC to find when $p'(x) = 0$.
$p''(x) = 6x - 20$ $p''(0.730) < 0 \Rightarrow x = 0.730$ is a maximum	Use the second derivative to determine the nature of these points.
Since x is in thousands of units, the production level necessary to maximize profits is 730 fishing poles.	Interpret your answer in the context of the problem.
c $p''(5.94) > 0 \Rightarrow x = 5.94$ is a minimum, so at $x = 5.94$, the profit function p has an absolute minimum in its domain. The production level that would maximize losses is 5940 fishing poles.	

Calculus

Continued on next page

d The company will break even at 296 fishing poles and at 1190 fishing poles. 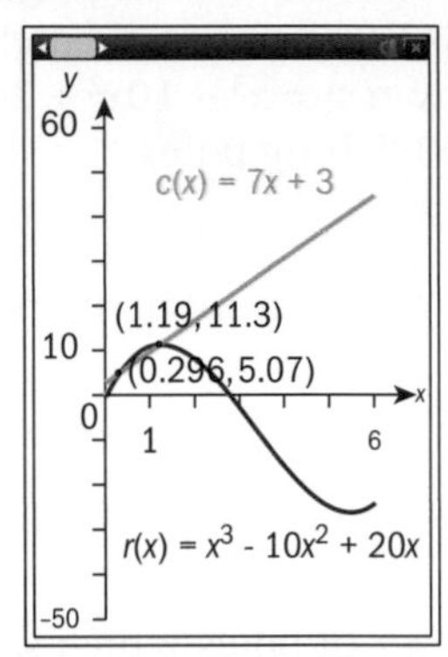	The company breaks even when costs equal revenues. From the graph that is at $x = 0.296$ (that is, 296 fishing poles) or at $x = 1.19$ (that is, 1190 fishing poles).
e The company's revenues decrease, despite the costs continuing to increase. This may be because of lack of demand for so many poles; which would mean they cannot sell them all.	

Exercise 5Q

1 A farmer wishes to enclose a rectangular field of area 100 m^2. The farmer must buy fencing material for three of the sides, but the fourth side (one of the longer sides) will be an existing fence.

The shorter sides of the rectangular enclosure are to have length x m.

a Write down the length of the longer sides of the enclosure, in terms of x.

b Find an expression, in terms of x, for the total length of new fencing the farmer must buy to enclose the rectangular field.

c Find the minimum length of fencing required, and the value of x for which this minimum occurs.

d Sketch a graph of fencing length against x to confirm your answer.

2 The profit, \$$y$, generated from the sale of x laptops is given by the formula $y = 600x + 15x^2 - x^3$. Find the number of laptops, x, which maximizes profit, and determine the maximum profit.

3 An open-top water tank is in the shape of a cuboid. The tank has a square base of side length s m, and has a volume of 216 m^3.

Find expressions, in terms of s, for:

a the height of the tank

b the surface area of the tank.

Given that the surface area of the open-top water tank is minimized:

c find the value of s.

4 A 150 cm length of wire is cut into two pieces. One of the pieces is bent to form a square with side length s cm, and the other piece is bent to form a rectangle with a side length that is twice its width.

Find the value of s which minimizes the sum of the areas enclosed by the square and rectangle.

5 An architect wants to design a rectangular water tank with an open top that can hold a capacity of 10 m^3. The materials needed for the tank cost \$10 per square metre for the base, and \$6 per square metre for the sides. The length of the base must be twice its width. Find the minimum cost of the material required to build the container. Give your answer to the nearest dollar.

Kinematics

Kinematics is the study of the movement of objects. A particle moving in a straight line is an example of the simplest type of motion. To describe simple linear motion, you need a starting point, a direction and a distance.

- A **path** is the set of points along which an object travels.
- The **distance travelled** by an object is the *total length* of the path followed by the object after it leaves its starting point.
- The **displacement** of an object is the *difference* between the object's position and a fixed point. This fixed point is sometimes called the *origin*, and it is not necessarily the same as the object's starting point.

If you run a complete lap around a standard athletics track, the *distance* you will travel is 400 m. However, after you have completed the lap, your displacement will be 0 m, since your initial and end positions are the same.

Displacement tells you the *shortest distance* and *direction* of the object from its starting point. Displacement can be a positive value or a negative value, depending on the direction of the object from its starting point. Distance, however, is a scalar quantity which will never be negative, as it only tells you the *length of the path* the object has travelled.

In the diagram on the right, the *path* a particle travels along is the purple curve joining A to B. The length of this path will tell you the *distance* travelled by the particle. The *displacement* of the particle from point A to point B is the length of the segment [AB]. The displacement would be positive, because point B is to the right of point A. Hence, the displacement is always less than or equal to the distance traveled.

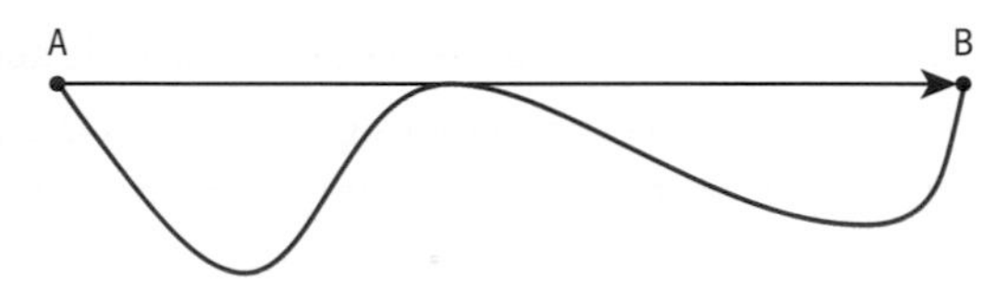

- Velocity, $v(t)$, is the derivative of the displacement function $s(t)$.

  $v(t) = \frac{\mathrm{d}s}{\mathrm{d}t}$ or $s'(t)$
- Acceleration, $a(t)$, is the derivative of the velocity.

  $a(t) = \frac{\mathrm{d}v}{\mathrm{d}t} = \frac{\mathrm{d}^2 s}{\mathrm{d}t^2}$ or $s''(t)$

Investigation 21

Ben runs from point A to point B along a straight line. When he reaches B, he then turns and runs back to point A. When he reaches A, he turns once again and runs back to point B.

Ben's position, s, from point A at any time, t after he leaves point A (where $0 \le t \le 4$), can be modelled by the function $s(t) = t^3 - 6t^2 + 9t$.

1 Use your GDC to plot the graph of Ben's displacement, $s(t)$, against t.

2 Use your GDC to plot the graph of Ben's velocity, $s'(t)$, against t.

Write down the time intervals where Ben's velocity is:

a positive **b** negative.

Which direction is Ben moving in each interval?

Continued on next page

3 **Conceptual** What does a positive or negative value for velocity represent?

4 **Conceptual** What is the connection between speed and velocity?

5 Find the times at which Ben's velocity is zero. What is Ben doing at each of these times?

6 At which two times did Ben's maximum velocity occur? Describe Ben's motion at each of these two times.

Investigation 22

A particle moves along a straight line such that its position (s metres) at any time (t seconds) is given by $s(t) = t^3 - 7t^2 + 11t - 2.5$.

1 Find expressions for the velocity and acceleration of the particle in terms of t.

2 Find the times at which:

a the particle is at rest **b** the particle is speeding up **c** the particle is slowing down.

Justify your answers.

3 **Conceptual** What must be true about the signs of the velocity and acceleration in order for a particle to be speeding up, and in order for a particle to be slowing down?

4 At what values of t does the particle change direction? How can you tell this from the graph of acceleration against time?

When v and a have the same sign, an object is speeding up (accelerating).
When v and a have different signs, an object is slowing down (decelerating).

Example 28

A particle moves in a horizontal line so that its position from a fixed point after t seconds, where $t \geq 0$, is s metres, where $s(t) = 5t^2 - t^4$.

a Find the position, velocity and acceleration of the particle after 1 second.

b Determine whether the particle is speeding up or slowing down at $t = 1$.

c Find the values of t when the particle is at rest.

d Find the time intervals on which the particle is speeding up, and the intervals on which it is slowing down.

e Find the total distance the particle travels in the first 3 seconds.

a $s(1) = 5(1) - 1 = 4\text{ m}$	Find the value of $s(1)$.
$v(t) = s'(t) = 10t - 4t^3\text{ m s}^{-1}$ $v(1) = 6\text{ m s}^{-1}$	Differentiate s to get an expression for v, and evaluate at $t = 1$.
$a(t) = v'(t) = 10 - 12t^2\text{ m s}^{-2}$ $a(1) = -2\text{ m s}^{-2}$	Differentiate v to get an expression for a, and evaluate at $t = 1$.
b Since v and a have opposite signs at $t = 1$, the particle is slowing down.	
c $10t - 4t^3 = 0$ $2t(5 - 2t^2) = 0$	The particle is at rest when $v = 0$.
$t = 0,\ t = \sqrt{\frac{5}{2}}$	Reject the negative value of the square root, since $t \geq 0$.

d $10-12t^2=0$

$t=\sqrt{\frac{5}{6}}$

Find the places at which the acceleration changes sign. Again, ignore the negative value for t.

t	$0<t<\sqrt{\frac{5}{6}}$	$\sqrt{\frac{5}{6}}<t<\sqrt{\frac{5}{2}}$	$t>\sqrt{\frac{5}{2}}$
Sign of v	+	+	−
Sign of a	+	−	−

Compare the signs of velocity and acceleration each time at each interval on which one of them changes.

The particle is speeding up when $0<t<\sqrt{\frac{5}{6}}$ and $t>\sqrt{\frac{5}{2}}$.

Both v and a have the same sign on these intervals.

The particle is slowing down when $\sqrt{\frac{5}{6}}<t<\sqrt{\frac{5}{2}}$.

v and a have different signs on this interval.

e Distance is always positive, so you must find the distance the particle travels in one direction, and then the distance it travels the other direction, and add these together.

$s(0)=0$

Initially the particle is at 0.

$$s\left(\sqrt{\frac{5}{2}}\right)=5\left(\sqrt{\frac{5}{2}}\right)^2-\left(\sqrt{\frac{5}{2}}\right)^4=6.25\text{ m}$$

Find the displacement of the particle when it changes direction at $t=\sqrt{\frac{5}{2}}$.

Particle travels $6.25-0=6.25$ m in the positive direction.

$s(3)=5(3)^2-(3)^4=-36$ m

Find the displacement of the particle after 3 seconds.

Particle travels $6.25-(-36)=42.25$ m in the negative direction.

Find distance particle travels in the negative direction.

So total distance travelled in first 3 seconds $=6.25+42.25=48.5$ m.

Add the two distances together.

Exercise 5R

1 A particle moves in a straight line such that its position in metres from a fixed origin is given by $s(t)=t^3-3t+1$, where t is the time in seconds, $t\geq 0$.

a Find expressions for the particle's velocity and acceleration at time t.

b Find the initial position, velocity and acceleration. Hence describe the particle's motion at this instant.

c Describe the motion of the particle at $t=2$ seconds.

d At what time(s) does the particle reverse direction?

e For what time interval is the particle speeding up?

f What is the total distance travelled by the particle between $t=0$ and $t=3$ seconds?

2 When a ball is thrown straight up in the air its height above the ground is given by $s(t) = 12t - t^3 + 1$ metres where t is in seconds, $t \geq 0$.

a From what height above the ground was the ball released?

b Find the maximum height reached by the ball.

c Find the ball's velocity:
i when released **ii** at $t = 1$ **iii** at $t = 3$

d Find the total distance travelled by the ball before it hits the ground.

3 A stone is projected vertically so that its position in metres above ground level after t seconds is given by $s(t) = 15t - 5t^2$. Calculate the time at which the stone reaches its maximum height, and find the maximum height.

4 A person jumps from a diving board above a swimming pool. At time t seconds after leaving the board, the person's height above the surface of the pool, s metres, can be modelled by the function $s(t) = 10 + 5t - t^2$. Find:

a the height of the diving board above the surface of the pool

b the time between the person leaving the board and hitting the water

c the velocity and acceleration of the diver upon impact with the water. Interpret your answers in the context of this problem.

5 A child's toy launches toy rockets into the air. The height, h m, of a toy rocket which is launched into the air with an initial velocity of v_0 and initial height of h_0 can be modelled by the function $h(t) = h_0 + v_0(t) - 4.9t^2$, where t is time in seconds that have passed since the rocket was launched.

A toy rocket is launched from the ground with an initial velocity of 50 m s^{-1}. Find the maximum height the rocket reaches, and the time that passes before it hits the ground again.

6 You begin walking eastward through a park, and your velocity graph is shown below.

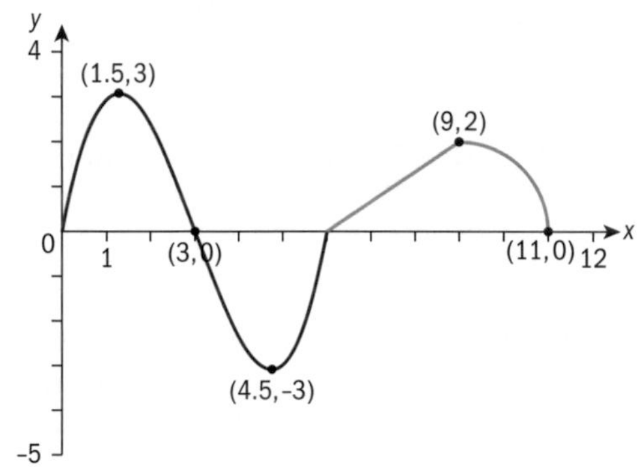

Find:

a the times when you are standing still

b the time intervals when you are moving **i** eastward **ii** westward

c the time intervals when you are moving most quickly **i** eastward **ii** westward

d the time(s) when you are moving most quickly

e the time intervals when you are **i** speeding up **ii** slowing down.

Developing your toolkit

Now do the Modelling and investigation activity on page 274.

Chapter summary

Limits

The limit, L, of a function f exists as x approaches a real value a if and only if $\lim_{x \to a^+} f(x) = L$ **and** $\lim_{x \to a^-} f(x) = L$.

We then write $\lim_{x \to a} f(x) = L$.

Limit of a sequence

If $\lim_{n\to\infty}\{a_n\} = L, L \in \mathbb{R}$, then the sequence is said to converge, otherwise it diverges.

Convergence of geometric series

For an infinite geometric series, when $|r| < 1$ the series converges to it sum, $S_\infty = \frac{u_1}{1-r}$.

Basic differentiation rules

- If $f(x) = x^n$, then $f'(x) = nx^{n-1}$, where $n \in \mathbb{R}$.
- If $f(x) = c,\ c \in \mathbb{R}$, then $f'(x) = 0$.
- For $c \in \mathbb{R}$, $(cf)'(x) = c(f'(x\,\mathbb{R}))$, provided $f'(x)$ exists.
- If $f(x) = u(x) \pm v(x)$, then $f'(x) = u'(x) \pm v'(x)$.

Equations of tangents and normals

Suppose $f(x)$ is a non-linear function and (x_1, y_1) is a point which lies on the curve $y=f(x)$. If $f'(x_1) = m$, where m is a constant, then m is the gradient of the tangent to $y=f(x)$ at (x_1, y_1), and the equation of the tangent line is $(y-y_1) = m(x-x_1)$.

If a tangent line has gradient m, the gradient of the normal is $\frac{-1}{m}$.

Chain rule

If $y=g(u)$ where $u=f(x)$ then $\frac{dy}{dx} = \frac{dy}{du} \times \frac{du}{dx}$.

The product rule

If $y = uv \Rightarrow \frac{dy}{dx} = u\frac{dv}{dx} + v\frac{dv}{dx}$.

Another way of writing this is:

If $f(x) = u(x)v(x)$, then $f'(x) = u(x)v'(x) + v(x)u'(x)$.

The quotient rule

If $y = \frac{u}{v}$, $v(x) \neq 0$ then $\frac{dy}{dx} = \frac{v\frac{du}{dx} - u\frac{dv}{dx}}{v^2}$

Another way of writing this is:

$y = \frac{u}{v}$, $v(x) \neq 0$ then $\frac{dy}{dx} = \frac{v(x)u'(x) - u(x)v'(x)}{(v(x))^2}$.

Higher derivatives

$$f'(x) = \frac{dy}{dx};\ f''(x) = \frac{d^2y}{dx^2};\ f'''(x) = \frac{d^3y}{dx^3};\ f^{(n)}(x) = \frac{d^ny}{dx^n}$$, where $n = 4, 5, \ldots$

Increasing or decreasing functions

If $f'(x) > 0$ for all $x \in]a, b[$ then f is increasing on $]a, b[$.

If $f'(x) < 0$ for all $x \in]a, b[$ then f is decreasing on $]a, b[$.

Calculus

Continued on next page

First derivative test

- If $f'(c) = 0$, test for turning points.
- If $f' > 0$ for $x < c$ and $f' < 0$ for $x > c$, then f has a local maximum at $x = c$.
- If $f' < 0$ for $x < c$ and $f' > 0$ for $x > c$, then f has a local minimum at $x = c$.

Second derivative test for maxima and minima

If $f'(c) = 0$ and $f''(c) < 0$ then $f(x)$ has a local maximum at $x = c$.

If $f'(c) = 0$ and $f''(c) > 0$ then $f(x)$ has a local minimum at $x = c$.

If $f'(c) = 0$ and $f''(c) = 0$ then the second derivative test is inconclusive. If the second derivative test is inconclusive, revert to the first derivative test for local extrema.

Points of inflexion

If f has a point of inflexion at $x = c$, then $f''(c) = 0$.

If $f''(c) = 0$ **and** f changes concavity at $x = c$, then f has a point of inflexion at $x = c$.

f is concave down in an open interval if, for all x in the interval, $f''(c) < 0$.

f is concave up in an open interval if, for all x in the interval, $f''(c) > 0$.

Optimization and kinematics

- Velocity, $v(t)$, is the derivative of the displacement function $s(t)$. $v(t) = \frac{ds}{dt}$ or $v(t) = s'(t)$.
- Acceleration, $a(t)$, is the derivative of the velocity. $a(t) = \frac{dv}{dt} = \frac{d^2s}{dt^2}$ or $a(t) = v'(t) = s''(t)$.
- When v and a have the same sign, the object is speeding up (accelerating).
- When v and a have different signs, the object is slowing down (decelerating).

Developing inquiry skills

The distance a sprinter has travelled at various times is given by the function $s(t) = -0.102t^3 + 1.98t^2 + 0.70t$ for $0 \le t \le 10$ seconds.

1 What is the fastest instantaneous speed of the sprinter in the race?

2 Draw the graph of $v(t)$ and state whether or not the sprinter is running at a constant speed throughout the race.

3 How could you determine the sprinter's speed at a particular instant $t = c$?

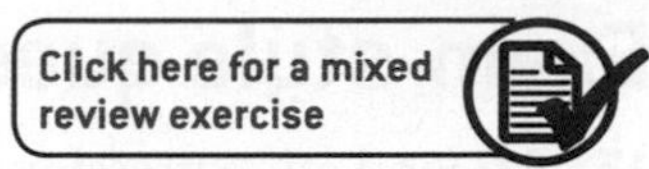

Chapter review

Click here for a mixed review exercise

1 Consider the function $f(x) = \frac{2x}{x-a}$, for $x \neq a$.

a Write down the equation of the horizontal asymptote.

b Find the value of a if the vertical and horizontal asymptotes to the graph of f intersect at the point $(2, 2)$.

2 **a** Sketch the graph of $y = \frac{5x + x^2}{x}, x \neq 0$.

b Find $\lim_{x\to 0} \frac{5x + x^2}{x}$ numerically.

3 Using technology, find the equations of any asymptotes of each function.

a $f(x) = \frac{6x^2 + 7}{x^2 - 9}$ **b** $f(x) = \frac{80x}{3x^3 + 81}$

4 Differentiate each function with respect to x:

a $y = (1 - 2x)^5(3x - 2)^6$

b $y = \frac{x-3}{x^2 - 3x}$

c $y = \sqrt{x} - 4\sqrt[3]{x}$

5 Let $y = \frac{x}{x^2 - 1}$

a Find the equations of the vertical and horizontal asymptotes of y.

b Show that $\frac{dy}{dx} < 0$ for all x.

c Sketch the function.

6 Find the equations of any horizontal tangents to the curve $y = x^3 - 3x^2 - 9x + 2$.

7 The tangent to the curve $y = \frac{2x}{x-1}, x \neq 1$ at the point $(2,4)$ and the normal to the same curve at the point $(3, 3)$ intersect at the point P. Find the coordinates of P.

8 Find the equation of the normal line to the curve $f(x) = 2x^3 - 3x + 1$ at the point $(1, 0)$.

9 Find the coordinates of the point(s) on the curve $y = -x^3 + 2x^2 + 1$, where the tangent to the curve is parallel to the line $y = -4x$.

10 Find the coordinates of the points where the first derivative is equal to zero on the curve $y = \frac{x^2}{x+1}, x \neq -1$ and determine their nature.

11 $f(x) = \frac{9(x+1)}{x^2}$

a Find the x-intercept(s) of $f(x)$.

b Find the equations of the asymptote(s) of $f(x)$.

c Find the coordinates of the local minimum of $f(x)$, and justify that it is a minimum.

d Find the interval(s) where $f(x)$is concave up.

e Sketch the graph of $f(x)$, clearly showing and labelling the features you found in parts **a–d**.

12 $f(x) = x - b\sqrt{x}; x \geq 0, b \in \mathbb{R}^+$. Find, in terms of b:

a the zeros of f

b the intervals where f is **i** increasing and **ii** decreasing.

c State the conditions on b for f to be **i** concave up or **ii** concave down.

13 A rocket is launched vertically upwards. Its height, h metres, t seconds after it was launched, is given by $h(t) = 49t - 2.45t^2$.

a Find an expression for the velocity v of the rocket at time t.

b Find the maximum height of the rocket.

14 A particle travels in a straight line such that its velocity at any time t is given by $v(t) = 1 + t - \sqrt{4t+9}$ m s^{-1}. Find:

a the particle's initial velocity

b the time when the particle is at rest

c the particle's acceleration at the instant it comes to rest

d the time interval(s) when the particle is **i** slowing down **ii** speeding up.

Exam-style questions

15 P1: Find an expression for derivative of each of the following functions.

a $f(x) = x^4 - 2x^3 - x^2 + 3x - 4$ (1 mark)

b $g(x) = \dfrac{-4x}{x^2+1}$ (3 marks)

c $h(x) = (x+2)(x-7)$ (2 marks)

d $i(x) = (2x+3)^3$ (2 marks)

16 P1: Consider the following graphs.

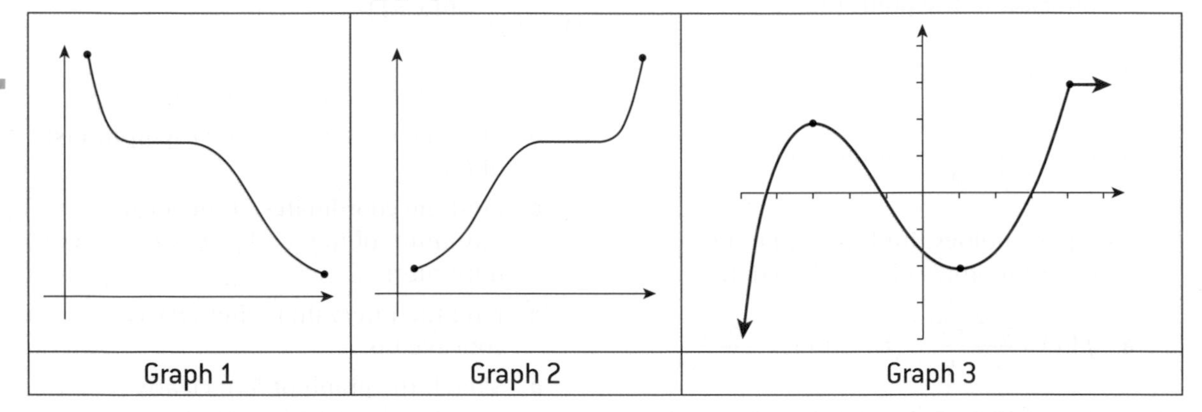

For each condition **a–d** state, with reasons, which graph satisfies that condition.

a The graph has no tangent parallel to $y = x$ (2 marks)

b The graph represents an increasing function. (2 marks)

c $\lim\limits_{x \to \infty} f(x) = 2$. (2 marks)

d The derivative of *f* is always negative. (2 marks)

17 P1: A particle *P* moves in a straight line. Its position, *s*, after *t* seconds is described by the following graph.

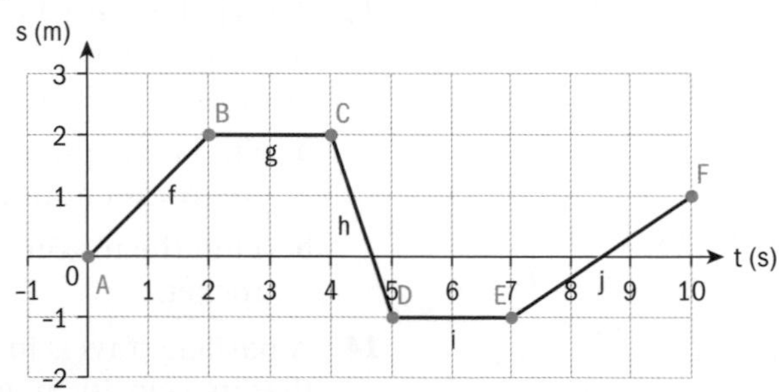

a State the intervals in which the particle

i is moving away from the origin (3 marks)

ii is at rest (2 marks)

iii is to the left of the origin. (1 mark)

b Find expressions for s in terms of t which describe position of the particle in each of the intervals *f*, *g*, *h*, *i* and *j* shown in the diagram. (4 marks)

c Draw a graph of the particle's velocity against time for $0 \le t \le 10$. (4 marks)

18 **P2:** A real estate agent manages 200 one-bedroom apartments. Currently, all apartments are rented for the price of \$320 per month. Market studies indicate that for each \$10 increase in monthly rent above \$320 will result in five unoccupied apartments.

a Given that the real estate agent wants to maximize the value of the total rental income gained from the 200 properties, determine the monthly rent that the agent should charge for a single apartment. (4 marks)

b If the agent charges this price for rent, find

i the number of flats rented (1 mark)

ii the total monthly rental income the agent gains from these properties. (2 marks)

19 **P2:** A projectile is launched vertically from the ground. After t seconds its height is given by $h(t) = 112t - 4.9t^2$ metres.

a Determine $h(4)$ and $h(5)$. (2 marks)

b Find an expression for the velocity of the projectile in terms of t. (2 marks)

c Find the time at which the projectile reaches its maximum height. (2 marks)

d Determine how long it takes for the projectile to hit the ground. (2 marks)

e Sketch the graph of $y = h(t)$, stating clearly the domain of validity of the model and its maximum point. (3 marks)

f Find the velocity of the projectile at the instant it hits the ground. (2 marks)

g Show that the acceleration of the projectile is constant, stating its value. (3 marks)

20 **P1:** Consider the information given in the following table.

x	$f(x)$	$f'(x)$	$g(x)$	$g'(x)$
0	1	−2	12	−12
1	2	4	6	−3
2	9	10	4	$-\frac{4}{3}$
3	22	16	3	$-\frac{3}{4}$

a Find

i $\left(\dfrac{f}{g}\right)'(2)$ (3 marks)

ii $(g \circ f)'(1)$ (2 marks)

b Determine whether or not the following statements are true, and justify your answer.

i f is an increasing function. (2 marks)

ii The tangents to the graph of g at the points $x = 2$ and $x = 3$ are perpendicular lines. (2 marks)

21 **P2:** An epidemic is spreading through a country. Authorities estimate that the number of people who will be affected by the disease can be modelled by the function $N(t) = 450t^2 - 30t^3$ where $0 \le t \le 30$ represents the time measured in days.

a Compare the average rate of change at which the disease is expected to spread between the 1st and 3rd day, with the average rate of change at which the disease is expected to spread between the 4th and 5th day. (3 marks)

b Find an expression for the instantaneous rate of change of the spread of the disease for $0 \le t \le 30$. (2 marks)

c Hence determine when the spread of the disease reaches its maximum. (2 marks)

d Find $\dfrac{d^2N}{dt^2}$ and comment on its meaning in the context of the question. (3 marks)

22 **P1:** The functions $g(x)$ and $h(x)$ are defined by $g(x) = \dfrac{x+2}{x-1}$ and $h(x) = x^2$.

a Show that g is a self-inverse. (4 marks)

b Show that $(h \circ g^{-1})'(x) \ne (h' \circ g^{-1})(x)$. (5 marks)

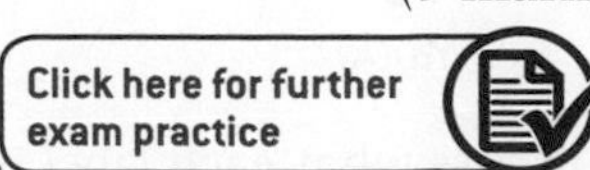

Calculus

River crossing

Approaches to learning: Thinking skills: Evaluate, Critiquing, Applying
Exploration criteria: Personal engagement (C), Reflection (D), Use of mathematics (E)
IB topic: Differentiation, Optimisation

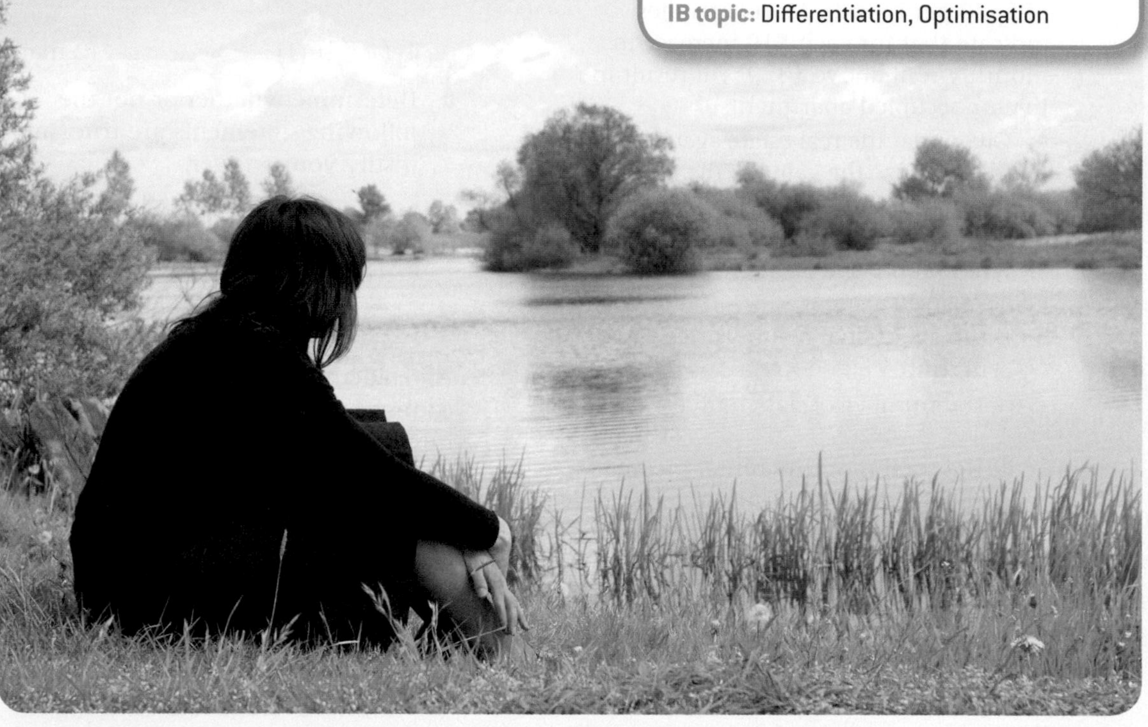

The problem

You are standing at the edge of a slow-moving river which is one kilometre wide. You want to return to your campground on the opposite side of the river. You can swim at 3 km/h and run at 8 km/h. You must first swim across the river to any point on the opposite bank. From there you must run to the campground, which is 2 km from the point directly across the river from where you start your swim.

What route will take the least amount of time?

Visualize the problem

Here is a diagram of this situation:

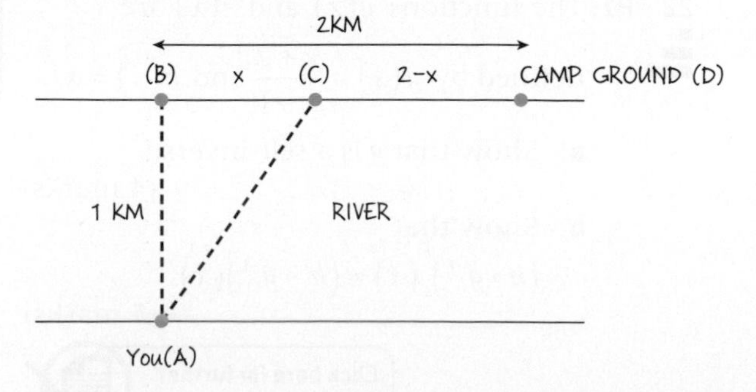

Discuss what each label in the diagram represents.

Solve the problem

What is the length of AC in terms of x?

Using the formula for time taken for travel at a constant rate of speed from this chapter, write down an expression in terms of x for:

1 the time taken to swim from A to C

2 the time taken to run from C to D.

Hence write down an expression for the total time taken, T, to travel from A to D in terms of x.

You want to minimize this expression (find the minimum time taken).

Find $\frac{dt}{dx}$.

Now solve $\frac{dt}{dx} = 0$ to determine the value of x that minimizes the time taken.

How do you know this is a valid value?

For this value of x, find the minimum time possible and describe the route.

Assumptions made in the problem

The problem is perhaps more accurately stated as:

You are standing at the edge of a river. You want to return to your campground which you can see further down the river on the other side. You must first swim across the river to any point on the opposite bank. From there you must run to the campground.

What route will take the least amount of time?

Look back at the original problem.

What additional assumptions have been made in the original question?

What information in the question are you unlikely to know when you are standing at the edge of the river?

What additional information would you need to know to determine the shortest time possible?

The original problem is a simplified version of a real-life situation. Criticize the original problem, and the information given, as much as possible.

Extension

In an exploration it is important to reflect critically on any assumptions made and the subsequent significance and limitations of the results.

Consider the open-box problem and the cylindrical can problem and/or some of the questions in Exercise 5R on page 267–268.

If you were writing an exploration using these problems as the basis or inspiration of that exploration, then:

- What assumptions have been made in the question?
- What information in the question are you unlikely to know in real-life?
- How could you find this missing information?
- What additional information would you need to know?
- Criticize the questions as much as possible!

Representing data: statistics for univariate data

Univariate data analysis involves handling data which has only one variable ("uni" means "one"). For example, the heights of all students in your class is univariate data. The main purpose of univariate analysis is to describe data. You can draw charts, find averages and analyse data in many other ways.

Univariate analysis *does not* deal with relationships between variables, such as the relationship between height and weight for students in your class. This is called bivariate analysis, and you will study this in chapter 7.

Concepts

- Representation
- Approximation

Microconcepts

- Population, sample, discrete and continuous data
- Convenience, simple random, systematic, quota and stratified sampling
- Frequency distributions (tables)
- Grouped data
- Histogram
- Central tendency. Mean, mode, median
- Spread (or dispersion). Cumulative frequency, cumulative frequency graphs.
- Median, quartiles, percentiles. Range. Interquartile range. Outliers
- Box-and-whisker plots
- Skew
- Standard deviation and variance

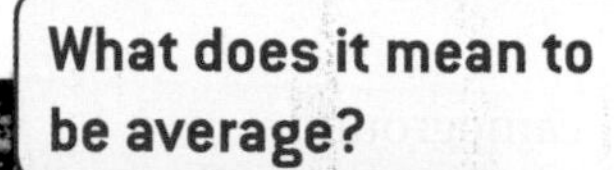

How could you choose a sample of people from your school to survey which would reliably tell you about the eating habits of students in the whole school?

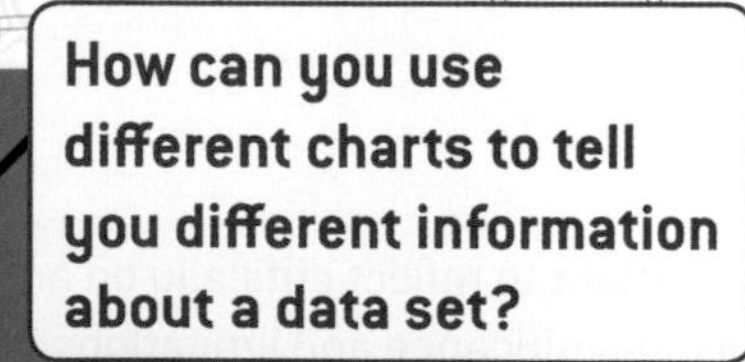

A group of 32 students took a test with maximum mark 10. Their scores are listed below:

0, 1, 1, 2, 2, 2, 3, 3, 4, 4, 4, 5, 5, 5, 5, 5, 6, 6, 6, 6, 7, 7, 7, 7, 7, 7, 8, 8, 8, 8, 9, 10.

The study of statistics allows mathematicians to **collect** a set of data, like this one, and **organize** it, **analyse** it, **represent** it and **interpret** it.

- What should you do next with this data?
- How can you organize the data to give a better picture of the scores?
- How should you display the scores?
- Should you use an average?
- What should you do about converting the scores to letter grades?
- Can you draw any conclusions from the scores?

Developing inquiry skills

Think of some other types of data that you collect. How would you collect, organize, analyse, represent and interpret the data?

Think about the questions in this opening problem and answer any you can. As you work through the chapter, you will gain mathematical knowledge and skills that will help you to answer them all.

Before you start

You should know how to:

1 Draw a bar chart.

eg Draw a bar chart for the number of cars in the families of 40 students in the frequency table.

Cars	Frequency
0	2
1	10
2	18
3	6
4	4

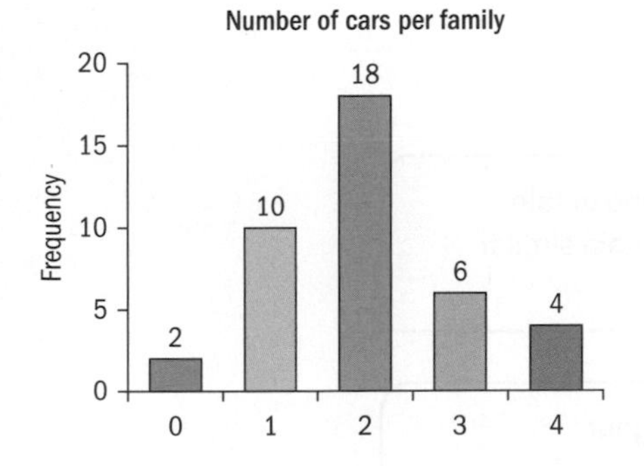

2 The mean is sometimes called the *average*. The mode is the value that occurs most often. The median is the middle value when the data is arranged in order least to greatest.

Find **a** the mode, **b** the median and **c** the mean of 1, 2, 2, 4, 5, 10, 11

a 2 **b** 4

c Mean $= \frac{1+2+2+4+5+10+11}{7} = \frac{35}{7} = 5$

Skills check

Click here for help with this skills check

1 Find the mean of each set of values:

a 2, 3, 4, 5, 6

b 13, 9, 7, 12, 15, 19, 2

2 Find the mode of each set of values:

a 2, 3, 5, 5, 7

b 4, 1, 1, 6, 7, 7

3 Find the median of each set of values:

a 4, 5, 6, 7, 8

b 9, 2, 3, 8, 5, 7

6.1 Sampling

Quantitative data are always numbers. Quantitative data describes information that can be counted, such as "How many people live in your house?" or "How long does it take you to get home after school?"

Quantitative data can be **discrete** or **continuous.**

Discrete data can only take particular values. Although there may be an infinite number of these values, each data value is distinct from the others and there are no "gaps" between data values. Examples of discrete data include:

- The number of pets in your house (there can only be a whole number of pets!).
- The result of rolling a dice (you can only get the values 1, 2, 3, 4, 5 or 6. There are no other possibilities).
- A person's shoe size (shoes are only made in discrete sizes, for example "size 39" or "size 40" – you *cannot* be "size 39.2").

Continuous data is not restricted to certain fixed values, but it can occupy any value within a continuous range. Its values are often expressed using fractions or decimals. Examples of continuous data include:

- The mass of an object in kg (for example, 24.16 kg).
- The time it takes you to travel to school (for example, 12.5 minutes).

International-mindedness

Ronald Fisher (1890–1962) lived in the UK and Australia and has been described as "a genius who almost single-handedly created the foundations for modern statistical science". He used statistics to analyse problems in medicine, agriculture and social sciences.

Reflect Is the taxiing speed of an airplane discrete or continuous?

Is the number of airplanes waiting to take off discrete or continuous?

Could data ever be classified as both discrete *and* continuous?

Why is it important to consider the nature of the variable, rather than just the data values themselves, when classifying whether data is discrete or continuous?

Reliability of data

Data is **reliable** if you can repeat the data collection process and obtain similar results. For example, could you repeat a survey and obtain similar findings?

Data is **sufficient** if there is enough data available to support your conclusions.

When collecting data, students often ask "How many data items should I have?" There is no fixed value, but you need to ensure that you collect enough data so that your results are reliable, and are **representative**: they represent the whole population well.

Reflect How many students would you need to survey to find a good estimate of the average time students at your school spend doing homework?

Two factors which can cause unreliable data are:

1 **Missing data**. This is a common problem in nearly all types of research. Missing data might be caused by:
   a a lack of response to questionnaires or surveys
   b it not being possible to record the data (for example, if you were surveying number of cars on a road at different times of day, it might not be possible to record data at night if you are asleep).

Missing data reduce the validity of a sample and can therefore distort inferences about the population. Missing values are automatically excluded from analysis. A few data values missing, like 10 in a sample of 1000 might not be a problem, but if the missing data makes up 20% or more of the sample it can be a serious issue.

In a survey:

- One way to minimize missing data in surveys is to avoid asking questions where "not available" or "don't know" answers are available for selection.
- You might be missing data because certain questions only applied to some of the people interviewed. For instance, if a question was only applicable to students who took the SAT, then data might appear to be missing for other students in the school.

2 **Errors in handling data**. Data might be entered incorrectly, or columns within a table might get muddled up; all of which affects the final results that you obtain from the data.

To avoid these issues, data collection should be closely monitored and checked to minimize errors.

What is the difference between a population and a sample?

Suppose you want to know how many people in your city, who drive a car to work, take one or more passengers with them. Here, the **population** is everyone who lives in the city and drives a car to work.

In this scenario, it's very impractical to ask every single person from the population about the number of passengers they take, and so you will probably ask a small selection of drivers. This selection of drivers is called a **sample**; it is a subset of the population.

The **population** consists of every member in the group that you want to find out about.

A **sample** is a subset of the population that will give you information about the population as a whole.

Choosing a sample which gives a good representation of the population

There are a number of ways in which you can draw a sample from the population. You should always try to choose a method which results in the sample giving the best approximation for the population as a whole.

Investigation 1

A group of IB students wish to investigate the mean time that each student in their school spends on homework per night. They would have liked to ask each of the 1500 students in their school, but realized that they do not have enough time to collect and analyse this amount of data, so they decide to ask a sample of students in the hope that the sample will give a good approximation for students in the whole school.

To decide how to choose the sample:

- Beth suggests they should interview their friends.
- Emily suggests they should interview two people from each year group.
- Natasha suggests they should pick 10 boys and 10 girls.
- Amanda suggests they should assign a number to each student in the school, and use a random number generator to choose a sample.
- Greg suggests they obtain a list of all students in the school, organized alphabetically by surname, and choose every 20th person in the list.

HINT

Most calculators have a random number generator.

On the TI-84 press MATH, PRB and randInt (5,...) press enter.

1 In a group, discuss the advantages and disadvantages of each of the five methods. For each suggestion, you should consider

- How easy would it be to obtain a sample?
- What sample size would it give? Is this sample size big enough?
- Would it give results which are representative of all students in the school?

Record your results in a table like this.

Sampling technique	Advantages	Disadvantages
Beth's suggestion: interview their friends		
Emily's suggestion: interview two people from each year group		

2 How could you combine the suggestions of two or more of the students in order to get results which are more representative of all students in the school?

3 Can you think of any other techniques to obtain a sample in this scenario?

4 **Conceptual** Why must you consider the context of a scenario in order to choose an appropriate sampling technique?

In investigation 1, you learned about choosing a sample which would give a good representation of the population as a whole. A sampling **bias** occurs when you take a sample from a population and some members of the population are not as likely to be chosen as others. When a sampling bias occurs, the sample might not give an accurate representation of the population, and there can be inappropriate conclusions drawn about the population.

In practice, an unbiased sampling scheme is very important, but in many cases it is also not easy to produce.

Sampling techniques

In order to produce an unbiased sample, there are a number of established sampling methods which statisticians have developed. The best method for any given scenario will depend on what information you want to find out, and the nature of the population from which you're taking the sample.

The combination of two or more of these sampling techniques might well result in the best sample.

Convenience sampling

- This is the easiest method by which you can generate a sample. You select those members of the population who are most easily accessible or readily available.
- **Example**: To conduct a convenience sample to find the mean length of time spent doing homework, you might survey those students who are in the same class as you.

Simple random sampling

- Each member of the population has an equal chance of being selected. A sample is chosen by drawing names from a hat, or assigning numbers to the population and using a random number generator.
- **Example**: To conduct a simple random sample to find the mean length of time spent doing homework, you might put the names of every student into a hat and draw out the names of 100 students to form a sample.

Systematic sampling

- Here, you list the members of the population and select a sample according to a random starting point and a fixed interval.
- **Example**: If you wanted to create a systematic sample of 100 students at a school with an enrolled population of 1000, you would choose every tenth person from a list of all students.

Stratified sampling

- This involves dividing the population into smaller groups known as strata. The strata are formed based on members' shared characteristics. You then choose a random sample from each stratum, and put them together to form your sample.

- **Example**: In a high school of 1000 students, you could choose 25 students from each of the four year-groups to form a sample of 100 students.

Quota sampling

- This is like stratified sampling, but involves taking a sample size from each stratum which is in proportion to the size of the stratum.
- **Example:** In a high school of 1000 students where 60% of the students are female and 40% are male, your sample should also be 60% female and 40% male.

Reflect In investigation 1, the five students suggested sampling techniques to help investigate the mean average time that students in their school spend doing homework.

Decide which category of sampling technique each of the five suggestions fits into.

HINT

Pair share: one partner describes a sampling method and the other gives an example. Then swap over.

Exercise 6A

1 Classify each of the following as either discrete or continuous data.

  a the number of computers you have owned

  b the length of a computer monitor

  c the weight of your computer

  d the number of laptop bags that you have.

TOK

Why have mathematics and statistics sometimes been treated as separate subjects?

2 Ben is studying the average height of students who attend his school. Choose from *convenience, simple random, systematic, stratified* or *quota* to classify each of the following sampling techniques that Ben might use.

  a A sample of 100 students is taken by organizing the students' names by classification (freshman, sophomore, junior or senior), and then selecting 25 students from each classification.

  b A random name is chosen. Starting with that student, every 50th student is chosen until 80 students are included in the sample.

  c A completely random method is used to select 100 students. Each student has the same probability of being chosen at any stage of the sampling process.

  d The population of the school consists of 70% mathematicians and 30% non-mathematicians. Seven mathematicians and three non-mathematicians are chosen from each grade.

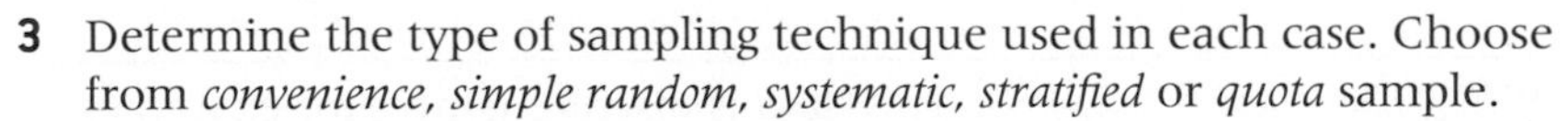

3 Determine the type of sampling technique used in each case. Choose from *convenience, simple random, systematic, stratified* or *quota* sample.

a A fishmonger selects six fish whose lengths are between 11 cm and 15 cm, seven fish whose lengths are between 11 cm and 15 cm, and three fish whose lengths are between 21 cm and 25 cm to test for taste.

b The school newspaper interviews 50 high school female teachers and 50 high school male teachers.

c A DJ chooses every fourth song on a playlist.

d The mathematics club uses a computer to generate 50 random numbers and then picks students whose names correspond to the numbers.

f A local newspaper selects to interview people at the seaside. It specifies that 90% of those interviewed must be local residents and 10% must be tourists.

Developing inquiry skills

In the opening problem for the chapter, you were given the test scores, out of 10, of 32 students.

- Are the test scores an example of discrete or continuous data?
- Before marking every student's test paper, the teacher wishes to choose a sample of eight papers to mark first that will give her an estimate of the mean average mark for the class. Describe a suitable sampling method the teacher could use.

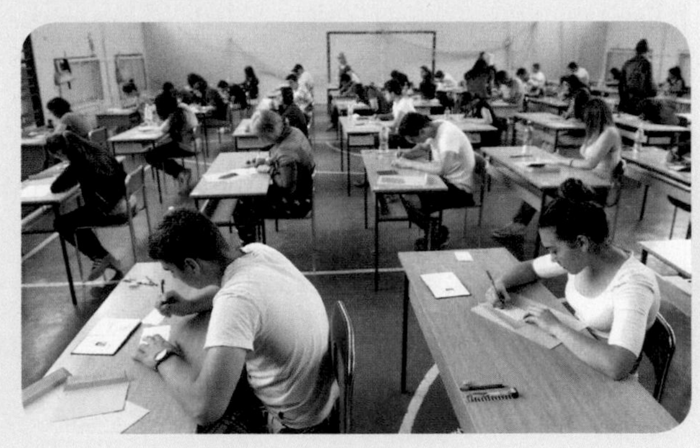

Statistics and probability

6.2 Presentation of data

When you have raw data (primary or secondary data that you have collected), a **frequency table** is an easy way to view your data quickly and look for patterns. From the frequency table, you can plot a histogram. This gives you a visual interpretation of the data.

Example 1

In the opening problem, you studied the marks of 32 students who took a test with maximum mark 10. Their scores are listed below:

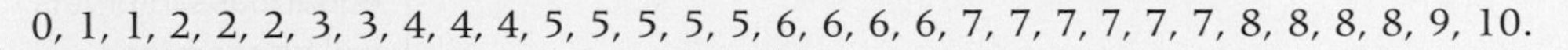

0, 1, 1, 2, 2, 2, 3, 3, 4, 4, 4, 5, 5, 5, 5, 5, 6, 6, 6, 6, 7, 7, 7, 7, 7, 7, 8, 8, 8, 8, 9, 10.

Construct a frequency table for this data and draw a bar chart for it.

Continued on next page

Score	f
0	1
1	2
2	3
3	2
4	3
5	5
6	4
7	6
8	4
9	1
10	1

This data is discrete (since, for example, you cannot have a score of 2.5). So the left-hand column of the table gives a separate row for each score.

The number of students who earned each score is shown in the right-hand column of the table. The f at the top of the column stands for *frequency*.

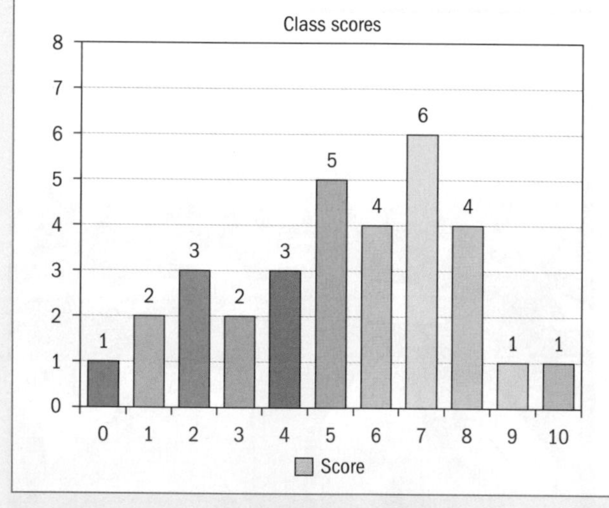

The bar chart for this data shows the test score on the x-axis, and the frequency (that is, the number of students who attained each score) on the y-axis.

Since this is discrete data, we use a bar chart rather than a histogram. You will learn about histograms after the example.

Reflect What do the frequency table and bar chart show you about the test scores that you didn't see from the raw data?

Bar charts and histograms

You should already be familiar with a bar chart. A histogram looks similar to a bar chart, but a bar chart is suitable for discrete data. When you are working with continuous data, you should use a histogram.

The main differences between a histogram and a bar chart are:

- A histogram has no gaps between the bars because it shows continuous data with no gaps. A bar chart has gaps between the bars.
- The horizontal axis on a histogram has a continuous scale, but the horizontal axis on a bar chart has discrete values or categories.

TOK

Can you justify using statistics to mislead others?

How easy is it to be misled by statistics?

HINT

Only frequency histograms with equal class intervals will be examined.

- For a bar chart, the width of each bar does not matter. For a histogram, the width of each bar should span the width of the data class it represents.

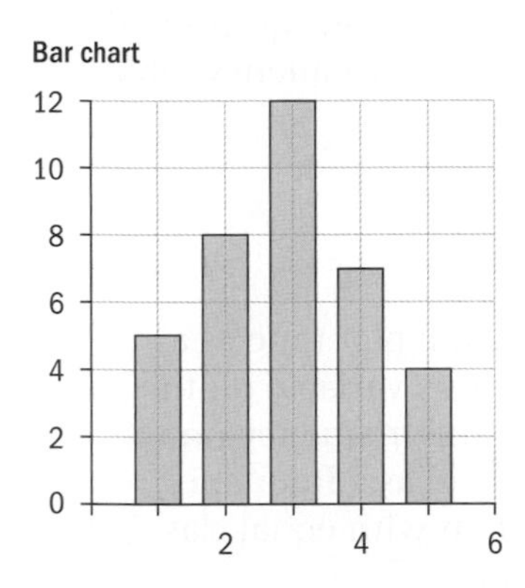

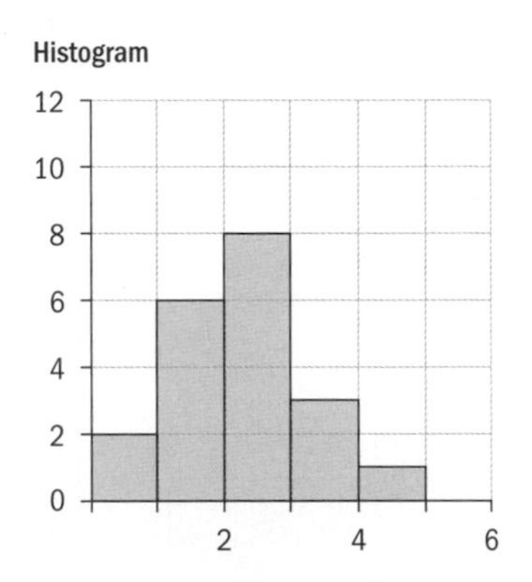

HINT

A bar chart is suitable for discrete data and may have gaps between the bars.

A histogram is suitable for continuous data and has no gaps between the bars.

Describing the shape of a histogram

Constructing a histogram is only the first step in your analysis. To analyse the data, you have to describe what the shape of the histogram tells you about the data. **Skew** is one thing to look for. When data are skewed, the majority of the data are located on one side of the histogram.

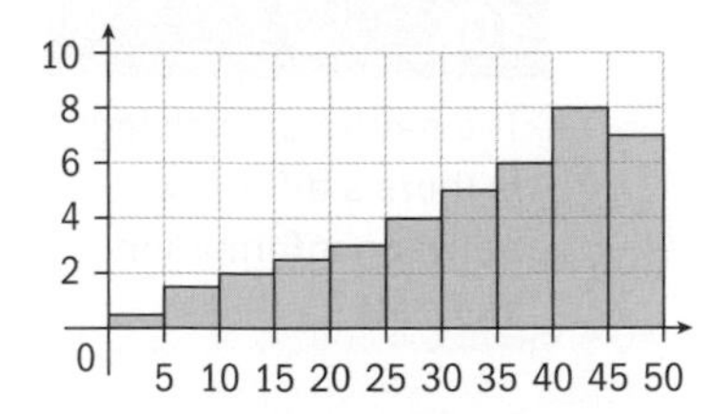

Left or negatively skewed. Most of the data are bunched on the right side of the histogram and the tail is longer on the left side.

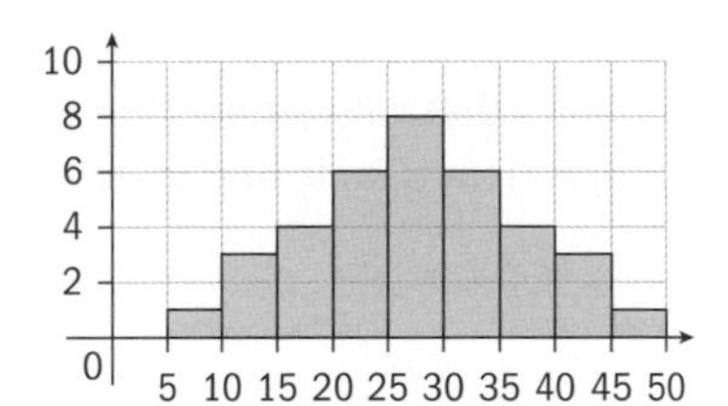

Normal distribution. Data is equally spread out on either side of the peak.

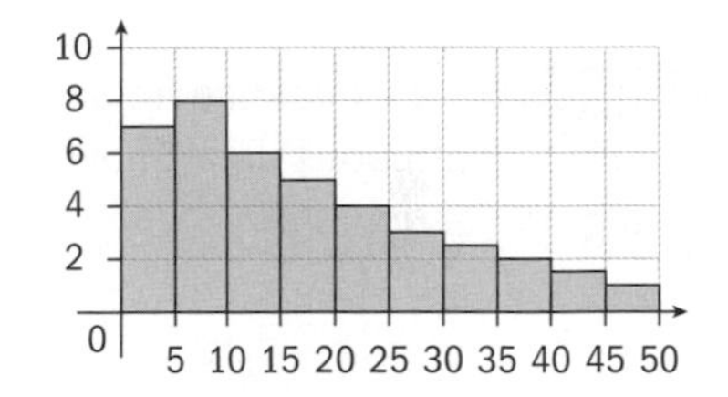

Right or positively skewed. Most of the data are bunched on the left side of the histogram and the tail goes to the right.

Example 2

A group of 37 teachers were asked how much time, in minutes, they spent eating lunch at school. The raw data is listed below:

1, 1, 1, 1, 2, 2, 3, 4, 5, 5, 5, 5, 5, 5, 5, 5, 5, 6, 6, 7, 8, 9, 9, 9, 9, 10, 10, 10, 11, 12, 12, 13, 13, 13, 13, 14, 15, 16.

a Construct a frequency table for this data with classes of $0 < t \leq 4$, $4 < t \leq 8$, etc.

b Construct a histogram for this data.

Continued on next page

a

Time (t, hours)	f
$0 < t \le 4$	8
$4 < t \le 8$	12
$8 < t \le 12$	10
$12 < t \le 16$	7

Having one line for each age would give us a table 16 lines deep! Time is a continuous variable and you can show this as a grouped frequency table.

b

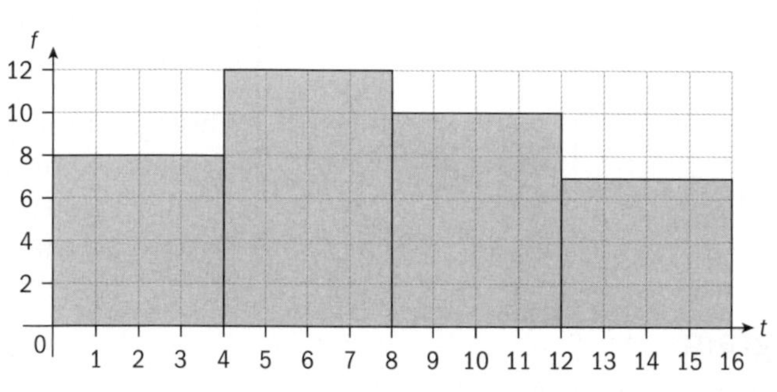

You should plot time as a continuous variable on the x-axis, and frequency on the y-axis. and as a frequency histogram with equal class intervals.

Reflect Write a question for your classmates that will require them to collect continuous data. Show this as a frequency table and draw a histogram for your data.

TOK

The nature of knowing: is there a difference between information and data?

Exercise 6B

1 This set of raw data shows the waiting time, in minutes, at a fast food restaurant.

13, 1.3, 8, 5, 9.25, 10, 7.1, 3, 2, 9, 18, 15, 6, 8, 2.25, 1, 7, 12, 9, 14, 16, 11, 7, 8, 6.8, 11, 20, 5, 4.2, 3.4

a Is the data discrete or continuous?

b Construct a frequency table with class intervals $0 < t \le 4$, $4 < t \le 8$, etc.

c Draw a histogram for this data.

d Describe the skew.

2 This list shows the number of hours spent studying, per student per month, for a class of IB Mathematics students.

50, 49, 15.25, 21.35, 12, 18, 34, 22.51, 45, 24, 30, 20, 52, 26, 23, 40, 15.75, 24, 28.8, 18.6, 35, 40, 38

a Is the data discrete or continuous?

b Construct a frequency table with class intervals $10 < t \le 20$, $20 < t \le 30$, etc.

c Draw a histogram for this data.

d Describe the skew.

3 The masses of a group of children are recorded in the grouped frequency table below.

Mass (m, kg)	f
$30 < m \leq 40$	6
$40 < m \leq 50$	10
$50 < m \leq 60$	12
$60 < m \leq 70$	18
$70 < m \leq 80$	13
$80 < m \leq 90$	9
$90 < m \leq 100$	5

a Is the data discrete or continuous?

b Draw a histogram for this data.

c Describe the skew.

4 A group of biology students measured the lengths of shellfish, in cm, they found at the beach. They produced this histogram to display their results.

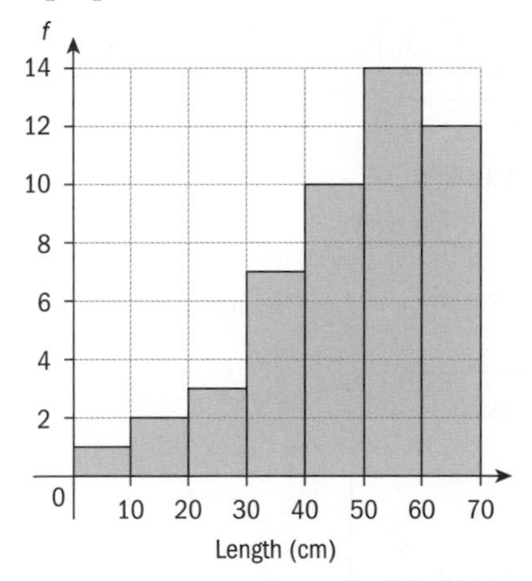

a Construct a frequency table with classes of $0 < l \leq 10$, $10 < l \leq 20$, etc. for this data.

b Describe the skew.

5 The number of hours that a professional tennis player trains each day in the month of May is represented in this histogram.

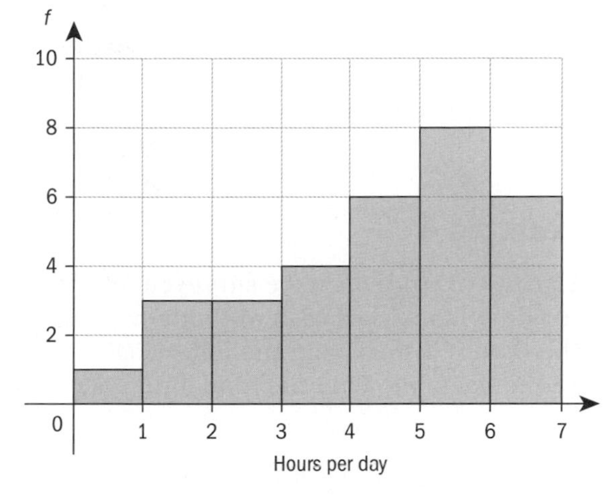

a Construct a frequency table with classes of $0 < h \leq 1$, $1 < h \leq 2$, etc. for this data.

b Describe the skew.

6 Write down everything that you know about histograms and share this with a classmate. Discuss your answers.

6.3 Measures of central tendency

In descriptive statistics, so far, you have seen how to collect and represent a data set. The next step is to choose a single value to represent the data. A measure of central tendency is a measure that tells us where the middle of a set of data lies. The three most common measures of central tendency are the mode, the mean and the median.

EXAM HINT

You will only use the population mean and standard deviation in examinations.

HINT

There can be more than one mode if two or more values occur most often.

If no number occurs more than once in the set, then there is no mode.

The mode

The mode is the value that occurs most frequently in a set of data.

The mode in a list of numbers is the number that occurs most often. A trick to remember this one is to remember that **mo**de starts with the same first two letters that **mo**st does.

Example 3

Find the mode of 4, 7, 3, 3, 1, 2, 7, 5, 7, 11.

The mode is 7.	7 occurs the most at three times.

When presented with a frequency table, the mode is the group with the greatest frequency. In a bar chart or histogram, the mode is the item from the data set with the tallest bar.

Example 4

A group of students were surveyed to find out how many hats each of them owned. The data is displayed in the following frequency table. Find the modal number of hats.

Hats	f
0	3
1	7
2	3
3	3
4	2

The mode is 1 hat.	"1" has a frequency of 7, which is the biggest frequency in the table.

HINT

In example 4, students often make two errors:

Error 1: "The mode is 7."
This is not the case, because 7 is the *frequency* of the modal number of hats, which is 1 hat.

Error 2: "The mode is 3."
This is not the case. 3 is the most common frequency—0 hats, 2 hats and 3 hats all have frequency 3, but the mode is the number of hats with the greatest frequency.

The modal class

When asked for the mode of grouped data, you should state the group that has the highest frequency. This is called the modal class.

You cannot tell which value within that group has the highest mode.

Example 5

This table shows the ages of students who exercise in the gym after school.

Find the modal age.

Age (y, years)	f
$13 < y \leq 14$	2
$14 < y \leq 15$	6
$15 < y \leq 16$	8
$16 < y \leq 17$	10
$17 < y \leq 18$	3

The modal age is $16 < y \leq 17$ years.	The $16 < y \leq 17$ class has the greatest frequency.

Exercise 6C

1 Find the mode of the following sets of data:

a 8, 6, 5, 9, 8, 2, 8

b 3, 4, 7, 1, 13, 4, 2, 11, 9, 4, 10

c 19, 13, 26, 31, 25, 19, 13, 18, 20, 13, 17

d 9, 2, 4, 3, 7, 1, 0, 5

e 5, 6, 2, 7, 4, 2, 1, 0, 4

> **HINT**
>
> A set of data is **bimodal** if it has two modes.
>
> If all of the numbers in a list appear only once, there is **no mode.**

2 Find the mode of each frequency table.

a

Shoe size	f
8	8
9	6
10	9
11	7
12	7

b

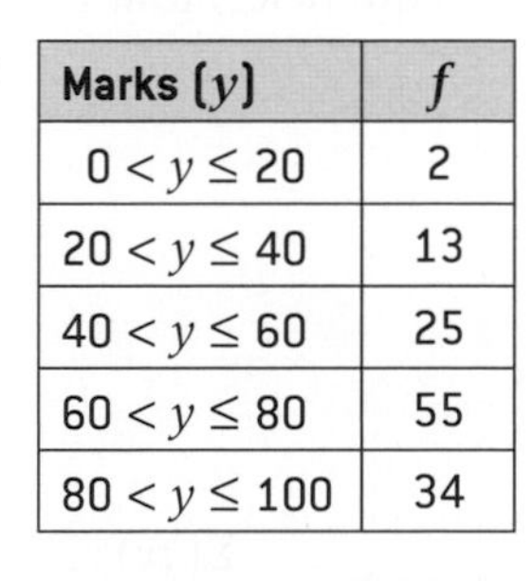

Marks (y)	f
$0 < y \leq 20$	2
$20 < y \leq 40$	13
$40 < y \leq 60$	25
$60 < y \leq 80$	55
$80 < y \leq 100$	34

3 a Find the mode from each histogram.

i

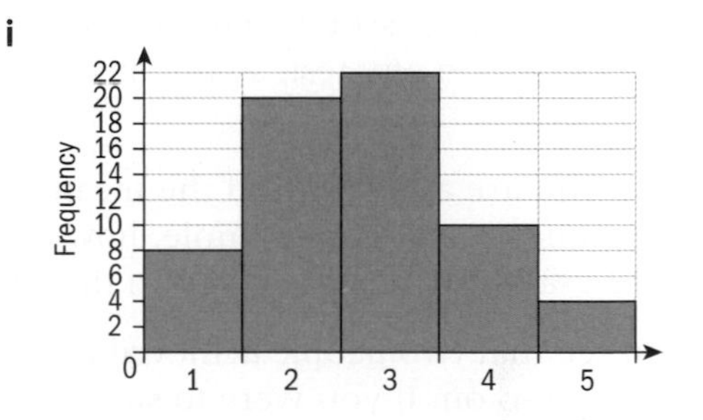

ii

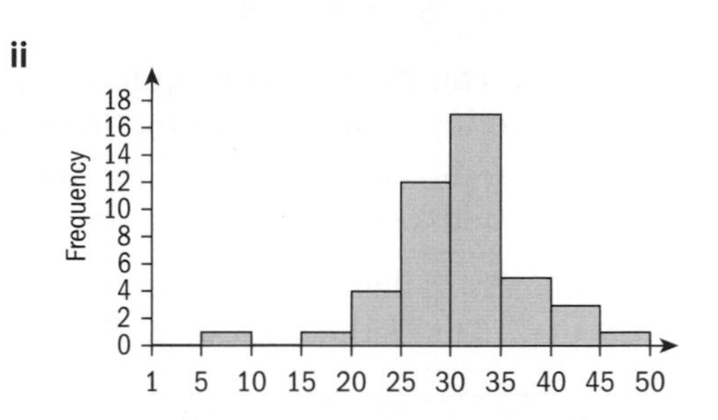

b Consider whether each histogram shows discrete or continuous data.

Reflect Write down what you now know about the mode. Share your answers with a classmate and discuss.

The mean

The arithmetic mean is sometimes called the "average", and is the commonest measure of central tendency. The mean is simply the sum of the values, divided by the number of values in a set of data. It is often written as μ or $\overline{x}$.

$$\text{Mean} = \frac{\text{sum of the data values}}{\text{number of data values}}$$

This can be written mathematically as:

$$\overline{x} = \frac{\Sigma x}{n}$$

where Σx is the sum of the data values and n is the number of data values in the population. Σ is the upper-case Greek letter "sigma", and means "to sum".

The mean gives you a single number that tells you the centre of the data set. It is a representative value: unlike the mode, the mean is often not a member of the raw data itself. For example, your average mathematics score for the year might be 85.73%, even though your teacher always gives scores that are whole numbers. You did not achieve 85.73% in any test, so the mean is not a member of the data itself.

Suppose you are asked to find the mean of data which is presented in a frequency table. For example, how could you find the mean of a class's IB grades, which are given in this table?

Grade (x)	f
7	2
6	4
5	8
4	3
3	2
2	1
1	0

You can see that two people achieved grade 7, four people achieved grade 6, and so on. If you were to sum the data values, you would add up $7 + 7 + 6 + 6 + 6 + 6 + \ldots$ etc.

However, you can do this more quickly by multiplying each grade, x, by its frequency, f, and summing these values, that is, sum of data values $(7 \times 2) + (6 \times 4) + \ldots$. This is the same as $\Sigma(fx)$. Usually, it is most convenient to add a third column to the table to write the values $f \times x$.

The total *number* of data values is the sum of all the frequencies, that is $2 + 4 + 8 + \ldots$. This is the same as $\Sigma(f)$.

Grade(x)	f	fx
7	2	14
6	4	24
5	8	40
4	3	12
3	2	6
2	1	2
1	0	0
	$\Sigma(f) = 20$	$\Sigma(fx) = 98$

So, the mean is $\overline{x} = \dfrac{\Sigma(fx)}{\Sigma(f)}$

$$= \frac{98}{20} = 4.9$$

When the data is given in a grouped frequency table, you have to assume that the data is equally spread out in each of the classes, and use the mid-interval value of each class as the representative value for that class. Because of this assumption, you can only find an estimate of the mean.

This process is shown in example 6.

TOK

What are the benefits of sharing and analysing data from different countries?

Example 6

This table shows the ages (in years) of 10 pet cats.

Age (x, years)	f
$0 < x \leq 2$	2
$2 < x \leq 4$	4
$4 < x \leq 6$	3
$6 < x \leq 8$	1

Find an estimate of the mean age of the cats.

Age (x, years)	f	Mid-interval value (m)	fm
$0 < x \leq 2$	2	1	2
$2 < x \leq 4$	4	3	12
$4 < x \leq 6$	3	5	15
$6 < x \leq 8$	1	7	7
	$\Sigma f = 10$		$\Sigma fm = 36$

Add a third column to the table and write in the mid-interval value m of each class.

To find m, add the first and value of the interval and divide by 2.

Add a fourth column to the table and multiply f by m for each class.

Find Σf and Σfm.

$$\bar{x} = \frac{\Sigma fm}{\Sigma f} = \frac{36}{10} = 3.6$$

Use the formula to find an estimate of the mean.

The average age of the cats is 3.6 years.

State your result in the context of the question.

Sometimes, you may be given an average of a group of data and asked to find the data value that is missing.

Example 7

Will has scored 85, 93, 92 and 84 on the first four mathematics tests of the term. Find the score that Will must get on the last test to finish with an average of 90%.

Let x be Will's final test score.

$$\frac{85+93+92+84+x}{5} = 90$$

The sum of all five test scores is 85 + 93 + 92 + 84 plus Will's final test.

$$\frac{354+x}{5} = 90$$

$$354 + x = 450$$

Use the formula for the mean.

$$x = 96$$

Solve for x.

Will must score 96 on the last test.

Exercise 6D

1 Use your GDC to help you find the mean of the data in each of the frequency tables by entering the data into the statistics function on your calculator.

a

x	f
6	12
5	11
4	8
3	12
2	7
1	10

b

x	f
10	2
11	8
12	4
13	9
14	7
15	6

c

x	f
2.5	4
3	5
3.5	6
4	3
4.5	2

2 a

x	f
$0 < x \le 10$	18
$10 < x \le 20$	14
$20 < x \le 30$	12
$30 < x \le 40$	9
$40 < x \le 50$	7

b

x	f
$0 < x \le 12$	4
$12 < x \le 24$	0
$24 < x \le 36$	8
$36 < x \le 48$	15
$48 < x \le 60$	13
$60 < x \le 72$	7

c

x	f
$1 < x \le 1.5$	4
$1.5 < x \le 2$	6
$2 < x \le 2.5$	7
$2.5 < x \le 3$	7
$3 < x \le 3.5$	5

3 A group of teachers were asked how many cups of coffee they drank in a typical day. The results are shown in this table.

Cups	0	1	2	3	4	5
f	6	5	4	7	10	4

Find the mean number of cups of coffee.

4 Phil grows tomato plants in his garden. He records the number of tomatoes on each plant, and presents his data using this table.

Tomatoes	3	4	5	6	7	8	9	10
f	2	4	4	6	10	15	4	5

a How many tomato plants did Phil have in total?

b What is the modal number of tomatoes per plant?

c What is the mean number of tomatoes per plant?

5 Noot records the number of fish that she catches each day throughout the month of January.

What is the mean number of fish that she catches per day?

Fish	0	1	2	3	4	5	6	7	8	9	10
f	1	5	4	2	3	5	3	2	3	1	2

6 Mo works as a barista in a coffee shop and records the total amount, per day, that he receives in tips.

Tip($)	f
$0 < \$ \le 10$	6
$10 < \$ \le 20$	14
$20 < \$ \le 30$	15
$30 < \$ \le 40$	8
$40 < \$ \le 50$	2

What is his average amount of tips per day, to the nearest cent?

7 The bar chart shows the number of children per family in a selection of families.

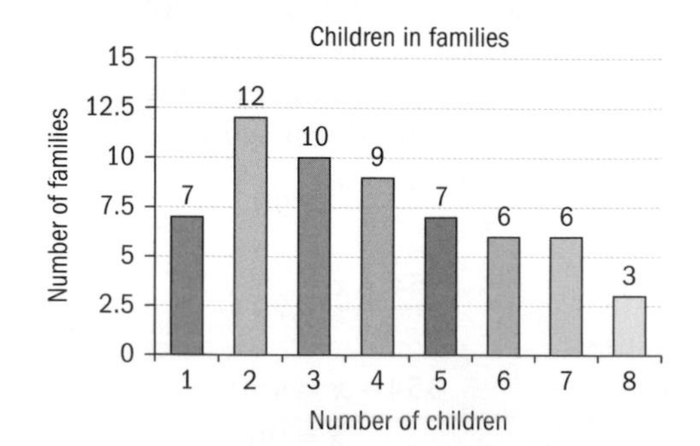

a How many families are represented?

b Describe the skew.

c Write down the mode of the distribution.

d Find the mean number of children per family, to the nearest tenth.

8 The histogram represents the ages of the people in a village.

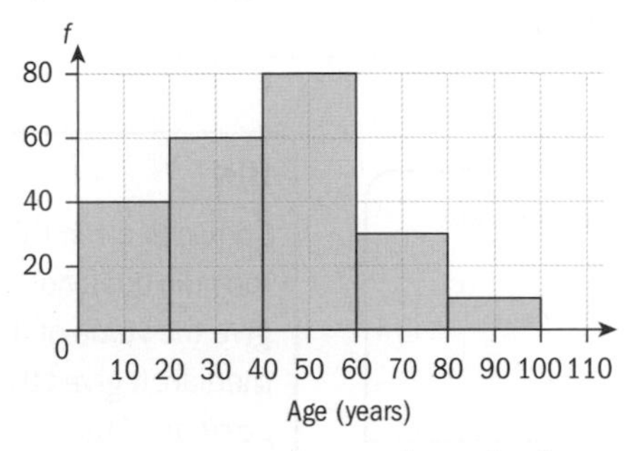

a How many people are there in the village?

b Write down the modal class of the distribution.

c Find the mean age.

9 If the mode of this set of numbers is 2 and the mean is 5, find the values of a and b.

$$1, 2, a, 4, 5, 6, b, 8, 10$$

10 The mean of $8, x, 17, 2x + 3$ and 45 is 23. Find the value of x.

11 Six positive integers, all less than 8, have a mean of 4 and a mode of 4. Find the numbers, given that 2 and 3 are the first two.

12 In an IB class, there are 8 boys and 12 girls.

The mean mass of the boys is 52 kg and the mean mass of the girls is 44 kg.

What is the mean mass of the whole class of 20 students?

Investigation 2

1 What is the total number of smiles per day that you would expect from a group of 19 adults?

One child joins the group.

2 What is the total number of smiles per day that you would now expect from the group?

3 What is the average number of smiles per person now that the group has 19 adults and 1 child?

4 **Conceptual** What happens to the mean when we add one extreme value to the data set?

5 What other measure of central tendency might be more appropriate to measure the average number of smiles per day in this scenario?

The median

The median is the middle data value when the data values are arranged in order of size. If the number of values in a data set is even, then the median is the mean of the two middle numbers.

International-mindedness

The 19th century German psychologist Gustav Fechner popularized the median although French mathematician Pierre Laplace had used it earlier.

Example 8

Find the median of 7, 12, 1, 4, 2, 17, 9, 11, 16, 10, 18.

1, 2, 4, 7, 9, 10, 11, 12, 16, 17, 18	Arrange the data in order of size.
	Find the data value in the middle
Median = 10	There are 11 numbers. The median will be the sixth data value.

If there are a lot of numbers and it is difficult to find the middle member, we can use the formula median $= \left(\frac{n+1}{2}\right)$th data value, where n is the number of data values in the set.

HINT

Common error: This formula does not give the *value* of the median. It gives the *position* of the median in the data set.

Exercise 6E

1 Find the median of each data set.

a 11, 14, 17, 18, 19, 20, 26

b 11, 14, 17, 18, 19, 20, 26, 28

c 4, 2, 5, 5, 1

d 9, 8, 8, 7, 4, 3, 3, 2, 0, 0

e 5, 10, 8, 6, 4, 8, 2, 5, 7, 7

f 2, 8, 5, 6, 3, 5

2 Craig keeps chickens and collects their eggs every day. Find the median number of eggs that he collects in a day.

Eggs	5	6	7	8	9	10	11	12
f	2	4	3	7	11	18	6	2

3 A factory owner and the factory workers in Indobodia are involved in a dispute over wages. They both start with the basic data of monthly income (in US$).

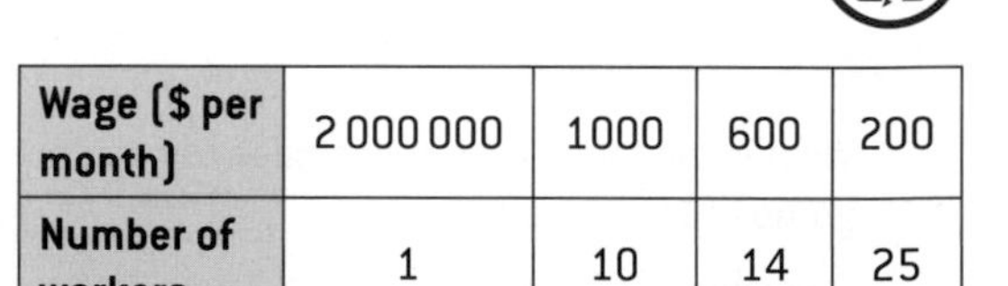

Wage ($ per month)	2 000 000	1000	600	200
Number of workers	1	10	14	25

Find

a the mean

b the mode

c the median.

Reflect **Discuss** your answer with a partner then share your **ideas** with your class.

- Which measure of central tendency would the workers choose and why?
- Which measure of central tendency would the owner choose and why?
- Which measure of central tendency do you think is most fair and why?

What will happen to the measures of central tendency if we add the same amount to all data values, or multiply each data value by the same amount?

TOK

Do different measures of central tendency express different properties of the data?

How reliable are mathematical measures?

Reflect Copy and complete the following table. You should use your GDC to calculate the mean, mode and median each time.

	Data	Mean	Mode	Median
Data set	6, 7, 8, 10, 12, 14, 14, 15, 16, 20			
Add 4 to each value				
Multiply each value by 2				

Now copy and complete the following sentences to explain what happens to the mean mode and median of the original data set.

a If you add 4 to each data value ....

b If you multiply each data value by 2 ....

Mode

The mode tells you the most frequently occurring value in a data set.

Advantages:

- Extreme values do not affect the mode.

Disadvantages:

- Does not use all members of the data set.
- Not necessarily unique—may be more than one answer.
- When no values repeat in the data set, the mode is every value and is useless.
- When there is more than one mode, it is difficult to interpret and/or compare.

Mean

The mean is the sum of all data values, divided by the number of values in the set.

Advantages:

- Most popular measure of "average" in fields such as business, engineering and computer science.
- Uses all members of the data set.
- It is unique—a data set has only one mean.
- Useful when comparing sets of data.

Disadvantages:

- Affected by extreme values.

Median

The median describes the middle of a set of data.

Advantages:

- Extreme values do not affect the median as strongly as they do the mean.
- Useful when comparing sets of data.
- It is unique—a data set has only one median.
- As the median is the middle value, 50% of the data is either side of it.

Disadvantages:

- Does not take into account every value in the data set.
- Less use in further calculations.

Developing inquiry skills

1. Find the mean, mode and median of the class test scores from the start of this chapter.
2. Which of mean, median or mode gives the best indication of the "averages" score? Are any two averages roughly equal?

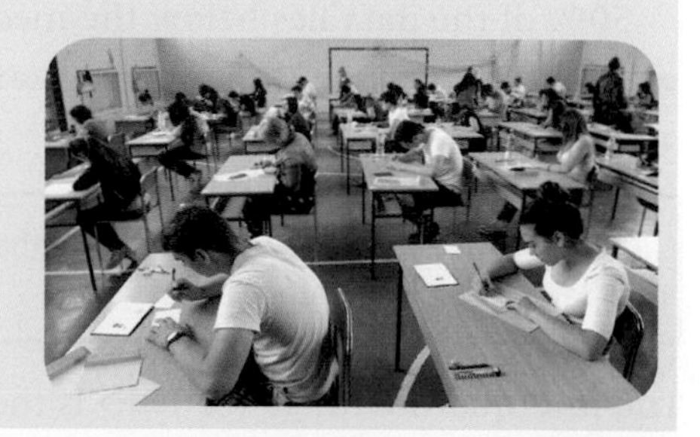

Statistics and probability

6.4 Measures of dispersion

Measures of central tendency (mean, median, mode) help you to explore the middle of a data set, but do not always give you a clear picture of the whole data set.

One of the measures of dispersion that will give you more information about the data is called the **range**.

The range is the difference between the maximum and minimum values of the data set.

$$\text{Range} = \max - \min$$

The range is a simple measure of how spread out the data is (the **dispersion**).

In the opening problem, you studied the test results of 32 students:

0, 1, 1, 2, 2, 2, 3, 3, 4, 4, 4, 5, 5, 5, 5, 5, 6, 6, 6, 6, 7, 7, 7, 7, 7, 7, 8, 8, 8, 8, 9, 10

Here, the minimum test score is 0 and the maximum score is 10.

Therefore, the range is 10 − 0 = 10.

Like the mean, the range can be affected by one very large or small value.

Suppose the test scores of ten students from one class were 0, 1, 1, 2, 2, 2, 3, 3, 4, 10.

Here, the range is still 10 − 0 = 10, but the data is much more compact than the range suggests. The value 10 influences the range, so that it is not representative of the spread of the whole data.

Knowing whether the data is widely spread out or bunched around the mean provides you with a better understanding of the distribution of the whole data. You need to consider more values than just the smallest and the largest.

Quartiles

Quartiles are values that divide the data into quarters.

- The first quartile (called the **lower quartile** or Q_1) has 25% of the data below it.
- The second quartile is the median; the middle value in the data set. 50% of the data lies below the median, and 50% lies above it.
- The third quartile (the **upper quartile** or Q_3) has 75% of the data below it.

Q_1 is the median of the lower 50% of the data, and Q_3 is the median of the upper 50% of the data.

The interquartile range (or IQR) is the difference between Q_3 and Q_1.

HINT
Algorithms for quartiles are implemented in various calculators, spreadsheets and statistical software packages in diverse ways that give rise to slightly different results, particularly for a small sample size.

Reflect The values from Q_1 to Q_3 are sometimes called the range of the "middle half" of the data. Can you explain why is this the case?

The important advantage of interquartile range is that it can be used to give a more accurate measure of the data spread if extreme values distort the range.

$$\text{IQR} = Q_3 - Q_1$$

Example 9

For the data set 1, 4, 9, 16, 25, 36, 49, 64, 81, 100, 121, 144, find:

a the median **b** the lower quartile
c the upper quartile **d** the interquartile range
e the range.

a Median $= \left(\frac{n+1}{2}\right)\text{th} = \left(\frac{12+1}{2}\right)\text{th}$ $= 6.5\text{th} = 42.5$	First, you should make sure the data values are written in ascending order! There are 12 numbers so $n = 12$. The median is the mean of 36 and 49. $\frac{36+49}{2} = 42.5$
b $Q_1 = 12.5$	Q_1 is the median of the lower half. It lies between 9 and 16. $\frac{9+16}{2} = 12.5$
c $Q_3 = 90.5$	Q_3 lies between 81 and 100. $\frac{81+100}{2} = 90.5$
d IQR $= Q_3 - Q_1 = 90.5 - 12.5 = 78$ **e** Range = max − min = 144 − 1 = 143	

You can visualize the four quarters of the data set from example 9:

First quarter	Second quarter	Third quarter	Fourth quarter
1, 4, 9	16, 25, 36	49, 64, 81	100, 121, 144

TOK

To what extent can we rely on technology to produce our results?

Exercise 6F

1 Find **a** the median, **b** the lower quartile, **c** the upper quartile, **d** the IQR, **e** the range of 3, 6, 7, 7, 8, 8, 9, 10, 12, 12, 15.

2 Find **a** the median, **b** the lower quartile, **c** the upper quartile, **d** the IQR, **e** the range of 4, 5, 5, 7, 9, 8, 15, 6, 2, 10, 6, 5.

3 Lincy records the number of sit-ups she does each day over a period of 16 days. These are her results:

12, 25, 16, 37, 10, 25, 30, 14, 16, 50, 45, 2, 20, 28, 10, 40

a Find the median number of sit-ups.

b Find the interquartile range.

Copy and complete the following sentences.

c On 8 of the 16 days, Lincy did more than .... sit-ups.

d The "middle half" of the number of sit-ups Lincy did was between .... and ....

e On 4 of the 16 days, Lincy did more than .... sit-ups.

4 The number of cars passing a school was recorded every hour for 11 hours. The results were:

20, 20, 35, 45, 50, 35, 35, 45, 25, 30, 35

Find the interquartile range.

5 This table shows the number of passengers in each of 16 cars.

People	2	3	4	5	6
Cars	6	3	3	2	2

Find the interquartile range.

6 The data below is listed in ascending order:

5, 6, 7, 7, 9, 9, r, 10, s, 13, 13, t

The median of the data is 9.5. The upper quartile Q_3 is 13.

a Write down the value of:

i r

ii s

b The mean of the data is 10. Find the value of t.

Box plots

It is possible to get a sense of a data set's distribution by examining a **five-number summary.** This is:

1 the minimum value

2 the first quartile

3 the median (or second quartile)

4 the third quartile

5 the maximum.

HINT

Make sure you know how to use your GDC to find a five-number summary and construct a box plot.

A five-number summary can be represented graphically as a **box-and-whisker diagram** (sometimes called a box plot).

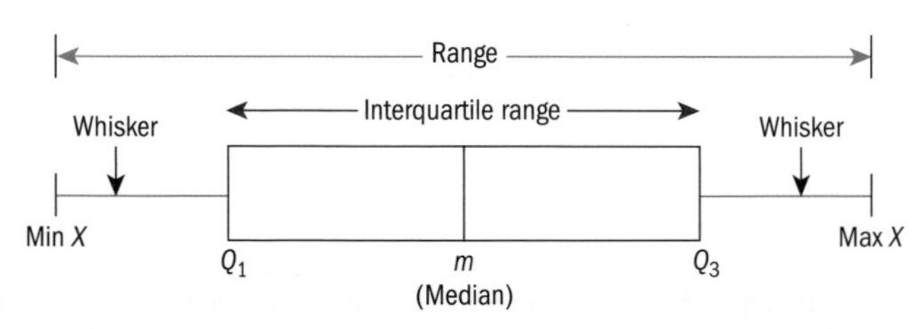

The first and third quartiles are at the ends of the box; the median is indicated with a vertical line in the interior of the box; and the maximum and minimum points are at the ends of the whiskers. The diagram should be drawn to scale along a single, horizontal axis. The test data from the class scores at the start of this chapter is displayed in this box-and-whisker plot below.

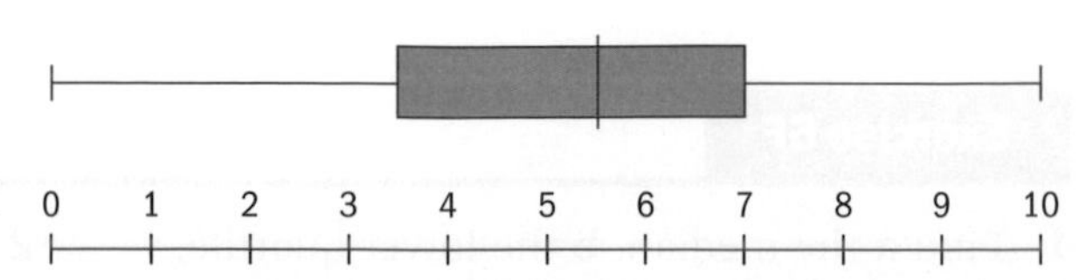

Outliers

Extreme values are called **outliers** and might be considered for removal from a sample.

> An outlier is any value at least $1.5 \times \text{IQR}$ above Q_3 or below Q_1.

Example 10

a Using your GDC, find the median and upper and lower quartiles for this data set:

37, 43, 43, 44, 44, 46, 46, 47, 47, 47, 47, 48, 51, 52, 53, 53, 54

b Show this data on a box plot.

c Determine whether 37 is an outlier.

a Median = 47 Lower quartile = 44 Upper quartile = 51.5	Use your GDC to calculate the median and quartiles.
b 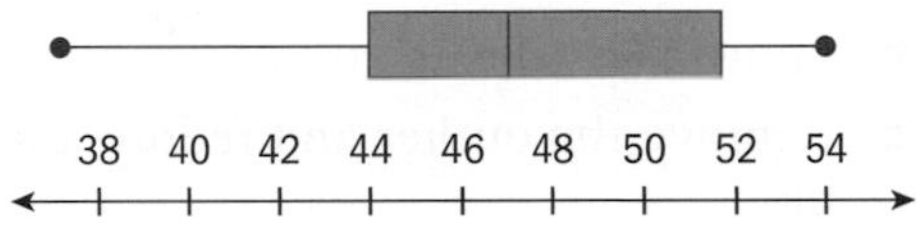	Display the information on a box-and-whisker plot.
c IQR = 51.5 − 44 = 7.5 $Q_1 - 1.5(\text{IQR}) = 44 - 1.5(7.5) = 32.75$ 32.75 < 37, so 37 is not an outlier.	Use $\text{IQR} = Q_3 - Q_1$. Determine the value that is 1.5 IQR below Q_1. Compare 37 with the boundary for outliers.

> The size of the box or length of the whisker does **not** tell you how many data values lie within that area. Each whisker, and each section of the box, all contain 25% of the data values. The wider the spread of the whisker or box, the more spread out the data is. In example 10, the lowest 25% are the most spread out, whilst the data values between Q_1 and Q_2 are the least spread out.

Deciding when to keep and when to reject outliers

You have seen how to identify an outlier, and now you must decide whether to reject or accept it. Outliers can have a distorting effect on statistical measures like the mean, but some of them are still valid data values and it is not acceptable just to reject an outlier without a reason.

There are occasions when an outlier appears due to an error in data collection or entry, and that outlier should be removed.

A case to remove an outlier

For example, the heights of 15 boys, in cm, are 174, 166, 182, 177, 164, 382, 175, 170, 181, 173, 168, 172, 166, 172, 176.

Clearly there is an error in measurement or recording: a boy cannot be 382 cm tall! This is an outlier that should be discarded.

A case to keep an outlier

Suppose the test results of seven students are 20%, 22%, 18%, 30%, 26%, 89%, and 21%.

If the data is checked and the student who scored 89% was not an error in recording, then this has significant effects on the conclusions that you draw: without the outlier, a teacher might conclude the test was too difficult. With the outlier, the teacher might conclude that the test was fine, but that six of the seven students had not studied properly!

Exercise 6G

1 Draw a box-and-whisker plot for this data:

Minimum	First quartile	Median	Third quartile	Maximum
59	62	66	69	72

2 Mario conducted a science experiment to see how many minutes it took cups of ice to turn into water in the classroom.

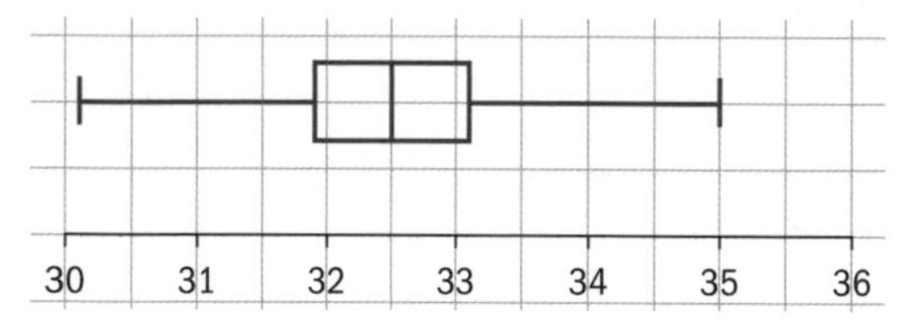

a What was the minimum time?

b What was the maximum time?

c What was the median time?

d What was the interquartile range of the times?

3 The number of students in international schools in the country of Portmany is shown below.

20, 600, 650, 710, 790, 800, 855, 878, 943, 990, 1000, 1500, 1700, 1850, 2000, 2150, 2400, 2500, 2135, 6000

a Show this information as a box-and-whisker plot.

b Show that 6000 is an outlier.

c Remove the outlier and redraw the box plot.

d State a reason to remove the outlier.

Reflect **Discuss** your answers with a partner and comment on which is a better chart and why.

4 Test scores for an IB morning exam are:

99, 56, 78, 55.5, 32, 90, 80, 81, 56, 59, 45, 77, 84.5, 84, 70, 72, 68, 32, 79, 90

Test scores for an IB afternoon exam are:

98, 78, 68, 83, 81, 89, 88, 76, 65, 45, 98, 90, 80, 84.5, 85, 79, 78, 98, 90, 79, 81, 25.5

a Draw a box-and-whisker plot for each data set. Use one number line for both plots, and draw one above the other.

Reflect **Discuss** your answers with a partner and decide which class has the better scores.

b State which examination has the greatest spread for the middle 50%.

c Describe what your answer to part **b** tells you about the range in performance for the two exams.

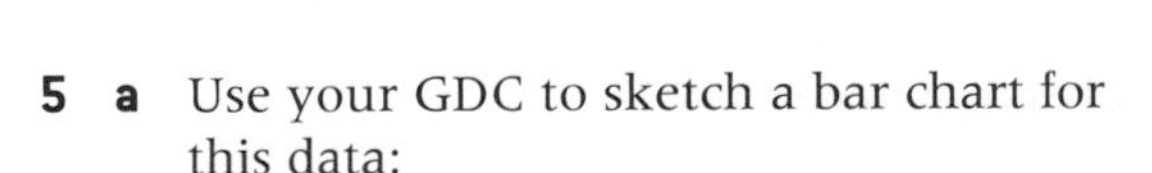

5 a Use your GDC to sketch a bar chart for this data:

x	0	1	2	3	4	5	6	7	8	9	10
f	0	5	10	8	6	4	2	2	1	1	1

b Use your GDC to sketch a box plot for the same data.

c Describe the skew of the data.

Reflect **Discuss** with a partner how skew is seen in a box plot, and share your answers with the class.

6 Match each histogram with its corresponding box-and-whisker plot.

1
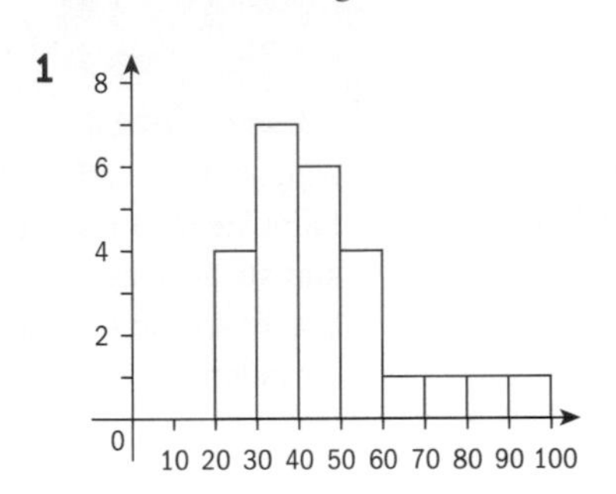

2
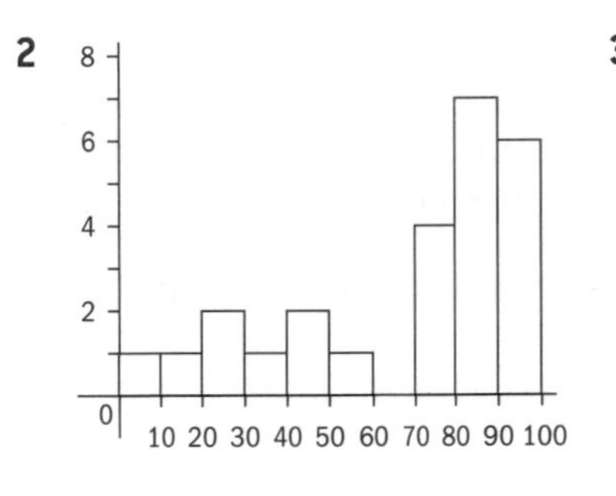

3
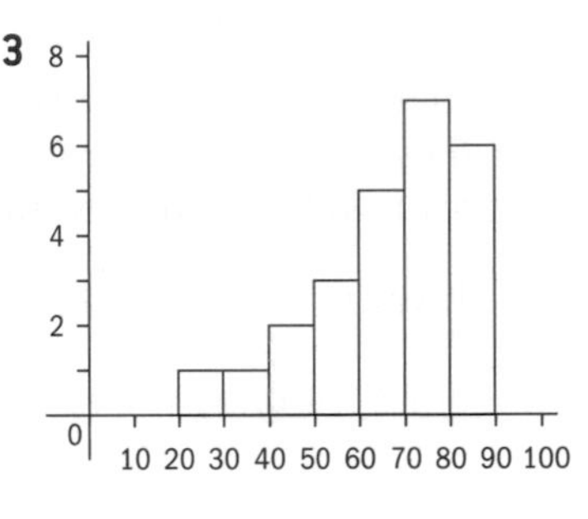

A
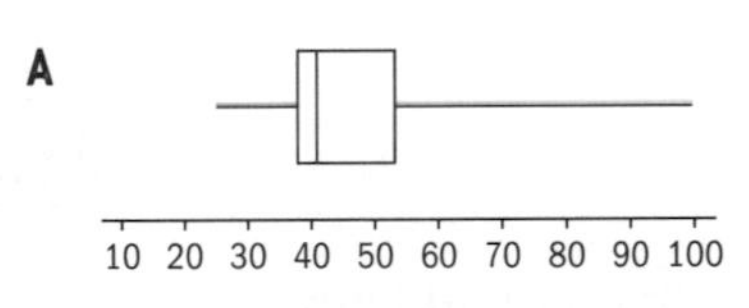

B
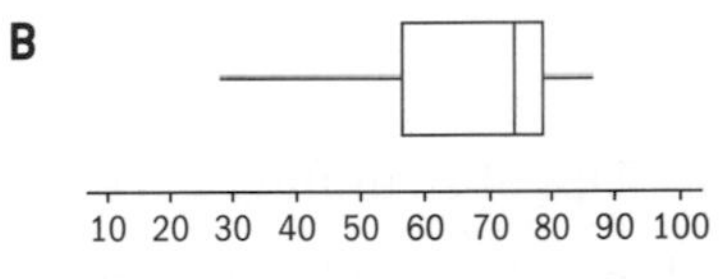

C
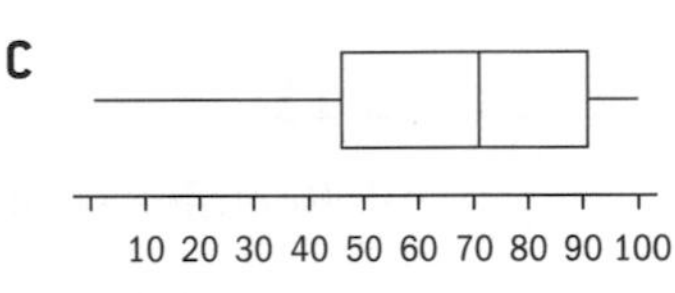

Cumulative frequency

Investigation 3

The data below shows the number of times that 50 students have lost a pencil this week:

5, 9, 10, 5, 9, 9, 8, 4, 9, 8, 5, 7, 3, 10, 7, 7, 8, 7, 6, 6, 9, 6, 4, 4, 10, 5, 6, 6, 3, 8, 7, 8, 3, 4, 6, 6, 5, 7, 5, 4, 3, 5, 2, 4, 2, 8, 1, 0, 3, 5

1 Construct a frequency table for this data.

2 Copy and complete this table:

Pencils	0	1	2	3	4	5	6	7	8	9	10
Number of students who lost this many or fewer (cumulative frequency)	1	2	4	9							

The **cumulative frequency** for x pencils lost is found by adding up the frequency of students who have lost "x pencils or fewer".

3 Plot the number of pencils on the x-axis and cumulative frequency on the y-axis. Join up the points to form a smooth curve.

4 Explain what this graph tells you about how the number of students losing x pencils changes as x increases? Discuss your answer with a classmate.

5 Explain to a classmate why the cumulative frequency curve cannot turn down or to the left.

When you have raw data (for example, a list of numbers), you can use the formula:

$$\text{median} = \left(\frac{n+1}{2}\right)\text{th value}$$

to find the median and quartiles when the data is arranged in ascending order.

However, when you have grouped data, it is hard to tell accurately what the median or quartile might be if the $\left(\frac{n+1}{2}\right)$th value lies in the middle of a group.

Cumulative frequency curves enable you to find the median and quartiles from a set of grouped data. To find the median, you draw a horizontal line across from the frequency axis to the curve at the $\left(\frac{n}{2}\right)$th value, and then down to the x-axis to determine the median data point. Finding the quartiles is similar.

International-mindedness

Why are there different formulae for the same statistical measures like mean and standard deviation.

Example 11

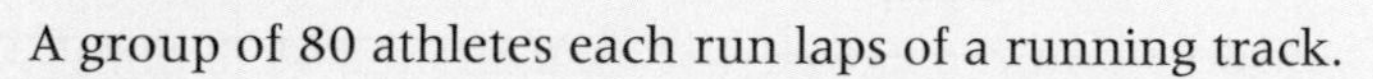

A group of 80 athletes each run laps of a running track.

The grouped frequency table shows the number of laps run by different athletes.

Laps (x)	f
$0 < x \leq 10$	1
$10 < x \leq 20$	20
$20 < x \leq 30$	31
$30 < x \leq 40$	21
$40 < x \leq 50$	7

Construct a cumulative frequency table for this data, and use it to draw a cumulative frequency curve for the number of laps run by different athletes.

Use the curve to estimate the median and interquartile range of number of laps.

Laps (x)	f	CF
$0 < x \leq 10$	1	1
$10 < x \leq 20$	20	21
$20 < x \leq 30$	31	52
$30 < x \leq 40$	21	73
$40 < x \leq 50$	7	80

CF stands for cumulative frequency

Add a CF column to the table.

f	CF
1	1
20	$1 + 20 = 21$
31	$1 + 20 + 31 = 52$
21	$1 + 20 + 31 + 21 = 73$
7	$1 + 20 + 31 + 21 + 7 = 80$

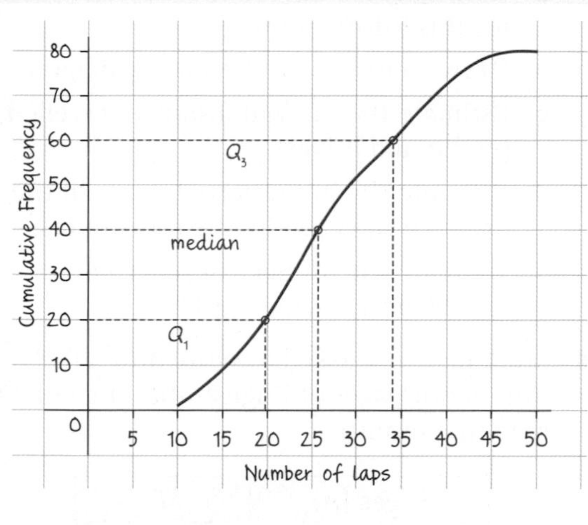

Median = 26 laps

$Q_1 = 20$, $Q_3 = 34$

IQR = 34 − 20 = 14 laps

You should plot the cumulative frequency against the end point of the data class. Then join your points up with a smooth curve.

Median $= \left(\frac{80}{2}\right)$th = 40th data value.

Draw a line across from 40 on the vertical axis until it hits the curve and then draw a line vertically down until it touches the horizontal axis. This is the median. Similarly,

$Q_1 = \left(\frac{80}{4}\right)$th = 20th data value.

$Q_1 = 3\left(\frac{80}{4}\right)$th = 60th data value.

You will notice that, for large data sets, there is often very little difference between using $\left(\frac{n+1}{2}\right)$ and $\left(\frac{n}{2}\right)$ to find the median.

Notice also that the example used "estimate" instead of "find". It is assumed that the number of laps run by different athletes are equally spread out through the class intervals. Since this is not necessarily the case in reality, the median and quartiles might be an approximation of the true values.

Percentiles

A percentile is a number where a certain percentage of scores fall below that percentile. If you score 64 out of 80 on a test, you do not know how well you have done unless you compare your score to the whole class. If you know that your score is in the 90th percentile, that means you scored better than 90% of people who took the test.

The mean is often called the 50th percentile, the lower quartile is the 25th percentile and the upper quartile is the 75th percentile.

Exercise 6H

1 The time, in minutes, taken by 100 students to reply to their friends on social media is shown as a cumulative frequency curve.

a Find the longest time taken to reply.

b Estimate the median time.

c Estimate the interquartile range in time taken to reply.

d 90% of the students replied in k minutes or less. Find k.

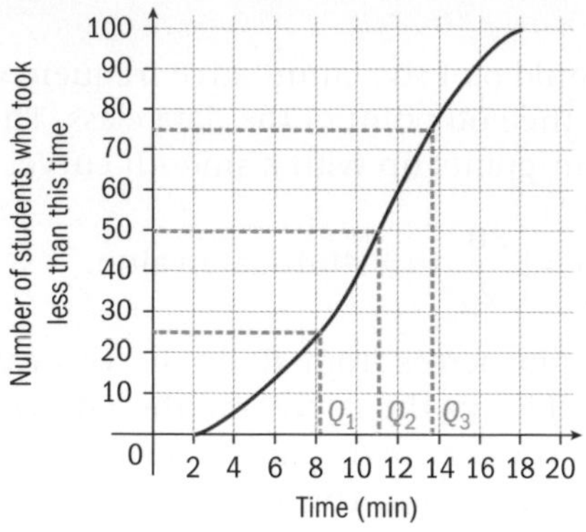

2 The marks obtained by 100 students are shown on this cumulative frequency curve. Estimate:

a the median

b the interquartile range

c the lowest mark needed to be in or above the 80th percentile

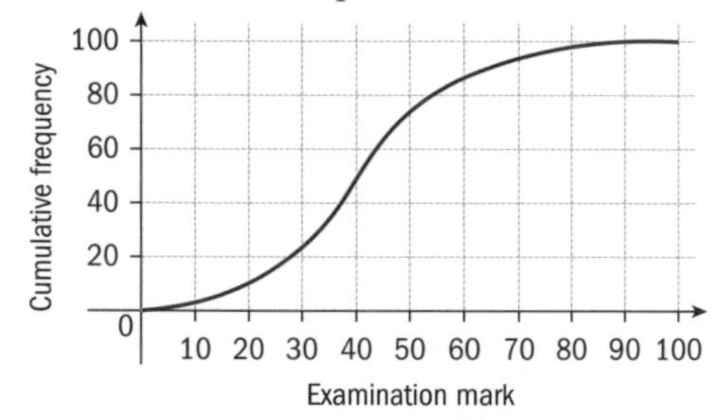

3 A taxi company recorded the distance (km) travelled by each of its drivers one Saturday evening.

Distance (d, km)	f
$0 < d \le 25$	0
$25 < d \le 50$	32
$50 < d \le 75$	102
$75 < d \le 100$	86
$100 < d \le 125$	16
$125 < d \le 150$	4

a Construct a cumulative frequency table for this information.

b Draw a cumulative frequency diagram.

c Estimate the median distance travelled by the taxi drivers.

d Estimate the interquartile range in the distance travelled by the taxi drivers.

e Estimate the number of cars that travelled more than 130 km.

4 Students were asked to write down the number of pages of English that they had to read in a month.

Pages (p)	f
$100 < p \le 200$	12
$200 < p \le 300$	36
$300 < p \le 400$	42
$400 < p \le 500$	53
$500 < p \le 600$	33
$600 < p \le 700$	20
$700 < p \le 800$	4

a Construct a cumulative frequency table for this information.

b Draw a cumulative frequency diagram.

c Estimate the median.

d Estimate the interquartile range.

e Estimate the number of students who read more than 450 pages.

5 Match each cumulative frequency curve to its corresponding box-and-whisker plot.

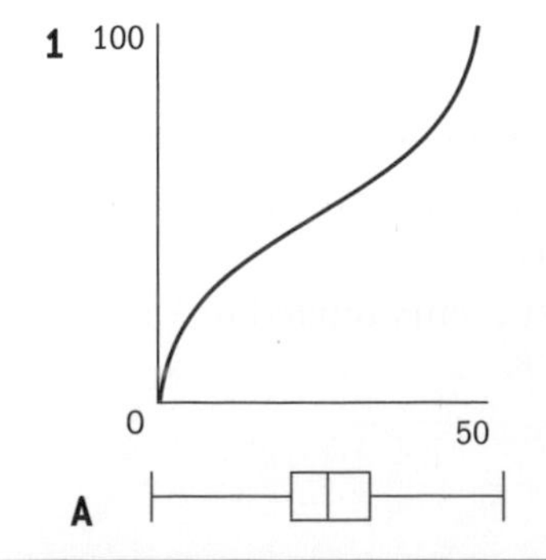

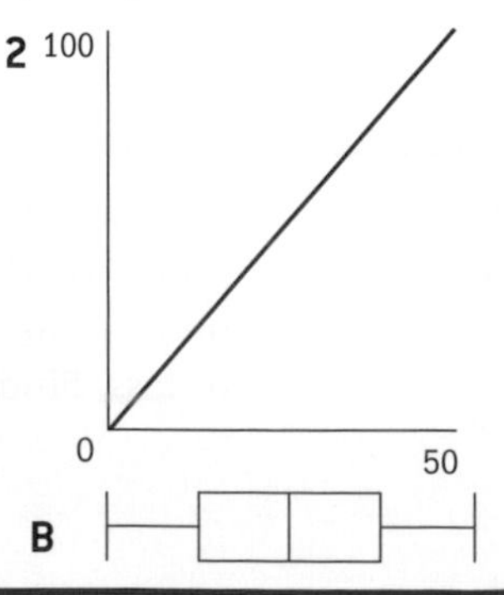

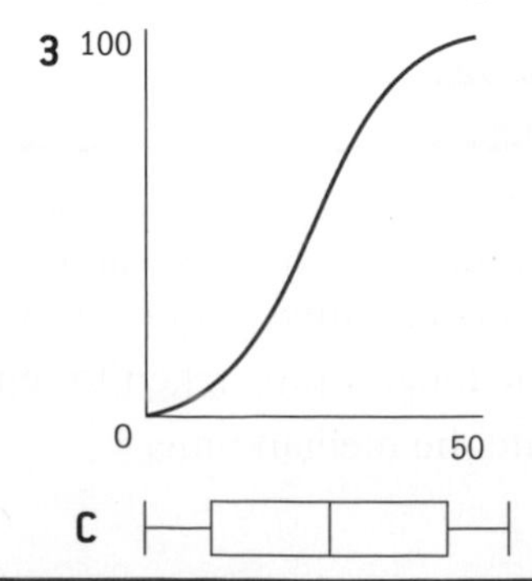

Variance and standard deviation

The range and quartiles are good measures of how spread out a data set is around the median, but they do not use all the data values to give an indication of the data's spread. The **variance**, on the other hand, combines *all* the data values in a data set to give a measure of a data's spread about its mean. It is a measure of how far, on average, each data value differs from the mean.

Consider the opening problem for the chapter. The variance would tell us whether students' test scores were, in general, close to the mean average score, or far away. The variance of a data set is calculated by taking the arithmetic mean of the squared differences between each data value and the mean value.

Suppose we need to find the variance of the 32 test scores from the opening problem:

0, 1, 1, 2, 2, 2, 3, 3, 4, 4, 4, 5, 5, 5, 5, 5, 6, 6, 6, 6, 7, 7, 7, 7, 7, 7, 8, 8, 8, 8, 9, 10

$$\bar{x} = \frac{\Sigma x}{n} = \frac{168}{32} = 5.25$$

x	$(x-\bar{x})$	$(x-\bar{x})^2$
0	−5.25	27.5625
1	−4.25	18.0625
1	−4.25	18.0625
2	−3.25	10.5625
2	−3.25	10.5625
2	−3.25	10.5625
3	−2.25	5.0625
3	−2.25	5.0625
4	−1.25	1.5625
4	−1.25	1.5625
4	−1.25	1.5625
5	−0.25	0.0625
5	−0.25	0.0625
5	−0.25	0.0625
5	−0.25	0.0625
5	−0.25	0.0625
6	0.75	0.5625
6	0.75	0.5625
6	0.75	0.5625
6	0.75	0.5625
7	1.75	3.0625

1 First, find the mean.

2 Find $(x - \bar{x})$ for each data value x. This is the distance between x and the mean $\bar{x}$.

3 Square each value of $(x - \bar{x})$. This does two things:

- **a** Squaring makes each term positive so that values above the mean do not cancel values below the mean, which would give you an inaccurate idea of the average distance from the mean.
- **b** Squaring adds more weighting to the larger differences, and in many cases this extra weighting is appropriate since points further from the mean may be more significant.

Continued on next page

7	1.75	3.0625
7	1.75	3.0625
7	1.75	3.0625
7	1.75	3.0625
7	1.75	3.0625
8	2.75	7.5625
8	2.75	7.5625
8	2.75	7.5625
8	2.75	7.5625
9	3.75	14.0625
10	4.75	22.5625

$$\sigma^2 = \frac{\Sigma\left(x-\bar{x}\right)^2}{n} = \frac{198}{32} = 6.19$$

Variance (σ^2) is the sum of the squared distances from the mean, divided by the number of data values.

Variance gives a *mean average* of the *squared distance* between each data point and the mean.

Because the differences are squared, the units of variance are not the same as the units of the data. Therefore, statisticians often find the **standard deviation**, which is the square root of the variance. The units of standard deviation are the same as those of the data set.

$$\sigma = \text{Population standard deviation} = \sqrt{\frac{\sum_{i=1}^{n}(x_i-\mu)^2}{n}}$$

$$\sigma^2 = \text{Population variance} = \frac{\sum_{i=1}^{n}(x_i-\mu)^2}{n}$$

HINT

It is expected that a GDC will be used to calculate the population standard deviation and variance.

The standard deviation shows how much variation there is from the mean and gives an idea of the shape of the distribution.

A low standard deviation (sd) shows that the data points tend to be closer to the mean than a larger standard deviation.

TOK

Is standard deviation a mathematical discovery or a creation of the human mind?

Properties of the standard deviation

- Standard deviation is only used to measure spread or dispersion around the mean of a data set.
- Variance (and therefore standard deviation) is never negative because it is an average of the *squared* distances $(x-\bar{x})^2$.
- Standard deviation is sensitive to outliers. A single outlier can raise the standard deviation and in turn, distort the picture of spread.
- For data with approximately the same mean, the greater the spread, the greater the standard deviation.

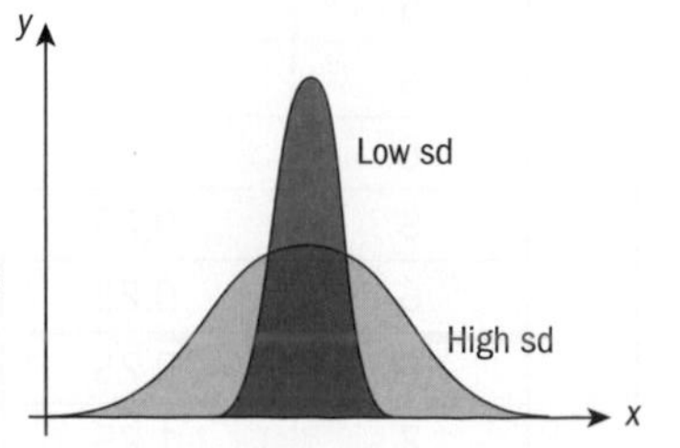

Investigation 4

A coach has two basketball players who, in the last 10 games, have scored the following points for their team:

Player A: 12, 13, 15, 15, 15, 15, 16, 16, 16, 17

Player B: 0, 1, 5, 8, 15, 15, 20, 22, 25, 39

1 Copy and complete the following table:

	Total points	Mean	Median	Range	Standard deviation
Player A					
Player B					

On the basis of their performances in the last 10 games, comment on which player the coach might pick in each of these two situations:

2 Situation 1. Coach needs a reliable points-scorer to start the game. Who should he choose and why?

3 Situation 2. The team is down by 17 points with 10 minutes to go. Who should he choose and why?

4 **Conceptual** What factors do you need to take into account to compare two data sets effectively?

Exercise 6I

1 Find the mean, variance and standard deviation of each data set.

a 4, 6, 7, 7, 5, 1, 2, 3

b 2, 5, 8, 7, 1, 3, 9, 11, 4, 2

c −4, −2, 0, 3, −5

d 1, 2, 3, 4, 5

e 1, 2, 3, 4, 5, 500

2 Find the mean and standard deviation of the data in each table.

a

x	f
1	3
2	8
3	6
4	6
5	7

b

x	f
1	5
3	12
5	16
7	22
9	27
11	30
13	18

c

x	f
$0 < x \le 10$	18
$10 < x \le 20$	14
$20 < x \le 30$	13
$30 < x \le 40$	11
$40 < x \le 50$	6

3 A real estate agent recorded the number of rooms in the houses that she was selling.

Rooms	1	2	3	4	5	6	7	18
Houses	2	2	4	10	12	2	2	1

a Find the mean and standard deviation.

b Remove the house with 18 rooms and find the new mean and standard deviation.

Reflect **Discuss** your answers with a partner then **share** your ideas with your class.

What difference does one large value make to the mean and standard deviation?

Would you remove this value if it was your data?

4 Here are the world's top ten athletes' earnings (in millions of dollars) in the year 2017.

Athlete	Earnings (mil US$)	Athlete	Earnings (mil US$)
Cristiano Ronaldo	93	Andrew Luck	50
LeBron James	86.2	Rory McIlroy	50
Lionel Messi	80	Stephen Curry	47.3
Roger Federer	64	James Harden	46.6
Kevin Durant	60.6	Lewis Hamilton	46

Find the mean amount of money earned by the top ten athletes, and the standard deviation.

5 Han's car dealership has 25 cars for sale. The prices are in thousands of Yen.

Price (x, Yen 1000s)	Cars
$100 < x \leq 200$	3
$200 < x \leq 300$	6
$300 < x \leq 400$	11
$400 < x \leq 500$	5

Find the mean price of a car and the standard deviation.

6 Sofia's security company recorded data for the total amount of overtime, in hours, her employees worked each month. She presented her data in this histogram.

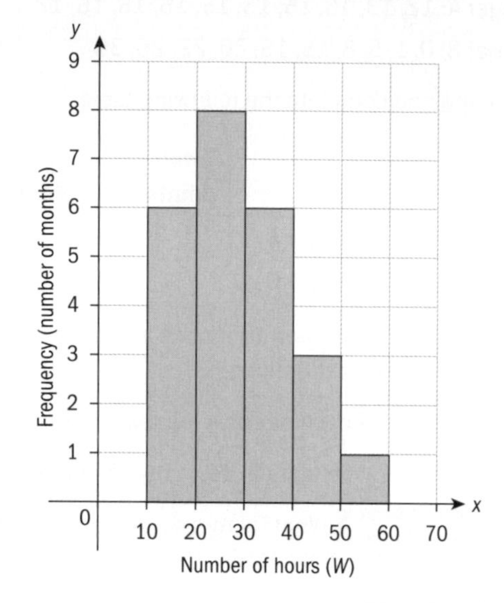

a For how many months did Sofia collect data?

b What was the modal number of hours of overtime?

c Estimate the mean number of hours of overtime.

d Estimate the standard deviation.

Investigation 5

Data set A is:
4, 2, 0, 9, 3, 5, 5, 1, 4, 6

1 Calculate the mean and standard deviation of data set A.

2 Data set B takes each of the values in set A and adds 100 to them. Write out the values in data set B.

3 **Factual** Find the mean of data set B. How does this differ from the mean of data set A? Explain why this is the case.

4 **Factual** Find the standard deviation of data set B. How does this differ from the standard deviation of data set A? Explain why this is the case.

5 **Conceptual** Using your answers to questions **3** and **4**, state what happens to the mean and standard deviation of a data set which has a constant c added to each value.

6 Data set C takes each of the values in set A and multiplies them by 2. Write out the values in data set C.

7 **Factual** Find the mean of data set C. How does this differ from the mean of data set A? Explain why this is the case.

8 **Factual** Find the standard deviation of data set C. How does this differ from the standard deviation of data set C? Explain why this is the case.

9 **Conceptual** Using your answers to questions **7** and **8**, state what happens to the mean and standard deviation of a data set when each value is multiplied by a constant d.

Reflect How will the variance of a data set be affected either by adding c to every value, or by multiplying each value by d? Explain why this is the case.

Example 12

A data set has a mean of 12 and a standard deviation of 2.

a Each value in the data set has 20 added to it. Find the new mean and standard deviation.

b Each value in the original data set is multiplied by 4. Find the new mean and variance.

a The mean is 32. The standard deviation is unchanged at 2.	The mean is increased by 12 but the standard deviation remains unchanged under addition or subtraction.
b The mean is 48. The variance is 64.	The mean is multiplied by 4. The standard deviation is multiplied by 4, and the variance is the square of the standard deviation.

Exercise 6J

1 **a** Find the mean, median and mode of 1, 3, 5, 5, 8.

b Add 4 to each value and find the mean, median and mode of the new data set.

c Describe the effect that adding 4 to each value has on the mean, median and mode.

2 **a** Find the standard deviation and variance of the set 7, 9, 3, 0, 1, 8, 6, 4, 10, 5, 5.

b Multiply each member of the data set by 3 and find the new standard deviation and variance.

c Describe the effect that multiplying each member by 3 has on the mean and standard deviation.

3 The mean age of a group of friends at school is 17.2 years, their median age is 17, and the standard deviation of their ages is 0.5 years. They meet again 4 years later at a school reunion. What are the new mean, median and standard deviation of their ages?

4 All of the items in a data set are doubled. State which of the following are also doubled: mean, median, standard deviation, variance.

5 If the variance of a data set is x and all of the numbers in a data set are multiplied by 9, state the new variance.

Developing your toolkit

Now do the Modelling and investigation activity on page 316.

Chapter summary

- Discrete data can take only particular values, and each value is distinct from the others.
- Continuous data is not restricted to certain fixed values, but it can occupy any value within a continuous range.
- Data is **reliable** if you can repeat the data collection process and obtain similar results.
- Data is **sufficient** if there is enough data available to support your conclusions.
- The **population** consists of every member in the group that you want to find out about.
- A **sample** is a subset of the population that will give you information about the population as a whole.
- Different sampling methods include ***convenience, simple random, systematic, stratified*** or ***quota*** sampling.
- A bar chart is used to display discrete data, whereas a histogram is used to display continuous data.

In examinations the data set will be treated as the population.

Measures of dispersion

- The **range** is the difference between the largest and smallest values.
- **Quartiles** divide a sample of data into four groups containing equal numbers of observations.
- Q_1 is the median of the lower 50% of the data, and Q_3 is the median of the upper 50% of the data.
- The **interquartile range** is the value of the third quartile minus the value of the first quartile.
- Different methods for finding quartiles exist and therefore the values obtained using technology and by hand may differ.
- A **box -and-whisker** plot is a way of summarizing a set of data.

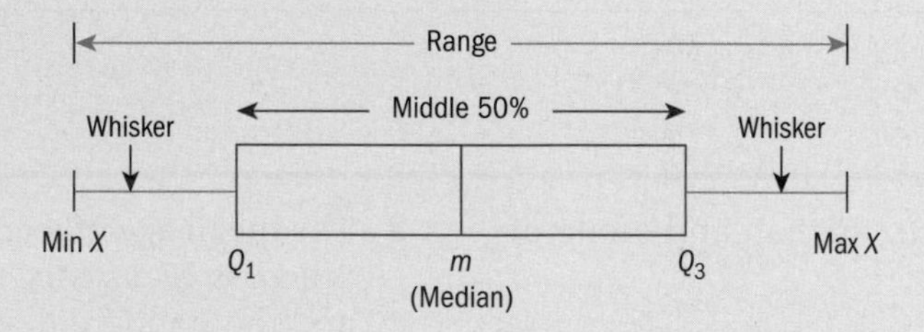

- Extreme values are called **outliers** and might be considered for removal from a sample. An outlier is any value at least 1.5 IQR above Q_3 or below Q_1
- Adding up the frequencies of the data values as we go along is called calculating the cumulative frequency (CF).

x	f	CF
$0 < x \leq 10$	0	0
$10 < x \leq 20$	19	19
$20 < x \leq 30$	30	49
$30 < x \leq 40$	20	69
$40 < x \leq 50$	6	75

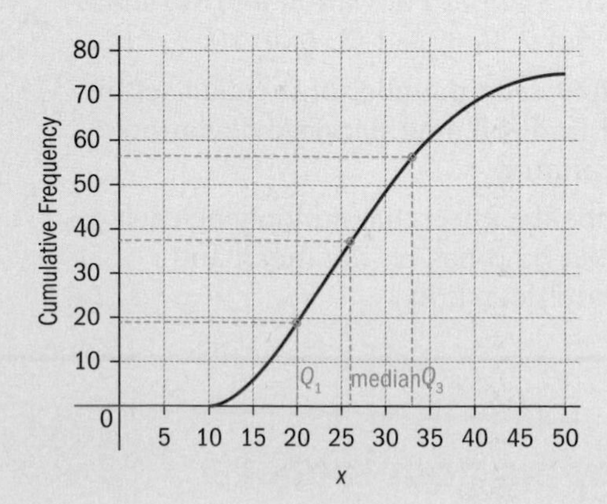

Variance and standard deviation

- Standard deviation is a measure of how spread out the items of a data set are from the mean.
- Standard deviation is the square root of the variance, where $\sigma^2 = \dfrac{\Sigma\left(x - \bar{x}\right)^2}{n}$.
- A GDC should be used to find both of these measures.

Effect of constant changes on the original data

- Adding a constant c to every value in a data set increases the mean and the median by c, and has no effect on the standard deviation.
- Multiplying every value in a data set by a constant d means that the mean and standard deviation will also be multiplied by d, and the variance will be multiplied by d^2.

Developing inquiry skills

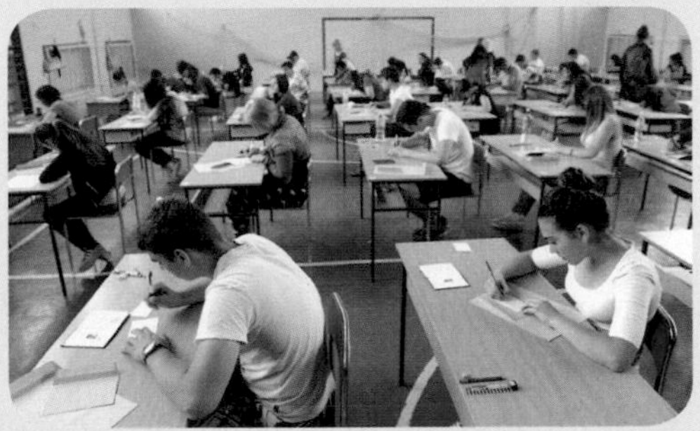

In the opening problem for the chapter, you were given the test scores, out of 10, of 32 students.

1 Represent the scores on a cumulative frequency graph.

2 Write a five-number summary for the data from the opening problem on class test scores.

3 Draw a box-and-whisker diagram for the data.

4 What does this tell you about the performance of the class? How would you allocate letter grades A, B, C and D?

5 What would happen to the mean, median and standard deviation if the teacher decided to multiply all of the scores by 10 to show them as a percentage?

Chapter review

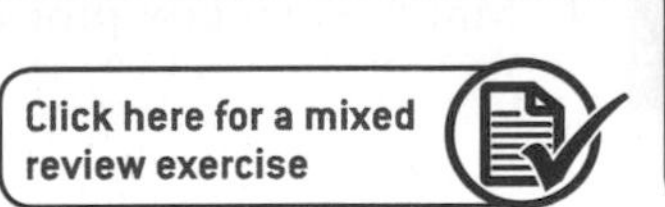

1 Find **a** the mode, **b** the median, **c** the mean and **d** the range of 2, 8, 1, 5, 0, 4, 4, 1, 1, 6.

2 From January to September, the mean number of car accidents in a city per month was 420. From October to December, the mean was 740 accidents per month.

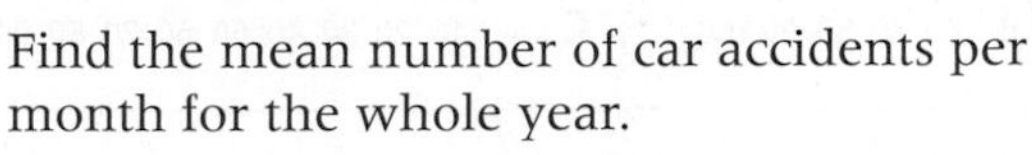

Find the mean number of car accidents per month for the whole year.

3 Students collected data on the number of ties that their fathers owned.

Ties	0	1	2	3	4	5	6
f	4	8	10	20	4	3	1

a Write down the mode.

b Find the median.

c Calculate the mean number of ties that a father owns.

4 The mean age of a group of students when they leave school is 17.9 years and the standard deviation is 1.1 years. If they all come back to see their old school 4 years later find the new mean and standard deviation of their ages.

5 On Monday, 23 students in a chemistry class spent a total of 736 minutes on an experiment.

Statistics and probability

a Find the mean number of minutes the students spent doing the experiment.

Two students forgot to report their times. One spent 24 minutes and the other spent 15 minutes.

b Calculate the new mean including these two students.

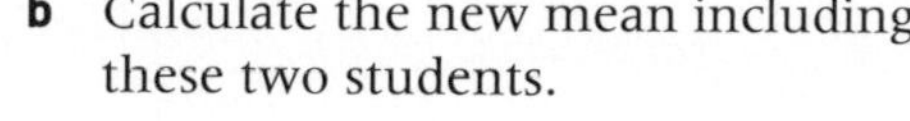

6 A data set has a mean of 48 and a standard deviation of 5.

a Each value in the data set has 10 added to it. Find the new mean and standard deviation.

b Each value in the original data set is multiplied by 10. Find the new mean and variance.

7 The box plot shows the heights, in cm, that a class of 8 year-olds could jump.

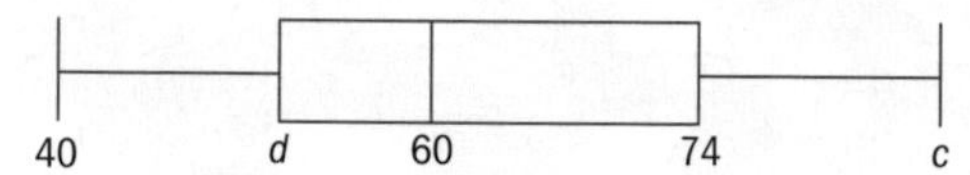

a What is the lowest height?

b Write down the median.

c If the range is 50 cm, find the value of c.

d Find the value of d if the interquartile range is 24 cm.

8 The test results for a group of children in a school district are shown on this cumulative frequency diagram.

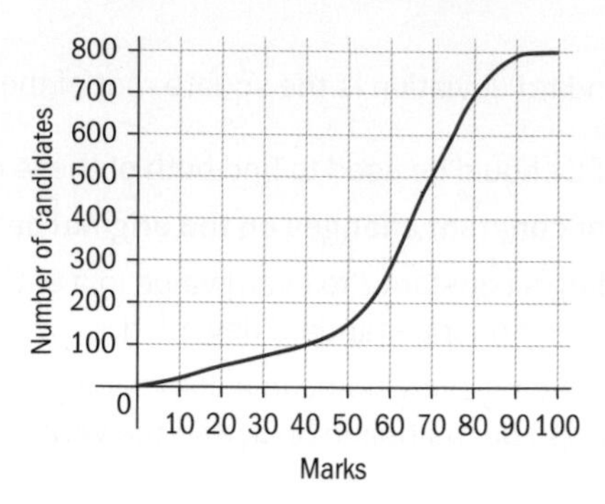

a How many students' test scores were recorded?

b What is the median score?

c Show that the interquartile range is 20 marks.

d How many students scored more than 80 marks on the test?

e If Calvin earned 80 marks, would he be in the 90th percentile? Give reasons for your answer.

f 100 students scored less than k marks. Find the value of k.

9 Match each box plot with the correct histogram:

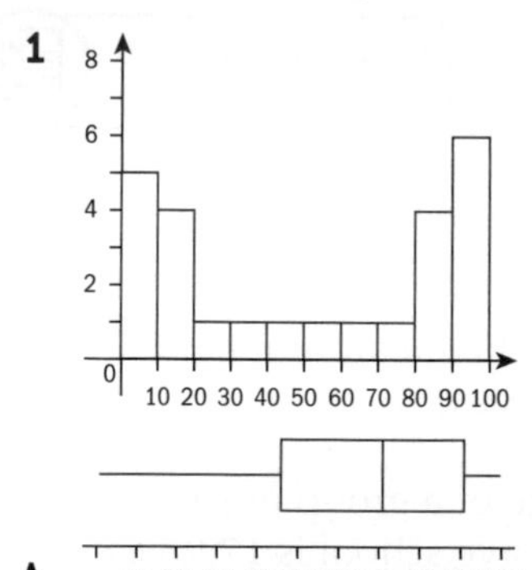

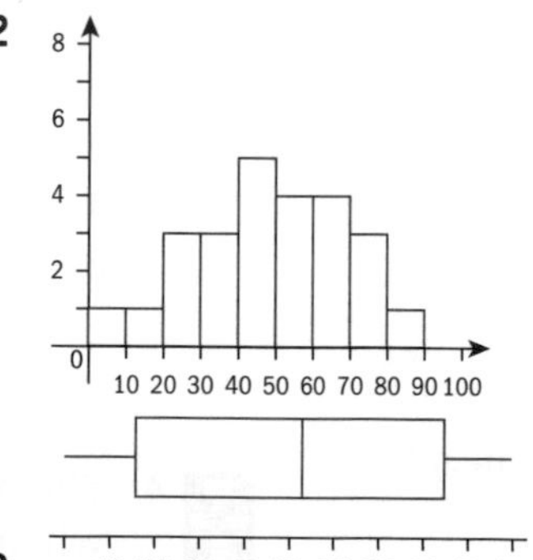

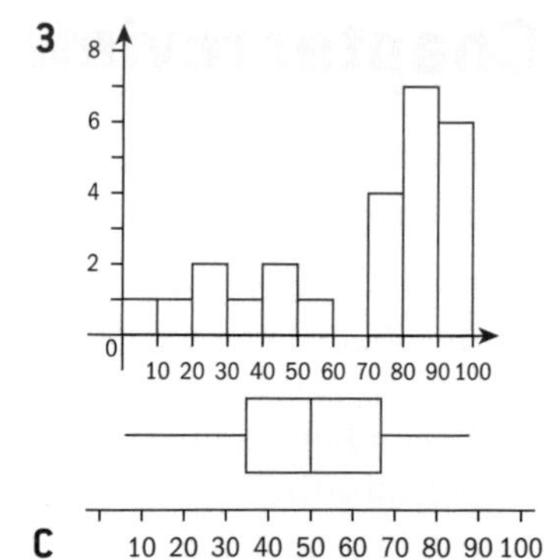

EXAM HINT

The command "Write down ..." means that marks are awarded for the answer only, but "find ..." means that marks will be awarded for working.

10 a Write down the mean and standard deviation and **b** find the interquartile range of 15, 12, 22, 30, 25, 7, 19, 33, 19, 41, 53, 12, 3, 8, 6, 17.

11 Mr Mateo's class were wearing rings for Geology Day. Mr Mateo recorded the number of rings per student in this table.

Rings	2	3	4	5	6	7
f	3	4	10	3	2	2

Write down the mode, mean, median and standard deviation of the number of rings.

12 Ode surveyed students for his internal assessment about the number of friends who attended each of their birthday parties last year. His results are shown in this grouped frequency table.

Friends (p)	f
$0 < p \le 5$	15
$5 < p \le 10$	11
$10 < p \le 15$	9
$15 < p \le 20$	12
$20 < p \le 25$	6

a Write down estimates for mean, median and standard deviation.

b Explain why you are only able to find estimates in part **a**.

13 This table shows the number of watches owned by each student in a group.

Watches	0	1	2	3	4	5
f	11	7	6	k	8	10

If the mean number of watches owned is 2.5, find the value of k.

14 A study to examine the weight of bats in a small cave produced this cumulative frequency graph.

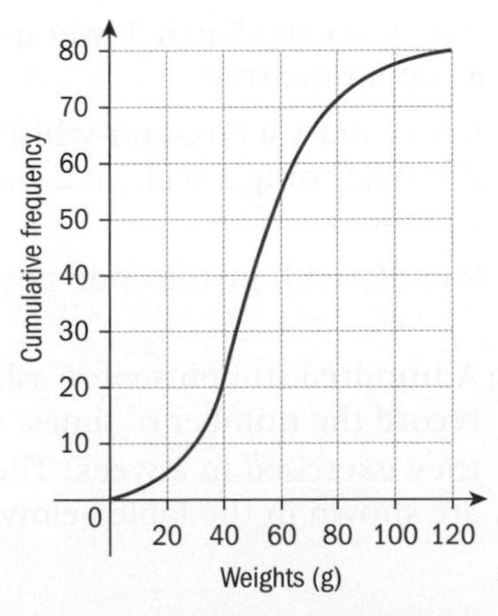

a How many bats were weighed?

b Write down the median weight.

c Find the percentage of bats that weighed 20 grams or less.

The data is also presented in this table.

Weight (w, g)	Bats
$0 < w \le 30$	a
$30 < w \le 60$	45
$60 < w \le 90$	b
$90 < w \le 120$	c

d Write down the values of a and c.

e Find the value of b.

f Use the values from the table to estimate the mean and standard deviation of the weights.

15 Consider this cumulative frequency table.

x	f	Cumulative frequency
10	3	3
15	11	14
20	16	n
25	m	42
30	8	50

a Find the values of m and n.

b Write down the value of the mean.

c Find the variance.

16 A survey was conducted of the number of mobile devices that families owned.

Mobile devices	1	2	3	4	5	6
f	41	60	52	32	15	8

a State whether the data is discrete or continuous.

b Write down the mean number of mobile devices per family.

c Write down the standard deviation.

d Find how many families have a number of mobile devices greater than one standard deviation above the mean.

Exam-style questions

17 P1: State whether each of the following descriptions would generate discrete or continuous data.

a A student's grade on an IB SL Mathematics exam. (1 mark)

b The volume of water that a person uses when having a shower. (1 mark)

c The length of time that a person spends having a shower. (1 mark)

d The number of emails that a person receives during a day. (1 mark)

18 P1: Consider the following set of data 3, 6, 1, 5, a, b where $a > b$.

The mode of this data is 5. The median of this data is 4.5.

a Find the value of a and the value of b. (5 marks)

b Find the mean of this data. (2 marks)

19 P1: a State the definition of an outlier for a set of statistical data. (1 mark)

b Given the data

1, 2, 3, 5, 6, 7, 8, 8, 9, 11, 19

find the

i mode

ii median

iii lower quartile

iv upper quartile. (4 marks)

c Identify, with a reason, any outliers for this set of data. (2 marks)

20 P2: a A set of 10 students have a mean mass of 70 kg. A new student joins the group. The mean mass of the 11 students is 72kg. Find the mass of the new student. (4 marks)

b The new lower quartile and upper quartile for the 11 students are 66 and 76 respectively. Determine with justification whether the mass of the new student is an outlier. (3 marks)

21 P2: Sue has collected continuous data on the heights of flowers and has represented it in the table below.

Height (h, cm)	$0 < h \le 10$	$10 < h \le 20$	$20 < h \le 30$	$30 < h \le 40$	$40 < h \le 50$
Frequency	40	45	50	60	5

a State how many flower heights were measured. (1 mark)

b Find the mid-point of the modal interval. (1 mark)

c Estimate the **i** mean **ii** standard deviation. (5 marks)

d Sue's calculator states that the median is 25. Find a better estimate than this by considering the median's position within the interval that it belongs to. Give your answer to the nearest integer. (3 marks)

22 P2: Data on IB SL Mathematics grades is represented in the table below.

Grade	1	2	3	4	5	6	7
Frequency	2	5	8	40	50	20	15

a State whether this data is discrete or continuous. (1 mark)

b Find the mode. (1 mark)

c Find the **i** mean **ii** standard deviation. (4 marks)

d Find the **i** median **ii** lower quartile **iii** upper quartile. (3 marks)

e Hence draw a box-and-whisker plot for this data using a scale of 2 cm for 1 grade. (3 marks)

f Identify with justification any outliers. (3 marks)

23 P2: A hundred students were asked to record the number of times, x, that they exercised in a week. The results are shown in the table below.

x	Frequency	Cumulative frequency
0	10	10
1	7	
2	11	28
3		41
4	15	56
5		71
6	12	
7	10	93
8		
9	2	99
10	1	100

a Copy this table and fill in the missing numbers. (4 marks)

b Find the **i** median **ii** lower quartile **iii** upper quartile. (3 marks)

c Find the **i** mean **ii** variance. (3 marks)

d State with a reason whether this data has a unique mode. (2 marks)

24 P1: Grouped, continuous data for the mass, w kg, of a group of adults is given in the table below.

Mass	$40<w\leq 50$	$50<w\leq 60$	$60<w\leq 70$	$70<w\leq 80$	$80<w\leq 90$	$90<w\leq 100$	$100<w\leq 110$	$110<w\leq 120$
Frequency	5	15	25	30	50	35	25	15

a State the modal interval. (1 mark)

b Construct a labelled cumulative frequency table for this data. (3 marks)

c On graph paper draw a cumulative frequency curve, with 1 cm representing 10 kg on the x-axis and 1 cm representing 10 adults on the y-axis. (5 marks)

d Hence, estimate values for the **i** median **ii** lower quartile **iii** upper quartile. Draw lines on your graph to indicate how you obtained these values. (4 marks)

25 P2: Sally and Rob each teach an IB SL Mathematics class. In a common test, Sally's students scored

1, 1, 4, 7, 8, 8, 10, 10

and Rob's students scored

4, 4, 4, 5, 6, 6, 10, 10, 10, 10.

a For the data from Sally's class find the **i** median **ii** mean. (3 marks)

b For the data from Rob's class find the **i** median **ii** mean. (3 marks)

c Give a reason why Sally could claim that her class did better than Rob's class. (1 mark)

d Give a reason why Rob could claim that his class did better than Sally's class. (1 mark)

26 P2: Discrete data on an IB exam taken by 76 students is given in the table below.

Grade	1	2	3	4	5	6	7
Frequency	4	8	16	20	16	8	4

a On graph paper, draw a bar chart to represent this data. Use 2 cm to represent 1 grade on the x-axis and use 2 cm to represent 5 students on the y-axis. (3 marks)

b Find the **i** mode **ii** median **iii** mean. (3 marks)

c Explain your answers to parts **bii** and **iii** by referring to a geometrical property of the bar chart drawn in part **a**. (2 marks)

What's the difference?

Approaches to learning: Thinking skills, Communicating, Collaborating, Research
Exploration criteria: Presentation (A), Mathematical communication (B), Personal engagement (C), Reflection (D), Use of mathematics (E)
IB topic: Statistics, Mean, Median, Mode, Range, Standard deviation, Box plots, Histograms

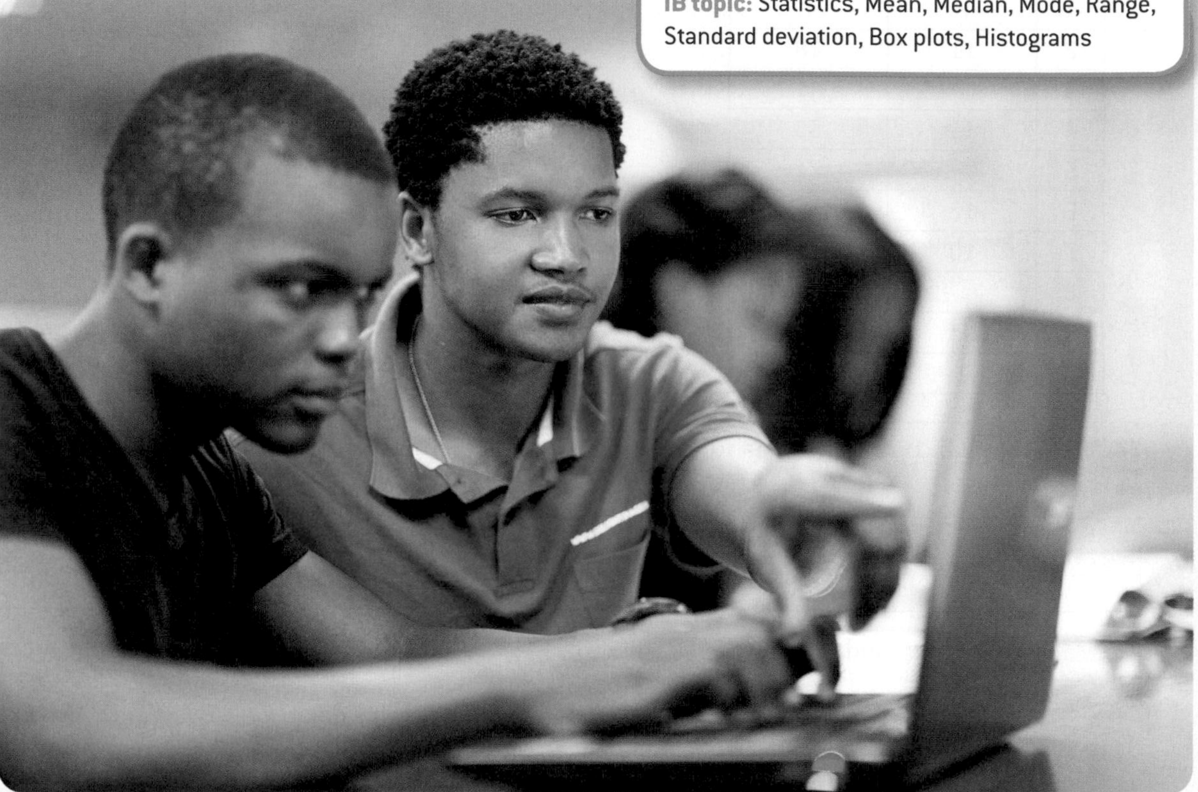

Example experiment

Raghu does an experiment with a group of 25 students.

Each member of the group does a reaction test and Raghu records their times.

Raghu wants to repeat the experiment, but with some change.

He then wants to then compare the reaction times in the two experiments.

Discuss:

How could Raghu change his experiment when he does it again?

With each change, is the performance in the group likely to improve/stay the same/get worse?

Alternatively, Raghu could use a different group when he repeats the experiment.

What different group could he use?

With each different group, is the performance likely to improve/stay the same/get worse?

Your experiment

Your task is to devise an experiment to test your own hypothesis.

You will need to do your experiment two times and compare your results.

Step 1: What are you going to test? State your aim and hypothesis.

Write down the aim of your experiment and your hypothesis about the result.

Why do you think this is important?

What are the implications of the results that you may find?

Make sure it is clear what you are testing for.

Step 2: How are you going to collect the data? Write a plan.

- What resources/sites will you need to use?
- How many people/students will you be able to/ need to collect data from to give statistically valid results?
- Exactly what data do you need to collect? How are you going to organize your data? Have you done a trial experiment?
- Are there any biases in the way you present the experiment? How can you ensure that everyone gets the same instructions?
- Is your experiment a justifiable way of testing your hypothesis? Justify this. What are the possible criticisms? Can you do anything about these?
- Is the experiment reliable? Is it likely that someone else would reach a similar conclusion to you if they used the same method?

Step 3: Do the experiment and collect the data.

Prepare a results sheet to collect the data.

Give clear, consistent instructions.

Step 4: Present the data for comparison and analysis.

How are you going to present the data so that the two sets can be easily compared?

How are you going to summarize the summary statistics of the two data sets so that you can compare them?

Do you need to find all of the summary statistics covered in this chapter?

Step 5: Compare and analyse.

Describe the differences between your two sets of data.

Make sure that your conclusion is relevant to your aim and hypothesis stated at the beginning.

Step 6: Conclusions and implications.

What are the conclusions from the experiment? Are they different to or the same as your hypothesis? To what extent? Why?

How confident are you in your results? How could you be more certain?

What is the scope of your conclusions?

How have your ideas changed since your original hypothesis?

Extension

- How could you test whether the spread of the data has changed significantly, rather than the average?
- How could you analyse changes in individual results, rather than whole class changes?
- Investigate the "difference in means test".

7 Modelling relationships between two data sets: statistics for bivariate data

Bivariate data analysis looks at the strength of the relationship between two variables. Mathematicians do this using two measures:

- correlation, which gives a numerical value for the strength of the relationship between the variables
- regression, which gives an equation to model the relationship between the variables.

You can then use this information to make predictions about the data.

For example, when a class of students take a mathematics test and a science test, you could use bivariate data analysis to determine whether students who are good at mathematics tend to be good at science as well, and regression to determine whether the marks in science can be predicted from marks in mathematics.

Concepts

- Representation
- Modelling

Microconcepts

- Scatter plots
- Correlation
- Pearson's correlation coefficient (r)
- The line of best fit, by eye, passing through the mean point
- y on x regression line
- x on y regression line
- Interpreting the parameters of a regression line, in context

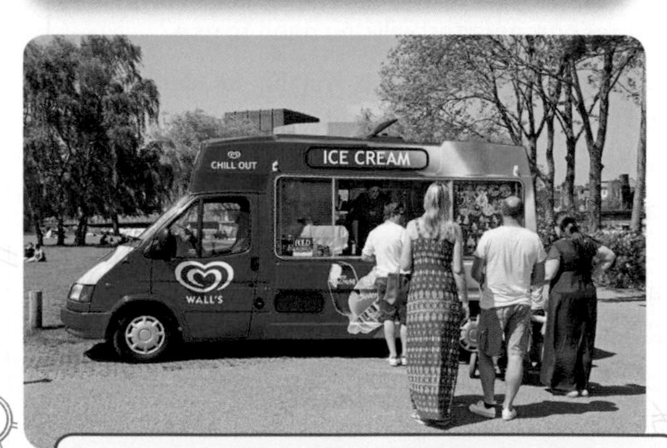

Are ice cream sales and number of sunshine hours linked? How strongly?

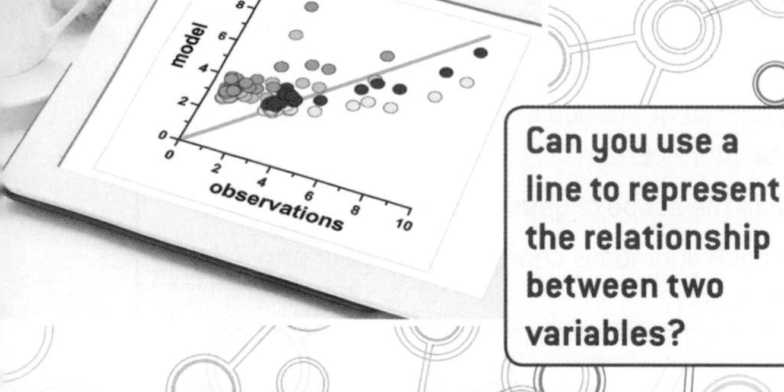

Can you use a line to represent the relationship between two variables?

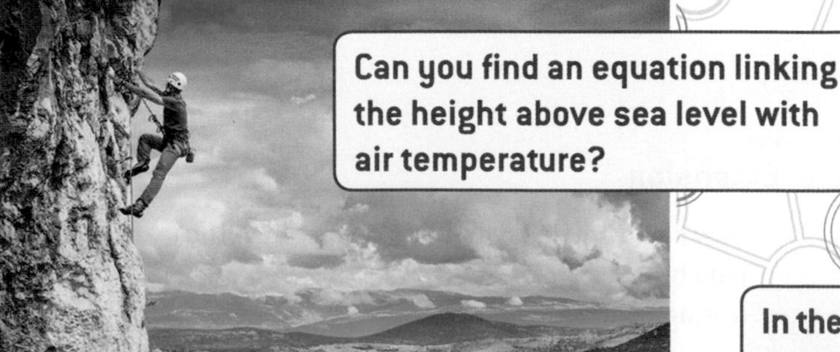

Can you find an equation linking the height above sea level with air temperature?

In the UK, the number of hot drinks consumed is linked with the sales in socks. But does one cause the other, or is there a third factor which influences them both?

A rice farmer records data on the amount of rainfall (in cm) and rice yield (in tonnes) from his farm for the last eleven years.

Rainfall (cm)	130	150	140	180	350	210	190	100	160	175	145
Rice yield (tonnes)	94	100	90	120	20	150	135	80	130	110	105

- Are the variables discrete or continuous?
- Does it look as though more rainfall gives a higher rice yield?
- Can you suggest better ways to view the data which would help you to see the connection between the variables?
- Can this to help predict the rice yield in a year when you know the rainfall forecast?

Developing inquiry skills

Think of some other variables that you think might be connected – for example, the number of Instagram posts and the number of followers. How do we know whether variables influence each other?

Look up the divorce rates over the twentieth century and the number of cars sold over the same period. Are cars causing divorces?

What is the difference between correlation and causation?

Before you start

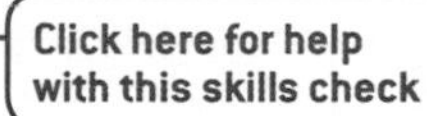

Click here for help with this skills check

You should know how to:	Skills check
1 Calculate simple positive exponents. eg Evaluate 3^4 $3^4 = 3 \times 3 \times 3 \times 3 = 81$ eg Evaluate $\left(\frac{2}{5}\right)^3$ $\left(\frac{2}{5}\right)^3 = \frac{2\times2\times2}{5\times5\times5} = \frac{8}{125}$	**1** Evaluate: **a** 6^4 **b** 2^6 **c** 7^3
2 Convert numbers to exponential form. eg Find n given $4^n = 64$ $4 \times 4 \times 4 = 64$ $4^3 = 64$ $n = 3$	**2** Find the value of n. **a** $2^n = 32$ **b** $5^n = 625$ **c** $(-4)^n = -64$ **d** $\left(\frac{2}{3}\right)^n = \frac{8}{27}$
3 Find the equation of a line through two given points. eg Find the equation of the line passing through $(1, 5)$ and $(4, 14)$, in gradient-intercept form. $m = \frac{y_2 - y_1}{x_2 - x_1} = \frac{14-5}{4-1} = 3$ $y - y_1 = m(x - x_1)$ $y - 5 = 3(x - 1)$ $y - 5 = 3x - 3$ $y = 3x + 2$	**3 a** Find the equation of the line passing through $(2, 6)$ and $(3, 10)$, in gradient-intercept form. **b** Find the equation of the line passing through $(5, -9)$ and $(2, -3)$, in gradient-intercept form.

7.1 Scatter diagrams

Bivariate data analysis is the study of the relationship between two data sets.

One way to determine whether there is a relationship between two data sets is to plot the bivariate data on a scatter diagram. A scatter diagram takes the two sets of data and plots one set on the x-axis and the other set on the y-axis.

Example 1

The height, and corresponding weight, of 10 students are shown in the table.

Height (cm)	182	155	174	166	158	171	179	180	168	160
Weight (kg)	90	60	84	73	68	77	84	86	74	73

Plot a scatter diagram to represent the data gathered for the 10 students.

Describe what the scatter diagram shows you about the relationship between students' height and weight.

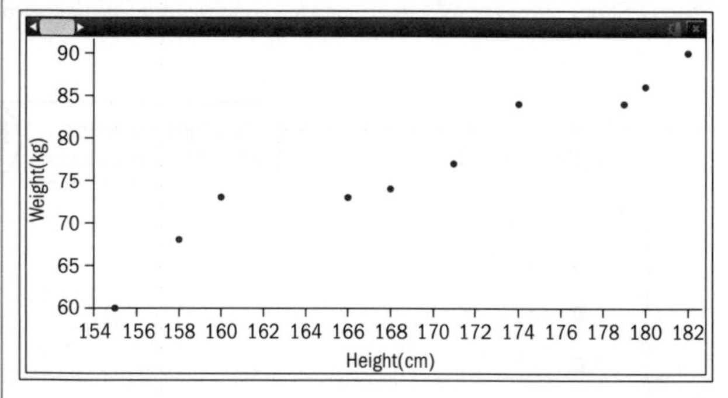

In general, as height increases, weight also increases.

Enter "height" and "weight" into separate columns of a spreadsheet on your GDC.

Plot a scatter diagram of this data on your GDC.

Correlation

A **correlation** exists between two variables, x and y, when a change in x corresponds to a change in y.

One of the benefits of a scatter diagram is that you can visually see whether there is any correlation between two variables and, if there is, how strong the correlation may be.

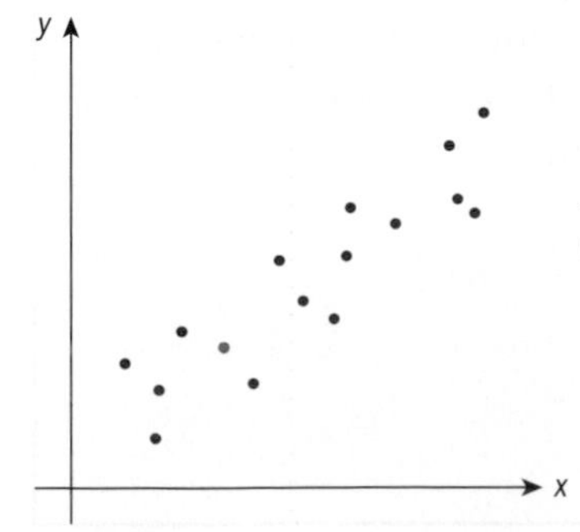

A general upward trend shows a **positive correlation**: y *increases* as x increases.

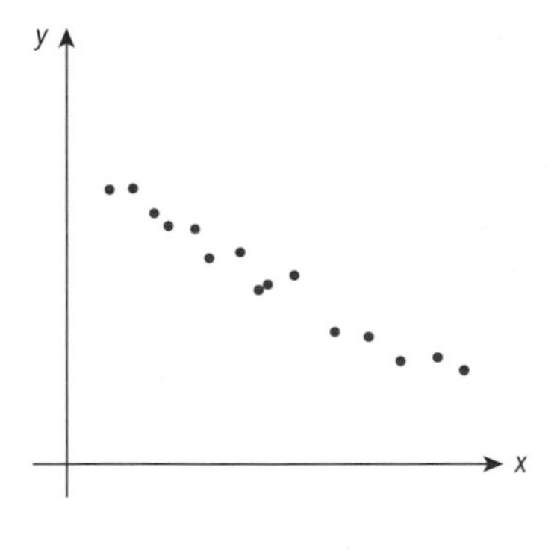

A general downward trend shows a **negative correlation**: *y decreases* as *x* increases.

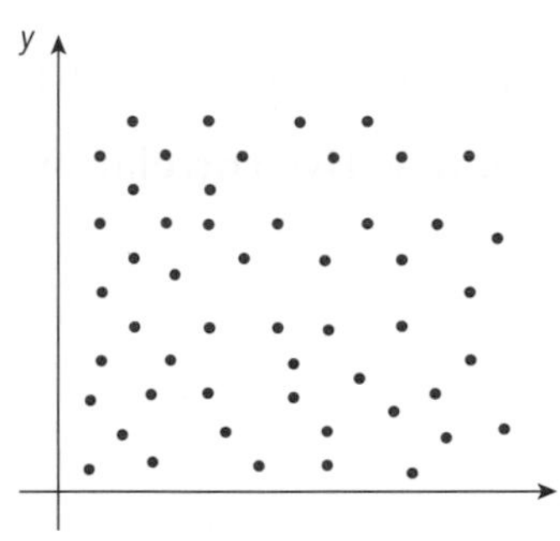

Scattered points with no trend may indicate **little or no correlation**. This suggests there is no relationship (or a very weak one) between the two variables.

International-mindedness

In 1956, Australian statistician Oliver Lancaster made the first convincing case for a link between exposure to sunlight and skin cancer using statistical tools including correlation and regression.

Reflect Can you suggest two real-life variables which have:

- a positive correlation?
- a negative correlation?
- no correlation?

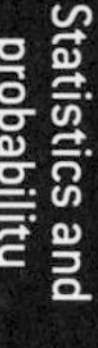

A scatter diagram not only shows you whether the correlation between two variables is positive or negative, but also tells you something about the **strength of the correlation**. The stronger the correlation, the more the points will look like a straight line.

The three graphs below all show variables which have *positive* linear correlation, but each has a different strength of correlation.

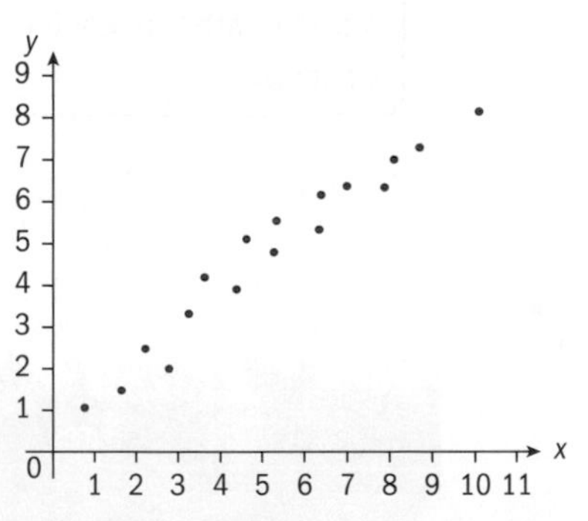

Strong positive correlation

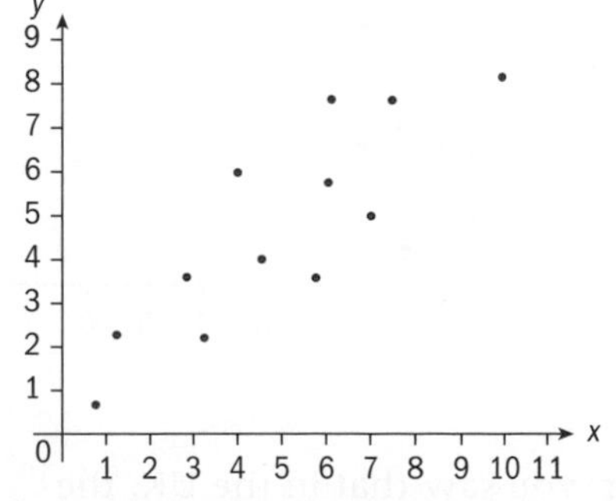

Moderate positive correlation

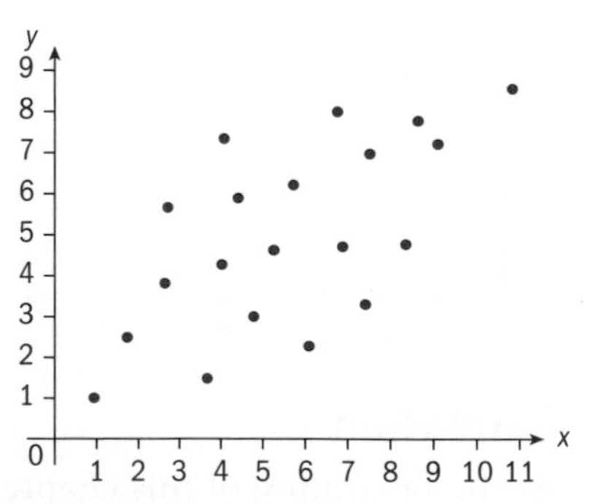

Weak positive correlation

Likewise, the three graphs below all show variables which have *negative* linear correlation, but each has a different strength of correlation.

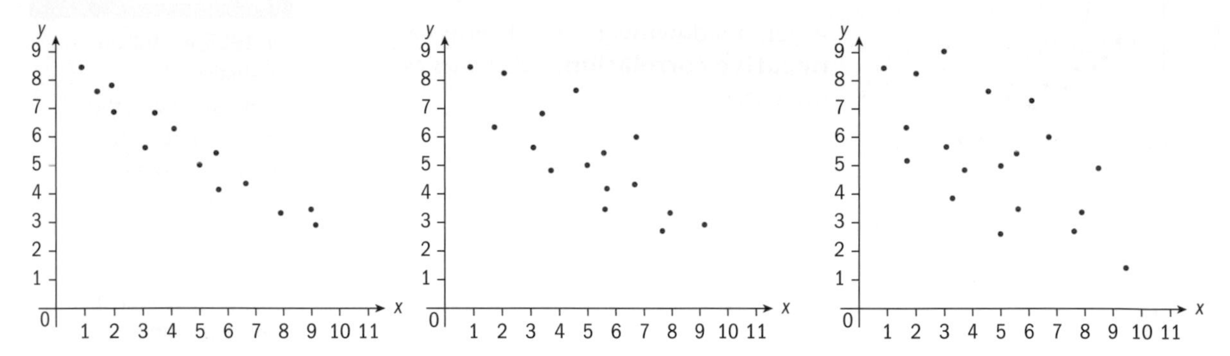

Strong negative correlation **Moderate negative correlation** **Weak negative correlation**

Just as you have already studied both linear and non-linear functions, some correlations will be linear (the scatter graph will follow a straight line) and other correlations will be non-linear (the scatter graph will follow a curve, or other non-linear relationship).

To describe the correlation between two variables, you must say whether the correlation is:

- positive, negative or zero
- strong or weak
- linear or non-linear.

Reflect How would you describe the correlation between variables x and y in these two scatter diagrams?

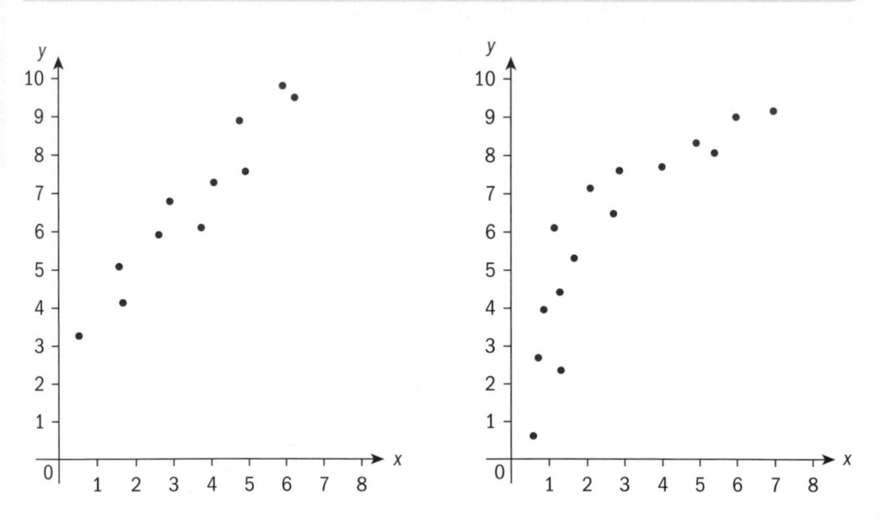

HINT

You would use a similar graph to these if you were to have a scatter diagram in your exploration but replace x and y with the real-life descriptions.

Causation

At the beginning of this chapter, you saw that in the UK, the number of hot drinks consumed is linked with the sales in socks. But does one cause the other, or is there a third factor which influences them both?

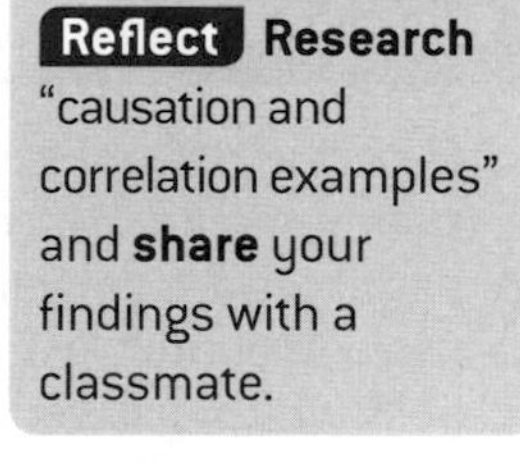

Adam saw this information. He thought his wife Sarah was spending too much money on socks. To try to stop this, he suggested she only drink one cup of tea per day.

Had Adam misinterpreted the information? The fact that hot drinks and sock sales are correlated does not mean that buying socks *causes* people to drink more hot drinks, or vice versa.

In fact, both the consumption of hot drinks *and* sales in socks change as a result of a third factor: temperature change.

Reflect **Research** "causation and correlation examples" and **share** your findings with a classmate.

A correlation between two data sets does not necessarily mean that change in one variable *causes* change in the other.

There are also some instances where completely random variables are correlated by chance.

Here is a real-life example: Using the data on the right, you could try to argue that the more lemons imported into the USA from Mexico, the fewer road fatalities!

It is clear that the number of lemons do not **cause** a decrease in the fatality rate, but there is a chance correlation between the data sets. You sometimes need to be aware of these chance correlations when analysing claims made as a result of misinterpretation of statistics.

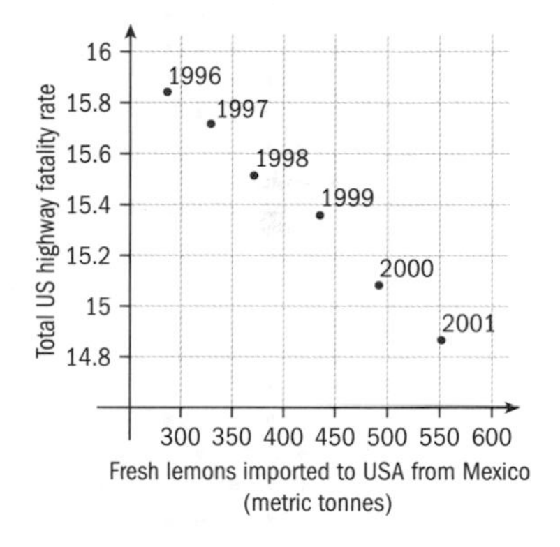

Example 2

a Represent this data on a scatter diagram.

x	1	2	3	4	5	6	7	8	9
y	1	3	2	6	6	7	9	9	10

b Describe the correlation between x and y.

a

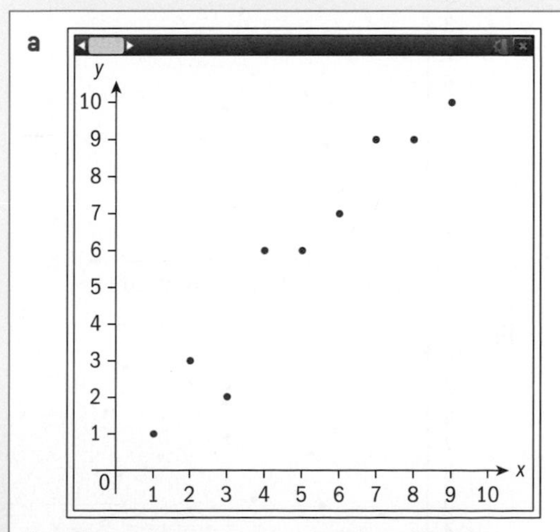

Use your GDC to plot a scatter diagram of the points.

b There is a strong, positive, linear correlation.

Remember that you need to give three pieces of information about the correlation to describe it accurately.

Exercise 7A

1 Describe whether each graph shows strong / weak / no correlation. If the variables are correlated, describe the correlation using the words positive / negative; linear / non-linear.

a

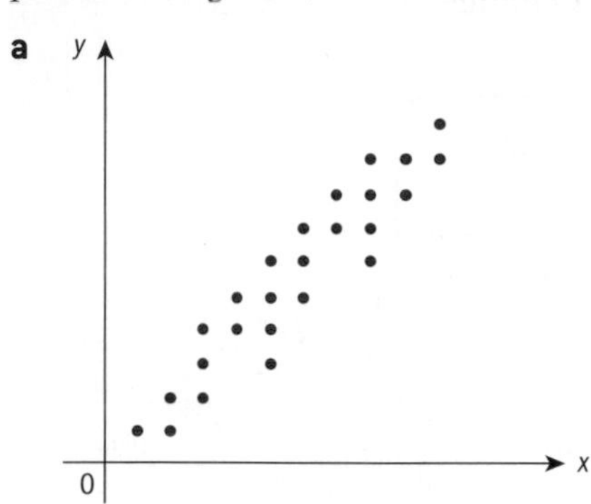

b

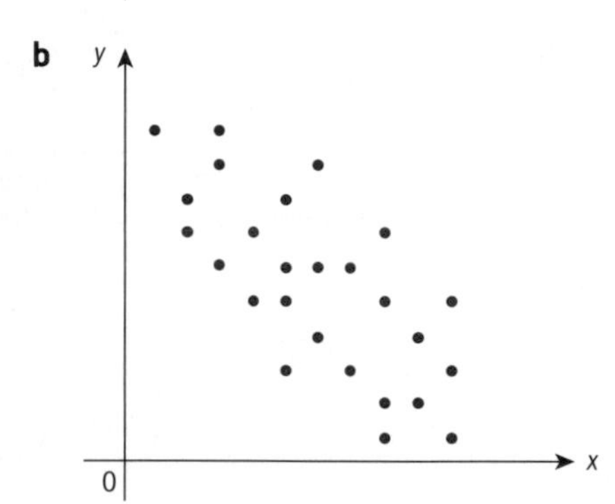

c

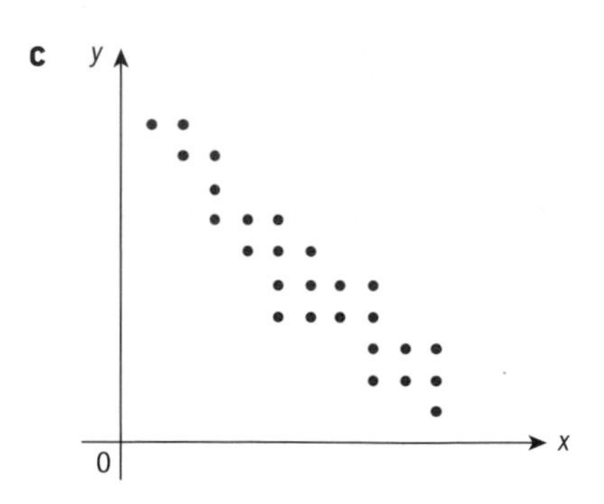

d

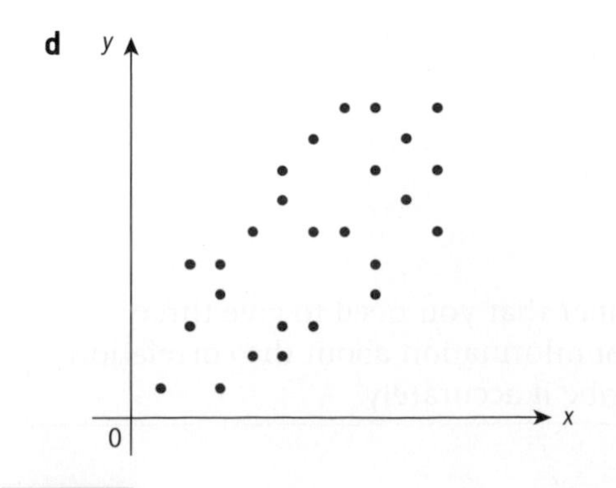

e

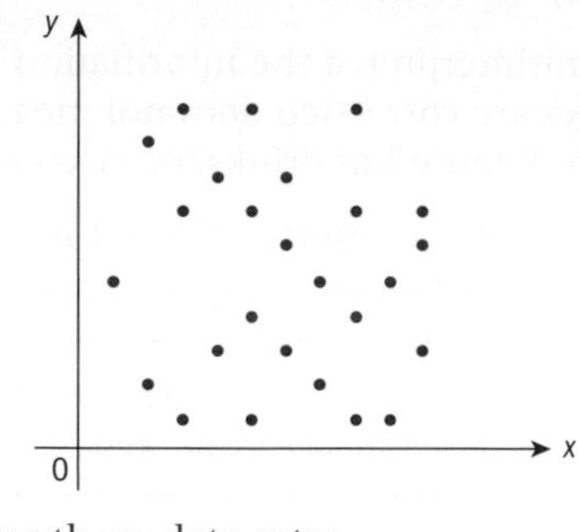
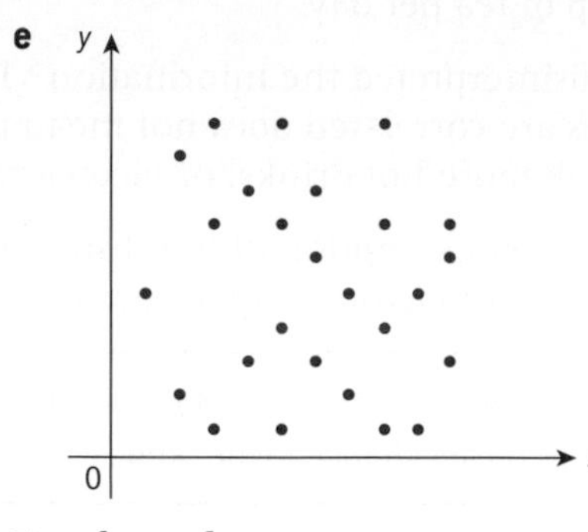
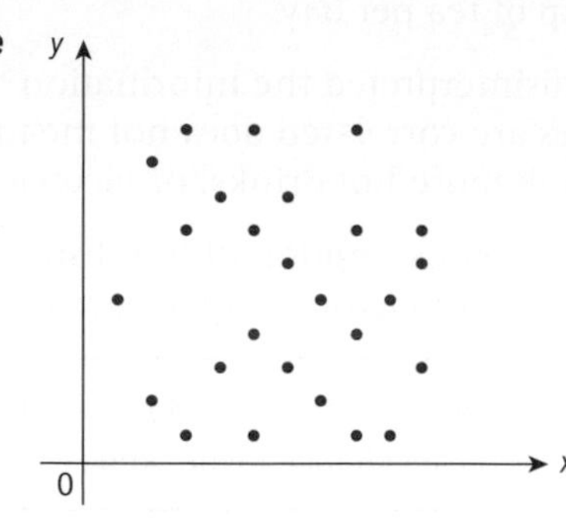

TOK

What is the difference between correlation and causation?

To what extent do these different processes affect the validity of the knowledge obtained?

2 For these data sets:

a Is the relationship positive, negative or is there no association?

b Is the relationship linear or non-linear?

c Is the association strong, moderate, weak or zero?

i

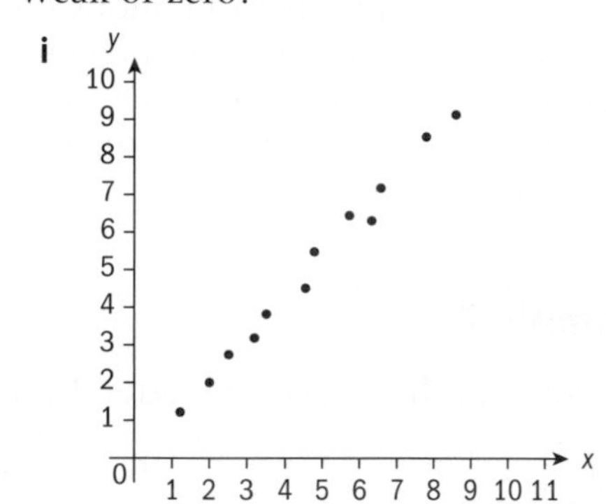

ii

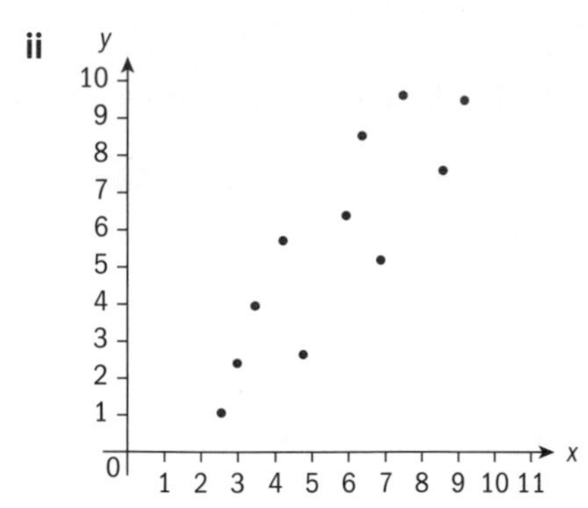

iii

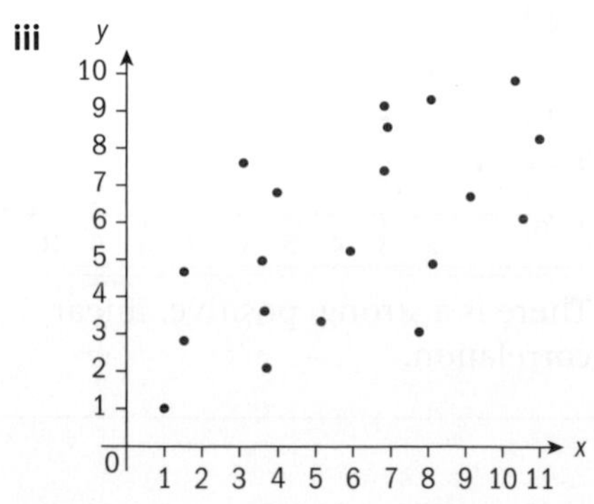

iv

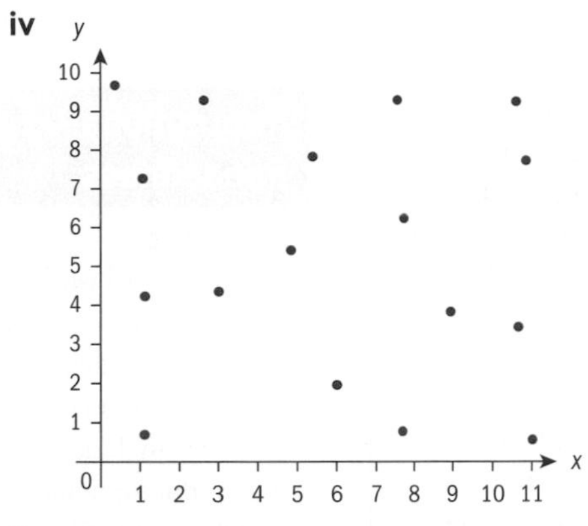

v

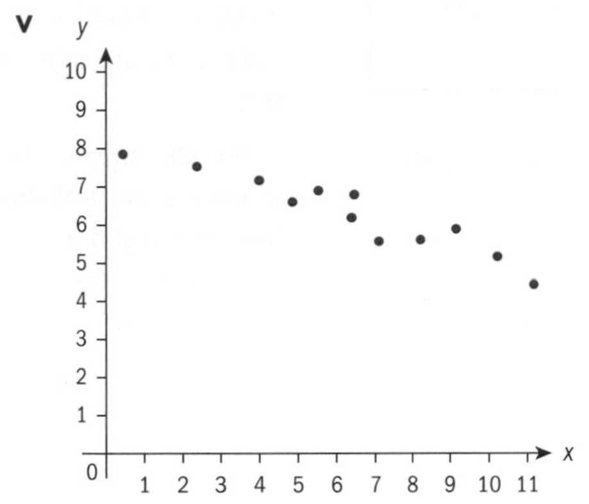

vi

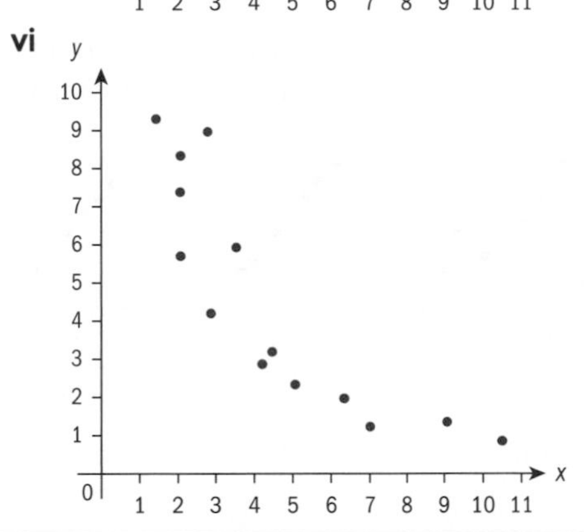

3 SCUBA divers have maximum dive times they cannot exceed when going to different depths. This table shows the times for certain depths.

Depth (m)	18	21	24	27	30	33
Time (min)	55	45	35	25	22	15

a Show this data on a scatter diagram.

b Describe the correlation.

c Copy and complete the sentence.
As the maximum depth .... the time at that depth ....

Reflect **Discuss** with a partner then **share** your ideas with your class.

- What is meant by independent and dependent variables?

4 The amount of fuel required for a car to travel certain distances is shown in this table:

Distance (km)	20	35	55	62	85	96	105	120
Fuel (l)	1.3	1.9	2.8	3.2	4.7	5.1	5.8	6.6

a Show this data on a scatter diagram.

b Describe the correlation.

5 Yumi has a new kitten. This table shows the age and weight of the kitten for its first eight weeks.

Week	1	2	3	4	5	6	7	8
Weight (kg)	0.25	0.35	0.5	0.66	0.73	1	1.2	1.3

a Show this data on a scatter diagram.

b Describe the correlation.

c What happens to the weight of the kitten as it gets older?

Statistics and probability

Developing inquiry skills

In the opening problem for the chapter, you saw how a rice farmer recorded the data of rainfall (in cm) and rice yield (in tonnes) from his farm for the last eleven years.

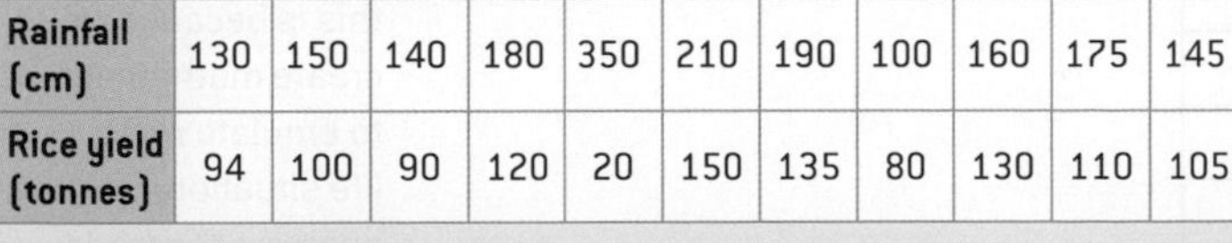

Rainfall (cm)	130	150	140	180	350	210	190	100	160	175	145
Rice yield (tonnes)	94	100	90	120	20	150	135	80	130	110	105

Using what you have learned so far in this chapter:

a draw a scatter diagram for the data

b describe the correlation.

7.2 Measuring correlation

So far, you have described the strength of correlation between two variables using words: very strong; strong; weak; very weak; or no correlation. However, statisticians need to be able to represent the strength of correlation as a numerical value. This allows them to make judgments about whether one correlation is stronger than another, say, if it is difficult to tell from a scatter graph.

> The **Pearson product-moment correlation coefficient** (denoted by r, where $-1 \le r \le 1$), is a measure of the correlation strength between two variables x and y. It is commonly used as a measure of the strength of **linear** correlation between two variables.

A stronger relationship between the two variables is shown by the r-value being closer to either +1 or −1, depending on whether the relationship is positive or negative. A correlation coefficient of zero indicates that no linear relationship exists between the two variables, and a correlation coefficient of −1 or +1 indicates a perfect linear relationship. The strength of relationship can be anywhere between −1 and +1.

International-mindedness

Karl Pearson (1857–1936) was an English lawyer and mathematician. His contributions to statistics include the product-moment correlation coefficient and the chi-squared test.

He founded the world's first university statistics department at the University College of London in 1911.

Correlation

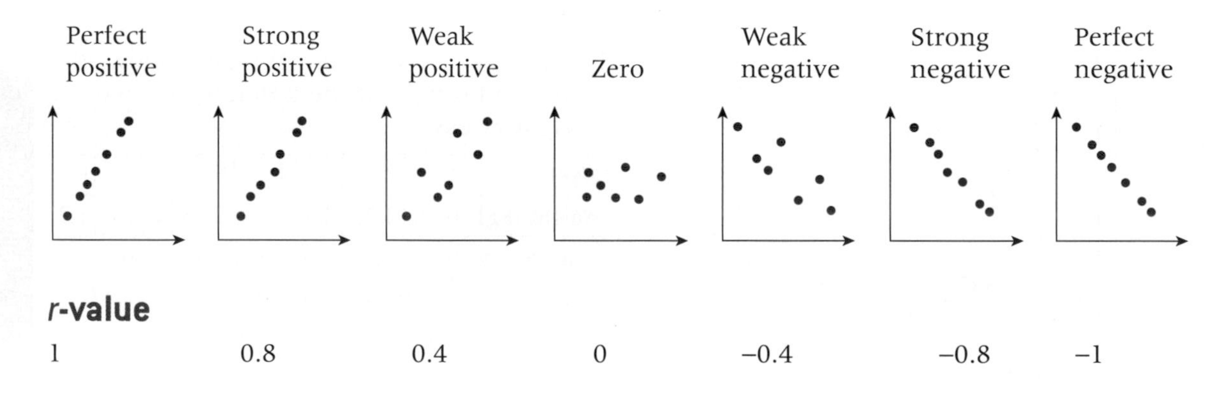

r-value

1 0.8 0.4 0 −0.4 −0.8 −1

In the external examinations, you will be expected to use a calculator to find the correlation coefficient.

Here is a quick way to interpret the r-value:

r-value	Correlation
$0 < r \le 0.25$	Very weak
$0.25 < r \le 0.5$	Weak
$0.5 < r \le 0.75$	Moderate
$0.75 < r \le 1$	Strong

TOK

We can often use mathematics to model everyday processes.

Do you think that this is because we create mathematics to emulate real-life situations or because the world is fundamentally mathematical?

Example 3

A school basketball coach kept a record of the number of games played (x) and the number of points scored (y) for seven basketball players.

Player	Games (x)	Points (y)
Ali	3	9
Mateo	4	10
Jerry	4	20
Poom	4	16
Ayo	5	20
Chen	6	29
Jimmy	10	43
Total	**36**	**147**

Use Pearson's correlation coefficient to determine the strength of the correlation between the number of games played and the number of points scored.

Player	x	y	x^2	y^2	xy
Ali	3	9	9	81	27
Mateo	4	10	16	100	40
Jerry	4	20	16	400	80
Poom	4	16	16	256	64
Ayo	5	20	25	400	100
Chen	6	29	36	841	174
Jimmy	10	43	100	1849	430
Total	**36**	**147**	**218**	**3927**	**915**

Enter the data in your GDC to find the value of r.

$r = 0.957$

Use the formula $r = \frac{S_{xy}}{S_x S_y}$.

There is a strong, positive, linear correlation between the number of games played and the number of points scored.

Since $0.75 < r \leq 1$, this is a strong correlation.

You can find Pearson's product-moment correlation coefficient directly from your GDC.

Exercise 7B

1 Kelly is a real-estate agent and makes a table to show the floor area and the prices of houses in her town.

Area (m^2)	20	24	30	40	50	75	80	90	100	120
Price ($1000)	250	300	350	360	480	580	750	840	900	1000

a Find the r-value for this data.

b Describe the correlation.

c Interpret the correlation between floor area and house price.

TOK

Can all data be modelled by a known mathematical function?

Statistics and probability

2 Uncle Mac insisted that his motorcycle would always be valuable, regardless of its age. His nephew, Jay, recorded its age and value in this table.

Age (years)	1	2	3	4	5	6	7	8	9	10
Price (euros)	40000	36500	31000	26658	24250	19540	19100	18750	15430	12600

a Find the value of r.

b Describe the correlation.

c Will the price of the motorcycle continue to decrease forever? Was Uncle Mac right? Discuss your answers with a classmate.

---

3 For a biology project, Luciana surveyed the heights and shoe sizes of her friends.

Height (cm)	148	153	165	142	155	141	171	154	170	168
Shoe size	34	38	42	36	42	32	40	34	40	38

a Find the value of r.

b Describe the correlation between height and shoe size for Luciana's friends.

4 The students' council took a sample of both the number of hours that its members studied on the night before a test and their test score.

Time (hr)	6	4	7	5	1	2	4	6	8	4
Score	78	80	86	88	66	70	78	95	97	76

a Find the r-value.

b Describe the correlation.

c Based on the results for the sample, would you say that more study time leads to a better test score?

5 Thierry had always been told that he spent too much time playing computer games and not enough time studying. He disagreed with this, and conducted a survey of his friends and their GPA.

GPA	3.9	2.7	3.8	2.4	1.7	2.6	4.0	3.7
Game time (hrs week^{-1})	10	14	5	8	24	17	21	7

a Find the value of r.

b Describe the correlation.

c Based on your results, determine whether Thierry was right, and give a reason for your answer.

6 A mathematics teacher wanted to see whether the number of extra practice questions attempted by his students had any effect on the students' course grade. He obtained the following results.

Questions	52	60	62	65	68	76	77	78	80	84	85	95
Grade	60	68	66	69	75	82	83	84	88	90	93	92

a Based on the results for the sample, would you say that more study time leads to a better test score?

b Describe the correlation.

c Based on these results, state the effect of doing more practice questions on the course grade.

Developing inquiry skills

In the opening problem for the chapter, you saw how a rice farmer recorded the data of rainfall (in cm) and rice yield (in tonnes) from his farm for the last eleven years.

Rainfall (cm)	130	150	140	180	350	210	190	100	160	175	145
Rice yield (tonnes)	94	100	90	120	20	150	135	80	130	110	105

1 Plot a scatter diagram for this data, and describe the correlation.
2 Use technology to find the value of r for this bivariate data. Comment on this r-value.
3 Observe the point at (350, 20). This distorts the data. Remove this outlier and redraw your scatter diagram.
4 Find the value of r without (350, 20). Is the correlation stronger or weaker when this point is removed?
5 Give a reason why the outlier may have occurred.

Reflect **Discuss** your answers with a partner then **share** your ideas with your class.

- Give a reason why the outlier may have occurred.

You should be aware that, in context, some outliers are a valid part of the sample, but some outlying data items may be an error in the sample.

7.3 The line of best fit

Investigation 1

Work with a partner or in a small group.

1 Draw a scatter diagram for this data set.

x	1	2	3	4	4	5	6
y	1	2	1.5	5	2	4	3

2 In words, describe the correlation between x and y, as it appears in your scatter diagram.
3 Draw a straight line on the graph which you think best approximates the relationship between x and y. Each person in the group should draw at least one line. Discuss why you drew the line of best fit in the way that you did.
4 Find the mean of the x-values ($\bar{x}$) and the mean of the y-values ($\bar{y}$).
5 Plot the point $(\bar{x}, \bar{y})$ on your scatter diagram. Can you adjust your line of best fit to pass through this point?
6 **Conceptual** Can you explain why your line of best fit should pass through the mean point $(\bar{x}, \bar{y})$?
7 **Conceptual** What does the line of best fit represent, and what point should it pass through?

You have been searching for a line of best fit in this investigation.

A **line of best fit** (sometimes called a trend line) is a straight line drawn through the centre of a group of data points plotted on a scatter diagram. The line of best fit lies as close as possible to all of the data points.

Drawing a line "by eye" is not the best way as it can lead to more than one line—like this set of data, which have several possible trend lines drawn.

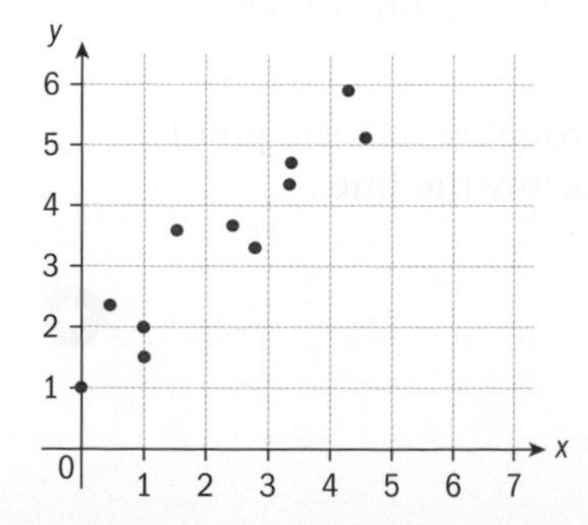

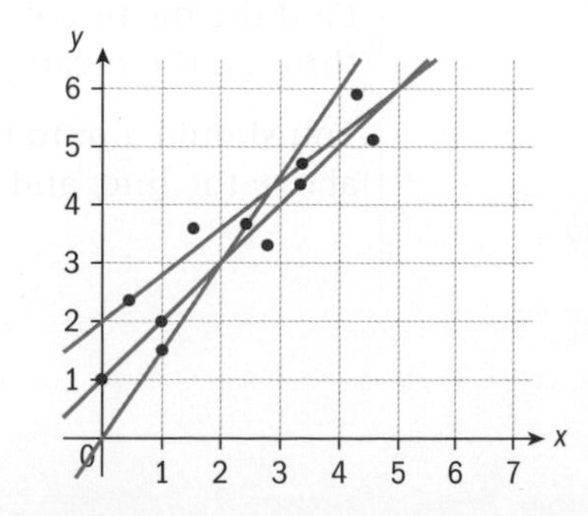

Statistics and probability

To draw a line of best fit by eye, draw a line that will balance the number of points above the line with the number of points below the line. An improvement is to have a reference point: one point which the line must pass through. This is the mean point, $(\bar{x}, \bar{y})$ and is calculated by finding the mean of the x-values and the mean of the y-values.

In the following example, you will plot a line of best fit to make an estimate about the connection between the total fat and the total calories in fast food.

Example 4

The table below shows the total fat and the total calories in fast food.

Meal	Total fat (g)	Total calories
Hamburger	9	260
Cheeseburger	13	320
Double burger	21	420
Double burger with cheese	30	530
Big burger	31	560
Toasted sandwich	31	550
Chicken wings	34	590
Crispy chicken	25	500
Fish fillet	28	560
Grilled chicken	20	440
Grilled chicken light	5	300

a Find the mean number of grams of fat.

b Find the mean number of calories.

c Construct a scatter plot for this data.

d Plot the mean point on your scatter plot and use it to draw a line of best fit.

e Find the equation of the line of best fit.

a Mean grams of fat $= \frac{247}{11} = 22.\dot{4}\dot{5}$

Divide the sum of the fat content by the number of items.

b Mean calories $= \frac{5030}{11} = 457.\dot{2}\dot{7}$

Divide the sum of the calories by the number of items.

c and **d**

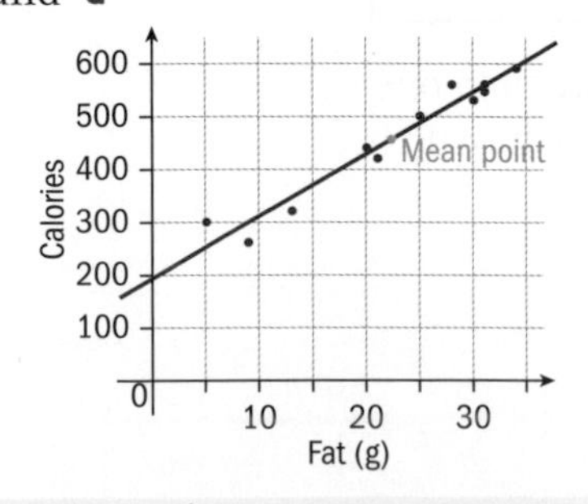

Choose suitable scales for your axes which allow you to see the spread of points well.

Find the mean point, and draw a line of best fit through the mean point.

You should aim to make roughly half the points above the line, and half below the line.

e $m = \frac{y_2 - y_1}{x_2 - x_1} = \frac{457.\dot{2}\dot{7} - 280}{22.\dot{4}\dot{5} - 10} = 14.2$	Find the gradient using the mean point and any other point on the line (which does not have to be a data point). Here, we use (10, 280).
$y - y_1 = m(x - x_1)$ $y - 457.\dot{2}\dot{7} = 14.2(x - 22.\dot{4}\dot{5})$	Use the point-gradient formula to find the equation of the line.
$y = 14.2x + 138$	Write in the form $y = mx + b$.

HINT

Remember, your line of best fit is not meant to pass through all of the points. It is meant to be a representative line for the data.

Interpolation and extrapolation

When you have a line of best fit, you can use it to make estimates about the data.

There are two types of predictions that can be made.

Interpolation is when you make predictions *inside* the domain which your data points span. Extrapolation is when you make predictions *outside* the domain which your data points span.

A good way to remember this is:

- **In**terpolation **in**side your data
- **Extra**polation for **extra** data.

The accuracy of interpolation depends on how strong the linear correlation was.

Extrapolation is often considered unreliable, as predicting becomes less accurate as we move further outside of the domain. It is difficult to know whether the data will continue in the same trend outside the domain of the data you have.

For example, Ayman is 17 years old. His mother has been recording his height since birth. This scatter graph shows his age and height.

Ayman's height chart

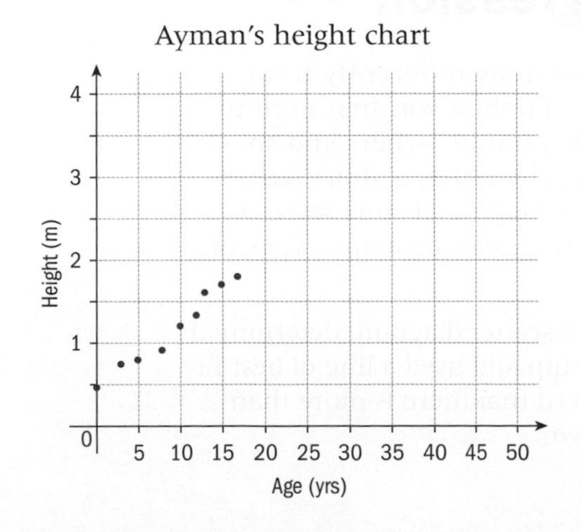

TOK

Is extrapolation knowledge gained using intuition and, possibly, emotion?

If so, how would you describe interpolation in terms of ways of knowing?

Reflect **Discuss** your answer with a partner then **share** your ideas with your class.

- Can you use the scatter diagram to estimate Ayman's height at any time during his first 17 years?
- What happens when you use the scatter diagram to estimate his height when he is 40 years old?
- Where can you use the scatter diagram to make an accurate estimation about Ayman's height?

Exercise 7C

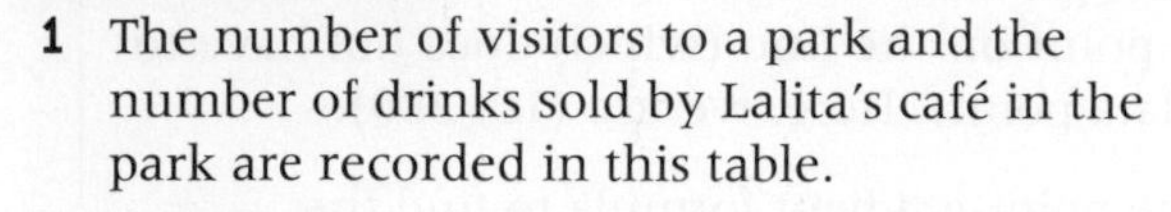

1 The number of visitors to a park and the number of drinks sold by Lalita's café in the park are recorded in this table.

Number of visitors	36	55	42	35	58	65
Drinks sold	17	30	23	11	44	51

a Find the mean number of visitors.

b Find the mean number of drinks sold.

c Construct a scatter diagram and draw a line of best fit through the mean point.

2 Giraffes are the world's tallest mammal and can grow to a height of 6 m. This table shows a baby giraffe's height during its first year.

Age (months)	1	2	3	4	6	8	10	12
Height (m)	1.78	1.98	2.17	2.40	2.82	3.26	3.71	4.14

a Find the mean age and height of the baby giraffe during its first year.

b Construct a scatter plot and draw a line of best fit through the mean point.

c Find the equation of your line of best fit.

d Use your equation to estimate the height of this giraffe at 9 months.

e Use your equation to estimate the height of this giraffe at 10 years.

f Explain whether your answer to part e is reliable.

3 Gayathri has found a job in the city and now has to find an apartment in which to live. She surveys the monthly rent of several places and their distance from the city centre.

Distance (km)	3	6	10	12	15	20
Monthly rent (thousands of rupees)	60	45	32	28	18	15

a Find the mean distance of the apartments from the city centre.

b Find the mean cost of these apartments.

c Construct a scatter plot and draw a line of best fit through the mean.

d Find the equation of your line of best fit.

e Estimate the cost of an apartment which is 8 km from the city centre.

f Gayathri can afford to pay 50 000 rupees per month. Find how close to the city centre she can live.

g Explain whether you can use your line of best fit to accurately calculate the cost of an apartment 30 km from the city centre.

7.4 Least squares regression

The term "regression" is used in statistics quite differently from the way it is used in other contexts. The method was first used to examine the relationship between the heights of fathers and sons. A tall father tended to have sons shorter than him; a short father tended to have sons taller than him. The heights of sons *regressed* to the mean. The term "regression" is now used for many types of curve fitting.

In this chapter, you have plotted data on a scatter diagram, determined the strength and direction of the relationship and used a line of best fit to represent the data. You may have noticed that there is more than one line of best fit by eye that can be drawn.

The line can be improved by using residuals.

A **residual** is the vertical distance between a data point and the graph of a regression line.

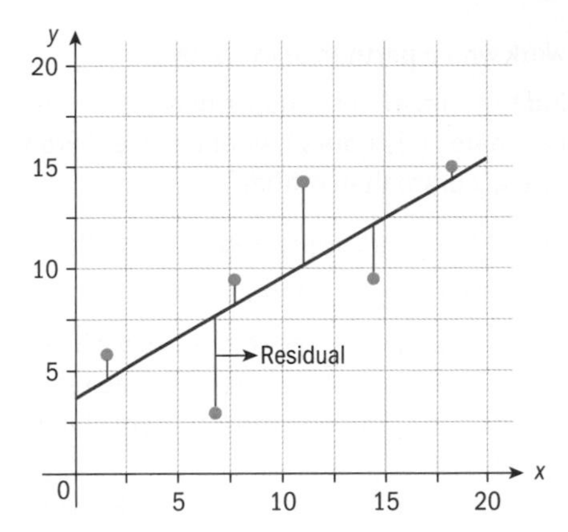

A method called **least squares regression** will help you to find the most accurate line of best fit, and therefore help you to make better predictions.

The **least squares regression line** is the one that has the smallest possible value for the sum of the **squares** of the residuals.

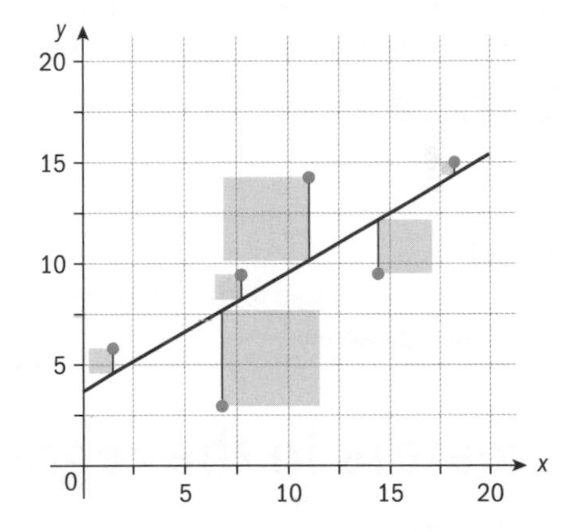

Investigation 2

In the opening problem for the chapter, you looked at the connection between rainfall and rice yield.

Rainfall (cm)	130	150	140	180	210	190	100	160	175	145
Rice yield (tonnes)	94	100	90	120	150	135	80	130	110	105

1. Find the mean rice yield and mean rainfall.
2. Plot a scatter diagram for this data with rainfall on the x-axis and rice yield on the y-axis. Draw a line of best fit, by eye, which passes through the mean point. Find the equation of your line of best fit.
3. On the scatter diagram, draw in the residuals between each data point and your line of best fit. Measure the length of each residual and square it. Record your results in a table like this.

Rainfall (cm)	130	150	140	180	210	190	100	160	175	145
Rice yield (tonnes)	94	100	90	120	150	135	80	130	110	105
Residual length										
Residual squared										

Continued on next page

Now work with a partner, or in a group.

4 Find the sum of the squares of all the residuals on your scatter graph. Compare this sum with your classmates—is yours higher or lower? From your results, do you think that some lines of best fit are more accurate than others?

5 Use the Geogebra applet found here https://www.geogebra.org/m/XUkhCJRj.
Without moving any of the points:
   a Tick the top three boxes to show a line of best fit and the squares of the residuals. The red line shows you the sum of the squares of the residuals.
   b Adjust the gradient and intercept sliders. Can you minimize the value of the red line, and hence find the best regression line?
   c Finally, tick the fourth box. This shows the actual regression line. How close were you?

6 Use technology to find the equation of the y on x regression line for the rice yield data. How do you think your calculator finds this equation?

7 **Conceptual** How does the regression line estimate the line of best fit?

EXAM HINT
You will be expected to use technology to find the equation of the line of regression during exams.

Parameters in the regression line equation

When you have a regression line $y = ax + b$ between two variables x and y, it is important that you are able to interpret the meaning of the parameters in the context of the problem.

The constant term b is the value at which the line crosses the y-axis (the value of y when $x = 0$). In the opening problem, b would represent the rice yield when no rain fell.

The gradient a of the regression line represents the rate of change in y as x changes. In the context of the opening problem, a gradient of a means rice yield increases by a tonnes for every 1 cm increase in rainfall.

HINT
Notice that the regression line found by your GDC passes through the **mean point** of the data.

Example 5

As a car's tires become worn, it uses more fuel to travel the same distance. This table shows the age of the tires and the number of kilometres the car travelled on one litre of diesel.

Age (years)	1	2	3	4	5	6
km l^{-1}	20	18.5	17.4	15.6	14.8	13.2

a Write down the regression equation.

b Interpret the meaning of the constant value.

c Interpret the meaning of the gradient of the regression line.

d Use the equation of the regression line to estimate the distance the car would have travelled on 1 litre of diesel when the tires were 3.5 years old.

a $y = -1.34x + 21.3$	Find the equation of the y on x regression line from your GDC.
b With brand new tires, the car would probably have travelled 21.3 km on 1 litre of diesel.	This is the distance the car travelled with tires age = 0.
c Every year, the car travels 1.34 km less far on 1 litre of fuel.	The gradient is negative, so the distance is decreasing as age increases.
d $y = -1.34(3.5) + 21.3 = 16.6$ km	Find y when $x = 3.5$.

Exercise 7D

1 A school principal collected data on the number of days per year (x) various students spend playing sports, and the number of hours of homework (y) they do per week. She came up with the equation of the regression line $y = 35 - 0.5x$. Interpret the meaning of the gradient and y-intercept in this context.

2 A police chief wants to investigate the relationship between the number of times a person has been convicted of a crime (y) and the number of criminals that person knows (x). The equation was found to be $y = 6x + 1$. Interpret the meaning of the gradient and y-intercept in this context.

3 The value of a speaker, $\$y$ dollars, is related to its age, x years, by the equation $y = 300 - 40x$. Interpret the meaning of the gradient and y-intercept.

HINT

The y-intercept may not be relevant to the relationship being studied, especially if it falls far outside the domain of the data.

TOK

To what extent can you reliably use the equation of the regression line to make predictions?

4 This table shows both the time, in seconds, which it takes for 10 cars to accelerate from 0 km h^{-1} to 90 km h^{-1}, and the maximum speed of the car.

0 to 90 time (seconds)	6	7	8	8.5	9	9.5	10	11	11.5	12
Max speed (km hr^{-1})	157	155	147	142	138	132	134	127	120	115

Statistics and probability

a Find Pearson's product-moment correlation coefficient (r) for this data.

b Find the equation of the y on x regression line.

c Use this equation to estimate the maximum speed of a car that takes 7.5 seconds to accelerate from 0 km h^{-1} to 90 km h^{-1}.

d Interpret the meaning of the gradient of the regression line in this context.

5 Ms Fung's class has 10 students. Their scores for both classwork and the final exam are shown below. Both scores are out of 100.

Student	Lin	Jung	Erika	Non	Chen	Park	Ploy	Jai	Fae	Nit
Classwork	80	73	95	84	67	88	69	92	75	90
Final	74	62	93	75	73	81	58	90	Abs	84

Fae was absent for the final exam. Do not include her grades in your calculations.

a Find the correlation coefficient (r) for this data.

b Find the equation of the y on x regression line.

c Use the equation of the regression line to estimate Fae's score, had she been present for the final exam.

6 Morgan records the length and height, in cm, of each of seven cats.

Length (cm)	35	38	42	45	47	48	50
Height (cm)	13	18	27	28	36	34	40

a Find the value of r for this data. Interpret the r-value in this context.

b Find the equation of the y on x regression line.

c Morgan's new cat has a length of 40 cm, but it would not stay still for long enough for its height to be measured. Use your regression equation to estimate its height.

d Interpret the meaning of the gradient in the equation of this regression line.

7 At low tide, mud banks are exposed on both sides of a river. Sonny recorded the number of mudbugs at various distances from the river's edge.

Distance (m)	0.5	1	1.5	2	2.5	3	4
Number of mudbugs	30	28	14	18	10	7	1

Sonny uses his GDC to model the relationship between the variables by a regression equation of the form $y = ax + b$.

a Write down the values of a and b.

b Hence, estimate the number of mudbugs, to the nearest whole number, that are 3.5 m from the river's edge.

c Find Pearson's product-moment correlation coefficient (r) for this data.

d State which two of these words best describe the correlation between Sonny's variables: strong, zero, positive, negative, no correlation, weak.

8 The table shows the Diploma score (x points) and first year university score (y%) for six IB Diploma students.

Diploma score (x points)	28	33	35	42	40	38
First year university score (y%)	66	70	85	94	96	80

a Find Pearson's product-moment correlation coefficient (r) for this data.

The relationship between the variables can be modelled by the regression line with equation $y = ax + b$.

b Write down the values of a and b.

c Interpret the meaning of the gradient in the regression line.

Glen scored a total of 30 in the IB Diploma.

d Use your regression line to estimate Glen's first year university score.

9 Tyler manages a pizza shop. Over a nine-day period, Tyler recorded the number of pizzas sold (x) and the total production cost ($\$y$). Tyler's results are shown in the table.

x	25	40	65	53	46	30	50	74	70
y	200	260	350	360	260	250	310	600	450

a Find the equation of the y on x regression line.

b Use your regression line as a model to find and interpret the meaning of:

i the gradient

ii the y-intercept.

c Estimate the cost of producing 60 pizzas.

d Comment on the appropriateness of using your model to:

i estimate the cost of producing 5000 pizzas

ii estimate the number of pizzas produced when the total production cost is \$100.

10 The value, y thousand Yen, of a used car changes as the age of the car (x years) increases. This table shows the age and value of Chan's car over 10 years.

x	1	2	3	4	5	8	9	10
y	115	110	92	89	80	63	59	54

The relationship between x and y can be modelled by the regression equation $y = ax + b$.

a Find the value of r for this data.

b Write down the values of a and b.

c Use the regression equation to estimate the value of Chan's car after six years. Give your answer to the nearest ¥1000.

Choosing a line of regression

You can draw two different regression lines depending upon which variable you wish to approximate. So far, you have used the y on x regression line. The x on y regression line is often a different line.

The y on x regression line is given by $y = a + bx$ and allows you to estimate the value of y from a given value of x.

The x on y regression line is given by $x = c + dy$ and allows you to estimate the value of x from a given value of y.

If there is a perfect correlation between the data—that is, if all the points lie on a straight line—then the two regression lines will be the same.

Your GDC can compute the equation of either line for you.

HINT

Pearson's product-moment correlation coefficient (the r-value) of x and y is the same, whether you compute using (x, y) or (y, x).

Example 6

The scores, marked out of 10, for a class of students in English and TOK are:

English (x)	2	3	4	4	5	6	6	7	7	8	10	10
TOK (y)	4	5	4	6	5	6	4	9	6	10	9	10

Use technology to help you:

a Find the regression equation of y on x.

b Find the regression equation of x on y.

c Show both lines on a scatter plot with the English score on the horizontal axis.

d Find the predicted TOK score, to the nearest whole number, for a student who scores 9 in English.

e Find the predicted English score, to the nearest whole number, for a student who scores 8 in TOK.

a $y = 1.92 + 0.764x$

b $x = 0.139 + 0.902y$

Find the equations of both lines of regression using your GDC.

c

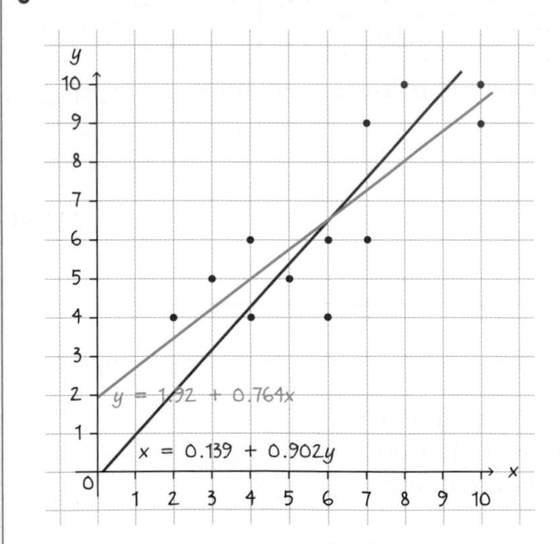

Plot a scatter diagram of the data, and then draw both regression lines using their equations. It can be helpful to show the two lines in different colours.

d $y = 1.92 + 0.764(9) = 9$ to the nearest whole number.

TOK is the y-variable, so use the y on x regression line.

e $x = 0.139 + 0.902(8) = 7$ to the nearest whole number.

English is the x-variable, so use the x on y regression line.

Exercise 7E

1 For seven different theatres, the table below shows the cost of theatre tickets and the number of tickets sold in that theatre over the course of a week.

Cost ($$x$)	12	15	18	18	22	25	30
Tickets sold (y)	45	44	45	42	40	34	26

a Find the regression line of y on x and use it to predict how many tickets, to the nearest whole number, would be sold at a theatre which charged $20 per ticket.

b Find the regression line of x on y and use it to predict the cost of a ticket, to the nearest Dollar, at a theatre which sold 35 tickets.

2 Mo is building a tower from wooden blocks. He measures the height of the tower at different times, in minutes, after he starts building.

Time (min)	2	3	5	7	8	10
Height (cm)	6	10	22	33	42	56

a Find the equation of the regression line of time on height.

b Use your equation from part **a** to approximate the time at which the tower is 50 cm tall. Give your answer to the nearest tenth of a minute.

3 Cherry scores 52 in a physics test but was absent for the mathematics test. The table shows the test results for the rest of the class.

Mathematics (x)	90	88	65	92	50	67	100	100	73	90	83	94	83
Physics (y)	87	57	52	76	30	67	96	74	65	87	78	89	78

Find a suitable regression equation and estimate Cherry's score in mathematics, to the nearest whole number.

4 The success of a shopping mall is related to its distance, in km, from the town centre.

For 10 different malls, the number of visitors per hour and the distance from the town centre were recorded.

Distance (km)	1	5	9	12	14	19	21	24	30	34
Visitors	180	164	148	120	118	90	85	82	65	60

a Estimate how many people per hour would visit a mall which is 7 km from the town centre.

b A developer wants to build a mall which will attract at least 100 visitors per hour. Estimate the furthest distance from the town centre that he should build it.

Piecewise linear functions

Some problems are not easily represented by a single linear function, but are better represented by two or more linear functions. These are known as piecewise linear functions.

TOK

"Everything that can be counted does not count. Everything that counts cannot be counted."
—Albert Einstein

What counts as understanding in mathematics?

Investigation 3

Robin starts a business making fishing rods and has to produce a report at the end of the year to show his stakeholders.

This table was included in his report, and shows the number of rods made (x) and the total production cost ($\$y$).

x	10	20	30	40	50	60	70	80	90	100
y	100	120	140	144	155	195	250	275	340	360

1 Show this data on a scatter diagram.

2 Draw a line of best fit on your scatter diagram.

3 **Factual** Comment on how well the line of best fits the data, and how accurately you could use it to make predictions.

Robin found that, to make more than 50 rods a day, he had to use two machines rather than one.

4 **Factual** Explain how this information might cause you to conclude that a single regression line is not enough to best represent the relationship between number of rods and production cost.

5 **Factual** What is a piecewise linear model?

6 **Conceptual** When can piecewise models be used to represent bivariate data?

7 Use technology to find one regression line for 0 to 50 rods and another for 50 to 100 rods. Plot the piecewise regression lines on your scatter diagram.

8 Comment on whether the piecewise model, or the linear model, better fits the shape of the data.

Piecewise linear models should not be used for data which is related by a non-linear equation, for example an exponential or quadratic relation.

Example 7

Draw the piecewise functions.

a $f(x) = \begin{cases} x+2, \text{when } x \le 1 \\ 4-x, \text{when } x > 1 \end{cases}$

b $f(x) = \begin{cases} 2x+1, \text{when } x \le 2 \\ 6, \text{when } x > 2 \end{cases}$

a

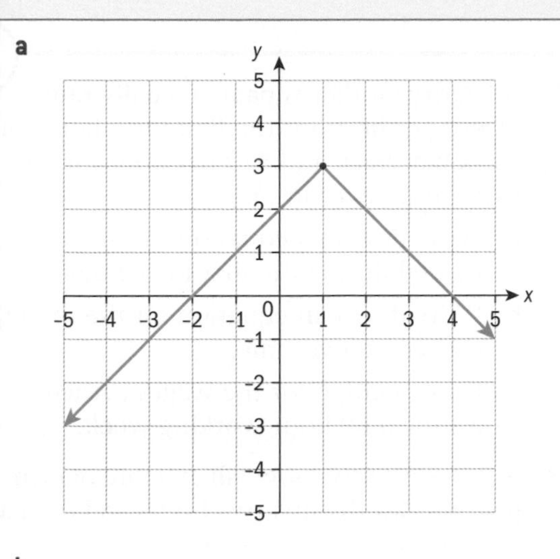

a The graph of $f(x) = x + 2$ should be drawn up $x = 1$.

After this point you should draw $f(x) = 4 - x$ from $x = 1$ onwards.

b

b The graph of $f(x) = 2x + 1$ should be drawn up to $x = 2$. A solid dot should be at $(2, 5)$ to indicate that this point is part of the function.

Now you should draw $f(x) = 6$ from $x = 2$ onwards, with a hollow dot at the point $(2, 6)$ as x is greater than but not equal to 2.

Example 8

In Thabodia, income is taxed at 20% for the first \$15 000 an individual earns, and then at 30% for anything over \$15000. Draw a graph of tax paid for people earning up to \$30 000.

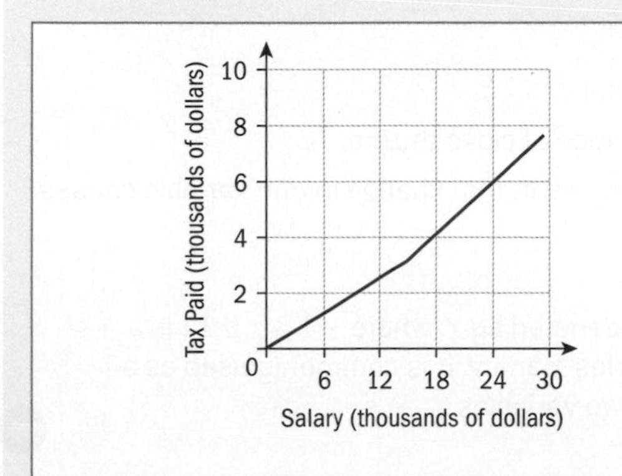

A 20% tax for the first \$15 000 gives $y = 0.2x, 0 \leq x \leq 15$ where the axes scales are in thousands of dollars. This means the first line stops at the point $(15, 3)$.

For any money earned above \$15 000, a 30% tax is implemented. Therefore, the second line has a gradient of 0.3. However, everything up to \$15 000 is still taxed at 20%, so we deduct 10% of 15.

This gives the line $y = 0.3x - 1.5$, $x > 15$

Exercise 7F

1 Draw the piecewise functions.

a $f(x)=\begin{cases}-x+12, \text{when } x\le 5\\ x+2, \text{ when } x>5\end{cases}$

b $f(x)=\begin{cases}-2x-1, \text{when } x\le 2\\ 4-x, \text{when } x>2\end{cases}$

c $f(x)=\begin{cases}x-2, \text{when } x\le 1\\ 6-x, \text{when } x>1\end{cases}$

d $f(x)=\begin{cases}4-2x, \text{when } x>2\\ x-2, \text{ when } -2\le x\le 2\\ -4, \text{ when } x<-2\end{cases}$

e $f(x)=\begin{cases}x+1, \text{when } x>2\\ x, \text{when } -2\le x\le 2\\ -4, \text{when } x<-2\end{cases}$

2 A factory worker is paid at a daily rate of \$10 per hour for the first forty hours of work per week and \$15 per hour for any extra time (overtime).

a Write down an equation for the money earned during the first forty hours.

b Write down an equation for the money earned on overtime.

c Draw a graph for the wages earned during a 60 hour working week.

3 A ski resort saw snow fall at a rate of 6cm per hour for the first two hours and then at 4cm per hour for the next three hours.

a Draw a graph to show the snowfall.

b Draw a line on your graph to estimate when there will be 20cm of snow. Write your answer under the graph.

Developing your toolkit

Now do the Modelling and investigation activity on page 350.

Chapter summary

- **Bivariate analysis** is concerned with the relationship between two data sets.
- A scatter diagram takes the two sets of data and plots one set on the x-axis and the other set on the y-axis.
- A **correlation** exists between two variables, x and y, when a change in x corresponds to a change in y.
  - A general upward trend shows a positive correlation.
  - A general downward trend shows a negative correlation.
  - Scattered points with no trend may indicate a correlation of close to zero.
- A correlation between two data sets does not necessarily mean that change in one variable *causes* change in the other.

Measuring correlation

- The **Pearson product-moment correlation coefficient** (denoted by r, where $-1\le r\le 1$) is a measure of the correlation strength between two variables x and y. It is commonly used as a measure of the strength of **linear** correlation between two variables.

r-value	Correlation
$0 < r \leq 0.25$	Very weak
$0.25 < r \leq 0.5$	Weak
$0.5 < r \leq 0.75$	Moderate
$0.74 < r \leq 1$	Strong

The line of best fit

- A **line of best fit** (sometimes called a trend line) is a straight line drawn through the centre of a group of data points plotted on a scatter diagram. The line of best fit lies as close as possible to all of the data points.
- To draw a line of best fit by eye, draw a line that will balance the number of points above the line with the number of points below the line. An improvement is to have a reference point: one point which the line must pass through. This is the mean point, $(\bar{x}, \bar{y})$ and is calculated by finding the mean of the x-values and the mean of the y-values.

Interpolation and extrapolation

- When you have a line of best fit, you can use it to estimate further data.
- There are two types of predictions that can be made. One is called interpolation and the other is called extrapolation.
- Interpolation means from within the given domain and extrapolation means from outside the given domain. It is not usually advisable to extrapolate outside the domain of the given data.

Least squares regression

- The **least squares regression line** is the one that has the smallest possible value for the sum of the **squares** of the residuals. You can find the equation of this line on your GDC.
- The y on x regression line is given by $y = a + bx$ and allows you to estimate the value of y from a given value of x.
- The x on y regression line is given by $x = c + dy$ and allows you to estimate the value of x from a given value of y.
- If there is a perfect correlation between the data—that is, if all the points lie on a straight line—then the two regression lines will be the same.
- Piecewise linear models display situations where one single equation of a line does not fit the entire set of data points.

Statistics and probability

Developing inquiry skills

Look back at the scatter graph you produced for the rice farmer's data on the amount of rainfall in cm and rice yield in tonnes from the last eleven years.

1. Find the mean point.
2. Draw a line of best fit by eye through the mean point.
3. Now find the regression line in the form $y = ax + b$ from your GDC and draw this on the same graph.
4. Comment on which is the best line to represent the data and why.

Chapter review

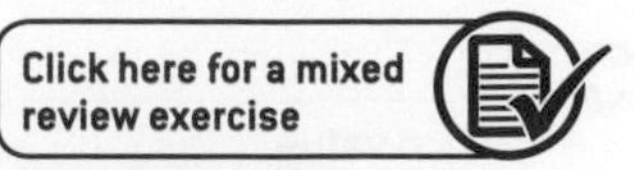

1 **a** Write down the maximum and minimum values that Pearson's product-moment correlation coefficient (r) can take.

b Match each scatter graph with the r-value from the list below which best describes the correlation between variables x and y.

r-values: –0.96, –0.6, 0, 0.5, 0.9

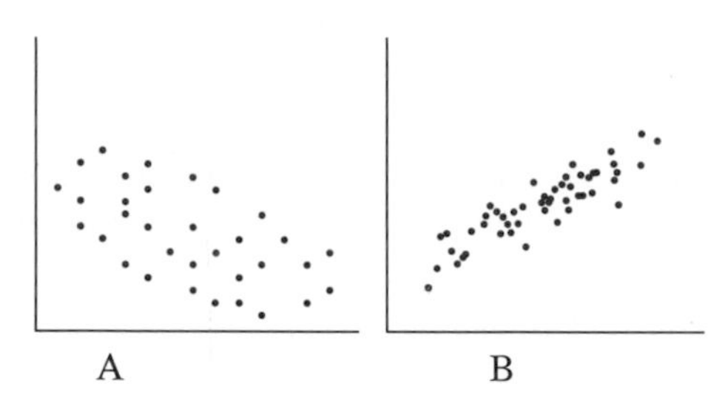

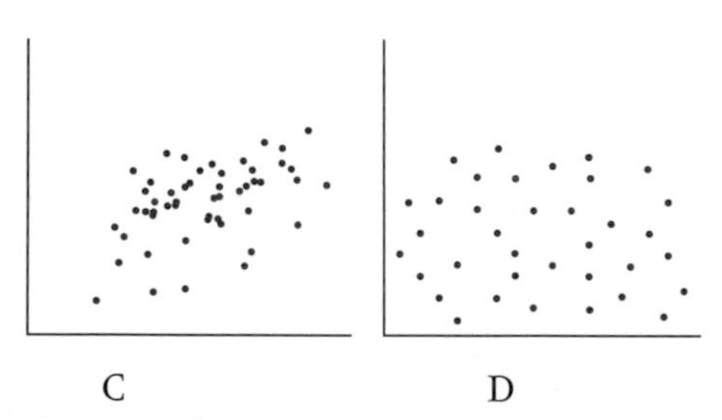

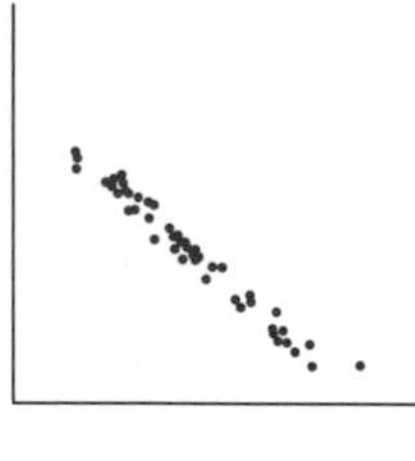

c Choose two words from the list below which best describe the correlation between x and y in scatter diagram E:

perfect, zero, linear, strong positive, strong negative, weak positive, weak negative

2 Kasem recorded the average monthly outside temperature (in °C) and the cost of his monthly electricity bill (in hundreds of dirhams) over the course of six months. His results are shown in the table.

Temperature (°C)	Electricity bill (dirhams 100s)
39	11
36	8
45	16
41	13
42	14
37	10

a Plot this data on a scatter diagram.

b Find the mean outside temperature over the six months.

c Find the mean cost of Kasem's electricity bill over the six months.

d Plot and label the mean point, M, on your scatter diagram.

The line of best fit passes through M and the point (45,16).

e Use these two points to find the equation of the line of best fit for this data.

f Draw the line of best fit on your scatter diagram.

3 An environmental group records the annual numbers of eggs (e) laid each year by eagles in a wildlife park, t years after 2010.

Years (t)	Eggs (e)
0	29
2	38
4	27
6	19
8	12

The relationship between the variables can be modelled by the regression equation $e = at + b$.

a Find the values of a and of b.

b Use the regression equation to estimate the number of eggs in the park when $t = 5$.

c Explain why it is not reliable to use the regression equation to estimate the number of eggs in the year 2050.

4 A furniture manufacturer makes tables to sell to shops. Over a six-week period, the production cost $\$y$ of manufacturing x chairs is given in this table.

Week	Number of chairs (x)	Production cost (y)
1	28	3600
2	46	5200
3	38	4400
4	34	3800
5	52	6000
6	50	5900

a Find the equation of the y on x regression line for this data.

b The chairs are sold at \$120 each. Find the least number of chairs which the factory must sell each week in order to make a profit.

5 The maximum daily temperature at the beach (x°C) and the number of bottles of water sold per day (y) at Marley's beach stall is given in this table.

x	24	23.5	23	22	21	20.3	20	18.2	17	26
y	260	199	174	162	149	135	118	115	102	246

a Write down the mean maximum temperature.

b Write down Pearson's product-moment correlation coefficient, r.

c Comment on what the value of r for this data tells you.

d Write down the equation of the y on x regression line.

e Estimate how many bottles of water Marley will sell on a day when the maximum temperature is 19.6°C.

f On a day when the temperature is forecast to be 36°C, Marley estimates that he will sell 429 bottles of water. Give one reason why his answer might be unreliable.

6 The annual advertising expenditures ($\$x$) and sales ($\$y$) for a small company are listed below, covering a period of eight years.

x	3500	2000	5000	6000	5000	3000	4000	8000
y	110 000	65 000	100 000	135 000	120 000	90 000	100 000	140 000

a Write down the equation of the y on x regression line for this data.

b Estimate the annual sales that would result when \$7000 is spent on advertising.
Give your answer to the nearest thousand Dollars.

c The annual advertising expenditures and annual sales figures above were converted into Japanese Yen (1 Dollar = 69.4017 Yen).
Which of the following quantities would change, and which would remain the same?

i the mean of the advertising expenditures

ii the standard deviation of the advertising expenditures

iii the correlation coefficient, r

7 The amount of fuel (y litres) used by a car to travel x km is shown in this table.

Distance (x km)	Amount of fuel (y litres)
30	3.2
65	7.5
110	8.4
140	15.1
185	16.5

This data can be modelled by the regression line with equation $y = ax + b$.

a Write down the values of a and of b.

b Explain what the gradient, a, represents.

c Use the model to estimate the amount of fuel the car will use when it is driven for 160 km.

d Explain why it would be unreliable to use this model to predict the amount of fuel used on a journey of 5 km.

8 Sukhila is a biologist investigating the relationship between the amount of growth hormone, x grams, supplied to an orchid plant, and the number of flowers (y) it produces.

x	1	1.5	2	2.5	3	3.5	4	4.5
y	6	7	10	15	9	17	20	18

Sukhila models the data using the y on x regression line with equation $y = ax + b$.

a Write down the values of a and of b.

b Explain what the gradient represents.

c Explain what the y-intercept represents.

d Estimate the number of flowers that Sukhila would expect from 1.75 grams of growth hormone.

e Sukhila wanted the plant to produce 12 flowers. Find how much growth hormone she should use.

f Explain why it might be inappropriate to use this model to predict the number of flowers when the plant is treated with 1 kg of growth hormone.

9 Joel is conducting a physics experiment. One end of a spring is attached to a hook, and masses of different sizes are attached to the other end. Joel measures the length of the spring for each of the different masses, and records his results in this table.

Mass (g)	Spring length (mm)
100	204
200	257
300	292
400	315
500	330
600	355
700	370

Joel represented the data with a linear regression equation $y = ax + b$.

a Find the values of a and b.

b Interpret the meaning of the gradient, a.

c Interpret the meaning of b.

d Estimate the length of the spring when a mass of 550 grams was attached.

e Explain whether Joel could reliably use this equation to predict the spring length for a mass of 2 kg.

f Joel wants the spring to stretch to 300 mm. Find the equation of the x on y regression line, and use it to determine the mass that should be attached to the spring.

10 Luther wishes to see whether there is any correlation between a person's age and the number decimal places of π which they can remember after spending five minutes memorizing them. His results are as follows.

Age (x years)	Number of decimal places (y)
15	26
25	30
35	25
45	26
55	20
65	41

a Luther wants to represent the data as a linear regression. Find the equation of the regression line of y on x.

b Use your equation to estimate the number of decimal places of π that a person aged 50 years would remember.

c Write down the correlation coefficient r.

d Use the value of r to describe the correlation between age and number of decimal places a person remembered.

Exam-style questions

11 **P2:** For 20 paired pieces of bivariate data (x, y), with a Pearson product moment correlation coefficient of $r = 0.93$, a line of best fit of y on x is calculated as $y = 0.51x + 7.5$.
The mean of the x data, $\bar{x} = 100$.

a If an x-value of 120 is known, estimate the corresponding y-value, assuming that this is interpolation. (2 marks)

b Find the mean of the y data, $\bar{y}$. (2 marks)

c State the type of linear correlation that is shown in this example. (2 marks)

d If a y-value was known state which linear regression line should be used to find an estimate for the corresponding x-value. (1 mark)

12 **P1:** Use the list below to find the correct description for the following values of Pearson product moment correlation coefficients.
perfect positive, strong positive, weak positive, zero, weak negative, strong negative, perfect negative

i 1 **ii** −0.8 **iii** 0.4
iv −0.4 **v** 0 (5 marks)

13 **P2:** Ten pieces of paired bivariate data are obtained from ten twins, one of which is female and the other is male, by giving them an intelligence test. The data is given in the table below.

Female	100	110	95	90	103	120	97	105	89	111
Male	98	107	95	89	100	112	99	101	89	109

a Find the Pearson product moment correlation coefficient, r. (2 marks)

b State the type of linear correlation that is shown in this example. (2 marks)

c Let the male score be represented by x and the female score by y. Find the equation of the

i y on x line of best fit

ii x on y line of best fit. (4 marks)

d Another pair of female–male twins is discovered. The male scored 105 on the test but the female was too ill to take it. Estimate the score that she would have obtained, giving your answer to the nearest integer. (1 mark)

e Another pair of female/male twins is discovered. The female scored 95 on the test but the male refused to take it. Estimate the score that he would have obtained, giving your answer to the nearest integer. (1 mark)

f If for a further pair of male/female twins the male scored 140 on the test, explain why it would be unreliable to use a line of best fit to estimate the females score. (1 mark)

14 **P2:** Eight pieces of paired bivariate data (x, y) are given in the table below.

x	1	2	3	4	5	6	7	9
y	9	8	7	6	5	5	5	2

a On graph paper, draw a scatter diagram to represent this data using a scale of 1 cm for 1 unit on both the x- and y-axis. (4 marks)

b By considering the scatter diagram, state the type of linear correlation that is shown in this example. (2 marks)

c **i** Find the mean of the x-values. (2 marks)

ii Find the mean of the y-values. (2 marks)

iii Mark the point $(\bar{x}, \bar{y})$ on the scatter diagram using the symbol ⊗. (1 mark)

d Sketch the y on x line of best fit on the scatter diagram. (2 marks)

e If the x value is known to be 8, use the diagram to estimate (to one decimal place) the y value. Show on the diagram how this estimate was obtained. (2 marks)

15 P2: Paired, bivariate data (x, y) that is strongly correlated has a y on x line of best fit of $y = mx + c$. When $x = 70$, an estimate for y is 100. When $x = 100$, an estimate for y is 140.

a Find the value of **i** m **ii** c. (3 marks)

b State if there is positive or negative correlation. (1 mark)

c The value of $\bar{x}$ is 90, find the value of $\bar{y}$. (3 marks)

d When $x = 60$ find an estimate for the value of y, given that this is interpolation. (2 marks)

16 P1: The temperature, T °C, of liquid in a chemical reaction is a function of time, t, in seconds, where t is the time from when the experiment started. The relationship is given by the equation:

$$f(x) = \begin{cases} 40 + 2t & 0 \le t \le 30 \\ 130 - t & 30 \le t \le 60. \end{cases}$$

Find the initial temperature of the liquid. (1 mark)

a Find the temperature of the liquid after 60 seconds. (1 mark)

b Find the maximum temperature that the liquid reaches. (1 mark)

c **i** Sketch a graph of T as a function of t.

ii Find the time interval during which the temperature of the liquid is greater or equal to 80°C. (4 marks)

17 P2: Ten teenagers were asked their age and the number of brothers and sisters they have. The data are shown in the scatter diagram below, where x represents the teenager's age and y represents the number of brothers and sisters that they have.

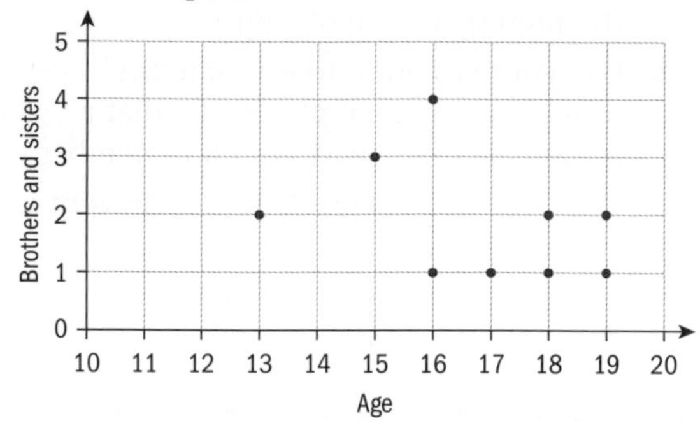

a Use the scatter diagram to copy and complete the following table.

x	13	14	15	16	16	17	18	18	19	19
y					4			2	1	

(3 marks)

b Calculate the Pearson product moment correlation coefficient for this data. (2 marks)

c Give two reasons why using this scatter diagram would not be valid for estimating the number of brothers and sisters a 25 year old would have. (2 marks)

18 P1: One of the ramps in a skateboard park is as shown in the diagram below. There is a horizontal part in the middle and the units shown are in metres.

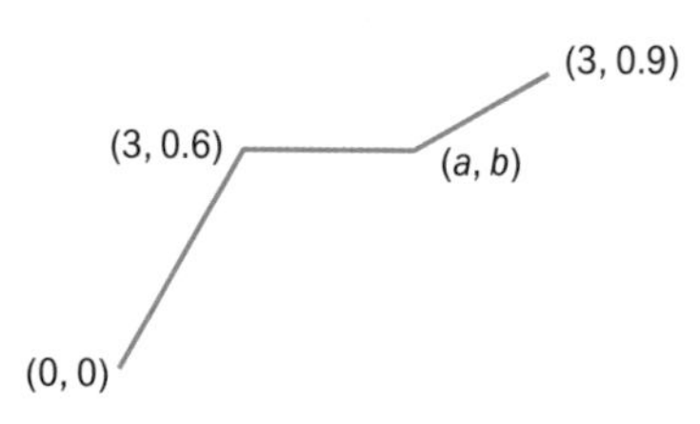

Diagram not to scale

The formula for this piecewise linear function is given by

$$f(x) = \begin{cases} mx & 0 \le x \le k \\ l & k \le x \le 5 \\ px + q & 5 \le x \le r \end{cases}$$

Find the values of

i m **ii** l **iii** k **iv** a

v b **vi** p **vii** q **viii** r. (11 marks)

19 P2: Paired bivariate data (x, y) is given in the table below. It represents the heights x, m and lengths y, m of a rare type of animal found on a small island.

x	2.4	3.6	2.8	1.8	2.0	2.2	3.0	3.4
y	3.0	4.0	3.0	1.7	2.0	2.3	3.1	2.7

a **i** Calculate the Pearson product moment correlation coefficient for these data.

ii State the type of linear correlation that is shown in this example.

iii Calculate the y on x line of best fit. (6 marks)

Another four examples of this rare animal are found on a nearby smaller island. These extra data are given in the table below.

x	2.3	2.7	3.0	3.5
y	4.1	1.5	4.2	1.5

b **i** Calculate the Pearson product moment correlation coefficient for the combined data of all twelve animals.

ii State the type of linear correlation that is shown in this example.

iii Suggest a reason why it would not be particularly valid to calculate the y on x line of best fit for the combined data. (5 marks)

20 P1: A set of bivariate data has a Pearson product moment correlation coefficient of $r = 0.87$ for 25 pairs (x, y). The y on x line of best fit is given by $y = 15x + 11$. Consider the value of r and line of best fit together with their definitions to answer the following.

a All the original x-values are increased by adding 5 and all the original y-values are decreased by subtracting 4.

i State the new value of r.

ii State the new value for the gradient of the y on x line of best fit.

iii Give a reason for your answers to **i** and **ii**.

iv State the type of linear correlation that is shown in this example.

(5 marks)

b All the original y-values are altered by multiplying by 2 and all the original x values remain unchanged.

i State the new value of r.

ii State the new value for the gradient of the y on x line of best fit.

iii Give a reason for your answers to **i** and **ii**. (3 marks)

c All the original x values are altered by multiplying by -3 and all the original y-values remain unchanged.

i State the new value of r.

ii State the new value for the gradient of the y on x line of best fit.

iii Give a reason for your answers to **i** and **ii**.

iv Describe in two words the linear correlation that exists for the new data. (6 marks)

Ranking

> **Approaches to learning:** Collaboration, Communication
>
> **Exploration criteria:** Personal engagement (C), Use of mathematics (E)
>
> **IB topic:** Bivariate Data, Correlation, Spearman's rank

Pearson product moment correlation coefficient

As you have seen in this chapter, the Pearson product moment correlation coefficient (PMCC) evaluates the strength of linear correlation between two variables.

What happens if there appears to be a relationship, but the correlation does not appear to be linear?

You can represent this data on a scatter diagram as shown:

x	4	5	6	7	8	9	10	11	12	13	14
y	4	4.1	4.2	4.4	4.7	5.1	5.5	6	7	9	20

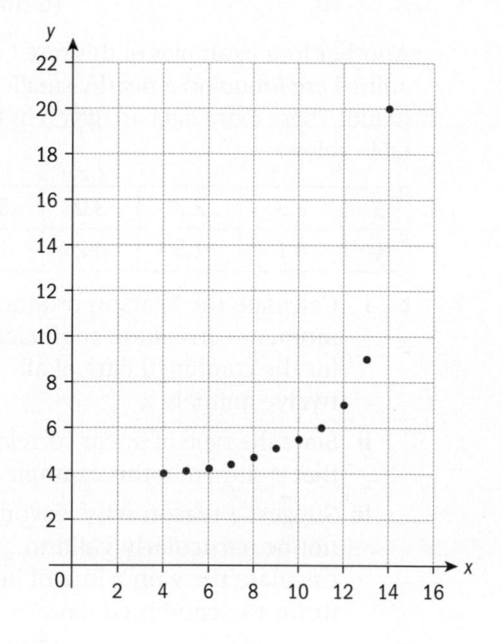

Do you think there is a relationship between the variables x and y?

Is the relationship strong? Describe the relationship.

Find the Pearson's correlation coefficient, r, for this data.

What does this tell you about the correlation for this data?

Explain why the relationship is very strong but the correlation coefficient does not reflect this.

Spearman rank correlation

You can also use **Spearman Rank Correlation** to analyse the relationship.

The Spearman correlation evaluates the relationship between variables which is not necessarily linear. The Spearman correlation coefficient is based on the **ranked values** for each variable rather than on the **raw data**.

What are the main differences between the Pearson correlation and the Spearman correlation?

The formula for Spearman's correlation, r_s, is:

$$r_s = 1 - \frac{6\sum D^2}{n(n^2-1)}$$

where D is the difference between a pair of scores and n is the number of pairs of ranks.

Research how to interpret Spearman's rank correlation.

What is the value of Spearman's rank for the above data?

What does this value of r_s tell you about these data?

Activity 1

Country	GDP/Capita in US$(2017)	Number of Children per woman(2017)	Country	GDP/Capita in US$(2017)	Number of Children per woman(2017)
Afghanistan	1981	5.12	Niger	1017	6.49
Argentina	20787	2.26	Paraguay	9691	1.90
Australia	47047	1.77	Peru	13434	2.12
Barbados	18640	1.68	Philippines	8342	3.02
Brazil	15848	1.75	Poland	29291	1.35
Cambodia	4002	2.52	Qatar	128378	1.90
China	16807	1.60	Russia	25533	1.61
Gambia, The	1715	3.52	Serbia	15090	1.44
Haiti	1815	2.72	Spain	38091	1.50
India	7056	2.43	Tunisia	11911	2.23
Japan	43876	1.41	United Kingdom	43877	1.88
Kenya	3286	2.98	United States	59532	1.87
Malawi	1202	5.49	Uzbekistan	6865	1.76
Moldova	5698	1.57	Zimbabwe	2086	3.98
Nepal	2682	2.12			

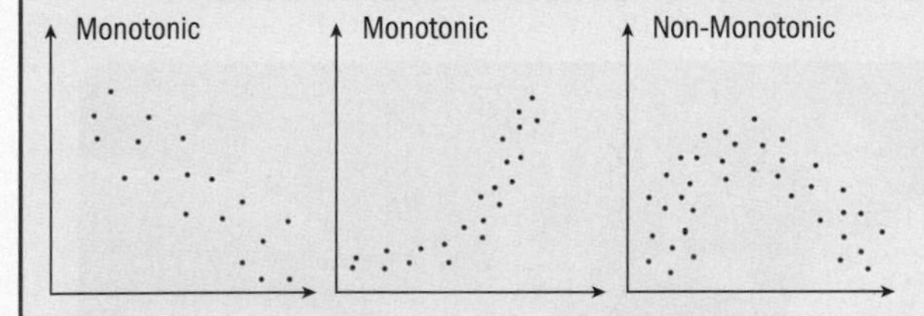

Describe the relationship between GDP/Capita and Number of children per woman.

Calculate the Spearman's rank correlation coefficient for this data.

Describe the relationship between GDP/Capita and Number of children per woman.

Activity 2

Individually rank the songs you have selected by number, with 1 being the favourite.

Record the rankings of everyone in your group. Do not collaborate or communicate with each other.

Find the Spearman rank correlation between each pair of students in your group.

Do any of the pairs of students display strong correlations?

Write a conclusion for this experiment based on the results you have found.

Extension

What else could you compare the ranking of using the same process as in Activity 2?

Brainstorm some possible ideas within your groups.

Design and run an experiment for one of your ideas.

Think carefully about what you would need to be aware of as you run this experiment to ensure that your results are accurate and unbiased.

8 Quantifying randomness: probability

Concepts

- Representation
- Quantity

Microconcepts

- Concepts of trial, outcome, equally likely outcomes, relative frequency, sample space (U) and event
- The probability of an event
- The complementary events A and A' (not A)
- Expected number of occurrences
- Combined events
- Mutually exclusive events
- Conditional probability P
- Probabilities with and without replacement
- Independent events
- Use of Venn diagrams, tree diagrams, sample space diagrams and tables of outcomes to calculate probabilities

You use the language of probability all the time:

- What is the chance it will rain tomorrow on my way to school?
- How likely do you think it is you will pass your driving test on your next attempt?
- What do you think the probability is that we will win the game this afternoon?
- Am I certain to get to school on time if I catch the bus rather than walk?

Uncertainty and randomness occur in many aspects of our daily life. Decision making in the worlds of business, investment, agriculture and industry, healthcare and all aspects of real life is based on expectation and prediction. Having a good knowledge of probability helps you make sense of these uncertainties, understand risk and make better decisions about the future.

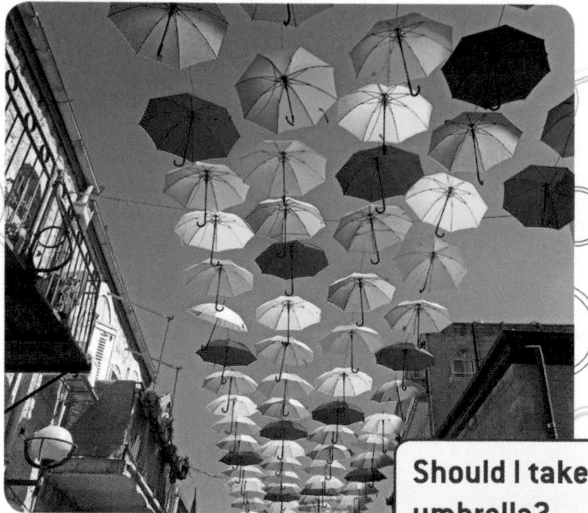

Should I take my umbrella?

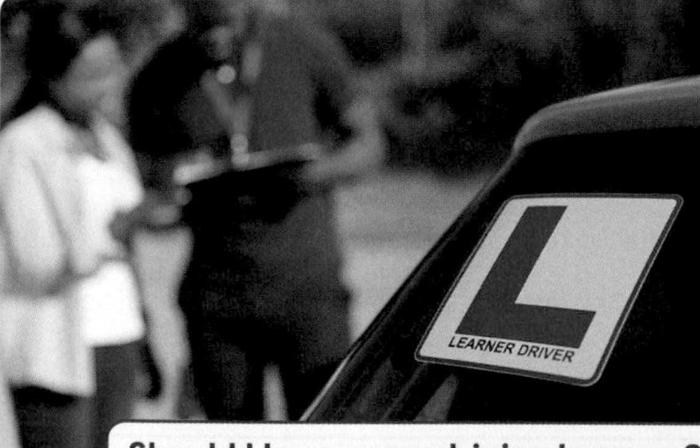

Should I have more driving lessons?

What transport should I use to make sure I am on time for my important exam?

Should we play our best team right from the start?

The Monty Hall problem is one of the most famous in probability. It is based on the 1980s American gameshow "Let's Make a Deal".

In the game, contestants are given a choice of three doors. Behind one door is a car and behind the other two doors are goats!

The contestant picks a door, which remains closed for the time being. Monty Hall, the presenter, knows what is behind the doors. He then opens one of the doors that has a goat behind.

Two doors now remain. The contestant is then given the choice of whether to stick with their original choice or switch to the remaining door.

What should they do?

- Stick with their first choice.
- Switch to the remaining door.
- It does not matter. Chances are even.

Developing inquiry skills

What other types of games and real-life situations can you use probability to help you make decisions?

Think about the questions in this opening problem and answer any you can. As you work through the chapter, you will gain mathematical knowledge and skills that will help you to answer them all.

Before you start

Click here for help with this skills check

You should know how to:	Skills check
1 Add, subtract, multiply and divide fractions both with and without a calculator. $\frac{2}{3}+\frac{1}{5}=\frac{10}{15}+\frac{3}{15}=\frac{13}{15}$ $1-\frac{2}{9}=\frac{9}{9}-\frac{2}{9}=\frac{7}{9}$ $\frac{3}{4}\times\frac{3}{5}=\frac{3\times 3}{4\times 5}=\frac{9}{20}$ $\frac{3}{7}\div\frac{4}{7}=\frac{3}{7}\times\frac{7}{4}=\frac{3}{4}$	**1** Evaluate: **a** $1-\frac{3}{7}$ **b** $\frac{2}{5}+\frac{5}{7}$ **c** $\frac{2}{5}\times\frac{2}{3}$ **d** $1-\left(\frac{1}{7}\times\frac{3}{8}\right)$ **e** $\frac{\frac{3}{20}}{\frac{7}{20}}$
2 Add, subtract, multiply and divide decimals both with and without a calculator. $0.2+0.7=0.9$ $1-0.08=0.92$ $0.2\times 0.34=0.1\times 2\times 0.01\times 34$ $=0.001\times 68=0.068$	**2** Evaluate: **a** $1-0.375$ **b** $0.65+0.05$ **c** 0.7×0.6 **d** 0.25×0.64 **e** 50% of 30 **f** 22% of 0.22
3 Calculate percentages both with and without a calculator: 52% of 60 = $0.52\times 60=31.2$	**3** Check your answers to questions **1** and **2** with a calculator.

8.1 Theoretical and experimental probability

Imagine a factory that produces electrical components. Some of the components produced are known to be faulty. How **likely** is it that a particular component chosen at random, say component A, is faulty? How could we determine this?

The study of probability helps us to understand the degree of uncertainty of something happening.

Investigation 1

At lunchtime every Monday, Andy and Bexultan play each other at chess. They only have time to play just two games in their regular lunchtime competition.

Andy is a little bit better at chess than Bexultan. Over the course of a school year (36 weeks), will Andy win more competitions than Bexultan? We can simulate this situation with a dice.

Factual What is a simulation?

Let 1, 2, 3 and 4 correspond to Andy winning a chess game, and 5 and 6 correspond to Bexultan winning a chess game.

Why does this choice of outcomes represent Andy being better at chess than Bexultan?

How much better is Andy than Bexultan in this situation?

What other choice of outcomes would also suggest that Andy is better than Bexultan?

What if you had a dice with 12 sides? What different combinations would represent Andy being better than Bexultan?

We will first investigate what happens one Monday:

1 Throw the dice. A role of 1, 2, 3 or 4 means Andy wins; a role of 5 or 6 means Bexultan wins. Record the result of the first game on a result sheet.

2 Throw the dice again, to see who wins the second game, and record it on the result sheet.

Competition number:	1st game won by:	2nd game won by:	Competition winner A (Andy), B (Bexultan) or D (Draw)
1			
2			
...			
36			

3 What was the result of the first lunchtime competition? Compare your result with others in your class. Are there any surprises?

4 Now repeat the experiment 35 more times, to get the results for a full year. Are you surprised by your results? How do they compare with what you would expect? Which player wins most often? Will this always be the case?

5 How often would you **expect** each player to win? How often would you **expect** a draw?

6 How would you expect the answers to these questions to change if we used 1, 2, 3, 4 and 5 to represent Andy winning? What if we used a dice with a different number of sides?

7 How can we get a more accurate representation for the model of Andy or Bexultan winning?

8 **Conceptual** How can a simulation be helpful in making predictions?

9 **Conceptual** How can we calculate an expected number of occurrences?

In the investigation, we considered the question, "Will Andy win more competitions than Bexultan?" In this and other simple question such as, "Will it rain today?" we can express probability using words such as "unlikely", "likely", "impossible" and "certain".

For each of these words, state an event in relation to your day today; for example, "It is likely I will have a physics lesson later today" (is it certain?).

A **probability experiment** (or just an "experiment") is something that has an uncertain result. An experiment or "**trial**" is the process by which we find out information about events by collecting data in a careful manner.

An event is a single result of an experiment; for example:

- Getting tails (Experiment: tossing a coin)
- Rolling a "5" (Experiment: rolling a six-sided dice)
- Choosing a "King" (Experiment: picking a card from a deck)
- Getting an "even number" (Experiment: spinning a numbered spinner)

The probability of the occurrence of an event can be represented using words, but more useful is a value from 0 to 1. On this scale, 0 represents an impossible event and 1 represents an event that is certain to happen.

Impossible Unlikely Even chance Likely Certain

0 $\frac{1}{2}$ 1

We write P(A) to represent the probability of an event A occurring.

Hence, $0 \leq P(A) \leq 1$.

There are two main ways of finding the probability of an event:

1 theoretical approach to probability

2 experimental approach to probability.

Theoretical approach

The theoretical approach to probability is based on a number of events where the likelihood of each event occurring is known.

A fair dice has six numbered sides, all of which are equally likely to occur. The list of equally likely possible outcomes is 1, 2, 3, 4, 5, 6.

The set of all outcomes in an experiment is called the **sample space** of that experiment. It is denoted by U.
The notation $n(U)$ shows how many elements of the sample space there are.

In this case, we would write $n(U) = 6$ as there are six elements of the sample space of rolling a fair six-sided dice {1, 2, 3, 4, 5, 6}. The sample space of flipping a fair coin is {heads, tails}. The sample space of picking a card from a pack of cards would be a list of all the cards in the pack.

TOK

A model is used to represent a mathematical situation in this case. In what ways may models help or hinder the search for knowledge?

HINT

The probability of an event cannot be greater than 1.

You can write a probability as a decimal, a fraction or a percentage.

International-mindedness

Probability theory was first studied to increase the chances of winning when gambling. The earliest work on the subject was by Italian mathematician Girolamo Cardano in the 16th century.

Now, let event A be defined as "rolling a 6".

In this sample space there is one 6, so $n(A) = 1$. This shows that there is one 6 in the sample space.

The probability of getting a 6 when you roll the dice once is therefore one out of six or $\frac{1}{6}$.

In probability notation, $P(A) = \frac{n(A)}{n(U)} = \frac{1}{6}$.

> Generalizing, this gives us the theoretical probability of any event A is
> $P(A) = \frac{n(A)}{n(U)}$, where $n(A)$ is the number of ways that event A can occur and $n(U)$ is the total number of possible outcomes.

Expected number of occurrences

So if you roll a six-sided dice six times you would **expect** to get a 6 once. For 12 rolls, we would expect $12 \times \frac{1}{6} = 2$ sixes.

> If the probability of an event is P, in n trials you would expect the event to occur $n \times P$ times.

But what would we expect to get for 10 rolls?

Here we would expect to get $10 \times \frac{1}{6} = 1\frac{2}{3}$ sixes. Clearly it is **not possible** to get $1\frac{2}{3}$ sixes, but it is what you would **expect**! We will revisit this idea when we look at probability distributions in chapter 14.

Example 1

A fair 20-sided dice with faces numbered from 1 to 20 is rolled.

The event M is defined as "the number obtained is a multiple of 3".

a Determine $P(M)$.

The dice is rolled 100 times.

b How many times would you expect to get a multiple of 3?

> **HINT**
> A 20-sided polyhedron is called an "icosahedron".

a $n(M) = 6$ and $n(U) = 20$ So $P(M) = \frac{n(M)}{n(U)} = \frac{6}{20} = \frac{3}{10} (= 0.3)$	There are 20 possible outcomes of which 6 belong to M {3, 6, 9, 12, 15, 18}.
b $\frac{3}{10} \times 100 = 30$	Expected number of occurrences = probability × number of trials

Exercise 8A

1 Find the probability of choosing an odd number from the set of numbers {1, 2, 3, 4, 5, 6, 7, 8, 9, 10}.

2 A used car dealer has 150 used cars on his lot. The dealer knows that 30 of the cars are defective. One of the 150 cars is selected at random. What is the probability that it is defective?

HINT

"At random" means that any car has an equal chance of being selected. One of the 30 defective cars is as equally likely to be chosen as one of the cars that are not defective.

3 Thys has signed up to do a drama production at school. He discovers that the director of the production is going to randomly allocate students to roles. There are 20 places in the chorus, 10 minor speaking roles and 5 main speaking roles. What is the probability that Thys will be in the chorus?

4 An octahedral (eight-sided) dice is thrown. The faces are numbered 1 to 8. Find the probability that the number thrown is:

a an even number

b a multiple of 3

c a multiple of 4

d not a multiple of 4

e less than 4

f a 9.

5 Each letter of the word STATISTICS is written on a separate card. The 10 cards are placed face down and a card is drawn at random.

What is the probability of picking a card with:

a the letter C

b the letter P

c a vowel.

6 A bag contains 50 discs numbered 1 to 50. A disc is selected at random. Find the probability that the number on the disc:

a is an even number

b has the digit 1 in it.

7 A school has five buses for transporting students. Four of them are minibuses with the same number of seats and there is a large coach which has three times the number of seats as a minibus. A student is allocated to a bus. What is the probability that the student is allocated to the coach given that seats are allocated at random until all of them are filled?

8 A spinner has sections that are coloured red, blue, green and yellow. The probabilities of getting a red and getting a blue are shown in the table. The probability of getting green is twice that of getting yellow.

Colour	Red	Yellow	Blue	Green
Frequency	0.4		0.3	

Find the probability of getting green.

9 At a school charity fundraising gala Sebastian decides to buy a ticket for the raffle. There are 360 people at the gala in total. Half of the people at the gala buy raffle tickets. Of those who buy raffle tickets, half buy two raffle tickets and the rest buy one ticket. There is only one winner of the raffle. What is the probability that Sebastian will win?

TOK

Play the game of the St Petersburg Paradox and decide how much would you pay to play the game.

Experimental (empirical) probability

Investigation 2

A **cuboctahedron** is a polyhedron with eight triangular faces and six square faces. It is formed by cutting equilateral triangles from the corner of a cube.

It is unlikely that you have access to a cuboctahedron dice but it is possible to make one using a net like the one shown on the right.

Number the square faces from 1 to 6 and the triangular faces from 7 to 14.

What is the probability that when you roll the dice the uppermost face will be square? As this is not a dice with equal-sized sides it is perhaps not possible to predict based on theory alone. You will need to conduct an experiment.

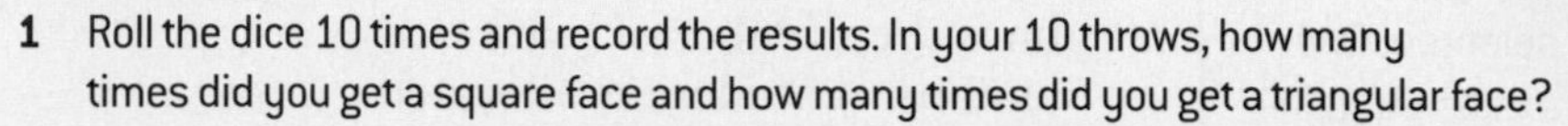

1 Roll the dice 10 times and record the results. In your 10 throws, how many times did you get a square face and how many times did you get a triangular face?

Based on these results, estimate the probability of getting a square face.

P(square face) = number of trials in which a square face was at the top ÷ total number of rolls.

For example, if you got seven square faces in 10 rolls then this would give an estimate of the probability of getting a square face as $\frac{7}{10} = 0.7$. This is called the **relative frequency** of getting a square face.

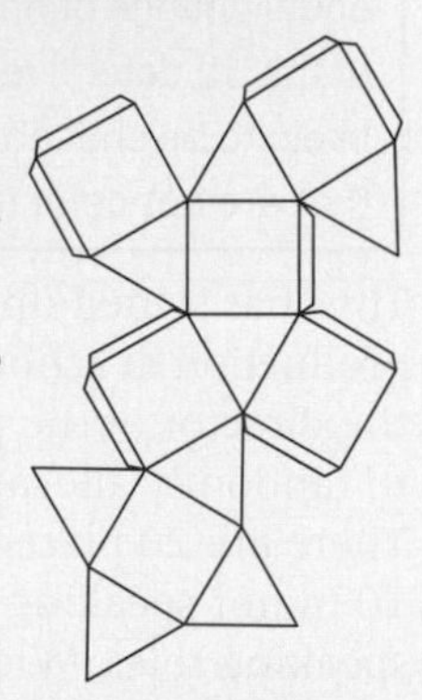

Net of a cuboctahedron

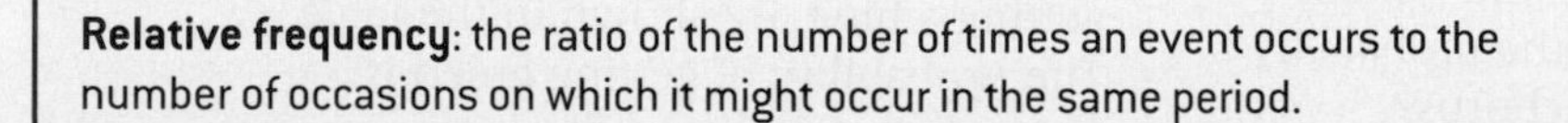

Relative frequency: the ratio of the number of times an event occurs to the number of occasions on which it might occur in the same period.

2 Repeat the experiment by rolling 10 more times. What is the new estimate of the probability of a square face based on 20 rolls? Which estimate for the probability do you think is more accurate? Why?

3 Repeat and estimate again after every 10 rolls, and add your findings to a copy of the table below. You will need to roll the dice several times! Let's go to 100. If there are other people in the class then you may like to combine results to make this quicker or to get even more results.

Number of rolls	Total number of times a square face is at the top	Estimate of the probability of getting a square face at the top (relative frequency)
10		
20		
30		
...		
100		

4 Are the short-run relative frequencies (those for 10 or 20 rolls) the same as the long-run probabilities (those for 100 rolls)?

5 **Conceptual question** Why does the number of trials of an experiment affect the accuracy of predictions?

6 Have you conducted enough trials here to get a good estimate? What would be the problem with doing more? When should you stop?

7 **Conceptual** How many trials of an experiment before we can be sure of the theoretical probability?

It is also possible to represent your series of estimates on a graph.

8 Plot the relative frequencies that you have found after every 10 rolls. You could do this by hand or by using your GDC.

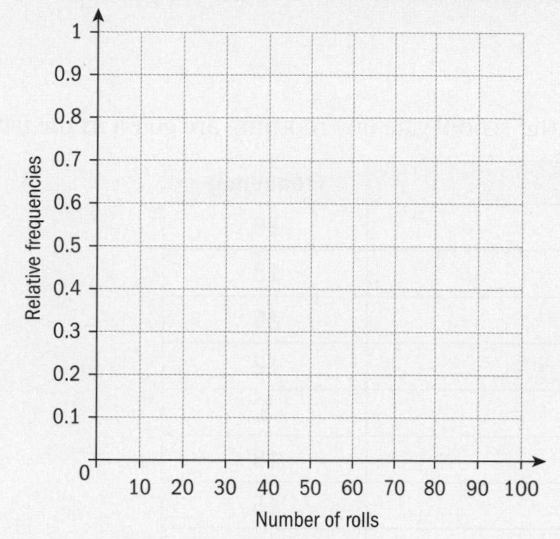

9 Join up the points on the graph. What do you notice?

10 **Factual** What happens to the relative frequency of a square face as the number of rolls increases?

11 Based on your experiment what is a good estimate for the probability of getting a square face at the top? How reliable is this?

12 **Conceptual** What happens to the relative frequency as the number of trials increases?

> In experimental probability, the probability of an event is based on the outcome of a series of trials. As the number of trials increases, the probability will get closer to the actual probability of the event occurring. This method can be useful when the outcomes are not equally likely and so estimation is required.

Let's return to the real-life situation of the factory that produces different electrical components, where some of the components produced are known to be faulty. We want to determine the probability that a particular component, say component A, is faulty so we need to test some. (Clearly we cannot test them all otherwise we will have no components left to sell!)

If the first component tested is faulty we could conclude that all of the components are faulty. However, this may not be the case. If the second component is not faulty, we could conclude the probability of a component being faulty is $\frac{1}{2}$ since (so far) half (one out of two) of all components are faulty.

Continuing this process a number of times and calculating the ratio:

$$\frac{\text{the number of faulty components}}{\text{the number of components tested}} = \text{the relative frequency of the component being faulty.}$$

Just as with the dice investigation, as the number of components tested increases, the relative frequency will get closer and closer to the true probability that a component is faulty.

You can use relative frequency as an estimate of probability. The larger the number of trials, the closer the relative frequency is to the probability.

TOK

When watching a crime series, reading a book or listening to the news, the evidence of DNA often closes a case. If it was that simple, no further detection or investigation would be needed. Research the "Prosecutor's fallacy".

How does reason contrast with emotion in making a decision based solely on DNA evidence?

Example 2

The colours of cars passing the school gate one morning are given in the table.

Colour	Frequency
Red	26
Black	18
White	20
Green	12
Yellow	3
Blue	16
Other	15
Total	**110**

a Estimate the probability that the next car to pass the school gates will be red.

b Using this information, if 150 cars were to pass the school gate the next day, estimate the number of red cars expected that morning.

The colour of cars is also recorded on the following day. The results are given in the table.

Colour	Frequency
Red	20
Black	21
White	12
Green	9
Yellow	4
Blue	16
Other	10
Total	**92**

c What is an improved estimate of the probability of the next car to pass the school gate being red?

a $P(\text{red}) = \frac{26}{110} = \frac{13}{55} \approx 0.236$	In reality, it is not possible to have a fraction of a car, but this is the number that is expected.
b $150 \times \frac{13}{55} \approx 35.5$ cars	
c $\frac{26+20}{110+92} = \frac{46}{202} = \frac{23}{101} \approx 0.228$	Total number of red cars in two days ÷ total number of cars in two days.

Exercise 8B

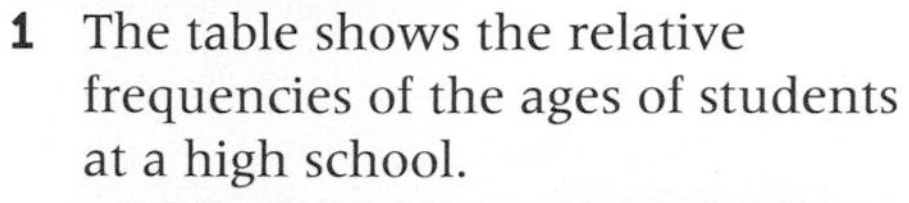

1 The table shows the relative frequencies of the ages of students at a high school.

Age (years)	Relative frequency
13	0.08
14	0.12
15	0.18
16	0.22
17	0.27
18	0.13
Total	1

a A student is randomly selected from this school. Find the probability that:

i the student is 15 years old

ii the student is 16 years of age or older.

There are 1200 students at this school.

b Calculate the number of 15-year-old students at school.

2 The sides of a six-sided spinner are numbered from 1 to 6. The table shows the results for 100 spins.

Number on spinner	1	2	3	4	5	6
Frequency	27	18	17	15	16	7

a What is the relative frequency of getting a 1?

b Do you think the spinner is fair? Give a reason for your answer.

The spinner is spun 3000 times.

c Estimate the number of times the result will be a 4.

3 An eight-sided dice numbered from 1 to 8 is rolled 80 times to determine whether it is fair.

a If the dice is fair, how many of each number would you expect to get?

The results obtained were:

Number on dice	1	2	3	4	5	6	7	8
Frequency	8	8	12	11	11	12	11	7
Relative frequency								

b Copy and complete the table with the relative frequency of each of the possible outcomes. Give your answers to 3 significant figures.

The dice is rolled 320 times more.

Number on dice	1	2	3	4	5	6	7	8
Frequency	29	41	43	39	45	46	32	45

c Using the **combined data from the two tables**, determine the relative frequency for each possible outcome.

Relative frequency								

d Conclude from the data whether the dice is fair.

International-mindedness

A well-known French gambler, Chevalier de Méré consulted Blaise Pascal in Paris in 1650 with questions about some games of chance. Pascal began to correspond with his friend Pierre Fermat about these problems which began their study of probability.

In this section we have considered experimental and theoretical probability. Use what you have learned to suggest which method would be best to use to predict the probability of:

a obtaining a run of five heads on a fair coin
b obtaining a run of five heads on a damaged coin
c rain tomorrow
d winning the lottery

Reflect How can you predict the probability that an event occurs?

Developing inquiry skills

In order to find a solution to the Monty Hall problem you could try to run a simulation in class.

You will need a partner and three cups.

- In your pair, decide which of you will be the host and which the contestant. The host should place an object under one of the cups while the second person (the contestant) closes their eyes.
- When ready, the contestant should choose a cup, but not lift it.
- The host then lifts one of the other two cups which is empty.
- The contestant either switches or does not switch their choice of cup.
- Record whether the player has won or lost.
- Calculate the win percentage for switching and the win percentage for not switching.
- What is the best strategy?

	Switch	Do not switch
Number of wins		
Number of loses		
Win percentage (%)		

What do your results suggest the probability of winning is if you switch or do not switch?

How many trials do you think will be necessary for this simulation to see more reliable results?

When will you know when you have done enough?

It is also possible to use an online simulator for this problem. There are many available if you search for them. Here is one http://www.mathwarehouse.com/monty-hall-simulation-online/. Alternatively you could write a program for a spreadsheet or a code to do the same.

What is the advantage of running a simulation using technology?

8.2 Representing probabilities: Venn diagrams and sample spaces

Investigation 3

Shapes are to be sorted into two sets, R and T.
Let R be the set of shapes containing a right angle and T be the set of shapes with three sides.

1 In which set would the following belong?

- **a** square
- **b** regular pentagon
- **c** isosceles triangle
- **d** right-angled triangle

2 What do you notice?

3 Sort the following shapes into the diagram below.

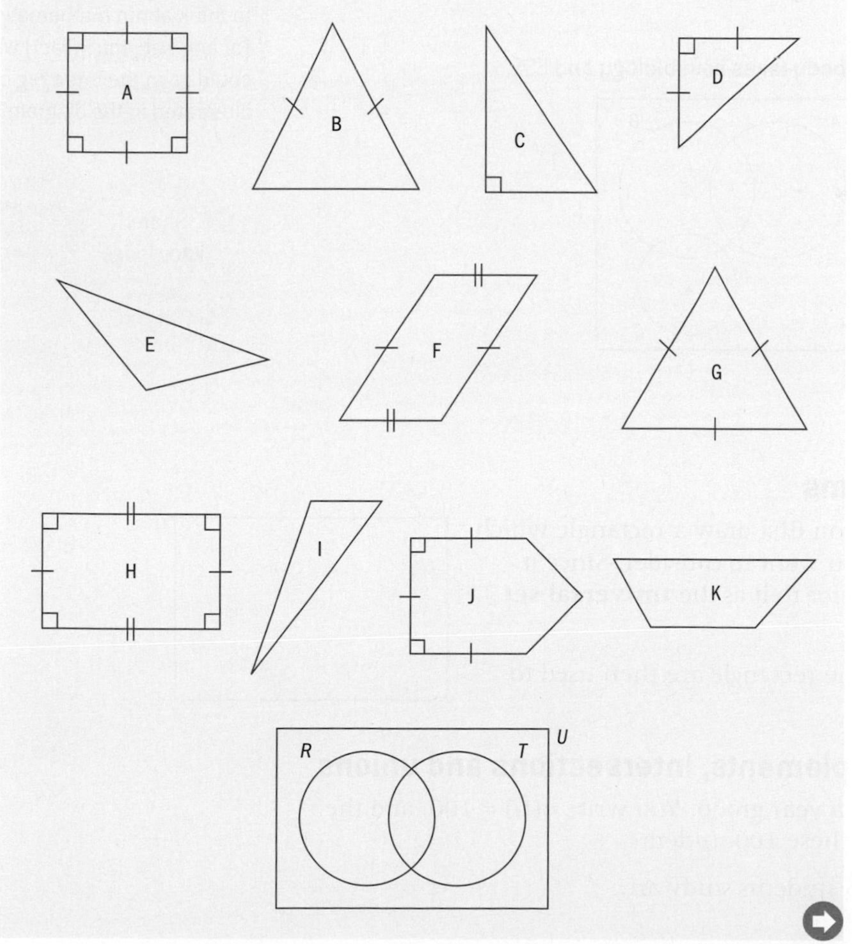

Continued on next page

4 Draw another shape that belongs in:

a set R but not set T

b neither set R nor T

c both sets

5 Where does each of these shapes belong in the diagram?

This diagram is called a **Venn diagram**. A Venn diagram is a diagram that shows the relationship between and among a finite collection of sets.

6 If a shape is chosen at random from the set of shapes at the beginning of this investigation what is the probability that it is in both sets R and T?

7 **Conceptual** What are Venn diagrams useful for?

Try to define sets A, B and C such that all regions (denoted by *) have some members in them.

You could perhaps try to use data from your classroom or school. For example, does the following work for your classroom?

A = students who take biology

B = students who take chemistry

C = students who take ESS

Note that in this example nobody takes both biology and ESS.

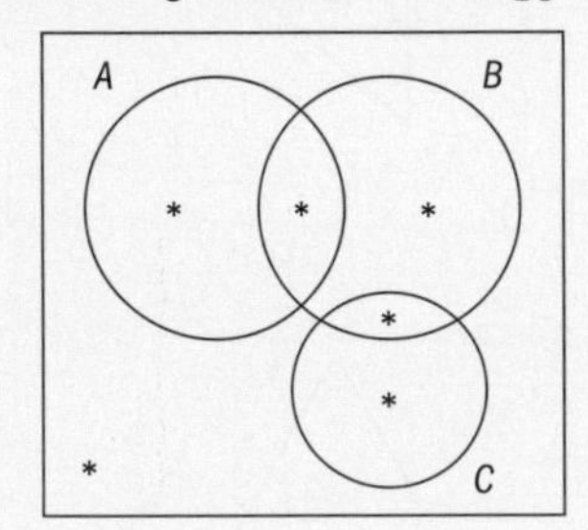

TOK

In TOK it can be useful to draw a distinction between shared knowledge and personal knowledge. The IB use a Venn diagram to represent these two types of knowledge. If you are to think about mathematics (or any subject, in fact) what could go in the three regions illustrated in the diagram?

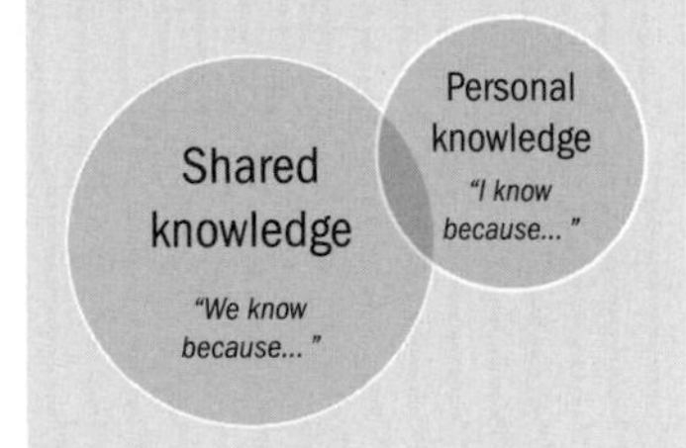

Drawing Venn diagrams

To draw a Venn diagram you first draw a rectangle which will contain every item you want to consider. Since it contains every item, we refer to it as the **universal set** and it is denoted by U.

Enclosed regions within the rectangle are then used to represent different sets.

Venn diagrams, complements, intersections and unions

There are 100 students in a year group. You write $n(U) = 100$, and the rectangle now represents these 100 students.

Within the year group, 36 students study art.

Francis and Hannah are two of the students who study art, and Grant and Iona are two students who do not study art. In the following diagram where would you place Francis, Grant, Hannah and Iona?

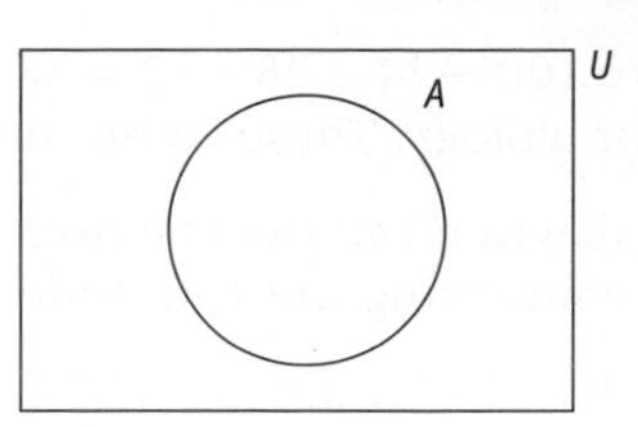

Within the circle there are 36 people. You do not need to name each student!

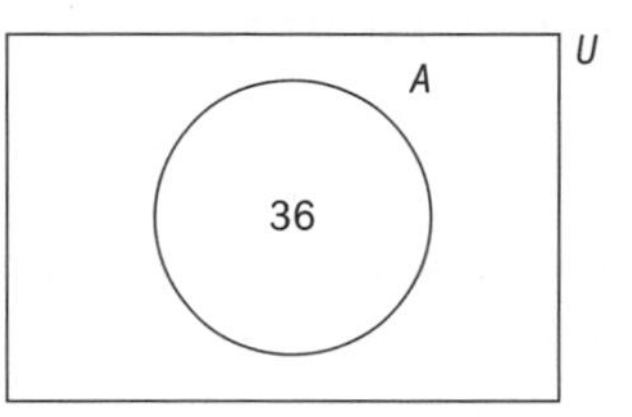

A student is chosen at random from the year group.

What is the probability that the student studies art?

The probability that a student does art is written as $P(A)$.

Here $P(A) = \frac{36}{100} = \frac{9}{25}$.

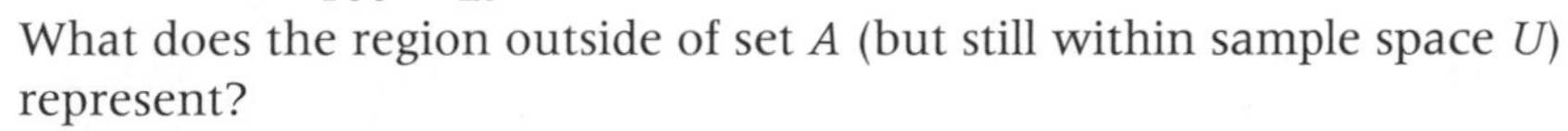

What does the region outside of set A (but still within sample space U) represent?

What do you know about students in this region?

How many students are there in this region?

This is A', the **complement** of set A, and represents those students who do not study art.

So $n(A') = n(U) - n(A)$.

In this example $n(A') = 100 - 36 = 64$.

What is the probability that a student chosen from the year group does not study art?

We require $P(A')$.

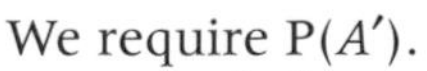

$P(A') = \frac{n(A')}{n(U)} = \frac{64}{100} = \frac{16}{25}$

What do you notice about $P(A) + P(A')$? Why is this the case?

So $P(A') + P(A) = \frac{9}{25} + \frac{16}{25} = 1$.

In general, if you consider event A and its complement A', then

$P(A') + P(A) = 1$

and

$P(A') = 1 - P(A)$.

Intersection of events

If 40 students study biology and, of these, 12 students study both art and biology, then how can this be represented on the Venn diagram? You can add another region, B, to the Venn diagram to represent students who do biology.

The 12 students who study both art and biology are placed in the **intersection** of A and B.

This region is written as $A \cap B$ and is represented by the shaded area.

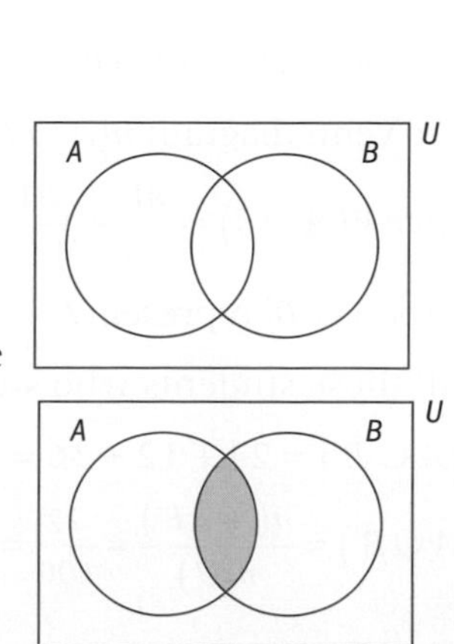

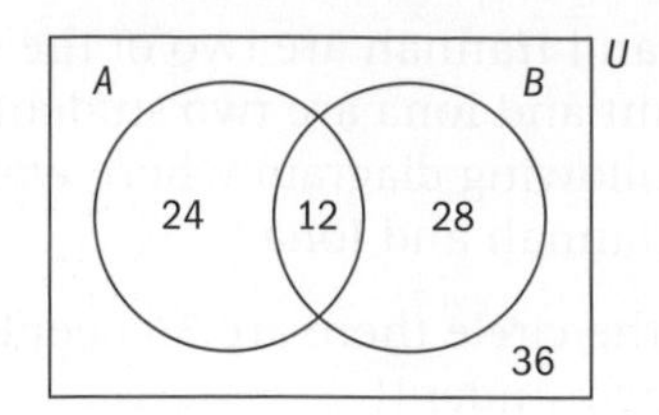

There were 100 – 24 – 28 – 12 = 36 students who do not study art or biology. These go on the outside of the circles.

The Venn diagram can now be used to answer probability questions concerning the 100 students.

For example, what is the probability that a student chosen at random studies both art and biology? $P(A \cap B)$

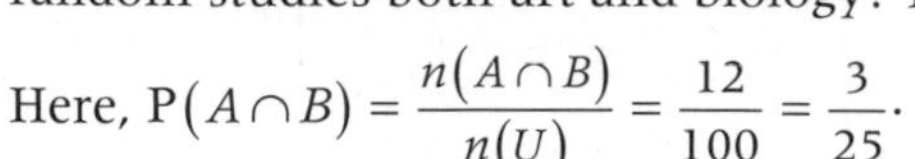

Here, $P(A \cap B) = \dfrac{n(A \cap B)}{n(U)} = \dfrac{12}{100} = \dfrac{3}{25}$.

What is the probability that a student chosen at random does not study biology but does study art?

This is written $P(A \cap B')$.

Here, $P(A \cap B') = \dfrac{n(A \cap B')}{n(U)} = \dfrac{24}{100} = \dfrac{6}{25}$.

What does $A' \cap B'$ represent?

This is the intersection of those students who do not do art and do not do biology.

Where is this region on the Venn diagram?

In this case, $n(A' \cap B') = 36$.

Union of events

The probability that a student chosen at random does either art **or** biology is written $P(A \cup B)$.

Reflect What do "and" and "or" mean in mathematics? How do they differ from regular usage in English (inclusive and exclusive or)?

In mathematics, "or" includes the possibility of "both". So "A or B" means "either in A or in B" or "in both A and B". This is known as the **"inclusive or"**.

In other contexts, we might use the "exclusive or". For example, the question, "Would you like a drink or a snack?" does not invite the possibility of "both".

The area $P(A \cup B)$ is shaded on this Venn diagram.

From our Venn diagram $n(A \cup B) = 24 + 12 + 28 = 64$.

And hence $P(A \cup B) = \dfrac{n(A \cup B)}{n(U)} = \dfrac{64}{100} = \dfrac{16}{25}$.

What does $A \cup B'$ represent?

This is all those students who study art or do not study biology.

Here $n(A \cup B') = 24 + 12 + 36 = 72$

and $P(A \cup B') = \dfrac{n(A \cup B')}{n(U)} = \dfrac{72}{100} = \dfrac{18}{25}$.

Example 3

In a group of 50 people, 29 like eating fish (F) and 38 like eating chips (C) and 9 people like neither fish nor chips. How many like both fish and chips?

1 Draw a Venn diagram to show this information.

2 Use your diagram to find the probability that a person chosen at random from the group likes:

a chips

b both fish and chips

c fish but not chips.

3 **a** Describe in words those people in the set $F' \cup C$.

b Find $n(F' \cup C)$.

1 Let F = likes fish C = likes chips	First define your sets.
Let $x = n(F \cap C)$. $n(F \cap C') = 29 - x$ $n(F' \cap C) = 38 - x$	You do not know how many like fish **and** chips so let's use x for this value.
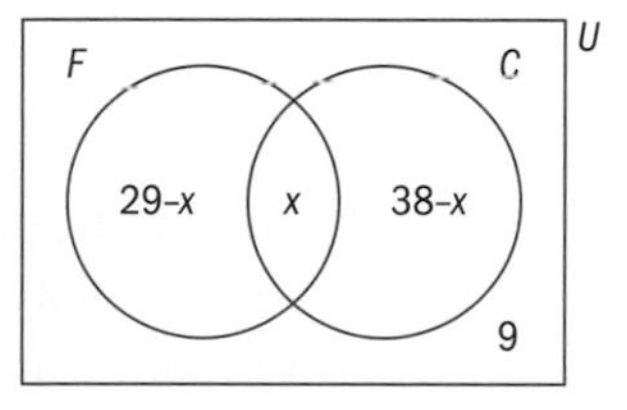	
$(29 - x) + x + (38 - x) + 9 = 50$ $76 - x = 50$ $x = 26$	The four regions of the diagram make up the universal set U and so must all add to give 50. Solve for x.
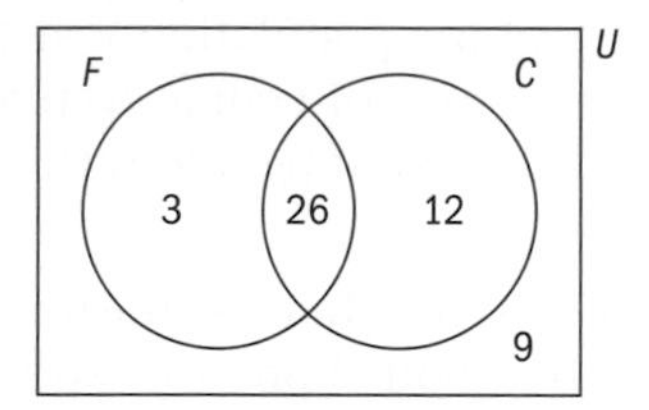	Substitute $x = 26$ to get the number in each section of the diagram.
a $P(C) = \frac{38}{50} = \frac{19}{25}$ **b** $P(F \cap C) = \frac{26}{50} = \frac{13}{25}$ **c** $P(F \cap C') = \frac{3}{50}$ $F' \cup C$ are those people who do not like fish or do like chips. **d** $n(F' \cup C) = 12 + 9 + 26 = 47$	

Statistics and probability

Exercise 8C

1 In a group of 38 students, 29 play computer games, 10 play board games and 9 play both.

a Draw a Venn diagram to represent this situation.

A student is selected at random.

b Find the probability that the student plays neither computer games nor board games.

2 In a sports club, 40 members play badminton, 37 members play squash, 21 play both and 7 play neither.

a Represent this information in a Venn diagram.

b Determine how many people are members of the club.

c Determine the probability that a member:

i plays badminton

ii plays both sports

iii plays neither sport

iv plays at least one sport.

3 A group of 50 people were asked whether they gave their partner a card or a present on their last birthday. The results were: 31 gave a card, 40 gave a present and 25 gave both a card and a present. If one of the people was chosen at random, determine the probability that they gave:

a a card or a present

b a card but not a present

c neither a card nor a present.

4 Set A contains letters needed to spell the word PROBABILITY and set B contains the letters needed to spell the word COMPLEMENTARY.

a Draw a Venn diagram for the two sets A and B.

b What is in the intersection of A and B?

c What is in the union of A and B?

5 If $U = \{1, 2, 3, 4, 5, 6, 7, 8, 9, 10\}$, $A = \{2, 4, 6, 8, 10\}$ and $B = \{3, 6, 9\}$, list the members of the following sets:

a $A \cap B$

b $A \cup B$

c A'

d $A' \cap B$

e $A \cup B'$

f $A' \cup B'$

6 The universal set U is defined as the set of positive integers less than or equal to 15.

M is the set of integers that are in set U and are multiples of 3.

F is the set of integers that are in set U and are factors of 30.

a List the elements of:

i M

ii F

b Place the elements of M and F in the appropriate regions of a Venn diagram.

c A number is chosen at random from U.

Find the probability that the number is:

i both a multiple of 3 and a factor of 30

ii neither a multiple of 3 nor a factor of 30.

7 In a town, 10% of the population watch the news at 1 pm, 30% of people watch the news at 6 pm and 40% of people watch the news at 9 pm.

It is found that 5% watch at both 6 pm and 9 pm, 4% watch at both 1 pm and 9 pm, 3% watch at 1 pm and 6 pm, and 2% of the people watch all three news shows.

a Complete a Venn diagram to show this information. For this Venn diagram, you will need three circles, one for each time the news is on.

b Find the probability that a person chosen at random from the town:

i watches only the news at 9 pm

ii watches only the news at 6 pm

iii does not watch the news.

The addition rule

Investigation 4

Let's return to the Venn diagram of the art and biology students.

1. Write down the following probabilities again:

   $\mathrm{P}(A)$, $\mathrm{P}(B)$, $\mathrm{P}(A \cap B)$ and $\mathrm{P}(A \cup B)$.

2. Explain why the probability that a randomly chosen student who studies both art and biology, $\mathrm{P}(A \cup B)$, does not equal the probability of a student studying art plus the probability of a student studying biology, $\mathrm{P}(A) + \mathrm{P}(B)$.

3. **Factual** Use the values of $\mathrm{P}(A)$, $\mathrm{P}(B)$, $\mathrm{P}(A \cap B)$ and $\mathrm{P}(A \cup B)$ to determine the correct rule for $\mathrm{P}(A \cup B)$ in terms of $\mathrm{P}(A)$, $\mathrm{P}(B)$ and $\mathrm{P}(A \cap B)$.

4. Explain this rule using Venn diagrams.

5. **Conceptual** Why might rules such as this be useful?

International-mindedness

The Dutch scientist Christian Huygens, a teacher of Leibniz, published the first book on probability in 1657.

The **addition rule** states that $\mathrm{P}(A \cup B) = \mathrm{P}(A) + \mathrm{P}(B) - \mathrm{P}(A \cap B)$.

Example 4

A and B are two events such that $\mathrm{P}(A) = \frac{9}{20}$ and $\mathrm{P}(B) = \frac{3}{10}$. It is known that $\mathrm{P}(A \cup B) = 2\mathrm{P}(A \cap B)$.

Find:

a $\mathrm{P}(A \cup B)$

b $\mathrm{P}(A \cup B)'$

c $\mathrm{P}(A \cap B')$

a Let $x = \mathrm{P}(A \cap B)$.	Use $\mathrm{P}(A \cup B) = \mathrm{P}(A) + \mathrm{P}(B) - \mathrm{P}(A \cap B)$.
$2x = \frac{9}{20} + \frac{3}{10} - x$	
$3x = \frac{15}{20}$	
$x = \frac{3}{4} \div 3$	
$x = \frac{1}{4} = \mathrm{P}(A \cap B)$	
$\mathrm{P}(A \cup B) = \frac{1}{2}$	Since $\mathrm{P}(A \cup B) = 2\mathrm{P}(A \cap B)$
b If $\mathrm{P}(A \cup B) = \frac{1}{2}$ then $\mathrm{P}(A \cup B)' = 1 - \frac{1}{2} = \frac{1}{2}$.	Since $\mathrm{P}(A') = 1 - \mathrm{P}(A)$
c If $\mathrm{P}(A \cap B') = \mathrm{P}(A) - \mathrm{P}(A \cap B) = \frac{9}{20} - \frac{1}{4} = \frac{1}{5}$.	Using the result from part **a**.

In some of the previous examples the question could have been answered by use of the Venn diagram alone.

HINT

The probability, $P(A)$ and the probability $P(B)$ each include the probability of both A and B, $P(A \cap B)$. For $P(A \cup B)$ we only wish to include this probability once so we subtract one of these probabilities.

Reflect When might a formula or a diagram be more appropriate? Which is the most efficient method? What do we mean by efficient here?

In the following questions, consider whether to use the formula or a diagram (or perhaps a combination of both).

TOK

Do ethics play a role in the use of mathematics?

Exercise 8D

1 A ten-sided dice, numbered 1 to 10, is rolled. Calculate the probability that:

a The number rolled is a prime number.

b The number rolled is a prime number or a multiple of 3.

c The number rolled is a multiple of 3 or a multiple of 4.

2 In a group of 55 tourists, 30 have cameras, 25 are female and 18 are females with cameras. Find the probability that a tourist picked at random from this group is either a camera owner or a female.

3 A letter is chosen at random from the 26-letter English alphabet. Find the probability that the letter is:

a in the word MATHEMATICS

b in the word TRIGONOMETRY

c in the word MATHEMATICS **and** in the word TRIGONOMETRY

d in the word MATHEMATICS **or** in the word TRIGONOMETRY.

4 Arnav goes to the library. The probability that he takes out a fiction book is 0.4, a non-fiction book is 0.3 and both a fiction and a non-fiction book is 0.2.

a What is the probability that Arnav takes out a fiction book, a non-fiction book or both?

b What is the probability that Arnav does not check out a book?

5 If X and Y are two events such that $P(X)=\frac{1}{4}$ and $P(Y)=\frac{1}{8}$ and $P(X \cap Y)=\frac{1}{8}$, find:

a $P(X \cup Y)$ **b** $P(X \cup Y)'$

6 If $P(A) = 0.2$ and $P(B) = 0.4$ and $P(A \cup B) = 0.5$, find:

a $P(A \cap B)$ **b** $P(A' \cup B)$

7 A and B are two events such that $P(A)=\frac{3}{16}$ and $P(B)=\frac{3}{8}$ and it is known that $P(A \cup B) = 3P(A \cap B)$.

Find:

a $P(A \cup B)$ **b** $P(A \cup B)'$ **c** $P(A \cap B')$

Mutually exclusive events

Of the 100 students, it is found that 28 students study chemistry. It is not possible to study both chemistry and art as the classes are at the same time.

The events A, studying art, and C, studying chemistry, are said to be mutually exclusive events.

Mutually exclusive (or disjoint) events are events whose outcomes cannot occur at the same time.

In general, if two events A and B are mutually exclusive then it follows that $P(A \cap B) = 0$ and the addition rule in these cases is $P(A \cup B) = P(A) + P(B)$.

Mutually exclusive events do not intersect if represented on a Venn diagram.

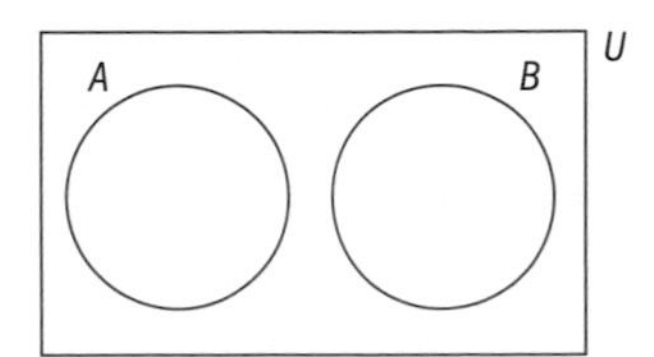

Example 5

Arthur has a drawer which contains ties. He picks a tie at random from the drawer. The probability that the tie is blue is $\frac{5}{12}$ and the probability it is red is $\frac{1}{3}$.

What is the probability of Arthur picking out a tie that is neither red nor blue?

Let B = blue tie selected R = red tie selected	First define your sets.
$P(B \cup R) = P(B) + P(R) = \frac{5}{12} + \frac{1}{3} = \frac{9}{12} = \frac{3}{4}$	B and R are mutually exclusive since you cannot select both a red and blue tie.
$P(B \cup R)' = 1 - P(B \cup R) = 1 - \frac{3}{4} = \frac{1}{4}$	Since $P(A') = 1 - P(A)$.

Exercise 8E

1 Here are some events relating to throwing two dice:

A: Both dice show a 4.

B: The total is 7 or more.

C: There is at least one 6.

D: The two dice show the same number.

E: Both dice are odd.

Which of these pairs of events are mutually exclusive?

a A and B **b** A and C **c** A and D
d A and E **e** B and E **f** C and D
g B and C

2 Two events N and M are such that $P(N) = \frac{1}{5}$ and $P(M) = \frac{1}{10}$ and $P(N \cup M) = \frac{3}{10}$. Are N and M mutually exclusive?

3 In an inter-school quiz, the probability of School A winning the competition is $\frac{1}{3}$, the probability of School B winning is $\frac{1}{4}$ and the probability of School C winning is $\frac{1}{5}$. Find the probability that:

a A or B wins the competition.

b A, B or C wins the competition.

c Are there any other schools in the competition? How do you know?

Representing outcomes: sample space diagrams

Investigation 5

Consider this series of three different problems.

First problem

In a group of 12 people, one person plays both archery and badminton, two people do only archery, and seven people do badminton.

Find the probability that a person chosen at random only plays badminton.

1 First represent this information on a Venn diagram.

This is similar to questions we have tackled previously.

2 Next copy and complete the following table with the same information. In each space put the number of people represented.

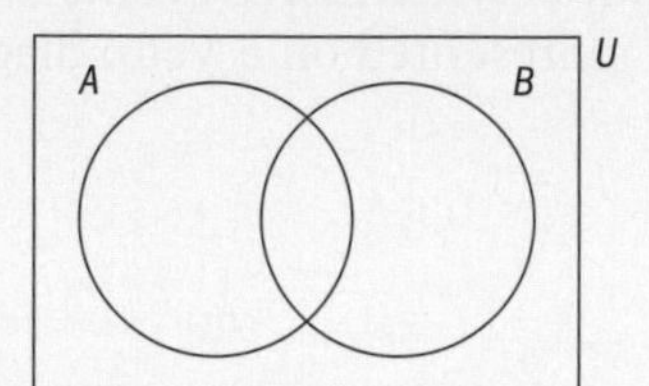

	Archery	Not archery	Totals
Badminton	1		
Not badminton			
Totals			

This is called a two-way table (or contingency table).

3 Now answer the question: Find the probability that a person chosen at random does only badminton.

4 In your opinion, which of these two methods of representing the 12 students in the sample space, Venn diagram or contingency table, is preferable? Why?

Second problem

What is the probability of scoring less than 6 when two fair six-sided dice are rolled?

To answer this question we need to know how many possible outcomes there are.

One possible outcome is a 3 on the first dice and a 5 on the second dice. We could represent this as (3, 5).

1 List all the possible outcomes. How many outcomes are there?

2 What is the problem with listing outcomes? (Think about if you had a dice with lots of sides!)

Another way of representing this information is in a sample space diagram also called a probability space diagram. A 3 on the first dice and a 5 on the second dice is represented like this:

		Dice 2					
		1	2	3	4	5	6
Dice 1	1						
	2						
	3					(3, 5)	
	4						
	5						
	6						

3 Copy and complete the diagram. Highlight all the ways of getting a total score of less than 6.

4 Now answer the question: What is the probability of scoring less than 6 when two fair six-sided dice are rolled?

5 What is the advantage of a sample space diagram over the list of outcomes?

6 When might a list of outcomes be more suitable than a sample space diagram?

Third problem

A fair spinner with the numbers 1, 2, 3 on it is spun three times.

Find the probability that the score on the last spin is greater than the scores on the first two spins.

1. Could you represent the same information in a Venn diagram or contingency table? Would that be helpful here?
2. List all the possible outcomes from this experiment.
3. How many possible outcomes are there?

When listing all outcomes, you need to be systematic so that you do not miss any.

4. Circle those sets of outcomes in your list in which the last spin is greater than the scores on the first two spins. How many are there?
5. What would the probability space diagram look like in this problem?
6. Why do we not draw a sample space diagram here?
7. Now answer the question: Find the probability that the score on the last spin is greater than the scores on the first two spins.
8. What are the different ways sample spaces can be represented?

Example 6

In an experiment a coin is tossed and a six-sided dice is rolled.

a Make a sample space diagram for this experiment.

b Find the probability that in a single experiment you obtain a head and a number less than 3 on the dice.

a

	1	2	3	4	5	6
H	(H, 1)	(H, 2)	(H, 3)	(H, 4)	(H, 5)	(H, 6)
T	(T, 1)	(T, 2)	(T, 3)	(T, 4)	(T, 5)	(T, 6)

The outcomes that give a head and a number less than 3 are highlighted.

b P(head and number less than 3) $= \frac{2}{12} = \frac{1}{6}$.

Exercise 8F

1 Three unbiased coins are tossed one at a time and the results are noted. One possible outcome is that all three coins are heads. This is written HHH. Another is that the first two coins are heads and the last one is a tail. This is written HHT.

List the complete sample space for this random experiment.

Find the probability that:

a The number of heads is greater than the number of tails.

b At least two heads are tossed consecutively.

c Heads and tails are tossed alternately.

2 Two tetrahedral dice, one blue and the other red, are each numbered 1 to 4. The two dice are rolled and the results noted.

a Draw a sample space diagram for this experiment.

b Find the probability that:

i the number on the red dice is greater than the number on the blue dice

ii the difference between the numbers of the dice is 1

iii the red dice shows an odd number and the blue dice shows an even number

iv the sum of the numbers on the dice is a prime number.

3 A box contains three cards bearing the numbers 1, 2, 3. A second box contains four cards with the numbers 2, 3, 4, 5. A card is chosen at random from each box.

a Draw the sample space diagram for the experiment.

b Find the probability that:

i the cards have the same number.

ii the larger of the two numbers drawn is a 3

iii the sum of the two numbers on the card is less than 7

iv the product of the numbers on the card is at least 8

v at least one even number is chosen.

4 Toby plays a game with a dice called "Come and Go".

He rolls the dice. If the score is 1 he moves forward 1 m. If the score is 2 he moves right 1 m. If the score is 3 he moves backwards 1 m. If the score is 4 he moves left 1 m. If the score is a 5 or 6 he stays where he is.

Toby rolls the dice twice. What is the probability that he is:

a at the same point where he started

b exactly 2 m away from his starting point

c more than one but less than 2 m away from his starting point.

Developing inquiry skills

Monty Hall problem: listing sample space solution

If we return to the Monty Hall problem, can any of these methods of representing sample spaces be used?

Copy and complete the following table for all the possible outcomes.

Contestant chooses	Prize door	Door Monty opens	Result if stay	Result if switch
1	3	2	LOSE	WIN
1	2	3	LOSE	WIN
1	1	2 or 3	WIN	LOSE
2	3	1	LOSE	WIN
...	...	...	...	...

How many different options are there in total?

In how many of these options do you win the prize when you switch?

In how many of them do you win a prize if you don't switch?

What is the probability of winning a prize if you switch?

What is the probability of winning a prize if you don't switch?

Should you switch or stay?

8.3 Independent and dependent events and conditional probability

Product rule for independent events

Investigation 6

Consider the example of a dice being rolled and a coin tossed in the previous section.

Here is the sample space diagram for the dice and coin again:

	1	2	3	4	5	6
H	(H, 1)	(H, 2)	(H, 3)	(H, 4)	(H, 5)	(H, 6)
T	(T, 1)	(T, 2)	(T, 3)	(T, 4)	(T, 5)	(T, 6)

1. Copy the diagram.
2. How many possible outcomes are there?
3. Shade in all the outcomes that determine the probability of H, getting a head. What is $P(H)$?
4. In a different colour, shade all the outcomes that determine the probability of L, getting a score of less than 6. What is $P(L)$?
5. What is the probability of getting a head and a score less than a 6? Which part of the diagram represents this probability? What is $P(H \cap L)$?
6. Look at your answers. What is the connection between $P(H)$, $P(L)$ and $P(H \cap L)$?
7. When conducting this experiment, does the outcome of getting a head affect the outcome of getting a score of less than 3?

Two events A and B are **independent** if the occurrence of one does not affect the chance that the other occurs.

Write down a rule to find $P(A \cap B)$ if A and B are independent events.

This is called the product rule or multiplication rule for independent events. This rule can also be used to test whether two events are independent.

If A and B are independent events, then $P(A \cap B) = P(A) \times P(B)$.

International-mindedness

During the mid-1600s, mathematicians Blaise Pascal, Pierre de Fermat and Antoine Gombaud puzzled over this simple gambling problem:

Which is more likely: rolling at least one six on four throws of one dice or rolling at least one double six on 24 throws with two dice?

HINT

In this investigation, the sample space diagram helped to visualize the number of possible outcomes in the intersection, but it is not always necessary. It may just be appropriate to just apply the rule if we know that A and B are independent events.

Example 7

One bag contains three red balls and two white balls. Another bag contains one red ball and four white balls. A ball is selected at random from each bag. Find the probability that:

a both balls are red **b** the balls are different colours **c** at least one ball is white.

a From the first bag $P(R_1) = \frac{3}{5}$	The events "picking a red from the first bag" (R_1) and "picking a red from the second bag" (R_2) are independent events.
From the second bag $P(R_2) = \frac{1}{5}$	
Therefore, $P(R_1 \cap R_2) = \frac{3}{5} \times \frac{1}{5} = \frac{3}{25}$	$P(R_1 \cap R_2) = P(R_1) \times P(R_2)$
b From the first bag $P(R_1) = \frac{3}{5}$	
From the second bag $P(W_2) = \frac{4}{5}$	
Therefore, $P(R_1 \cap W_2) = \frac{3}{5} \times \frac{4}{5} = \frac{12}{25}$	
From the first bag $P(W_1) = \frac{2}{5}$	
From the second bag $P(R_2) = \frac{1}{5}$	
Therefore, $P(W_1 \cap R_2) = \frac{2}{5} \times \frac{1}{5} = \frac{2}{25}$	If the balls are different colours, then either the first one is red and the second one is white or the first one is white and the second one is red.
$P(R_1 \cap W_2) + P(W_1 \cap R_2) = \frac{12}{25} + \frac{2}{25} = \frac{14}{25}$	
P(different colours) = $P(R_1 \cap W_2) + P(W_1 \cap R_2) = \frac{12}{25} + \frac{2}{25} = \frac{14}{25}$	These are mutually exclusive events.
c P(at least one white) = 1 − P(both red) $= 1 - P(R_1 \cap R_2)$ $= 1 - \frac{3}{25} = \frac{22}{25}$	We **could** calculate the probability that both are white, the probability the first is white and the second is red, and the probability the first is red and the second is white and add these up. **or** If at least one is white then it means that both cannot be red.

Exercise 8G

1 Erland's wardrobe contains five shirts—one blue, one brown, one white, one black and one purple. He reaches into the wardrobe and chooses a shirt without looking. He replaces this shirt and chooses again. What is the probability Erland chooses the purple shirt both times?

2 A large school conducts a survey of the food provided by the school cafeteria. It is found that $\frac{4}{5}$ of the students like pasta. Three students are chosen at random. What is the probability that all three students like pasta?

TOK

Do you think that mathematics is a useful way to measure risks?

To what extent do emotion and faith play a part in taking risks?

3 Millie is playing in a cricket match and a game of hockey at the weekend. The probability that her team will win the cricket match is 0.75 and the probability of her team winning the hockey match is 0.85. What is the probability that Millies' team loses both matches?

4 Three events A, B and C are such that A and B are mutually exclusive and $P(A) = 0.2$, $P(C) = 0.3$, $P(A \cup B) = 0.4$ and $P(B \cup C) = 0.34$.

   a Calculate $P(B)$ and $P(B \cap C)$.

   b Determine whether B and C are independent.

5 Muamar tosses a coin and rolls a six-sided dice. Find the probability that Muamar gets a head on the coin, and does not get a 6 on the dice.

6 A missile has a $\frac{8}{9}$ chance of hitting its target. If four missiles are launched what is the probability that the target is not hit?

7 Given that $P(E') = P(F) = 0.6$ and $P(E \cap F) = 0.24$:

   a Write down $P(E)$.

   b Explain how you know E and F:

      i are independent

      ii are not mutually exclusive.

   c Find $P(E \cup F')$.

8 A six-sided dice is numbered 1, 2, 2, 5, 6, 6. It is thrown three times.

What is the probability that the scores add up to 6?

9 A and B are independent events such that $P(A) = 0.9$ and $P(B) = 0.3$.

Find:

   a $P(A \cap B)$ b $P(A \cap B')$ c $P(A \cup B')$

Conditional probability

Investigation 7

Here is the diagram showing the 100 students from the previous section. The diagram represents the number of students who study art and biology.

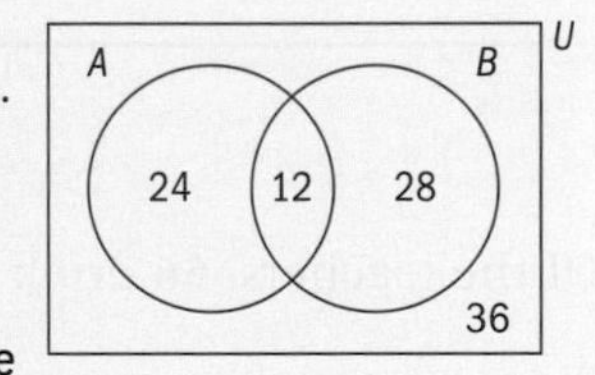

If we know that a particular student studies biology, how does this affect the probability that they also study art?

1 How many students study biology? (What is $n(B)$?) Where is this region in the Venn diagram?

2 Of those students who study biology how many also study art? Where is this region in the Venn diagram? What notation is used to represent this region?

3 What is the probability that a student studies art given that they study biology?

We write the probability that a student does art **given that** they study biology as $P(A|B)$.

Continued on next page

TOK

Consider the question posed at the beginning of this section. Which is more likely: rolling at least one six on four throws of one dice or rolling at least one double six on 24 throws with two dice?

International-mindedness

In 1933 Russian mathematician Andrey Kolmogorov built up probability theory from fundamental axioms in a way comparable with Euclid's treatment of geometry that forms the basis for the modern theory of probability. Kolmogorov's work is available in an English translation titled "The Foundations of Probability Theory".

4 Write down a rule that connects $P(A|B)$, $n(B)$ and $n(A \cap B)$.

5 Write down a rule that connects $P(A|B)$, $P(B)$ and $P(A \cap B)$

The second of these rules is the one that appears in the formula book. Check you have the correct answer!

What would the rule look like if you wanted to find $P(B|A)$ or $P(A|B')$?

Recall that, for independent events, $P(A \cap B) = P(A) \times P(B)$.

6 Rearrange your formula for $P(A|B)$ to make $P(A \cap B)$ the subject.

7 **Conceptual** What does this imply about $P(A|B)$ and $P(A)$ for independent events?

8 **Conceptual** Which situations describe conditional probability?

9 **Factual** What is conditional probability and how may it be represented?

$P(A|B)$ is known as **conditional probability**. We say *the probability of A given B*.

For two events A and B, the probability of A occurring given that B has occurred can be found using $P(A|B) = \frac{P(A \cap B)}{P(B)}$.

For independent events $P(A|B) = P(A)$.

It also follows that for independent events:

$$P(B|A) = P(B)$$
$$P(A|B') = P(A)$$
$$P(B|A') = P(B)$$

Example 8

There are 84 teachers in a school. Of the teachers, 56 drink tea, 37 drink coffee and 12 drink neither tea nor coffee.

a How many teachers drink both tea and coffee?

One member of the teaching staff is chosen at random.

Find the probability that:

b she drinks tea but not coffee

c if she is a tea drinker she drinks coffee as well

d if she is a tea drinker she does not drink coffee.

a T, C, U; 56–x, x, 37–x, 12	A Venn diagram may be useful here to display this information.

Let $n(T \cap C) = x$

so

$$(56 - x) + x + (37 - x) + 12 = 84$$
$$105 - x = 84$$
$$x = 21$$

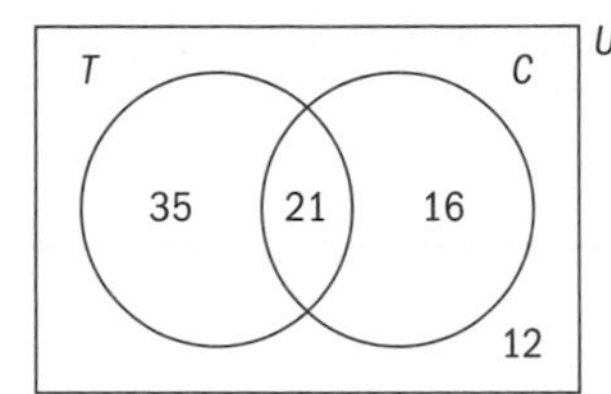

b $\mathrm{P}(T \cap C') = \dfrac{35}{84}$

c $\mathrm{P}(C|T) = \dfrac{\mathrm{P}(C \cap T)}{\mathrm{P}(T)} = \dfrac{\frac{21}{84}}{\frac{56}{84}} = \dfrac{21}{56}$

d $\mathrm{P}(C'|T) = \dfrac{\mathrm{P}(C' \cap T)}{\mathrm{P}(T)} = \dfrac{\frac{35}{84}}{\frac{56}{84}} = \dfrac{35}{56}$

$\mathrm{P}(C' \cap T) = \mathrm{P}(T \cap C')$ = answer from part **b**

Exercise 8H

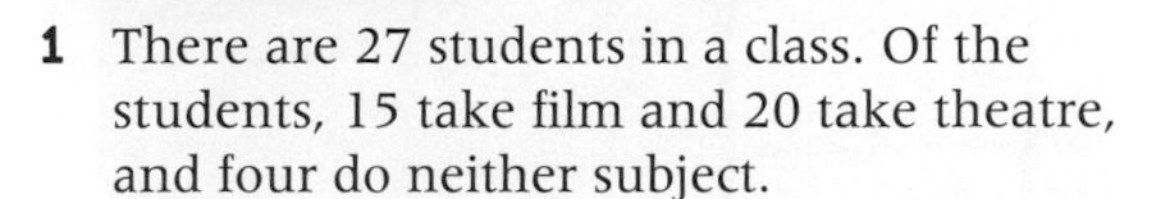

1 There are 27 students in a class. Of the students, 15 take film and 20 take theatre, and four do neither subject.

a How many students take both subjects?

b One person is chosen at random. Find the probability that the person:

i takes theatre but not film

ii takes at least one of the two subjects

iii takes theatre given that they take film.

2 A number is chosen at random from this list of eight numbers:

1 2 6 7 11 14 24 29

Find:

a P(it is even | it is not a multiple of 4)

b P(it is less than 15 | it is greater than 5)

c P(It is less than 5 | it is less than 15)

d P(it lies between 1 and 10 | it lies between 5 and 25)

3 V and W are mutually exclusive events where $\mathrm{P}(V) = 0.26$ and $\mathrm{P}(W) = 0.37$.

Find:

a $\mathrm{P}(V \cap W)$ **b** $\mathrm{P}(V|W)$

c $\mathrm{P}(V \cup W)$

4 The table below shows the number of left-handed and right-handed table tennis players in a sample of 50 males and females at a club.

	Left-handed	Right-handed	Total
Male	5	32	37
Female	2	11	13
Total	7	43	50

A table tennis player was selected at random from the group. Find the probability that the player was:

a male and left-handed

b right-handed

c right-handed, given that the player selected was female.

5 J and K are independent events. Given that $\mathrm{P}(J|K) = 0.3$ and $\mathrm{P}(K) = 0.5$, find $\mathrm{P}(J)$.

6 Your neighbour has two children. You learn that he has a son, Samuel. What is the probability that Samuel's sibling is a brother?

HINT

This is not as obvious as it might initially seem!

Reflect How can conditional probability be used to make predictions?

Developing inquiry skills

The Monty Hall problem: conditional probability solution

Let's take a typical situation in the game. Suppose the contestant has chosen door 3 and Monty Hall reveals that there is an unwanted prize behind door 2.

What is the conditional probability that the prize is behind door 1?

Let A be the condition that there is a car behind door 1 and the contestant has chosen door 3.

Let B be the condition that there is a dud behind door 2 given that the choice was door 3.

What is $\mathrm{P}(A \cap B)$? This is the probability that there is a car behind door 1, the contestant has chosen door 3 and there is a dud behind door 2.

Here the car is behind door 1 (probability $\frac{1}{3}$) and the contestant has chosen door 3 (probability $\frac{1}{3}$).

Monty Hall has to show what is behind door 2 (probability 1),

so $\mathrm{P}(A \cap B) = \frac{1}{3} \times \frac{1}{3} = \frac{1}{9}$.

What is the probability of being shown a dud behind door 2 given that the choice was door 3?

This situation can arise in two ways:

1 When the car is behind door 1:

The probability is $\frac{1}{9}$ as shown above.

2 When the car is behind door 3:

Here Monty could reveal either what is behind door 1 or door 2. He is equally likely to choose either of these doors, so the probability of showing what is behind door 2 is $\frac{1}{2} \times \frac{1}{9} = \frac{1}{18}$.

Therefore the probability of there being revealed an unwanted prize behind door 2 given that the contestant has chosen door 3 is the sum of these: $\frac{1}{18} + \frac{1}{9} = \frac{3}{18}$.

This is $\mathrm{P}(B)$.

We want the conditional probability $\mathrm{P}(A|B)$ as this relates to the contestant winning the prize. This is given by $\mathrm{P}(A|B) = \frac{\mathrm{P}(A \cap B)}{\mathrm{P}(B)} = \frac{\frac{1}{9}}{\frac{3}{18}} = \frac{2}{3}$.

This means that the conditional probability that the car is behind door 3 given that the contestant has chosen door 3 and has been shown that there is an unwanted prize behind door 2 is only $\frac{1}{3}$.

It is therefore worthwhile switching!

8.4 Probability tree diagrams

Investigation 8

A bag contains five blue balls and two green balls. Three balls are picked, one at a time, from the bag.

Consider two different situations:

1 A ball is selected and the colour is noted, then the ball is placed back in the bag.

2 A ball is selected and the colour is noted, then the ball is not placed back in the bag.

A single trial in these experiments consists of selecting a sample of three balls, one at a time, from the bag.

1 Make a list of all the possible outcomes in each situation; for example, you could get BBG (a blue followed by a blue followed by a green) in each case.

	Situation 1	Situation 2
List of possible outcomes		
Number of possible outcomes		

Continued on next page

Statistics and probability

2 Which outcome is impossible in situation 2 that is possible in situation 1? Why?

3 What is the probability of getting a blue on the first pick in each situation?

4 If you get a blue on the first pick are you more likely to get a blue on the second pick in situation 1 or situation 2?

5 What is the probability of getting a blue on the second pick in each situation?

6 Is the probability of getting a blue on the second pick dependent on or independent of the outcome on the first pick?

7 What is the probability of getting a blue on the third pick in each situation?

	Situation 1	Situation 2
Probability of blue on first pick		
Probability of blue on second pick		
Probability of blue on third pick		
Dependent or independent?		

8 Is getting three blues more likely in situation 1 or in situation 2?

9 In which situation is BBG most likely? How can you calculate it?

	Situation 1	Situation 2
Probability of blue followed by blue followed by blue (BBB)		
Probability of BBG		

10 Calculate the probabilities associated with each possible outcome in your list.

11 **Conceptual** How does "with" or "without" replacement affect the analysis and representation of probability?

Situations in probability may involve either "with replacement" or "without replacement" and there is a difference in the probabilities that are calculated in each case.

Creating a list of the possible outcomes and the probabilities is not necessarily the most efficient method of representing probabilities.

We can use probability tree diagrams to represent probabilities where more than one event occurs. We will consider two additional examples before returning to problem above.

"With replacement" and repeated events

The following example works through the construction of a tree diagram for a situation that is similar to the "with replacement" problem. What aspect of this situation makes it similar to the "with replacement" problem?

Example 9

The probability that Raghav, a keen member of the school archery club, hits the bullseye is 0.8. Raghav takes two shots. Assume that success with each shot is independent from the previous shot.

a Represent this information on a tree diagram.

b Find the probability that Raghav:

i hits two bullseyes

ii hits only one bullseye

iii hits at least one bullseye.

a The first section of the tree diagram represents Raghav's first shot.

How many possible outcomes are there for this shot?

In terms of bullseyes, there are two possible outcomes—either hitting the bullseye (H) or missing the bullseye (M).

$\mathrm{P}(H) = 0.8$

$\mathrm{P}(M) = 0.2$

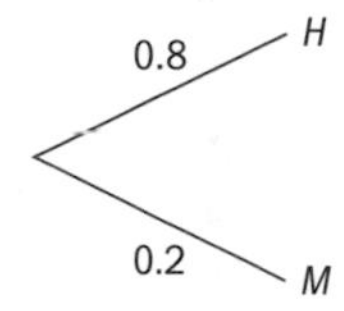

The outcome is on the end of the branch; the probability associated with it is beside the branch.

The second shot will also either hit or miss the bullseye.

There are therefore four possible outcomes of the experiment:

- a hit followed by a hit (H and H)
- a hit followed by a miss (H and M)
- a miss followed by a hit (M and H)
- a miss followed by a miss (M and M)

These four outcomes can be seen on the tree diagram.

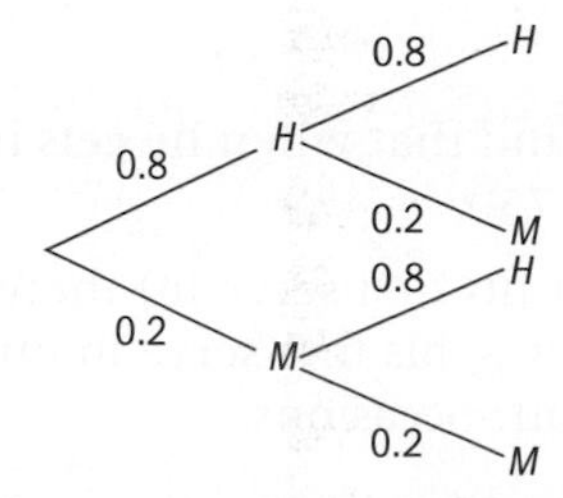

Continued on next page

Statistics and probability

b **i** We require $P(H \cap H)$.

Since a hit with the first shot is independent of getting a hit with the second shot, we multiply the probabilities together.

Multiplying along the top two branches:

$P(H \cap H) = 0.8 \times 0.8 = 0.64$

ii We require $P(H \cap M) + P(M \cap H)$ as only one hit could mean a hit on the first shot, or a miss on the first shot and a hit on the second shot.

The events $(H \cap M)$ and $(M \cap H)$ are mutually exclusive; that is, they cannot both happen at the same time. So here we multiply along the branches for $P(H \cap M)$ and $P(M \cap H)$ as the events are independent and then add the two values as they are mutually exclusive.

$P(H \cap M) + P(M \cap H)$
$= (0.8 \times 0.2) + (0.2 \times 0.8) = 0.32$

iii P(at least one bullseye) $= 1 - P(M \cap M)$
$= 1 - (0.2 \times 0.2)$
$= 1 - 0.04 = 0.96$

Now return to the problem in investigation 8. Draw a tree diagram to represent this problem. How many different sections of the tree diagram will there be? How many different outcomes are there in each section? Add the relevant probabilities to the diagram.

TOK

What do we mean by a "fair" game? Is it fair that casinos should make a profit?

"Without replacement" and conditional probability

The following example works through the construction of a tree diagram for a situation that is similar to the "without replacement" problem. What aspect of this situation makes it similar to a "without replacement" problem?

Example 10

Riley is a rising star in the school tennis club. He has found that when he gets his first serve in the probability that he wins that point is 0.75.

When he uses his second serve (because he does not get his first serve in) there is a 0.45 chance of him winning the point. He is successful at getting his first serve in on three out of five occasions and the second serve in on three out of four occasions.

a Find the probability that the next time Riley serves he wins the point.

b Given that Riley wins the point, what is the probability that he got his first serve in?

Start by drawing a tree diagram to show the probabilities of getting the first serve in or out. If the serve is out, then Riley must serve again. If the serve is in, then we show the probability of winning the point.

It is not necessary to continue the branches once the point has been won or lost.

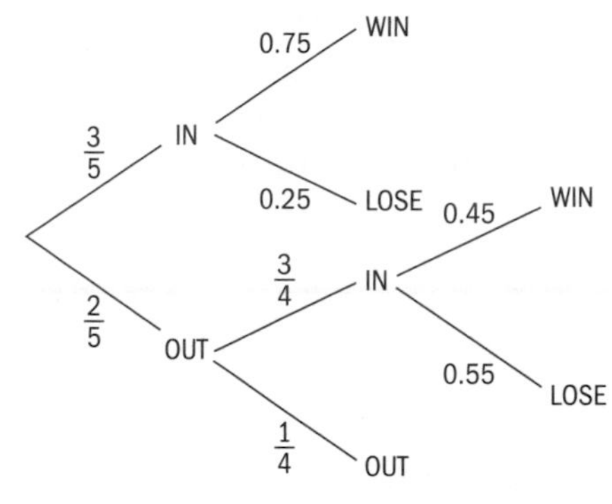

a P(win) = P(get first serve in and win) + P(miss first serve, get second serve in and win)

$$= \left(\frac{3}{5} \times 0.75\right) + \left(\frac{2}{5} \times \frac{3}{4} \times 0.45\right)$$

$$= 0.45 + 0.135$$

$$= 0.585$$

b P(First serve in | win point)

$$= \frac{\text{P(first serve in and win point)}}{\text{P(win point)}}$$

(Both of these values have been found in part **a**.)

$$= \frac{\left(\frac{3}{5} \times 0.75\right)}{0.585} = 0.769 (3 \text{ s.f.})$$

Exercise 8I

1 Three cards are drawn at random from a deck of playing cards. Each card is not replaced.

Find the probability of obtaining:

a three picture cards

b two picture cards.

2 A box of colouring pens contains five pens that are broken and nine pens that work. Two children, a girl and a boy, each need to take a pen.

a What is the probability that two broken pens are chosen?

b What is the probability that at least one broken pen is chosen?

c If exactly one broken pen is chosen, what is the probability that the girl chose it?

3 The yearbook team has 10 members, of whom seven are female and three are male. One of the members is chosen at random to be the lead editor of the book.

a Find the probability that the chosen person is male.

b Two people are chosen to take photographs at the school sports day. Find the probability that one is male and the other is female.

4 On average, Luca answers five problems correctly out of seven. Ian's average is five out of nine. They both attempt the same problem.

a What is the probability that at least one of the students answers the question correctly?

b If the question is answered correctly, what is the probability that Luca got the correct answer?

c If the question is answered correctly, what is the probability that Ian got the correct answer?

d If there was at least one correct answer what is the probability that there were two correct answers?

Developing your toolkit

Now do the Modelling and investigation activity on page 392.

Chapter summary

Theoretical and experimental probability

- The study of probability helps us to understand the degree of uncertainty of something happening.
- The theoretical approach to probability is based on a number of events where the likelihood of each event occurring is known.
- The set of all outcomes in an experiment is called the **sample space** of that experiment. It is denoted by U.

  The notation $n(U)$ shows how many members of the sample space there are.
- The theoretical probability of any event A is $\mathrm{P}(A) = \frac{n(A)}{n(U)}$, where $n(A)$ is the number of ways that event A can occur and $n(U)$ is the total number of possible outcomes.
- If the probability of an event is P, in n trials, you would expect the event to occur $n \times \mathrm{P}$ times.
- **Relative frequency** : the ratio of the number of times an event occurs to the number of occasions on which it might occur in the same period.
- In experimental probability, the probability of an event is based on the outcome of a series of trials. As the number of trials increases, the probability will get closer to the actual probability of the event occurring. This method can be useful when the outcomes are not equally likely and so estimation is required.

Representing probabilities: Venn diagrams and sample spaces

- In mathematics, "or" includes the possibility of "both". So "A or B" means "either in A or in B" or "in both A and B". This is known as the **"inclusive or"**.
- The **addition rule** states that $\mathrm{P}(A \cup B) = \mathrm{P}(A) + \mathrm{P}(B) - \mathrm{P}(A \cup B)$.

  The probability $\mathrm{P}(A)$ and the probability $\mathrm{P}(B)$ each include the probability of both A and B $\mathrm{P}(A \cap B)$. For $\mathrm{P}(A \cup B)$ we only wish to include this probability once so we subtract one of these probabilities.

- **Mutually exclusive (or disjoint) events** are events where two outcomes cannot occur at the same time.
- In general, if two events A and B are mutually exclusive then it follows that $\mathrm{P}(A \cap B) = 0$ and the addition rule in these cases is $\mathrm{P}(A \cup B) = \mathrm{P}(A) + \mathrm{P}(B)$.

Independent and dependent events and conditional probability

- Two events A and B are **independent** if the occurrence of one does not affect the chance that the other occurs.
- If A and B are independent events, then $\mathrm{P}(A \cap B) = \mathrm{P}(A) \times \mathrm{P}(B)$.
- $\mathrm{P}(A|B)$ is known as **conditional probability**. We say *the probability of A given B.*
- For two events A and B, the probability of A occurring given that B has occurred can be found using $\mathrm{P}(A|B) = \frac{\mathrm{P}(A \cap B)}{\mathrm{P}(B)}$.
- For independent events, $\mathrm{P}(A|B) = \mathrm{P}(A)$.

  It also follows that for independent events:

$$\mathrm{P}(B|A) = \mathrm{P}(B)$$
$$\mathrm{P}(A|B') = \mathrm{P}(A)$$
$$\mathrm{P}(B|A') = \mathrm{P}(B)$$

Probability tree diagrams

- Situations in probability may involve either "with replacement" or "without replacement" and there is a difference in the probabilities that are calculated in each case.

Developing inquiry skills

The Monty Hall problem: tree diagram solution

Represent this problem in a tree diagram. The diagram has been partially started here.

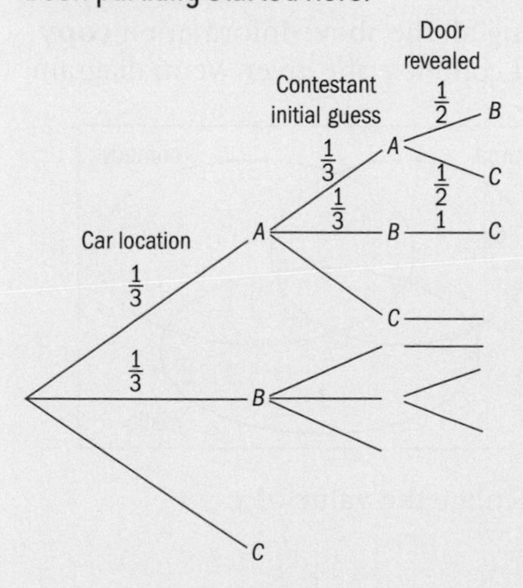

Continued on next page

Once the tree diagram has been completed you can find the probability of winning when switching (or losing when not switching) and losing when switching (or winning when not switching) by adding up the relevant results in the last column.

What is the probability of winning when switching? Is it worthwhile switching?

We have seen four different solutions to the Monty Hall problem provided in this chapter. There are many other possible methods of solving the problem.

How could you extend this problem for an exploration?

What about changing the number of doors? What effect will this have on the situation?

Chapter review

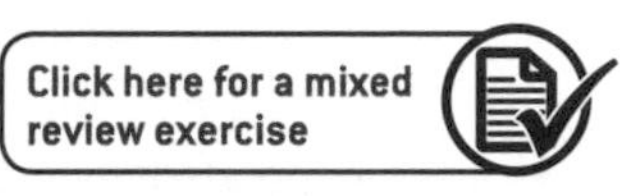

1 A two digit number between 10 and 99 inclusive is written down at random. Find the probability that it is:

a divisible by 5

b divisible by 3

c greater than 50

d a square number.

2 In a class of 30 pupils, 18 have a dog, 20 have a cat and 3 have neither. If a member is selected at random, find the probability that this pupil has both a cat and a dog.

3 For events C and D it is known that:
$P(C) = 0.7 \quad P(C' \cap D') = 0.25 \quad P(D) = 0.2$

a Find $P(C \cap D')$.

b Explain why C and D are not independent events.

4 The two events A and B are such that $P(A) = 0.6$, $P(B) = 0.2$ and $P(A \mid B) = 0.1$. Calculate the probabilities that:

a both of the events occur

b at least one of the events occur

c exactly one of the events occur

d B occurs given that A has occurred.

5 A group of 100 students are asked which of three types of TV programme, Drama, Comedy and Reality, they watch regularly.

They provide the following information:

15 watch all three types;

18 watch drama and comedy;

22 watch comedy and reality TV;

35 watch drama and reality TV;

10 watch of none these programmes regularly.

There are three times as many students who watch drama only than comedy only, and two times as many who watch comedy only than reality TV only.

a If x is taken as the number of students who watch reality TV only, write an expression for the number of students who play drama only.

b Using all the above information **copy** and complete the given Venn diagram.

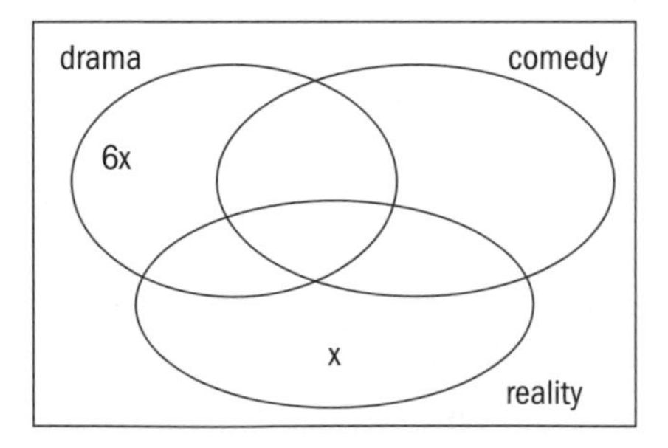

c Calculate the value of x.

6 Let P(C) = 0.4; P(D) = 0.5; P(C|D) = 0.6.

a Find P(C and D).

b Are C and D mutually exclusive? Give a reason for your answer

c Are C and D independent events? Give a reason for your answer

d Find P(C or D).

e Find P(D|C).

7 Jack does $\frac{3}{5}$ of the jobs around the house and Jill does the rest. Jack finishes 35% of his jobs properly and Jill finishes 55% of her jobs properly. Find the probability that a job done around the house will be done:

a properly

b by Jill if it was not done properly.

8 Yannick travels to school each day either by bicycle, by bus or by car. The probability that he travels by bus on any day is 0.6. The probability that he travels by bicycle on any day is 0.3.

a Draw a tree diagram showing the possible outcomes for Yannick's journeys on Monday and Tuesday.

b Find the probability that Yannick travels

i by bicycle on Monday and Tuesday

ii by bicycle on Monday and by bus on Tuesday

iii by the same method of travel on Monday and Tuesday.

c Yannick travelled to school by bicycle on Monday and Tuesday. Find the probability that he does not travel to school by bicycle on Wednesday and Thursday and Friday.

d Find the probability that in any three days Yannick travels twice by car and once by bus or twice by bicycle and once by car.

9 A bag contains 6 red apples and 10 green apples. Without looking into the bag, Lauren randomly selects one apple.

a Find the probability that it is red.

The apple is red and Lauren eats it. The bag is then passed to Oscar. Without looking into the bag, he randomly selects one apple.

b Find the probability that it is green.

The apple is green and Oscar replaces it in the bag. The bag is then passed to Darius. Without looking into the bag, he randomly selects two apples.

c Find the probability that they are both red.

10 On a walk Paul counts 70 rabbits. Of these rabbits, 42 are female, 34 are not eating carrots, and 23 are female and not eating carrots.

Draw a Venn diagram and hence find the number that are both female and eating carrots.

a Find the probability that a rabbit is male and not eating carrots.

b Find the probability that a rabbit is female given that it is eating carrots.

c Is being female independent of eating carrots? Justify your answer.

Exam-style questions

11 P1: An ordinary, fair, six-sided dice is thrown. Find the probability of obtaining

a a five (1 mark)

b an even number (1 mark)

c a prime number (2 marks)

d a number greater than 3 and smaller than 6 (2 marks)

e a number exactly divisible by 7. (1 mark)

12 P1: Two ordinary, fair, six-sided dice are thrown. One dice is red and the other is blue. Find the probability that

a both dice are a 5 (1 mark)

b both dice are the same number (1 mark)

c the red dice is a 5 and the blue dice is a 4 (1 mark)

d one dice is a 5 and the other a 4 (1 mark)

e the total of the two dice is 7 (2 marks)

f the red dice is a 5 given that the blue dice is a 4 (1 mark)

g one of the dice is a 5 (2 marks)

h one of the dice is a 5 given that the total of the two dice is 7. (2 marks)

13 P1: The probability that James is selected for the school football team is $\frac{1}{3}$. The probability that he is selected for the school rugby team is $\frac{1}{4}$. The probability that he is selected for the both teams is $\frac{1}{6}$.

a State, with a reason, whether the events of him being selected for football and being selected for rugby are independent. (2 marks)

b Find the probability that he is selected for at least one team. (2 marks)

c Find the probability that he is selected for exactly one team. (2 marks)

d Find the probability that he is selected for both teams given that he is selected for football. (2 marks)

14 P1: A control tower sends a message to a ship. The probability that the ship hears the message is $\frac{3}{4}$. The ship will reply if and only if it receives the message. If the ship replies, the probability that the tower hears the reply is $\frac{3}{5}$.

a Sketch a probability tree to represent this information. (2 marks)

b Hence find the probability that the tower hears a reply to the message it sends. (2 marks)

c Write down the probability that it does not hear a reply. (2 marks)

d Given that the tower did not hear a reply, find the probability that the ship did not hear the original message. (2 marks)

e Events are defined as follows:

A: The ship does not hear the original message.

B: The tower receives a reply.

State, with a reason, whether the two events A and B are mutually exclusive or not. (2 marks)

15 P1: On average in Wales it rains on 3 days out of every 5 days.

a Over a month, that has 30 days in it, calculate how many of these days would be expected to have rain. (2 marks)

b Over a period of 50 days, calculate how many days would not be expected to have rain. (2 marks)

c Calculate what length of time in days, would be expected to have 30 days of rain in it. (2 marks)

16 P1: At an international school with 100 students, 60 students speak English, 40 speak French and 10 students speak neither of these two languages.

a Calculate how many students speak both English and French. (2 marks)

b Sketch a Venn diagram to represent this information. In each enclosed region put in the number of students corresponding to this region. (3 marks)

c Calculate the probability that a student chosen at random

i speaks English but not French

ii speaks either English or French

iii does not speak English. (3 marks)

d Given that a student speaks French, find the probability that they also speak English. (2 marks)

e Let E be the event that a student speaks English and F be the event that a student speaks French. Determine with a reason if events E and F are independent or not. (2 marks)

17 P2: A disease that spreads rapidly is about to arrive at a city with a very large number of inhabitants. If a person does not do anything, then the probability that they will catch the disease is 90%. If they have taken a special injection then the probability that they will catch the disease reduces to 20%. Sadly only 30% of the population could afford the injection and they were all injected.

a Sketch a probability tree that represents all this information. (4 marks)

b Find the probability that a random person is not injected and catches the disease. (1 mark)

c Find the probability that a random person is injected and does not catch the disease. (1 mark)

d Find the probability that a random person does catch the disease. (2 marks)

e Given that a random person has caught the disease find the probability that they were not injected. (2 marks)

f Given that a random person has not caught the disease find the probability that they were injected. (2 marks)

18 P1: A college with 200 students offers rowing, kayaking and surfing as activities. Of the students, 8 do all three activities, 30 do rowing and kayaking, 20 do rowing and surfing, 10 do surfing and kayaking but not rowing, 40 do only rowing, 70 do kayaking and 48 do surfing.

a Sketch a Venn diagram to represent all this information. In each enclosed space in the diagram put the number of students that this space specifically refers to. (4 marks)

b Calculate how many students do not do any of these three activities. (2 marks)

c If a student is chosen at random find the probability they

i do only kayaking

ii do rowing or kayaking

iii do rowing or kayaking but not surfing

iv do not do rowing. (5 marks)

d Given that a student does surfing find the probability that they also do rowing. (2 marks)

19 P2: Adrian has a box with 5 red and 3 green apples in it. He takes out an apple at random and eats it. He then takes out another apple at random and also eats it.

a Find the probability that he eats

i two red apples

ii two apples of different colours. (4 marks)

Sally also has a box with 5 red and 3 green apples in it. Sally is on a diet. She takes out an apple at random and then puts it back in the box. She then takes out another apple at random.

b Find the probability that she takes

i two red apples

ii two apples of different colours. (4 marks)

20 P1: a For two events, A and B, $P(B) = 0.5$, $P(A|B) = 0.4$ and $P(A \cap B') = 0.4$.

Calculate

i $P(A \cap B)$

ii $P(A)$

iii $P(A \cup B)$

iv $P(A|B')$ (8 marks)

b Determine, with a reason, whether events A and B are independent or not. (2 marks)

Random walking

Approaches to learning: Critical thinking
Exploration criteria: Mathematical communication (B), Personal engagement (C), Use of mathematics (E)
IB topic: Probability, Discrete Distributions

The problem

A man walks down a long, straight road. With each step he either moves left or right with equal probability. He starts in the middle of the road. If he moves 3 steps to the left or 3 steps to the right, he will fall into a ditch on either side of the road. The aim is to find probabilities related to the man falling into the ditch, and in particular to **find the average number of steps he takes before inevitably falling into the ditch.**

Explore the problem

Use a counter to represent the man and a 'board' to represent the scenario:

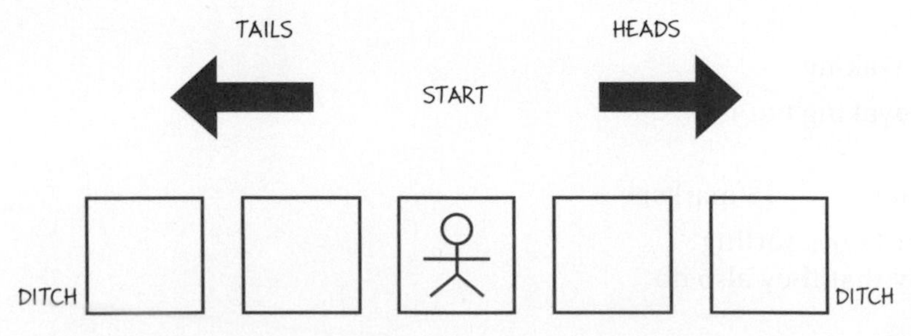

Toss a coin.

Let a tail (T) represent a left step and a head (H) represent a right step.

Write down the number of tosses/steps it takes for the man to fall into the ditch.

Do this a total of 10 times.

Calculate the average number of steps taken.

Construct a spreadsheet with the results from the whole class.

Calculate the average number of steps taken from these results.

How has this changed the result?

Do you know the actual average number of steps required?

How could you be certain what the average is?

Calculate probabilities

Construct a tree diagram that illustrates the probabilities of falling into the ditch within 5 steps.

Use your tree diagram to answer these questions:

What is the probability associated with each sequence in which the man falls into the ditch after a total of exactly 5 steps?

What is the probability that the man falls into the ditch after a total of exactly 5 steps?

What is the minimum number of steps to fall into the ditch?

What is the maximum number?

What is the probability that the man falls into the ditch after a total of exactly 3 steps?

Explain why all the paths have an odd number of steps.

Let x be the number of the steps taken to fall into the ditch.

Copy and complete this table of probabilities:

x	1	2	3	4	5	6	7	8	9	10	11	12	....
$P(X=x)$													

Look at the numbers in your table.

Can you see a pattern?

Could you predict the next few entries?

Simulation

Since there is an infinite number of values of x, calculating the expected number of steps to fall into the ditch would be very complicated.

An alternative approach is to run a computer simulation to generate more results, and to calculate an average from these results.

You can write a code in any computer language available that will run this simulation as many times as needed.

This will allow you to improve on the average calculated individually and as a class.

Although this would not be a proof, it is convincing if enough simulations are recorded.

Extension

Once you have a code written you could easily vary the problem.

What variations of the problem can you think of?

You may also be able to devise your own probability question which you could answer using simulation.

9 Representing equivalent quantities: exponentials and logarithms

Concepts

- Equivalence
- Quantity

Microconcepts

- Laws of exponents
- The exponential function *e* and its graph
- Logarithmic functions and their graphs
- Laws of logarithms including change of base
- Transformations of exponential and logarithmic graphs
- Solving exponential equations using logarithms
- Derivatives of exponential and logarithmic functions

You have already studied linear, quadratic and rational functions. In this chapter, you will look at a new class of function: one involving powers, where the independent variable x is used as an exponent (or index). The simplest of these functions is $f(x) = a^x$ where a is a positive real number.

This class of functions is used to model many real-life scenarios where the dependent variable (y) increases or decreases *exponentially* with the independent variable (x).

Exponential models enable scientists to measure very small changes in y at one end of the domain, and very large changes in y at the other end.

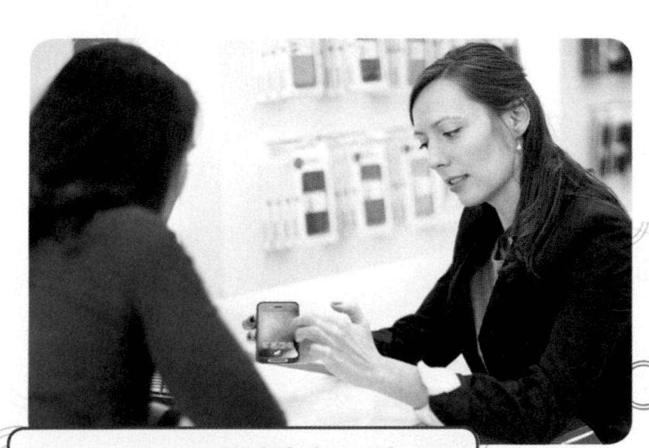

How can you model the sales of smartphones between 2003 and now? Can this model be used indefinitely?

Earthquakes are measured in the Richter scale, sound is measured using a decibel scale. What do these two scales have in common? Why is this scale more convenient to use?

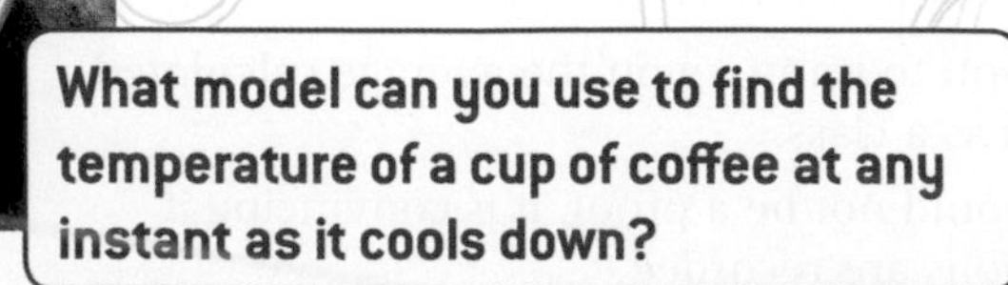

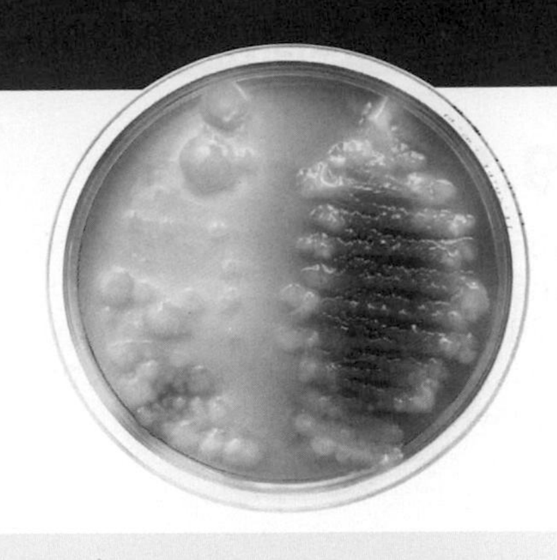

The number of bacteria in a Petri dish doubles every hour.

At 9am there are N_0 bacteria cells present in a Petri dish.

- How many bacteria cells will be in the Petri dish at 10am?
- How many bacteria cells will be in the Petri dish at 1pm?

Could you write down a rule which links the number of bacteria cells N in the dish, with the number of hours t which have passed since 9am?

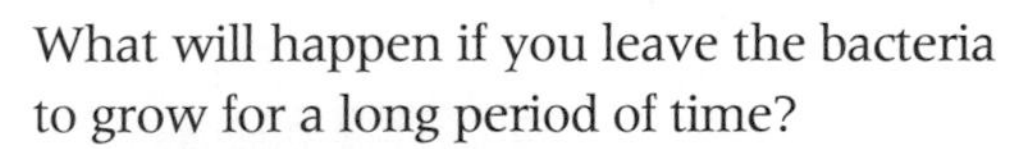

What will happen if you leave the bacteria to grow for a long period of time?

To answer these questions, think about:

i What must you multiply N_0 by to find the number of bacteria cells at 10am?

ii What must you, in turn, multiply this by to find the number of bacteria at 11am?

iii Do you always multiply by the same number?

iv How can you make a connection between this and the variable t?

Developing inquiry skills

Think of similar scenarios in which the rate of change is exponential.

Growth is often described as "exponential" – for example, a newspaper reporting "exponential rise in property prices" – but is it really?

What is the difference in the relationship between the variables in the newspaper headline and the bacteria example?

Think about the questions in this opening problem and answer any you can. As you work through the chapter, you will gain mathematical knowledge and skills that will help you to answer them all.

Before you start

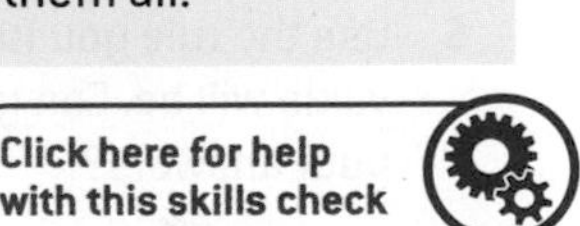

You should know how to:	Skills check
1 Evaluate simple exponents. eg $2^4 = 2 \times 2 \times 2 \times 2 = 16$ eg $\left(\frac{3}{4}\right)^3 = \frac{3^3}{4^3} = \frac{27}{64}$	**1** Evaluate: **a** 2^5 **b** 10^3 **c** $\left(\frac{1}{3}\right)^5$ **d** $\left(\frac{5}{6}\right)^3$
2 Convert simple numbers into exponential form. eg Find x when $5^x = 125$. $5^3 = 125$, so $x = 3$	**2** Solve: **a** $2^x = 8$ **b** $10^x = 10\,000$ **c** $4^x = 256$
3 Transform functions. eg Given the graph of $y = x^2$, sketch the graph of $y = (x - 2)^2 + 3$. 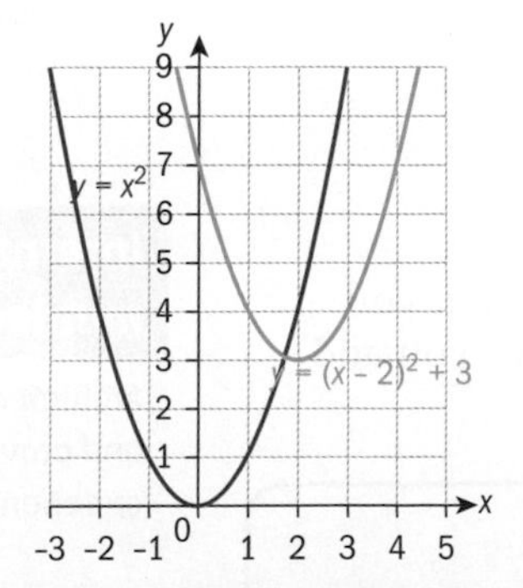	**3** Sketch the graph of $y = (x + 1)^2 - 2$.

9.1 Exponents

Exponents (or powers, or indices) are a quick way of writing repeated multiplication. Instead of writing $2 \times 2 \times 2 \times 2$, we write 2^4.

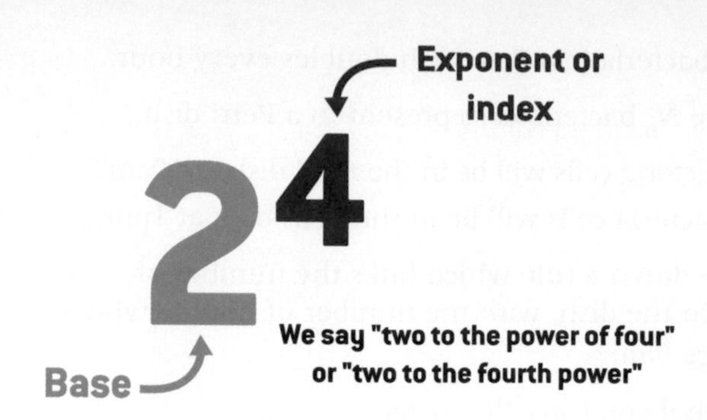

Investigation 1

1 Copy the following and fill in the blanks:

$2^1 =$ ___ $2^2 =$ ___ $2^3 =$ ___ $2^4 =$ ___

$2^5 =$ ___ $2^6 =$ ___ $2^7 =$ ___ $2^8 =$ ___

$2^9 =$ ___ $2^{10} =$ ___ $2^{11} =$ ___ $2^{12} =$ ___

2 Look at the last digit in each of your answers to question **1**. What do you notice about the last digit as the power of 2 increases?

3 Find the next four terms in the sequence. Does the pattern you identified in question **2** continue?

4 Using your answers so far, write down a method for finding the last digit of 2^n, where n is a positive integer.

5 Use the rule you found in question **4** to predict what the last digit of each value will be. Can you use your calculator, or other technology, to check your answers?

a 2^{20} **b** 2^{37} **c** 2^{99}

6 What is an exponent?

7 How can powers be used to represent real numbers?

Reflect Discuss your answer with a classmate then **share** your ideas with your class.

- What do you know about exponents?
- What other names are there for exponents?

Laws of exponents

Multiplication

Suppose you are asked to simplify $x^6 \times x^2$.

You could do it the following way:

$x^6 \times x^2 = (x \times x \times x \times x \times x \times x) \times (x \times x)$

$= x \times x \times x \times x \times x \times x \times x \times x$
$= x^8$

So, $x^6 \times x^2 = x^{(6+2)} = x^8$.

Instead of writing it all out in this way, you can see that when you multiply powers which have the same base (in this case, x), you add the exponents together.

$$a^m \times a^n = a^{m+n}$$

International-mindedness

Archimedes discovered and proved this law of exponents.

Division

Could you find a similar rule to simplify $x^6 \div x^2$?

$$x^6 \div x^2 = \frac{x \times x \times x \times x \times x \times x}{x \times x} = x \times x \times x \times x = x^4$$

So, $x^6 \div x^2 = x^{6-2} = x^4$.

When you divide powers which have the same base, you subtract the exponents.

$$a^m \div a^n = \frac{a^m}{a^n} = a^{m-n}$$

Reflect You have probably heard that any number to the power of zero is 1, and zero to any power is 0. But if that is the case, what is zero to the power of zero?

Research and discuss your answer with a classmate.

A power of zero

Let's try to find the value of $x^2 \div x^2$.

First, perform the division: $x^2 \div x^2 = \frac{x^2}{x^2} = \frac{x \times x}{x \times x} = 1$.

Second, use the rule for dividing exponents which have the same base:

$$x^2 \div x^2 = x^{2-2} = x^0$$

So, $x^0 = 1$.

$$a^0 = 1,\ a \neq 0$$

International-mindedness

French mathematician Nicolas Chuquet created his own notation for exponentials and might have been the first to recognize zero and negative exponents.

Negative exponents

Reflect How could you use both $\frac{a^m}{a^n} = a^{m-n}$ to write x^{-p} in terms of x^p?

You could also go back to the definition of division in order to find an expression for x^{-p}.

Consider $x^2 \div x^6$.

$$x^2 \div x^6 = \frac{x^2}{x^6} = \frac{x \times x}{x \times x \times x \times x \times x \times x} = \frac{1}{x \times x \times x \times x} = \frac{1}{x^4}.$$

But also, using the rule for dividing powers with the same base, $x^2 \div x^6 = x^{2-6} = x^{-4}$.

So, $x^{-4} = \frac{1}{x^4}$.

$$a^{-m} = \frac{1}{a^m}$$

Raising a power to a power

Simplify $(x^6)^2$.

$(x^6)^2 = (x \times x \times x \times x \times x \times x) \times (x \times x \times x \times x \times x \times x) = x^{12}$

So, $(x^6)^2 = x^{6\times 2} = x^{12}$.

$$(a^m)^n = a^{mn}$$

International-mindedness

The term power was used by the Greek mathematician Euclid for the square of a line.

Example 1

Simplify **a** $3x^2 \times 5x^4$ **b** $\frac{12a^7c^2}{2a^3c^5}$ **c** $(-4pq^2)^3$

a $3x^2 \times 5x^4 = (3 \times 5) \times (x^2 \times x^4)$ $= 15x^{2+4}$ $= 15x^6$	First, multiply the numbers together: $3 \times 5 = 15$ Then use the multiplication rule for exponents: $x^2 \times x^4 = x^{2+4}$
b $\frac{12a^7c^2}{2a^3c^5} = \left(\frac{12}{2}\right) \times \left(\frac{a^7}{a^3}\right) \times \left(\frac{c^2}{c^5}\right)$ $= 6a^{7-3}c^{2-5}$ $= 6a^4c^{-3}$ $= \frac{6a^4}{c^3}$	First, divide the number in the numerator by the number in the denominator: $12 \div 2 = 6$ Then use the division rule to simplify exponents with the same base: $\frac{a^7}{a^3} = a^{7-3}$, $\frac{c^2}{c^5} = c^{2-5}$ Finally, use $c^{-3} = \frac{1}{c^3}$.
c $(-4pq^2)^3 = (-4pq^2) \times (-4pq^2) \times (-4pq^2)$ $= (-4)^3 \times (p)^3 \times (q^2)^3$ $= -64p^3q^{2\times 3}$ $= -64p^3q^6$	Expand the cubed expression. Use the rule $(a^m)^n = a^{mn}$.

Exercise 9A

In questions **1–10**, use exponent laws to simplify the expression as far as possible.

1 $a^5 \times a^3 \times a^7$

2 $2x^3y^2 \times 7x^4y^6$

3 $4ab^3 \times 0.5a^6c$

4 $\frac{8m^5}{4m^3}$

5 $\frac{6u^5v^2}{9u^3v^3}$

6 $(3rs^3)^3$

7 $(-2x^4yz^5)^3$

8 $\left(\frac{x^{12}y^8}{x^5y^6}\right)^2$

9 $\frac{(5x)^2(5y^3)}{(5x^3y^4)^3}$

10 $\frac{9x^3(y^3)^3}{-81(x^{-2})^4y^{11}}$

11 Find an expression for the area of a square with side length $3x^2y$. Write your answer in its simplest form.

12 Find the area of a rectangle with width $4a^3b^2$ and length $\frac{5a}{2b^3}$. Write your answer in its simplest form.

Fractional exponents

The exponents that you have used so far have been integers. However, exponents can also be fractions.

To understand what $x^{\frac{1}{n}}$ means, we will use the multiplication law for indices:

$$x^{\frac{1}{2}} \times x^{\frac{1}{2}} = x^{\frac{1}{2}+\frac{1}{2}} = x^1 = x$$

However, you also know that, by definition:

$$\sqrt{x} \times \sqrt{x} = x$$

This implies that:

$$x^{\frac{1}{2}} = \sqrt{x}\cdot$$

Hence, raising any number x to the power of $\frac{1}{n}$ is equivalent to finding the nth root of x.

You can use this to understand what $x^{\frac{n}{m}}$ means.

Example 2

Write an expression for $x^{\frac{3}{2}}$ in a form involving a radical term.

$x^{\frac{3}{2}} = x^{\left(\frac{1}{2}\times 3\right)}$	
$= \left(x^{\frac{1}{2}}\right)^3$	Raising a "power to a power" rule.
$= \left(\sqrt{x}\right)^3$	Since $x^{\frac{1}{2}} = \sqrt{x}$.
$= \sqrt{x^3}$	By laws of surds.

$$a^{\frac{m}{n}} = \left(\sqrt[n]{a}\right)^m = \sqrt[n]{a^m}$$

International-mindedness

In the 9th century, the Persian mathematician Muhammad al-Kwarizmi used the terms "mal" for a square and "kahb" for a cube.

Example 3

a Without the use of technology, find the value of $8^{\frac{2}{3}}$.

b Write $\sqrt[4]{x^3}$ in exponential form.

a $8^{\frac{2}{3}} = \left(\sqrt[3]{8}\right)^2$	Write $8^{\frac{2}{3}}$ in radical form.
$= 2^2$	The cube root of 8 is 2.
$= 4$	2 squared gives 4.

Continued on next page

b $\sqrt[4]{x^3} = \left(\sqrt[4]{x}\right)^3$ $= \left(x^{\frac{1}{4}}\right)^3$ $= x^{\frac{1}{4}\times 3}$ $= x^{\frac{3}{4}}$	You can use the rule that $a^{\frac{m}{n}} = \sqrt[n]{a^m}$, or you can work it through step by step using the definitions. Using $x^{\frac{1}{n}} = \sqrt[n]{x}$ "Raising a power to a power" rule.

Exercise 9B

1 Write each expression in radical form:

a $7^{\frac{1}{2}}$ **b** $2^{\frac{3}{5}}$ **c** $6^{\frac{3}{2}}$

d $2^{\frac{5}{4}}$ **e** $5^{-\frac{1}{2}}$ **f** $(3x)^{-\frac{3}{2}}$

g $3x^{-\frac{3}{2}}$

2 Write each expression in exponential form:

a $\sqrt{10}^3$ **b** $\sqrt[5]{a^6}$ **c** $\sqrt[3]{m^7}$

d $\frac{1}{\sqrt{5x}}$ **e** $\frac{1}{\sqrt[4]{(2d)^5}}$ **f** $3\sqrt{x}$

g $\frac{3}{\sqrt{x}}$

Exponential equations

You can solve equations in which the unknown is in the exponent, such as $2^x = 16$ and $3^x = 9^{x+2}$.

Solving these equations can be done simply if all the terms in the equation have the same base.

International-mindedness

The word *exponent* was first used in the 16th century by German monk and mathematician Michael Stifel. Samuel Jeake introduced the term *indices* in 1696.

Example 4

Solve **a** $3^{x+2} = 27$ **b** $4^{2x+1} = 8^{x+2}$

a $3^{x+2} = 27$ $3^{x+2} = 3^3$ $x + 2 = 3$ $x = 1$	Convert 27 to a power of 3. (Note $27 = 3^3$.) Equate the exponents.
b $4^{2x+1} = 8^{x+2}$ $(2^2)^{2x+1} = (2^3)^{x+2}$ $2^{4x+2} = 2^{3x+6}$ $4x + 2 = 3x + 6$ $x = 4$	Convert the bases of 4 and 8 to base 2. (Note $4 = 2^2$ and $8 = 2^3$.) "Raising a power to a power" rule. Equate the exponents.

Exercise 9C

1 Solve each equation:

a $2^x = 16$ b $10^x = 1\,000\,000$

c $2^{x+1} = 64$ d $3^{2x-1} = 27$

e $3^{1-2x} = 1$ f $3 \times 2^x = 48$

g $4^{x+2} = \frac{1}{64}$ h $\sqrt[4]{3} = 9^x$

i $\left(\frac{1}{5}\right)^x = 25$ j $2^x = 2\sqrt{2}$

2 Convert to the same base and solve each equation:

a $2^{x+3} = 4^{x-2}$ b $5^{x-3} = 25^{x-4}$

c $6^{2x-6} = 36^{3x-5}$ d $9^{5x+2} = \left(\frac{1}{3}\right)^{11-x}$

Functions

Exponential functions

Investigation 2

The bacterium *Streptococcal pharyngitis* can double its number in 30 minutes. At 9am there are 1000 bacteria cells in a sample.

1 Copy and complete the table showing the number of bacteria cell present at each time.

Time (x)	9am	9.30am	10am	10.30am	11am	11.30am	12pm
Thousands of cells (y)	1			8			

2 How many bacteria cells will there be at 12pm?

3 Use graph paper to plot a graph of the number of bacteria cells in the sample from 9am to 12pm You should let 9am be zero on the x-axis, 9.30am be 1 on the x-axis and so on. Join up the points to sketch a smooth curve.

4 By drawing the line $y = 50$ on the same set of axes and finding the x-value at the point where the line intersects the curve, find the time when there are 50 000 cells.

6 **Conceptual** What is exponential growth or decay?

7 **Conceptual** How do exponential functions model real-world problems and their solutions?

TOK

The phrase "exponential growth" is used popularly to describe a number of phenomena.

Do you think that using mathematical language can distort understanding?

In investigation 2, you modelled the growth of bacteria in a sample using an exponential function.

An exponential function is a function of the form $f(x) = a^x$, where $a > 0, a \neq 1$

Reflect Find the meaning of exponential growth and decay and discuss your answer with a classmate about how this relates to your functions.

Investigation 3

1 Use your GDC to help you sketch, on the same set of axes, the graphs of:

a $y = 2^x$ b $y = 3^x$ c $y = 4^x$

Continued on next page

2 For each of these functions, can you find the equation of the asymptote?

3 Where do the functions cross the y-axis? Can you explain why this is the case?

4 What is the domain and range of each function?

5 **Conceptual** What effect does the value of the base have on the graph of an exponential function?

6 Use your GDC to help you sketch the graphs of:
a $y = 2^x + 2$ **b** $y = 2^{x+2}$

7 Describe the transformations that map the graph of $y = 2^x$ onto each of the two graphs in question **6**.

8 Use your GDC to help you sketch, on the same axes, the graphs of:
a $y = 2^{-x}$ **b** $y = 3^{-x}$ **c** $y = 4^{-x}$

9 How is the graph of $y = a^{-x}$ the same as the graph of $y = a^x$?

10 How is the graph of $y = a^{-x}$ different from the graph of $y = a^x$?

11 **Factual** If $y = 2^x$ is the parent function, copy and complete this table to summarize your findings about the parameters of an exponential function.

Function	Transformation of the parent function
$y = a2^x$	
$y = 2^{ax}$	
$y = 2^x + b$	
$y = 2^{x+c}$	
$y = 2^{-x}$	
$y = -2^x$	

12 **Conceptual** How do the parameters of an exponential functions affect the graph of the function?

Example 5

Consider the function $y = 2^x + 3$.

a Find **i** the y-intercept **ii** the equation of the asymptote.

b State the domain and range of the function.

c Sketch the graph of the function, showing the asymptote as a dotted line.

d Using your GDC, solve the equation $2^x + 3 = 2 + 3^{-x}$.

a i $y(0) = 2^0 + 3 = 1 + 3 = 4$ Hence, y-intercept is $(0, 4)$	The y-intercept occurs when $x = 0$.
ii $y = 3$	As $x \to -\infty$, $2^x \to 0$ so $y \to 3$.
b $x \in \mathbb{R}$, $y > 3$	The function is defined for all values of x, and $2^x + 3 > 3$ for all x.

c 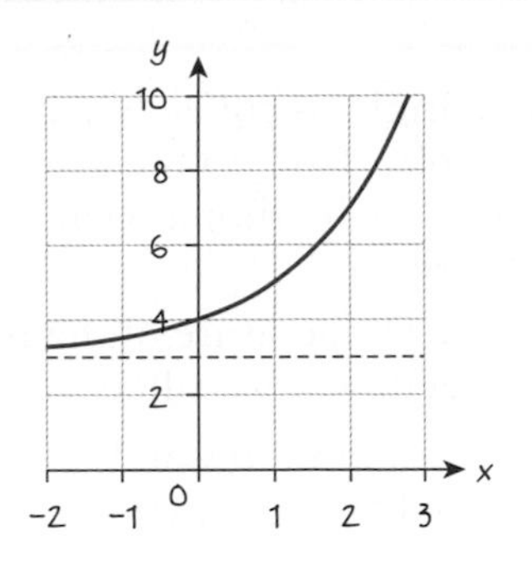	The function crosses the y-axis at $(0, 4)$. Mark the asymptote $y = 3$ as a dashed line. y should tend towards the asymptote as $x \to -\infty$.
d 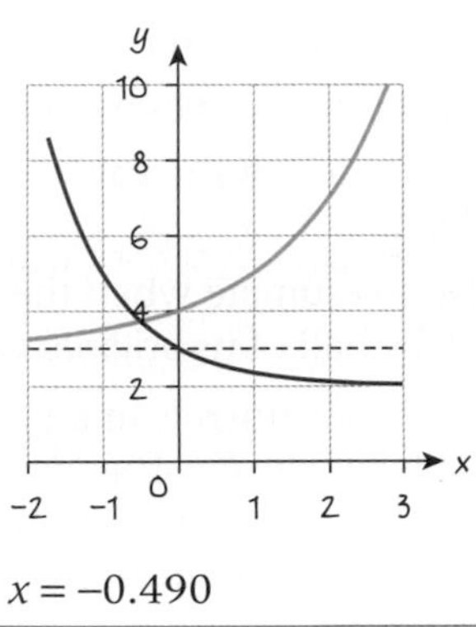 $x = -0.490$	Plot the graphs of $y = 2^x + 3$ and $y = 2 + 3^{-x}$ on the same axes and find the x-value of the intersection point.

Example 6

The value of a boat, y, in thousands of UK pounds (£) is modelled by the function $y = 20(0.85)^x$, where x is the number of years since the boat was manufactured.

a Find the value of the boat when it was brand new.

b Estimate the value of the boat when it is 3 years old. Give your answer to the nearest pound.

c Use your GDC to estimate when the value of the boat will be worth half its original value.

a $y(0) = 20(0.85)^0 = £20$ thousand	The value when the boat was new is found by letting $x = 0$.
b $y(3) = 20(0.85)^3 = 12.2825$ thousand $= £\,12\,283$	Let $x = 3$ and find y.
c 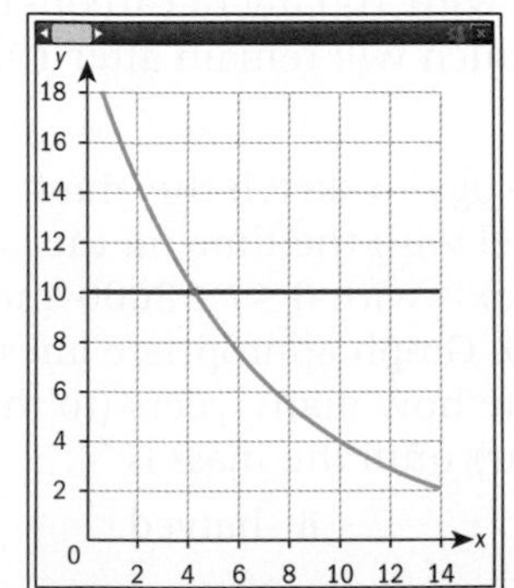 The boat will be worth half its original value in approximately 4.27 years.	The value of the boat will have halved when it is £10 thousand. On your GDC, plot $y = 20(0.85)^x$ and the line $y = 10$ on the same graph. Find the x-coordinate of the intersection point. The functions intersect at $(4.265, 10)$.

Exercise 9D

1 State whether each function represents a relationship of exponential growth or exponential decay.

a $y = 10^x$ **b** $y = 6^{-x}$ **c** $y = \left(\frac{3}{5}\right)^x$

d $y = (0.45)^x$ **e** $y = (1.5)^x$

2 Match each of these functions with its graph:

$f(x) = 2^x$

$g(x) = 2^{x+1}$

$h(x) = 2^{-x}$

$i(x) = 3(2^x)$

$j(x) = 2^{x-2}$

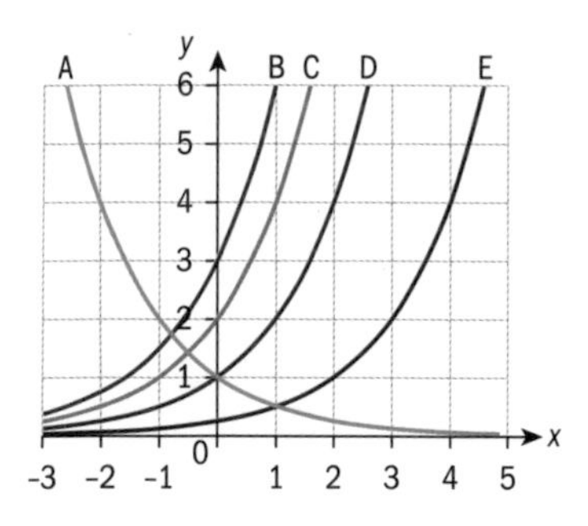

3 For each function:

a $y = 3^x + 4$ **b** $y = 2^{-x} - 3$

c $y = 2 - 2^x$

i Sketch the graph and mark the asymptote as a dashed line (if the asymptote is not the x-axis).

ii Label the coordinates of the y-intercept on your graph.

iii Write down the equation of the asymptote.

iv Label the coordinates of one additional point on the curve (other than the y-intercept).

v State the domain and range.

4 Teacher Tom makes a cup of coffee and leaves it on his table.

Its temperature (H °C) is modelled by the function $H = (65)2^{-\frac{t}{2}} + 25$, where t is the time in minutes after Tom makes the coffee.

a State the initial temperature of the coffee.

b Find the temperature of the coffee after 3 minutes.

c Tom likes to drink his coffee when it is 40°C or cooler. Find, to the nearest minute, how long he should wait to drink his coffee.

d Determine the temperature of the room where Tom set down his coffee.

5 The value, y, of a car, in thousands of dollars, is modelled by the function $y = 30(0.9)^x$, where x is the number of years since the car was manufactured.

a Find the value of the car when it was new.

b Determine the value of the car when it is 3 years old.

c Use your GDC to estimate when the value of the car will be half of its original value.

6 The population (P) of squirrels in a park is modelled by the function $P = 40(1.5)^t$, where t is the number of years that have elapsed since recording the squirrel population began.

a State how many squirrels there were initially.

b Estimate the population of squirrels in the park after two years.

c By plotting a graph on your GDC, find how long will it take the population of squirrels to reach 200.

7 The rate of decay of the radioactive substance carbon-14 can be modelled by the equation $A = A_0 (2)^{-\frac{t}{5730}}$, where A is the mass of carbon-14 at time t (measured in years) and A_0 is the initial mass.

a If you start with 100 mg of carbon-14, find how much will remain after 1000 years.

b Use technology to sketch the graph of the model with the time on the horizontal axis with $0 \le t \le 8000$ and $0 \le A \le 100$. Graph appropriate lines to approximate how many years (to the nearest year) until the mass is:

i 75 grams **ii** halved.

The natural base e

Mathematics has many special constants, such as π. Another important mathematical constant is Euler's constant e. We will study the constant e in investigation 4.

Investigation 4

1 Copy and complete the table by entering the value of $\left(1+\frac{1}{x}\right)^x$ into the table for each x. Give your answers to correct to five decimal places.

x	$\left(1+\frac{1}{x}\right)^x$
1	
2	
3	
4	
5	
10	
100	
1000	
10 000	
100 000	

2 Use technology to plot a graph of $y=\left(1+\frac{1}{x}\right)^x$ for $0 \le x \le 10$.

3 Can you find an approximate equation for the asymptote of the function $y=\left(1+\frac{1}{x}\right)^x$?

You have been introduced to the limit notation in earlier chapters of this book.

4 Use your table to approximate $\lim_{x\to\infty}\left(1+\frac{1}{x}\right)^x$ correct to three decimal places.

The number you have just found is called Euler's number, e.

5 What is the value of $y = e^0$?

6 Use your GDC to plot the graphs of $f(x) = a^x$ for different values of a between 2.5 and 3. You will find that they all have an asymptote of $y = 0$ and cross the y-axis at 1, but only one value of a has a gradient of 1 as it crosses the y-axis. Can you use your GDC to find what value of a this is?

Euler's number, e, is a constant approximately equal to 2.718.

Exercise 9E

1 Find the value of each expression to 3 decimal places:

a e^1 **b** e^2 **c** e^{-2} **d** $3e$ **e** $\frac{e}{2}$

f $5\sqrt{e}$ **g** $4e - 5$.

2 On the same set of axes sketch $f(x) = 1^x$, $g(x) = 2^x$, $h(x) = e^x$ and $i(x) = 3^x$.

a Describe the graph of $y = f(x)$.

b Compare the graph of $h(x) = e^x$ to the graphs of $g(x)$ and $i(x)$. How are they the same? How are they different?

3 Sketch the graphs of $f(x) = e^x$ and $g(x) = e^{-x}$.

Describe the transformation that maps f onto g.

4 The growth of a fungus on a tree is modelled by $G(t) = 4500e^{0.3t}$, where G cm^2 is the area covered by the fungus after t days.

a Find the initial area covered by the fungus.

b Determine the area the fungus will cover after 10 days have passed.

5 When interest is compounded continuously, the amount of money in an account is modelled by the function:

$FV = PVe^{rt}$

where FV is the future value of the account, PV is the present value, r is the annual rate of interest as a decimal and t is the number of years that have elapsed since the time of the present value.

If you invest \$5000 at 5% compound interest, how much money will there be in the account after six years? Give your answer to the nearest dollar.

6 When an amount (a) increases or decreases by a fixed percent (r) each time period, the value (y) of the amount after t years can be modelled by one of these equations:

exponential growth of $y = a(1 + r)^t$ or exponential decay of $y = a(1 - r)^t$

In 2011, the population of the world was approximately 7 billion, and it is growing at a rate of 1.1% a year.

a Using the information given at the start of the question, write down a model for the population of the world t years after 2011.

b Estimate the world population in 2025.

c In what year will the world population be estimated to reach 10 billion?

7 The value of a car decreases by 15% every year. Tatiana buys a new car for \$25 000.

a Write down a function to model the value of the car after t years.

b Find the value of Tatiana's car after three years, to the nearest \$100.

c Tatiana will sell the car when its value decreases to \$10 000. Use a graph to determine how old the car will be when Tatiana sells it.

Developing inquiry skills

In Investigation 2 earlier in the chapter you looked at bacteria growing in a Petri dish. At 9 am there were 1000 bacteria present.

Using the mathematics you have learned in this chapter, write down an expression for N, the number of bacteria cells present t hours after 9 am.

Use this to find the time at which the number of bacteria cells reaches 64 000.

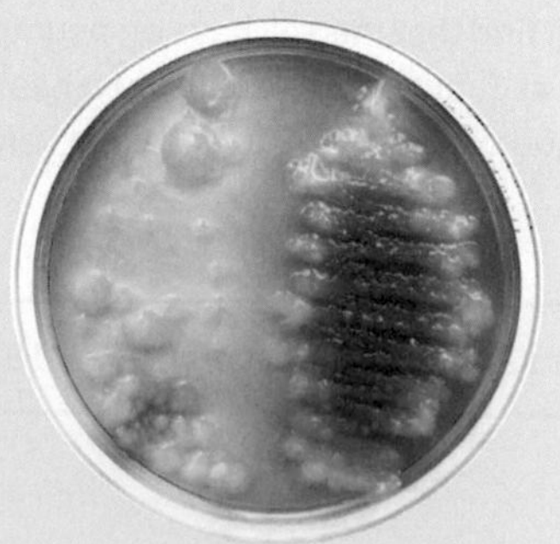

9.2 Logarithms

Logarithms can be used to help us deal with the manipulation of large numbers. The word "logarithm" was coined by Scottish mathematician John Napier in the 16th century from the Greek word *logos* which has many meanings, one of which is *ratio*. Logarithms are used by mathematicians and scientists in many fields, including finance and astronomy.

International-mindedness

Logarithms do not have units but many measurements use a log scale such as earthquakes, the pH scale and human hearing.

Investigation 5

1 On your GDC, plot the graphs of $f(x) = 10^x$ and $g(x) = \log_{10}x$ on the same set of axes.

2 Find the domain, range and asymptote for $f(x) = 10^x$.

3 Find the domain, range and asymptote for $g(x) = \log_{10}x$.

4 Now plot the line $y = x$ on the same set of axes.

5 **Factual** What does the line $y = x$ tell you about the functions f and g? Explain your reasoning.

6 **Conceptual** What is the relationship between exponents and logarithms?

7 **Conceptual** How could you use the domain and range of an exponential function to state the domain and range of a logarithmic function?

8 **Factual** What graph do you get when you reflect the graph of $f(x) = a^x$ in the line $y = x$?

If $f(x) = a^x$, $f^{-1}(x) = \log_a x$

Logarithmic functions

The inverse of $y = a^x$ is $y = \log_a x$. So, the graph of $y = \log_a x$ is a reflection of $y = a^x$ in the line $y = x$.

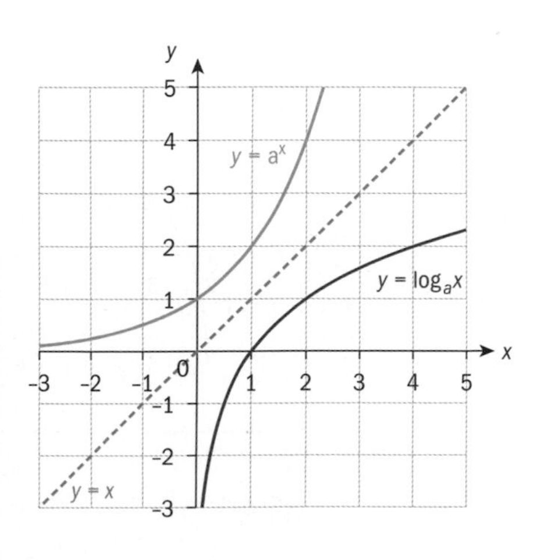

Properties of the graph of the logarithmic function $y = \log_a x$, where $a > 0, a \neq 1$:

$x = 0$ is an asymptote.

The curve crosses the x-axis at $(1, 0)$.

The graph continually increases for $a > 1$, and decreases for $0 < a < 1$.

The **domain** is the set of all *positive* real numbers.

The **range** is the set of all real numbers.

Functions

You can now transform the logarithmic function using the skills from earlier chapters.

Example 7

The function $f(x) = \log_{10}(x - 2) + 1$ is given.

a Describe the transformations necessary to transform the graph of $g(x) = \log_{10}x$ onto f.

b Write down the equation of the asymptote of f.

c Find the domain and range of f.

d Sketch $y = f(x)$ and the line $y = x$ on the same graph. Hence, sketch $y = f^{-1}(x)$ on the same axes.

a Translation of 2 units to the right, followed by translation of 1 unit up.	First transform $\log_{10}x$ to $\log_{10}(x - 2)$, and then transform $\log_{10}(x - 2)$ to $\log_{10}(x - 2) + 1$.
b $x = 2$	The asymptote of g was $x = 0$ and this is translated 2 units to the right by the first transformation. Since this is a vertical asymptote, the translation of 1 unit up does not affect the asymptote.
c $x > 2$, $y \in \mathbb{R}$	The translations do not affect the range.
d 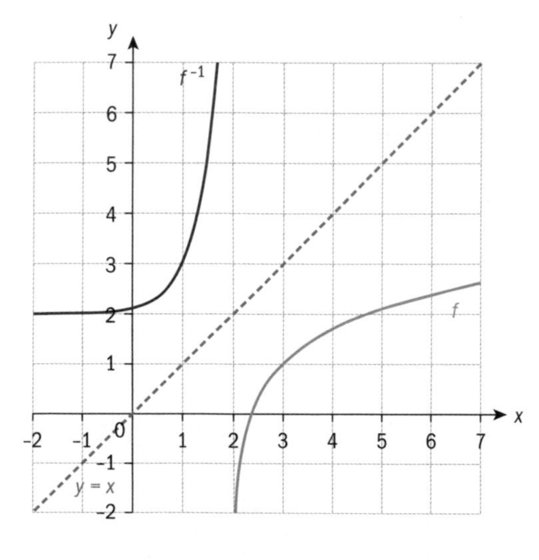	$y = f^{-1}(x)$ is a reflection of $y = f(x)$ in the line $y = x$.

Exercise 9F

1 Describe the transformations required to map the graph of $f(x) = \log_3 x$ onto the graph of:

a $g(x) = \log_3 x + 4$

b $h(x) = \log_3 (x - 3)$

c $i(x) = 2\log_3 x$

2 a On separate axes, sketch the graph of each function.

Show the asymptote as a dashed line (if it is not the y-axis), and label the coordinates of the x-intercept in each case.

a $f(x) = \log x$

b $g(x) = -\log x$

c $h(x) = 3 + \log x$

d $i(x) = 2\log(x + 2)$

3 Solve the equation $e^{x-3} = \log(x + 2)$ by plotting graphs of two appropriate functions on your GDC and finding any points intersection.

Investigation 6

Logarithm laws provide the means to find inverses of exponential functions which model real-life situations.

1 Copy and complete this table with the use of your calculator.

$10^0 =$	$\log_{10} 1 =$
$10^1 =$	$\log_{10} 10 =$
$10^2 =$	$\log_{10} 100 =$
$10^3 =$	$\log_{10} 1000 =$
$10^4 =$	$\log_{10} 10\,000 =$
$10^5 =$	$\log_{10} 100\,000 =$
$10^6 =$	$\log_{10} 1\,000\,000 =$

2 **Conceptual** Using the table above, describe how the same information can be conveyed in both exponential form and logarithmic form.

3 Write $2^3 = 8$ in logarithmic form.

4 Write $a^b = c$ in logarithmic form.

5 Write $\log_5 25 = 2$ in exponential form.

6 **Conceptual** How are logarithms and exponents different representations of the same quantities? why is it useful to be able to move from one form to another?

For $a > 0$, $a \neq 1$, and $c > 0$, $a^b = c$ is equivalent to $\log_a c = b$.

TOK

Mathematics is all around us in patterns, shapes, time and space.

What does this tell you about mathematical knowledge?

When we write $\log_2 8 = 3$, the number 2 is the *base* of the logarithm, and the number 8 is the *argument*. We say "the logarithm of 8, to base 2, is 3", or simply "log base 2 of 8 equals 3". The word "logarithm" is often abbreviated to just "log".

HINT

Since 10 is a very common base, it is sometimes left out of the expression. Hence, $\log x$ is understood to mean $\log_{10} x$.

Investigation 7

1 Given that $\log_a c = b$ is equivalent to $a^b = c$, copy and complete the table.

Logarithmic form	Exponential form	Solve for x
$\log_2 2 = x$	$2^x = 2$	$x = 1$
$\log_3 3 = x$		
$\log_4 4 = x$		
$\log_5 5 = x$		
$\log_a a = x$		

2 What does the table tell you about the value of the exponent when the base and the argument of a logarithm are the same?

3 Use your result from question **2** to solve each of these for the unknown variable:

a $\log_{11} 11 = x$ **b** $\log_5 p = 1$ **c** $\log_q 3 = 1$

4 Now use $\log_a c = b$ is equivalent to $a^b = c$ to solve:

a $\log_6 1 = x$ **b** $\log_7 1 = x$ **c** $\log_8 1 = x$

5 What does this tell you about the value of a logarithm when the argument is 1?

6 **Factual** How can you write 1 in logarithmic form?

7 **Conceptual** How can equivalent values be represented in different ways by using logarithms?

TOK

How does mathematical proof differ from reasoning in everyday life?

For any base $a > 0, a \neq 1$

$$\log_a 1 = 0$$

And

$$\log_a a = 1$$

Reflect You can extend this result to give $\log_a(a^x) = x$. Using the result above, and the definition of a logarithm, can you explain why this is the case?

For any base $a > 0, a \neq 1$

$$\log_a(a^x) = x$$

Example 8

a Write $\log_7 x = 2$ in exponential form.

b Write $x = \sqrt{25}$ in logarithmic form.

c Find the value of $\log_2 32$.

1 $x = 7^2$	Using $a^b = c$ is equivalent to $\log_a c = b$.
2 $x = \sqrt{25} = 25^{\frac{1}{2}}$ $\log_{25} x = \frac{1}{2}$	First write $\sqrt{25}$ in exponential form, then use the identity above.
3 $\log_2 32 \Rightarrow 2^x = 32$ $2^x = 2^5$ $x = 5$ So, $\log_2 32 = 5$	Write in exponential form, and then write 32 as a power of 2 in order to solve.

Functions

Exercise 9G

1 Write each equation in exponential form:

a $\log_p q = r$ **b** $\log_3 5 = r$

c $\log_7 q = 6$ **d** $\log_p 5 = 3$

e $\log 11 = x$

2 Write each equation in logarithmic form.

a $r^s = t$ **b** $8^2 = 64$

c $10^x = 25$ **d** $3^{-2} = \frac{1}{9}$

e $27^{\frac{2}{3}} = 9$

3 Find the value of each expression:

a $\log_7 1$ **b** $\log_8 1$

c $\log_9 1$ **d** $\log_x 1$

4 Find the value of each expression:

a $\log_3 3$ **b** $\log_4 4$

c $\log_5 5$ **d** $\log_x x$

5 Find the value of each expression:

a $\log_3 9$ **b** $\log_2 32$

c $\log_5 125$ **d** $\log_4 256$

e $\log_{25} 5$ **f** $\log_8 2$

g $\log_3 \frac{1}{27}$

Investigation 8

Part one:

1 Copy and complete these equations using your calculator. Give your answers in exact form, or rounded to 3 significant figures.

$\log 2 + \log 3 =$
$\log 3 + \log 4 =$
$\log 5 + \log 6 =$

Continued on next page

$\log 5 + \log 3 =$ $\quad$ $\log 6 + \log 3 =$
$\log 4 + \log 5 =$ $\quad$ $\log 6 =$
$\log 12 =$ $\quad$ $\log 15 =$
$\log 18 =$ $\quad$ $\log 20 =$
$\log 30 =$

2 Work with a classmate. Share your answers and discuss any patterns that you notice.

3 **Factual** Can you work together to suggest a general rule for writing $\log A + \log B$ as a single logarithm, for all positive values of A and B?

4 Test your rule using several other values for A and B.

Part two:

Now you will investigate what happens when you subtract two logarithms.

5 Copy and complete these equations using your calculator:

$\log 6 - \log 3 =$ $\quad$ $\log 16 - \log 4 =$
$\log 15 - \log 5 =$ $\quad$ $\log 100 - \log 10 =$
$\log 15 - \log 3 =$ $\quad$ $\log 2 =$
$\log 3 =$ $\quad$ $\log 4 =$
$\log 5 =$ $\quad$ $\log 10 =$

6 Work with a classmate. Share you answers and discuss any patterns that you notice.

7 **Factual** Can you work together to suggest a general rule for writing $\log A - \log B$ as a single logarithm, for all positive values of A and B?

8 Test your general rule on several other values for A and B.

9 **Conceptual** How do properties of logarithms help to simplify expressions?

The laws of logarithms

We will now formalize the general rules that you found in the investigation above.

Suppose that $\log_a x = m$ and $\log_a y = n$. Then, by definition, $x = a^m$ and $y = a^n$.

So

$$\begin{aligned}
\log_a(xy) &= \log_a(a^m a^n) && \text{by substitution} \\
&= \log_a(a^{m+n}) && \text{by properties of exponents from section 9.1} \\
&= m + n && \text{because } \log_a(a^b) = b \\
&= \log_a x + \log_a y && \text{by substitution}
\end{aligned}$$

Reflect Can you use the same approach to show, step-by-step, that $\log_a \frac{x}{y} = \log_a x - \log_a y$?

$$\log_a xy = \log_a x + \log_a y$$

$$\log_a \frac{x}{y} = \log_a x - \log_a y$$

Remember that if $\log_a x = m$, then $a^m = x$.

Then, given a constant p:

$\log_a(x^p) = \log_a((a^m)^p)$	because $x = a^m$
$= \log_a(a^{mp})$	by rules of exponents
$= mp$	because $\log_a(a^b) = b$
$= p\log_a x$	substituting $m = \log_a x$

$$\log_a(x^p) = p\log_a x$$

Consider now $a^{\log_a x}$:

$a^{\log_a x} = a^m$	because $m = \log_a x$
$= x$	because $a^m = x$

$$a^{\log_a x} = x$$

TOK

"One reason why mathematics enjoys special esteem, above all other sciences, is that its propositions are absolutely certain and indisputable"—Albert Einstein.

How can mathematics, being after all a product of human thought which is independent of experience, be appropriate to the objects of the real world?

Functions

Example 9

Use the laws of logarithms to write each expression as a single logarithm:

a $\log_8 6 + \log_8 3$ **b** $\log_5 9 - \log_5 3$ **c** $\log_7 6 - 1$ **d** $2\log_{10} 5 + 2$

a $\log_8 6 + \log_8 3$ $= \log_8(6 \times 3) = \log_8 18$	Using $\log_a xy = \log_a x + \log_a y$
b $\log_5 9 - \log_5 3$ $= \log_5 \frac{9}{3} = \log_5 3$	Using $\log_a \frac{x}{y} = \log_a x - \log_a y$

Continued on next page

c $\log_7 6 - 1$ $= \log_7 6 - \log_7 7 = \log_7 \frac{6}{7}$	Using $\log_a a = 1$ for any base, hence $1 = \log_7 7$
d $2\log_{10} 5 + 2$ $= \log_{10}(5^2) + \log_{10}(10^2)$ $= \log_{10}(25) + \log_{10}(100)$ $= \log_{10}(25 \times 100)$ $= \log_{10}(2500)$	$p\log_a x = \log_a(x^p) \Rightarrow 2\log_{10} 5 = \log_{10} 5^2$ Also, $\log_a a = 1 \Rightarrow 2 = 2\log_{10} 10 = \log_{10}(10^2)$. Using $\log_a xy = \log_a x + \log_a y$

Exercise 9H

1 Write each expression as a single logarithm:

a $\log 3 + \log 5$ **b** $\log 16 - \log 2$ **c** $3 \log 5$ **d** $3 \log 4 - 4 \log 3$ **e** $\log x + \log 1$ **f** $\log 236 - \log 1$ **g** $5 \log 2 + 2 \log 5$ **h** $\log 128 - 6 \log 2$ **i** $\log 2 + \log 3 + \log 4$ **j** $\log 12 - 2 \log 2 + \log 3$ **k** $5 \log 2 + 4 \log 3$ **l** $\log 6 + 3 \log 3 - \log 2$

Example 10

If $x = \log 2$ and $y = \log 6$, write each of these expressions in terms of x and y:

a $\log 12$ **b** $\log 3$ **c** $\log 48$

a $\log 12 = \log(2 \times 6) = \log 2 + \log 6 = x + y$	Write 12 as a product of 6 and 2. Use $\log(ab) = \log a + \log b$.
b $\log 3 = \log\frac{6}{2} = \log 6 - \log 2 = y - x$	Write 3 as the quotient of 6 and 2. Use $\log\left(\frac{a}{b}\right) = \log a - \log b$.
c $\log 48 = \log(8 \times 6)$ $= \log 8 + \log 6$ $= \log 2^3 + \log 6$ $= 3\log 2 + \log 6$ $= 3x + y$	Write 48 as a product of 6 and 8. $8 = 2^3$

Example 11

Let $x = \log_3 P$, $y = \log_3 Q$ and $z = \log_3 R$.

Write each of these expressions in terms of x, y and z:

a $\log_3\left(\frac{PQ}{R}\right)$ **b** $\log_3\left(\frac{P^2R}{Q^3}\right)$

a $\log_3\left(\frac{PQ}{R}\right)$ $= \log_3 P + \log_3 Q - \log_3 R$ $= x + y - z$	Expand the brackets using both $\log(ab) = \log a + \log b$ and $\log\left(\frac{a}{b}\right) = \log a - \log b$. Substitute x, y and z.
b $\log_3\left(\frac{P^2R}{Q^3}\right)$ $= \log_3 P^2 + \log_3 R - \log_3 Q^3$ $= 2\log_3 P + \log_3 R - 3\log_3 Q$ $= 2x + z - 3y$	Expand the brackets using the multiplication and division log rules. Use $\log_a(x^p) = p\log_a x$. Substitute x, y and z.

Exercise 9I

1 If $x = \log 3$ and $y = \log 6$, write each expression in terms of x and y:

a $\log 18$ **b** $\log 2$ **c** $\log 9$

d $\log 27$ **e** $\log 36$ **f** $\log\frac{1}{2}$

2 If $m = \log_5 7$ and $n = \log_5 4$, write each expression in terms of m and n:

a $\log_5 28$ **b** $\log_5\frac{7}{4}$ **c** $\log_5 49$

d $\log_5 64$ **e** $\log_5\frac{49}{4}$ **f** $\log_5\frac{7}{16}$

g $\log_5 112$ **h** $\frac{\log_5 7}{\log_5 4}$ **i** $\frac{\log_5 49}{\log_5 64}$

j $\log_5 100$

3 Let $x = \log_2 A$, $y = \log_2 B$ and $z = \log_2 C$.

Express $\log_2\left(\frac{A}{BC^3}\right)^4$ in terms of x, y and z.

4 If $\log_3 P = x$ and $\log_3 Q = y$, write each expression in terms of x and y:

a $\log_3 P^3Q$ **b** $\log_3\frac{\sqrt{P}}{Q}$

5 a Given that $\log x - \log(x-5) = \log M$, express M in terms of x.

b Hence, or otherwise, solve the equation $\log x - \log(x-5) = 1$

Natural logarithms

In section 9.1 you learned about powers with the natural base: $e \approx 2.718$.

The natural logarithm is a logarithm with the base of e, but instead of writing $\log_e x$, it can be abbreviated to $\ln x$, where ln was derived from the French expression "Logarithme Naturel".

The laws of logarithms are the same for natural logarithms as for logarithms with any other base.

Reflect Can you use what you have learned about logarithms so far to write down the values of $\ln 1$ and $\ln e$? Justify your answers.

Example 12

1 Find the **exact** values of:

a $\ln(e^2)$ **b** $e^{\ln 5}$ **c** $\ln\left(\frac{e^x}{e^y}\right)$

a $\ln(e^2) = 2\ln e$ $= 2$	$\ln e = 1$
b $e^{\ln 5} = 5$	Using $a^{\log_a x} = x$
c $\ln\left(\frac{e^x}{e^y}\right) = \ln(e^{x-y})$	Using $\frac{e^x}{e^y} = e^{x-y}$
$= x - y$	Using $\log_a(x^p) = p\log_a x$

Exercise 9J

1 Simplify each expression:

a $\ln e^3$ **b** $\ln e^4$ **c** $\ln\sqrt{e}$

d $\ln\sqrt[3]{e}$ **e** $\ln\frac{1}{e}$ **f** $\ln\frac{1}{e^2}$

2 Simplify each expression:

a $e^{\ln 2}$ **b** $e^{\ln 3}$ **c** $e^{\ln x}$

d $e^{2\ln 4}$ **e** $e^{3\ln x}$ **f** $e^{-\ln 3}$

3 Solve each of these equations:

a $\ln x = 2.7$ **b** $\ln(x + 1) = 1.86$

4 Simplify the expression $e^{x\ln a}$ as far as possible.

Change of base

It can be convenient to change the base of a logarithm. You may want to do this for a number of reasons, including:

- You may need to simplify an expression which involves logarithms with different bases. Changing to the same base allows you to use the laws of logarithms.
- You may need to find the value of a logarithmic expression using a scientific calculator which, unlike many GDCs, may only be able to calculate with terms in $\log_{10}$ and ln.

Suppose you want to express $\log_a x$ in logarithms in terms of base b.

Let $y = \log_a x$.

Then $a^y = x$.

TOK

Is mathematics invented or discovered?

Make the expression on each side of the equation the argument of a logarithm with base b.

$\log_b a^y = \log_b x$

$y\log_b a = \log_b x$ (bringing the power down on the left-hand side)

$y = \dfrac{\log_b x}{\log_b a}$ (dividing both sides by $\log_b a$)

Therefore, $\log_a x = \dfrac{\log_b x}{\log_b a}$ when $a, b, x > 0$.

> $\log_a x = \dfrac{\log_b x}{\log_b a}$ where $a, b, x > 0$.

Example 13

Find the value of $\log_3 20$ to 3 significant figures by converting to **a** base 10, **b** base e. Comment on your answers to parts **a** and **b**.

a $\log_3 20 = \dfrac{\log 20}{\log 3} = 2.73$	Using $\log_a x = \dfrac{\log_b x}{\log_b a}$
b $\log_3 20 = \dfrac{\ln 20}{\ln 3} = 2.73$	
Changing to base 10 or base e still gives the same value for $\log_3 20$.	You can use the "change of base formula" to change a logarithm to an expression which involves logarithmic terms in any base you choose.

Example 14

If $\log_4 7 = p$ and $\log_4 9 = q$, find $\log_7 9$ in terms of p and q.

$\log_7 9 = \dfrac{\log_4 9}{\log_4 7} = \dfrac{q}{p}$	Using $\log_a x = \dfrac{\log_b x}{\log_b a}$ and changing to base 4

Exercise 9K

1 Find the value of each of these expressions.

a $\log_3 8$ **b** $\log_6 24$ **c** $\log_5 8$

d $\log_3 30$ **e** $\log_7 \dfrac{1}{4}$ **f** $\log_2 \dfrac{3}{5}$

2 If $p = \log_3 A$ and $q = \log_3 B$, find $\log_A B$ in terms of p and q.

3 If $\log_x 3 = r$ and $\log_x 6 = s$, write each of these expressions in terms of r and s:

a $\log_3 6$ **b** $\log_6 3$ **c** $\log_3 36$ **d** $\log_3 54$

e $\log_9 6$ **f** $\log_6 18$ **g** $\log_3 2$

4 Prove that $\log_x y = \dfrac{1}{\log_y x}$.

5 Show that $\ln 10 = \dfrac{1}{\log e}$.

6 Explain the difference between the word "prove" and "show that" in the last two questions.

Using logarithms to solve equations

A basic rule for manipulating equations is that whatever you do to the left-hand side of the equation, you should also do to the right-hand side. For example, you can add a constant to both sides of an equation, or you can square both sides of an equation.

When solving equations where the variable is an exponent, it is often helpful to use logarithms.

Reflect In deriving the change of base formula, you saw an example of how using logarithms in an equation allowed you to isolate the variable. Look back at this, and explain what rule of logarithms we used in order to isolate the variable.

In exercise 9C, you solved equations where the variable was an exponent, such as $2^x = 16$.

You were only able to do this because, for each of the equations you solved in exercise 9C, you could easily convert both sides of the equation to exponential expressions with the same base.

For example, with $2^x = 16$, you saw that $2^x = 2^4$, and equating the exponents gave $x = 4$.

However, you cannot use this technique with many equations in which the variable is an exponent because it may not be as easy to convert the terms of the equation to exponential expressions with the same base.

For example, suppose you are asked to solve $2^x = 10$. Since 10 is not a rational power of 2, we need a different technique to find x. Using logarithms can help us with this.

Example 15

Use logarithms to solve each equation.

a $2^x = 10$ **b** $3^{2x} = 20$ **c** $5^{1-2x} = 50$

a $2^x = 10$ $\log 2^x = \log 10$ $x\log 2 = \log 10$	When converting to logarithms, it may be easiest to use base 10 or base e.
$x = \dfrac{\log 10}{\log 2}$	Make x the subject of the equation.
$= 3.32$	Use your calculator to find the value of $\dfrac{\log 10}{\log 2}$.
b $3^{2x} = 20$ $\ln 3^{2x} = \ln 20$ $2x \ln 3 = \ln 20$	Convert to ln on both sides of the equation (you could also use log).
$2x = \dfrac{\ln 20}{\ln 3}$	Make x the subject of the equation.
$x = \dfrac{1}{2}\left(\dfrac{\ln 20}{\ln 3}\right) = 1.36$	Use your calculator to find the value of the expression. (Note: It is better to leave the calculation until the end so you do not continue working with a rounded figure.)

c $5^{1-2x} = 50$ $\log 5^{1-2x} = \log 50$ $(1-2x)\log 5 = \log 50$ $1-2x = \frac{\log 50}{\log 5}$ $-2x = \frac{\log 50}{\log 5} - 1$ $x = -\frac{1}{2}\left(\frac{\log 50}{\log 5} - 1\right)$ $= -0.715$	Convert to logarithmic expressions on both sides. Make x the subject of the equation. Use your calculator to find the answer.

In some equations, the variable occurs more than once. In these cases, you should still convert to logarithms. However, you must also rearrange your equation so that all terms involving the variable are on one side of the equation. You can then factorize the expression and solve.

Example 16

Solve for x:

a $2 \times 3^x = 5 \times 2^x$ **b** $e^{2x} - 3e^x + 2 = 0$

a $2 \times 3^x = 5 \times 2^x$ $\log(2 \times 3^x) = \log(5 \times 2^x)$ $\log 2 + \log 3^x = \log 5 + \log 2^x$ $\log 3^x - \log 2^x = \log 5 - \log 2$ $x\log 3 - x\log 2 = \log 5 - \log 2$ $x(\log 3 - \log 2) = \log 5 - \log 2$ $x = \frac{\log 5 - \log 2}{\log 3 - \log 2} = 2.26$	Rewrite the equation using logarithms Use laws of logarithms to write as separate terms. Take terms in x to the left-hand side. Bring the exponents in front of the logarithms. Factorize the left-hand side. Make x the subject. Use your calculator to find the value of $\frac{\log 5 - \log 2}{\log 3 - \log 2}$
b $e^{2x} - 3e^x + 2 = 0$ $(e^x)^2 - 3e^x + 2 = 0$ $(e^x - 1)(e^x - 2) = 0$ $e^x = 1$ or $e^x = 2$ $x = \ln 1, \ln 2$ $x = 0, \ln 2$	Notice that $e^{2x} = (e^x)^2$. Factorize and solve the quadratic. Using the definition of a logarithm. Remember $\ln 1 = 0$.

Example 17

The radius, R cm, of a fungus on a tree can be modelled by $R(t) = 5(2.4)^{kt}$, where t is the time in days since the fungus was first observed.

a State the radius of the fungus when it was first observed.

b After four days, the radius is 8 cm. Find the value of k.

c Calculate how long it will take the for the radius of the fungus to be 10 cm.

a $R(0) = 5(2.4)^{k\times 0} = 5$ cm	This is the initial radius; when the time equals zero.
b $8 = 5(2.4)^{k\times 4}$ $1.6 = (2.4)^{4k}$	Let $t = 4$ and $R = 8$. Divide both sides by 5.
$\log 1.6 = \log(2.4)^{4k}$	Rewrite the equation using logarithms.
$\log 1.6 = 4k\log(2.4)$ $k = \frac{1}{4}\left(\frac{\log 1.6}{\log 2.4}\right) = 0.134215... = 0.134$	Use the logarithm law $\log_a(x^p) = p\log_a x$. Make k the subject. Solve.
c $10 = 5(2.4)^{0.134t}$ $2 = (2.4)^{0.134t}$ $\log 2 = \log(2.4)^{0.134t}$ $\log 2 = (0.134t)\log(2.4)$	Let $R = 10$ and $k = 0.134$, but use the **unrounded** value for k in your calculations. Divide through by 5.
$t = \frac{1}{0.134}\left(\frac{\log 2}{\log 2.4}\right) = 5.90$ days	Make t the subject and solve.

Exercise 9L

1 Solve each equation:

a $2^x = 5$ **b** $3^x = 17$ **c** $9^x = 49$

d $3^x = 69$ **e** $16^x = 67$ **f** $12^x = 5$

g $7^x = 4$ **h** $19^x = 2$ **i** $e^x = 5$

j $e^x = 10$

2 Solve each equation:

a $2^{4x} = 9$ **b** $6^{3x} = 4$ **c** $5^{\frac{1}{2}x} = 79$

d $2^{x+1} = 15$ **e** $6^{x-2} = 4$ **f** $e^{x-1} - 4 = 6$

g $2^{3x-2} = 53$ **h** $4^{2x+1} = 10$

i $11^{x-8} - 11 = 48$ **j** $9^{x+10} + 22 = 100$

3 Solve each equation:

a $6 \times 2^x = 14$ **b** $4 \times 6^{3x} = 16$

c $3 \times 4e^{2-2x} + 1 = 4$ **d** $10 - 2e^{7x+5} = 3$

e $2^{x-1} = 3^{x+1}$ **f** $3^{2x-1} = 5^x$

g $4^{3x+1} = 6^{1-2x}$ **h** $e^{x+1} = 5^{x-2}$

4 a Write the expression $3\ln 2 + \ln 3$ in the form $\ln a$.

b Hence, solve $6\ln 2 - \ln 4 = -\ln x$.

5 Solve each equation:

a $e^{2x} - 5e^x + 4 = 0$ **b** $e^{2x} - 2e^x - 3 = 0$

c $e^{4x} + 4e^{2x} - 12 = 0$

6 The height, h cm, of Sonny's tomato vine is modelled by the equation $h(t) = 10(1.075)^{kt}$, where t is the time measured in days after Sonny plants it in the garden.

a Find the height of Sonny's vine when he first plants it in the garden.
After four days, Sonny's tomato vine is 12 cm tall.

b Find the value of k.

c Calculate when the tomato vine will have doubled in height from when it was planted.

7 Mo bought a new car from a showroom. The value, $\$P$, of Mo's car is modelled by $P(t) = 20\,000(0.9)^{kt} + 1000$, where t is the number of years since Mo bought the car.

a Find the price that Mo paid for the car.

After three years, the car was valued at \$16 000.

b Find k.

Mo will sell his car when it is worth \$5000.

c Determine the number of years, after buying the car, that Mo sells it.

8 For their homework, a class of students was asked to memorize 100 Spanish words.

The average number of words remembered, W, is modelled by $W(t) = 84 - 10\ln(t + 1)$, where t is the number of days that have passed since the homework was set.

a State the average number of words remembered on the day the homework was set.

b Find the average number of words remembered ten days after the homework was set.

c Find the number of days that pass before students will, on average, remember fewer than half the words.

9 The radioactive substance plutonium has a half-life of 25 000 years.

A sample of 500 kg of plutonium is put into storage. After t years have passed, the amount of plutonium remaining in a sample, A, is modelled by:

$$A(t) = A_0(0.5)^{\frac{t}{25\,000}}$$

where A_0 is the initial amount of plutonium in the sample.

a If the sample is stored for 100 years find, to the nearest kg, how much plutonium remains when the sample is taken out of storage.

b Find, to the nearest year, how long the sample was stored if 100 kg remained when the sample was taken out.

10 Solve $2^x = 3e^{4x}$.

Developing inquiry skills

In Investigation 2 earlier in the chapter you looked at bacteria growing in a Petri dish.

Using the mathematics you have learned in this chapter, and your results from the end of section 9.1, use logarithms to determine the time at which the number of bacteria cells equals 10 000. Give your answer to the nearest minute.

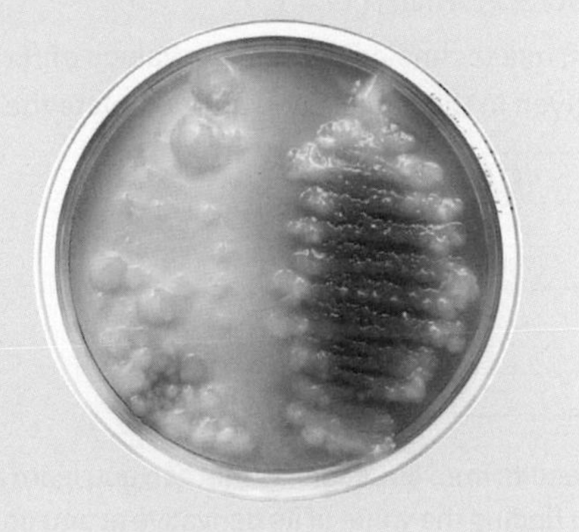

9.3 Derivatives of exponential functions and the natural logarithmic function

Now that you have studied exponential and logarithmic functions, it is time to explore the derivatives of these functions.

Investigation 9

1 Use technology to help you copy and complete the table. For each function $f(x)$, you should enter values of $f(1)$ and $f'(1)$. Round your answers to 3 decimal places.

$f(x)$	$f(1)$	$f'(1)$
2^x		
$(2.1)^x$		
$(2.2)^x$		
$(2.3)^x$		
$(2.4)^x$		
$(2.5)^x$		
$(2.6)^x$		
$(2.7)^x$		
e^x		
$(2.8)^x$		
$(2.9)^x$		
3^x		

2 What did you notice about the values of $f'(1)$ as you moved down the table?

3 What did you notice about $f'(1)$ when $f(x) = e^x$?

4 For the function $f(x) = e^x$, use technology to find the values of $f(x)$ and $f'(x)$ for the values of x given in this table. Copy and complete the table.

x	$f(x)$	$f'(x)$
1		
2		
3		
4		
5		

5 **Conceptual** Use your results from question **4**. What can you learn about the slope of the graph of e^x by finding the value of its derivative at any point?

6 **Conceptual** With a classmate, discuss what your results from questions **4** and **5** tell you about the derivative of $f(x) = e^x$.

Copy and complete this sentence: When $f(x) = e^x$, $f'(x) =$ ....

7 **Factual** How is the derivative of e^x different to the derivative of other functions?

TOK

Why is proof important in mathematics?

If $f(x) = e^x$, $f'(x) = e^x$.

Investigation 10

1 Use technology to find the gradient of $f(x) = \ln x$ for the values of x in this table.

x	1	2	3	4	5	10	$\frac{1}{2}$	$\frac{1}{3}$	$\frac{1}{4}$
$f'(x)$									

2 **Conceptual** With a classmate, look at your results from question **1** and suggest a function which is the derivative of $f(x) = \ln x$.

If $f(x) = \ln x$, $f'(x) = \frac{1}{x}$.

Example 18

Find the derivative of each function.

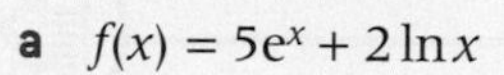

a $f(x) = 5e^x + 2\ln x$ **b** $g(x) = e^{2x}$ **c** $h(x) = e^x \ln x$ **d** $j(x) = \frac{e^x}{\ln x}$

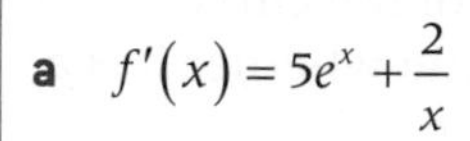

a $f'(x) = 5e^x + \frac{2}{x}$

The derivative of e^x is e^x and the derivative of $\ln x$ is $\frac{1}{x}$.

b Let $u(x) = 2x$ and $g(u) = e^u$

Then $g'(u) = e^u$ and $u'(x) = 2$

Then by the chain rule,

$g'(x) = g'(u) \times u'(x)$

$= e^u \times 2$

$= 2e^{2x}$

Set up to use the chain rule.

c Let $u(x) = e^x$ and $v(x) = \ln x$

Then $u'(x) = e^x$ and $v'(x) = \frac{1}{x}$

Then by the product rule

$h'(x) = u(x)v'(x) + u'(x)v(x)$

$= e^x \frac{1}{x} + e^x \ln x$

$= e^x\left(\frac{1}{x} + \ln x\right)$

Use the product rule.

Factorize e^x as a common factor.

d $u(x) = e^x$ and $v(x) = \ln x$

Then $u'(x) = e^x$ and $v'(x) = \frac{1}{x}$

Use the quotient rule

Continued on next page

Then by the quotient rule $j'(x)=\frac{v(x)u'(x)-u(x)v'(x)}{[v(x)]^2}$ $=\frac{\ln x\cdot(e^x)-e^x\cdot\frac{1}{x}}{(\ln x)^2}$ $=\frac{e^x\left(\ln x-\frac{1}{x}\right)}{(\ln x)^2}$	Factorize e^x as a common factor.

You will see from the use of the chain rule that if $f(x) = e^{kx}$ where k is constant, then $f'(x) = ke^{kx}$.
Also, if $f(x) = \ln kx$, then $f'(x) = \frac{1}{x}$.

TOK

In what ways might a pragmatist view the differentiation of exponentials and logarithms?

Exercise 9M

1 Find the derivative of each function:

a $7e^x$ b $-\frac{1}{4}e^x$ c $9\ln x$
d $\pi\ln x$ e $\ln 5x$ f $\ln 6x$
g $\ln 7x$ h e^{2x} i e^{4x} j e^{5x}

2 Find the derivative of each function:

a $5\ln x-2e^x$ b $x^2-e^{-\frac{1}{2}x}+\ln x$
c $4-\ln 9x+e^{-5x}+x^3$
d $\ln 7x+\ln 7+e^{7x}-7x$ e $e^{10}-5\ln x+6e^{4x}$
f $\ln(e^x x^3)$ g $\ln\left(\frac{x^2+1}{x^3-x}\right)$

3 Find the derivative of each function:

a e^{2x^3} b $e^{(4x^3+5)^2}$ c $\ln(3x^5)$
d $(\ln x)^3$ e xe^{2x} f $2x^3e^{-3x}$
g $(x^2+1)e^{3x}$ h xe^{ax^2+1}
i $x\ln x$ j $x^3\ln x$ k $x^2\ln(2x+3)$
l $\frac{e^{3x}}{x^2}$ m $\frac{2e^{4x}}{1-e^x}$ n $\frac{e^x+1}{e^x-1}$
o $\frac{x}{\ln x}$ p $\frac{2-\ln x}{x}$ q $\frac{1+\ln x}{x^2}$

4 Find the turning point of the curve $y=\ln x-x$ and determine whether it is a maximum or a minimum.

5 Find the equation of the tangent to the curve $f(x)=2e^{2x}$ at the point where $x=0$.

6 Find the equation of the normal to the curve $y=\ln x$ at the point where $x=1$.

7 Find the exact value of the gradient of the tangent to the curve $f(x)=e^{-x}+4$ at the point where $x=6$.

8 Find the exact values of the coordinates of the points on the curve $f(x)=x^2+\ln x$ where the derivative is 3.

9 Find the exact values of the coordinates of the points on the curve $f(x)=\ln(e^x+e^{-x})$ where the derivative is 0.6.

10 Find the equations of the tangent and normal lines to the curve $f(x)=xe^x-e^x$ at the point (1, 0).

Developing your toolkit

Now do the Modelling and investigation activity on page 430.

Chapter summary

- Laws of exponents:
  - $a^m \times a^n = a^{m+n}$
  - $a^m \div a^n = \dfrac{a^m}{a^n} = a^{m-n}$
  - $a^0 = 1,\ a \neq 0$
  - $a^{-m} = \dfrac{1}{a^m}$
  - $(a^m)^n = a^{mn}$
- Roots and radicals:
  - $a^{\frac{m}{n}} = \sqrt[n]{a^m}$
- An exponential function is a function of the form $f(x) = a^x$, where $a > 0,\ a \neq 1$.
- If $f(x) = a^x$, then $f^{-1}(x) = \log_a x$
- For $a > 0,\ a \neq 1$, and $c > 0$, $a^b = c$ is equivalent to $\log_a c = b$
- The graphs of exponential and logarithmic functions:

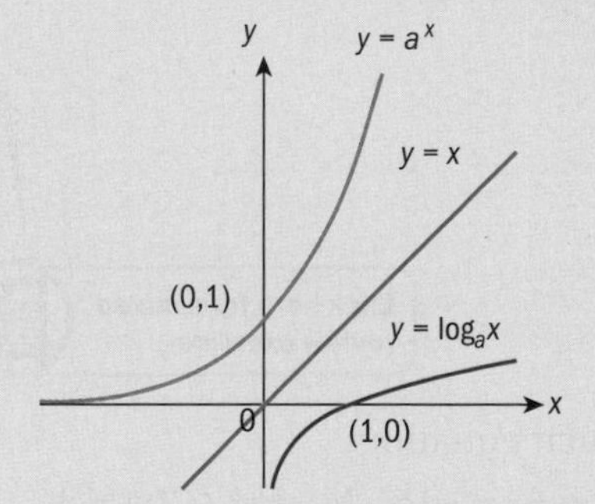

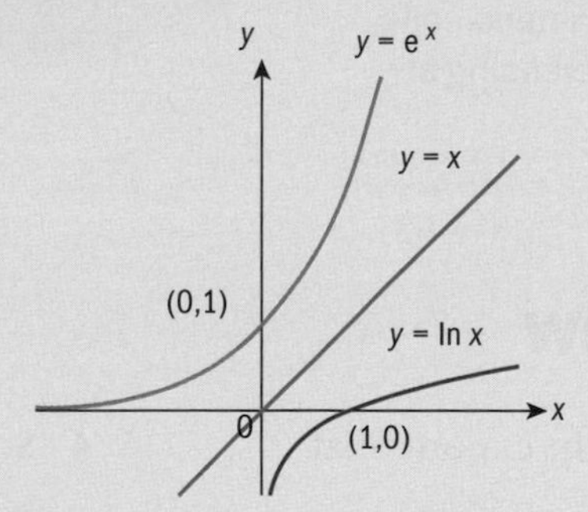

- Properties of the logarithmic function $y = \log_a x$
  - $x = 0$ is an asymptote.
  - The curve crosses the x-axis at $(1, 0)$.
  - The graph continually increases if $a > 1$, and decreases if $0 < a < 1$
  - The **domain** is the set of all *positive* real numbers.
  - The **range** is the set of all real numbers.
- Laws of logarithms:
  - $\log_a xy = \log_a x + \log_a y$
  - $\log_a \dfrac{x}{y} = \log_a x - \log_a y$
  - $\log_a x^n = n \log_a x$

Continued on next page

- $a^{\log_a x} = x$
- $\log_a(a^x) = x$

- $\log_a a = 1$
- Change of base:
  - $\log_a x = \dfrac{\log_b x}{\log_b a}$ where $a, b, x > 0$.
- The derivatives of exponential and logarithmic functions:
  - If $f(x) = e^x$, then $f'(x) = e^x$
  - If $f(x) = \ln x$, then $f'(x) = \dfrac{1}{x}$

Developing inquiry skills

In investigation 2 earlier in the chapter you looked at bacteria multiplying in a Petri dish.

The number of bacteria cells in a second Petri dish after 9 am that day is modelled by the equation $B = 24\, e^{2t-3} + \ln(3t)$.

Using the mathematics you have learned in this chapter, find the rate at which the number of bacteria cells in the second dish are increasing at 10 pm that same day.

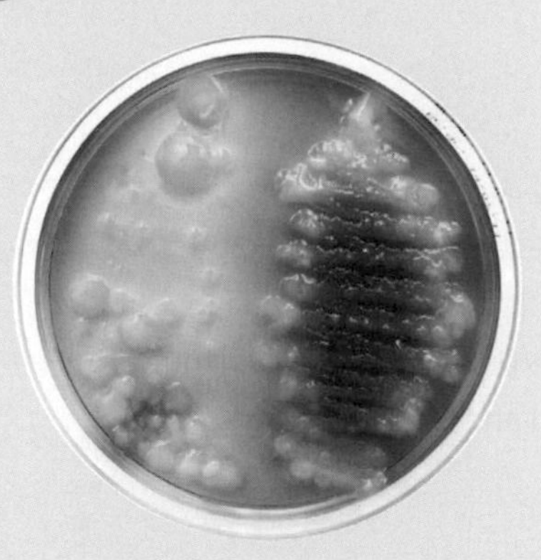

Chapter review

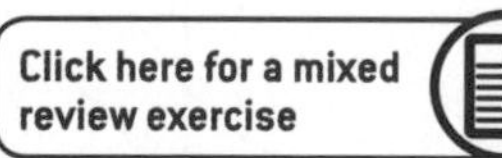

Click here for a mixed review exercise

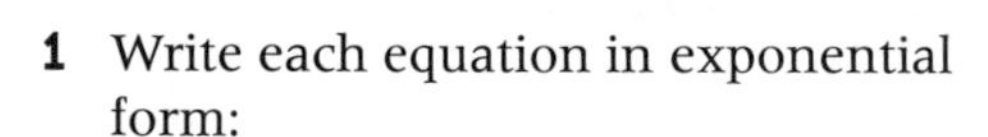

1 Write each equation in exponential form:

a $\log_2 16 = 4$ **b** $\log_5 125 = 3$

c $\log_9 81 = 2$ **d** $\log_{12} 144 = 2$

e $\log 10\,000 = 4$

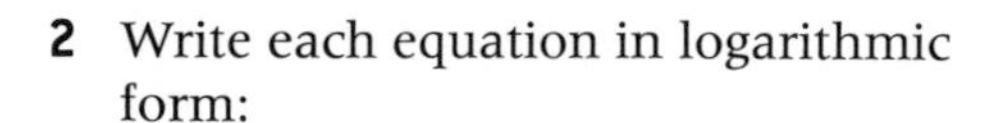

2 Write each equation in logarithmic form:

a $3^4 = 81$ **b** $15^2 = 225$

c $a^9 = \dfrac{1}{2}$ **d** $a^{14} = c$

e $e^4 = x$

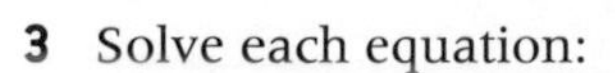

3 Solve each equation:

a $2^{2x-2} = 16$ **b** $9^{3x-1} = \dfrac{1}{27}$

c $3^{2-x} = 243$ **d** $4^{1-2x} = \dfrac{1}{64}$

4 Solve each equation:

a $15 \log 6x = -15$ **b** $\log(-7x) = 3$

c $3 \log 10x = -6$ **d** $-\log 4x = -2$

5 Without the use of a calculator, find the value of each expression:

a $\log_5 25$ **b** $\log_2 128$

c $\log_{21} 21$ **d** $\log_a a$

e $\log_6 1$ **f** $3^{\log_3(79)}$

g $\ln e^{19}$ **h** $e^{\ln 7}$

6 Solve each equation:

a $2^x = 17$ **b** $6^{6x+3} = 19$

c $2 \times 12^{3x} = 11$ **d** $6 \times 8^{-5x} = 18$

e $e^{x+5} = 13$ **f** $2e^{6x+8} = 10$

g $4e^{6x+9} - 3 = 30$ **h** $5 - 4e^{-5x-4} = -16$

7 Write each expression as a single logarithm:

a $\log 3 + \log 4$ b $\log 15 - \log 5$

c $2\log x - 5\log y$ d $8\log_5 x + 2\log_5 y$

e $\ln x + \frac{1}{2}\ln y + \ln\frac{1}{2}z$ f $4\ln x - 3\ln y - 2\ln z$

8 If $\ln a = p$ and $\ln b = q$, write each expression in terms of p and q:

a $\ln(ab)$ b $\ln a^3$

c $\ln a^2b^3$ d $\ln\frac{b^5}{a^4}$

9 Use the change of base formula to find the value of each expression:

a $\log_3 17$ b $\log_5 0.5$ c $\log_8 200$

10 Use your GDC, and a sketch, to solve $e^{x-4} = 3\ln x$.

11 a Sketch $f(x) = \log x$ and $g(x) = -2 + 5\log x$ on the same axes for $-2 \leq x \leq 8$.

b Write down the equation of the asymptote of $g(x)$.

c Describe a series of transformations which map the graph of $f(x)$ onto the graph of $g(x)$.

12 a Sketch $f(x) = \ln x$ and $g(x) = \ln(2 - x)$ on the same axes for $-4 \leq x \leq 6$.

b Write down equation of the asymptote of $g(x)$.

c Describe a transformation which maps the graph of $f(x)$ onto the graph of $g(x)$.

d Use your sketch to solve the equation $f(x) = g(x)$.

13 The graph of $f(x) = 2^x$ is reflected in both the x- and y-axes, and then translated 2 units down to become $g(x)$. Find the equation of the function $g(x)$.

14 $f(x) = \ln x$ is translated 5 units to the left and then vertically stretched by a scale factor of 3 to become $g(x)$. Find the equation of the function $g(x)$.

15 Differentiate each expression with respect to x:

a $8e^x + 7\ln x$ b e^{3x}

c $x\ln x - x$ d e^{6x^2+5x}

e $\ln(x^2 + 8)$ f $\frac{9e^x + 1}{2e^x + 1}$

g $\ln\sqrt{3x - 2}$ h $e^x \ln x$

16 Find the equation of the normal to the curve $f(x) = 4xe^{x^2-1}$ at the point (1, 4). Give your answer in the form $ax + by + c = 0$ where a, b and c are integers.

17 Find $f'(2)$ when $f(x) = x^2 + \ln x$.

18 The projected population P, in thousands, that will live in Chilltown can be modelled by $P(t) = 30e^{0.032t}$, where t is the number of years after 2020.

a State the population of Chilltown in 2020.

b Find the projected population of Chilltown in 2025.

c Determine in what year is the population of Chilltown is expected to reach 40 000.

19 A house in Chilltown was purchased for \$150 000 in 2020. The value V of the house is modelled by $V(t) = 150\,000e^{0.05875t}$, where t is the number of years after 2020. Find the year in which the value of the house will reach \$200 000.

20 Lionel is given a 500 mg dose of a medication that decays by 25% every hour. Find the amount of the drug which remains in his system after one day has passed.

Exam-style questions

21 **P1:** Consider the graph of the exponential function f defined by $f(x) = 16 - 4 \times 2^{-x}$.

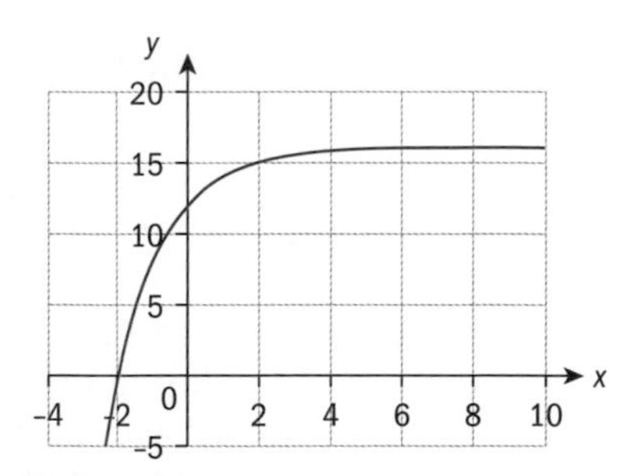

State

a the domain of f (1 mark)

b the zero of f (1 mark)

c the y-intercept of the graph of f (1 mark)

d the range of f (1 mark)

e the interval where f is positive (1 mark)

f the equation of the horizontal asymptote of f. (1 mark)

22 P2: In a research laboratory, biologists studied the growth of a culture of bacteria. From the data collected hourly, they concluded that the culture increases in number according to the formula

$$N(t) = 35 \times 1.85^t$$

where N is the number of bacteria present and t is the number of hours since the experiment began.

Use the model to calculate

a the number of bacteria present at the start of the experiment (1 mark)

b the number of bacteria present after 4 hours, giving your answer correct to the nearest whole number of bacteria (2 marks)

c the time it would take for the number of bacteria to exceed 1000. (2 marks)

Due to lack of nutrients, the culture cannot exceed 1000 bacteria.

d State the domain of validity of the model. (1 mark)

23 P2: In a controlled experiment, the temperature T ° C of a liquid, t hours after the start of the experiment, is $T = 25 + e^{0.4t}$, $0 \le t \le 12$.

a Sketch the graph of the temperature $T = T(t)$ for $0 \le t \le 12$. (2 marks)

b State the temperature halfway through the experiment, to the nearest 0.1 °C. (1 mark)

c Find the time at which the temperature of the liquid reaches 100 °C. Give your answer in hours and minutes, to the nearest minute. (3 marks)

24 P1: Let $f(x) = 3\ln(x-2)+1$.

a State the largest possible domain of f. (1 mark)

b State the equation of the vertical asymptote to the graph of f. (1 mark)

c Find the x-intercept of the graph of f. (2 marks)

d Find the equation of the tangent to the graph of f that is parallel to the line $y = x$. (4 marks)

25 P1: Simplify the following expressions, giving your answers as a single logarithm.

a $\log_2 3 + \log_2 x - \log_2(x-3)$ (2 marks)

b $3\ln x - 2\ln(x-1) + 1$ (4 marks)

26 P2: A large city is concerned about pollution and decides to look at the number of people using taxis.

At the end of the year 2010, there were 50 thousand taxis in the city. At the end of the n th year after 2010, the number of thousands of taxis, T, is given by $T_n = 50 \times 1.1^{n-1}$.

a **i** Find the number of taxis in the city at the end of 2015. (2 marks)

ii Assuming that this model remains valid, find the year in which the number of taxis is double the number of taxis at the end of 2010. (2 marks)

At the end of 2010 there were 2 million people in the city who used taxis. n years later, the number of million people P_n who use taxis is given by $P_n = 2 \times 1.01^{n-1}$.

b Assuming that this model remains valid, find the value of P at the end of 2020, giving your answer to the nearest thousand. (3 marks)

Let R be the ratio of the number of people using taxis to the number of taxis. At the end of 2010, R was 43.6 : 1.

c Find the ratio R at the end of

i 2015 (2 marks)

ii 2020. (2 marks)

d Comment on the values found in part **c**. (1 mark)

27 P1: a Show that $\ln(x^2) \neq (\ln x)^2$ for all $x > 0$. (3 marks)

b Solve the equation $(\ln x)^2 - \ln(x^2) - 15 = 0$. (5 marks)

28 P2: A cup of tea is left on a table for several hours. Initially its temperature was 94 °C. After 20 minutes, the temperature had dropped to 29°.

The temperature of the tea in the cup is modelled by $T(t) = 25 + ae^{bt}$, where t represents the time in minutes that the cup of tea has been on the table.

a Find the value of

i a (2 marks)

ii b (2 marks)

b Use the model to estimate the temperature of the tea after it has been sitting on the table for 30 minutes. (2 marks)

c Write down the equation of the horizontal asymptote of the graph of T. (1 mark)

d State the meaning of the asymptote found in **b**. (1 mark)

e Find an expression for the rate of change of the temperature of the tea after t minutes. (2 marks)

f Find an expression for $\frac{d^2T}{dt^2}$. (2 marks)

g Hence explain how the temperature of the tea will vary if the tea is left in the cup for a long time. (2 marks)

29 P2: The diagram shows three graphs.

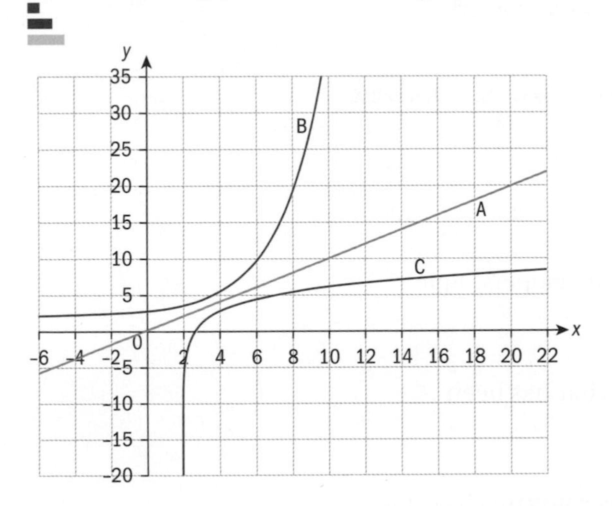

A is part of the line $y = x$.

B is part of the graph of

$$f(x) = \left(\frac{3}{2}\right)^{x-1} + 2.$$

C is the reflection of graph of

$f(x) = \left(\frac{3}{2}\right)^{x-1} + 2$ in the line $y = x$.

a State

i the y-intercept of $y = f(x)$ (1 mark)

ii the coordinates of the point where C cuts the x-axis. (1 mark)

b Determine an equation for C in the form $y = g(x)$ (4 marks)

c Write down

i the equation of the horizontal asymptote to the graph of $y = f(x)$ (1 mark)

ii the equation of the vertical asymptote to the graph of $y = g(x)$ (1 mark)

d Hence state the domain of g. (1 mark)

e Solve the equation $f(x) + g(x) = 0$ (2 marks)

30 P1: Given $f(x) = \log_2 x + \log_2 (x^2 - 1) - \log_2 (x + 1)$,

a Verify that $f(4) = \log_2 12$. (2 marks)

b Prove that $f(x) = \log_2 (x^2 - x)$ for all $x > 1$. (4 marks)

A passing fad?

Approaches to learning: Communication, Research
Exploration criteria: Presentation (A), Reflection (D), Use of mathematics (E)
IB topic: Exponentials, Logarithms

Look at the data

Fortnite was released by Epic games in July 2017 and quickly grew in popularity.

Here are some data for the total number of registered players worldwide from August 2017 to June 2018:

Date	August 2017	November 2017	December 2017	January 2018	June 2018
Months since launch, t	1	4	5	6	11
Number of registered players, P (million)	1	20	30	45	125

The data is taken from the press releases of the developers, Epic Games.

Are these data reliable?

Are there any potential problems with the data that has been collected?

What other data might be useful?

Plotting these data on your GDC or other graphing software gives this graph:

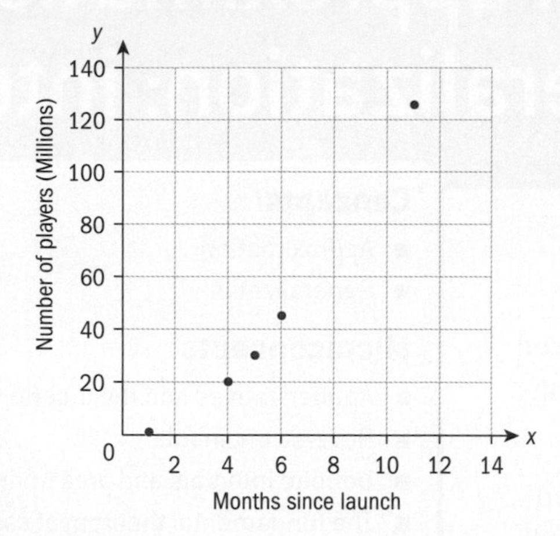

Describe the shape of the graph.

What could be some possible explanations for this growth?

Model the data

Why might it be useful to find a model that links the number of players of Fortnite against the number of months since the launch of the game?

Who might this be useful for?

Let t be the number of months since July 2017.

Let P be the number of players in millions.

Assume that the data is modelled by an exponential function of the form $p = a \cdot b^t$ where a and b are constants to be found.

Use the techniques from the chapter or previous tasks to help you.

Consider different models **by hand** and **using technology.**

How do these models differ? Why?

Which one is preferable? Why?

What alterations could be made to the model?

Use the model to predict the number of users for the current month.

Do you think this is likely to be a reliable prediction?

How reliable do you think the models are at predicting how many Fortnite players there are now? Justify your answer.

Research the number of players there are in this current month who play Fortnite.

Compare this figure with your prediction based on your model above. How big is the error?

What does this tell you about the reliability of your previous model?

Plot a new graph with the updated data and try to fit another function to this data.

Will a modified exponential model be a good fit?

If not, what other function would be a better model that could be used to predict the number of users now?

Extension

Think of another example of data that you think may currently display a similar exponential trend (or exponential decay).

Can you collect reliable and relevant data for your example?

Find data and present it in a table and a graph.

Develop a model or models for the data (ensure that your notation is consistent and your variables are defined). You could use technology or calculate by hand.

For how long do you think your model will be useful for making predictions?

Explain.

10 From approximation to generalization: integration

You can find the velocity of a moving object by taking the derivative of the displacement function. Now consider the reverse process. You can find a displacement function for a moving object, if you know the velocity function, by finding an antiderivative. The process of finding an antiderivative of a function is called integration. In this chapter you will learn about integration and about definite integrals, which can be used to find the area between two curves.

Concepts

- Approximation
- Generalization

Microconcepts

- Antiderivatives and the indefinite integral
- Reverse chain rule
- Definite integrals and area under curves
- The fundamental theorem of calculus
- Area between two curves

How can you find how much paint is needed to cover a parabolic-shaped door?

How much work is needed to stretch a spring a fixed amount?

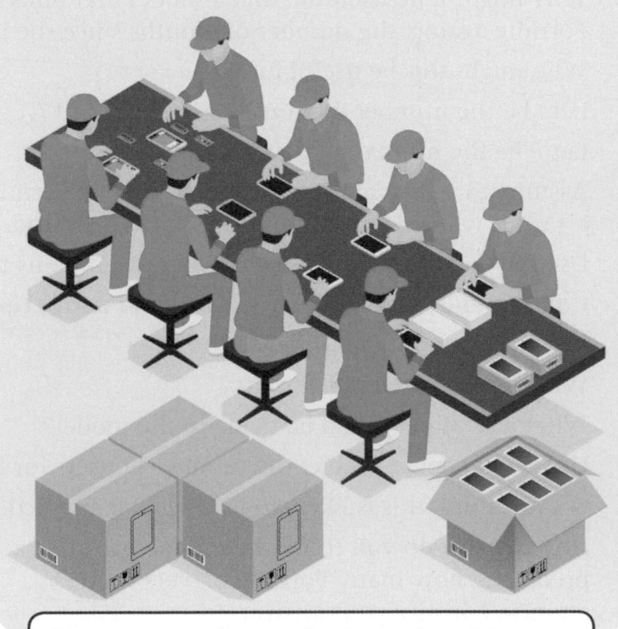

Suppose you know the function that models the rate at which a given item moves along an assembly line in metres per second. How can you find a function that models the number of metres the manual is from the start of the assembly line t seconds after it begins moving?

Piyakan wants to plant a flowering shrub in her garden, but does not want the shrub to grow higher than 1.5 m. Her best friend who is a mathematician says that a hydrangea would be ideal, based on the following functions.

The height of a particular hydrangea shrub is modelled by the function $h(t) = \frac{1}{5}e^{\frac{t}{10}-0.9}$, where h is in metres and t is measured in days for $0 \le t \le 20$. After 20 days the rate of growth of the shrub is modelled by the function $g(t) = \frac{1}{50}e^{-\frac{t}{10}+3.09}$, where g is measured in meters per day and t is measured in days since the shrub was planted.

1 According to the model, what is the height of the hydrangea shrub when it is first planted? What is its height 20 days after the hydrangea shrub is planted?

2 Find $h'(t)$. Then use $h'(t)$ to show the hydrangea bush does not exceed a height of 1.5 m for $0 \le t \le 20$.

3 Find the antiderivatives of $g(t)$.

4 The function h models the height of the hydrangea plant for $0 \le t \le 20$. Suppose that the function f models the height for $t \ge 20$. Find the function f.

5 Use the function f to show that the height of the hydrangea plant does not exceed 1.5 m for $t \ge 20$.

Developing inquiry skills

Think about the questions in this opening problem and answer any you can. As you work through the chapter, you will gain mathematical knowledge and skills that will help you to answer them all.

Before you start

Click here for help with this skills check

You should know how to:

1 Use geometric formulae to find area.

a 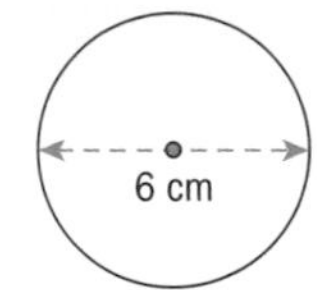

b 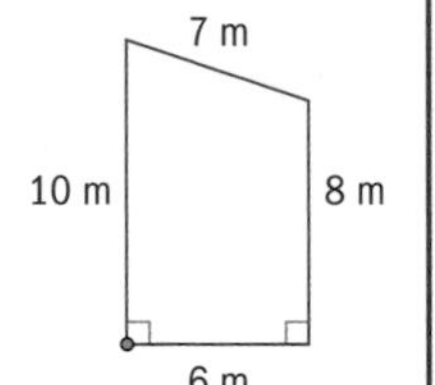

Area of a circle:

$A = \pi r^2 = \pi(3^2) = 9\pi \text{ cm}^2$

Area of a trapezium:

$A = \frac{1}{2}(a+b)h = \frac{1}{2}(10+8)(6) = 54$

2 Expand mathematical expressions.

$4x(x-3) = 4x(x) - 4x(3)$
$= 4x^2 - 12x$

3 Express radicals with rational exponents.

a $\sqrt{x} = x^{\frac{1}{2}}$ b $\sqrt[5]{x^3} = x^{\frac{3}{5}}$

Skills check

1 Find the area of each figure.

a 

b 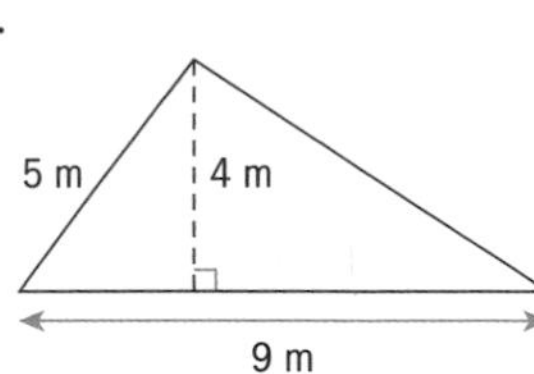

c 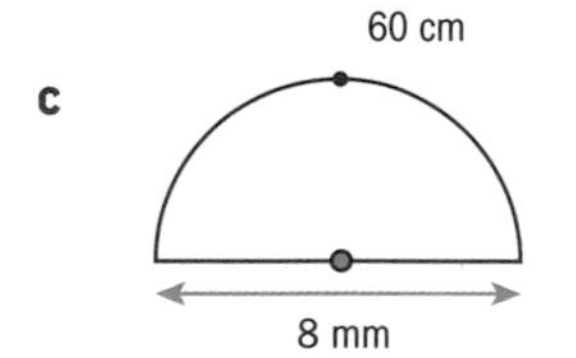

2 Expand each expression.

a $3x(x+5)$ b $(x-5)(x+5)$

c $(3x+1)^2$ d $(2x+1)(x-5)$

3 Rewrite each expression using rational exponents.

a $\sqrt[3]{x}$ b $\sqrt[7]{x^4}$

c $\sqrt{2x+5}$ d $\sqrt[3]{(x-4)^2}$

10.1 Antiderivatives and the indefinite integral

Think about the use of the term *"inverse"* in mathematics.

- Addition and subtraction are *inverse operations*. Can you name other pairs of inverse operations?
- The functions $y = 2x + 4$ and $y = \frac{1}{2}x - 2$ are *inverse functions*. What is the inverse of the function $y = e^x$?
- Suppose you know that $f(x) = 4x - 1$ and would like to find a function F so that $F'(x) = f(x)$. You are looking for an **antiderivative** of $f(x)$. To find the antiderivative of a function use the *inverse process* of finding the derivative (that is, you work backwards).

Investigation 1

1 Work backwards from $f(x) = 4x - 1$ to find a function F so that $F'(x) = f(x)$.

2 Find: **a** $\frac{d}{dx}(2x^2 - x)$ **b** $\frac{d}{dx}(2x^2 - x + 8)$ **c** $\frac{d}{dx}(2x^2 - x - 3)$

3 The functions $F(x) = 2x^2 - x$, $F(x) = 2x^2 - x + 8$ and $F(x) = 2x^2 - x - 3$ are each an **antiderivative** of $4x - 1$. Write a function to represent all possible antiderivatives of $f(x) = 4x - 1$.

4 **Conceptual** The process of finding the antiderivatives of a function leads to a family of functions. What can you say about this family of functions?

5 Consider the antiderivatives of $f(x) = x^n$, where $n \in \mathbb{Z}$. Complete a table like the one below. The first entry has been done for you.

$f(x)$	**Antiderivatives of f (Let C represent an arbitrary constant.)**
x	$\frac{1}{2}x^2 + C$
x^2	
x^3	
x^4	

6 **Factual** What is a general expression (rule) for the antiderivatives of x^n?

7 Consider the antiderivatives of $f(x) = x^n$, where $n \in \mathbb{Q}$. Show whether your rule gives correct antiderivatives for x^{-4} and $x^{\frac{1}{2}}$.

8 Are there any rational values of n for which your rule does not apply? Why?

HINT

The notation $\frac{d}{dx}$ followed by the expression for a function means to find the derivative of that function.

The function F is called an **antiderivative** of f if $F'(x) = f(x)$.

The antiderivatives of $f(x) = x^n$ are given by $F(x) = \frac{1}{n+1}x^{n+1} + C$, where $n \neq -1$ and C is an arbitrary constant.

HINT

Just as the process of finding a derivative is called **differentiation**, the process of finding an antiderivative is called **antidifferentiation.**

Example 1

Find the antiderivatives of each function.

a $f(x) = x^8$ **b** $f(x) = \frac{1}{x^3}$ **c** $f(x) = \sqrt[5]{x^3}$

a $F(x) = \frac{1}{8+1}x^{8+1} + C$	Apply the rule $\frac{1}{n+1}x^{n+1} + C$, where $n = 8$.
$= \frac{1}{9}x^9 + C$	
b $F(x) = \frac{1}{x^3} = x^{-3}$	Rewrite using a rational exponent.
$F(x) = \frac{1}{-3+1}x^{-3+1} + C$	Apply the rule $\frac{1}{n+1}x^{n+1} + C$, where $n = -3$.
$= -\frac{1}{2}x^{-2} + C$ or $-\frac{1}{2x^2} + C$	Simplify.
c $F(x) = \sqrt[5]{x^3} = x^{\frac{3}{5}}$	Rewrite using a rational exponent.
$F(x) = \frac{1}{\frac{3}{5}+1}x^{\frac{3}{5}+1} + C$	Apply the rule $\frac{1}{n+1}x^{n+1} + C$, where $n = \frac{3}{5}$.
$= \frac{1}{\left(\frac{8}{5}\right)}x^{\frac{8}{5}} + C$	Simplify.
$= \frac{5}{8}x^{\frac{8}{5}} + C$	

Exercise 10A

Find the antiderivatives of each function.

1 $f(x) = x^{10}$

2 $f(x) = x^5$

3 $f(x) = x^{25}$

4 $f(x) = x^{-6}$

5 $f(x) = \frac{1}{x^8}$

6 $f(x) = \frac{1}{x^2}$

7 $f(x) = x^{\frac{2}{3}}$

8 $f(x) = x^{\frac{1}{10}}$

9 $f(x) = x^{-\frac{1}{4}}$

10 $f(x) = \sqrt{x}$

11 $f(x) = \sqrt[4]{x^3}$

12 $f(x) = \frac{1}{\sqrt[7]{x}}$

Antidifferentiation is also known as **indefinite integration** and is denoted with the integral symbol, $\int dx$. For example, $\int x^8 dx = \frac{1}{9}x^9 + C$ means that the indefinite integral (or any antiderivative) of x^8 is $\frac{1}{9}x^9 + C$.

If $F'(x) = f(x)$, we write $\int f(x)dx = F(x) + C$.
The expression $\int f(x)dx$ is called an **indefinite integral**.
$\int f(x)dx$ is read as "the antiderivative of f with respect to x" or "the integral of f with respect to x".

Variable of integration ↓

$$\int f(x)dx = F(x) + C$$

↑ **Integrand** ↑ **Constant of integration**

These basic differentiation rules have a corresponding integration rule.

	Derivative rules	Integral rules
Power	$\frac{d}{dx}(x^n) = nx^{n-1}$	$\int x^n dx = \frac{1}{n+1}x^{n+1} + C,\ n \neq -1$
Constant multiple	$\frac{d}{dx}[kf(x)] = kf'(x)$	$\int kf(x)dx = k\int f(x)dx$
Sum or difference	$\frac{d}{dx}[f(x) \pm g(x)] = f'(x) \pm g'(x)$	$\int[f(x) \pm g(x)]dx = \int f(x)dx \pm \int g(x)dx$
Constant function	$\frac{d}{dx}(k) = 0$	$\int k\,dx = kx + C$

HINT

To think about the rule for the integral of a constant function, rewrite k as kx^0.

$$\int k\,dx = \int kx^0\,dx = k\int x^0\,dx = k \cdot \frac{1}{0+1}x^{0+1} + C = kx + C$$

TOK

Where does mathematics come from?

Galileo said that the universe is a grand book written in the language of mathematics.

Does mathematics start in our brains or is it part of the universe?

Example 2

Find each indefinite integral.

a $\int x^7 dx$ **b** $\int 6\,du$ **c** $\int 2x^3 dx$ **d** $\int (4t^3 + 6t^2 + 3)\,dt$ **e** $\int \left(x^2 + \sqrt{x}\right) dx$

a $\int x^7 dx = \frac{1}{7+1}x^{7+1} + C$ $= \frac{1}{8}x^8 + C$	Apply the power rule with $n = 7$.
b $\int 6\,du = 6u + C$	Apply the constant function rule. The du tells you the variable of integration is u.
c $\int 2x^3 dx = 2\int x^3 dx$	Apply the constant multiple rule.
$= 2\left(\frac{1}{3+1}x^{3+1} + C_1\right)$	Apply the power rule with $n = 3$.
$= \frac{1}{2}x^4 + 2C_1$ $= \frac{1}{2}x^4 + C$	$2C_1$ is some arbitrary constant C and usually only the final arbitrary constant is shown.
d $\int (4t^3 + 6t^2 + 3)\,dt = \int 4t^3 dt + \int 6t^2 dt + \int 3\,dt$	Apply the sum rule.
$= 4\int t^3 dt + 6\int t^2 dt + \int 3\,dt$	Apply the constant multiple rule.
$= 4\left(\frac{1}{3+1}t^{3+1}\right) + 6\left(\frac{1}{2+1}t^{2+1}\right) + 3t + C$ $= t^4 + 2t^3 + 3t + C$	Apply the power rule and constant rule, with variable of integration t.
e $\int \left(x^2 + \sqrt{x}\right) dx = \int \left(x^2 + x^{\frac{1}{2}}\right) dx$	Rewrite x using rational exponents.
$= \frac{1}{2+1}x^{2+1} + \frac{1}{\frac{1}{2}+1}x^{\frac{1}{2}+1} + C$ $= \frac{1}{3}x^3 + \frac{2}{3}x^{\frac{3}{2}} + C$	Using the sum rule, apply the power rule to each term.

Exercise 10B

Find the indefinite integral in questions **1–10**. You can check answers to definite integrals by differentiating your answer and checking to see that the derivative is the same as the integrand.

1 $\int x^4 dx$

2 $\int (6x^2 + 4x + 5)\,dx$

3 $\int (15t^4 + 12t^3 + 2t + 5)\,dt$

4 $\int 8\,dx$

5 $\int \frac{1}{u^7} du$

HINT

$\int du = \int 1\,du = \int u^0 du$

6 $\int \frac{2}{x^5} dx$

7 $\int \left(w^3 + \sqrt[3]{w}\right) dw$

8 $\int \left(4\sqrt{x} + 3\right) dx$

9 $\int \sqrt[9]{x^5} dx$

10 $\int du$

11 Given that $f(x) = x^5 + \frac{3}{x^2}$, find:

a $f'(x)$ **b** $\int f(x)\,dx$

12 Given that $\int (3x^2 + px + q)\,dx = x^3 + 8x^2 + 7x + C$, find the values of p and q.

Calculus

Investigation 2

1 In investigation 1, question **8**, you saw that there is a value of n for which the power rule, $\int x^n \mathrm{d}x = \frac{1}{n+1}x^{n+1} + C$, does not apply. What value of n is this?

2 For the value of n you identified in question **1**, x^n is the derivative of a certain function you have already met. What function is this?

3 Use question **2** to find $\int x^n \mathrm{d}x$ for this value of n.

4 Think about a function that is its own derivative. For what function does $\int f(x)\,\mathrm{d}x = f(x) + C$? (Hint: Think about a function that is its own derivative.)

HINT

To answer the questions in investigation 2, refer to the derivative rules in the Chapter 9 summary.

HINT

For $x > 0$, $\frac{\mathrm{d}}{\mathrm{d}x}(\ln x) = \frac{1}{x}$. So, for $x > 0$, $\int \frac{1}{x}\mathrm{d}x = \ln x + C$.

For $x < 0$, $-x > 0$ and $\frac{\mathrm{d}}{\mathrm{d}x}(\ln(-x)) = \frac{1}{-x}\cdot -1 = \frac{1}{x}$. So, for $x < 0$, $\int \frac{1}{x}\mathrm{d}x = \ln(-x) + C$.

These two statements are combined as $\int \frac{1}{x}\mathrm{d}x = \ln|x| + C$.

$$\int \frac{1}{x}\mathrm{d}x = \ln|x| + C$$

$$\int \mathrm{e}^x \mathrm{d}x = \mathrm{e}^x + C$$

Example 3

Find the indefinite integral.

a $\int \frac{3}{x}\mathrm{d}x$ **b** $\int \frac{\mathrm{e}^x}{5}\mathrm{d}x$

a $\int \frac{3}{x}\mathrm{d}x = 3\int \frac{1}{x}\mathrm{d}x$	Apply the constant multiple rule.
$= 3\ln\|x\| + C$	Use the rule $\int \frac{1}{x}\mathrm{d}x = \ln\|x\| + C$.
b $\int \frac{\mathrm{e}^x}{5}\mathrm{d}x = \frac{1}{5}\int \mathrm{e}^x \mathrm{d}x$	Apply the constant multiple rule.
$= \frac{1}{5}\mathrm{e}^x + C$	Use the rule $\int \mathrm{e}^x \mathrm{d}x = \mathrm{e}^x + C$.

To work out some integrals, you may need to rewrite the integrand by expanding brackets, separating terms or simplifying.

Example 4

Find the indefinite integral.

a $\int (2x^2 + 3)^2 \mathrm{d}x$ **b** $\int \frac{4x^2 + 2x + 3}{x}\mathrm{d}x$ **c** $\int \ln(\mathrm{e}^{4t})\,\mathrm{d}t$

a $\int (2x^2+3)^2 \mathrm{d}x = \int (4x^4 + 12x^2 + 9)\mathrm{d}x$	Expand and integrate each term.
$= \frac{4}{5}x^5 + 4x^3 + 9x + C$	

b $\int \frac{4x^2+2x+3}{x}\,\mathrm{d}x = \int\left(\frac{4x^2}{x}+\frac{2x}{x}+\frac{3}{x}\right)\mathrm{d}x$ $= \int\left(4x+2+\frac{3}{x}\right)\mathrm{d}x$ $= 2x^2+2x+3\ln\lvert x\rvert + C$	Separate the terms. Simplify and then integrate each term.
c $\int \ln\left(\mathrm{e}^{4t}\right)\mathrm{d}t = \int 4t\,\mathrm{d}t$ $= 2t^2 + C$	Simplify $\ln(\mathrm{e}^{4t})$, using the fact that e^x and $\ln x$ are inverses.

Exercise 10C

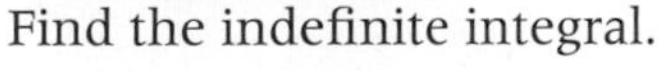

Find the indefinite integral.

1 $\int \frac{6}{x}\,\mathrm{d}x$

2 $\int 5\mathrm{e}^u\,\mathrm{d}u$

3 $\int \frac{1}{2x}\,\mathrm{d}x$

4 $\int \frac{\mathrm{e}^x}{3}\,\mathrm{d}x$

5 $\int (3x+2)^2\,\mathrm{d}x$

6 $\int \ln(\mathrm{e}^{x+1})\,\mathrm{d}x$

7 $\int (t^2(t+3))\,\mathrm{d}t$

8 $\int \mathrm{e}^{\ln(3x)}\,\mathrm{d}x$

9 $\int \frac{x^4+3x^2+2x}{x}\,\mathrm{d}x$

10 $\int \frac{\mathrm{e}^u-4}{2}\,\mathrm{d}u$

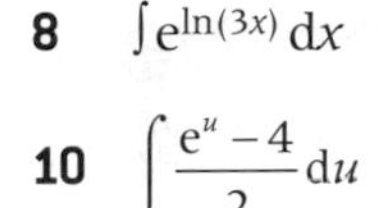

Developing inquiry skills

In the opening scenario for this chapter, the function $g(t) = \frac{1}{50}e^{-\frac{t}{10}+3.09}$ models the rate of growth of the hydrangea bush in meters per day for $t \geq 20$ days. Find the antiderivatives of the function g.

10.2 More on indefinite integrals

You have already learned how to find the integrals of x^n, e^x and $\frac{1}{x}$.

In this section you will consider integrals of the composites of these functions with the linear function $ax+b$, that is, $(ax+b)^n$, e^{ax+b} and $\frac{1}{ax+b}$.

Investigation 3

1 Show that $\int x^4\,\mathrm{d}x = \frac{1}{5}x^5 + C$, by finding $\frac{\mathrm{d}}{\mathrm{d}x}\left(\frac{1}{5}x^5+C\right)$.

2 Show that $\int (2x+3)^4\,\mathrm{d}x$ is NOT equal to $\frac{1}{5}(2x+3)^5 + C$ by using the chain rule to find the derivative of $\frac{\mathrm{d}}{\mathrm{d}x}\left[\frac{1}{5}(2x+3)^5+C\right]$.

3 Think of reversing the chain rule to find the following. Verify your answers by finding the derivative.

a $\int (2x+3)^4\,\mathrm{d}x$ **b** $\int (8x+4)^3\,\mathrm{d}x$ **c** $\int (-5x+7)^9\,\mathrm{d}x$

Continued on next page

Calculus

4 a **Factual** What is $\int (ax+b)^n \mathrm{d}x$?

b Verify your answer by differentiating it. Do you get $(ax+b)^n$?

If you need help to answer questions **5** and **6**, choose some of your own values for a and b. Then you will have specific functions to work with, as you did in question **3**.

5 a **Factual** What is $\int \mathrm{e}^{ax+b} \mathrm{d}x$?

b Verify your answer by differentiating it. Do you get e^{ax+b}?

6 a **Factual** What is $\int \frac{1}{ax+b} \mathrm{d}x$?

b You know that $\frac{\mathrm{d}}{\mathrm{d}x}(\ln x) = \frac{1}{x}$. It is also true that $\frac{\mathrm{d}}{\mathrm{d}x}(\ln|x|) = \frac{1}{x}$. Use this fact and the chain rule to verify your answer by differentiating it. Do you get $\frac{1}{ax+b}$?

7 **Conceptual** What rule of differentiation can be reversed to help you integrate composite functions of the form $f(ax+b)$?

$\int f(ax+b)\,\mathrm{d}x = \frac{1}{a}F(ax+b)+C$, where $\int f(x)\,\mathrm{d}x = F(x)$.

- For $f(x) = x^n$: $\int (ax+b)^n \,\mathrm{d}x = \frac{1}{a(n+1)}(ax+b)^{n+1} + C$
- For $f(x) = \mathrm{e}^x$: $\int \mathrm{e}^{ax+b}\mathrm{d}x = \frac{1}{a}\mathrm{e}^{ax+b} + C$
- For $f(x) = \frac{1}{x}$: $\int \frac{1}{ax+b}\mathrm{d}x = \frac{1}{a}\ln|ax+b| + C$

TOK

Consider

$f(x) = \frac{1}{x}$, $1 \le x \le \infty$

An infinite area sweeps out a finite volume. Can this be reconciled with our intuition? What does this tell us about mathematical knowledge?

Example 5

Find the indefinite integral.

a $\int (4x-5)^6 \mathrm{d}x$ b $\int \frac{1}{3x+2}\mathrm{d}x$ c $\int 2\mathrm{e}^{5x-1}\mathrm{d}x$ d $\int \frac{1}{(5x+2)^4}\mathrm{d}x$

a $\int (4x-5)^6\,\mathrm{d}x = \frac{1}{4\cdot 7}(4x-5)^7 + C$ $= \frac{1}{28}(4x-5)^7 + C$	Find $\frac{1}{a(n+1)}(ax+b)^{n+1} + C$ for $a=4$, $b=-5$ and $n=6$. Check by differentiating. $\frac{\mathrm{d}}{\mathrm{d}x}\left[\frac{1}{28}(4x-5)^7\right] = \frac{1}{28}\left(7(4x-5)^6(4)\right) = (4x-5)^6$
b $\int \frac{1}{3x+2}\mathrm{d}x = \frac{1}{3}\ln|3x+2| + C$	Find $\int \frac{1}{ax+b}\mathrm{d}x = \frac{1}{a}\ln|ax+b| + C$ for $a=3$ and $b=2$. Check by differentiating. $\frac{\mathrm{d}}{\mathrm{d}x}\left[\frac{1}{3}\ln|3x+2|\right] = \frac{1}{3}\left(\frac{1}{3x+2}(3)\right) = \frac{1}{3x+2}$

c $\int 2e^{5x-1}dx = 2\int e^{5x-1}dx$ $= 2\left(\frac{1}{5}e^{5x-1}\right) + C$ $= \frac{2}{5}e^{5x-1} + C$	Apply the constant multiple rule. Find $\int e^{ax+b}dx = \frac{1}{a}e^{ax+b} + C$ for $a = 5$ and $b = -1$. Check by differentiating. $\frac{d}{dx}\left[\frac{2}{5}e^{5x-1}\right] = \frac{2}{5}\left(e^{5x-1} \cdot 5\right) = 2e^{5x-1}$
d $\int \frac{1}{(5x+2)^4}dx = \int (5x+2)^{-4}dx$ $= \frac{1}{5\cdot(-4+1)}(5x+2)^{-4+1} + C$ $= -\frac{1}{15(5x+2)^3} + C$	Rewrite using rational exponents. Find $\frac{1}{a(n+1)}(ax+b)^{n+1} + C$ for $a = 5$, $b = 2$ and $n = -4$. Check by differentiating. $\frac{d}{dx}\left[-\frac{1}{15(5x+2)^3}\right] = \frac{d}{dx}\left[-\frac{1}{15}(5x+2)^{-3}\right]$ $= -\frac{1}{15}\left(-3(5x+2)^{-4}(5)\right) = \frac{1}{(5x+2)^4}$

Exercise 10D

Find the indefinite integral in questions **1–12**.

1 $\int (7x-5)^4 dx$

2 $\int (-3x+7)^6 dx$

3 $\int \frac{1}{10x+13}dx$

4 $\int e^{-4x+3}dx$

5 $\int 4(5x+1)^3 dx$

6 $\int \frac{2}{3x+8}dx$

7 $\int 3e^{4-2x}dx$

8 $\int 7(2x-9)^4 dx$

9 $\int \frac{1}{(4x+3)^2}dx$

10 $\int (2x+1)^{\frac{1}{3}}dx$

11 $\int \left(e^{5x} + \frac{8}{5x-3}\right)dx$

12 $\int \frac{1}{\sqrt[3]{4x+7}}dx$

13 Given that $f(x) = (3x+10)^5$, find the following:

a $f'(x)$ **b** $\int f(x)dx$

14 Given that $f(x) = \frac{1}{12x+7}$, find the following:

a $f'(x)$ **b** $\int f(x)\,dx$

HINT

You can check answers to indefinite integrals by differentiating your answer and checking that the derivative equals the integrand.

Investigation 4

Suppose that an object is moving in a straight line.

1 Given the displacement function for the object, how do you find the velocity function? (Hint: look back at chapter 5, section 5.5.)

2 **Factual** Given the velocity function, how would you find the displacement function?

3 Suppose that an object is travelling with velocity $v(t) = 6t + 3$. Find the displacement function, s, for the object.

Continued on next page

4 Is knowing the derivative of a certain function enough information to find that particular function? Why or why not?

5 Suppose at $t = 2$, the displacement of the object with velocity $v(t) = 6t + 3$ is $s = 4$. Find the displacement function for the object.

6 **Conceptual** In general, given a derivative of a certain function, what else do you need to know to find that particular function?

Suppose that $f'(x) = 4x + 3$ and you want to find f.

- You know that $f(x) = \int (4x + 3)\,dx = 2x^2 + 3x + C$.
  So, a general solution is $f(x) = 2x^2 + 3x + C$.
- To find a particular solution for f, you need to know a **boundary condition,** such as $f(1) = 9$. This allows you to find C.
  $f(1) = 2(1)^2 + 3(1) + C = 9 \Rightarrow C = 4$ and thus $f(x) = 2x^2 + 3x + 4$

Sometimes a boundary condition is given as an **initial condition**. An initial condition tells you the value of the dependent variable when the independent variable is zero.

HINT

If you are given a velocity function and a boundary condition, you can integrate to find the displacement function.

In the same way, if you are given an acceleration function and a boundary condition, you can integrate to find the velocity function.

Example 6

a If $f'(x) = (2x - 3)^3$ and $f\left(\frac{1}{2}\right) = 4$, find $f(x)$.

b The curve $y = g(x)$ passes through the point $(2, 5e^8)$. The gradient of the curve is given by $g'(x) = 8e^{4x}$. Find the equation of the curve.

c The rate of change of the population of fish in a lake is given by $\frac{dP}{dt} = 60\sqrt{t}$ for $0 \le t \le 4$ years. The initial population was 140 fish. Find the number of fish at $t = 4$ years.

a $f'(x) = (2x - 3)^3$	
$f(x) = \int (2x - 3)^3\,dx$	Integrate to find the general solution.
$f(x) = \frac{1}{8}(2x-3)^4 + C$	$\int (2x-3)^3\,dx = \frac{1}{2\cdot 4}(2x-3)^4 + C$
$4 = \frac{1}{8}\left(2\cdot\frac{1}{2} - 3\right)^4 + C$	Use the fact that $f\left(\frac{1}{2}\right) = 4$ to find C.
$C = 2$	
$f(x) = \frac{1}{8}(2x-3)^4 + 2$	Substitute the value of C into the general solution.
b $g'(x) = 8e^{4x}$	
$g(x) = \int 8e^{4x}\,dx$	Integrate to find the general solution.
$g(x) = 2e^{4x} + C$	$\int 8e^{4x}\,dx = 8\cdot\frac{1}{4}e^{4x} + C$
$5e^8 = 2\cdot e^{4\cdot 2} + C$	Use the point $(2, 5e^8)$ to find C.
$C = 3e^8$	
$g(x) = 2e^{4x} + 3e^8$	Substitute the value of C into the general solution.

c $\frac{dP}{dt} = 60\sqrt{t} = 60t^{\frac{1}{2}}$	Write in exponent form.
$P(t) = \int 60t^{\frac{1}{2}}dt$ $P(t) = 40t^{\frac{3}{2}} + C$	Integrate to find the general solution. $\int 60t^{\frac{1}{2}}\,dt = 60\cdot\frac{2}{3}t^{\frac{3}{2}} + C$
$140 = 40(0)^{\frac{3}{2}} + C$ $C = 140$ $P(t) = 40t^{\frac{3}{2}} + 140$	The initial population was 140 fish, so $P(0) = 140$. Use this to find C.
$P(4) = 40(4)^{\frac{3}{2}} + 140$ $= 460$ There are 460 fish when $t = 4$ years.	Find P when $t = 4$.

Exercise 10E

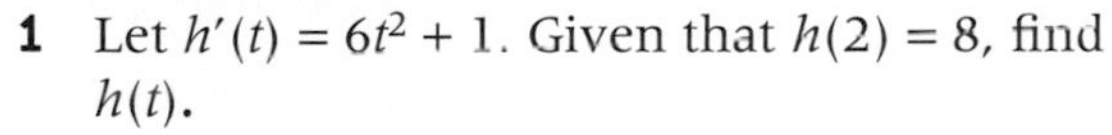

1 Let $h'(t) = 6t^2 + 1$. Given that $h(2) = 8$, find $h(t)$.

2 It is given that $\frac{dy}{dx} = 8(2x-3)^3$ and that $y = 6$ when $x = 2$. Find y in terms of x.

3 The acceleration, a m s^{-2}, of particle moving in a straight line is given by $a(t) = 4t + 1$ at time t seconds. The initial velocity of the particle is 2 m s^{-1} and its initial displacement is 8 m.

a Find an expression for the velocity of the particle.

b Find an expression for the displacement of the particle.

4 A particle moving in a straight line has velocity v m s^{-1} and displacement s m at time t seconds. The particle's velocity is given by $v(t) = 4e^{2t} + t$.

a Find the acceleration when $t = 3$.

When $t = 0$, the displacement, s, of the particle is 4 m.

b Find an expression for the displacement of the particle in terms of t.

5 The derivative of the function f is given by $f'(x) = \frac{1}{8x-7}$. The graph of f passes through the point $\left(1, \frac{7e}{8}\right)$. Find an expression for $f(x)$.

Developing inquiry skills

In the opening scenario for this chapter, you were asked to find the function f that models the height or the hydrangea bush for $t \geq 20$ and to show that the height of the hydrangea bush does not exceed 1.5 m. You should now be able to answer these questions.

10.3 Area and definite integrals

You know that an indefinite integral, $\int f(x)\mathrm{d}x$, gives a family of functions that differ by a constant.

In this section you will learn about the **definite** integral, $\int_a^b f(x)\mathrm{d}x$. Although these integrals share a common symbol, the definite integral gives you a **number** rather than a function. You will learn about the connection between these two types of integrals in section 10.4.

$\int_a^b f(x)\mathrm{d}x$ is read as "the integral from a to b of f with respect to x", where a and b are the limits of integration. You will use your GDC to evaluate definite integrals in this section.

Investigation 5

1 Use a geometric formula to find the area of the shaded regions shown in the following diagrams.

a

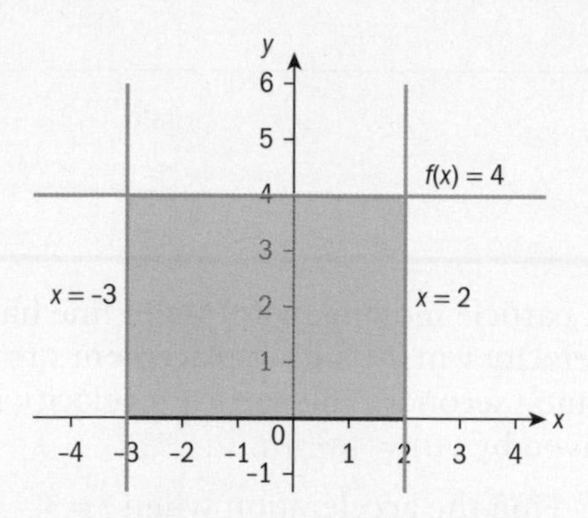

b

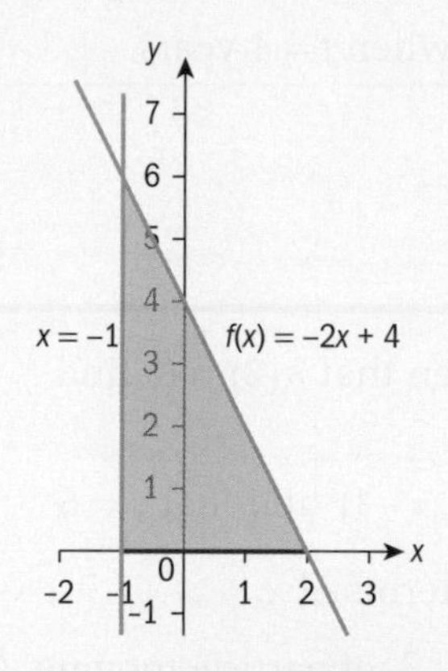

c

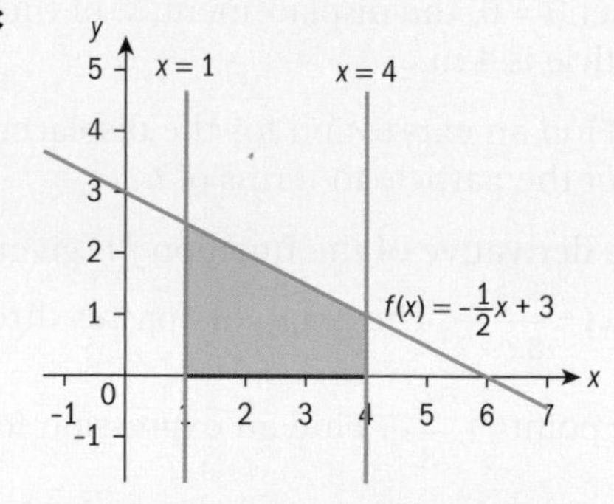

d

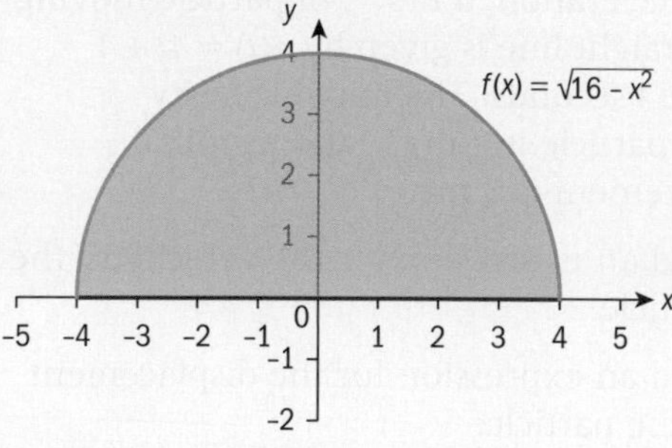

2 Use a GDC to evaluate the following definite integrals.

a $\int_{-3}^{2} 4\,\mathrm{d}x$ **b** $\int_{-1}^{2} (-2x+4)\,\mathrm{d}x$ **c** $\int_{1}^{4}\left(-\frac{1}{2}x+3\right)\mathrm{d}x$ **d** $\int_{-4}^{4}\left(\sqrt{16-x^2}\right)\mathrm{d}x$

3 What do you notice about your answers to questions **1** and **2**?

> **HINT**
>
> The area between the graph of a function f and the x-axis is called the **area under the curve**. In mathematics, a **curve** is a graph on a coordinate plane, so curves include straight lines. In question **1a**, the shaded region is described as the area under the curve $f(x)=4$ between $x=-3$ and $x=2$.

4 **Conceptual** What does a definite integral represent?

Example 7

Consider the shaded region in the diagram on the right.

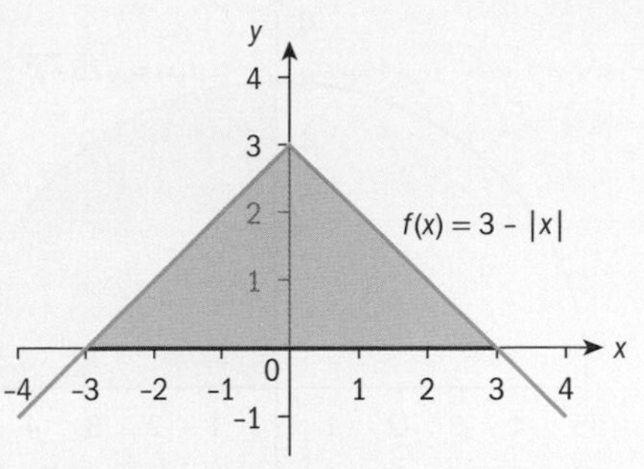

a Write down a definite integral that gives the area of the shaded region, and evaluate it using your GDC.

b Find the area using a geometric formula to verify your answer in part **a**.

a $\int_{-3}^{3}(3-\lvert x\rvert)\,dx = 9$	The shaded region is under the curve $f(x)=3-\lvert x\rvert$, between $x=-3$ and $x=3$.
b Area $=\frac{1}{2}(6\times 3)=9$	Use the area formula for a triangle with a base of 6 and a height of 3.

Exercise 10F

Consider the shaded region in the following diagrams.

a Write down a definite integral that gives the area of the shaded region, and evaluate your definite integral using a GDC.

b Find the shaded area using geometric formulae to verify your answers to part **a**.

1

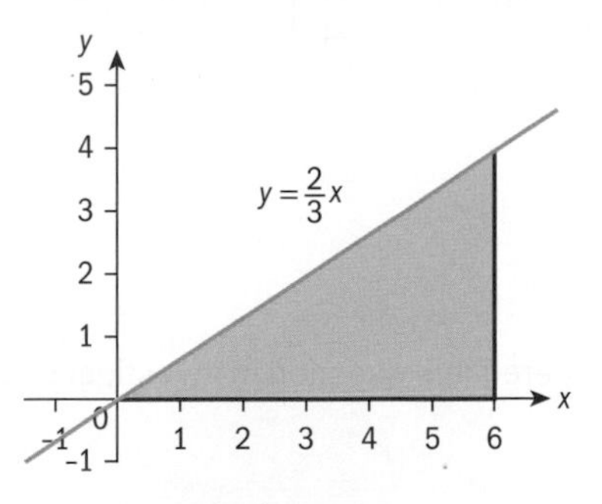

2

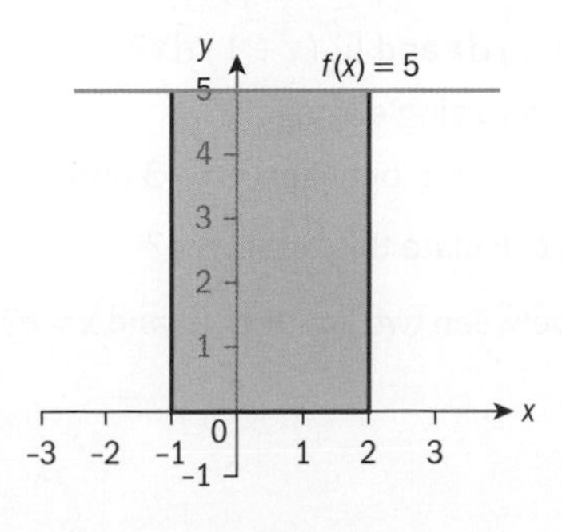

3

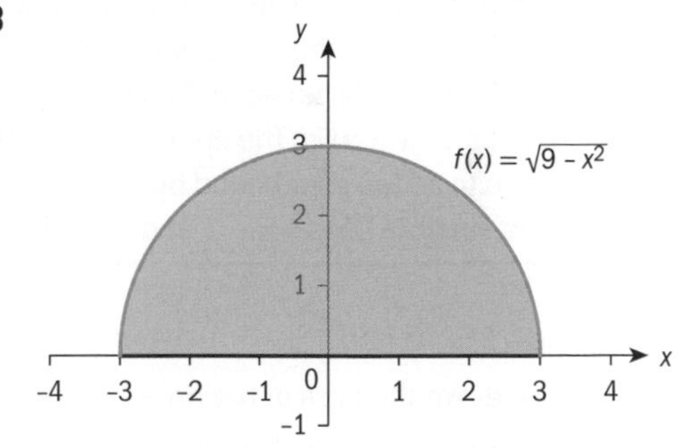

4

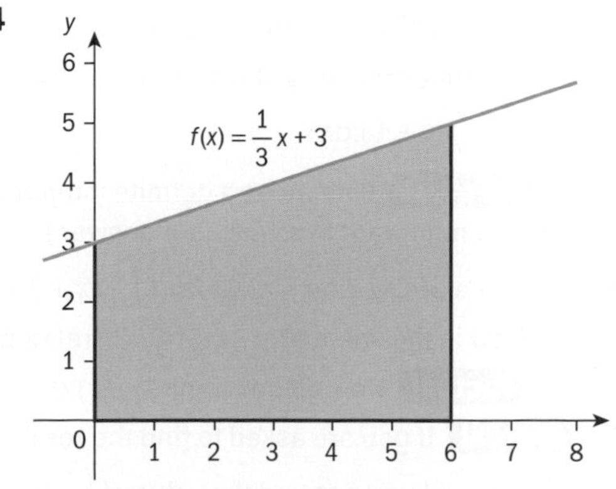

5

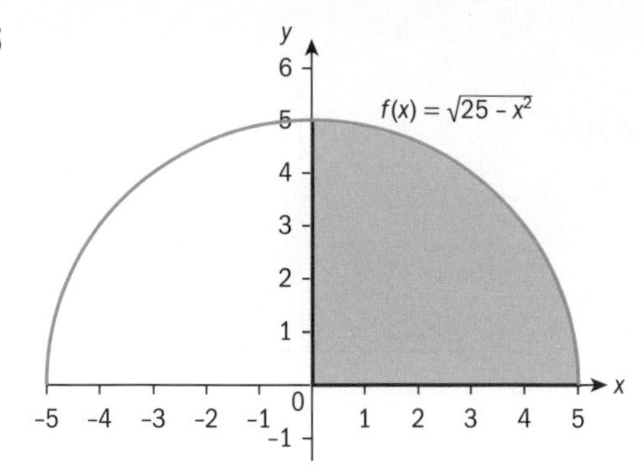

6

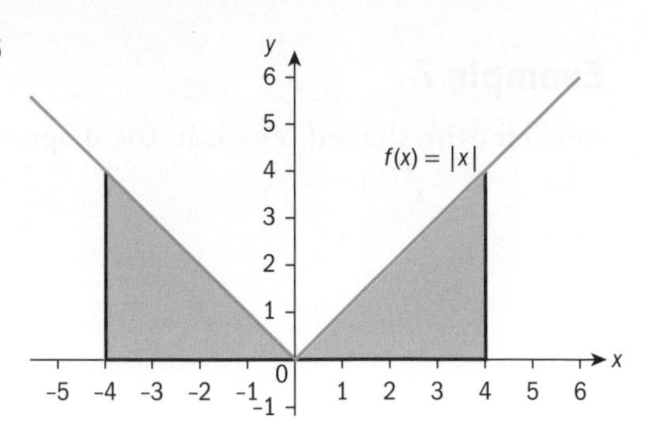

Investigation 6

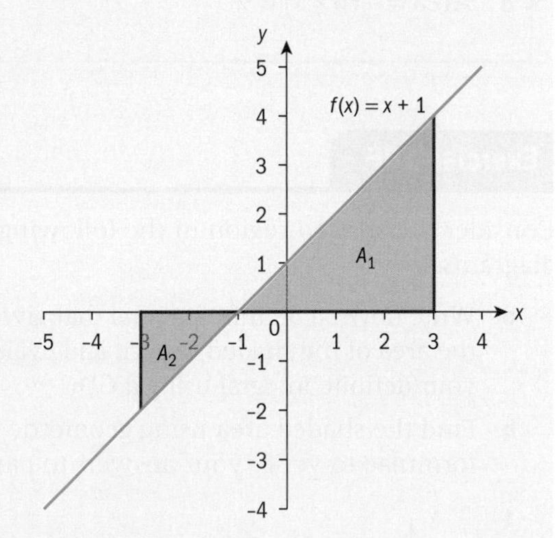

The diagram on the right shows part of the graph of $f(x) = x + 1$ and two triangular regions. Let A_1 represent the area under f from $x = -1$ to $x = 3$, and A_2 represent the area under f from $x = -3$ to $x = -1$.

HINT

Even though the region with area A_2 is below the x-axis, it is still called the area under the curve because the region is between the graph of the function f and the x-axis. The area A_2 could also be referred to as the area bound by the graph of f, the x-axis, and the lines $x = -3$ and $x = -1$.

1 a Use a GDC to find $\int_{-1}^{3} (x+1)\,dx$.

b Write down the area of region A_1.

2 a Use geometric formula to find the area of triangular region A_2.

b Use a GDC to find the value of $\int_{-3}^{-1} (x+1)\,dx$.

c How does the area of region A_2 compare to the value of $\int_{-3}^{-1} (x+1)\,dx$?

3 **Conceptual** What does a definite integral represent? (You answered this question in investigation 5, but you may want to refine your answer.)

4 a Use a GDC to find the value of $\int_{-3}^{3} (x+1)\,dx$.

b How is the value of $\int_{-3}^{3} (x+1)\,dx$ related to the values of $\int_{-3}^{-1} (x+1)\,dx$ and $\int_{-1}^{3} (x+1)\,dx$?

c **Factual** How do you think $\int_{a}^{c} f(x)\,dx + \int_{c}^{b} f(x)\,dx$ can be written as a single integral?

5 **Factual** If you are asked to find the **total area** under the graph $f(x) = x + 1$ between $x = -3$ and $x = 3$, why should you **not** find the value of $\int_{-3}^{3} (x+1)\,dx$? How would you calculate this total area?

6 **Conceptual** How could you find the total area under a curve $f(x)$ between two limits $x = a$ and $x = b$?

If f is a non-negative function for $a \le x \le b$, then $\int_a^b f(x)\,dx$ represents the area under the curve f from $x = a$ to $x = b$

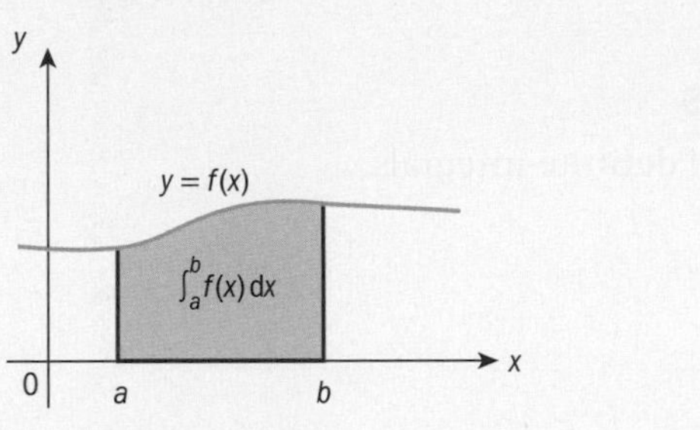

HINT

In the IB formula booklet, the formula for the area under a curve between $x = a$ and $x = b$ is given as $A = \int_a^b |y|\,dx$. The absolute value reflects any points below the x-axis on the graph of $y = f(x)$ in the x-axis. The non-negative absolute value function preserves the area of the region enclosed by the curve and the x-axis.

Example 8

The graph of $y = f(x)$ consists of four line segments as shown in the following diagram.

Evaluate $\int_0^{12} f(x)\,dx$ using geometric formulae.

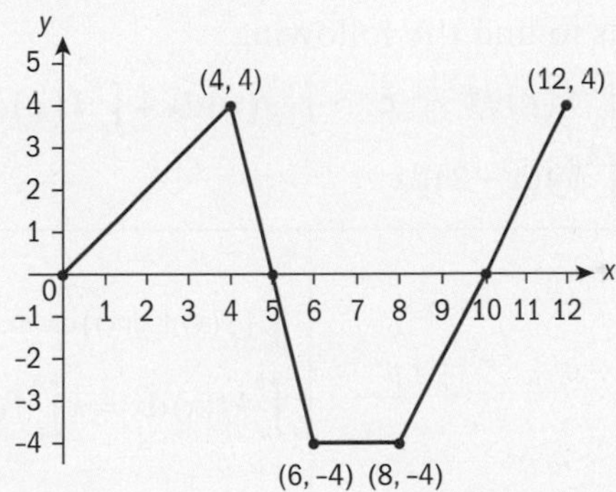

a

$$\int_0^{12} f(x)\,dx = \int_0^5 f(x)\,dx + \int_5^{10} f(x)\,dx + \int_{10}^{12} f(x)\,dx$$

$$= A_1 - A_2 + A_3$$

$$= \frac{1}{2}(5)(4) - \frac{1}{2}(5+2)(4) + \frac{1}{2}(2)(4)$$

$$= 10 - 14 + 4$$

$$= 0$$

Find the area of triangle A_1, minus the area of trapezium A_2, plus the area of triangle A_3.

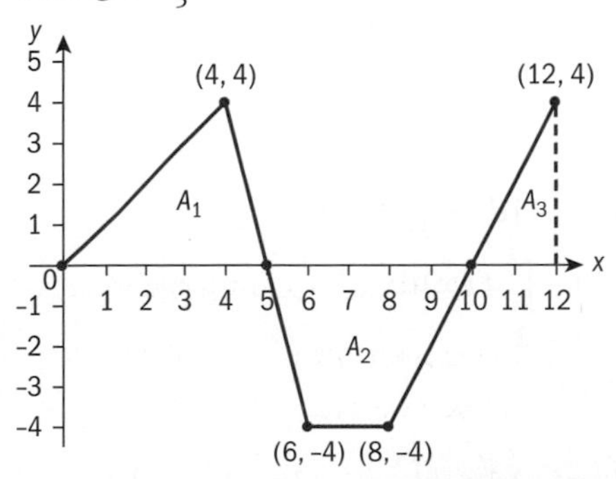

In the previous example, you used the property $\int_a^b f(x)\,dx = \int_a^c f(x)\,dx + \int_c^b f(x)\,dx$ and the relationship between definite integrals and area.

TOK

Is imagination more important than knowledge?

Calculus

Reflect In Example 8, the total area under the curve $y=f(x)$ between $x=0$ and $x=12$ is **not** zero. How could you sum three definite integrals to find this total area?

Properties of definite integrals

Here are some additional properties of definite integrals.

$$\int_a^b kf(x)\,dx = k\int_a^b f(x)\,dx$$

$$\int_a^b (f(x) \pm g(x))\,dx = \int_a^b f(x)\,dx \pm \int_a^b g(x)\,dx$$

$$\int_a^a f(x)\,dx = 0$$

$$\int_a^b f(x)\,dx = -\int_b^a f(x)\,dx$$

$$\int_a^b f(x)\,dx = \int_a^c f(x)\,dx + \int_c^b f(x)\,dx$$

Example 9

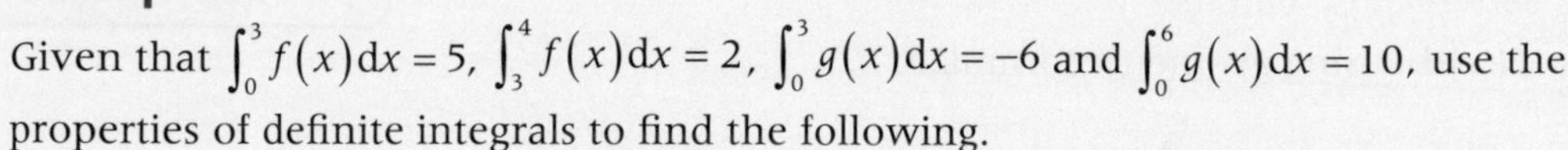

Given that $\int_0^3 f(x)\,dx = 5$, $\int_3^4 f(x)\,dx = 2$, $\int_0^3 g(x)\,dx = -6$ and $\int_0^6 g(x)\,dx = 10$, use the properties of definite integrals to find the following.

a $\int_0^3 (2f(x)+g(x))\,dx$ **b** $\int_0^4 f(x)\,dx$ **c** $\int_3^3 g(x)\,dx + \int_4^3 f(x)\,dx$

d $\int_3^6 g(x)\,dx$ **e** $\int_2^8 3g(x-2)\,dx$

a $\int_0^3 (2f(x)+g(x))\,dx$	
$=\int_0^3 2f(x)\,dx + \int_0^3 g(x)\,dx$	$\int_a^b (f(x) \pm g(x))\,dx = \int_a^b f(x)\,dx \pm \int_a^b g(x)\,dx$
$=2\int_0^3 f(x)\,dx + \int_0^3 g(x)\,dx$	$\int_a^b kf(x)\,dx = k\int_a^b f(x)\,dx$
$=2(5)+-6$	Substitute and evaluate.
$=4$	
b $\int_0^4 f(x)\,dx$	
$=\int_0^3 f(x)\,dx + \int_3^4 f(x)\,dx$	$\int_a^b f(x)\,dx = \int_a^c f(x)\,dx + \int_c^b f(x)\,dx$
$=5+2$	
$=7$	Substitute and evaluate.
c $\int_3^3 g(x)\,dx + \int_4^3 f(x)\,dx$	$\int_a^a f(x)\,dx = 0$
$=0-\int_3^4 f(x)\,dx$	and $\int_a^b f(x)\,dx = -\int_b^a f(x)\,dx$
$=0-2$	Substitute and evaluate.
$=-2$	
d $\int_0^6 g(x)\,dx = \int_0^3 g(x)\,dx + \int_3^6 g(x)\,dx$	$\int_a^b f(x)\,dx = \int_a^c f(x)\,dx + \int_c^b f(x)\,dx$
$\int_3^6 g(x)\,dx = \int_0^6 g(x)\,dx - \int_0^3 g(x)\,dx$	Rearrange terms.
$=10-(-6)$	Substitute and evaluate.
$=16$	

e $\int_2^8 3g(x-2)\,dx$	
$= 3\int_2^8 g(x-2)\,dx$	$\int_a^b kf(x)\,dx = k\int_a^b f(x)\,dx$
$= 3\int_0^6 g(x)\,dx$	The graph of $g(x-2)$ is a result of translating the graph of $g(x)$ to the right 2 units. So the limits of integration were shifted from $x = 0$ and $x = 6$ to $x = 2$ and $x = 8$.
$= 3(10)$ $= 30$	Substitute and evaluate.

Exercise 10G

The graph of $y = f(x)$ consists of three line segments as shown in the diagram.

Find the definite integral in questions **1–3** using geometric formulae and properties of integrals.

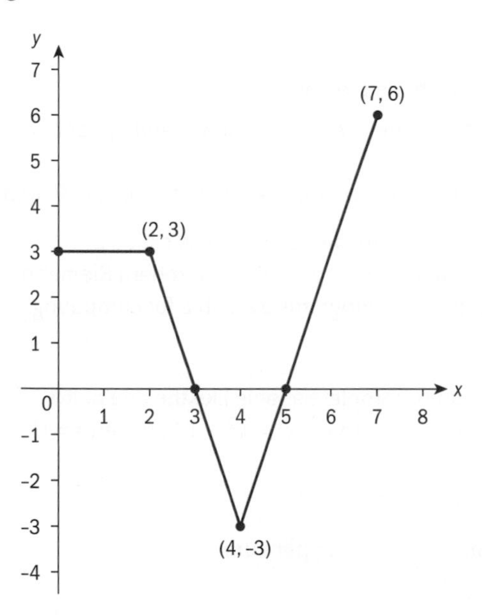

1 $\int_0^3 f(x)\,dx$ **2** $\int_3^5 f(x)\,dx$ **3** $\int_0^7 f(x)\,dx$

Given that $\int_1^6 f(x)\,dx = -4$, $\int_1^{10} f(x)\,dx = 12$, $\int_1^6 g(x)\,dx = 6$ and $\int_6^{10} g(x)\,dx = 14$, find the definite integral in questions **4–11**.

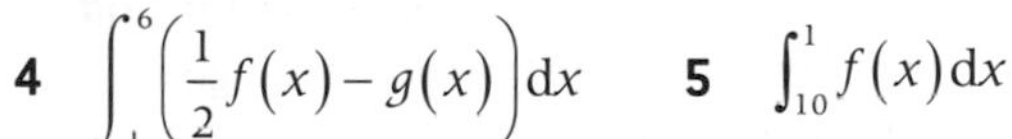

4 $\int_1^6 \left(\frac{1}{2}f(x) - g(x)\right)dx$ **5** $\int_{10}^1 f(x)\,dx$

6 $\int_1^{10} g(x)\,dx$ **7** $\int_6^6 g(x)\,dx$

8 $\int_6^{10} f(x)\,dx$ **9** $\int_{-2}^7 f(x+3)\,dx$

10 $\int_3^{12} \left(\frac{1}{2}g(x-2)\right)dx$ **11** $\int_1^{10} (f(x)+4)\,dx$

12 Let f be a function such that $\int_{-3}^3 f(x)\,dx = 20$.

a Find $\int_{-3}^3 \frac{1}{2}f(x)\,dx$.

b Given that $\int_{-3}^1 f(x)\,dx = 6$, find $\int_1^3 f(x)\,dx$.

c Given that $\int_a^b f(x-4)\,dx = 20$, find a possible value of a and of b.

d Given that $\int_{-3}^3 (f(x)+k)\,dx = 32$, find k.

Calculus

10.4 Fundamental theorem of calculus

In section 10.3, you saw that the definite integral is related to area by looking at geometric regions with known area formulae. Now you will examine a region where you cannot use geometry to find the exact area.

Investigation 7

Consider the shaded area under the curve $f(x) = x^2 + 1$ between $x = 0$ and $x = 2$. The sum of the areas of the four rectangles in Figure 1 gives a lower bound on the shaded area. The sum of the areas of the four rectangles in Figure 2 gives an upper bound on the shaded area.

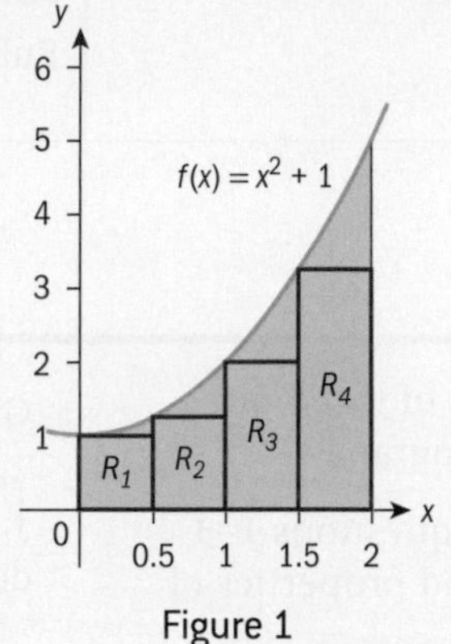

Figure 1

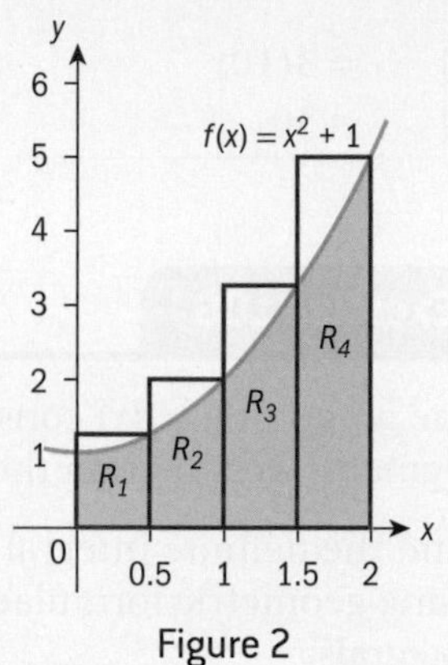

Figure 2

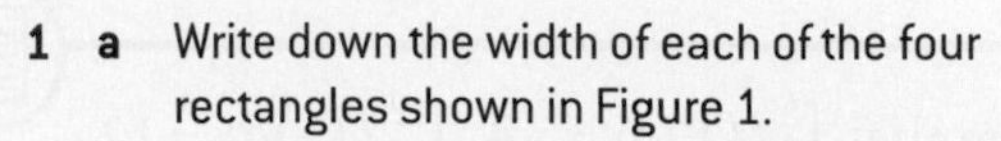

1 **a** Write down the width of each of the four rectangles shown in Figure 1.

b Find the height of each rectangle.

c Find the sum of the areas of the four rectangles.

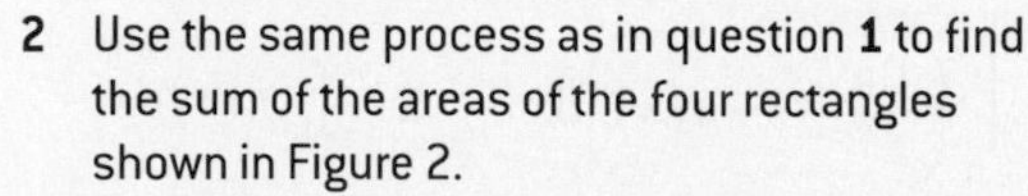

2 Use the same process as in question **1** to find the sum of the areas of the four rectangles shown in Figure 2.

3 **a** Write down a definite integral that represents the area of shaded region.

b Evaluate the definite integral with a GDC, to show that the value lies between lower and upper bounds found in questions **1** and **2**.

4 How can you find upper and lower bounds for the area under the curve that are closer to the actual area?

The process of using sums of areas of rectangles to approximate the area under a curve is called a Riemann sum. This method is named after the German mathematician Georg Friedrich Bernhard Riemann (1826–1866), who generalized the process. There are many software programs available for computing Riemann sums, using areas of rectangles.

5 Using software which enables you to calculate a Riemann sum, complete a table like the one below for the lower and upper sum approximations of the area under the curve $f(x) = x^2 + 1$ between $x = 0$ and $x = 2$.

Number of rectangles	Lower sum	Upper sum
4		
10		
50		
100		

6 **Conceptual** What major concept of calculus can be applied to the process of summing areas of rectangles, to give the exact value of the area under a curve and of the definite integral representing that area?

Consider the region under the curve $y = f(x)$ from $x = a$ to $x = b$, as shown in the diagram on the right. Let n represent the number of rectangles in a Riemann sum approximation of the area of the shaded region.

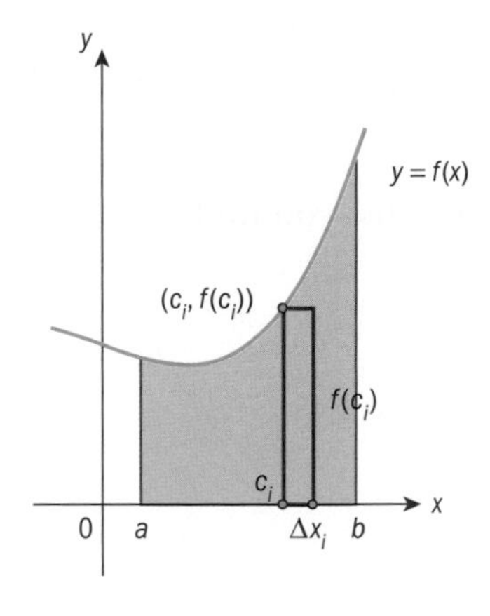

The expression $\sum_{i=1}^{n} f(c_i)\Delta x_i$ represents the sum of the areas of n rectangles, where $f(c_i)$ is the height of the rectangle which begins at $x = c_i$ and Δx_i is the width of this rectangle. As the number of rectangles approaches infinity, the widths of the rectangles will approach 0.

$\lim_{\Delta x_i \to 0} \sum_{i=1}^{n} f(c_i)\Delta x_i$ gives the exact area of the shaded region and is equal to $\int_a^b f(x)\,dx$. Notice the similarities in the limit notation and the definite integral notation.

- $f(c_i)$ and $f(x)$ represent the height of a rectangle.
- Δx_i and dx represent the width of a rectangle.
- $\sum_{i=1}^{n}$ and $\int_a^b$ represent a sum.

You have seen that the product $(\Delta y)(\Delta x)$, the area of a rectangle, gives an approximation for the area under a curve, much like the quotient $\frac{\Delta y}{\Delta x}$, the gradient of a secant line, gives an approximation for the gradient of a tangent line. In much the same sense that multiplication and division are inverse operations, Isaac Newton and Gottfried Leibniz independently came to realize that differentiation and definite integrals are inverse processes. This fact established the following important theorem, that allows you to evaluate a definite integral analytically.

Fundamental theorem of calculus

If f is a continuous function on the interval $a \le x \le b$ and F is an antiderivative of f on $a \le x \le b$, then $\int_a^b f(x)\,dx = [F(x)]_a^b = F(b) - F(a)$.

This theorem provides a method for evaluating definite integrals and is still considered one of the major advances in modern mathematics.

International-mindedness

The definite integral notation was introduced by the German mathematician Gottfried Wilhelm von Leibniz towards the end of the 17th century.

HINT

The notation $[F(x)]_a^b$ means $F(b) - F(a)$.

International-mindedness

Augustin-Lois Cauchy (1789–1857) was the first mathematician to formalize and prove the fundamental theorem of calculus in its current form.

In the investigation at the beginning of this section, you evaluated the definite integral $\int_0^2 (x^2+1)\,dx$ with a GDC. You found $\int_0^2 (x^2+1)\,dx \approx 4.67$.

Using the fundamental theorem of calculus, you find:

$$\int_0^2 (x^2+1)\,dx = \left[\frac{1}{3}x^3 + x\right]_0^2$$
$$= \left(\frac{1}{3}(2^3)+2\right) - \left(\frac{1}{3}(0^3)+0\right)$$
$$= \frac{14}{3} \approx 4.67$$

HINT

When applying the fundamental theorem of calculus, although F can be any antiderivative of f, the "simplest" one is chosen, that is, the one where the constant of integration is $C = 0$.

You can do this because, for any C,

$$[F(x)+C]_a^b = [F(b)+C]-[F(a)+C] = F(b)-F(a)$$

Example 10

Evaluate the definite integral without using a GDC.

a $\int_0^3 (x-2)\,dx$ **b** $\int_2^6 \frac{1}{t}\,dt$ **c** $\int_0^2 4e^u\,du$

a $\int_0^3 (x-2)\,dx = \left[\frac{1}{2}x^2 - 2x\right]_0^3$	Find the "simplest" antiderivative of $x - 2$.
$= \left(\frac{1}{2}(3)^2 - 2(3)\right) - \left(\frac{1}{2}(0)^2 - 2(0)\right)$	Evaluate $\frac{1}{2}x^2 - 2x$ at $x = 3$ and $x = 0$, and find the difference.
$= \left(\frac{9}{2} - 6\right) - (0-0) = -\frac{3}{2}$	
b $\int_2^6 \frac{1}{t}\,dt = [\ln\lvert t\rvert]_2^6$	Find the "simplest" antiderivative of $\frac{1}{t}$.
$= \ln 6 - \ln 2 = \ln 3$	Recall that $\ln a - \ln b = \ln\frac{a}{b}$.
c $\int_0^2 4e^u\,du = [4e^u]_0^2$	Find the "simplest" antiderivative of $4e^u$.
$= 4e^2 - 4e^0 = 4e^2 - 4$	

Exercise 10H

1 $\int_{-2}^3 6x\,dx$

2 $\int_1^4 \frac{e^x}{2}\,dx$

3 $\int_3^4 \frac{5}{u}\,du$

4 $\int_{-1}^2 (3x^2+4x-2)\,dx$

5 $\int_1^2 \left(\frac{4}{x^2}+1\right)dx$

6 $\int_1^3 \frac{1}{x^3}\,dx$

7 $\int_0^{16} \left(\frac{1}{t^{\frac{1}{4}}} - t^{\frac{1}{2}}\right)dt$

8 $\int_{e^2}^{e^5} \frac{1}{x}\,dx$

9 Write down a definite integral that represents the area of the shaded region. Then find the area.

a $f(x) = -x^2 + 2x$

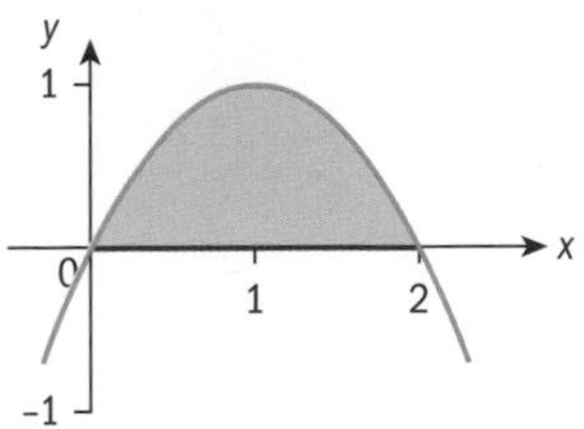

b $f(x) = \frac{1}{x^2}$

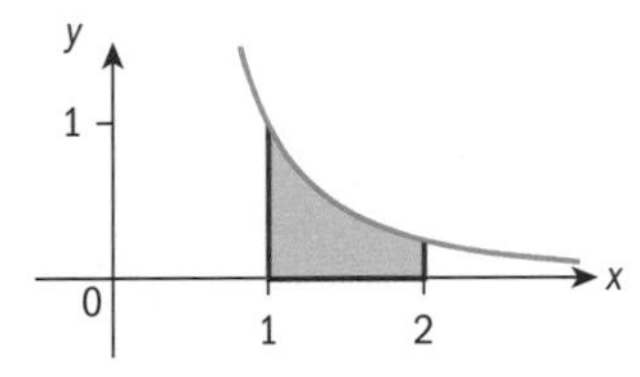

10 It is given that $\int_0^4 f(x)\,dx = 10$.

a Write down the value of $\int_0^4 2f(x)\,dx$.

b Find the value of $\int_0^4 (2f(x) + x)\,dx$.

11 a Find $\int \frac{2}{x}\,dx$.

b Given that $\int_2^k \frac{2}{x}\,dx = \ln 9$ and $k > 0$, find the value of k.

Example 11

Evaluate the definite integral without using a GDC.

a $\int_{-2}^{2} 2x^2(x-3)\,dx$ **b** $\int_{-1}^{0} (2x+1)^3\,dx$ **c** $\int_{-\frac{5}{4}}^{1} \sqrt{4x+5}\,dx$ **d** $\int_1^3 \frac{e^{4x}+e^{2x}}{e^{2x}}\,dx$

a $\int_{-2}^{2} 2x^2(x-3)\,dx = \int_{-2}^{2}(2x^3 - 6x^2)\,dx$	Expand $2x^2(x-3)$.
$= \left[\frac{1}{2}x^4 - 2x^3\right]_{-2}^{2}$	
$= \left(\frac{1}{2}(2)^4 - 2(2)^3\right) - \left(\frac{1}{2}(-2)^4 - 2(-2)^3\right)$	
$= (8-16) - (8+16) = -32$	
b $\int_{-1}^{0}(2x+1)^3\,dx = \left[\frac{1}{2\cdot 4}(2x+1)^4\right]_{-1}^{0}$	$\int (ax+b)^n\,dx = \frac{1}{a(n+1)}(ax+b)^{n+1} + C$
$= \left(\frac{1}{8}(2(0)+1)^4\right) - \left(\frac{1}{8}(2(-1)+1)^4\right)$	
$= \frac{1}{8} - \frac{1}{8} = 0$	
c $\int_{-\frac{5}{4}}^{1}\sqrt{4x+5}\,dx = \int_{-\frac{5}{4}}^{1}(4x+5)^{\frac{1}{2}}\,dx$	Rewrite using rational exponents.
$= \left[\frac{1}{4\cdot\frac{3}{2}}(4x+5)^{\frac{3}{2}}\right]_{-\frac{5}{4}}^{1}$	$\int (ax+b)^n\,dx = \frac{1}{a(n+1)}(ax+b)^{n+1} + C$

Continued on next page

Calculus

$= \left(\frac{1}{6}(4(1)+5)^{\frac{3}{2}} - \frac{1}{6}\left(4\left(-\frac{5}{4}\right)+5\right)^{\frac{3}{2}}\right)$	
$= \left(\frac{1}{6}(9)^{\frac{3}{2}} - \frac{1}{6}(0)^{\frac{3}{2}}\right) = \frac{27}{6} = \frac{9}{2}$	
d $\int_1^3 \frac{e^{4x}+e^{2x}}{e^{2x}}dx = \int_1^3 \left(\frac{e^{4x}}{e^{2x}} + \frac{e^{2x}}{e^{2x}}\right)dx$	Separate the terms.
$= \int_1^3 (e^{2x}+1)dx$	$\frac{e^{4x}}{e^{2x}} = e^{4x-2x} = e^{2x}$
$= \left[\frac{1}{2}e^{2x} + x\right]_1^3$	$\int e^{ax+b}dx = \frac{1}{a}e^{ax+b} + C$
$= \left(\frac{1}{2}e^{2(3)} + 3\right) - \left(\frac{1}{2}e^{2(1)} + 1\right) = \frac{1}{2}e^6 - \frac{1}{2}e^2 + 2$	

Exercise 10I

In questions **1–8**, evaluate the definite integral.

1 $\int_1^4 \frac{1}{4x-2}dx$ **2** $\int_0^1 (2t-1)^2 dt$

3 $\int_{-1}^2 e^{3x+4}dx$ **4** $\int_0^2 (x+2)(x-1)dx$

5 $\int_1^2 (3x-3)^3 dx$ **6** $\int_0^4 \sqrt{4x+9}\,dx$

7 $\int_{-2}^2 (e^t - e^{-t})dt$ **8** $\int_4^9 \frac{3\sqrt{t}+2}{\sqrt{t}}dt$

9 The diagram shows part of the graph of $f(x) = -2x(x^2-4)$ and a shaded region bounded by the graph of f and the x-axis.

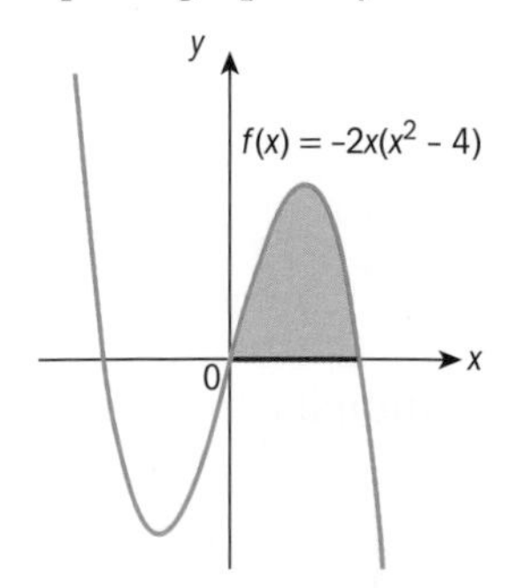

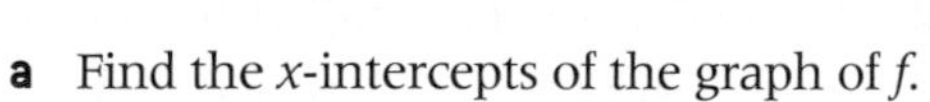

a Find the x-intercepts of the graph of f.

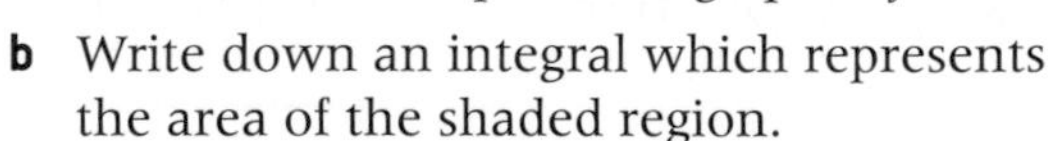

b Write down an integral which represents the area of the shaded region.

c Find the area of the shaded region.

10 The diagram shows part of the graph of $y = \frac{1}{2x+1}$. The area of the shaded region under the graph of f from $x = 1$ to $x = k$ is $\ln 3$.

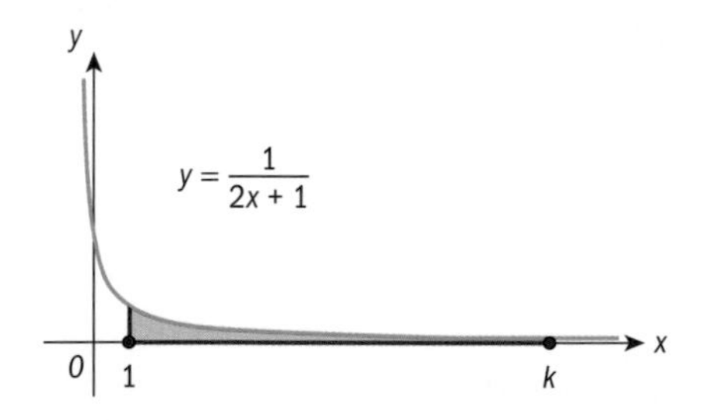

a Write down an integral which represents the area of the shaded region.

b Find the value of k.

10.5 Area between two curves

In section 10.4, you approximated the area *under* a curve using sums of areas of rectangles, to show the connection between area and the definite integral. Now you will consider the area *between* two curves.

TOK

Why do we study mathematics?

What's the point?

Can we do without it?

Investigation 8

The diagram shows four regions bounded by the curves $y=f(x)$ and $y=g(x)$. The area of each region can be approximated using areas of rectangles. A representative rectangle is shown for each region.

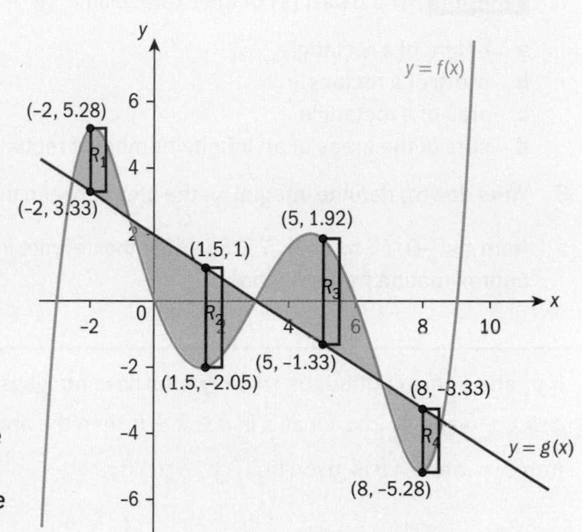

1 The height of each rectangle can be expressed as either $f(x)-g(x)$ or $g(x)-f(x)$, for some value of x. For example, the height of the rectangle shown in region R_1 can be expressed as $h=f(-2)-g(-2)$ $=5.28-3.33=1.95$. Express the heights of the rectangles shown in the other three regions in the same manner.

2 **Factual** When using rectangles to approximate the area of a region between the curves $y=f(x)$ and $y=g(x)$, how do you choose whether to represent the height of the rectangle as $f(x)-g(x)$ or $g(x)-f(x)$?

Consider the area between the graphs of the two curves $f(x)=x^2-2x-1$ and $g(x)=\frac{1}{2}x-7$ from $x=-0.75$ to $x=1.75$.

The five rectangles shown in the diagram are used to approximate the area.

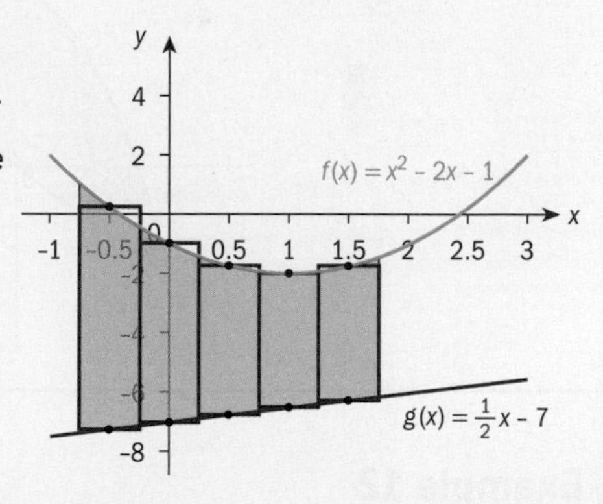

3 Would you use $f(x)-g(x)$ or $g(x)-f(x)$ to represent the heights of the rectangles? Why?

4 Complete a table like the one below, to find the area of each of the five rectangles. The first row in the table is given for you.

Interval	Width	Height	Area
$-0.75 \le x \le -0.25$	0.5	$f(-0.5)-g(-0.5)$ $=0.25-(-7.25)=7.5$	$0.5(7.5)=3.75$
$-0.25 \le x \le 0.25$			
$0.25 \le x \le 0.75$			
$0.75 \le x \le 1.25$			
$1.25 \le x \le 1.75$			

Continued on next page

5 Approximate the area between the curves $f(x) = x^2 - 2x - 1$ and $g(x) = \frac{1}{2}x - 7$ from $x = -0.75$ to $x = 1.75$, by finding the sum of the areas of the rectangles.

The area *under* a curve $y = f(x)$ between $x = a$ and $x = b$ is given as $A = \int_a^b |y| dx$.

6 **Conceptual** How can you modify the formula for the area *under* a curve to find the area *between* two curves? Explain your reasoning.

7 **Factual** Which part (s) of the expression $\int_a^b (y_1 - y_2)\, dx$ represent(s) each of the following?

a height of a rectangle
b width of a rectangle
c area of a rectangle
d sum of the areas of an infinite number of rectangles from $x = a$ to $x = b$

8 Write down a definite integral for the area between the two curves $f(x) = x^2 - 2x - 1$ and $g(x) = \frac{1}{2}x - 7$, from $x = -0.75$ to $x = 1.75$. Evaluate the definite integral on a GDC and compare the value to your approximation from question **5**.

If y_1 and y_2 are continuous (their graphs have no "gaps" or "breaks") on $a \le x \le b$ and $y_1 \ge y_2$ for all x in $a \le x \le b$, then the area between y_1 and y_2 from $x = a$ to $x = b$ is given by $\int_a^b (y_1 - y_2)\, dx$.

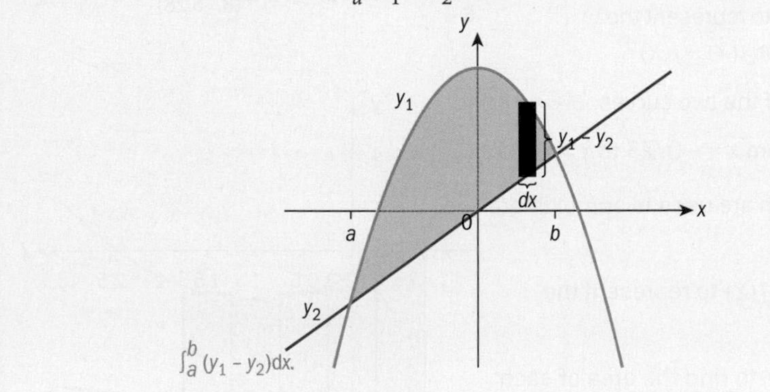

HINT

Since y_1 must be greater than or equal to y_2, you can always think of the height of the rectangle as the "top curve" minus the "bottom curve".

Example 12

Find the area of the region bounded by the curves $y = 4 - x^2$ and $y = x + 2$.

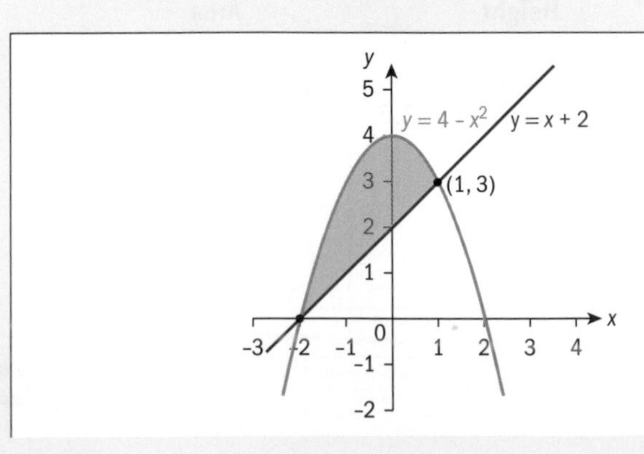

Graph the region bounded by the curves.

$4 - x^2 = x + 2$ $x^2 + x - 2 = 0$ $(x+2)(x-1) = 0$ $x = -2, 1$	Find the x-coordinates of the points of intersection of the two functions. Notice that for $x = -2$ to $x = 1$, $4 - x^2 \geq x + 2$. You can think of $y = 4 - x^2$ as the "top curve" and $y = x + 2$ as the "bottom curve".
$\text{Area} = \int_{-2}^{1} \left[\left(4 - x^2\right) - (x+2)\right] \mathrm{d}x$	Write down a definite integral for the area between the two functions.
$= \int_{-2}^{1} \left(2 - x - x^2\right) \mathrm{d}x$ $= \left[2x - \frac{1}{2}x^2 - \frac{1}{3}x^3\right]_{-2}^{1}$ $= \left(2(1) - \frac{1}{2}(1)^2 - \frac{1}{3}(1)^3\right) - \left(2(-2) - \frac{1}{2}(-2)^2 - \frac{1}{3}(-2)^3\right)$ $= \left(2 - \frac{1}{2} - \frac{1}{3}\right) - \left(-4 - 2 + \frac{8}{3}\right)$ $\text{Area} = \frac{9}{2}$	Evaluate the integral.

Example 13

Find the area of the region bounded by the curves $f(x) = e^{2x} + 1$ and $g(x) = x + 4$.

	Use a GDC to help sketch the graphs and to find the x-coordinates of the points of intersection.
$e^{2x} + 1 = x + 4$ $x = -2.998, 0.6469$	From $x = -2.998$ to $x = 0.6469$, $g(x) \geq f(x)$. (g is the "top curve")
$\int_{-2.997}^{0.6469} \left[(x+4) - \left(e^{2x} + 1\right)\right] \mathrm{d}x = 4.83$	Write down a definite integral for the area between the two functions. Evaluate with a GDC.

Exercise 10J

In questions **1** and **2**, sketch the graphs of the given functions on the same set of axes, and shade the region bound by the two curves. Write down an expression that gives the area of the region and then find the area.

1 $f(x) = x^2$ and $g(x) = -2x + 3$

2 $y = x^2 - 5x + 4$ and $y = 4$

In questions **3–8**, sketch the graphs of the given functions on the same set of axes, and shade the region bound by the two curves. Write down an expression that gives the area of the region. Find the area using a GDC.

> **HINT**
>
> You can store the x-coordinates of the points of intersection of the two functions in your GDC and use the stored values to evaluate the definite integral.

3 $y = -0.5x^2 + 8$ and $y = 0.5x^2 - 8$

4 $g(x) = x^2 - 2x$ and $h(x) = 4\sqrt{x}$

5 $y = \ln x$ and $y = \frac{1}{2}x - 2$

6 $f(x) = x^2 - 5x + 1$ and $g(x) = 6 - x^2$

7 $f(x) = 2e^x$ and $g(x) = 3x + 4$

8 $y = \frac{x+4}{x-2}$ and $y = -x - 7$

9 The diagram shows a shaded region, bounded by the curves $y = \frac{2}{x}$, $y = -2$, $x = 1$ and $x = 4$.

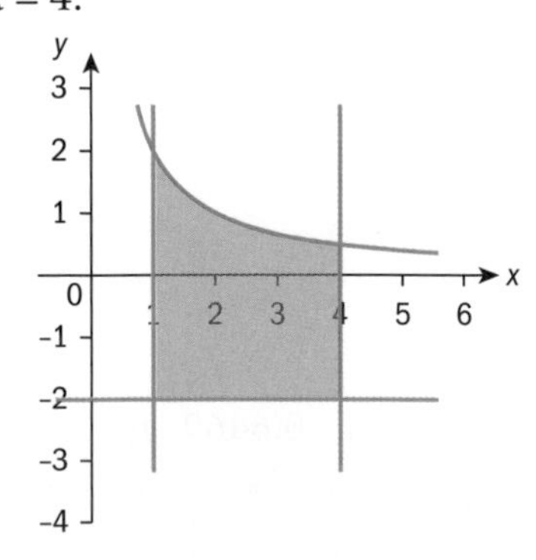

The area of the region can be written in the form $p \ln(2) + q$. Find the values of p and of q.

10 Consider the functions $f(x) = \frac{1}{2}x$ and $g(x) = \sqrt{x}$. Part of the graphs of f and g are shown in the diagram.

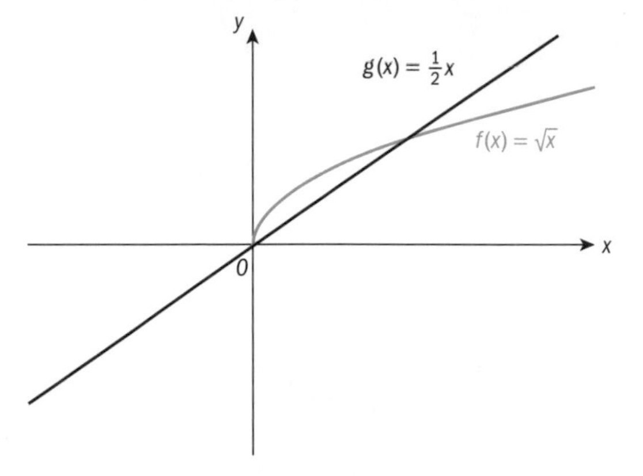

a Find the coordinates of the points of intersection of the graphs of f and g.

b **i** Write an integral expression for the area enclosed by the graphs of f and g.

ii Find this area.

c The line $x = k$ divides the area of the region from part **b** in half.

i Write down an integral expression for half the area of the region from part **b**.

ii Find an expression, not involving an integral, for half the area of the region.

iii Using your GDC, find the value of k.

> **TOK**
>
> How do "believing that" and "believing in" differ?
>
> How does belief differ from knowledge?

Suppose that y_1 and y_2 are continuous on $a \le x \le b$ and y_1 is not greater than or equal to y_2 for all x in $a \le x \le b$. In this case, you need to set up more than one integral to find the area between the curves.

Example 14

Find the area of the region between the curves $f(x) = x^3 - 3x^2 + 3x$ and $g(x) = x^2$.

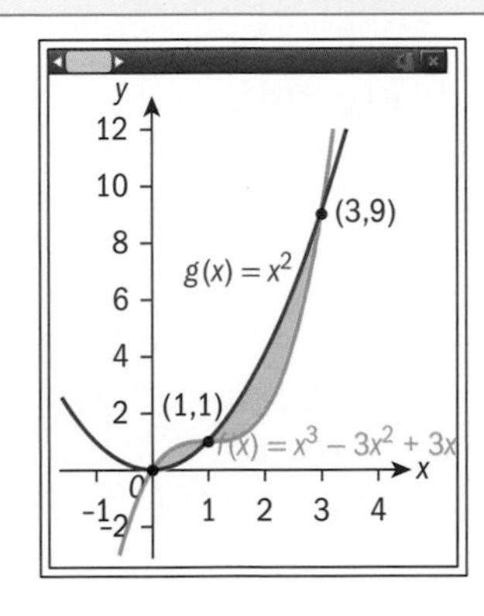

Use a GDC to help sketch the graphs and to find points of intersection.

From $x = 0$ to $x = 1$:

$f(x) \ge g(x)$

(f is the "top curve")

Area $= \int_0^1 (f(x) - g(x))\,dx$

From $x = 1$ to $x = 3$:

$g(x) \ge f(x)$

(g is the "top curve")

Area $= \int_1^3 (g(x) - f(x))$

Use a GDC to evaluate the sum of the integrals.

Area $= \int_0^1 (f(x) - g(x))\,dx + \int_1^3 (g(x) - f(x))\,dx$

$= \int_0^1 \left[(x^3 - 3x^2 + 3x) - x^2\right]dx + \int_1^3 \left[x^2 - (x^3 - 3x^2 + 3x)\right]dx$

$= \frac{37}{12} \approx 3.083$

Exercise 10K

In questions **1–4**, write an integral expression to find the area of the region bounded by the two curves, and then find this area. Use your GDC to examine the graphs and evaluate any definite integrals.

1. $f(x) = x^2 - 2x$ and $g(x) = 10x + x^2 - 3x^3$
2. $f(x) = x^3 - 3x^2 + 3x + 1$ and $g(x) = x + 1$
3. $f(x) = 3xe^{-x^2}$ and $g(x) = x^3 - 2x$

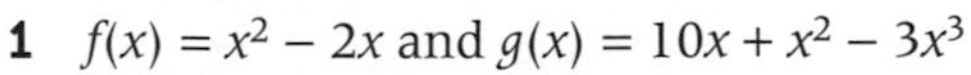

4. $f(x) = x^4 - 20x^2 + 64$ and $g(x) = -x^4 + 16x^2$
5. The graph shows a shaded region bounded by the functions $f(x) = \frac{1}{2}x^2 - 1$, $g(x) = -x^2 - 1$ and $h(x) = -3x - 5.5$. Point P is the intersection of the graphs of f and h and Q is the intersection of the graphs of g and h.

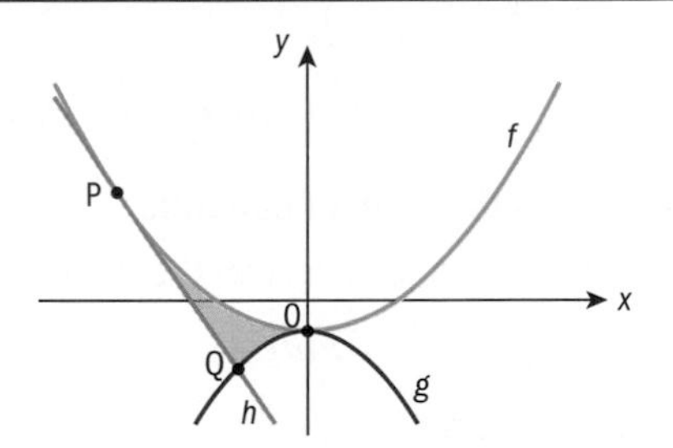

   a. Find the coordinates of P.
   b. Show that the graph of h is tangential to the graph of f at P.
   c. Using your GDC, find the coordinates of Q to 2 d.p.
   d. i. Write down an integral expression for the shaded area.
      ii. Find the area of the shaded region.

Developing your toolkit

Now do the Modelling and investigation activity on page 464.

Calculus

Chapter summary

Antiderivatives and the indefinite integral

Power rule: $\int x^n \mathrm{d}x = \frac{1}{n+1}x^{n+1} + C,\ n \neq 1$

Constant rule: $\int k\,\mathrm{d}x = kx + C$

Constant multiple rule: $\int kf(x)\,\mathrm{d}x = k\int f(x)\,\mathrm{d}x$

Sum or difference rule: $\int [f(x) \pm g(x)]\,\mathrm{d}x = \int f(x)\,\mathrm{d}x \pm \int g(x)\,\mathrm{d}x$

More on indefinite integrals

$$\int \frac{1}{x}\mathrm{d}x = \ln|x| + C$$

$$\int \mathrm{e}^x \mathrm{d}x = \mathrm{e}^x + C$$

$$\int (ax+b)^n\,\mathrm{d}x = \frac{1}{a(n+1)}(ax+b)^{n+1} + C$$

$$\int \mathrm{e}^{ax+b}\mathrm{d}x = \frac{1}{a}\mathrm{e}^{ax+b} + C$$

$$\int \frac{1}{ax+b}\mathrm{d}x = \frac{1}{a}\ln|ax+b| + C$$

Properties of definite integrals

$$\int_a^b kf(x)\,\mathrm{d}x = k\int_a^b f(x)\,\mathrm{d}x$$

$$\int_a^b (f(x) \pm g(x))\,\mathrm{d}x = \int_a^b f(x)\,\mathrm{d}x \pm \int_a^b g(x)\,\mathrm{d}x$$

$$\int_a^a f(x)\,\mathrm{d}x = 0$$

$$\int_a^b f(x)\,\mathrm{d}x = -\int_b^a f(x)\,\mathrm{d}x$$

$$\int_a^b f(x)\,\mathrm{d}x = \int_a^c f(x)\,\mathrm{d}x + \int_c^b f(x)\,\mathrm{d}x$$

Fundamental theorem of calculus

If f is a continuous function on the interval $a \leq x \leq b$ and F is an antiderivative of f on $a \leq x \leq b$, then $\int_a^b f(x)\,\mathrm{d}x = [F(x)]_a^b = F(b) - F(a)$.

Area under a curve

If f is a non-negative function for $a \leq x \leq b$, then $\int_a^b f(x)\,\mathrm{d}x$ gives the area under the curve f from $x = a$ to $x = b$.

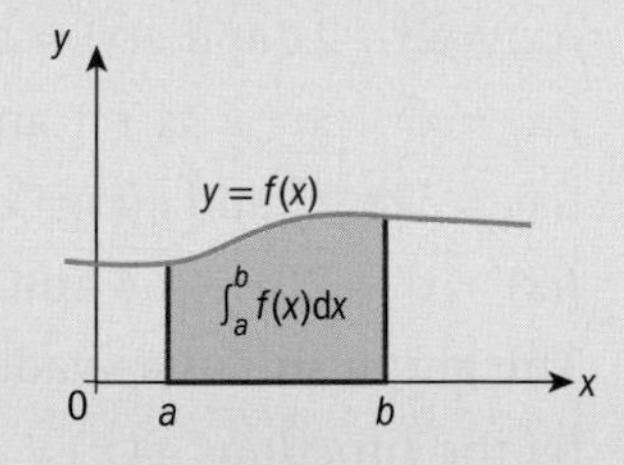

Area between curves

If y_1 and y_2 are continuous on $a \leq x \leq b$ and $y_1 \geq y_2$ for all x in $a \leq x \leq b$, then the area between y_1 and y_2 from $x = a$ to $x = b$ is given by $\int_a^b (y_1 - y_2)\,\mathrm{d}x$.

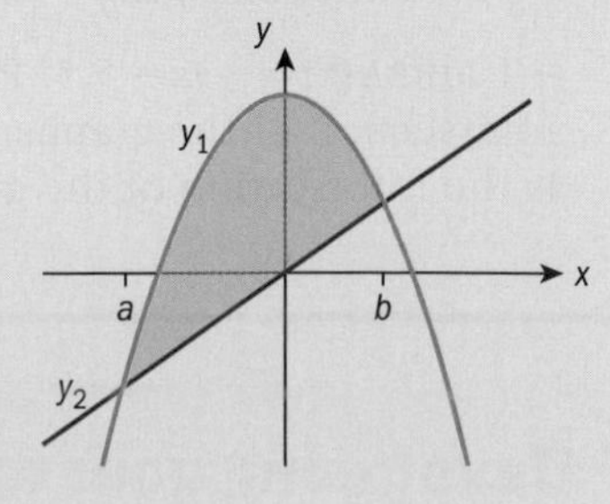

Developing inquiry skills

Return to the opening problem. How have the skills you have learned in this chapter helped you to solve the problem?

Chapter review

Click here for a mixed review exercise

1 Find the indefinite integral.

a $\int x^8\,dx$ **b** $\int (5x^4 - 6x^2 + 7)\,dx$

c $\int \sqrt[10]{x^3}\,dx$ **d** $\int \frac{4}{x^9}\,dx$

e $\int \frac{8x^5 + 4x}{2x^2}\,dx$ **f** $\int 4e^x\,dx$

g $\int \left(6\sqrt{x} + 2\right)dx$ **h** $\int (x^2 + 3)^2\,dx$

i $\int (4x + 5)^6\,dx$ **j** $\int 6e^{3x+2}\,dx$

k $\int \frac{1}{6x - 7}\,dx$ **l** $\int \ln(e^{3x})\,dx$

2 Find the definite integral.

a $\int_{-2}^{3} (6x - 1)\,dx$ **b** $\int_{-1}^{3} x^2\,dx$

c $\int_{9}^{25} \frac{3}{\sqrt{x}}\,dx$ **d** $\int_{1}^{e^4} \frac{5}{x}\,dx$

e $\int_{-2}^{0} 8(2x + 3)^3\,dx$ **f** $\int_{3}^{5} e^{4x}\,dx$

3 Given that $\int_1^4 f(x)\,dx = 10$ and $\int_1^3 f(x)\,dx = 6$, find:

a $\int_1^4 2f(x)\,dx$ **b** $\int_3^4 f(x)\,dx$

c $\int_1^4 \left(f(x) + 4\right)dx$

4 Let $f'(x) = 4x^3 + 2$. Given that $f(2) = 24$, find $f(x)$.

5 The following diagram shows the graph of $y = f(x)$, for $-8 \le x \le 8$.

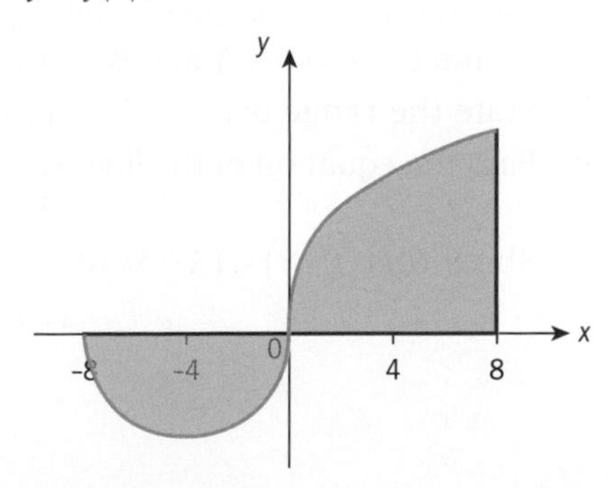

The first part of the graph is a semicircle with centre $(-4, 0)$.

a Find $\int_{-8}^{0} f(x)\,dx$.

b The shaded region is enclosed by the graph of f, the x-axis and the line $x = 8$. The area of this region is 21π. Find $\int_{-8}^{8} f(x)\,dx$.

6 Find the area under the graph of $f(x) = -x^2 - 4x$.

7 Find the area of the region enclosed by the graphs of the given functions.

a $f(x) = x^4 - 5x^2$ and $g(x) = 0.5e^x + 1$

b $f(x) = x^3 - 9x$ and $g(x) = x^2 - 3x$

8 **a** Find the equation of the tangent line to the graph of $f(x) = x^3$ at $x = 1$.

b The tangent line to the graph of $f(x) = x^3$ at $x = 1$ intersects f at one other point. Find the coordinates of this point.

c Find the area of the region enclosed by the graph of f and the tangent line.

Exam-style questions

9 **P1:** A function f is defined by

$$f(x) = \frac{x^2 + 1}{2x}.$$

Find expressions for:

a $f'(x)$

b $\int f(x)\,dx$ (3 marks)

10 **P1:** The velocity, $v\ ms^{-1}$, of a particle at time t seconds is given by $v = 40 - 3t$.

a Find an expression for the acceleration of the particle in ms^{-2}. (2 marks)

b Let s represent the displacement, in metres, of the particle from the origin at a time t.

Given that $s = 10$ when $t = 1$, find an expression for s as a function of t. (4 marks)

11 P1: Consider the two regions represented on the diagram below. Region 1 is the triangle with vertices $(-2, 1)$, $(0, 0)$ and $(0, 2)$; Region 2 is the triangle with vertices $(3, 1)$, $(0, 0)$ and $(0, 2)$.

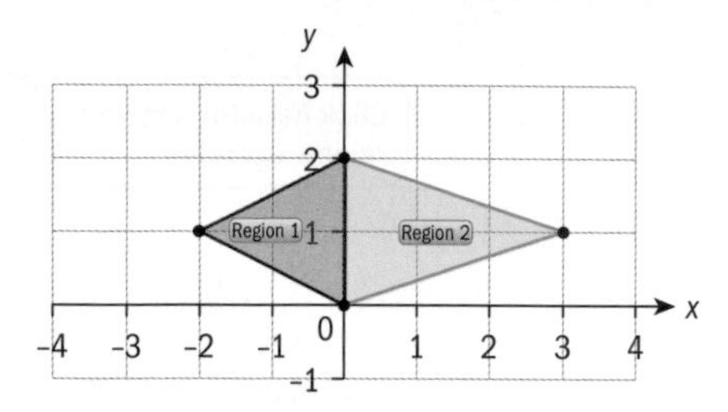

a Determine the area of:

i Region 1

ii Region 2. (4 marks)

b Hence determine the area of Region 3, shown on the diagram below (1 mark)

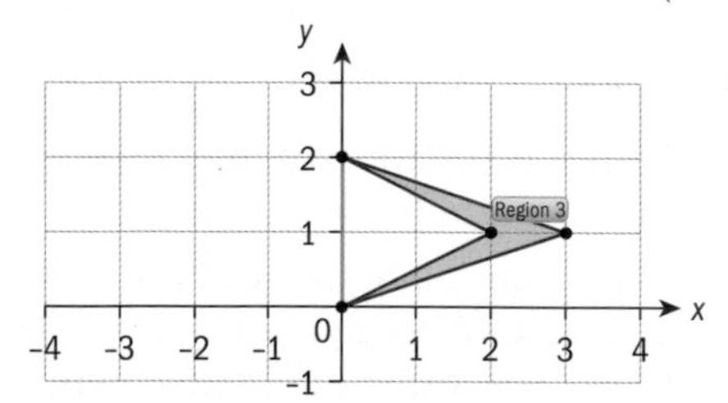

12 P2: The diagram shows the graph of the function defined by $f(x) = \frac{3e^{-x^2}}{2}$.

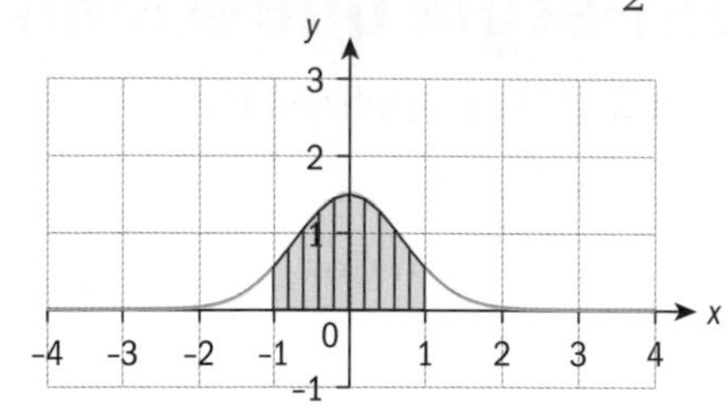

a Find $A = \int_0^1 f(x)\, dx$ correct to 5 significant figures. (2 marks)

b Hence find the value of

i the area of the shaded region on the diagram. (1 mark)

ii $\int_1^2 2f(x-1)\, dx$ (2 marks)

13 P2: Consider the function f where $f(x) = x(x^2 - 1)$, $x \in \mathbb{R}$.

a Find the coordinates of the points where the graph of f intersects the axes. (2 marks)

b Determine an expression for the derivative of f. (2 marks)

c Hence find the x-coordinates of the turning points on the graph of f. (2 marks)

d Find the value of $\int_{-1}^{1} f(x)\, dx$. (2 marks)

e Explain why the value of the integral found in part **d** does not represent the area of the region S enclosed by the graph of f and the x-axis between $x = \pm 1$. (1 mark)

f Find the area of the region S. (2 marks)

14 P1: Consider the functions f and g defined by $f(x) = x^2$ and $g(x) = 3 - 2x$.

a Show that the graphs of f and g intersect at points with x- coordinates -3 and 1. (3 marks)

b Hence find the area enclosed by the graphs of f and g. (4 marks)

15 P1: a Expand $(x - 2)^4$. (3 marks)

b Hence find $\int (x-2)^4 dx$ (3 marks)

16 P1: Use the fundamental theorem of calculus to show that $\int_{-1}^{2} |x|\, dx = 2.5$ (4 marks)

17 P2: Consider the function f defined by $f(x) = 3xe^{-x}$ on the interval $I = [0, 5]$.

a Sketch the graph of f, showing clearly the coordinates of the maximum point M and end-points A and B. (4 marks)

b State the range of f. (1 mark)

c Find the equation of the line AB. (3 marks)

d Show that $f'(x) = (3 - 3x)e^{-x}$. (2 marks)

e Hence find the equation of the tangent to the graph of f that is parallel to AB. Give all the coefficients in your equation correct to 3 significant figures. (3 marks)

f Find the area of the region enclosed by the line AB and the graph of f. (3 marks)

18 P1: Consider the function defined by $f(x) = 4x - x^3$ for $-2 \le x \le 2$.

a Complete the table

x	$\frac{1}{2}$	1	$\frac{3}{2}$
$f(x)$			

(2 marks)

b Sketch the graph of f, showing clearly the axes intercepts. (3 marks)

c Use four rectangles with equal widths to find an approximation of $\int_0^2 f(x)\,dx$. (4 marks)

d Find the exact value of $\int_0^2 f(x)\,dx$. (3 marks)

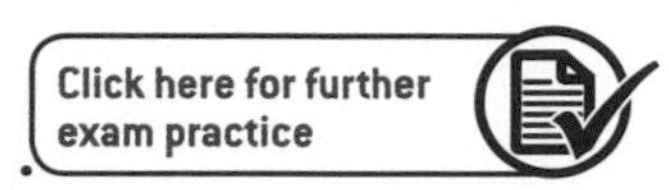

In the footsteps of Archimedes

Approaches to learning: Research, Critical thinking

Exploration criteria: Mathematical communication (B), Personal engagement (C), Use of mathematics (E)

IB topic: Integration, Proof, Coordinate Geometry

The area of a parabolic segment

A **parabolic segment** is a region bounded by a parabola and a line.

Consider this shaded region which is the area bounded by the line $y = x + 6$ and the curve $y = x^2$:

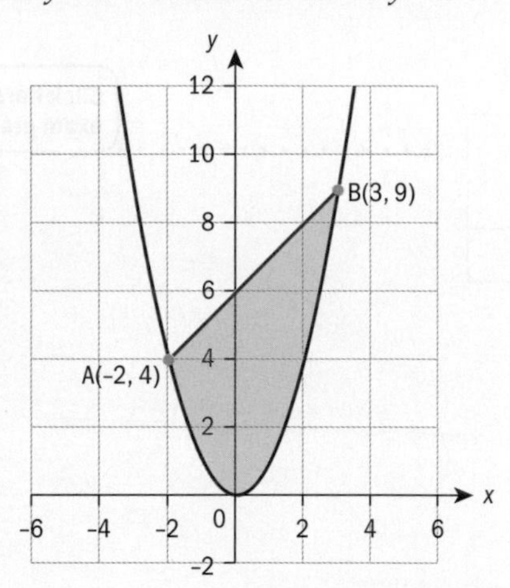

From this chapter you know that you can calculate the shaded area using integration.

On the diagram points A(−2, 4) and B(3, 9) are marked.

Point C is such that the x-value of C is halfway between the x-values of points A and B.

What are the coordinates of point C on the curve?

Triangle ABC is constructed as shown:

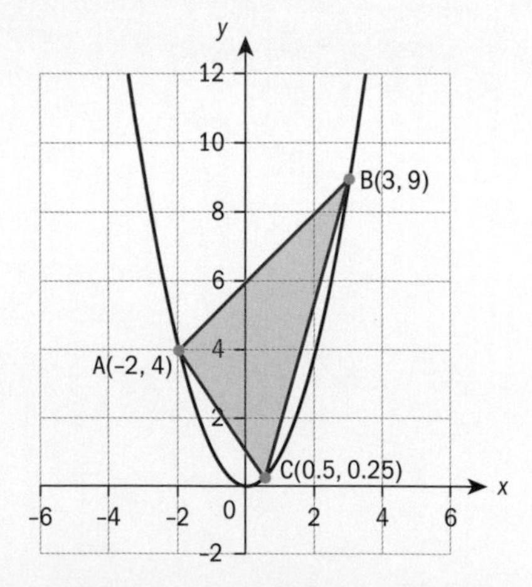

Archimedes showed that the area of the parabolic segment is $\frac{4}{3}$ of the area of triangle ABC.

Calculate the area of the triangle shown.

What methods are available to calculate the area of the triangle?

Use integration to calculate the area between the two curves.

Hence verify that Archimedes' result is correct for this parabolic segment.

You can show that this result is true for any parabola and for any starting points A and B on the parabola.

Consider another triangle by choosing point D on the parabola such that its x-value is halfway between the x-values of A and C, similar to before.

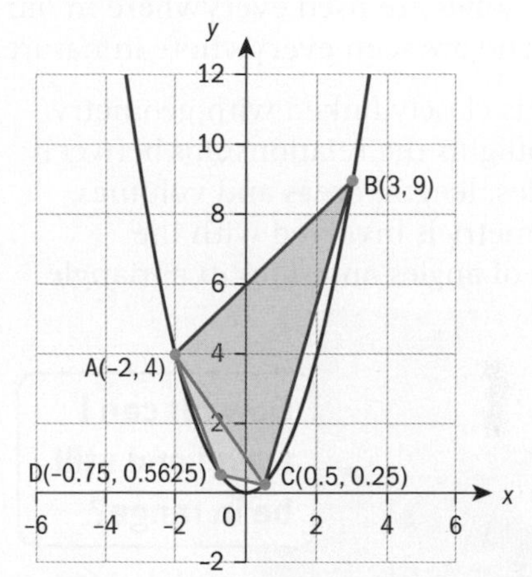

What are the coordinates of point D?

Calculate the area of triangle ACD.

Similarly, for line BC, find E such that its x-value is half-way between C and B.

What are the coordinates of point E?

Hence calculate the area of triangle BCE.

Calculate the ratio between the areas of the new triangles and original triangle ABC.

What do you notice?

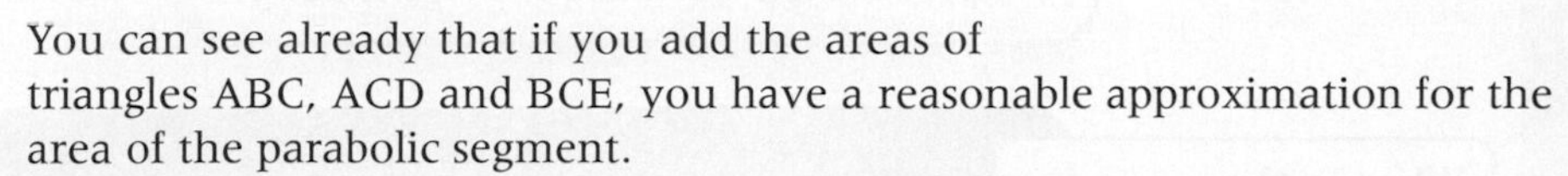

You can see already that if you add the areas of triangles ABC, ACD and BCE, you have a reasonable approximation for the area of the parabolic segment.

You can improve this approximation by continuing the process and forming four more triangles from sides AD, CD, CE and BE.

If you add the areas of these seven triangles, you have an even better approximation.

How could the approximation be improved?

Generalise the problem

Let the area of the first triangle be X.

What is the total area of the next two, four and eight triangles in terms of X?

If you continued adding the areas of an *infinite* number of such triangles, you would have the *exact* area for the parabolic segment.

By summing the areas of all the triangles, you can show that they form a geometric series.

What is the common ratio?

What is the first term?

What is the sum of the series?

What has this shown?

Extension

This task demonstrates the part of the historical development of the topic of limits which has led to the development of the concept of calculus.

Look at another area of mathematics that you have studied on this course so far or one that interests you looking forward in the book.

- What is the history of this particular area of mathematics.
- How does it fit into the development of the whole of mathematics?
- How significant is it?
- Who are the main contributors to this branch of mathematics?

Relationships in space: geometry and trigonometry in 2D and 3D

From the geometry required to build the pyramids of Egypt to the Louvre pyramid in Paris, Platonic solids are used everywhere in our everyday life and are seen everywhere in nature.

Trigonometry is closely linked with geometry. Geometry spotlights the relationships between angles and sides, length, areas and volumes, while trigonometry is involved with the measurement of angles and sides in a triangle.

Concepts

- Space
- Relationships

Microconcepts

- Distance and midpoint formulas
- 3D coordinates
- Volume and surface area of 3D shapes
- Right-angled triangle trigonometry
- Angles between lines and a line and a plane
- Elevation, depression, bearings
- Area of triangle ABC as $A = \frac{1}{2}ab\sin C$
- Sine rule
- The ambiguous case
- Cosine rule
- Applications of geometry and trigonometry

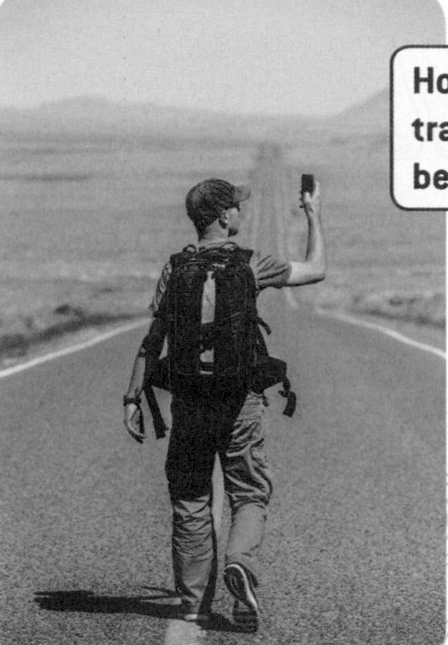

How far can I travel and still be in range?

What is the distance between the Sun and Venus?

Will this hold enough water for a month?

The pyramids of Giza were originally covered with a layer of white "casing stone", six feet deep. How much casing stone is needed to cover the pyramid?

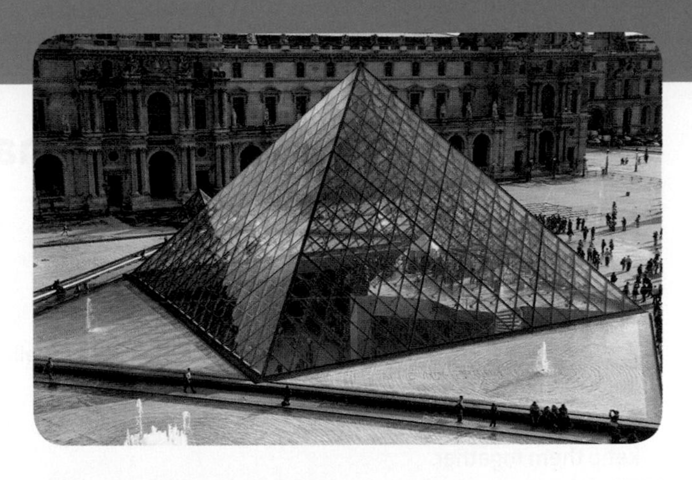

The world-famous Louvre Museum in Paris was commissioned by the President of France in 1984 and designed by the architect I. M. Pei. The structure, which is constructed entirely of glass sections and metal rods, reaches a height of 21.6 m. Its square base has sides of 35 m.

Imagine you are the architect working on the Louvre pyramid. Make a list of five questions that need to be answered for you to build this pyramid.

- Find the total volume of the pyramid.
- Find the angle between a face and a base of the pyramid.
- Find the angle between an edge and the base of the pyramid.

Developing inquiry skills

Write down any similar inquiry questions you might ask if you were building the Louvre pyramid.

Research some examples of similar structures from around the world, for example, the step-pyramid at Koh Ker in Cambodia. What different questions would you ask if you were building these?

Think about the questions in this opening problem and answer any you can. As you work through the chapter, you will gain mathematical knowledge and skills that will help you to answer them all.

Before you start

You should know how to:

Use the Pythagorean theorem to find the side lengths of right triangles.

1 Find the length of AB in this triangle.

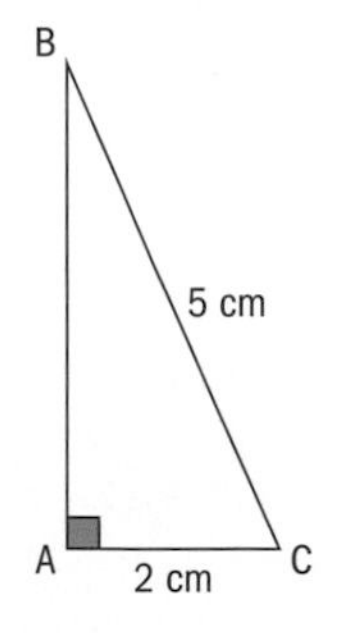

$AB^2 = BC^2 - AC^2$

$AB^2 = 5^2 - 2^2$

$AB^2 = 21$

$AB = \sqrt{21} = 4.58$

2 Convert units for length, area and volume.

$1\,\text{m} = 100\,\text{cm}$

$(1\,\text{m})^2 = (100\,\text{cm})^2 \quad (1\,\text{m})^3 = (100\,\text{cm})^3$

$1\,\text{m}^2 = 10\,000\,\text{cm}^2 \quad 1\,\text{m}^3 = 1\,000\,000\,\text{cm}^3$

$1000\,\text{cm}^3 = 1000\,\text{cc} = 1$ litre

Skills check

Click here for help with this skills check

1 Find the value of x.

a **b**

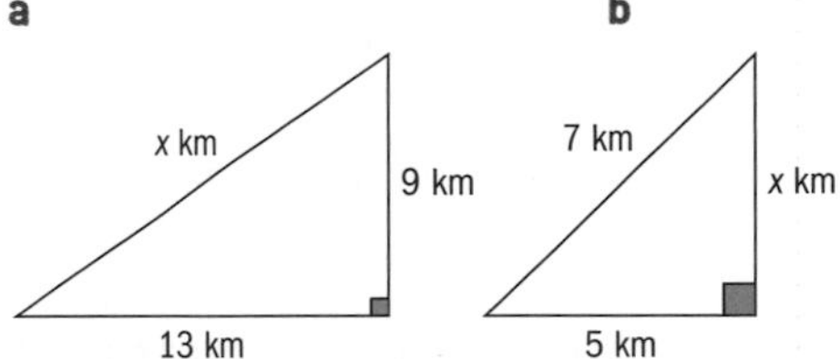

2 a Find the area of a rectangle of base 90 cm and height 1.4 m in:

i cm^2 **ii** m^2

b Find the number of litres of water in a cylindrical tank of base radius 1 m and height 4 m.

11.1 The geometry of 3D shapes

Investigation 1

Work with a partner for this investigation.

You will need a piece of plain paper. Draw this shape (it is called a net), make each side 4 cm and cut it out.

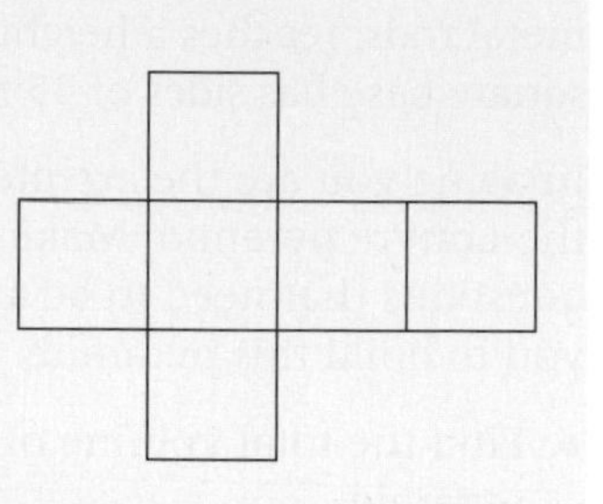

Fold along all of the lines to make a cube. You might want to tape the edges to keep them together.

Make one corner (vertex) the origin and label that O and the others ABCDEFG as in the diagram.

Now put dots or lines to mark 1, 2, 3, 4 on the x, y and z axes at 1 cm intervals.

You now have cube OABCDEFG with sides 4 cm.

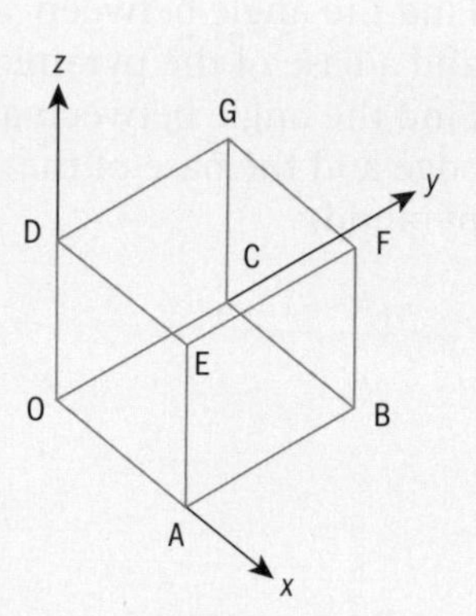

The rule for our game is that you have to start at O, and you can only move your finger in the x, y or z direction along the outside of your cube.

A lies on the x-axis, C lies on the y-axis and D lies on the z-axis.

From O you move 4 units along the x-axis to get to A. A has coordinates (4, 0, 0).

From O you move 4 units along the x-axis and 4 units along the y-axis to get to B. B has coordinates (4, 4, 0).

From O you move 4 units along the x-axis, 4 along the y-axis and 4 along the z-axis to get to F. F has coordinates (4, 4, 4).

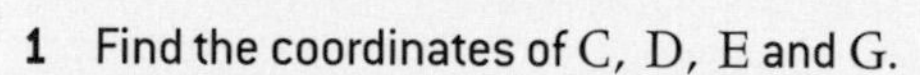

1 Find the coordinates of C, D, E and G.

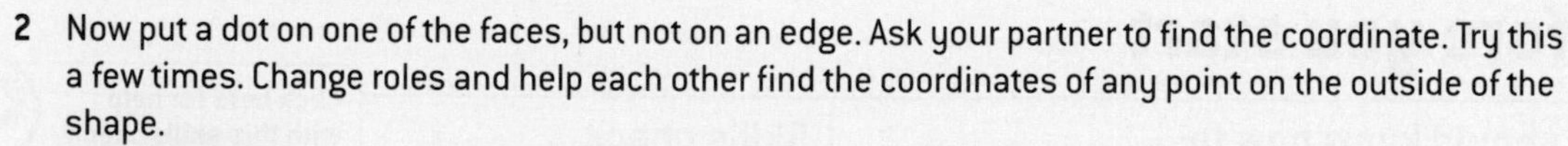

2 Now put a dot on one of the faces, but not on an edge. Ask your partner to find the coordinate. Try this a few times. Change roles and help each other find the coordinates of any point on the outside of the shape.

3 **Conceptual** How are 3D coordinates used to find measures?

4 **Conceptual** How is the Pythagorean theorem related to the distance formula?

5 Can you find why the coordinate of the point at the centre of the cube, halfway from O to F, is (2,2,2)?

6 Can you find the length of OB using the Pythagorean theorem? What about the length of OF?

The midpoint and distance formulas

In investigation 1, you will have noticed that the midpoint of OF was half of OA, half of OC and half of OD. The coordinates of the midpoint of a line segment are the averages of the x-, y- and z-coordinates of the endpoints.

The midpoint of (x_1, y_1, z_1) and (x_2, y_2, z_2) is $\left(\frac{x_1 + x_2}{2}, \frac{y_1 + y_2}{2}, \frac{z_1 + z_2}{2}\right)$.

In our cube, the length OF can be measured using the Pythagorean theorem. When you deal with 3D figures, it is easier to show the individual 2D shapes that you are working with.

In triangle OAB

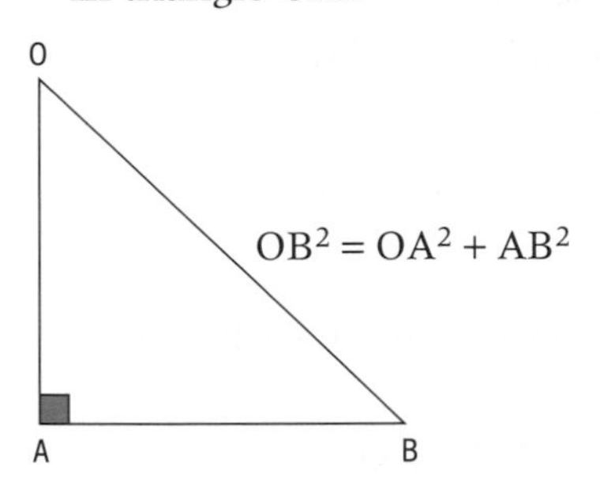

In triangle OBF

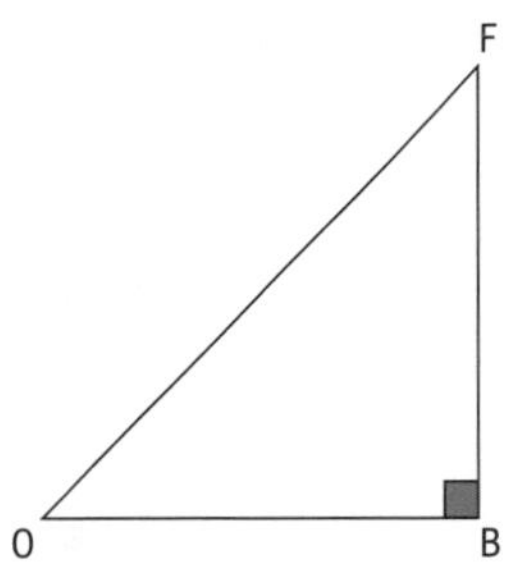

$OF^2 = OB^2 + BF^2$

$OF^2 = OA^2 + AB^2 + BF^2$

$OF = \sqrt{OA^2 + AB^2 + BF^2}$

OA is the difference in x, $x_2 - x_1$, AB is the difference in y, $y_2 - y_1$ and BF is the difference in z, $z_2 - z_1$.

The distance (d) between (x_1, y_1, z_1) and (x_2, y_2, z_2) is

$$d = \sqrt{(x_2 - x_1)^2 + (y_2 - y_1)^2 + (z_2 - z_1)^2}.$$

HINT

[AB] is used to denote the line segment with endpoints A and B.

AB is used to denote the length of [AB].

So when we say "Find AB.", we are asking for the length of [AB]. Or we could say "Find the length of [AB].", or "Find the length AB."

Example 1

Find the midpoint and length of [AB] when A is (1, 2, 5) and B is (3, 3, 7).

Midpoint of AB

$$\left(\frac{x_1 + x_2}{2}, \frac{y_1 + y_2}{2}, \frac{z_1 + z_2}{2}\right)$$

$$= \left(\frac{1+3}{2}, \frac{2+3}{2}, \frac{5+7}{2}\right) = (2, 2.5, 6)$$

Length AB

$$AB = \sqrt{(x_2 - x_1)^2 + (y_2 - y_1)^2 + (z_2 - z_1)^2}$$

$$AB = \sqrt{(3-1)^2 + (3-2)^2 + (7-5)^2}$$

$$AB = \sqrt{4+1+4}$$

$$AB = \sqrt{9}$$

$$AB = 3$$

One of the best ways to show working and gain credit in IB exams is to write down a formula, substitute and evaluate, and good mathematical layout is to keep the = under, or on the same line as, the = as you go down the page.

Exercise 11A

1 The cuboid, OABCDEFG is such that OA = 3, OC = 4, and OD = 2.

A lies on the *x*-axis, C lies on the *y*-axis and D lies on the *z*-axis.

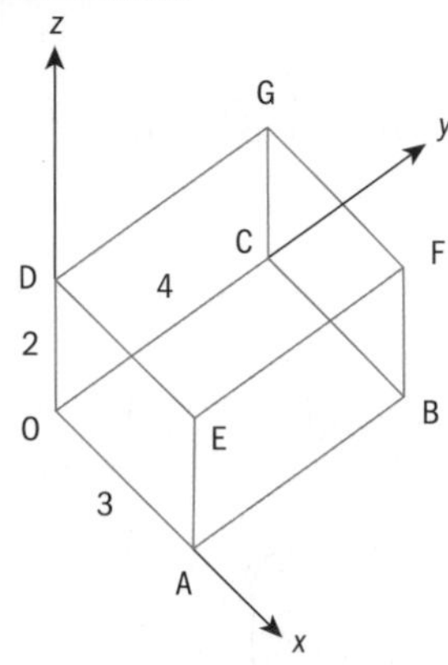

Write down the coordinates of:

a A **b** B

c E **d** F

e Find the midpoint of [OF].

f Find OF.

2 Find the midpoint of the line segment with the given endpoints.

a (−4, 4, 3) and (5, −1, 3)

b (−4, 4, 5) and (−2, 2, 9)

c (5, 2, −4) and (−4, −3, −8)

d (−5.1, −2, 9) and (1.4, 1.7, 11)

3 Find the distance between each pair of points.

a (2, 3, 5) and (4, 3, 1)

b (−3, 7, 2) and (2, 4, −1)

c (−1, 3, −4) and (1, −3, 4)

d (2, −1, 3) and (−2, 1, 3)

4 Find the distance between each pair of points.

a (1, 2, 3) and (−5, −6, −7)

b (0, −4, 2) and (4, 0, 5)

c (−1, −1, −1) and (1, 2, 3)

d (4, 1, −3) and (1, 1, 1)

Investigation 2

You should be familiar with the names, surface area and volume of prisms.

1 **a** Draw four prisms.

b Write down five things that you know about prisms.

You might want to research online if you do not remember facts about prisms.

2 **a** Draw a cylinder.

b Write down five things that you know about a cylinder.

3 Share your thoughts with the class to generate a pool of information. Had you forgotten anything that you now remember? Are there any unfamiliar things that you will have to go back and review?

4 **Conceptual** How can you classify or describe a three-dimensional shape?

Pyramids, cones and spheres

> The base of a **pyramid** is a polygon, and the three or more triangular faces of the pyramid meet at a point known as the apex. In a right-angled pyramid, the apex is vertically above the centre of the base.

The figure on the left is a square-based pyramid, but there are also tetrahedrons (triangular-based pyramids), hexagonal-based pyramids and so on.

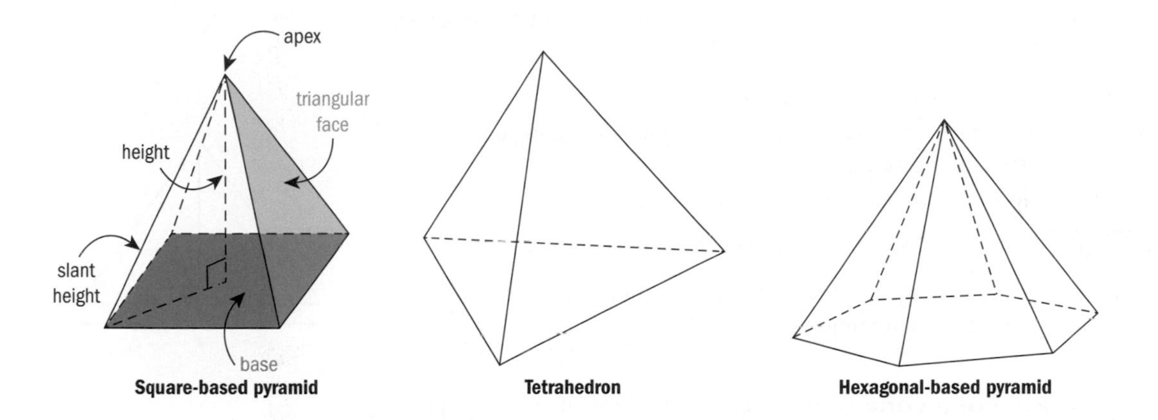

There is a close relationship between pyramids and cones. The base of a cone is a circle rather than a polygon, so it does not have triangular faces.

A sphere is defined as the set of all points in three-dimensional space that are equidistant from a central point. Half of a sphere is called a hemisphere.

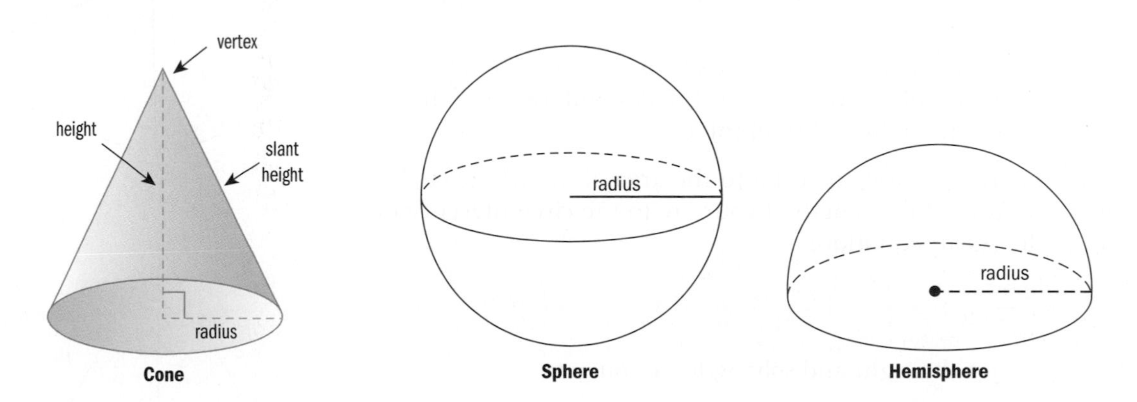

Surface area

The surface area of a pyramid is the sum of the areas of all of its faces.

Investigation 3

Square-based pyramids have a total of five faces. You can see that the faces are a square and four triangles.

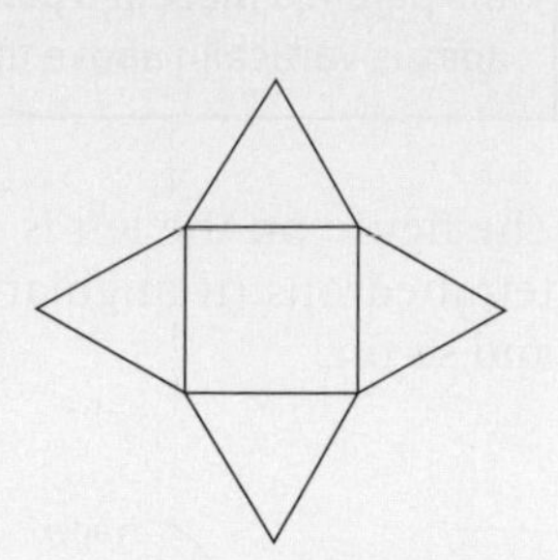

1 **Conceptual** How do you find the surface area of 3D shapes?

2 **Factual** What is a net?

3 Draw this net on plain paper, the size is not important, but a larger shape is easier to work with. It has a square in the middle and identical isosceles triangles on each side of the square.

   a Fold up your shape to make a pyramid.

   b Label the sides of the square x and the height of the triangles, l.

   c Find the area of the square.

   d Find the area of one triangle and multiply it by 4.

4 **Factual** What is the surface area of a right-angled pyramid?

5 Add your two results together to make a formula for the surface area of a square-based pyramid.

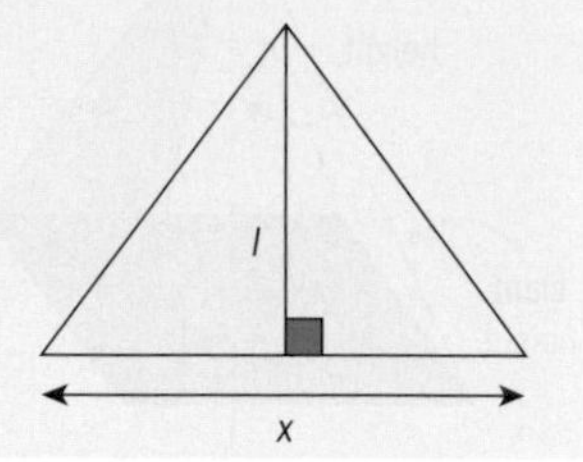

The surface area of a cone is the sum of the areas of the circular base and the curved side.

You can make a cone by cutting a circle out of plain paper, then cutting a sector out of the circle, and then folding the shape so that the straight edges touch.

l is the slant height of the cone.

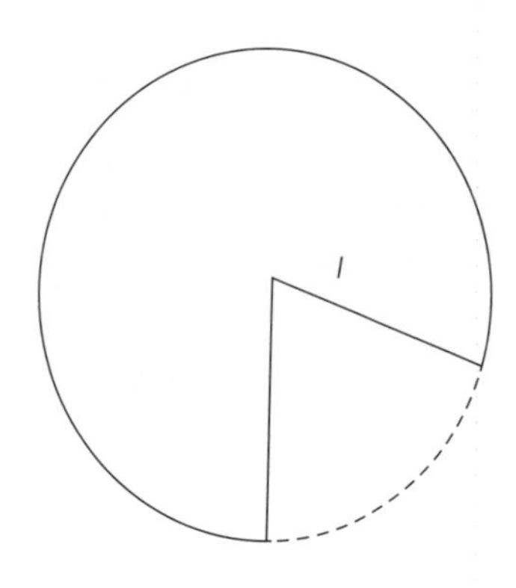

The area of the circle that you started with was πl^2 and its circumference was $2\pi l$.

After you cut your shape, the arc of your sector that you are left with wraps around the base of the cone, and is the circumference of the base, $2\pi r$, where r is the base radius of the cone.

The ratio of the area (x) of your sector to the area of the whole circle is the same as the ratio of the length of your arc to the circumference of the whole circle. As an equation:

$$\frac{x}{\pi l^2} = \frac{2\pi r}{2\pi l}$$

Cancelling by 2π on the right and solving for x you get:

$$x = \pi r l$$

This is called the area of the curved surface of the cone.

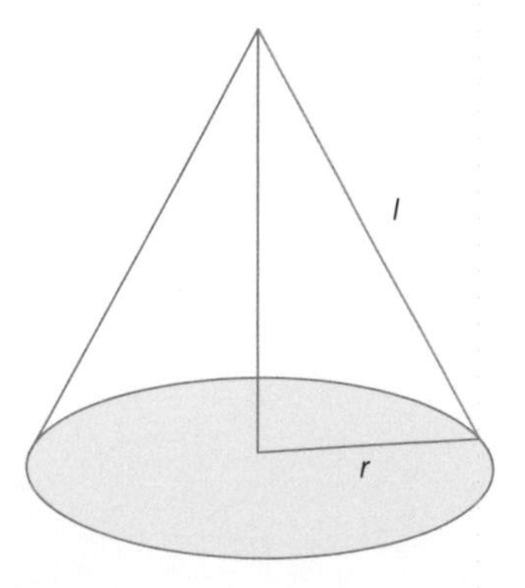

To get the total surface area of the cone, you add the area of the base to get $\pi r^2 + \pi r l$.

Here is a summary of the surface area formulas along with the volume formulas.

Shape	Surface area	Volume
Pyramid	The sum of all of the faces. $x^2 + 2xl$ for a squared-based pyramid, where x is the length of the sides of the square base and l is the slant height.	$V = \frac{1}{3}(\text{base area} \times \text{height})$
Cone	$\pi r^2 + \pi rl$, where r is the base radius and l is the slant height.	$V = \frac{1}{3}\pi r^2 h$
Sphere	$4\pi r^2$	$V = \frac{4}{3}\pi r^3$

Example 2

Find the volume and surface area of the cone.

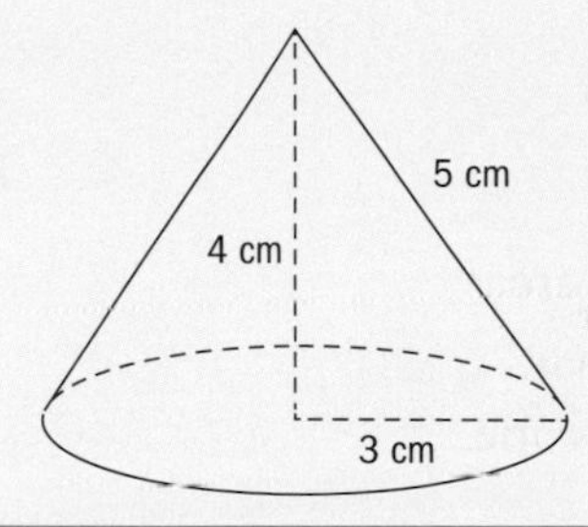

$V = \frac{1}{3}\pi r^2 h = \frac{1}{3}\pi(3)^2 4 = 37.7\,\text{cm}^3$ $\text{SA} = \pi r^2 + \pi rl = \pi(3)^2 + \pi(3)(5) = 75.4\,\text{cm}^2.$	Substitute into the selected formulas.

Exercise 11B

1 Find the surface area of each figure.

a

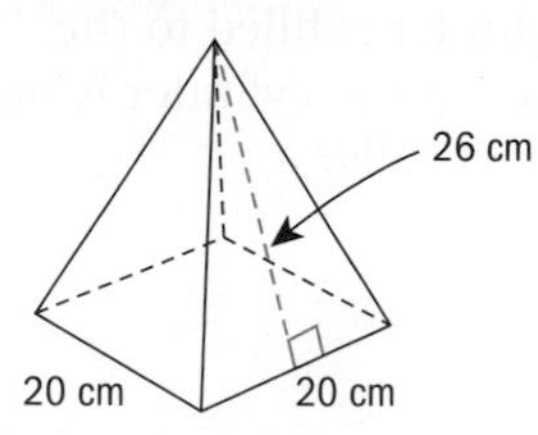

b

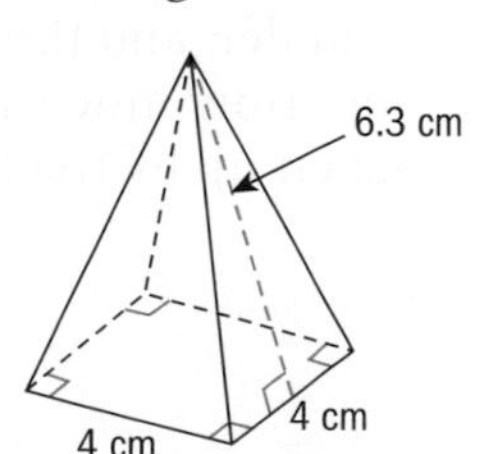

c

d

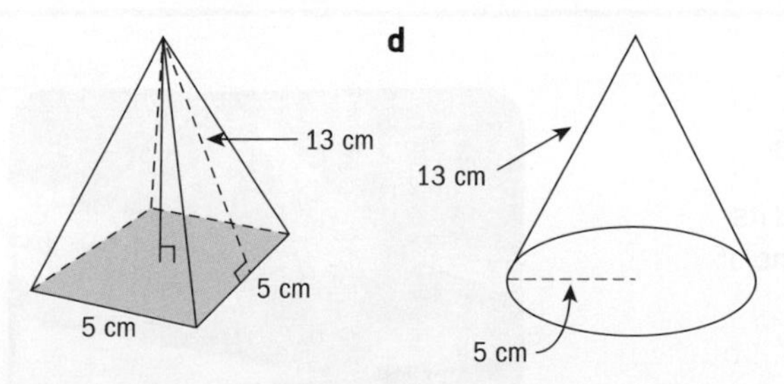

e

f

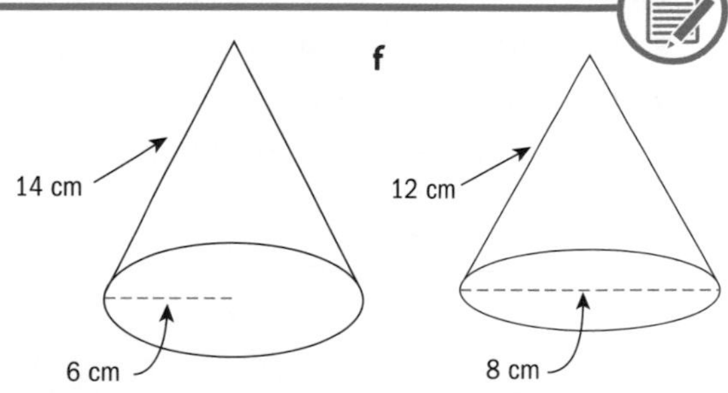

2 Find the surface area and volume of each sphere.

a

b

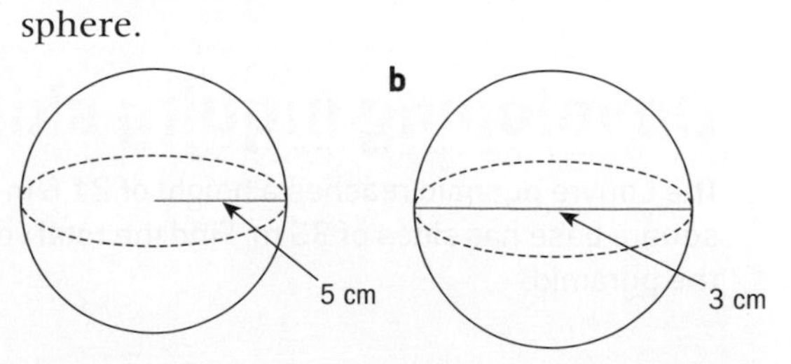

3 Find the volume of each figure.

a

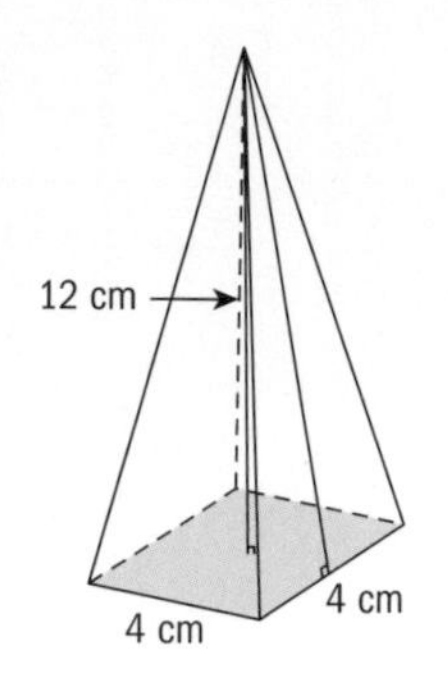

b

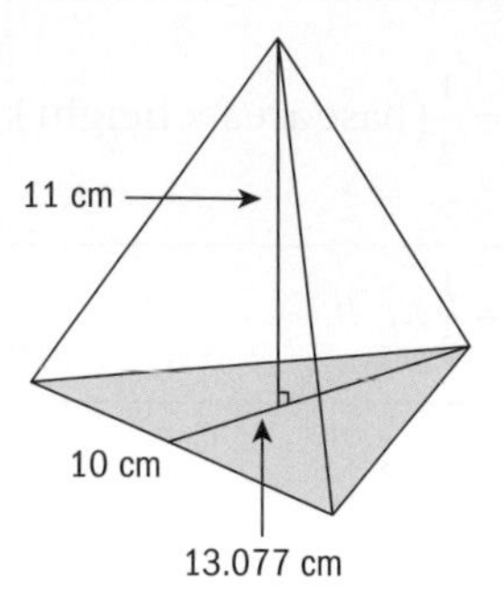

c

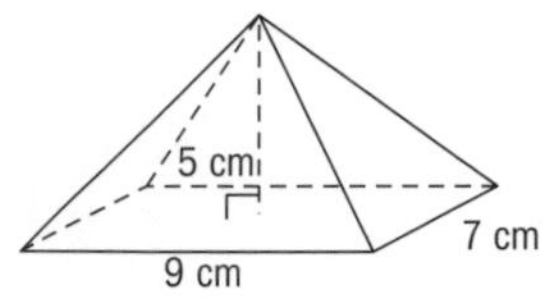

4 Find:

a the curved surface area

b the total surface area

c the volume of this cone.

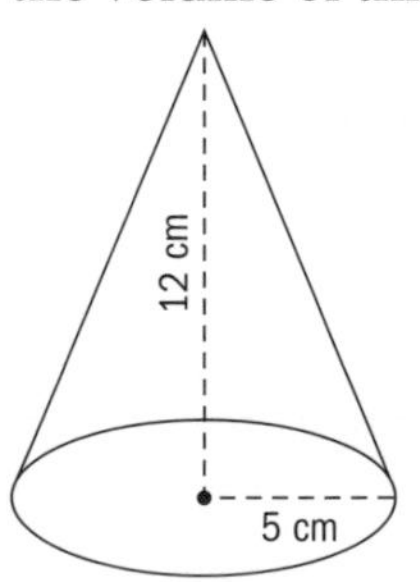

5 Find:

a the curved surface area

b the total surface area

c the volume of this hemisphere.

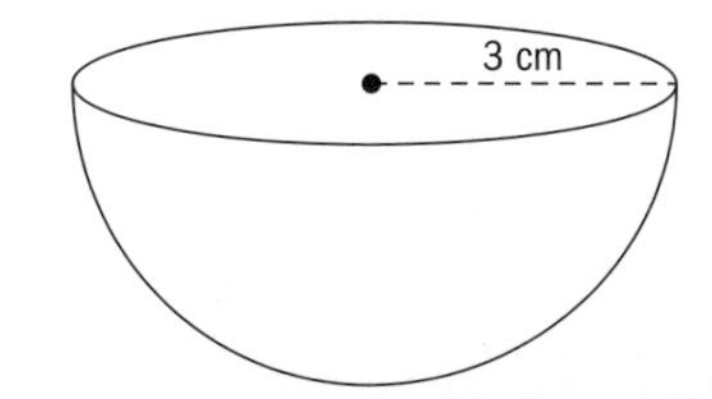

6 Benji has an ice cream cone. The ice cream cone is in the shape of a hemisphere sitting on top of a cone.

a What is the volume of the ice cream cone?

b What is the surface area?

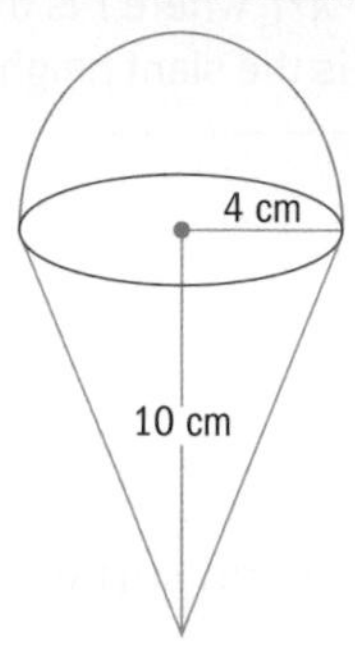

7 A water tank is in the shape of a cone sitting on top of a cylinder. How many litres of water can it contain?

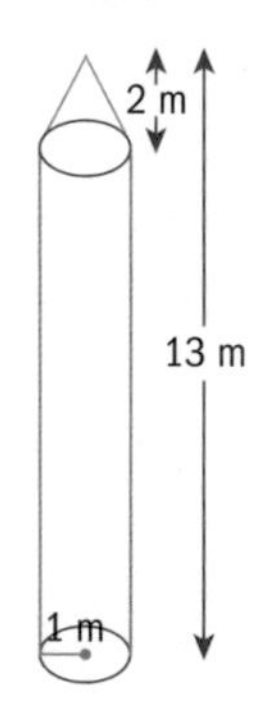

8 A cylindrical can holds three tennis balls. Each ball has a diameter of 6 cm, which is the same diameter as the interior of the cylinder, and the cylinder is filled to the top. How much space in the cylinder is not taken up by the tennis balls?

Developing inquiry skills

The Louvre pyramid reaches a height of 21.6 m and its square base has sides of 35 m. Find the total volume of the pyramid.

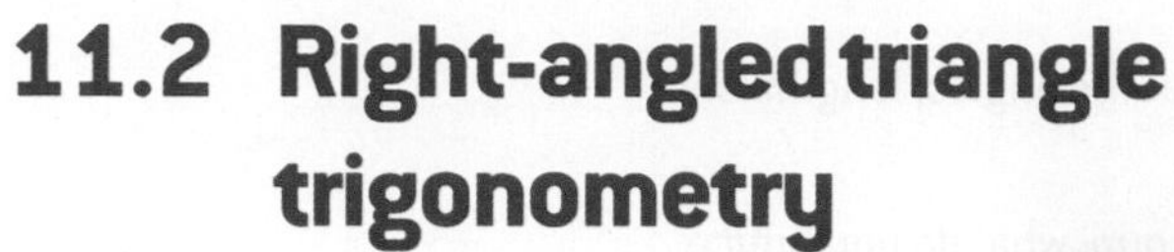

11.2 Right-angled triangle trigonometry

Trigonometry is a branch of mathematics that uses triangles to help solve problems. Initially, trigonometry dealt with relationships among the sides and angles of triangles and was used in the development of astronomy, navigation, building and surveying.

Investigation 4

1 Write down what you already know about right-angled triangles.

Did you include the Pythagorean theorem?

2 Measure the sides of each of the three triangles on the right, then copy and complete the table below for the angle of 20°. For each triangle, form ratios using the required lengths and round to three significant figures.

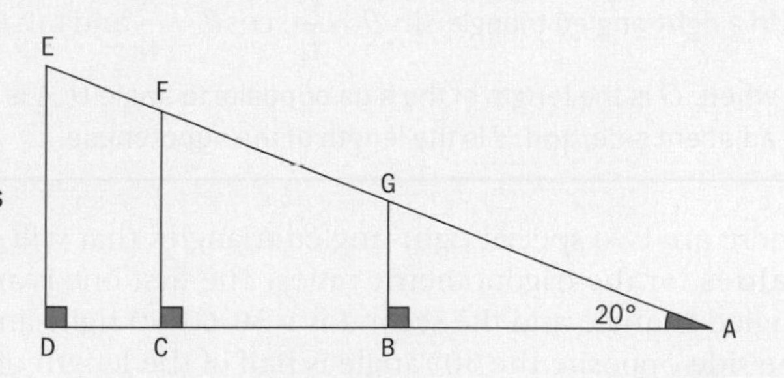

ΔEAD	ΔFAC	ΔGAB
$\frac{EA}{AD} \approx$	$\frac{FA}{AC} \approx$	$\frac{GA}{AB} \approx$
$\frac{AD}{EA} \approx$	$\frac{AC}{FA} \approx$	$\frac{AB}{GA} \approx$
$\frac{EA}{DE} \approx$	$\frac{FA}{CF} \approx$	$\frac{GA}{BG} \approx$
$\frac{AD}{DE} \approx$	$\frac{AC}{CF} \approx$	$\frac{AB}{BG} \approx$
$\frac{DE}{EA} \approx$	$\frac{CF}{FA} \approx$	$\frac{BG}{GA} \approx$
$\frac{DE}{AD} \approx$	$\frac{CF}{AC} \approx$	$\frac{BG}{AB} \approx$

Continued on next page

3 Write a sentence describing what you notice about the angles, lengths and ratios of your data.

4 After comparing your data with the data of your group, what do you notice about your group's data?

5 Write a hypothesis about the relationships among the length of the sides of the right-angled triangles based on the information that your group gathered and discussed. Get ready to share this with the class.

6 Using your calculator, find sine, cosine and tangent of 20°. What do you notice?

7 **Factual** What are the three trigonometric ratios?

8 **Factual** What are sine, cosine and tangent?

9 **Conceptual** How do trigonometric ratios allow us to solve problems involving right-angled triangles?

The Greek letter θ, pronounced "theta", is used to denote an angle.

From the investigation above, you should have seen that:

In a right-angled triangle $\sin\theta = \frac{\text{O}}{\text{H}}$, $\cos\theta = \frac{\text{A}}{\text{H}}$ and $\tan\theta = \frac{\text{O}}{\text{A}}$ where O is the length of the side opposite to angle θ, A is the length of the adjacent side, and H is the length of the hypotenuse.

HINT

The phrase **SOH-CAH-TOA** can help you remember how to use these trigonometric ratios.

There are two special right-angled triangles that will give you **exact values** for the trigonometric ratios. The first one is an isosceles right-angled triangle and the second is a 30-60-90 right-angled triangle where the side opposite the 30° angle is half of the length of the hypotenuse.

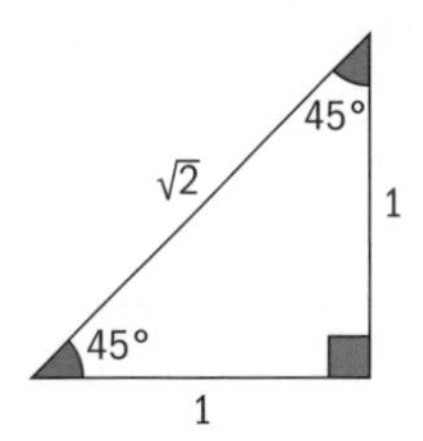

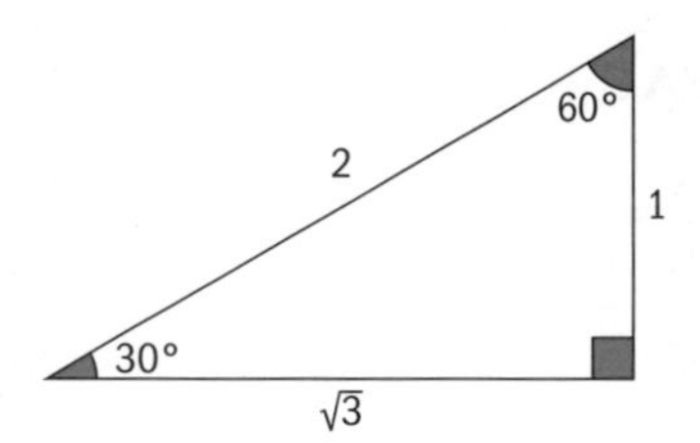

These results are shown in the table below.

θ	30°	45°	60°
$\sin\theta$	$\frac{1}{2}$	$\frac{\sqrt{2}}{2}$	$\frac{\sqrt{3}}{2}$
$\cos\theta$	$\frac{\sqrt{3}}{2}$	$\frac{\sqrt{2}}{2}$	$\frac{1}{2}$
$\tan\theta$	$\frac{\sqrt{3}}{3}$	1	$\sqrt{3}$

EXAM HINT

You will need to remember these values for your exams, as they may be required for questions in which a calculator is not used.

Investigation 5

1 Copy and complete the table for the given values of H

$H =$	2	3	4	5
$O =$		$3\sin\theta$		
$A =$	$2\cos\theta$			

2 Write down $\sin\theta$ in terms of O and H.

3 Rearrange this to make O the subject.

4 Write down $\cos\theta$ in terms of A and H.

5 Rearrange this to make A the subject.

6 Explain to a classmate, what you have created.

7 **Conceptual** How do trigonometric ratios allow us to solve problems involving right-angled triangles?

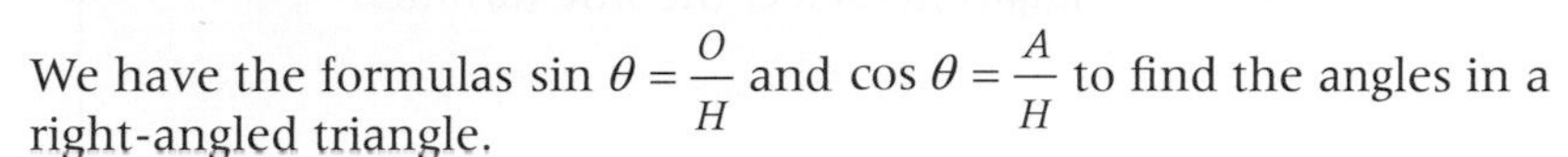
We have the formulas $\sin\theta = \frac{O}{H}$ and $\cos\theta = \frac{A}{H}$ to find the angles in a right-angled triangle.

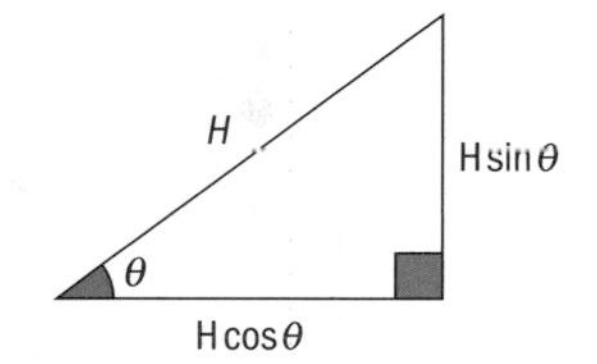

The investigation has seen you find two formulas for the sides of a right-angled triangle.

$O = H\sin\theta$ and $A = H\cos\theta$

The angle between a line and a plane

The angle between the line EF and the plane ABCD is defined as the angle between EF and its projection on the plane ABCD.

To find the angle between the line and the plane:

- Draw a perpendicular line, EG, from E to the plane. (G is called the foot of the perpendicular.)
- Join E to F.
- The required angle is $E\hat{F}G$.

The angle between an edge or a face and the base of a 3D shape

Using the method from a line and a plane, you can find angles in a 3D shape. In this pyramid:

- Angle A is an angle between a face and the base.
- Angle B is an angle between an edge and the base.

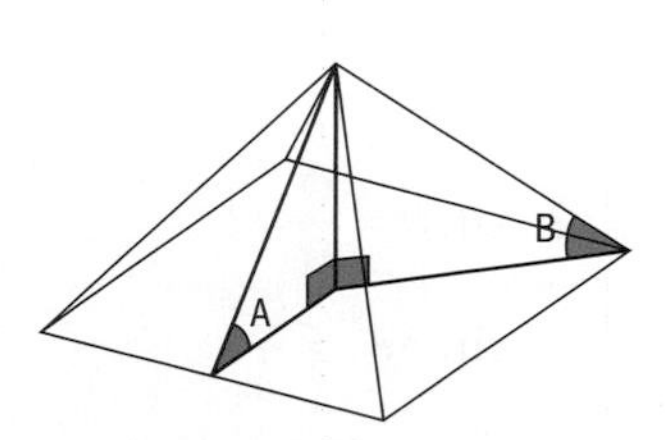

Example 3

1 Find the value of θ.

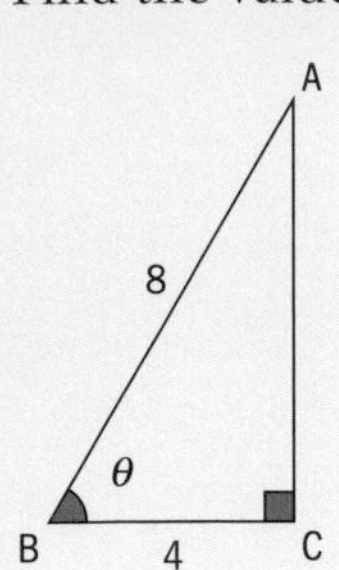

2 Find the value of x.

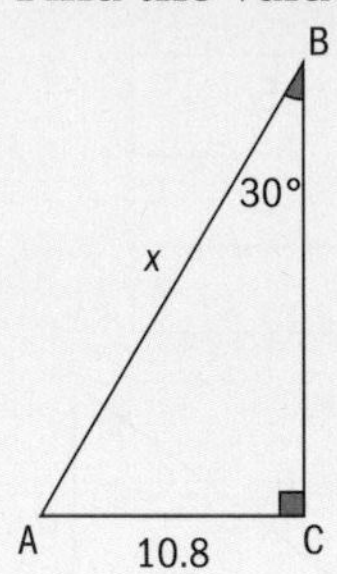

1 $\cos\theta = \frac{A}{H} = \frac{4}{8}$ $\theta = 60°$	AB is the hypotenuse and BC is the adjacent. Select the cosine formula.
2 $\sin\theta = \frac{O}{H}$ $\sin 30 = \frac{10.8}{x}$	AB is the hypotenuse and AC is the adjacent. Select the sine formula.
$x = \frac{10.8}{\frac{1}{2}} = 21.6$	Rearrange to make x the subject.

Example 4

1 Find $\hat{K}$.

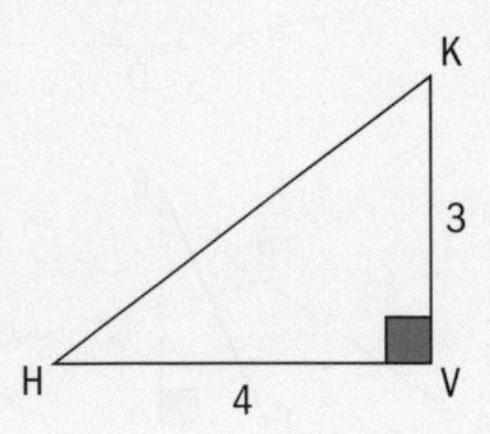

2 Find the value of r.

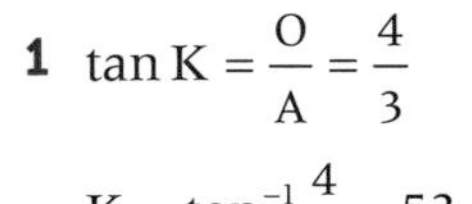

1 $\tan K = \frac{O}{A} = \frac{4}{3}$ $K = \tan^{-1}\frac{4}{3} = 53.1°$	HV is the side opposite angle K, and KV is the adjacent side. Select the tangent formula. Find the inverse tangent.
2 $\sin\theta = \frac{O}{H}$ $\sin 38° = \frac{21}{r}$ $r = \frac{21}{\sin 38°} = 34.1$	r is the hypotenuse and 21 is the opposite side. Select the sine formula. Rearrange to make r the subject and evaluate.

Exercise 11C

1 Find the value of θ:

a

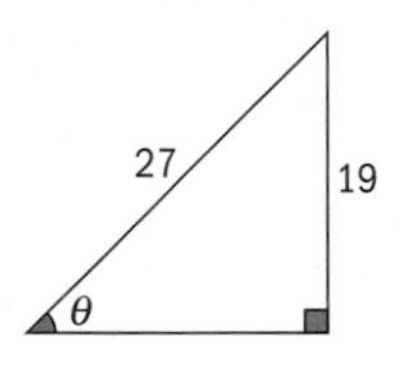

b

c

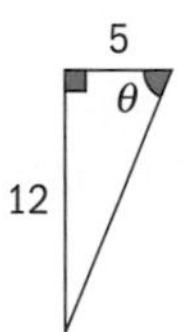

d

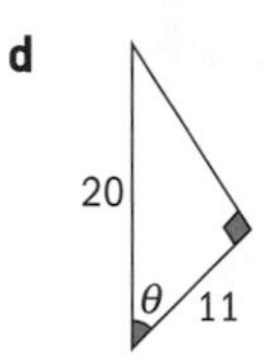

2 Find the value of x:

a

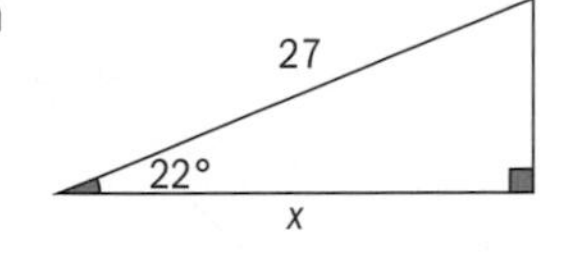

b

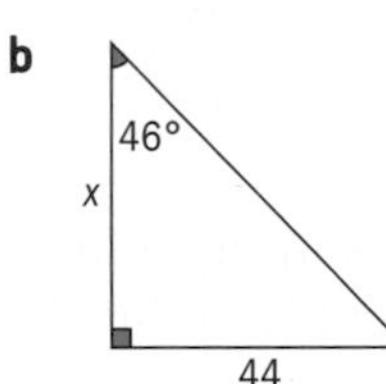

c

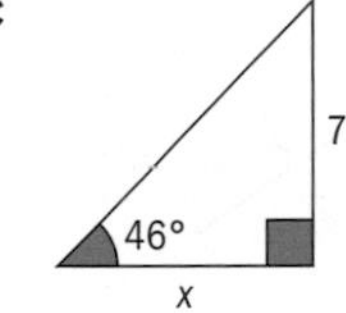

d

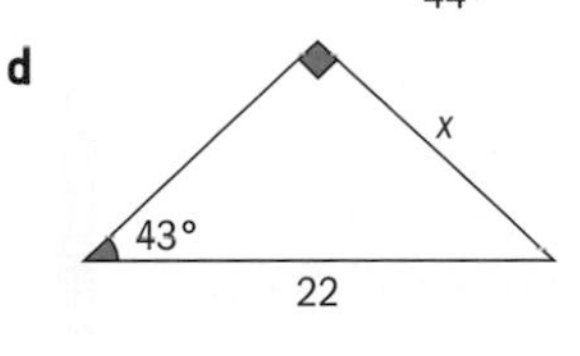

3 Find the value of h and $\hat{a}$:

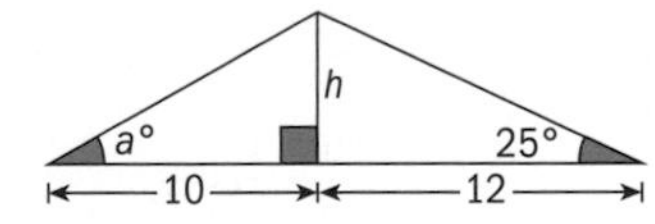

4 Find the value of p and q:

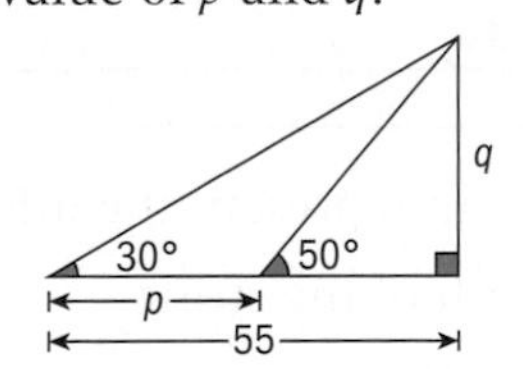

5 Find the value of x:

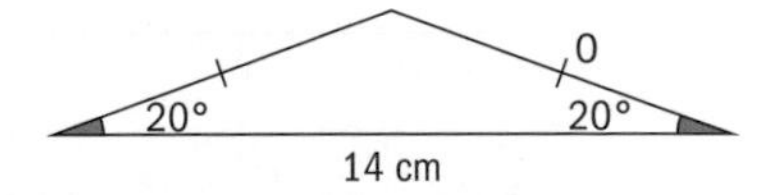

6 Find the height of a cone of base radius 6 cm and $\theta = 30°$.

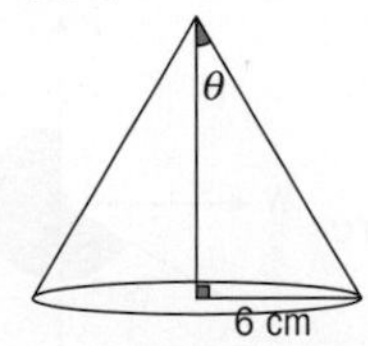

7 The diagram shows a rectangular-based right-angled pyramid ABCDT in which AD = 20 cm, CD = 15 cm, TN = 30 cm.

a Find the volume of the pyramid.

b Find TC.

c Find the angle between the edge [TC] and the base ABCD.

M is the midpoint of [DC].

d Find TM.

e Find the angle between the face TCD and the base ABCD, $T\hat{M}N$.

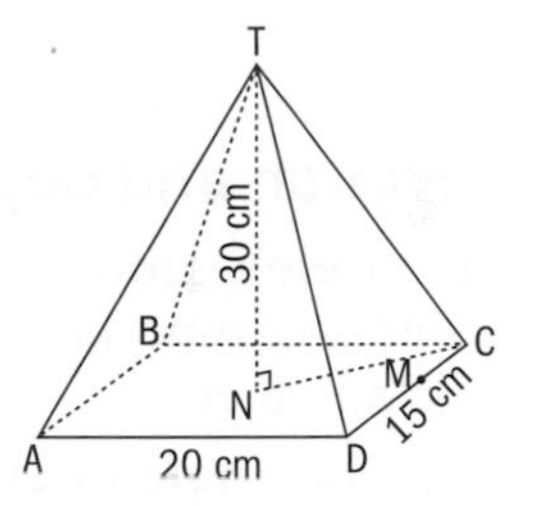

8 The Great Pyramid of Giza is a square-based pyramid. The base of the pyramid has side length 230.4 m and the vertical height is 146.5 m. The Great Pyramid is represented in the diagram below as ABCDE. The vertex E is directly above the centre O of the base. M is the midpoint of BC.

a Calculate the volume of the pyramid.

b Find EM.

c Show that the angle between [EM] and the base of the pyramid is 52° correct to 2 significant figures.

d Find the total surface area of the pyramid including the base.

e Find the angle between an edge and the base.

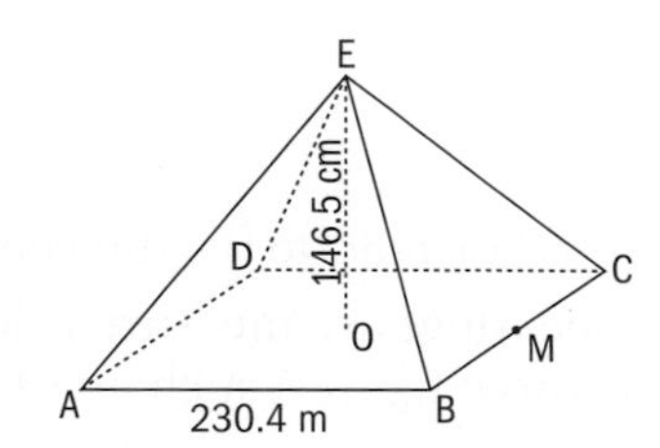

Solving problems with right-angled triangles

Example 5

A tree casts a shadow of 10 m when the sun is at an angle of 40° to the ground. How tall is the tree?

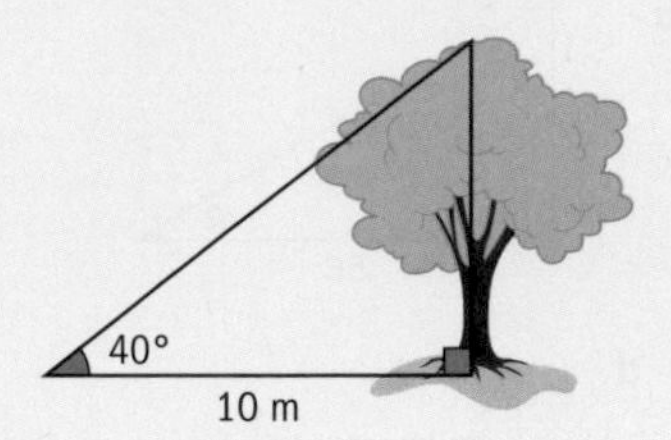

$\tan 40° = \frac{\text{height}}{10}$ height $= 10 \tan 40° = 8.39$ m	Draw a diagram to represent the information. Select the tangent formula.

Angles of elevation and depression

The terms **angle of elevation** and **angle of depression** are often used in word problems, especially those involving people as they look up at or down on an object.

When looking up, the **angle of elevation** is the angle up from the horizontal and the line from the object to the person's eye.

When looking down, the **angle of depression** is the angle down from the horizontal and the line from the object to the person's eye.

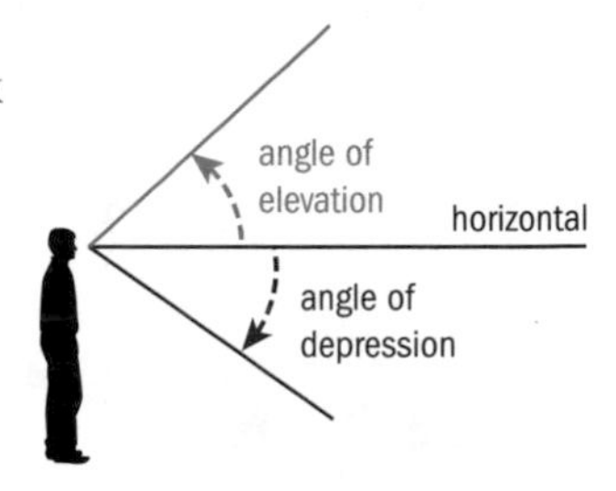

Example 6

Rondo is flying a kite. The kite string is at an angle of elevation of 60°. If Rondo is standing 30 m from the point on the ground directly below the kite, find the length of the kite string.

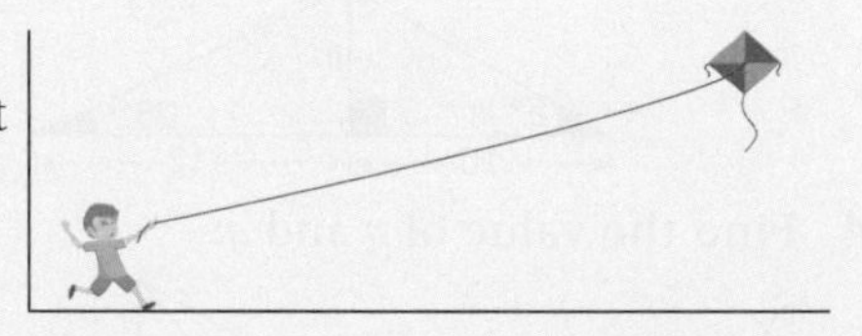

$\cos 60° = \frac{30}{l}$ $l = \frac{30}{\cos 60°} = 60$ m	Draw a diagram to represent the information. Let l be the length of the string. Select the cosine formula.

Bearings

A bearing is used to indicate the direction of an object from a given point.

Three-figure bearings are measured clockwise from North, and can be written as three figures, such as 120°, 057°, or 314°.

Compass bearings are measured from either North or from South, and can be written such as N20°E, N55°W, S36°E, or S67°W, where the angle between the compass directions is less than 90°.

The bearing of A in the diagram may be written as 240° or as S60°W.

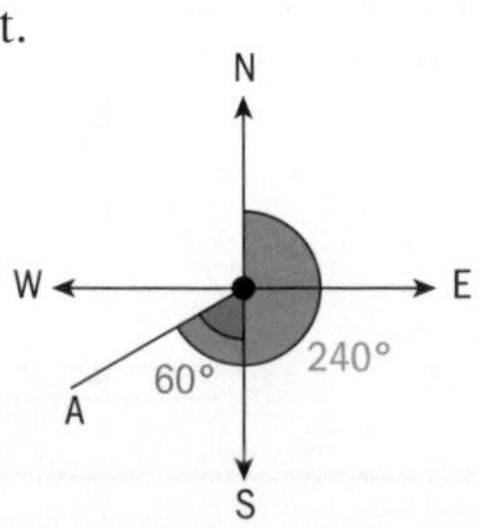

Example 7

Nina runs 5 km on a bearing of 120° from the start. How far south has she travelled?

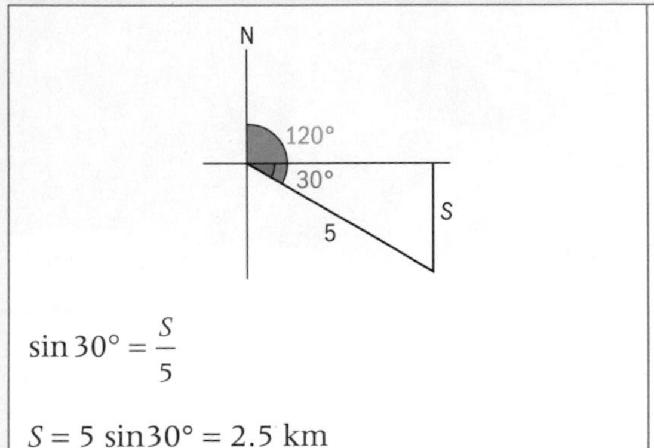	Draw a diagram to represent the information. Note that a bearing of 120° is the same direction as S60°E, or 30° south of east.
$\sin 30° = \frac{S}{5}$ $S = 5 \sin 30° = 2.5$ km	Let S be the distance south. Select the sine formula.

Exercise 11D

1 A tree casts a 12-metre shadow when the angle of elevation of the sun is 25°. How tall is the tree?

2 A kite in the air has its string tied to the ground. The length of the string is 50 m. Find the height of the kite above the ground when the angle of elevation of the string to the ground is 55°.

3 A 3 m ladder rests against a vertical wall so that the distance between the base of the ladder and the bottom of the wall is 0.8 m.

a Find the angle that the ladder makes with the wall.

b Find the height above the ground where the ladder meets the wall.

4 A man standing on a cliff looks down with an angle of depression of 40° to a boat in the water below the cliff. The cliff is 80 m high and the man's eyes are 1.5 m above the cliff. How far is the boat from the cliff?

5 Romeo stands 3 m away from the wall of Juliet's house. Juliet is standing on the balcony. Romeo raises his eyes 70° to look into Juliet's eyes. Romeo and Juliet are both the same height. How high is Juliet's balcony?

6 A boat travels on a bearing of N36°W for 25 km.

a How many miles north has the boat travelled?

b How many miles west has the boat travelled?

7 A boat is due south of a lighthouse and sails on a bearing of 292° for 51 km until it is due west of the lighthouse. How far away is it from the lighthouse now?

8 Wally looks out of his office window and sees the top and bottom of an apartment block 300 m away. The angle of elevation to the top of the office building is 23° and the angle of depression to the base of the building is 30°. How tall is the apartment block?

9 Pierre is standing on the viewing deck of the Eiffel Tower, 300 m above the ground. He looks down with an angle of depression of 40°, to see Alain, at point A. He looks down a further 32° to see Bailee, at point B.What is the distance from Alain to Bailee?

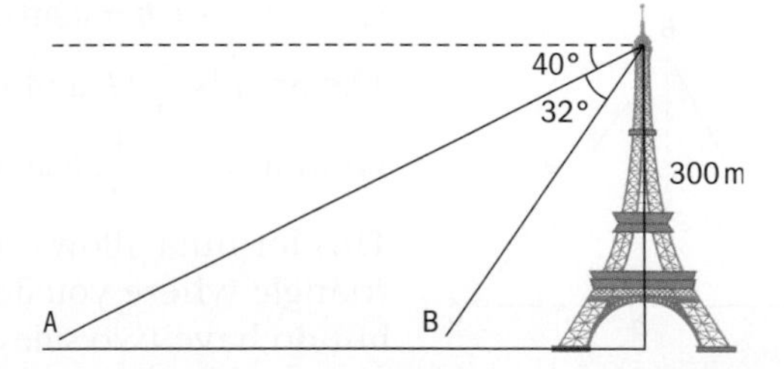

10 When you stand at the edge of the Grand Canyon at its deepest point and look horizontally across, the canyon is 498 m wide. To look down to the bottom of the canyon from the cliff edge above, you need to move your eyes through an angle of depression of 75°. When your eyes are 1.5 m above the cliff edge, how deep is the Grand Canyon at this point? Answer to the nearest metre.

Investigation 6

In a right-angled triangle there is a right angle and two other angles that add up to 90°. These are called complementary angles.

1 Use technology to copy and complete this table:

θ	$\sin\theta$	$\cos(90-\theta)$	$\cos\theta$	$\sin(90-\theta)$
30				
50				
60				
75				
80				

2 Write down, in words, two things that you notice.

3 Share these with a partner.

4 Now write your findings as mathematical equations.

5 Copy and complete the following sentences:

The sine of 40° is equal to the cosine …

The cosine of 70° is equal to the sine …

$\sin A = \cos B$ when …

The area of a triangle

You have used the formula $\frac{1}{2}$ base × height to find the area of a triangle previously. You can use that formula along with right-angled triangle trigonometry to produce another formula.

For this triangle, using the right-hand side of the shape,

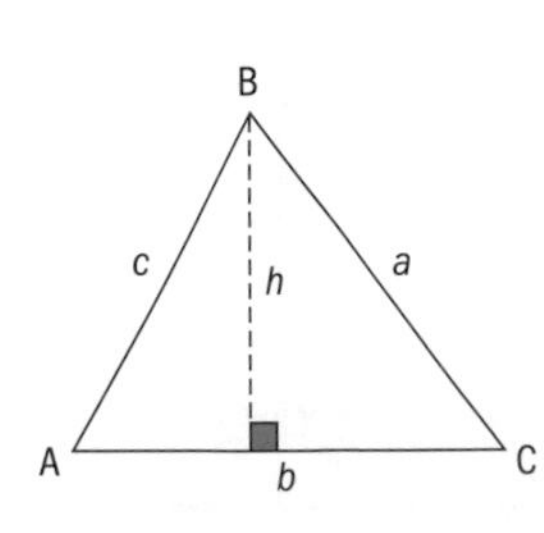

$\sin C = \frac{h}{a}$ or $h = a\sin C$

The area is $\frac{1}{2}bh$ and you substitute $h = a\sin C$ you have $A = \frac{1}{2}ab\sin C$.

This formula allows you to find the area of a triangle where you do not have a right angle but do have two sides and the angle between them (the included angle).

HINT

a is across from angle A, b is across from angle B, and so on.

> **The area of a triangle** can be found using $A = \frac{1}{2}ab\sin C$ where a and b are two sides and C is their included angle.

Example 8

Find the area of the triangle.

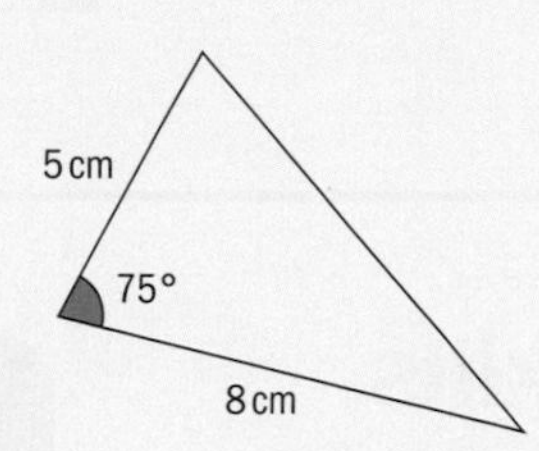

$\text{Area} = \frac{1}{2}ab\sin C$	Substitute the two sides and the included angle into the area formula.
$\text{Area} = \frac{1}{2}\times 5\times 8\times \sin 75°$	
$\text{Area} = 20\sin 75°$	$20\sin 75°$ is the exact answer.
$\text{Area} = 19.3\text{ cm}^2$	19.3 is the answer rounded to 3 s.f.

Exercise 11E

1 Find the area of each triangle.

a 8 cm, 80°, 6 cm

b 10 m, 125°, 15 m

c **d**

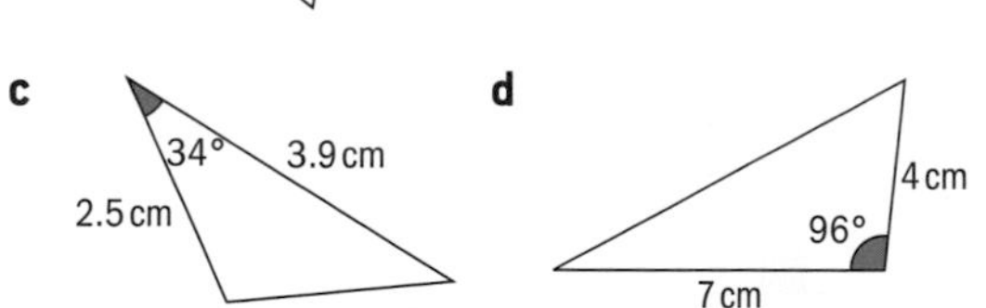

e **f**

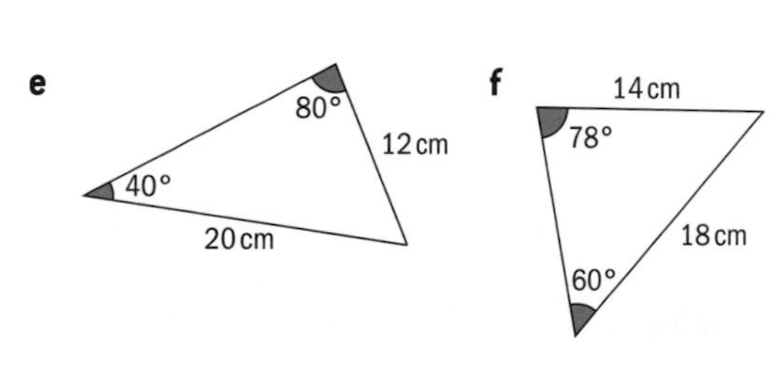

g

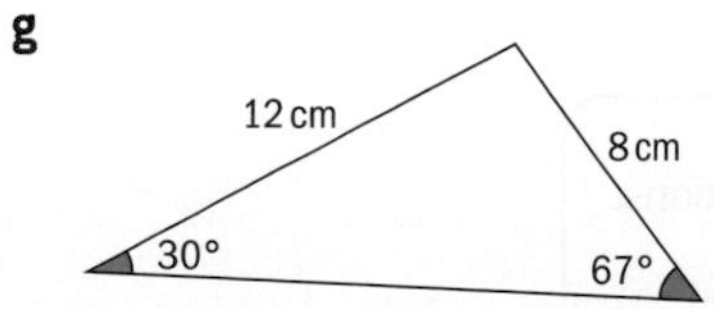

2 a Find the area of a triangle with two sides that measure 5 cm and 8 cm with an included angle of 39°.

b In triangle ABC, $a = 8$ cm, $b = 8$ cm and the area of triangle ABC is 16 cm^2. Find the measure of angle C.

3 Find the area of this parallelogram.

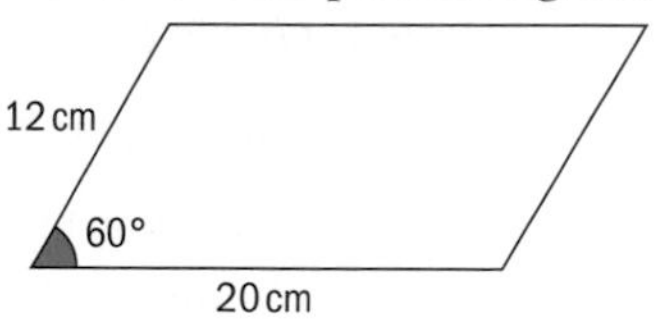

4 Find the surface area of tetrahedron ABCD when all of the faces are equilateral triangles with side length 10 cm.

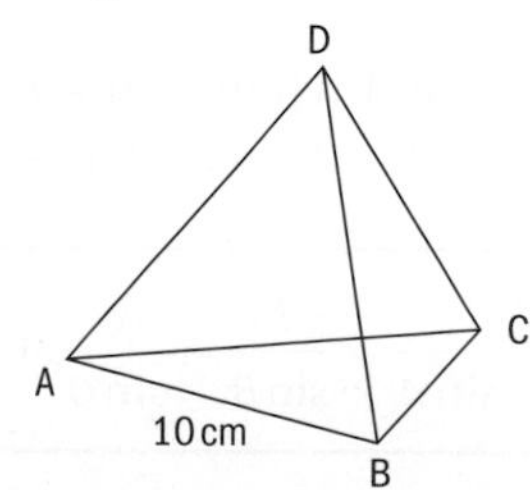

5 Find the area of a regular pentagon of radius 4 m.

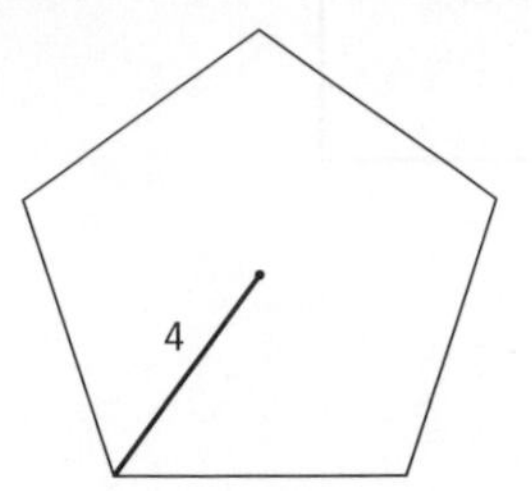

6 Find the value of x when the area of this triangle is $5\sqrt{3}$.

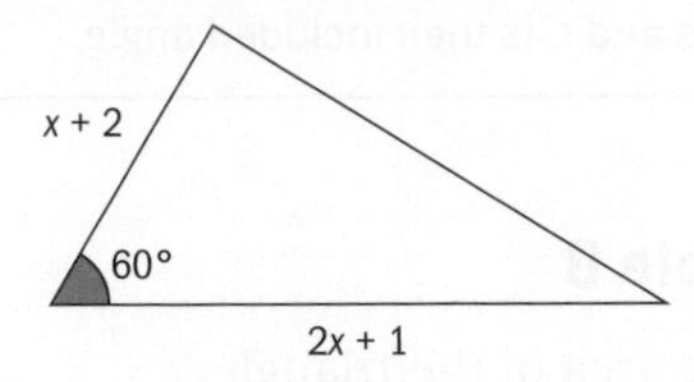

Developing inquiry skills

Find the angle between a face and a base of the Louvre pyramid.

Find the angle between an edge and the base.

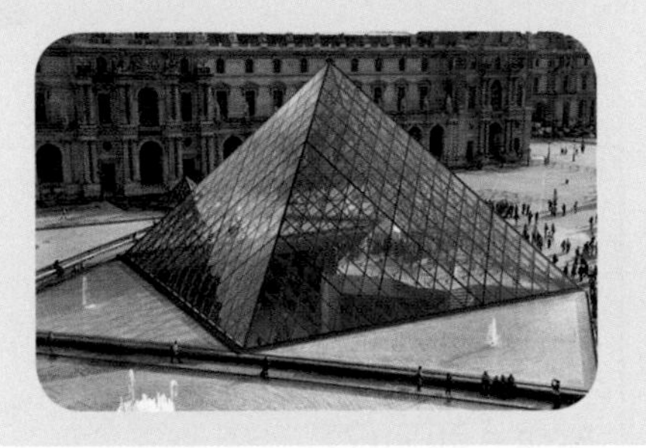

11.3 The sine rule

In section 11.2 you saw how to find the sides and angles of a right-angled triangle: given two sides, or one side and one acute angle. Now we will look at non-right-angled triangles.

Investigation 7

Let h be height of triangle ABC.

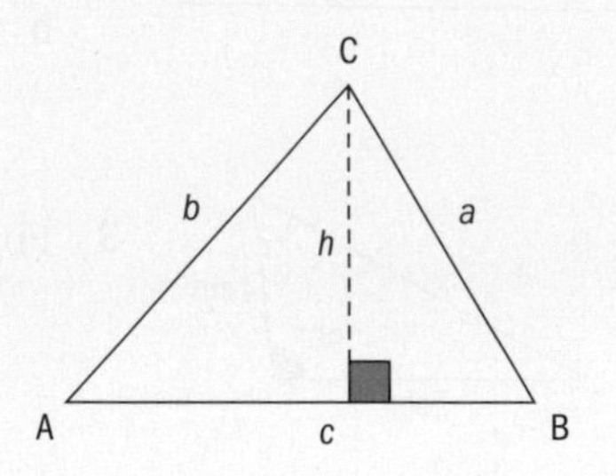

1 Write $\sin B$ in terms of h and a.

2 Now make h the subject.

3 Write $\sin A$ in terms of h and b.

4 Replace h with your answer to part **2**.

5 Divide both side of your equation by a.

You should now have $\frac{\sin A}{a} = \frac{\sin B}{b}$.

6 **Factual** What is the sine rule?

The sine rule is used when you are given either two angles and one side, or two sides and a non-included angle.

The sine rule is $\frac{a}{\sin A} = \frac{b}{\sin B} = \frac{c}{\sin C}$. It is usual to use only two of these fractions.

Example 9

1 Find BC.

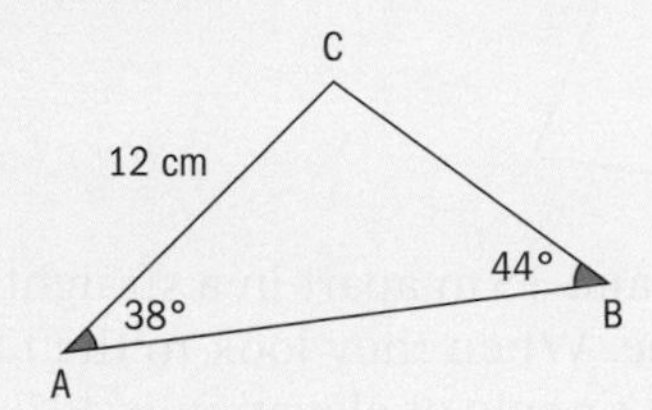

2 Find $\hat{C}$.

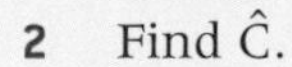

1 $\dfrac{a}{\sin A} = \dfrac{b}{\sin B}$	We have angles A and B and side b, so choose the two fractions with these letters.
$\dfrac{BC}{\sin 38°} = \dfrac{12}{\sin 44°}$	Substitute and solve.
$BC = \dfrac{12 \sin 38°}{\sin 44°} = 10.6\,\text{cm}$	
2 $\dfrac{\sin C}{c} = \dfrac{\sin A}{a}$	We have angle A and sides a and c, so you choose the two fractions with these letters.
$\dfrac{\sin C}{12} = \dfrac{\sin 88°}{20}$	Note that the sine rule can be set up with either the side lengths or the sine values in the numerator. When looking for an angle, it may be easier for you to write the sine values in the numerator.
$\sin C = \dfrac{12 \sin 88°}{20}$	Substitute and solve.
$C = \sin^{-1}\left(\dfrac{12 \sin 88°}{20}\right) = 36.8°$	It is better to leave all calculations until the last step to avoid inaccuracy and early-rounding errors.

Exercise 11F

1 Find the value of θ in each triangle.

a

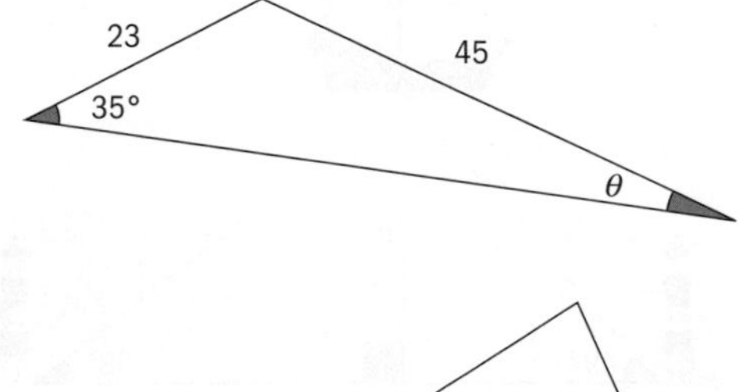

b

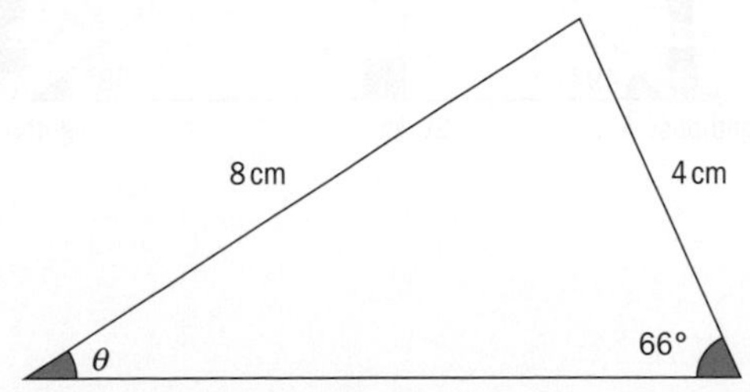

c

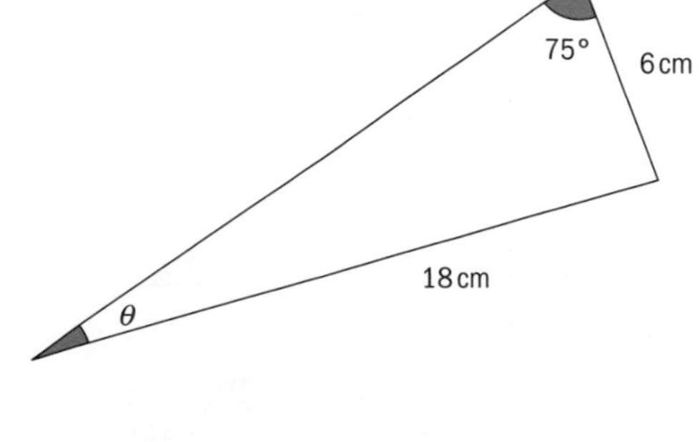

d

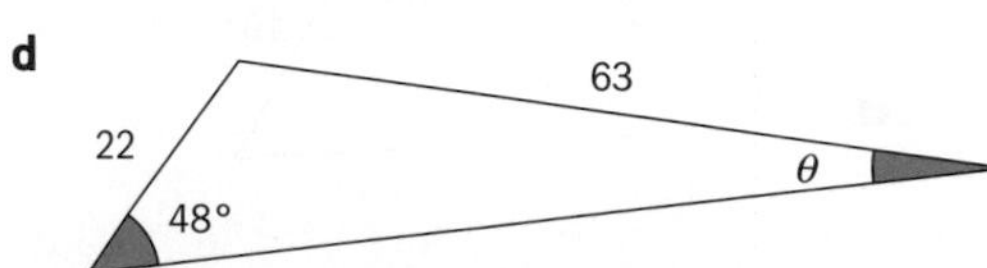

Geometry and trigonometry

e

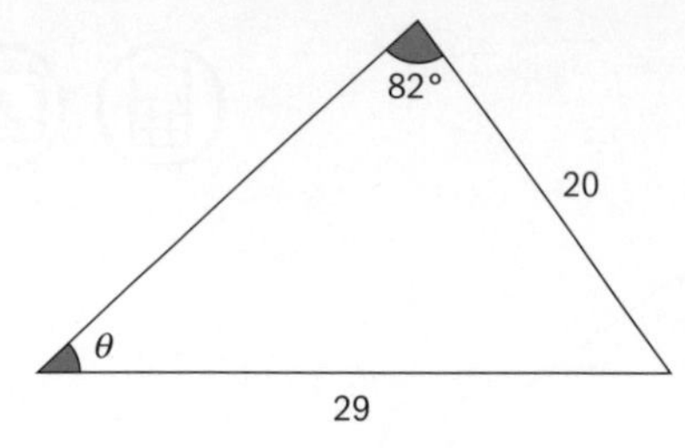

f

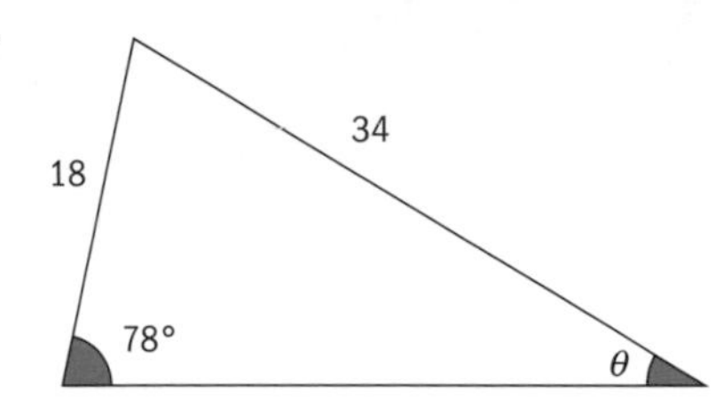

2 Find the value of x in each triangle.

a

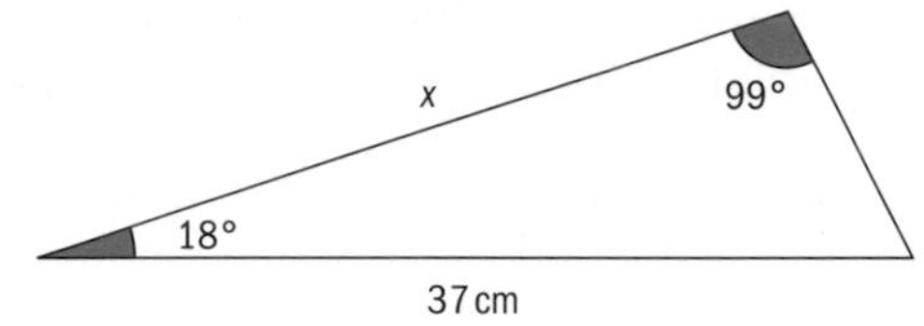

b

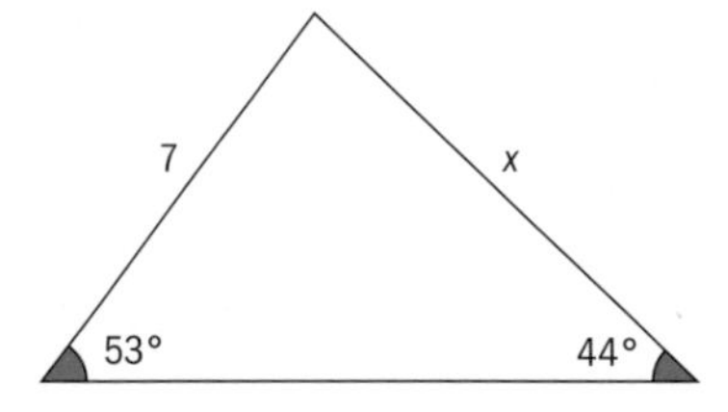

c

d

e

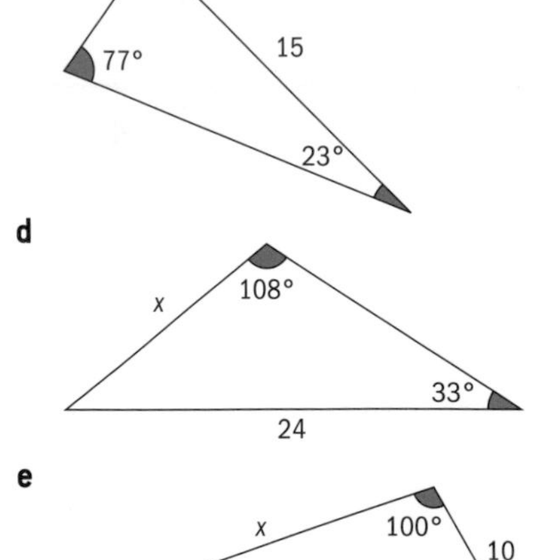

f

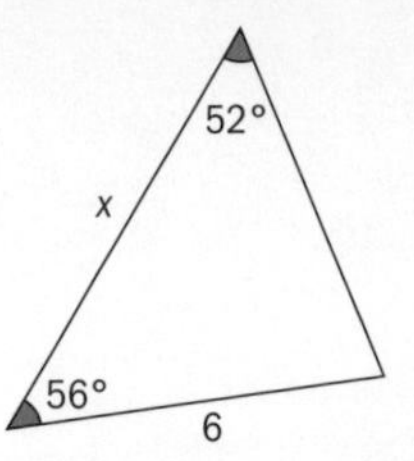

3 Nit and Pon stand 15 m apart in a straight line from a tree. When they look to the top of the tree, Nit's angle of elevation is 40° and Pon's is 70°. How tall is the tree?

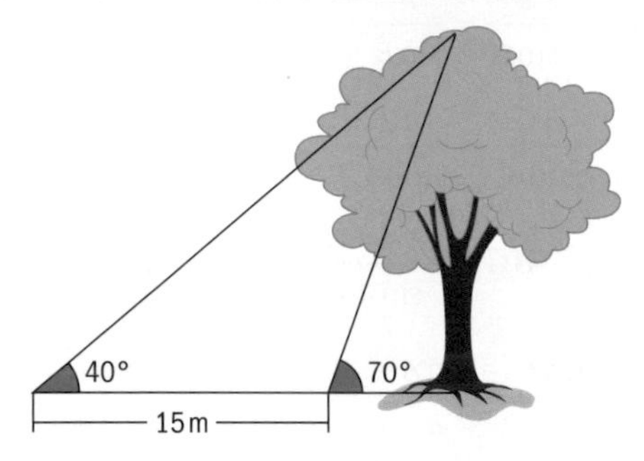

4 Fire towers A and B are located 10 km apart, with B due East of A. Rangers at fire tower A spot a fire on a bearing of 048°, and rangers at fire tower B spot the same fire at 334°. How far from tower A is the fire?

5 Triangulation can be used to find the location of an object by measuring the angles to the object from two points. Two lighthouses, 20 km apart on the shore spot a ship at sea. Using this figure, find the distance, d, the ship is from shore to the nearest tenth of a km.

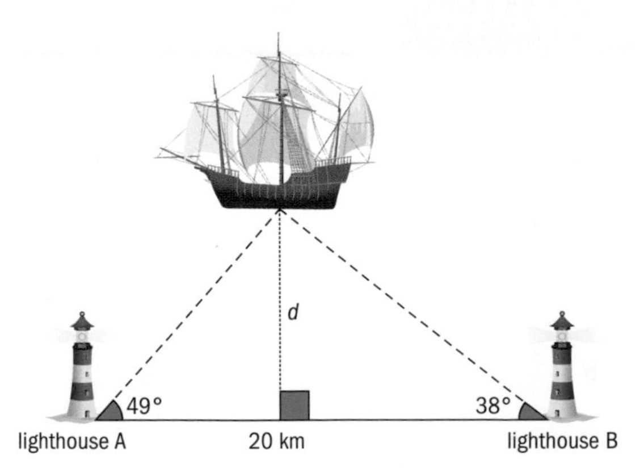

6 A pole, BD, is attached to point C by two wires as shown. How far is point C from the base of the pole?

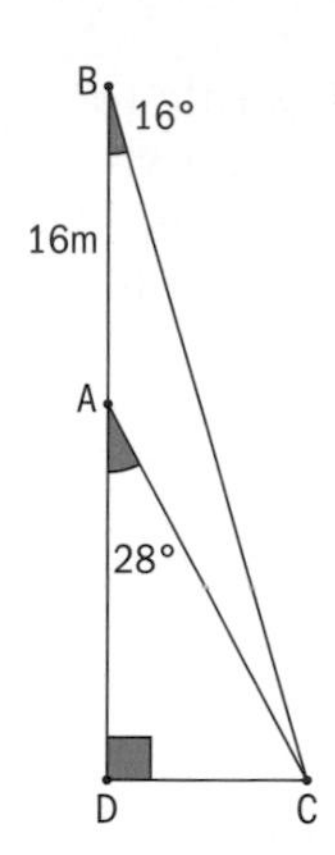

7 While exploring the Hawa Mahal, "The Palace of Winds", in Jaipur, Taylor estimated an angle of elevation of 15° from his position on the ground to the top of the tower. Moving 10 m closer, Taylor now estimates the angle of elevation to be 18°. How high does Taylor estimate the tower of the Hawa Mahal to be?

8 Find the area of this triangle.

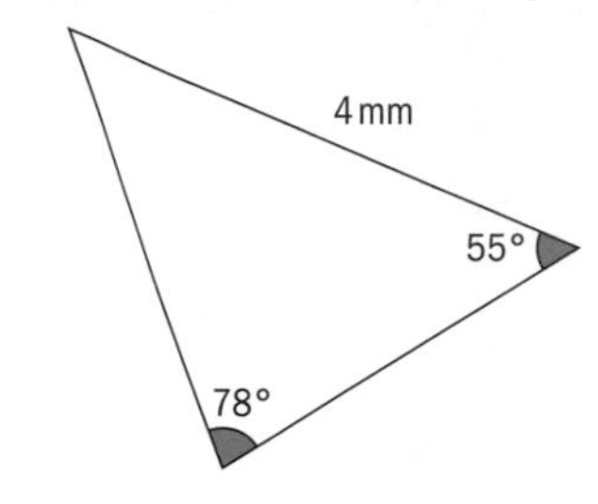

Investigation 8

You will need a compass, protractor, pencil and ruler for this investigation.

First, you need to know some basic facts.

1 Working with a partner or in a small group, copy and complete these sentences.

In a triangle, the longest side is opposite the .... angle.

The shortest distance from a point to a line is ....

The angles in a triangle add up to ....

In an acute triangle, all of the angles ....

In an obtuse triangle, one angle is .... and the other two are ....

Now it is time to construct some triangles. Follow these instructions.

2 On plain paper, draw a line segment of length 10 cm.

Label the endpoints A and B.

3 Draw another line segment, starting from A at 30° to AB. Make it go all of the way to the edge of your paper.

4 Now open your compass out to 7 cm.

Put the point of the compass on B and draw an arc.

You should notice that the arc crosses the line from A two times.

5 Mark one of those intersection points C_1 and the other C_2.

6 Join B with both C points.

7 Do you notice that you have two triangles., ABC_1 and ABC_2 for the same information of one side 10 cm, one side 7 cm and an angle of 30°?

8 How can a given situation appear as two different diagrams?

The **ambiguous case** occurs when you use the sine rule to determine an angle when you are given the lengths of two sides of a triangle and an acute angle opposite the shorter of the two sides. This can lead to two possible triangles just like the ones that you found in the investigation.

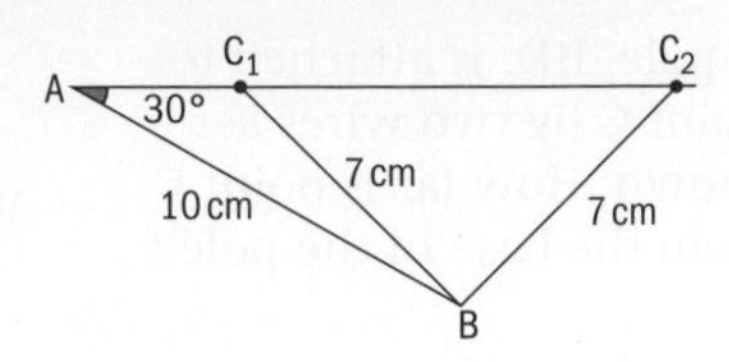

Investigation 9

1 Use technology to copy and complete this table.

θ	$\sin\theta$	$\sin(180-\theta)$
10		
30		
45		
60		
80		

2 Explain what you notice.

3 Complete this sentence: $\sin\theta$ ..... $\sin(180-\theta)$.

4 **Factual** What is the relationship between the sines of supplementary angles?

5 Copy and complete these sentences:

The sine of 50° is equal to the sine ...

The sine of 120° is equal to the sine ...

$\sin A = \sin B$ when ...

6 Write two answers for:

a $\sin^{-1}0.5 =$ ... and ...

b $\sin^{-1}0.7 =$ ... and ...

Example 10

ΔABC has $\hat{B} = 34^\circ$, $b = 15$ cm and $c = 20$ cm. Find two possible values for $\hat{C}$.

$\frac{\sin C}{c} = \frac{\sin B}{b}$	
$\frac{\sin C}{20} = \frac{\sin 34}{15}$	Solve for the angle as before.
$C = \sin^{-1}\frac{20\sin 34}{15} = 48.2^\circ \text{ or } 132^\circ$	To find the second angle, subtract the first answer from 180°. Write your answers correct to 3 s.f.

$\sin A = \sin B$ when $A + B = 180^\circ$

Exercise 11G

It might help you to draw a rough sketch to view these triangles. You know that this triangle might have an ambiguous case since you are given two sides and an angle that is not in between them. In exams, it is common to be asked for two possible values.

1 Find two possible values for $\hat{A}$ in triangle ABC when $\hat{C} = 64°$, $a = 8$ and $c = 10$.

2 Find two possible values for $\hat{A}$ in triangle ABC when $\hat{B} = 20°$, $b = 3$ and $a = 5$.

3 Find two possible values for $\hat{B}$ in triangle ABC when $\hat{A} = 45°$, $a = 8$ and $b = 10$.

4 Find two possible values for $\hat{C}$ in triangle ABC when $\hat{B} = 40°$, $b = 24$ and $c = 30$.

5 When the area of this triangle is $20\,cm^2$, find the size of **obtuse** angle ABC.

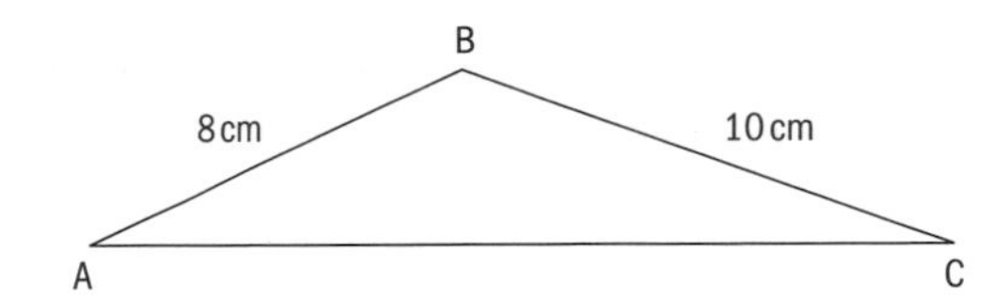

11.4 The cosine rule

The Pythagorean theorem and right-angle triangle relationships are the foundations of trigonometry. In the previous section, you saw how the sine rule could be derived using right-angle triangles. The sine rule can be used when you have an angle, its opposite side, and one more piece of information, but not when you have two sides and the included angle. The Pythagorean theorem and right-angle triangle trigonometry come to the rescue again, to find a suitable method for this situation.

Investigation 10

You will be using the triangle. You might want to work with a friend to compare results.

1 **Conceptual** How does the relationship between the sides and the angles of triangles lead you to the cosine rule?

2 In triangle ADC, apply the Pythagorean theorem to complete the equation $b^2 = \ldots$

3 Still using triangle ADC, complete $\cos A = \ldots$ and then rewrite this equation to make x the subject.

4 Now use the Pythagorean theorem in triangle BCD to complete the equation $a^2 = \ldots$

5 Expand the expression $(c - x)^2$.

You should now have $a^2 = c^2 - 2cx + x^2 + h^2$.

6 Substitute your answer from question **2**.

7 Substitute your answer from question **3**.

8 Simplify.

You now have the cosine rule.

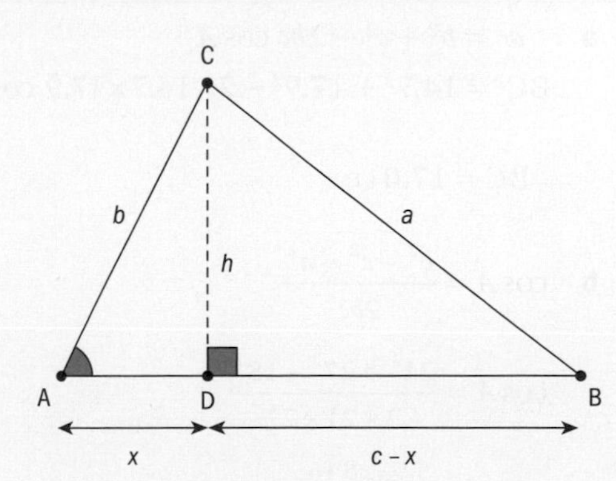

Geometry and trigonometry

The cosine rule to find a side in ΔABC is

$$a^2 = b^2 + c^2 - 2bc\cos A$$

or $b^2 = a^2 + c^2 - 2ac\cos B$ or $c^2 = a^2 + b^2 - 2ab\cos C$

The cosine rule may also be used **to find an angle** when you are **given three sides**.

The cosine rule to find an angle in ΔABC is

$$\cos A = \frac{b^2 + c^2 - a^2}{2bc}$$

or $$\cos B = \frac{a^2 + c^2 - b^2}{2ac}$$

or $$\cos C = \frac{a^2 + b^2 - c^2}{2ab}$$

Example 11

a Find BC:

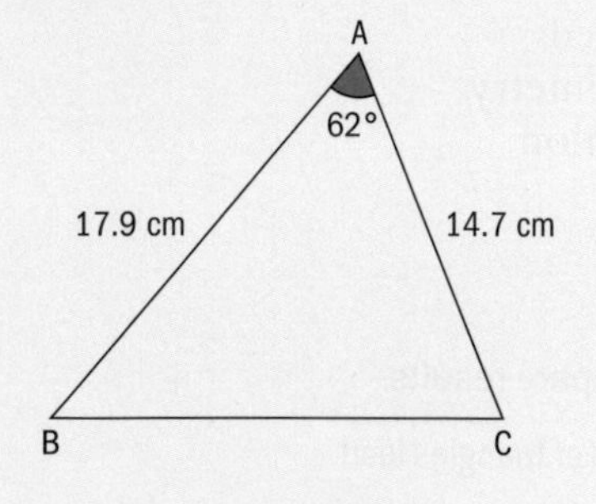

b Find $\hat{A}$.

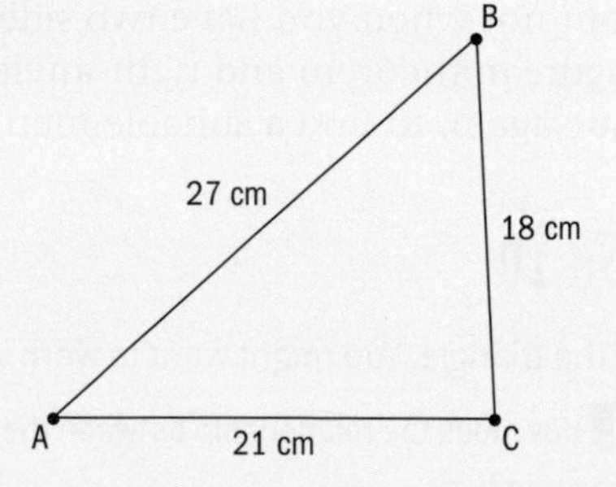

a $a^2 = b^2 + c^2 - 2bc\cos A$	$b = 14.7,\ c = 17.9,\ \hat{A} = 62°$
$BC^2 = 14.7^2 + 17.9^2 - 2\times 14.7\times 17.9\cos 62$	Substitute into the cosine rule.
	Evaluate.
$BC = 17.0\,\text{cm}$	
	$a = 18,\ b = 21,\ c = 27$
b $\cos A = \dfrac{b^2 + c^2 - a^2}{2bc}$	
$\cos A = \dfrac{21^2 + 27^2 - 18^2}{2\times 21\times 27}$	Substitute into the cosine rule.
	Evaluate.
$A = \cos^{-1}\dfrac{846}{1134} = 41.8°$	

Exercise 11H

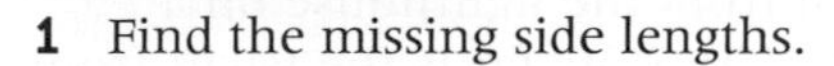

1 Find the missing side lengths.

a

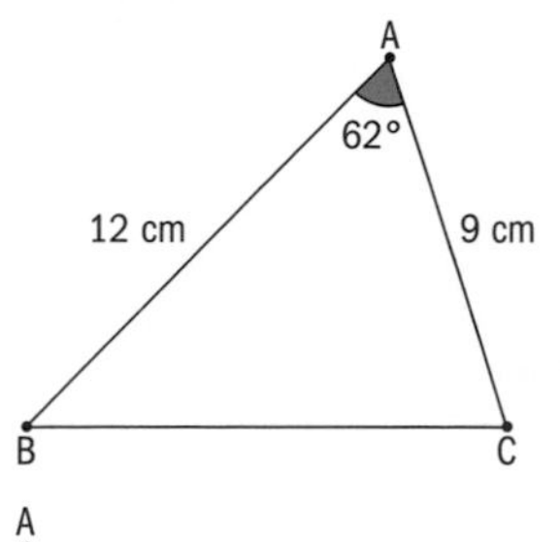

b

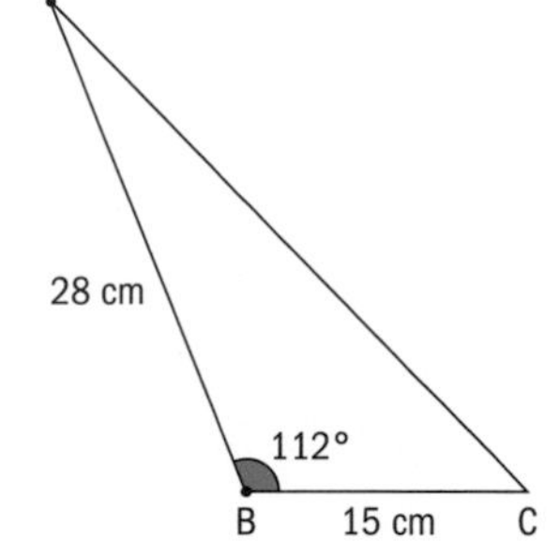

c

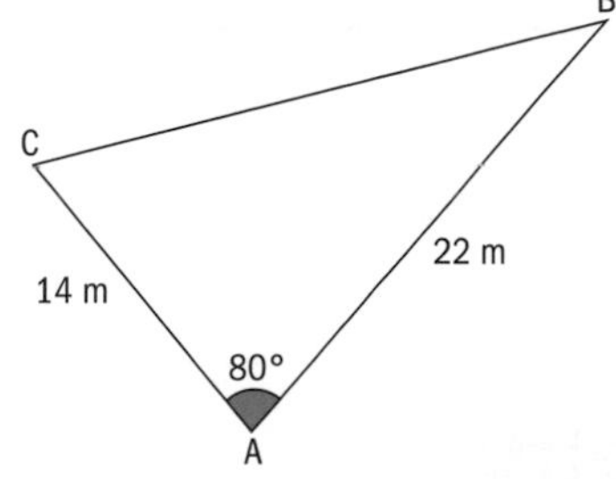

d

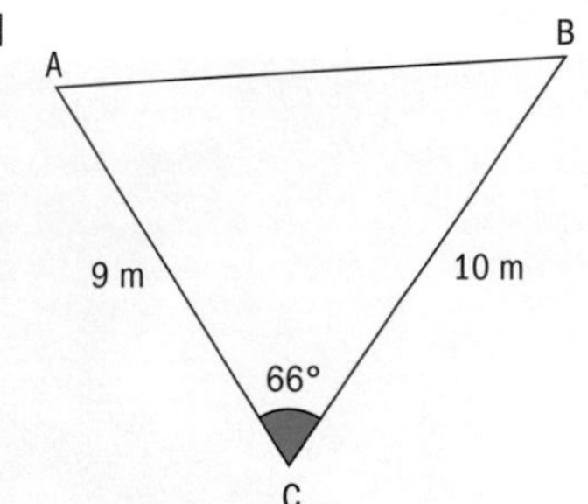

e

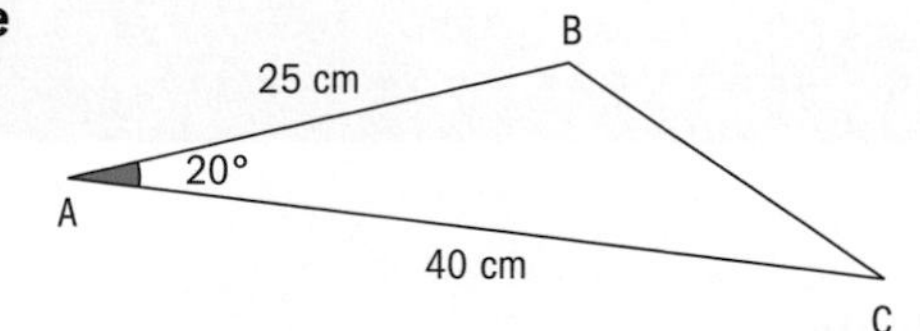

f

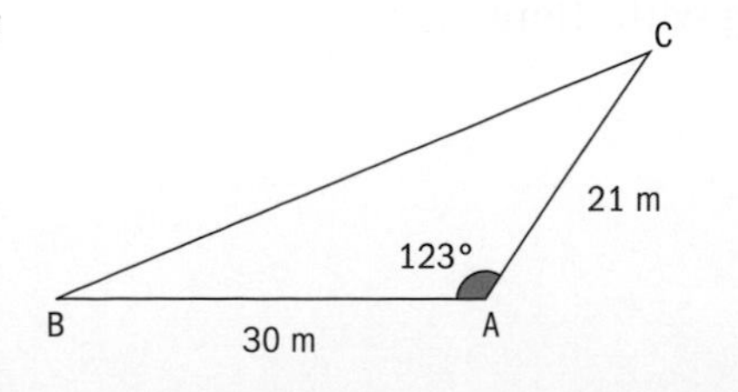

2 Find θ:

a

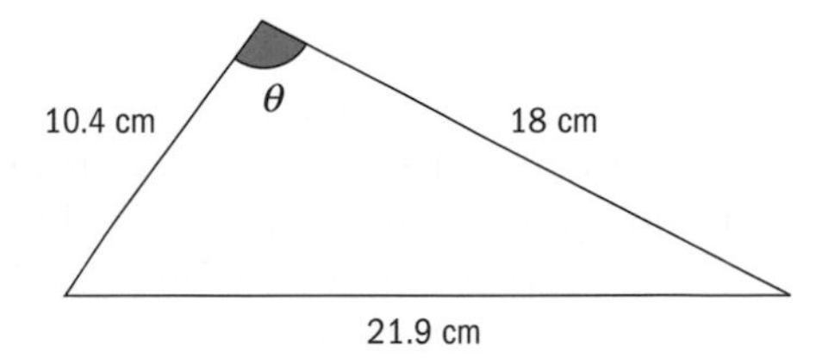

b

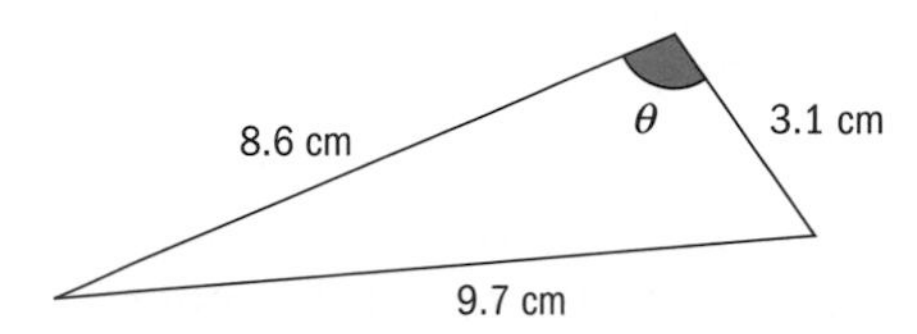

c

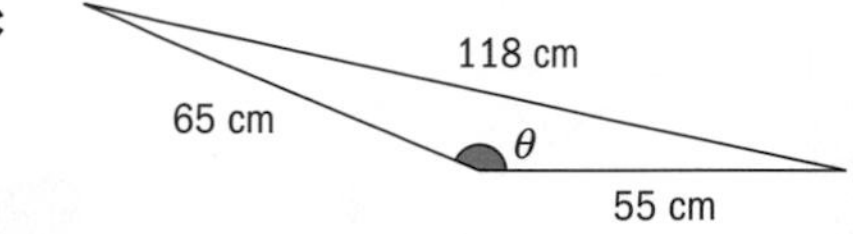

d

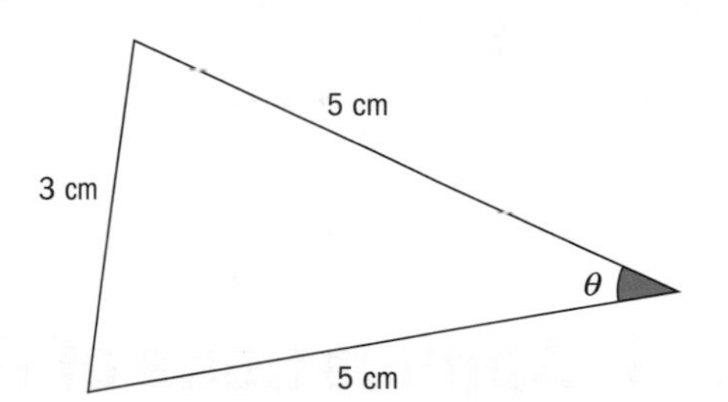

e

f

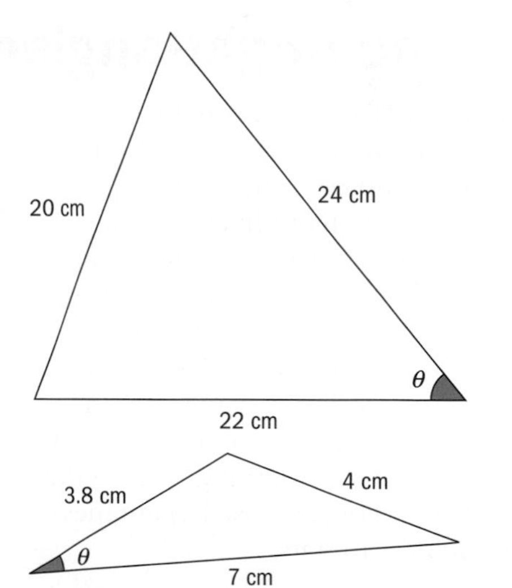

3 The lengths of the three sides of a triangle are 6 cm, 9 cm and 12 cm.

a Find the measure of the smallest angle in the triangle.

b Find the area of the triangle.

4 A ship leaves port at 2pm traveling north at a speed of 30 km h^{-1}. At 4pm, the ship adjusts its course 20° to the east. How far, to the nearest km, is the ship from the port at 5pm?

5 Joon stands between two buildings of different heights. He is 46 m away from the building on his left, and the angle of elevation to the top of the building is 33°. The building on his right is 28 m away, with an angle of elevation of 17° to the top of the building.

a Find the heights of the two buildings.

b Find the distance from the top of one building to the top of the other.

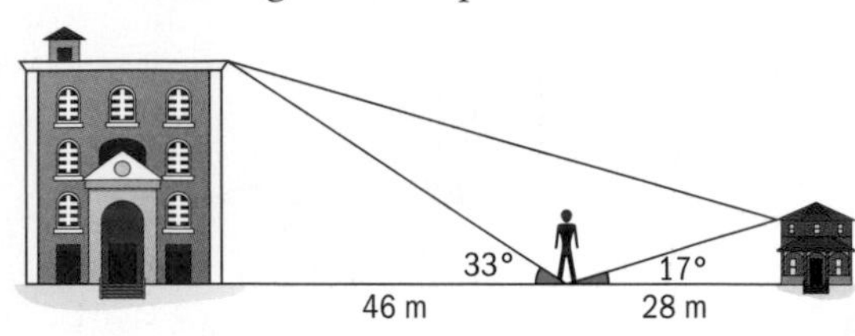

6 A lighthouse keeper spots a yacht 15 km away on bearing 070°. There is also a rock 9 km from the lighthouse on a bearing of 210°. How far is the yacht from the rock?

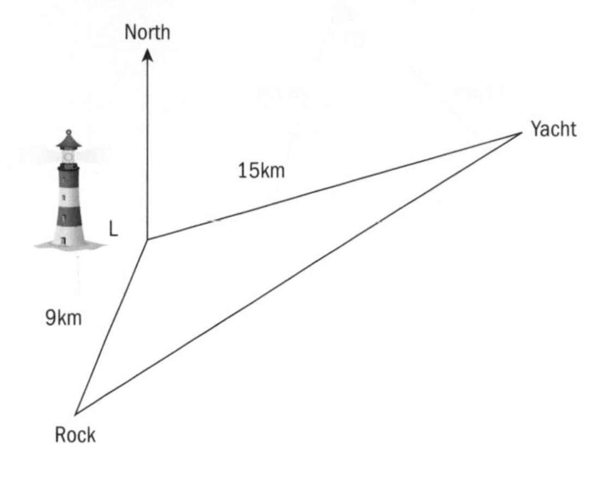

11.5 Applications of right and non-right-angled trigonometry

Trigonometry is a tool for solving problems that involve triangles. There are many examples of, and uses for, trigonometry in our lives—from astronomy and navigation to surveying and sports. Sometimes we need to know measurements that we cannot find directly, such as the height of a building or a distance to a ship at sea. Architects, scientists and engineers use trigonometry to determine all sorts of things—from the sizes and angles of parts used in engines to the distances to stars.

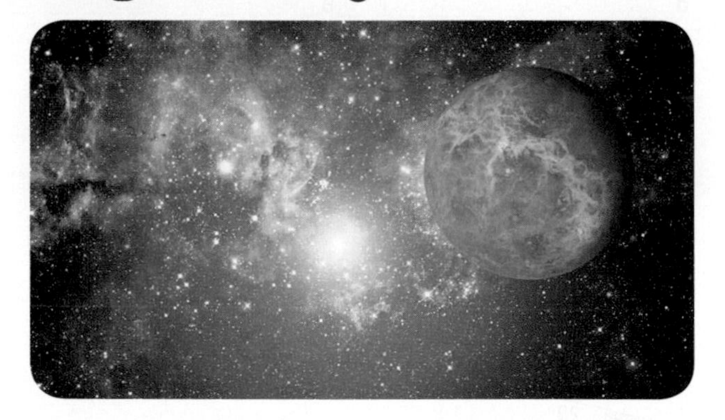

In order to make use of trigonometry, you have to apply the appropriate formula for the situation that you are working with. Here is a guide:

Use the **Pythagorean theorem** in a right-angled triangle when you know the lengths of two sides of the triangle, and you want to find the length of the third side.
Use **SOH-CAH-TOA** in a right-angled triangle when you have two sides or one side and one angle.
Use the **sine rule** when you have the measures of an angle and its opposite side, along with one more piece of information.
Use the **cosine rule** when you know the measures of two sides and the included angle, or when you know the measures of all three sides.

Example 12

Jon and Jan have two gardens. Jon has garden PQR and Jan has PRS.

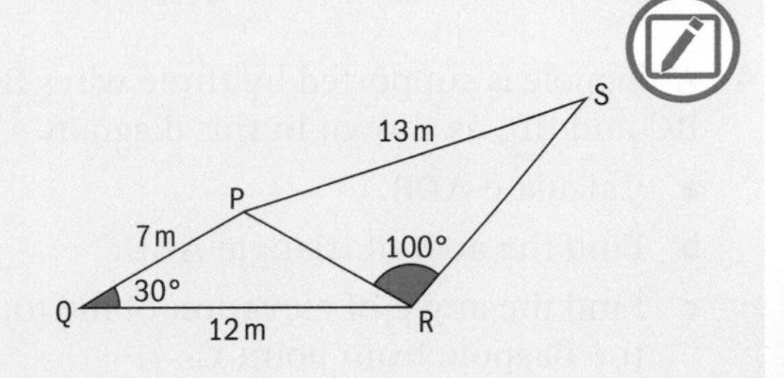

a Find the length of the fence, PR, between their gardens.

b Find the area of Jon's garden.

c Find $P\hat{S}R$ in Jan's garden.

a $q^2 = p^2 + r^2 - 2\,pr\cos Q$ $q^2 = 12^2 + 7^2 - 2 \times 12 \times 7 \times \cos 30°$ $q^2 = 47.507\ \ldots$ $q = 6.89\,\text{m}$	You have two sides, p and r and the included angle, Q. Select the cosine rule.
b $A = \frac{1}{2}pr\sin Q$ $A = \frac{1}{2} \times 12 \times 7 \times \sin 30° = 21\,\text{m}^2$	Use $\frac{1}{2}ab\sin C$ to find the area.
c $\frac{\sin S}{s} = \frac{\sin R}{r}$ $\frac{\sin S}{6.89} = \frac{\sin 100}{13}$ $\sin S = \frac{6.89\sin 100}{13}$ $S = 31.5°$	You have side r and angle R opposite side s. Select the sine rule.

Exercise 11I

1 Calculate the area of the triangle with sides measuring 12 cm, 14 cm and 20 cm.

2 The radius of the Earth's orbit around the Sun is approximately 150 million km. When the parallax angle to a nearby star is 7°, find the distance from the Earth to the star.

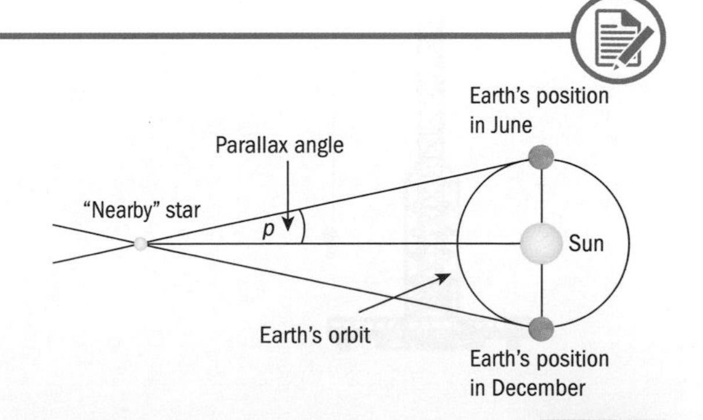

3 a Find the length of [PR].
b Find the length of [QR].
c Calculate the area of triangle PQS.

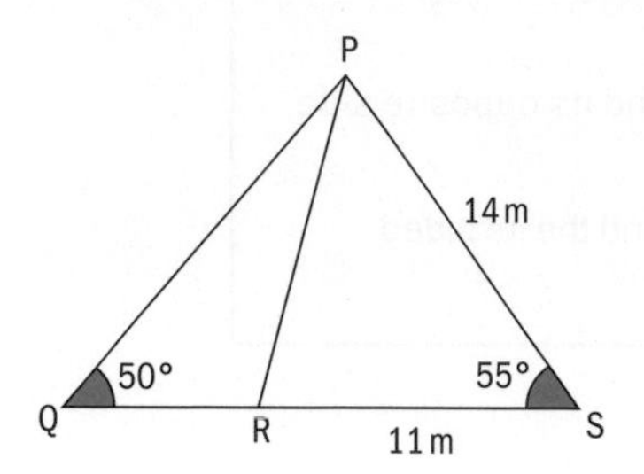

4 A flagpole is supported by three wires BA, BC and BD, as shown in this diagram.
a Calculate $A\hat{D}B$.
b Find the area of triangle ABD.
c Find the angle of elevation of the top of the flagpole from point C.
d Show that triangle ABC is not right angled.

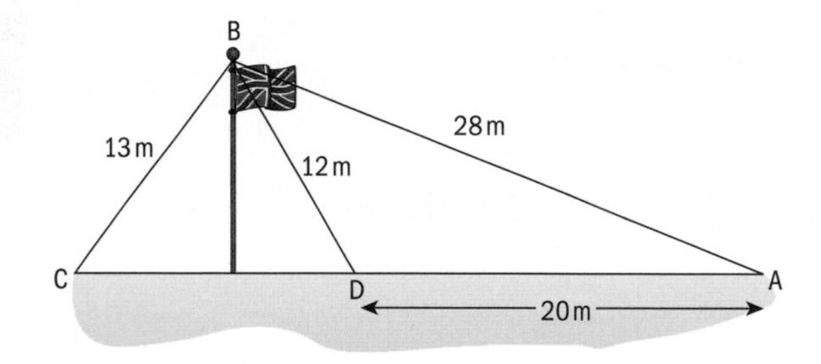

5 A dockyard crane ABC, is used to unload cargo from ships. AB = 22.5 m, AC = 48 m and angle ABC = 46°.
a Find $A\hat{C}B$.
b Find the length of the lifting arm, BC.

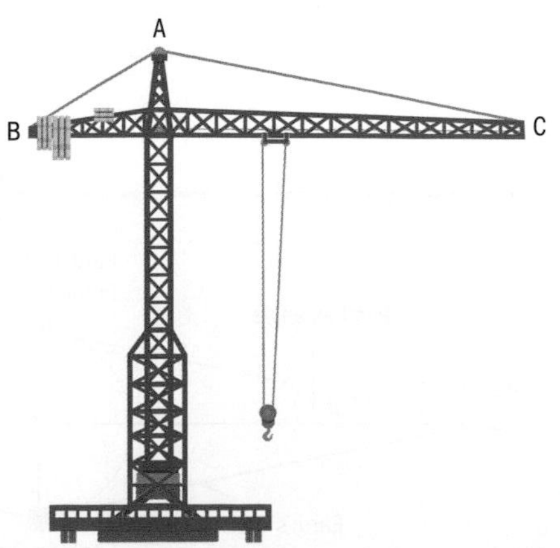

6 A toy bicycle frame is constructed from five plastic pieces.
a Calculate the length of [AC].
b Calculate the size of $B\hat{A}C$.
c Write down the size of $A\hat{C}D$.
d Calculate the length of [AD].
e Calculate the area of the quadrilateral ABCD.

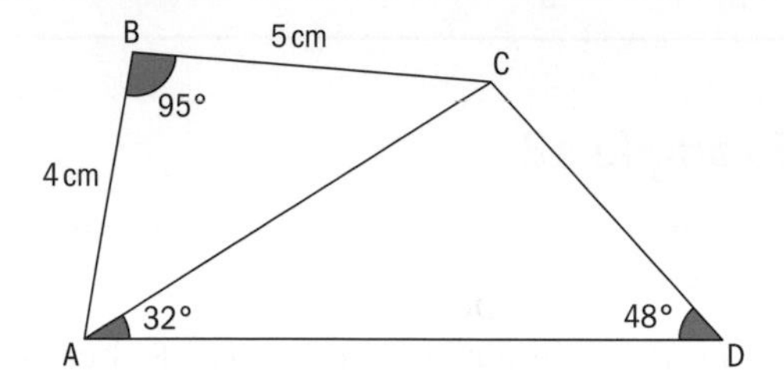

7 Two security cameras are positioned on the ceiling of a gym, 10 m apart.
One camera has an angle of depression of 50° to a point on the floor, and the other camera has an angle of depression of 60° to the same point. Calculate the height, h metres, of the gym.

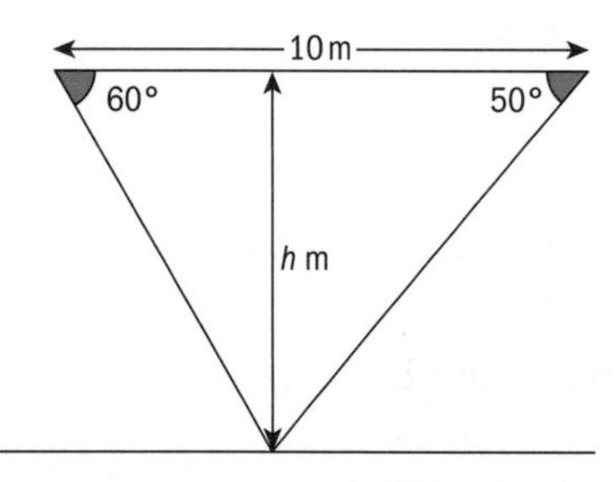

HINT

In bearing problems we often encounter two parallel north lines. The angles $\hat{A}$ and $\hat{B}$ are supplementary angles, they add up to 180°. For example, when angle $\hat{A}$ is 60°, angle $\hat{B}$ is 120°.

North
North
B
A

8 Porpor flies from A to B on a bearing of 067° for a distance of 80 km and from there to C on a bearing of 123°.
a Find $A\hat{B}C$.
b Calculate the distance for Porpor to fly from C back to A.

c Find the bearing that Porpor must fly on to return to A from C.

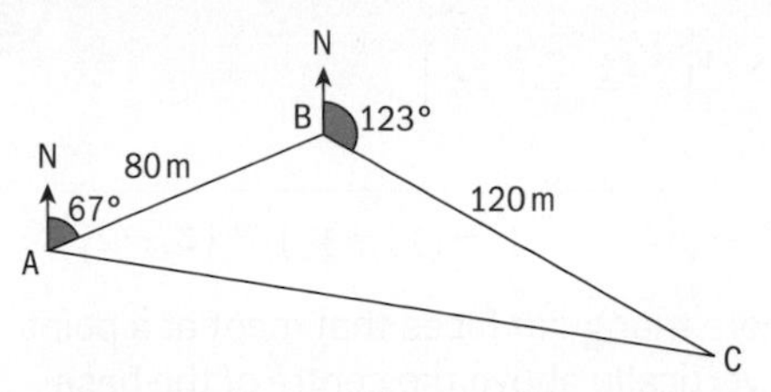

9 A ship sails from harbour H on a bearing of 084° for 340 km until it reaches point P. It then sails on a bearing of 210° for 160 km until it reaches point Q.

a Calculate the distance between point Q and the harbour.

b On what bearing must the ship sail to return directly to the harbour from Q?

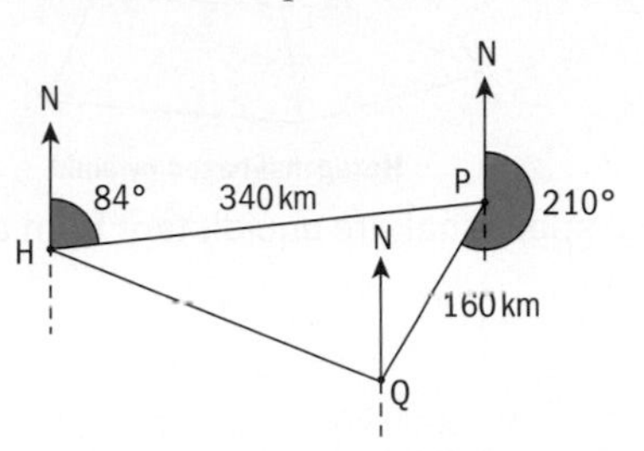

10 A ship leaves Shanghai on a bearing of 030°. It sails a distance of 500 km to Jeju.

In Jeju, the ship changes direction to a bearing of 100°. It sails a distance of 320 km to reach Nagasaki.

Calculate the bearing and distance for the ship to return from Nagasaki to Shanghai.

11 A right-angled ABCD with height OV = 20 cm has a square base of side 10 cm. Find

a the length of the slant edge

b the angle between a sloping face and the base

c the angle between two sloping faces.

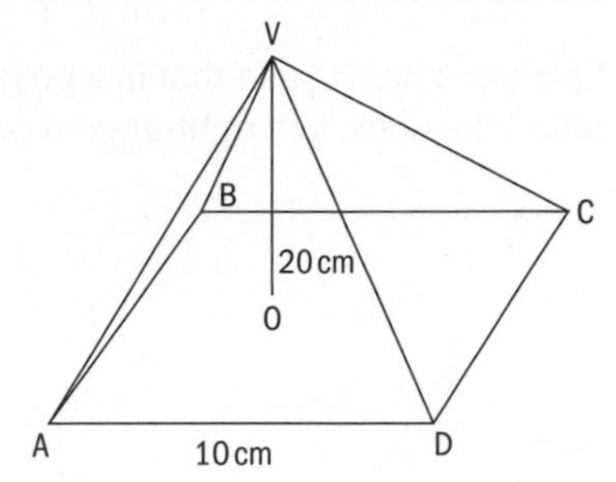

12 a Calculate the value of cos $A\hat{B}C$ in this triangle.

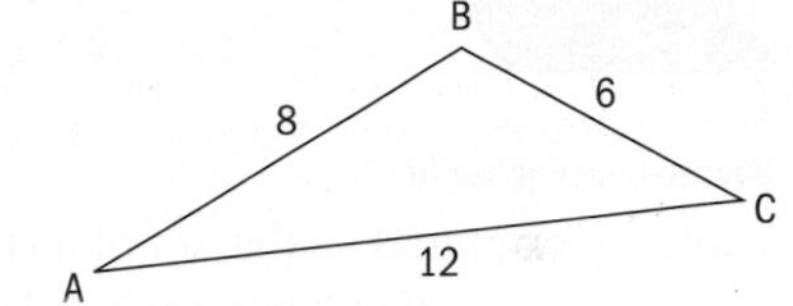

b Without actually calculating the size of the angle you will be able to say that $A\hat{B}C$ is obtuse. Using your answer in part **a**, explain why you are able to do this.

Developing your toolkit

Now do the Modelling and investigation activity on page 504.

Chapter summary

- The **midpoint** of (x_1, y_1, z_1) and (x_2, y_2, z_2) is $\left(\frac{x_1 + x_2}{2}, \frac{y_1 + y_2}{2}, \frac{z_1 + z_2}{2}\right)$.
- The **distance** (d) from (x_1, y_1, z_1) and (x_2, y_2, z_2) is $d = \sqrt{(x_2 - x_1)^2 + (y_2 - y_1)^2 + (z_2 - z_1)^2}$.
- A **pyramid** has a base that is a polygon and three or more triangular faces that meet at a point called the apex. In a right-angled pyramid, the apex is vertically above the centre of the base.

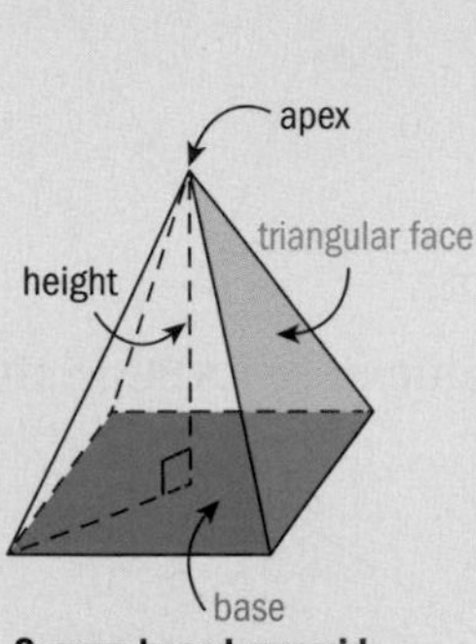

Square-based pyramid

Tetrahedron

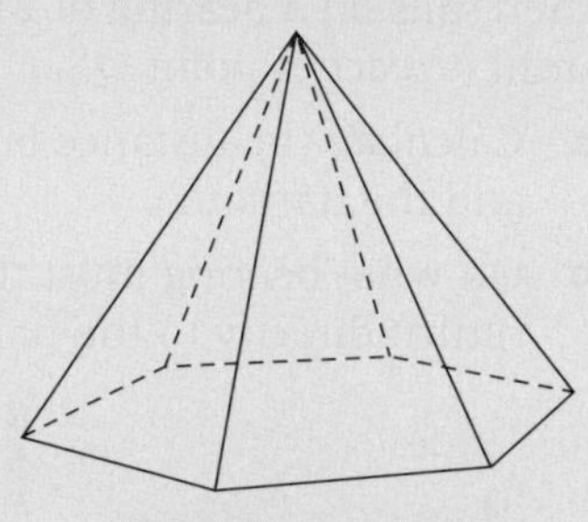
Hexagonal-based pyramid

- A sphere is defined as the set of all points in three-dimensional space that are equidistant from a central point. Half of a sphere is called a hemisphere.

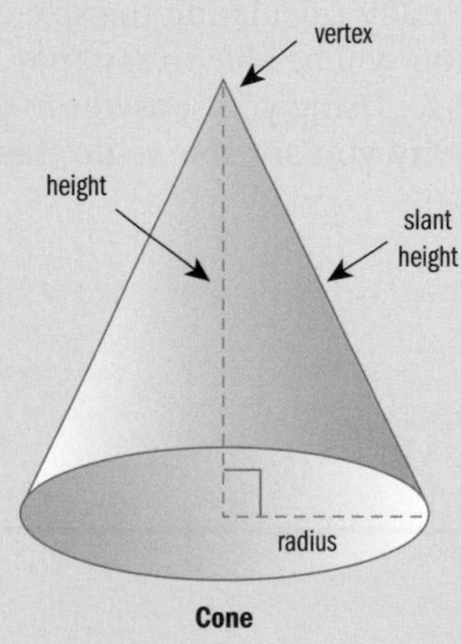

Cone

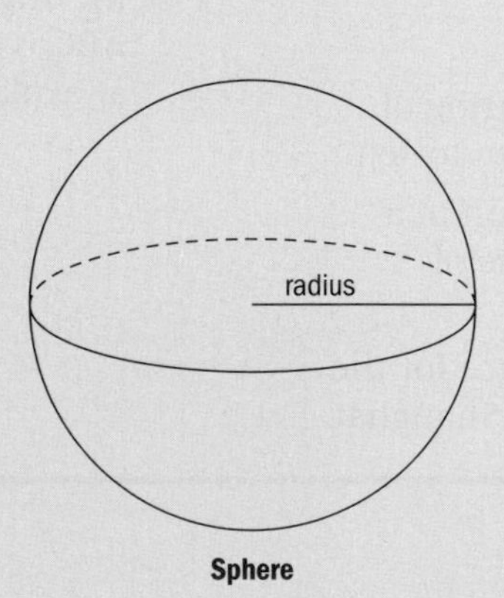

Sphere

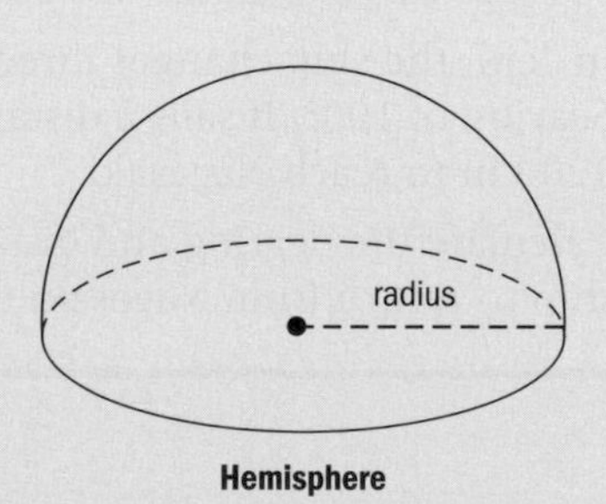

Hemisphere

Surface area

- The **surface area of a pyramid** is the sum of the areas of all of its faces.
- The **surface area of a cone** is the sum of the areas of the circular base and the curved side.

Shape	Surface area	Volume
Pyramid	The sum of all of the faces. $x^2 + 2xl$ for a squared-based pyramid where x is the side length of the base and l is the slant height.	$V = \frac{1}{3}(\text{base area} \times \text{height})$
Cone	$\pi r^2 + \pi rl$, where r is the base radius and l is the slant height.	$V = \frac{1}{3}\pi r^2 h$
Sphere	$4\pi r^2$	$V = \frac{4}{3}\pi r^3$

Right-angled triangle trigonometry

- In a right-angled triangle:

$$\sin\theta = \frac{O}{H}, \cos\theta = \frac{A}{H} \text{ and } \tan\theta = \frac{O}{A}$$

where O is the opposite side to angle θ, A is the adjacent side and H is the hypotenuse.

There are two special right-angled triangles that will give you **exact values** for the trigonometric ratios. The first one is an isosceles right-angled triangle and the second is a 30-60-90 right-angled triangle where the side opposite the 30° angle is half of the length of the hypotenuse.

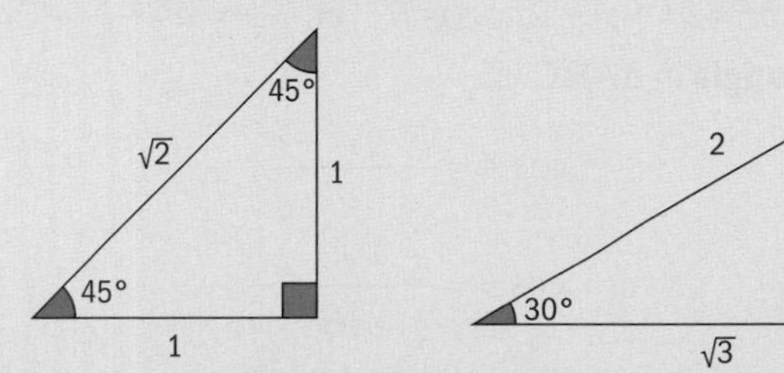

θ	30°	45°	60°
$\sin\theta$	$\frac{1}{2}$	$\frac{\sqrt{2}}{2}$	$\frac{\sqrt{3}}{2}$
$\cos\theta$	$\frac{\sqrt{3}}{2}$	$\frac{\sqrt{2}}{2}$	$\frac{1}{2}$
$\tan\theta$	$\frac{\sqrt{3}}{3}$	1	$\sqrt{3}$

Angles of elevation and depression

- When looking up, the **angle of elevation** is the angle up from the horizontal to the line from the object to the person's eye.
- When looking down, the **angle of depression** is the angle down from the horizontal to the line from the object to the person's eye.

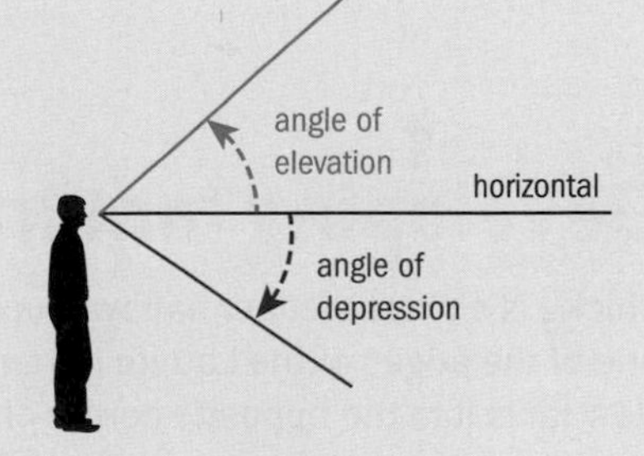

Bearings

- A bearing is used to indicate the direction of an object from a given point. Three-figure bearings are measured clockwise from North, and can be written as three figures, such as 120°, 057°, or 314°. Compass bearings are measured from either North or from South, and can be written such as N20°E, N55°W, S36°E, or S67°W, where the angle between the compass directions is less than 90°. The bearing of A in the diagram may be written as 240° or as S60°W.

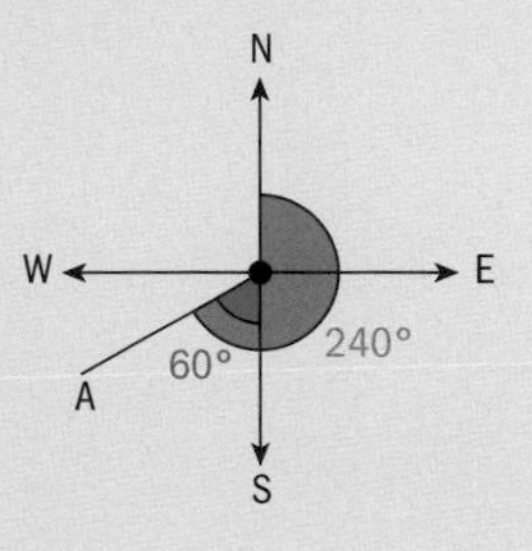

Continued on next page

Geometry and trigonometry

Non-right-angled triangles

- The **area of a triangle can be found using** $A = \frac{1}{2}ab \sin C$ where a and b are two sides and C is their included angle.
- The sine rule is $\frac{\sin A}{a} = \frac{\sin B}{b} = \frac{\sin C}{c}$. It is usual to only use two of these fractions.
- $\sin A = \sin B$ when $A + B = 180°$.
- The **cosine rule to find a side** in ΔABC is:

$$a^2 = b^2 + c^2 - 2bc\cos A$$

or $b^2 = a^2 + c^2 - 2ac\cos B$ or $c^2 = a^2 + b^2 - 2ab\cos C$

- The **cosine rule to find an angle** in ΔABC is:

$$\cos A = \frac{b^2 + c^2 - a^2}{2bc}$$

or $$\cos B = \frac{a^2 + c^2 - b^2}{2ac}$$

or $$\cos C = \frac{a^2 + b^2 - c^2}{2ab}$$

- Use the **Pythagorean theorem** in a right-angled triangle when you know the lengths of two sides, and you want to find the length of the third side.
- Use **SOH-CAH-TOA** in a right-angled triangle when you have two pieces of information.
- Use the **sine rule** when you have the measures of an angle and its opposite side, along with one more piece of information.
- Use the **cosine rule** when you know the measures of two sides and the included angle, or when you know the measures of all three sides.

Developing inquiry skills

Chicky is a glass cleaner half way up of one of the edges of the Louvre pyramid. How far is it to the opposite corner of the base?

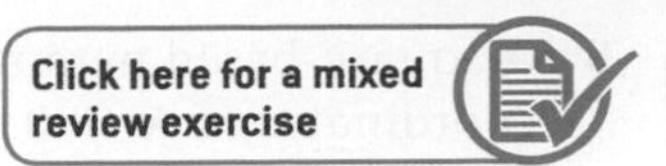

Chapter review

1 Find the volume and the surface area of a square-based pyramid with a base side length of 8m and a height of 3 m.

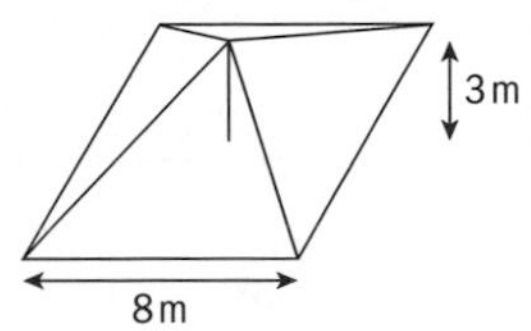

2 Find the volume and the total surface area of a cone with radius 6 cm and height 8 cm. Write your answer as an exact value (this means in terms of π).

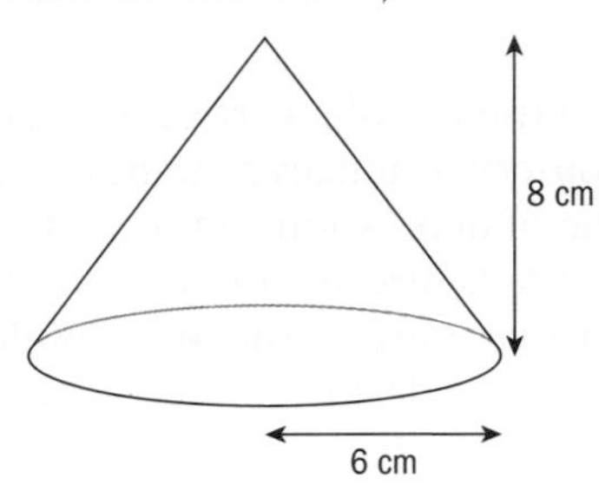

3 A sphere has a volume of $\frac{32}{3}\pi$ m^3. Find the exact value of its surface area.

4 **a** Find the volume of a cone with height 10 cm and base radius 4 cm.

b A truncated cone is formed when the top of the cone is removed with a horizontal cut, leaving a height of 6 cm and top radius of 2 cm. Find the volume of the truncated cone

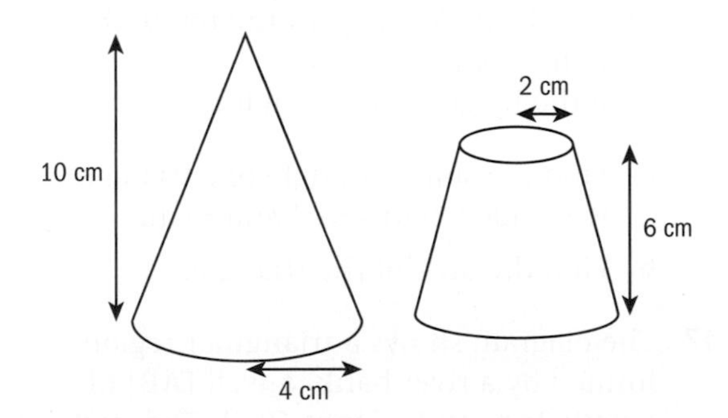

5 Jamie has a circular cylinder with a lid. The cylinder has a height of 39 cm and diameter 65 mm.

a Calculate the volume of the cylinder in cm^3, correct to 2 decimal places.

The cylinder is used for storing tennis balls. Each ball has a radius of 3.25 cm.

b Calculate how many balls Jamie can fit in the cylinder when it is filled to the top.

c Jamie fills the cylinder with the number of balls found in part **b** and puts the lid on. Calculate the volume of air inside the cylinder in the spaces between the tennis balls.

d Convert your answer to part **c** into cubic metres.

6 A metal fuel tank is made of a cylinder of length 8.5 m and diameter 3 m with a hemisphere welded to each end.

a Find the volume of the fuel tank.

b Find the surface area of the fuel tank.

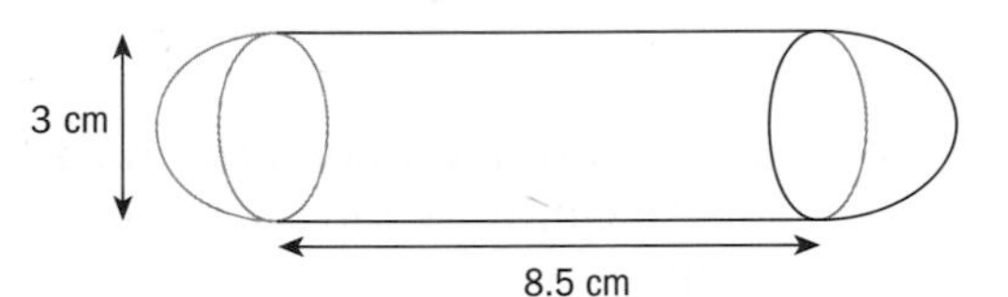

7 ABCDV is a solid stone pyramid. The base of the pyramid is a square with side length 6 cm. The vertical height is 3 cm. The vertex V is directly above the centre O of the base.

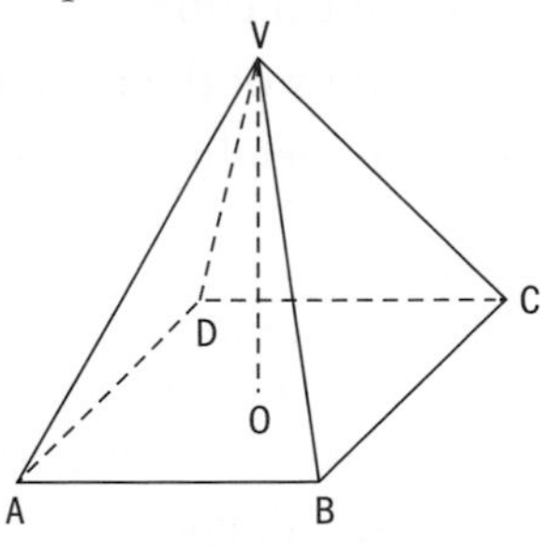

a Calculate the volume of the pyramid.

b The stone weighs 12 grams per cm^3. Calculate the weight of the pyramid.

c Show that the length of the sloping edge [VC] of the pyramid is 5.20 cm.

d Find the angle at the vertex, $B\hat{V}C$.

e Calculate the total surface area of the pyramid.

Geometry and trigonometry

8 A square-based pyramid, with the coordinates of A $(1, 0, 3)$, B $(1, 5, 3)$ and E $(7, \sqrt{15}, 10)$ is shown. Find:

a the length AB

b the midpoint of [AB]

c the length AE.

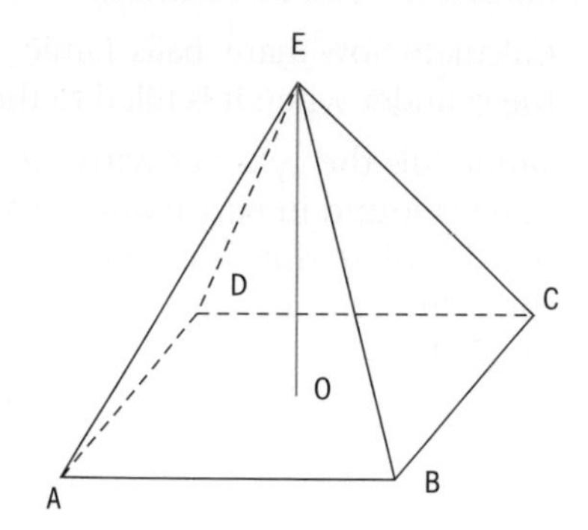

9 Find the area of the triangle:

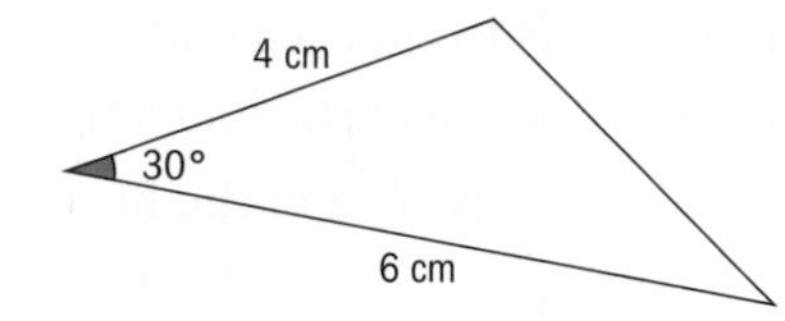

10 80 metal, spherical cannon balls, each of diameter 16 cm, were salvaged from a pirate ship.

a Calculate the total volume of all 80 cannon balls to the nearest whole number.

The cannon balls are to be melted down to make a cone.

The base radius of the cone is 40 cm.

b Calculate the height of the cone.

11 Your teacher asks you to measure the height of a flagpole. You stand at a point 10 metres from the base and measure the angle of elevation to the top of the flagpole as 60°.

a How tall is the flagpole?

Your friend stands directly behind you and measures the angle of elevation to the top of the flagpole as 30°.

b What is the distance between you and your friend?

12 Find the surface area of a tetrahedron with all faces equilateral triangles of side 6 cm.

13 The length of the base of a cuboid is three times the width x, and its height is h cm. Its total surface area, A, is 600 cm^2.

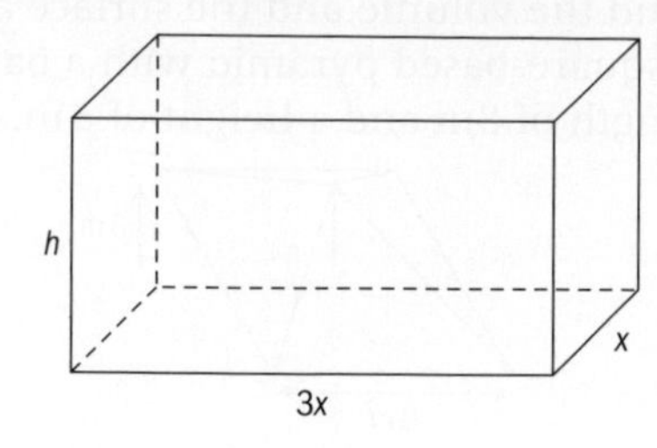

a Show that $A = 6x^2 + 8xh$.

b When $A = 600$, find an expression for h, in terms of x.

c Show that the volume is $\frac{9}{4}(100 - x^2)$.

14 From the top of a vertical cliff 70 m high, an observer notices a yacht at sea. The angle of depression to the yacht is 45°. The yacht sails directly away from the cliff, and after five minutes the angle of depression is 10°. How fast does the yacht sail in $km\,h^{-1}$?

15 A radio antenna, [AB], 80 metres tall, is on the roof of a communications building.

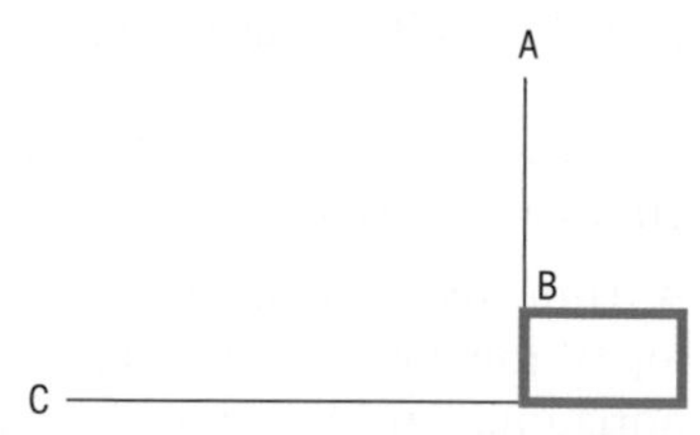

The angle of depression from A to a point C on the horizontal ground is 32°.

The angle of elevation of the top of the building from C is 15°.
Find the height of the building.

16 **a** Find the smallest angle of a triangle with side lengths 6, 7 and 8 cm.

b Find the area of the triangle.

17 The diagram shows a triangular region formed by a river bank, a wall [AB] of length 34 m and a fence [BC]. The angle between the wall and the river bank is 25°. The end of the fence, point C, can be positioned anywhere along the river bank.

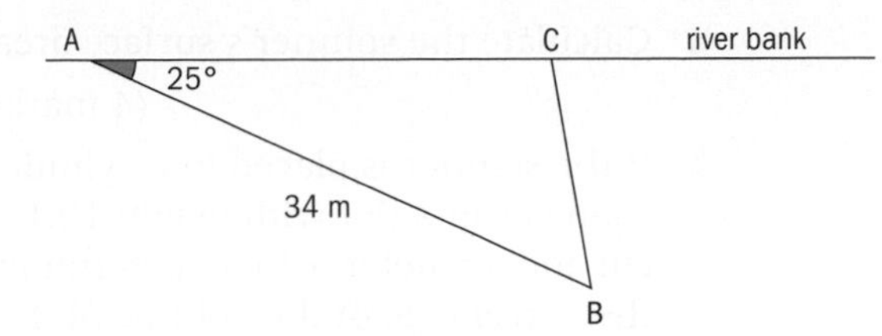

a Find the length of the fence [BC], when point C is 15 m from A.

The farmer has another, longer fence. It is possible for him to enclose two different triangular regions with this fence. He places the fence so that $A\hat{B}C = 85°$.

b Find the distance from A to C.

c Find the area of the region ABC with the fence in this position.

d To form the second region, he moves the fencing so that point C is closer to point A. Find the new distance from A to C.

18 This plan shows a plot of land PQRS crossed by a path QS.

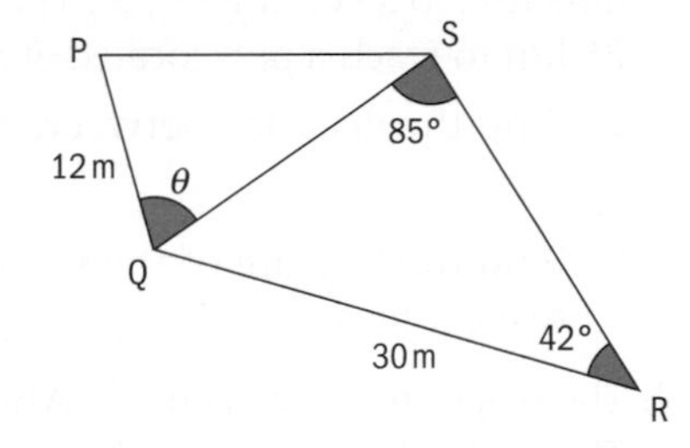

a Find QS.

b Find the area of triangle QRS.

The area of triangle PQS is half the area of triangle QRS.

c Find the possible values of θ.

d Given that θ is obtuse, find PS.

19 The diagram below, which is not drawn to scale, represents the positions of three mobile phone masts.

The bearing of mast B from mast A is 100° and they are 70 km apart.

The bearing of mast C from mast B is 150°.

Masts A and C are 100 km apart.

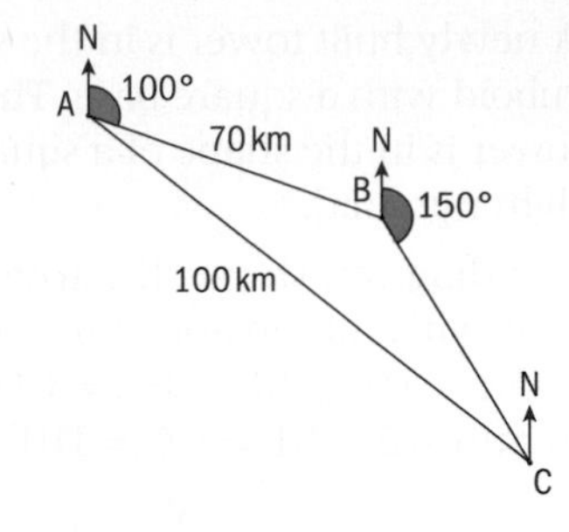

a Find $A\hat{B}C$.

b Hence find the bearing of mast A from mast C.

c Find the distance from mast B to mast C.

20 The Leaning Tower of Pisa, [PQ], leans at 5° away from the vertical.

Galli stands at point A on flat ground 119 m away from Q in the direction of the lean.

He measures the angle of elevation to the top of the tower P to be 26.5°.

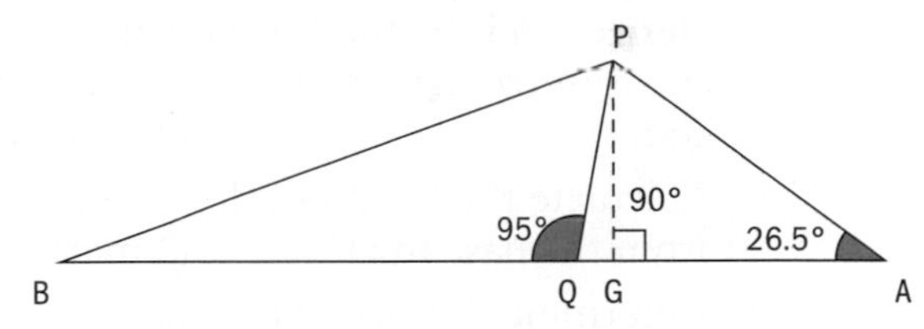

a Find the measure of angle $Q\hat{P}A$.

b Find the length, PQ, of the Leaning Tower of Pisa.

c Calculate the vertical height PG of the top of the tower.

Exam-style questions

21 P2: A squared-based pyramid has slant height 7 centimetres. The edges of the base are 5 centimetres long and the apex is located vertically above the centre of its base.

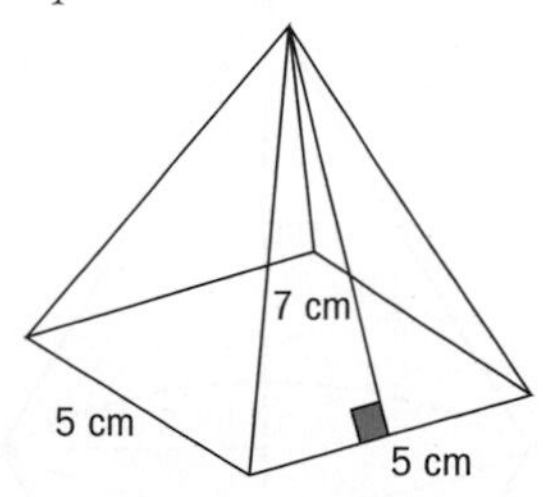

a Find the total surface area of the pyramid. (2 marks)

b Calculate the volume of the pyramid. (4 marks)

22 P2: A newly built tower is in the shape of a cuboid with a square base. The roof of the tower is in the shape of a square-based right pyramid.

The diagram shows the tower and its roof with dimensions OE = OF = OG = OH = 10m, AB = BC = CD = AD = 6 m and AE = BF = CG = DH = 42 m.

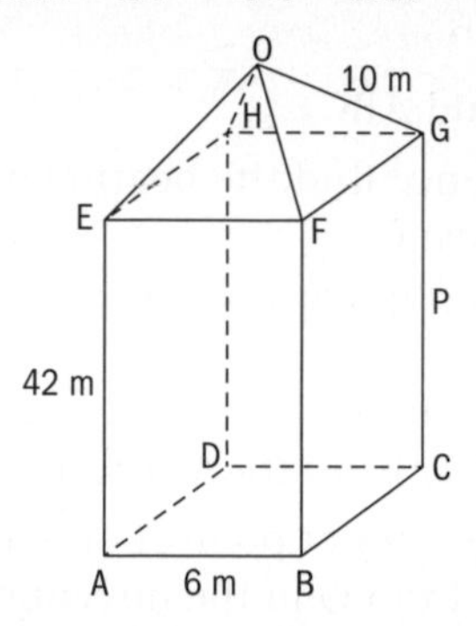

a Calculate the shortest distance from O to EF. (2 marks)

b Hence, find the total surface area of the four triangular sections of the roof. (2 marks)

c Calculate the height of the tower from the base to O. (2 marks)

d Determine the size of the angle between OE and EF. (2 marks)

A bird nest is perched at a point P on the edge, CG, of the tower. A person at the point B, outside the building, measures the angle of elevation to point P to be 60°.

e Find the height of the nest from the base of the tower. (2 marks)

23 P2: A spinner is made of two equal cones with heights 5 cm and radii 3 cm, as illustrated in the diagram.

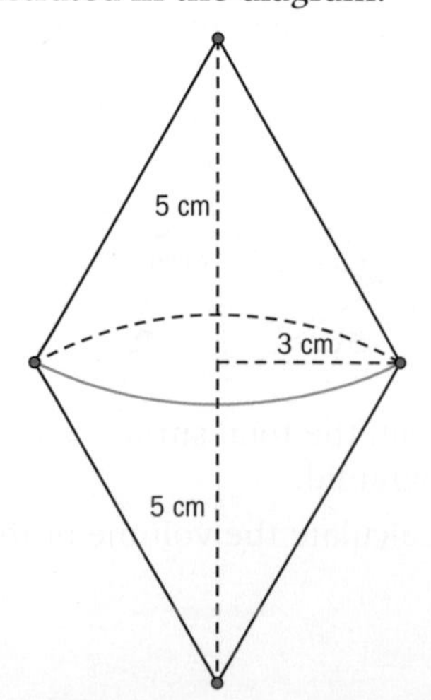

a Calculate the spinner´s surface area. (4 marks)

b If the spinner is placed in a cylinder-shaped container with height 10.1 cm and diameter 6.1 cm, determine the percentage of the volume of the container occupied by the spinner. (2 marks)

24 P2: Consider a trapezium ABCD with AB = CD and BC parallel to AD. Given that AB = 13 cm, BC = 12 cm, and AD = 22 cm,

a show that the trapezium has height 13 centimetres. (3 marks)

b Hence, find the area of the trapezium. (2 marks)

c Find the size of the angle D. (3 marks)

d Calculate the length of the diagonal AC. (3 marks)

25 P1: A ship leaves a port located at point A on a bearing of 030°. It sails a distance of 20 km to a point B where it changes direction to a bearing of 75°. Then it sails 25 km to reach a port located at a point C.

a Find the distance between A and C. (4 marks)

b Find the bearing of the point C with respect to A. (3 marks)

26 P2: The diagram shows a cuboid ABCDEFGH. The vertex F is located at the origin of a plane and the vertex C has coordinates (8, 10, 6). The faces of the cuboid are parallel to the coordinate planes.

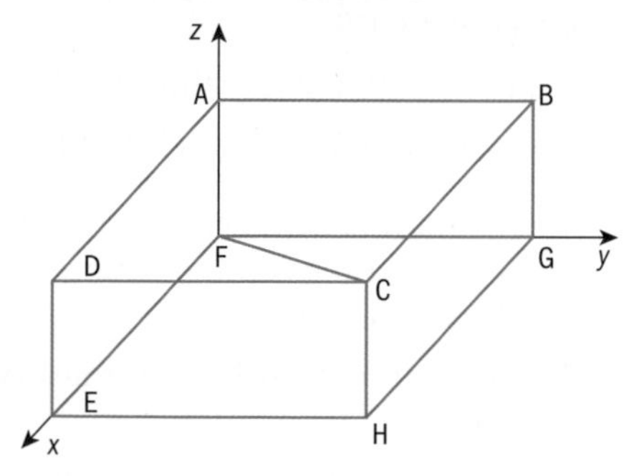

a Show that the length of the diagonal of the cuboid is $10\sqrt{2}$ cm. (2 marks)

b Determine the coordinates of the midpoint M of the segment AD. (2 marks)

c Find the distance between the points F and M. (2 marks)

d Find the perimeter of the triangle CFM, giving your answer exactly. (3 marks)

e Show that cosine of the angle AMF is equal to $\frac{2\sqrt{13}}{13}$. (3 marks)

27 P1: The diagram below shows a quadrilateral ABCD such that

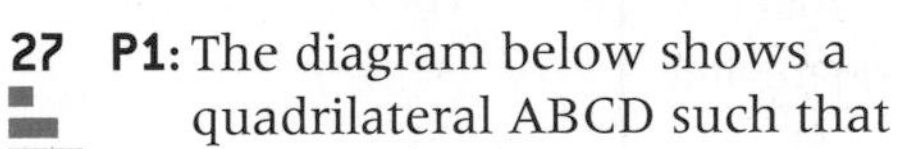

AB= 10 cm, AD =5 cm, BC=13 cm, $B\hat{C}D = 45°$ and $B\hat{A}D = 30°$.

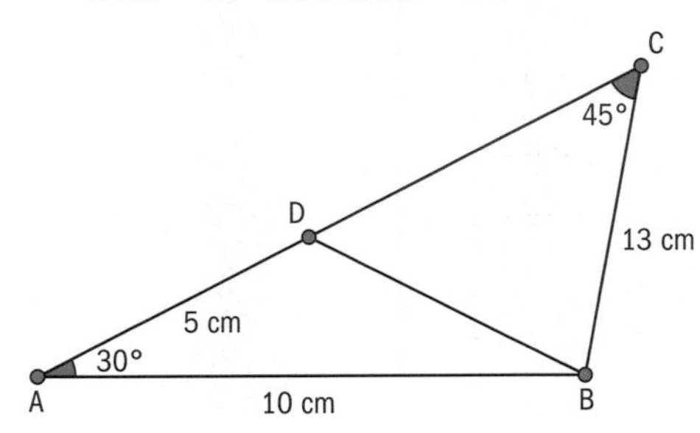

a Find the exact area of the triangle ABD. (2 marks)

b Show that $BD = 5\sqrt{5 - 2\sqrt{3}}$. (4 marks)

c Find the exact value of $\sin C\hat{D}B$. (3 marks)

d Hence, explain why there are two possible values for the size of the angle $C\hat{B}D$, stating the relationship between the two possible values. (2 marks)

28 P2: The figure shows a container in the shape of a circular cylinder topped by a hemisphere. The radii of both cylinder and hemisphere is 3 centimetres and the height of the cylinder is 7 centimetres.

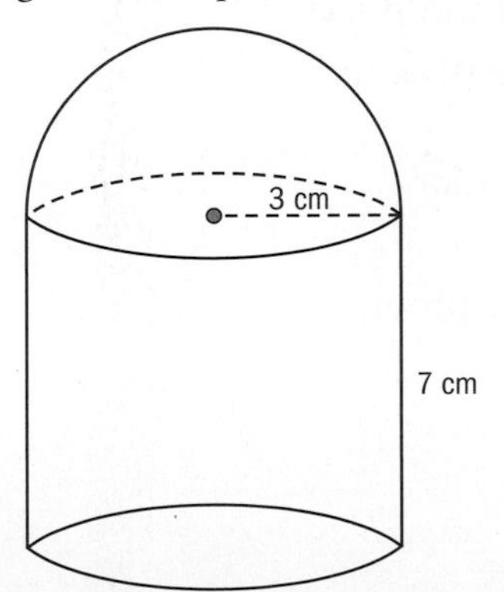

Calculate:

a the volume of the container (4 marks)

b the total surface area of the container. (3 marks)

29 P2: The diagram shows a water tower with height 30 meters and width 3 meters. The tower stands on a horizontal platform.

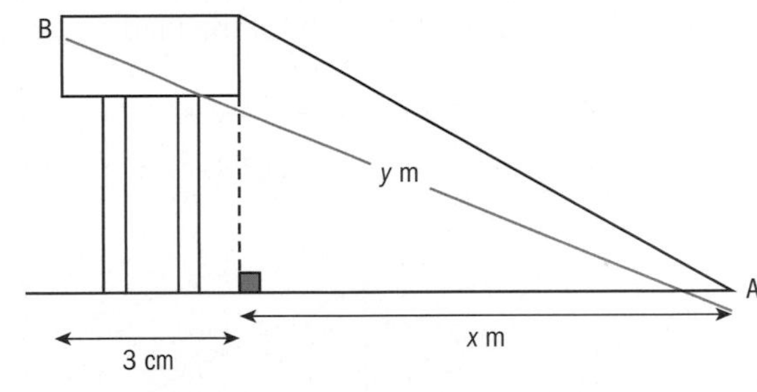

From a point A on the ground the angle of elevation to the top of the tower is 32°.

a Calculate the distance x, giving your answer correct to the nearest metre. (3 marks)

b Hence, determine the distance y between A and B. (2 marks)

c Find the angle of depression of A from B. (2 marks)

30 P2: The diagram shows an obtuse triangle ABC with AB = 57 cm, AC = 48 cm and $B\hat{A}C = 117°$.

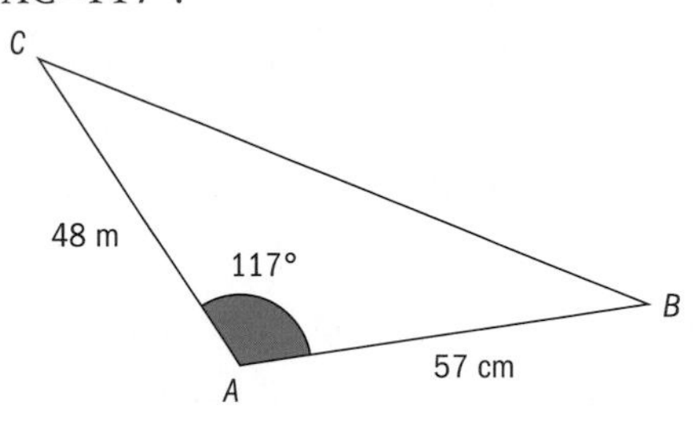

Calculate:

a the length BC (3 marks)

b the area of triangle ABC (2 marks)

c the size of the angle ABC. (3 marks)

Geometry and trigonometry

Three squares

Approaches to learning: Research, Critical Thinking
Exploration criteria: Personal engagement (C), Use of mathematics (E)
IB topic: Proof, Geometry, Trigonometry

The problem

Three identical squares with length of 1 are adjacent to one another. A line is connected from one corner of the first square to the opposite corner of the same square, another to the opposite corner of the second square and another to the opposite corner of the third square:

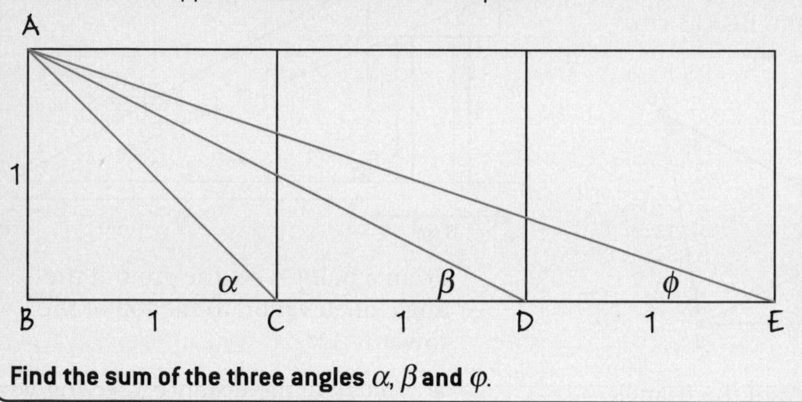

Find the sum of the three angles α, β and φ.

Exploring the problem

Look at the diagram

What do you think the answer may be?

Use a protractor if that helps.

How did you come to this conjecture?

Is it convincing?

This is not an accepted mathematical truth. It is a conjecture, based on observation.

You now have the conjecture $\alpha + \beta + \varphi = 90°$ to be proved mathematically.

Direct proof

What is the value of α?

Given that $\alpha + \beta + \varphi = 90°$, what does this tell you about α and $\beta + \varphi$?

What are the lengths of the three hypotenuses of ΔABC, ΔABD and ΔABE?

Hence explain how you know that ΔACD and ΔACE are similar.

What can you therefore conclude about $C\hat{A}D$ and $C\hat{E}A$?

Hence determine why $A\hat{C}B = C\hat{A}D + A\hat{D}C$ and conclude the proof.

Proof using an auxiliary line

An additional diagonal line, CF, is drawn in the second square and the intersection point between CF and AE is labelled G:

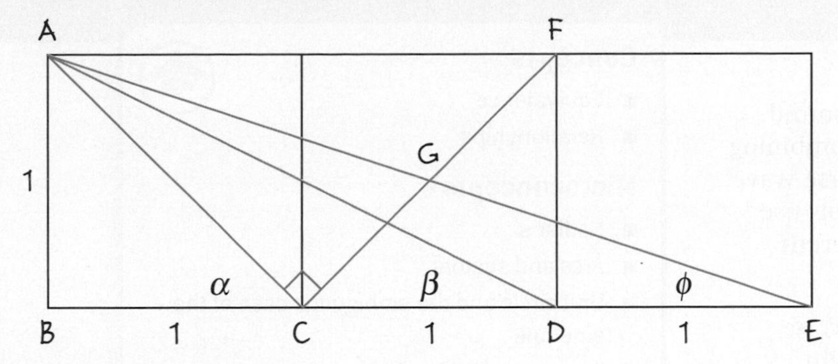

Explain why $B\hat{A}C = \alpha$.

Explain why $E\hat{A}F = \varphi$.

If you show that $G\hat{A}C = \beta$, how will this complete the proof?

Explain how you know that ΔGAC and ΔABD are similar.

Hence explain how you know that $G\hat{A}C = B\hat{D}A = \beta$.

Hence complete the proof.

Proof using the cosine rule

The diagram is extended and the additional vertices of the large rectangle are labelled X and Y and the angle is labelled θ:

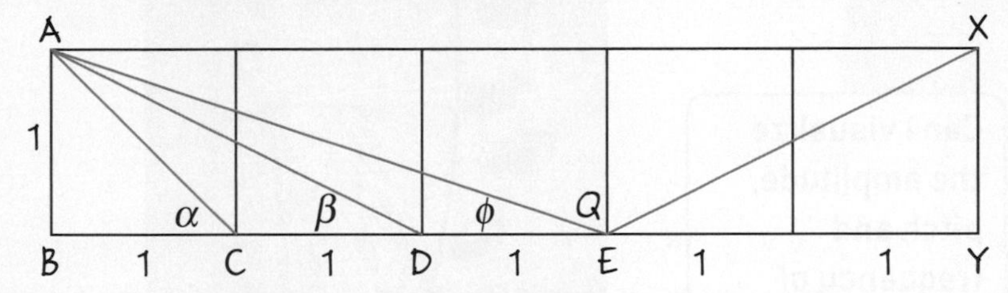

Explain why $X\hat{E}Y = \beta$.

Calculate the lengths of AE and AY.

Now calculate $A\hat{E}Y$ (θ) using the cosine rule.

Hence explain how you know that $\beta + \varphi = 45°$.

Hence complete the proof.

Extension

Research other proofs on the internet.

You could also try to produce a proof yourself.

You do not have to stop working when you have the proof.

What could you do next?

You could use the methods devised in the task in Chapter 7 on Spearman's Rank to rank the proofs and discuss the results.

12 Periodic relationships: trigonometric functions

Sine waves are the building blocks of sound. In fact, **all** sounds can be created by combining sine waves. Sound travels as a transverse wave. Musical instruments and electronic tools use these waves to produce sounds of different frequencies, volume and pitch.

Many phenomena that have a periodic, or repeating, pattern, such as tides, sunlight and Ferris wheels, may be modelled with trigonometric functions.

Concepts

- Equivalence
- Relationships

Microconcepts

- Radians
- Arcs and sectors
- Unit circle and the ambiguous case of the sine rule
- Trigonometric identities and equations
- Trigonometric periodic functions

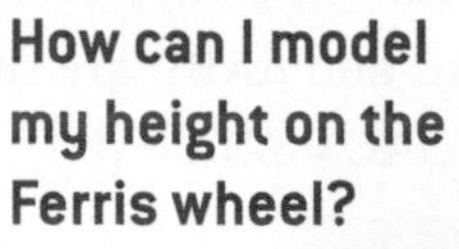

Can I visualize the amplitude, pitch and frequency of musical notes?

What time is sunset in Bali on my birthday?

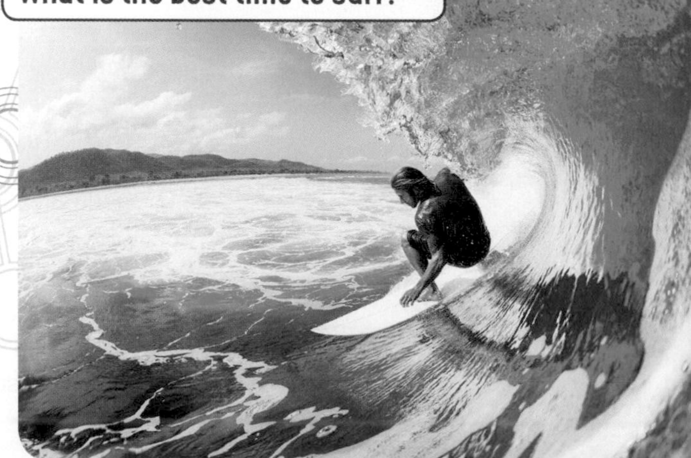

The sounds we hear are caused by vibrations that send pressure waves through the air. Our ears respond to these pressure waves and send information to the brain about the amplitude and frequency of these waves. The brain then interprets these signals as sound.

A speaker usually consists of a paper cone attached to an electromagnet. By sending an oscillating electric current through the electromagnet, the paper cone can be made to oscillate back and forth. When you make a speaker cone oscillate 440 times per second, it will sound like a pure A note.

We need to describe oscillations that occur many times per second. The graph of a sine function that oscillates through one cycle in a second looks like this:

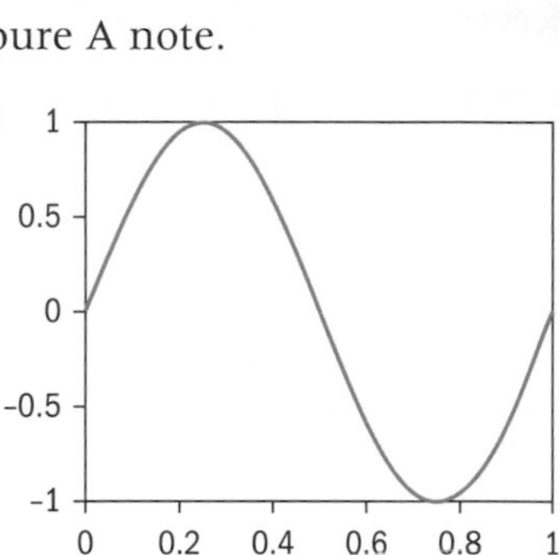

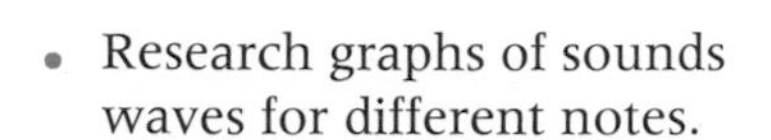

- Research graphs of sounds waves for different notes.
- What is the same about all these graphs?
- What is different?
- What is the relationship between a graph of a sine function that oscillates through one cycle per second (above) and the graph of a sine function that oscillates through two cycles per second?

Developing inquiry skills

Research how Bach and other composers used mathematics in their musical compositions.

You might start by watching the Ted Talk on music and maths, available on YouTube.

Think about the questions in this opening problem and answer any you can. As you work through the chapter, you will gain mathematical knowledge and skills that will help you to answer them all.

Before you start

Click here for help with this skills check

You should know how to:	Skills check
1 Find the exact values of certain trigonometric ratios. eg Find the exact value of $\sin 30°$. $\sin 30° = 0.5$	**1** Find the exact value of each. **a** $\sin 45°$ **b** $\tan 60°$ **c** $\cos 30°$
2 Work with the graphing functions of your GDC. eg Use the graphing functions of your GDC to find the x-intercepts of the graph of $f(x) = x^2 - 3x - 6$. $x \approx -1.37, 4.37$. eg Use the graphing functions of your GDC to solve the equation, $3\ln x = 2 - x^2$ $x \approx 1.20$	**2** Use the graphing functions of your GDC to find the x-intercepts of the graph of each function. **a** $f(x) = x^3 - 2x^2 + 1$ **b** $f(x) = e^x - 2$ **3** Use the graphing functions of your GDC to solve each equation. **a** $e^x = 2x^2 - 5$ **b** $\ln x = 3 - x$

12.1 Radian measure, arcs, sectors and segments

Investigation 1

For this investigation you should be in a group of two or three.

You will need: compass, protractor, string, scissors, poster paper, pencil and glue.

1 At one end of the paper, construct a circle on a set of axes with radius 10 cm.

2 Use a protractor to measure and mark every $10°$ around the circle.

3 To the right of the circle draw an x-axis about 65 cm long, and draw a y-axis about 20 cm tall.

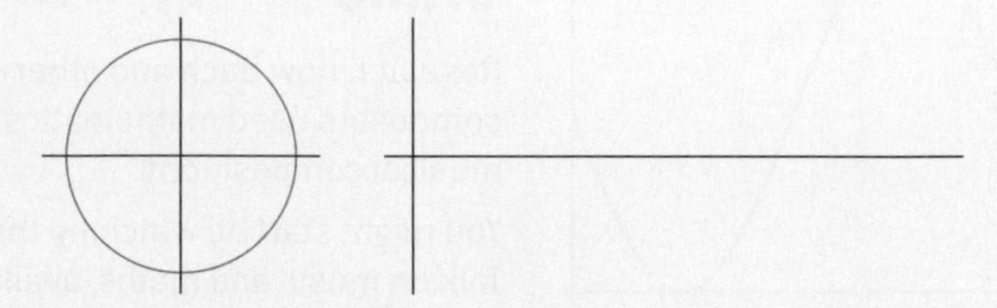

4 Place the string around the circle with the end at zero. Mark the $10°$ marks onto the string. Then put the string on the x-axis of the right-hand set of axes and transfer the marks labelling every $10°$.

5 Measure the vertical distance from the $10°$ mark to the x-axis. At your mark of $10°$ on the other set of axes, draw a line segment the same height as the one you just measured.

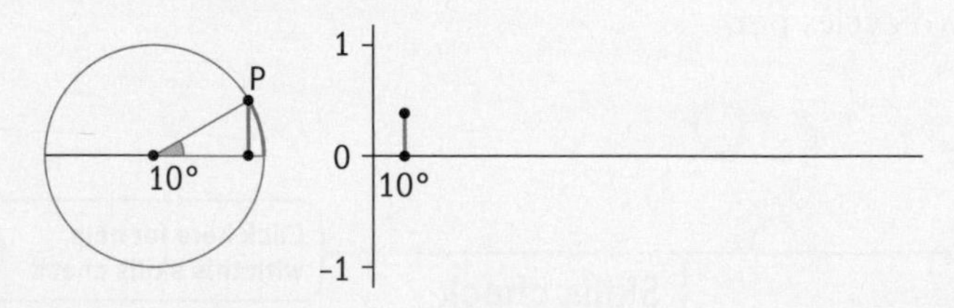

6 Now repeat for $20°$, $30°$ and so on, until you have gone completely around the circle to $360°$.

7 Take your string and glue it to your poster from zero degrees, along the ends of the line segments to form a smooth curve.

This is the sine curve.

8 Research periodic functions. Is the sine function periodic? Why?

9 **Factual** What is a periodic function?

10 **Factual** What shape is the graph of a periodic function?

11 Use your calculator to graph $y = \sin x$. Does it look like your string graph?

12 **Conceptual** How does the shape of a periodic function show that there will always be multiple x values that give the same values of y?

13 Write an explanation about why $\sin 60° = \sin 120°$.

14 **Conceptual** How do we extend trigonometric ratios beyond right-angled triangles?

15 **Discuss** your answer to the following questions with your group and then **share** your ideas with your class.

- a What is the radius of the circle in string units?
- b What is the circumference in string units?
- c What happens after $360°$?
- d What are the zeros of your sine function?
- e Where is the highest point above the x-axis? At what angle is this?
- f Where is the lowest point below the x-axis? At what angle is this?
- g How would you do this investigation differently for a cosine curve?

You already know how to use **degrees** when measuring angles, but most science and engineering applications use **radians**. Radians are an alternative way of measuring the size of an angle.

Investigation 2

You will need: string, a compass, a ruler, a protractor and a pencil.

1 Draw an x- and y-axis.

2 Use your compass to construct a circle with the origin as the centre.

3 Label the point where the circle cuts the positive x-axis $0°$, the point where the circle cuts the positive y-axis $90°$ and keep going counter-clockwise round the circle to label $180°$, $270°$ and $360°$. You may notice that 0 and 360 are in the same place.

4 Take your string and put it around the outside of your circle. Cut off any excess string so that you have the exact circumference in string.

5 Put one end of your string at the origin and mark the point where it touches the edge of the circle. You should have marked the length of one radius on the string.

6 Mark any point on the circle A. Put the end of the string here, draw it out along the circumference of the circle and mark the length of one radius. Mark this point B.

7 Mark the centre O and draw segments [OA] and [OB].

8 Repeat this for several starting points on your circle.

9 Use your protractor to measure angle A$\hat{O}$B for each shape and average those answers.

10 Share your results with the class by writing everybody's averages on the board and averaging all of those values.

11 Use the Internet to search "What is a radian?"

12 About how many radians is a straight line? Can you think of a mathematical value or constant that is close to this number? We say that an angle which forms a straight line measures π radians.

13 **Factual** What is the relationship between radians and degrees?

14 **Factual** What are some uses of the radian measure?

15 **Conceptual** Why are radians dimensionless?

16 With a partner, discuss what you have discovered, what you now know about radian measure and what you wonder.

Radians

A **radian** is the measure of the angle with its vertex at the centre of the circle between two radii with endpoints that are one radius length apart on the circumference of the circle.

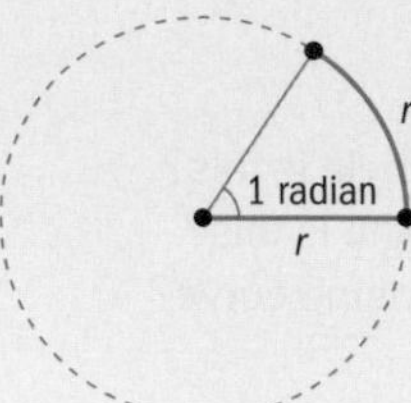

You might recall that the circumference of a circle is $2\pi r$, so that means there are 2π, or approximately 6.28 radians, in a full circle. This gives you the conversion:

$$2\pi \text{ radians} = 360° \text{ or } 1 \text{ radian} \approx 57.3°$$

HINT
Make sure you know how to use your GDC to work with both degrees and radians.

From this you can deduce that:

π radians = 180°

$\frac{\pi}{2}$ radians = 90°

$\frac{\pi}{3}$ radians = 60°

$\frac{\pi}{4}$ radians = 45°

$\frac{\pi}{6}$ radians = 30°

You can use radians in the place of degrees on the sine curve from Investigation 1.

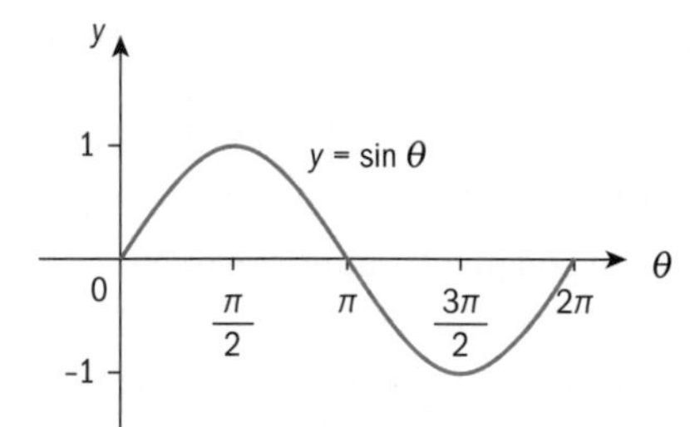

Converting between degrees and radians

Degrees to radians:

Multiply the degree measure by $\frac{\pi}{180}$

Radians to degrees:

Multiply the radian measure by $\frac{180°}{\pi}$

The symbol "c" or the abbreviation "rad" are used for radians, but they are often omitted, particularly when π is present. The symbol for degrees is never omitted.

Example 1

a Convert 20° to radians. **b** Convert 56.5° to radians. **c** Convert $\frac{4\pi}{3}$ to degrees.

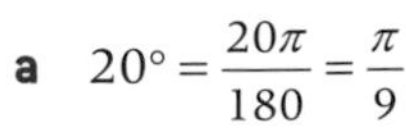 **a** $20° = \frac{20\pi}{180} = \frac{\pi}{9}$	The 20 and the 180 cancel. Leave your answer in terms of π.
b $56.5° = \frac{56.5\pi}{180} \approx 0.986\,\text{rad}$	The numerator and denominator do not cancel. Give your answer to 3 s.f.
c $\frac{4\pi}{3} = \frac{4\pi}{3} \times \frac{180°}{\pi} = 240°$	The π cancels. Do not forget the degree symbol.

Exercise 12A

1 Express each angle in radians.

a 45° **b** 60° **c** 270° **d** 360°

e 18° **f** 225° **g** 80°

h 200° **i** 120° **j** 135°

2 Express each angle in degrees.

a $\frac{\pi}{6}$ **b** $\frac{\pi}{10}$ **c** $\frac{5\pi}{6}$ **d** 3π

e $\frac{7\pi}{20}$ **f** $\frac{4\pi}{5}$ **g** $\frac{7\pi}{4}$

h $\frac{14\pi}{9}$ **i** $\frac{5\pi}{3}$ **j** $\frac{13\pi}{4}$

3 Express each angle in radians, correct to 3 significant figures.

a 10° **b** 40° **c** 25° **d** 300°

e 110° **f** 75° **g** 85° **h** 12.8°

i 37.5° **j** 1°

4 Express each angle in degrees, correct to 3 significant figures.

a 1 rad **b** 2 rad **c** 0.63 rad

d 1.41 rad **e** 1.55 rad **f** 3 rad

g 0.36 rad **h** 1.28 rad **i** 0.01 rad

j 2.15 rad

Reflect **Discuss** your answer with a friend then **share** your ideas with your class.

- Who "invented" the symbol for pi?
- Write a sentence stating who you think it could have been.
- Research to find the correct answer.

Sectors and segments

An arc is part of the circumference of a circle. Minor arc AB is the shortest distance between two points A and B on the circumference of a circle, major arc AB is the longer distance around the circumference from A to B. Unless otherwise indicated in a question, we assume arc AB refers to the minor arc.

A sector is the region enclosed by two radii and an arc. Unless otherwise indicated in a question, we assume sector AOC refers to the minor sector.

A chord is the line segment joining two points on the circumference of a circle.

A segment is the region between a chord and an arc.

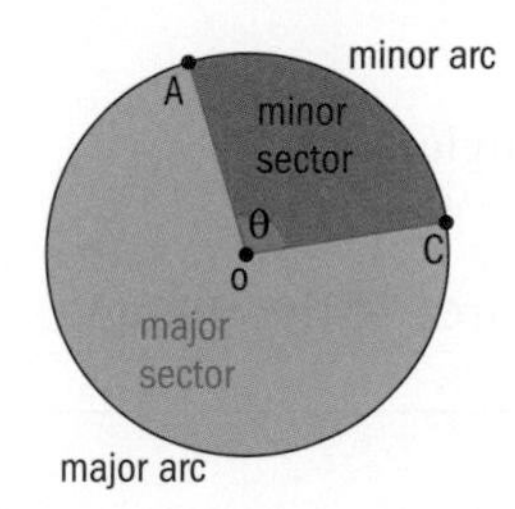

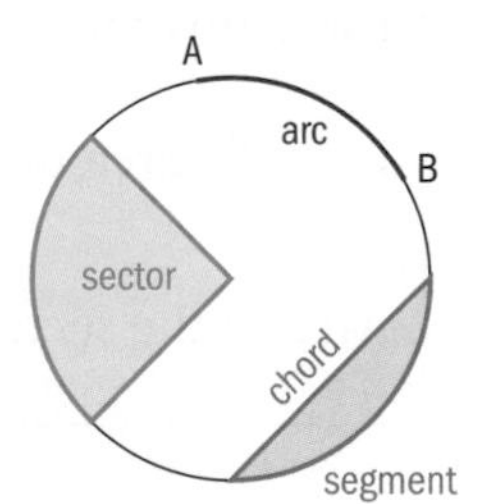

The length of arc, l, can be found by using the circumference formula for a circle:

$$l = \frac{\theta}{2\pi} \times 2\pi r = r\theta$$

The area of a sector, A, can be found by using the area formula for a circle:

$$A = \frac{\theta}{2\pi} \times \pi r^2 = \frac{1}{2} r^2 \theta$$

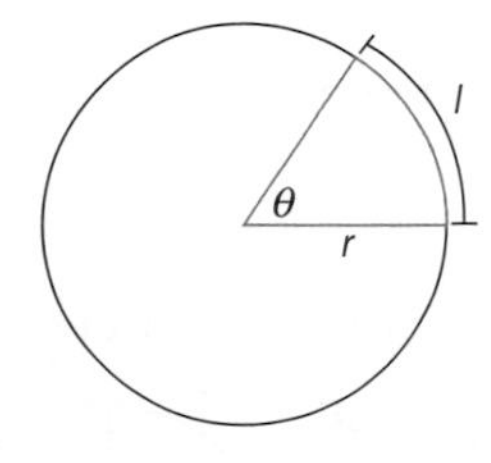

The length of an arc is given by

$$l = r\theta$$

and the area of a sector by

$$A = \frac{1}{2} r^2 \theta$$

where θ is measured in radians.

Example 2

Find:

a the arc length and

b the area of the sector of this circle.

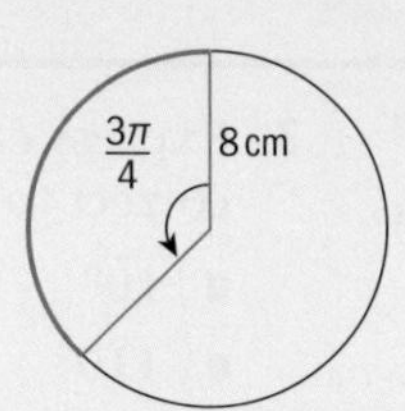

a $l = r\theta = 8 \times \frac{3\pi}{4} = 6\pi$ cm	The exact answer would leave your arc length in terms of π. You can also round to 3 s.f. to give 18.8 cm.
b $A = \frac{1}{2}r^2\theta = \frac{1}{2}(8)^2\left(\frac{3\pi}{4}\right) = 24\pi$ cm	You can also round to 3 s.f. to give 75.4 cm^2.

Example 3

Find the shaded area.

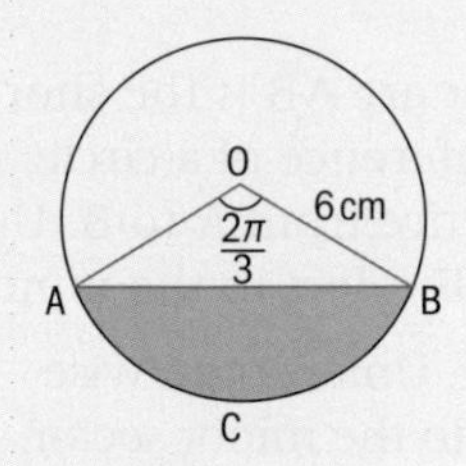

$A = \frac{1}{2}r^2\theta = \frac{1}{2}(6)^2\left(\frac{2\pi}{3}\right) = 12\pi$	Find the area of sector OACB. Leave your answers as an exact value as you will be using this again.
$A = \frac{1}{2} \times 6 \times 6 \times \sin\frac{2\pi}{3} = 18 \sin\frac{2\pi}{3}$	Find the area of triangle OAB.
Shaded area $= 12\pi - 18 \sin\frac{2\pi}{3} = 22.1$ cm^2	Area of the sector – area of the triangle

Exercise 12B

1 Find:

a the arc length

b the area of the sector in each circle.

i

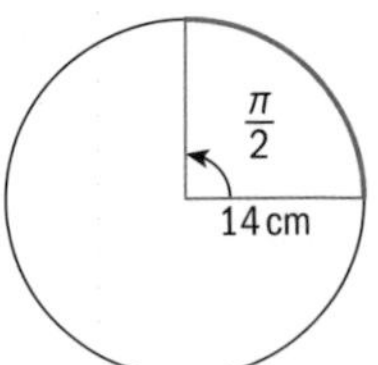

ii

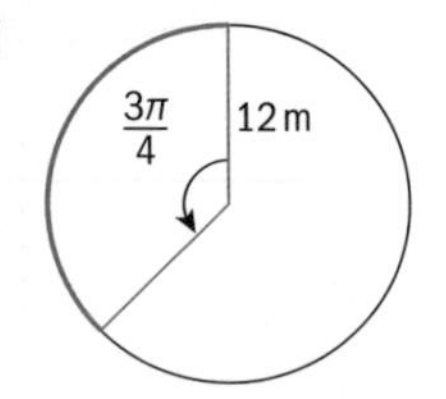

iii

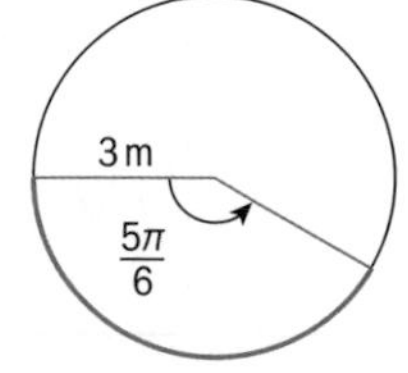

iv

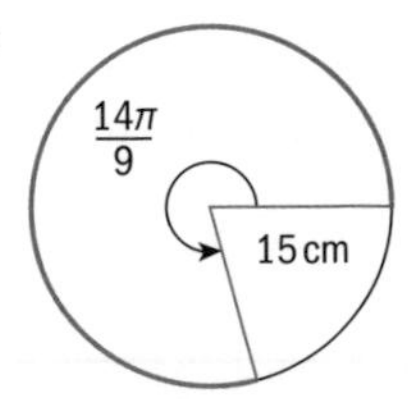

2 The sector formed by a central angle of $\frac{\pi}{12}$ has an area of 3π cm^2. Find the radius of the circle.

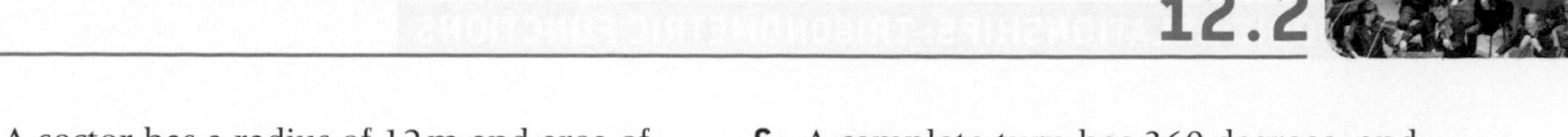

3 A sector has a radius of 12 m and area of 36π m^2. Find:

a the central angle which forms the sector

b the perimeter of the sector.

4 Find the shaded area when $\theta = 1.5$ rad.

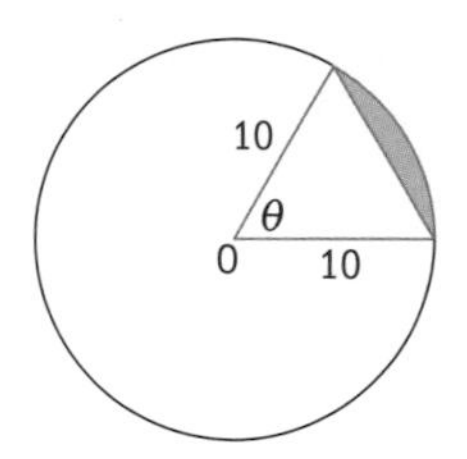

5 The pendulum on a large grandfather clock swings from one side to the other once every second. The length of the pendulum is 4 m and the angle through which it swings is $\frac{\pi}{12}$. Find the total distance the tip of the pendulum travels in one minute.

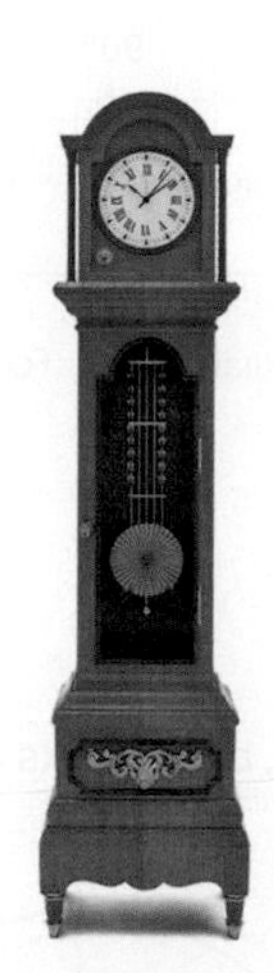

6 A complete turn has 360 degrees, and each degree is divided into 60 parts called minutes.

When the central angle with its vertex at the centre of the earth has a measure of 1 minute, the arc on the surface of the earth that is formed by this angle (known as the great-circle distance) has a measure of 1 nautical mile.

The radius of the earth is approximately 6371 km. Find how many km there are in 1 nautical mile.

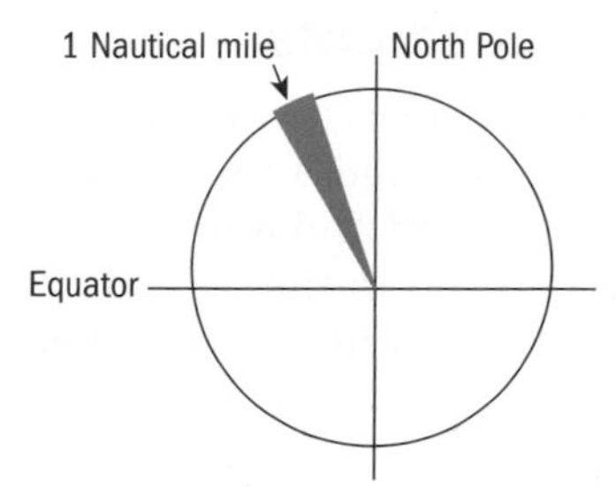

Developing inquiry skills

How would you describe the shape of the function in the opening problem?

Is this periodic function? Explain your reasoning.

12.2 Trigonometric ratios in the unit circle

The **unit circle** is a circle with its centre at the origin and of radius 1 unit.

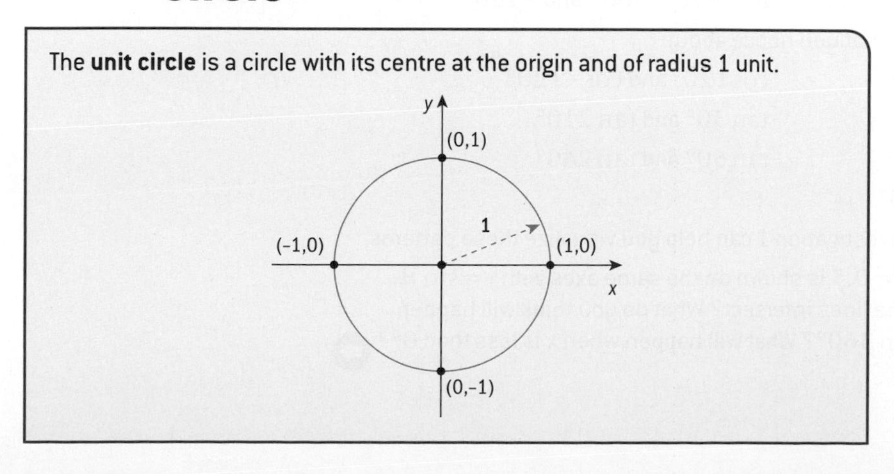

Angles in **standard position** in the unit circle are measured starting on the positive x-axis and turning anticlockwise (positive angles) or clockwise (negative angles).

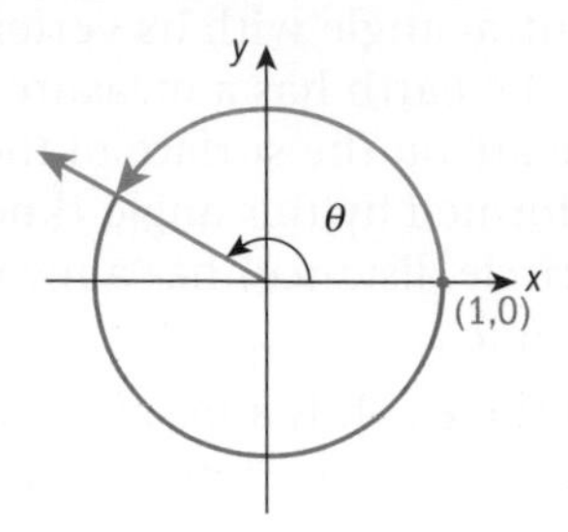

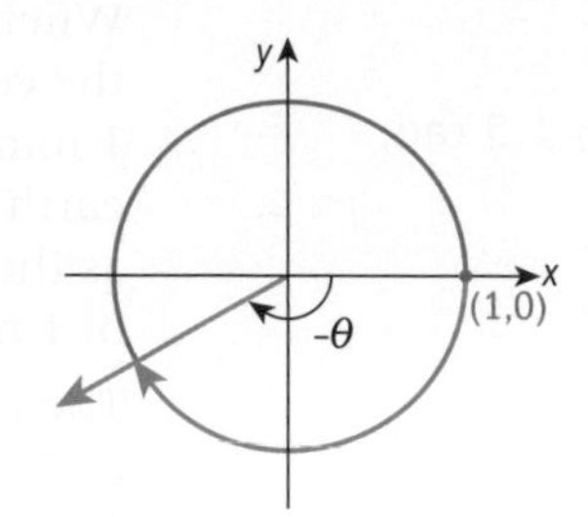

The four quadrants

The coordinate axes divide the plane into four quadrants, labelled first, second, third and fourth, as shown. Angles in the second quadrant, for example, lie between 90° and 180°.

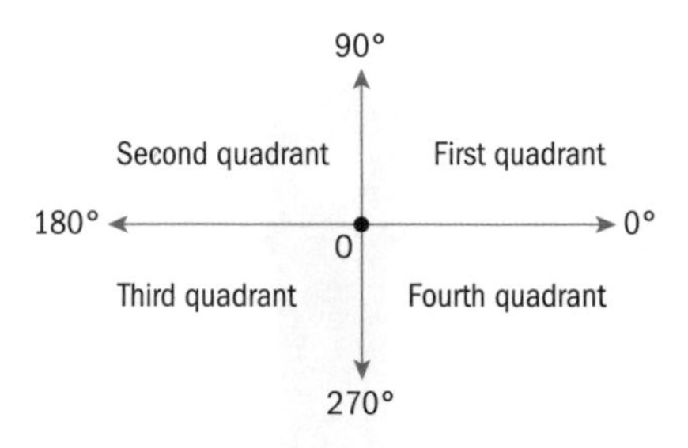

Investigation 3

Work in small groups. This investigation uses degrees, but it works equally well if you want to convert the degrees into radians.

1 Draw a unit circle on graph paper.

2 Now draw point A on your circle with $\theta = 30°$ and draw a line vertically to the x-axis to make point B.

3 Use trigonometry to find OB. This is the x-coordinate of point A.

4 Use trigonometry to find AB. This is the y-coordinate of A.

5 Write down the coordinates of A.

Notice that the coordinates of A at any angle θ are $(\cos\theta, \sin\theta)$.

6 Repeat your drawing of A and find its coordinates when $\theta = 60°, 120°, 150°, 210°, 240°, -30°, -60°$ and $-120°$.

7 Explain to your group what you notice about:

$\sin 30°$ and $\sin 150°$

$\sin 60°$ and $\sin 120°$

$\cos 30°$ and $\cos -30°$

$\cos 60°$ and $\cos -60°$

$\cos 120°$ and $\cos -120°$

$\tan 30°$ and $\tan 210°$

$\tan 60°$ and $\tan 240°$

The sine graph from investigation 1 can help you visualize these patterns.

8 In the graph below, $y = 0.5$ is shown on the same axes with $y = \sin\theta$. Can you see where the lines intersect? What do you think will happen when x is greater than 360°? What will happen when x is less than 0°?

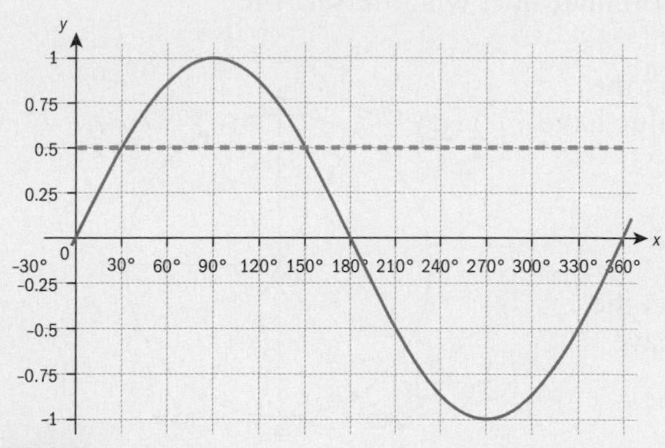

9 **Conceptual** Using symmetry, state the different relationships between angles in different quadrants for each of the trigonometric values.

10 How are positive and negative angles of all sizes represented on a unit circle?

11 How can you find different angles with the same trigonometric values?

12 How can you find the exact values of the trigonometric ratios of $0, \frac{\pi}{6}, \frac{\pi}{4}, \frac{\pi}{3}, \frac{\pi}{2}, \pi$ and their multiples?

13 Did you notice any other patterns?

14 What happens when you add $360°$ to any of those angles?

From your investigation, you should see that:

$$\sin\theta = \sin(180 - \theta)$$
$$\cos\theta = \cos(-\theta)$$
$$\tan\theta = \tan(180 + \theta)$$

Looking at your results, the x- and y-coordinates of any point that lies in each of the four quadrants, we can identify the sign of each of the trigonometric ratios in a given quadrant.

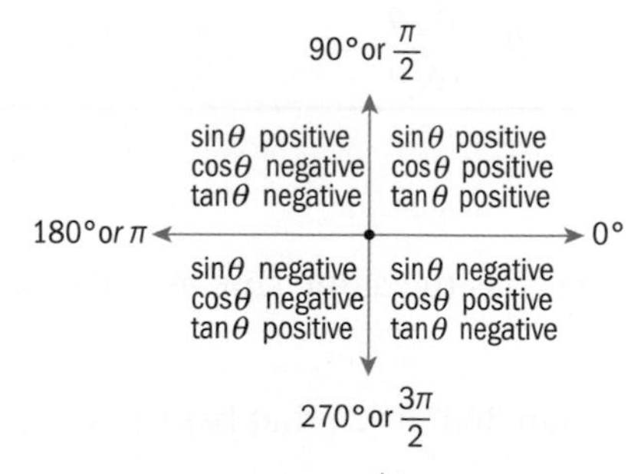

To help remember the signs of the three trigonometric ratios in the quadrants, you can see that all of the ratios are positive in the first quadrant, then only one of the ratios is **positive** in the other three quadrants.

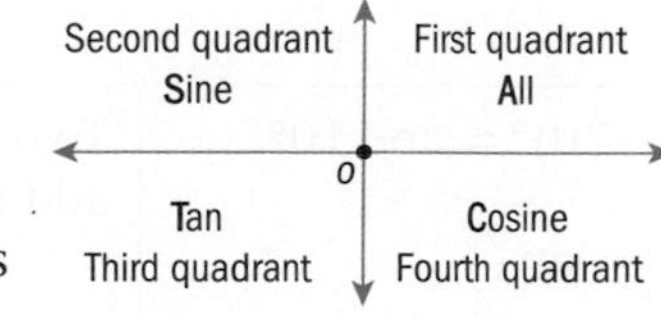

HINT

The mnemonic **All Students Take Calculus** is a good way to remember this.

Remember that the **standard position** of an angle is the position of an angle with its vertex at the origin of a unit circle and one side fixed on the positive x-axis. This side is called the initial side of the angle.

The other side of the angle, called the terminal side, will intersect the circle at a point.

Angles with terminal sides that intersect the circle at points with with the same y value have equal sines. For example:

$\sin 60° = \sin 120°$

$\sin 240° = \sin 300°$

Angles with terminal sides that intersect the circle at points with the same x value have equal cosines. For example:

$\cos 60° = \cos 300°$

$\cos 120° = \cos 240°$

Angles with terminal sides that are directly opposite each other on a line drawn through the origin have the same tangents:

$\tan 60° = \tan 240°$

$\tan 120° = \tan 300°$

In the unit circle, an angle can be used to form a right triangle with a hypotenuse of length one and two shorter sides of $\sin\theta$ and $\cos\theta$.

The Pythagorean identity gives you:

$\sin^2\theta + \cos^2\theta = 1$

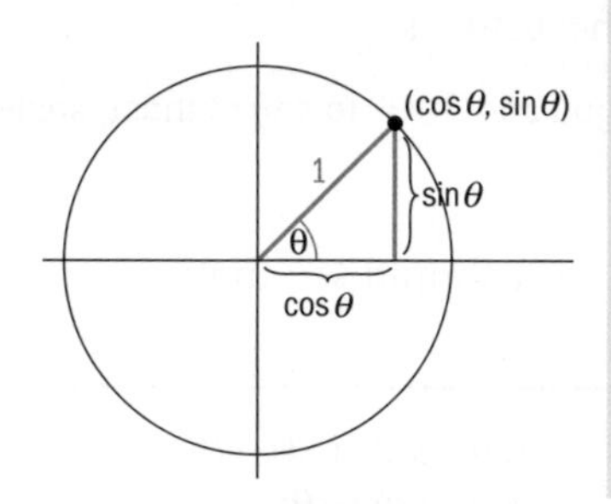

HINT

Notice that $(\sin\theta)^2$ is written $\sin^2\theta$.

In the same unit circle diagram:

$\tan\theta = \dfrac{\text{opposite side}}{\text{adjacent side}} = \dfrac{\sin\theta}{\cos\theta}$

$$\sin^2\theta + \cos^2\theta = 1$$

$$\tan\theta = \frac{\sin\theta}{\cos\theta}$$

Reflect **Discuss** your answer with a partner then **share** your ideas with your class.

- What facts do you know about the unit circle?
- What can the unit circle help you to find?
- Research why we use radians instead of degrees.

Example 4

1 Find an obtuse angle that has the same sine as the given acute angle.

a $70°$ **b** $\dfrac{\pi}{4}$

2 Find another angle less than $360°$ or 2π that has the same cosine as the given angle.

a $50°$ **b** $\dfrac{2\pi}{3}$

1 a $\sin 70° = \sin(180 - 70)° = \sin 110°$ **b** $\sin\dfrac{\pi}{4} = \sin\left(\pi - \dfrac{\pi}{4}\right) = \sin\dfrac{3\pi}{4}$	Two angles have the same sine if the angles add up to $180°$ or π radians.
2 a $\cos 50° = \cos(360 - 50)° = \cos 310°$ **b** $\cos\dfrac{2\pi}{3} = \cos\left(2\pi - \dfrac{2\pi}{3}\right) = \cos\dfrac{4\pi}{3}$	Two angles have the same cosine if the angles add up to $360°$ or 2π radians.

Example 5

Without finding θ, find $\cos\theta$, when $\sin\theta = \frac{3}{5}$ and θ is acute.

$\sin^2\theta + \cos^2\theta = 1$	Use the Pythagorean identity.
$\left(\frac{3}{5}\right)^2 + \cos^2\theta = 1$	
$\cos^2\theta = 1 - \left(\frac{3}{5}\right)^2 = \frac{16}{25}$	
$\cos\theta = \frac{4}{5}$	Use the positive value, as the angle is acute.

Exercise 12C

1 Find the sign of:

a $\cos 130°$ b $\sin 320°$

c $\tan 225°$

2 Find an obtuse angle that has the same sine as:

a $36°$ b $50°$ c $85°$

d $460°$ e $\frac{\pi}{3}$ f $\frac{\pi}{5}$

g $\frac{2\pi}{7}$ h $\frac{8\pi}{3}$

3 Find an angle less than $360°$ or 2π that has the same cosine as:

a $40°$ b $110°$ c $300°$ d $500°$

e $\frac{\pi}{8}$ f $\frac{\pi}{10}$ g $\frac{3\pi}{2}$ h $\frac{9\pi}{4}$

4 Solve these equations for $0 \le \theta \le 2\pi$.

a $\sin\theta = \frac{1}{2}$ b $\cos\theta = \frac{\sqrt{2}}{2}$ b $\tan\theta = \sqrt{3}$

5 a Find $\sin\theta$, when $\cos\theta = \frac{8}{17}$ and θ is acute.

b Find $\tan\theta$.

Developing inquiry skills

In the graph on the opening page, what do you think will happen after one second? Can you describe the next few seconds? What do you think could have happened before zero seconds?

12.3 Trigonometric identities and equations

Solving trigonometric equations

During this course, you have learned to solve many types of equations, such as quadratic, exponential and logarithmic. Now we will look at trigonometric equations.

When solving trigonometric equations, it is important to know whether your answer should be exact, in radians or degrees, within what interval, or domain. "Exact" will often mean radians in terms of π.

TOK

It is fascinating to see how mathematics is passed from nation to nation, irrespective of race or religion.

Research the etymology of the word "sine" and write a paragraph explaining when it was first used and why.

Share your ideas with a partner and compare answers.

Investigation 4

1. Solve $\sin\,\theta = \frac{1}{2}$ using graphing technology.
2. Now find $\sin^{-1}0.5$ on your calculator.

Note that your calculator will give you $30°$ (or $\frac{\pi}{6}$ radians).

3. Graph $y = \sin\,x$ and $y = 0.5$ on the same set of axes.
4. Notice that the graph has more than one intersection, but your calculator only gives one value. Show this answer on a unit circle. This is called the primary value.
5. Now show $390°$ on your unit circle.
6. Show $750°, 1110°, -330°$ and $-690°$. What do you notice?
7. Enter the sine of each of these angles on your calculator.
8. Now find $\sin\,750°$ and then $\sin^{-1}$ of the answer. What do you notice?
9. Discuss with a partner how to write all of the solutions to $\sin\theta = \frac{1}{2}$.

   Show this angle on a unit circle. This is called a secondary value.
10. Go back to your unit circle and find another value for $\sin\,\theta = \frac{1}{2}, 0 \le \theta \le 360°$.
11. Now show $510°$ on your unit circle.
12. Show $870°, 1230°, -210°$ and $-570°$. What do you notice?
13. Enter the sine of each of these angles on your calculator.
14. Now find $\sin\,510°$ and then $\sin^{-1}$ of the answer. What do you notice?
15. **Factual** How can you use these patterns to solve equations?
16. **Conceptual** How is trigonometry used to find all possible angles for unknown values? How is trigonometry used to find unknown values?
17. Discuss with a friend how to improve your solution to $\sin\theta = \frac{1}{2}$.
18. Share your final results with the class.

You will know from chapter 11 that one solution to $\sin\,\theta = \frac{1}{2}$ is $30°$ (or $\frac{\pi}{6}$), and from the unit circle you saw that angles with terminal sides that intersect the circle at points with the same y-value have equal sines, and angles with the same x values have equal cosines.

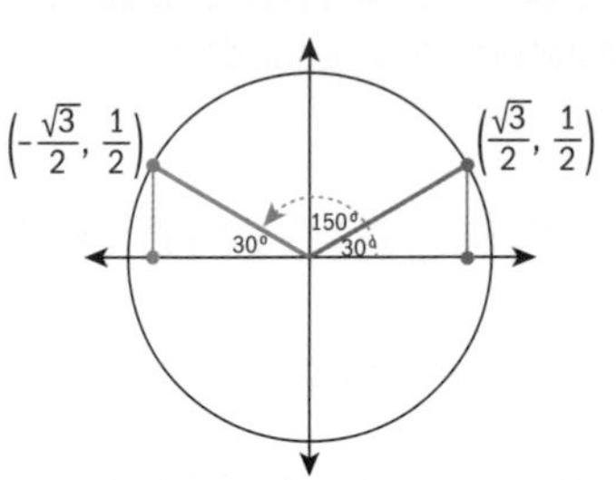

You also saw the identity $\sin\theta = \sin(180 - \theta)$, which can help you to find the second angle.

Now you have $\theta = 30°$ and $150°$, or $\frac{\pi}{6}$ and $\frac{5\pi}{6}$.

When you do a whole rotation, after each angle you can add or subtract 360° or 2π and so on. This will give you an infinite number of solutions and so you are usually asked to solve an equation within a specific domain.

$\theta = 30°$ or $\frac{\pi}{6}$ is called the primary value and $\theta = 150°$ or $\frac{5\pi}{6}$ is the secondary value and you can add or subtract as many complete turns to both as you wish to find further angles in the given domain.

Example 6

1 Solve $2\cos\theta - 1 = 0$ for $0 \le \theta \le 2\pi$.

2 Solve $-2\cos^2 x - \sin x + 1 = 0 \quad 0 \le x \le 4\pi$.

1 $2\cos\theta - 1 = 0$	Rearrange the equation.
$2\cos\theta = 1$ $\cos\theta = \frac{1}{2}$	Two angles have the same cosine when the angles add up to 2π radians.
$\theta = \cos^{-1}\frac{1}{2} = \frac{\pi}{3}, \frac{5\pi}{3}$	Answers for $0 \le \theta \le 2\pi$.
2 $-2\cos^2 x - \sin x + 1 = 0$ $-2(1 - \sin^2 x) - \sin x + 1 = 0$	From the Pythagorean identity, $\cos^2\theta = 1 - \sin^2\theta$
$-2 + 2\sin^2 x - \sin x + 1 = 0$ $2\sin^2 x - \sin x - 1 = 0$ $(2\sin x + 1)(\sin x - 1) = 0$	Form and solve by factorization, as you would with a quadratic equation.
$2\sin x + 1 = 0 \quad \sin x - 1 = 0$ $\sin x = -\frac{1}{2} \quad \sin x = 1$	
$x = \sin^{-1} -\frac{1}{2} \quad x = \sin^{-1} 1$	Find the inverse sine of the values.
$x = \frac{7\pi}{6}, \frac{11\pi}{6} \quad x = \frac{\pi}{2}$	
$x = \frac{\pi}{2}, \frac{7\pi}{6}, \frac{11\pi}{6}, \frac{5\pi}{2}, \frac{19\pi}{6}, \frac{23\pi}{6}$	Generate angles from the unit circle. Add 2π to give angles up to 4π.

Exercise 12D

1 Solve each equation for $0 \le \theta < 360°$, giving your answers to 1 decimal place.

a $\cos\theta = 0.6$ **b** $\sin\theta = 0.15$

c $\tan\theta = 0.2$ **d** $\tan\theta = -0.76$

e $\cos\theta = -0.43$

2 Solve each equation for $0 \le \theta \le 2\pi$.

a $\sin\theta = 0.82$ **b** $\tan\theta = -0.94$

c $\cos\theta = -0.94$ **d** $\cos\theta = 0.77$

e $\sin\theta = -0.23$

3 Solve the equation $2\sin^2\theta + 5\sin\theta = 3$ for $0 \le \theta \le 2\pi$.

4 Solve each equation for $0 \le \theta \le 2\pi$, giving your answers to 2 decimal places.

a $4\cos x - 3\sin x = 0$

b $2\sin x + \cos x = 0$

c $\tan^2 x - \tan x - 2 = 0$

d $2\cos^2 x + \sin x = 1$

5 Solve for θ in the given domain.

a $\cos\theta = 0.3, -\pi \le \theta \le 3\pi$

b $\tan\theta = 1.61, 0 \le \theta \le 4\pi$

c $\sin\theta = -2\cos\theta, -\pi \le \theta \le \pi$

d $2\tan^2\theta + 5\tan\theta = 3, -2\pi \le \theta \le 0$

6 Solve the equation $3\cos x = 5\sin x$, for $0 \le x \le 360°$, giving your answer to the nearest degree.

Example 7

1 Solve $\sin 2x = \frac{\sqrt{2}}{2}$ for $0 \le x \le 2\pi$.

1 $\sin 2x = \frac{\sqrt{2}}{2}$	
$2x = \sin^{-1}\frac{\sqrt{2}}{2}$	Take the inverse sine.
$2x = \frac{\pi}{4}, \frac{3\pi}{4}, \frac{9\pi}{4}, \frac{11\pi}{4}$	Because we have $2x$, and $0 \le x \le 2\pi$, then we use $0 \le 2x \le 4\pi$. When we divide these values by 2 the answers we require will fall in the domain $0 \le \theta \le 2\pi$.
$x = \frac{\pi}{8}, \frac{3\pi}{8}, \frac{9\pi}{8}, \frac{11\pi}{8}$	

Exercise 12E

1 Solve each equation for $-180° \le x \le 180°$.

a $\cos 2x = \frac{\sqrt{3}}{2}$ **b** $\cos 3x = \frac{\sqrt{3}}{2}$

c $2\cos 3x - 1 = 0$ **d** $3\tan\frac{x}{2} + 3 = 0$

2 Solve each equation for $0 \le \theta \le 2\pi$:

a $\sin 3\theta = \frac{\sqrt{3}}{2}$ **b** $\cos 3\theta - 1 = 0$

c $\sin\frac{\theta}{2} = \frac{\sqrt{2}}{2}$ **d** $\sin^2\left(\frac{2\theta}{3}\right) - 1 = 0$

Double angle identities

Now we will look at the trigonometric formulas known as the double angle formulas. They are called this because they involve trigonometric ratios involving double angles, that is, $\sin 2\theta$ and $\cos 2\theta$. **Trigonometric identities** are equations relating the trigonometric functions that are true for any value of the variable. An **equation** is an equality that is true only for certain values of the variable.

You have already seen some trigonometric identities such as $\tan\theta = \frac{\sin\theta}{\cos\theta}$ and $\sin^2\theta + \cos^2\theta = 1$.

This unit circle shows the angles θ and $-\theta$. This makes angle $B\hat{A}C = 2\theta$.

BD has length $\sin\theta$ and is the same length as CD, so $BC = 2\sin\theta$.

Using the cosine rule in triangle ABC:

$BC^2 = AB^2 + AC^2 - 2(AB)(AC)\cos 2\theta$

$BC^2 = 1^2 + 1^2 - 2(1)(1)\cos 2\theta = 2 - 2\cos 2\theta$

$BC = \sqrt{2 - 2\cos 2\theta}$

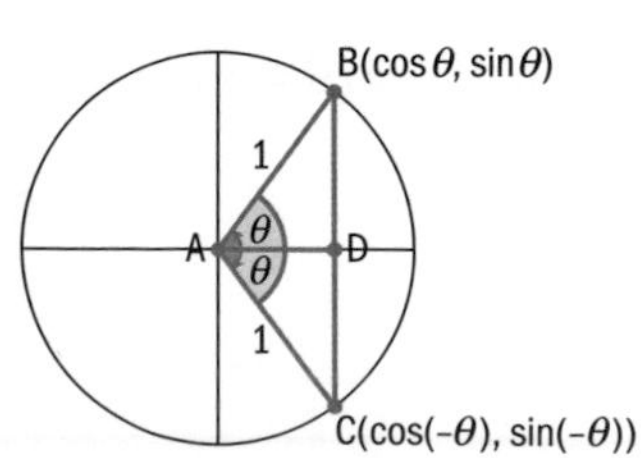

Equating the two expressions for BC gives $2\sin\theta = \sqrt{2-2\cos 2\theta}$.

Square both sides $4\sin^2\theta = 2 - 2\cos 2\theta$.

Rearranging this equation $2\cos 2\theta = 2 - 4\sin^2\theta$.

Divide both sides by 2 $\cos 2\theta = 1 - 2\sin^2\theta$.

$\cos 2\theta = 1 - 2\sin^2\theta$ is called the double angle identity for cosine.

There are two further identities for $\cos 2\theta$ that you can work out using the Pythagorean identity.

We know $\sin^2\theta + \cos^2\theta = 1$, so $\sin^2\theta = 1 - \cos^2\theta$

Using substitution, we would have $\cos 2\theta = 1 - 2(1 - \cos^2\theta)$.

Rearranging this equation gives us $\cos 2\theta = 2\cos^2\theta - 1$.

You can substitute $\sin^2\theta + \cos^2\theta = 1$ into this equation to get:

$\cos 2\theta = 2\cos^2\theta - (\sin^2\theta + \cos^2\theta)$

$$\cos 2\theta = \cos^2\theta - \sin^2\theta.$$

You can also use the Pythagorean identity to derive an identity for $\sin 2\theta$.

You know that $\sin^2 A + \cos^2 A = 1$

When you let $A = 2\theta$, you get $\sin^2 2\theta + \cos^2 2\theta = 1$

substitute $\cos 2\theta = 1 - 2\sin^2\theta$ $\sin^2 2\theta + (1 - 2\sin^2\theta)^2 = 1$

Rearrange $\sin^2 2\theta = 1 - (1 - 2\sin^2\theta)^2$

Expand the bracket and simplify $\sin^2 2\theta = 4\sin^2\theta - 4\sin^4\theta$

Factorize the right-hand side $\sin^2 2\theta = 4\sin^2\theta(1 - \sin^2\theta)$

$\cos^2\theta = 1 - \sin^2\theta$ $\sin^2 2\theta = 4\sin^2\theta(\cos^2\theta)$

Take the square root of both sides $\sin 2\theta = 2\sin\theta\cos\theta$

This is the double angle identity for sine.

The **double angle identities for sine and cosine** are:

$\sin 2\theta = 2\sin\theta\cos\theta$

$\cos 2\theta = 1 - 2\sin^2\theta$

$\cos 2\theta = 2\cos^2\theta - 1$

$\cos 2\theta = \cos^2\theta - \sin^2\theta$.

Example 8

1 Given that $\sin\theta = \frac{3}{5}$ and $0 \le \theta \le \frac{\pi}{2}$, find exact values for:

a $\cos\theta$ **b** $\sin 2\theta$ **c** $\cos 2\theta$ **d** $\tan 2\theta$.

Continued on next page

Geometry and trigonometry

a $\sin^2\theta + \cos^2\theta = 1$ $\left(\frac{3}{5}\right)^2 + \cos^2\theta = 1$	Use the Pythagorean identity.
$\cos^2\theta = 1 - \left(\frac{3}{5}\right)^2 = 1 - \frac{9}{25} = \frac{16}{25}$ $\cos\theta = \frac{4}{5}$	Take the positive answer only, as $0 \le \theta \le \frac{\pi}{2}$.
b $\sin 2\theta = 2\sin\theta\cos\theta$ $\sin 2\theta = 2\left(\frac{3}{5}\right)\left(\frac{4}{5}\right) = \frac{24}{25}$	Use the double angle formula for sine and substitute.
c $\cos 2\theta = 1 - 2\sin^2\theta$ $\cos 2\theta = 1 - 2\left(\frac{3}{5}\right)^2 = 1 - \frac{18}{25} = \frac{7}{25}$	Choose any one of the three double angle identities for cosine.
d $\tan 2\theta = \frac{\sin 2\theta}{\cos 2\theta} = \frac{24}{25} \times \frac{25}{7} = \frac{24}{7}$	Use the tangent identity.

Exercise 12F

1 Write each expression as a single trigonometric ratio.

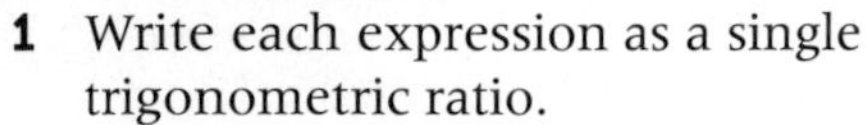

a $2\sin 5\cos 5$ **b** $2\sin\frac{\pi}{2}\cos\frac{\pi}{2}$

c $2\sin 4\pi\cos 4\pi$ **d** $\cos^2 0.4 - \sin^2 0.4$

e $2\cos^2(6) - 1$ **f** $1 - 2\sin^2\frac{\pi}{4}$

2 Given that $\sin\theta = \frac{1}{3}$ and θ is acute, evaluate each expression.

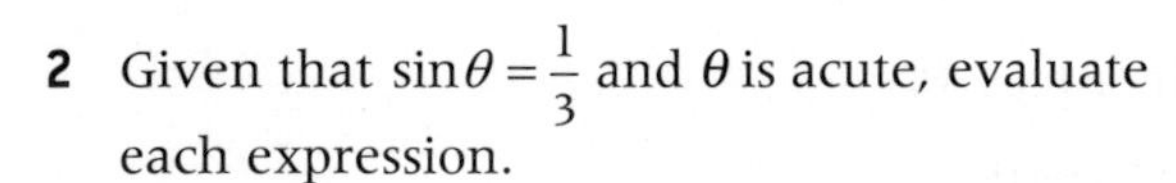

a $\cos\theta$ **b** $\sin 2\theta$

c $\cos 2\theta$ **d** $\tan 2\theta$

3 Given that $\cos\theta = -\frac{1}{2}$, and θ is obtuse, find:

a $\sin\theta$ **b** $\sin 2\theta$

c $\cos 2\theta$ **d** $\tan 2\theta$

4 Given that $\sin\theta = -\frac{1}{8}$, and $\pi \le \theta \le \frac{3\pi}{2}$, find:

a $\sin 2\theta$ **b** $\cos 2\theta$

c $\tan 2\theta$ **d** $\sin 4\theta$

5 Solve each equation for $0 \le \theta \le 2\pi$ giving your answers in terms of π.

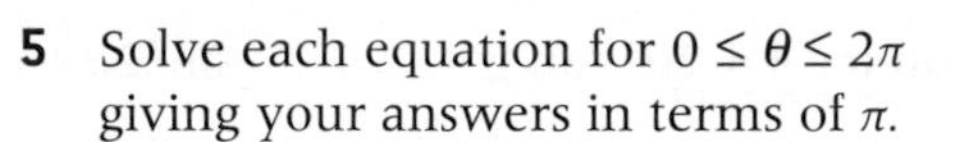

a $\sin 2\theta = \sin\theta$ **b** $\cos 2\theta + \sin\theta = 0$

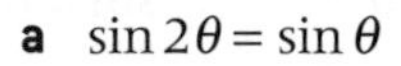

c $\sin 2\theta = \sqrt{3}\cos\theta$ **d** $\cos\theta = \sin\theta\sin 2\theta$

e $\cos 2\theta = \cos\theta$

6 The expression $32\sin x\cos x$ can be expressed in the form $a\sin bx$.

a Find the value of a and of b.

b Hence or otherwise, solve the equation $32\sin x\cos x = 8$, for $0 \le x \le \pi$.

7 Given that triangle ABC has an area of $10\,\text{cm}^2$ when $\sin\theta = \frac{1}{4}$, find x.

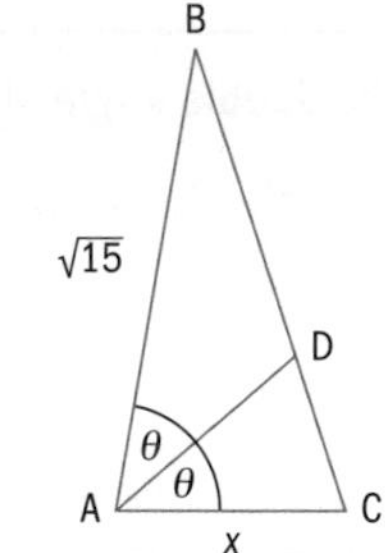

Developing inquiry skills

Notice that the function can have more than one intersection when you draw horizontal lines on the opening graph. What would those lines tell you?

Where would there be only one point of intersection?

Is there anywhere that would be more than 2?

12.4 Trigonometric functions

The graph of a function provides a useful image of its behaviour and allows you to see patterns. A graph shows us a function's behaviour and makes it easier to see the properties of a function. Data analysis and problem solving frequently involve the use of graphs and the understanding they provide to allow for calculation and prediction. The graphs of trigonometric functions have clearly visible patterns.

The London Eye is a Ferris wheel with diameter 120 m. The wheel completes one turn every 30 minutes. After one cycle, it repeats its journey again and again. This is an example of a **periodic function**. A periodic function is a function that repeats its values in regular intervals or periods. Tides, planetary orbits, daylight hours, biorhythms, heart beats and musical rhythms are some examples of phenomena that can be modelled by periodic functions.

The London Eye repeats its revolution, or cycle, every 30 minutes, and so we say it has a period of 30 minutes.

The sine and cosine curves

You can see many similarities between the sine and cosine curves in this diagram:

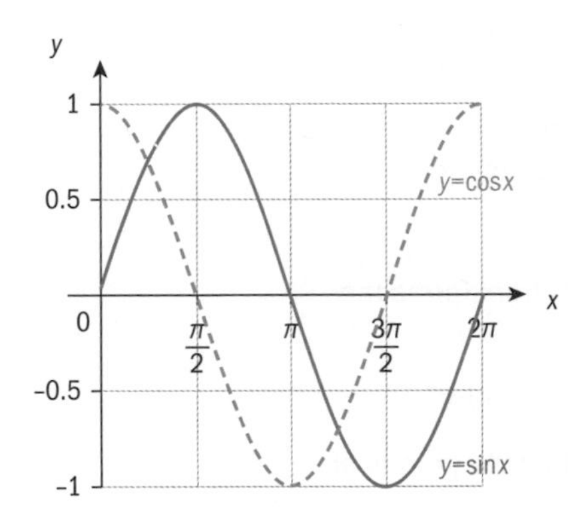

- The curves are the same shape.
- The curves are the same height. Both functions have a maximum value of 1 and a minimum value of –1.
- Translating the cosine curve (the dotted line) $\frac{\pi}{2}$, horizontally to the right, makes it identical to the sine curve.
- Both functions complete one wavelength in 2π radians.
- The functions are **periodic**, meaning they repeat the same cycle of values.

The characteristics of sine and cosine functions

This graph shows the function $y = \sin x$.

The **amplitude** is half of the difference of the maximum and minimum values. That is, the y value at point A minus the y value at point B, divided by 2.

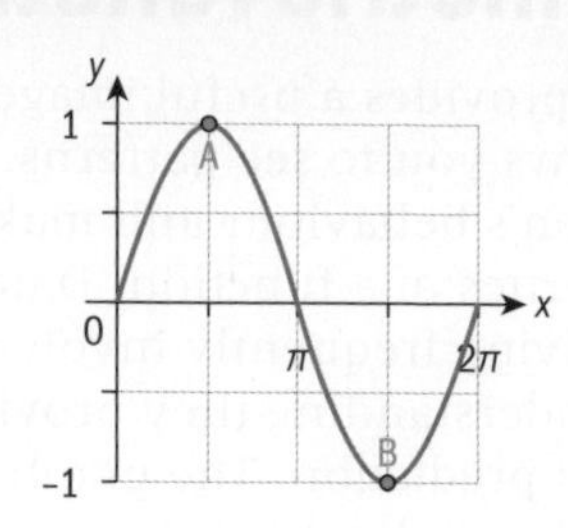

Investigation 5

1 Use technology to help you sketch the graph of each function for $0 \le x \le 2\pi$.

a $y = \sin x$ b $y = 2\sin x$ c $y = 3\sin x$

d Find the amplitude of each graph.

e How could you have found the amplitude by looking at the equation?

f Could the amplitude be negative?

2 Sketch the graph of each function for $0 \le x \le 2\pi$.

a $y = -\sin x$ b $y = -2\sin x$ c $y = -3\sin x$

d Explain to a partner what you notice and explain the effect of the negative sign.

3 Sketch the graph of each function for $-2\pi \le x \le 2\pi$.

a $y = \sin(-x)$ b $y = 2\sin(-x)$ c $y = 3\sin(-x)$

d Explain to a partner what you notice and explain the effect of the negative sign.

4 Use technology to help you sketch the graph of each function for $0 \le x \le 2\pi$.

a $y = \sin x$ b $y = \sin 2x$ c $y = \sin 3x$ d $y = 0.5\sin x$

The **period** of a trigonometric function is the length of one full cycle along the x-axis, ie the angle measure after which the graph begins to repeat.

5 Explain to a partner what you notice.

6 Write an explanation of how the value of b in $y = \sin bx$ transforms the graph of $y = \sin x$

7 **Factual** What do the parameters represent in $f(x) = a\sin bx$?

8 **Conceptual** How does changing the parameters affect the graphs of trigonometric functions?

9 **Conceptual** How can you transform trigonometric graphs?

For the curves $y = a\sin bx$ and $y = a\cos bx$:

$|a|$ is the amplitude, and the period is $\frac{2\pi}{b}$.

$y = -\sin x$ or $y = -\cos x$ is a **reflection in the x-axis.**

$y = \sin(-x)$ or $y = \cos(-x)$ is a **reflection in the y-axis.**

Example 9

Sketch the graph of $y = 3\cos 2x$, $0 \le x \le 2\pi$.

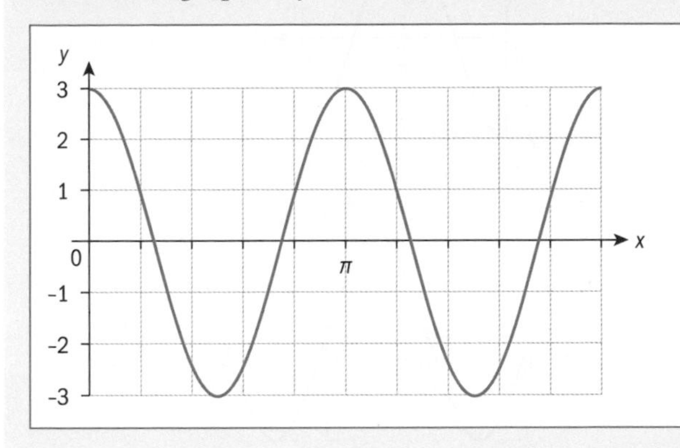

The function has amplitude 3 and period $= \frac{2\pi}{b} = \pi$.

Example 10

Write down the period and amplitude, and find the equation of this function.

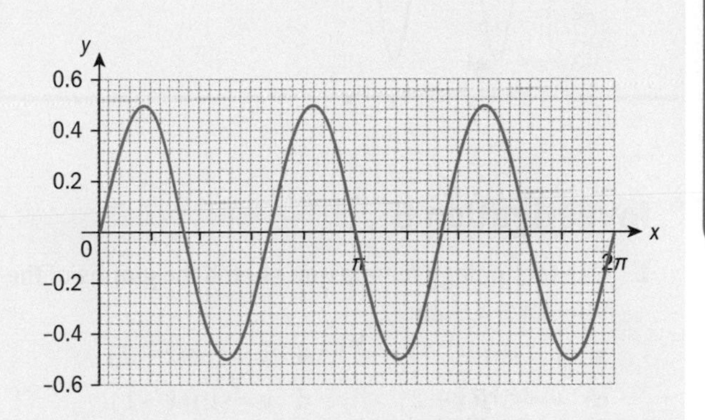

Amplitude: 0.5

Amplitude $= \frac{\max - \min}{2} = \frac{0.5-(-0.5)}{2} = 0.5$.

Period: $\frac{2\pi}{3}$

Period $= \frac{2\pi}{\left(\frac{2\pi}{3}\right)}$ =3 or 3 cycles in 2π radians.

$y = 0.5\sin 3x$

The graph is the shape of a sine curve.

Note that you could also use a cosine function, if you include a horizontal translation in your equation.

Exercise 12G

1 Sketch the following graphs for $0 \le x \le 2\pi$.

a $y = 2\sin x$ **b** $y = -3\cos x$

c $y = 2\cos 2x$ **d** $y = 3\sin 4x$

2 Find the amplitude and period of each function.

a $y = \sin 3x$ **b** $y = 0.5\cos 2x$

c $y = -4\cos 3x$ **d** $y = -\frac{1}{2}\sin\frac{x}{3}$

Geometry and trigonometry

3 Write down the period and amplitude, and find the equations of each graph.

a

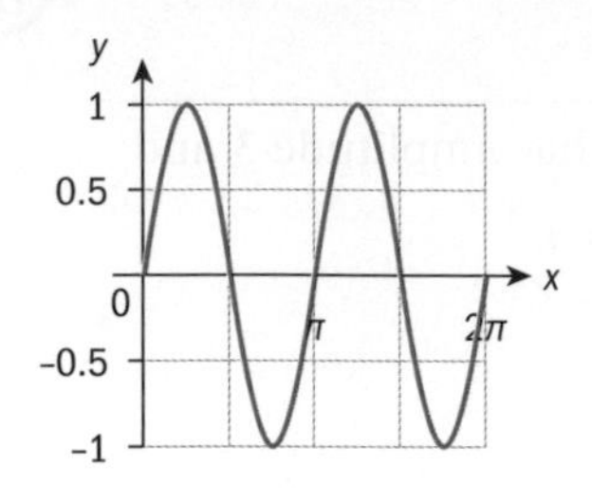

b

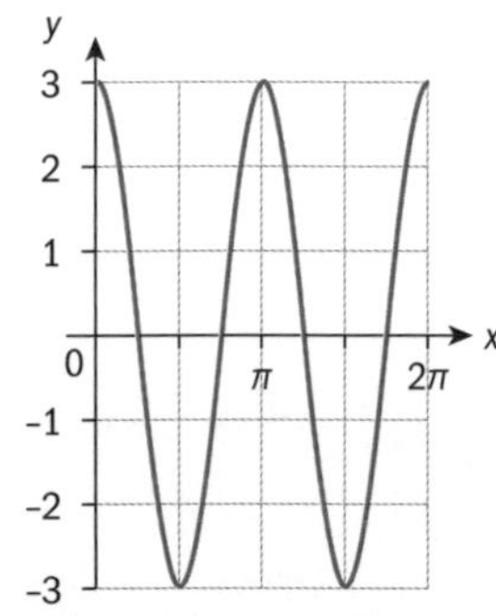

c

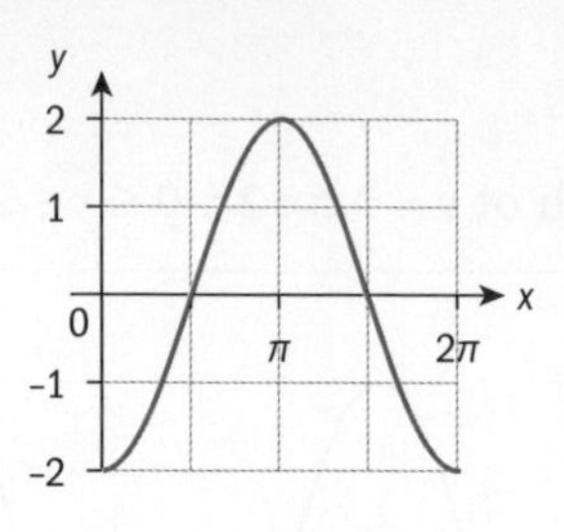

d

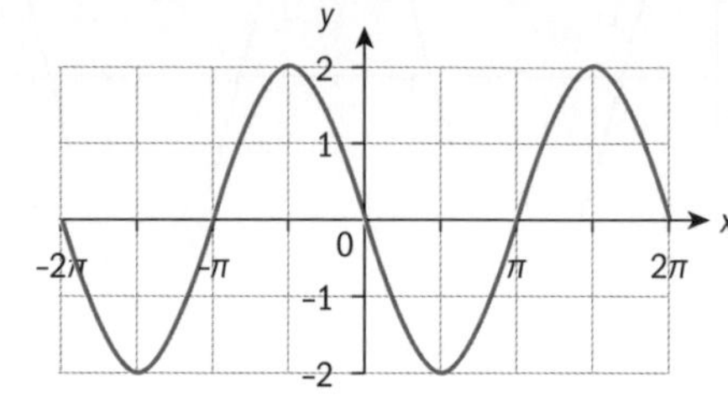

4 Let $f(x) = 6\sin\left(\frac{\pi}{2}x\right)$ for $0 \le x \le 4$.

a Write down the amplitude of f.

b Find the period of f.

Investigation 6

1 Use technology to help you sketch the graphs of the following functions for $-2\pi \le x \le 2\pi$.

a $y = \sin x$ **b** $y = \sin(x-2)$

c $y = \sin(x+2)$ **d** $y = \sin(x-1)$

e How does the value of c in $y = \sin(x-c)$ transform the graph of $y = \sin x$?

2 Sketch the graphs of the following functions for $0 \le x \le 2\pi$.

a $y = \sin x$ **b** $y = 2 + \sin x$

c $y = 3 + \sin x$ **d** $y = -1 + \sin x$

e With a partner, discuss what you notice and explain the effect of the constant.

3 **Factual** What do the parameters represent in $f(x) = \sin(x-c) + d$?

4 **Conceptual** How does changing the parameters affect the graphs of trigonometric functions?

5 How can you transform trigonometric graphs?

$y = \sin(x-c)$ or $y = \cos(x-c)$ translates the function c units horizontally.
$y = \sin x + d$ or $y = \cos x + d$ translates the function d units vertically.

This leads you to the general sine (or cosine) function.

$y = a\sin(b(x - c)) + d$

where $|a|$ is the amplitude. Amplitude $= \dfrac{\max - \min}{2}$

Period $= \dfrac{2\pi}{b}$

c is the horizontal shift and

d is the vertical shift. Vertical shift $= \dfrac{\max + \min}{2}$

Reflect **Discuss** your answer with a partner then **share** your ideas with your class.

- What do you see?
- What do you think?
- What do you wonder?

Example 11

Sketch the graph of $y = 2\sin(3(x + \pi)) - 1$, $0 \le x \le 2\pi$.

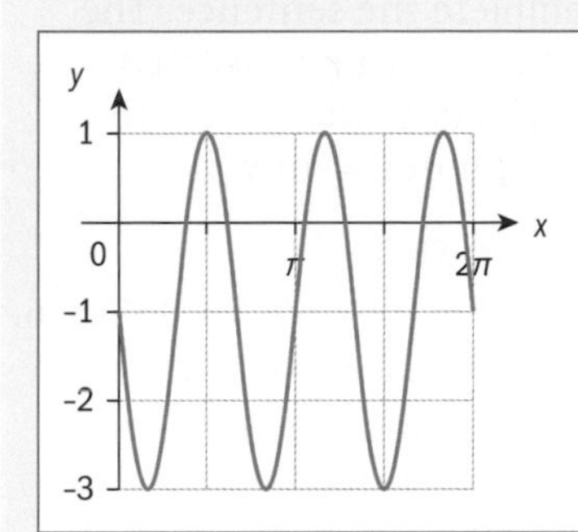

Shape of a sine curve

Amplitude: 2

$b = 3$; Period $= \dfrac{2\pi}{b} = \dfrac{2\pi}{3}$

Horizontal shift: $-\pi$ (or π units left)

Vertical shift: -1

Exercise 12H

1 Match each function with the correct graph: **i, ii, iii** or **iv.**

a $y = 3\sin 2x$ **b** $y = -2\cos x + 1$

c $y = \dfrac{1}{2}\sin\left(x + \dfrac{\pi}{2}\right)$ **d** $y = \sin\left(\dfrac{1}{2}\right)x + 2$

i

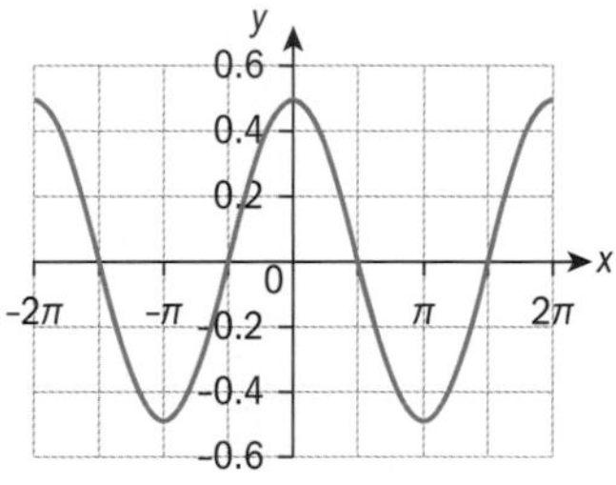

ii

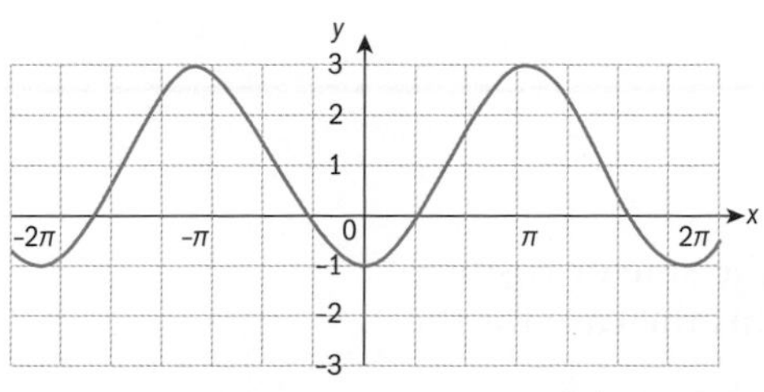

iii

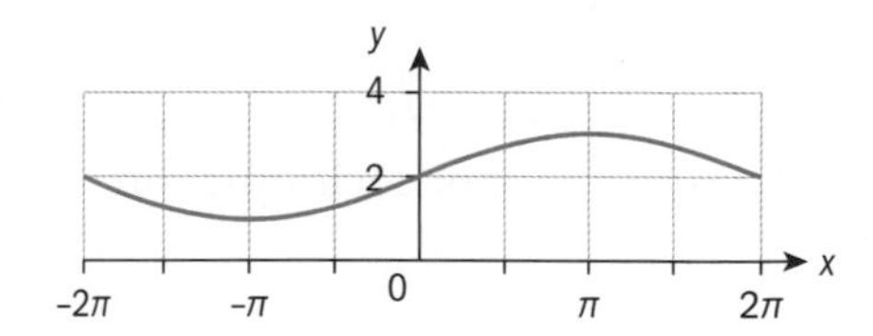

iv

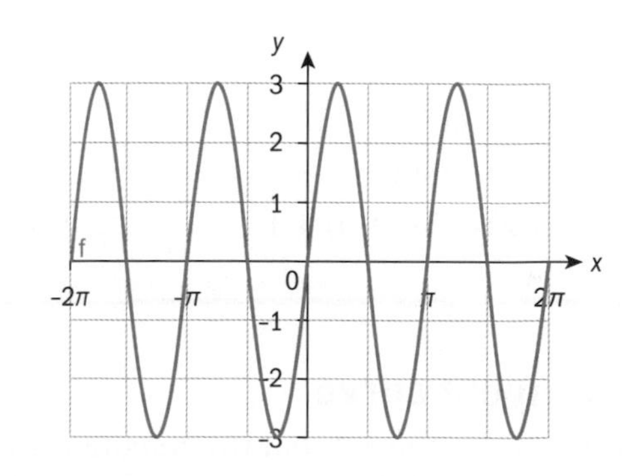

Geometry and trigonometry

2 Find the equations of each function.

a

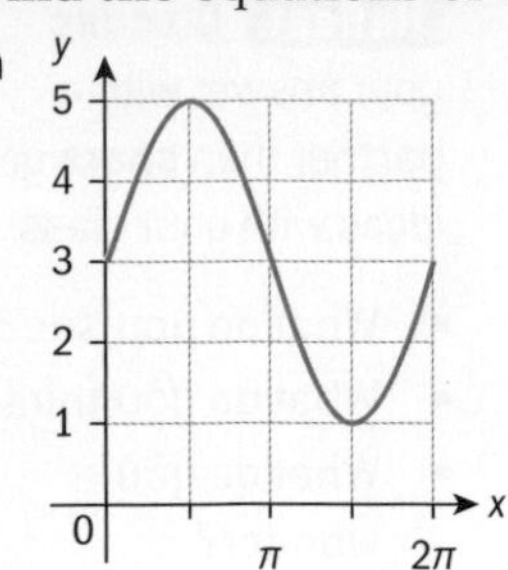

b

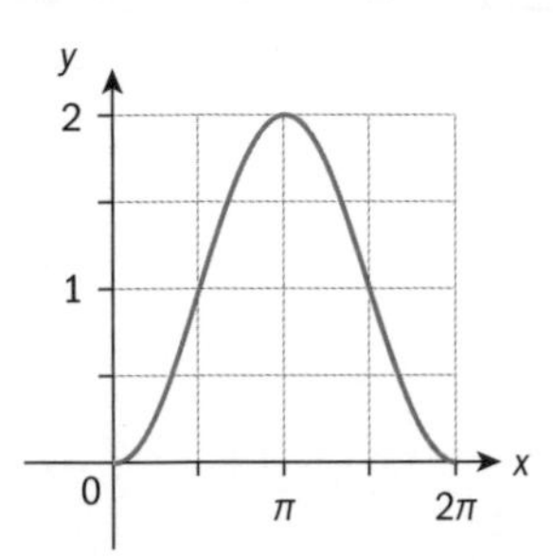

c

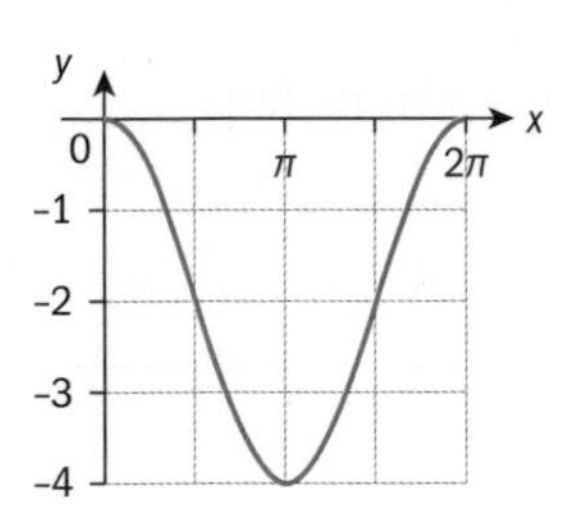

d

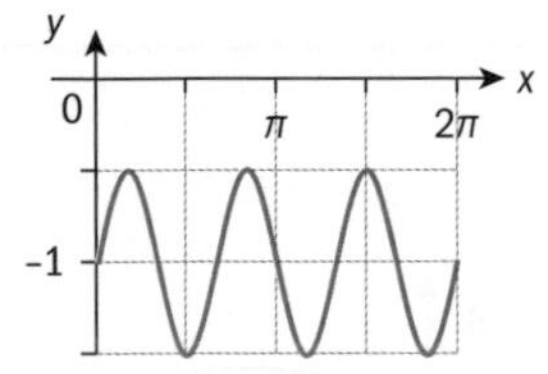

3 Sketch the graph of each function for $0 \le x \le 2\pi$.

a $y = 2 + 3\sin 2x$

b $y = 0.5\sin\left(x + \frac{\pi}{3}\right)$

c $y = \cos(x + \pi) - 1$

d $y = 2 - 2\sin 2x$

e $y = 2\cos 3(x + \pi) + 1$

4 Solve the equation $\cos(x) - x^2 = 0$ by sketching $y = \cos x$ and $y = x^2$ on the same axes and finding their intersection points.

5 Solve $2\sin x = x + 1$ graphically.

6 Solve $\sin x = \frac{1}{2}$, $0 \le x \le 4\pi$ graphically.

7 The equation $e^x = \cos x$ has a solution between –2 and –1. Find this solution.

8 Sketch $y = \sin x$ and $y = \cos x$ on the same axes.

a Use your sketch to help you find the value of c in $\sin x = \cos(x - c)$.

b Copy and complete the sentence. The graph of $y = \cos x$ may be translated .... horizontally to the .... to become the graph of $y = \sin x$.

9 Let $f(x) = \sin(1 + \sin x)$.

a Sketch the graph of $y = f(x)$, for $0 \le x \le 6$.

b Write down the x-coordinates of all minimum and maximum points of f. Give your answers correct to 4 significant figures.

10 Let $f(x) = \sin 2x$ and $g(x) = \sin(0.5x)$.

Find:

a the minimum value of the function f for $0 \le x \le 2\pi$

b the period of the function g

c Solve $f(x) = g(x)$ for $0 \le x \le \frac{3\pi}{2}$.

The tangent curve

You will have noticed that the maximum value of the sine or cosine of an angle is 1 and the minimum value is –1, but is this the case for the tangent function?

Investigation 7

1 Copy and complete this table with the use of technology.

$x°$	0	20	40	60	80	90	100	120	140	160	180
$\tan x$											

$x°$	200	220	240	260	270	280	300	320	340	360
$\tan x$										

2 Plot this data as points on a set of axes to give $y = \tan x$.

3 Where do they cross the x-axis?

4 What happens at 90° and 270°?

5 Connect your points to form smooth curves.

6 Is this a periodic function? Why?

7 Explain what you notice about the tangent function to a classmate.

8 How can you compare the graphs of the sine, cosine and tangent functions?

In radians, the tangent curve looks like this:

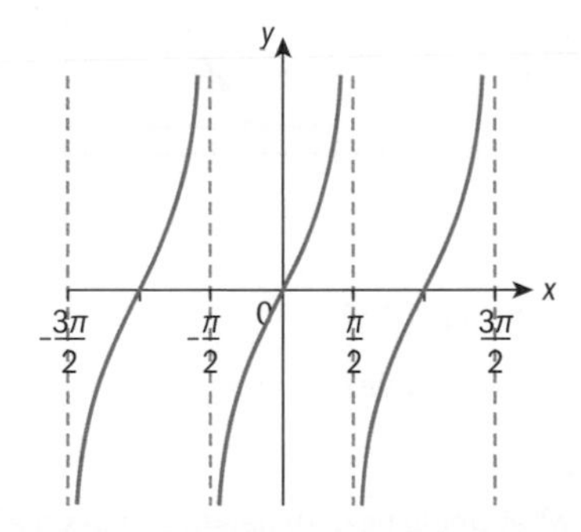

There are some similarities to the sine and cosine curves. All three curves are periodic functions. The transformations of horizontal and vertical stretches, reflections and translations all follow the same rules, though the period of the tangent function is 180° or π radians.

The tangent function does not have an amplitude since there are no local extrema points.

Exercise 12I

1 Sketch the graph of each function for $0 \le \theta \le 2\pi$.

a $y = \tan 2\theta$

b $y = \tan\frac{\theta}{2}$

c $y = -1 + \tan\theta$

d $y = \tan\theta + 2$

e $y = \tan\left(\theta - \frac{\pi}{2}\right) - 2$

f $y = 1 + \tan\left(\theta - \frac{3\pi}{4}\right)$

2 Use technology to help you solve each equation.

a $\tan\theta = 3,\ 0 \le \theta \le 2\pi$

b $\tan\theta = x + 1,\ 0 \le \theta \le 2\pi$

c $\tan\theta = \sin\theta,\ -\pi \le \theta \le \pi$

d $\tan 2\theta = 5 - x^2,\ -\pi \le \theta \le \pi$

Modelling periodic functions

Periodic occurrences happen in repeated and predicatable cycles, such as lunar orbits and high tides. However, many real-life situations are more complicated than the simple rotation about a unit circle. We often need to transform the sine and cosine curves, so we can use these functions to model an event.

One major difference with these graphs is that the original data might not use angles as the independent variable. For example, our opening problem with the London Eye will use time on the horizontal axis and height on the vertical axis.

Investigation 8

The London Eye

The London Eye was the fourth largest Ferris wheel in the world when it was built, with a diameter of 120 m. One full rotation takes 30 minutes, and you board from a platform which is 2 m above the ground. Can you find your height above ground as a function of time in minutes?

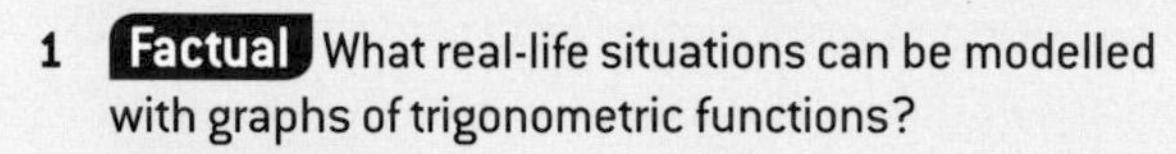

1 **Factual** What real-life situations can be modelled with graphs of trigonometric functions?

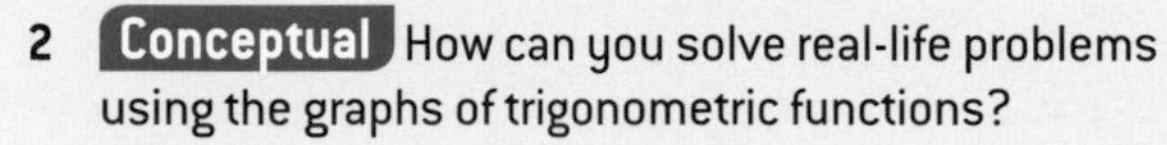

2 **Conceptual** How can you solve real-life problems using the graphs of trigonometric functions?

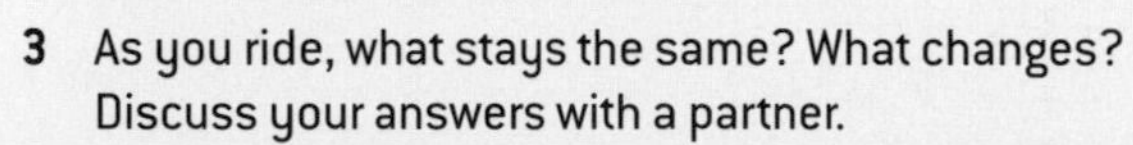

3 As you ride, what stays the same? What changes? Discuss your answers with a partner.

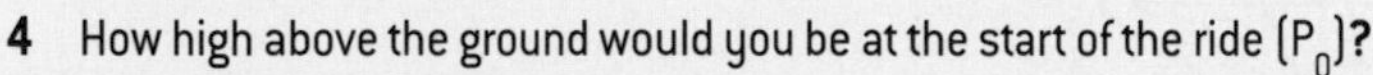

4 How high above the ground would you be at the start of the ride (P_0)?

5 What is your maximum height (P_6) above the ground during the ride?

6 How many minutes would it take you to get to point P_6?

7 How many minutes would it take you to get to P_3 and what would be your height at that time?

8 How many minutes would it take you to get to P_9 and what would be your height at that time?

9 Copy and complete this table.

Position	P_0	P_1	P_2	P_3	P_4	P_5	P_6	P_7	P_8	P_9	P_{10}	P_{11}
Time (mins)	0		5						20			
Height (m)	2		35.75						103.25			

10 **Conceptual** How can you model periodic behaviour?

11 Use graph paper to plot these points and draw a periodic function to model your journey on the London Eye.

12 Find the amplitude, period and vertical shift of the curve.

13 Model the function as a cosine function.

14 Check your model by finding your height after 15 mins. This should be the maximum height.

15 **Factual** What kinds of information can be modelled using periodic graphs?

16 The best pictures of Big Ben are taken after 18 mins on the London Eye. What will be the height at this time?

17 **Factual** What equations can be solved with trigonometric functions?

Example 12

Ben is a fisherman with a boat in Rhyl harbour. This is part of a graph of the depth of water at the mouth of the harbour.

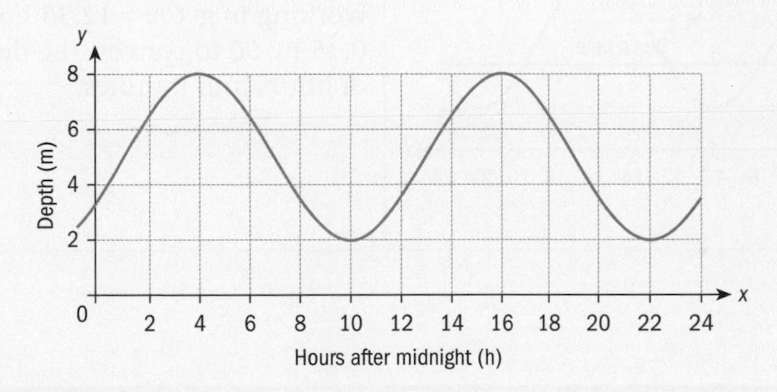

a Find the amplitude of the function.

b Find the period of the function.

Model this as a cosine function.

c Find the horizontal shift.

d Find the vertical shift.

e Write down the equation of the function for the depth of the water in terms of m and h.

f Calculate the depth of the water at 9:30am.

Ben's boat can only get in and out of the harbour when there is at least 4 m of water at the mouth of the harbour. Ben sleeps until 10am.

g Find the next time that he can go fishing.

Continued on next page

Geometry and trigonometry

a Amplitude $= \frac{\max - \min}{2} = \frac{8-2}{2} = 3$	
b Period = 12	Take the horizontal distance between maximum points as your period: $16 - 4 = 12$
c 4	The function $y = \cos x$ would have a maximum where $x = 0$. This function has a maximum at $x = 4$. Hence a horizontal shift of 4 to the right.
d Vertical shift $= \frac{\max + \min}{2} = \frac{8+2}{2} = 5$	You can find the vertical shift by looking at the graph or by using the formula $\frac{\max + \min}{2}$.
e $m = 3\cos\left(\frac{\pi}{6}(h-4)\right) + 5$	Put all of the answers together using $m = a\cos(b(h-c)) + d$
f $m = 3\cos\left(\frac{\pi}{6}(9.5-4)\right) + 5 = 2.10\text{ m}$	Substitute $x = 9.5$ hours into your equation.
g 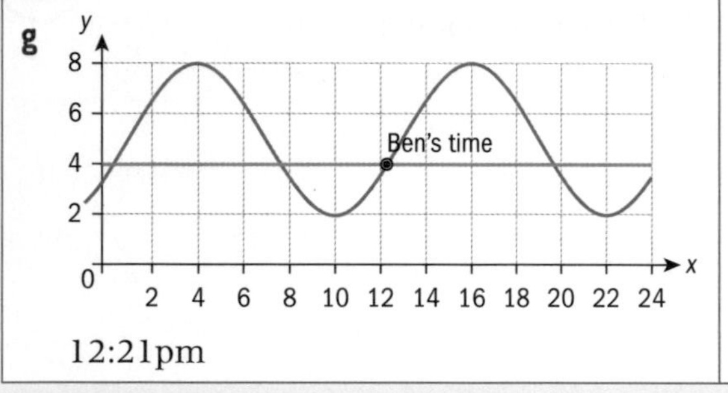 12:21pm	Graph your function and $m = 4$ on the same axes using technology. Show a sketch as working to get $m = 12.35$ hours. Multiply 0.35 by 60 to convert the decimal number of hours into minutes.

Exercise 12J

1 The height, h metres, of the tide on New Year's Day in Perth, Australia can be modelled using the function $h(t) = 5\sin\left(\frac{\pi}{6}(t-5)\right) + 7$, where t is the number of hours after midnight.

a Find the maximum and minimum values for the depth of water.

b What time is high tide?

c What time is low tide?

d What is the depth of the water at 9.00 am?

e Find all the times during a 24-hour period when the depth of the water is 3 m.

2 The number of visitors (y) to a ski resort in Switzerland is modelled by the function $y = 3000\cos(0.5(x-1)) + 10\,000$, where x is the number of months after 1 January.

a What is the maximum number of visitors and when does this occur?

b What is the minimum number of visitors and when does this occur?

c How many visitors are at the resort on 1 May?

3 This graph shows the number of fish in a lagoon over a 24-hour period starting at midnight.

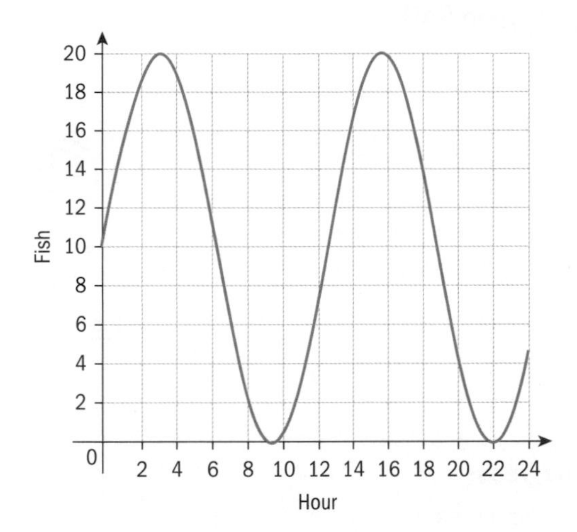

a What was the maximum number of fish in the lagoon?

b How many fish were in the lagoon at midnight?

c Find an equation for this function.

d How many fish were in the lagoon at 2pm?

4 The height, $h(t)$, in metres, of the end of a windmill blade above the ground at time, t seconds, is a cosine function of the form $h(t) = a\cos bt + c$. The following graph shows one rotation of the blade.

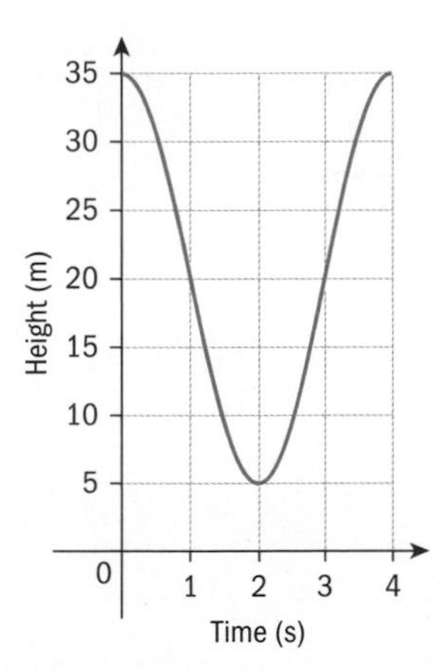

a Write down the maximum height of the end of blade above the ground.

b Write down the minimum height of the end of blade above the ground.

c Find the amplitude of the function.

d Hence or otherwise, find the value of a and of c.

e Write down the period of the function.

f Hence, find the value of b.

g Use technology to determine the first value of t for which the end of the blade is 30 metres above the ground.

5 A Ferris wheel moves with constant speed and completes one rotation every 40 seconds. The wheel has a radius of 12 m and its lowest point is 2 m above the ground.

The height of a chair on the Ferris wheel above ground can be modelled by the function, $h(t) = a\cos bt + c$, where t is the time in seconds.

a Find the value of a, b and c.

The chair first reaches a height of 20 m above the ground after p seconds.

b Find the angle that the chair has rotated through to reach this position.

c Find the value of p.

6 A Ferris wheel with a radius of 20 m rotates once every 40 seconds. Passengers get on 1 m above level ground.

a Find a cosine equation for the height of a seat on the Ferris wheel.

b Find the length of time during one rotation that a passenger on the Ferris wheel will be at least 23 m above the ground.

7 In a physics experiment, a weight is attached to the end of a long spring and bounces up and down. As it bounces, its distance from the floor can be modelled with a cosine function. The weight is released from a high point 60 cm above the floor at 0.3 seconds and first reaches its low point, 40 cm above the floor at 1.8 seconds.

a Write an equation to model the distance of the weight from the floor in seconds.

b Find the distance of the weight from the floor at 17.2 seconds.

c Find the time when the weight was first 59 cm above the floor.

Developing your toolkit

Now do the Modelling and investigation activity on page 540.

Chapter summary

- A **radian** is the measure of the angle with its vertex at the centre of the circle between two radii with endpoints that are one radius length apart on the circumference of the circle.
- 2π radians = 360° or 1 radian ≈ 57.3°.
- Converting between degrees and radians:

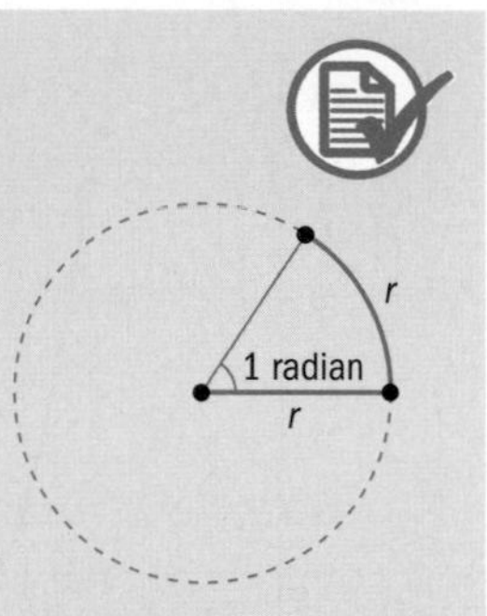

Degrees to radians:

Multiply the degree measure by:

$$\frac{\pi}{180°}$$

Radians to degrees:

Multiply the radian measure by:

$$\frac{180°}{\pi}$$

- The length of an arc is given by:

  $l = r\theta$

  and the area of a sector by:

  $A = \frac{1}{2}r^2\theta$

  where θ is measured in radians.

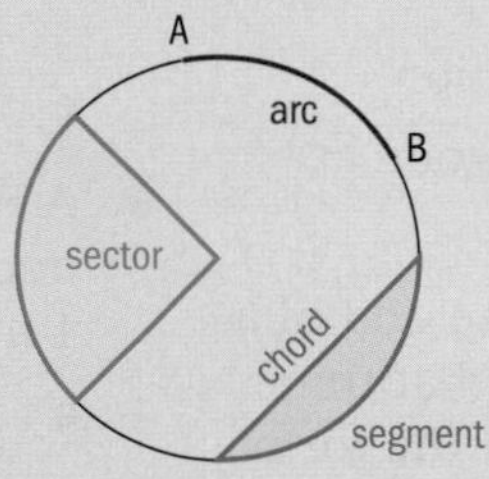

- The **unit circle** is a circle with its centre at the origin and of radius 1 unit.

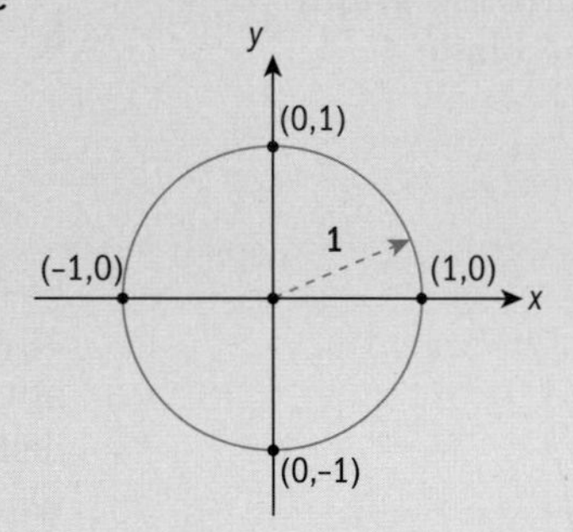

- **Trigonometric ratios**

  $\sin\theta = \sin(180 - \theta)$

  $\cos\theta = \cos(-\theta)$

  $\cos\theta = \cos(360 - \theta)$

  $\tan\theta = \tan(180 + \theta)$

- **Pythagorean identity**

  $\sin^2\theta + \cos^2\theta = 1$

- **Tangent identity**

  $\tan\theta = \dfrac{\sin\theta}{\cos\theta}$

- **The double angle identities**

  $\sin 2\theta = 2\sin\theta\cos\theta$

  $\cos 2\theta = 1 - 2\sin^2\theta$

  $\cos 2\theta = 2\cos^2\theta - 1$

  $\cos 2\theta = \cos^2\theta - \sin^2\theta$

HINT

You should keep the formula booklet on your table when you are working to see what is in there to help you, like trigonometric identities, and what is not there that you will have to remember, like the exact values of trigonometric ratios of $0, \frac{\pi}{6}, \frac{\pi}{4}, \frac{\pi}{3}, \frac{\pi}{2}, \pi$ and their multiples, including the relationship between angles in different quadrants.

- **The sine and cosine curves**

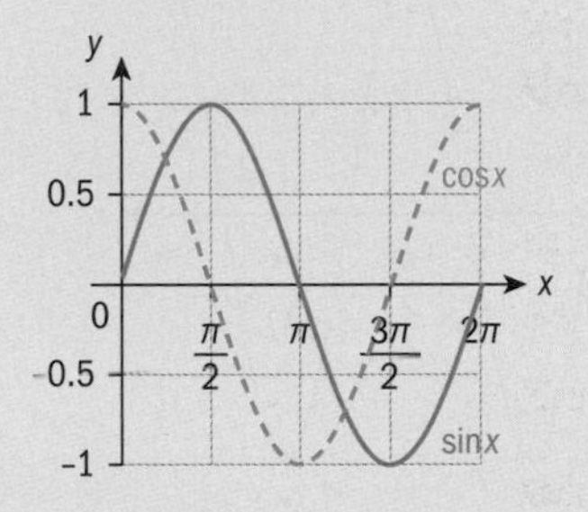

The tangent curve

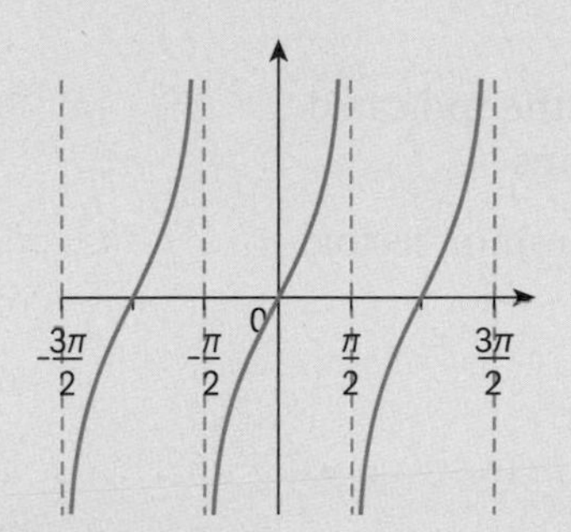

- $y = -\sin x$ or $y = -\cos x$ is a **reflection in the x-axis.**

  $y = \sin(-x)$ or $y = \cos(-x)$ is a **reflection in the y-axis.**

- $y = \sin(x - c)$ or $y = \cos(x - c)$ translates the function c units horizontally.

  $y = \sin x + d$ or $y = \cos x + d$ translates the function d units vertically.

- $y = a\sin b(x - c) + d$

  $|a|$ is the amplitude. Amplitude $= \dfrac{\max - \min}{2}$

  period $= \dfrac{2\pi}{b}$

  c is the horizontal shift and

  d is the vertical shift. Vertical shift $= \dfrac{\max + \min}{2}$

Developing inquiry skills

Can you find the equation of the function from the opening problem?

Would you use sine or cosine?

What would be the difference if you used sine and your friend used cosine?

Chapter review

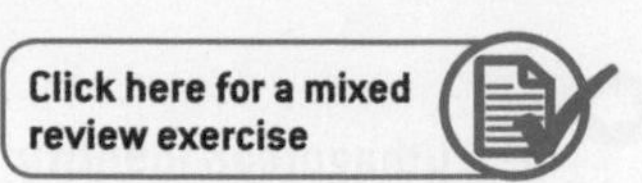

1 Rewrite each angle in radians as a multiple of π.

a $30°$ **b** $150°$ **c** $315°$

d $120°$ **e** $-20°$ **f** $-240°$

g $-270°$ **h** $144°$

2 Rewrite each angle in degrees.

a $\frac{3\pi}{2}$ **b** $\frac{7\pi}{6}$ **c** $-\frac{7\pi}{12}$ **d** $\frac{\pi}{9}$

e $\frac{7\pi}{3}$ **f** $-\frac{11\pi}{30}$ **g** $\frac{11\pi}{6}$ **h** $\frac{34\pi}{15}$

3 a Find the measure of the indicated central angle in radians

b Find the area of the minor sector.

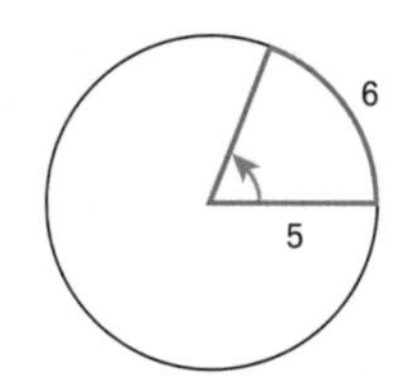

4 a Given that $\sin\theta = 0.6$, and θ is acute, find $\cos\theta$ and $\tan\theta$.

b Solve $2\sin x = \tan x$ for $-\frac{\pi}{2} \le x \le \frac{\pi}{2}$.

5 Solve for $0 \le \theta \le 2\pi$, giving your answer in radians as multiples of π.

a $\sin\theta = -\frac{1}{2}$ **b** $\cos\theta = \frac{\sqrt{2}}{2}$

c $\tan\theta = 1$

6 The expression $8\sin x\cos x$ can be expressed in the form $a\sin bx$.

a Find the values of a and of b.

b Hence or otherwise, solve the equation $6\sin x\cos x = 2$, $0 \le x \le 2\pi$

7 Given that $\frac{\pi}{2} \le \theta \le \pi$ and $\cos\theta = -\frac{12}{13}$, find each value.

a $\sin\theta$ **b** $\cos 2\theta$ **c** $\sin(\theta + \pi)$

8 a Given that $2\sin^2\theta + \sin\theta - 1 = 0$, find the two values for $\sin\theta$.

b Given that $0 \le \theta \le 2\pi$, find three possible values for θ.

9 Given that $\sin\theta = \frac{3}{4}$, where θ is an obtuse angle, find:

a $\cos\theta$ **b** $\cos 2\theta$

10 This graph is the function $y = a\cos bx$.

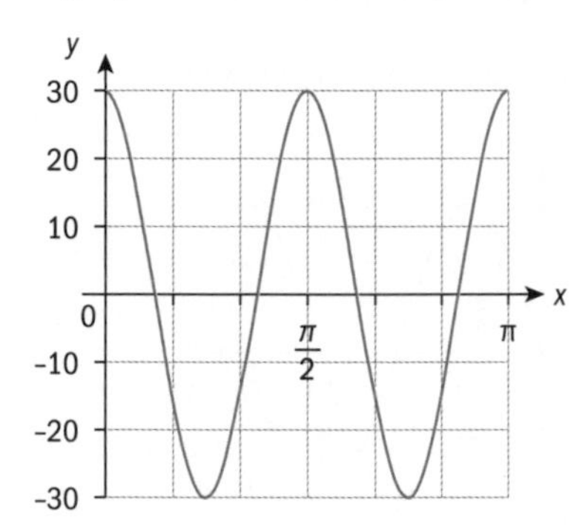

Find the values of a and b.

11 This graph shows the function $f(x) = a\sin bx$, where $b > 0$. Find

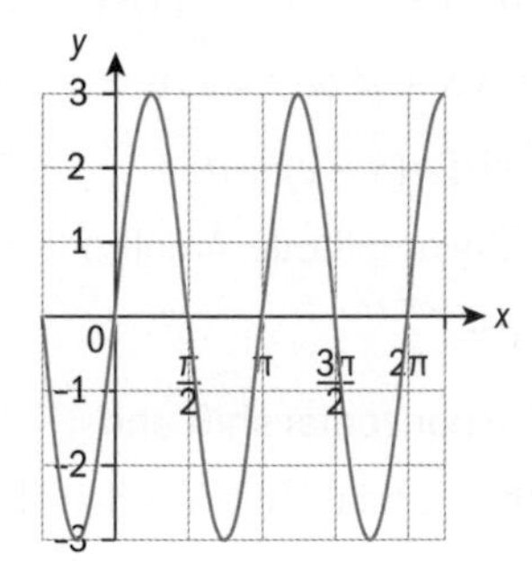

a Find the period of f.

b Write down the amplitude of f.

c Write down the value of a.

d Find the value of b.

12 $f(x) = 5\sin\left(\frac{x\pi}{2}\right)$ for $0 \le x \le 4$.

a Write down the amplitude of f.

b Find the period of f.

c Sketch the function.

13 OABC is a sector of a circle with radius 8 cm.

Find the shaded area when $\theta = \frac{\pi}{6}$

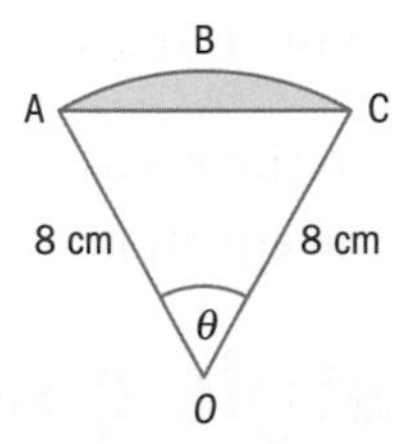

14 This diagram shows a triangle ABC and a sector BDC of a circle with centre B and radius 4 cm. The points A, B and D are on the same line and AB = $\sqrt{2}$ cm, BC = 4 cm, the area of triangle ABC is 2 cm^2 and angle ABC is obtuse. Find:

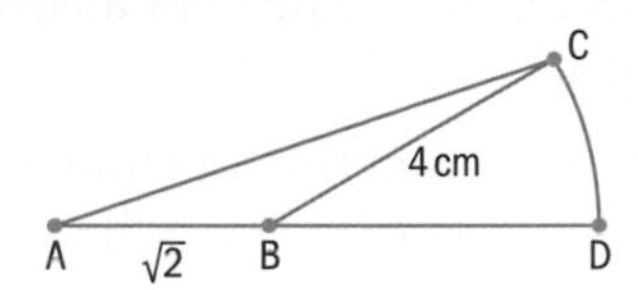

a the angle ABC

b the exact area of the sector BDC.

15 Use a sketch to help you solve

a $\cos x = x^2$ for $0 \le x \le \frac{\pi}{2}$

b $4\sin \pi x = 4e^{-x} - 3$ for $0.5 \le x \le 1.5$.

16 In triangle ABC, AB is 10 cm, BC is 8 cm and angle ABC is obtuse. The area of the triangle is 10 cm^2.

Find the size of angle ABC in radians.

17 This circle has a radius 10 cm with centre O. The points A and B are on the circle, and angle AOB is 0.8 radians. Angle ONA is a right angle. Find the area of the shaded region.

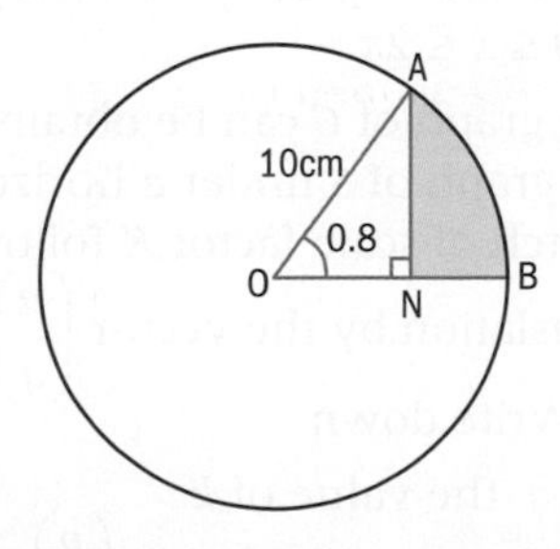

18 This circle has centre O and radius 8 cm. The points A, B and C lie on the circle. O, C, and D lie on a straight line. Angle ADC = 0.4 radians and angle AOC = 0.8 radians. Find:

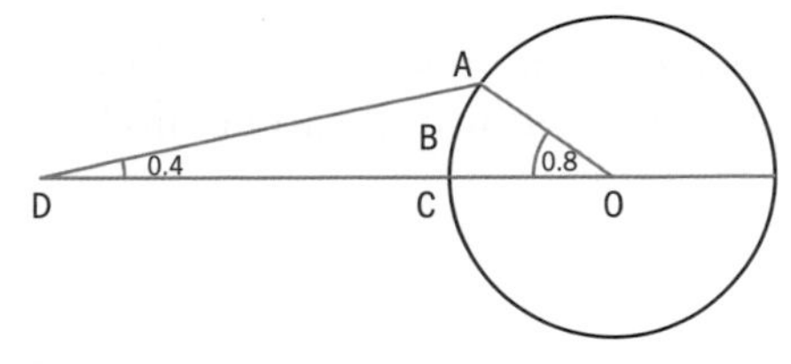

a AD

b OD

c the area of sector OABC

d the area of region ABCD.

19 The function f is defined by

$f(x) = 20\sin 3x \cos 3x,\ 0 \le x \le \frac{\pi}{3}$

a Write down an expression for $f(x)$ in the form $a \sin bx$.

b Solve $f(x) = 0$, giving your answers in terms of π.

20 Given $f(\theta) = 4\cos^2\theta + \cos\theta + 1$, for $-360° \le \theta \le 360°$,

a show that this function may be written as $f(\theta) = 4\cos^2\theta + \cos\theta + 1$.

b how many distinct values of $\cos\theta$ satisfy this equation?

c find all values of θ which satisfy this equation.

21 A Ferris wheel with centre O and a radius of 15 m is represented in the diagram. When seat A is at ground level, another seat is at B, where angle AOB = $\frac{\pi}{6}$.

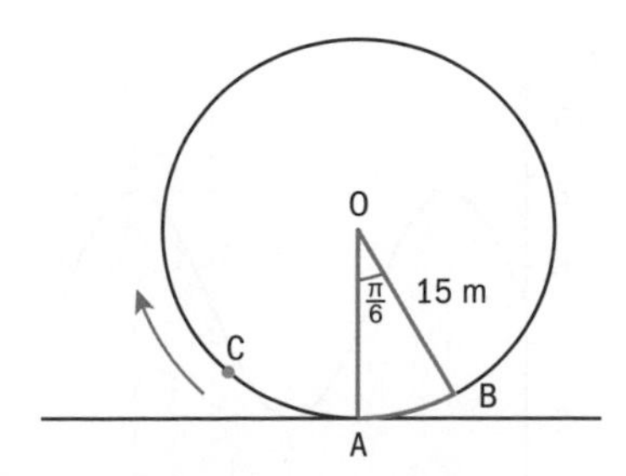

a Find the length of the arc AB.

Geometry and trigonometry

b Find the area of the sector AOB

c Find the height of A above the ground when the wheel turns clockwise through an angle of $\frac{2\pi}{3}$.

The height, h metres, of seat C above the ground after t minutes, can be modelled by the function

$$h(t) = 15 - 15\cos 2\left(t + \frac{\pi}{8}\right)$$

d Find the height of seat C when $t = \frac{\pi}{4}$

e Find the initial height of seat C

f Find the time at which seat C first reaches its highest point.

22 This graph shows the functions $f(x) = p\cos(qx)$ and $g(x) = \sin(bx) - d$.

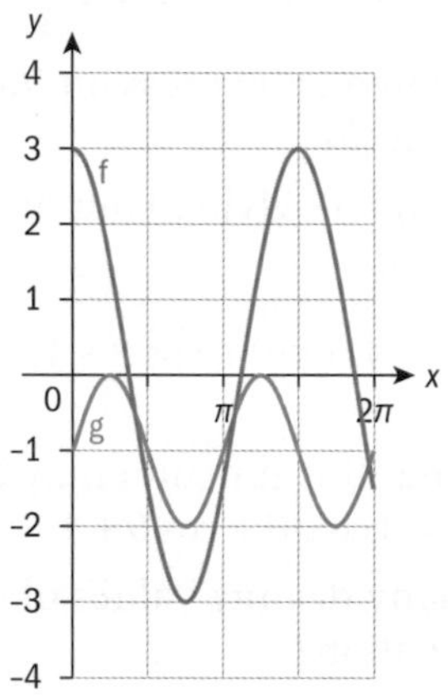

a Find the values of p, q, b and d.

b Solve the equation $f(x) = g(x)$ for $0 < x < 2\pi$

23 This graph shows the function $f(x) = a\sin(bx) + d$ for $0 \le x \le 12$. There is a maximum point at $(2, 8)$ and a minimum point at $(6, 2)$.

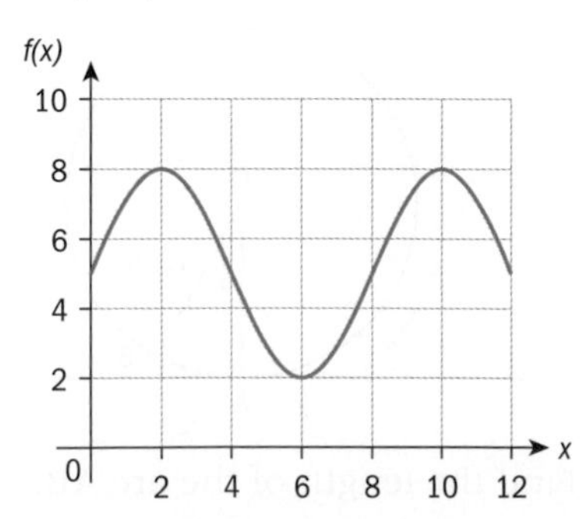

a Find the value of a.

b Show that $b = \frac{\pi}{4}$.

c Find the value of d.

The graph of f is translated by $\begin{pmatrix} 3 \\ 0 \end{pmatrix}$ and then reflected in the x-axis to give the graph of g.

d Find the equation of $g(x)$.

Exam-style questions

24 P1: Let $A(t) = 2\cos^2 t - 3\cos t + 1$, $0 \le t \le 2\pi$.

a Factorize $A(t)$. (2 marks)

b Hence, solve the equation $A(t) = 0$ for $0 \le t \le 2\pi$. (4 marks)

25 P1: Solve the equation $2\cos^2 x = \sin 2x$ for $0 \le t \le 2\pi$, giving your answer in terms of π. (5 marks)

26 P1: A trigonometric function defined by $f(x) = a\sin(bx) + c$, where $a, b, c \in \mathbb{R}$. The period of the function f is π, the maximum of f is 14 and the minimum is 8.

a Find the values of the parameters $a, b, c \in \mathbb{R}$. (4 marks)

b Hence, sketch the graph of f. (3 marks)

27 P1: Let $S(x) = (\sin 2x + \cos 2x)^2$.

a Show that $S(x) = \sin(4x) + 1$, for all $x \in \mathbb{R}$. (3 marks)

b Hence, sketch the graph of S for $0 \le x \le \pi$. (3 marks)

c State

i the period of S

ii the range of S. (2 marks)

d Sketch the graph of the function C defined by $C(x) = \cos(2x) - 1$, for $0 \le x \le 2\pi$. (3 marks)

The graph of C can be obtained from the graph of S under a horizontal stretch of scale factor K followed by a translation by the vector $\begin{pmatrix} p \\ q \end{pmatrix}$.

e Write down

i the value of k

ii a possible vector $\begin{pmatrix} p \\ q \end{pmatrix}$. (3 marks)

28 **P1:** The diagram shows a circle with centre O and radius r. The points A and B lie on the circumference of the circle.

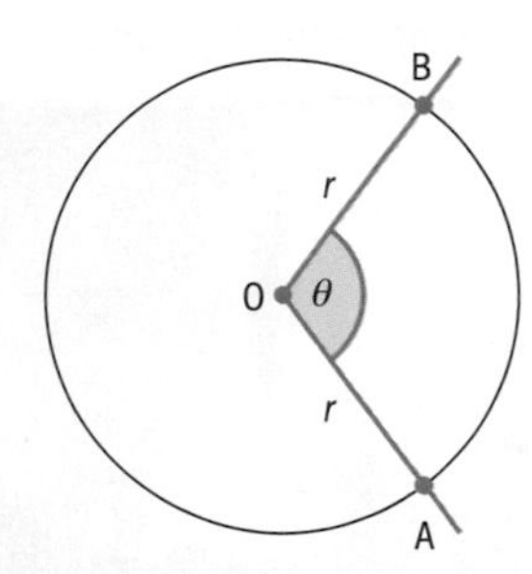

Let $\theta = A\hat{O}B$.

a If $\theta = \frac{2\pi}{3}$ and $r = 2$ cm, state, in terms of π

i the area of the sector AOB

ii the length of the arc AB. (2 marks)

b Given that the area of the sector AOB is πcm^2 and the length of the arc ABC is $\frac{\pi}{3}$ cm, find the exact value of

i r **ii** θ. (5 marks)

29 **P1:** The graph of a function defined by $f(x) = a\cos bx + c$ for $-1 \le x \le 3$ is shown below. The graph of f contains the points $(1, 3)$ and $(2, -1)$.

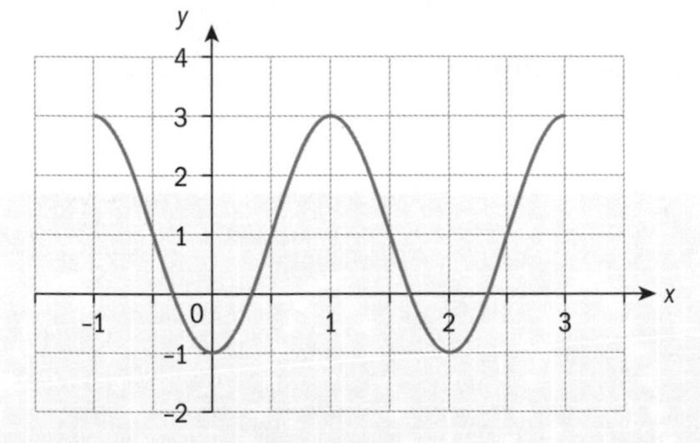

a State

i the range of f

ii the period of f. (2 marks)

b Hence, find the value of the parameters a, b and c. (4 marks)

c Find the zeros of f (2 marks)

30 **P1:** The graph below shows the graph of $f(x) = a\tan\left(b(x-c)\right) + d, -\frac{\pi}{4} < x < \frac{5\pi}{4}$, $x \ne 0, x \ne \frac{\pi}{2}, x \ne \pi$.

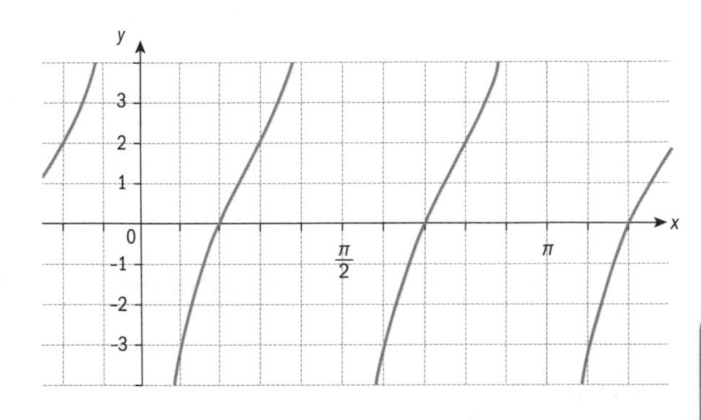

a State

i the equations of the asymptotes to the graph of f

ii the period of f

iii the range of f. (3 marks)

b Write down the value of the parameters b, and d. (2 marks)

c Show that a possible value of c is $\frac{\pi}{4}$. (1 mark)

Given that the graph contains the point $\left(\frac{\pi}{8}, -2\right)$,

d show that $a = 3$ (2 marks)

e write down the solutions to the equation $f(x) = -2$. (2 marks)

31 **P1: a** Write the expression $-2\cos^2 x + \sin x + 3$ in the form $a\sin^2 x + b\sin x + c$. (2 marks)

b Hence, solve the equation $-2\cos^2 x + \sin x + 3 = 2$ for $-2\pi \le x \le 2\pi$. (5 marks)

The sound of mathematics

Approaches to learning: Research, Critical thinking, Using technology

Exploration criteria: Mathematical communication (B), Personal engagement (C), Use of mathematics (E)

IB topic: Trigonometric functions

Brainstorm

In small groups, brainstorm some ideas that link music and mathematics.

Construct a **mindmap** from your discussion with the topic "MUSIC" in the centre.

Share your mindmaps with the whole class and discuss.

This task concentrates on the relationship between **sound waves** and **trigonometric functions.**

Research

Research how sound waves and trigonometric functions are linked.

Think about:

- How do vibrations cause sound waves?
- How do sound waves travel?
- What curve can be used to model a sound wave?

The fundamental properties of a basic sound wave are its **frequency** and its **amplitude**.

- What is the frequency of a sound?
- What units is it measured in?
- What is the amplitude of a sound?

Use what you studied in this chapter to answer these questions:

If you have a sine wave with the basic form $y = a\sin(bt)$, where t is measured in seconds, how do you determine its period, frequency and its amplitude?

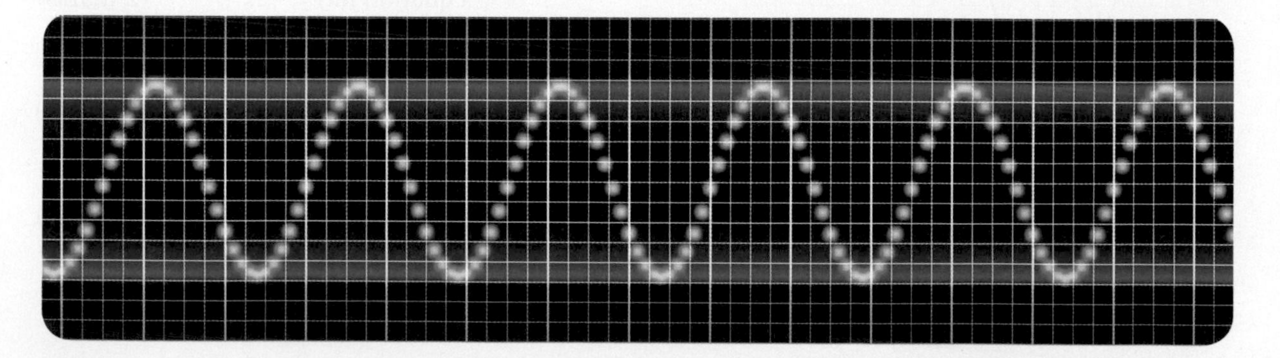

What do the values of *a* and *b* represent? What does *y* represent in this function?

With this information, determine the equivalent sine wave for a sound of 440Hz.

Technology

There are a large number of useful programmes that you can use to consider sound waves.

Using these programmes It is possible to record or generate a sound and view a graphical representation of the soundwave with respect to time.

If you have a music department in your school and/or access to such a programme, you could try this.

Design an investigation

Using the available technology and the information provided here, what could you investigate and explore further?

What experiment could you design regarding sounds?

Disucss your ideas with your group.

What exactly would be the aim of each investigation/exploration/experiment that your group thought of?

Select one of the ideas in your group and plan further.

What will you need to think about as you conduct this exploration?

How will you ensure that your results are reliable?

How will you know that you have completed the exploration and answered the aim?

Extension

Trigonometric functions also occur in many other areas that aid your understanding of the physical world.

Some examples are:

- Temperature modelling
- Tidal measurements
- The motion of springs and pendulums
- The electromagnetic spectrum

They can be thought of in similar terms of waves with different frequencies, periods, amplitudes and phase shifts too.

Think about the task that you have completed here and consider how you could collect other data that can be modelled using a trigonometric function. You could use one of the examples here or research your own idea.

13 Modelling change: more calculus

Concepts
- Change
- Relationships

Microconcepts
- Derivatives of trigonometric functions, including sums and multiples
- Optimization and kinematic problems
- Integration of trigonometric functions
- Integration by inspection or substitution (reverse chain rule)

Phenomena in many fields – the sciences, engineering, business, for example – can be modelled by an elementary function. An elementary function is a function that is algebraic (polynomials, rational functions, functions involving radicals), transcendental (logarithmic, exponential, trigonometric) or a sum, difference, product, quotient or composition of algebraic and transcendental functions. After studying the derivatives of sine and cosine in this chapter, you will be able to differentiate almost any elementary function. You will also study the integrals of sine and cosine, an integration technique called substitution, and the real-world applications of differentiation and integration.

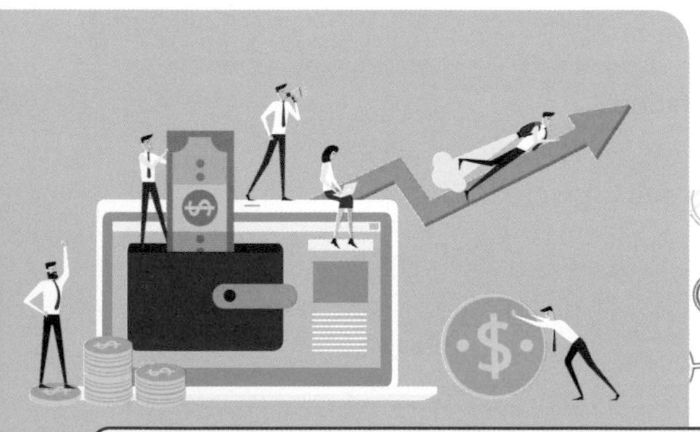

Marginal profit is the profit earned when one additional unit is produced and sold. Given a profit function, how can you use calculus to find the marginal profit function?

How can integration be used to find the center of mass of a thin plate with uniform density?

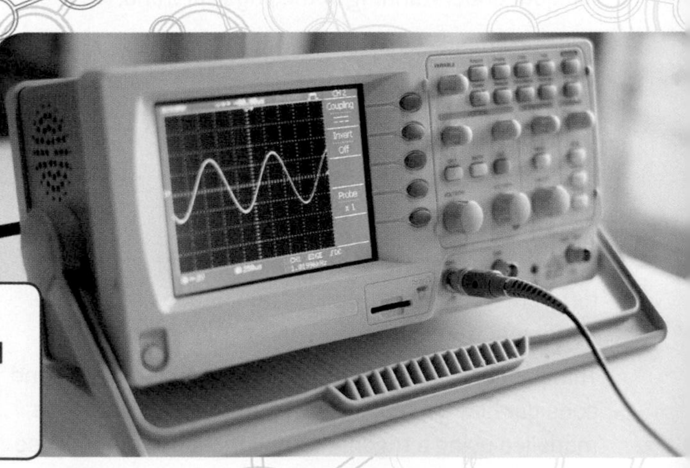

For an alternating current, how can you find the amount of charge stored in a capacitor given the relationship between current and time?

The number of gallons of water in storage tank A is given by $V(t) = -500\sin(0.3t) + 150t + 1450$, where $0 \le t \le 24$ hours.

1 Find the number gallons of water in the tank A at $t = 0$ hours and at $t = 15$ hours.

2 Find the rate at which the amount of water in the tank A is changing at time t hours. Then find the rate at $t = 15$ hours.

At time $t = 0$ hours, storage tank B contains 1800 gallons of water. During the time interval $0 \le t \le 24$ hours, water is pumped into tank B at the rate of $E(t) = 585\sin^2\left(\frac{t}{4}\right)$ gallons per hour. For the same time interval, water is removed from tank B at a rate of $E(t) = 480 + 300\cos(\sqrt{x})$ gallons per hour.

3 Determine whether the amount of water in tank B is increasing or decreasing at time $t = 12$ hours.

4 Find the amount of water in storage tank B at time $t = 12$ hours.

Developing inquiry skills

Think about the information that can be found directly from the given functions. Then think about whether differentiation or integration might be helpful in finding other information. State any requested information that you know how to find now. What further knowledge do you need to find the rest of the requested information?

Before you start

You should know:

1 Find the exact values of sine and cosine for angles on the unit circle.

You should memorize these values, use special right triangles, or use the unit circle to help you find them.

2 Use trigonometric identities to simply expressions

eg:

a $\sin^2(3x) + \cos^2(3x) = 1$ by the Pythagorean identity.

b $1 - 2\sin^2 5x = \cos(10x)$ by the double angle identity for cosine.

3 Use the power, product, quotient and chain rules to find derivatives

eg Find the derivative of $f(x) = x^3\ln(2x)$

$$f'(x) = x^3\left(\frac{1}{2x}\cdot 2\right) + (\ln(2x))(3x^2)$$
$$= x^2 + 3x^2\ln(2x)$$

Skills check

Click here for help with this skills check

1 Find the exact value of:

a $\sin\frac{5\pi}{4}$ **b** $\cos\pi$ **c** $\cos\frac{7\pi}{6}$

d $\sin\frac{\pi}{3}$ **e** $\sin 2\pi$ **f** $\cos\frac{2\pi}{3}$

2 Use an identity to simplify each expression:

a $\cos^2 2x - \sin^2 2x$

b $6\sin x\cos x$

c $e^x\sin^2 x + e^x\cos^2 x$

3 Find the derivative of each function.

a $f(x) = \sqrt[3]{4x^3 + 7x}$

b $f(x) = 3x^2e^{2x}$

c $f(x) = \frac{\ln x}{x^2}$

13.1 Derivatives with sine and cosine

You can use the graphs of the sine and cosine functions to investigate their derivatives.

Investigation 1

Use the graph of $f(x) = \sin x$, for $-2\pi \leq x \leq 2\pi$, to answer the following questions.

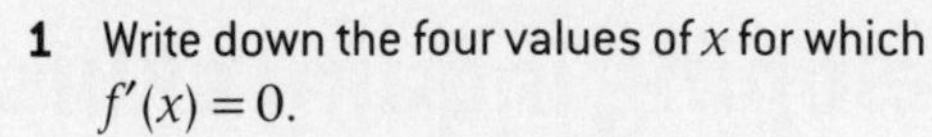

1 Write down the four values of x for which $f'(x) = 0$.

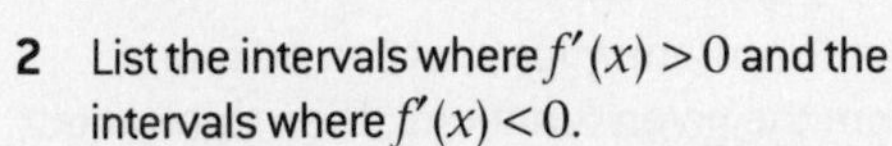

2 List the intervals where $f'(x) > 0$ and the intervals where $f'(x) < 0$.

3 Use the values you found in question **1** to plot four points that lie on the graph of $y = f'(x)$. Then use your answers to question **2** to help you sketch a possible graph of $y = f'(x)$.

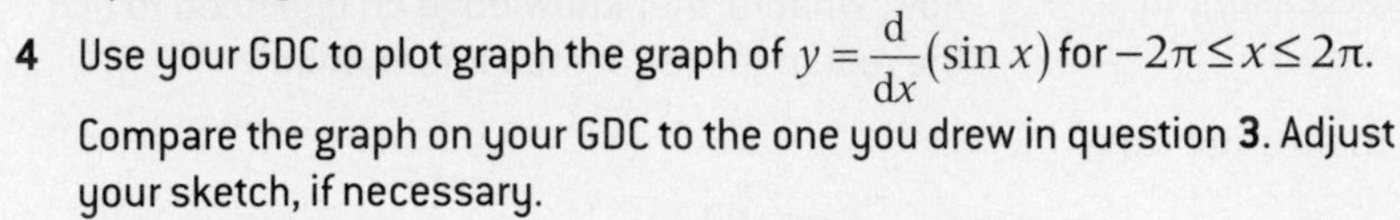

4 Use your GDC to plot graph the graph of $y = \frac{d}{dx}(\sin x)$ for $-2\pi \leq x \leq 2\pi$.

Compare the graph on your GDC to the one you drew in question **3**. Adjust your sketch, if necessary.

HINT

Be sure your GDC is in the radian mode.

5 Based on your graph of $y = \frac{d}{dx}(\sin x)$ from question **4**, can you suggest a function which is the derivative of $y = \sin x$?

To validate your suggestion about the derivative of $y = \sin x$, use your GDC to make a table of values for $y = \frac{d}{dx}(\sin x)$, and another table of values for the function you believe is the derivative. Are the values in these tables the same?

6 **Factual** What is the derivative of $f(x) = \sin x$?

7 **Conceptual** What does the cosine function tell you about the graph of the sine function for any given value in the domain of the sine function?

Sketch a graph of $f(x) = \cos x$, for $-2\pi \leq x \leq 2\pi$ and repeat the process you followed in questions **1–5**. Hence, make a suggestion about the function which is the derivative of the $f(x) = \cos x$.

8 **Factual** What is the derivative of $f(x) = \cos x$?

HINT

$f'(x) = 0$ when the tangent lines to the graph of f are horizontal.

Derivatives of sine and cosine

In investigation 1, you made conjectures about the derivatives of $y = \sin x$ and $y = \cos x$ by examining these functions both graphically and numerically.

The proof that $\frac{d}{dx}(\sin x) = \cos x$ is beyond the scope of this course. However, you can use the fact that $\frac{d}{dx}(\sin x) = \cos x$ to prove that $\frac{d}{dx}(\cos x) = -\sin x$. First, consider a translation of the graphs of the sine and cosine functions.

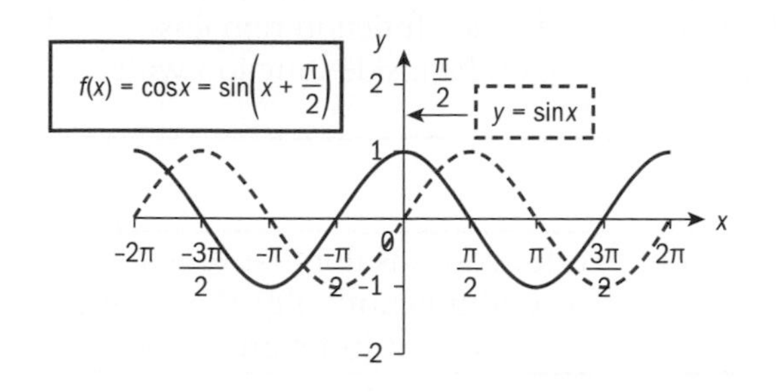

If you translate the graph of $y = \sin x$ horizontally to the left by $\frac{\pi}{2}$, you get the graph of $y = \cos x$.

So, $\cos x = \sin\left(x + \frac{\pi}{2}\right)$.

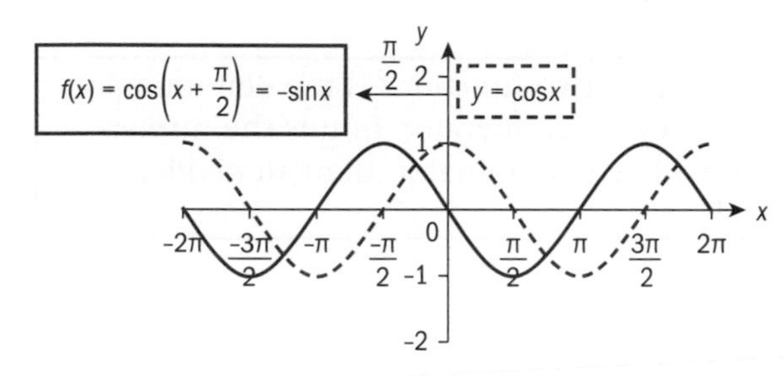

If you translate the graph of $y = \cos x$ horizontally to the left by $\frac{\pi}{2}$, you get a graph which is a reflection of $y = \sin x$ in the x-axis.

So, $\cos\left(x + \frac{\pi}{2}\right) = -\sin x$.

Now you can prove that $\frac{d}{dx}(\cos x) = -\sin x$.

$$\frac{d}{dx}(\cos x) = \frac{d}{dx}\left[\sin\left(x + \frac{\pi}{2}\right)\right]$$ Since $\cos x = \sin\left(x + \frac{\pi}{2}\right)$.

$$= \left[\cos\left(x + \frac{\pi}{2}\right)\right]\left[\frac{d}{dx}\left(x + \frac{\pi}{2}\right)\right]$$

$$= \left[\cos\left(x + \frac{\pi}{2}\right)\right] \times 1$$ Apply the chain rule and the fact $\frac{d}{dx}(\sin x) = \cos x$.

$$= -\sin x \times 1$$ Since $\cos\left(x + \frac{\pi}{2}\right) = -\sin x$.

Therefore, $\frac{d}{dx}(\cos x) = -\sin x$.

$$\frac{d}{dx}(\sin x) = \cos x \qquad \frac{d}{dx}(\cos x) = -\sin x$$

HINT
$\frac{d}{dx}(\sin x) = \cos x$ and $\frac{d}{dx}(\cos x) = -\sin x$ are only valid when x is measured in radians.

Recall the following rules for derivatives which you learned in chapter 5. Remember that u and v are functions of x. It may be helpful to express each rule verbally.

Chain rule

$y = g(u)$ and

$u = f(x) \Rightarrow \frac{dy}{dx} = \frac{dy}{du} \times \frac{du}{dx}$

The derivative of $y = g(f(x))$ is the derivative of the outside function with respect to the inside function (the inside function remains the same), multiplied by the derivative of the inside function with respect to x.

Product rule

$y = uv \Rightarrow \frac{dy}{dx} = u\frac{dv}{dx} + v\frac{du}{dx}$

The derivative of the product of two functions is the first function multiplied by the derivative of the second function, plus the second function multiplied by the derivative of the first function.

Quotient rule

$y = \frac{u}{v} \Rightarrow \frac{dy}{dx} = \frac{v\frac{du}{dx} - u\frac{dv}{dx}}{v^2}$

The derivative of the quotient of two functions is the denominator multiplied by the derivative of the numerator, minus the numerator multiplied by the derivative denominator, then all divided by the denominator squared.

Example 1

Find the derivative of each function.

a $f(x) = 2\sin x - 3\cos x$ **b** $y = \sin x \cos x$ **c** $g(t) = \sin\sqrt{t}$ **d** $f(x) = \cos^2 x$

a $f(x) = 2\sin x - 3\cos x$

$f'(x) = 2\cos x - 3(-\sin x)$

$= 2\cos x + 3\sin x$

Recall that when k is a real number, $\frac{d}{dx}(kf(x)) = kf'(x)$, and find the derivative of each term.

b $y = \sin x \cos x$

$y' = \underbrace{\sin x}_{\text{first factor}} \times \underbrace{(-\sin x)}_{\text{derivative of second factor}} + \underbrace{\cos x}_{\text{second factor}} \times \underbrace{\cos x}_{\text{derivative of first factor}}$

$= -\sin^2 x + \cos^2 x$ or $\cos(2x)$

Apply the product rule with $u = \sin x$ and $v = \cos x$.

Recall the double angle identity $\cos 2x = \cos^2 x - \sin^2 x$

c $g(t) = \sin\sqrt{t}$

$= \sin\left(t^{\frac{1}{2}}\right)$

Rewrite the function using rational exponents.

$g'(t) = \overbrace{\left[\cos\left(t^{\frac{1}{2}}\right)\right]}^{\text{derivative of outside function with respect to inside function}} \times \overbrace{\left[\frac{1}{2}t^{-\frac{1}{2}}\right]}^{\text{derivative of inside function with respect to } t}$ $= \frac{\cos\sqrt{t}}{2\sqrt{t}}$	Apply the chain rule, where the outside function is $y = \sin u$ and the inside function is $u = t^{\frac{1}{2}}$.
d $f(x) = \cos^2 x$ $= (\cos x)^2$	Rewrite $\cos^2 x$ as $(\cos x)^2$.
$f'(x) = \overbrace{[2(\cos x)]}^{\text{derivative of outside function with respect to inside function}} \times \overbrace{[-\sin x]}^{\text{derivative of inside function with respect to } x}$ $= -2\sin x\cos x$ or $-\sin 2x$	Apply the chain rule: the outside function is $y = u^2$ and the inside function is $u = \cos x$. Recall the double angle identity $\sin 2x = 2\sin x \cos x$

In example 1 parts **b** and **d**, you used double angle identities to help simplify your answers.

Trigonometric identities are often helpful when finding derivatives involving trigonometric functions.

Example 2

Find the derivative of each function.

a $y = \tan x$ **b** $y = 2\sin x\cos x$

a $y = \tan x$ $= \frac{\sin x}{\cos x}$	Rewrite the tangent function in terms of sine and cosine.
$y' = \frac{\overbrace{\cos x}^{\text{denominator}} \times \overbrace{\cos x}^{\text{derivative of numerator}} - \overbrace{\sin x}^{\text{numerator}} \times \overbrace{(-\sin x)}^{\text{derivative of denominator}}}{\underbrace{\cos^2 x}_{\text{denominator squared}}}$	Apply the quotient rule, where $u = \sin x$ and $v = \cos x$.
$= \frac{\cos^2 x + \sin^2 x}{\cos^2 x}$	Simplify.
$= \frac{1}{\cos^2 x}$	Use the Pythagorean identity $(\sin^2 x + \cos^2 x = 1)$ to further simplify the numerator.
b $y = 2\sin x\cos x$ $= \sin(2x)$	Instead of using the product rule like you did in example 1 **b**, you can rewrite the function using the double angle identity $\sin 2x = 2\sin x \cos x$.
$y' = \overbrace{[\cos(2x)]}^{\text{derivative of outside function with respect to inside function}} \times \overbrace{[2]}^{\text{derivative of inside function with respect to } x}$ $= 2\cos(2x)$	Apply the chain rule where $y = \sin u$ and $u = 2x$.

Calculus

Exercise 13A

In questions **1–10**, find the derivative of each function.

1 $f(x) = 4\sin x + 3\cos x$

2 $y = \sin(5x)$

3 $y = \sin\frac{x}{3} - \cos(3x)$

4 $f(x) = \frac{3}{\cos x}$

5 $h(t) = \sin^3 t$

6 $y = \sin\sqrt[3]{x}$

7 $y = \cos\frac{\pi}{x}$

8 $g(x) = \frac{1}{\tan x}$

9 $f(x) = \sin^2 x + \cos^2 x$

10 $y = \sin(2x)\cos(2x)$

11 Let $f(x) = 4x^3$ and $g(x) = \cos x$.

a Find $(g \circ f)(x)$. b Find $\frac{d}{dx}[(g \circ f)(x)]$.

Suppose that $r(x) = ((g \circ f)(x)) \times h(x)$ and $h(x) = x^2$.

c Find $r'(x)$.

12 Let $f(x) = \sin x$.

a Find the following:

i $f'(x)$ ii $f''(x)$

iii $f'''(x)$ iv $f^{(4)}(x)$

b Suppose that $f^{(n)}(x) = \sin x$, where n is a positive integer. Write down all possible values of n that are less than 15.

c Find the following:

i $f^{(80)}(x)$ ii $f^{(42)}(x)$

Exercise 13B

In questions **1** and **2**, find the equations of the tangent line and the normal line to the curve at the given value of x.

1 $f(x) = \cos x + 2\sin x$; $x = \pi$

2 $f(x) = \cos 6x$; $x = \frac{\pi}{3}$

3 Let $f(x) = \sin(\pi x)$.

a Find $f\left(\frac{1}{4}\right)$. b Find $f'(x)$.

c Find the equation of the tangent line to the graph of f at $x = \frac{1}{4}$.

4 Consider the function $f(x) = 2\cos x$ for $0 \le x \le 2\pi$. Find the value(s) of x for which the tangent lines to the graph of f are parallel to the line $y = -\sqrt{2}x - 6$.

HINT

Recall that the gradient of the tangent line to the graph of $y = f(x)$ at $x = a$ is equal to $f'(a)$ and that the normal line to the graph of a function at a given point is the line perpendicular to the tangent line at that point.

Example 3

Find the derivative of each function.

a $f(x) = 3e^{5x} + \sin(2x - 7)$ b $y = (\cos x)(\ln x)$ c $y = \cos(\ln x)$

a $f(x) = 3e^{5x} + \sin(2x - 7)$ $f'(x) = 3(e^{5x})(5) + [\cos(2x - 7)](2)$ $= 15e^{5x} + 2\cos(2x - 7)$	To differentiate the first term, use the constant multiple rule and chain rule. For the second term, use the chain rule.

b $y = (\cos x)(\ln x)$ $y' = (\cos x)\left(\frac{1}{x}\right) + (\ln x)(-\sin x)$ $= \frac{\cos x}{x} - \sin x \ln x$	Use the product rule.
c $y = \cos(\ln x)$ $y' = [-\sin(\ln x)]\left(\frac{1}{x}\right)$ $= -\frac{\sin(\ln x)}{x}$	Use the chain rule.

You can use the first and second derivatives of a function to analyse the graph of the function.

Example 4

Consider the function $f(x) = \cos x - \sin x$ for $0 \le x \le 2\pi$.

a Find the x- and y-intercepts.

b Find the intervals on which f is increasing and decreasing, and find the local maximum and minimum points.

c Find the intervals on which f is concave up and concave down, and find the inflexion points.

d Use the information from parts **a**, **b** and **c** to sketch the graph of f. Then use a GDC to plot the graph of the function and verify your answers.

HINT

Local extrema may also be called relative extrema.

a $\cos x - \sin x = 0$ $\sin x = \cos x$ $\tan x = 1$ $x = \frac{\pi}{4}, \frac{5\pi}{4}$ x-intercepts: $\left(\frac{\pi}{4}, 0\right), \left(\frac{5\pi}{4}, 0\right)$	To find the x-intercepts, set the function equal to 0, and solve for x in the given domain. Use your knowledge of special angles in the unit circle to find the solutions.
$f(0) = \cos 0 - \sin 0 = 1 - 0 = 1$ y-intercept: $(0,1)$	To find the y-intercept, evaluate the function when $x = 0$.
b $f(x) = \cos x - \sin x$ $f'(x) = -\sin x - \cos x$ $-\sin x - \cos x = 0$	Find the derivative of f and the points where $f'(x) = 0$.
$\sin x = -\cos x$ $\tan x = -1$	Test the values to the left and right of the roots of $f'(x)$.

Continued on next page

Calculus

$f'(x) = 0$ at $x = \frac{3\pi}{4}, \frac{7\pi}{4}$

Increasing: $\frac{3\pi}{4} < x < \frac{7\pi}{4}$

Decreasing: $0 < x < \frac{3\pi}{4}$ and $\frac{7\pi}{4} < x < 2\pi$

x	$0 < x < \frac{3\pi}{4}$	$\frac{3\pi}{4} < x < \frac{7\pi}{4}$	$\frac{7\pi}{4} < x < 2\pi$
Sign of $f'(x)$	$f'\left(\frac{\pi}{2}\right) < 0$ $-$	$f'(\pi) > 0$ $+$	$f'\left(\frac{11\pi}{6}\right) = \frac{1}{2} - \frac{\sqrt{3}}{2} < 0$ $-$
$f(x)$	decreasing	increasing	decreasing

Local maximum point: $\left(\frac{7\pi}{4}, \sqrt{2}\right)$

Local minimum point: $\left(\frac{3\pi}{4}, -\sqrt{2}\right)$

Use the first derivative test to determine the local maximum point occurs at $x = \frac{7\pi}{4}$ and the local minimum point occurs at $x = \frac{3\pi}{4}$. Evaluate f at $x = \frac{7\pi}{4}$ and $x = \frac{3\pi}{4}$ to find the maximum and minimum values.

c $f''(x) = -\cos x + \sin x$

$-\cos x + \sin x = 0$

$\sin x = \cos x$

$\tan x = 1$

$x = \frac{\pi}{4}, \frac{5\pi}{4}$

Concave up: $\frac{\pi}{4} < x < \frac{5\pi}{4}$

Concave down: $0 < x < \frac{\pi}{4}$ and $\frac{5\pi}{4} < x < 2\pi$

Find the second derivative of f and where $f''(x) = 0$.

Test the values to the left and right of the roots of $f''(x)$.

x	$0 < x < \frac{\pi}{4}$	$\frac{\pi}{4} < x < \frac{5\pi}{4}$	$\frac{5\pi}{4} < x < 2\pi$
Sign of $f''(x)$	$f''\left(\frac{\pi}{6}\right) < 0$ $-$	$f'(\pi) > 0$ $+$	$f'\left(\frac{3\pi}{2}\right) < 0$ $-$
$f(x)$	concave down	concave up	concave down

Inflexion points: $\left(\frac{\pi}{4}, 0\right)$ and $\left(\frac{5\pi}{4}, 0\right)$

Inflexion points occur when at points c where $f''(c) = 0$ and the sign of $f''(x)$ changes as x passes through c. Evaluate f at $x = \frac{\pi}{4}$ and $x = \frac{5\pi}{4}$ to find the y-coordinates of the points of inflexion.

d

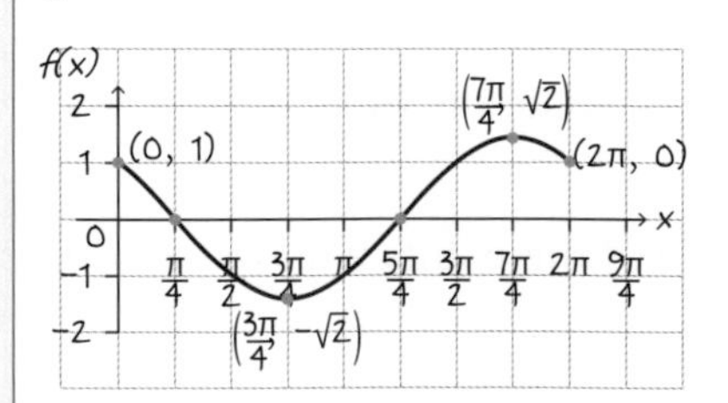

HINT

Do not extend the graph beyond the given domain. Check your work by graphing f on your GDC

Exercise 13C

In questions **1–10**, find the derivative of each function.

1 $f(x) = 3\sin\left(4x - \frac{\pi}{6}\right) + 5x$

2 $y = e^{4x^3+2x^2+x}$

3 $y = \frac{\sin x}{1-\cos x}$

4 $f(x) = x^2e^x + e^x$

5 $h(t) = 2e^{\cos t}$

6 $y = e^{5x}\sin 3x$

7 $y = \cos x + x\sin x$

8 $f(x) = \cos(e^{x^2})$

9 $f(x) = (\ln 3x)(\sin 3x)$

10 $f(x) = \ln(\sin 3x)$

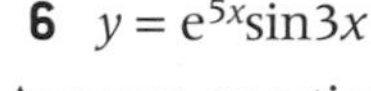

Answer questions **11–13 without** using your GDC. Then plot a graph of each function on your GDC to verify your answers.

11 Consider the function $f(x) = e^{\cos x}$, for $0 \le x \le 2\pi$.

a Find $f'(x)$.

b Find the intervals where the graph of f is increasing and the intervals where it is decreasing.

c Find the coordinates of any local minimum points and of any local maximum points on the graph of f.

12 Consider the function $f(x) = x + \cos x$, for $0 \le x \le 2\pi$.

a Find $f'(x)$ and $f''(x)$.

b Find the intervals where the graph of f is concave up and the intervals where it is concave down.

c Find the coordinates of any inflexion points on the graph of f.

13 Consider the function $f(x) = \cos 2x + \cos^2 x$, for $0 \le x \le \pi$.

a Show that $f'(x) = -3\sin 2x$.

b Find the coordinates of any relative extrema.

c Find the coordinates of any inflexion points.

d Sketch a graph of $f(x) = \cos 2x + \cos^2 x$ for $0 \le x \le \pi$.

Developing inquiry skills

In the opening scenario, you were given that the number of gallons of water in storage tank A at time t hours is modelled by $V(t) = -500\sin(0.3t) + 150t + 1450$. You should now have the knowledge to find the function that gives the rate at which the amount of water in the tank A is changing at time t hours and to find that rate of change at $t = 15$ hours.

13.2 Applications of derivatives

As you work through this section, think about how derivatives can be used to solve real-life problems.

Investigation 2

The total cost, \$$C$, for a bicycle manufacturer to produce x bicycles is given by $C(x) = -0.1x^2 + 150x + 3000$ for where $0 \leq x \leq 600$.

1 Find the cost of producing the following numbers of bicycles:
a 250 bicycles **b** 251 bicycles **c** 550 bicycles **d** 551 bicycles

Consider the actual cost of producing one additional bicycle. For example, the actual cost incurred in producing the 251st bicycle is equal to $C(251) - C(250)$.

2 Find the actual cost of manufacturing the:
a 251st bicycle **b** 551st bicycle

3 **Conceptual** Why is the cost of producing one additional bicycle not constant?

4 Differentiate C and evaluate $C'(x)$ at the following values:
a $C'(250)$ **b** $C'(550)$
Explain what the value of the derivative at each point tells you, in the context of this problem.

5 How does the actual cost incurred for producing an additional bicycle found in question **2** compare to the value you found in question **4**?

The derivative of the cost function, $C'(x)$, is called the **marginal cost** function. The marginal cost is the cost of producing one additional unit. In this situation, $C'(250)$ represents the cost of producing the 251st bicycle and $C'(550)$ represents the cost of producing the 551st bicycle.

6 **Conceptual** Why can the marginal cost function—the derivative of the cost function—be used to approximate the cost incurred for producing one additional item?

For a function $y = f(x)$, $\frac{f(b) - f(a)}{b - a}$ is the *average* rate of change of f over $[a, b]$. It is also possible to represent the *instantaneous* rate of change of a function, which is the rate of change at a certain point.

7 **Conceptual** What represents the instantaneous rate of change of a function at a given point?

Derivatives are useful in situations involving the rate of change of one variable with respect to another variable.

The economic theory of marginal analysis is the study of change in cost, revenue and profit for each increase of one unit in production or sales. The marginal cost, marginal revenue and marginal profit functions are the derivatives of the cost, revenue and profit functions.

$C(x)$ = total cost of producing x amount of units

$R(x)$ = total revenue for selling x amount of units

$P(x)$ = profit in selling x amount of units

$P(x) = R(x) - C(x)$

Example 5

A company uses the function $C(x) = 100 + x - 0.01x^2 + 0.00006x^3$ to estimate the cost, in euros, of producing x items. The revenue, in euros, of selling x items is modelled by $R(x) = 22.8x - 0.001x^2$.

a Find the cost of producing 300 items.

b Find the marginal cost of producing 300 items and explain what this means, in context.

c Find the marginal profit of selling 300 items and explain what this means, in context.

a $C(300) = 1120$ euros	Use a GDC to find $C(300)$.
b $C'(300) = 11.2$ This mean the cost of producing the 301st item is 11.20 euros.	The marginal cost function is the derivative of the cost function. Use a GDC to find $C'(300)$.
c $P(x) = -100 + 21.8x + 0.0009x^2 - 0.00006x^3$	First find the profit function using $P(x) = R(x) - C(x)$.
$P'(300) = 11$ This means that an additional profit of 11 euros is earned by selling the 301st item.	The marginal profit function is the derivative of the profit function. Use a GDC to find $P'(300)$.

The average rate of change of a function f on the interval $[a, b]$ is the slope of the secant line, $\frac{f(b)-f(a)}{b-a}$.

The instantaneous rate of change of f at $x = a$ is the slope of the tangent line at $x = a$, that is, $f'(a)$.

Example 6

During one month, the temperature of the water in a pond is modelled by the function $T(n) = 18 + 12ne^{-\frac{n}{3}}$, where n is measured in days and T is measured in degrees Celsius.

a Find the average rate of change in temperature during the first 18 days of the month. Include units in your answer.

b Find $T'(n)$.

c Find the exact value of the instantaneous rate of change in temperature on day 18. Check your answer using a GDC.

a average rate of change $= \frac{T(18)-T(0)}{18-0} = 0.0297$ °C day^{-1}	Find the slope of the secant line on the interval [0, 18]. The answer is given correct to 3 significant figures. The units for $\frac{\text{change in temperature}}{\text{change in time}}$ are °C day^{-1}.

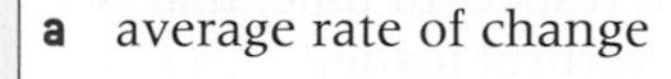

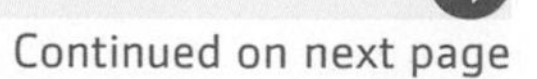

Continued on next page

b $T'(n) = 0 + \left[12n\left(-\frac{1}{3}e^{-\frac{n}{3}}\right) + 12e^{-\frac{n}{3}}\right]$ $= -4ne^{-\frac{n}{3}} + 12e^{-\frac{n}{3}}$	The derivative of the constant term is zero. Use the product rule and the chain rule to find the derivative of the second term.
c instantaneous rate of change $T'(18) = -4(18)e^{-\frac{18}{3}} + 12e^{-\frac{18}{3}}$ $= -60e^{-6}\ °\text{C day}^{-1}$ $-60e^{-6}\ °\text{C day}^{-1} \approx -0.149\ °\text{C day}^{-1}$	Find the slope of the tangent line to T at $n = 18$. Check your answer using your GDC.

You can see from your answers to example 6 parts **a** and **c** that the average rate of change and the instantaneous rate of change are often very different, and may even have different signs.

Example 7

A diver jumps from a platform at time $t = 0$ seconds. The distance of the diver above the water level at time t seconds is given by $s(t) = -4.9t^2 + 4.9t + 10$, where s is in metres.

a Find the average velocity of the diver during the dive.

b Find the velocity of the diver at the instant the diver hits the water.

c Explain why the answer to part **b** is negative.

d Find the speed of the diver at the instant the diver hits the water.

a $-4.9t^2 + 4.9t + 10 = 0,\ t \geq 0$ $t = 2.01354$	Find when the diver hits the water by solving $s(t) = 0$ for $t \geq 0$. Do not round too early in your calculations.
average velocity $= \frac{s(2.01354) - s(0)}{2.01354 - 0}$ $= -4.97\ \text{m s}^{-1}$	Find the average velocity over the interval $[0, 2.01354]$ using average velocity $= \frac{\text{total displacement}}{\text{time taken}}$
b instantaneous velocity when diver hits water  $= s'(2.01354) = -14.8\ \text{m s}^{-1}$	Use your GDC to find $s'(2.01354)$.
c The diver is moving downward towards the water, so the velocity of the diver in the vertical plane is negative.	Velocity gives both magnitude and direction. This model gives the rate of change of vertical height with respect to time, that is the velocity in the vertical plane. A positive velocity indicates upward movement and a negative velocity indicates downward movement.

d speed = |−14.8| = 14.8 m s^{-1}	Speed gives only magnitude; it does not indicate direction. Speed = $\|v(t)\|$

Exercise 13D

Answer questions **1–3** without using a GDC.

1 The cost C, in euros, of producing a certain table is modelled by $C(x) = 0.5x^2 - 50x + 2500$, where x is the number of tables produced. Find $C'(120)$, the marginal cost when $x = 120$, and explain its meaning.

2 The displacement of an object from a fixed point O is given by $s(t) = \frac{t}{e^t}$, for $0 \le t \le 4$.

a Show that the velocity of the object is $v(t) = \frac{1-t}{e^t}$.

b Find the velocity and the speed of the object when $t = 2$.

c Find the time when the particle changes direction.

3 The displacement d m (from the equilibrium position) of a weight moving in harmonic motion at the end of a spring is modelled by $d(t) = \frac{1}{6}\sin 18t - \frac{1}{3}\cos 18t$, where t is measured in seconds. Find the velocity of the object when $t = \frac{\pi}{27}$ s.

Use a GDC to help you answer questions **4–6**.

4 The number of bacteria in a science experiment on day t is modelled by $P(t) = 120e^{0.2t}$.

a Find the average rate of change of the number of bacteria over the interval 0 to 10 days.

b Find the instantaneous rate of change of the number of bacteria at any time t.

c Find the instantaneous rate of change of the number of bacteria on day 10 and explain the meaning of your answer.

5 The profit, in euros, obtained from selling x units of a certain chemical is modelled by $P(x) = -0.00005x^3 + 12x - 200$. The revenue from the sales of x units of the chemical is modelled by $R(x) = 10x - 4$.

a Find $P'(200)$, the marginal profit when $x = 200$, and explain its meaning.

b Use the fact that $P(x) = R(x) - C(x)$ to find the cost function C.

c Find the marginal cost of producing the 201st unit of the chemical.

6 A ball is thrown vertically upwards. Its height in metres above the ground t seconds after it is thrown is modelled by $s(t) = -4.9t^2 + 15.2t + 1.3$. Find:

a the time at which the ball hits the ground

b the average velocity of the ball during its flight

c the velocity of the ball at time t

d the value of t for which $v'(t) = 0$ and explain the significance of this value of t.

Derivatives are useful in optimization problems, such as the maximizing or minimizing profits, costs, areas, volumes or distances.

Example 8

The cost per item, in euros, to produce x items is $C(x) = x^3 - 3x^2 - 9x + 30$. Find the number of items that must be produced to minimize the cost per item. Justify your answer.

$C(x) = x^3 - 3x^2 - 9x + 30$ $C'(x) = 3x^2 - 6x - 9$	Find the derivative of the cost per item function.
$3x^2 - 6x - 9 = 0$ $3(x-3)(x+1) = 0$	Find the zeros of the derivative.
$x = 3, -1$ $x = 3$	It is not feasible to produce −1 items, so the only feasible zero in this context is $x = 3$.
$C''(x) = 6x - 6$ $C''(3) = 12 > 0$ Since the second derivative is greater than 0 at $x = 3$, there is a minimum at $x = 3$ items.	You can use either the first or second derivative test to justify that a minimum occurs at $x = 3$. The second derivative test is shown here.

Example 9

A rectangle is inscribed under the curve $f(x) = 2\cos\left(\frac{1}{2}x\right)$, for $-\pi \le x \le \pi$. Points A and B lie on the x-axis, and points C and D lie on the curve, as shown in the diagram.

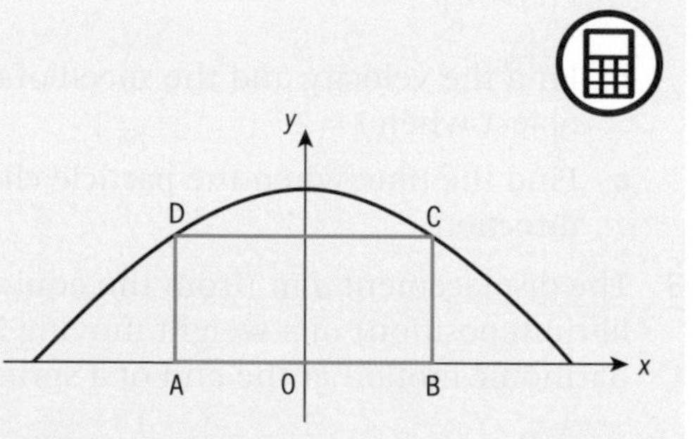

The coordinates of B are $(x, 0)$ and the coordinates of C are $\left(x, 2\cos\left(\frac{1}{2}x\right)\right)$.

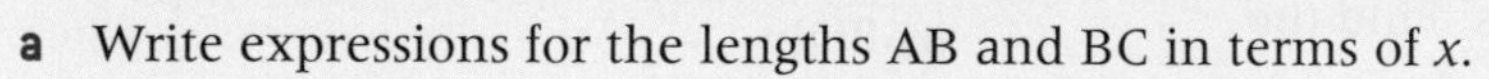

a Write expressions for the lengths AB and BC in terms of x.

b Write an expression for the area of the rectangle, $A(x)$, in terms of x.

c Find $A'(x)$.

d Use your answer from part **c** to find the value of x for which the area is a maximum.

e Use your GDC to plot a graph of $y = A(x)$ and verify your answer from part **d**. Find the maximum area of the rectangle.

a $AB = 2x$ and $BC = 2\cos\left(\frac{1}{2}x\right)$	
b $A(x) = 4x\cos\left(\frac{1}{2}x\right)$	Area = length × width = $2x \times 2\cos\left(\frac{1}{2}x\right)$

c $A'(x)=(4x)\left(-\sin\left(\frac{x}{2}\right)\right)\left(\frac{1}{2}\right)+\left(\cos\left(\frac{x}{2}\right)\right)(4)$ $=-2x\sin\left(\frac{x}{2}\right)+4\cos\left(\frac{x}{2}\right)$	Use the product and chain rules.
d 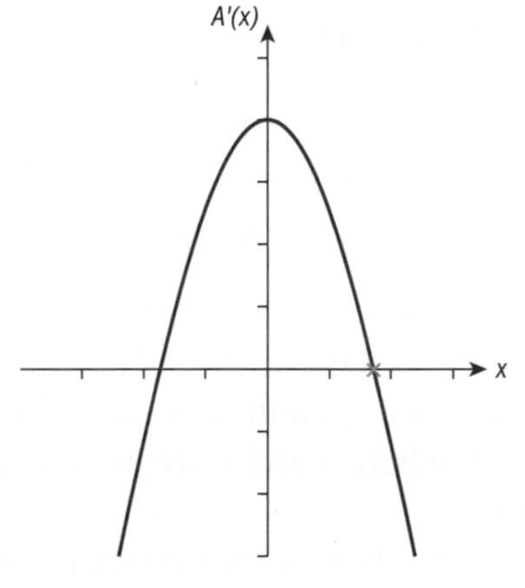 $x = 1.72$	By the 1st derivative test, $A(x)$ is maximized where $A'(x) = 0$ and $A'(x)$ changes sign from positive to negative. Graph $A'(x)=-2x\sin\left(\frac{x}{2}\right)+4\cos\left(\frac{x}{2}\right)$ on your GDC and find where the graph changes from positive to negative in $0<x\le\pi$. The answer is given correct to 3 s.f.
e 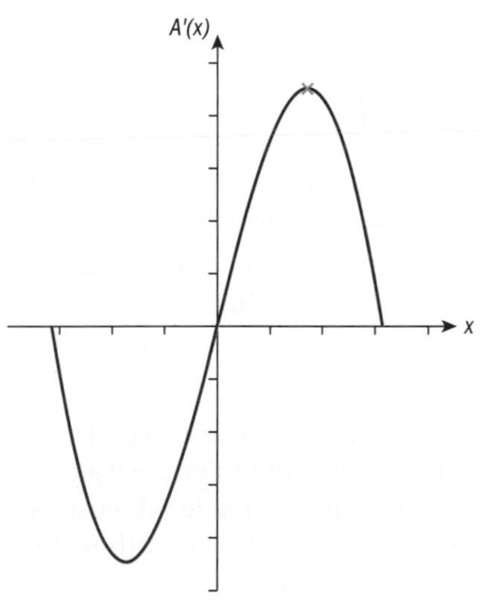 This graph verifies that the maximum occurs when $x = 1.72$. The maximum area is 4.49.	Graph $A(x)=4x\cos\left(\frac{x}{2}\right)$ and find the maximum.

Reflect What types of real-life problems can be solved using derivatives?

Exercise 13E

1 The cost per item, in euros, to produce x items is $C(x) = x + \dfrac{10\,000}{x}$. Items are produced in batches according to the size of the order.

Find the order size that will minimize costs. Justify your answer.

2 A rectangle PQRS is inscribed in a circle of radius 4 cm with centre O, as shown below.

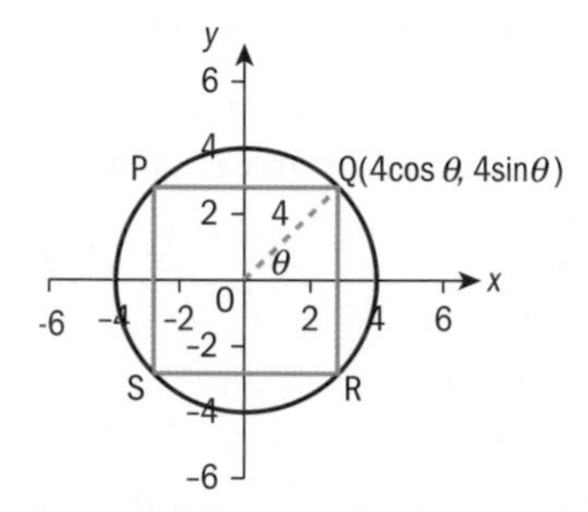

The angle between radius OQ and the x-axis is θ radians, where $0 \le \theta \le \dfrac{\pi}{2}$. Point Q has coordinates $(4\cos\theta, 4\sin\theta)$.

Let A be the area of rectangle PQRS.

a Show that $A = 32\sin 2\theta$.

b Find $\dfrac{dA}{d\theta}$ and use it to find the exact value of θ that maximizes the area of the rectangle.

c Find $\dfrac{d^2A}{d\theta^2}$ and use it to justify that the value you found for θ in part **b** gives the maximum area.

3 A cone has radius 4 cm and height 6 cm. A cylinder with radius r and height h is inscribed in the cone.

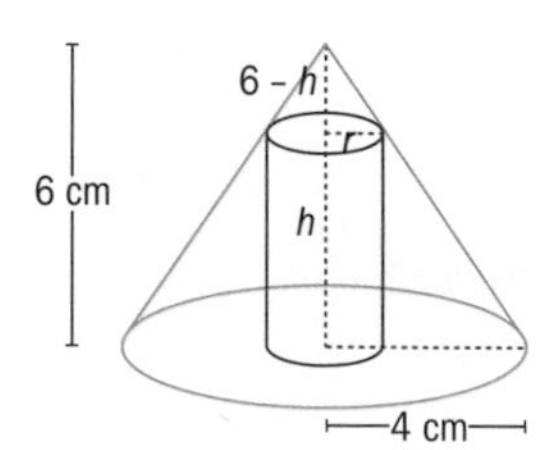

a Find an expression for r in terms of h.

b Let V be the volume of the cylinder. Show that the volume of the cylinder is $V = \dfrac{4\pi}{9}\left(36h - 12h^2 + h^3\right)$.

c Find $\dfrac{dV}{dh}$ and $\dfrac{d^2V}{dh^2}$.

d Find the radius and height of the cylinder with maximum volume. Use $\dfrac{d^2V}{dh^2}$ to justify your answer.

4 The revenue for selling x units of an item is modelled by $R(x) = 4\sqrt{x}$ and the cost of producing x units is modelled by $C(x) = 2x^2$, where R and C are in thousands of dollars.

a Write an expression for $P(x)$, the profit for producing and selling x units.

b Find $\dfrac{dP}{dx}$.

c Use $\dfrac{dP}{dx}$ to find the number of units that should be produced in order to maximize the profit.

d Use your GDC to graph the profit function. Verify that your answer to part **c** gives a maximum, and find the maximum profit.

5 A rectangular enclosure, ABCD, of area 675 m^2, is to be built to enclose a playground. The fencing used for side AB costs \$10 per metre. The fencing for the other three sides costs \$4 per metre.

a Find an expression for the total cost of the fencing in terms of a single variable.

b The fence is to be built so that the cost is a minimum. Find this minimum cost.

> **HINT**
>
> Note that in question **5b**, there are no demands on the work to be shown, and since the GDC is allowed, once you find a function for the minimum cost you may use a graph of the function on your GDC to find the minimum.

13.3 Integration with sine, cosine and substitution

In this section you will find integrals involving sine and cosine and learn about integration with the substitution method.

Integrals of sine and cosine

Investigation 3

1 For each derivative of a function there is a corresponding indefinite integral. For example:

$$\frac{d}{dx}(x^3) = 3x^2 \Rightarrow \int 3x^2 dx = x^3 + C$$

Use this fact to copy and complete the following blanks.

$\frac{d}{dx}(x^3) = 3x^2 \Rightarrow \int 3x^2 dx = x^3 + C$

$\frac{d}{dx}(e^{-2x}) =$ .... $\Rightarrow$ ....

$\frac{d}{dx}(\sin x) =$ .... $\Rightarrow$ ....

$\frac{d}{dx}(-\cos x) =$ .... $\Rightarrow$ ....

2 **Factual** What is $\int \sin x dx$?

3 **Factual** What is $\int \cos x dx$?

4 **Conceptual** What can you use as a basis for determining the anti-derivatives of many functions?

Integrals of sine and cosine

$$\int \sin x dx = -\cos x + C$$
$$\int \cos x dx = \sin x + C$$

Integrals of composition of sine or cosine with a linear function

$$\int \sin(ax+b)dx = -\frac{1}{a}\cos(ax+b)+C$$

$$\int \cos(ax+b)dx = \frac{1}{a}\sin(ax+b)+C$$

HINT

In chapter 10, you learned about integrals of compositions of functions with a linear function. The same concept applies to $\int \sin(ax+b)\,dx$ and $\int \cos(ax+b)\,dx$.

In general,

$$\int f(ax+b)dx = \frac{1}{a}F(ax+b)+C,$$

where $F'(x) = f(x)$.

Calculus

Example 10

Find these indefinite integrals.

a $\int 2\cos x\,dx$ **b** $\int \sin(3x-5)\,dx$ **c** $\int 4\cos(4x)\,dx$

a $\int 2\cos x\,dx = 2\int \cos x\,dx$ $= 2\sin x + C$	Use the constant multiple rule and then integrate cosine.
b $\int \sin(3x-5)\,dx = -\frac{1}{3}\cos(3x-5)+C$	$\int \sin(ax+b)\,dx = -\frac{1}{a}\cos(ax+b)+C$
c $\int 4\cos(4x)\,dx = 4\int \cos(4x)\,dx$ $= 4\cdot\frac{1}{4}\sin(4x)+C$ $= \sin(4x)+C$	Use the constant multiple rule and $\int \cos(ax+b)\,dx = \frac{1}{a}\sin(ax+b)+C.$

Exercise 13F

Find each indefinite integral.

1 $\int 5\sin x\,dx$

2 $\int (4\cos x - 2\sin x)\,dx$

3 $\int \cos(7x)\,dx$

4 $\int 6\cos(2x)\,dx$

5 $\int \sin(5x+3)\,dx$

6 $\int (x^3 + \sin(2x))\,dx$

7 $\int \cos\left(\frac{x}{2}\right)dx$

8 $\int 2\pi\sin(2\pi x)\,dx$

HINT

Check your answers by making sure the derivative of your answer is the same as the integrand.

Indefinite integrals and the substitution method

You can integrate a function of the form $f(ax+b)$ by recognizing it is the composite of a function with a linear function, and knowing that $\int f(ax+b)\,dx = \frac{1}{a}F(ax+b)+C$, where $F'(x) = f(x)$.

You will now consider integrands that are composites of more complicated functions: those which are not just linear.

Investigation 4

1 Suppose $y = \frac{1}{3}(x^2-4x)^3$ and let $u = x^2 - 4x$.

a Find $\frac{du}{dx}$.

b Write y in terms of u and find $\frac{dy}{du}$.

c Find $\frac{dy}{dx}$. What rule are you using here?

2 Based on your answer to question **1c**, what is $\int (x^2-4x)^2(2x-4)\,dx$?

3 Show *how* to find $\int (x^2-4x)^2(2x-4)\,dx$ by reversing the process in question **1**. Copy and fill in the blanks below.

$\int (x^2-4x)^2(2x-4)\,dx$

Let $u = x^2 - 4x$.

$\frac{du}{dx} = $ .... Find $\frac{du}{dx}$.

$du = $ .... dx Express du in terms of x.

$\int (x^2 - 4x)^2(2x - 4)dx = \int$ .... du Express the integrand in terms of u.

$\int$ .... $du = $ .... $+ C$ Find the integral in terms of u.

$\int (x^2 - 4x)^2(2x - 4)dx = $ .... $+ C$ Rewrite the answer in terms of x.

The method you used in question **3** is called the **substitution method** of integration. Notice that you rewrote an integrand with two factors as an integrand with one factor, using a new variable.

4 **Conceptual** The integration by substitution method is the reverse process of which method of differentiation?

In $\int (x^2 - 4x)^2(2x - 4)dx$, the first factor, $(x^2 - 4x)^2$, is a composite function. Notice that there is a relationship between the composite function and the second factor, $(2x - 4)$.

5 **Conceptual** When may the substitution method of integration apply?

HINT

Recall from chapter 5 that $\frac{du}{dx}$ is not a fraction. However, since $\frac{du}{dx}$ and $f'(x)$ represent rates of change, you can express $\frac{du}{dx} = f'(x)$ as $du = f'(x)\,dx$, or even $dx = \frac{du}{f'(x)}$

The substitution method of integration applies to integrals of the form $\int kg'(x)f(g(x))\,dx$. Sometimes you may be able to perform the substitution mentally and this is called **integration by inspection**.

HINT

The substitution method is sometimes called a u-substitution, because the variable u is commonly used to make the substitution.

Example 11

Each indefinite integral is in the form $\int kg'(x)f(g(x))\,dx$. Find each definite integral.

a $\int (3x^2 + 7x)^5(6x + 7)dx$ **b** $\int x^4 \cos(x^5 + 3)dx$ **c** $\int 2xe^{3x^2}dx$ **d** $\int \frac{8x^3 - 15x^2}{2x^4 - 5x^3}dx$

a $\int (3x^2 + 7x)^5(6x + 7)dx$	$f(g(x)) = (3x^2 + 7x)^5$, $g(x) = 3x^2 + 7x$ and $g'(x) = 6x + 7$
Let $u = 3x^2 + 7x$, then $(3x^2 + 7x)^5 = u^5$.	Define u.
$\frac{du}{dx} = 6x + 7$ and $du = (6x + 7)dx$.	Differentiate u with respect to x, and find du in terms of x.
$\int (3x^2 + 7x)^5(6x + 7)dx = \int u^5 du$	Substitute to write an integral in terms of u.
$= \frac{1}{6}u^6 + C$	Integrate with respect to u.
$= \frac{1}{6}(3x^2 + 7x)^6 + C$	Substitute $3x^2 + 7x$ for u to give your final answer in terms of x.
	Check your answer by finding $\frac{d}{dx}\left[\frac{1}{6}(3x^2 + 7x)^6 + C\right]$.
b $\int x^4 \cos(x^5 + 3)dx$	$f(g(x)) = \cos(x^5 + 3)$, $g(x) = x^5 + 3$ and $g'(x) = 5x^4$, which is a constant multiple of x^4.

Continued on next page

Calculus

Let $u = x^5 + 3$, then $\cos(x^5+3) = \cos u$.	Define u.
$\frac{du}{dx} = 5x^4$ and $\frac{1}{5}du = x^4 dx$.	Differentiate u with respect to x, and find du in terms of x.
$\int x^4 \cos(x^5+3)\,dx = \frac{1}{5}\int \cos u\,du$	Substitute to write an integral in terms of u.
$= \frac{1}{5}\sin u + C$	Integrate with respect to u.
$= \frac{1}{5}\sin(x^5+3) + C$	Substitute $x^5 + 3$ for u to give your final answer in terms of x. Check your answer by finding $\frac{d}{dx}\left[\frac{1}{5}\sin(x^5+3)+C\right]$.
c $\int 2xe^{3x^2}dx$	$f(g(x)) = e^{3x^2}$, $g(x) = e^{3x^2}$ and $g'(x) = 6x$, which is a constant multiple of $2x$.
Let $u = 3x^2$, then $e^{3x^2} = e^u$,	Define u.
$\frac{du}{dx} = 6x$ and $\frac{1}{3}du = 2x\,dx$.	Differentiate u with respect to x, and find du in terms of x.
$\int 2xe^{3x^2}dx = \frac{1}{3}\int e^u du$	Substitute to write an integral in terms of u.
$= \frac{1}{3}e^u + C$	Integrate with respect to u.
$= \frac{1}{3}e^{3x^2} + C$	Substitute $3x^2$ for u to give your final answer in terms of x. Check your answer by finding $\frac{d}{dx}\left[\frac{1}{3}e^{3x^2}+C\right]$.
d $\int \frac{8x^3-15x^2}{2x^4-5x^3}dx = \int \frac{1}{2x^4-5x^3}\cdot(8x^3-15x^2)\,dx$	$f(g(x)) = \frac{1}{2x^4-5x^3}$, $g(x) = 2x^4 - 5x^3$ and $g'(x) = 8x^3 - 15x^2$.
Let $u = 2x^4 - 5x^3$, then $\frac{1}{2x^4-5x^3} = \frac{1}{u}$,	Define u
$\frac{du}{dx} = 8x^3 - 15x^2$ and $du = (8x^3 - 15x^2)\,dx$.	Differentiate u with respect to x, and find du in terms of x.
$\int \frac{8x^3-15x^2}{2x^4-5x^3}dx = \int \frac{1}{u}du$	Substitute to write an integral in terms of u. Integrate with respect to u.
$= \ln\lvert u\rvert + C$	Substitute $2x^4 - 5x^3$ for u to give your final answer in terms of x.
$= \ln\lvert 2x^4 - 5x^3\rvert + C$	Check your answer by finding $\frac{d}{dx}\left[\ln\lvert 2x^4-5x^3\rvert + C\right]$. Note: $\frac{d}{dx}\left[\ln\lvert x\rvert\right] = \frac{1}{x}$, the same as if the absolute value sign were not present.

The more you practise the substitution method, the easier it becomes to identity the function to select for u.

Exercise 13G

Find the indefinite integrals in questions **1–10**.

1 $\int (5x^3 + 4x)^2(15x^2 + 4)\,dx$

2 $\int 15x^4 \sin(3x^5)\,dx$

3 $\int \frac{4x+3}{(2x^2+3x+1)^2}\,dx$

4 $\int (2x+7)e^{x^2+7x}\,dx$

5 $\int (8x^3 - 12x)(x^4 - 3x^2)^3\,dx$

6 $\int \frac{e^{\sqrt{x}}}{\sqrt{x}}\,dx$

7 $\int (2x-1)\cos(2x^2 - 2x)\,dx$

8 $\int \sin x \cos^4(x)\,dx$

9 $\int \frac{\sin(\ln x)}{x}\,dx$

10 $\int x^2 e^{x^3} \sqrt{e^{x^3} + 5}\,dx$

11 The gradient of a curve is given by $f'(x) = e^{\sin x}\cos x$. The curve passes though the point $(\pi, 12)$.

Find an expression for $f(x)$.

12 Let $f'(x) = \frac{4x}{2x^2 + e^2}$.

Given that $f(0) = 5$, find $f(x)$.

Definite integrals and the substitution method

Now consider some definite integrals. If you use a u-substitution with a definite integral, it is often easier to change the limits of integration into limits in u, rather than to change the integrated expression in u back to the variable x and use the original limits.

Example 12

Find the exact value of the definite integral without using a GDC. When necessary, first rewrite the definite integral using a u-substitution.

Check your answers by finding the value of the definite integral on your GDC.

a $\int_0^{\frac{\pi}{6}} 3\sin x\,dx$ **b** $\int_0^1 (3x^2+1)^3(6x)\,dx$ **c** $\int_{\frac{\pi}{4}}^{\frac{\pi}{2}} \cos(x)\sin^2(x)\,dx$

a $\int_0^{\frac{\pi}{6}} 3\sin x\,dx = 3\int_0^{\frac{\pi}{6}} \sin x\,dx$	
$= -3[\cos x]_0^{\frac{\pi}{6}}$	Find the antiderivative.
$= -3\left[\cos\frac{\pi}{6} - \cos 0\right]$	Apply the fundamental theorem of calculus.
$= -3\left[\frac{\sqrt{3}}{2} - 1\right]$	Use unit circle values to evaluate.
$= \frac{6-3\sqrt{3}}{2}$	

Continued on next page

Using your GDC gives the same value

$\frac{6-3\sqrt{3}}{2} = 0.4019237886$

Check your answer using your GDC.

b $\int_0^1 (3x^2+1)^3(6x)\mathrm{d}x$

Use the substitution method.

$f(g(x)) = (3x^2+1)^3$, $g(x) = 3x^2+1$ and $g'(x) = 6x$.

Let $u = 3x^2+1$, then $(3x^2+1)^3 = u^3$ and

Define u.

$\frac{\mathrm{d}u}{\mathrm{d}x} = 6x$ so $\mathrm{d}u = (6x)\mathrm{d}x$.

Differentiate u with respect to x, and find $\mathrm{d}u$ in terms of x.

When $x = 0$, $u = 3(0)^2 + 1 = 1$.

Find the limits of integration in terms of u.

When $x = 1$, $u = 3(1)^2 + 1 = 4$.

$$\begin{aligned}\int_0^1 (3x^2+1)^3(6x)\mathrm{d}x &= \int_{x=0}^{x=1} u^3\mathrm{d}u \\ &= \int_{u=1}^{u=4} u^3\mathrm{d}u \\ &= \int_1^4 u^3\mathrm{d}u \\ &= \left[\frac{1}{4}u^4\right]_1^4 \\ &= \left(\frac{1}{4}\cdot 4^4\right) - \left(\frac{1}{4}\cdot 1^4\right) \\ &= \frac{255}{4}\end{aligned}$$

The limits of integration, 0 and 1, are in terms of x. If you express the limits in terms of u, you do not need to express the antiderivative in terms of x before evaluating.

Integrate and evaluate with respect to u.

Using your GDC gives the same value

$\frac{255}{4} = 63.75$

Check your answer using your GDC.

HINT

You will get the same answer if you change the antiderivative back to terms of x and evaluate with the original limits given in terms of x.

$$\left[\frac{1}{4}(3x^2+1)^4\right]_0^1 = \frac{1}{4}\left(3(1)^2+1\right)^4 - \frac{1}{4}\left(3(0)^2+1\right)^4 = \frac{255}{4}$$

c

$$\int_{\frac{\pi}{4}}^{\frac{\pi}{2}} \cos(x)\sin^2(x)\mathrm{d}x = \int_{\frac{\pi}{4}}^{\frac{\pi}{2}} \cos(x)(\sin(x))^2\mathrm{d}x$$

Use the substitution method where $f(g(x)) = (\sin(x))^2$, $g(x) = \sin x$, and $g'(x) = \cos x$.

Let $u = \sin x$, then $(\sin x)^2 = u^2$, and

Define u.

$\frac{\mathrm{d}u}{\mathrm{d}x} = \cos x$ so $\mathrm{d}u = (\cos x)\mathrm{d}x$.

Differentiate u with respect to x, and find $\mathrm{d}u$ in terms of x.

When $x = \frac{\pi}{4}$, $u = \sin\frac{\pi}{4} = \frac{\sqrt{2}}{2}$. When $x = \frac{\pi}{2}$, $u = \sin\frac{\pi}{2} = 1$.	Find the limits of integration in terms of u.
$\int_{\frac{\pi}{4}}^{\frac{\pi}{2}} \cos(x)(\sin(x))^2\,dx = \int_{\frac{\sqrt{2}}{2}}^{1} u^2 du$	The limits in the new integral are in terms of u.
$= \left[\frac{1}{3}u^3\right]_{\frac{\sqrt{2}}{2}}^{1} = \left(\frac{1}{3}\cdot 1^3\right) - \left(\frac{1}{3}\cdot\left(\frac{\sqrt{2}}{2}\right)^3\right) = \frac{4-\sqrt{2}}{12}$	Integrate with respect to u, and evaluate.
Using your GDC gives the same value $\frac{4-\sqrt{2}}{12} = 0.2154822031$	Check your answer using your GDC.

Recall that you can use definite integrals to find the area under and between curves.

HINT

There are techniques of integration beyond the scope of this course, so there are many functions you cannot yet integrate. You can, however, use technology to find the values of definite integrals of these functions.

Example 13

a Find the area of the region bounded by the curve $f(x) = e^{-x^2}$, the lines $x = -1$ and $x = 1$ and the x-axis.

b Find the area of the region bounded by the curves $f(x) = \sin x$ and $g(x) = 0.5x$.

a 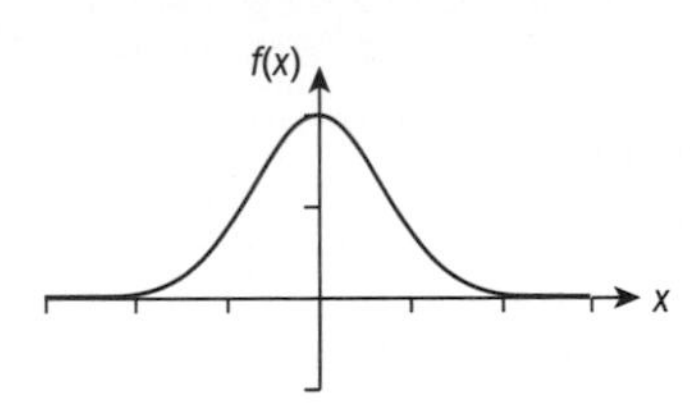 	Use a GDC to graph $f(x) = e^{-x^2}$. Notice that $f(x) > 0$ between $x = -1$ and $x = 1$. 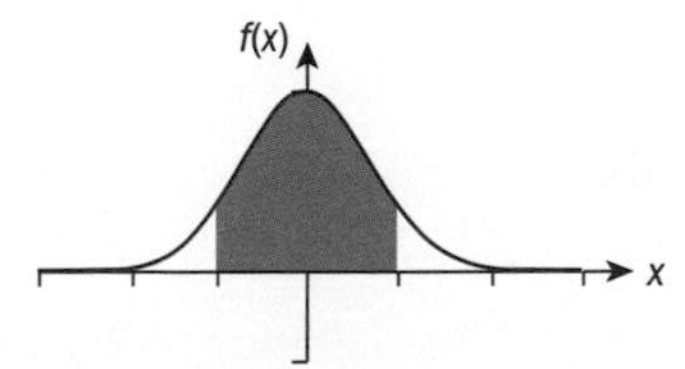
Area $= \int_{-1}^{1} e^{-x^2} dx$ Area $= 1.49$	Set up the definite integral for the area under the curve. Use a GDC to evaluate the integral (1.493648266). The answer is given correct to 3 significant figures.

Continued on next page

Calculus

b 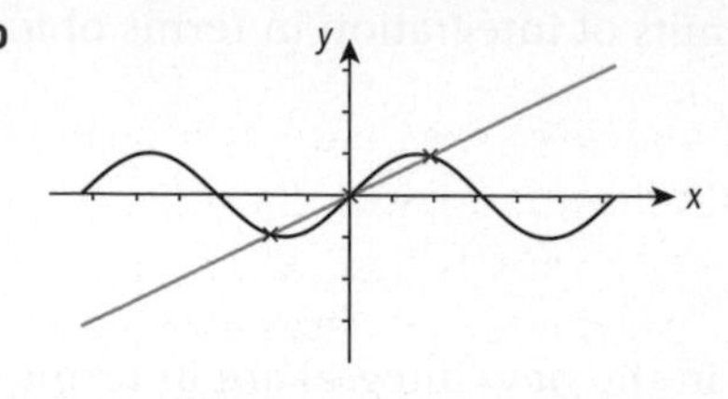	Use a GDC to graph $f(x) = \sin x$ and $g(x) = 0.5x$. Find the points of intersection, where $\sin x = 0.5x$. These occur at $x = -1.89549, 0, 1.89549$ and are the limits of integration. **HINT** Do not round the limits of integration too soon. You may be able to store them in your GDC.
$\text{Area} = \int_{-1.89549}^{0}\left(g(x) - f(x)\right)\text{d}x + \int_{0}^{1.89549}\left(f(x) - g(x)\right)\text{d}x$ or $\text{Area} = 2\int_{0}^{1.89549}(f(x) - g(x))\,\text{d}x$	Notice that in the third quadrant, $g(x) \geq f(x)$ and in the first quadrant $f(x) \geq g(x)$. Due to the symmetry of these graphs, you could use $2\int_{0}^{1.89549}(f(x) - g(x))\,\text{d}x$ to find the area. Use a GDC to evaluate the integral.
Your GDC gives the value 0.8415957901. Area = 0.842	The answer is given correct to 3 significant figures.

Exercise 13H

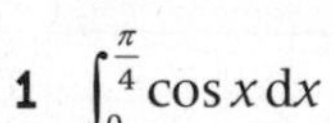

In questions **1** and **2**, find the exact value of the definite integral.

1 $\int_{0}^{\frac{\pi}{4}}\cos x\,\text{d}x$ **2** $\int_{\frac{\pi}{3}}^{\frac{5\pi}{6}}2\sin x\,\text{d}x$

In questions **3–6**, rewrite each definite integral using a u-substitution. Then find the exact value of the definite integral without using a GDC. Check your answer with a GDC.

3 $\int_{0}^{2}(x^2 + x)^3(2x + 1)\text{d}x$

4 $\int_{\frac{\pi}{6}}^{\frac{\pi}{3}}\sin x\cos^3 x\,\text{d}x$

5 $\int_{1}^{2}\frac{3x^2}{\sqrt{x^3+1}}\text{d}x$

6 $\int_{\ln\frac{\pi}{4}}^{\ln\frac{\pi}{3}}\text{e}^x\sin\left(\text{e}^x\right)\text{d}x$

7 A portion of the graph of $f(x) = e^x \sin x$ is shown in the diagram. The graph has x-intercepts at $x = 0$ and $x = k$.

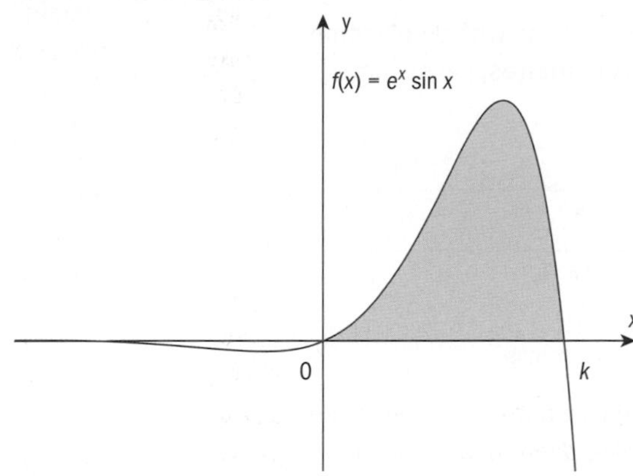

a Find the value of k.

b Find the area of the shaded region.

8 Find the area of the region bounded by the graphs of $f(x) = \sin x$ and $g(x) = -x^3 + 5x^2 - 4x$.

13.4 Kinematics and accumulating change

In chapter 5, you studied the relationship between derivatives and the kinematics of objects moving along a straight line.

Recall the following.

- s is the **displacement** of the particle **from a fixed origin** at time t.
- v is the **velocity** of the particle at time t and is given by $v(t) = s'(t)$.
- **Speed** is a scalar quantity, and is equal to $|v(t)|$.
- a is the **acceleration** of the particle at time t and is given by $a(t) = v'(t) = s''(t)$.

You also know that:

- Given the velocity function and a boundary condition, you can use an indefinite integral to find the displacement function.
- Given the acceleration function and a boundary condition, you can use an indefinite integral to find the velocity function.

In this chapter, you will learn how the **definite** integral is related to kinematics.

HINT

The displacement of the particle from a fixed origin is sometimes called the position of the particle.

Investigation 5

The motion of an object moving along a horizontal line with displacement $s(t) = 2t^3 - 21t^2 + 60t + 3$, for $0 \leq t \leq 7$ (s in metres, t in seconds), can be modelled by the following diagram.

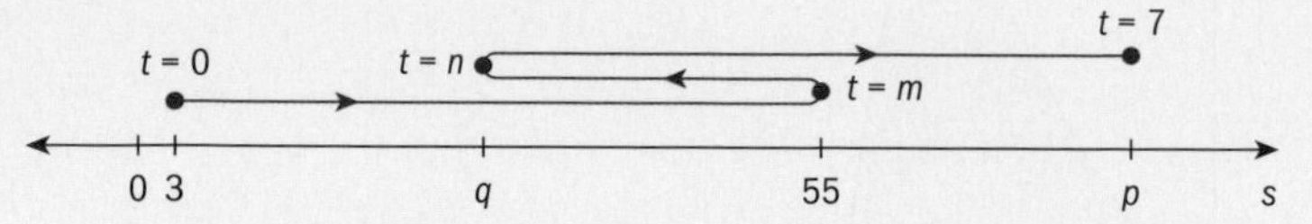

Motion of an object with displacement $s(t) = 2t^3 - 21t^2 + 60t + 3$, for $0 \leq t \leq 7$, where s is in metres and t is in seconds

The object is moving along line s. The purple curve above the line is used to show the direction the particle is moving and its position at key times.

1 Find the velocity function for the moving object.

2 Find the values of the following constants, shown in the motion diagram:

- **a** p, the displacement of the object from the origin **at** 7 seconds
- **b** m and n, the times when the object changes direction
- **c** q, the displacement of the object from the origin at n seconds

The displacement of the object over a time interval, $[t_1, t_2]$ is equal to $s(t_1) - s(t_2)$. This displacement describes the change is position from time t_1 to time t_2 in terms of direction and magnitude.

3 **a** Find $s(7) - s(0)$, the displacement of the object **after** 7 seconds. Describe what this value tells you about the position of the object at 7 seconds, relative to the position of the object at 0 seconds.

b Find the value of $\int_0^7 v(t)\,dt$ on your GDC and explain why this gives you the displacement of the object after 7 seconds.

4 **Factual** What does $\int_a^b v(t)\,dt$ represent?

5 Use the motion diagram to find the total distance the object travels over the time $0 \leq t \leq 7$.

6 **Conceptual** Is displacement the same as distance travelled? Why or why not?

7 Use your GDC to find $\int_0^7 |v(t)|\,dt$ and compare it to our answer in question **5**.

8 **Factual** What does $\int_a^b |v(t)|\,dt$ represent?

HINT

It might be helpful to look at the graph of $s(t)$ on your GDC. Compare the key features of the graph to the values you see in the motion diagram in this investigation.

The motion of an object moving along a horizontal straight line with position $s(t) = 2t^3 - 21t^2 + 60t + 3$, for $0 \leq t \leq 7$, where s is in metres and t is in seconds, is shown in the diagram below.

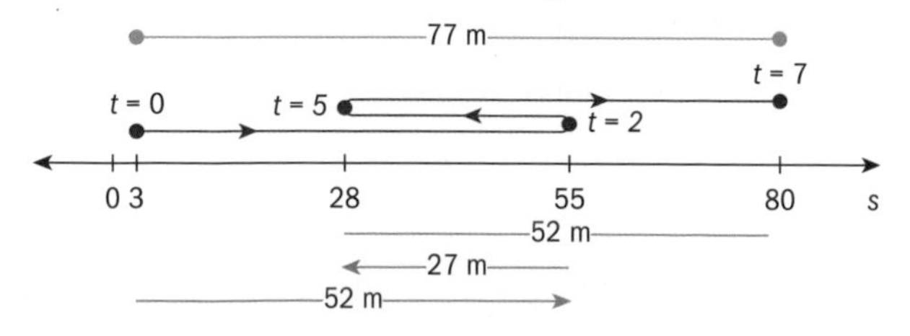

The displacement after 7 seconds is 77 m. This is the amount of change in position from 0 seconds to 7 seconds. The displacement is positive, so at 7 seconds the object is to the right of its initial position.

The total distance is the sum of 52 m travelled to the right, plus 27 m travelled to the left, plus 52 m travelled to the right; in total 131 m.

You can see that the displacement of the object over the time $0 \le t \le 7$ is not the same as the total distance travelled. Consider this in terms of area under the curve $v(t) = 6t^2 - 42t + 60$, where A_1, A_2 and A_3 represent the areas of the regions enclosed by the t-axis and the graph of v (to the left) or the graph of $|v|$ (to the right).

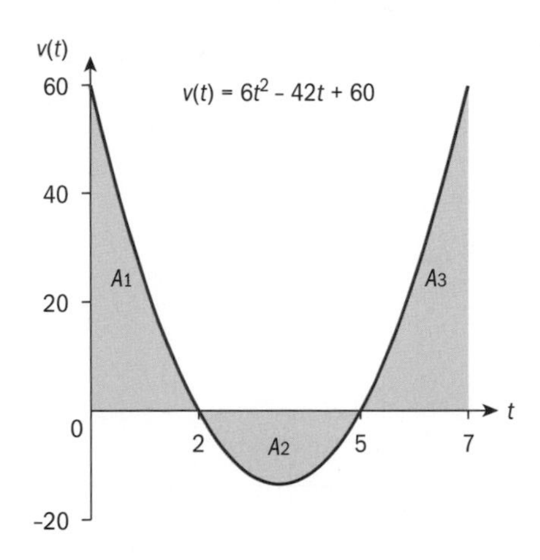

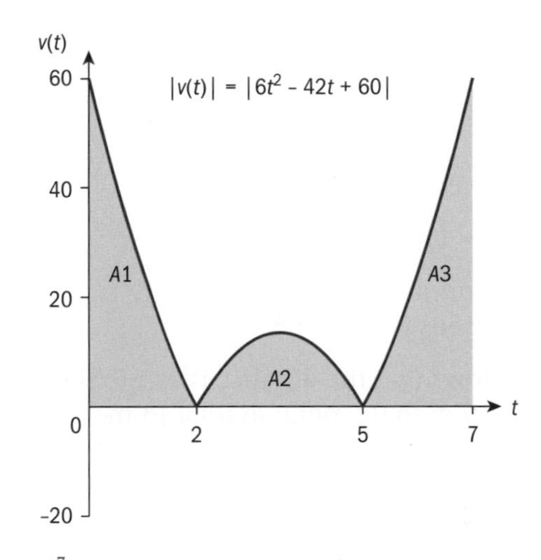

$\int_0^7 v(t)\,dt = s(7) - s(0) = A_1 - A_2 + A_3$

$=$ displacement from time 0 to time 7

$\int_0^7 |v(t)|\,dt = A_1 + A_2 + A_3$

$=$ total distance travelled from time 0 to time 7

If v is the velocity function for a particle moving along a straight line, then:

$\int_{t_1}^{t_2} v(t)\,dt = s(t_2) - s(t_1)$ is the **displacement** from t_1 to t_2.

$\int_{t_1}^{t_2} |v(t)|\,dt$ is the total **distance** travelled from t_1 to t_2.

Example 14

A particle moves along a straight line such that its displacement s in metres from an origin O is given by $s(t) = t^2 - 4t + 3$, for $0 \le t \le 5$, where t is time in seconds.

a Find the velocity of the particle at time t.

b Find when the particle is moving to the right and when it is moving to the left.

c Draw a motion diagram for the particle.

d Write definite integrals for the particle's displacement on the interval $0 \le t \le 5$ and for the distance travelled on the interval $0 \le t \le 5$. Use a GDC to find the value of the integrals. Use the motion diagram to verify the results.

a $v(t) = 2t - 4$	$v(t) = s'(t)$
b $2t - 4 = 0$ $t = 2$	Find when velocity equals 0. Then find when it is positive and when it is negative.

Continued on next page

Moving left for $0 < t < 2$ | $v(t) < 0$ on $[2, 5]$

Moving right for $2 < t < 5$ | $v(t) > 0$

c $s(0) = 3 \quad s(2) = -1 \quad s(5) = 8$

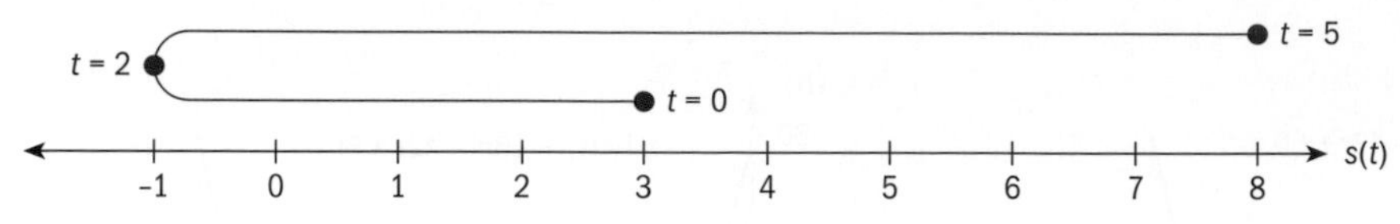

d displacement $= \int_0^5 (2t - 4)\,dt = 5\,\text{m}$

distance $= \int_0^5 |2t - 4|\,dt = 13$ m

The motion diagram shows that displacement is $8 - 3 = 5$ m.

It also shows that the distance is 4 m to the left, plus 9 m to the right; in total 13 m.

Use your GDC to evaluate the integrals $\int_0^5 (2x - 4)\,dx = 5$ and $\int_0^5 (|2x - 4|)\,dx =$ 12.99999796.

Example 15

The velocity, $v\,\text{m}\,\text{s}^{-1}$, of a particle moving along a line is shown in the figure on the right. Find the particle's displacement and the distance travelled over the interval $0 \le t \le 11$ seconds.

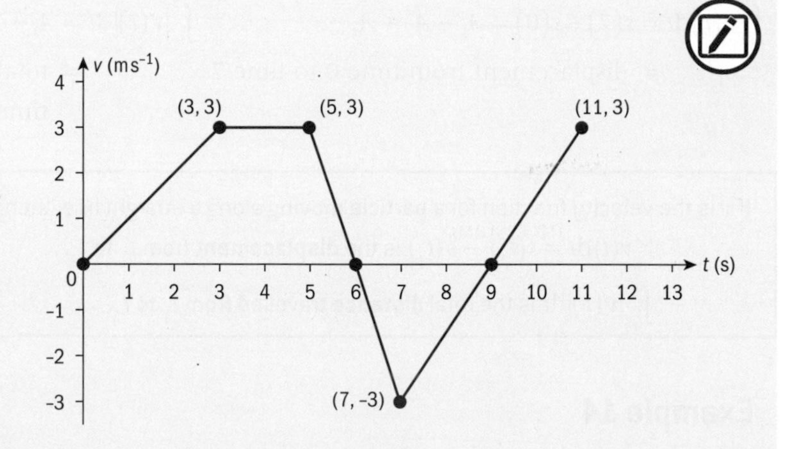

Let A_1, A_2 and A_3 represent the areas of the trapezoid and the two triangles.

Displacement

$= \int_0^{11} v(t)\,dt$

$= A_1 - A_2 + A_3$

$= \frac{1}{2}(6+2)(3) - \frac{1}{2}(3)(3) + \frac{1}{2}(2)(3)$

$= 10.5$ m

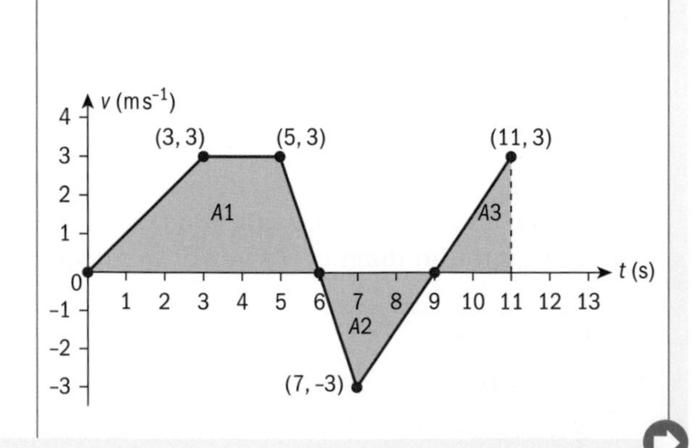

Distance

$= \int_0^{11} |v(t)| dt$

$= A_1 + A_2 + A_3$

$= \frac{1}{2}(6+2)(3) + \frac{1}{2}(3)(3) + \frac{1}{2}(2)(3)$

$= 19.5 \text{ m}$

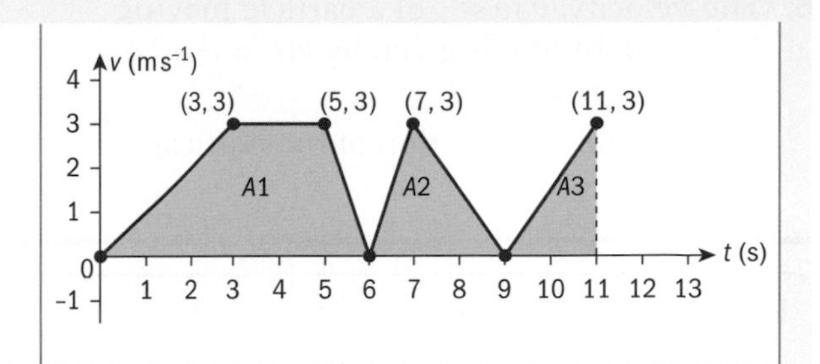

Reflect When will the total distance travelled be equal to the displacement for a given time interval?

Exercise 13I

Each of questions **1–3** gives a displacement function from a fixed origin for a particle moving along a straight line and a time interval, where t is measured in seconds and s is measured in metres.

a Find the velocity of the particle at time t.

b Write down a definite integral for the particle's displacement over the interval. Find the displacement.

c Write down a definite integral for the particle's distance travelled over the interval. Find the distance.

Use a GDC to evaluate definite integrals. You can also use a motion diagram to verify your results.

1 $s(t) = -\frac{1}{3}t^3 + 4t^2 - 12t + 2;\ 0 \le t \le 9$

2 $s(t) = t^2 - 6t + 8;\ 0 \le t \le 6$

3 $s(t) = (t-1)^3;\ 0 \le t \le 3$

4 The velocity, v m s^{-1}, of a particle moving along a line, for $0 \le t \le 14$, is shown in the following diagram.

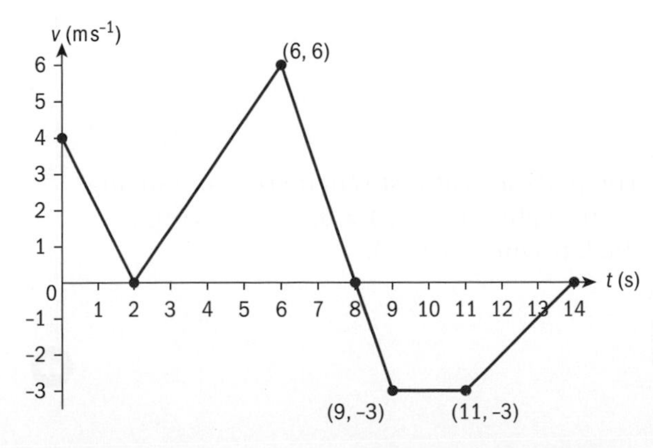

Write down a definite integral that represents the displacement and a definite integral that represents the total distance travelled for each of the following intervals. Use the diagram and concept of area to evaluate each integral.

a $0 \le t \le 8$

b $2 \le t \le 14$

c $0 \le t \le 14$

5 The velocity, v m s^{-1}, of a particle moving along a line, for $0 \le t \le 7$, is shown in the following diagram.

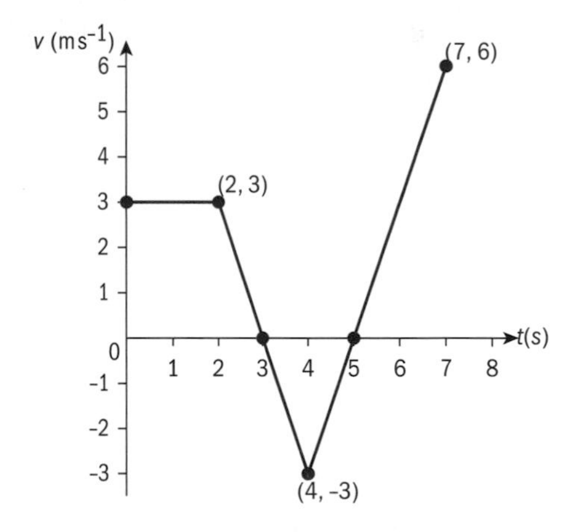

a Find the acceleration of the particle when $t = 3$.

b Write down the interval(s) on which the particle is travelling to the right.

c Write down a definite integral that represents the total distance travelled for $0 \le t \le 7$. Find this total distance.

6 The velocity, v m s^{-1}, of a particle moving in a straight line is given by $v(t) = t^2 - 16$, where $t \geq 0$ seconds.

a Find the acceleration of the particle at $t = 2$.

b The initial displacement of the particle is 10 m. Find an expression, s, for the displacement of the particle at time t.

c Find the distance travelled between times 2 seconds and 6 seconds.

HINT

When tackling questions, remember that:

- "Initially" $\Rightarrow$ at time 0
- "At rest" $\Rightarrow v(t) = 0$
- "Moving right" or "up" $\Rightarrow v(t) > 0$
- "Moving left" or "down" $\Rightarrow v(t) < 0$
- $v(t)$ and $a(t)$ have same signs $\Rightarrow$ speeding up
- $v(t)$ and $a(t)$ have different signs $\Rightarrow$ slowing down

The next examples of kinematics involve trigonometric functions.

Example 16

A particle moves along a horizontal line. The particle's displacement, in metres, from an origin at time t seconds is given by $s(t) = 4 - 2\sin 2t$.

a Find the particle's velocity and acceleration at any time t.

b Find the particle's initial displacement, velocity and acceleration.

c Find when the particle is at rest, moving to the right and moving to the left during the time $0 \leq t \leq \pi$.

d Find the displacement of the particle after π seconds.

a $v(t) = -4\cos 2t$ $a(t) = 8\sin 2t$	$v(t) = s'(t) = 0 - 2(\cos 2t)(2)$ $a(t) = v'(t) = -4(-\sin 2t)(2)$
b $s(0) = 4 - 2\sin(2(0))$ $= 4 - 2(0) = 4$ m $v(0) = -4\cos(2(0))$ $= -4(1) = -4$ m s^{-1} $a(0) = 8\sin(2(0))$ $= 8(0) = 0$ m s^{-2}	Evaluate each function at $t = 0$.
c $-4\cos 2t = 0$ $\cos 2t = 0$ $2t = \frac{\pi}{2}, \frac{3\pi}{2}$ $t = \frac{\pi}{4}, \frac{3\pi}{4}$	The particle is at rest when $v(t) = 0$, moving to the right when $v(t) > 0$, and moving to the left when $v(t) < 0$.

Working	Explanation
The particle is at rest at $\frac{\pi}{4}$ and $\frac{3\pi}{4}$ seconds, moving to the left for $0 < t < \frac{\pi}{4}$ and $\frac{3\pi}{4} < t < \pi$ seconds, and moving to the right for $\frac{\pi}{4} < t < \frac{3\pi}{4}$ seconds.	(see sign table below)
d $\int_0^{\pi} -4\cos(2t)\,dt = -4\int_0^{\pi}\cos(2t)\,dt$ $= \left[-4\left(\frac{1}{2}\right)\sin(2t)\right]_0^{\pi}$ $= -2\sin\pi - (-2)\sin 0$ $= 0\,\text{m}$	The displacement from time 0 seconds to time π seconds is equal to $\int_0^{\pi} v(t)\,dt$. $\int\cos(ax+b)\,dx = \frac{1}{a}\sin(ax+b)+C$ Note that a displacement of 0 m means the particle has returned to its initial position at π seconds.

t	$0 < t < \frac{\pi}{4}$	$\frac{\pi}{4} < t < \frac{3\pi}{4}$	$\frac{3\pi}{4} < t < \pi$
Sign of $v(t)$	negative	positive	negative

Example 17

A particle moves along a straight line so that its velocity, $v\,\text{m}\,\text{s}^{-1}$, at time t seconds is given by $v(t) = 4\sin t\cos^2 t$.

a When $t = 0$, the displacement, s, of the particle from a fixed origin is 5 m. Find an expression for s in terms of t.

b Find an expression for the acceleration, a, of the particle in terms of t.

c Find the acceleration, velocity and speed of the particle when $t = 6$.

d Determine whether the particle is speeding up or slowing down when $t = 6$.

e Find the displacement of the particle and the distance travelled over the interval $0 \le t \le 6$.

Working	Explanation
a Let $u = \cos t$, then $\frac{du}{dt} = -\sin t \Rightarrow -\sin t\,dt = du$ and $\cos^2 t = u^2$.	Integrate by substitution.
$\int 4\sin t\cos^2 t\,dt = -4\int u^2\,du$ $= -4\left(\frac{1}{3}u^3\right) + C$ $= -\frac{4}{3}\cos^3 t + C$	Integrate velocity to get displacement.
$5 = -\frac{4}{3}\cos^3(0) + C$ $5 = -\frac{4}{3}(1) + C$ $C = \frac{19}{3}$ So $s(t) = -\frac{4}{3}\cos^3 t + \frac{19}{3}$.	Use the fact that $s(0) = 5$ to find C.

Continued on next page

Calculus

b $a(t) = v'(t)$ $= 4\sin t[2\cos t(-\sin t)] + \cos^2 t(4\cos t)$ $= -8\sin^2 t\cos t + 4\cos^3 t$	Use the product and chain rules to find the derivative of velocity.
c $a(6) = 2.94\ \text{m s}^{-2}$ $v(6) = -1.03\ \text{m s}^{-1}$ speed $= \lvert v(6)\rvert = 1.03\ \text{m s}^{-1}$	Use a GDC to evaluate acceleration and velocity when $t = 6$. speed $= \lvert v(t)\rvert$
d The particle is slowing down when $t = 6$.	When velocity and acceleration have opposite signs, the particle is slowing down.
e displacement $= \int_0^6 v(t)\,dt$ $= 0.153$ m distance $= \int_0^6 \lvert v(t)\rvert\,dt$ $= 5.18$ m	Write down the definite integral for the displacement, $\int_{t_1}^{t_2} v(t)\,dt$, and for the distance, $\int_{t_1}^{t_2} \lvert v(t)\rvert\,dt$. Use a GDC to evaluate the definite integrals.

Exercise 13J

1 A particle moves along a straight line. Its displacement from an origin, in metres, is given by $s(t) = e^t \cos t$, for time t seconds.

- **a** Write down an expression for the velocity, v, in terms of t.
- **b** Write down an expression for the acceleration, a, in terms of t.

2 A particle moves along a straight line. Its displacement from an origin, in metres, is given by $s(t) = 6 - 2\sin t$, for time $t \geq 0$ seconds.

- **a** Calculate the velocity when $t = 0$.
- **b** Find the first time the particle is at rest.
- **c** Calculate the displacement of the particle from O the first time it is at rest.

3 The velocity, v m s^{-1}, of an object moving along a horizontal line at time t seconds is given by $v(t) = e^{\cos t}\sin t$.

- **a** **i** Find when the object is at rest during the interval $0 \leq t \leq 2\pi$.
  - **ii** Find when the object is moving to the left during the interval $0 \leq t \leq 2\pi$.
- **b** Find the object's acceleration, a, in terms of t.
- **c** The initial displacement s is 3e metres. Find an expression for s in terms of t.

4 An object moves along a straight line. Its velocity, v m s^{-1}, is given by $v(t) = 5\sin t + 2\cos t,\ t \geq 0$.

- **a** Find the displacement of the particle after 4 seconds.
- **b** Find the distance travelled by the particle after 4 seconds.

5 A particle moves along a straight line with velocity, v m s^{-1}, after t seconds given by

$$v(t) = -(t+2)\sin\left(\frac{t^2}{4}\right), \text{ where } 0 \leq t \leq 6.$$

- **a** **i** Find the acceleration when $t = 1.3$.
  - **ii** Determine whether the particle is speeding up or slowing down when $t = 1.3$.
- **b** Find all the times in the interval $0 < t < 6$ that the particle changes direction.
- **c** **i** Find the displacement of the particle at the end of the time period $0 \leq t \leq 6$.
  - **ii** Explain what this displacement tells you about the particle.
- **d** Find the total distance travelled by the particle during the time $0 \leq t \leq 6$.

6 The velocity, v m s^{-1}, of an object moving along a straight line at time t seconds is given by $v(t) = e^{3\cos t} - 2$, where t is measured in seconds, for $0 \le t \le 6$.

a Find the acceleration of the object at $t = 1$.

b Determine the value(s) of t when the particle has a velocity of 8 m s^{-1}.

c Find the displacement of the particle after 6 seconds.

d Find the distance travelled by the particle during the time interval $0 \le t \le 6$.

Accumulating rates of change

The velocity function, v, is the rate of change of the displacement function, s. Since $\int_{t_1}^{t_2} v(t)\,dt = s(t_2) - s(t_1)$, $\int_{t_1}^{t_2} v(t)\,dt$ represents the net change in displacement over the interval $[t_1, t_2]$. This concept can be extended to other functions that represent the rate of change of a quantity.

Reflect What does the definite integral of the rate of change of a quantity represent?

The definite integral of a quantity's rate of change between times t_1 and t_2 is equal to the net change in the quantity from t_1 to t_2.
This is, $\int_{t_1}^{t_2} r(t)\,dt = R(t_2) - R(t_1)$, where $r(t) = R'(t)$.

Example 18

A culture of bacteria is started with an inital population of 150 bacteria. The rate at which the number of bacteria changes over a one-month period can be modelled by the function $r(t) = e^{0.256t}$, where r is the rate of change in the number of bacteria on day t.

Find the population of bacteria 15 days after the culture was started.

First find the net change from day 0 to day 15. $\int_0^{15} e^{0.256t}\,dt = 178$ bacteria	The function $r(t) = e^{0.273t}$ is a rate of change, measured in bacteria per day. Let $R(t)$ be the number of bacteria on day t, then $R'(t) = r(t)$ Therefore $\int_0^{15} e^{0.273t}\,dt = R(15) - R(0)$ and is the change in the number of bacteria from day 0 to day 15.
$150 + \int_0^{15} e^{0.256t}\,dt = 328$ bacteria	The initial population of 150 bacteria is added to the net change in population over the 15 days to get the number of bacteria on day 15.

Calculus

In the previous example, you could have found the indefinite integral of $\int e^{0.256t}\,dt$, using the initial condition of 150 bacteria. This would have resulted in the function $R(t)$, where R is the number of bacteria on day t. Then evaluating $R(15)$ would have given the number of bacteria on day 15.

If you try this method, you will appreciate how much more convenient it is to use the definite integral of a rate of change to determine an accumulated amount. In these problems, it may even be necessary for you to use your calculator to find the value of the definite integral when you are faced with an indefinite integral that you do not know how to integrate.

Exercise 13K

Write an expression involving a definite integral that can be used to answer each question. Use a GDC to evaluate the expression.

1 The rate at which the number of spectators enter a stadium for a football game is modelled by the function $r(t) = 1025t^2 - t^3$ for $0 \le t \le 1.5$ hours. The function $r(t)$ is measured in people per hour. There are no spectators in the stadium when the gates open at $t = 0$ hours. The game begins at time $t = 1.5$ hours. How many spectators are in the stadium when the game begins?

2 There is 33.4 cubic feet of snow on a driveway at midnight. From midnight, when $t = 0$, to 8am snow accumulates on the driveway at a rate modelled by the function $s(t) = 5.2te^{(-0.05t^3 + 2.3)}$, where s is the number of cubic feet of snow accumulating per hour and t is the number of hours after midnight. How many cubic feet of snow are on the driveway at 8am?

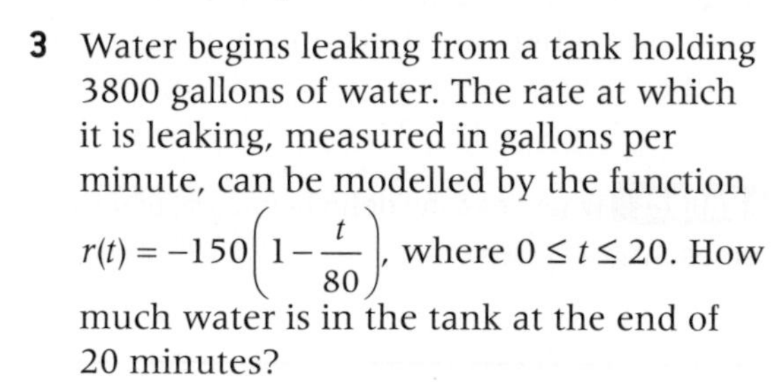

3 Water begins leaking from a tank holding 3800 gallons of water. The rate at which it is leaking, measured in gallons per minute, can be modelled by the function $r(t) = -150\left(1 - \frac{t}{80}\right)$, where $0 \le t \le 20$. How much water is in the tank at the end of 20 minutes?

4 The rate of consumption of oil in a certain country from 1 January 2008, to 1 January 2018 (in billions of barrels per year) is modelled by the function $C'(t) = 20.4e^{\frac{t}{18}}$, where t is the number of years since 1 January 2008. Find the country's total consumption of oil over the 10-year period.

Developing your toolkit

Now do the Modelling and investigation activity on page 580.

Chapter summary

- **Derivatives of sine and cosine**

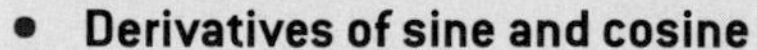

$f(x) = \sin x \Rightarrow f'(x) = \cos x$

$f(x) = \cos x \Rightarrow f'(x) = -\sin x$

- **Integrals of sine and cosine**

  $\int \sin x\,dx = -\cos x + C$

  $\int \cos x\,dx = \sin x + C$

  $\int \sin(ax+b)\,dx = -\frac{1}{a}\cos(ax+b) + C$

  $\int \cos(ax+b)\,dx = \frac{1}{a}\sin(ax+b) + C$

- **Kinematics**

  If v is the velocity function in terms of t, then over the time $[t_1, t_2]$:

  $\text{displacement} = \int_{t_1}^{t_2} v(t)\,dt$

  $\text{distance} = \int_{t_1}^{t_2} |v(t)|\,dt$

Developing inquiry skills

You have seen how the definite integral of a rate of change over a given time is equal to the net change in a quantity over that time period. Return to the opening scenario for this chapter and use this knowledge to find the amount of water in storage tank B at $t = 12$ hours. How could you find the minimum and maximum amounts of water in tank B during the time $0 \le t \le 24$ hours?

Chapter review

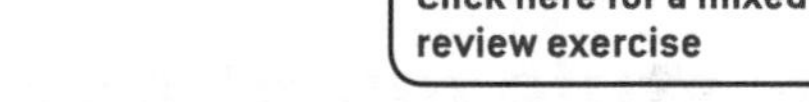

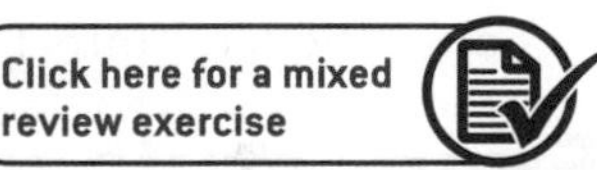

1 Find the derivative of each function.

a $f(x) = 3\sin x + 4\cos x$ **b** $y = \cos(3x - 4)$

c $h(t) = \sin^4 x$ **d** $f(x) = \sqrt{\cos x}$

e $y = (\sin x)(\ln x)$ **f** $y = \sin(\ln x)$

g $s(t) = \frac{\cos t}{e^t}$ **h** $f(x) = e^{2x}\sin 2x$

2 Find each indefinite integral.

a $\int (3x^4 + \cos x)\,dx$ **b** $\int \sin 4x\,dx$

c $\int \cos(2x + 3)\,dx$

d $\int (6x^2 + 5)(2x^3 + 5x)^4\,dx$

e $\int 3x^2\cos(x^3)\,dx$ **f** $\int (4x + 10)e^{x^2+5x}\,dx$

g $\int \frac{\cos(\ln x)}{x}\,dx$ **h** $\int \frac{2e^{4x}}{e^{4x} + 5}\,dx$

3 Evaluate each definite integral.

a $\int_{\frac{\pi}{2}}^{\frac{3\pi}{4}} \sin x\,dx$

b $\int_{-1}^{2} (3x^2 - 2)(x^3 - 2x)^3\,dx$

c $\int_0^{\frac{\pi}{6}} (1 + \cos x)\,dx$ **d** $\int_0^1 8xe^{4x^2+1}\,dx$

4 An open box has a height of h centimetres and a square base with length x centimetres. The surface area of the open box is 432 cm^2.

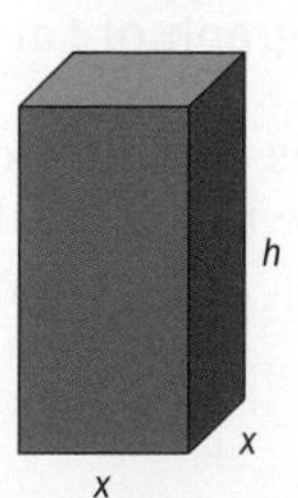

Calculus

a Show that $h = \frac{432 - x^2}{4x}$.

b Write an expression for the volume of the box in terms of x.

c Find the dimensions of the box, so that the volume will be a maximum.

5 A particle moves along a straight line. The particle's displacement, in metres, from an origin O is given by $s(t) = 3e^{\cos t} - 2$ for time t seconds.

a Find the particle's velocity and acceleration at any time t.

b Find when the particle is moving to the left during the time $0 \le t \le 2\pi$.

c Find when the velocity is increasing.

d Find the distance the particle travels for the time $0 \le t \le 2\pi$.

6 Water is dripping into a container over a ten-hour period. The height of the water in the container increases at a rate of $r(t) = 7t^2e^{-1.2t}$ centimetres per hour, where t is time in hours. If the height of the water in the container was 4.5 cm at the beginning of the ten-hour period, what was the height of the water after the ten hours?

Exam-style questions

7 **P1:** Consider the function g defined by $g(x) = \sin^2 x - 5x$.

a Find an expression for $g'(x)$. (2 marks)

b Hence, show that g is decreasing on all its domain. (3 marks)

8 **P2:** Consider the function f defined by

$f(x) = 5x - \tan x,\ 0 \le x < \frac{\pi}{2}$.

a Find the equation of the tangent to the graph of f at $x = \frac{\pi}{4}$. (5 marks)

b There is a point A on the graph of f where the normal to the graph is vertical. Determine the coordinates of A. (3 marks)

9 **P2:** Two particles, P_1 and P_2, are moving along a straight line. The position of the particles in relation to the origin t seconds after the start of the observation of the movement is given by $s_1(t) = \sin 2t$ and $s_2(t) = \sin(t - 0.24),\ 0 \le t \le 3$.

a Find an expression for the distance between the particles. (2 marks)

b Hence, determine the time when the particles will be furthest apart. (2 marks)

c Determine whether the particles will ever collide in the first 3 seconds of the movement. (2 marks)

10 **P1:** Consider the function f defined by $f(x) = x - \sin x,\ 0 \le x \le \pi$

a Find an expression for $\int f(x)\mathrm{d}x$. (3 marks)

b Hence, find the area of the region enclosed by the graph of f, the x-axis and the line $x = \pi$. (3 marks)

11 **P1:** Let $a(x) = e^x \cos x$.

a Use product rule to find $\frac{\mathrm{d}a}{\mathrm{d}x}$. (3 marks)

b Hence write an equation for the tangent to the graph at $x = 0$ (4 marks)

12 **P2:** Let $s(x) = \sin(1 + \sin 2x)$.

a Sketch the graph of $y = s(x)$ for $0 \le x \le \pi$. (3 marks)

b Hence,

i find the coordinates of all minimum and maximum points of s

ii state the range of s

iii state the y-intercept of $y = s(x)$. (6 marks)

c Let R be the region enclosed by both axes and the graph of $y = s(x)$.

i Write down an expression for the area of R.

ii Hence determine the area of the region R. (4 marks)

13 **P2:** Consider the function f defined by $f(x) = \sin 2x + \cos x$ for $-\frac{\pi}{2} \le x \le \frac{\pi}{2}$.

a Find

i the zero of f

ii the range of f

iii an expression for $f'(x)$.

(5 marks)

The diagram below shows the graph of f.

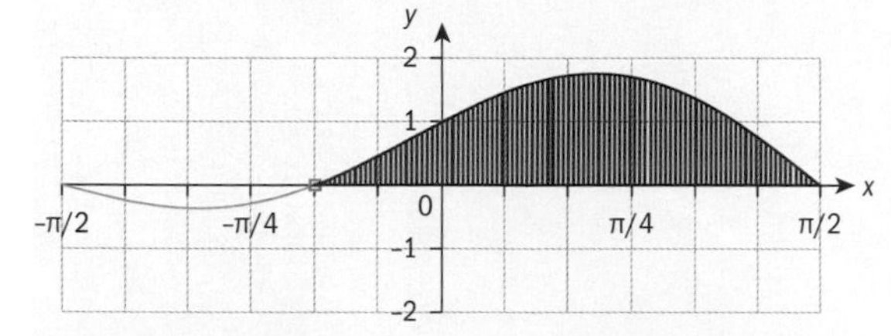

b Find the area of the region shaded.

(3 marks)

14 **P2:** A particle is moving along a straight line. The displacement from the origin, in millimetres, t seconds after the start of the movement is given by

$$s = 5\sin 3t + t^2 + 2,\ 0 \le t \le 3.$$

a State the initial displacement from the origin. (1 mark)

b Find an expression, in terms of t, for

i the velocity

ii the acceleration. (4 marks)

c Hence, find the times at which the displacement from the origin is at a maximum. (3 marks)

15 **P2:** Consider the function f defined by $f(x) = 5 + e^x \sin 2x$ for $0 \le x \le \pi$.

The graph has a maximum point at $A(a, f(a))$ and a minimum point at $B(b, f(b))$.

a Find **i** $f(0)$ **ii** $f(\pi)$.

(2 marks)

b Sketch the graph of f, showing clearly the points A and B.

(4 marks)

c Find $\int_a^b f(x)\,dx$. (2 marks)

d Hence, explain why the value found in part **c** does not represent the area of the region enclosed by the graph of f and the x-axis for $a \le x \le b$.

(2 marks)

e Hence, find the area of the region enclosed by the graph of f and the x-axis for $a \le x \le b$. (2 marks)

16 **P2:** Consider the function f defined by $f(x) = \sin\frac{x}{2}$ for $0 \le x \le 2\pi$.

a Show that $f(x) \ge 0$ for all $0 \le x \le 2\pi$. (2 marks)

Two points $A(\pi - x, a)$ and $B(\pi + x, b)$ are placed on the graph of f.

b Justify that $a = b$. (2 marks)

A rectangle of base AB is drawn such that the other base lies on the x-axis.

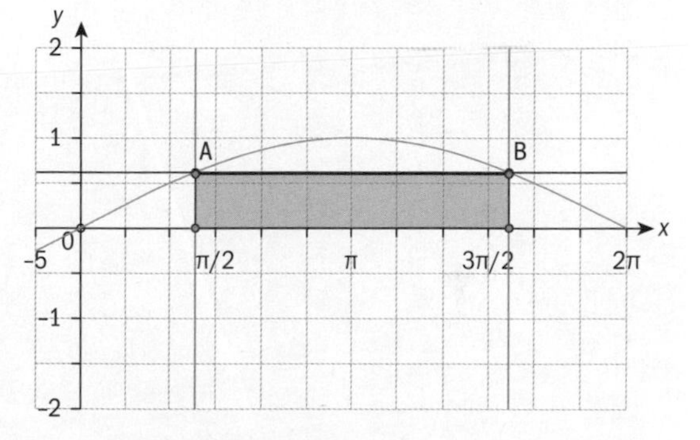

c Find the coordinates of A and B such that the rectangle has maximum area.

(6 marks)

d Hence, show that the perimeter of the rectangle found in part **c** is 8.19.

(2 marks)

Be the particle!

Approaches to learning: Collaboration, Communication
Exploration criteria: Personal engagement (C), Use of mathematics (E)
IB topic: Calculus, Kinematics–motion in a straight line

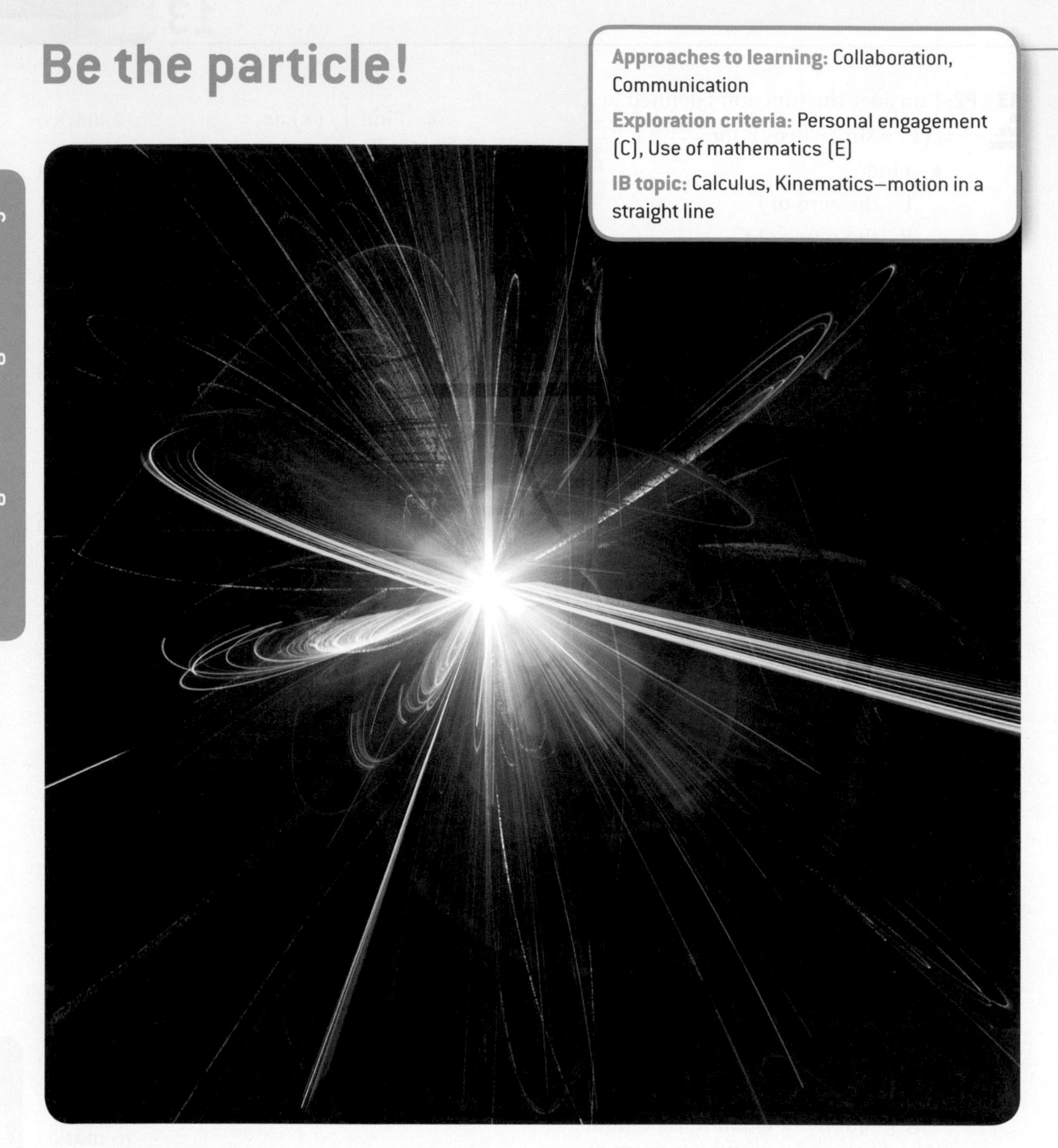

Motion of a particle in a straight line

In this task all times are in seconds and all distances are in "units".

A particle, P, moves with velocity function $v(t) = t^2 - 7t + 10$, $0 \leq t \leq 6$.

In your groups discuss:

- What is the initial speed of the particle?
- When is the particle stationary?
- At what velocity is the particle moving at the end of the journey?

The **displacement function** is $s(t)$.

In your groups discuss:

- At time $t = 0$, $s(t) = -5$. What does $s(t) < 0$ mean?
- What does $s(1)$ mean?
- What is the displacement function $s(t)$ of the particle?
- How far is the particle away from its starting point at the end of the journey?
- What is the particle's displacement at each second of the journey?

Find the **acceleration function** $a(t)$ of the particle.

In your groups discuss:

- What is the initial acceleration?
- When does the particle change direction?
- Find total distance travelled by the particle in the 6 second "journey".

Produce graphs of the displacement, velocity and acceleration against time for the particle's journey.

Be the particle for six seconds!

How could you walk to model the journey of the particle?

Discuss.

One member of your group should walk the "path of the particle" using the scale given on the board.

The other members should use a timer to advise the walker when to change direction, etc.

How could you check whether this is an accurate attempt?

Use a motion detector and/or a graphing progamme to fit a cubic curve to the displacement graph.

Compare this to the actual cubic curve for the displacement.

How similar is it to the cubic you produced previously?

What can you do to improve the model found?

Extension

Repeat the above experiment with another velocity function and initial displacement.

For example, try $v(t) = 2\sin\left(\frac{1}{2}t\right) + 1,\ 0 \le t \le 10$ and $s(0) = -2$.

Ensure here that all calculations are completed in radians.

You could also devise your own problem similar to the one in this task.

14 Valid comparisons and informed decisions: probability distributions

Concepts

- Representation
- Quantity

Microconcepts

- Discrete random variables and their probability distributions
- Expected value $\mathrm{E}(X)$ for discrete data
- Binomial distribution $X \sim \mathrm{B}(n, p)$
- Mean and variance of binomial distribution
- Normal distribution and curves $X \sim \mathrm{N}(\mu, \sigma^2)$
- Understanding the nature and occurrence of the normal distribution, properties of the normal distribution, diagrammatic representation
- Normal probability calculations
- Inverse normal calculations (with known mean or standard deviation, or with both mean and standard deviation unknown)
- Standardization of normal variables

Between 2008 and 2010, Paul the Octopus correctly predicted the outcome of 12 out of 14 football matches by approaching one of two boxes marked with the flag of the national teams playing. This is a success rate of almost 86%. Is this convincing? What is the probability that this would happen purely by chance?

Paul is not the only famous predicting animal. Punxsutawney Phil is a groundhog, immortalized in the film *Groundhog Day*. The legend is that if he sees his shadow on 2 February, six more weeks of winter weather lie ahead; no shadow indicates an early spring. Phil has been forecasting the weather on Groundhog Day for more than 130 years (although we suspect it is not the same groundhog!). Phil has a 39% success rate. Is this low? What success rate would you expect?

This chapter will be looking at situations such as these and will aim to determine what the probability of an event like this would be, since they really are a study of pure chance!

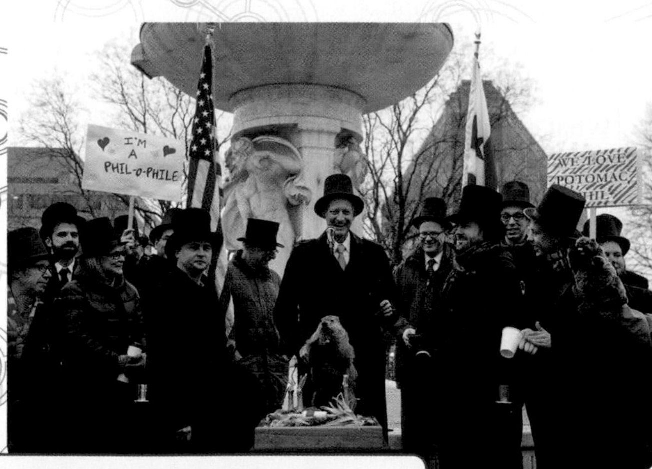

What would persuade you that an octopus could predict the outcomes of soccer matches? Is it just guessing? How about a groundhog predicting the weather?

Consider the following games.

You are initially given 30 tokens. The aim of the game is to collect as many tokens as possible. Which game should you play?

Developing inquiry skills

How can an understanding of games such as those above help us in real-world situations?

Research some common "cognitive biases" that can be overcome through an understanding of statistics and probability.

Before you start

You should know how to:

1 Find the mean of a set of numbers
eg Calculate the mean of the following frequency distribution of x.

x	0	1	2	3
Frequency	3	6	9	2

$$\bar{x} = \frac{\sum fx}{\sum f} = \frac{(0\times3)+(1\times6)+(2\times9)+(3\times2)}{3+6+9+2}$$

$$= \frac{30}{20} = 1.5$$

2 Use $\binom{n}{r}$ notation eg evaluate $\binom{5}{2}$

$\binom{5}{2} = 10$

3 Solve equations

eg Given $\frac{4}{x} = 3$, solve for x

$\frac{4}{x} = 3 \quad 4 = 3x \quad x = \frac{4}{3}$

Skills check

Click here for help with this skills check

1 Find the mean of the following frequency distributions of x:

a

x	3	4	5	6	7	8
Frequency	3	5	7	9	6	2

b

x	10	12	15	17	20
Frequency	3	10	15	9	2

2 Repeat question **1** above using your GDC.

3 Evaluate:

a $\binom{6}{2}$ **b** $\binom{8}{5}$ **c** $\binom{9}{6}(0.3)^3(0.7)^6$

4 Solve for x:

a $\frac{5.5}{x} = 3.2$ **b** $\frac{x-2.5}{1.2} = 0.4$

c $\frac{9-x}{0.2} = 1.6$

14.1 Random variables

Investigation 1

1 Write down four possible outcomes for each of these probability experiments.

  i A fair dice is rolled three times and the number of sixes is recorded.

  ii A researcher visits a prenatal clinic and records the number of babies that women are pregnant with.

  iii You measure the heights of a sample of students in your school.

  iv The time taken for a runner to complete the 100 m sprint is measured.

2 For each of the four probability experiments, write down one outcome that is impossible to obtain.

3 a What are the similarities between the types of possible outcome in experiments **i** and **ii**?

  b What are the similarities between the types of possible outcome in experiments **iii** and **iv**?

  c What are the differences between the types of possible outcome in experiments **i** and **ii**?

  d What are the differences between the types of possible outcome in experiments **i** and **iii**

In situations **i–iv** above, we could write:

X = the number of sixes obtained when a dice is rolled three times
B = the number of babies in a pregnancy
H = the height of students in a school
T = the time taken for a runner to complete the 100 m

This investigation will show you that there are two basic types of random variables:

Discrete random variables—These have a finite or countable number of possible values (such as X and B above).

A discrete random variable does not necessarily need to take just positive integer values. (For example, shoes sizes of a set of students could have possible values of ... 4, 4.5, 5, 5.5, 6, 6.5, ...)

Continuous random variables—These can take on any value in some interval (such as H and T above).

4 Categorize each of the following random variables as continuous or discrete:

  a A is "the age in completed years of the next person to call me on my phone"

  b B is "the length of the next banana I buy when shopping"

  c C is "how many cats I will see before the first white one"

  d D is "the diameter of the donuts in the cafeteria"

5 Discuss any issues you might have if actually collecting this data in each case.

TOK

"Those who have knowledge, don't predict. Those who predict, don't have knowledge."—Lao Tzu.

Why do you think that people want to believe that an outside influence such as an octopus or a groundhog can predict the future?

6 **Factual** What are specific examples of continuous and discrete random variables?

7 **Conceptual** What are the distinguishing features of continuous and discrete random variables?

A **random variable** is a quantity whose value depends on the outcome of a probability experiment. Random variables are represented using letters.

Random variables are represented by capital letters. The actual measured values which the random variable can take are represented by lower case letters.

For example, if X represents the number of sixes obtained when a dice is rolled 3 times,we would therefore write $P(X = x)$ to represent "the probability that the number of sixes is x" where x can take the values 0, 1, 2 and 3.

TOK

Is it possible to reduce all human behaviour to a set of statistical data?

Investigation 2

Consider a game in which we are going to need the difference between two dice when they are rolled in order to determine the number of spaces we move forward on the game board.

1 What possible values are there for the difference in dice scores?

The rules of the dice game are:

Roll two dice. If the scores on the two dice:

- are equal, then move forward 4 spaces
- differ by one or two, then move forward 2 spaces
- differ by three or more, then move forward 1 space

We will let M be the random variable representing the number of places moved per roll of the dice.

2 What possible values can M take?

We want to calculate the following probabilities.

$P(M=1), P(M=2), P(M=4)$

3 **Factual** Which of the following methods, which you learned in chapter 8 (Probability), would be best to use in order to calculate these probabilities?

Venn diagram, tree diagram, list of outcomes, sample space diagram

4 Using the method you identified in question **3**, determine

$P(M=1)$
$P(M=2)$
$P(M=4)$

Continued on next page

Statistics and probability

5 Complete the following table which represents the possible number of moves and the associated probabilities.

m	1	2	4
$P(M=m)$			

This table is called the **probability distribution** for the discrete random variable, M.

Note we often replace $P(M=m)$ in the table with just $P(m)$.

A probability distribution is a list of each possible value the random variable can take, together with the probability that when the experiment is run it will have that value.

6 Add up the probabilities in the table in question **5**. What do you notice?

7 **Conceptual** What does the probability distribution of a discrete random variable add up to? Why is this the case?

Example 1

Let X be the random variable that represents the number of sixes obtained when a dice is rolled three times.

Tabulate the probability distribution for X.

The four possible values that X can take are 0, 1, 2 and 3.

Using a tree diagram to help, we can find the values of $P(X=0)$, $P(X=1)$, $P(X=2)$ and $P(X=3)$.

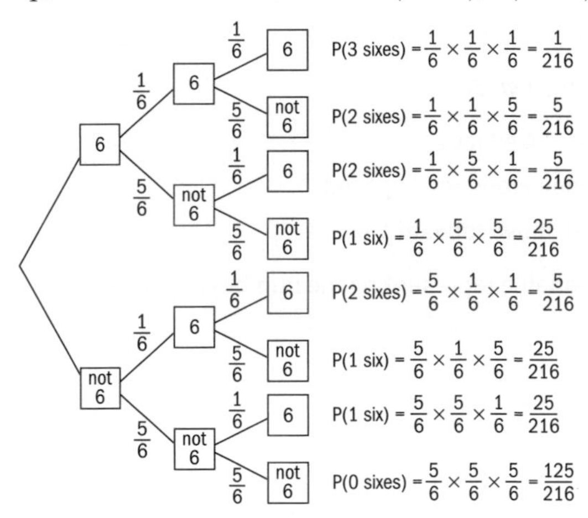

These probabilities can be represented in the following table giving the probability distribution for X:

x	0	1	2	3
$P(X=x)$	$\frac{125}{216}$	$\frac{25}{72}$	$\frac{5}{72}$	$\frac{1}{216}$

Again we can observe that $\frac{125}{216}+\frac{25}{72}+\frac{5}{72}+\frac{1}{216}=1$.

For any random variable X, note these two important facts.

$$0 \le P(X=x) \le 1$$
$$\sum P(X=x) = 1$$

The first of these states that a probability will always be a value from 0 to 1 (inclusive).

The second states that the sum of the probabilities will always be 1. This important fact is used in the next example.

Example 2

The random variable X has the probability distribution shown.

x	1	2	3	4	5
$P(X=x)$	$7c$	$5c$	$4c$	$3c$	c

a Find the value of c.

b Find $P(X \ge 4)$.

a $7c + 5c + 4c + 3c + c = 1$ $20c = 1$ $c = \frac{1}{20}$	Using $\Sigma P(X = x) = 1$
b $P(X \ge 4) = P(X = 4) + P(X = 5)$ $= \frac{3}{20} + \frac{1}{20}$ $= \frac{4}{20} = \frac{1}{5}$	

Exercise 14A

1 Tabulate the probability distribution for each random variables:

a the sum of the faces when two ordinary dice are thrown

b the number of twos obtained when two ordinary dice are thrown

c the smaller number when two ordinary dice are thrown

d the product of the faces when two ordinary dice are thrown.

2 A fair six-sided dice has a "1" on one face, a "2" on two of its faces and a "3" on the remaining three faces.

The dice is thrown twice, and T is the random variable "the sum of scores from both throws".

Find:

a the probability distribution of T

b the probability that the total score is more than 4.

3 A board game is played by moving a counter S spaces forward at a time, where S is determined by the following rule:

A fair six-sided dice is thrown once. S is half the number shown on the dice if that number is even; otherwise S is twice the number shown on the dice.

a Write out a table showing the possible value of S and their probabilities.

b Find the probability that, in a single turn, a player moves their counter forward more than 2 spaces.

4 The random variable X has the probability distribution shown.

x	1	2	3	4
$P(X=x)$	$\frac{1}{3}$	$\frac{1}{3}$	c	c

a Find the value of c.

b Find $P(1 < X < 4)$.

5 The probability distribution of a random variable Y is given by:

$$P(Y = y) = cy^3 \text{ for } y = 1, 2, 3$$

Given that c is a constant, find the value of c.

6 The random variable X has the probability distribution shown.

x	−1	0	1	2
$P(X=x)$	$2k$	$4k^2$	$6k^2$	k

Find the value of k.

7 The random variable X has the probability distribution given by $P(X = x) = k\left(\frac{1}{3}\right)^{x-1}$ for $x = 1, 2, 3, 4$, where k is a constant.

Find the exact value of k.

8 The discrete random variable X can take only the values 0, 1, 2, 3, 4, 5. The probability distribution of X is given by the following

$P(X = 0) = P(X = 1) = P(X = 2) = a$

$P(X = 3) = P(X = 4) = P(X = 5) = b$

$P(X \geq 2) = 3P(X < 2)$

where a and b are constants.

a Determine the values of a and b.

b Determine the probability that the sum of two independent observations from this distribution exceeds 7.

9 The discrete random variables A and B are independent and have the following distributions.

a	1	2	3
$P(A=a)$	$\frac{1}{3}$	$\frac{1}{3}$	$\frac{1}{3}$

b	1	2	3
$P(B=b)$	$\frac{1}{6}$	$\frac{2}{3}$	$\frac{1}{6}$

The random variable C is the sum of one observation from A and one observation from B.

a Show that $P(C = 3) = \frac{5}{18}$.

b Tabulate the probability distribution for C.

Expectation

Investigation 3

Recall the game involving rolling two dice from investigation 2:

Roll two dice. If the scores on the two dice:

- are equal, then move forward 4 spaces
- differ by one or two, then move forward 2 spaces
- differ by three or more, then move forward 1 space

M represents the number of places moved per roll of the dice.

TOK

What does it mean to say that mathematics can be regarded as a formal game lacking in essential meaning?

1 Use your results from investigation 2 to complete the following probability distribution for M.

m	1	2	4
$P(M=m)$			

2 The dice are rolled 36 times. Complete the following table to show the expected frequency of obtaining each of the different values of m.

m	1	2	4
Expected frequency			

3 Calculate the mean of this frequency distribution.

4 **Factual** What does the mean of this frequency distribution tell you?

5 The dice are now rolled 100 times. Complete the following table to show the expected frequency of obtaining each of the different values of m.

m	1	2	4
Expected frequency		$\frac{250}{9}$	

6 Calculate the mean again of the frequency distribution now.

7 **Factual** What do you notice from your answers to questions **3** and **6**?

8 **Factual** What would the mean value of M be if the experiment were repeated 10 times? Or 1000 times? Or just once?

9 **Conceptual** What does the mean of a frequency distribution (for multiple events) or the mean of a probability distribution (for just one event) tell you about the random variable?

Investigation 3 shows you that the mean value of M is the same, regardless of the number of trials. Therefore, we can find the mean or **expected value** $E(M)$ of the random variable M, by finding the sum of each value m, multiplied by its respective probability $P(M = m)$. This is equivalent to conducting the experiment just once.

$$E(M) = \sum_m mP(M = m)$$

The **mean** or **expected value** of a random variable X is the average value that we should expect for X over many trials of the experiment.

Expectation is actually the mean of the underlying distribution (the parent population). So it is often denoted by μ.

Example 3

In example 1, X was the random variable that represented the number of sixes obtained when a dice was rolled three times (which is also the number of sixes when three dice are rolled once).

Here is the probability distribution from example 1.

x	0	1	2	3
P(X=x)	$\frac{125}{216}$	$\frac{25}{72}$	$\frac{5}{72}$	$\frac{1}{216}$

Find the expected value of X.

Now

$$E(X) = \left(0 \times \frac{125}{216}\right) + \left(1 \times \frac{25}{72}\right) + \left(2 \times \frac{5}{72}\right) + \left(3 \times \frac{1}{216}\right)$$

$$E(X) = \frac{1}{2}$$

Therefore if three dice are rolled a large number of times, we should expect the mean number of sixes to be 0.5.

Note that the expected value of X does not need to be a value of X that is actually obtainable.

Using the calculator

The calculator can be used to find the expected value of a random variable.

Exercise 14B

1 When throwing a normal dice, let X be the random variable defined by

X = the square of the score shown on the dice.

What is the expectation of X?

2 A "Fibonacci dice" is unbiased, six-sided and labelled with these numbers: 1, 2, 3, 5, 8, 13.

What is the expected score when the dice is rolled?

3 The discrete random variable X has probability distribution $P(x) = \frac{x}{36}$ for $x = 1$, 2, 3, …, 8.

Find E(X).

4 For the discrete random variable X, the probability distribution is given by

$$P(X = x) = \begin{cases} kx & x = 1,2,3,4,5 \\ k(10 - x) & x = 6,7,8,9 \end{cases}$$

Find:

a the value of the constant k

b E(X).

5 Complete this probability distribution, in terms of k, for a discrete random variable X.

X	1	2	3
P(X=x)	0.2	$1 - k$	

a What range of values can k take? Give your answer in the form $a \le k \le b$, $a, b \in \mathbb{Q}$.

b Find the mean of the distribution in terms of k.

6 Ten balls of identical size are in a bag. Two of the balls are red and the rest are blue. Balls are picked out at random from the bag and are not replaced.

Let R be the number of balls drawn out, up to and including the first red one.

- **a** List the possible values of R and their associated probabilities.
- **b** Calculate the mean value of R.
- **c** What is the most likely value of R?

7 Consider the bag of balls in question **8**. Suppose now that each ball is replaced before the next is drawn.

- **a** Show that the probability of the first red being drawn out on the second go is $\frac{4}{25}$.
- **b** Calculate the probability that the first red is drawn after the third go.
- **c** Derive a formula to find the probability that the first red is drawn on the nth go.
- **d** What is the most likely value of R?

8 Suppose an instant lottery ticket is purchased for \$2. The possible prizes are \$0, \$2, \$20, \$200 and \$1000. Let Z be the random variable representing the amount won on the ticket, and suppose Z has the following distribution.

z	0	2	20	200	1000
$P(Z = z)$		0.2	0.05	0.001	0.0001

- **a** Determine $P(Z = 0)$.
- **b** Determine $E(Z)$ and interpret its meaning.
- **c** How much should you expect to gain or lose on average per ticket?

Developing inquiry skills

In the opening problem, you were faced with a number of games. Consider the first one.

10 tokens to roll 1 dice

If you get a 6 win **30 tokens**
If you get a 5 or 4 get your tokens back
If you get a 1, 2 or 3 you get nothing!

If you played this game a few times, do you think you would end up with more money than you started with?

How could we tell?

Define a random variable for the winnings from playing the game once.

Draw a probability distribution for your random variable.

Calculated the expected winnings from the game.

How much would you expect to win or lose if you played the game 10 times? 100 times?

Is the game "fair"?

How could you define a fair game?

What adjustment could you make to the prizes that would ensure that this is a fair game?

TOK

Do you think that people from very different backgrounds are able to follow mathematical arguments, as they possess deductive ability?

14.2 The binomial distribution

Investigation 4

1 Work with a classmate.

- Take five different cards—they could be five different cards from a deck of cards, or five cards with letters A–E, or five cards with five different shapes or pictures.
- Sit back to back with your partner. One of the pair will choose a card from the five, and the other person must try, without looking, to guess which card they have chosen.
- Replace the card, and repeat this process five more times.
- For each of the six cards chosen, record whether or not the guesser chose correctly using a table like this.

Trial number	Card chosen	Card guessed	Correct?
1			
2			
3			
4			
5			
6			

- Afterwards, swap roles so that the other person has a chance to guess.

For a single guess:

- Let C be the event that the person guesses the card correctly.
- Let W be the event that the person guesses the card incorrectly.

2 State the probabilities $\mathrm{P}(C)$ and $\mathrm{P}(W)$.

Let X be the discrete random variable "the number of cards predicted correctly out of the six attempts".

3 Write down the values that the random variable X can take.

4 What is the probability that, over the six attempts, the person guessing would get none of the predictions correct? (That is, what is the probability that $\mathrm{P}(X=0)$?)

$X=0$ can be written as the chain of guesses $WWWWWW$, as there are six incorrect guesses.

5 Ali plays this game and guesses the correct card in one of the six attempts. Write down six possible chains of guesses, involving C and W, that could represent Ali's game.

6 Find the probability for each possibility you listed in the previous question. Are they all the same?

7 Hence, state the probability $\mathrm{P}(X=1)$.

8 In a similar way, can you calculate $\mathrm{P}(X=2)$?

You might notice that $P(X=1)=6\times(P(C))^1\times(P(W))^5$

9 Can you write a similar expression for $P(X=2)$ in terms of $P(C)$ and $P(W)$?

10 Complete the following table.

- In column 1: Write each of the values, x, that X can take.
- In column 2: Write an expression to help you calculate the probability $P(X=x)$, such as $P(X=1)=6\times(P(C))^1\times(P(W))^5$
- In column 3: Write the numerical value of $P(X=x)$.

x	$P(X=x)$	
0		
1		
2		
3		
4		
5		
6		

11 In each row of the table, what do the powers in your expression for $P(X=x)$ (in column 2) add up to? How does this correspond to the total number of guesses a student makes?

12 Can you think of a theorem, which you have already learned in this course, that would help you to find the number of combinations of guessing r cards correctly out of six guesses?

The game you studied in the previous experiment can be modelled by what is known as a **binomial distribution.**

In every binomial distribution:

a There is a fixed number (n) of trials.

(In this case there were six guesses made.)

b Each trial has two possible outcomes—a "success" or a "failure".

(Here, "success" was predicting the correct card and "failure" was predicting the wrong card.)

c The probability of a success (p) is constant from trial to trial.

(In this case, the probability of success was $\frac{1}{5}$ *because there were 5 cards.)*

d Trials are independent of each other.

(In this case, if you get one prediction right it does not mean that you are more or less likely to get the next prediction right.)

Reflect How can you tell if a real-life situation can be modelled using the binomial distribution?

TOK

A model might not be a perfect fit for a real-life situation, and the results of any calculations will not necessarily give a completely accurate depiction. Does this make it any less useful?

You performed a binomial experiment in investigation 4. The outcomes of a **binomial experiment,** and the corresponding probabilities of these outcomes, are called a **binomial distribution.**

Since the parameters that define a unique binomial distribution are the values of n (the number of trials) and p (the probability of a success), the binomial distribution for a variable X is represented as $X \sim \mathrm{B}(n, p)$.

Investigation 5

Determine the probability of getting two heads in three tosses of a biased coin for which $\mathrm{P(Head)} = \frac{2}{3}$.

You may recall the following problem from chapter 8, where you answered this question using a tree diagram.

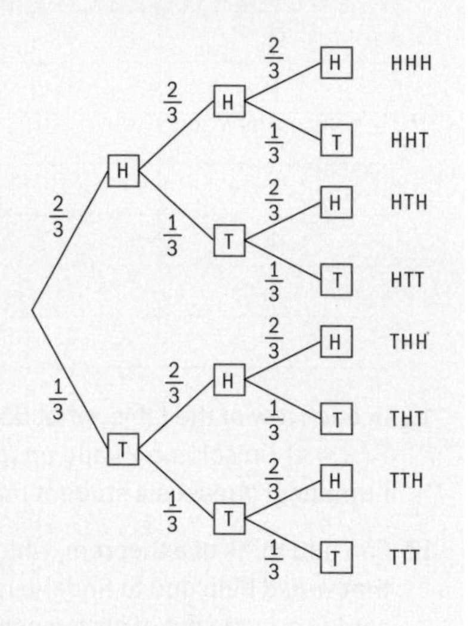

There are three ways of getting two heads in three tosses: HHT, HTH, THH.

So, P(Two heads in three tosses) = P(HHT) + P(HTH) + P(THH).

Each of the three probabilities are the same:

$$\mathrm{P(HHT)} + \mathrm{P(HTH)} + \mathrm{P(THH)} = \left(\frac{2}{3}\right)^2\left(\frac{1}{3}\right) = \frac{4}{27}$$

And so $\mathrm{P(Two\ heads\ in\ three\ tosses)} = 3 \cdot \left(\frac{2}{3}\right)^2\left(\frac{1}{3}\right) = \frac{12}{27} = \frac{4}{9}$.

Suppose, now, that you are asked to find the probability of getting two heads in six tosses of the coin.

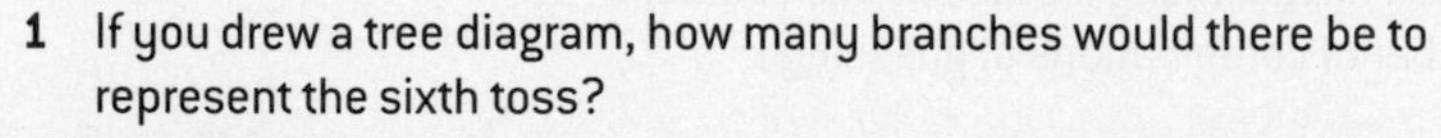

1 If you drew a tree diagram, how many branches would there be to represent the sixth toss?

2 **Factual** What would be the issue with drawing a tree diagram to represent this situation?

In this investigation, you will look for a formula that will allow you to calculate probabilities for multiple tosses of a coin.

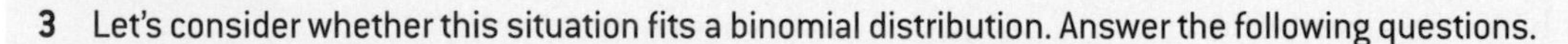

3 Let's consider whether this situation fits a binomial distribution. Answer the following questions.

   a Are there a fixed number (n) of trials? If so, how many?

   b Does each trial have two possible outcomes? What is a "success" and what is a "failure" for this experiment?

   c Is the probability (p) of a "success" constant from trial to trial? What is the value of p?

   d Are outcomes of trials independent from each other?

   Hence, can this experiment be modelled by a binomial distribution?

One combination of Hs and Ts that will produce two heads and six tails is HHTTTT.

4 What is P(HHTTTT)?

5 List the other combinations of Hs and Ts that will produce two heads and four tails. How many possible combinations are there?

6 Explain why every possible combination of two heads and four tails has the same probability.

In Chapter 1 you looked at the binomial expansion and were introduced to $\binom{n}{r} = {}_nC_r$.

Recall that $\binom{n}{r}$ represents the number of ways of choosing r items out of n items.

Recall that to calculate $\binom{n}{r}$ you can use:

1 your GDC, or

2 the formula $\binom{n}{r} = \frac{n!}{r!(n-r)!}$, or

3 the rth entry on the nth row of Pascal's triangle.

In this situation, n can represent the number of throws of the coin and r can represent the number of heads obtained.

7 Find how many different combinations of two heads you can get in six throws. Is your answer the same as that in question **5**?

8 Use your answers to questions **6** and **7** to find P (2 heads in 6 tosses)

9 **Conceptual** For a discrete random variable which is binomially distributed, how does the binomial theorem help to calculate probabilities?

10 **Conceptual** Why is using a tree diagram not always efficient?

Generalizing the method you used in the previous investigation for other values of n and r gives us the **binomial distribution function**, which is defined as follows.

If X is a discrete random variable which is binomially distributed, $X \sim B(n, p)$, then the probability of obtaining r successes out of n independent trials, when p is the probability of success for each trial, is

$$P(X = r) = \binom{n}{r} p^r (1-p)^{n-r}.$$

This is often shortened to $P(X = r) = \binom{n}{r} p^r q^{n-r}$ where $q = 1 - p$.

International-mindedness

The Galton board, also known as a quincunx or bean machine, is a device for statistical experiments named after English scientist Sir Francis Galton. It consists of an upright board with evenly spaced nails or pegs driven into its upper half, where the nails are arranged in staggered order, and a lower half divided into a number of evenly-spaced rectangular slots. In the middle of the upper edge, there is a funnel into which balls can be poured. Each time a ball hits one of the nails, it can bounce right or left with the same probability. This process gives rise to a binomial distribution of in the heights of heaps of balls in the lower slots and the shape of a normal or bell curve.

Investigation 6

1 Let $r = 2$ and $p = 0.5$.

Choose different values of n.

What happens to the probability $P(X = r)$ as value of the parameter, n, changes? Explore this and record your results in a suitable table.

2 Let $r = 2$ and $n = 5$.

Choose different values of p.

What happens to the probability $P(X = r)$ as the value of the parameter, p, changes? Explore this and record your results in a suitable table.

3 **Conceptual** How do the parameters of the binomial distribution affect the resulting probabilities?

4 **Conceptual** How do the parameters affect the graph of the binomial distribution?

Example 4

X is a binomially distributed discrete random variable which represents the number of successes in six trials. The probability of success in each trial is $\frac{1}{5}$.

What is the probability of

a exactly four successes **b** at least one success?

$X \sim \text{B}\left(6, \frac{1}{5}\right)$	Use the information given to write the distribution of X.
a $\text{P}(X=4) = \binom{6}{4}\left(\frac{1}{5}\right)^4\left(\frac{4}{5}\right)^2$ $= 15 \times \frac{1}{625} \times \frac{16}{25}$ $= \frac{48}{3125}$ $= 0.01536$ $= 0.0154$ (3 s.f.)	In part **a**, you need to find $\text{P}(X=4)$. Use the formula for binomial probability and simplify.
b $\text{P}(X \ge 1) = 1 - \text{P}(X=0)$ $= 1 - \binom{6}{0} \times \left(\frac{1}{5}\right)^0\left(\frac{4}{5}\right)^6$ $= 1 - \left(\frac{4}{5}\right)^6$ $= 1 - \frac{4096}{15\,625}$ $= \frac{11\,529}{15\,625}$ $= 0.738$ (3 s.f.)	$\text{P}(X \ge 1)$? To have at least one success, we require $\text{P}(X \ge 1)$. It is quicker to calculate $1 - \text{P}(X=0)$ than it is to calculate $\text{P}(X=1) + \text{P}(X=2) + \ldots + \text{P}(X+6)$. Use the formula for binomial probability and simplify. **EXAM HINT** A question like example 4 would most likely be on paper 2, and it would therefore be possible to use the GDC to speed up calculations.

Exercise 14C

1 X is a binomially distributed discrete random variable which represents the number of successes in four trials. The probability of success in each trial is $\frac{1}{2}$.

Without the use of a GDC, find the probabilities

a $\text{P}(X=1)$ **b** $\text{P}(X<1)$

c $\text{P}(X \le 1)$ **d** $\text{P}(X \ge 1)$

2 If $X \sim \mathrm{B}\left(6, \frac{1}{3}\right)$ find, correct to 3 significant figures, the values of:

a $\mathrm{P}(X = 2)$

b $\mathrm{P}(X < 2)$

c $\mathrm{P}(X \le 2)$

d $\mathrm{P}(X \ge 2)$

3 X is a binomially distributed discrete random variable which represents the number of successes in eight trials. The probability of success in each trial is $\frac{2}{7}$. Find the probability of:

a exactly five successes

b less than five successes

c more than five successes

d at least one success.

Example 5

On any morning, the probability that Sam takes a bus to work is 0.4. Find the probability that, in a working week of five days, Sam takes the bus to work twice.

$\mathrm{P}(X = 2) = \binom{5}{2}(0.4)^2(0.6)^3$	Let X be the number of days Sam takes the bus to work.
$= 10 \times 0.16 \times 0.216$	Because there are 5 working days, $n = 5$.
$= 0.3456$	The probability of "success" on any day is 0.4.
$= 0.346$ (3 s.f.)	Therefore, $X \sim \mathrm{B}(5, 0.4)$
	We require $\mathrm{P}(X = 2)$, so $r = 2$.

Example 6

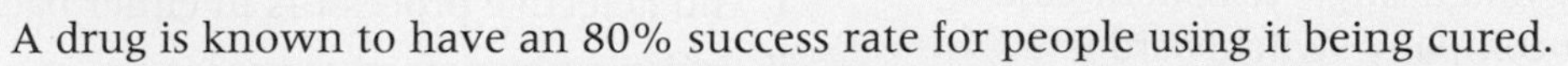

A drug is known to have an 80% success rate for people using it being cured.

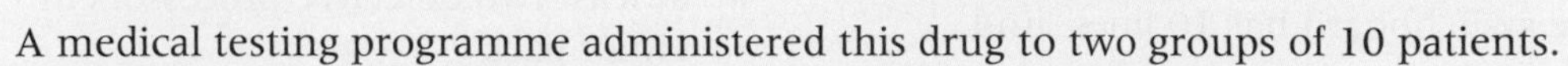

A medical testing programme administered this drug to two groups of 10 patients.

Find the probability that all 10 patients were cured in both groups.

You can assume that X is binomially distributed random variable since there are two outcomes: a "success" is a cure; and a "failure" is "not a cure".

We can assume that the trial results from patient to patient are independent, with a fixed probability of success of 0.8.

Let X be "the number of patients cured in a group of 10".	
Now $\mathrm{P}(X = 10) = \binom{10}{10}(0.8)^{10}(0.2)^0$	$n = 10$ because there are 10 trials of the drug.
$= 0.8^{10}$	$r = 10$ because we require all 10 patients to be cured.
$= 0.10737...$	

Continued on next page

So for all 10 patients to be cured in both groups, we require: $[P(X=10)]^2 = (0.10737...)^2$ $= 0.0115$ (3 s.f.)	We can multiply the probabilities $P(X=10)$ and $P(X=10)$ here because the two events ("the patients being cured in group A" and "the patients being cured in group B") are independent.

Exercise 14D

1 A regular tetrahedron has three white faces and one red face. It is rolled four times and the colour of the bottom face is noted. Find the most likely number of times that the red face will end downwards. Calculate the probability of this value occurring.

2 The probability that a marksman scores a bull's eye when he shoots at a target is 0.55.

Find the probability that in eight attempts:

a He **hits** the bull's eye five times.

b He **misses** the bull's eye at least five times.

3 A factory has four machines making the same type of component. The probability that any machine will produce a faulty component is 0.01. Find the probability that, in a sample of four components from each machine:

a None will be faulty.

b Exactly 13 will be not be faulty.

c At least two will be faulty.

4 The probability that a single telephone line is engaged at a company switchboard is 0.25. If the switchboard has 10 lines, find the probability that:

a One half of the lines are engaged.

b At least three lines are free.

5 The probability that Nicole goes to bed at 7.30pm on a given day is 0.4. Calculate the probability that, on five consecutive days, Nicole goes to bed at 7.30pm on at most three days.

6 In an examination hall, it is known that 15% of desks are wobbly.

a What is the probability that, in a row of six desks, more than one will be wobbly?

b What is the probability that exactly one desk will be wobbly in a row of six desks?

TOK

How can we trust the data collected from humans?

7 In the mass production of computer processors, it is found that 5% are defective. Processors are selected at random and put into packs of 15.

a If a pack is selected at random, find the probability that it will contain:

i three defective processors

ii no defective processors

iii at least two defective processors.

b Two packs are selected at random. Find the probability that there are:

i no defective processors in either pack

ii at least two defective processors in either pack

iii no defective processors in one pack and at least two in the other.

Example 7

A box contains a large number of carnations, $\frac{1}{4}$ of which are red. The rest are white.

Carnations are picked at random from the box. How many carnations must be picked so that the probability that there is at least one red carnation among them is greater than 0.95?

Let X be the random variable "the number of red carnations".	
Then $X \sim B(n, 0.25)$	State the distribution of X. n is the number of carnations picked. This is what we need to find.
$P(X \geq 1) > 0.95$	We require $P(X \geq 1) > 0.95$ in order that probability of *at least one* red carnation is greater than 0.95.
$1 - P(X = 0) > 0.95$	Since $P(X \geq 1) = 1 - P(X = 0)$
$1 - \binom{n}{0}(0.25)^0 (0.75)^n > 0.95$	Apply the formula for binomial probability.
$1 - (0.75)^n > 0.95$ $0.05 > (0.75)^n$	
$\log 0.05 > n \log 0.75$	Solve using logarithms, as in Chapter 9.
$\frac{\log 0.05}{\log 0.75} < n$ $10.4 < n$	Since $\log 0.75 < 0$, we divide through by a negative and the inequality sign changes direction.
At least 11 carnations must be picked out of the box to ensure that the probability that there is at least one red carnation among them is greater than 0.95.	$n > 10.4$, so the least value of n is 11. Note: You can also solve this type of problem by using a graph or table on your GDC.

Exercise 14E

1. If $X \sim B(n, 0.6)$ and $P(X < 1) = 0.0256$, find n.
2. 1% of fuses in a large box of fuses are faulty. What is the largest sample size that can be taken if it is required that the probability that there are no faulty fuses in the sample is greater than 0.5?
3. If $X \sim B(n, 0.2)$and $P(X \geq 1) > 0.75$ find the least possible value of n.
4. The probability that Anna scores a penalty goal in a hockey competition is 0.3. Find the least number of penalty attempts that she would need to make if the probability that she scores a goal at least once is greater than 0.95.
5. Find the least number of times an unbiased coin must be tossed so that the probability that at least one tail will occur is at greater than or equal to 0.99.

Statistics and probability

Expectation of a binomial distribution

Investigation 7

Consider again the biased coin you studied in investigation 5. In this example, $P(H) = \frac{2}{3}$.

Continued on next page

1 Suppose you toss the coin three times. How many times would you expect to get a head?

Previously, you learned that the expectation of any discrete random variable is $E(X) = \sum_x xP(X = x)$. Let X be the random variable "the number of heads obtained in three tosses of this biased coin".

2 Use the probabilities you calculated in investigation 5 to complete the table below.

x	$P(X=x)$	$x \cdot P(X=x)$
0		
1		
2		
3		

3 Now calculate $\sum_x xP(X = x)$ and hence state $E(X)$, the expected value of X.

4 **Conceptual** What is the relationship between the number of trials, the probability of success and the expected value for a binomial distribution?

For the binomial distribution where $X \sim B(n, p)$, the expectation of X is $E(X) = np$.

Example 8

A biased dice is thrown 30 times and the number of sixes seen is eight.

If the dice is thrown a further 12 times find the expected number of sixes.

Let X be random variable "the number of sixes seen in 12 throws". Then $X \sim B(12, p)$ where $p = \frac{8}{30} = \frac{4}{15}$	X is a binomially distributed random variable with $n = 12$, because you want the expected number of sixes in the next 12 throws. Since you are not told the value of p, you must calculate it using the experimental data given.
So $E(X) = np = 12 \times \frac{4}{15} = 3.2$.	Use the formula for the expectation of a binomially distributed random variable.

Exercise 14F

1 **a** A fair coin is tossed 40 times. Find the expected number of heads.

b A fair dice is rolled 40 times. Find the expected number of sixes.

c A card is drawn from a pack of 52 cards, noted and returned. This is repeated until 40 cards have been drawn. Find the expected number of hearts.

2 X is a random variable such that $X \sim B(n, p)$. Given that the mean of the distribution is 10 and $p = 0.4$, find n.

3 A multiple choice test has 15 questions, and each question has four possible answers. There is only one correct answer per question.

If X is "the number of questions a student guesses correctly", find:

a the distribution of X

b the expectation of X

c the probability that, if a student answers the test purely by guessing, they will achieve the pass mark of 10 or more.

4 A group of 100 families, each with three children, are found to have the following number of girls.

Number of girls	0	1	2	3
Frequency	13	34	40	13

a Using this sample of 100 families to represent the population, estimate the probability that a new baby born is a girl.

Another sample of 100 families, each with three children, is taken from the same population.

b Calculate the number of families in this sample that you would expect to have two girls.

The variance of a binomial distribution

Remember that, when handling data, the variance and standard deviation are a measure of how spread out the data is around the mean.

In the same way, you can calculate the variance of a discrete random variable. The full method to do this is beyond the scope of this course, but you will be expected to calculate the variance of a binomially distributed random variable.

The formula for the variance of a binomial distribution with $X \sim B(n, p)$ is given by $\text{Var}(X) = \sigma^2 = npq = np(1 - p)$.

Statistics and probability

Example 9

A biased dice is thrown 50 times and the number of sixes seen is 12.

If the dice is thrown a further 12 times, find:

a the probability that a six will occur exactly twice

b the expected number of sixes

c the variance of the number of sixes.

Let X be defined as "the number of sixes in 12 throws". $p = \frac{12}{50} = 0.24$ So $X \sim B(12, 0.24)$	X is a binomially distributed random variable with $n = 12$, because you want the expected number of sixes in the next 12 throws. Since you are not told the value of p, you must calculate it using the experimental data given.

Continued on next page

a $P(X=2)=\binom{12}{2}(0.24)^2(0.76)^{10}$ $=0.244$ (3 s.f.)	Using $P(X=r)=\binom{n}{r}p^r(1-p)^{n-r}$
b $E(X) = 12 \times 0.24 = 2.88$	Using $E(X) = np$
c $\text{Var}(X) = 12 \times 0.24 \times 0.76$ $= 2.19$ (3 s.f.)	Using $\text{Var}(X) = np(1-p)$

Exercise 14G

1. X is a discrete random variable where $X \sim B(n, p)$, $E(X) = 12$ and $\text{Var}(X) = 3$.

   Find n and p.

2. A multiple choice test has 20 questions. There are five possible answers to each question, where only one is correct.

   If X is "the number of questions a student guesses correctly", find:

   a the distribution of X

   b the mean and variance of X

   c the probability that, if a student answers the test purely by guessing, they will achieve the pass mark of 10 or more.

3. A random variable X has the distribution $B(12, p)$.

   Given that the variance of X is 1.92, find the possible values of p.

Developing inquiry skills

Consider now the second option from the original problem.

If you played this game a few times, do you think you would end up with more money than you started with?

How could we tell?

Define a random variable for the winnings from playing the game once.

Does this experiment fit a binomial distribution?

What are the parameters?

What is the probability that you will win 100 tokens? Your tokens back? 10 tokens? 0 tokens?

Draw a probability distribution table for your random variable representing the winnings.

Calculated the expected winnings from the game.

How much would you expect to win or lose if you played the game 10 times? 100 times?

Is the game "fair"? Explain why, or why not.

What adjustment could you make to the prizes that would ensure that this is a fair game?

14.3 The normal distribution

Investigation 8

The heights, in cm, of 70 grade 12 students at a school are recorded.

152	160	164	166	169	171	173	175	178	183
155	160	165	166	170	171	173	176	179	184
157	162	165	168	170	171	173	177	180	185
157	163	165	168	170	172	174	177	181	186
158	163	165	168	170	172	174	178	181	188
158	163	166	168	170	172	174	178	183	188
159	164	166	169	171	173	175	178	183	191

1. **Factual** Using words, describe how you would expect the heights of the 70 students to be distributed.
2. Represent this data using a histogram.
3. **Factual** Where is the peak of the histogram? What does the peak of the histogram represent?
4. Is the histogram roughly symmetrical? Where is the line of symmetry?
5. Join the midpoints of the tops of the bars of your histogram with a smooth curve.
6. Describe the shape of the curve that you have drawn.
7. Think of other data you could collect that is likely to have this shape.
8. Collect at least 50 pieces of data from students in your school (for example, weights, maximum hand span, length of foot, circumference of wrist) or some other set of data that you could represent in the same way. Repeat questions **1** to **5** above.
9. **Conceptual** How can you approximate the distribution of many real-life events that involve continuous data?

If more measurements of heights were taken in the previous investigation, a histogram plotted and the midpoints of the tops of the bars joined with a curve, it would likely become more symmetrical and bell-shaped until it would look like the curve below.

A **normal distribution** is roughly symmetrical about a central value, and the curve is bell-shaped, with the majority of measurements grouped around the central value.

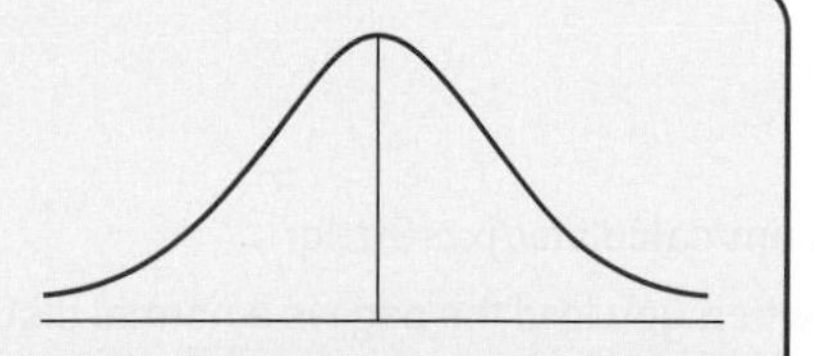

The **normal distribution** is one of the most important distributions in statistics since it is found to be a suitable model to represent many naturally occurring random variables which are distributed in this way. Such variables include the physical attributes of people, animals and plants; and even the specifications of mass-produced items from factories. The distribution could also be applied as an approximation of, for example, student exam scores, time to complete a piece of work, reaction times or IQ scores.

International-mindedness

The normal curve is also known as the Gaussian curve, and is named after the German mathematician, Carl Friedrich Gauss (1777–1855), who used it to analyse astronomical data. This is seen on the old 10 Deutsche Mark notes.

Investigation 9

Consider the data set of student heights given in investigation 8 (or, instead, you could use the data you collected in question **8** of investigation 8).

1. Calculate the mean, median and mode of this data set.
2. **Factual** What do you notice about the values of the mean, median and mode?
3. **Conceptual** Can you explain why your observation from question **2** is the case? Will this always be the case for a normally distributed data set?
4. **Conceptual** What are the characteristics of a perfect normal distribution?

For a perfect normal distribution:

- The curve is bell-shaped.
- The data is symmetrical about the mean (μ).
- The mean, mode and median are the same.

The characteristics of any normal distribution

There is no single normal curve, but a **family of curves.** Each normal curve is defined by its mean, μ, and its standard deviation, σ.

Recall that the mean, μ, is the "average" of the data, and the standard deviation, σ, is a measure of the data's spread.

μ and σ are called the **parameters** of the distribution. We can write that $X \sim \mathrm{N}(\mu, \sigma^2)$ to say that a random variable X is normally distributed with a mean of μ, and a standard deviation of σ.

Note that in the expression $X \sim \mathrm{N}(\mu, \sigma^2)$, it is actually the variance that is written (since variance is the square of standard deviation).

Investigation 10

Go to https://www.desmos.com/calculator/jxzs8fz9qr

The first curve you will see when you load the page is a normal distribution curve with a mean ($\mu = b$) of 0 and a standard deviation ($\sigma = a$) of 1.

This is called the **standard normal distribution curve.** The standard normal distribution is the normal distribution where $\mu = 0$ and $\sigma = 1$, and we write $Z \sim N(0, 1)$.

1 Where is the line of symmetry on the graph?

2 Change the value of b (the mean) and keep the value of the standard deviation the same. What happens to the line of symmetry of the graph?

The mean represents the central point of the distribution.

These diagram shows $X_1 \sim N(5, 2^2)$, $X_2 \sim N(10, 2^2)$ and $X_3 \sim N(15, 2^2)$.

Here, the standard deviations are all the same, but $\mu_1 < \mu_2 < \mu_3$.

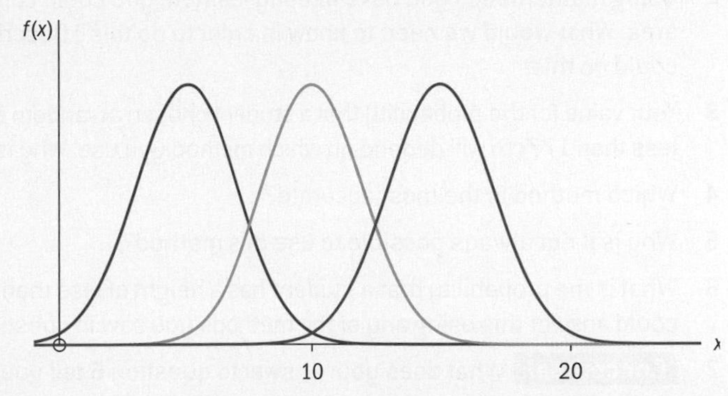

3 Label the curves X_1, X_3 and X_3.

The standard deviation describes the spread of the distribution. The higher the standard deviation, the wider the normal curve will be.

Go back to the online Desmos normal distribution curve. Change the value of a (the standard deviation) and keep the value of the mean the same.

4 Describe what happens to the shape of the graph.

These three graphs show $X_1 \sim N(5, 1^2)$, $X_2 \sim N(5, 2^2)$ and $X_3 \sim N(5, 3^2)$.

Here , the means are all the same, but $\sigma_1 < \sigma_2 < \sigma_3$.

5 Label the curves X_1, X_3 and X_3.

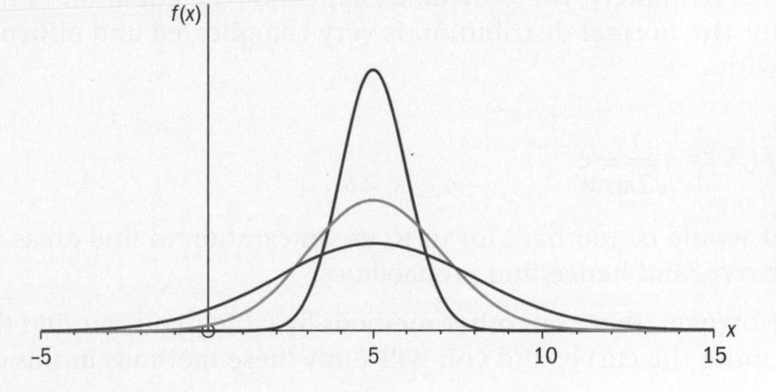

6 **Conceptual** How do the parameters of a normal distribution affect the shape and position of the curve?

The area beneath the normal distribution curve

Investigation 11

Consider the data set of student heights in investigation 8.

1 Calculate the probability that a student chosen at random has a height less than 177cm. Do this in each of these three different ways:

You should try could calculate this by either:

a Count the number of students less than 177 cm in the raw data list, and use this to find the probability.

Continued on next page

Statistics and probability

b Using your histogram, add up the areas of the rectangles that represent the heights of students who are less than 177 cm.

If you know the area under the normal distribution curve, you can also use this to calculate the probability that a student chosen at random has a height less than 177 cm.

2 Using mathematics you have already learned, you could calculate this area. What would we need to know in order to do this? Describe how you could do this.

3 Your value for the probability that a student chosen at random has a height less than 177cm will depend on which method you use. Why is this the case?

4 Which method is the most accurate?

5 Why is it not always possible to use this method?

6 What is the probability that a student has a height of less than 200 cm? (You could answer this using any of the methods you saw in questions **1** and **2**.)

7 **Conceptual** What does your answer to question **6** tell you about the probability of a random variable taking a value in a given range?

In any normal distribution, no matter what the values of μ and σ are, the total area under the curve is always equal to **one**. We can therefore consider partial areas under the curve as representing probabilities.

Unfortunately, the probability function (the equation of the curve) for the normal distribution is very complicated and difficult to work with.

$$f(X)=\frac{1}{\sqrt{2\pi}\sigma}e^{\left(\frac{-(X-\mu)^2}{2\sigma^2}\right)} \quad -\infty < X < \infty$$

It would be too hard for us to use integration to find areas under this curve, and hence find probabilities.

However, there are other methods by which you can find the area under the curve, and you will study these methods in this chapter.

The standard normal distribution

The **standard normal distribution** is the normal distribution where $\mu = 0$ and $\sigma = 1$.

Mathematicians always use Z to describe a random variable with a standard normal distribution, and we write $Z \sim N(0, 1)$.

We can use a GDC to calculate the areas under the standard normal distribution curve for values of Z between $Z = a$ and $Z = b$. This allows us to calculate $P(a < Z < b)$.

HINT

Note that the $P(Z = a) = 0$. You can think of this as a line having no width and therefore no area.

This means that $P(a<Z<b)=P(a\le Z\le b)=P(a<Z\le b)=P(a\le Z<b)$

Example 10

Given that $Z \sim N(0, 1)$, sketch the required area under the standard normal curve, then find the probability using your GDC:

a $P(-2 < Z < 1)$ **b** $P(Z < 1)$ **c** $P(Z > -1.5)$

d $P(Z < 0)$ **e** $P(|Z| > 0.8)$

a

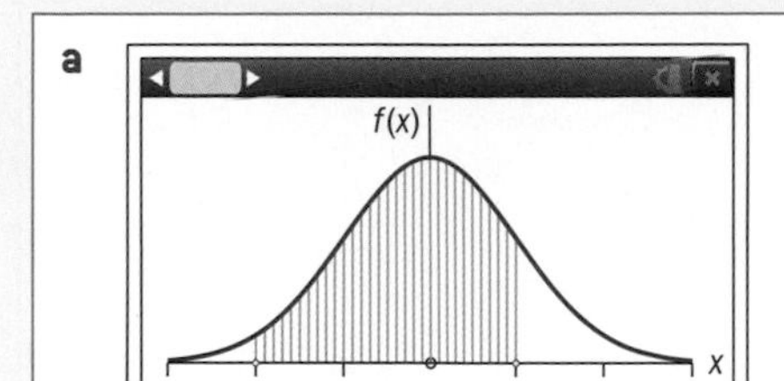

$P(-2 < Z < 1) = 0.819$

You will need to use your GDC to find the required probability (area) in each case.

b

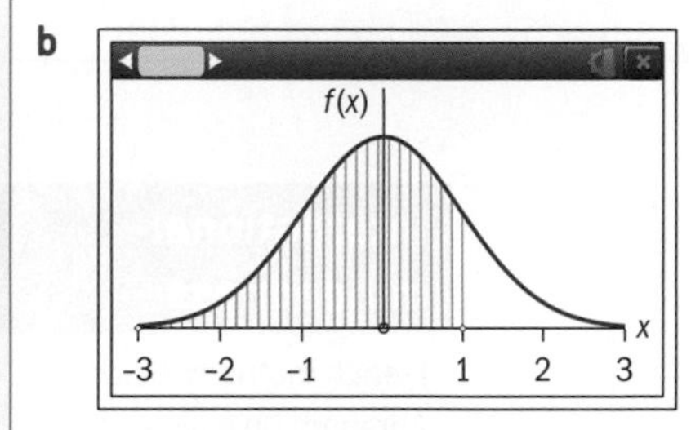

$P(Z < 1) = 0.841$

c

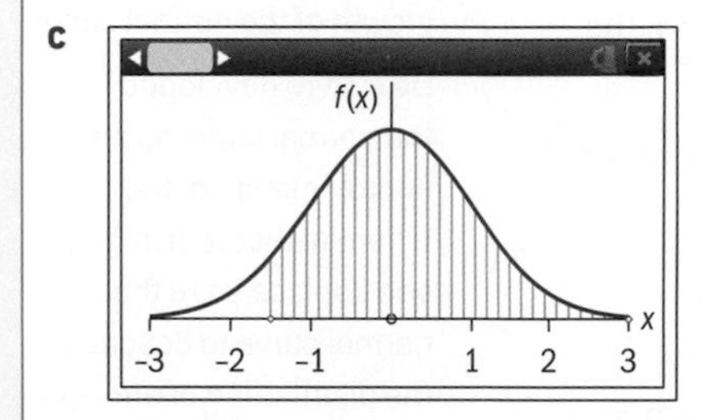

$P(Z > -1.5) = 0.933$

d $P(Z < 0) = 0.5$

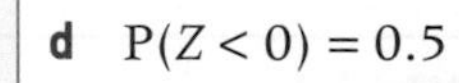

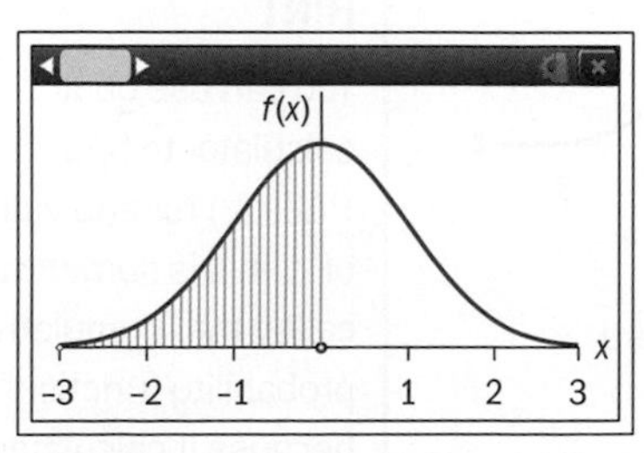

In this case we do not need to use the calculator due to the symmetry of the graph about the mean $\mu = 0$.

Continued on next page

e $P(|Z| > 0.8) = P(Z > 0.8) + P(Z < -0.8)$

Since $|Z| > 0.8 \Rightarrow Z < -0.8$ and $Z > 0.8$

Because of the symmetry of the curve about $Z = 0$, you can calculate either $P(Z > 0.8)$ or $P(Z < -0.8)$ and multiply that value by 2.

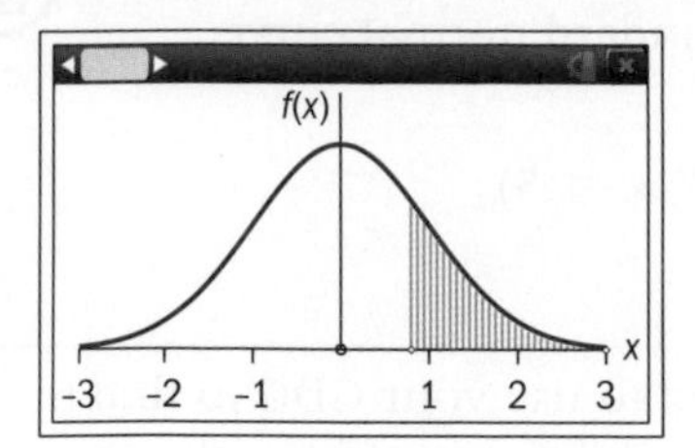

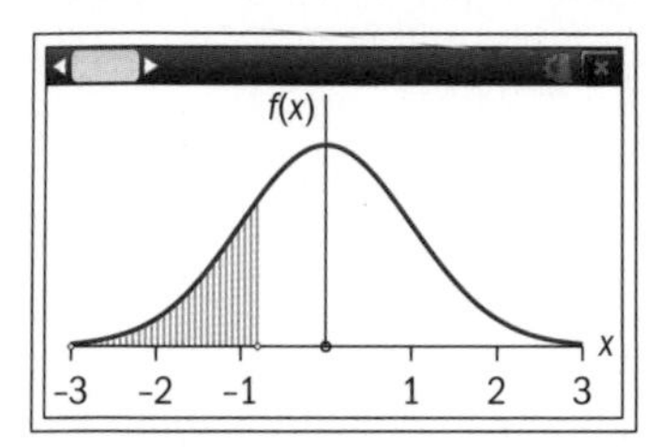

f $P(|Z| > 0.8) = 2P(Z < -0.8) = 0.424$

Investigation 12

1. Given that $Z \sim N(0, 1)$, use your GDC to find:

   a $P(-1 < Z < 1)$ **b** $P(-2 < Z < 2)$ **c** $P(-3 < Z < 3)$

2. Since the standard deviation of the standardized normal distribution is $\sigma = 1$, explain what the values you calculated in question **1** tell you about the probability that Z falls within 1, 2 or 3 standard deviations from the mean.

3. Using your answer to question **2**, express the percentage of data which:

   a lies within 1 standard deviation of the mean

   b lies within 2 standard deviations of the mean

   c lies within 3 standard deviations of the mean.

We can see that most of the data for a normal distribution will lie within 3 standard deviations of the mean.

4. Copy this diagram of the standard normal distribution curve. Draw appropriate vertical lines and shade, using different colours, the regions where data falls within 1, 2 or 3 standard deviations from the mean.

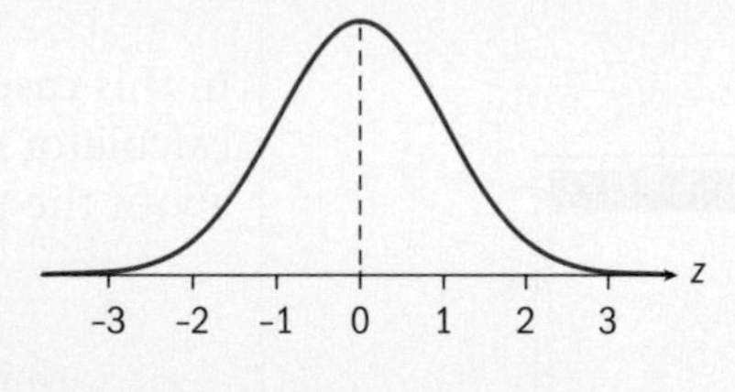

5. **Conceptual** For a standard normal distribution, how do standard deviations from the mean represent the data?

6. **Conceptual** How could we calculate how much data falls within 1.5 standard deviations of the mean?

International-mindedness

French mathematicians Abraham De Moivre and Pierre Laplace were involved in the early work of the growth of the normal curve. De Moivre developed the normal curve as an approximation of the binomial theorem in 1733 and Laplace used the normal curve to describe the distribution of errors on 1783 and in 1810 to prove the Central Limit Theorem

HINT

You can use your calculator to find $P(Z < z)$ for any value of z. This is sometimes called the "cumulative probability function" because it calculates the probability from $Z = -\infty$ up to $Z = z$.

Exercise 14H

1 Find the area under the standard normal distribution curve:

a between 1 and 2 standard deviations from the mean

b between 0.5 and 1.5 standard deviations from the mean.

2 Find the area under the standard normal distribution curve which is:

a more than 1 standard deviation above the mean

b more than 2.4 standard deviations above the mean

c less than1 standard deviation below the mean

d less than 1.75 standard deviations below the mean.

3 Given that $Z \sim N(0, 1)$ use your GDC to find:

a $P(Z < 0.65)$ b $P(Z > 0.72)$

c $P(Z \geq 1.8)$ d $P(Z \leq -0.28)$

4 Given that $Z \sim N(0, 1)$ use your GDC to find:

a $P(0.2 < Z < 1.2)$

b $P(-2 < Z \leq 0.3)$

c $P(-1.3 \leq Z \leq -0.3)$

5 Given that $Z \sim N(0, 1)$ use your GDC to find:

a $P(|Z| < 0.4)$ b $P(|Z| > 1.24)$

Probabilities for other normal distributions

So far, you have used the standardized normal random variable $Z \sim N(0, 1)$ in your calculations.

Investigation 13

Suppose Millie has a wrist circumference of 13.75 cm.

1 Is this a small wrist circumference? What might you need to know in order to answer this question?

2 The average wrist size for girls of Millie's age is 16 cm ($\mu = 16$). How far is Millie from the mean?

3 Using your answer to question **2**, would you conclude that Millie has a small wrist circumference? What else would you want to know?

4 The standard deviation of wrist sizes for girls of Millie's age is 1.5 cm ($\sigma = 1.5$).

How can you use this fact, and your findings from investigation 12, to make a judgment as to whether Millie has a small wrist circumference?

Clearly the circumference of the wrist relative to other people her age is important. However, it is also useful to know the distance of the value from the mean of the data in terms of the standard deviation of the data.

5 If Millie's wrist size is x, the mean wrist size μ is and the standard deviation is σ, how could you calculate the exact number of standard deviations Millie is from the mean?

Since Millie's wrist circumference was below the mean, the number of standard deviations from the mean could be given as a score: in this case, −1.5. This tells you that her wrist is 1.5 standard deviations below the mean.

This is known as the **standardized value** or z-value for Millie's circumference.

Most real-life random variables which are normally distributed do not have a mean of 0 and a standard deviation of 1. It is possible, however, to transform any normally distributed variable $X \sim N(\mu, \sigma^2)$ to a standard normally distributed variable as above. This is possible because all normal distribution curves have the same basic bell-shape, and so can be obtained through translations and stretches.

> Therefore, in order to transform any given value x of $X \sim N(\mu, \sigma^2)$ to its equivalent z-value of $Z \sim N(0, 1)$, we use the formula:
>
> $$z = \frac{x - \mu}{\sigma}$$
>
> This process is called **standardization.**

You can then use calculate probabilities using your GDC, as before.

Example 11

Given that $X \sim N(10, 2^2)$, find:

a $P(X < 13)$ **b** $P(X > 9)$ **c** $P(9.1 < X < 10.3)$.

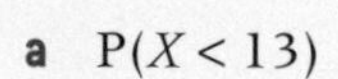

a 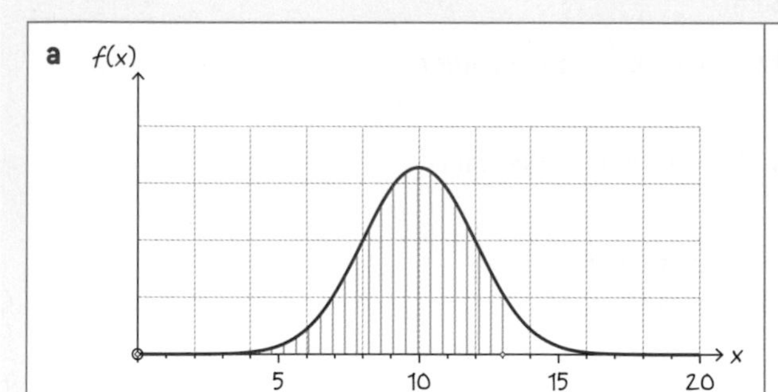	A sketch will help by showing the area required and will also give us an idea of the approximate value of the answer.
$z = \frac{13-10}{2}$ $= 1.5$	Standardize the value of x using $z = \frac{x - \mu}{\sigma}$
$P(X < 13) = P(Z < 1.5)$ $= 0.933$	$P(X < 13)$ is equivalent to $P(Z < 1.5)$. Calculate $P(Z < 1.5)$ using your GDC.

b

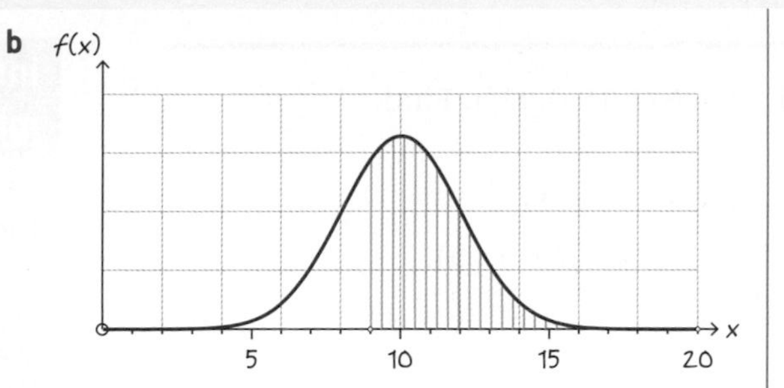

$z = \frac{9-10}{2}$
$= -0.5$

And so $P(X > 9) = P(Z > -0.5)$
$= 0.691$

Standardize the value of x using $z = \frac{x-\mu}{\sigma}$

$P(X > 9)$ is equivalent to $P(Z > -0.5)$.

Calculate $P(Z > -0.5)$ using your GDC.

c

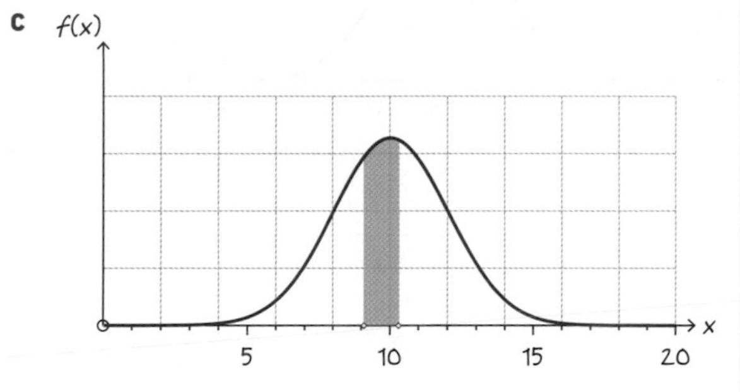

$z = \frac{9.1-10}{2}$
$= -0.45$

$z = \frac{10.3-10}{2}$
$= 0.15$

And so

$P(9.1 < X < 10.3) = P(-0.45 < Z < 0.15)$
$= 0.233$

Standardize each value of x using $z = \frac{x-\mu}{\sigma}$

Calculate $P(-0.45 < Z < 0.15)$ using your GDC.

HINT

Some GDCs allow you to calculate the probabilities of other normally distributed random variables directly. You must enter the mean and standard deviation into your GDC.

If you can do this, it is the quickest and most efficient method to use for this type of question. However, you need to know how to use standardization in order to answer other types of questions later in this chapter.

Exercise 14I

1 $X \sim N(14, 5^2)$. Find:

a $P(X < 16)$ b $P(X > 9)$

c $P(9 \leq X < 12)$ d $P(X < 14)$.

2 $X \sim N(48, 81)$. Find:

a $P(X < 52)$ b $P(X \geq 42)$

c $P(37 < X < 47)$.

3 $X \sim N(3.15, 0.02^2)$. Find:

a $P(X < 3.2)$

b $P(X \geq 3.11)$

c $P(3.1 < X < 3.15)$.

International-mindedness

Belgian scientist Lambert Quetelet applied the normal distribution to human characteristics (l'homme moyen) in the 19th century. He noted that characteristics such as height, weight and strength were normally distributed.

Example 12

Eggs laid by a chicken are known to have masses which are normally distributed with mean 55 g and standard deviation 2.5 g.

Find the probability that a single egg laid by this chicken has mass:

a greater than 59 g

b less than 53 g

c between 52 g and 54 g.

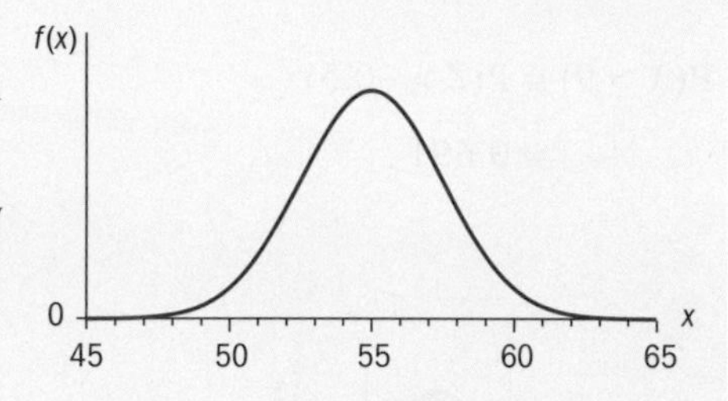

a $P(M > 59) = 0.0548$ (3 s.f.) b $P(M < 53) = 0.212$ (3 s.f.) c $P(52 < M < 54) = 0.230$ (3 s.f.)	

Exercise 14J

1 Households in Portugal spend a mean of €100 per week on groceries with a standard deviation of €20. Assuming that the distribution of grocery expenditure follows a normal distribution, find the probability of a household spending:

a less than €130 per week

b more than €90 per week

c between €80 and €125 per week.

2 A machine produces bolts with diameters that are normally distributed with a mean of 4 mm and a standard deviation of 0.25 mm.

Bolts are measured accurately, and any bolts which are smaller than 3.5 mm or bigger than 4.5 mm are rejected.

Out of a batch of 500 bolts, find the number bolts you would expect to be accepted.

3 The length of time patients have to wait in Dr Barrett's waiting room is known to be normally distributed with a mean of 14 minutes and a standard deviation of 4 minutes.

a Find the probability that a patient will have to wait more than 20 minutes to see Dr Barrett.

b Find the percentage of patients who wait for less than 10 minutes.

4 A company that produces breakfast cereal claims that packets of "flakey flakes" have a net mass of 550 g. The production of packets of "flakey flakes" is such that the masses are normally distributed with a mean of 551.3 g, and standard deviation of 15 g. Find the percentage of packets which are heavier than the stated mass.

5 The masses of packets of washing powder are normally distributed with a mean of 500 g and standard deviation of 20 g.

a Find the probability that a packet chosen at random has a mass less than 475 g.

b Three packets are chosen at random. Determine the probability that all three packets have masses less than 475 g.

The inverse normal distribution

Here we are given a cumulative probability, $P(Z < z)$ for unknown z, and would like to find the z-value that has this cumulative probability.

For example, the owner of a company that fills cartons of juice to a mean volume of 150 ml may wish to find the volume below which 5% of the cartons are, in order to reject them.

Again we can use the calculator to help us find this value. The calculator has an Inverse Normal function which will do this.

We will look first at inverse normal calculations using the standard normal distribution $Z \sim N(0, 1)$.

Example 13

Given that $Z \sim N(0, 1)$, use your GDC to find a when:

a $P(Z < a) = 0.877$ **b** $P(Z > a) = 0.2$

c $P(Z > a) = 0.55$ **d** $P(-a < Z < a) = 0.42$

a $P(Z < a) = 0.877$

a is the value for which the area under the curve, to the left of a, is 0.877.

Use the inverse normal function on your GDC to find a.

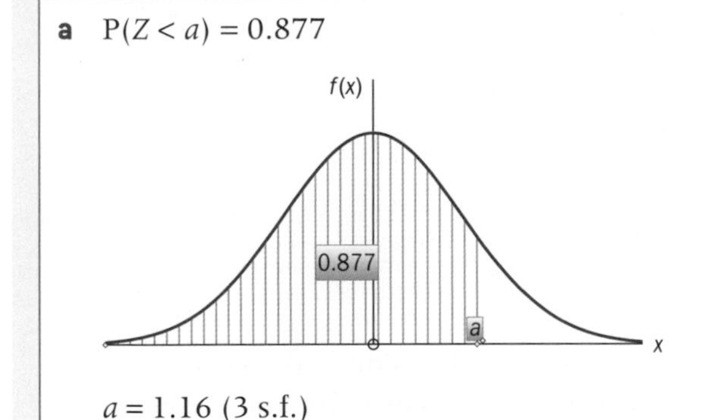

$a = 1.16$ (3 s.f.)

b $P(Z > a) = 0.2$

Notice that to find the value of a for which $P(Z > a) = 0.2$ we can more easily find $P(Z < a) = 0.8$.

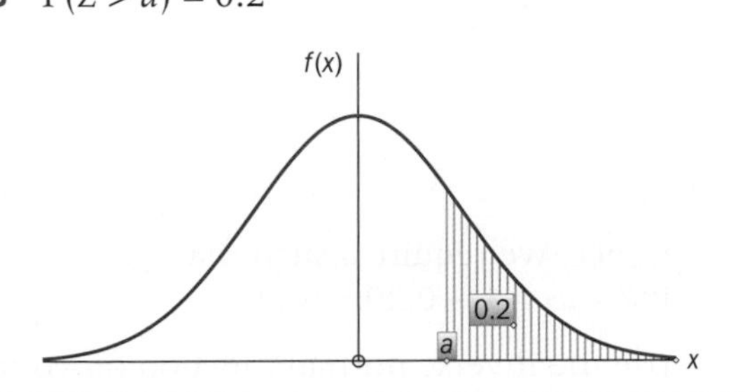

Continued on next page

Statistics and probability

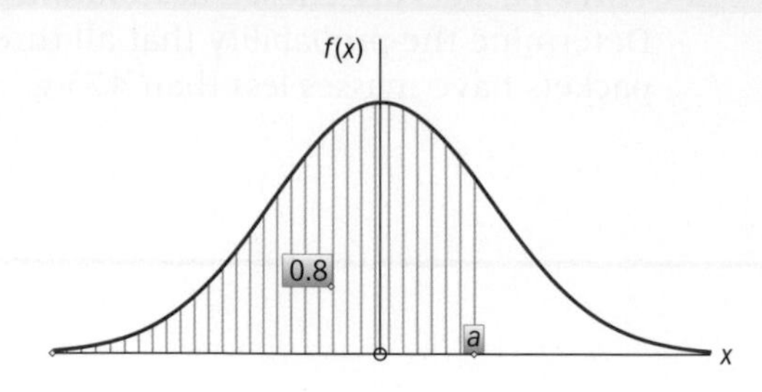

$P(Z > a) = 0.2 \Rightarrow P(Z < a) = 0.8$

$a = 0.842$ (3 s.f.)

Use the inverse normal function on your GDC to find a.

c $P(Z > a) = 0.55$

Notice that to find the value of a for which $P(Z > a) = 0.55$ we can more easily find $P(Z < a) = 0.45$.

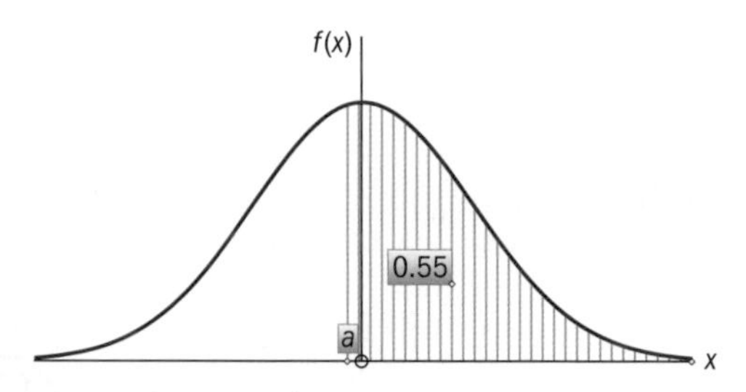

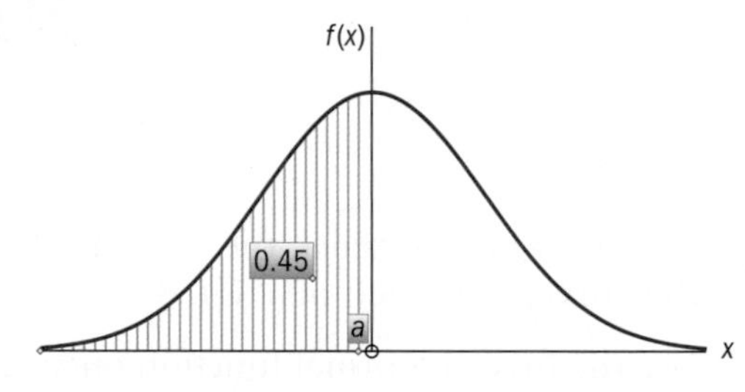

Use the inverse normal function on your GDC to find a.

$P(Z > a) = 0.55 \Rightarrow P(Z < a) = 0.45$

$a = -0.126$ (3 s.f.)

d $P(-a < Z < a) = 0.42$

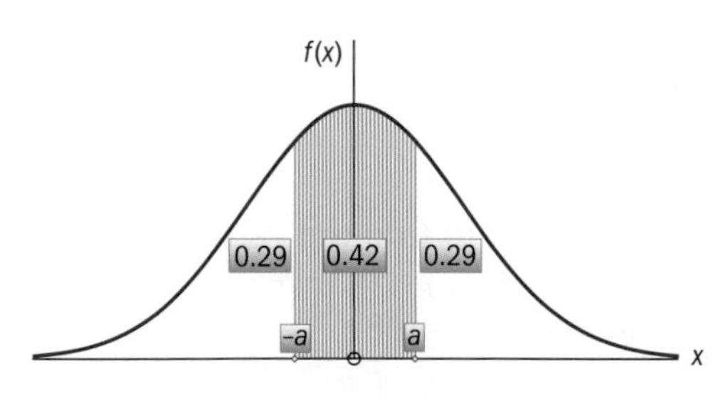

The areas either side of the shaded region are the same size and are equal to

$\frac{1}{2}(1 - 0.42) = 0.29.$

and hence $P(Z < a) = 0.71$

$a = 0.553$

Hence, we require a such that $P(Z < a) = 1 - 0.29 = 0.71.$

Use the inverse normal function on your GDC to find a.

Exercise 14K

1 Find a such that:

a $P(Z < a) = 0.922$

b $P(Z > a) = 0.342$

c $P(Z > a) = 0.005$.

2 Find a such that:

a $P(1 < Z < a) = 0.12$

b $P(a < Z < 1.6) = 0.787$

c $P(a < Z < -0.3) = 0.182$.

3 Find a such that:

a $P(-a < Z < a) = 0.3$

b $P(|Z| > a) = 0.1096$.

4 The diagrams below each show a standard normal distribution. Find the values of z.

a

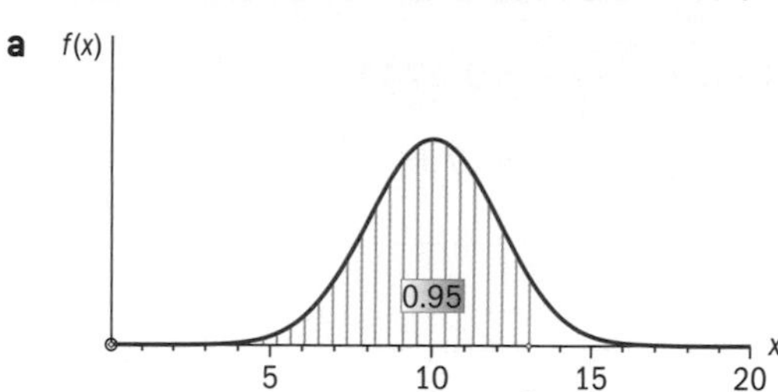

b

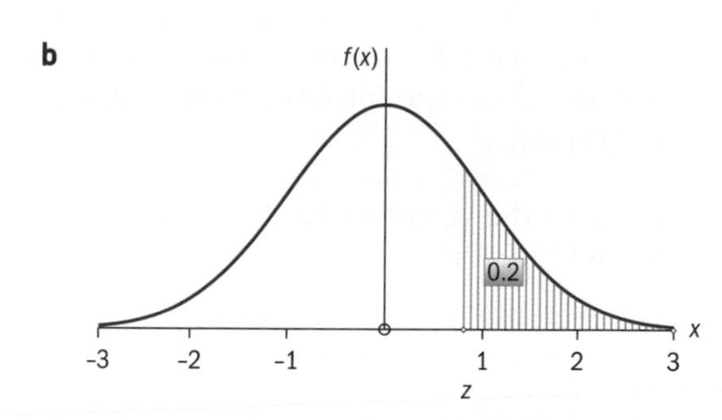

Once again, when solving real-life problems involving the normal distribution, the mean and standard deviation are unlikely to be 0 and 1, respectively. As such, you also have to transform variables to fit the standardized inverse normal distribution.

Example 14

Given that $X \sim N(15, 3^2)$, find the value of x for which $P(X < x) = 0.75$.

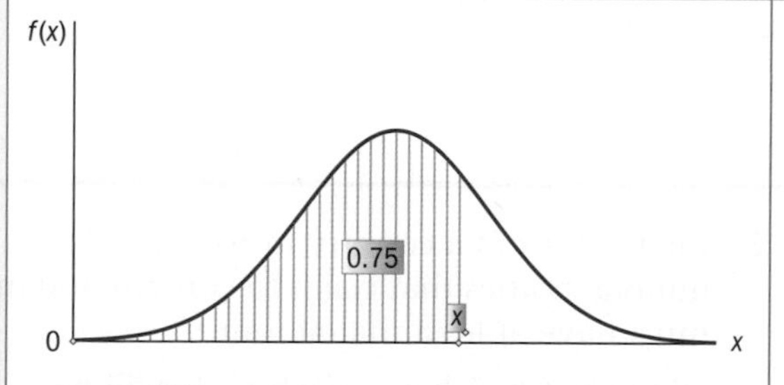

A sketch will help by showing the value of x required.

$$z = \frac{x - 15}{3}$$

First, standardize the value of x using $z = \frac{x - \mu}{\sigma}$

HINT

It is also possible to do example 14 entirely on your GDC, without standardizing the value of x.

Continued on next page

We require $P(Z < \frac{x-15}{3}) = 0.75$. Hence $\frac{x-15}{3} = 0.6744$ $x = 17.0$	Since $P(X < x) = 0.75 \Rightarrow$ $P(Z < \frac{x-15}{3}) = 0.75$ Use the inverse normal function on your calculator.

Example 15

Cartons of juice are such that their volumes are normally distributed with a mean of 150 ml and a standard deviation of 5 ml. Given that 5% of cartons are rejected for containing too little juice, find the minimum volume, to the nearest ml, that a carton must contain if it is to be accepted.

Let V be the volume of a carton, then $V \sim N(150, 5^2)$. Let m be the minimum volume that a carton must be to be accepted.	
We require $P(V < m) = 0.05$.	5% of cartons are rejected for containing too little juice, so $P(V < m) = 0.05$.
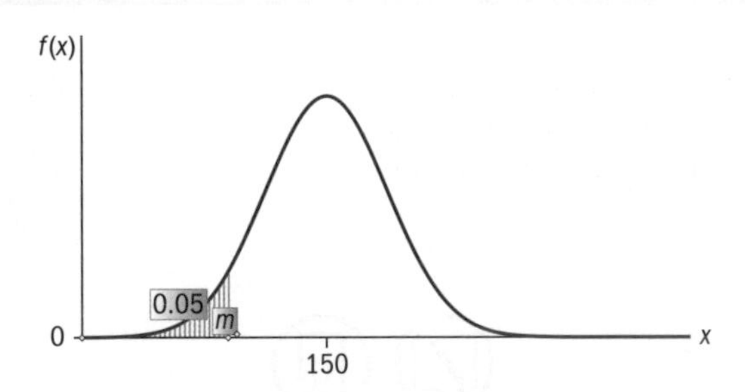	You could standardize the value m, so that $z = \frac{m-150}{5}$, or you could use the inverse normal function on your GDC with the given mean and standard deviation and find m directly.
The minimum volume is 142 to the nearest ml.	

Exercise 14L

1 $X \sim N(5.5, 0.2^2)$ and $P(X > a) = 0.235$. Find the value of a.

2 The masses, M, of tins of dog food is such that $M \sim N(420, 10^2)$. Find:

a the first quartile

b the 90th percentile.

3 Regulations in a country insist that all mineral bottles that claim to contain 500 ml must have at least that amount.

"Yummy Cola" has a machine for filling bottles, which puts a mean of 502 ml into each bottle, with a standard deviation of 1.6 ml. This follows a normal distribution.

a An inspector randomly selects a bottle of "Yummy Cola". Find the probability that the selected bottle violates the regulations.

b Find the proportion of bottles which contain between 500 ml and 505 ml.

c 95% of bottles contain between a ml and b ml of liquid where a and b are symmetrical about the mean. Find the values of a and b.

4 The masses of heads of lettuce sold at a hypermarket are normally distributed with a mean mass of 550 g and standard deviation of 25 g.

a If a head of lettuce is chosen at random, find the probability that its mass lies between 520 g and 570 g.

b Find the mass exceeded by 10% of the heads of lettuce.

5 The marks of 500 candidates in an examination are normally distributed with a mean of 55 marks and a standard deviation of 15 marks.

a If 5% of the candidates obtain a distinction by scoring d marks or more, find the value of d.

b If 10% of students fail by scoring f marks or less, find the value of f.

We may also be given cumulative probabilities and could be required to find either the mean μ (if σ is known) or the standard deviation σ (if μ is known), or both μ and σ if both are unknown. These questions can only partially be done using a GDC.

Example 16

Sacks of potatoes are packed by an automatic loader. The mean mass of a sack is 5 kg. In a test it was found that 10% of bags were over 5.2 kg. Given that the masses of the sacks are normally distributed, find the standard deviation.

Let M be the mass of a sack of potatoes.
$M \sim \mathrm{N}(5, \sigma^2)$

We require $\mathrm{P}(M > 5.2) = 0.1$.

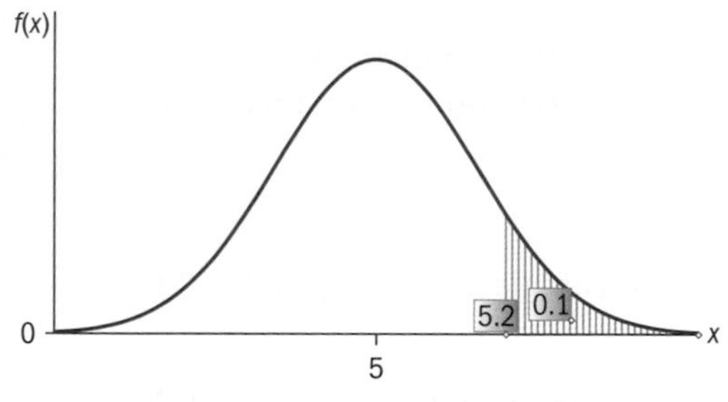

Standardizing gives $z = \frac{5.2-5}{\sigma} = \frac{0.2}{\sigma}$

You have not been told the standard deviation of M. However, you can standardize the m-value and use this to find σ.

We require $\mathrm{P}Z > \frac{0.2}{\sigma} = 0.1$.

You need to find z-value such that $\mathrm{P}(Z > z) = 0.1$.

This is the same as $\mathrm{P}Z < \frac{0.2}{\sigma} = 0.9$.

Continued on next page

Using the GDC gives $P(Z < 1.28155...) = 0.9$ And so $\frac{0.2}{\sigma} = 1.28155....$	Remember, you can store the z-value on your calculator and avoid using a rounded value in your calculations. Solve for σ.
$\sigma = 0.156$ (3 s.f.)	

Example 17

A manufacturer does not know the mean and standard deviation of the diameters of ball bearings he is producing. However, a sieving system rejects all ball bearings with diameter greater than 2.4 cm or less than 1.8 cm. It is found that 8% of the ball bearings are rejected as too small, and 5.5% are rejected because they are too big.

Assuming the size of the ball bearings are normally distributed, calculate the mean and standard deviation of the diameters of the ball bearings produced.

Let D be the diameters of ball bearings produced. So $D \sim N(\mu, \sigma^2)$.	We do not know either the mean or standard deviation of D.
We require $P(D < 1.8) = 0.08$ and $P(D > 2.4) = 0.055$.	8% are rejected as too small.
Standardizing each gives: $z_1 = \frac{1.8 - \mu}{\sigma}$ and $z_2 = \frac{2.4 - \mu}{\sigma}$	5.5% are rejected because they are as too big. Standardize to a normal distribution with mean 0 and standard deviation 1. Find the z-values.
$P(Z < \frac{1.8 - \mu}{\sigma}) = 0.08$ and $P(Z > \frac{2.4 - \mu}{\sigma}) = 0.055$ or	You need z_1 and $z_2$which give $P(Z < z_1) = 0.08$ and $P(Z > z_2) = 0.055$.
$P(Z < \frac{2.4 - \mu}{\sigma}) = 0.945$	Easier to calculate $P(Z < z_2) = 1 - 0.055$.
Now from the GDC we know that $P(Z < -1.40507...) = 0.08$ and $P(Z < 1.59819...) = 0.945$.	Store the z-values on your calculator to avoid using rounded values in your calculations.
Hence $\frac{1.8 - \mu}{\sigma} = -1.40507...$	

and $\frac{2.4-\mu}{\sigma}=1.59819...$ $\mu = 2.08$ and $\sigma = 0.200$	Solve these two equations simultaneously for μ and σ.

HINT

You must be able to standardize a normally distributed random variable in order to find unknown values of μ and σ. There is no way to find these values directly from your GDC.

Exercise 14M

1 $X \sim \mathrm{N}(30, \sigma^2)$ and $\mathrm{P}(X > 40) = 0.115$. Find the value of σ.

2 $X \sim \mathrm{N}(\mu, 4^2)$ and $\mathrm{P}(X < 20.5) = 0.9$. Find the value of μ.

3 $X \sim \mathrm{N}(\mu, \sigma^2)$. Given that $\mathrm{P}(X > 58.39) = 0.0217$ and $\mathrm{P}(X < 41.82) = 0.0287$, find μ and σ.

4 A random variable X is normally distributed with mean μ and standard deviation σ, such that $\mathrm{P}(X < 89) = 0.90$ and $\mathrm{P}(X < 94) = 0.95$.

Find μ and σ.

5 The mean height of children of a certain age is 136 cm. 12% of children have a height of 145 cm or more. Find the standard deviation of the heights.

6 The standard deviation of masses of loaves of bread is 20 g. Only 1% of loaves have mass less than 500 g. Find the mean mass of the loaves.

7 The masses of cauliflowers are normally distributed with mean 0.85 kg. 74% of cauliflowers have mass less than 1.1kg. Find:

a the standard deviation of masses of cauliflowers

b the percentage of cauliflowers with mass greater than 1 kg.

8 The lengths of some boxes are normally distributed with mean μ and standard deviation 7 cm. If 2.5% of the boxes measure more than 68 cm, find the value of μ.

9 A roll of wrapping paper is sold as "3m long". It is found that actually only 35% of rolls are over 3 m long and that the average length of the rolls of wrapping paper is 2.9 m. Find the value of the standard deviation of the lengths of rolls of wrapping paper, assuming that the lengths of rolls follow a normal distribution.

10 It is suspected that the scores in a test are normally distributed. 30% of students scored less than 108 marks on the test, and 20% scored more than 154 marks.

a Find the mean and standard deviation of the scores, if they are normally distributed.

60% of students score more than 117 marks.

b Explain whether this fact appears to be reasonably consistent with the claim that the scores are normally distributed.

11 Due to variations in manufacturing, the length of wool in a randomly chosen ball of wool can be modelled by a normal distribution. Find the mean and standard deviation in the lengths of wool, given that 95% of balls of wool have lengths exceeding 495 m, and 99% have lengths exceeding 490 m.

Developing your toolkit

Now do the Modelling and investigation activity on page 626.

Chapter summary

- $0 \leq \mathrm{P}(X=x) \leq 1$
  $\sum \mathrm{P}(X=x)=1$
- $\mathrm{E}(M)=\sum_{m} m\mathrm{P}(M=m)$
- If X is a discrete random variable which is binomially distributed, $X \sim \mathrm{B}(n, p)$, then the probability of obtaining r successes out of n independent trials, when p is the probability of success for each trial, is $\mathrm{P}(X=r)=\binom{n}{r} p^{r}(1-p)^{n-r}$.
  This is often shortened to $\mathrm{P}(X=r)=\binom{n}{r} p^{r} q^{n-r}$ where $q=1-p$.
- For the binomial distribution where $X \sim \mathrm{B}(n, p)$, the expectation of X is $\mathrm{E}(X)=np$.
- The formula for the variance of a binomial distribution with $X \sim \mathrm{B}(n, p)$ is given by $\mathrm{Var}(X)=\sigma^{2}=npq=np(1-p)$.
- A **normal distribution** is roughly symmetrical about a central value, and the curve is bell-shaped, with the majority of measurements grouped around the central value.

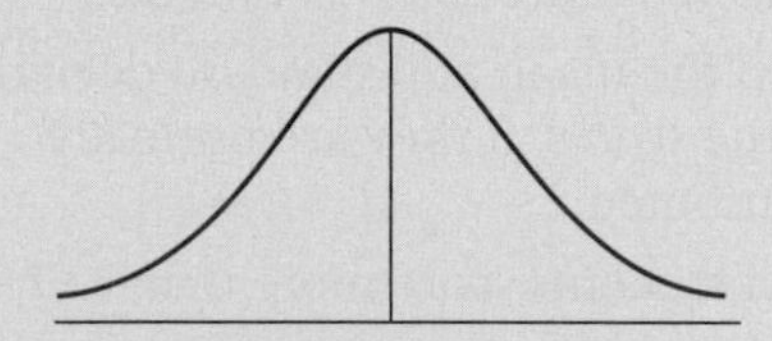

- For a perfect normal distribution:
  - The curve is bell-shaped.
  - The data is symmetrical about the mean (μ).
  - The mean, mode and median are the same.
- Therefore, in order to transform any given value x of $X \sim \mathrm{N}(\mu, \sigma^{2})$ to its equivalent z-value of
  $Z \sim \mathrm{N}(0, 1)$, we use the formula: $z=\frac{x-\mu}{\sigma}$
  This process is called **standardization**.

Developing inquiry skills

Consider now the third option from the original problem.

If you played this game a few times, do you think you would end up with more money than you started with?

How could we tell?

When a ball is dropped and it hits one of the pegs it has an equal chance of falling either side. Let "L" and "R" denote whether a ball goes left or right at any given peg.

What path does the ball need to take in order to go in to box number 1?

What is the probability of this happening?

What possible paths can the ball take in order to go into box number 2?

The "game" is called a quincunx or Galton board or bean machine. Sir Edward Galton, a British scientist, originally devised it for probability experiments.

What is the relationship between the quincunx and the binomial distribution?

What is the probability that you will win 500 tokens? 50? 10? 0?

Draw a probability distribution table for your random variable.

Calculate the expected winnings from the game.

How much would you expect to win or lose if you played the game 10 times? 100 times?

Is the game "fair"?

Interestingly if a large number of balls are dropped into the quincunx then the shape of the resulting balls approximates a normal distribution.

This can be seen in the simulation at this site:

https://www.mathsisfun.com/data/quincunx.html

You can alter the number of boxes and the speed that the balls drop. It is also possible to change the probability that the ball falls to one side or the other.

What effect does this have on the distribution of the balls?

Statistics and probability

Chapter review

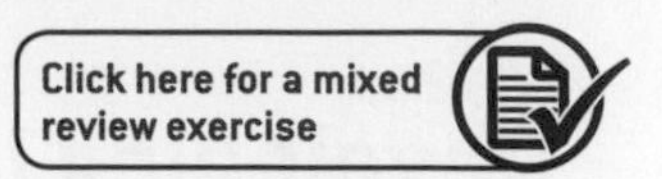

1 The following table shows the probability distribution of a discrete random variable X.

x	$P(X = x)$
−2	0.3
−1	$\frac{1}{k}$
0	$\frac{2}{k}$
1	0.1
2	0.1

a Find the value of k

b Find the expected value of X.

2 The probability distribution of a discrete random variable X is defined by $P(X = x) = cx(6 - x)$, $x = 1, 2, 3, 4, 5$.

a Find the value of c.

b Find $E(X)$.

3 In a game a player rolls a biased tetrahedral (four-faced) die. The probability of each possible score is shown below.

Score	Probability
1	$\frac{1}{4}$
2	$\frac{1}{4}$
3	$\frac{1}{8}$
4	x

Find the probability of a total score of six after two rolls.

4 A game involves spinning two spinners. One is numbered 1, 2, 3, 4. The other is numbered 2, 2, 4, 4. Each spinner is spun once and the number on each is recorded.

Let P be the product of the numbers on the spinners.

a Write down all the possible values for P.

b Find the probability of each value of P.

c Find the expected value of P.

d A mathematician determines the amount of pocket money to give his son each week by getting him to play the game on Monday morning. If the son spins and the product is greater than 10, then the boy gets £10. Otherwise the boy gets £5. Find how much in total should the boy expect to get after 10 weeks of playing the game.

5 On a train, $\frac{1}{3}$ of the passengers are listening to music. Five passengers are chosen at random. Find the probability that exactly three passangers are listening to music.

6 When Abhinav plays a game at a fair, the probability that he wins a prize is 0.1. He plays the game twice. Let X denote the total number of prizes that he wins. Assuming that the games are independent, find $E(X)$

7 Let X be normally distributed with mean 75 and standard deviation 5.

a Given that $P(X < 65) = P(X > a)$, find the value of a.

b Given that $P(65 < X < a) = 0.954$, Find $P(X > a)$.

8 Three dice are thrown. If a 1 or a 6 is rolled somewhere on these three dice, you will be paid \$1, but if neither are rolled you will pay \$5. You play the game.

a Find the probability that you will win \$1.

b Complete the table below that shows the probability distribution of X, 'the number of dollars won in a game'.

x	$P(X = x)$
−5	
1	

c How much would you expect to gain (or lose) in

i 1 game **ii** 9 games?

9 I like 30% of the songs on my friends MP3 player. If I choose eight songs at random

a find the probability that I like exactly three songs

b find the probability that I like at least three songs.

10 Find the probability of throwing three sixes twice in five throws of six dice.

11 In a large school one person in five is left-handed.

a A random sample of ten people is taken. Find the probability that

i exactly four will be left-handed

ii more than half will be left-handed.

b Find the most likely number of left-handed people in the sample of ten people.

c How large must a random sample be if the probability that it contains at least one left-handed person is greater than 0.95?

12 Z is the standardized normal random variable with mean 0 and variance 1. Find the value of a such that $P(|Z| \le a) = 0.85$.

13 The results of a test given to a group of students are normally distributed with a mean of 71. In the test, 85% of students scored less than 80.

a Find the standard deviation of the scores.

To pass the test a student must score at least 65.

b Find the probability that a student chosen at random passes the test.

14 The lifespans of certain batteries are normally distributed. It is found that 15% of batteries last less than 30 hours and 10% of batteries last more than 50 hours. Find the mean and standard deviation of the lifespans of the batteries.

15 The time taken for Samuel to get to school each morning is normally distributed with a mean of μ minutes and a standard deviation of 2 minutes

The probability that the journey takes more than 35 minutes is 0.2.

a Find the value of μ.

Samuel should be at school at 08.45 each morning and so on five consecutive days he sets out at 08.10.

b Find the probability that he arrives before 08.45 on all five days.

c Find the probability he is late on at least two days.

Exam-style questions

16 **P1:** There are ten black and eight white socks in a drawer. Jeannie, without looking, takes out a sock at random, looks at its colour and puts it back in the drawer. Then, again without looking, takes another sock out.

Let X be the random variable that represents the number of black socks taken out by Jeannie.

a State the possible values of the random variable X. (1 mark)

b Find $P(X = 2)$. (2 marks)

c Construct the probability distribution table for the random variable X. (2 marks)

17 **P1:** The table shows the probability distribution of a discrete random variable X.

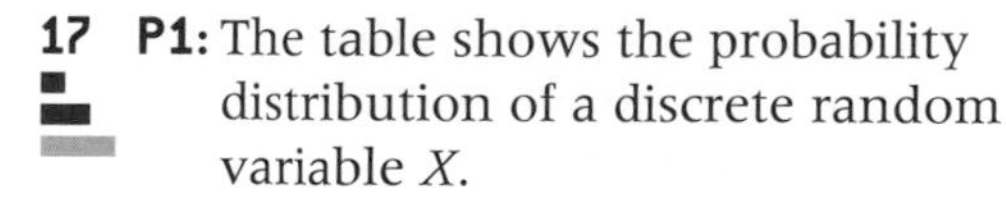

x	$P(X = x)$
0	0.2
1	$k + 0.25$
2	$k - 0.05$
3	0.3

a Find the value of k. (3 marks)

b Hence calculate $E(X)$. (2 marks)

Statistics and probability

18 **P2:** An emergency call centre has four service lines that operate 24 hours every day of the week. The random variable X represents the number of phone lines in use during any five-minute period. Based on data collected over a long period of time, the centre manager has determined the following:

x	$P(X = x)$
0	0.05
1	0.22
2	0.27
3	a
4	b

a State a condition on a and b. (2 marks)

b Given that $E(X) = 2.46$, determine the value of a and the value of b. (5 marks)

19 **P2:** In a mass production of batteries, the probability that one battery is defective is 0.5%. Given that the batteries are packed randomly in packs of 10 and quality control inspect batteries from randomly selected packs, find the probability that a selected pack contains

a one defective battery (2 marks)

b not more than one defective battery (2 marks)

c exactly 1 defective battery, given that it has not more than one defective battery. (3 marks)

20 **P1:** A random variable X follows a binomial distribution with mean 3 and variance 1.2. Find the value of the parameters n and p of the distribution of X. (5 marks)

21 **P2:** A packing machine produces bags of rice whose weights are normally distributed with mean 50.1 kg and standard deviation 0.4 kg. If a bag produced by this machine is selected at random, find the probability that its weight is:

a less than 49.5 kg (2 marks)

b between 49.5 kg and 50.5 kg (2 marks)

c more than 49 kg given that it is less than 49.5 kg. (3 marks)

22 **P2:** A random normal variable X has mean μ and standard deviation 3. Given that $P(X < 5) = 0.754$

a find the value of μ. (3 marks)

b Hence find $P(4 < X < 5)$. (2 marks)

23 **P2:** An exam consists of 20 true/false questions. John knows the correct answers to 8 of the questions and decides to choose at random the answers to the remaining questions.

a Find the probability that John answers

i 10 questions correctly

ii all the 20 questions correctly. (4 marks)

A student gains 2 marks for each correct answer, has one mark subtracted for each incorrect answer, and gains no marks if the question is left unanswered.

b Compare the expected number of marks when the student answers all the questions with the marks she will gain if she just answers the questions to which she knows the correct answer. (5 marks)

24 **P2:** A survey is conducted in a large factory. It is found that 27% of the factory workers weigh less than 65 kg and that 25% of the factory workers weigh more than 96 kg.

Assuming that the weights of the factory workers is modelled by a normal distribution with mean μ and standard deviation σ,

a **i** determine two simultaneous linear equations satisfied by μ and σ

ii find the values of μ and σ. (6 marks)

b Find the probability that a factory worker chosen at random weighs more than 100 kg. (2 marks)

The factory has 1000 workers, 630 of which are males. The weights of the males are normally distributed with mean 80.5 and standard deviation 10.1.

c If a male factory worker is selected at random, find the probability that he weights between 75 and 85 kilograms. (2 marks)

d Find the expected number of male factory workers that weight more than 85 kilograms. (2 marks)

The female workers weights have mean m.

e Find the value of m. (2 marks)

25 **P2:** A national park houses a large number of tall trees. The heights of the trees are normally distributed with mean of 45 metres and a standard deviation of 9 metres.

a One tree is selected at random from the park.

i Find the probability that this tree is at least 55 metres high.

ii Given that this tree is at least 55 metres, find the probability that it is taller than 65 metres. (5 marks)

b Three trees are selected at random. Find the probability that they are all at least 55 metres high. (2 marks)

c A selection of 50 trees is made at random.

i Find the expected number of these trees that taller than 55 metres.

ii Find the probability that at least 5 of these trees are taller than 55 metres. (5 marks)

26 **P2:** The marks M of a mathematics exam at a large school were normally distributed. Let μ be the mean and σ be the standard deviation of the distribution of these marks.

Given that 10% of the candidates that sat the exam scored at least 82 marks and 20% scored less than 40, find the value of

a the mean, μ

b the standard deviation, σ.

Click here for further exam practice

Fair game!

Approaches to learning: Collaboration, Communication, Self-management
Exploration criteria: Presentation (A), Mathematical communication (B), Personal engagement (C), Reflection (D), Use of mathematics (E)
IB topic: Probability, Expected Value, Probability Distributions

The three different problems in the Developing inquiry skills section of this chapter involved probability distributions and understanding **expected value** and the meaning of a **fair game**.

The task

In your pairs or groups of 3, your task is to design your own fully functioning game for other students to play.

In the task you will need to discuss:

- What equipment can you use to create probabilities?
- How can you make your game exciting/appealing?
- How can you make sure you make a profit from your game?

Your game must be unique, and **not** copied from an existing game!

Part 1: Design and understand the probabilities involved in your game

In your groups decide on the game you will produce.

Brainstorm some ideas first.

You may need to trial the game in your group to check it works before you play it with other students.

Provide a brief overview of your game and make sure that you are able to explain the probabilities involved for your teacher to check through.

Think about:

- What equipment do players need to play the game?
- How much will the game cost to play?
- What will the prizes be?

Create a set of clear, step-by-step **instructions** to explain your game to players.

A player should be able to read the instructions and be able to start playing your game immediately.

Part 2: Set up the game in a "class fair"

You will need to provide any required equipment for your game.

Your game needs to be attractive and eye-catching.

Play the games in class.

Part 3: Reflect on the success or otherwise of your game

Either in a written report, interview or in a video reflect on the success or otherwise of your game.

Think about, for example:

- Explain how your game worked. What went well?
- What did you do as the game supervisor and what did your participants do as game players?
- Provide accurate analysis of the mathematics behind your game.
- Was your game popular? Why? Why not?
- What was your expected profit per game? Is this what you actually received? Why? Why not? Did the experimental probability match the theoretical probability?
- If the game was not fair, how could you change the game to make it fair?
- If anything didn't go as planned, what went wrong?
- What did you learn from other games that would improve your game? What would you improve or change if you did this task again?

Extension

As you reflect on this task and on this chapter consider one, some or all of these questions regarding mathematics, probability and gambling.

You can approach each of these questions from a combination of mathematics and TOK concepts.

- What does "the house always wins" mean? Is it **fair** that casinos should make a profit? Is there such a thing as "ethical gambling"?
- Could and/or should mathematics and mathematicians help increase incomes in gambling? What is the ethical responsibility of a mathematician?
- What is luck? How would a mathematician explain luck?

15 Exploration

All IB Diploma subjects have an Internal Assessment (IA). The IA in Mathematics is called an exploration. The exploration will be assessed internally by your teacher and externally moderated by the IB and counts for 20% of your final grade.

This chapter gives you advice on planning your exploration, as well as hints and tips to help you to achieve a good grade by making sure that your exploration satisfies the assessment criteria. There are also suggestions on choosing a topic and how to get started on your exploration.

About the exploration

The exploration is an opportunity for you to show that you can apply mathematics to an area that interests you. It is a piece of written work investigating an area of mathematics.

There are 30 hours in the syllabus for developing your mathematical toolkit and your exploration. The "toolkit" is the inquiry, investigative, problem solving and modelling skills you need to write a good exploration. You can build these skills throughout this book—in particular, in the Investigations, Developing Inquiry Skills and Modelling activities in each chapter.

You should expect to spend around 10–15 hours of class time on your exploration and up to 10 hours of your own time.

During **class time** you will:

- go through the assessment criteria with your teacher
- brainstorm to come up with suitable topics/titles
- look at previous explorations and the grading
- meet with your teacher to discuss your choice of topic and your progress.

During **your own time** you will:

- research the topic you have chosen, to make sure that it is appropriate for an exploration (if not, you will have to conduct further research to help you select a suitable topic)
- collect and organize your information/data and decide which mathematical processes to apply
- write your exploration
- submit a draft exploration to your teacher (your teacher will set a deadline for this)
- present your draft exploration to some of your peers, for their feedback.
- submit the final exploration (your teacher will set a deadline for this). If you do not submit an exploration then you receive a grade of "N" and will not receive your IB Diploma.

How the exploration is marked

After you have submitted the final version of your exploration your teacher will mark it. This is "internal assessment" (in school). Your teacher submits these marks to the IB, from which a random sample of explorations is selected automatically. Your teacher uploads these sample explorations to be marked by an external moderator. This external moderation of IA ensures that all teachers in all schools are marking students' work to the same standards.

To begin with, the external moderator will mark three of your school's explorations. If the moderator's mark is within 2 marks of your teacher's mark, then all your teacher's marks stay the same.

If the moderator's mark is more than 2 marks higher or lower than your teacher's mark, the external moderator will mark the remaining explorations in the sample. This may increase the mark if the teacher marked too harshly or decrease the mark if the teacher marked too leniently. The moderator sends a report to the school to explain the reason for any change in the marks.

IA tip

When coloured graphs are uploaded in black and white, they are very difficult to follow, as information such as colour-coded keys is lost. Make sure your exploration is uploaded in colour if it contains colour diagrams.

Internal assessment criteria

Your exploration will be assessed by your teacher, against the criteria given below. The IB external moderator will use the same assessment criteria.

The final mark for each exploration is the sum of the scores for each criterion. The maximum possible final mark is 20. This is 20% of your final mark for Mathematics: analysis and approaches Standard level.

The criteria cover five areas, A to E

Criterion A	Presentation
Criterion B	Mathematical communication
Criterion C	Personal engagement
Criterion D	Reflection
Criterion E	Use of mathematics

IA tip

Make sure that **all** the pages are uploaded. It is almost impossible to mark an exploration with pages missing.

IA tip

Make sure you understand these criteria. Check your exploration against the criteria frequently as you write it.

Criterion A: Presentation

This criterion assesses the organization, coherence and conciseness of your exploration.

Achievement level	Descriptor
0	The exploration does not reach the standard described by the descriptors below.
1	The exploration has some coherence or some organization.
2	The exploration has some coherence and shows some organization.
3	The exploration is coherent and well organized.
4	The exploration is coherent, well organized and concise.

To get a good mark for Criterion A: Presentation

- A **well organized** exploration has:
  - a **rationale** which includes an explanation of why you chose this topic
  - an **introduction** in which you discuss the context of the exploration
  - a statement of the **aim** of the exploration, which should be clearly identifiable
  - a **conclusion**.
- A **coherent** exploration:
  - is logically developed and easy to follow
  - should "read well" and express ideas clearly
  - includes any graphs, tables and diagrams where they are needed—not attached as appendices to the document.
- A **concise** exploration:
  - focuses on the aim and avoids irrelevancies
  - achieves the aim you stated at the beginning
  - explains all stages in the exploration clearly and concisely.
- References must be cited where appropriate. Failure to do so could be considered academic malpractice.

IA tip

For more on citing references, academic honesty and malpractice, see page 633.

Criterion B: Mathematical communication

This criterion assesses how you:

- use appropriate mathematical language (notation, symbols, terminology)
- define key terms, where required
- use multiple forms of mathematical representation, such as formulas, diagrams, tables, charts, graphs and models, where appropriate.

Achievement level	Descriptor
0	The exploration does not reach the standard described by the descriptors below.
1	There is some relevant mathematical communication which is partially appropriate.
2	The exploration contains some relevant, appropriate mathematical communication.
3	The mathematical communication is relevant, appropriate and is mostly consistent.
4	The mathematical communication is relevant, appropriate and consistent throughout.

IA tip

Only include forms of representation that are relevant to the topic, for example, don't draw a bar chart and pie chart for the same data. If you include a mathematical process or diagram without using or commenting on it, then it is irrelevant.

To get a good mark for Criterion B: Mathematical communication

- Use appropriate mathematical language and representation when communicating mathematical ideas, reasoning and findings.
- Choose and use appropriate mathematical and ICT tools such as graphic display calculators, screenshots, mathematical software, spreadsheets, databases, drawing and word-processing software, as appropriate, to enhance mathematical communication.

IA tip

Use technology to enhance the development of the exploration—for example, by reducing laborious and repetitive calculations.

- Define key terms that you use.
- Express results to an appropriate degree of accuracy.
- Label scales and axes clearly in graphs.
- Set out proofs clearly and logically.
- Define variables.
- Do not use calculator or computer notation.

Criterion C: Personal engagement

This criterion assesses how you engage with the exploration and make it your own.

Achievement level	Descriptor
0	The exploration does not reach the standard described by the descriptors below.
1	There is evidence of some personal engagement.
2	There is evidence of significant personal engagement.
3	There is evidence of outstanding personal engagement.

IA tip

Students often copy their GDC display, which makes it unlikely they will reach the higher levels in this criterion. You need to express results in proper mathematical notation, for example, use 2^x and not 2^x

use × not *

use 0.028 and not 2.8E-2

To get a good mark for Criterion C: Personal engagement

- Choose a topic for your exploration that you are interested in, as this makes it easier to display personal engagement
- Find a topic that interests you and ask yourself "What if...?"
- Demonstrate personal engagement by using some of these skills and practices from the mathematician's toolkit.
  - Creating mathematical models for real-life situations.
  - Designing and implementing surveys.
  - Running experiments to collect data.
  - Running simulations.
  - Thinking and working independently.
  - Thinking creatively.
  - Addressing your personal interests.
  - Presenting mathematical ideas in your own way.
  - Asking questions, making conjectures and investigating mathematical ideas.
  - Considering historical and global perspectives.
  - Exploring unfamiliar mathematics.

IA tip

Just showing personal interest in a topic is not enough to gain the top marks in this criterion. You need to write in your own voice and demonstrate your own experience with the mathematics in the topic.

Criterion D: Reflection

This criterion assesses how you review, analyse and evaluate your exploration.

Achievement level	Descriptor
0	The exploration does not reach the standard described by the descriptors below.
1	There is evidence of limited reflection.
2	There is evidence of meaningful reflection.
3	There is substantial evidence of critical reflection.

To get a good mark for Criterion D: Reflection

- Include reflection in the conclusion to the exploration, but also throughout the exploration. Ask yourself "What next?"
- Show reflection in your exploration by:
  - discussing the implications of your results
  - considering the significance of your findings and results
  - stating possible limitations and/or extensions to your results
  - making links to different fields and/or areas of mathematics
  - considering the limitations of the methods you have used
  - explaining why you chose this method rather than another.

IA tip

Discussing your results without analysing them is not meaningful or critical reflection. You need to do more than just describe your results. Do they lead to further exploration?

Criterion E: Use of mathematics

This criterion assesses how you use mathematics in your exploration.

Achievement level	Descriptor
0	The exploration does not reach the standard described by the descriptors below.
1	Some relevant mathematics is used.
2	Some relevant mathematics is used. Limited understanding is demonstrated.
3	Relevant mathematics commensurate with the level of the course is used. Limited understanding is demonstrated.
4	Relevant mathematics commensurate with the level of the course is used. The mathematics explored is partially correct. Some knowledge and understanding are demonstrated.
5	Relevant mathematics commensurate with the level of the course is used. The mathematics explored is mostly correct. Good knowledge and understanding are demonstrated.
6	Relevant mathematics commensurate with the level of the course is used. The mathematics explored is correct. Thorough knowledge and understanding are demonstrated.

To get a good mark for Criterion E: Use of mathematics

- Produce work that is commensurate with the level of the course you are studying. The mathematics you explore should either be part of the syllabus, at a similar level or beyond.
- If the level of mathematics is not commensurate with the level of the course you can only get a maximum of two marks for this criterion.
- Only use mathematics relevant to the topic of your exploration. Do not just do mathematics for the sake of it.
- Demonstrate that you fully understand the mathematics used in your exploration.

IA tip

Make sure the mathematics in your exploration is not only based on the prior learning for the syllabus. When you are deciding on a topic, consider what mathematics will be involved and whether it is commensurate with the level of the course.

- Justify **why** you are using a particular mathematical technique (do not just use it).
- Generalize and justify conclusions.
- Apply mathematics in different contexts where appropriate.
- Apply problem-solving techniques where appropriate.
- Recognize and explain patterns where appropriate.

Academic honesty

This is very important in all your work. Your school will have an Academic Honesty Policy which you should be given to discuss in class, to make sure that you understand what malpractice is and the consequences of committing malpractice.

According to the IB Learner Profile for Integrity:

"We act with integrity and honesty, with a strong sense of fairness and justice, and with respect for the dignity and rights of people everywhere. We take responsibility for our actions and their consequences."

Academic Honesty means:

- that your work is authentic
- that your work is your own intellectual property
- that you conduct yourself properly during examinations
- that any work taken from another source is properly cited.

Authentic work:

- is work based on your own original ideas
- can draw on the work of others, but this must be fully acknowledged in footnotes and bibliography
- must use your own language and expression
- must acknowledge all sources fully and appropriately in a bibliography.

IA tip

Reference any photographs you use in your exploration, including to decorate the front page.

Malpractice

The IB Organization defines malpractice as "behaviour that results in, or may result in, the candidate or any other candidate gaining an unfair advantage in one or more assessment components."

Malpractice includes:

- plagiarism—copying from others' work, published or otherwise, whether intentional or not, without the proper acknowledgement
- collusion—working together with at least one other person in order to gain an undue advantage (this includes having someone else write your exploration)
- duplication of work—presenting the same work for different assessment components

IA tip

Plagiarism detection software identifies text copied from online sources. The probability that a 16-word phrase match is "just a coincidence" is $\frac{1}{10^{12}}$.[1]

[1]Words & Ideas. The Turnitin Blog. Top 15 misconceptions about Turnitin. Misconception 11: matched text is likely to be completely coincidental or common knowledge (posted by Katie P., March 09, 2010).

- any other behaviour that gains an unfair advantage such as taking unauthorized materials into an examination room, stealing examination materials, disruptive behaviour during examinations, falsifying CAS records or impersonation.

Collaboration and collusion

It is important to understand the distinction between collaboration (which is allowed) and collusion (which is not).

Collaboration

In several subjects, including mathematics, you will be expected to participate in group work. It is important in everyday life that you are able to work well in a group situation. Working in a group entails talking to others and sharing ideas. Every member of the group is expected to participate equally and it is expected that all members of the group will benefit from this collaboration. However, the end result must be your own work, even if it is based on the same data as the rest of your group.

Collusion

This is when two or more people work together to intentionally deceive others. Collusion is a type of plagiarism. This could be working with someone else and presenting the work as your own or allowing a friend to copy your work.

IA tip

Discussing individual exploration proposals with your peers or in class before submission is collaboration.

Individually collecting data and then pooling it to create a large data set is collaboration. If you use this data for your own calculations and write your own exploration, that is collaboration. If you write the exploration as a group, that is collusion.

References and acknowledging sources

The IB does not tell you which style of referencing you should use—this is left to your school.

The main reasons for citing references are:

- to acknowledge the work of others
- to allow your teacher and moderator to check your sources.

To refer to someone else's work:

- include a brief reference to the source in the main body of your exploration—either as part of the exploration or as a footnote on the appropriate page
- include a full reference in your bibliography.

The bibliography should include a list with full details of **all** the sources that you have used.

IA tip

Be consistent and use the same style of referencing throughout your exploration.

IA tip

Cite references to others' work even if you paraphrase or rewrite the original text.

You do not need to cite references to formulas taken from mathematics textbooks.

Choosing a topic

You need to choose a topic that interests you, as then you will enjoy working on the exploration and you will be able to demonstrate personal engagement by using your own voice and demonstrating your own experience.

Discuss the topic you choose with your teacher and your peers before you put too much time and effort into developing the exploration. Remember that the work does not need to go beyond the level of the course which you are taking, but you can choose a topic that is outside the syllabus and is at a commensurate level. You should avoid choosing topics that are too ambitious, or below the level of your course.

These questions may help you to find a topic for your exploration:

- What areas of the syllabus are you enjoying most?
- What areas of the syllabus are you performing best in?
- Would you prefer to work on purely analytical work or on modelling problems?
- Have you discovered, through reading or talking to peers on other mathematics courses, areas of mathematics that might be interesting to look into?
- What mathematics is important for the career that you eventually hope to follow?
- What are your special interests or hobbies? Where can mathematics be applied in this area?

One way of choosing a topic is to start with a general area of interest and create a mind map. This can lead to some interesting ideas on applications of mathematics to explore. The mind map on the following pages shows how the broad topic "Transport" can lead to suggestions for explorations into such diverse topics as baby carriage design, depletion of fossil fuels and queuing theory.

On page 636 there is an incomplete mind map for you to continue, either on your own or by working with other mathematics students.

IA tip

You must include a brief reference in the exploration as well as in the bibliography. It is not sufficient just to include a reference in the bibliography.

IA tip

Your exploration should contain a substantial amount of mathematics at the level of your course, and should not just be descriptive. Although the history of mathematics can be very interesting it is not a good exploration topic.

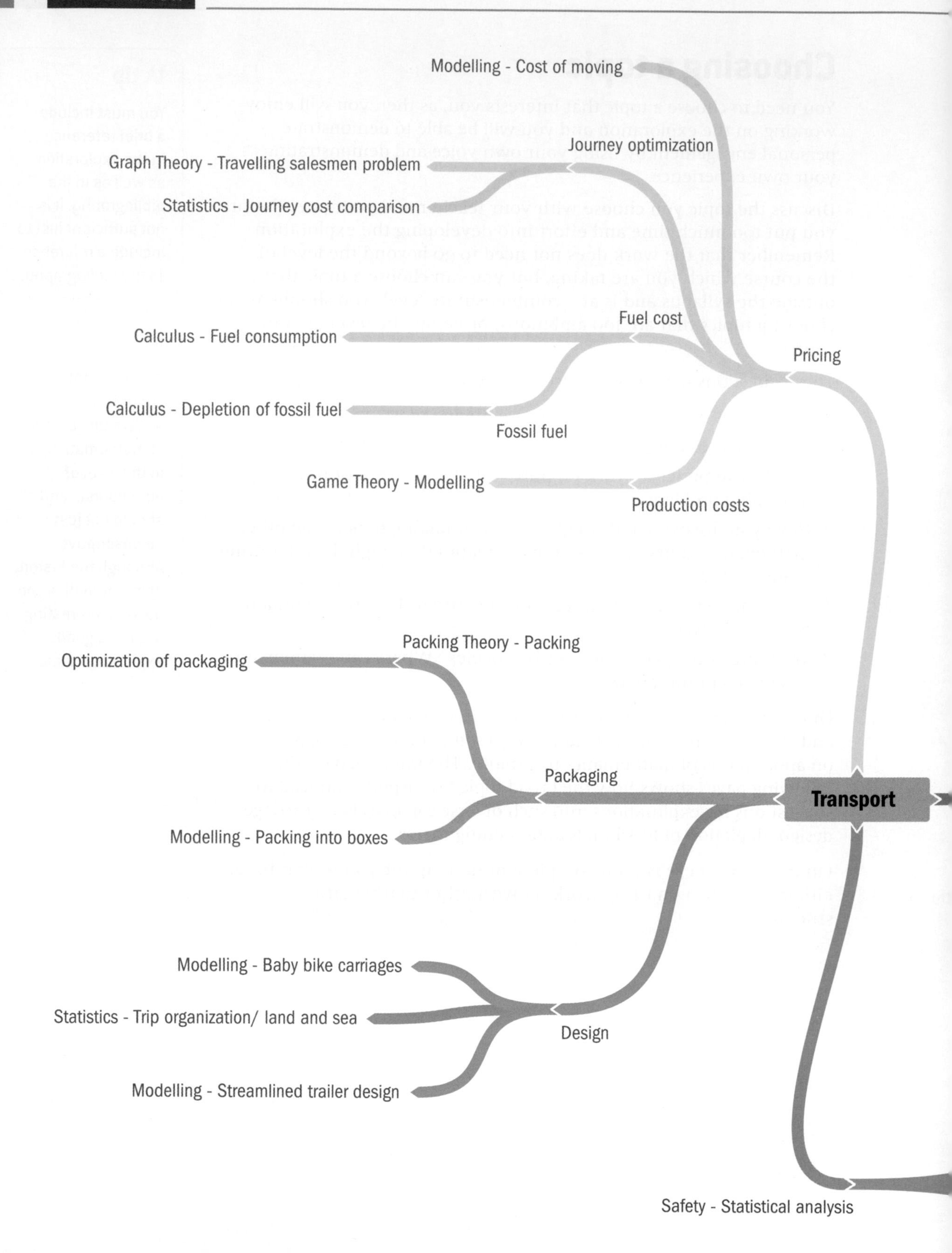
Modelling - Cost of moving
Journey optimization
Graph Theory - Travelling salesmen problem
Statistics - Journey cost comparison
Fuel cost
Calculus - Fuel consumption
Pricing
Calculus - Depletion of fossil fuel
Fossil fuel
Game Theory - Modelling
Production costs
Packing Theory - Packing
Optimization of packaging
Packaging
Transport
Modelling - Packing into boxes
Modelling - Baby bike carriages
Statistics - Trip organization/ land and sea
Design
Modelling - Streamlined trailer design
Safety - Statistical analysis

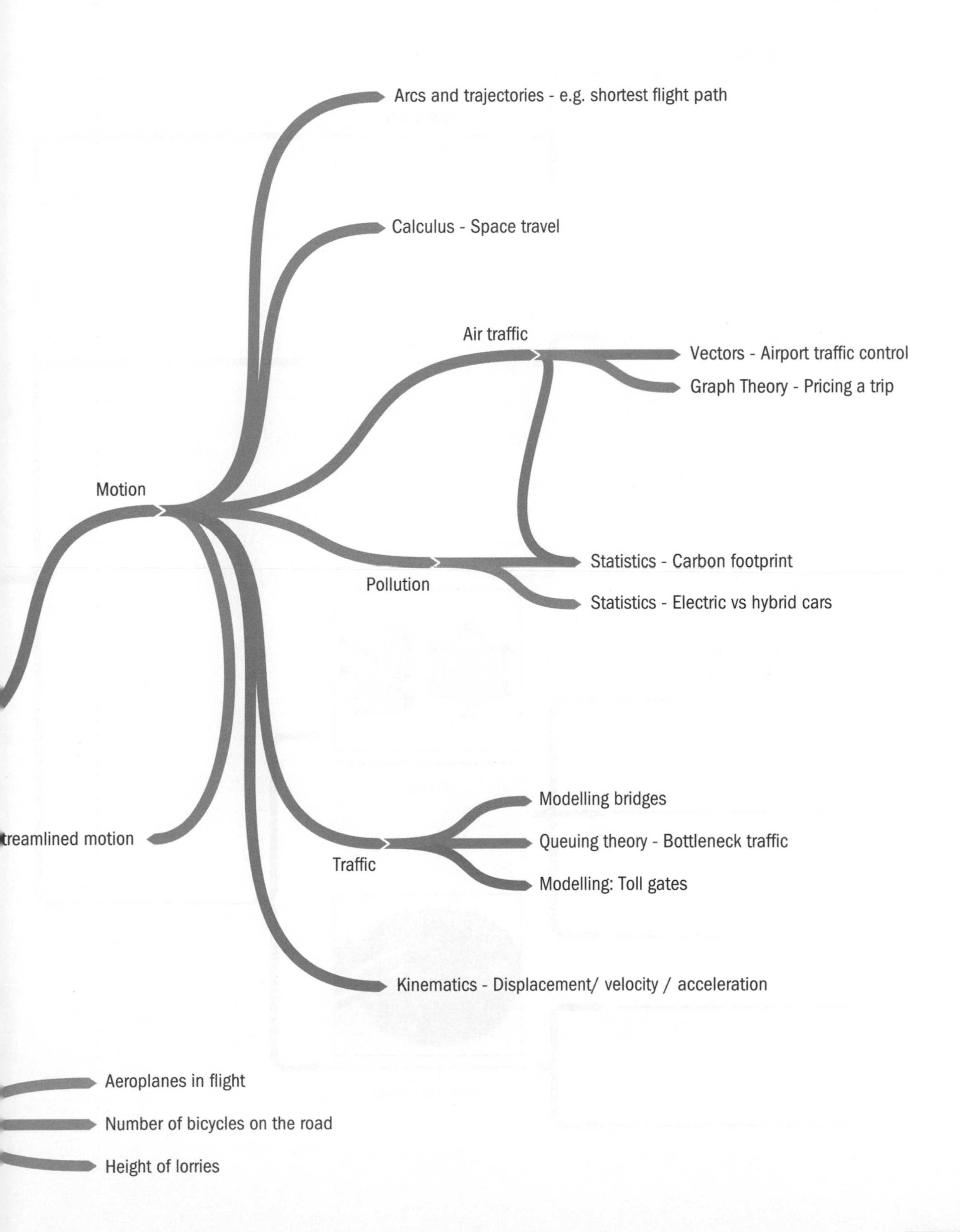

Arcs and trajectories - e.g. shortest flight path
Calculus - Space travel
Air traffic
Vectors - Airport traffic control
Graph Theory - Pricing a trip
Motion
Statistics - Carbon footprint
Pollution
Statistics - Electric vs hybrid cars
Modelling bridges
treamlined motion
Queuing theory - Bottleneck traffic
Traffic
Modelling: Toll gates
Kinematics - Displacement/ velocity / acceleration
Aeroplanes in flight
Number of bicycles on the road
Height of lorries

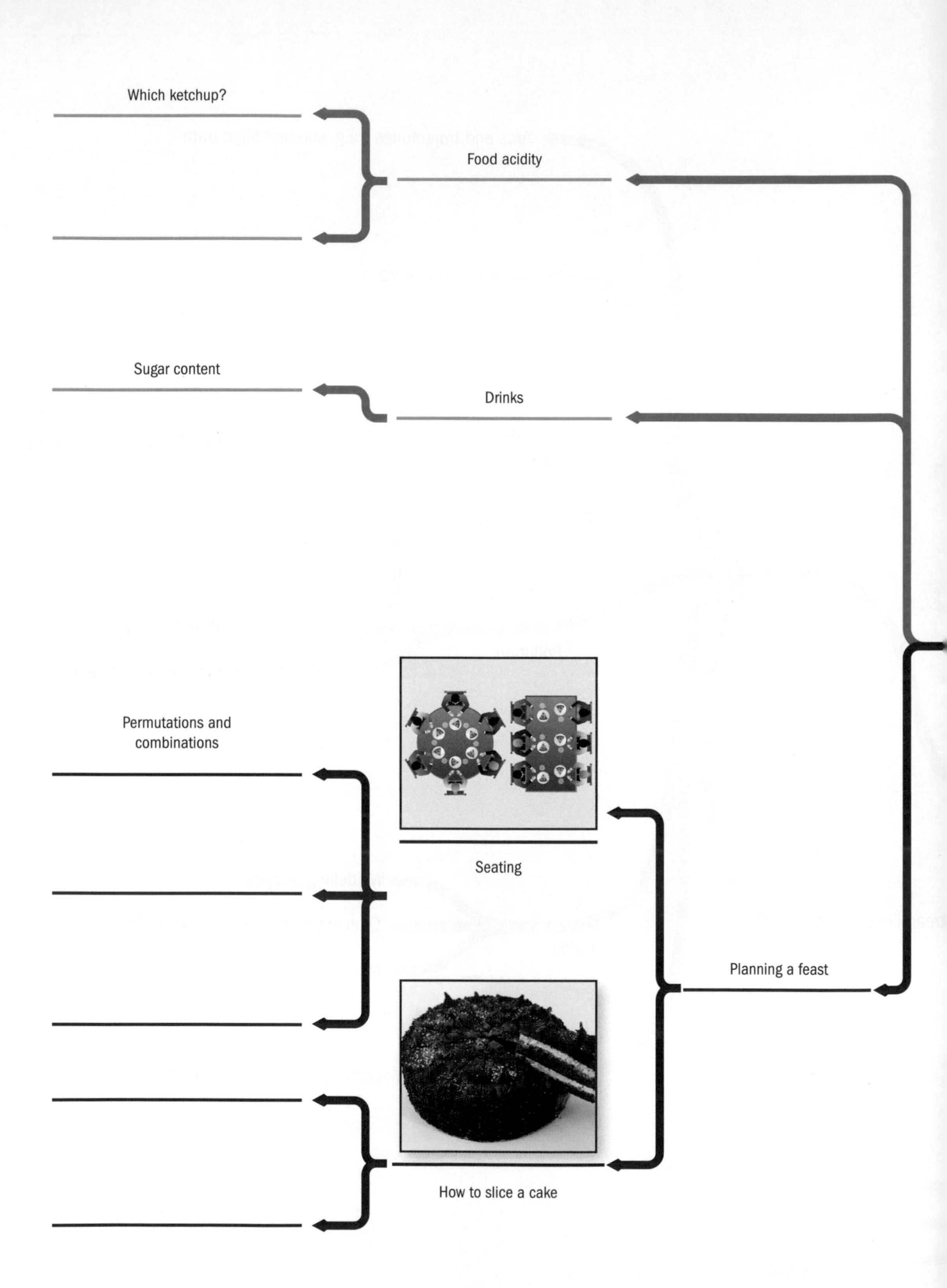
Which ketchup?
Food acidity
Sugar content
Drinks
Permutations and combinations
Seating
Planning a feast
How to slice a cake

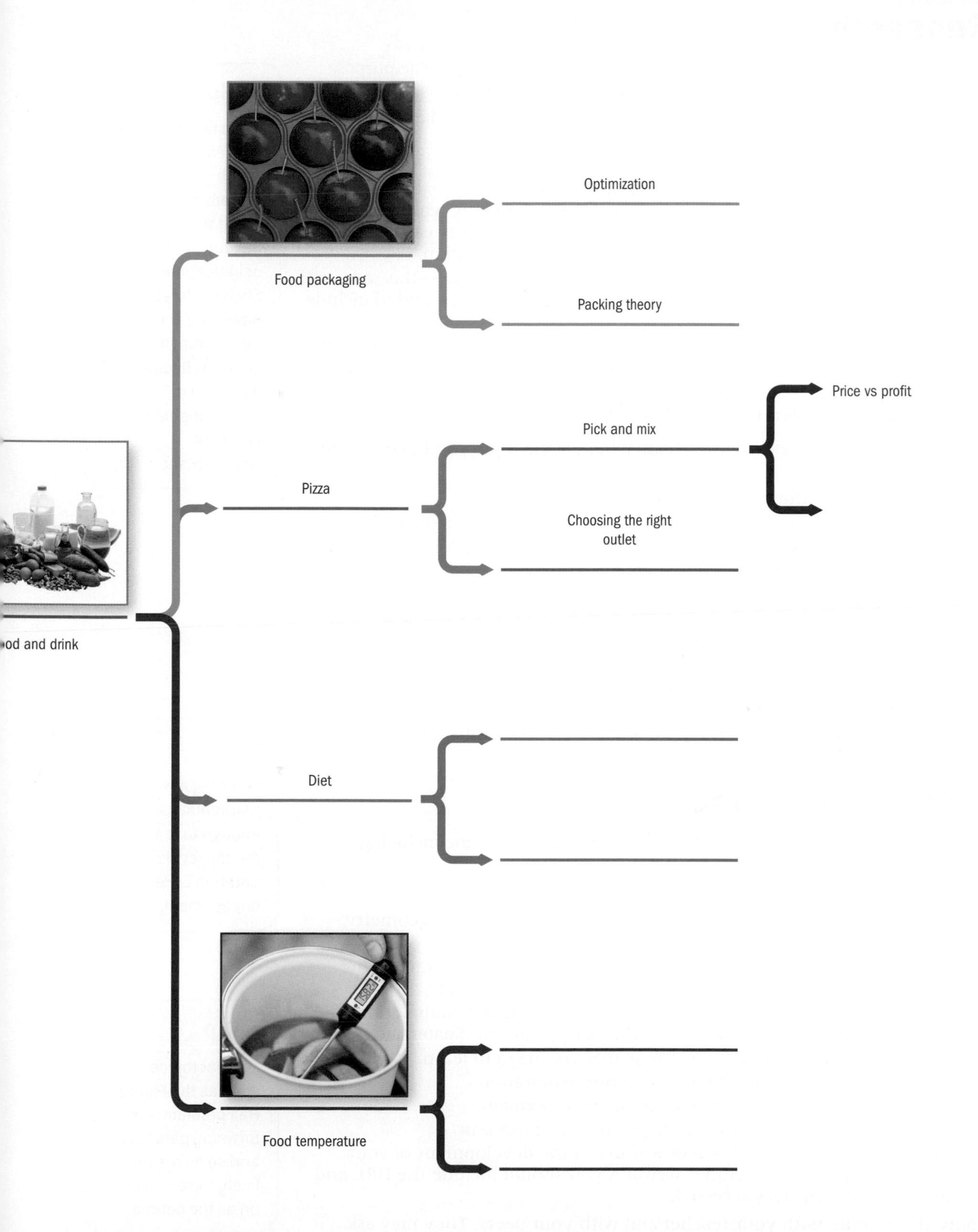

Optimization
Food packaging
Packing theory
Price vs profit
Pick and mix
Pizza
Choosing the right outlet
od and drink
Diet
Food temperature

Research

Once you have chosen a topic, you will need to do some research. The purpose of this research is to help you determine how suitable your topic is.

- Don't rely on the Internet for all your research—you should also make use of books and academic publications.
- Plan your time wisely—make sure that you are organized.
- Don't put it off—start your research in good time.
- For Internet research: refine your topic so you know exactly what information you are looking for, and use multiple-word searches. It is very easy to spend hours on the Internet without finding any relevant information.
- Make sure that you keep a record of all the websites you use—this saves so much time afterwards. You will need to cite them as sources, and to include them in your bibliography.
- Make sure that the sources are reliable—who wrote the article? Are they qualified? Is the information accurate? Check the information against another source.
- Research in your own language if you find this easier.

These questions will help you to decide whether the topic you have chosen is suitable.

- What areas of mathematics are contained in the topic?
- Which of these areas are contained in the syllabus that you are following?
- Which of these areas are not in the syllabus that you are following but are contained in the other IB mathematics course?
- Which of these areas are in none of the IB mathematics courses? How accessible is this mathematics to you?
- Would you be able to understand the mathematics and write an exploration in such a way that a peer is able to understand it all?
- How can you demonstrate personal engagement in your topic?
- Will you manage to complete an exploration on this topic and meet all the top criterion descriptors within the recommended length of 12 to 20 pages (double spaced and font size 12)?

> **IA tip**
>
> Try to avoid writing a research report in which you merely explain a well-known result that can easily be found online or in textbooks. Such explorations have little scope for meaningful and critical reflection and it may be difficult to demonstrate personal engagement

Writing an outline

Once you think you have a workable topic, write a brief outline including:

- Why you chose this topic.
- How your topic relates to mathematics.
- The mathematical areas in your topic, for example, algebra, geometry, calculus, etc.
- The key mathematical concepts covered in your topic, for example, modelling data, areas of irregular shapes, analysing data, etc.
- The mathematical skills you will use in the exploration, for example, integration by parts, working with complex numbers, using polar coordinates, etc.
- Any mathematics outside the syllabus that you need to learn.
- Technology you could use to develop your exploration.
- New key terms that you will need to define or explain.
- How you are going to demonstrate personal engagement.
- A list of any resources you have/ will use in the development of your exploration. If this list includes websites you should include the URL and the date when this was accessed.

Share this outline with your teacher and with your peers. They may ask questions that lead you to improve your outline.

> **IA tip**
>
> Learning new mathematics is not enough to reach the top levels in Criterion C: Personal engagement.

> **IA tip**
>
> Popular topics such as the Monty Hall problem, the Birthday paradox, and so on are not likely to score well on all the criteria.

This template may help you write the outline for the exploration when presenting a formal proposal to your teacher.

Mathematics exploration outline

Topic:
Exploration title:
Exploration aim:
Exploration outline:
Resources used:
Personal engagement:

IA tip

As you write your exploration, remember to refer to the criteria below.

Writing your exploration

Now you should be ready to start writing your exploration in detail.

You could ask one of your classmates to read the exploration and give you feedback before you submit the draft to your teacher. If your exploration is related to another discipline, for example, Economics, it would be better if the peer reading your exploration is someone who does not study economics.

IA tip

Remember that your peers should be able to read and understand your work.

Mathematical exploration checklist

Work through this checklist to confirm that you have done everything that you can to make your exploration successful.

- ☐ Does your exploration have a title?
- ☐ Have you given a rationale for your choice of exploration?
- ☐ Have you ensured your exploration does not include any identifying features—for example, your name, candidate number, school name?
- ☐ Does your exploration start with an introduction?
- ☐ Have you clearly stated your aim?
- ☐ Does your exploration answer the stated aim?
- ☐ Have you used double line spacing and 12-point font?
- ☐ Is your exploration 12–20 pages long?
- ☐ Have you cut out anything that is irrelevant to your exploration?
- ☐ Have you checked that you have not repeated lots of calculations?
- ☐ Have you checked that tables only contain relevant information and are not too long?
- ☐ Is your exploration easy for a peer to read and understand?
- ☐ Is your exploration logically organized?
- ☐ Are all your graphs, tables and diagrams correctly labelled and titled?
- ☐ Are any graphs, tables and diagrams placed appropriately and not all attached at the end?
- ☐ Have you used appropriate mathematical language and representation (not computer notation, eg *, ^, etc.)?
- ☐ Have you used notation consistently through your exploration?
- ☐ Have you defined key terms (mathematical and subject specific) where necessary?
- ☐ Have you used appropriate technology?
- ☐ Have you used an appropriate degree of accuracy for your topic/exploration?
- ☐ Have you shown interest in the topic?
- ☐ Have you used original analysis for your exploration (eg simulation, modelling, surveys, experiments)?
- ☐ Have you expressed the mathematical ideas in your exploration in your own way (not just copy-and-pasted someone else's)?
- ☐ Does your exploration have a conclusion that refers back to the introduction and the aim?
- ☐ Do you discuss the implications and significance of your results?
- ☐ Do you state any possible limitations and/or extensions?
- ☐ Do you critically reflect on the processes you have used?
- ☐ Have you explored mathematics that is commensurate with the level of the course?
- ☐ Have you checked that your results are correct?
- ☐ Have you clearly demonstrated understanding of why you have used the mathematical processes you have used?
- ☐ Have you acknowledged direct quotes appropriately?
- ☐ Have you cited all references in a bibliography?
- ☐ Do you have an appendix if one is needed?

Paper 1

Time allowed: 1 hour 30 minutes

Answer all the questions.

All numerical answers must be given exactly or correct to three significant figures, unless otherwise stated in the question.

Answers should be supported by working and/or explanations. Where an answer is incorrect, some marks may be awarded for a correct method, provided this is shown clearly.

You are not allowed to use a calculator for this paper.

Section A (Short answers)

1 [Maximum mark: 5]

Find the derivatives of the following functions.

a $y = x^4$ [1 mark]

b $y = \frac{1}{x^2}$ [1 mark]

c $y = x\sin x$ [2 marks]

d $y = e^{4x}$. [1 mark]

2 [Maximum mark: 4]

A single piece of wire has been bent to form the perimeter of a square, with side length of 2 cm. The wire is straightened out and then bent again to form an arc of a circle of radius 4 cm. The wire just forms the arc, not the radii. Calculate the angle, in radians, that the arc subtends at the centre of the circle. [4 marks]

3 [Maximum mark: 7]

Functions from $\mathbb{R} \to \mathbb{R}$ are defined by

$f(x) = 2x + 1$, $g(x) = 3x - 2$.

Find **a** $(f \circ g)(x)$ [2 marks]

b $(g \circ f)(x)$ [2 marks]

c $f^{-1}(x)$. [3 marks]

4 [Maximum mark: 8]

Let $\log x = p$ and $\log y = q$. Find

a $\log(xy)$ [2 marks]

b $\log\left(\frac{x^2}{y}\right)$ [2 marks]

c $\log\sqrt{x}$ [2 marks]

d $\log(100y)$. [2 marks]

5 [Maximum mark: 5]

Find the equation of the straight line that goes through the point (4, 3) and is perpendicular to the line $y = 2x + 11$. Give the answer in the form $y = mx + c$. [5 marks]

6 [Maximum mark: 6]

A piece of wood is 290 cm long. It is going to be cut into 10 pieces, each of which is 2 cm longer than the piece cut off before it. All the wood will be used up. Find

a the length of the first piece [4 marks]

b the length of the last piece. [2 marks]

7 [Maximum mark: 5]

Find $\int_0^4 \frac{1}{2x+1}\,dx$ giving the answer in the form In k, where $k \in \mathbb{Z}$. [5 marks]

Section B (Long answers)

8 [Maximum mark: 11]

a The probability distribution function for a discrete random variable, X, is given by

x	0	1	2	3
$P(X = x)$	a	0.2	0.3	0.1

i Find the value of a.

ii Find $E(X)$. [4 marks]

b Toby takes part in a sports match. A random variable, Y, represents Toby's winnings in euros. The probability distribution function for Y is given by

y	−2	1	2	3
$P(Y = y)$	b	c	0.1	0.2

If it is known that the game is a fair game, find the values of b and c. [7 marks]

9 [Maximum mark: 13]

Let $f(x) = x^2 + 6x + (8 + k)$.

a Find the discriminant, D, of this quadratic, in terms of k. [2 marks]

b Given that $f(x) > 0$ for all values of x, state, with reasons, the set of values that can be taken by **i** D **ii** k. [5 marks]

c Write $f(x)$ in the form $f(x) = (x + p)^2 + q$, where $p \in \mathbb{Z}$ and q is an expression involving k. [2 marks]

d Explain how the form obtained in **c** confirms the answer to **b ii**. [2 marks]

e If $k = 4$, write down the minimum point of the quadratic. [2 marks]

10 [Maximum mark: 16]

A farmer has a tractor in a field at point A, 2 km due south of point B, on a road which runs from west to east. The farmer wishes to get to point C, which is 2 km east of B, on the road. The tractor travels at 20 km/h on the road and 12 km/h in the field. The tractor will travel in a straight line across the field, to join the road at point, P, x km east of B.

a Sketch a diagram to represent this information. [1 mark]

b Show that the total time T for the tractor to travel from A to C, is given by

$$T = \frac{\sqrt{4 + x^2}}{12} + \frac{2 - x}{20}$$

[4 marks]

c Use calculus to find the value of x that minimises the journey time, T. [11 marks]

Paper 2

Time allowed: 1 hour 30 minutes

Answer all the questions.

All numerical answers must be given exactly or correct to three significant figures, unless otherwise stated in the question.

Answers should be supported by working and/or explanations. Where an answer is incorrect, some marks may be awarded for a correct method, provided this is shown clearly.

You need a graphic display calculator for this paper.

Section A (Short answers)

1 [Maximum mark: 7]

a **i** Factorize $x^2+9x+14$.

ii Hence solve $x^2+9x+14=0$. [3 marks]

b Solve $x^2+4x+2=0$, giving your answers

i exactly

ii to 4 significant figures. [4 marks]

2 [Maximum mark: 6]

The random variable X is normally distributed, with mean equal to 8. Given that $P(X>7)=0.69146$, find the value of the standard deviation of X. [6 marks]

3 [Maximum mark: 9]

Abi plants ten seeds. The probability that a seed grows into a plant is always 0.8 (this is independent of what the other seeds do). Let the random variable, X, be the number of seeds that grow into plants.

a State the distribution that X satisfies, including the parameters. [2 marks]

b Write down the mean of X. [1 mark]

c Calculate $P(X=7)$. [1 mark]

d Calculate $P(X>6)$. [2 marks]

e Given that $X>6$ find the probability that $X=7$. [3 marks]

4 [Maximum mark: 7]

Maria sees and is frightened by a huge monster. The angle of elevation from Maria's feet to the top of the monster's head is 40°. Maria runs (on horizontal ground) 5 m further away from the monster. The angle of elevation from Maria's feet to the top of the monster's head is now 35°.

a Sketch a diagram to represent this story. [1 mark]

b Find the height of the monster. [6 marks]

5 [Maximum mark: 7]

An object is moving in a straight line from the origin with velocity v m/s given by $v = 5\sin\left(t^2\right)$, where t is the time in seconds from when the experiment started.

a Write down the initial velocity of the object. [1 mark]

b Find the total distance that the object travels in the first 3 seconds. [3 marks]

c Find the displacement of the object from the origin, when $t = 3$. [3 marks]

6 [Maximum mark: 4]

Find the constant term in the expansion of $\left(x + \frac{2}{x}\right)^6$ [4 marks]

Section B (Long answers)

7 [Maximum mark: 16]

Data is collected from ten football teams in the same league. The amount that each team spent on players' wages, w, in millions of pounds and the number of goals, g, that the team scored is recorded in the following table.

Wages, w	200	185	190	150	160	120	50	100	40	10
Goals, g	92	87	85	81	74	69	62	54	41	32

a For the amounts spent on wages calculate

i the mean **ii** the standard deviation. [3 marks]

b For the goals scored calculate

i the mean **ii** the standard deviation. [3 marks]

c **i** Calculate the Pearson product moment correlation coefficient.

ii Describe the linear correlation, using two words.

iii Write down the line of best fit g on w. [6 marks]

d Another club in the same league spent £80 million on players' wages. Estimate, to the nearest integer, the number of goals that they scored. [2 marks]

e Yet another club in the same league scored 17 goals. Give two comments on the validity of using the answer from **c iii** to estimate the club's wage bill. [2 marks]

8 [Maximum mark: 13]

A function is defined by $f(x) = \dfrac{6x+3}{2x-10}, x \in \mathbb{R}, x \neq 5.$

a **i** Write down the equation of the vertical asymptote.

ii Write down the equation of the horizontal asymptote. [2 marks]

b **i** Write down the intercept on the y-axis.

ii Write down the intercept on the x-axis. [2 marks]

c **i** Find $f'(x)$.

ii State what this indicates about the graph of $f(x)$. [4 marks]

d Sketch the graph of $f(x)$, showing the information found in **a**, **b** and **c**. [3 marks]

e Write down the equation of the tangent to the curve at the point where $x = 2$. [2 marks]

9 [Maximum mark: 11]

Anna invests $200 in a bank that gives a compound interest rate of 5% per year.

a **i** Calculate how much money will she have after 3 complete years. Give the answer to two decimal places.

ii Calculate how many complete years she would have to wait until her money has more than doubled. [4 marks]

Anna wishes to buy a mountain bike which costs $350 but is depreciating in value by 6% per year.

b **i** Calculate how many complete years it will take for the bike to cost less than $300.

ii Calculate how many complete years it will take until Anna can afford the bike. [4 marks]

c If after two years the interest rate changes to only 2% and the depreciation rate changes to 4%, calculate how many complete years (from when she first invested the $200) it will now take until Anna can afford the bike. [3 marks]

Answers

Chapter 1

Skills check

1 a $x=-3$ b $a=2$ c $x=4$

2 a $\frac{5}{12}$ b $\frac{53}{48}$ c $-\frac{6}{5}$

3 a 128 b 9 c −81

Exercise 1A

1 a −20, −23, −26

b 49, 64, 81

c 30, 36, 42

d $\frac{125}{2}, \frac{125}{4}, \frac{125}{8}$, or 62.5, 31.25, 15.625

e $\frac{5}{6}, \frac{6}{7}, \frac{7}{8}$

f $\frac{5}{243}, \frac{6}{729}, \frac{7}{2187}$

2 a $u_n = 10 \times 5^{n-1}$, geometric

b $u_n = -6n + 47$, arithmetic

c $u_n = (-1)^{n+1}\frac{1}{3^n}$, geometric

d $u_n = u_{n-1} + u_{n-2}$, neither

e $u_n = \frac{2n-1}{2n}$, neither

f $u_n = -4 \times 3^n$, geometric

3 a 100, 200, 300, …, $u_n = 100n$, arithmetic

b $6, 3, \frac{3}{2}, \ldots, u_n = 6\left(\frac{1}{2}\right)^{n-1}$, geometric

c 70, 77, 84.7,…, $u_n = 70(1.1)^{n-1}$, geometric

Exercise 1B

1 a 1, −4, 16, −64, 256

b $3, -\frac{2}{3}, 3, -\frac{2}{3}, 3$

c −1, 2, 8, 128, 32 768

d m, $3m + 5$, $9m + 20$, $27m + 65$, $81m + 200$

2 a $u_n = u_{n-1} - 2$, $u_1 = -2$

b $u_n = 4u_{n-1}$, $u_1 = 1$

c $u_n = \frac{u_{n-1}}{10}$, $u_1 = 52$

d $u_n = u_{n-1} + 5$, $u_1 = 14$

e $u_1 = 2$, $u_2 = 3$, $u_{n+1} = u_n \times u_{n-1}$

f $u_n = (n+1)(u_{n-1})$, $u_1 = 1$

Exercise 1C

1 a

$$\sum_{n=1}^{4}(-1)^n(n+1) = -2+3-4+5 = 2$$

b

$$\sum_{n=2}^{6}4n-3 = 5+9+13+17+21 = 65$$

c $\sum_{n=1}^{3}n(n+1) = 2+6+12 = 20$

d $\sum_{n=3}^{5}\frac{(-1)^{n+1}}{n-2} = 1+\frac{-1}{2}+\frac{1}{3} = \frac{5}{6}$

2 a $\sum_{n=1}^{\infty}4^n$ b $\sum_{n=3}^{5}\frac{n}{n+1}$

c $\sum_{n=1}^{100}(-1)^n\frac{1}{n}$ d $\sum_{n=1}^{8}-2$

e $\sum_{n=2}^{\infty}n^2+1$

f $\sum_{n=7}^{11}n^2m^{n-1}$ or $\sum_{n=6}^{10}(n+1)^2m^n$

Exercise 1D

1 $u_9 = 69$

2 $u_{11} = -40$

3 $u_7 = 6.97$

4 $u_6 = \frac{13}{6}$

5 $u_9 = x + 26$

6 $u_{12} = 36a$

Exercise 1E

1 $u_1 = 105$

2 $d = -4.86$

$u_{15} = -71.74$

3 $n = 4$

4 $d = 60$

$u_{14} = 664$

5 $n = 7$;

the 7th term.

6 $u_{12} = 52$ seats

7 $n = 11$

Since they are held in 2050 (since n is a natural number), the next year they will be held is 2054.

8 $n = 8$;

in 7 weeks

Exercise 1F

1 a $u_6 = 2187$

b Not geometric

c $u_7 \approx 1.068$ d $u_8 \approx 68.3$

e $u_{13} = \frac{4}{1\,953\,125}$

f Not geometric

g $u_{12} = 3m^{11}$

2 $u_{30} = 536\,870\,912$ cents or \$5 368 709.12

3 Use an r value that is a factor of 64. For example, $r = 2$:

2, 4, 8, 16, 32,...

Exercise 1G

1 $r = 1.5$

$u_{15} \approx 2310$ (3 sf)

2 $u_{20} = -56.3$ (3 sf)

3 $r = \pm\frac{1}{4}$

$x = 2$

or

$x = -6$

4 $n = 9$

5 $r = 2$

6 $u_{10} = 303$ students

7 a $u_{30} = 536\ 870\ 912$ grains

b $n = 10$th square

8 $r = 2$

8, 16, 32, 64, 128

Exercise 1H

1 This is geometric because you are multiplying each previous height by $\frac{1}{2}$.

$u_{10} = \frac{1}{512}$ metres

2 This is arithmetic because you are adding more money to your account every month.

$x = \$123.57$

3 This is arithmetic because you are adding from year to year.

$n = 5.58\bar{3}$

Finland did not gain independence in the year of the tiger.

4 This is geometric because a rate implies you are multiplying.

$r \approx 122\%$

Exercise 1I

1 a $d = \frac{1}{3}$

$S_7 = \frac{42}{5}$

b $u_1 = -\frac{3}{2}$

$u_2 = \frac{9}{2}$

$r = -3$

$S_8 = 2460$

c $r = 0.5$

$S_8 \approx 0.199$

d $d = 6$

$n = 48$

$S_{48} = 7056$

e $u_1 = 4,\ u_2 = 8,\ d = 8 - 4 = 4$

As 1000 is a multiple of 4, the largest multiple of 4 less than 999 would be 996.

$n = 249$

$S_{249} = 124\ 500$

f $u_1 = 2$

$u_2 = -4$

$r = -2$

$S_6 = -42$

2 $d = 4$

$S_{30} = 2400$ seats

3 The series is $1 + 2 + 4 + \ldots$

$S_6 = 63$ family members

4 The series is $1 + 2 + 3 + \ldots$

$d = 1$

$S_{12} = 78$

But since there are two 12-hour cycles in a 24-hour day:

$S_{24} = 78 \times 2 = 156$ chimes

5 The series is $5 + 9 + 13 + \ldots$

$S_{48} = 4752$ line segments

Exercise 1J

1 a Not converging as $r = 1.5$.

b $r = -\frac{3}{4}$

$S_\infty = \frac{-3}{14}$

c $r = \frac{u_n}{u_{n-1}} = \frac{\frac{-5}{4}}{\frac{-5}{2}} = \frac{-5}{4} \times \frac{2}{-5} = \frac{1}{2}$

$S_\infty = \frac{u_1}{1-r} = \frac{-\frac{5}{2}}{1-\frac{1}{2}} = \frac{-\frac{5}{2}}{\frac{1}{2}}$

$= \frac{-5}{2} \times \frac{2}{1} = -5$

d Not converging as $r = -2$

e $r = \frac{1}{3}$

$S_\infty = \frac{81}{2} x - \frac{81}{2}$

f Not converging as $r = 2$

g $r = \frac{1}{2}$

$S_\infty = \frac{2}{\sqrt{2}}$

2 Any infinite geometric series where $|r| < 1,\ r \neq 0$

3 Any infinite geometric series where $r < -1,\ r > 1$

4 $S_\infty = 30$

Total distance = 48 ft.

5 17 892 000 gallons

Exercise 1K

1 a $d = 5$

b i $u_{28} = 127$

ii $S_{28} = 1666$

c $n = 31$

2 $d = 3,\ u_1 = -11$

3 $u_1 = 16$

4 $r = \frac{2}{3}$

5 Choose any r value $-1 < r < 1,\ r \neq 0$.

Example:

if $r = \frac{1}{2}$

$\therefore$ The series is $4 + 2 + 1 + \ldots$

6 $u_{10} = 1280$

7 By GDC, $n = 4.69116\ldots$

A minimum of 5 rounds are required.

8 a Since the geometric series has only positive terms,

$r = \frac{2}{3}$

b $u_1 = 9$

9 a $r = \frac{m+8}{6}$

b i $m = -11\ m = 4$

ii $r = 2$

c i Since the sum of an infinite series can onlybe found when $-1 < r < 1,\ r \neq 0$,

$r = -\frac{1}{2}$.

ii $S_\infty = -8$

Exercise 1L

1 a $A = 2400$

$I = \$900$

b $A = 35\,200$

$I = 3200$ GBP

c $A = 15\,043\,348.839\,948\ldots$

$I \approx 875\,348.84$ Yen

d $A = P(1 + r)^n$

$$A = 300\,000\left(1+\frac{0.04}{365}\right)^{2\times 365}$$

$$A = 300\,000\left(1+\frac{0.04}{365}\right)^{730}$$

$A = 324\,984.695\,81\ldots$

$I = 324\,984.70 - 300\,000 \approx 24\,984.70$ Mexican Pesos

e $A = 438\,532.634627\ldots$

$I \approx 188\,532.63$ Swiss Francs

2 i $r = 0.000\,649\,855\ldots$

Fernando will pay an annual simple interest rate of 3.38%.

ii 23846 Columbian Pesos.

3 $A \approx \$100\,705.89$

4 $P \approx \$26\,410.17$

5 $n \approx 21.4$ years

6 $P \approx 24\,500$ Brazilian Reals

7 Oliver:

$A \approx 425.78$ GBP

Harry:

$A \approx 436.25$ GBP

Harry earned more than Oliver.

8 Savings account:

$A = 20\,000(1 + 0.012n)$

GIC:

$n \approx 6.03$ years

Exercise 1M

1 a $r = \frac{11000}{12\,500} = 0.88$

$C = 12\,500 \times 0.88^t$, where C represents the white blood cell count and t is the time every 12 hours.

b 3 days = 72 hours = 6 12-hour periods

$C = 12\,500 \times 0.88^6$
$= 5805.0510\ldots$

$C \approx 5805$ cells mcL^{-1}

c The limitation of the general formula is that white blood cell count does not continue to decrease infinitely. Once the antibiotics killed the infection, the patient's white blood cell count would return to normal.

2 a This is an arithmetic sequence since the rate decreases by −0.2% each month.

b $U = 7.9 - 0.2(t - 1)$, where U represents the unemployment rate and t is the month starting with January.

c $U = 7.9 - 0.2(12 - 1)$

$U = 7.9 - 0.2(11)$

$U = 5.7\%$

d It is not realistic. There will always be people you are not capable of working, or are switching jobs, or looking for jobs.

3 a This means that it takes 1.23 years for the substance to decrease to half of the original mass.

b $A = A_0\left(\frac{1}{2}\right)^{\frac{t-1}{h}}$ where A is the amount remaining after t years, A_0 is the original mass, and h is the half-life.

d $$A = 52\left(\frac{1}{2}\right)^{\frac{7.2-1}{1.23}}$$

$$A = 52\left(\frac{1}{2}\right)^{5.040\,650\,4\ldots}$$

$A = 1.579\,85\ldots$

$A \approx 1.58$ g

Exercise 1N

1 $x^4 + 20x^3 + 150x^2 + 500x + 625$

2 $-b^5 + 10b^4 - 40b^3 + 80b^2 - 80b + 32$

3 $64x^6 - 192x^5 + 240x^4 - 160x^3 + 60x^2 - 12x + 1$

4 $256x^4 + 256x^3y + 96x^2y^2 + 16xy^3 + y^4$

5 $x^3 - 9x^2y + 27xy^2 - 27y^3$

6 $243x^5 + 1620x^4y + 4320x^3y^2 + 5760x^2y^3 + 3840xy^4 + 1024y^5$

Exercise 1O

1 a $451\,068\,750x^7$

b $75\,582\,720x^2y^8$

c $-4320y^3$

d $-19\,683$

e $-\dfrac{2187}{x^7}$

2 a 1, 4, 6, 4, 1

b $-96x$

3 a $-13\,608x^3$

b $-11\,340x^5$

4 $336\,798x^9$

5 $r = 5$

$k = 2$

6 $k = 6$

7 $-20.$

8 16.003

9 a $x^5 - 25x^4 + 250x^3 - 1250x^2 + 3125x - 3125$

b $525x^4$

10 a $(3 - 2x)^4 = 16x^4 - 96x^3 + 216x^2 - 216x + 81$

$(-2x + 3)^4 = 16x^4 - 96x^3 + 216x^2 - 216x + 81$

b No, when the exponent is odd, the expansions will not be the same.

11 $(a + b)^n = (3x - 4y)^3$

12 $2^n = (1 + 1)^n$

$$2^n = \binom{n}{0}(1)^n(1)^0 + \binom{n}{1}(1)^{n-1}(1)^1 + \binom{n}{2}(1)^{n-2}(1)^2 + \ldots$$

$$\binom{n}{n-1}(1)^1(1)^{n-1} + \binom{n}{n}(1)^0(1)^n$$

Since $1^x = 1$ for any $x \in \mathbb{R}$,

$$2^n = \binom{n}{0} + \binom{n}{1} + \binom{n}{2} + \ldots$$

$$\binom{n}{n-1} + \binom{n}{n}$$

Exercise 1P

1

LHS	RHS
$-2(a-4)+3(2a+6)-6(a-5)$	$-2(a-28)$
$-2a+8+6a+18-6a+30$	
$-2a+56$	
$-2(a-28)$	
LHS ≡ RHS	

2

LHS	RHS
$(x-3)^2+5$	$x^2-6x+14$
$x^2-6x+9+5$	
$x^2-6x+14$	
QED	

3

LHS	RHS
$\dfrac{1}{m}$	$\dfrac{1}{m+1}+\dfrac{1}{m^2+m}$
	$\dfrac{1}{m+1}+\dfrac{1}{m(m+1)}$
	$\dfrac{m}{m(m+1)}+\dfrac{1}{m(m+1)}$
	$\dfrac{m+1}{m(m+1)}$
	$\dfrac{1}{m}$
	QED

4 a

LHS	RHS
$\dfrac{x-2}{x} \div \dfrac{3x-6}{x^2+x}$	$\dfrac{x+1}{3}$
$\dfrac{x-2}{x} \cdot \dfrac{x^2+x}{3x-6}$	
$\dfrac{x-2}{x} \cdot \dfrac{x(x+1)}{3(x-2)}$	
$\dfrac{\cancel{x-2}}{\cancel{x}} \cdot \dfrac{\cancel{x}(x+1)}{3\cancel{(x-2)}}$	
$\dfrac{x+1}{3}$	
LHS ≡ RHS	

b $x \neq -1, 0, 2$

Chapter review

1 i a This sequence is not arithmetic since $18 - 6 \neq 6 - 3$.
This sequence is not geometric since $\dfrac{18}{6} \neq \dfrac{6}{3}$.

ii a This sequence is arithmetic since $-12 - -14 = 2$.

b $u_n = 2n - 18$

c $u_{10} = 2$

d $S_8 = -72$

iii a This sequence is geometric since $\dfrac{500}{1000} = \dfrac{1}{2}$.

b $u_n = u_1 r^{n-1}$

$u_n = 2000 \times \left(\dfrac{1}{2}\right)^{n-1}$

c $u_9 \approx 7.81$

d $S_7 = 3968.75$

iv a This is a geometric sequence since $\dfrac{12}{6} = \dfrac{6}{3} = 2$.

b $u_n = 3 \times (2)^{n-1}$

c $u_5 = 48$

d $S_{10} = 3069$

v a This sequence is arithmetic since $115-110=110-105=5$.

b $u_n = 5n + 100$

c $u_7 = 135$

d $S_9 = 1125$

2 $S_{16} = -550$

3 The first five terms are $-4, 11, -19, 41, -79$.

4 $n = 4$

5 $r = 1.5$

$u_{-1} = 2$

6 Sequence A. An infinite sum can only be found for a converging geometric sequence.

$$r=-0.5;\ S_\infty = \frac{\frac{1}{4}}{\frac{3}{2}} = \frac{1}{6}$$

7 $n = 20$

8 $r = 2$

$u_9 = 768$

9 $x = 7$ or $x = -6$

10 a 55, 51.15, 47.5695, 44.239635…

b It is a geometric sequence because $\frac{47.5695}{51.15} = 0.93$

c $u_{11} \approx 26.6$ litres left in the tank

d 36.5 litres drained from the tank

e $S_\infty \approx 785$ minutes or 13 hours and 5 minutes

11 a $u_n = 4n + 41$

b $u_{10} = 81$ cm

c Eventually the spring will hit the ground or the surface it is sitting on, so the length will become constant. Also, the spring could break from too much weight.

d 1500 g or 1.5 kg

12 $r \approx 2.248\%$

13 a 1, 13, 78, 286, 715, 1287, 1716, 1716, 1287, 715, 286, 78, 13, 1

Each row in Pascal's triangle is symmetric.

b 1, 14, 91, 364, 1001, 2002, 3003, 3432, 3003, 2002, 1001, 364, 91, 14, 1

14 $729x^6 - 1458x^5y + 1215x^4y^2 - 540x^3y^3 + 135x^2y^4 - 18xy^5 + y^6$

15 The coefficient is $-870\,912$.

16 $n = 5$

17 a $10\,206x^5$ b $-20412x^6$

18 $a = 9$

$k = 2$

19 $(2x-1)(x-3) - 3(x-4)^2 = -x^2 + 17x - 45$

$2x^2 - 7x + 3 - 3(x^2 - 8x + 16) = -x^2 + 17x - 45$

$2x^2 - 7x + 3 - 3x^2 + 24x - 48 = -x^2 + 17x - 45$

$-x^2 + 17x - 45 = -x^2 + 17x - 45$

QED

20 a

$$\frac{x^2-x-6}{x+4}\cdot\frac{x^2-16}{x^2+2} = \frac{x^2-7x+12}{x}$$

$$\frac{(x-3)(x+2)}{x+4}\cdot\frac{(x-4)(x+4)}{x(x+2)} = \frac{x^2-7x+12}{x}$$

$$\frac{(x-3)\cancel{(x+2)}}{\cancel{x+4}}\cdot\frac{(x-4)\cancel{(x+4)}}{x\cancel{(x+2)}} = \frac{x^2-7x+12}{x}$$

$$(x-3)\cdot\frac{(x-4)}{x} = \frac{x^2-7x+12}{x}$$

$$\frac{x^2-7x+12}{x} = \frac{x^2-7x+12}{x}$$

QED

b $x \neq -4, -2, 0$

Exam-style questions

21 a

$$\left(1-\frac{x}{4}\right)^5 = \binom{5}{0}1^5 + \binom{5}{1}1^4\left(-\frac{x}{4}\right) + \binom{5}{2}1^3\left(-\frac{x}{4}\right)^2 + \binom{5}{3}1^2\left(-\frac{x}{4}\right)^3 + \binom{5}{4}1^1\left(-\frac{x}{4}\right)^4 + \binom{5}{5}\left(-\frac{x}{4}\right)^5$$ (2 marks)

$$= 1 - \frac{5x}{4} + \frac{5x^2}{8} - \frac{5x^3}{32} + \frac{5x^4}{256} - \frac{x^5}{1024}$$ (1 mark)

b Substituting $x = 0.1$ (1 mark)

$$0.975^5 \approx 1 - \frac{5\times 0.1}{4} + \frac{5\times 0.1^2}{8}$$

$$= 1 - \frac{1}{8} + \frac{5}{800}$$ (1 mark)

$$= \frac{800}{800} - \frac{100}{800} + \frac{5}{800}$$

$$= \frac{705}{800}\left(=\frac{141}{160}\right)$$ (1 mark)

22 a Using $u_n = a+(n-1)d$ (1 mark)

$143 = a+14d$

$183 = a+30d$ (1 mark)

Solving simultaneously (1 mark)

$a = 108$ (1 mark)

$d = \frac{5}{2}$ (1 mark)

b 100th term is $a+99d$ (1 mark)

$= 108+99\times\frac{5}{2}$

$= 355.5$ (1 mark)

23 a Money in the account would be

3000×1.015^{10} $(= \$3482)$ (2 marks)

Therefore, interest gained is

$3000\times 1.015^{10} - 3000 = \$482$ (1 mark)

b Total amount is

$3000\times 1.015^{11} + \left(1200\times 1.015+1200\times 1.015^2+\ldots +1200\times 1.015^{10}\right)$ (1 mark)

$= 3000\times 1.015^{11}+(1200\times 1.015)\left(\frac{1.015^{10}-1}{1.015-1}\right)$ (2 marks)

$= \$16\,570$ (1 mark)

24 a 5500×1.0275^4 (2 marks)

$= \$6130.42$ (1 mark)

b Consider

$5500\times 1.0275^n = 12\,000$ (2 marks)

$1.0275^n = \frac{12\,000}{5500}$

Using GDC: (1 mark)

$n = 29.76$ (1 mark)

So Brad must wait 30 years (1 mark)

25 Require ($3\times$ coefficient of term in x^5) $+(1\times$ coefficient of term in $x^4)$

$3\times\binom{8}{5}4^3(-2x)^5+1\times\binom{8}{4}4^4(-2x)^4$ (3 marks)

$= 3\times(-114688)+1\times 286720$

$= -57344$ (1 mark)

26 $\binom{n}{2}(1^{n-2})(3x)^2 = 495x^2$ (2 marks)

$\frac{9n(n-1)}{2} = 495$

$n(n-1) = 110$

$n^2-n-110 = 0$ (1 mark)

$(n-11)(n+10) = 0$

$n = 11$ (2 marks)

27 Require $\binom{8}{2}(x^3)^2\left(-\frac{2}{x}\right)^6$ (2 marks)

$= 28\times(-2)^6$ (1 mark)

$= 28\times 64$

$= 1792$ (1 mark)

---

28 a $\left(\frac{1}{2x}-x\right)^4 = \binom{4}{0}\left(\frac{1}{2x}\right)^4(-x)^0+\binom{4}{1}\left(\frac{1}{2x}\right)^3(-x)^1+\binom{4}{2}\left(\frac{1}{2x}\right)^2(-x)^2+\binom{4}{3}\left(\frac{1}{2x}\right)^1(-x)^3+\binom{4}{4}\left(\frac{1}{2x}\right)^0(-x)^1$ (2 marks)

$= x^4+\frac{1}{16x^4}-2x^2-\frac{1}{2x^2}+\frac{3}{2}$ (1 mark)

b $(3-x)^3 = 27-27x+9x^2-x^3$ (2 marks)

$(3-x)^3\left(\frac{1}{2x}-x\right)^4 = (27-27x+9x^2-x^3)\left(x^4+\frac{1}{16x^4}-2x^2-\frac{1}{2x^2}+\frac{3}{2}\right)$

Therefore required term is $\left(27\times\frac{3}{2}\right)-\frac{9}{2}$ (1 mark)

$= 36$ (1 mark)

---

29 $120 = \frac{a}{1-0.2}$ (2 marks)

$120 = \frac{5a}{4}$

$a = 96$ (1 mark)

The 6th term is therefore

$96\times 0.2^5 = 0.03072$ (2 marks)

30 $ar = 180$ and $ar^5 = \frac{20}{9}$ (2 marks)

Solving simultaneously (1 mark)

$\frac{ar^5}{ar} = \frac{20}{9\times 180} = \frac{1}{81} = r^4$

Therefore $r = \pm\frac{1}{3}$ (1 mark)

So $a = \frac{180}{r} = \frac{180}{\frac{1}{3}} = 540$ (1 mark)

Using $r = \frac{1}{3}$,

$$S_\infty = \frac{a}{1-r} = \frac{540}{1-\frac{1}{3}} = \frac{3 \times 540}{2} = 810$$

$S_\infty = -405$

$r = \frac{1}{3}$ (2 marks)

31 First part is geometric sum, $a = 1$, $r = 1.6$, $n = 16$ (1 mark)

Second part is arithmetic sum,

$a = 0$, $d = -12$, $n = 16$ (1 mark)

Third part is $16 \times 1 = 16$ (1 mark)

Geometric sum:

$$S_{16} = \frac{1.6^{16} - 1}{1.6 - 1} = 3072.791$$ (1 mark)

Arithmetic sum:

$$S_{16} = \frac{16}{2}\left(2 \times 0 + 15 \times (-12)\right)$$
$= -1440$ (1 mark)

So $\sum_{n=0}^{n=15}\left(1.6^n - 12n + 1\right)$

$= 3072.791 - 1440 + 16$

$= 1648.8$ (1 mark)

32 Required distance

$$= 20 + \left(2 \times \frac{5}{6} \times 20\right) + \left(2 \times \frac{5}{6} \times \frac{5}{6} \times 20\right) + \cdots$$ (3 marks)

$$= 20 + \frac{\frac{100}{3}}{1 - \frac{5}{6}}$$ (1 mark)

$$= 20 + \frac{\frac{100}{3}}{\frac{1}{6}}$$

$= 20 + 200$ (1 mark)

$= 220$ m

33 $\binom{n-1}{k} + \binom{n-1}{k-1}$

$$= \frac{(n-1)!}{k!(n-k-1)!} + \frac{(n-1)!}{(k-1)!(n-k)!}$$ (3 marks)

$$= \frac{(n-k)(n-1)! + k(n-1)!}{k!(n-k)!}$$ (1 mark)

$$= \frac{n(n-1)! - k(n-1)! + k(n-1)!}{k!(n-k)!}$$ (1 mark)

$$= \frac{n(n-1)!}{k!(n-k)!}$$ (1 mark)

$$\left(= \frac{n!}{k!(n-k)!}\right)$$

$$= \binom{n}{k}$$

34 $1400 = 504 + 7(n-1)$

$\Rightarrow n = 129$ (1 mark)

So the sum of the multiples of 7 is

$$S_{129} = \frac{129}{2}\left(2 \times 504 + 7 \times (129 - 1)\right)$$
$= 122\,808$ (2 marks)

Sum of the integers from 500 to 1400 (inclusive) is

$$S_{901} = \frac{901}{2}\left(2 \times 500 + 1 \times (901 - 1)\right)$$
$= 855\,950$ (2 marks)

Therefore require

$855\,950 - 122\,808 = 733\,142$ (1 mark)

Chapter 2

Skills check

1

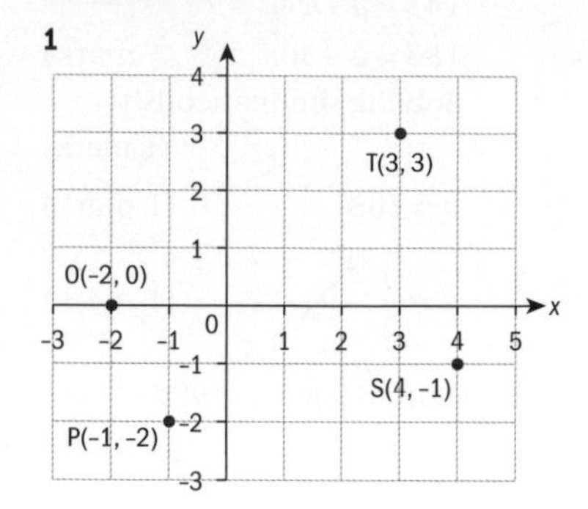

2 $A(3, 0)$, $B(-2, 4)$, $E(-1, 1)$, $R(2, -1)$

3 **a** 17 **b** -5 **c** $-\frac{3}{2}$ **d** $-\frac{3}{2}$

4 **a** 2 **b** $\frac{1}{2}$ **c** 13 **d** -1

5 **a**

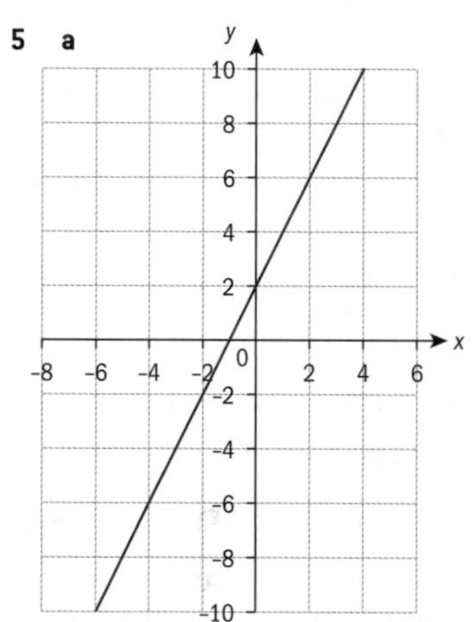

b

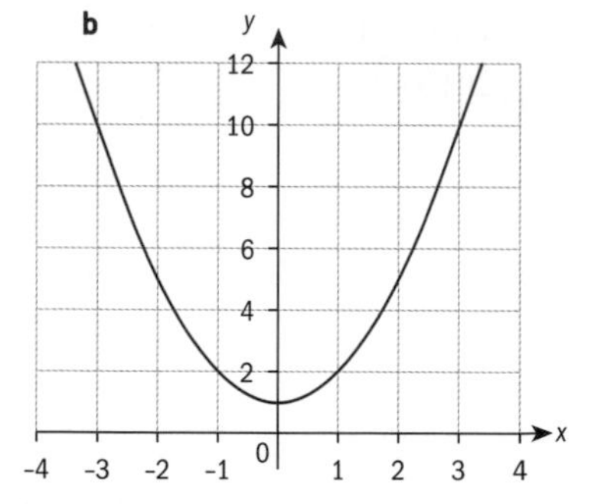

c

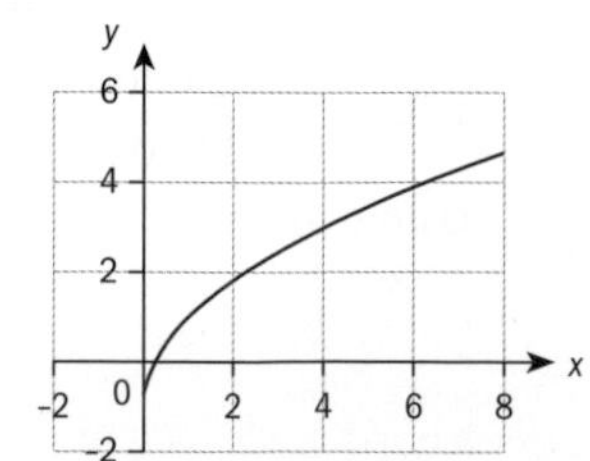

d

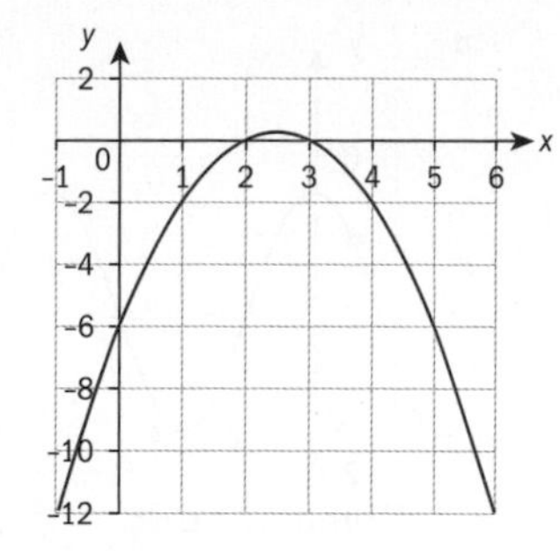

Exercise 2A

1 **a** Functon

b Function

c Not a function as there are tickets with different costs (adults vs child), so the same number of tickets could a different total cost.

d Function

e Not a function

f Function

g Function

h Function

i Not a function

j Function

k Function

l Function

m Not a function

2 Answers will vary

3 "All functions are relations, but not all relations are functions."

Exercise 2B

1 **a** $g(-4) = -14$

b $f: \to -9 = -46$

c $C(100) = 2250$

d $h(5) = -4$

e 3 **f** 5

g 1

2 **a** $f(-3) = -28$

b $g(15) = -53$

c $f(1) + g(-1) = 7$

d 6

e $f(x-2) = -3x^2 + 12x - 13$

f $g(n) = -4n + 7$

g -24

h $f(x+1) = -3x^2 - 6x - 4$

$g(x-2) = -4x + 15$

$f(x+1) \times g(x-2)$
$= 12x^3 - 21x^2 - 74x - 60$

3 **a** Yes, it is a function. Every value of t will yield only one value of d.

b $d(t) = -75t + 275$

c $d(0) = 275\,\text{km}$

d $0 < t < n$, where n is the amount of time it takes to drive to Perth.

4 **a** Yes, it is a function. Every temperature in Celsius will only yield one temperature in Fahrenheit.

b $F(17)$ is asking what temperature in °F is equivalent to 17 °C.

$F(17) = 62.6$°F.

c $F(C) = 100$ is asking what temperature in °C is equivalent to 100°F.

$C = 37.\bar{7} \approx 37.8$°C

d $F(0) = 32$°F

e $F(100) = 212$°F

f $F(38.75) = 101.75$ °F

g $C = 176.\bar{6} \approx 177$°C

5 **a** $C(g) = 10g + 25$

b $g > 0$

c $C(14) = 165$

d $g = 7.5$ gigs

Exercise 2C

1 **a**

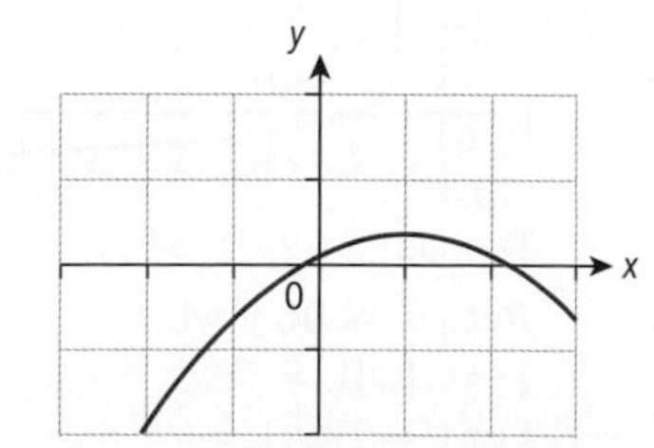

b

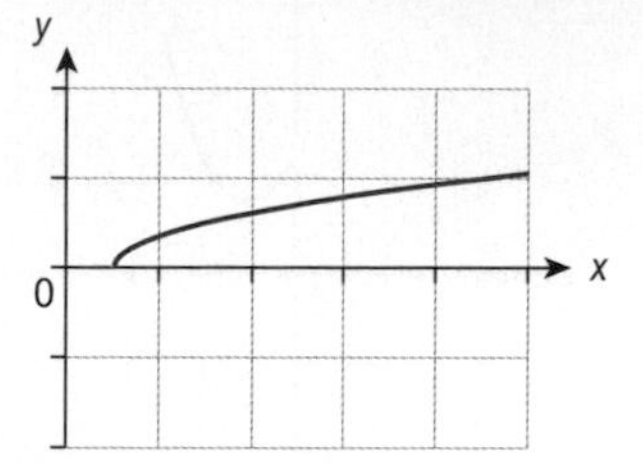

c

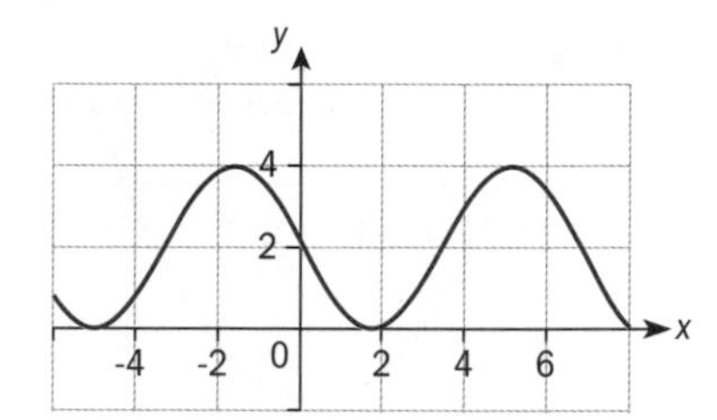

2 **a**

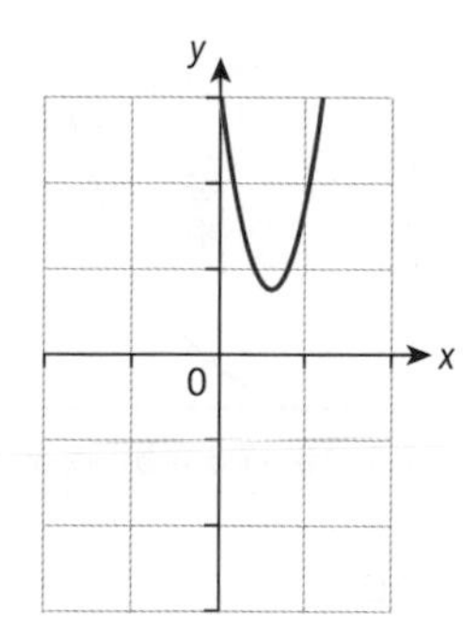

b

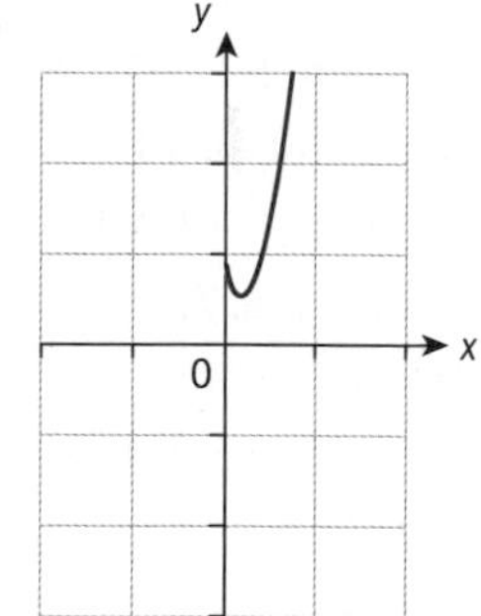

c

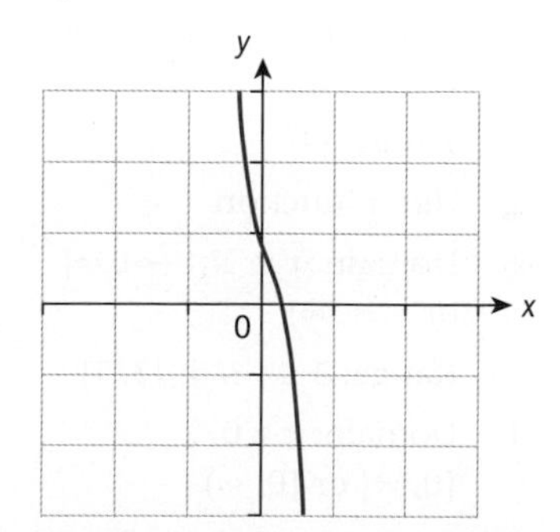

d

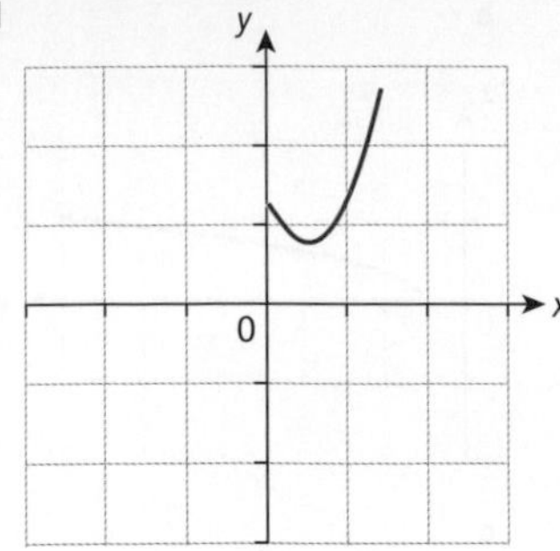

3 a

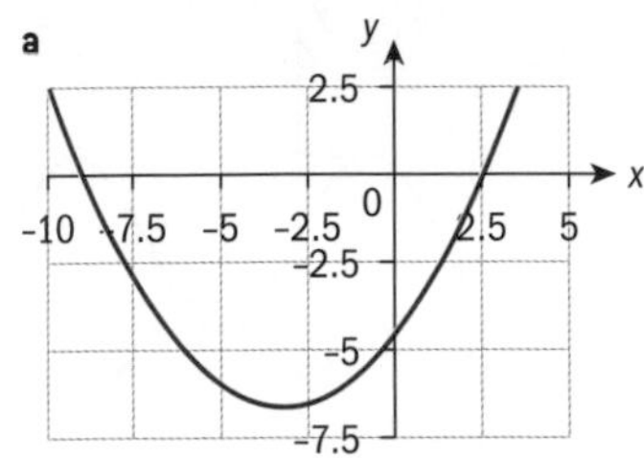

b

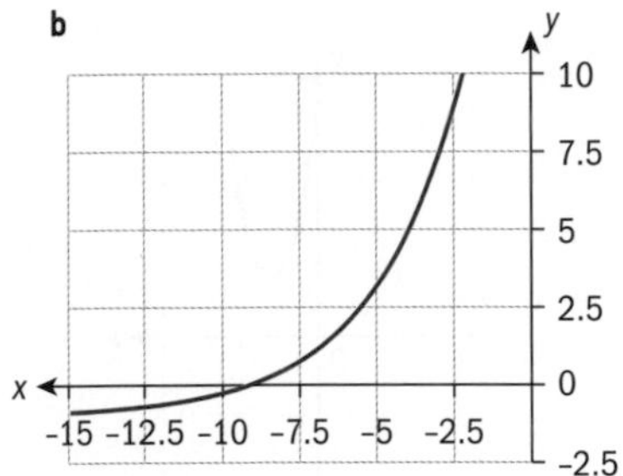

Exercise 2D

1 a Not a function

b Domain: {−5, −2, 3}

Range: {4, 6, 14}

c Domain: {−12, −8, −5}

Range: {−8, 7}

d Not a function

e Domain: $x \in \mathbb{R},]-\infty, \infty[$ or $(-\infty, \infty)$

Range: $y \in \mathbb{R},]-\infty, \infty[$ or$(-\infty, \infty)$

f Domain: $x \in \mathbb{R},]-\infty, \infty[$ or $(-\infty, \infty)$

Range: {4}

g Not a function

h Domain: $x \in \mathbb{R},]-\infty, \infty[$ or $(-\infty, \infty)$

Range: $3 \le y \le 5$, $[3, 5]$

i Domain: $x \ge 0$, $[0, \infty[$ or $[0, \infty)$

Range: $y \le 0$, $]-\infty, 0]$ or $(-\infty, 0]$

j Not a function

k Domain: $x <= -1$ or $x >= 1$, $]-\infty, -1] \cup [1, \infty[$ or $(-\infty, -1] \cup [1, \infty)$

Range: $y \ge 3$, $[3, \infty[$ or $[3, \infty)$

l Not a function.

2 a

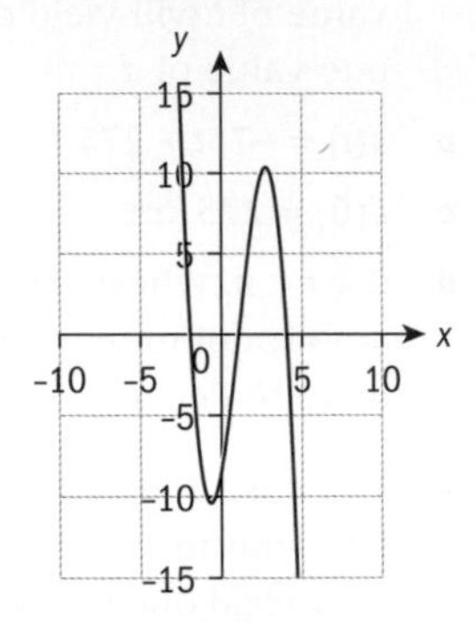

Domain: $x \in \mathbb{R}$, $]-\infty,\infty[$ or$(-\infty,\infty)$

Range: $y \in \mathbb{R},]-\infty,\infty[$ or $(-\infty,\infty)$

b

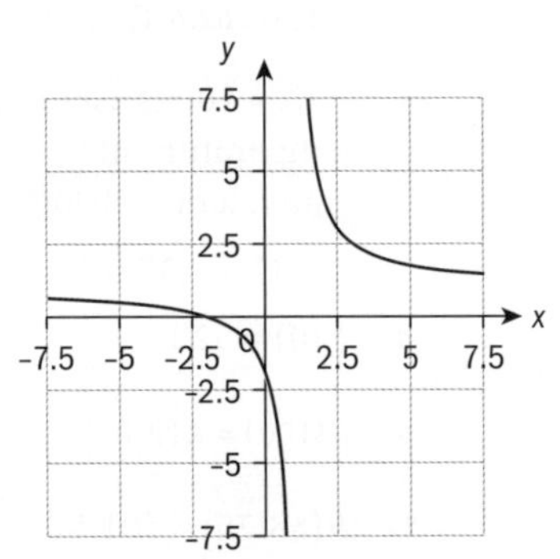

Domain: $x \in \mathbb{R}, x \ne 1$, $]-\infty, 1[\cup]1, \infty[$ or $(-\infty, 1) \cup (1, \infty)$

Range: $y \in \mathbb{R}, y \ne 1$, $]-\infty, 1[\cup]1, \infty[$or $(-\infty, 1) \cup (1, \infty)$

c

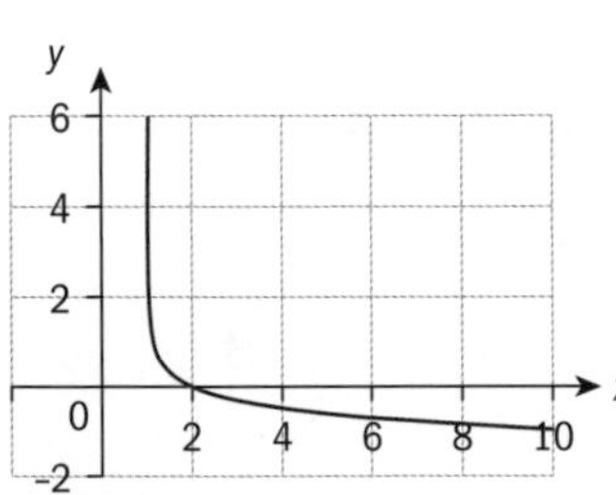

Domain: $x > 1$, $]1, \infty[$ or $(1, \infty)$

Range: $y \in \mathbb{R}$, $y \ne 1$, $]-\infty,1[\cup]1, \infty$ [or $(-\infty,1) \cup (1, \infty)$

d

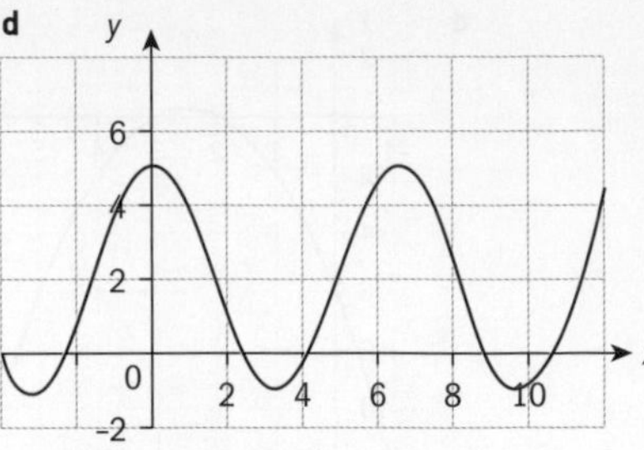

Domain: $x \in \mathbb{R}$, $]-\infty,\infty$ [or$(-\infty, \infty)$

Range: $-1 \le y \le 5$, $[-1, 5]$

3 Answers will vary.

a

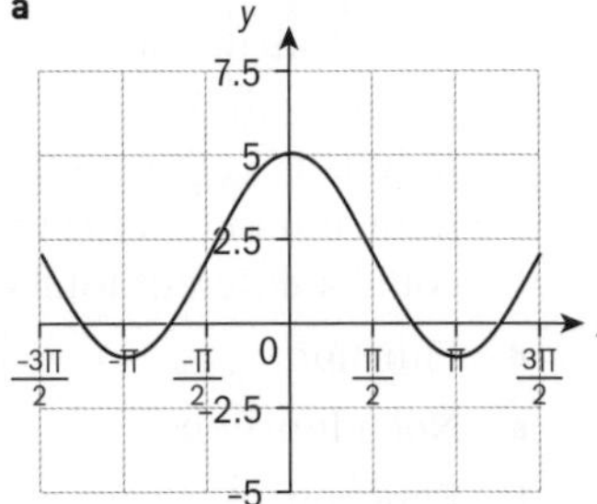

b

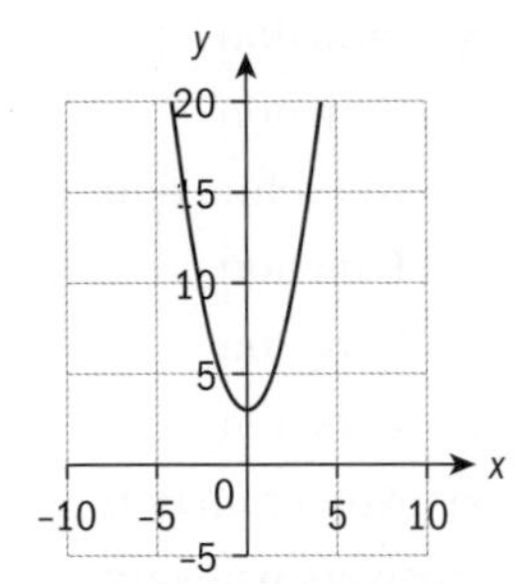

c

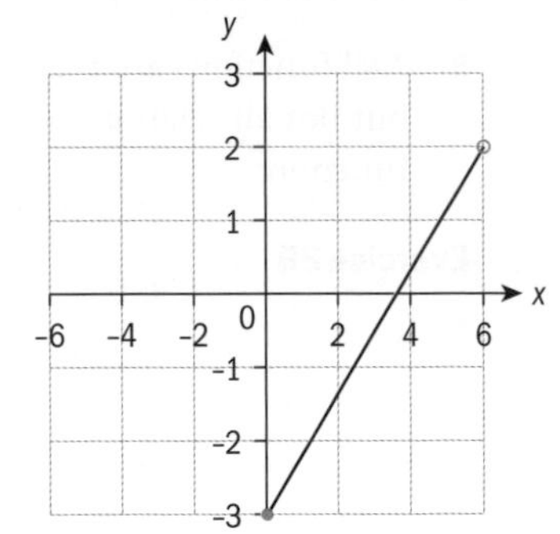

d

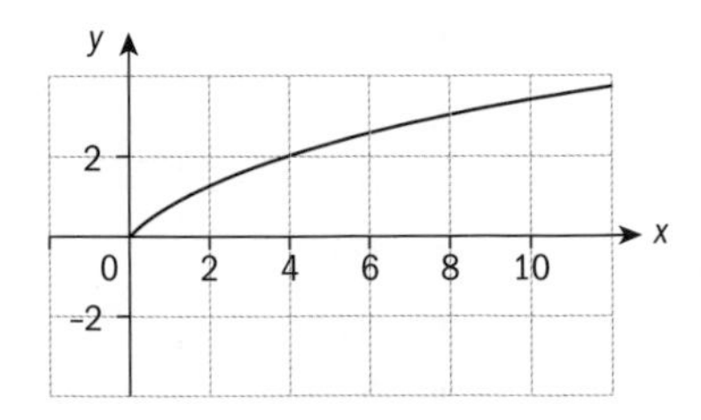

4 a i 4 ii 2

b

c $0 \le x \le 10$; $0 \le y \le 4$

5 $f(x) = \begin{cases} -x, & -3 \le x \le -1 \\ 2x+3, & -1 < x \le 2 \\ 7, & 2 < x \le 6 \end{cases}$

It is also possible to include −1 in the second interval rather than the first and 2 in the third interval rather than the second.

Exercise 2E

1 a $C(n) = 40 + 21n$, where C is the cost and n is the number of hours.

b Domain: $n \ge 0$, $[0, \infty[$ or $[0, \infty)$

Range: $C(n) \ge 40$, $[40, \infty[$ or $[40, \infty)$

c $C(4) = \$124$

2 a

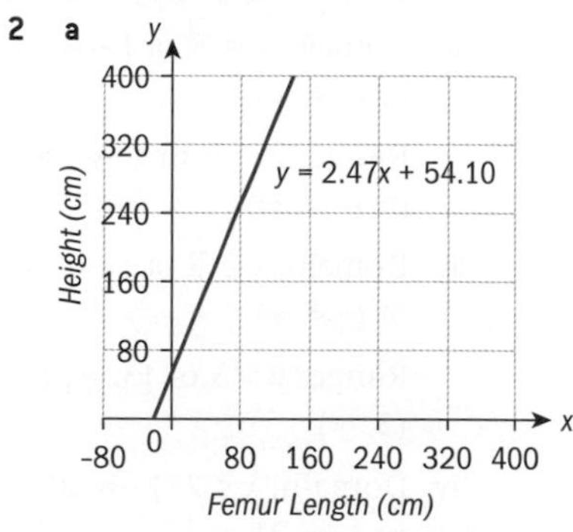

b Domain eg: $10 \le f \le 80$

Range eg: $75.8 \le h \le 228$

c $h(51) \approx 180$ cm

d $f \approx 43.3$ cm

3 a $I(t) = 10000\left(1 + \frac{0.025}{12}\right)^{12t}$

b The equation satisfies the vertical line test.

c Domain: $t \in \mathbb{R}, t \ge 0$. Range: $[10000, \infty)$.

d Javier needs 27 years and 10 months to double his money.

Exercise 2F

1 a i $f(g(x)) = -16x^2 + 36x - 14$

ii $f(f(x)) = -x^4 + 10x^3 - 30x^2 + 25x$

iii $f(h(x)) = -x + 3\sqrt{x} + 4$

iv $g \circ h(x) = 4\sqrt{x} + 2$

v $f(-1) = -6$

$f \circ f(-1) = -66$

$f \circ f \circ f(-1) = 4686$

vi $g(h(9)) = 14$

vii $g \circ f(2) = 22$

$f \circ g(2) = -6$

$g \circ f(2) + f \circ g(2) = 22 - 6 = 16$

b i $x \in \mathbb{R}$ or $]-\infty, \infty[$ or $(-\infty, \infty)$

ii $x \in \mathbb{R}$ or $]-\infty, \infty[$ or $(-\infty, \infty)$

iii $x \ge 0$ or $[0, \infty[$ or $[0, \infty)$

iv $x \ge 0$ or $[0, \infty[$ or $[0, \infty)$

2 Answers will vary.

3 a $f(g(x)) = -8x + 7$

b $x = -\frac{5}{8}$

4 a $f(g(x)) = 3x^2 - 24x + 42$

b

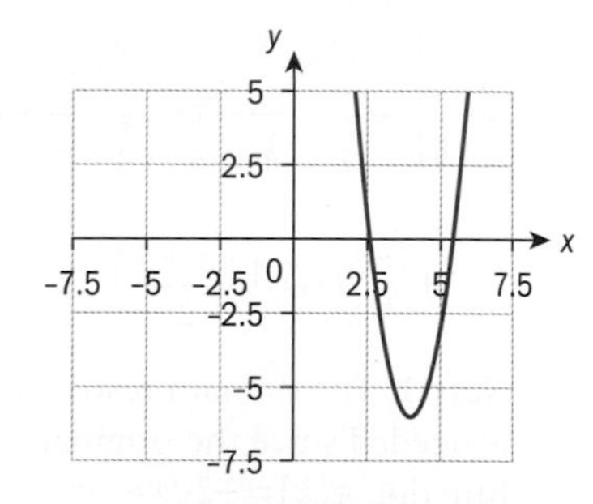

c Domain: $x \in \mathbb{R}$ or $]-\infty, \infty[$ or $(-\infty, \infty)$

Range: $y \ge -6$ or $[-6, \infty[$ or $[-6, \infty)$

5 a $f(x) = x + 25$

b $g(x) = 1.06x$

c $f(g(x)) = 1.06x + 25$; this represents only paying tax on the price of the fridge.

$g(f(x)) = 1.06(x + 25)$; this represents paying tax on both the price of the fridge and the delivery fee.

d $f(g(x))$

Exercise 2G

1 a i $g(-2) = 3$

$f(g(-2)) = f(3) = 2$

ii $f(5) = 6$

$g(f(5)) = g(6) = -3$

iii $g(3) = 11$

$g(g(3)) = g(11) = 0$

b For $f(x)$:

Domain: {2, 3, 5, 10}

Range: {−4, 1, 2, 6}

For $g(x)$:

Domain: {−2, 3, 6, 11}

Range: {−3, 0, 3, 11}

2 a $g(3) = 7$

$f \circ g(3) = f(7) = -2$

b $f(-1) = 9$

$g \circ f(-1) = g(9) = -4$

c $f(9) = -1$

$f \circ f(9) = f(-1) = 9$

3 a $f(0) = 0$

$g(f(0)) = g(0) = -4$

b $f(1) = 1$

$g(f(1)) = g(1) = -3$

c $g(-2) = 0$

$f(g(-2)) = f(0) = 0$

d $g(-0) = -4$

$f(g(0)) = f(-4) = 4$

4 Answers will vary; $g(x)$ contains a point with an x-coordinate of −1; $f(x)$ must have a point with a y-coordinate of 2.

Exercise 2H

1 **a** $f(n) = n - 100$

$g(n) = 2.20n$

b $f(n)$ represents that you receive commission on every new person who signs up after the first 100 people.

$g(n)$ represents that you receive 2.20 GBP for each person (after the first 100) who sign up.

c **i** $f(224) = 124$

ii $g(124) = 272.80$ GBP

d **i** $S(276) = 387.20$ GBP

ii $x = 152$ people

2 **a** $f(x) = x - 25$; this could represent \$25 off the price of the TV.

$g(x) = 1.10x$; this could represent a tax of 10%.

b **i** You paid $699.99 - 25 = \$674.99$

ii After tax, the TV cost $1.10 \times 674.99 = 742.489 \approx \$742.49$

c **i** $P(x) = 1.10(x - 25) + 49.99$

ii $P(525.99) \approx \$601.08$

3 Answer will vary.

Exercise 2I

1 **a** $f(g(x)) = -x - 2$

Since $f(g(x)) \neq x$, these are not inverses.

b $f(g(x)) = x$

$g(f(x)) = x$

Since $f(g(x)) = g(f(x)) = x$, these are inverses.

c $f(g(x)) = x$

$g(f(x)) = x$

Since $f(g(x)) = g(f(x)) = x$, these are inverses.

d $g(h(x)) = x$

$h(g(x)) = x$

Since $f(g(x)) = g(f(x)) = x$, these are inverses.

2 x-intercept: $\left(\frac{1}{2}, 0\right)$

y-intercept: $(0, 2)$

Since you only need two points to graph a line, you can switch the coordinates to find two point that the inverse passes through: $\left(0, \frac{1}{2}\right)$ and $(2, 0)$.

3 **a** **i** & **ii**

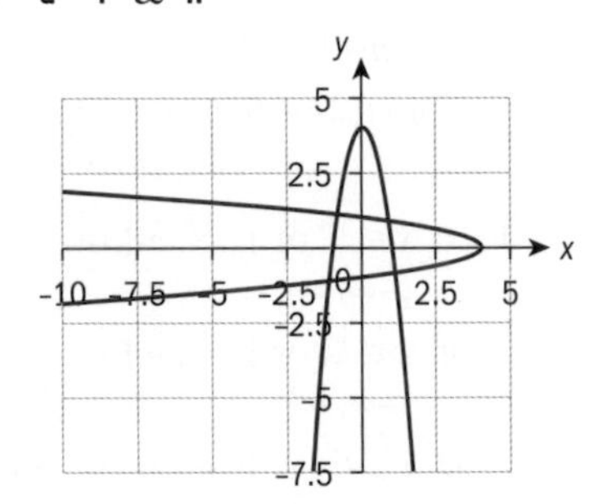

iii $f^{-1}(x) = \pm\sqrt{\frac{-x}{4} + 1}$

b **i** & **ii**

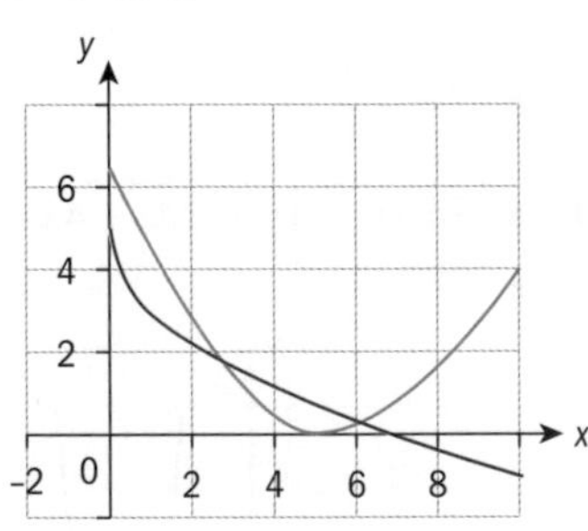

iii $g^{-1}(x) = \left(\frac{-x+5}{2}\right)^2, x \geq 0$

Note 1: The domain restriction is needed since the original function $g(x) = -2\sqrt{x} + 5$ would have the same restriction.

Note 2: The inverse, $g^{-1}(x)$, can be simplified further if desired:

$$y = \frac{1}{4}x^2 - \frac{5}{2}x + \frac{25}{4}, x \geq 0$$

c **i** & **ii**

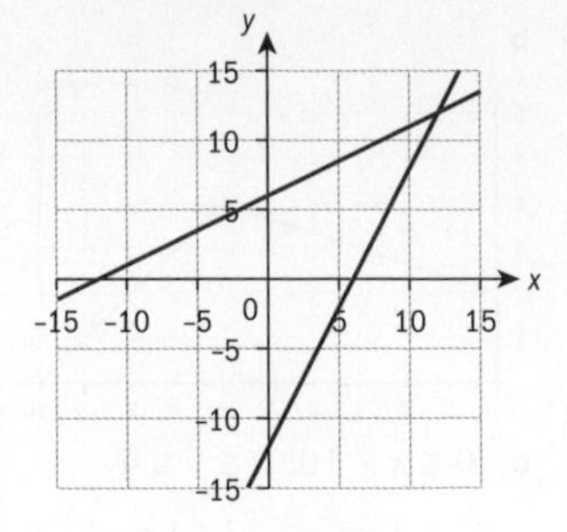

iii $g : x \to \frac{1}{2}x + 6$

$$y = \frac{1}{2}x + 6$$

$$x = \frac{1}{2}y + 6$$

$$\frac{1}{2}y = x - 6$$

$$y = 2x - 12$$

$$g^{-1}(x) = 2x - 12$$

4 **a** **i** Domain: $x \geq 1.25$ or $[1.25, \infty[$ or $[1.25, \infty)$

Range: $y \geq 2.875$, $[2.875, \infty[$ or $[2.875, \infty)$

ii Domain: $x \in \mathbb{R}$ or $]-\infty, \infty[$ or $(-\infty, \infty)$

Range: $y \in \mathbb{R}$ or $]-\infty, \infty[$ or $(-\infty, \infty)$

iii Domain: $x \in \mathbb{R}$ or $]-\infty, \infty[$ or $(-\infty, \infty)$

Range: $y > 3$ or $]3, \infty[$ or $(3, \infty)$

iv Domain: $x \leq 2$, $]-\infty, 2]$ or $(-\infty, 2]$

Range: $y \leq 1$, $]-\infty, 1]$ or $(-\infty, 1]$ or $[-\infty, 2]$ or $(-\infty, 2]$

b The domain of the function becomes the range of its inverse, and the range of the function becomes the domain of its inverse.

5 Answers will vary. In order for the function to be a one-to-one function, the inverse must be a function.

6 **a** $x = 8$

b $2y = x + 5$

$y = \frac{x+5}{2}$

$f^{-1}(x) = \frac{x+5}{2}$

c $f^{-1}(11) = \frac{11+5}{2}$

$f^{-1}(11) = \frac{16}{2}$

$f^{-1}(11) = 8$

d $f(x) = 11$ gives the same answer as $f^{-1}(11)$.

e $f(x) = y = f^{-1}(y)$

7 $f(x) = -2x - 1$

$y = -2x - 1$

$x = -2y - 1$

$2y = -x - 1$

$y = \frac{-x-1}{2}$

$f^{-1}(x) = \frac{-x-1}{2}$

$g \circ f^{-1}(x) = -3\left(\frac{-x-1}{2}\right)^2$

$g \circ f^{-1}(x) = -3\frac{(-x-1)^2}{4}$

$g \circ f^{-1}(x) = -3\frac{(x^2+2x+1)}{4}$

$g \circ f^{-1}(x) = \frac{-3x^2-6x-3}{4}$

$g \circ f^{-1}(-1) = \frac{-3(-1)^2-6(-1)-3}{4}$

$g \circ f^{-1}(-1) = \frac{-3(1)-6(-1)-3}{4}$

$g \circ f^{-1}(-1) = \frac{-3+6-3}{4}$

$g \circ f^{-1}(-1) = \frac{0}{4}$

$g \circ f^{-1}(-1) = 0$

Exercise 2J

1

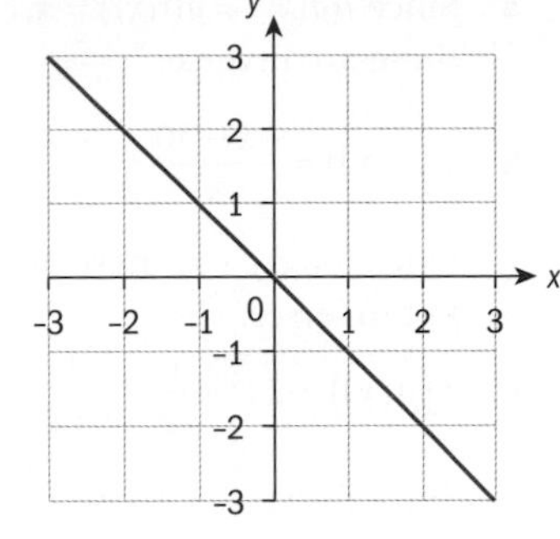

2 **a** $f(f(x)) = x$

$3 - (3 - x) = x$

$3 - 3 + x = x$

$x = x$

b $f(f(x)) = x$

$-2 - (-2 - x) = x$

$-2 + 2 + x = x$

$x = x$

c $f(f(x)) = x$

$\frac{1}{2} - \left(\frac{1}{2} - x\right) = x$

$\frac{1}{2} - \frac{1}{2} - x = x$

$x = x$

d $f(x) = n - x$, $n \in \mathbb{R}$ is a self-inverse.

3 $f(f(x)) = x$

$\frac{-\left(\frac{-x-2}{5x+1}\right)-2}{5\left(\frac{-x-2}{5x+1}\right)+1} = x$

$\frac{\frac{(x+2)}{(5x+1)} - \frac{2(5x+1)}{(5x+1)}}{\frac{(-5x-10)}{(5x+1)} + \frac{(5x+1)}{(5x+1)}} = x$

$\frac{\frac{x+2-10x-2}{5x+1}}{\frac{-9}{5x+1}} = x$

$\frac{-9x}{-9} = x$

$x = x$

4 $f(f(x)) = x$

$\frac{2\left(\frac{2x-4}{x+m}\right)-4}{\left(\frac{2x-4}{x+m}\right)+m} = x$

$\frac{\left(\frac{4x-8}{x+m}\right)-4\left(\frac{x+m}{x+m}\right)}{\left(\frac{2x-4}{x+m}\right)+m\left(\frac{x+m}{x+m}\right)} = x$

$\frac{\frac{4x-8-4x-4m}{x+m}}{\frac{2x-4+xm+m^2}{x+m}} = x$

$\frac{-4m-8}{2x-4+xm+m^2} = x$

$-4m - 8 = x(2x - 4 + xm + m^2)$

$-4m - 8 = 2x^2 - 4x + x^2m + m^2$

$-4m - 8 = (2 + m)x^2 - 4x + m^2$

$0 = 2 + m$

$m = -2$ No solution because for $m = -2$ we get the constant function $y = 2$, which has no inverse.

Chapter review

1

a	Yes	**b**	No	**c**	Yes
d	No	**e**	Yes	**f**	Yes
g	Yes	**h**	No	**i**	Yes
j	Yes	**k**	Yes	**l**	No
m	No				

2 **a** Domain: $\{-5, -1, 0, 1, 4, 9\}$

Range: $\{-8, -1, 0, 1, 6, 9\}$

b Domain: $\{0, 2, 3, 4\}$

Range: $\{-2, 2, 3\}$

c Domain: $\{-8, -5, 0, 1\}$

Range: $\{-2, 2, 3\}$

d Domain: $x \in \mathbb{R}$ or $(-\infty, \infty)$ or $]-\infty, \infty[$

Range: $y \in \mathbb{R}$ or $(-\infty, \infty)$ or $]-\infty, \infty[$

e Domain: $-3 < x \le 3$ or $(-3, 3]$ or $]-3, 3]$

Range: $-3 \le x \le -1$ or $[-3, -1]$

f Domain: $x \in \mathbb{R}$ or $(-\infty, \infty)$ or $]-\infty, \infty[$

Range: $x \ge -12.25$ or $[12.25, \infty)$ or $[12.25, \infty[$

g Domain: $x \geq 0$ or $[0, \infty)$ or $[0, \infty[$

Range: $y \leq 1$ or $(-\infty, 1]$ or $]-\infty, 1]$

h Domain: $x \in \mathbb{R}$ or $(-\infty,\infty)$ or $]-\infty,\infty[$

Range: $x \geq 5$ or $[5, \infty)$ or $[5, \infty[$

3 a $f(3) = 3$

b $f(-2) = -2$

c $g(-6) = 12$

d $f(1) + h(2) = -9$

e $2f(0) - 2g(-1) = -16$

f $h(0) \times f(-1) = 20$

g $g^{-1}(-3) = \dfrac{3}{2}$

h $f(g(x)) = 4x^2 - 6$

i $f \circ g^{-1}(x) = \dfrac{x^2 - 24}{4}$

4 a

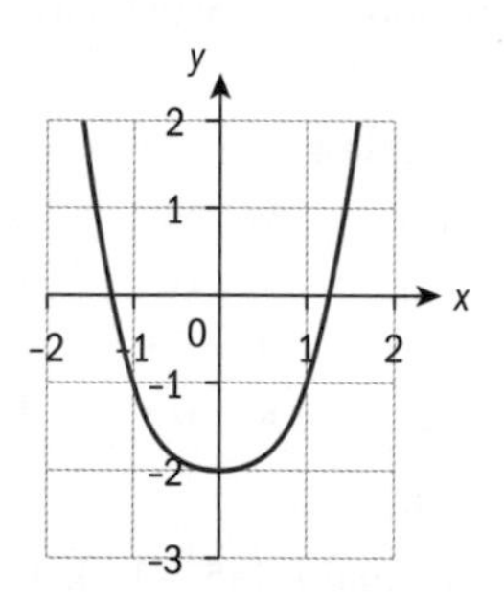

Domain: $x \in \mathbb{R}$ or $(-\infty, \infty)$ or $]-\infty, \infty[$

Range: $y \geq -2$ or $[-2, \infty)$ or $[-2, \infty[$

b

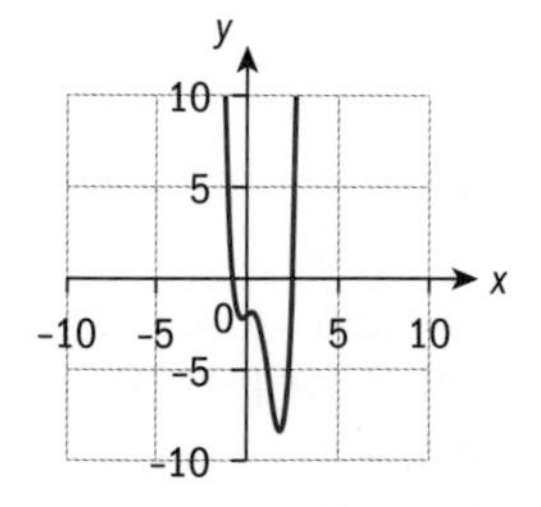

Domain: $x \in \mathbb{R}$ or $(-\infty,\infty)$ or $]-\infty, \infty[$

Range: $y \geq -8.38$ or $[-8.38, \infty)$ or $[-8.38, \infty[$

5 a Since $f(g(x)) = g(f(x)) = x$, these are inverses.

b $f(g(x)) = \dfrac{-x - 30}{8}$

Since $f(g(x)) \neq x$, these are not inverses.

c $f(g(x)) = 2x^2 + \dfrac{1}{2}x + \dfrac{1}{32} + 4$

Since $f(g(x)) \neq x$, these are not inverses.

d $g(f(x)) = x$

Since $f(g(x)) = g(f(x)) = x$, these are inverses.

6 a i -4

ii 4

b -4 or $\leq x \geq 4$ or $[-4, 4]$

c

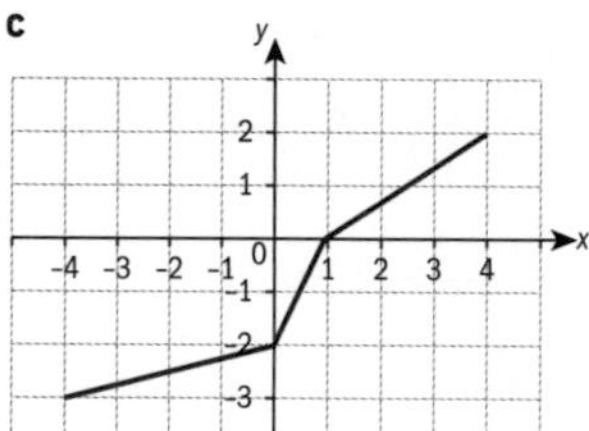

7 $(f \circ g)(x) = (g(x) + 2)^3$

$\sqrt[3]{-8x^6} = \sqrt[3]{(g(x)+2)^3}$

$-2x^2 = g(x) + 2$

$g(x) = -2x^2 - 2$

8 $f(h^{-1}(2)) = 264$

9 $f(f(x)) = -\dfrac{3}{\left(-\dfrac{3}{x}\right)}$

$f(f(x)) = -3 \times \dfrac{-x}{3}$

$f(f(x)) = x$

Exam-style questions

10 a $-24 \leq f(x) \leq 26$ (2 marks)

b $f(x) = \{-4, -2, 0, 2, 4, 6\}$ (2 marks)

c $0 \leq f(x) \leq 100$ (2 marks)

d $125 \leq f(x) \leq 250$ (2 marks)

11 a $f(-2) = 4 \times (-2) - 2$
$= -8 - 2 = -10$ (2 marks)

b $g(-2) = (-2)^2 - 8(-2) + 15$
$= 4 + 16 + 15 = 35$ (2 marks)

c $y = 4x - 2$

$x = 4y - 2$

$y = \dfrac{x+2}{4}$ (1 mark)

$f^{-1}(x) = \dfrac{x+2}{4}$ (1 mark)

d $x^2 - 8x + 15 = 35$ (1 mark)

$x^2 - 8x - 20 = 0$

$(x-10)(x+2) = 0$

(1 mark)

$x = 10$ or $x = -2$

(2 marks)

12 a $f(x) = 128\left(\frac{3}{2}\right) - 15 = 177$

(2 marks)

b $f(-3) = 128(-3) - 15 = -399$

(2 marks)

$f(15) = 128(15) - 15 = 1905$

(1 mark)

Range is $-399 < f(x) < 1905$

(1 mark)

c Solving $128a - 15 = 1162.6$

(1 mark)

$a = 9.2$ (1 mark)

13 a Domain is $-3 \le x \le 3$

(2 marks)

Range is $-1 \le f(x) \le 1$

(2 marks)

b Domain is $-1.5 \le x \le 5$

(2 marks)

Range is $-5 \le f(x) \le 4$

(2 marks)

c Domain is $0 \le x \le 24$

(2 marks)

Range is $0 \le f(x) \le 12$

(2 marks)

d Domain is $-3 \le x \le 3$

(2 marks)

Range is $0 \le f(x) \le 9$

(2 marks)

14 a Solving $3x - 10 = 5$ and

$3x - 10 = 50$ (3 marks)

Domain is $5 < x < 20$

(1 mark)

b $ff(10) = f(f(10))$

(1 mark)

$= f(20)$ (1 mark)

$= 50$ (1 mark)

c $y = 3x - 10$

$x = 3y - 10$

$y = \frac{x+10}{3}$ (1 mark)

$f^{-1}(x) = \frac{x+10}{3}$ (1 mark)

Range is $5 < f^{-1}(x) < 20$

(2 marks)

15 a NOT a function, since e.g. the value of $x = 5$ is related to more than one coordinate on the y-axis.

(2 marks)

b This is a function. Each value of x is related to only one value for y.

(2 marks)

c This is a function. Each value of x is related to only one value for y.

(2 marks)

d This is a function. Each value of x is related to only one value for y.

(2 marks)

16 a $y = \frac{k}{x-1} + 1$

$x = \frac{k}{y-1} + 1$

$x(y-1) = k + y - 1$

(1 mark)

$xy - x = k + y - 1$ (1 mark)

$y = \frac{k}{x-1} + 1$ (1 mark)

$f^{-1}(x) = \frac{k}{x-1} + 1$

So f is self-inverse

b Range is $f(x) > 1$,

$f(x) \in \mathbb{R}$ (2 marks)

c

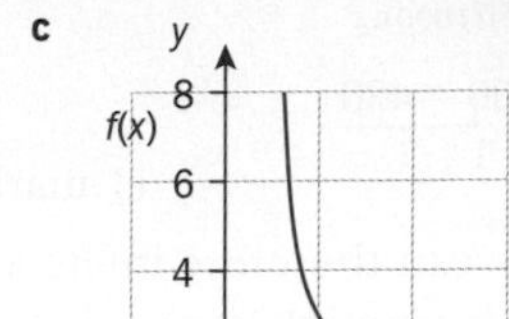

(2 marks)

17 a Range is $f(x) \ne -\frac{2}{3}$,

$(f(x) \in \mathbb{R})$ (1 mark)

b $x(3y+6) = 1 - 2y$ (1 mark)

$3xy + 6x = 1 - 2y$ (1 mark)

$f^{-1}(x) = \frac{1-6x}{2+3x}$ (1 mark)

c Domain is, $x \ne -\frac{2}{3}$, $(x \in \mathbb{R})$

(1 mark)

Range is $f(x) \ne -2$,

$(f(x) \in \mathbb{R})$ (1 mark)

18 a $x^2 = 2x - 1$ (1 mark)

$(x-1)^2 = 0$ (1 mark)

$x = 1$ (1 mark)

b $fg(x) = (2x-1)^2$ (1 mark)

$gf(x) = 2x^2 - 1$ (1 mark)

$(2x-1)^2 = 2x^2 - 1$ (1 mark)

$(x-1)^2 = 0$ (1 mark)

$x = 1$ (1 mark)

19 a $C = 430 + 14.5p$ (2 marks)

b $f(p)$ is a function since there is only one value of C which corresponds to each value of p in the domain.

(1 mark)

c $C = 430 + 14.5p$ (1 mark)

$f^{(-1)}(p) = \frac{P - 430}{14.5}$

(1 mark)

c $f^{(-1)}(1000) = \frac{1000-430}{14.5}$ (1 mark)

She can therefore invite a maximum of 39 people. (1 mark)

d $C = 430 + 14.5 \times 16 = \$662$ (1 mark)

$\frac{662}{16} = 41.375$ (1 mark)

Katie will therefore need to charge a minimum of \$41.38 per guest. (1 mark)

20 a $h(x) \geq 2, (h(x) \in \mathbb{R})$ (1 mark)

b $y = \frac{x}{3} + 2$

$3x = y + 6$ (2 marks)

$h^{-1}(x) = 3x - 6$ (1 mark)

c $hh(x) = \frac{\frac{x}{3}+2}{3} + 2$ (2 marks)

$= \frac{x}{9} + \frac{8}{3}$ (1 mark)

d $\frac{x}{3} + 2 = 3x - 6$ (1 mark)

$\frac{8x}{3} = 8$

$x = 3$ (1 mark)

e Because $h(x)$ and $h^{-1}(x)$ both intersect on the line $y = x$. (1 mark)

21 $x^2 + 4x - 11 = (x+2)^2 - 15$ (2 marks)

Therefore $h(x) = x + 2$ (1 mark)

$g(x) = x^2$ (1 mark)

$f(x) = x - 15$ (1 mark)

22 a $(f(x) \in \mathbb{R}), f(x) \geq -4$ (1 mark)

b $(g(x) \in \mathbb{R}), g(x) \neq 0$ (1 mark)

c $(h(x) \in \mathbb{R}), h(x) > 0$ (1 mark)

d $gf(x) = \frac{1}{(x^2-4)+1}$ (1 mark)

$= \frac{1}{x^2-3}$ (1 mark)

e $\frac{1}{x^2-3} = 9$ (1 mark)

$x = \pm\frac{2\sqrt{7}}{3}$ (1 mark)

f $gh(x) = \frac{1}{2^x+1}$ (2 marks)

$\frac{1}{2^x+1} > \frac{1}{17}$ (1 mark)

$2^x < 16$ (1 mark)

$x < 4$ (1 mark)

23 a $-8 \leq p(x) \leq 8$ (2 marks)

b $p^{-1}(x) = \sqrt[3]{x}, -8 \leq x \leq 8,$

$(x \in \mathbb{R})$ (2 marks)

c Using GDC, or otherwise, solving $x^3 = x$ (1 mark)

$x = -1, x = 0, x = 1$ (1 mark)

d

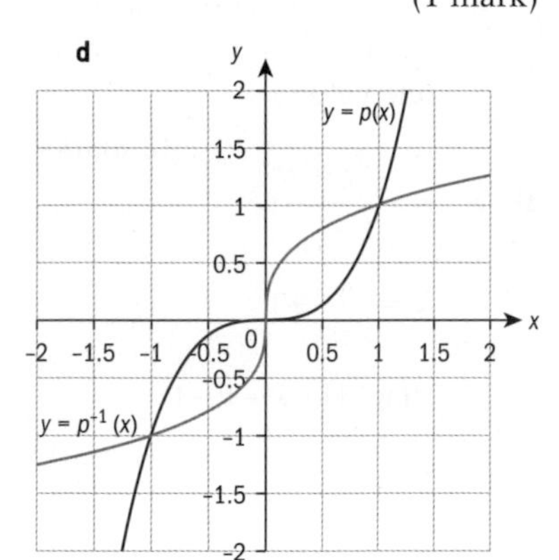

(2 marks)

24 a $x(4y-3) = 3y+5$ (1 mark)

$4xy - 3x = 3y + 5$ (1 mark)

$y = \frac{3x+5}{4x-3}$ (1 mark)

So $r(x)$ is self-inverse.

b $rrrrrr(5) = rrrr(5) = rr(5) = 5$ (2 marks)

25 a $x^2 - 6x + 13 = (x-3)^2 + 4$ (2 marks)

Therefore $k = 3$ (1 mark)

b $(y-3)^2 = x - 4$ (2 marks)

$f^{-1}(x) = 3 + \sqrt{x-4}$ (1 mark)

c The domain of $f^{-1}(x)$ is $x \geq 4, (x \in \mathbb{R})$ (1 mark)

The range of $f^{-1}(x)$ is $f(x) \geq 3, (f(x) \in \mathbb{R})$ (1 mark)

26 $(x-3)^2 = x^2 - 6x + 9$ (1 mark)

$2(x-3)^2 = 2x^2 - 12x + 18$ (2 marks)

$2(x-3)^2 + 12x = 2x^2 + 18$

Therefore $g(x) = 2x^2 + 12x$ (1 mark)

Chapter 3

Skills check

1 a -3 **b** $\pm\sqrt{7}$ **c** $-\frac{9}{2}$

2 a $3m(m-5)$ **b** $(x+6)(x-6)$

c $(n+1)(n+7)$

d $(4x-3)(x+1)$

e $9x(x+2)$

f $(2a-5)(a+1)$

g $(3x+2)(4x-1)$

h $(4a-7b)(4a+7b)$

Exercise 3A

1 a $-\frac{1}{2}$ **b** $\frac{2}{5}$

2 a $\frac{3}{4}$ **b** -1 **c** $\frac{1}{2}$

3 Rate of change $= -0.015$

The height of the candle decreases 0.015 cm each additional second it burns.

4 a $(342.93, 100)$ **b** 0.20

c 20%

Exercise 3B

1 **a** $m_1 = \frac{5}{3}$ $m_2 = \frac{3}{5}$ neither

b $m_1 = -4$ $m_2 = -4$ parallel

2 -1

3 **a** 8, 12

b For the first 40 hours Liam works in a week he is paid \$8 per hour. For each hour over 40 hours that Liam works, he earns \$12 per hour.

Exercise 3C

1 **a** $m = 3$; $(0, -7)$

b $m = -\frac{2}{3}$; $(0, 4)$

c $m = 0$; $(0, -2)$

2 $y = \frac{1}{5}x + 1$

3 **a** $y = 4x - 1$ **b** $y = 3x + 7$

4 **a** $x = 8$ **b** $y = -10$

c $x = 9$ **d** $(-2, 7)$

Exercise 3D

1 **a**

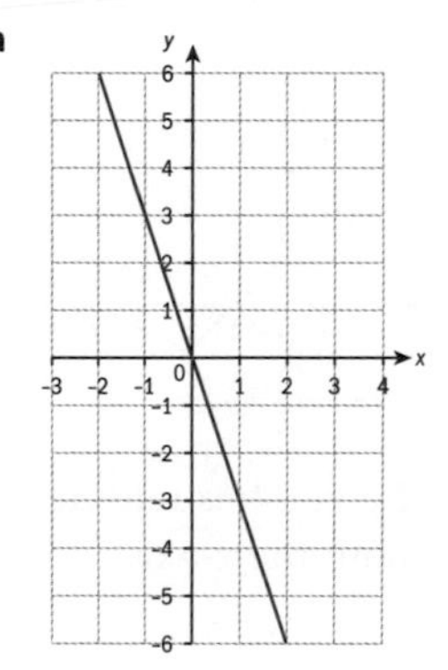

b

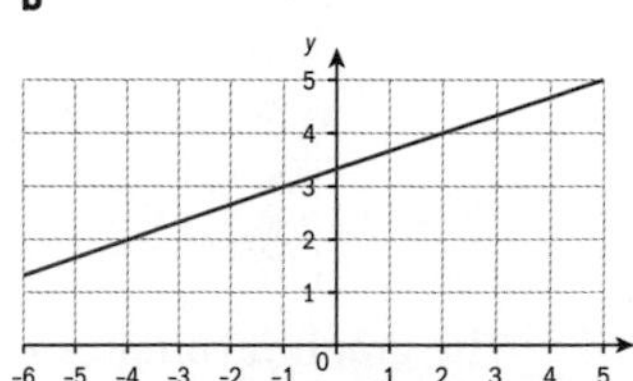

c

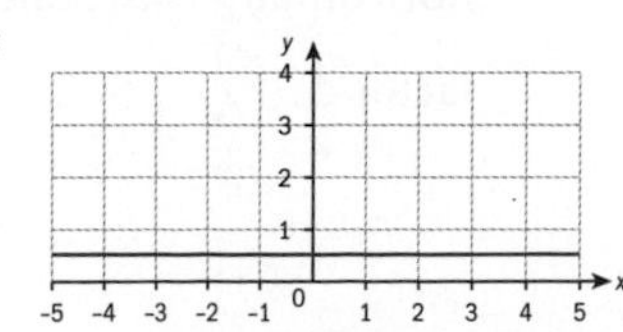

d

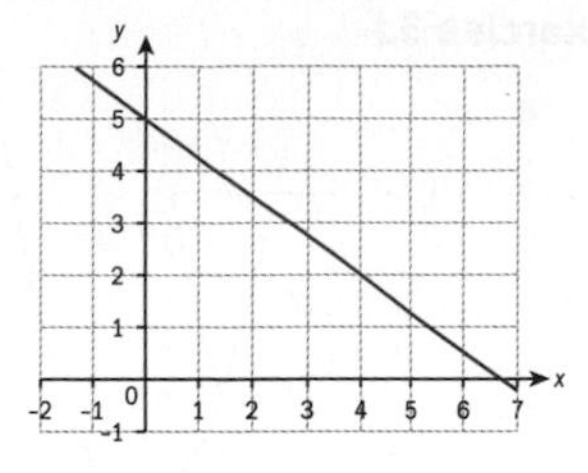

2 $y - 6 = -3(x - 2)$

3 **a** -3

b $y + 4 = -3(x + 3)$ or $y - 2 = -3(x + 5)$

c Change the equations to slope-gradient form: $y = -3x - 13$

Exercise 3E

1 **a** $x - 6y - 18 = 0$

b $2x + 3y - 12 = 0$

c $x + y + 1 = 0$

2 **a** $y = -3x + 5$ **b** $y = \frac{1}{2}x + 2$

c $y = -\frac{5}{2}x - \frac{7}{2}$

3 **a**

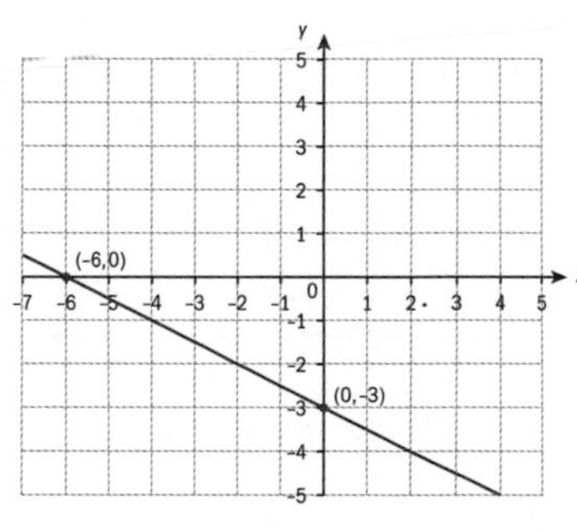

b

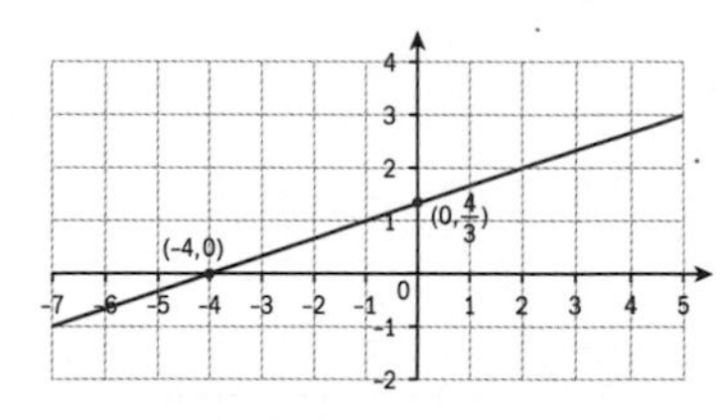

Exercise 3F

1 **a** $(-2, -5)$ **b** $(0.75, 2.5)$

c $(-3.58, -8.19)$

d $(1.18, 1.12)$

2 **a** 0.9

b -5.05

3 \$1666.67

Exercise 3G

1 **a** 2 **b** 3 **c** -7

d 4 **e** -6 **f** 0

g $-2x + 2$ **h** $-\frac{1}{3}x - \frac{7}{3}$

2 **a** **Domain:** all real numbers
Range: all real numbers

b **Domain:** all real numbers
Range: all real numbers

3 **a**

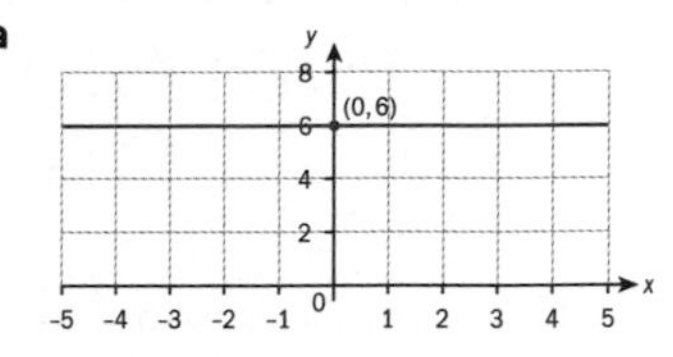

b

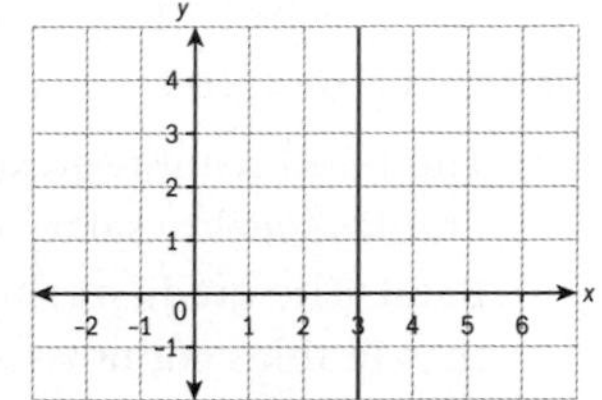

No vertical line is a function as the y corresponding to the x-coordinate of the x-intercept is not unique (in fact, any real number y corresponds to it).

4 **a** $f^{-1}(x) = 2x - 8$

b $f^{-1}(x) = -\frac{1}{3}x + 3$

Exercise 3H

1 **a** $f^{-1}(x) = \frac{1}{4}x + \frac{5}{4}$

b $f^{-1}(x) = -6x + 18$

c $f^{-1}(x) = 4x - 7$

2

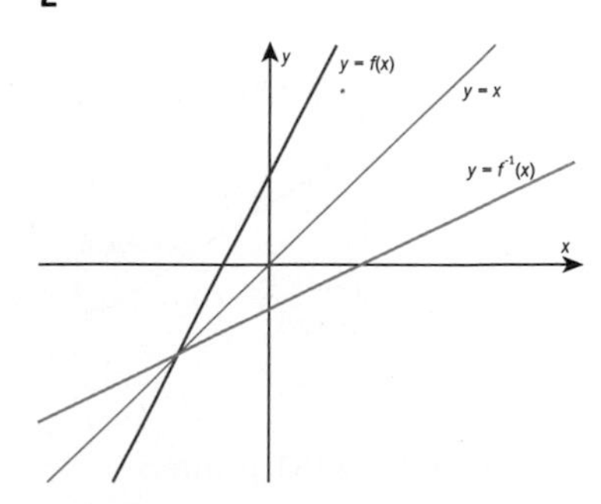

3 a \$615

b $f^{-1}(x) = \frac{1}{10}x - 6.5$, where x is the total cost of a t-shirt order in dollars and $f^{-1}(x)$ is the number of t-shirts in the order

c 500 t-shirts

Exercise 3I

1 a $d = \frac{1}{32}F$ or $F = 32d$, where d = distance in cm and F = force in newtons

b 11.5625 cm (exact) or 11.6 cm to 3 s.f.

2 a $y = 0.16x + 360$

b The y-intercept represents Frank's weekly salary of £360. The gradient shows that Frank's commission is 16% of his sales.

c £504

3 a Plan A: $c(m) = 9.99m + 79.99$ and Plan B: $c(m) = 20m$, where c is the cost for m months at the gym

b Month 8

4 a

$$p(h) = \begin{cases} 8h, & 0 \le h \le 40 \\ 12h - 160, & 40 < h \le 60 \end{cases}$$

b i £176 **ii** £404

5 a 1700

b The model predicts that raising the price €20 will result in 130 fewer printers sold.

c 75 Euro

d

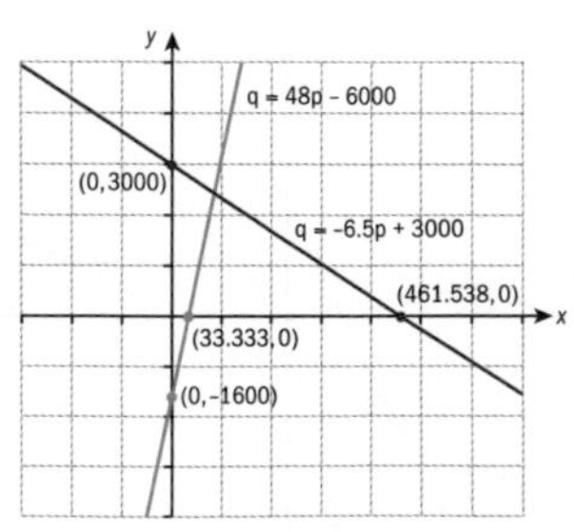

e €84.40; 2451 printers

Exercise 3J

1 a

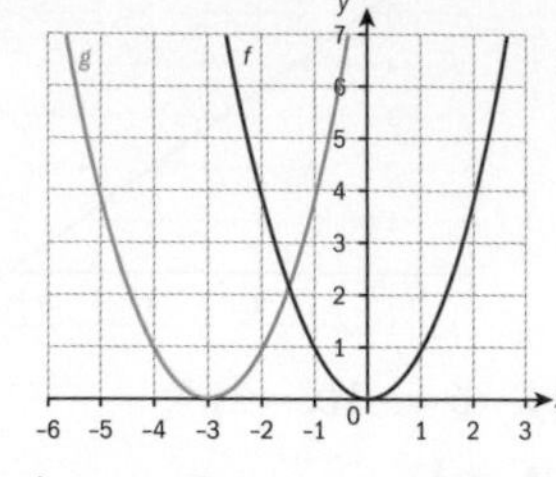

Axis: $x = -3$ vertex: $(-3, 0)$

b

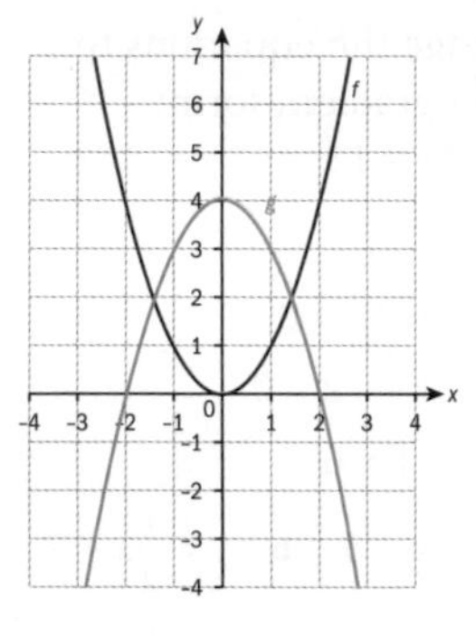

Axis: $x = 0$ vertex: $(0, 4)$

c

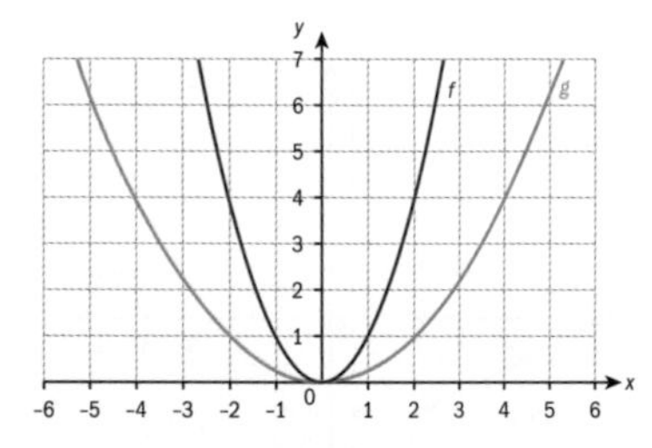

Axis: $x = 0$ vertex: $(0, 0)$

d

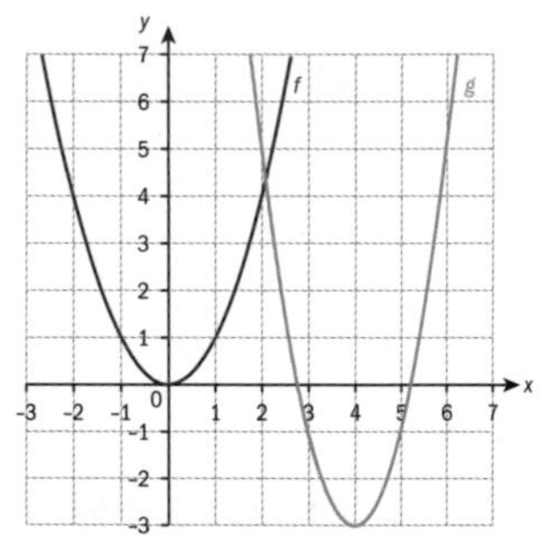

Axis: $x = 4$ vertex: $(4, -3)$

2 a Vertical compression with scale factor $\frac{1}{4}$; $g(x) = \frac{1}{4}x^2$

b Reflection in x-axis, vertical stretch with scale factor 2; $g(x) = -2x^2$

c Horizontal translation right 3, vertical translation up 2; $g(x) = (x - 3)^2 + 2$

d Horizontal translation left 3, vertical stretch with scale factor $\frac{3}{2}$, vertical translation down 5; $g(x) = \frac{3}{2}(x + 3)^2 - 5$

Exercise 3K

1 a

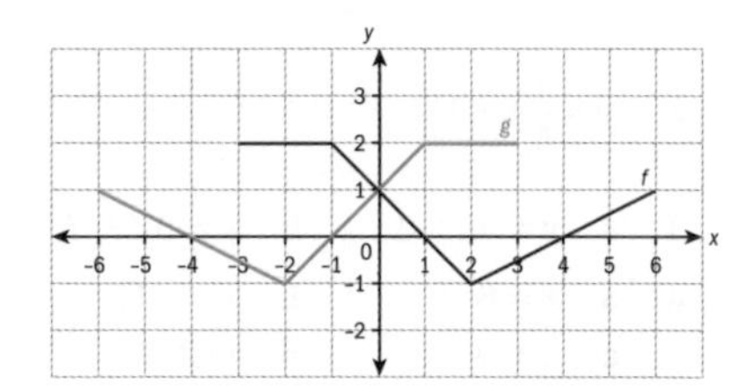

The graph is reflected about the y-axis.

b

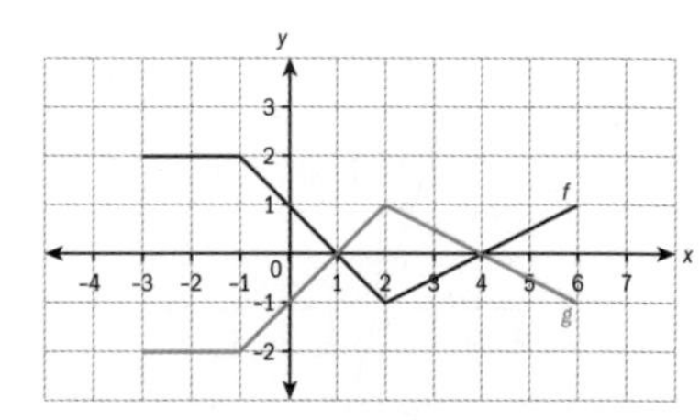

The graph is reflected about the x-axis.

c

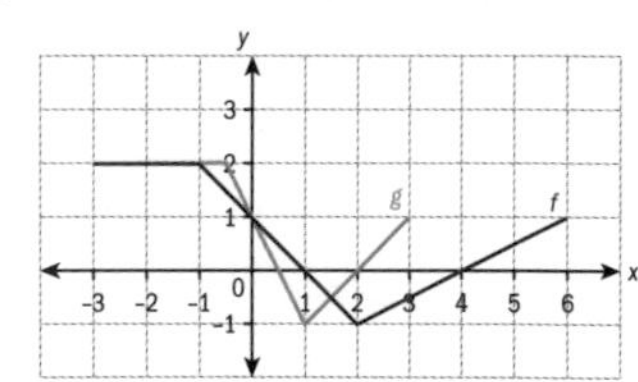

The graph is compressed horizontally with scale factor $\frac{1}{2}$.

d

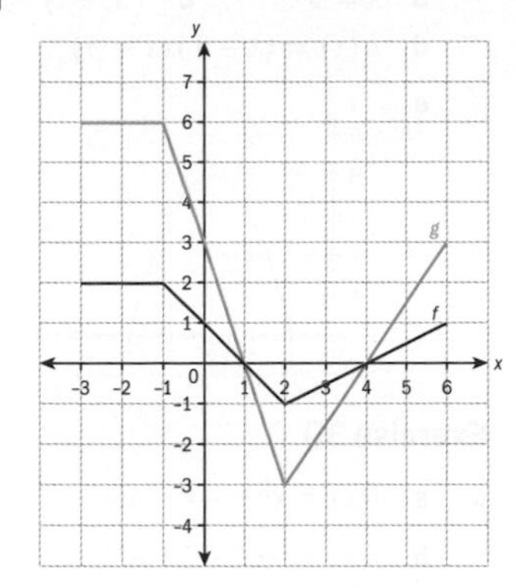

The graph is stretched vertically with scale factor 3.

e

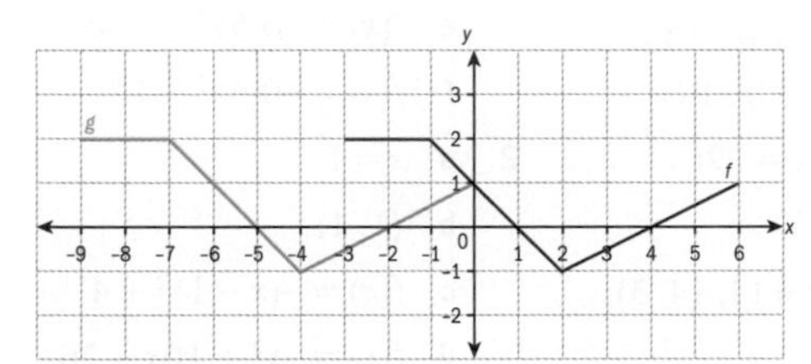

The graph is translated to the left by 6 units.

---

f

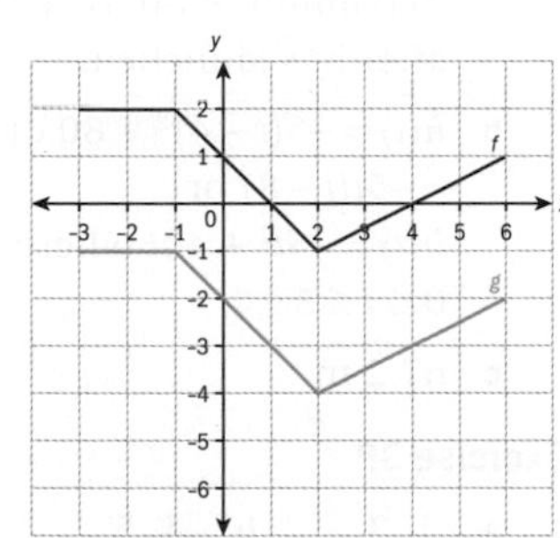

The graph is translated downwards by 3 units.

2 **a** $r(x) = 2f(x)$; $s(x) = -f(x - 3)$

b $r(x) = f(-x)$;

$s(x) = f\left(\frac{1}{2}x\right) - 4$

3 **a** $0 \leq y \leq 6$

b

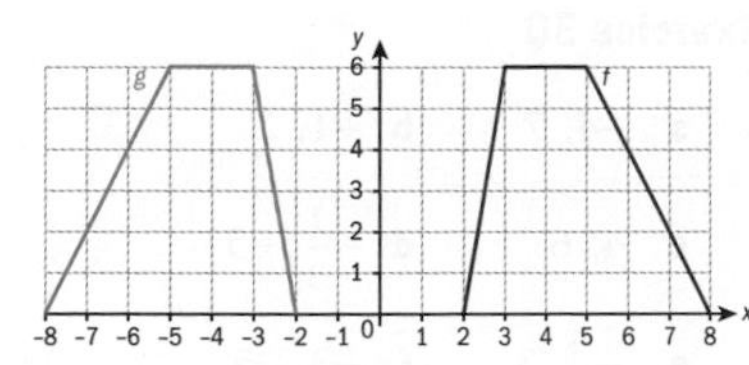

c $-8 \leq x \leq -2$

d $h(x) = g(x) - 4$

e $h(x) = f(-x) - 4$

Exercise 3L

1 x-intercepts: $(-2.81, 0)$, $(0.475, 0)$; y-intercept: $(0, -4)$; vertex: $(-1.17, -8.08)$

2 x-intercepts: none; y-intercept: $(0, -3)$; vertex: $(0.726, -0.785)$

3 **Domain:** $x \in \mathbb{R}$
Range: $f(x) \leq 9.125$

4 **Domain:** $x \in \mathbb{R}$
Range: $f(x) \geq -30.752$

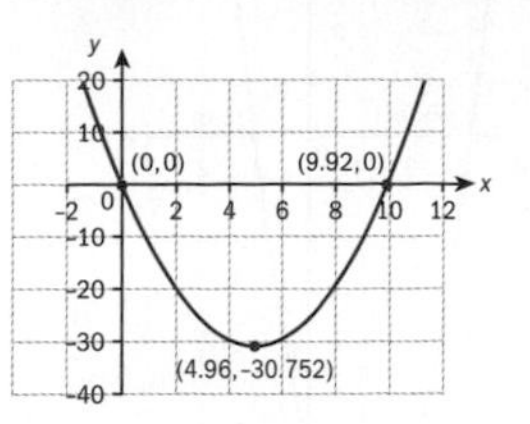

5 **Range:** $0 \leq f(x) \leq 8.1$

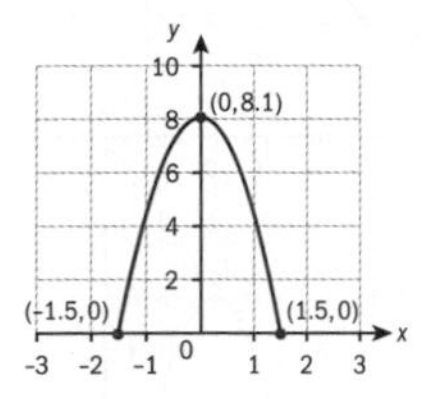

6 **Range:** $-6 \leq f(x) \leq 3$

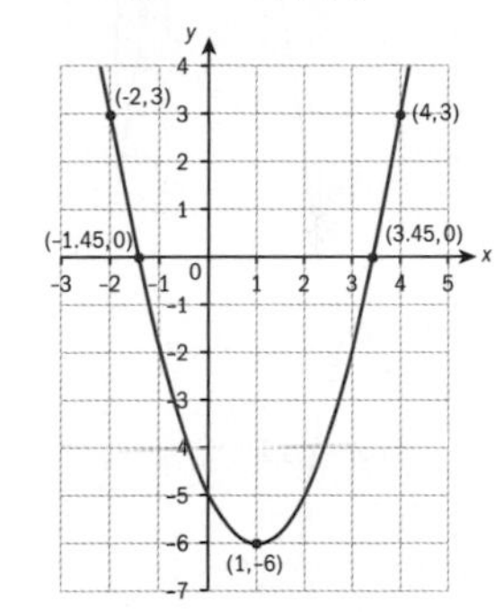

Exercise 3M

1 **a** $x = 3$; $(3, 4)$

b $x = 1$; $(1, -5)$

c $x = -3$; $(-3, 2)$

d $x = -6$; $(-6, -5)$

2 **a** $(0, 5)$; $x = 4$; $(4, -11)$

b $(0, 2)$; $x = 1$; $(1, -1)$

c $(0, -11)$; $x = -2$; $(-2, -3)$

d $(0, 3)$; $x = -\frac{3}{2}$; $\left(-\frac{3}{2}, -\frac{3}{2}\right)$

3 **a** $(2, 0)$, $(4, 0)$; $x = 3$; $(3, -1)$

b $(-3, 0)$, $(1, 0)$; $x = -1$; $(-1, -16)$

c $(-5, 0)$, $(3, 0)$; $x = -1$; $(-1, 16)$

d $(-3, 0)$, $(-2, 0)$; $x = -\frac{5}{2}$; $\left(-\frac{5}{2}, -\frac{1}{2}\right)$

4 **a**

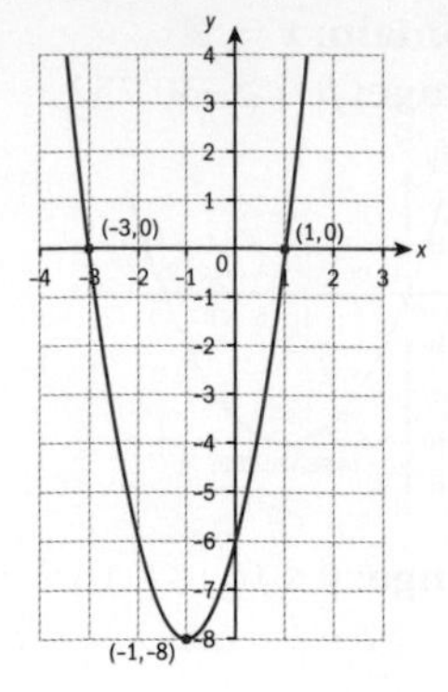

b

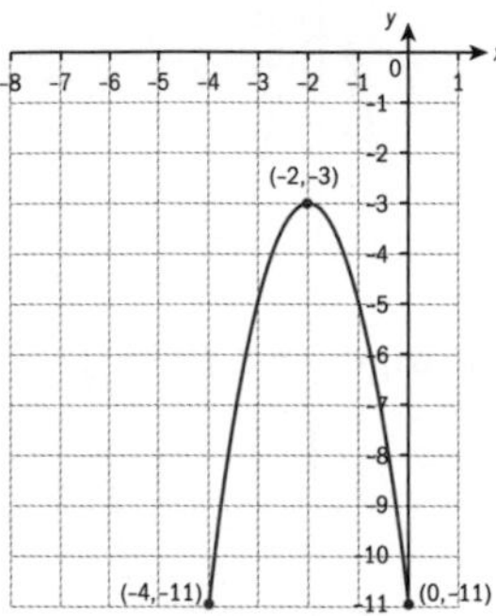

c

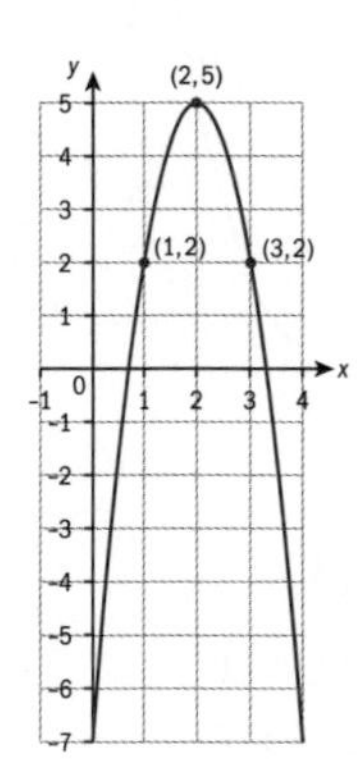

d

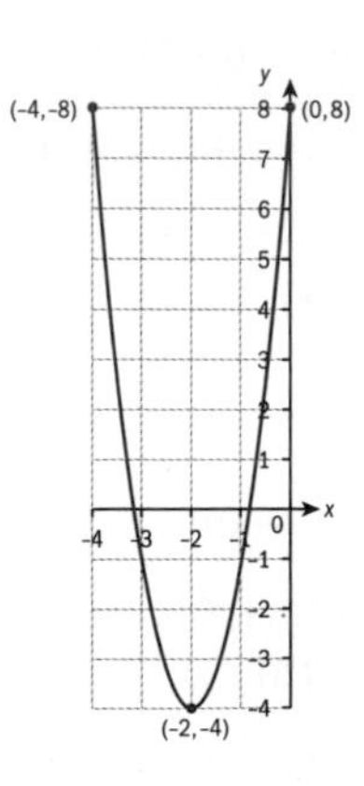

Exercise 3N

1 **a** $a = 1, p = 2, q = -9$;
$(-9, 0), (2, 0); (0, -18)$

b $a = 3, p = 2, q = \frac{5}{3}$;
$\left(\frac{5}{3}, 0\right), (2, 0); (0, 10)$

c $a = 0.5, p = -2, q = -4$;
$(-2, 0), (-4, 0); (0, 4)$

d $a = -4, p = 4, q = \frac{1}{2}$;
$\left(\frac{1}{2}, 0\right), (4, 0); (0, -8)$

2 **a** $a = 4, b = 16, c = -20$;
$(1, 0), (-5, 0); (0, -20)$

b $a = -2, b = -16, c = -14$;
$(-7, 0), (-1, 0); (0, -14)$

3 **a** $a = -3, b = -6, c = -9$;
$(-1, -6); (0, -9)$

b $a = \frac{1}{2}, b = -4, c = 11; (4, 3)$;
$(0, 11)$

4 **a** $a = 1, p = 4, q = -2$
$f(x) = (x - 4)(x + 2)$

b **i** $(4, 0), (-2, 0)$

ii $(0, -8)$

c $(1, -9)$

d

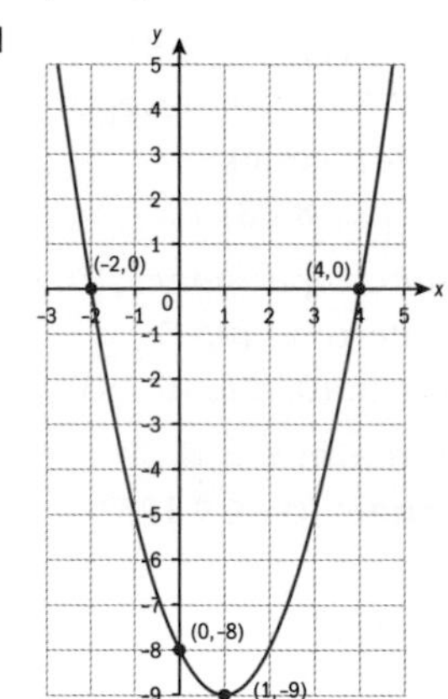

5 **a** **i** $(3, -2)$ **ii** $x = 3$

b $f(x) = x^2 - 6x + 7$

c 7 **d** 6

6 **a**

$$h(x) = (f \circ g)(x)$$
$$= (x-2)^2 - 2(x-2) - 3$$
$$= (x^2 - 4x + 4) - 2x + 4 - 3$$
$$= x^2 - 6x + 5$$

b $x = 3$ **c** $(3, -4)$

d $h(x) = (x - 1)(x - 5)$

e

Exercise 3O

1 **a** $f(x) = x^2 - 4x - 12$

b $f(x) = -x^2 - 2x + 3$

c $f(x) = 4x^2 - 24x + 20$

d $f(x) = 3x^2 - 12x + 6$

e $f(x) = -0.5x^2 - 1.5x + 5$

f $f(x) = -0.6x^2 - 12x$

2 **a** $x = 1$

b $(1, 4)$

c $f(x) = -(x - 1)^2 + 4$

d $f(x) = -x^2 + 10x - 26$

3 **a** $(4, 80)$; The model rocket is predicted to reach a maximum of 80 m, 4 s after it is launched.

b $h(t) = -5(t - 4)^2 + 80$ or $h(t) = -5t(t - 8)$ or $h(t) = -5t^2 + 40t$ where $0 \le t \le 8$

c 67.2 m

Exercise 3P

1 **a** 1, 3 **b** −4, 5

c 2, 6 **d** −11, 11

e −7, 6 **f** 4

2 **a** $1, -\frac{3}{2}$ **b** $-3, \frac{4}{3}$

c $-2, -\frac{3}{4}$ **d** $-\frac{7}{3}, \frac{7}{3}$

e $\frac{1}{2}, \frac{7}{2}$ **f** $-\frac{5}{4}, \frac{1}{3}$

Exercise 3Q

1 **a** −4, 7 **b** $-1, \frac{8}{3}$

c 4, 6 **d** $-\frac{1}{3}, -5$

e $-\frac{4}{3}, 3$ **f** −3, −5

2 a $(f \circ g)(x) = f(2x+1)$
$= (2x+1)^2 - 2$
$= (4x^2 + 4x + 1) - 2$
$= 4x^2 + 4x - 1$

b $-1, \frac{4}{3}$

Exercise 3R

1 $4 \pm \sqrt{10}$; 0.838, 7.16

2 $-10 \pm \sqrt{15}$; −6.13, −13.9

3 $-6 \pm 2\sqrt{3}$; −2.54, −9.46

4 $5 \pm 3\sqrt{3}$; −0.196, 10.2

5 (0.5, 4.5), (−1.5, 6.5)

6 (1.18, 7.35), (−1.96, 1.07)

7 (1, 5)

8 (2.72, 7.64), (0.613, −0.0872)

9 (−0.802, −8.64); $x = -0.802, 1.80$

10 (−2.91, −4.50); $x = -2.91, 0.915$

Exercise 3S

1 $-6 \pm \sqrt{38}$ 2 $\frac{3 \pm \sqrt{17}}{2}$

3 $3 \pm \sqrt{5}$ 4 $6 \pm 4\sqrt{2}$

5 $\frac{-5 \pm \sqrt{41}}{2}$ 6 $\frac{-1 \pm 3\sqrt{5}}{2}$

Exercise 3T

1 $-4 \pm \sqrt{21}$ 2 $3 \pm \sqrt{11}$

3 $1 \pm \sqrt{\frac{3}{2}}$ 4 $-4 \pm 3\sqrt{2}$

5 $\frac{3}{2}, -2$ 6 $-4 \pm \sqrt{10}$

7 a $35x - 0.25x^2 = 300 + 15x$

b $0.25(x - 40)^2 = 100$; $x = 20, 60$

c 20 or 60

d 40

e 100 000 Euros

Exercise 3U

1 a $-2 \pm \sqrt{6}$ b $1, \frac{5}{3}$

c $\frac{5 \pm \sqrt{41}}{4}$

2 a $\frac{-3 \pm 3\sqrt{5}}{2}$ b $\frac{2 \pm \sqrt{10}}{3}$

c $1 \pm \sqrt{3}$ d No real solution

e $\frac{5 \pm \sqrt{7}}{2}$ f $3, \frac{3}{2}$ g $3 \pm 2\sqrt{3}$

3 a $\frac{2}{3}, -\frac{3}{2}$ b $\frac{2 \pm \sqrt{2}}{2}$

c $1 \pm \sqrt{5}$

4 a −2 b (1, −4)

c $r = 1$; $s = 2$

Exercise 3V

1 a −11; no real roots

b 121; two distinct real roots

c −44; no real roots

d 112; two distinct real roots

e 0; two equal roots

f −199; no real roots

2 a $k < \frac{9}{4}$ b $k < 20$

3 a $p = \frac{25}{4}$ b $p = 12$

c $p = \pm 2\sqrt{2}$ d $p = 0, -\frac{8}{9}$

4 a $m > 1$ b $m > 3$ c $m > \frac{33}{4}$

Exercise 3W

1 a $x \le -2$ or $x \ge \frac{1}{3}$

b $-\sqrt{5} \le x \le \sqrt{5}$

c $-2 - \sqrt{10} < x < -2 + \sqrt{10}$

2 a $x \le -0.245$ or $x \ge 12.2$

b $-\frac{2}{3} < x < 3$

c $-0.890 < x < 1.26$

3 a $k > 4$ or $k < -4$

b $k > \sqrt{3}$ or $k < -\sqrt{3}$

4 $0 < m < \frac{1}{9}$

5 $k < 0$ or $k > \frac{1}{4}$

6 a $p^2 - 48$ b $-4\sqrt{3} < p < 4\sqrt{3}$

c $m = 6$ d $a = 3$; $h = -1$; $k = 1$

Exercise 3X

1 4 m, 12 m

2 a 17.9 m

b 0.211 s, 3.87 s

c 22.4 m

3 a Fare $= 5.50 - 0.05x$

b Number of riders $= 800 + 10x$

c Revenue $= (5.50 - 0.05x)(800 + 10x) = 4400 + 15x - 0.5x^2$

d 10 or 20 decreases

e $0 \le x \le 110$, where x is an integer

4 a $y = -(x-2)^2 + 4$ or $y = -x(x-4)$ or $y = -x^2 + 4x$

b If the centre of the object is aligned with the centre of the archway, it spans form $x = 0.5$ to $x = 3.5$. Evaluating the function at $x = 0.5$ and $x = 3.5$ gives 1.75. Since 1.6 < 1.75, the object will fit through the archway.

5 a $A(x) = x(155 - x)$ or $A(x) = 155x - x^2$

b 77.5 m by 77.5 m

c No; the touchline would not be longer than the goal line and 77.5 m is less than the minimum of 90 m for the touchline.

d $90 \le x \le 120$ (If the goal line restrictions are also taken into consideration the answer is $90 \le x \le 110$.)

e 5850 m²

Chapter review

1 **a**

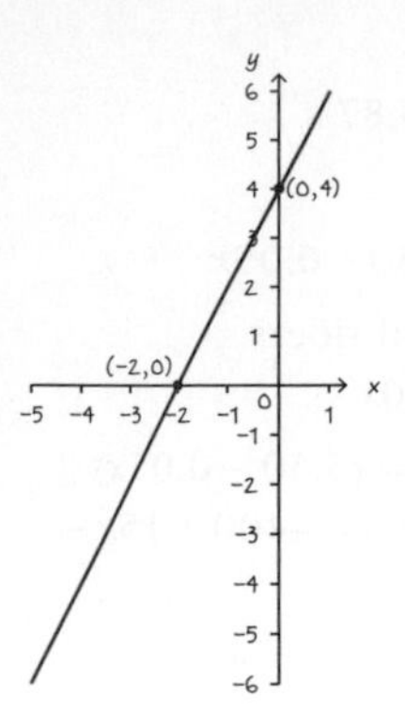

b

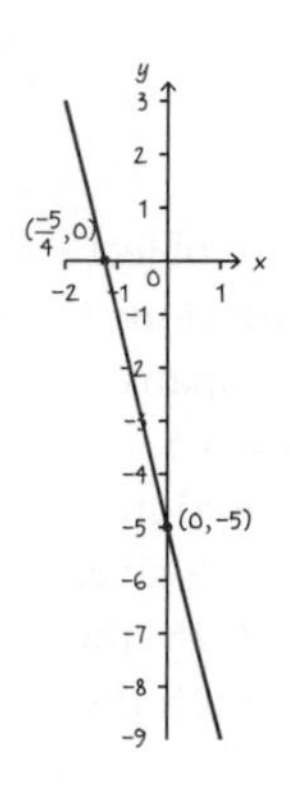

c

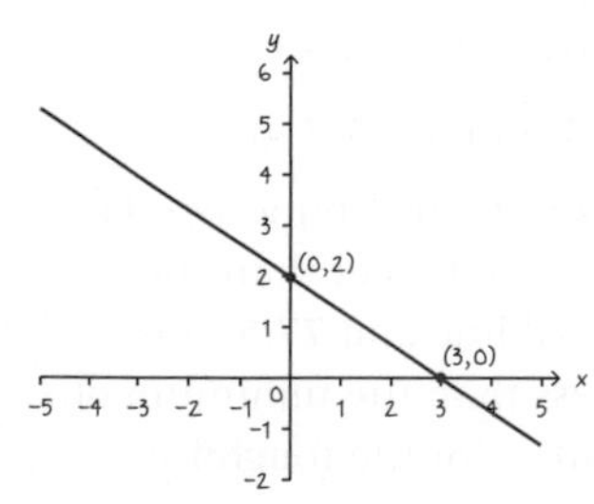

2 **a** $y=-\frac{1}{4}x+1$

b $y=\frac{1}{2}x-5$

c $y-4=\frac{3}{2}(x-2)$ or $y=\frac{3}{2}x+1$

d $y=-4$

3 **a** $f(1)=3$; $f(2)=3$

b

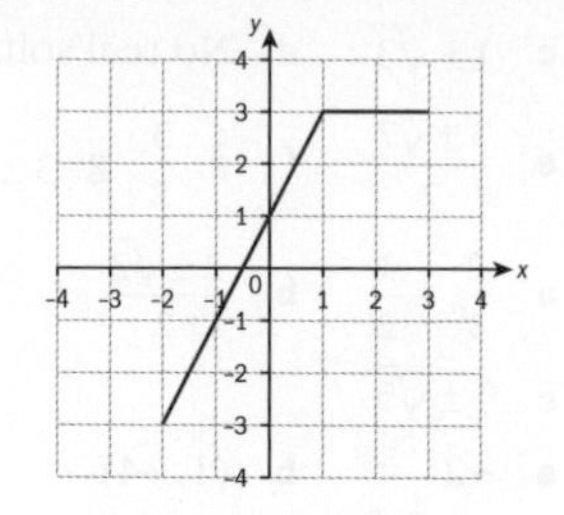

4 **a** Vertical stretch with scale factor 2, horizontal translation right 3

b Horizontal compression with scale factor $\frac{1}{2}$, vertical translation up 5

c Reflection in the x-axis, horizontal translation left 2, vertical translation down 1

d Horizontal compression with scale factor 3

e Reflection in the y-axis, vertical translation up 6

5 **a** $(3,0)$ and $(-7,0)$; $x=-2$

b $x=4$; $(4,2)$

c $x=-2$; $(0, 6)$

6 **a** **i** $a=3$ **ii** $h=-3$ **iii** $k=-7$

b $(-3,-7)$ **c** $(2,-10)$

7 **a** $11, -5$ **b** $-2\pm\sqrt{7}$ **c** -7

d $-4, 3$ **e** $\frac{7}{3}, 1$

8 $k=\frac{4}{3}, -\frac{4}{3}$

9 $f(x)=2x^2+4x-16$

10 **a** $-0.679, 3.68$

b $-4.92, 1.42$

11 **a** 18 m **b** 26.6 m **c** 3.66 s

d $0\le t\le 3.66$ **e** 1.72 s

12 **a** $A(-4,0)$; $B(0,7)$; $C(4,0)$

b $y=-1.75x+7$

c $2p$ cm by $-1.75p+7$ cm

d Area $=2p(-1.75p+7)$

e 4 cm by 3.5 cm

f 14 cm²

Exam-style questions

13 **a** $12y=-7x+168$ (1 mark)

$y=-\frac{7x}{12}+14$ (1 mark)

b $A(24,0)$ and $B(0,14)$ (2 marks)

c Area

$=\frac{1}{2}\times 24\times 14=168$ units² (2 marks)

14 **a**

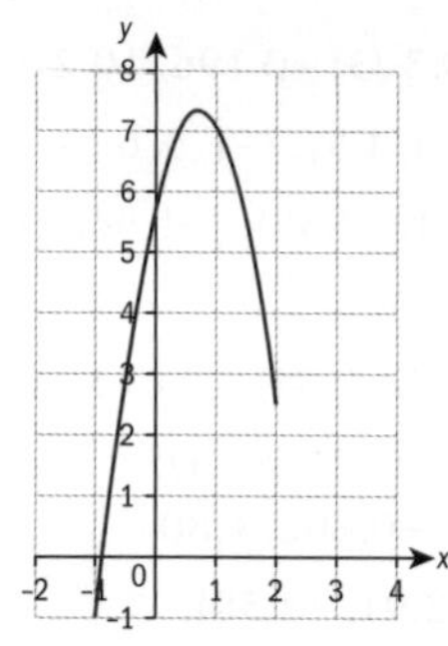

(2 marks)

b $(0,5.9)$ and $(-0.885,0)$ (2 marks)

c $-1.1\le f(x)\le 7.35$ (2 marks)

15 **a** $x=\frac{-b\pm\sqrt{b^2-4ac}}{2a}$ (1 mark)

$x=\frac{6\pm\sqrt{208}}{2}$ (1 mark)

$x=3\pm\sqrt{52}$

$x=3\pm 2\sqrt{13}$ (2 marks)

b Using GDC

$3-2\sqrt{13}\le x\le 3+2\sqrt{13}$ (2 marks)

16 **a** $3(x-1)^2-18$

$=3(x^2-2x+1)-18$ (1 mark)

$=3x^2-6x-15$ (1 mark)

b $(1,-18)$ (1 mark)

c $x=1$ (1 mark)

d $f(x)\in\mathbb{R}$, $f(x)\ge -18$ (2 marks)

e $g(x)=3((x-2)-1)^2-18-1$ (2 marks)

$=3x^2-18x+8$ (1 mark)

17 a $(4x+5)(2x-1)=0$ (2 marks)

$4x+5=0 \Rightarrow x=-\frac{5}{4}$ (1 mark)

$2x-1=0 \Rightarrow x=\frac{1}{2}$ (1 mark)

b $8x^2+6x-5-k=0$

No real solutions

$\Rightarrow b^2-4ac<0$ (1 mark)

$36-4\times8\times(-5-k)<0$ (1 mark)

$k<-\frac{49}{8}$ (1 mark)

18 a $x^2-10x+27=(x-5)^2-25+27$ (2 marks)

$=(x-5)^2+2$ (1 mark)

b Coordinates of the vertex are $(5, 2)$. (1 mark)

c Equation of symmetry is $x=5$. (1 mark)

19 a At $(10,0)$, $0=10^2+10b+c$,

so $10b+c=-100$ (2 marks)

Line of symmetry is $x=-\frac{b}{2}$,

so $b=-5$ (1 mark)

Solving simultaneously gives

$-50+c=-100$

So $c=-50$ (1 mark)

Therefore, the equation is

$y=x^2-5x-50$

b One coordinate is

$(0,-50)$ (1 mark)

The other coordinate is

$(-5,0)$ (1 mark)

20 a $f(x)=2\left[x^2-2x-4\right]$ (1 mark)

$=2\left[(x-1)^2-1-4\right]$ (1 mark)

$=2(x-1)^2-10$ (1 mark)

b A horizontal translation right 1 unit. (1 mark)

A vertical stretch with scale factor 2. (1 mark)

A vertical translation down 10 units. (1 mark)

21 a Two equal roots

$\Rightarrow b^2-4ac=0$ (1 mark)

$36-4(2k)(k)=0$ (1 mark)

$k=\pm\frac{\sqrt{3}}{2}$ (2 marks)

b Equation of line of symmetry is

$x=-\frac{b}{2a}=-\frac{6}{4k}=-\frac{3}{2k}$ (2 marks)

Therefore $\frac{3}{2k}=1$

$k=\pm\frac{3}{\sqrt{2}}$ or $=\pm\frac{3\sqrt{2}}{2}$ (1 mark)

c $k=2 \Rightarrow 4x^2+6x+2=0$

$2x^2+3x+1=0$

$(2x+1)(x+1)=0$ (1 mark)

$x=-\frac{1}{2}$ or $x=-1$ (2 marks)

22

a $A'(-6,10)$, $B'(0,-16)$, $C'(1,9)$ and $D'(7,-10)$ (4 marks)

b $A(12,13)$, $B(0,-13)$, $C(-2,12)$ and $D(-14,-7)$ (4 marks)

Chapter 4

Skills check

1 a -5 **b** 6 **c** $\frac{5}{2}$

2

y = 4
x = −2
x = 3
y = −3

Exercise 4A

1 a $\frac{1}{3}$ **b** $\frac{1}{5}$ **c** $-\frac{1}{2}$ **d** -1

e $\frac{5}{3}$ **f** $\frac{7}{22}$ **g** $-\frac{9}{8}$ **h** $\frac{4}{11}$

2 a $\frac{2}{3}$ **b** $\frac{1}{x}$ **c** $\frac{1}{2x}$

d $\frac{1}{4y}$ **e** $\frac{4}{3x}$ **f** $\frac{t}{d}$

g $\frac{4d}{3}$ **h** $\frac{x-3}{x+2}$

3 a $4\times\frac{1}{4}=1$

b $\frac{7}{11}\times\frac{11}{7}=1$

c $\frac{2}{x}\times\frac{x}{2}=1$

d $\frac{x-1}{x-2}\times\frac{x-2}{x-1}=1$

Exercise 4B

1 a

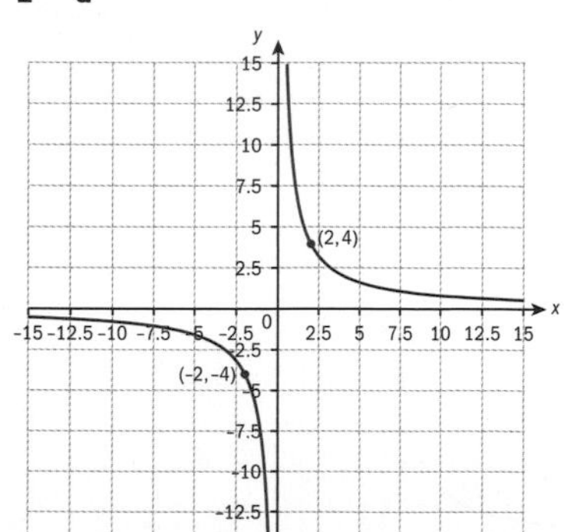

b

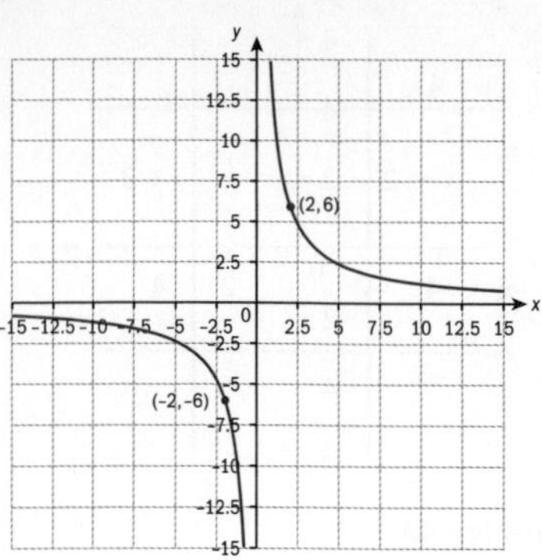

c

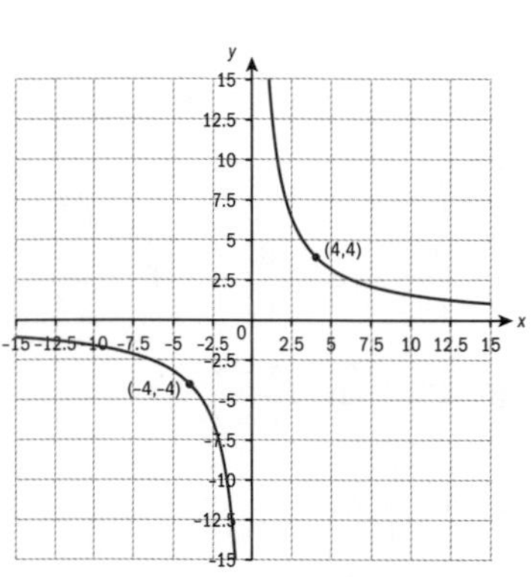

2

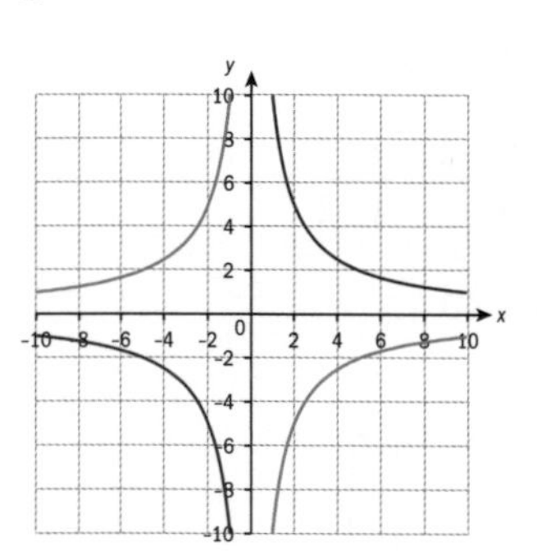

3 **a**

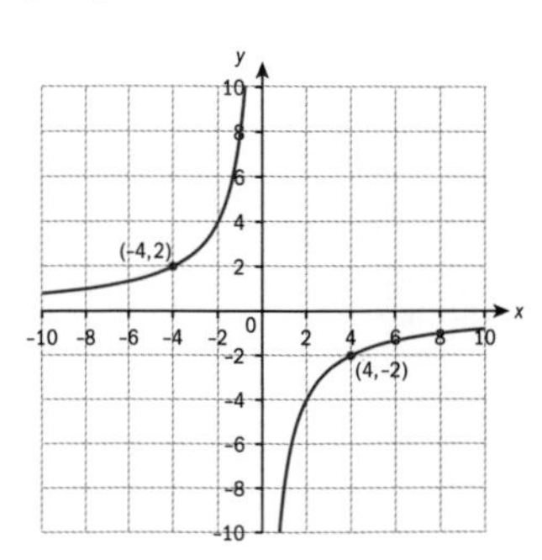

b

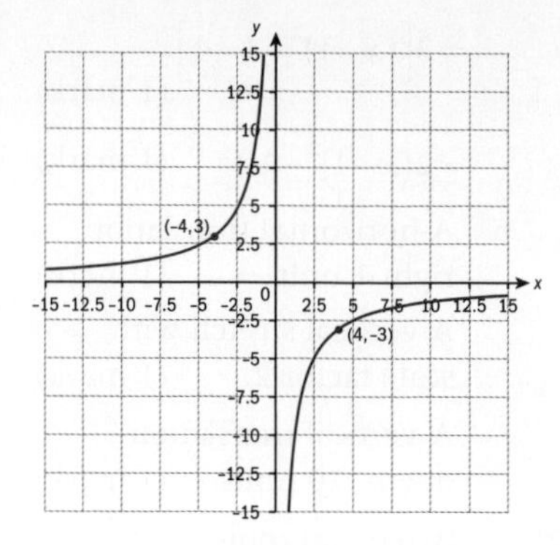

c

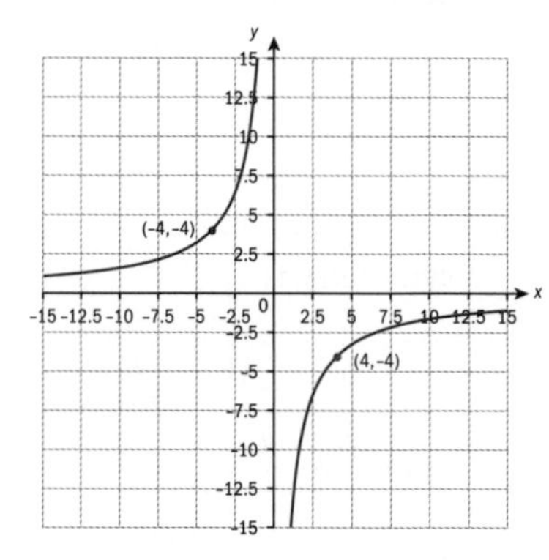

d The curves are in the opposite quadrants. The negative reflects the function in the x-axis.

4 $x = 0, y = 0$

Domain $x \in \mathbb{R}, x \neq 0$

Range $y \in \mathbb{R}, y \neq 0$

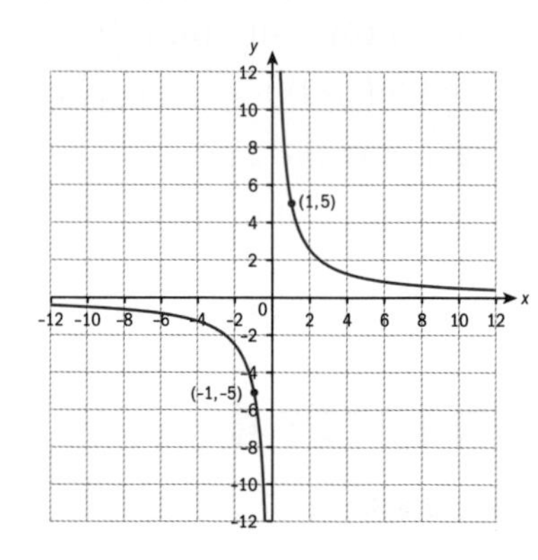

Exercise 4C

1 **a** 1

b 30 s

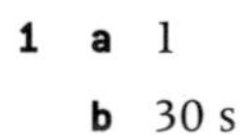

2 **a** and **c**

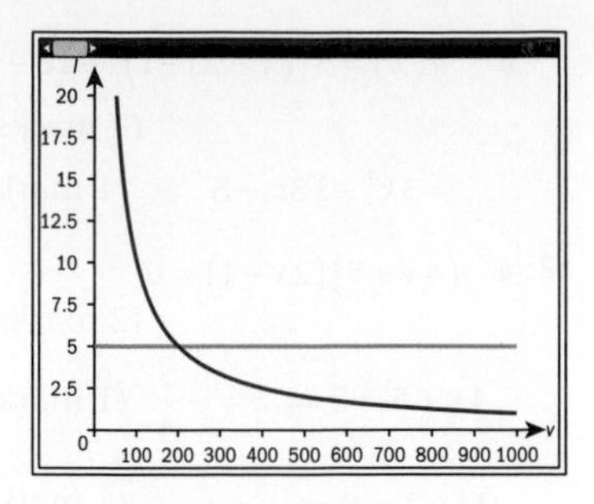

b 100

c (200, 5); string 5 cm long has vibrations of frequency 200 Hz.

3 **a** 4 **b** $y = \dfrac{64}{x}$

c and **d**

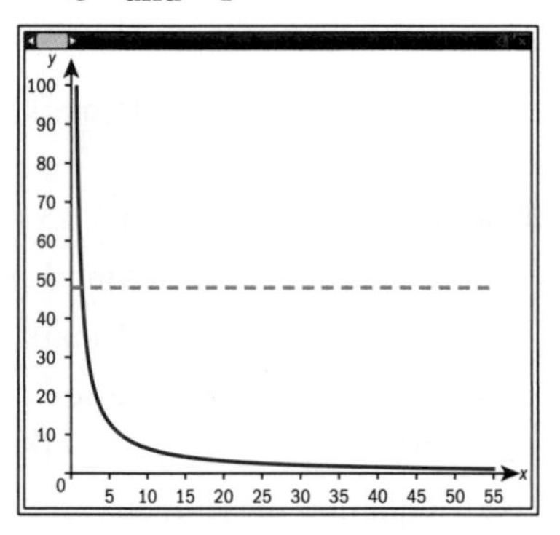

1.33 mins

Exercise 4D

1 **a** $x = -1, y = 0$
Domain $x \in \mathbb{R}, x \neq -1$
Range $y \in \mathbb{R}, y \neq 0$

b $x = 5, y = 0$
Domain $x \in \mathbb{R}, x \neq 5$
Range $y \in \mathbb{R}, y \neq 0$

c $x = 4, y = 0$
Domain $x \in \mathbb{R}, x \neq 4$
Range $y \in \mathbb{R}, y \neq 0$

d $x = -5, y = 0$
Domain $x \in \mathbb{R}, x \neq -5$
Range $y \in \mathbb{R}, y \neq 0$

e $x = -1, y = 2$
Domain $x \in \mathbb{R}, x \neq -1$
Range $y \in \mathbb{R}, y \neq 2$

f $x = -1, y = -2$
Domain $x \in \mathbb{R}, x \neq -1$
Range $y \in \mathbb{R}, y \neq -2$

g $x = 3, y = 2$
Domain $x \in \mathbb{R}, x \neq 3$
Range $y \in \mathbb{R}, y \neq 2$

h $x = 4, y = -4$

Domain $x \in \mathbb{R}, x \neq 4$

Range $y \in \mathbb{R}, y \neq -4$

2 **a** $x \in \mathbb{R}, x \neq -4 \quad y \in \mathbb{R}, y \neq 0$

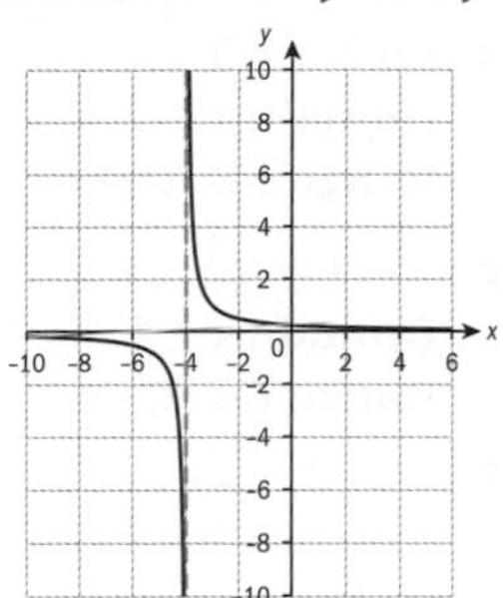

b $x \in \mathbb{R}, x \neq -4 \quad y \in \mathbb{R}, y \neq 0$

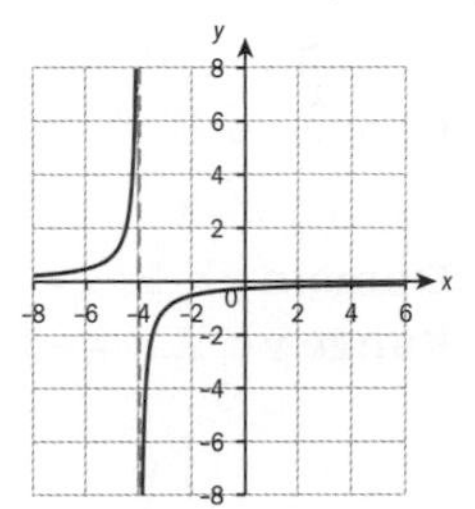

c $x \in \mathbb{R}, x \neq -4 \quad y \in \mathbb{R}, y \neq 1$

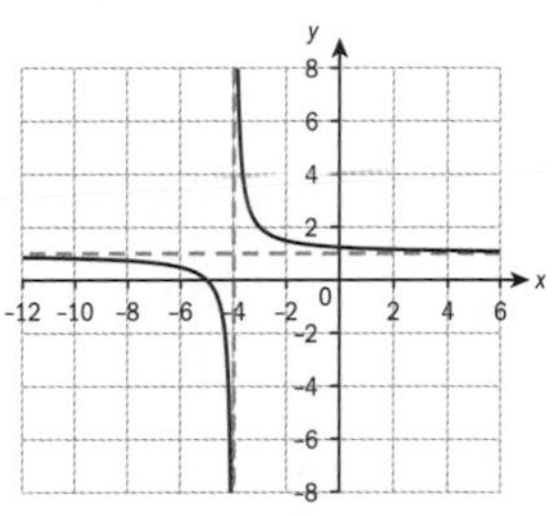

d $x \in \mathbb{R}, x \neq -5 \quad y \in \mathbb{R}, y \neq -1$

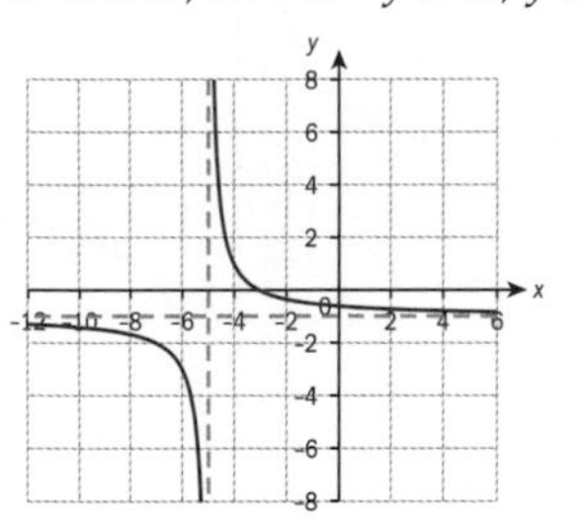

e $x \in \mathbb{R}, x \neq -0.5 \quad y \in \mathbb{R}, y \neq 0$

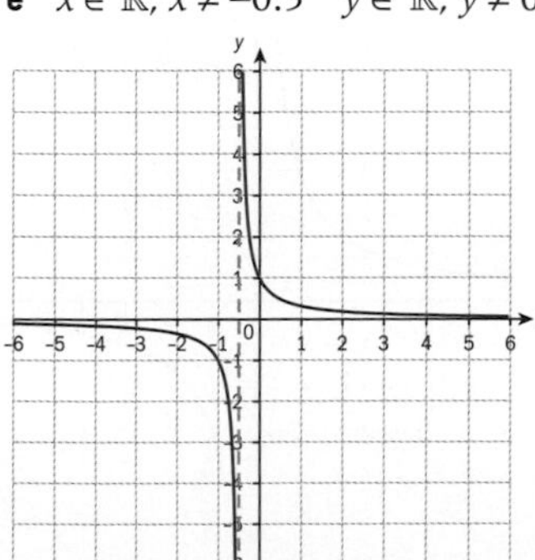

f $x \in \mathbb{R}, x \neq -2 \quad y \in \mathbb{R}, y \neq 0$

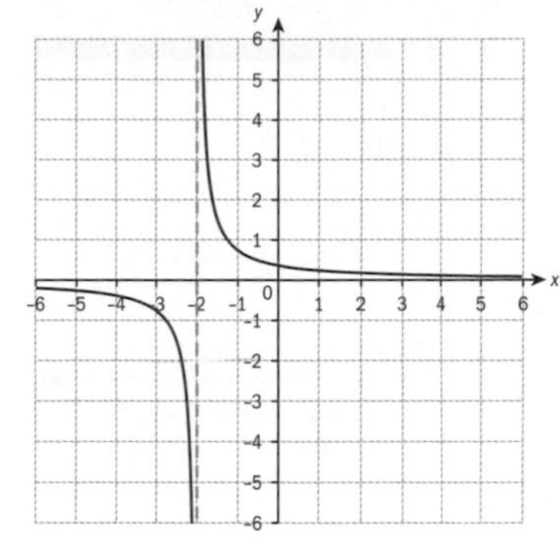

g $x \in \mathbb{R}, x \neq 2 \quad y \in \mathbb{R}, y \neq 2$

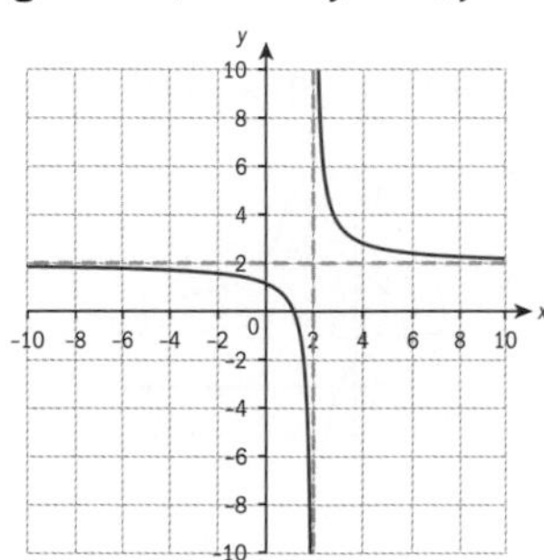

h $x \in \mathbb{R}, x \neq 2 \quad y \in \mathbb{R}, y \neq 1$

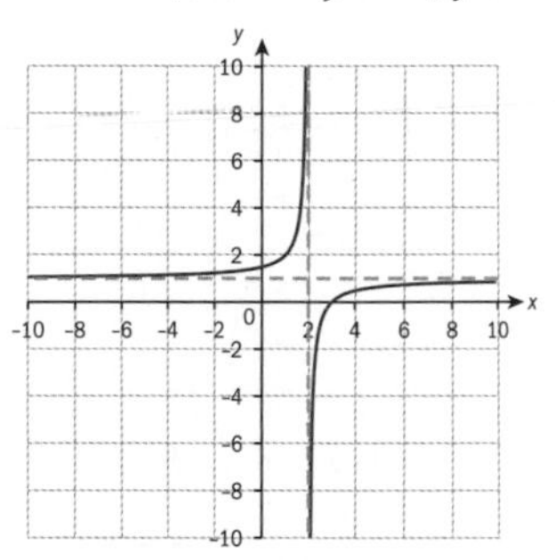

3 **a** **2:** Translation of 2 units right

b **5:** Reflection in $y = 0$ and a translation of 2 units right

c **1:** Translation of 2 units right and 2 units up

d **4:** Translation of 2 units right and 2 units down

e **3:** Translation of 2 units right and vertical stretch by a factor of 3

4 **a**

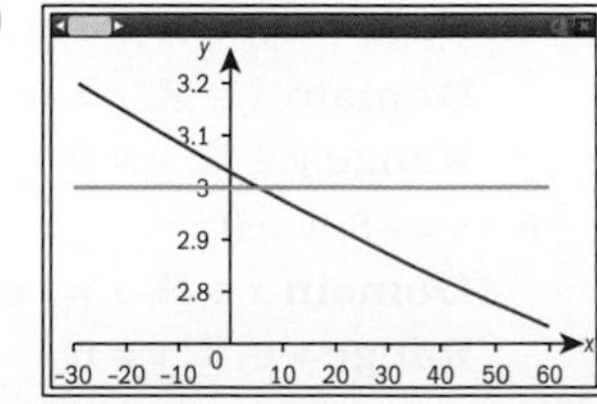

b 5.56 **c** −272.22°

5 **a** Asymptotes $s = 5$, $c = 0$

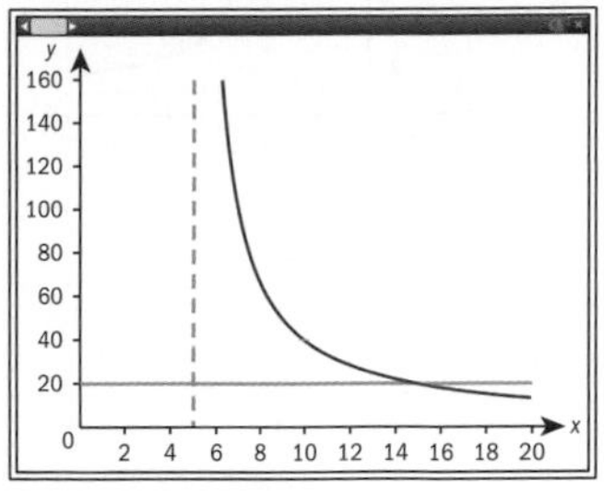

b 15 sessions

6

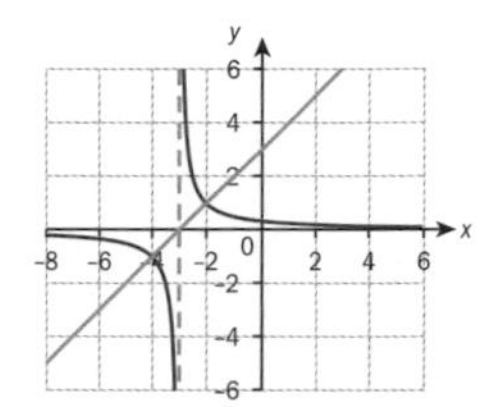

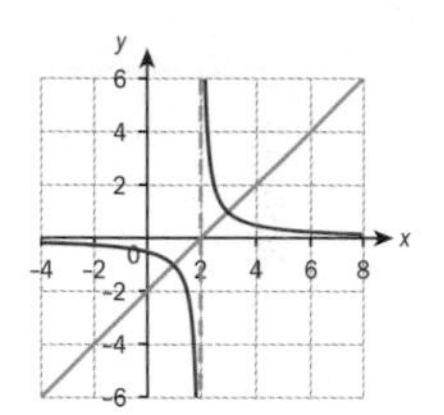

The linear function is a line of symmetry for the rational function. The linear function crosses the x-axis at the same place as the vertical asymptote.

Exercise 4E

1 **a** $x = 1, y = 1$

Domain $x \in \mathbb{R}, x \neq 1$

Range $y \in \mathbb{R}, y \neq 1$

b $x = -1, y = 2$

Domain $x \in \mathbb{R}, x \neq -1$

Range $y \in \mathbb{R}, y \neq 2$

c $x = -2, y = 3$

Domain $x \in \mathbb{R}, x \neq -2$

Range $y \in \mathbb{R}, y \neq 3$

d $x = 1.25, y = 0.75$

Domain $x \in \mathbb{R}, x \neq 1.25$

Range $y \in \mathbb{R}, y \neq 0.75$

e $x = 2, y = -3$

Domain $x \in \mathbb{R}, x \neq 2$

Range $y \in \mathbb{R}, y \neq -3$

2 **i** B; **ii** A; **iii** D; **iv** C

3 $x = q, y = 1$

Domain $x \in \mathbb{R}, x \neq q$

Range $y \in \mathbb{R}, y \neq 1$

4 **a**

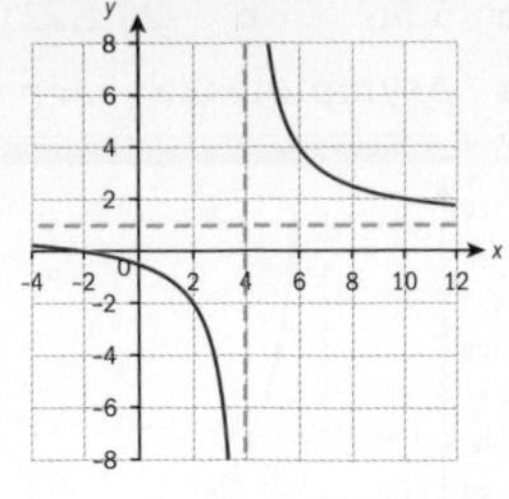

b

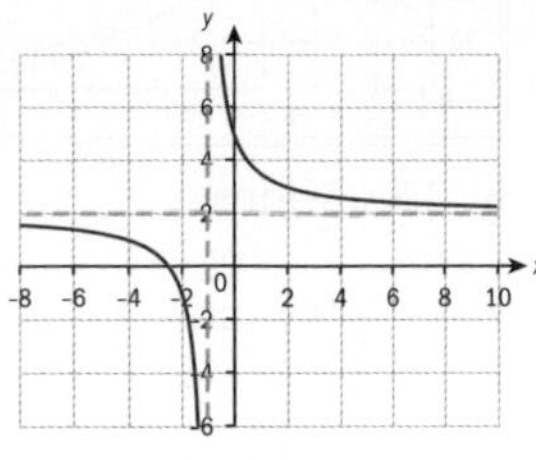

c

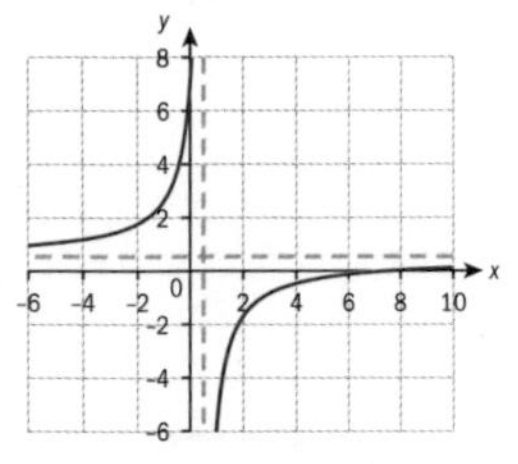

d

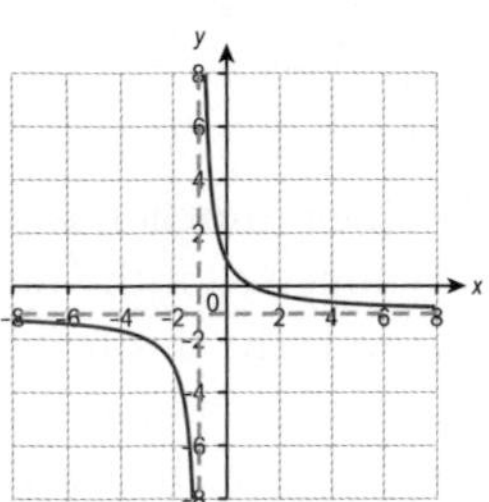

5 **a** $-\dfrac{5}{2}, 4$ **b** 0, 14

c $-\dfrac{6}{7}, 3$ **d** $-\dfrac{17}{3}$

6 2

7 **a** $\dfrac{2x+3}{x-1}$ **b** $\dfrac{7}{x+2}$

c $\dfrac{9x-1}{x+7}$ **d** $\dfrac{5-6x}{x+11}$

8 **a**

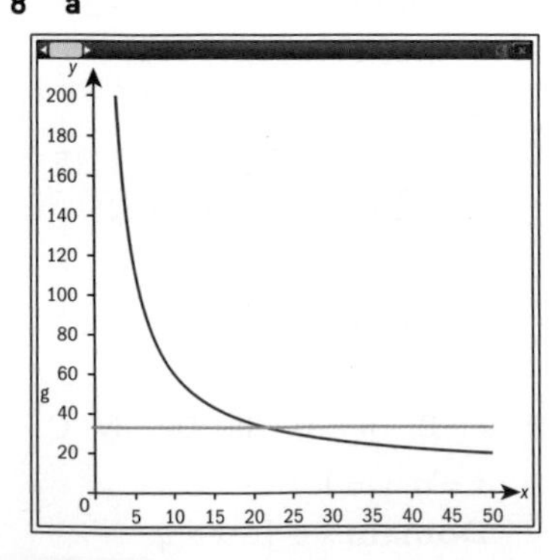

b 20 **c** 50

9 **a** $C(m) = \dfrac{20+10m}{m}$

b

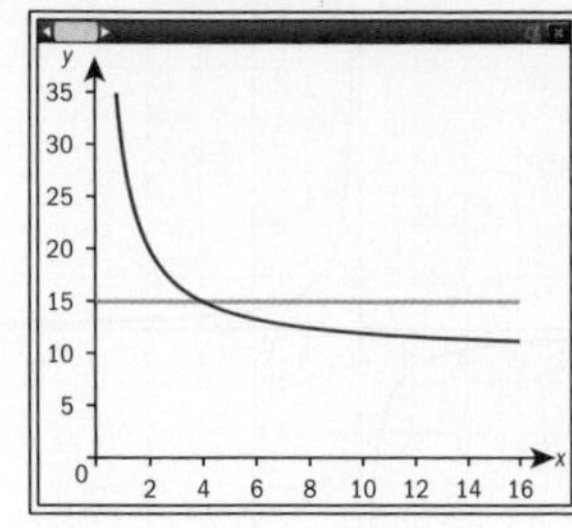

c 4 months **d** 10AUD

10 **a** $n = 5$ **b** $m = 4$

c $y = 4$

11 **a** $y = 3$ **b** $x = 2$

c (0.667. 0) and (0,1)

d

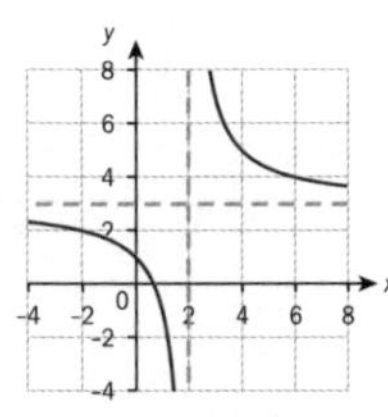

12 Let $f(x) = \dfrac{2x+1}{x-1}$

a

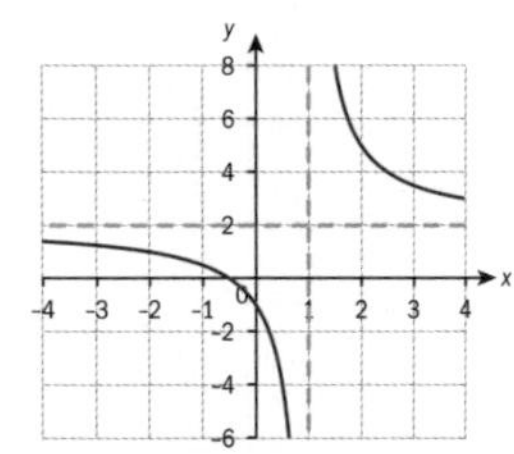

b $x = 1, y = 2$ **c** $(-0.5, 0)$

13 **a** $(g \circ f)(x) = \dfrac{x+3}{x+2}$

b

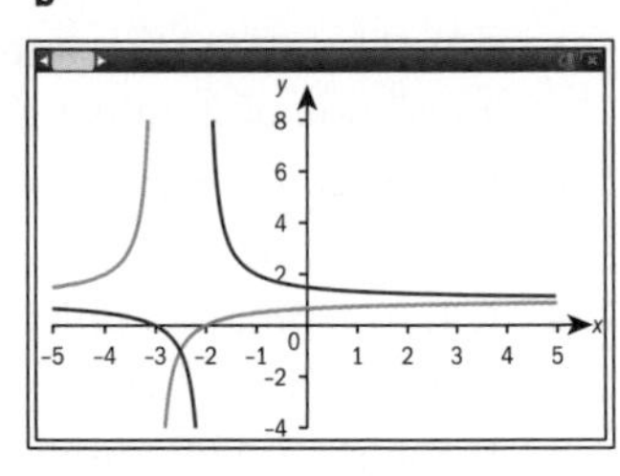

$x = -2.5$

Chapter review

1 **a** $x = 0, y = 0$

Domain $x \in \mathbb{R}, x \neq 0$

Range $y \in \mathbb{R}, y \neq 0$

b $x = -8, y = 0$

Domain $x \in \mathbb{R}, x \neq -8$

Range $y \in \mathbb{R}, y \neq 0$

c $x = 5, y = \dfrac{1}{2}$

Domain $x \in \mathbb{R}, x \neq 5$

Range $y \in \mathbb{R}, y \neq \dfrac{1}{2}$

d $x = 2, y = 3$

Domain $x \in \mathbb{R}, x \neq 2$

Range $y \in \mathbb{R}, y \neq 3$

e $x = 9, y = 2$

Domain $x \in \mathbb{R}, x \neq 9$

Range $y \in \mathbb{R}, y \neq 2$

f $x = -2, y = 4$

Domain $x \in \mathbb{R}, x \neq -2$

Range $y \in \mathbb{R}, y \neq 4$

g $x = -4, y = -1$

Domain $x \in \mathbb{R}, x \neq -4$

Range $y \in \mathbb{R}, y \neq -1$

h $x = -3, y = -3$

Domain $x \in \mathbb{R}, x \neq -3$

Range $y \in \mathbb{R}, y \neq -3$

2 **a**

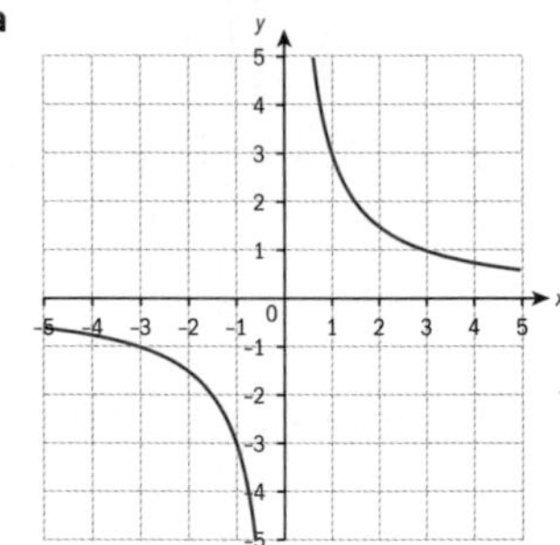

b

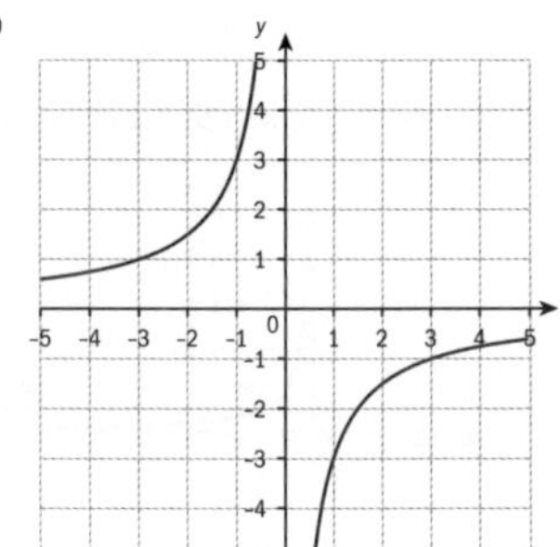

c

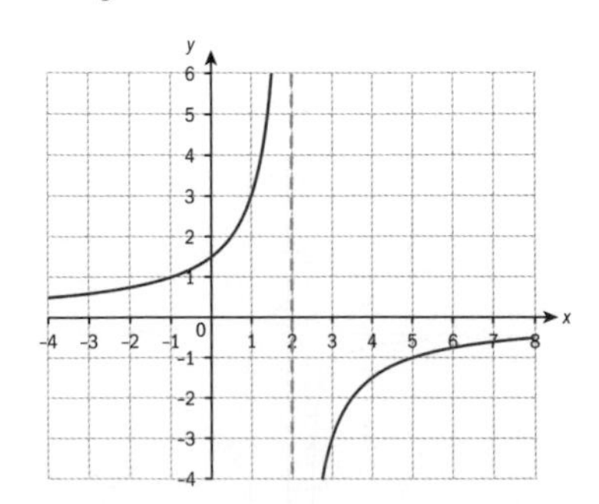

d

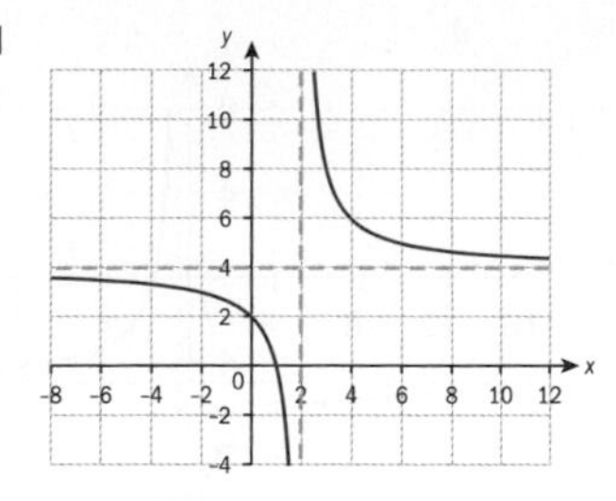

3 a $x = 1, y = 2$

b $(0.5, 0)$ and $(0, 1)$

c

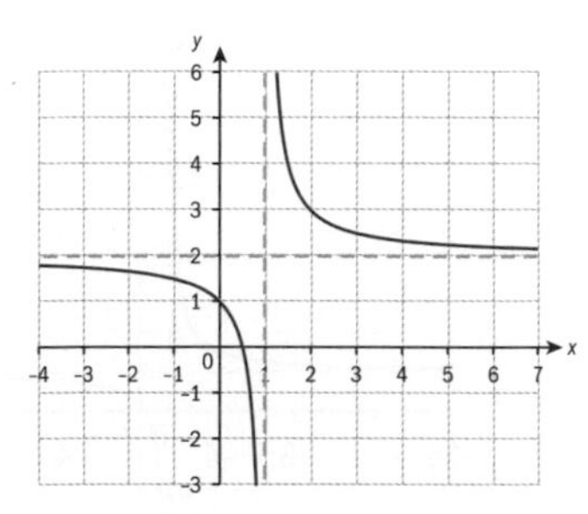

4

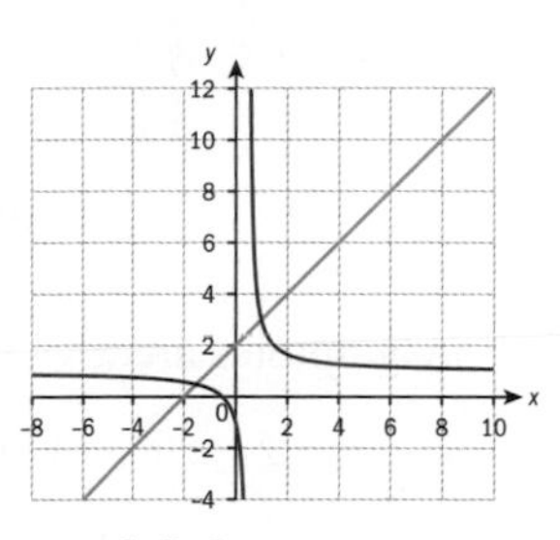

$x = -1.5, 1$

5 a 1,4 b 2 c 1.27

6 a $(4, 0)$ b $x = 1$ c $y = -2$

7 a $a = 2, d = 3$ b $b = 6$

8 a $m = 4$

b $n = \frac{33}{4}$

c $y = \frac{33}{4}$

9 a The x-intercept is $\left(-\frac{1}{2}, 0\right)$.

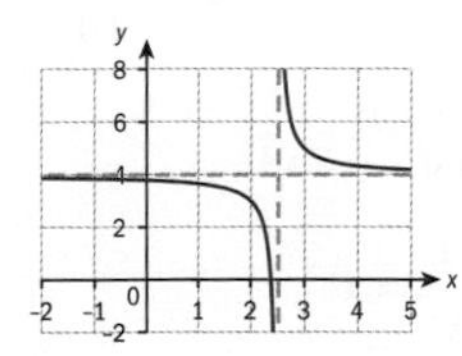

b $x = 2.5, y = 4$

c 2.375 d 3.8

10 a $f^{-1}(x) = \frac{x+1}{x-2}$

b

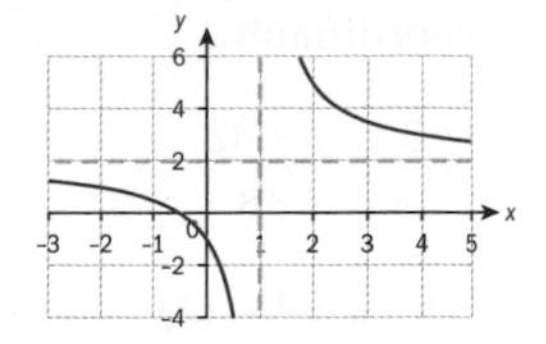

c $x = 1, y = 2$ d $(-0.5, 0)$

e $-0.303, 3.30$

11 a $f^{-1}(x) = \frac{1}{x} + 2$

b

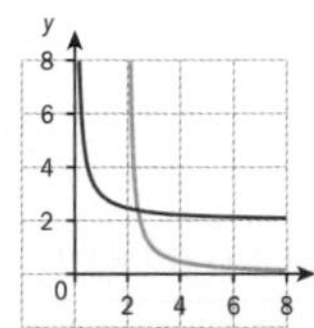

c 2.41

12 a

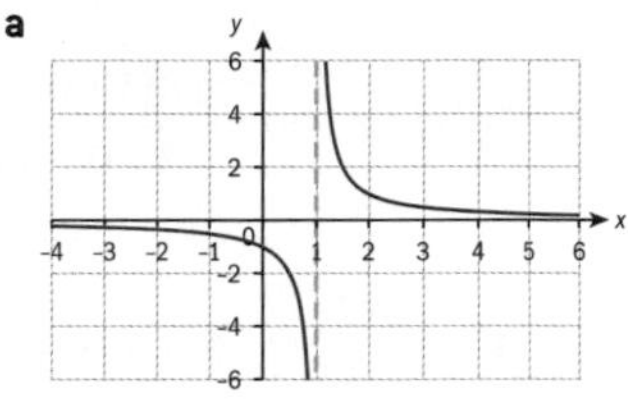

b $g(x) = \frac{1}{x-3} + 3$

c $(2.67, 0)$ and $(0, 2.67)$

d $x = 3, y = 3$

e

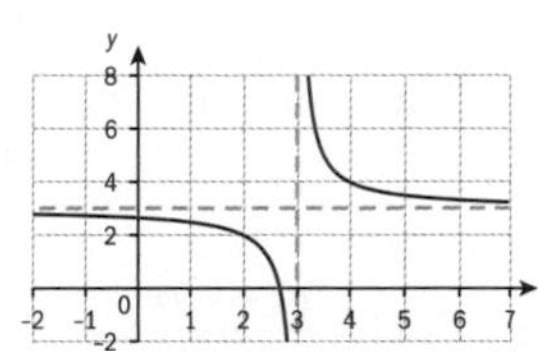

13 a $\frac{x-3}{2}$

b $(g \circ f^{-1})x = \frac{5}{4\left(\frac{x-3}{2}\right)} = \frac{5}{2x-6}$

c $(0, -0.833)$

d

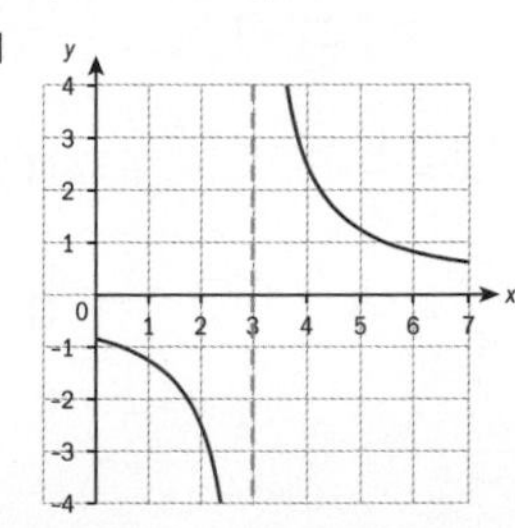

e $(-0.833, 0)$

f $x = 0$

14 Consider the function $f(x) = 2 + \frac{10}{x-4}$

a $x = 4$ and $y = 2$

b **Domain** $x \in \mathbb{R}, x \neq 4$
Range $y \in \mathbb{R}, y \neq 2$

c $(-1, 0)$ and $(0, -0.5)$

d

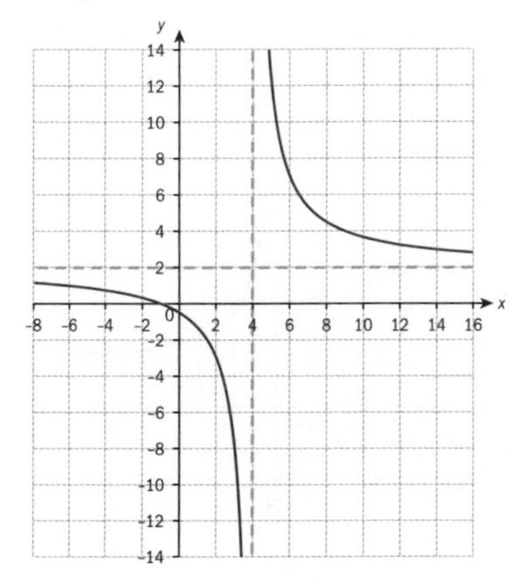

e Horizontal shift of 4 units right and a vertical shift of 2 units up

Exam-style questions

15 a $x \in \mathbb{R}, x \neq -2$ (1 mark)

b $f(x) \in \mathbb{R}, f(x) \neq \frac{3}{2}$ (1 mark)

c When $x = 0$,

$f(x) = -\frac{20}{4} = -5$

So one coordinate is $(0, -5)$ (1 mark)

When $y = 0$, $x = \frac{20}{3}$

So the other coordinate is $\left(\frac{20}{3}, 0\right)$ (1 mark)

16 a Domain is $x \in \mathbb{R}, x \neq -2$

Range is $f(x) \in \mathbb{R}, f(x) \neq 0$

(2 marks)

b Domain is $x \in \mathbb{R},\ x \neq -2$

Range is $f(x) \in \mathbb{R},\ f(x) \neq 4$ (2 marks)

c Domain is $x \in \mathbb{R},\ x \neq 0$

Range is $f(x) \in \mathbb{R},\ f(x) \neq 4$ (2 marks)

d Domain is $x \in \mathbb{R},\ x \neq 0$

Range is $f(x) \in \mathbb{R},\ f(x) \neq 0$ (2 marks)

17 a $x = 1$ (1 mark)

b $y = 3$ (1 mark)

c

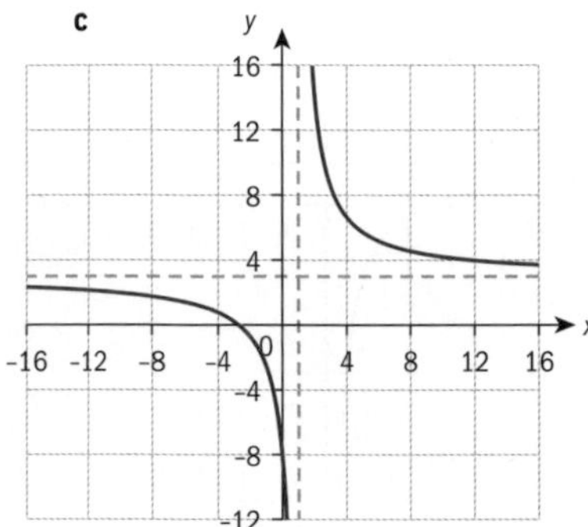

(3 marks)

18 a $y = 10$ (1 mark)

b $x = 2$ (1 mark)

c $f(x) = 10 + \frac{3}{2-x}$

$= \frac{10(2-x)+3}{2-x}$ (2 marks)

$= \frac{-10x+23}{-x+2}$ (1 mark)

19 a Vertical asymptote occurs when $c + 8x = 0$ (1 mark)

$c + 8\left(-\frac{3}{4}\right) = 0$

$c = 6$ (1 mark)

b $y = \frac{a+bx}{6+8x}$

Substituting the first coordinate:

$\frac{2}{5} = \frac{a+\frac{1}{2}b}{10}$ (1 mark)

$4 = a + \frac{1}{2}b$

$8 = 2a + b$ **(1)** (1 mark)

Substituting the second coordinate:

$-\frac{3}{38} = \frac{a+4b}{38}$

$-3 = a + 4b$ **(2)** (1 mark)

Solving **(1)** and **(2)** simultaneously:

$a = 5$ (1 mark)

$b = -2$ (1 mark)

20 a 6 (1 mark)

b $P = \frac{18(1+0.82\times 12)}{3+(0.034\times 12)} \approx 57$ (2 marks)

c Solving $100 = \frac{18(1+0.82t)}{3+0.034t}$ (1 mark)

$t = \frac{282}{11.36} = 24.8$ months (1 mark)

d Horizontal asymptote at $P = \frac{18\times 0.82}{0.034} = 434.12$ (2 marks)

Therefore for $t \geq 0,\ P \leq 434$ (1 mark)

21 a $f(x) = \frac{17-10x}{2x-1}$

$= \frac{12+5-10x}{2x-1}$ (2 marks)

$= \frac{12+5(1-2x)}{2x-1}$ (1 mark)

$= \frac{12}{2x-1} - 5$ (1 mark)

b $x = \frac{1}{2}$ (1 mark)

c $y = -5$ (1 mark)

d

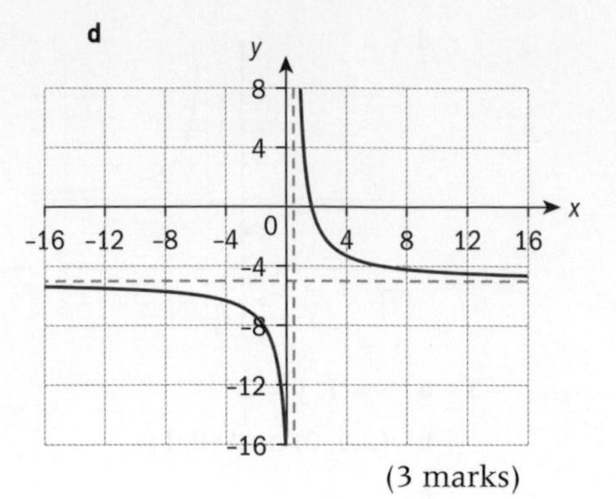

(3 marks)

22

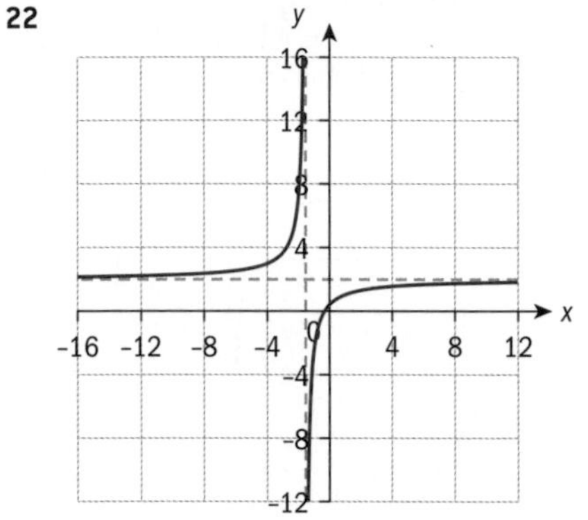

(2 marks)

Asymptotes are $x = -\frac{3}{2}$ and $y = 2$ (2 marks)

Intersections with axes are at $\left(0, \frac{1}{3}\right)$ and $\left(-\frac{1}{4}, 0\right)$ (2 marks)

Chapter 5

Skills check

1 a $\frac{3}{4}$ **b** $-\frac{12}{19}$

2 a $7\sqrt{x} = 7x^{\frac{1}{2}}$

b $\frac{1}{x^2} = x^{-2}$

c $\frac{8}{5\sqrt{x^3}} = \frac{8}{5}x^{-\frac{3}{2}}$

3 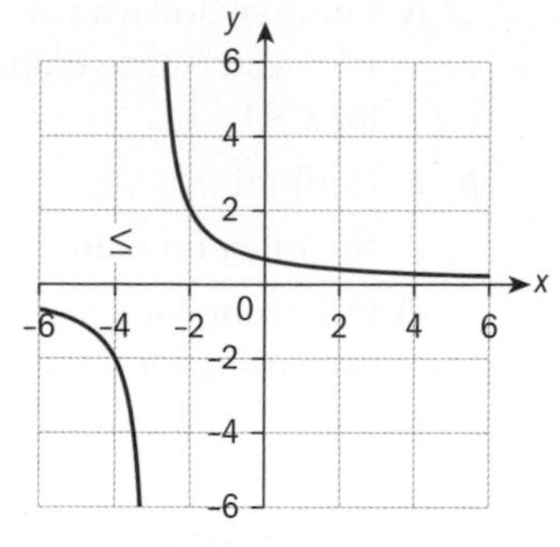

4 Since $\left|\frac{1}{2}\right| < 1$,

$\sum_{n=0}^{\infty} 5\left(\frac{1}{2}\right)^n = 10$

Exercise 5A

1 10 2 3 3 −1 4 1

Exercise 5B

1 Vertical asymptote at $x = \frac{1}{6}$

Horizontal asymptote at $y = \frac{1}{2}$

2 Vertical asymptotes at $x = \pm\sqrt{3}$

Horizontal asymptote at $y = -1$

3 Vertical asymptote at $x = 1$

Horizontal asymptote at $y = -1$

4 Vertical asymptotes at $x = \pm\sqrt{2}$

Horizontal asymptote at $y = 0$

Exercise 5C

1 $f'(x) = 7x^6$ 2 $f'(x) = 18x^{17}$

3 $f'(x) = -\frac{1}{2}x^{-\frac{3}{2}}$ 4 $f(x) = \frac{1}{5}x^{-\frac{4}{5}}$

5 $f(x) = -\frac{1}{2}x^{-\frac{3}{2}}$ 6 $f(x) = \frac{3}{4}x^{-\frac{1}{4}}$

Exercise 5D

1 a $4x^3 - x$ b $15x^2 - 5$

c $24x^3 - 6x$ d $4t + 3$

e -9.8 f 24

2 a $3x^{\frac{-1}{2}}$ b $3x^{\frac{-2}{5}}$

c $-2x^{-2} - \frac{3}{2}x^{\frac{-1}{2}}$

3 a $-3x^{-3}$ b $-\frac{3x^{-3}}{2}$

c $12\pi x^2$ d $2x + 2$

e $2x + 3x^{-2}$ f $6x^2 - 2x + 6$

4 a $\frac{3}{2}x^{\frac{1}{2}}$ b $-14x^{-3} + \frac{1}{2}x^{\frac{-3}{2}}$

c $\frac{1}{3}x^{\frac{-2}{3}} + \frac{1}{4}x^{\frac{-3}{4}}$

Exercise 5E

1 a -6 b 13 c $-\frac{15}{4}$

2 $\left(-\frac{1}{2}, \frac{199}{24}\right)$ and $\left(5, -\frac{217}{6}\right)$

Exercise 5F

1 $y_N = -4x + \frac{29}{4}$

2 $y = -7x + \frac{419}{27}$

3 $y = \frac{1}{3}x + \frac{7}{3}$ and $y = \frac{1}{3}x - \frac{7}{3}$

4 −2

5 Coordinates are

$\left(\frac{1+\sqrt{7}}{3}, \frac{7-14\sqrt{7}}{27}\right)$,

$\left(\frac{1-\sqrt{7}}{3}, \frac{7+14\sqrt{7}}{27}\right)$

6 $g(x) = \frac{1}{x^n} = x^{-n}$

$\therefore g'(x) = -nx^{-n-1}$

$\Rightarrow xg'(x) + ng(x)$

$= x(-nx^{-n-1}) + nx^{-n}$

$= -nx^{-n} + nx^{-n} = 0$

7 a $f'(x) = 15ax^2 - 4bx + 4c$

b $f'(x) \geq 0$

$\Rightarrow 15ax^2 - 4bx + 4c \geq 0$

$\Rightarrow x^2 - \frac{4b}{15a}x + \frac{4c}{15a} \geq 0$

$\Rightarrow \left(x - \frac{2b}{15a}\right)^2 - \frac{4b^2}{125a^2} + \frac{4c}{15a} \geq 0$

The LHS is valid for all real x and attains its minimum at $x = \frac{2b}{15a}$ so

$-\frac{4b^2}{125a^2} + \frac{4c}{15a} \geq 0$

$\Rightarrow b^2 \geq 15ac$

(A solution correctly using the discriminant of a quadratic is equally permissible)

8 $x = 2$

Exercise 5G

1 a $10(2x + 3)^4$ b $\frac{-1}{\sqrt{1-2x}}$

c $\frac{6x}{\sqrt{(2x^2-1)^3}}$

d $12\left(x^2 - \frac{2}{x}\right)^2\left(x + \frac{1}{x^2}\right)$

2 $y = 6 - 4x$

3 $a = 2, b = 3$ 4 $y = \frac{-4}{3}x + \frac{11}{6}$

5 $x = \pm\frac{\sqrt{2}}{3}$

Exercise 5H

1 a $2x(3x - 1)$

b $(x + 3)^2(8x - 3)$

c $\frac{4-9x}{2\sqrt{2-3x}}$

d $2x(5x - 1)(x^2 - x + 1)$

e $\frac{-10-9x}{2\sqrt{(x+2)}}$

2 b $x = 3$ 3 $y = -x$

$x = \frac{-1}{5}$

Exercise 5I

1 a $\frac{16}{(5-x)^2}$ b $\frac{2+x}{2(2-x)^2\sqrt{x}}$

c $\frac{x+2}{(1-x^2)^{\frac{3}{2}}}$ d $\frac{-3x^2-2x+3}{(x^2+1)^2}$

2 $y = -\frac{1}{3}x - 2$

3 $x = \frac{1}{\sqrt[3]{2}}$

Exercise 5J

1 **a** $(x+3)(3x+1)$

b $-\dfrac{3x}{\sqrt{1-2x}}$

c $\dfrac{-2}{(x-1)^2}$ **d** $\dfrac{-4(2x^3-1)}{(x^4-2x+1)^2}$

2 equation of normal:
$y = 24x - \dfrac{431}{2}$
equation of tangent:
$y = -\dfrac{1}{24}x + \dfrac{7}{8}$

Exercise 5K

1 **a** **i** $x > 0$ **ii** Nowhere

b **i** $x \in (-\infty, -1) \cup (-1, 0)$

ii $x \in (0, 1) \cup (1, \infty)$

c **i** $x \in (-\infty, -0.215) \cup (1.55, \infty)$

ii $x \in (-0.215, 1.55)$

d **i** $x \in (-\infty, -1) \cup (1, \infty)$

ii $x \in (-1, 1)$

2 **a** Increasing: nowhere
Decreasing: $\forall x \in \mathbb{R}$

b Increasing: $x > 0$
Decreasing: $x < 0$

c Increasing: nowhere
(note the function is only valid here for $x > 1$)
Decreasing: $x \in (1, \infty)$

d Increasing: $x \in \left(0, \dfrac{1}{16}\right)$
(note the function is only valid here for $x > 0$)
Decreasing: $x \in \left(\dfrac{1}{16}, \infty\right)$

Exercise 5L

1 **a** $(0, -2)$ min **b** $(1, -1)$ min

c $(0, 2)$ max; $(4, -30)$ min

2 $a = -\dfrac{4}{3}$

3 $a = -5$, $b = -6$, $c = 3$, $d = 1$

4 $(0, 21)$

Exercise 5M

1 $\dfrac{15}{2}x^{\frac{1}{2}}$ **2** $x = \pm 2$

3 $f''(x) = -4(5-4x)^{-\frac{3}{2}}$

4 $2a^2 = 8 \Rightarrow a = \pm 2$

Exercise 5N

1 **a** $(0, 0)$ **b** $]0, \infty[$

c $]-\infty, 0[$

2 **a** No points of inflexion

b f is concave up throughout its domain.

c f is never concave down.

3 **a** $(2, -38)$ **b** $]2, \infty[$

c $]-\infty, 2[$

4 **a** $\left(-\dfrac{1}{3}, -\dfrac{25}{27}\right)$ **b** $\left]-\dfrac{1}{3}, \infty\right[$

c $\left]-\infty, -\dfrac{1}{3}\right[$

5 **a** $(0, 0)$, $(2, 16)$ **b** $0 < x < 2$

c $x < 0$; $x > 2$

6 **a** $(1, 0)$ **b** $x > 1$ **c** $x < 1$

7 **a** $(-0.25, 0.992)$, $(0, 1)$

b $x < -0.25$; $x > 0$

c $-0.25 < x < 0$

8 **a** Coordinates of point of inflexion are $(2, 0)$ and $(0, -16)$

b Function is concave up when $x < 0$, $x > 2$

c Function is concave down when $0 < x < 2$

9 **a** $\left(\dfrac{-2}{3}, \dfrac{43}{27}\right)$-non-horizontal

b $(1, 0)$ = horizontal inflexion

c $\left(\dfrac{-4}{3}, \dfrac{310}{27}\right)$ = non-horizontal inflexion and $(0, 2)$ = horizontal inflexion point

d No inflexion points

10 **a** **i** $(-0.732, 3.39)$ max
$(2.73, -17.4)$ min

ii $(1, -7)$ = non-horizontal inflexion point

iii Increasing: $x < -0.732$ or $x > 2.73$ decreasing for $-0.732 < x < 2.73$

iv Concave downward $x < 1$ and concave upward for $x > 1$

b **i** $(1, 0)$ min

ii No inflexion point

iii Increasing for $x > 1$
decreasing for $x < 1$

iv Concave upward for $x \in \mathbb{R}$

c **i** $(-1, -3)$ min

ii $(0, -2)$ = horizontal inflexion point
$\left(\dfrac{-2}{3}, -\dfrac{70}{27}\right)$ = non-horizontal inflexion point

iii Increasing $x > -1$
Decreasing $x < -1$

iv Concave upward $x < \dfrac{-2}{3}$ or $x > 0$; concave downward $\dfrac{-2}{3} < x < 0$

Exercise 5O

1 **a**

b

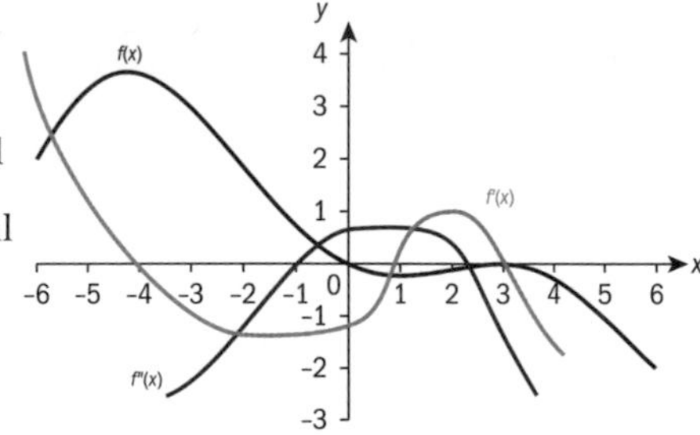

c

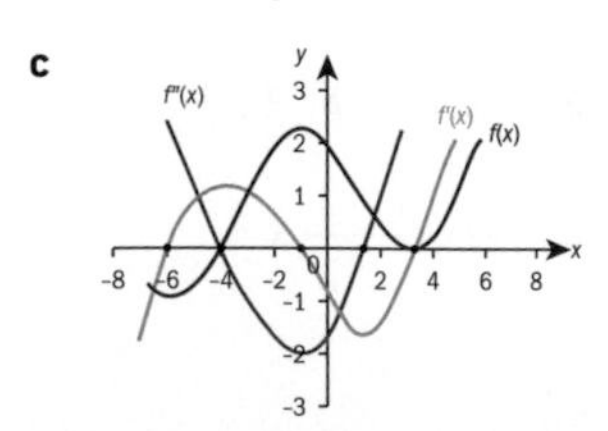

d

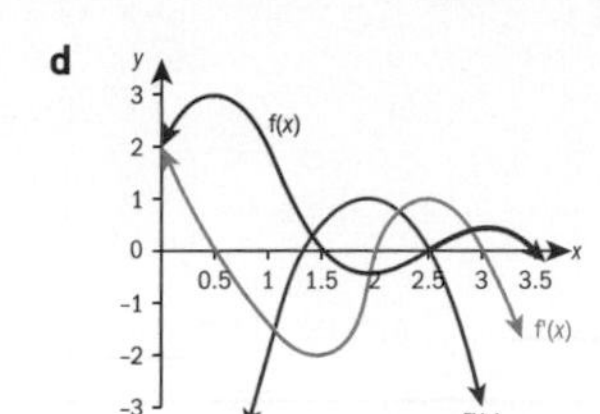

Exercise 5P

1 a

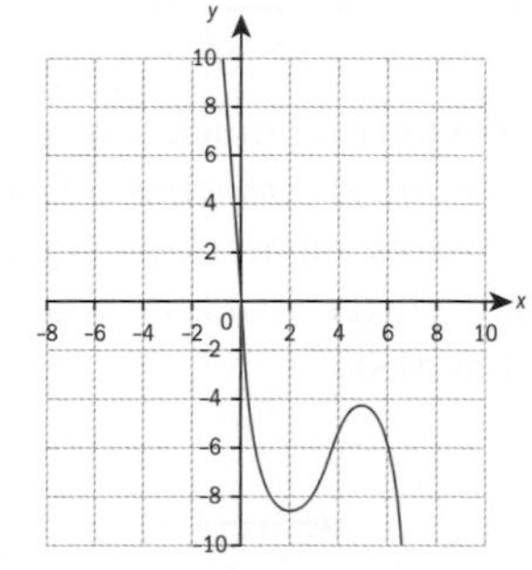

b

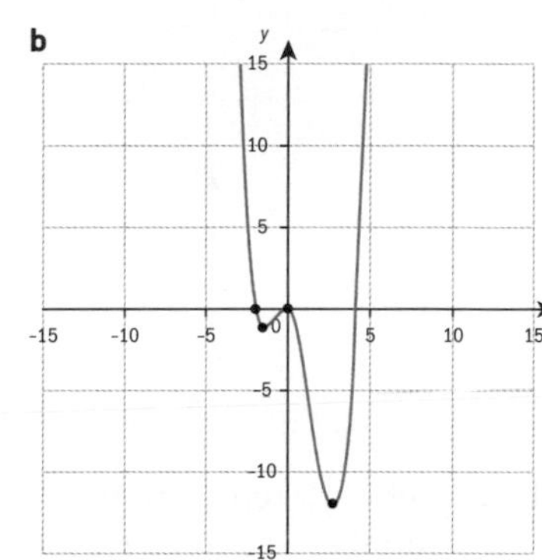

2 a

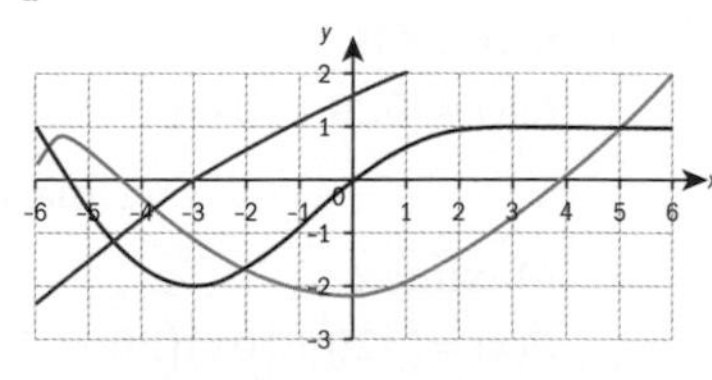

b

Exercise 5Q

1 a $L = \dfrac{100}{x}$ **b** $P = 2x + \dfrac{100}{x}$

c $x = 5\sqrt{2}$

$\therefore P\left(5\sqrt{2}\right) = 20\sqrt{2}$ (metres)

d

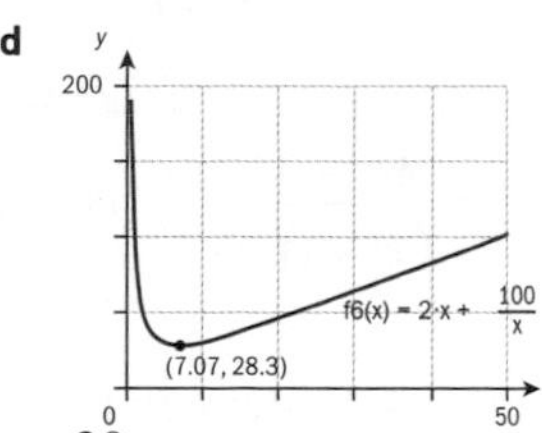

2 $x = 20$

$y = \$10000$

3 a $h = \dfrac{216}{s^2}$

b $A = s^2 + \dfrac{864}{s}$

c $s = \sqrt[3]{432}$

4 $s = \dfrac{300}{17}$

5 $C \simeq \$164$

Exercise 5R

1 a $v(t) = 3t^2 - 3$ $a(t) = 6t$

b $s(0) = 1$ m, $v(0) = -3$ m s^{-1}, $a(0) = 0$ At this instant, the particle is 1 metre from the origin in the positive direction, traveling towards the origin at 3 ms^{-1}, and is not accelerating.

c The particle is moving away from the origin at 9 m s^{-1} and is accelerating away from the origin at 12 m s^{-2}

d $t = 1$ s **e** $t > 1$

f $d = 22$ m

2 a 1 m **b** 17 m

c i $v(0) = 12$ m s^{-1}

ii $v(1) = 9$ m s^{-1}

iii $v(3) = -15$ m s^{-1}

d $d = 33$ m

3 $t = 1.5$ s $s(t) = 11.25$

4 a 10 m **b** 6.53 s

c The diver hits the water with a velocity of -8.06 ms^{-1}, and a constant vertical acceleration of -2 ms^{-2}. Since both velocity and acceleration are negative, the diver is speeding up as he/she approaches the water.

5 127.6 m; 10.2 s

6 a $t = 0, 3, 6, 11$

b i $0 < t < 3$; $6 < t < 11$

ii $3 < t < 6$

c i $0 < t < 1.5$ eastward;

ii $3 < t < 4.5$ westward

d $t = 1.5, 4.5$

e Speeding up: $t \in (0, 1.5)$; $t \in (3, 4.5)$; $t \in (6, 9)$

Slowing down: $t \in (1.5, 3)$; $t \in (4.5, 6)$; $t \in (9, 11)$

Chapter review

1 a $y = 2$ **b** $x = 2$

2 a

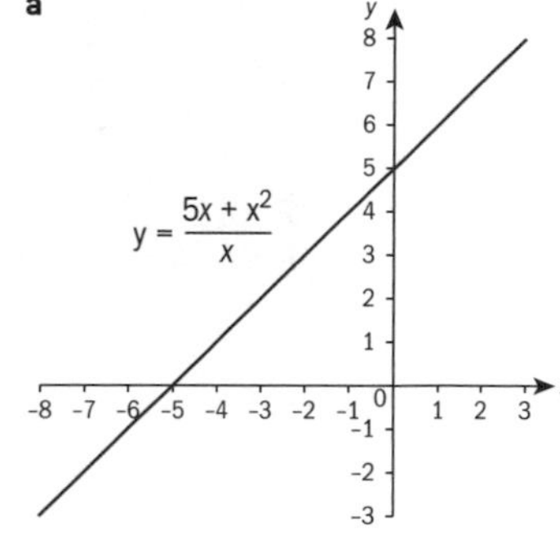

b 5

3 a $x = \pm 3, y = 6$ **b** $x = -3, y = 0$

4 a $y' = 2(1 - 2x)^4(3x - 2)^5(19 - 33x)$

b $y' = \dfrac{-1}{x^2}$ **c** $y' = \dfrac{1}{2}x^{\frac{-1}{2}} - \dfrac{4}{3}x^{\frac{-2}{3}}$

5 a $x = \pm 1, y = 0$

b $\dfrac{dy}{dx} = \dfrac{-\left(x^2+1\right)}{\left(x^2-1\right)^2} < 0$

c

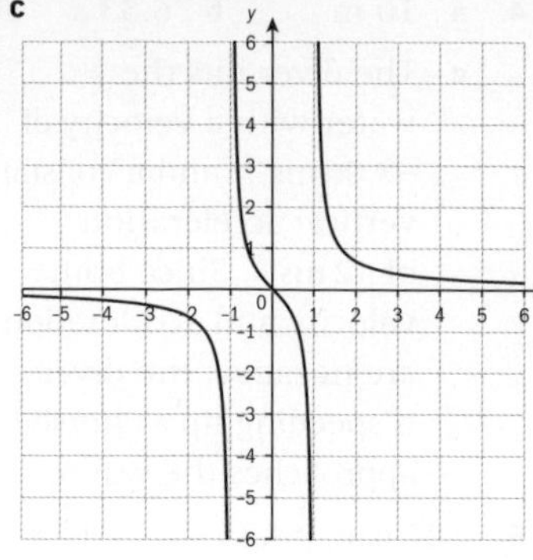

6 $y = 7, y = -25$

7 $\left(\frac{7}{3}, \frac{10}{3}\right)$ **8** $y = \frac{-1}{3}x + \frac{1}{3}$

9 $(2, 1)$ and $\left(\frac{-2}{3}, \frac{59}{27}\right)$

10 $(0, 0)$ minimum
$(-2, -4)$ maximum

11 a $(-1, 0)$ **b** $x = 0, y = 0$

c $\left(-2, \frac{-9}{4}\right)$ **d** $x > -3$

e

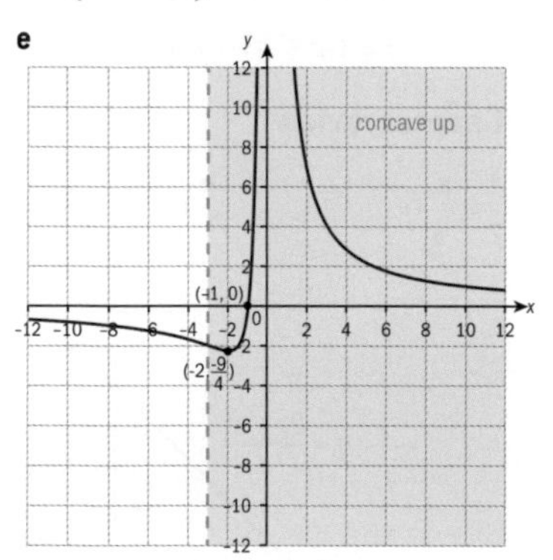

12 a $x = 0, x = b^2$

b $x > \frac{b^2}{4}$ increasing,

$0 < x < \frac{b^2}{4}$ decreasing

c If $b > 0$ concave upward,

If $b < 0$ concave downward

13 a $v(t) = 49 - 4.9t$

b $h = 245\text{ m}$

14 a -2 ms^{-1} **b** $t = 4\text{ s}$

c $a = \frac{3}{5}\text{ m s}^{-2}$

d Always speeding up since acceleration is always positive

Exam-style questions

15 a $f'(x) = 4x^3 - 6x^2 - 2x + 3$ (1 mark)

b

$$g'(x) = \frac{-4(x^2+1) - (-4x)\cdot 2x}{(x^2+1)^2}$$ (2 marks)

$$= \frac{4x^2 - 4}{(x^2+1)^2}$$ (1 mark)

c

$h'(x) = 1\cdot(x-7) + (x+2)\cdot 1 = 2x - 5$ (2 marks)

d

$i'(x) = 3\cdot 2\cdot(2x+3)^2 = 6(2x+3)^2$ (2 marks)

16 a Graph 1 (1 mark)

The gradient of the tangent to the graph at any point is negative or zero, whereas $y = x$ has gradient 1. (1 mark)

b Graph 2 (1 mark)

y increases as x increases (1 mark)

c Graph 3 (1 mark)
Since the functions represented by graphs 1 and 2 are not defined at infinity. (1 mark)

d Graph 1 (1 mark)
as the function is decreasing. (1 mark)

17 a i $0 \le t \le 2$, $4 \le t \le 4.6$, $8.5 \le t \le 10$ (3 marks)

ii $2 \le t \le 4$ and $5 \le t \le 7$ (2 marks)

iii $4.6 \le t \le 8.5$ (1 mark)

b $f(t) = 2t$, $f(t) = t$, $g(t) = 2$, $h(t) = -3t + 14$, $i(t) = -1$ and

$f(t) = \frac{1}{3}(2t - 17)$ (5 marks)

c

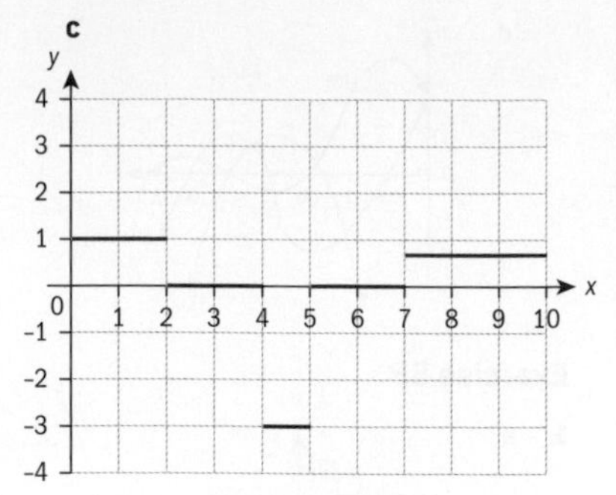

(Up to two branches correct: 1 mark; all branches correct: 2 marks; all branches correct and correct labels & scale: 3 marks)

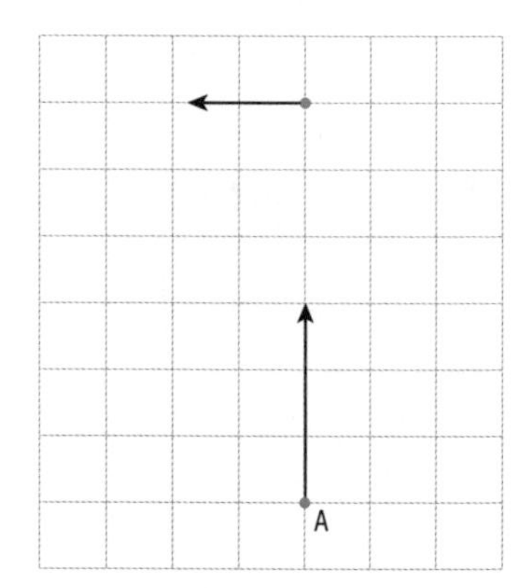

18 a Letting x represent the number of \$10 increases above \$320, then rental income is
$R(x) = (320 + 10x)(200 - 5x)$ (1 mark)

Maximize
$R(x) = (320 + 10x)(200 - 5x)$ (1 mark)

$x = 4$ (1 mark)

Which corresponds to \$360 rent (1 mark)

b

i $200 - 5\times 4 = 180$ (1 mark)

ii $360\times 180 = \$64\,800$ (2 marks)

19

a $h(4) = 370$ and

$h(5) = 438$ (3 s.f.) (2 marks)

b $v(t) = h'(t) = 112 - 9.8t$ (2 marks)

c $v(t) = 0 \Rightarrow 112 - 9.8t = 0$ (1 mark)

$t = 11.4$ s (3 s.f.) (1 mark)

d Either: double x-coordinate of maximum height

Or: solve $h(t) = 112t - 4.9t^2 = 0$ (1 mark)

Gives $t = 22.8$ s (3 s.f.) (1 mark)

e

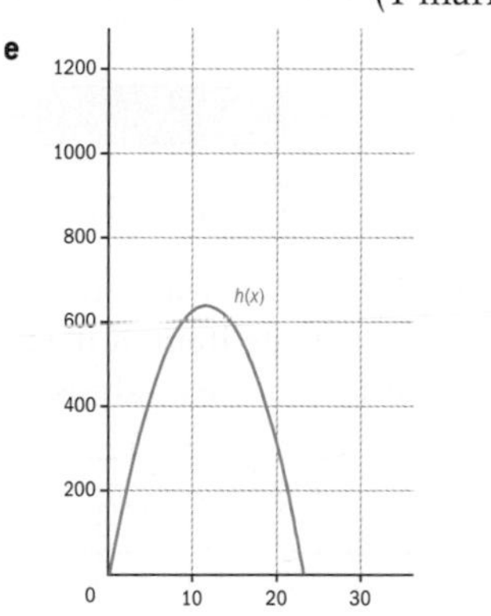

(Shape: 1 mark; domain: $0 \le x \le 22.9$ (3 sf) 1 mark; maximum: 640 (3 sf) 1 mark)

f $v(22.9 = -112$ m s^{-1} (2 marks)

g $a(t) = v'(t)$ (1 mark)

$= -9.8$ (1 mark)

which is constant. (1 mark)

20 a i

$$\left(\frac{f}{g}\right)'(2) = \frac{f'(2)g(2) - f(2)g'(2)}{(g(2))^2}$$ (1 mark)

$$= \frac{10\times4 - 9\times\left(-\frac{4}{3}\right)}{4^2}$$ (1 mark)

$$= \frac{52}{16}\left(= \frac{13}{4} = 3.25\right)$$ (1 mark)

ii $(g \circ f)'(1) = g'(f(1))f'(1)$ (1 mark)

$= -\frac{4}{3}\times 4 = -\frac{16}{3}$ (1 mark)

b i False (1 mark)

as derivative changes sign. (1 mark)

ii False (1 mark)

Since $g'(2)\times g'(3) = -\frac{4}{3}\times -\frac{3}{4} = 1 \ne -1$. (1 mark)

21 a $\frac{N(3) - N(1)}{3-1} = 1410$ (1 mark)

$\frac{N(5) - N(4)}{5-4} = 2220$ (1 mark)

Between day 1 and day 3, the number of people affected is increasing on average by 1410 per day. Between day 4 and day 5, the number of people affected is increasing on average by 2220 per day. (1 mark)

b $\frac{dN}{dt} = 900t - 90t^2$ (2 marks)

c After 10 days (reaches 15000 cases) (2 marks)

d $\frac{d^2N}{dt^2} = 900 - 180t$ (2 marks)

This tells you how the rate at which the disease spreads varies. (1 mark)

22 a $x = \frac{y+2}{y-1}$ (1 mark)

$x(y-1) = y + 2$

$xy - y = x + 2$ (1 mark)

$g^{-1}(x) = \frac{x+2}{x-1} = g(x)$ (2 marks)

b

$(h \circ g^{-1})'(x) = \frac{d}{dx}\left[h(g^{-1}(x))\right]$

$= h'(g^{-1}(x))(g^{-1})'(x)$ (1 mark)

$= 2\cdot\left(\frac{x+2}{x-1}\right)\cdot\left(-\frac{3}{(x-1)^2}\right)$

$= \left(-\frac{6(x+2)}{(x-1)^3}\right)$ (1 mark)

$(h' \circ g^{-1})(x) = h'(g^{-1}(x))$ (1 mark)

$= 2\cdot\left(\frac{x+2}{x-1}\right) = \frac{2x+4}{x-1}$ (1 mark)

$(h \circ g^{-1})'(x) \ne (h' \circ g^{-1})(x)$ (1 mark)

Chapter 6

Skills check

1 a 4 **b** 11

2 a 5 **b** 1 and 7

3 a 6 **b** 6

Exercise 6A

1 a Discrete **b** Continuous **c** Continuous **d** Discrete

2 a Stratified **b** Systematic **c** Simple random **d** Quota

3 a Stratified **b** Stratified **c** Systematic **d** Simple random **e** Quota

Exercise 6B

1 a Continuous

b

Time (mins)	f
$0 < t \le 4$	6
$4 < t \le 8$	11
$8 < t \le 12$	7
$12 < t \le 16$	4
$16 < t \le 20$	2

c

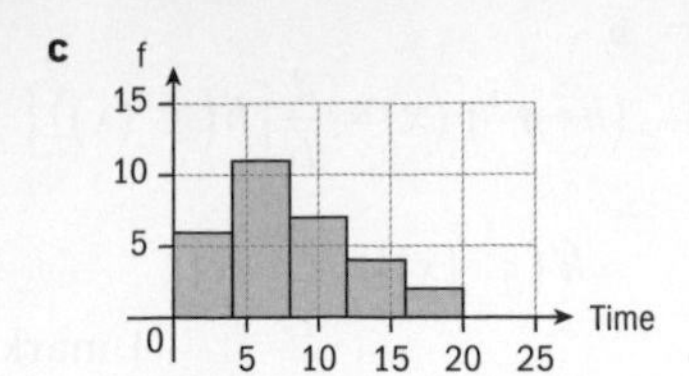

d Positive skew

2 **a** Continuous

b

Time (mins)	f
$10 < t \le 20$	6
$20 < t \le 30$	8
$30 < t \le 40$	5
$40 < t \le 50$	3
$50 < t \le 60$	1

c

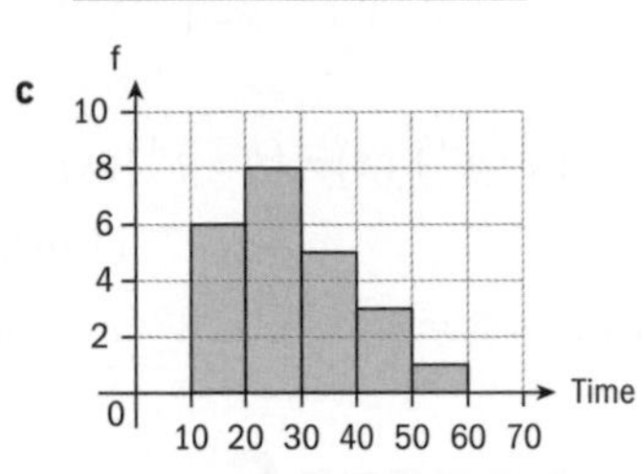

d Positive skew

3 **a** Continuous

b

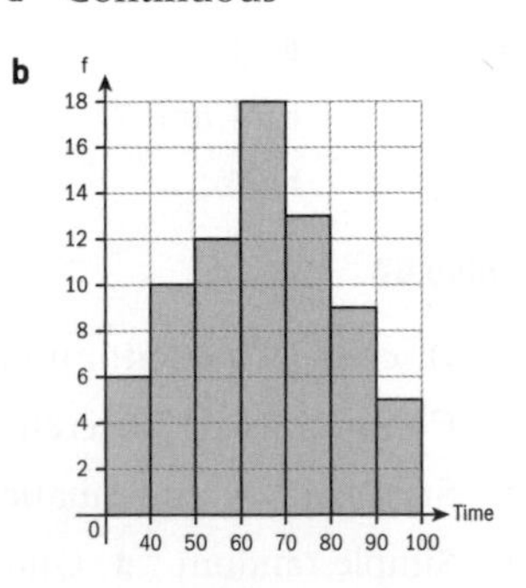

c Normal distribution

4 **a**

Time (mins)	f
$0 < l \le 10$	1
$10 < l \le 20$	2
$20 < l \le 30$	3
$30 < l \le 40$	7
$40 < l \le 50$	10
$50 < l \le 60$	14
$60 < l \le 70$	12

b Negative skew

5 **a**

Hours	Days
$0 < h \le 1$	1
$1 < h \le 2$	2
$2 < h \le 3$	3
$3 < h \le 4$	4
$4 < h \le 5$	6
$5 < h \le 6$	8
$6 < h \le 7$	6

b Negative skew

Exercise 6C

1 **a** 8 **b** 4 **c** 13
d No mode **e** 2 and 4

2 **a** 10 **b** $60 < y \le 80$

3 **ai** 3 **ii** $30 < x \le 35$
bi Discrete **ii** Continuous

Exercise 6D

1 **a** 3.65 **b** 12.8056
c 3.35

2 **a** 20.5 **b** 43.8 **c** 2.30

3 2.6

4 **a** 50 **b** 8 **c** 7.08

5 4.55 **6** \$21.89

7 **a** 60 **b** Right skewed
c 2 **d** 3.9

8 **a** 220 **b** $40 < a \le 60$
c 41.8

9 2 and 7 **10** 14

11 2, 3, 4, 4, 5, 6.

12 47.2 kg

Exercise 6E

1 **a** 18 **b** 18.5 **c** 4
d 3.5 **e** 6.5 **f** 5

2 9

3 **a** \$40 468 **b** \$200 **c** \$400

Exercise 6F

1 **a** 8 **b** 7 **c** 12
d 5 **e** 12

2 **a** 6 **b** 5 **c** 8.5
d 3.5 **e** 13

3 **a** 22.5 **b** 20.5
c 22.5 **d** 13 and 33.5
e More than 33.5

4 20 **5** 2.5

6 **a** $r = 10$, $s = 13$,
b $t = 18$

Exercise 6G

1

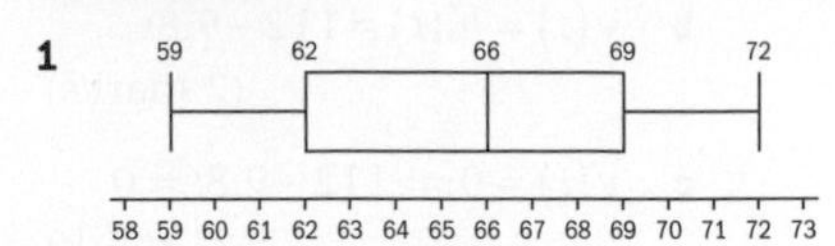

2 **a** 30.1 **b** 35 **c** 32.5 **d** 1.2

3 **a**

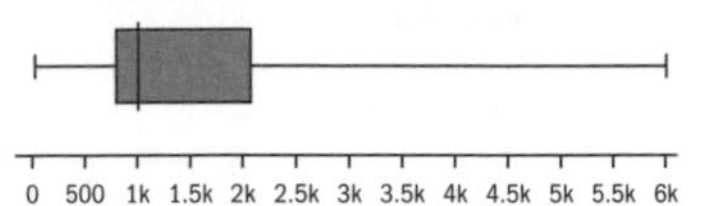

b $2067.5 + 1.5(1272.5) =$
3976.25
$6000 > 3976.25$

c

d The outlier was removed as a single item of data distorted the analysis.

4 **a**

b The morning exam

c This means that there is a bigger difference between the 25% and the 75% of the scores

5 **a**

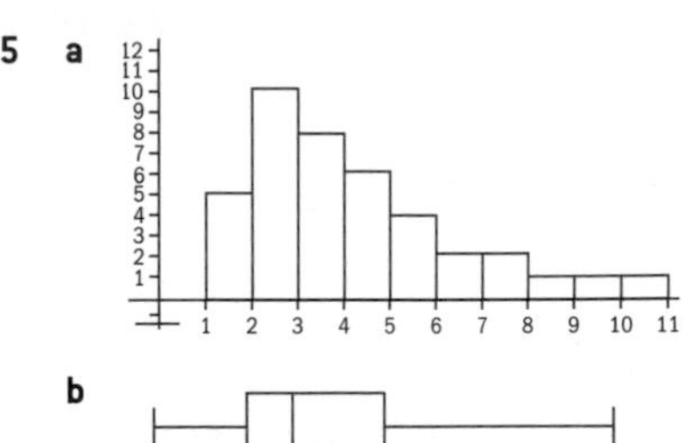

b

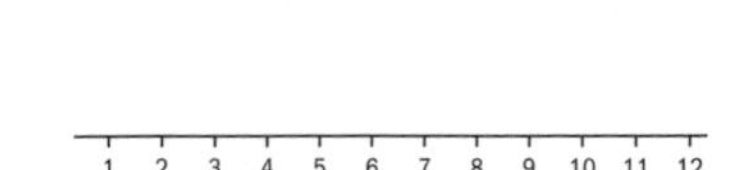

c Right skew

6 1A, 2C, 3B.

Exercise 6H

1 **a** 18 min **b** 11 min **c** $13.6 - 8.2 = 5.4$ min **d** 15.6 min

2 **a** 40 **b** $50 - 30 = 20$ **c** 53

3 **a**

Distance	$0 < d \le 25$	$25 < d \le 50$	$50 < d \le 75$	$75 < d \le 100$	$100 < d \le 125$	$125 < d \le 150$
CF	0	32	134	220	236	240

b

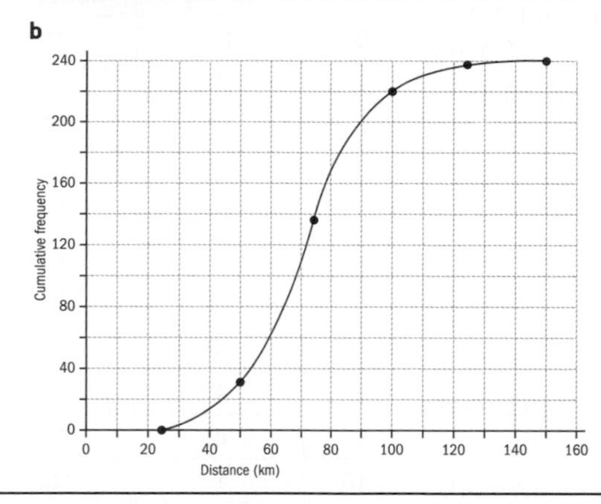

c 73

d $82 - 60 = 22$ **e** 3

4 **a**

Pages	f
$100 < p \le 200$	12
$200 < p \le 300$	48
$300 < p \le 400$	90
$400 < p \le 500$	143
$500 < p \le 600$	176
$600 < p \le 700$	196
$700 < p \le 800$	200

b

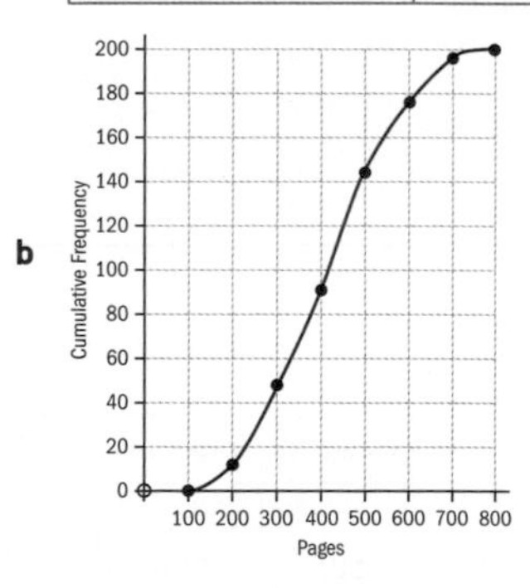

c 420 **d** 210 **e** 80

5 1C 2B 3A

Exercise 6I

1 **a** 4.375, 4.48, 2.12

b 5.2, 10.4, 3.22

c −1.6, 8.24, 2.87

d 3, 2, 1.41

e 85.8, 34 314, 185.2

2 **a** 3.2, 1.33 **b** 8.32, 3.34

c 20.6, 13.3

3 **a** 4.63, 2.67 **b** 4.24, 1.38

4 62.37, 16.9 **5** 322, 91.7

6 **a** 24 months

b 20 to 30 hours

c 28.75 **d** 11.1

Exercise 6J

1 **a** Mean = 4.4, median = 5, mode = 5.

b Mean = 8.4, median = 9, mode = 9.

c Adding 4 to each value increases the mean, median and mode by 4.

2 **a** Standard deviation = 3.02, variance = 9.108

b Standard deviation = 9.054, variance = 82.0

c The standard deviation is multiplied by 3 and the mean is multiplied by 3.

3 Mean = 21.2, median = 21, standard deviation = 0.5.

4 The mean, median and standard deviation are doubled.

5 The variance is multiplied by 81

Chapter review

1 **a** 1 **b** 3 **c** 3.2 **d** 8

2 500

3 **a** 3 **b** 3 **c** 2.5

4 Mean 21.9, standard deviation 1.1

5 **a** 32 min **b** 31 min

6 **a** Mean 58, standard deviation 5

b Mean 480, variance 2500

7 **a** 40 **b** 60 **c** 90 **d** 50

8 **a** 800 **b** 65

c $75 - 55 = 20$ **d** 100

e No. The 90th percentile is more than 80 marks

f 40

9 1B 2C 3A

10 **a** 20.125, 13.2

b $27.5 - 10 = 17.5$

11 4, 4.125, 4, 1.36

12 **a** 10.9, 12.5, 6.91

b We are assuming that the number of items is equally spread throughout the class interval.

13 8

14 **a** 80 **b** 50 **c** 6.25%

d $a = 10$, $c = 5$ **e** 20

f 52.5, 22.5

15 **a** $m = 12$, $n = 30$

b 21.1 **c** 32.3

16 **a** Discrete **b** 2.73

c 1.34 **d** 23

Exam-style questions

17 **a** Discrete (1 mark)

b Continuous (1 mark)

c Continuous (1 mark)

d Discrete (1 mark)

18 **a** As the mode is 5 there must be at least another 5. (1 mark)

So we have 1, 3, 5, 5, 6 with another number to be placed in order (1 mark)

The median will be the average of the 3rd and 4th pieces of data. (1 mark)

For this to be 4.5 the missing piece of data must be a 4.

Thus $a = 5, b = 4$ (2 marks)

b $\bar{x} = \dfrac{1+3+4+5+5+6}{6}$

$= \dfrac{24}{6} = 4$ (2 marks)

19 a An outlier is further than 1.5 times the IQR below the lower quartile or above the upper quartile. (1 mark)

b **i** Mode = 8 (1 mark)

ii Median = 7 (1 mark)

iii Lower quartile = 3 (1 mark)

iv Upper quartile = 9 (1 mark)

c IQR = 6 1.5 × IQR = 9

19 − 9 = 10 (1 mark)

19 is the (only) outlier (1 mark)

20 a $\dfrac{\sum x}{10} = 70 \Rightarrow \sum x = 700$ (1 mark)

Let the mass of the new student be s. $\dfrac{\sum x + s}{11} = 72$ (1 mark)

$700 + s = 792$ (1 mark)

So $s = 92$ kg (1 mark)

b IQR = 10 (1 mark)

$76 + 1.5 \times IQR = 76 + 15 = 91$ (1 mark)

So the new student's mass of 92 is an outlier. (1 mark)

21 a 200 (1 mark)

b 35 (1 mark)

c Using mid-points 5, 15, 25... as estimates for each interval (1 mark)

i Estimate for mean is 22.25 (2 marks)

ii Estimate for standard deviation is 11.6 (3 s.f.) (2 marks)

d Median is approximately the 100th piece of data which lies in the interval $20 < h \le 30$. (1 mark)

Will be 15 pieces of data into this interval

Estimate is

$20 + \dfrac{15}{50} \times 10 = 23$

(2 marks)

22 a Discrete (1 mark)

b 5 (1 mark)

c **i** 4.79 (3 s.f.)
ii 1.62 (3 s.f.) (2 marks)

d **i** 5 **ii** 4 **iii** 5.5 (1 mark each)

e

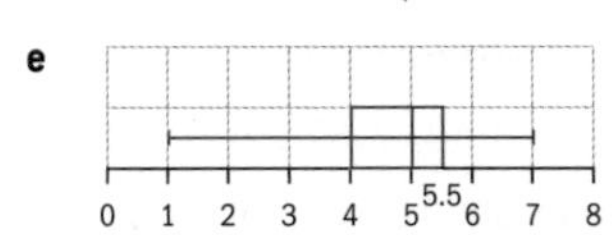

(1 mark) general shape
(1 mark) median
(1 mark) quartiles

f IQR = 1.5 $1.5 \times 1.5 = 2.25$ (1 mark)

5.5 + 2.25 = 7.75
4 − 2.25 = 1.75 (1 mark)

So the 2 (unhappy) candidates with grade 1 are outliers. (1 mark)

23 a

x	Frequency	Cumulative frequency
0	10	10
1	7	17
2	11	28
3	13	41
4	15	56
5	15	71
6	12	83
7	10	93
8	4	97
9	2	99
10	1	100

(4 marks for 6 correct, 3 marks for 4 or 5, 2 marks for 2 or 3, 1 mark for 1)

b **i** 4 **ii** 2 **iii** 6 (1 mark each)

c **i** 4.05 (3 s.f.)

ii $(2.4140...)^2 = 5.83$ (3 s.f.) (3 marks)

d No. It is bimodal at $x = 4$ and 5. (2 marks)

24 a $80 < w \le 90$ (1 mark)

b

Mass	$0<w\le 50$	$50<w\le 60$	$60<w\le 70$	$70<w\le 80$	$80<w\le 90$	$90<w\le 100$	$100<w\le 110$	$110<w\le 120$
Cumulative frequency	5	20	45	75	125	160	185	200

(2 marks) numbers
(1 mark) labelling

c

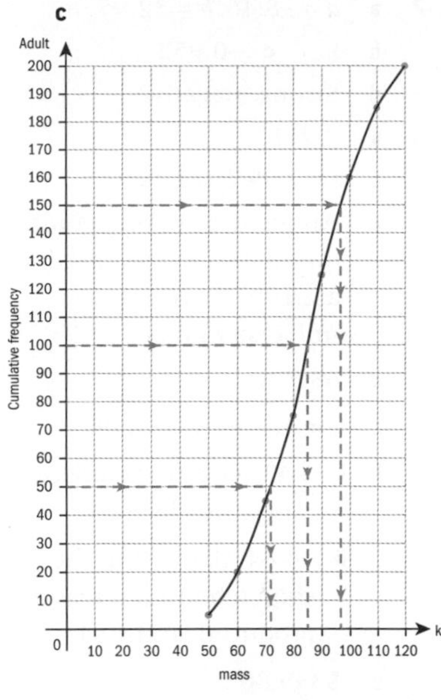

(2 marks) scales (1 mark) points and curve

i 85 **ii** 73 **iii** 97

(1 mark each) (1 mark) lines

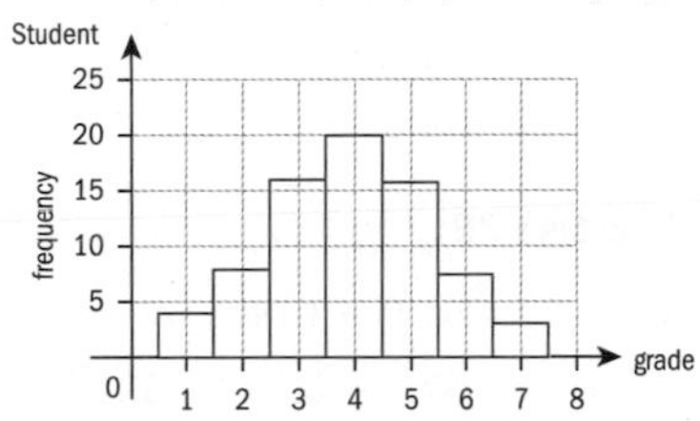

25 a i 7.5 **ii** 6.125 (2 marks)

b i 6 **ii** 6.9 (2 marks)

c Sally's had the greater median. (1 mark)

d Rob's had the greater mean. (1 mark)

26 a not to scale

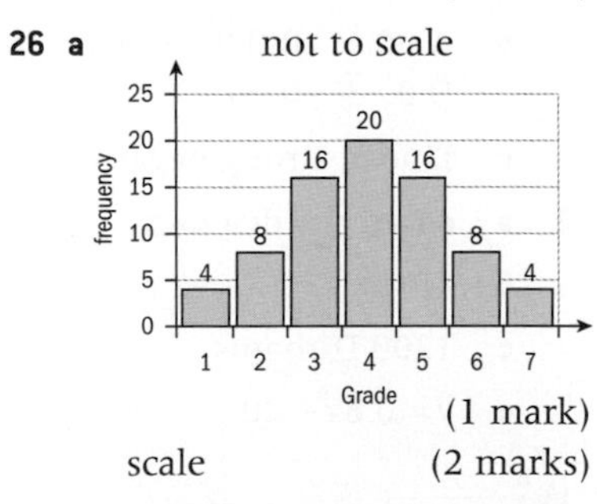

(1 mark)

scale (2 marks)

b i 4 **ii** 4 **iii** 4 (3 marks)

c The values of the median and the mean are the same due to the symmetry of the bar chart. (2 marks)

Chapter 7

Skills check

1 a 1296 **b** 64 **c** 343

2 a 5 **b** 4 **c** 3 **d** 3

3 a $y = 4x - 2$

b $y = 1 - 2x$

Exercise 7A

1 a Strong, positive, linear

b Weak, negative, linear

c Strong, negative, linear

d Weak, positive, linear

e No correlation

2 i a Positive

b Linear **c** Strong

ii a Positive

b Linear **c** Moderate

iii a Positive

b Linear **c** Weak

iv a No correlation

b Non-linear

c Zero

v a Negative **b** Linear

c Strong

vi a Negative **b** Non-linear

c Strong

3 a

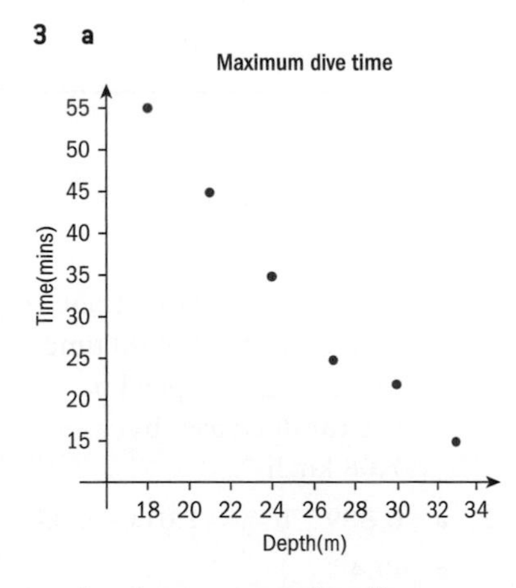

b Strong, negative, linear

c As the maximum depth increases the time at that depth decreases.

4 a

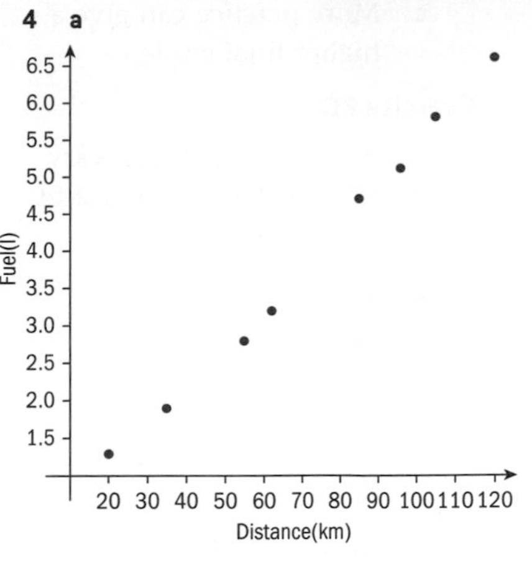

b Strong, positive, linear correlation

5 a

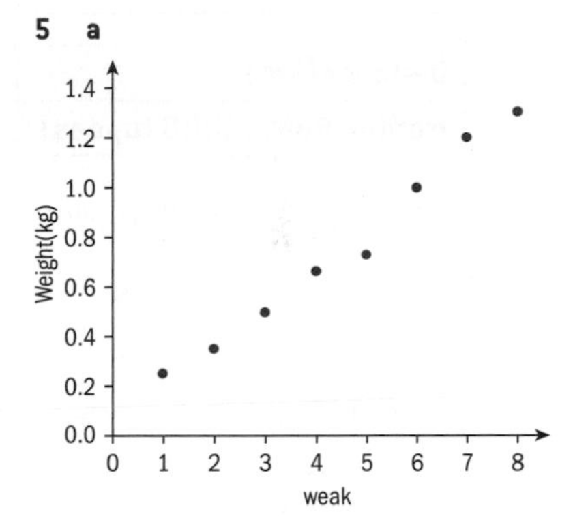

b Strong, positive, linear correlation

c The weight has increased.

Exercise 7B

1 a 0.987 **b** Strong positive

c As the floor area increases, the house price increases.

2 a −0.976

b Strong negative

c The price of the motorbike can never fall below 0.

3 a 0.698

b Moderate positive

4 a 0.878 **b** Strong positive

c Yes

5 a −0.449 **b** Weak negative

c Yes. The correlation between GPA and game time is weak.

6 a 0.970

b Strong positive

c More practice can give a higher final grade.

Exercise 7C

Note that answers may vary slightly for the line of best fit by eye.

1 **a** 48.5 **b** 29.3

c

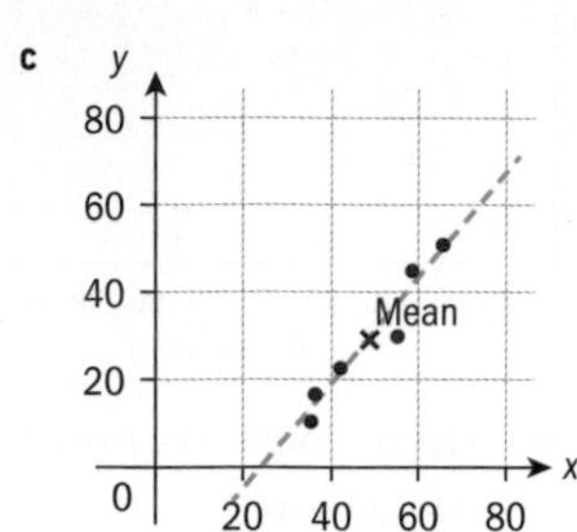

2 **a** 5.75, 2.7825

b

c $y = 0.208x + 1.587$

d 3.46 m **e** 26.5 m

f No. Giraffes only grow to 6 m. Extrapolation is unreliable.

3

Distance (km)	3	6	10	12	15	20
Monthy Rent (1000 rupees)	60	45	32	28	18	15

a 11 km **b** 33 000 rupees

c

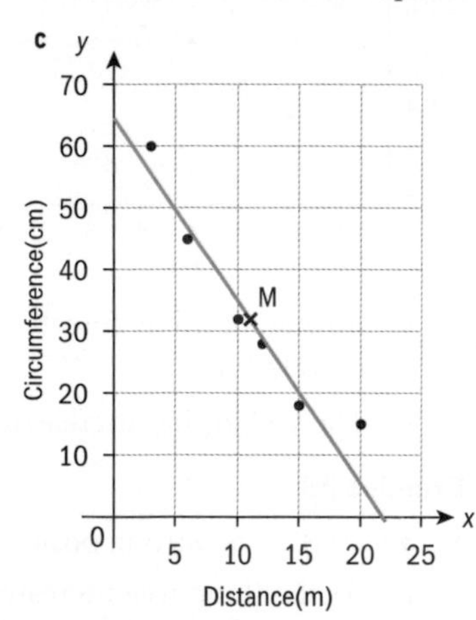

d $y = -2.83x + 64.2$

e 41 500 rupees

f 5 km

g No. Extrapolation is inaccurate. The rent is negative.

Exercise 7D

1 A student who does not play sports does 35 hours of homework per week.

Each sport played decreases the hours of homework by 30 min.

2 A person with no criminal friends has 1 conviction.

Having one criminal friend increases the number of conviction by 6.

3 A new speaker costs $300.

Each year its cost goes down by $40.

4 **a** −0.988

b $y = -7.08x + 202$

c 149 km h^{-1}

d For every increase of one second in the 0 to 90 time, the maximum speed of the car decreases by 7.08 km h^{-1}.

5 **a** 0.889 **b** $y = 1.01x - 6.32$

c 69.4

6 **a** 0.985 **b** $y = 1.74x - 48.0$

c 21.9 cm

d For every cm that the cat grows in length, it grows 1.74 cm in height.

7 **a** $a = -8.46, b = 32.95$

b 3 **c** −0.952

d Strong, negative.

8 **a** 0.911

b $a = 2.99, b = 2.19$

c An increase of one point on the IB diploma produces a 2.19% increase in university grade

d 68.7%

9 **a** $y = 6.416x + 14.839$

b **i** 6.42; each pizza produced raises costs by $6.42.

ii 14.84; the cost to the shop when no pizzas are produced is $14.84.

c $399.80

d **i** Unreliable as it is too far from the domain.

ii 13 pizzas

10 **a** −0.984 **b** $a = -6.71$, $b = 117.98$

c ¥78 000

Exercise 7E

1 **a** $y = 61.0 - 1.08x$, 39 tickets.

b $x = 52.1 - 0.814y$, $24

2 $x = 1.35 + 0.159y$, 9.3 min

3 $x = 31.7 + 0.708y$, 69.

4 **a** 149 **b** 19.7 km

Chapter review

1 **a** −1 and 1

b A −0.6 B 0.9 C 0.5 D 0 E −0.96

c Linear, strong negative

2 **a** **d f** on the diagram

b 40 °C

c 1200 Dirhams

e $y = 0.8x - 20$

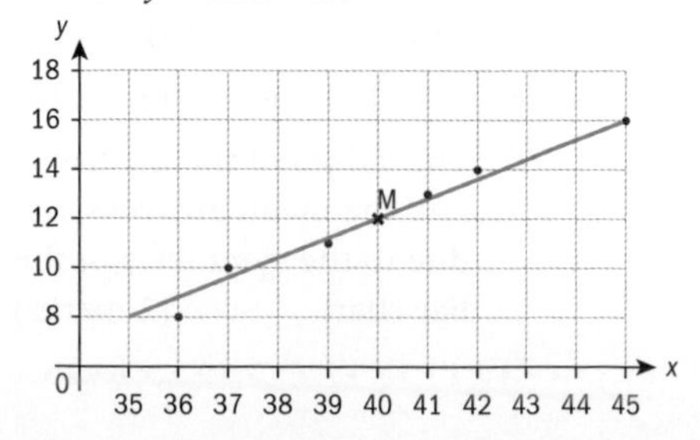

3 a $a = -2.65,\ b = 35.6$

b 22

c 40 is too far outside of the domain.

4 a $y = 108x + 361$ **b** 30

5 a 21.5°C **b** 0.924

c There is a strong positive correlation: the hotter the day, the more water sold.

d $y = 18.3x - 227$ **e** 132

f 36 °C is outside of the domain. Extrapolation is unreliable.

6 a $y = 11.8x + 53754$

b \$136 354 **c** i and ii

7 a $a = 0.0874,\ b = 0.876$

b The car travels 1 km on 0.0874 litres of fuel

c 14.9 litres

d 5 km is outside of the domain and extrapolation is unreliable.

8 a $a = 3.90,\ b = 2.03$

b An increase of 1 g of growth hormone produces nearly 4 extra flowers.

c A plant with no growth hormone would have 2 flowers

d An average of 8.86 flowers

e 2.56 g

f 1 kg is outside of the domain and extrapolation is unreliable.

9 a $a = 0.261,\ b = 199$

b Every time 1 g is added the spring expands 0.261 mm

c The spring was originally 199 mm long

d 343 mm

e 2 kg is beyond the domain and extrapolation is unreliable

f 388 g

10 a $y = -0.254x + 33.7$

b 21 **c** −0.842

d There is a strong, negative correlation.

Exam-style questions

11 a $0.51 \times 120 + 7.5 = 68.7$ (2 marks)

b The line of best fit goes through $(\bar{x}, \bar{y})$ (1 mark)

$\bar{y} = 0.51 \times 100 + 7.5 = 58.5$ (1 mark)

c Strong, positive (2 marks)

d x on y (1 mark)

12 i Perfect positive (1 mark)

ii Strong negative (1 mark)

iii Weak positive (1 mark)

iv Weak negative (1 mark)

v Zero (1 mark)

13 a $r = 0.979$ (3 s.) (2 marks)

b Strong, positive (2 marks)

c i $x = 1.23y - 20.9$ (2 marks)

ii $y = 0.77x - 20.8$ (2 marks)

d $1.23 \times 105 - 20.9 = 108$ (1 mark)

e $0.776 \times 95 + 20.8 = 95$ (1 mark)

f It is extrapolation (1 mark)

14 a

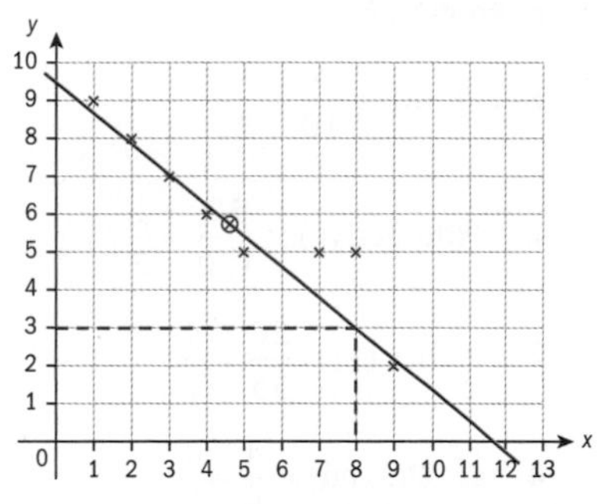

(1 mark scales 3 marks points (2 marks) 6 points, 1 mark 3 points)

b Strong, negative (2 marks)

c i $\bar{x} = 4.625$

ii $\bar{y} = 5.875$

iii See above (5 marks)

d See above

M1 thru average (1 mark)

a 3.2 see above for lines drawn on (2 marks)

15 a $100 = 70m + c$

$140 = 100m + c$

$40 = 30m \quad m = \frac{4}{3} \quad c = \frac{20}{3}$ (3 marks)

b Positive (1 mark)

c Line goes through $(\bar{x}, \bar{y})$ (1 mark)

$\bar{y} = \frac{4}{3}90 + 6\frac{2}{3} = 126\frac{2}{3}$ (2 marks)

d Estimate is

$\frac{4}{3}60 + 6\frac{2}{3} = 86\frac{2}{3}$ (2 marks)

16 a 40°C (1 mark)

b 70°C (1 mark)

c 100°C (1 mark)

d i (30,100), (60,70), (0,40) (1 mark)

ii $T \geq 80$

$40 + 2t = 80 \Rightarrow t = 20$

$130 - t = 80 \Rightarrow t = 50$ (1 mark)

Interval is $20 \leq t \leq 50$. (2 marks)

17 a

x	13	14	15	16	16	17	18	18	19	19
y	2	0	3	1	4	1	1	2	1	2

(3 marks) (2 marks for 5, 1 mark for 3)

b $r = -0.0695(3\text{s.f.})$ (2 marks)

c Very weak (negative) correlation so line of best fit is meaningless. (1 mark)
25-year-old would be extrapolation. (1 mark)

18 i Gradient $m = \frac{0.6}{3} = 0.2$ (2 marks)

ii $l = 0.6$ (1 mark)

iii $k = 3$ (1 mark)

iv $a = 5$ (1 mark)

v $b = 0.6$ (1 mark)

vi Gradient $p = \frac{0.9-0.6}{8-5} = 0.1$ (2 marks)

vii $0.6 = 0.1\times5+q \Rightarrow q = 0.1$ (2 marks)

viii $r = 8$ (1 mark)

19 a i 0.849 (3 s.f.) (2 marks)

ii Strong, positive (2 marks)

iii $y = 0.937x + 0.242$ (2 marks)

b i 0.267 (3 s.f.) (2 marks)

ii Weak, positive (2 marks)

iii The r value is too small for this to be particularly meaningful (1 mark)

20 a i No change $r = 0.87$ (1 mark)

ii No change 15 (1 mark)

iii The scatter diagram has been moved down by 4 and to the right by 5. (1 mark)

iv Strong, positive (2 marks)

b i No change $r = 0.87$ (1 mark)

ii $2\times15 = 30$ (1 mark)

iii The scatter diagram has been stretched vertically. (1 mark)

c i $r = -0.87$ (1 mark)

ii $\frac{15}{-3} = -5$ (1 mark)

iii The scatter diagram has been stretched horizontally and reflected in the y-axis. (2 marks)

iv Strong, negative (2 marks)

Chapter 8

Skills check

1 a $\frac{4}{7}$ **b** $1\frac{4}{35}$ **c** $\frac{4}{15}$ **d** $\frac{53}{56}$ **e** $\frac{3}{7}$

3 a 0.625 **b** 0.7 **c** 0.42 **d** 0.16 **e** 15 **f** 0.0484

Exercise 8A

1 $\text{P(odd)} = \frac{5}{10} = \frac{1}{2}$

2 $\text{P(defective)} = \frac{30}{150} = \frac{1}{5}$

3 $\text{P(chorus)} = \frac{20}{35} = \frac{4}{7}$

4 a $\text{P(even)} = \frac{4}{8} = \frac{1}{2}$

b $\text{P(multiple of 3)} = \frac{2}{8} = \frac{1}{4}$

c $\text{P(multiple of 4)} = \frac{2}{8} = \frac{1}{4}$

d $\text{P}(\text{not a multiple of 4}) = 1 - \frac{1}{4} = \frac{3}{4}$

e $\text{P}(\text{less than 4}) = \frac{3}{8}$

f $\text{P}(\text{nine}) = \frac{0}{8} = 0$

5 a $\text{P}(\text{C}) = \frac{1}{10}$

b $\text{P}(\text{P}) = \frac{0}{10} = 0$

c $\text{P}(\text{Vowel}) = \frac{3}{10}$

6 a $\text{P}(\text{Even}) = \frac{1}{2}$

b $\text{P(Contains digit 1)} = \frac{7}{25}$

7 a $\text{P}(\text{coach}) = \frac{3}{7}$

8 a P(green) = 0.2

9 $\text{P(Sebastian wins)} = \frac{1}{360}$

Exercise 8B

1 a i P(15 years old) = 0.18

ii P(16 or older) = 0.22 + 0.27 + 0.13 = 0.62

b 1200 × 0.18 = 216 students

2 a 0.27

b Probably not. The relative frequencies are very different.

c 450

3 a 10 of each

b

Relative frequency	0.1	0.1	0.15	0.138	0.138	0.15	0.138	0.0875

c

Relative frequency	0.0925	0.1225	0.1375	0.125	0.14	0.145	0.1075	0.13

d There is a big difference between relative frequency of getting a 1 and getting a 6. This suggests the dice is not fair.

7 a

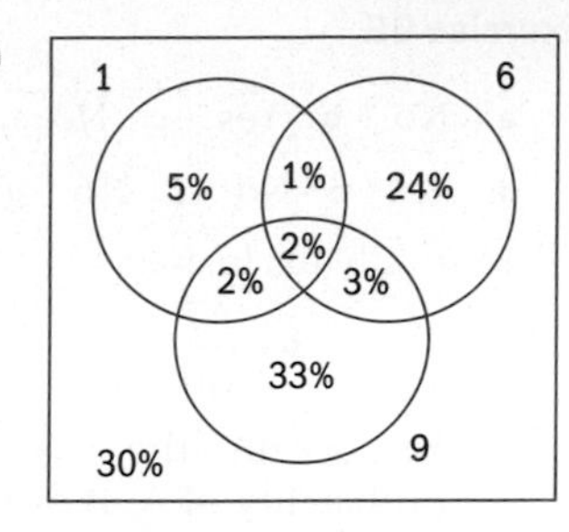

b i P(News at 9 only) = 0.33

ii P(News at 6 only) = 0.24

iii P(Does not watch news) = 0.30

Exercise 8C

1 a

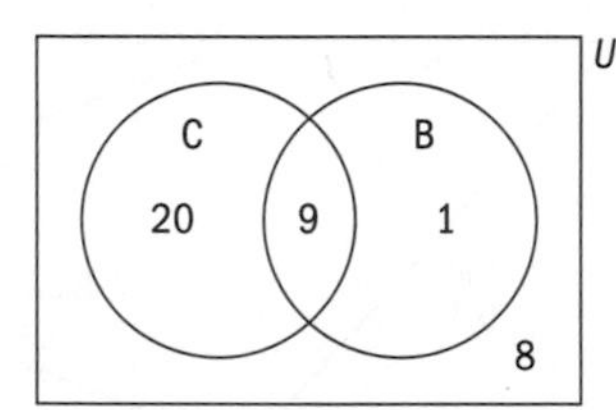

b P(neither) = $\frac{8}{38} = \frac{4}{19}$

2 a

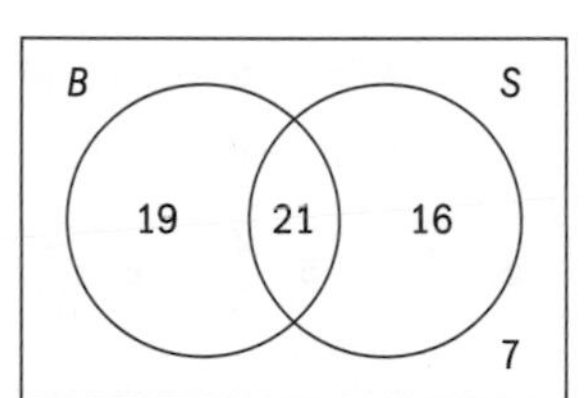

b 63 members of the club

c i P(Badminton) = $\frac{40}{63}$

ii P(Both) = $\frac{21}{63} = \frac{1}{3}$

iii P(Neither) = $\frac{7}{63} = \frac{1}{9}$

iv

P(At least one sport) = $\frac{56}{63} = \frac{8}{9}$

3 a

P(Card or present) = $\frac{46}{50} = \frac{23}{25}$

b P(Card but not a present)

$= \frac{6}{50} = \frac{3}{25}$

c P(Neither a card nor a present) = $\frac{4}{50} = \frac{2}{25}$

4 a

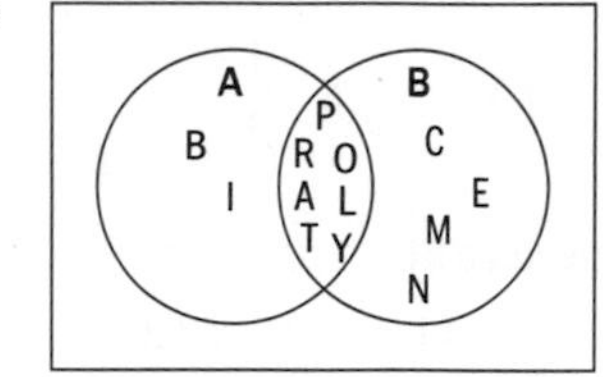

b {P, R, O, A, L, T, Y}

c {P, R, O, A, L, T, Y, B, I, C, M, E, N}

5 a {6}

b {2, 3, 4, 6, 8, 9, 10}

c {1, 3, 5, 7, 9}

d {3, 9}

e {1, 2, 4, 5, 6, 7, 8, 10}

f {1, 2, 3, 4, 5, 7, 8, 9, 10}

6 a i $M = \{3, 6, 9, 12, 15\}$

ii $F = \{1, 2, 3, 5, 6, 10, 15\}$

b

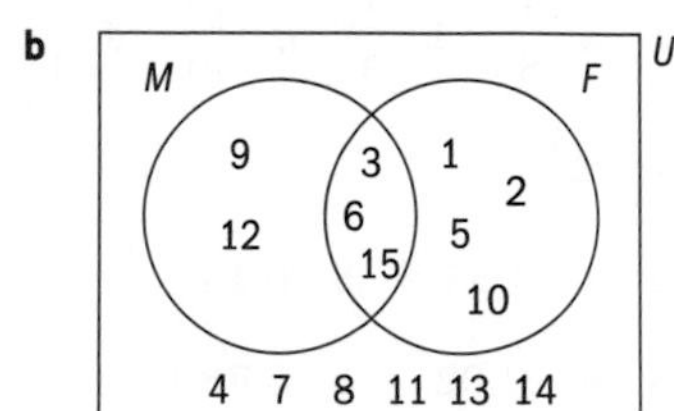

c i $P(M \cap F) = \frac{3}{15} = \frac{1}{5}$

ii $P(M' \cap F') = \frac{6}{15} = \frac{2}{5}$

Exercise 8D

1 a P(Prime) = $\frac{4}{10} = \frac{2}{5}$

b P(Prime or multiple of 3)

$= \frac{3}{5}$

c P(multiple of 3 or 4)

$= \frac{1}{2}$

2 P(Have camera or female)

$= \frac{12+18+7}{55} = \frac{37}{55}$

3 a P(M, A, T, H, E, I, C, S)

$= \frac{8}{26} = \frac{4}{13}$

b

P(T, R, I, G, O, N, M, E, Y) = $\frac{9}{26}$

c P(T, I, M, E) = $\frac{2}{13}$

d

P(M, A, T, H, E, I, C, S, R, G, O, N, Y)

$= \frac{13}{26} = \frac{1}{2}$

4 a 0.5 b 0.5

5 a $\frac{1}{4}$ b $\frac{3}{4}$

6 a 0.1 b 0.9

7 a $\frac{27}{64}$ b $\frac{37}{64}$ c $\frac{3}{64}$

Exercise 8E

1 a No b Yes c No
d Yes e No f No g No

2 Mutually exclusive

3 a $\frac{7}{12}$ b $\frac{47}{60}$

c Yes, because the probability of A, B or C winning is not equal to 1.

Exercise 8F

1 a $\frac{1}{2}$ b $\frac{3}{8}$ c $\frac{1}{4}$

2 a

	1	2	3	4
1	1, 1	1, 2	1, 3	1, 4
2	2, 1	2, 2	2, 3	2, 4
3	3, 1	3, 2	3, 3	3, 4
4	4, 1	4, 2	4, 3	4, 4

b i $\frac{3}{8}$ ii $\frac{3}{8}$ iii $\frac{1}{4}$ iv $\frac{9}{16}$

3 a

	1	2	3
2	2, 1	2, 2	2, 3
3	3, 1	3, 2	3, 3
4	4, 1	4, 2	4, 3
5	5, 1	5, 2	5, 3

b i $\frac{1}{6}$ ii $\frac{1}{4}$ iii $\frac{3}{4}$

iv $\frac{5}{12}$ v $\frac{2}{3}$

4 a $\frac{2}{9}$ b $\frac{1}{9}$ c $\frac{2}{9}$

Exercise 8G

1 $\frac{1}{25}$ 2 $\frac{64}{125}$ 3 0.0375

4 a 0.2, 0.16

b Not independent

5 $\frac{5}{12}$ 6 $\frac{1}{6561}$

7 a $P(E) = 0.4$

b i $P(E) \times P(F) = P(E \cap F)$
$0.4 \times 0.6 = 0.24$

ii $P(E \cap F) \neq 0$

c 0.64

8 $\frac{1}{27}$

9 a 0.27 b 0.63 c 0.97

Exercise 8H

1 a 12

b i $\frac{8}{27}$ ii $\frac{23}{27}$ iii $\frac{4}{5}$

2 a $\frac{3}{7}$ b $\frac{2}{3}$ c $\frac{1}{3}$ d $\frac{4}{5}$

3 a 0 b 0 c 0.63

4 a $\frac{1}{10}$ b $\frac{43}{50}$ c $\frac{11}{13}$

5 0.3 6 $\frac{1}{3}$

Exercise 8I

1 a $\frac{11}{1105}$ b $\frac{132}{1105}$

2 a $\frac{10}{91}$ b $\frac{55}{91}$ c $\frac{1}{4}$

3 a $\frac{3}{10}$ b $\frac{7}{15}$

4 a $\frac{55}{63}$ b $\frac{9}{11}$

c $\frac{7}{11}$ d $\frac{5}{11}$

Chapter review

1 a $\frac{1}{5}$ b $\frac{1}{3}$ c $\frac{49}{90}$ d $\frac{1}{15}$

2 $\frac{11}{30}$

3 a 0.55 b 0.14

4 a 0.02 b 0.78

c 0.76 d $\frac{1}{30}$

5 a $6x$

b

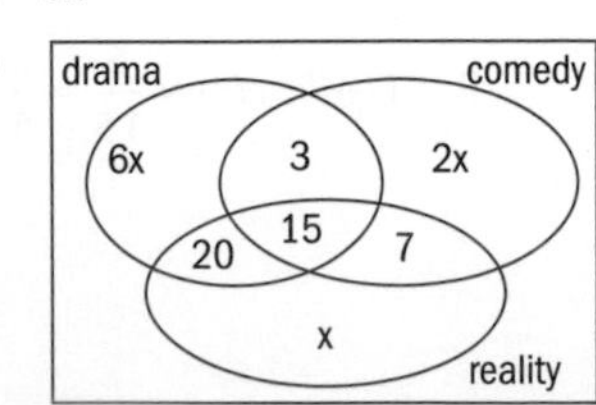

c $x = 5$.

6 a 0.3 b $P(C \text{ and } D) \neq 0$

c No $P(C) \times P(D) \neq P(C \text{ and } D)$

d 0.6 e 0.75

7 a 0.43 b 0.316

8

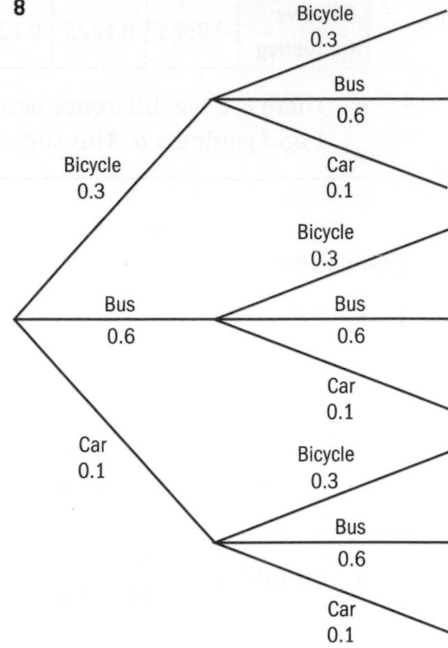

b i 0.09 ii 0.18 iii 0.46

c 0.343 d 0.045

9 a $\frac{3}{8}$ b $\frac{2}{3}$ c $\frac{2}{21}$

10

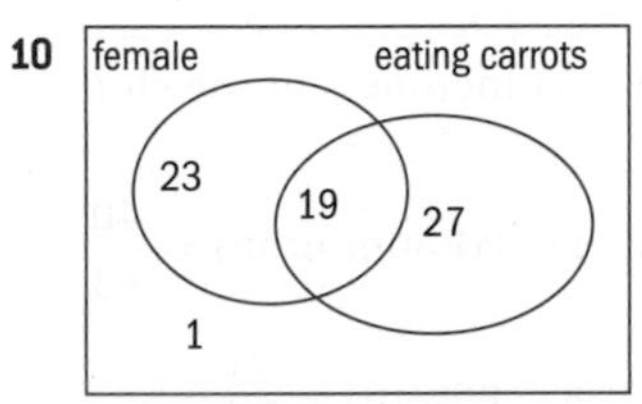

Both female and eating carrots = 19.

a $\frac{1}{70}$ b $\frac{19}{46}$

c No $P(F) \times P(C) \neq P(F \text{ and } C)$

Exam-style questions

11 a $\frac{1}{6}$ (1 mark)

b 2,4 or 6: $\frac{3}{6} = \frac{1}{2}$ (1 mark)

c Primes are 2,3,5: $\frac{3}{6} = \frac{1}{2}$ (2 marks)

d 4 or 5: $\frac{2}{6} = \frac{1}{3}$ (2 marks)

e Impossible: 0 (1 mark)

12 a $\frac{1}{36}$ (1 mark)

b $\frac{6}{36}=\frac{1}{6}$ (1 mark)

c $\frac{1}{36}$ (1 mark)

d $2\times\frac{1}{36}=\frac{1}{18}$ (1 mark)

e $(1,6),(2,5),(3,4),(4,3),$
$(5,2),(6,1)$
or using a lattice diagram
$\frac{6}{36}=\frac{1}{6}$ (2 marks)

f $\frac{1}{6}$ since independent (1 mark)

g $P(R5\cup B5)=$
$P(R5)+P(B5)-P(R5\cap B5)$

$=\frac{1}{6}+\frac{1}{6}-\frac{1}{36}=\frac{11}{36}$ or using a lattice diagram (2 marks)

h Considering the list in **e** $\frac{2}{6}=\frac{1}{3}$ or using conditional probability formula (2 marks)

13 a Independent
$\Leftrightarrow P(F\cap R)=P(F)\times P(R)$ (1 mark)

$\frac{1}{6}\neq\frac{1}{3}\times\frac{1}{4}$ so not independent (1 mark)

b
$P(F\cup R)=P(F)+P(R)-P(F\cap R)$ (1 mark)

$\frac{1}{3}+\frac{1}{4}-\frac{1}{6}=\frac{5}{12}$ (1 mark)

c P(exactly one team) =
$\frac{5}{12}-\frac{1}{6}=\frac{1}{4}$ (2 marks)
Could also use a Venn diagram in (b) and (c).

d $P(F\cap R|F)=\frac{P(F\cap R)}{P(F)}=\frac{\frac{1}{6}}{\frac{1}{3}}=\frac{1}{2}$ (2 marks)

14 a

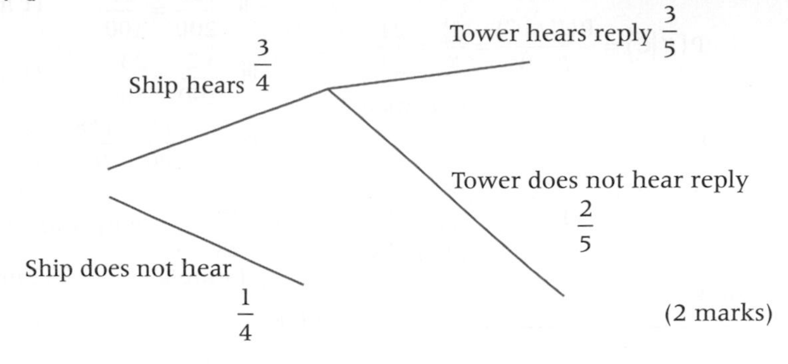

(2 marks)

b $\frac{3}{4}\times\frac{3}{5}=\frac{9}{20}$ (2 marks)

c $1-\frac{9}{20}=\frac{11}{20}$ (2 marks)

d
$P(\text{ship not hear}|\text{tower has no reply})=$

$\frac{P(\text{ship not hear}\cap\text{tower has no reply})}{P(\text{tower has no reply})}$

$=\frac{\frac{1}{4}}{\frac{11}{20}}=\frac{5}{11}$ (2 marks)

e $P(A\cap B)=0$ so events are mutually exclusive. (2 marks)

15 a $30\times\frac{3}{5}=18$ (2 marks)

b $50\times\frac{2}{5}=20$ (2 marks)

c $T\times\frac{3}{5}=30\Rightarrow T=50$ (2 marks)

16 a Let x be the number speaking both English and French.
$(60-x)+x+(40-x)+10=100$
$\Rightarrow 110-x=100\Rightarrow x=10$ (2 marks)

b

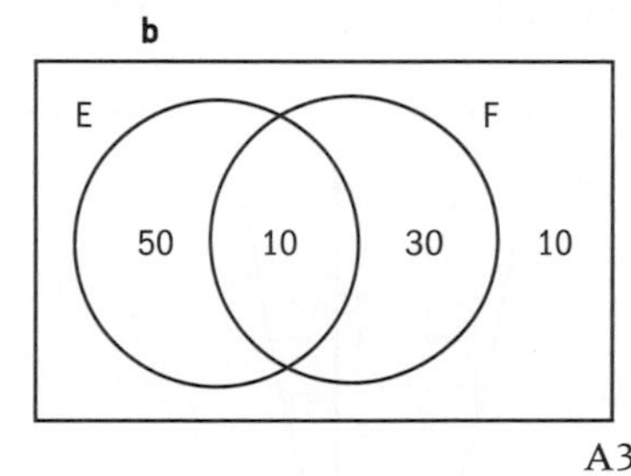

A3

(1 mark shape 2 marks numbers)

c i $\frac{50}{100}=\frac{1}{2}$ (1 mark)

ii $\frac{90}{100}=\frac{9}{10}$ (1 mark)

iii $\frac{40}{100}=\frac{2}{5}$ (1 mark)

d $P(E|F)=\frac{P(E\cap F)}{P(F)}=\frac{10}{40}=\frac{1}{4}$ (2 marks)

e If independent then
$P(E|F)=P(E)$ (1 mark)

$\frac{1}{4}\neq\frac{60}{100}$ so not independent (1 mark)

17 a

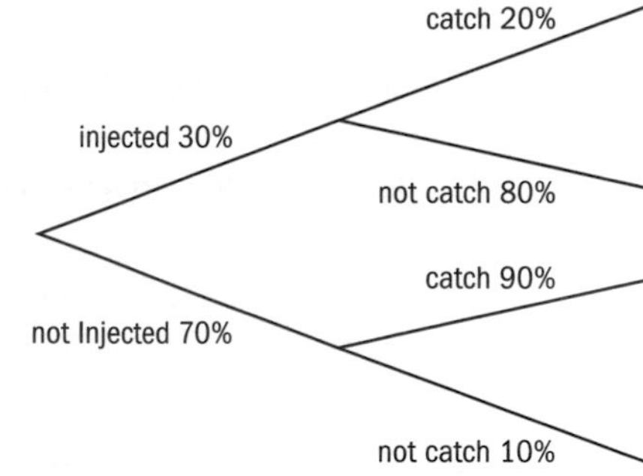

(4 marks) (2 marks layout 2 marks numbers)

b $\frac{70}{100}\times\frac{90}{100}=\frac{63}{100}$ (1 mark)

c $\frac{30}{100}\times\frac{80}{100}=\frac{6}{25}$ (1 mark)

d $\frac{30}{100}\times\frac{20}{100}+\frac{70}{100}\times\frac{90}{100}=\frac{69}{100}$ (2 marks)

e

$$P(I'|C) = \frac{P(I' \cap C)}{P(C)} = \frac{\frac{63}{100}}{\frac{69}{100}} = \frac{21}{23}$$

(2 marks)

f

$$P(I|C') = \frac{P(I \cap C')}{P(C')} = \frac{\frac{24}{100}}{\frac{31}{100}} = \frac{24}{31}$$

(2 marks)

18 a

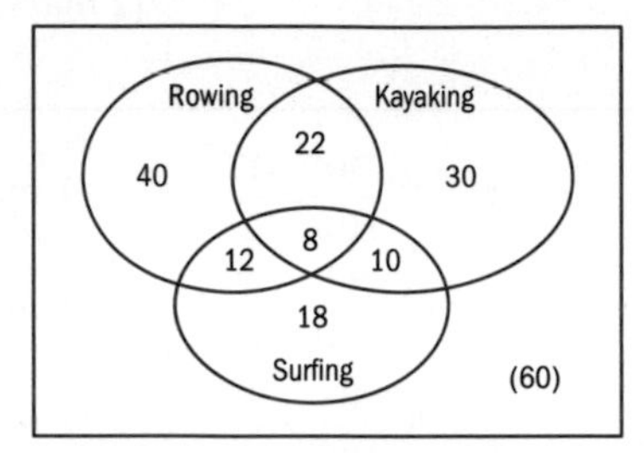

(4 marks) (A1 shape, 3 marks, 7 numbers, (2 marks, 4 numbers, 1 mark 2 numbers)

b $200 - 140 = 60$ (2 marks)

c i $\frac{30}{200} = \frac{3}{20}$ (1 mark)

ii $\frac{122}{200} = \frac{61}{100}$ (1 mark)

iii $\frac{92}{200} = \frac{23}{50}$ (1 mark)

iv $1 - \frac{82}{200} = \frac{118}{200} = \frac{59}{100}$ (2 marks)

d $\frac{20}{48} = \frac{5}{12}$ or by using the formula (2 marks)

19 a

i $\frac{5}{8} \times \frac{4}{7} = \frac{5}{14}$ (2 marks)

ii RG or GR $\frac{5}{8} \times \frac{3}{7} + \frac{3}{8} \times \frac{5}{7} = \frac{15}{28}$ (2 marks)

b i $\frac{5}{8} \times \frac{5}{8} = \frac{25}{64}$ (2 marks)

ii RG or GR $\frac{5}{8} \times \frac{3}{8} + \frac{3}{8} \times \frac{5}{8} = \frac{15}{32}$ (2 marks)

20 a i $P(A|B) = \frac{P(A \cap B)}{P(B)} \Rightarrow 0.4 = \frac{P(A \cap B)}{0.5} \Rightarrow P(A \cap B) = 0.2$ (2 marks)

ii $P(A) = P(A \cap B) + P(A \cap B') = 0.2 + 0.4 = 0.6$ (2 marks)

iii $P(A \cup B) = P(A) + P(B) - P(A \cap B) = 0.6 + 0.5 - 0.2 = 0.9$ (2 marks)

iv $P(A|B') = \frac{P(A \cap B')}{P(B')} = \frac{0.4}{0.5} = 0.8$ (2 marks)

b $P(A|B) \neq P(A|B')$ so not independent (2 marks)

Skills check

1 a 32 **b** 1000

c $\frac{1}{243}$ **d** $\frac{125}{216}$

2 a 3 **b** 4 **c** 4

3

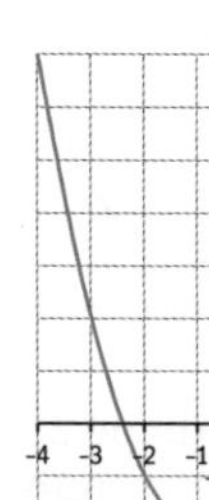

Exercise 9A

1 a^{15} **2** $14x^7y^8$

3 $2a^7b^3c$ **4** $2m^2$

5 $\frac{2u^2}{3v}$ **6** $27r^3s^9$

7 $-8x^{12}y^3z^{15}$ **8** $x^{14}y^4$

9 $\frac{1}{x^7y^9}$ **10** $-\frac{x^{11}}{9y^2}$

11 $9x^4y^2$ **12** $\frac{10a^4}{b}$

Exercise 9B

1 a $\sqrt{7}$ **b** $\sqrt[5]{2^3}$ **c** $\sqrt[2]{6^3}$

d $\sqrt[4]{2^5}$ **e** $\frac{1}{\sqrt{5}}$ **f** $\frac{1}{\sqrt{(3x)^3}}$

g $\frac{3}{\sqrt{x^3}}$

2 a $10^{\frac{3}{2}}$ **b** $a^{\frac{6}{5}}$

c $m^{\frac{7}{3}}$ **d** $(5x)^{-\frac{1}{2}}$

e $(2d)^{-\frac{5}{4}}$ **f** $3x^{\frac{1}{2}}$ **g** $3x^{-\frac{1}{2}}$

Exercise 9C

1 a 4 **b** 6 **c** 5

d 2 **e** $\frac{1}{2}$ **f** 4

g −5 **h** $\frac{1}{8}$ **i** −2

j $\frac{3}{2}$

2 a 7 **b** 5 **c** $\frac{1}{2}$ **d** $-\frac{5}{3}$

Exercise 9D

1 a Growth **b** Decay

c Decay **d** Decay

e Growth

2 $f(x)$ = D $g(x)$ = C $h(x)$ = A
$i(x)$ = B $j(x)$ = E

3 a i, ii, iv

iii $y = 4$ **v** $x \in \mathbb{R}, y > 4$

b **i, ii, iv**

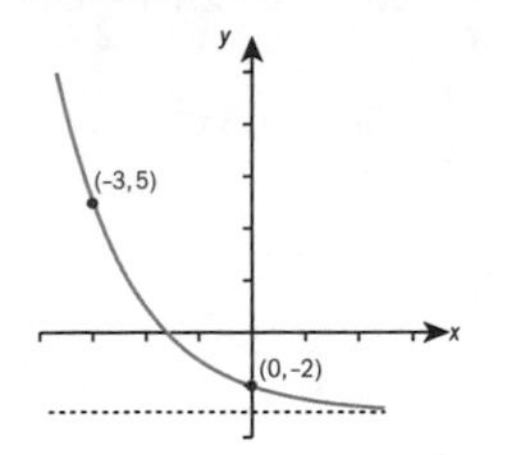

iii $y = -3$ **v** $x \in \mathbb{R}, y > -3$

c **i, ii, iv**

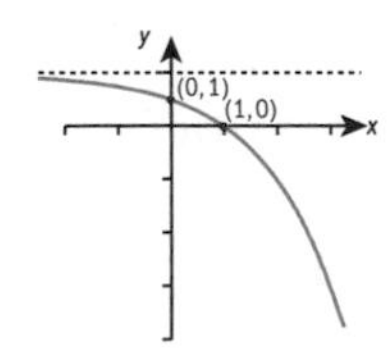

iii $y = 2$ **v** $x \in \mathbb{R}, y < 2$

4 **a** 90 °C **b** 48.0 °C
c 4 min **d** 25 °C

5 **a** \$30 000 **b** \$21 870
c 6.58 years

6 **a** 40 **b** 90

c

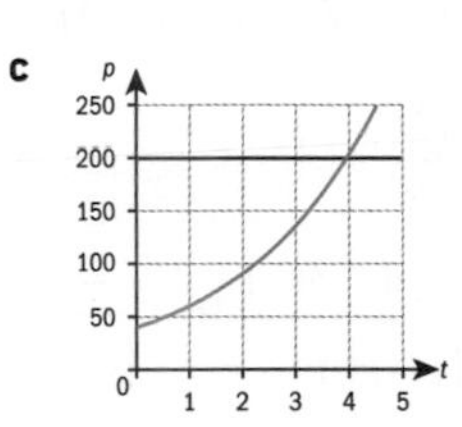

The population will reach 200 in 3.97 years.

7 **a** 88.6 g

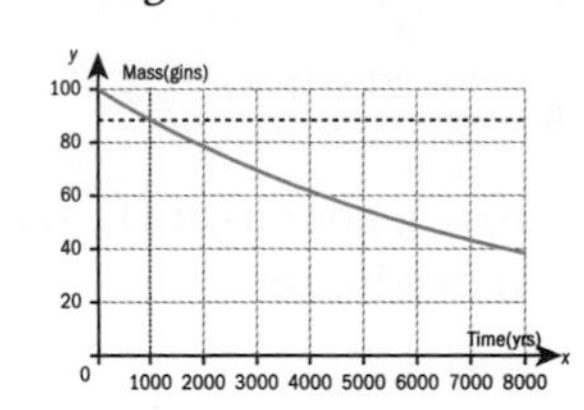

b **i** 2378 years
ii 5730 years

Exercise 9E

1 **a** 2.718 **b** 7.389
c 0.135 **d** 8.155
e 1.359 **f** 8.244
g 5.873

2

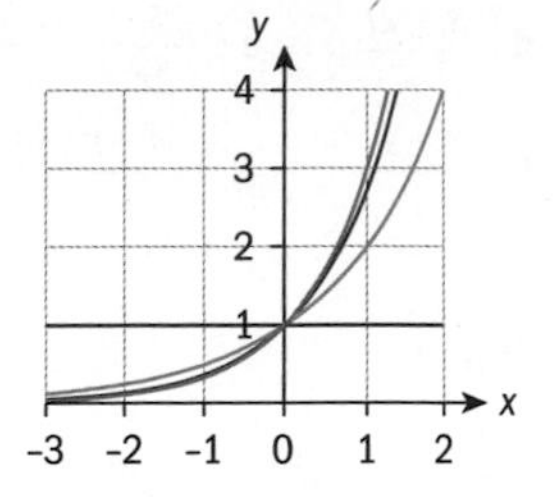

a It is a straight line.

b In between $g(x) = 2^x$ and $i(x) = 3^x$

3

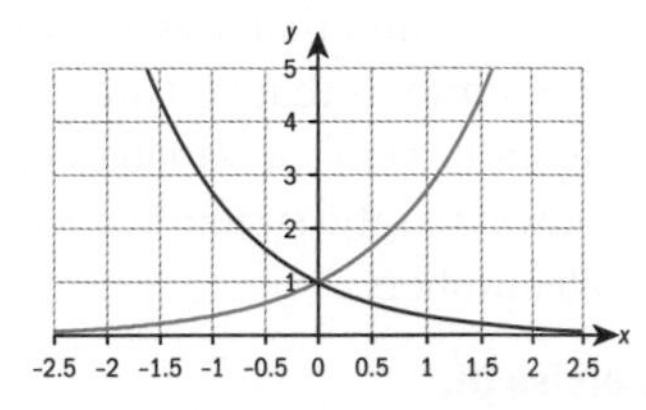

A reflection in the y-axis

4 **a** 4500 cm^2 **b** $90\,385 \text{ cm}^2$.

5 \$6749

6 In 2011 the population of the world was approximately 7 billion, and it is growing at a rate of 1.1%

a $y = 7(1 + 0.011)^t$

b 8.16 billion

c During the year 2044

7 The value of a car decreases by 15% every year. If Tatiana bought a new car for \$25 000,

a $y = 25(1 - 0.15)^t$, in thousands of dollars

b \$15 400

c After 5.64 years

Exercise 9F

1 **a** A vertical translation of 4 units up

b A horizontal translation of 3 units right

c A vertical stretch of scale factor 2

2 **a**

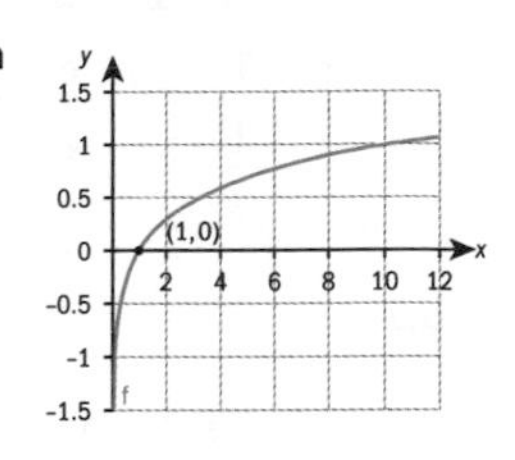

b

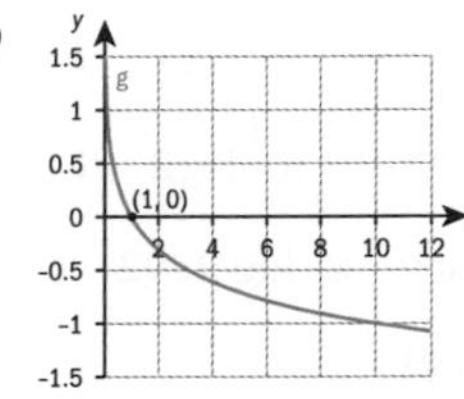

c

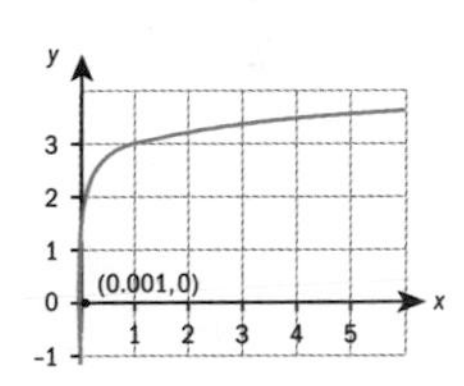

d

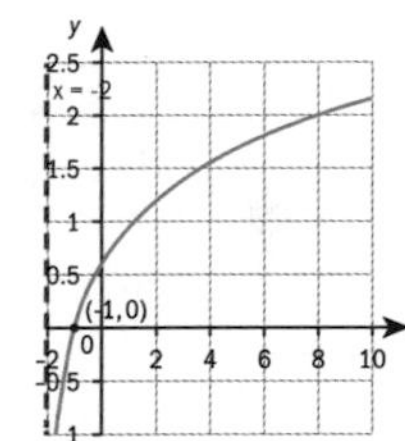

3

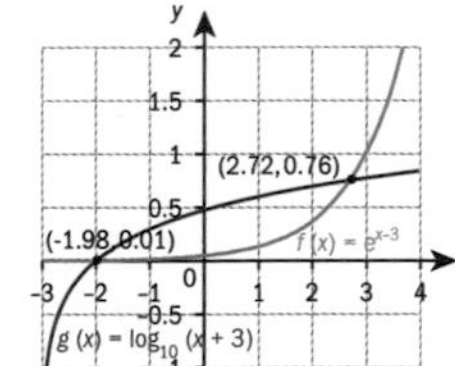

$x = -1.98, 2.72$

Exercise 9G

1 **a** $p^r = q$

b $3^r = 5$

c $7^6 = q$

d $p^3 = 5$

e $10^x = 11$

2 **a** $\log_r t = s$ **b** $\log_8 64 = 2$

c $\log_{10} 25 = x$ **d** $\log_3 \frac{1}{9} = -2$

e $\log_{27} 9 = \frac{2}{3}$

3 **a** 0 **b** 0 **c** 0 **d** 0

4 **a** 1 **b** 1 **c** 1 **d** 1

5 **a** 2 **b** 5 **c** 3

d 4 **e** $\frac{1}{2}$ **f** $\frac{1}{3}$

g −3

Exercise 9H

1 **a** $\log 15$ **b** $\log 8$ **c** $\log 125$

d $\log \frac{64}{81}$ **e** $\log x$ **f** $\log 236$

g $\log 800$ **h** $\log 2$ **i** $\log 24$

j $\log 9$ **k** $\log 2592$ **l** $\log 81$

Exercise 9I

1 **a** $x + y$ **b** $y - x$ **c** $2x$

d $3x$ **e** $2y$ **f** $x - y$

2 **a** $m + n$ **b** $m - n$ **c** $2m$

d $3n$ **e** $2m - n$ **f** $m - 2n$

g $m + 2n$ **h** $\frac{m}{n}$ **i** $\frac{2m}{3n}$

j $2 + n$

3 $4(x - y - 3z)$

4 **a** $3x + y$ **b** $\frac{1}{2}x - y$

5 **a** $M = \frac{x}{x-5}$ **b** $\frac{50}{9}$

Exercise 9J

1 **a** 3 **b** 4 **c** $\frac{1}{2}$

d $\frac{1}{3}$ **e** −1 **f** −2

2 **a** 2 **b** 3 **c** x

d 16 **e** x^3 **f** $\frac{1}{3}$

3 **a** 14.9 **b** 5.42

4 a^x

Exercise 9K

1 **a** 1.89 **b** 1.77

c 1.29 **d** 3.10

e −0.712 **f** −0.737

2 $\frac{q}{p}$

3 **a** $\frac{s}{r}$ **b** $\frac{r}{s}$ **c** $\frac{2s}{r}$

d $\frac{s+2r}{r}$ **e** $\frac{s}{2r}$

f $\frac{r+s}{s}$ **g** $\frac{s-r}{r}$

4 $\frac{1}{\log_y x}$

5 $\frac{1}{\log e}$

6 "Show that" is to use numbers to demonstrate a certain property and that it works for the numbers that you are using. To prove is to use variables to prove that the system works for all numbers.

Exercise 9L

1 **a** 2.32 **b** 2.58 **c** 1.77

d 3.85 **e** 1.52 **f** 0.65

g 0.712 **h** 0.235 **i** 1.61

j 2.30

2 **a** 0.792 **b** 0.258 **c** 5.43

d 2.91 **e** 2.77 **f** 3.30

g 2.58 **h** 0.330 **i** 9.70

j −8.02

3 **a** 1.22 **b** 0.258 **c** 1.69

d −0.535 **e** −4.42 **f** 1.87

g 0.0524 **h** 6.92

4 **a** $\ln 24$ **b** $\frac{1}{16}$

5 **a** 0, $\ln 4$ **b** $\ln 3$ **c** $\frac{\ln 2}{2}$

6 **a** 10 cm **b** 0.630

c 15.2 days

7 **a** \$21 000 **b** 0.910

c 16.8 years

8 **a** 84 **b** 60 **c** 29 days

9 **a** 499 kg **b** 58 048 years

10 $\frac{\ln 3}{\ln 2 - 4}$

Exercise 9M

1 **a** $7e^x$ **b** $-\frac{1}{4}e^x$ **c** $\frac{9}{x}$

d $\frac{\pi}{x}$ **e** $\frac{1}{x}$ **f** $\frac{1}{x}$

g $\frac{1}{x}$ **h** $2e^{2x}$ **i** $4e^{4x}$

j $5e^{5x}$

2 **a** $\frac{5}{x} - 2e^x$ **b** $2x + \frac{1}{2}e^{-\frac{1}{2}x} + \frac{1}{x}$

c $-\frac{1}{x} - 5e^{-5x} + 3x^2$

d $\frac{1}{x} + 7e^{7x} - 7$

e $-\frac{5}{x} + 24e^{4x}$ **f** $1 + \frac{3}{x}$

g $\frac{2x}{x^2+1} - \frac{3x^2-1}{x^3-x}$

3 **a** $6x^2e^{2x^3}$

b $24x^2(4x^3+5)e^{(4x^3+5)^2}$

c $\frac{5}{x}$ **d** $\frac{3(\ln x)^2}{x}$

e $(1 + 2x)e^{2x}$

f $6x^2(1 - x)e^{-3x}$

g $2xe^{3x} + 3e^{3x}(x^2 + 1)$

h $e^{ax^2+1} + 2ax^2e^{ax^2+1}$

i $\ln x + 1$ **j** $x^2(1 + 3\ln x)$

k $\frac{2x^2}{2x+3} + 2x\ln(2x+3)$

l $\frac{e^{3x}(3x-2)}{x^3}$

m $\frac{2e^{4x}(4-3e^x)}{(1-e^x)^2}$ **n** $\frac{-2e^x}{(e^x-1)^2}$

o $\frac{\ln x - 1}{(\ln x)^2}$

p $\frac{-3+\ln x}{x^2}$

q $\frac{1-2(1+\ln x)}{x^3}$

4 Turning point is at (1, −1) and is a maximum

5 $y = 4x + 2$

6 $y = -x + 1$

7 $-e^6$

8 $\left(\frac{1}{2}, \frac{1}{4} - \ln 2\right)$, (1,1)

9 $\left(\ln 2, \ln\frac{5}{2}\right)$

10 Tangent $f(x) = ex - e$. Normal $f(x) = -\frac{1}{e}x + \frac{1}{e}$

Chapter review

1 **a** $2^4 = 16$ **b** $5^3 = 125$
c $9^2 = 81$ **d** $12^2 = 144$
e $10^4 = 10\,000$

2 **a** $\log_3 81 = 4$ **b** $\log_{15} 225 = 2$
c $\log_{81} 9 = \frac{1}{2}$ **d** $\log_a c = 14$
e $\ln x = 4$

3 **a** 3 **b** $-\frac{1}{6}$ **c** -3 **d** 2

4 **a** $\frac{1}{60}$ **b** $-\frac{1000}{7}$
c $\frac{1}{1000}$ **d** 25

5 **a** 2 **b** 7 **c** 1 **d** 1
e 0 **f** 79 **g** 19 **h** 7

6 **a** 4.09 **b** -0.226
c 0.229 **d** -0.106
e -2.44 **f** -1.07
g -1.15 **h** -1.13

7 **a** $\log 12$ **b** $\log 3$
c $\log \frac{x^2}{y^5}$ **d** $\log_5 x^8y^2$
e $\ln x\sqrt{y}\frac{1}{2}z$ **f** $\ln \frac{x^4}{y^3z^2}$

8 **a** $p + q$ **b** $3p$
c $2p + 3q$ **d** $5q - 4p$

9 **a** 2.58 **b** -0.431
c 2.55

10 1.02, 5.65

11 **a**

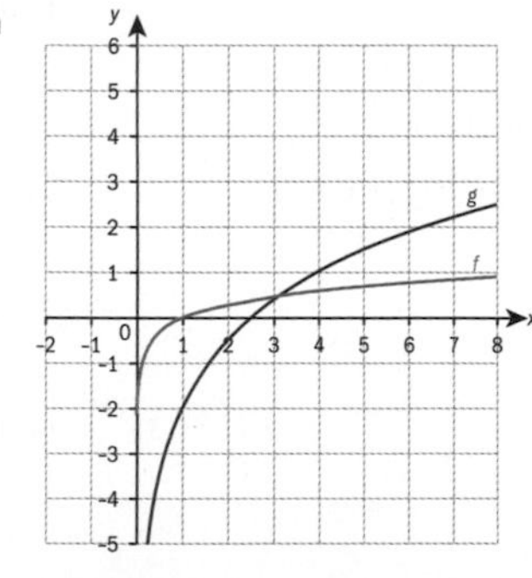

b $x = 0$

c Vertical stretch of scale factor 5 and a vertical translation of 2 units down.

12 **a**

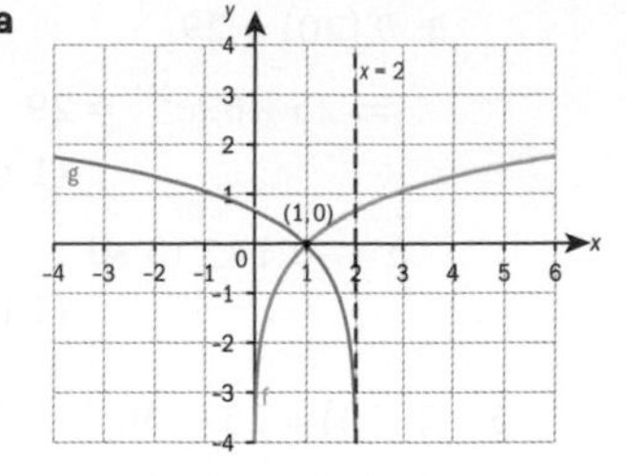

b $x = 2$

c A reflection in the y-axis and a horizontal translation of 2 units to the right.

d $x = 1$

13 $g(x) = -(2^{-x}) - 2$

14 $g(x) = 3\ln(x + 5)$

15 **a** $8e^x + \frac{7}{x}$ **b** $3e^{3x}$ **c** $\ln x$
d $(12x + 5)e^{6x^2+5x}$ **e** $\frac{2x}{x^2+8}$
f $\frac{7e^x}{\left(2e^x+1\right)^2}$ **g** $\frac{3}{2(3x-2)}$
h $\frac{e^x}{x} + e^x \ln x$

16 $x + 12y - 49 = 0$

17 4.5

18 **a** 30 000 **b** 35 205 **c** 2029

19 2025

20 0.502 mg

Exam-style questions

21 **a** $\mathbb{R}$ (1 mark)
b -2 (1 mark)
c 12 (1 mark)
d $y < 16$ (1 mark)
e $x > -2$ (1 mark)
f $y = 16$ (1 mark)

22 **a** $N(0) = 35$ (1 mark)
b $N(4) = 409.97$; 410 bacteria (2 marks)
c $N(t) > 1000$ (1 mark)
$t > 5.449...$ (1 mark)
d $0 \le t < 5.45$ (1 mark)

23 **a**

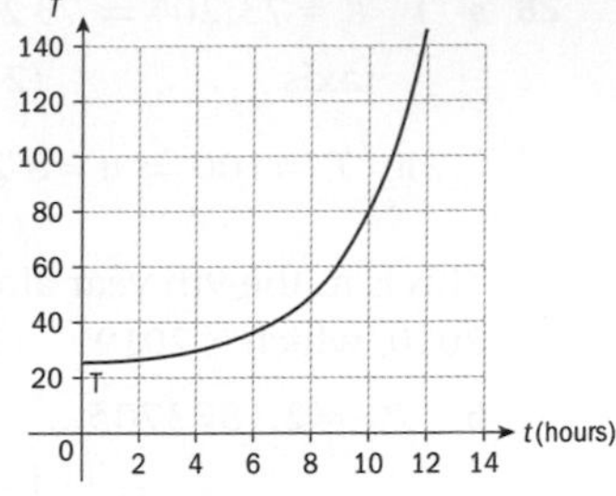

(shape: 1 mark; domain: 1 mark)

b $T(6) = 25 + e^{0.4\times6}$
36.0 (1 d.p.) (1 mark)

c Solve $25 + e^{0.4t} = 100$ (1 mark)
$t = 10.793...$ (1 mark)
10 hours 48 minutes (1 mark)

24 **a** $x > 2$ (1 mark)
b $x = 2$ (1 mark)
c $3\ln(x-2) + 1 = 0$
$\Rightarrow x = 2 + e^{-\frac{1}{3}}$ (2 marks)
d $f'(x) = \frac{3}{x-2}$ (1 mark)
$f'(x) = 1$ for tangent parallel to $y = x$
$\Rightarrow \frac{3}{x-2} = 1 \Rightarrow x = 5$ (1 mark)
$f(5) = 3\ln 3 + 1$ so tangent is at the point $(5, 3\ln 3 + 1)$ (1 mark)
Tangent line is
$y - (3\ln 3 + 1) = x - 5$ (1 mark)
(or $y = x + 3\ln 3 - 4$)

25 **a** Combining any two of the three terms, e.g.
$\log_2 3x - \log_2(x-3)$ (1 mark)
Combining the remaining terms, e.g.
$\log_2 \frac{3x}{x-3}$ (1 mark)
b $\ln\left(\frac{ex^3}{(x-1)^2}\right)$ (3 marks)

26 **a** **i** $T_5 = 73.205 \Rightarrow 73\,205$ taxis (2 marks)

ii $T_n = 100 \Rightarrow n = 8.27$ (1 mark)

This is in the 9th year after 2010, which is 2019 (1 mark)

b $P_{10} = 2.1873705...$ (2 marks)

2 187 000 people (nearest thousand) (1 mark)

c Adjusting units in (i) or (ii) (1 mark)

i $\frac{P_5 \times 10^6}{T_5 \times 10^3} = 28.4$ people per taxi; (2 marks)

ii $\frac{P_{10} \times 10^6}{T_{10} \times 10^3} = 18.6$ people per taxi. (1 mark)

d The model predicts a reduction in the number of people per taxi which may mean that the taxis are in use for less hours, or less taxis are used everyday. (1 mark)

27 **a** Choosing a counterexample; e.g. for $x = \text{e}$, (1 mark)

$$\ln(x^2) = \ln(\text{e}^2) = 2 \neq (\ln \text{e})^2 = 1$$

(2 marks)

b $(\ln x)^2 - \ln(x^2) - 15 = 0$

$(\ln x)^2 - 2\ln(x) - 15 = 0$ (1 mark)

$(\ln x - 5)(\ln x + 3) = 0$ (1 mark)

$\ln x - 5 = 0,\ \ \ln x + 3 = 0$

$\ln x = 5,\ \ \ln x = -3$ (1 mark)

$x = \text{e}^5,\ \ x = \text{e}^{-3}$ (2 marks)

28 **a** **i** $T(0) = 94 \Rightarrow 25 + a = 94$ (1 mark)

$a = 69$ (1 mark)

ii $T(20) = 29$

$\Rightarrow 25 + 69\text{e}^{20b} = 29$ (1 mark)

$b = -0.142$ (3 s.f.) (1 mark)

b $T(30) = 25 + 69\text{e}^{-0.142 \times 30}$

$= 26.0\,°\text{C}$ (3 s.f.) (2 marks)

c $y = 25$ (1 mark)

d The temperature of the room. (1 mark)

e $\frac{\text{d}T}{\text{d}t} = -9.79...\text{e}^{-0.142...t}$ (2 marks)

f $\frac{\text{d}^2T}{\text{d}t^2} = 1.40\,\text{e}^{-0.142...t}$ (1 mark)

g The rate of change is always negative which means the temperature is decreasing; (1 mark)

as the second derivative is always positive, the temperature will not have a minimum but will approach the value 25 given by the horizontal asymptote. (1 mark)

29 **a**

i $f(0) = \left(\frac{3}{2}\right)^{-1} + 2$

$= 2\frac{2}{3} = 2.67$ (3 s.f) (2 marks)

ii $(2.67, 0)$ (1 mark)

b $x = \left(\frac{3}{2}\right)^{y-1} + 2$ (1 mark)

$\left(\frac{3}{2}\right)^{y-1} = x - 2$ (1 mark)

$y = \log_{\frac{3}{2}}(x-2) + 1$ (2 marks)

$g(x) = \log_{\frac{3}{2}}(x-2) + 1$

c **i** $y = 2$ (1 mark)

ii $x = 2$ (1 mark)

d $x > 2$ (1 mark)

e Use GDC solver or intersection of graphs (1 mark)

$x = 2.16$ (1 mark)

---

30 **a** $f(4) = \log_2 4 + \log_2(15) - \log_2(5)$ (1 mark)

$= \log_2 \frac{4 \times 15}{5} = \log_2 12$ (1 mark)

b $f(x) = \log_2 x + \log_2(x^2 - 1) - \log_2(x+1)$

$f(x) = \log_2 x + \log_2(x-1)(x+1) - \log_2(x+1)$ (1 mark)

$f(x) = \log_2 \frac{x(x-1)(x+1)}{x+1}$ (2 marks)

$f(x) = \log_2 x(x-1)$ (1 mark)

$f(x) = \log_2(x^2 - x)$

Chapter 10

Skills check

1 a $1800\,\text{cm}^2$ b $18\,\text{m}^2$
c $8\pi\,\text{mm}^2$

2 a $3x^2+15x$ b x^2-25
c $9x^2+6x+1$ d $2x^2-9x-5$

3 a $x^{\frac{1}{3}}$ b $x^{\frac{4}{7}}$
c $(2x+5)^{\frac{1}{2}}$ d $(x-3)^{\frac{2}{3}}$

Exercise 10A

Note: throughout this chapter, denotes an arbitrary constant.

1 $F(x)=\frac{1}{11}x^{11}+C$

2 $F(x)=\frac{1}{6}x^6+C$

3 $F(x)=\frac{1}{26}x^{26}+C$

4 $F(x)=-\frac{1}{5x^5}+C$

5 $F(x)=-\frac{1}{7x^7}+C$

6 $F(x)=-\frac{1}{x}+C$

7 $F(x)=\frac{3}{5}x^{\frac{5}{3}}+C$

8 $F(x)=\frac{10}{11}x^{\frac{11}{10}}+C$

9 $F(x)=\frac{4}{3}x^{\frac{3}{4}}+C$

10 $F(x)=\frac{2}{3}x^{\frac{3}{2}}+C$

11 $F(x)=\frac{4}{7}x^{\frac{7}{4}}+C$

12 $F(x)=\frac{7}{6}x^{\frac{6}{7}}+C$

Exercise 10B

1 $\frac{1}{5}x^5+C$

2 $2x^3+2x^2+5x+C$

3 $3t^5+3t^4+t^2+5t+C$

4 $8x+C$ 5 $-\frac{1}{6u^6}+C$

6 $-\frac{1}{2x^4}+C$ 7 $\frac{1}{4}w^4+\frac{3}{4}w^{\frac{4}{3}}+C$

8 $\frac{8}{3}x^{\frac{3}{2}}+3x+C$ 9 $\frac{9}{14}x^{\frac{14}{9}}+C$

10 $u+C$

11 a $5x^4-\frac{6}{x^3}$ b $\frac{x^6}{6}-\frac{3}{x}+C$

12 $p=16;\ q=7$

Exercise 10C

1 $6\ln|x|+C$ 2 $5e^u+C$

3 $\frac{1}{2}\ln|x|+C$ 4 $\frac{1}{3}e^x+C$

5 $3x^3+6x^2+4x+C$

6 $\frac{1}{2}x^2+x+C$ 7 $\frac{1}{4}t^4+t^3+C$

8 $\frac{3}{2}x^2+C$ 9 $\frac{1}{4}x^4+\frac{3}{2}x^2+2x+C$

10 $\frac{1}{2}e^u-2u+C$

Exercise 10D

1 $\frac{1}{35}(7x-5)^5+C$

2 $-\frac{1}{21}(-3x+7)^7+C$

3 $\frac{1}{10}\ln|10x+13|+C$

4 $-\frac{1}{4}e^{-4x+3}+C$

5 $\frac{1}{5}(5x+1)^4+C$

6 $\frac{2}{3}\ln|3x+8|+C$

7 $-\frac{3}{2}e^{4-2x}+C$

8 $\frac{7}{10}(2x-9)^5+C$

9 $-\frac{1}{4(4x+3)}+C$

10 $\frac{3}{8}(2x+1)^{\frac{4}{3}}+C$

11 $\frac{1}{5}e^{5x}+\frac{8}{5}\ln|5x-3|+C$

12 $\frac{3}{8}(4x+7)^{\frac{2}{3}}+C$

13 a $15(3x+10)^4$
b $\frac{1}{18}(3x+10)^6+C$

14 a $-\frac{12}{(12x+7)^2}$
b $\frac{1}{12}\ln|12x+7|+C$

Exercise 10E

1 $h(t)=2t^3+t-10$

2 $y=(2x-3)^4+5$

3 a $v(t)=2t^2+t+2$
b $s(t)=\frac{2}{3}t^3+\frac{1}{2}t^2+2t+8$

4 a $8e^6+1$ b $2e^{2t}+\frac{1}{2}t^2+2$

5 $f(x)=\frac{1}{8}\ln|8x-7|+\frac{7e}{8}$

Exercise 10F

1 a $\int_0^6 \frac{2}{3}x\,dx=12$
b $\frac{1}{2}(6\times4)=12$

2 a $\int_{-1}^{2}5\,dx=15$ b $3(5)=15$

3 a $A=\int_{-3}^{3}\sqrt{9-x^2}\,dx=14.1$
b $A=\frac{\pi r^2}{2}=\frac{\pi(3)^2}{2}\approx14.1$

4 a $\int_0^6\left(\frac{1}{3}x+3\right)dx=24$
b $\frac{1}{2}(3+5)(6)=24$

5 a $A=\int_0^5\sqrt{25-x^2}\,dx=19.6$
b $A=\frac{1}{4}\pi r^2=\frac{1}{4}\pi(5)^2\approx19.6$

6 a $\int_{-4}^{4}|x|\,dx=16$
b $2\left[\frac{1}{2}(4\times4)\right]=16$

Exercise 10G

1 $\frac{15}{2}$ 2 -3 3 $\frac{21}{2}$ 4 -8

5 -12 6 20 7 0 8 16

9 12 10 10 11 48

12 a 10 b 14
c $a=1;\ b=7$ d $k=2$

Exercise 10H

1 15 2 $\frac{e^4-e}{2}$ 3 $5\ln\frac{4}{3}$ 4 9

5 3 6 $\frac{4}{9}$ 7 -32 8 3

9 **a** $\int_0^2 \left(-x^2+2x\right)\mathrm{d}x = \frac{4}{3}$

b $\int_1^2 \frac{1}{x^2}\mathrm{d}x = \frac{1}{2}$

10 **a** 20 **b** 28

11 **a** $2\ln|x| + C$ **b** $k = 6$

Exercise 10I

1 $\frac{\ln 7}{4}$ **2** $\frac{1}{3}$ **3** $\frac{e^{10}-e}{3}$ **4** $\frac{2}{3}$

5 $\frac{27}{4}$ **6** $\frac{49}{3}$ **7** 0 **8** 19

9 **a** $(-2, 0), (0, 0), (2, 0)$

b $\int_0^2 -2x(x^2-4)\mathrm{d}x$ **c** 8

10 **a** $\int_1^k \frac{1}{2x+1}\mathrm{d}x$ **b** $k = 13$

Exercise 10J

1 $\int_{-3}^{1}\left(\left(-2x+3\right)-x^2\right)\mathrm{d}x = \frac{32}{3}$

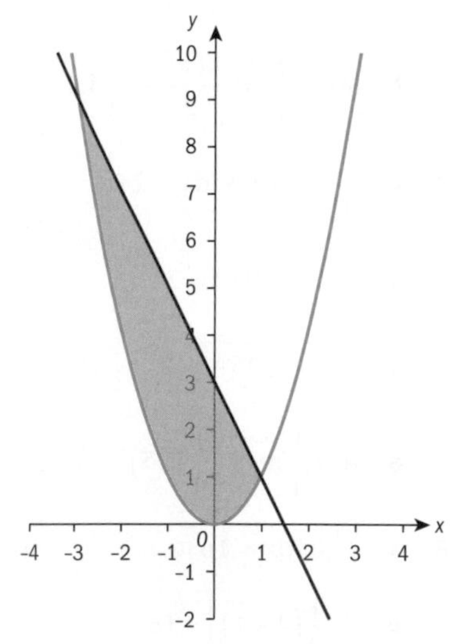

2 $\int_0^5\left(4-\left(x^2-5x+4\right)\right)\mathrm{d}x = \frac{125}{6}$

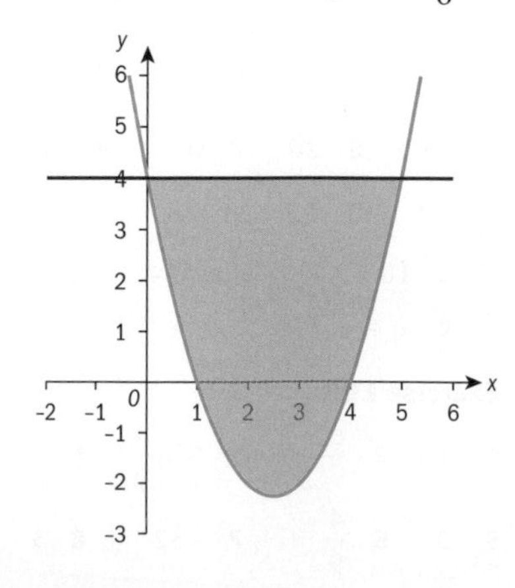

3 $\int_{-4}^{4}\left(\left(-0.5x^2+8\right)-\left(0.5x^2-8\right)\right)\mathrm{d}x \approx 85.3$

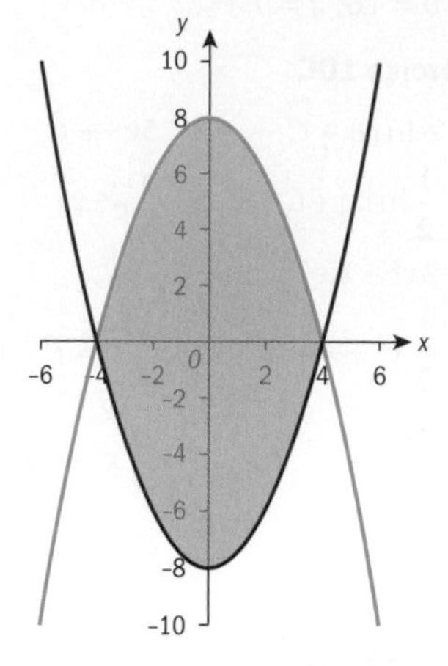

4 $\int_{-4}^{4}\left(4\sqrt{x}-\left(x^2-2x\right)\right)\mathrm{d}x = 16$

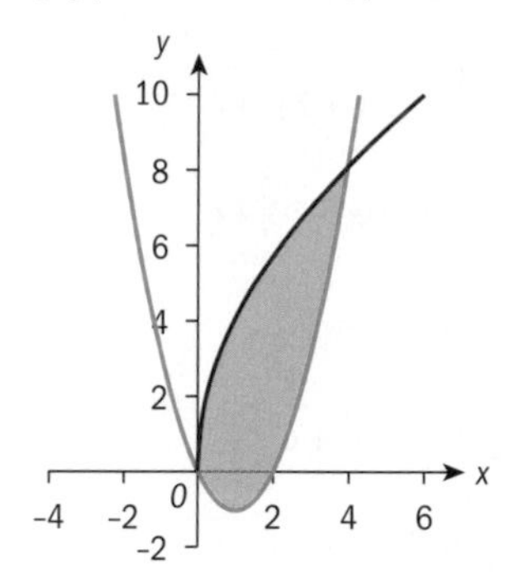

5 $\int_{0.15}^{8.21}\left(\ln(x)-\left(\frac{1}{2}x-2\right)\right)\mathrm{d}x \approx 8.78$

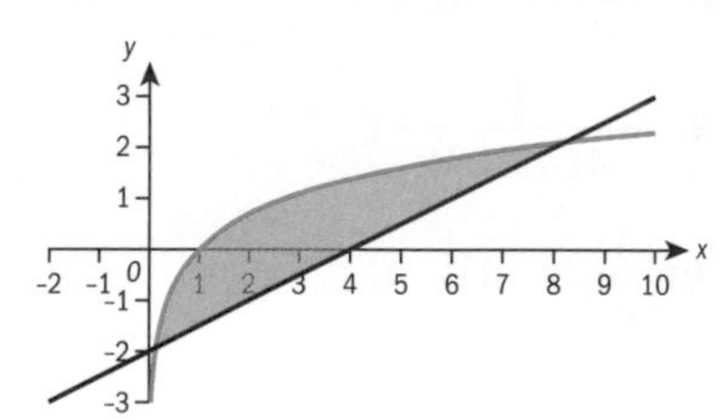

6 $\int_{-0.77}^{3.27}\left(\left(6-x^2\right)-\left(x^2-5x+1\right)\right)\mathrm{d}x \approx 21.8$

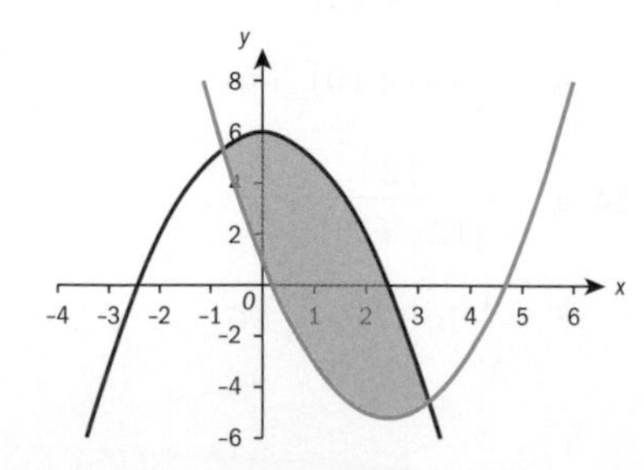

7 $\int_{-1.11}^{2}\left(\left(3x+4\right)-2e^x\right)\mathrm{d}x \approx 3.68$

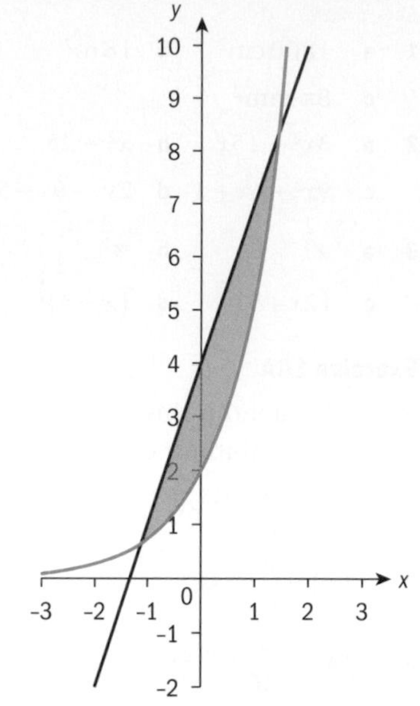

8 $\int_{-7.36}^{1.36}\left(\frac{x+4}{x-2}-\left(-x-7\right)\right)\mathrm{d}x \approx 27.5$

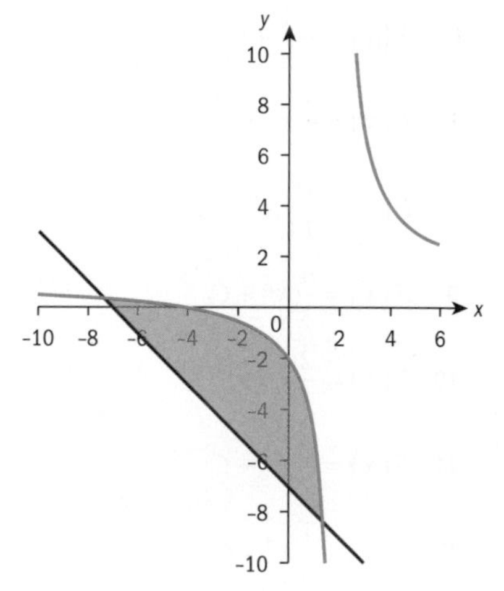

9 $p = 4, q = 6$

10 **a** $(0, 0), (4, 2)$

b **i** $\int_0^4\left(\sqrt{x}-\frac{1}{2}x\right)dx$

ii $\frac{4}{3}$ or 1.33

c **i** $\int_0^k\left(\sqrt{x}-\frac{1}{2}x\right)\mathrm{d}x$

ii $\frac{2}{3}k^{\frac{3}{2}}-\frac{1}{4}k^2$

iii $k = 1.51$

Exercise 10K

1 $\int_{-2}^{0}\left(\left(x^2-2x\right)-\left(10x+x^2-3x^3\right)\right)dx+\int_{0}^{2}\left(\left(10x+x^2-3x^3\right)-\left(x^2-2x\right)\right)dx=24$

2 $\int_{0}^{1}\left(\left(x^3-3x^2+3x+1\right)-\left(x+1\right)\right)dx+\int_{1}^{2}\left(\left(x+1\right)-\left(x^3-3x^2+3x+1\right)\right)dx=0.5$

3 $\int_{-1.51677}^{0}\left(\left(x^3-2x\right)-3xe^{-x^2}\right)dx+\int_{0}^{1.51677}\left(3xe^{-x^2}-\left(x^3-2x\right)\right)dx\approx 4.65$

4 $\int_{-4}^{-1.41421}\left(\left(-x^4+16x^2\right)-\left(x^4-20x^2+64\right)\right)dx$

$+\int_{-1.41421}^{1.41421}\left(\left(x^4-20x^2+64\right)-\left(-x^4+16x^2\right)\right)dx$

$\int_{1.41421}^{4}\left(\left(-x^4+16x^2\right)-\left(x^4-20x^2+64\right)\right)dx\approx 440$

5 a $P(-3, 3.5)$

b $f'(x)=x$

$m=f'(-3)=-3$

$y-3.5=-3\left(x-(-3)\right)$

$y-3.5=-3x-9$

$h(x)=-3x-5.5$

c $Q(-1.10, -2.21)$

d i

$\int_{-3}^{-1.09808}\left(\left(\frac{1}{2}x^2-1\right)-(-3x-5.5)\right)dx+\int_{-1.09808}^{0}\left(\left(\frac{1}{2}x^2-1\right)-\left(-x^2-1\right)\right)dx$

ii 1.81

Chapter review

1 a $\frac{1}{9}x^9+C$

b x^5-2x^3+7x+C

c $\frac{10}{13}x^{\frac{13}{10}}+C$

d $-\frac{1}{2x^8}+C$

e $x^4+2\ln|x|+C$

f $4e^x+C$

g $4x^{\frac{3}{2}}+2x+C$

h $\frac{1}{5}x^5+2x^3+9x+C$

i $\frac{1}{28}(4x+5)^7+C$

j $2e^{3x+2}+C$

k $\frac{1}{6}\ln|6x-7|+C$

l $\frac{3}{2}x^2+C$

2 a 10 b $\frac{28}{3}$ c 12

d 20 e 80 f $\frac{e^{20}-e^{12}}{4}$

3 a 20 b 4 c 22

4 $f(x)=x^4+2x+4$

5 a -8π b 5π

6 $\frac{32}{3}$

7 a 7530 b 21.1

8 a $y=3x-2$ b $(-2, -8)$

c 6.75

Exam-style questions

9 a $f'(x)=\frac{2x(2x)-(x^2+1)(2)}{(2x)^2}$

$=\frac{2x^2-2}{4x^2}=\frac{x^2-1}{2x}$ (3 marks)

b

$\int\frac{x^2+1}{2x}dx=\int\left(\frac{1}{2}x+\frac{1}{2x}\right)dx$

$=\frac{1}{4}x^2+\frac{1}{2}\ln|x|+C$

(3 marks)

10 a $a=\frac{dv}{dt}=-3\,m\,s^{-2}$ (2 marks)

b

$s(t)=\int(40-3t)\,dt=40t-\frac{3t^2}{2}+C$

(2 marks)

$s(1)=10\Rightarrow 40-\frac{3}{2}+C$

$=10\Rightarrow C=-\frac{57}{2}$ (1 mark)

$s(t)=40t-\frac{3t^2}{2}-\frac{57}{2}$ (1 mark)

11 a $A_{region1}=\frac{2\times 2}{2}=2$ (2 marks)

$A_{region2}=\frac{2\times 3}{2}=3$ (2 marks)

b $A_{region3}=3-2=1$ (1 mark)

12 a Use GDC to obtain value of definite integral (1 mark)

$A=\int_0^1 f(x)\,dx=1.12$ (3 sf)

(1 mark)

b i 2.24 (3 s.f.) (1 mark)

ii

$\int_1^2 2f(x-1)\,dx=2\int_0^1 f(x)\,dx$

$=2.24$ (3 s.f.)

(2 marks)

13 a (0,0), (−1.0), (1,0) (2 marks)

b Either:

$f'(x)=1\cdot\left(x^2-1\right)+2x^2=3x^2-1$

(2 marks)

OR

$f(x)=x^3-x$ (1 mark)

$f'(x)=3x^2-1$ (1 mark)

c $f'(x)=0\Rightarrow x=\pm\sqrt{\frac{1}{3}}$

$=\pm 0.577$

(2 marks)

d $\int_{-1}^{1} f(x)\,dx = 0$ (2 marks)

e The function changes sign in the interval $[-1,1]$, so the areas above and below the x-axis cancel out.

(1 mark)

f Either: $S = \int_{-1}^{1} |f(x)|\,dx = 0.5$

(2 marks)

Or (by symmetry)

$S = 2\int_{-1}^{0} f(x)\,dx = 0.5$

(2 marks)

14 a $f(1) = 1^2 = 3 - 2\times 1 = g(1)$

(2 marks)

$f(-3) = (-3)^2 = 3 - 2\times(-3)$
$= g(-3)$

(1 mark)

b

$$\int_{-3}^{1} (3 - 2x - x^2)\,dx = \left[3x - x^2 - \frac{x^3}{3}\right]_{-3}^{1}$$

(3 marks)

$$= \left(3 - 1 - \frac{1}{3}\right) - \left(-9 - 9 + \frac{27}{3}\right) = \frac{32}{3}$$

(1 mark)

15 a $(x-2)^4 = x^4 + 4\times(-2)x^3 + 6\times(-2)^2 x^2 + 4\times(-2)^3 x + (-2)^4$

(3 marks)

$= x^4 - 8x^3 + 24x^2 - 32x + 16$

b $\int (x-2)^4 dx = \int x^4 - 8x^3 + 24x^2 - 32x + 16\,dx$ (1 mark)

$= \frac{x^5}{5} - 2x^4 + 8x^3 - 16x^2 + 16x + C$ (2 marks)

16 Use GDC to obtain graph of $y = |x|$ (2 marks)

Attempt to calculate area of both triangles (1 mark)

$\int_{-1}^{2} |x|\,dx = \frac{1\times 1}{2} + \frac{2\times 2}{2} = 2.5$ (1 mark)

Another solution is:

$$\int_{-1}^{2} |x|\,dx = \int_{-1}^{0}(-x)\,dx + \int_{0}^{2} x\,dx = \frac{5}{2}$$

$$= \left[-\frac{1}{2}x^2\right]_{-1}^{0} + \left[\frac{1}{2}x^2\right]_{0}^{2} = \left(0 + \frac{1}{2}\right) + (2 - 0) = \frac{5}{2}$$

17 a

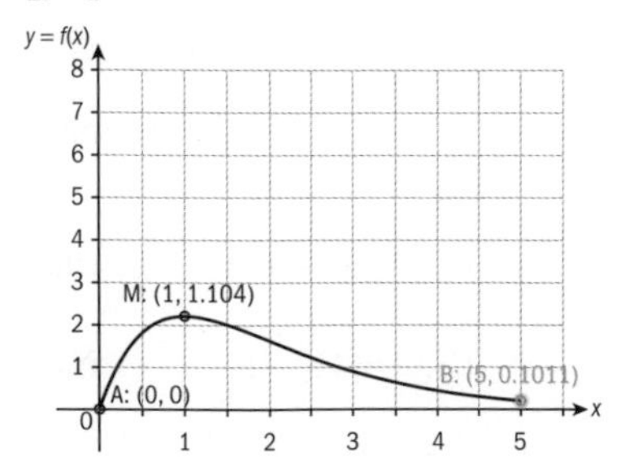

(1 mark for shape; 1 mark for domain; 1 mark for end-points coordinates; 1 mark for maximum point and its coordinates.)

b $0 \le y \le 1.104$ (1 mark)

c $m = \frac{0.1011}{5} = 0.0202...$

(1 mark)

AB contains the origin

(1 mark)

$y = 0.0202x$ (1 mark)

d $f'(x) = 3e^{-x} + 3x(-e^{-x})$
$= (3 - 3x)e^{-x}$

(2 marks)

e Solve

$f'(x) = 0.0202... \Rightarrow x = 0.98201...$

(1 mark)

$f(0.98201...) = 1.10345...$

(1 mark)

$y - 1.10 = 0.0202(x - 0.982)$

(1 mark)

f Use of GDC to calculate

(1 mark)

$\int_{0}^{5} (3xe^{x} - 0.0202x)\,dx = 2.63$

(2 marks)

18 a

x	$\frac{1}{2}$	1	$\frac{3}{2}$
$f(x)$	$\frac{15}{8}$	3	$\frac{21}{8}$

(1 correct: 1 mark; all correct: 2 marks)

b

(1 mark for shape; 1 mark for domain; 1 mark for intercepts)

c Either

$$A = \left(0\times\frac{1}{2}\right) + \left(\frac{15}{8}\times\frac{1}{2}\right) + \left(3\times\frac{1}{2}\right) + \left(\frac{21}{8}\times\frac{1}{2}\right) = \frac{15}{4}(= 3.75)$$

(4 marks)

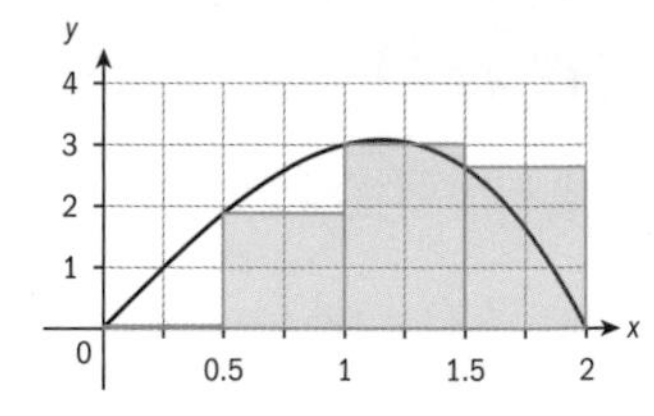

Or

$$A = \left(\frac{15}{8}\times\frac{1}{2}\right) + \left(3\times\frac{1}{2}\right) + \left(\frac{21}{8}\times\frac{1}{2}\right) + \left(0\times\frac{1}{2}\right) = \frac{15}{4}(= 3.75)$$

(4 marks)

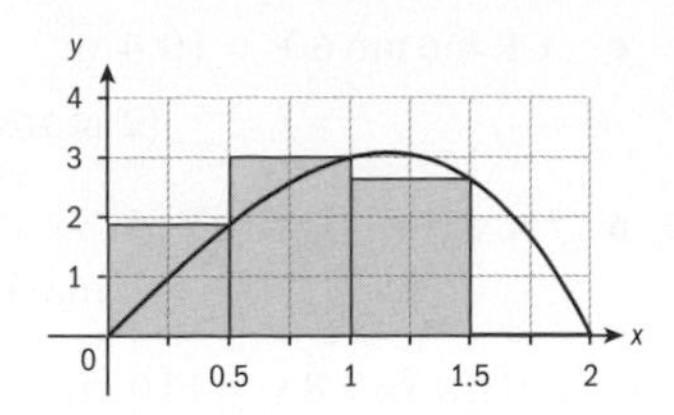

d $\int_0^2 (4x - x^3)\,dx = \left[2x^2 - \frac{x^4}{4}\right]_0^2$ (2 marks)

$= 8 - 4 = 4$ (1 mark)

Chapter 11

Skills check

1 a 15.8 b 4.90

2 a i 12 600 cm²
ii 1.26 m²
b 12 566 litres

Exercise 11A

1 a (3, 0, 0) b (3, 4, 0)
c (3, 0, 2) d (3, 4, 2)
e (1.5, 2, 1) f 5.4

2 a (0.5, 1.5, 3) b (−3, 3, 7)
c (0.5, −0.5, −6)
d (−1.85, −0.15, 10)

3 a 4.47 b 6.56
c 10.2 d 4.47

4 a 14.1 b 6.40
c 5.39 d 5

Exercise 11B

1 a 1440 cm² b 66.4 cm²
c 155 cm² d 283 cm²
e 377 cm² f 201 cm²

2 a SA = 314 cm²
Volume = 524 cm³
b SA = 28.3 cm²
Volume = 14.1 cm³

3 a 64.0 cm³ b 240 cm³
c 105 cm³

4 a 204 cm² b 283 cm²
c 314 cm³

5 a 56.5 cm² b 84.8 cm²
c 56.5 cm³

6 a 302 cm³
b 236 cm²

7 36 700 litres

8 170 cm³

Exercise 11C

1 a 44.7° b 30.5°
c 67.4° d 56.6°

2 a 25.0 b 42.5
c 6.76 d 15.0

3 $h = 5.60$ $a = 29.2°$

4 $p = 28.3$ $q = 31.8$

5 7.45

6 10.4 cm

7 a 3000 cm³ b 32.5 cm
c 67.4° d 31.6
e 71.7°

8 a 2 592 274.5 m³
b 186 m
c 52°
d 139 000 cm² e 42.0°

Exercise 11D

1 5.60 m 2 41.0 m

3 a 15.5° b 2.89 m

4 97.1 m 5 8.24 m

6 a 20.2 km b 14.7 km

7 47.3 km 8 301 m

9 260 m 10 1857 m

Exercise 11E

1 a 23.6 cm² b 61.4 m²
c 2.74 cm² d 13.9 cm²
e 104 cm² f 84.3 cm²
g 47.6 cm²

2 a 12.6 cm² b 30°

3 208 cm²

4 173 cm²

5 38.0 m².

6 2

Exercise 11F

1 a 17.0° b 27.2°
c 18.8° d 15.0°
e 43.1° f 31.2°

2 a 33.4 cm b 8.05
c 6.02 d 13.7
e 22.6 f 7.24

3 18.1 m 4 9.35 km

5 9.31 km 6 9.96 m

7 15.4 m 8 4.90 mm²

Exercise 11G

1 46.0° and 134°

2 34.8° and 145.2°

3 62.1° and 117.9°

4 53.5° and 126.5°

5 150°

Exercise 11H

1 a 11.1 cm b 36.4 cm
c 23.9 m d 10.4 m
e 18.6 cm f 45.0 m

2 a 97.3° b 101°
c 159° d 34.9°
e 51.3° f 26.9°

3 a 29.0 b 26.2 cm²

4 89 km

5 a 29.9 m and 8.56 m
b 77.0 m

6 22.6 km

Exercise 11I

1 82.6 cm²

2 1230.8 million km

3 a 11.8 m b 6.85 m
c 102 m²

4 a 120° b 104 m²
c 53.1°
d Angle $ABC = 105.1°$

5 a 19.7° b 60.8 m

6 a 6.67 cm b 48.3°
c 100° d 8.80 cm
e 25.6 cm²

7 7.06 m

8 **a** 124° **b** 178 km

c 281°

9 **a** 278 km **b** 292°

10 680 km, 236°

11 **a** 21.2 cm **b** 76.0°

c 95.6°

12 **a** $-\frac{11}{24}$

b The cosine is negative

Chapter review

1 volume = $64\,m^3$
surface area = $144\,m^2$

2 volume = $96\pi\,cm^3$
surface area = $96\pi\,cm^2$

3 $16\pi\,m^2$

4 **a** $168\,cm^3$ **b** $151\,cm^2$

5 **a** $1294.14\,cm^3$ **b** 6

c $431\,cm^3$ **d** $0.000\,431\,m^3$

6 **a** $74.2\,m^3$ **b** $108\,m^2$

7 **a** $36\,cm^3$ **b** 432 g

c $OC = \frac{1}{2}AC$

$AC = \sqrt{AB^2 + BC^2}$
$= \sqrt{6^2 + 6^2} = 8.49$ cm

Then $OC = \frac{1}{2}AC = \frac{1}{2} \times 8.49$
$= 4.24$ cm

Then $VC = \sqrt{VO^2 + OC^2}$
$= \sqrt{3^2 + 4.24^2}$
$= 5.20$ cm

d 70.5° **e** $87.0\,cm^2$

8 **a** 5 cm **b** (1, 2.5, 3)

b 10 cm

9 $6\,cm^2$

10 **a** $171\,573\,cm^3$ **b** 102 cm

11 **a** $10\sqrt{3}$ m **b** 20 m

12 $36\sqrt{3}\,cm^2$

13 **a** $A = 2(x)(3x) + 2xh + 2(3x)h$
$= 6x^2 + 8xh$

b $h = \frac{300 - 3x^2}{4x}$

c

V = base area × height $= (3x)(x)h$

$V = 3x^2\,\frac{300 - 3x^2}{4x}$

$= \frac{3}{4}x(300 - 3x^2)$

$= \frac{9}{4}x(100 - x^2)$

14 $3.9\,km\,h^{-1}$

15 60.1 m

16 **a** 46.6° **b** $20.3\,cm^2$

17 **a** 21.4° **b** 36.0°

c $259\,m^2$ **b** 25.6 m

18 **a** 20.2 m **b** 242 m

c 86.7° or 93.3°

d 24.1 m

19 **a** 130° **b** 298°

c 39.4 km

20 **a** 68.5° **b** 57.1 m

c 56.8 m

Exam-style questions

21 **a** $S = 4 \times \frac{5 \times 7}{2} + 5^2 = 95\ cm^2$ (2 marks)

b $h = \sqrt{7^2 - 2.5^2} = 6.54$ cm (2 marks)

$V = \frac{1}{3} \times 5^2 \times 6.6.538... = 54.5\ cm^3$ (2 marks)

22 **a** $l = \sqrt{10^2 - 3^2} = 9.54$ m (2 marks)

b $S = 4 \times \frac{6 \times 9.54}{2}$
$= 114.48 = 114\ m^2$ (2 marks)

c

$h = 42 + \sqrt{9.54^2 - 3^2} = 51.1$ m (2 marks)

d $\arccos\left(\frac{3}{10}\right) = 72.5°$ (2 marks)

e $CP = 6\tan 60° = 10.4$ m (2 marks)

23 **a** $l = \sqrt{5^2 + 3^2} = 5.83$ cm (2 marks)

$S = 2 \times (\pi \times 3 \times 5.83) = 110\ cm^2$ (2 marks)

b $\frac{2 \times \frac{1}{3} \times \pi \times 3^2 \times 5}{\pi \times 3.05^2 \times 10.1} \times 100\%$
$= 31.9\%$ (2 marks)

24 **a** $x = \frac{22 - 12}{2} = 5$ (1 mark)

$h = \sqrt{13^2 - 5^2} = 12$ (2 marks)

b $A = \frac{22 + 12}{2} \times 12 = 204\ cm^2$ (2 marks)

c $D = \cos^{-1}\left(\frac{5}{13}\right) = 67.4°$ (2 marks)

d $AC = \sqrt{17^2 + 12^2} = 20.8$ cm (2 marks)

25 **a** $A\hat{B}C = 135°$ (1 mark)

$AC = \sqrt{20^2 + 25^2 - 2 \times 20 \times 25 \times \cos 135°}$ (2 marks)

$AC = 40.8$ km (1 mark)

b $\frac{\sin\hat{C}}{20} = \frac{\sin 135°}{40.8\ ...}$ (1 mark)

$\hat{C} = 20.3°$ (1 mark)

Therefore the bearing of C with respect to A is 234° (1 mark)

26 **a** $FC = \sqrt{8^2 + 10^2 + 6^2} = 10\sqrt{2}$ (3 marks)

b $M\left(\frac{0+8}{2}, \frac{0+0}{2}, \frac{6+6}{2}\right) = (4, 0, 6)$ (2 marks)

c $FM = \sqrt{4^2 + 0^2 + 6^2} = 2\sqrt{13}$ (2 marks)

d $CM = \sqrt{4^2+10^2+0^2} = 2\sqrt{29}$ (1 mark)

$p = 2\sqrt{13} + 2\sqrt{29} + 10\sqrt{2}$ cm (2 marks)

e $\tan M = \frac{6}{4} = \frac{3}{2}$ (1 mark)

$\cos M = \frac{1}{\sqrt{\frac{9}{4}+1}} = \frac{2\sqrt{13}}{13}$ (3 marks)

27 a $A = \frac{1}{2} \times 5 \times 10 \sin 30° = \frac{25}{2}$ (2 marks)

b

$BD^2 = 5^2 + 10^2 - 2 \times 5 \times 10 \cos 30°$ (2 marks)

$BD = \sqrt{125 - 50\sqrt{3}}$ (1 mark)

$BD = \sqrt{25(5 - 2\sqrt{3})}$ (1 mark)

$BD = 5\sqrt{5 - 2\sqrt{3}}$ (1 mark)

c $\frac{\sin C\hat{D}B}{13} = \frac{\sin 45°}{5\sqrt{5-2\sqrt{3}}}$ (2 marks)

$\sin C\hat{D}B = \frac{13\sqrt{2}}{10\sqrt{5-2\sqrt{3}}}$ (1 mark)

d The angle $C\hat{D}B$ can either be acute or obtuse (1 mark)

and the two possible values add up to 180°. (1 mark)

28 a $V = \frac{1}{2} \times \frac{4\pi}{3} \times 3^3 + \pi \times 3^2 \times 7$ (3 marks)

$= 81\pi = 254 \text{ cm}^3$ (3 s.f.) (1 mark)

b

$S = \frac{1}{2} \times 2\pi \times 3^2 + 2\pi \times 3 \times 7 + \pi \times 3^2$ (2 marks)

$= 60\pi = 188 \text{ cm}^2$ (3 s.f.) (1 mark)

29 a $\tan 32° = \frac{30}{x} \Rightarrow x = \frac{30}{\tan 32°} = 48.0$ metres (3 marks)

b $AB = y = \sqrt{(3 + 48.0...)^2 + 30^2} = 59.2$ metres (2 marks)

c $\arctan\left(\frac{30}{51.0\ ...}\right) = 30.5°$ (2 marks)

30 a $BC = \sqrt{48^2 + 57^2 - 2 \times 48 \times 57 \cos 117°} = 89.7$ metres (3 marks)

b $A = \frac{1}{2} \times 48 \times 57 \sin(117°) = 1219$ sq metres (3 s.f.) (2 marks)

c $\frac{\sin B}{48} = \frac{\sin 117°}{89.7} \Rightarrow B = 28.5°$ (3 marks)

Chapter 12

Skills check

1 a $\frac{\sqrt{2}}{2}$ **b** $\sqrt{3}$ **c** $\frac{\sqrt{3}}{2}$

2 a $(-0.618, 0), (1, 0), (1.62, 0)$

b $(0.633, 0)$

3 a $(-1.61, 0.199)$

b $(2.21, 0.792)$

Exercise 12A

1 a $\frac{\pi}{4}$ **b** $\frac{\pi}{3}$ **c** $\frac{3\pi}{2}$

d 2π **e** $\frac{\pi}{10}$ **f** $\frac{5\pi}{4}$

g $\frac{4\pi}{9}$ **h** $\frac{10\pi}{9}$ **i** $\frac{2\pi}{3}$

j $\frac{3\pi}{4}$

2 a 30° **b** 18° **c** 150°

d 540° **e** 63° **f** 144°

g 315° **h** 280° **i** 300°

j 585°

3 a 0.175 **b** 0.698 **c** 0.436

d 5.24 **e** 1.92 **f** 1.31

g 1.48 **h** 0.223 **i** 0.654

j 0.0175

4 a 57.3° **b** 115° **c** 36.1°

d 80.8° **e** 88.8° **f** 172°

g 20.6° **h** 73.3° **i** 0.573°

j 123°

Exercise 12B

1 a i 7π cm **ii** 9π cm

iii $\frac{5\pi}{2}$ cm **iv** $\frac{70\pi}{3}$ cm

b i $49\pi \text{ cm}^2$ **ii** $54\pi \text{ m}^2$

iii $\frac{15\pi}{4} \text{ m}^2$ **iv** $175\pi \text{ cm}^2$

2 $6\sqrt{2}$ cm

3 a $\frac{\pi}{2}$ **b** 42.8 m

4 25.1 units^2

5 20π m

6 1.85 km

Exercise 12C

1 a Negative **b** Negative

c Positive

2 a 144° **b** 130° **c** 95°

d 100° **e** $\frac{2\pi}{3}$ **f** $\frac{4\pi}{5}$

g $\frac{5\pi}{7}$ **h** $\frac{2\pi}{3}$

3 **a** 320° **b** 250° **c** 60°

d 140° **e** $\frac{15\pi}{8}$ **f** $\frac{19\pi}{10}$

g $\frac{\pi}{2}$ **h** $\frac{\pi}{4}$

4 **a** $\frac{\pi}{6}, \frac{5\pi}{6}$ **b** $\frac{\pi}{4}, \frac{7\pi}{4}$

c $\frac{\pi}{3}, \frac{4\pi}{3}$

5 **a** $\frac{15}{17}$ **b** $\frac{15}{8}$

Exercise 12D

1 **a** 53.1°, 306.9°

b 8.63°, 171.4°

c 11.3°, 191.3°

d 142.8°, 322.8°

e 115.5°, 244.5°

2 **a** 0.96, 2.18

b 2.39, 5.53

c 2.79, 3.49

d 0.69, 5.59

e 6.05, 3.37

3 $\frac{\pi}{6}, \frac{5\pi}{6}$

4 **a** 0.93, 4.07

b 2.68, 5.82

c 1.11, 2.36, 4.25, 5.50

d 1.57, 3.67, 5.76

5 **a** −1.27, 1.27, 5.02, 7.55

b 1.01, 4.16, 7.30, 10.4

c −1.11, 2.03

d −5.82, −4.39, −2.68, −1.25

6 31°, 211°

Exercise 12E

1 **a** −165°, −15°, 15°, 165°

b −130°, −110°, −10°, 10°, 110°, 130°

c −140°, −100°, −20°, 20°, 100°, 140°

d −90°

2 **a** $\frac{\pi}{9}, \frac{2\pi}{9}, \frac{7\pi}{9}, \frac{8\pi}{9}, \frac{13\pi}{9}, \frac{14\pi}{9}$

b $0, \frac{2\pi}{3}, \frac{4\pi}{3}, 2\pi$ **c** $\frac{\pi}{2}, \frac{3\pi}{2}$

d $\frac{3\pi}{4}$

Exercise 12F

1 **a** $\sin 10°$ **b** $\sin \pi$

c $\sin 8\pi$ **d** $\cos 0.8$

e $\cos 12$ **f** $\frac{\pi}{2}$

2 **a** $\frac{2\sqrt{2}}{3}$ **b** $\frac{4\sqrt{2}}{9}$

c $\frac{7}{9}$ **d** $\frac{4\sqrt{2}}{7}$

3 **a** $\frac{\sqrt{3}}{2}$ **b** $-\frac{\sqrt{3}}{2}$

c $-\frac{1}{2}$ **d** $\sqrt{3}$

4 **a** $\frac{\sqrt{63}}{32}$ **b** $\frac{31}{32}$

c $\frac{\sqrt{63}}{31}$ **d** $\frac{31\sqrt{63}}{512}$

5 **a** $0, \frac{\pi}{3}, \pi, \frac{5\pi}{3}, 2\pi$

b $\frac{\pi}{2}, \frac{7\pi}{6}, \frac{11\pi}{6}$

c $\frac{\pi}{3}, \frac{\pi}{2}, \frac{2\pi}{3}, \frac{3\pi}{2}$

d $\frac{\pi}{4}, \frac{\pi}{2}, \frac{3\pi}{4}, \frac{5\pi}{4}, \frac{3\pi}{2}, \frac{7\pi}{4}$

e $0, \frac{2\pi}{3}, \frac{4\pi}{3}, 2\pi$

6 **a** $a = 16, b = 2$

b $\frac{\pi}{12}, \frac{5\pi}{12}$

7 $\frac{32}{3}$

Exercise 12G

1 **a**

b

c

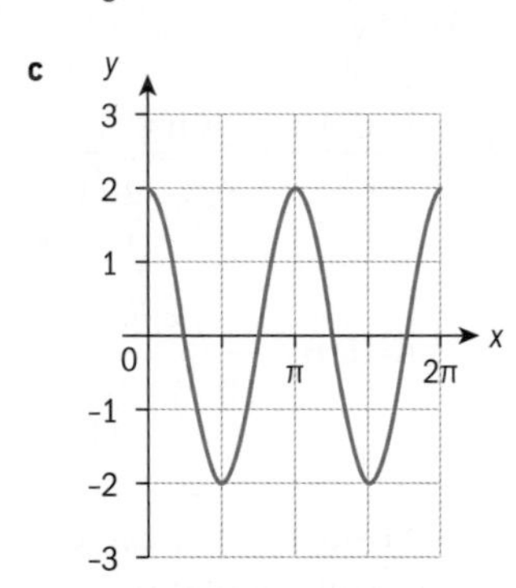

d

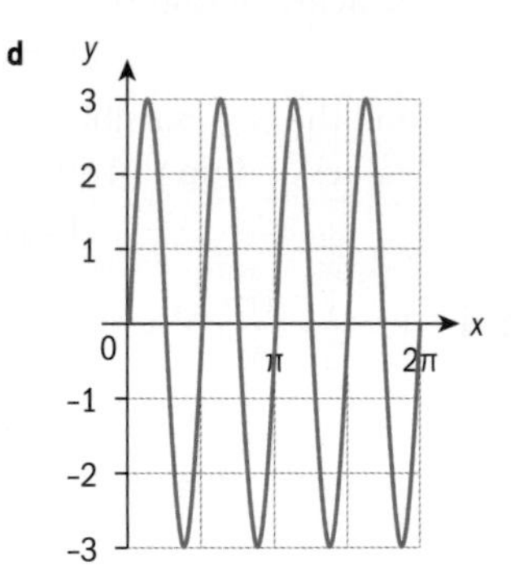

2 **a** Amplitude = 1, period = $\frac{2\pi}{3}$

b Amplitude = 0.5, period = π

c Amplitude = 4, period = $\frac{2\pi}{3}$

d Amplitude = $\frac{1}{2}$, period = 6π

3 a Amplitude = 1, period = π, $y = \sin 2x$

b Amplitude = 3, period = π $y = 3\cos 2x$

c Amplitude = 2, period = 2π $y = -2\cos x$

d Amplitude = 2, period = 2π $y = 2\sin(-x)$

4 a 6

b 4

Exercise 12H

1 a iv b ii

c i d iii

2 a $y = 2\sin x + 3$

b $y = \cos(x - \pi) + 1$

c $y = 2\cos(x) - 2$

d $y = 0.5\sin(3x) - 1$

3 a

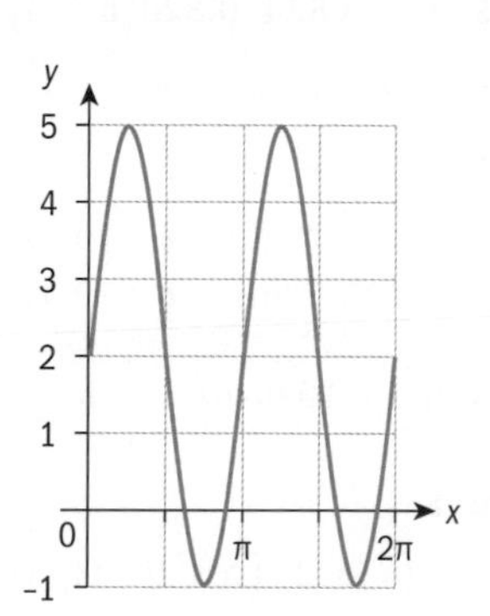

b

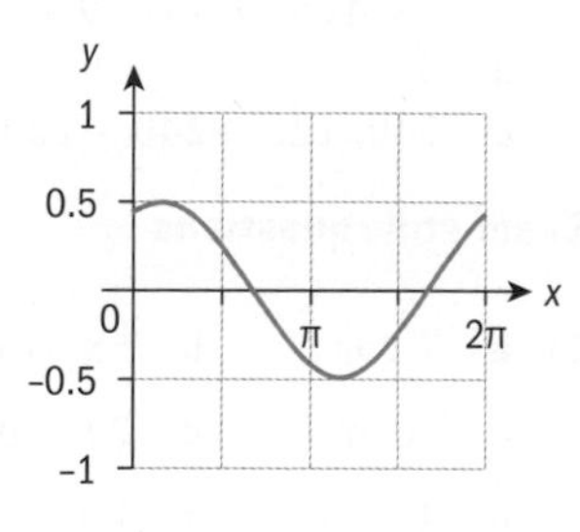

c

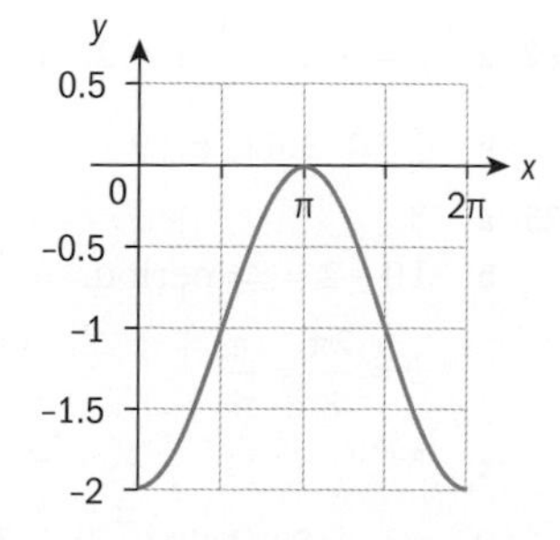

d

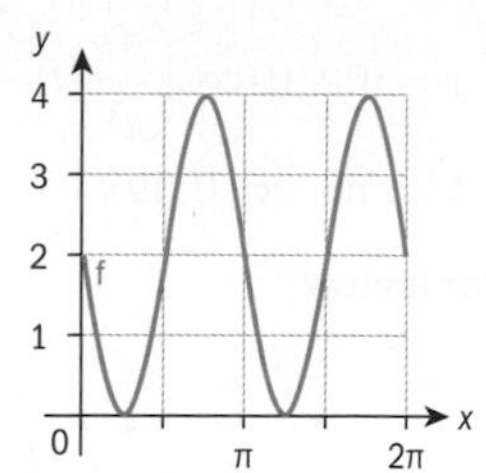

e

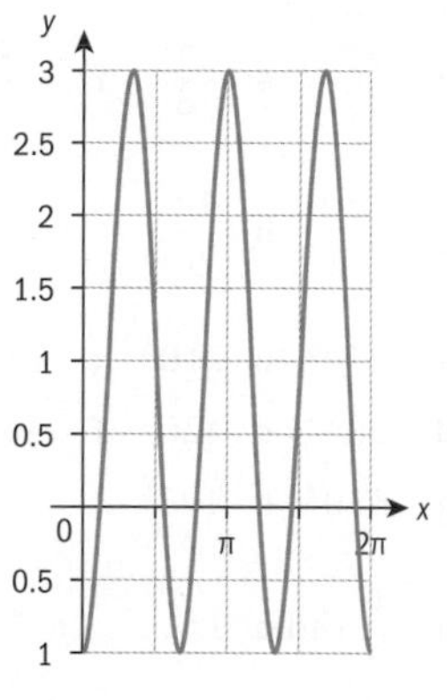

4 $(-0.824, 0)$, $(0.824, 0)$

5 $(-2.38, 0)$

6 $\frac{\pi}{6}, \frac{5\pi}{6}, \frac{13\pi}{6}, \frac{17\pi}{6}$ 7 $(-1.29, 0)$

8 a $\frac{\pi}{2}$

b The graph of $y = \cos x$ may be translated $\frac{\pi}{2}$ to the right to become the graph of $y = \sin x$.

9 a

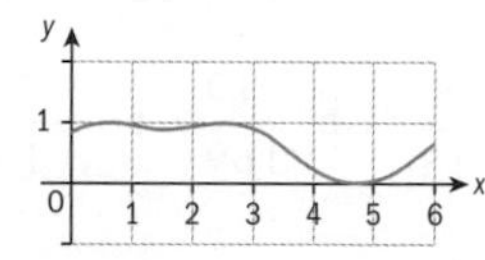

b 0.6075, 1.571, 2.534, 4.712

10 a $(2.36, -1)$ b 4π

c 0, 1.26, 3.77, 4.19

Exercise 12I

1 a

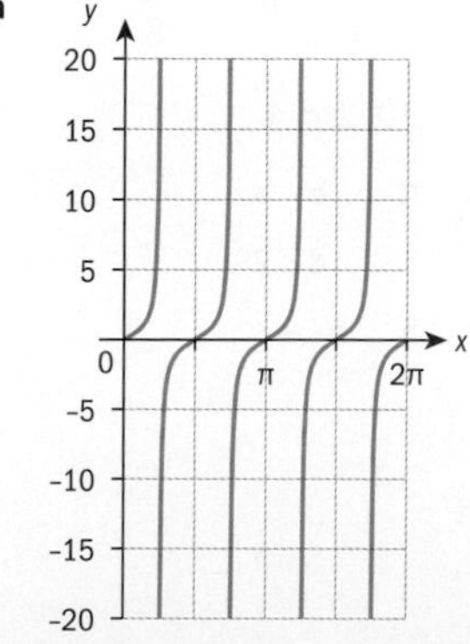

b

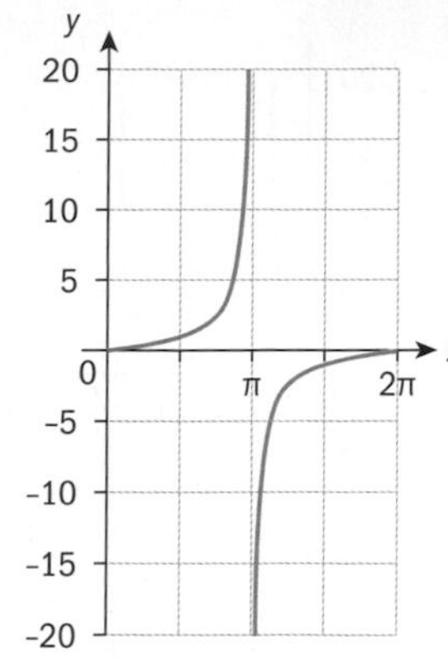

c

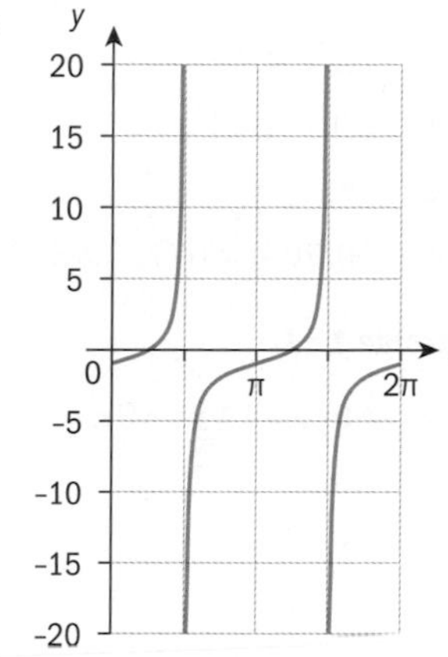

d

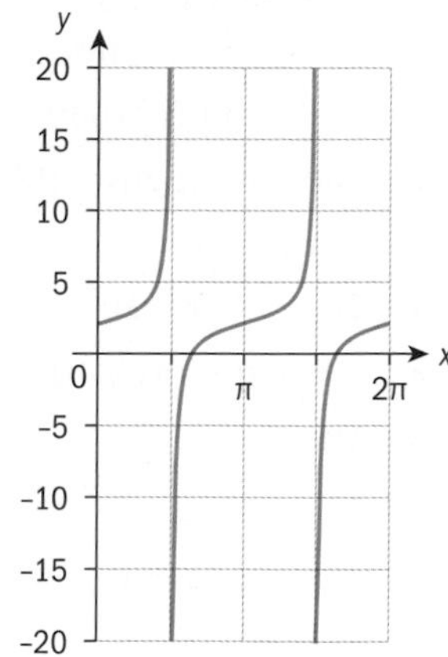

e

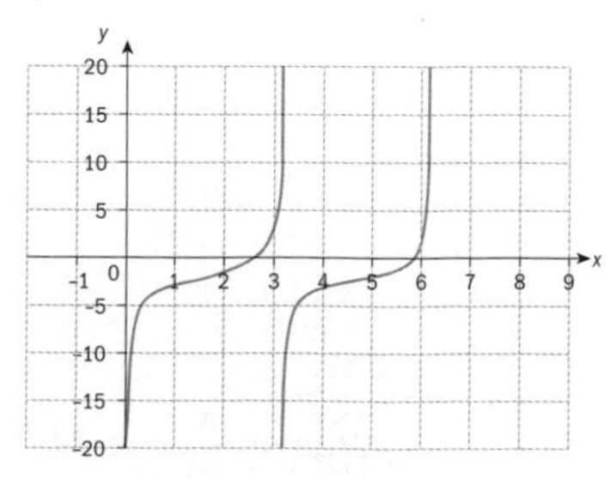

f

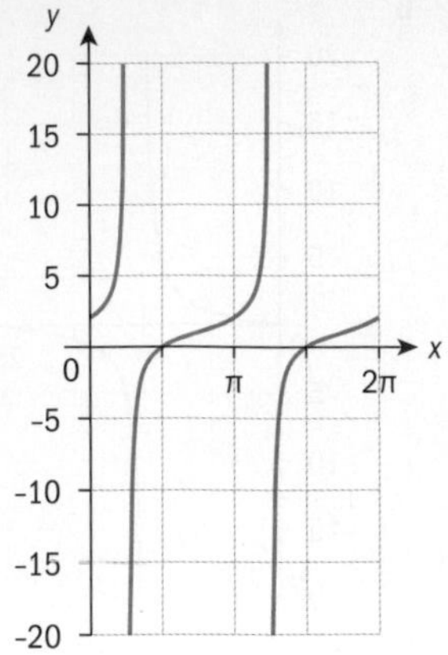

2 **a** 1.25, 4.39

b 1.13, 4.53

c $-\pi$, 0, π

d −0.903, 0.677, 1.98, 2.61

Exercise 12J

1 **a** Max = 12 m Min = 2 m

b 8 am

c 2 am

d 11.3 m

e 0:46am 3:14 am
12:46 pm 3:14 pm

2 **a** 13 000 on February 1st

b 7000 on August 9th

c 10 212

3 **a** 20 **b** 10

c $y = 10\sin 0.5x + 10$

d 18

4 **a** 35 **b** 5

c 15 **d** $a = 15, c = 20$

e 4 **f** $\frac{\pi}{2}$

g 0.535

5 **a** $a = -12$ $b = \frac{\pi}{20}$ $c = 14$

b $\frac{2\pi}{3}$ **c** 13.3 s

6 **a** $y = -20\cos\left(\frac{\pi}{20}x\right) + 21$

b 10.64 s

7 **a** $y = 50 + 10\cos\left(\frac{2\pi}{3}(t - 0.3)\right)$

b 43.3 m **c** 0.09 s

Chapter review

1 **a** $\frac{\pi}{6}$ **b** $\frac{5\pi}{6}$ **c** $\frac{7\pi}{4}$

d $\frac{2\pi}{3}$ **e** $-\frac{\pi}{9}$ **f** $-\frac{4\pi}{3}$

g $-\frac{3\pi}{2}$ **h** $\frac{4\pi}{5}$

2 **a** 270° **b** 210° **c** −105°

d 20° **e** 420° **f** −66°

g 330° **h** 408°

3 **a** $\frac{6}{5}$ rad **b** 15

4 **a** $\cos\theta = 0.8$ and $\tan\theta = 0.75$

b $-\frac{\pi}{3}, 0, \frac{\pi}{3}$

5 **a** $\frac{7\pi}{6}, \frac{11\pi}{6}$ **b** $\frac{\pi}{4}, \frac{7\pi}{4}$

c $\frac{\pi}{4}, \frac{5\pi}{4}$

6 **a** $a = 4$, $b = 2$

b $\frac{\pi}{12}, \frac{5\pi}{12}, \frac{13\pi}{12}, \frac{17\pi}{12}$

7 **a** $\frac{5}{13}$ **b** $\frac{119}{169}$ **c** $-\frac{5}{13}$

8 **a** $\frac{1}{2}, -1$

b $\sin^{-1}\frac{1}{2} = \frac{\pi}{6}, \frac{5\pi}{6}$

9 **a** $-\frac{\sqrt{7}}{4}$ **b** $-\frac{1}{8}$

10 $a = 30$, $b = 4$

11 **a** π **b** 3

c 3 **d** 2

12 **a** 5 **b** 4

c

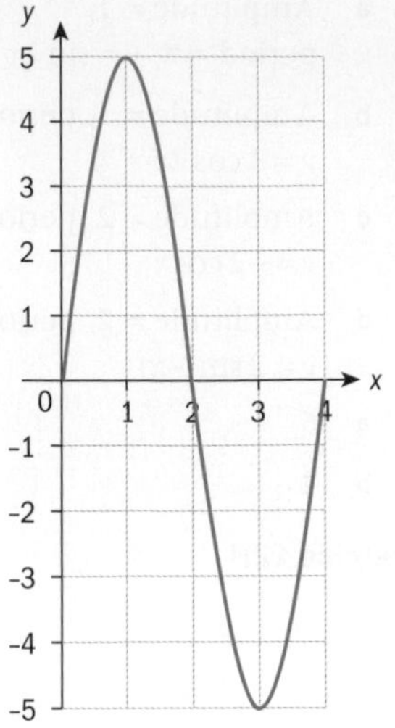

13 $\frac{16\pi}{3} - 16$

14 **a** $\frac{3\pi}{4}$ **b** 2π

15 **a** −0.824, 0.824 **b** 1.14

16 2.89 rad

17 15.0 cm^2

18 **a** 14.7 cm **b** 19.2 cm
c 25.6 cm^2 **d** 29.4 cm^2

19 $f(x) = 10\sin 6x$ **b** $0, \frac{\pi}{6}, \frac{\pi}{3}$

20 **a**

$$f(\theta) = 2\times(\cos^2\theta - 1) + \cos\theta + 3$$
$$= 4\cos^2\theta + \cos\theta + 1$$

b 4

c 240, 120, −240, −120

Exam-style questions

21 **a** 7.85 m **b** 58.9 m^2

c 20 m **d** 25.6 m

e 4.39 m **f** 1.18 s

22 **a** $p = 3$, $q = \frac{4}{3}$, $b = 2$, $d = 1$

b 1.30, 3.41, 6.19

23 **a** 3

b $10 - 2 = 8 =$ period,

$b = \frac{2\pi}{8} = \frac{\pi}{4}$

c 5

d $y = -(5 + 3\sin(\frac{\pi}{4}(x - 3))$

24 **a** $A(t) = (2\cos t - 1)(\cos t - 1)$ (2 marks)

b $(2\cos t - 1)(\cos t - 1) = 0$

$2\cos t - 1 = 0 \Rightarrow \cos t = \frac{1}{2} \Rightarrow t = \frac{\pi}{3}, \frac{5\pi}{3}$ (3 marks)

$\cos t - 1 = 0 \Rightarrow \cos t = 1 \Rightarrow t = 0, 2\pi$ (1 mark)

25 $2\cos^2 x = \sin 2x \Rightarrow$

$2\cos^2 x - 2\sin x \cos x = 0$ (1 mark)

$2\cos x(\cos x - \sin x) = 0$ (1 mark)

$\cos x = 0, \cos x = \sin x$ (1 mark)

$x = \frac{\pi}{2}, x = \frac{3\pi}{2}$ (1 mark)

$x = \frac{\pi}{4}, x = \frac{5\pi}{4}$ (1 mark)

26 **a** Correct attempt to at least one parameter (1 mark)

$a = \frac{14-8}{2} = 3$ (1 mark)

$b = \frac{2\pi}{\pi} = 2$ (1 mark)

$c = \frac{14+8}{2} = 11$ (1 mark)

b

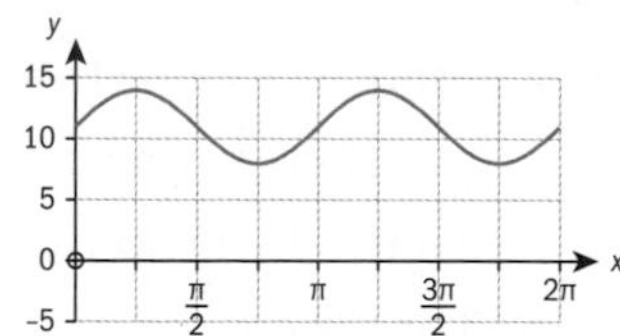

(1 mark for trigonometric scale and correct domain, 1 mark for correct max/min, 1 mark for two complete cycles)

27 **a** $S(x) = \underbrace{\sin^2 2x + \cos^2 2x} + \underbrace{2\sin 2x \cos 2x}$ (3 marks)

$= 1 + \sin 4x$ (1 mark)

b (1 mark for correct shape, 1 mark for 2 cycles, 1 mark for correct max/min)

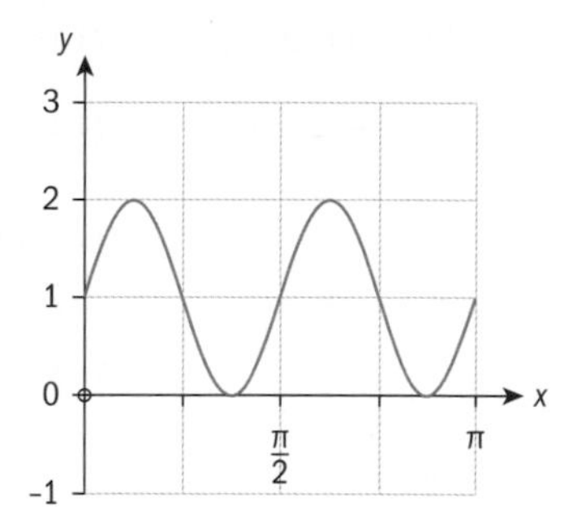

c **i** $\frac{\pi}{2}$ (1 mark)

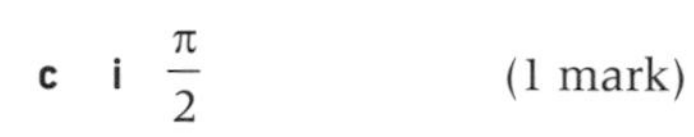

ii $0 \le y \le 2$ (1 mark)

d

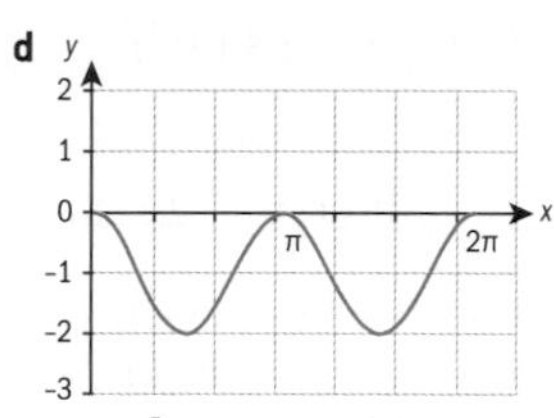

e **i** $\frac{1}{2}$ (1 mark)

ii $p = \pi$ (or any $\pi + 2n\pi, n \in \mathbb{Z}$) (1 mark)

$q = -2$ (1 mark)

28 **a** **i** $A = 2^2 \times \frac{\pi}{3} = \frac{8\pi}{3}$ (1 mark)

ii $l = 2 \times \frac{2\pi}{3} = \frac{4\pi}{3}$ (1 mark)

b **i** $r\theta = \frac{\pi}{3}$ (1 mark)

$r^2 \frac{\theta}{2} = \pi$ (1 mark)

Solve simultaneously (1 mark)

$r = 6$ cm (1 mark)

ii $\theta = \frac{\pi}{18}$ (1 mark)

29 **a** **i** $-1 \le y \le 3$ (1 mark)

ii 2 (1 mark)

b $a = -2$ (1 mark)

$b = \frac{2\pi}{2} = \pi$ (2 marks)

$c = 1$ (1 mark)

c $-2\cos \pi x + 1 = 0 \Rightarrow \cos \pi x = \frac{1}{2}$ (1 mark)

$\pi x \in \left\{-\frac{\pi}{3}, \frac{\pi}{3}, \frac{5\pi}{3}, \frac{7\pi}{3}, \right\}$ (1 mark)

$x \in \left\{-\frac{1}{3}, \frac{1}{3}, \frac{5}{3}, \frac{7}{3}, \right\}$ (1 mark)

30 **a** **i** $x = 0, x = \frac{\pi}{2}, x = \pi$ (1 mark)

ii $\frac{\pi}{2}$ (1 mark)

iii IR (1 mark)

b $b = 2$ (1 mark)

$d = 1$ (1 mark)

c The first point of inflexion occurs at $x = \frac{\pi}{4}$ (1 mark)

d $f\left(\frac{\pi}{8}\right) = -2 \Rightarrow f(x) = a\tan\left(2\left(\frac{\pi}{8} - \frac{\pi}{4}\right)\right) + 1 = -2$ (1 mark)

$a \underbrace{\tan\left(-\frac{\pi}{4}\right)}_{-1} = -3$ (1 mark)

$a = 3$ (1 mark)

e $\frac{\pi}{8}$ (1 mark)

$\frac{5\pi}{8}$ and $\frac{9\pi}{8}$ (1 mark)

31 a $-2\cos^2 x+\sin x+3=-2(1-\sin^2 x)+\sin x+3$ (1 mark)

$=2\sin^2 x+\sin x+1$ (1 mark)

b $-2\cos^2 x+\sin x+3=2 \Rightarrow 2\sin^2 x+\sin x+1=2$ (1 mark)

$2\sin^2 x+\sin x-1=0 \Rightarrow (2\sin x-1)(\sin x+1)=0$ (1 mark)

$\sin x=\frac{1}{2}, \sin x=-1$ (1 mark)

$x\in\left\{-\frac{11\pi}{6}, -\frac{\pi}{2}, -\frac{7\pi}{6}, \frac{\pi}{6}, \frac{3\pi}{2}, \frac{5\pi}{6}\right\}$ (2 marks)

Award (1 mark for two correct solutions)

Chapter 13

Skills check

1 a $-\frac{\sqrt{2}}{2}$ **b** -1 **c** $-\frac{\sqrt{3}}{2}$

d $\frac{\sqrt{3}}{2}$ **e** 0 **f** $-\frac{1}{2}$

2 a $\cos 4x$ **b** $3\sin 2x$ **c** e^x

3 a $\frac{12x^2+7}{3(4x^3+7x)^{\frac{2}{3}}}$

b $6x^2e^{2x}+6xe^{2x}$

c $\frac{1-2\ln x}{x^3}$

Exercise 13A

1 $f'(x)=4\cos x-3\sin x$

2 $y'=5\cos(5x)$

3 $y'=\frac{1}{3}\cos\left(\frac{x}{3}\right)+3\sin(3x)$

4 $f'(x)=\frac{3\sin x}{\cos^2 x}$

5 $h'(t)=3\sin^2 t\cos t$

6 $y'=\frac{\cos\sqrt[3]{x}}{3\sqrt[3]{x^2}}$

7 $y'=\frac{\pi}{x^2}\sin\frac{\pi}{x}$

8 $g'(x)=-\frac{1}{\sin^2 x}$

9 $f'(x)=0$

10 $y'=2\cos(4x)$ or $2\cos^2(2x)-2\sin^2(2x)$

11 a $(g\circ f)(x)=\cos(4x^3)$

b $\frac{d}{dx}[(g\circ f)(x)]=-12x^2\sin(4x^3)$

c $r'(x)=2x\cos(4x^3)-12x^4\sin(4x^3)$

12 a i $f'(x)=\cos x$

ii $f''(x)=-\sin x$

iii $f'''(x)=-\cos x$

iv $f^{(4)}(x)=\sin x$

b 4, 8, 12

c i $f^{(80)}(x)=\sin x$

ii $f^{(42)}(x)=-\sin x$

Exercise 13B

1 Tangent: $y+1=-2(x-\pi)$

Normal: $y+1=\frac{1}{2}(x-\pi)$

2 Tangent: $y=1$; normal: $x=\frac{\pi}{3}$

3 a $\frac{\sqrt{2}}{2}$ **b** $f'(x)=\pi\cos(\pi x)$

c $y-\frac{\sqrt{2}}{2}=\frac{\pi\sqrt{2}}{2}\left(x-\frac{1}{4}\right)$

4 $x=\frac{\pi}{4}, \frac{3\pi}{4}$

Exercise 13C

1 $f'(x)=12\cos\left(4x-\frac{\pi}{6}\right)+5$

2 $y'=(12x^2+4x+1)e^{4x^3+2x^2+x}$

3 $y'=-\frac{1}{\cos x-1}$

4 $f'(x)=e^x(x+1)^2$

5 $h'(t)=-2(\sin t)(e^{\cos t})$

6 $y'=3e^{5x}\cos 3x+5e^{5x}\sin 3x$

7 $y'=x\cos x$

8 $f(x)=-2xe^{x^2}\sin(e^{x^2})$

9 $f'(x)=3\ln(3x)\cos(3x)+\frac{\sin 3x}{x}$

10 $f'(x)=\frac{3\cos 3x}{\sin 3x}$ or $\frac{3}{\tan 3x}$

11 a $f'(x)=-\sin x(e^{\cos x})$

b increasing: $\pi<x<2\pi$
decreasing: $0<x<\pi$

c local minimum at (π, e^{-1})

12 a $f'(x)=1-\sin x$; $f''(x)=-\cos x$

b Concave up when

$f''(x)>0 \Rightarrow x\in\left(\frac{\pi}{2}, \frac{3\pi}{2}\right)$

Concave down when $f''(x)<0$

$\Rightarrow x\in\left(0, \frac{\pi}{2}\right)\cup\left(\frac{3\pi}{2}, 2\pi\right)$

c Inflexion points: $\left(\frac{\pi}{2}, \frac{\pi}{2}\right)$, $\left(\frac{3\pi}{2}, \frac{3\pi}{2}\right)$

13 a

$f'(x)=-2\sin(2x)+2\cos x(-\sin x)$
$=-2\sin(2x)-(2\sin x\cos x)$
$=-2\sin(2x)-\sin(2x)$
$=-3\sin(2x)$

b $(\frac{\pi}{2}, -1)$

c Inflexion points: $\left(\frac{\pi}{4}, \frac{1}{2}\right)$, $\left(\frac{3\pi}{4}, \frac{1}{2}\right)$

d

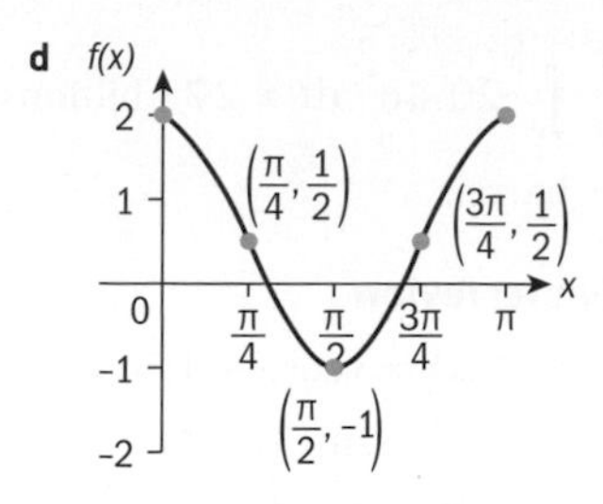

Exercise 13D

1 $C'(120) = 70$; this means it cost 70 Euros to produce the 121st table.

2 **a** $v(t) = s'(t) = \dfrac{e^t(1) - t(e^t)}{(e^t)^2}$

$= \dfrac{e^t(1-t)}{e^{2t}} = \dfrac{1-t}{e^t}$

b Velocity $= -\dfrac{1}{e^2}$; speed $= \dfrac{1}{e^2}$

c $t = 1$; the particle changes direction at 1 second, because velocity changes from positive to negative when $t = 1$.

3 $\dfrac{6\sqrt{3}-3}{2}$

4 **a** 76.7 bacteria per day

b $P'(t) = 24e^{0.2t}$

c 177 bacteria per day; at day 10 the number of bacteria are increasing at a rate of 177 bacteria per day.

5 **a** $P'(200) = 6$; the profit gained by selling the 201st unit of the chemical is 6 Euros.

b $C(x) = 0.00005x^3 - 2x + 196$

c 4 Euros

6 **a** 3.19 s

b $-0.408\ \text{m s}^{-1}$ (3sf)

c $v(t) = -9.8t + 15.2$

d 1.55 s; the ball reaches its maximum height and changes direction at 1.55 s.

Exercise 13E

1 The order size is 100 items; justification: $x \geq 0$ and $C'(x) = 0 \Rightarrow x = 100$

$C''(100) = \dfrac{20000}{100^3} > 0$

$\Rightarrow$ minimum

2 **a** $b = 2x = 8\cos\theta$

$h = 2y = 8\sin\theta$

$A = bh = (8\cos\theta)(8\sin\theta)$

$= 32(2\sin\theta\cos\theta) = 32\sin 2\theta$

b $\dfrac{dA}{dt} = 64\cos 2\theta$; maximum when $\theta = \dfrac{\pi}{4}$

c $\dfrac{d^2A}{d\theta^2} = -128\sin 2\theta$;

$\left.\dfrac{d^2A}{d\theta^2}\right|_{\theta=\frac{\pi}{4}} = -128\sin\left(\dfrac{\pi}{2}\right)$

$= -128 < 0 \Rightarrow$ maximum

3 **a** $r = \dfrac{2}{3}(6-h)$

b $V = \pi r^2 h = \pi\left(\dfrac{2}{3}(6-h)\right)^2(h)$

$= \dfrac{4\pi}{9}(36 - 12h + h^2)(h)$

$= \dfrac{4\pi}{9}(36h - 12h^2 + h^3)$

c $\dfrac{dV}{dh} = \dfrac{4\pi}{9}(36 - 24h + 3h^2)$

$= \dfrac{4\pi}{3}(12 - 8h + h^2)$

$\dfrac{d^2V}{dh^2} = \dfrac{4\pi}{3}(-8 + 2h)$

$= \dfrac{8\pi}{3}(h-4)$

d $h = 2, r = \dfrac{8}{3}$;

$\left.\dfrac{d^2V}{dh^2}\right|_{h=2} = -\dfrac{16\pi}{3} < 0 \Rightarrow$

maximum

4 **a** $P(x) = 4\sqrt{x} - 2x^2$

b $P'(x) = \dfrac{2}{\sqrt{x}} - 4x$

c $x = \sqrt[3]{\dfrac{1}{4}}$ or 0.630

d 2.38 thousand dollars or $2381.10

5 **a** $C = 14x + \dfrac{5400}{x}$ **b** $550

Exercise 13F

1 $-5\cos x + C$

2 $4\sin x + 2\cos x + C$

3 $\dfrac{1}{7}\sin(7x) + C$

4 $3\sin(2x) + C$

5 $-\dfrac{1}{5}\cos(5x+3) + C$

6 $\dfrac{1}{4}x^4 - \dfrac{1}{2}\cos(2x) + C$

7 $2\sin\left(\dfrac{x}{2}\right) + C$

8 $-\cos(2\pi x) + C$

Exercise 13G

1 $\dfrac{1}{3}(5x^3 + 4x)^3 + C$

2 $-\cos(3x^5) + C$

3 $-\dfrac{1}{2x^2 + 3x + 1} + C$

4 $e^{x^2 + 7x} + C$

5 $\dfrac{1}{2}(x^4 - 3x^2)^4 + C$

6 $2e^{\sqrt{x}} + C$

7 $\dfrac{1}{2}\sin(2x^2 - 2x) + C$

8 $-\dfrac{1}{5}\cos^5(x) + C$

9 $-\cos(\ln x) + C$

10 $\dfrac{2}{9}\left(e^{x^3} + 5\right)^{\frac{3}{2}} + C$

11 $f(x) = e^{\sin x} + 11$

12 $f(x) = \ln(2x^2 + e^2) + 3$

Exercise 13H

1 $\dfrac{\sqrt{2}}{2}$ **2** $\sqrt{3} + 1$

3 $\int_0^6 u^3 du = 324$

4 $-\int_{\frac{\sqrt{3}}{2}}^{\frac{1}{2}} u^3\, du = \dfrac{1}{8}$

5 $\int_2^9 u^{-\frac{1}{2}} du = 6 - 2\sqrt{2}$

6 $\int_{\frac{\pi}{4}}^{\frac{\pi}{3}} \sin u \, du = \frac{\sqrt{2}-1}{2}$

7 **a** $k = \pi$ or 3.14 **b** 12.1

8 11.4

Exercise 13I

1 **a** $v(t) = -t^2 + 8t - 12$ **b** $\int_0^9 (-t^2 + 8t - 12)\,dt = -27\,\text{m}$

c $\int_0^9 |-t^2 + 8t - 12|\,dt = 48.3\,\text{m}$

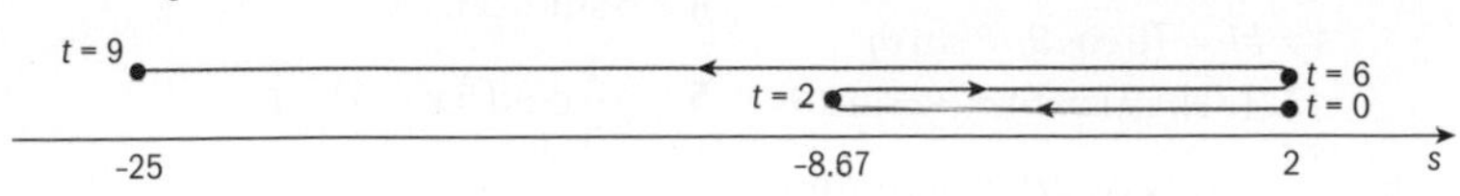

2 **a** $v(t) = 2t - 6$ **b** $\int_0^6 (2t - 6)\,dt = 0\,\text{m}$ **c** $\int_0^6 |2t - 6|\,dt = 18\,\text{m}$

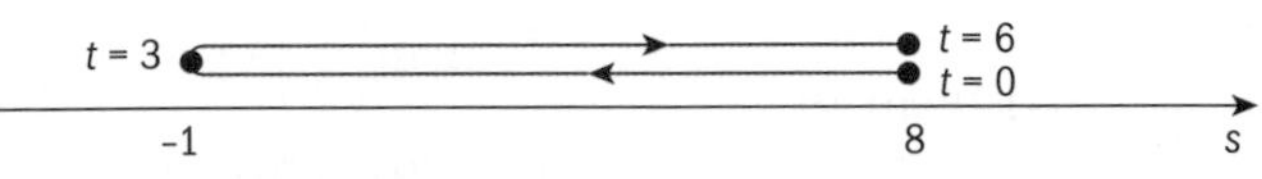

3 **a** $v(t) = 3(t-1)^2$ **b** $\int_0^3 (3(t-1)^2)\,dt = 9\,\text{m}$ **c** $\int_0^3 |3(t-1)^2|\,dt = 9\,\text{m}$

t = 0, t = 1, t = 3; -1, 0, 8, s

4 **a** Displacement $= \int_0^8 v(t)\,dt = 22\,\text{m}$; distance $= \int_0^8 |v(t)|\,dt = 22\,\text{m}$

b Displacement $= \int_2^{14} v(t)\,dt = 6\,\text{m}$; distance $= \int_2^{14} |v(t)|\,dt = 30\,\text{m}$

c Displacement $= \int_0^{14} v(t)\,dt = 10\,\text{m}$; distance $= \int_0^{14} |v(t)|\,dt = 34\,\text{m}$

5 **a** $-3\,\text{m}\,\text{s}^{-2}$ **b** $(0, 3)$ or $(5, 7)$

c $\int_0^7 |v(t)|\,dt = 16.5\,\text{m}$

6 **a** $4\,\text{m}\,\text{s}^{-2}$ **b** $s(t) = \frac{1}{3}t^3 - 16t + 10$

c 32 m

Exercise 13J

1 **a** $v(t) = e^t \cos t - e^t \sin t$

b $a(t) = -2e^t \sin t$

2 **a** $-2\,\text{m}\,\text{s}^{-1}$

b $t = \frac{\pi}{2}$ **c** 4 m

3 **a** **i** $t = 0, \pi, 2\pi$

ii $\pi < t < 2\pi$

b $a(t) = e^{\cos t}(\cos t - \sin^2 t)$

c $s(t) = -e^{\cos t} + 4e$

4 **a** 6.75 m **b** 14.0 m

5 **a** **i** $-2.37\,\text{m}\,\text{s}^{-2}$

ii Speeding up

b 3.54 s and 5.01 s

c **i** $-6.92\,\text{m}$

ii At 6 s, the particle is 6.92 m to the left of its initial position.

d 18.6 m

6 **a** $-12.8\,\text{m}\,\text{s}^{-2}$

b $t = 0.696, 5.59$

c 13.2 m

d 24.8 m

Exercise 13K

1 $\int_0^{1.5} (1025t^2 - t^3)\,dt$

≈ 1152 spectators

2 $33.4 + \int_0^8 5.2te^{(-0.05t^3 + 2.3)}\,dt$

$= 206\,\text{cm}^3$

3 $3800 + \int_0^{20} -150\left(1 - \frac{t}{80}\right)dt \approx 1175$ gallons

4 $\int_0^{10} 20.4e^{\frac{t}{18}}\,dt \approx 273$ billions of barrels

Chapter review

1 **a** $f'(x) = 3\cos x - 4\sin x$

b $y' = -3\sin(3x - 4)$

c $h'(t) = 4\sin^3 x \cos x$

d $f'(x) = -\frac{\sin x}{2\sqrt{\cos x}}$

e $y' = \frac{\sin x}{x} + (\ln x)(\cos x)$

f $y' = \frac{\cos(\ln x)}{x}$

g $s'(t) = \frac{-\sin t - \cos t}{e^t}$

h $f'(x) = 2e^{2x}\cos 2x + 2e^{2x}\sin 2x$

2 **a** $\frac{3}{5}x^5 + \sin x + C$

b $-\frac{1}{4}\cos 4x + C$

c $\frac{1}{2}\sin(2x + 3) + C$

d $\frac{1}{5}(2x^3 + 5x)^5 + C$

e $\sin(x^3) + C$

f $2e^{x^2 + 5x} + C$

g $\sin(\ln x) + C$

h $\frac{1}{2}\ln(e^{4x} + 5) + C$

3 **a** $\frac{\sqrt{2}}{2}$

b $\frac{255}{4}$

c $\frac{\pi + 3}{6}$

d $e^5 - e$

4 **a**

$x^2 + 4xh = 432 \Rightarrow 4xh = 432 - x^2$

$\Rightarrow h = \frac{432 - x^2}{4x}$

b

$V = x^2\left(\frac{432 - x^2}{4x}\right) = 108x - \frac{1}{4}x^3$

c 12 cm by 12 cm by 6 cm

5 a $v(t) = -3e^{\cos t}\sin t$;
$a(t) = 3e^{\cos t}\sin^2 t - 3e^{\cos t}\cos t$

b $0 < t < \pi$

c $0.905 < t < 5.38$

d 14.1 m

6 12.6 cm

Exam-style questions

7 a $g'(x) = 2\sin x\cos x - 5$ (2 marks)

b

$g'(x) = \sin 2x - 5 \le 1 - 5 = -4 < 0$ (3 marks)

Therefore g is decreasing on all its domain. (1 mark)

8 a $f'(x) = 5 - \frac{1}{\cos^2 x}$ (2 marks)

$f'\left(\frac{\pi}{4}\right) = 5 - \frac{1}{\cos^2\left(\frac{\pi}{4}\right)} = 3$ (1 mark)

$f\left(\frac{\pi}{4}\right) = \frac{5\pi}{4} - 1$ (1 mark)

$y - \left(\frac{5\pi}{4} - 1\right) = 3\left(x - \frac{\pi}{5}\right)$ (1 mark)

b The normal to the graph is vertical when the tangent is horizontal. (1 mark)

$f'(x) = 0 \Rightarrow x = 1.1071$ (2 marks)

$f'(1.1071) = 3.5357$
The coordinates of A are (1.11, 3.54) (2 marks)

9 a $d(t) = |\sin 2t - \sin(t - 0.24)|$ (2 marks)

b Use GDC to find the maximum (1 mark)

$t = 2.25$ (1 mark)

c Find intersection of graphs (1 mark)

$t = 1.13$ (1 mark)

10 a $\int (x - \sin x)\, dx = \frac{x^2}{2} + \cos x + C$ (3 marks)

b $\int_0^{\pi} (x - \sin x)dx = \left[\frac{x^2}{2} + \cos x\right]$

$= \frac{\pi^2}{2} - 2$ (3 marks)

11 a $\frac{da}{dx} = (e^x)'\cos x + e^x(\cos x)'$

$= e^x\cos x - e^x\sin x$ (3 marks)

b $a(0) = 1$ (1 mark)

$a'(0) = 1$ (1 mark)

$y = x + 1$ (2 marks)

12 a 1 mark for shape, 1 mark for domain, 1 mark for scale on axes

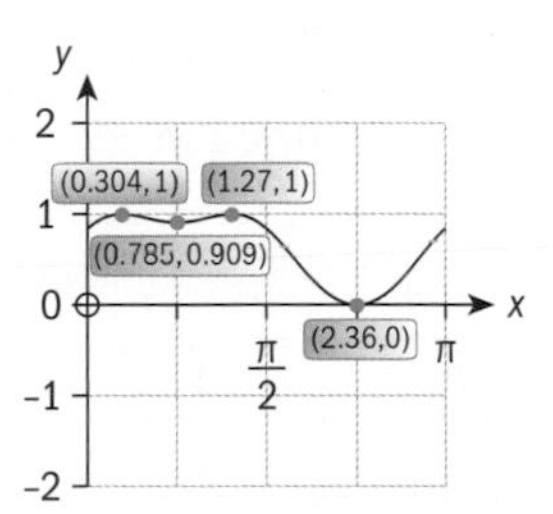

b i Minimum points: (0.785, 0.909) and (2.36, 0) (2 marks)

Maximum points: (0.304, 1) and (1.27, 1) (2 marks)

ii $0 \le s(x) \le 1$ (1 mark)

iii sin 1, 0.841 (1 mark)

c i $\int_0^{2.36} \sin(1 + \sin 2x)\, dx$ (2 marks)

ii 1.76 sin 2x) (2 marks)

d $-1 \le \sin(2x) \le 1$ $1 - 1 \le 1 + \sin(2x) \le 1 + 1$
$0 \le 1 + \sin(2x) \le 2$
So, $\sin(1 + \sin(2x) \ge 0$ for all x real.

13 a i -0.524 (1 mark)

ii $-0.369 \le y \le 1.76$ (2 marks)

iii $f'(x) = 2\cos x - \sin x$ (2 marks)

b $\int_{-0.524}^{\frac{\pi}{2}} (\sin 2x + \cos x)\, dx = 2.37$ (3 marks)

14 a $s(0) = 2$ mm (1 mark)

b i $v = s' = 15\cos 3t + 2t$ (2 marks)

ii $a = v' = -45\sin 3t + 2$ (2 marks)

c $v = 0 \Rightarrow t = 0.548, t = 1.50, t = 2.74$ (1 mark)

$a < 0 \Rightarrow t = 0.548, t = 2.74$ (2 marks)

15

a i $f(0) = 5$ (1 mark)

ii $f(\pi) = 5$ (1 mark)

b (1 mark for coordinates of A, 1 mark for coordinates of B,

1 mark for zeros, 1 mark for shape and domain

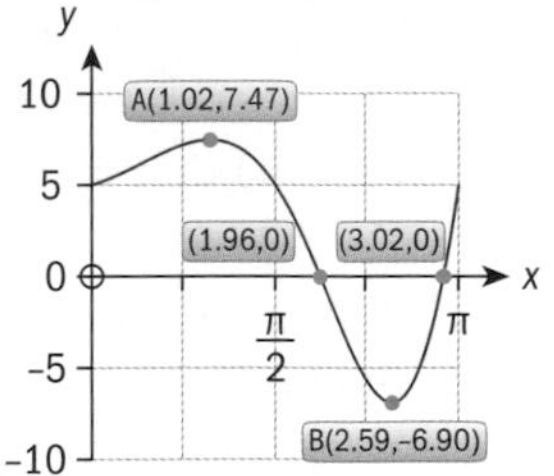

c $\int_a^b f(x)\, dx = 2.11$ or 2.07 using 3 sf for a and b. (2 marks)

d The graph crosses the x-axis between a and b (1 mark)

e Either

$\int_a^b |f(x)|\, dx = 7.39$ (2 marks)

Or

$\int_{1.02}^{1.96} f(x)\, dx - \int_{1.96}^{2.59} f(x)\, dx$

$= 7.39$ (2 marks)

16 a $0 \le x \le 2\pi \Rightarrow 0 \le \frac{x}{2} \le \pi$

(1 mark)

In the 1st and 2nd quadrants sine is positive

(1 mark)

Therefore $f(x) \ge 0$ for all $0 \le x \le 2\pi$. (1 mark)

b

$$f(\pi - x) = \sin\left(\frac{\pi}{2} - \frac{x}{2}\right)$$
$$= \sin\left(\pi - \left(\frac{\pi}{2} - \frac{x}{2}\right)\right)$$
$$= \sin\left(\frac{\pi}{2} + \frac{x}{2}\right) = f(\pi + x)$$

(2 marks)

Therefore

$a = f(\pi - x) = f(\pi + x) = b$

(1 mark)

c $A(x) = 2x\sin\left(\frac{\pi - x}{2}\right)$

(2 marks)

Find maximum point (1.72,2.24) (2 marks)

A(1.42,0.652) and B(4.86,0.652) (2 marks)

d $p = 2AB + 2 \times 0.652 = 8.19$

(3 marks)

Chapter 14

Skills check

1 a 5.5 **b** 14.6 (3 s.f.)

2 a 5.5 **b** 14.6 (3 s.f.)

3 a 15 **b** 56 **c** 0.267

4 a $x = 1.719$ **b** $x = 2.98$ **c** $x = 8.68$

Exercise 14A

1 a

s	2	3	4	5	6	7	8	9	10	11	12
$P(S=s)$	$\frac{1}{36}$	$\frac{2}{36}$	$\frac{3}{36}$	$\frac{4}{36}$	$\frac{5}{36}$	$\frac{6}{36}$	$\frac{5}{36}$	$\frac{4}{36}$	$\frac{3}{36}$	$\frac{2}{36}$	$\frac{1}{36}$

b

n	0	1	2
$P(N=n)$	$\frac{25}{36}$	$\frac{10}{36}$	$\frac{1}{36}$

c

n	1	2	3	4	5	6
$P(N=n)$	$\frac{11}{36}$	$\frac{9}{36}$	$\frac{7}{36}$	$\frac{5}{36}$	$\frac{3}{36}$	$\frac{1}{36}$

d

P	1	2	3	4	5	6	8	9	10
$P(P=p)$	$\frac{1}{36}$	$\frac{2}{36}$	$\frac{2}{36}$	$\frac{3}{36}$	$\frac{2}{36}$	$\frac{4}{36}$	$\frac{2}{36}$	$\frac{1}{36}$	$\frac{2}{36}$
P	12	15	16	18	20	24	25	30	36
$P(P=p)$	$\frac{4}{36}$	$\frac{2}{36}$	$\frac{1}{36}$	$\frac{2}{36}$	$\frac{2}{36}$	$\frac{2}{36}$	$\frac{1}{36}$	$\frac{2}{36}$	$\frac{1}{36}$

2 a

t	2	3	4	5	6
$P(T=t)$	$\frac{1}{36}$	$\frac{4}{36}$	$\frac{10}{36}$	$\frac{12}{36}$	$\frac{9}{36}$

b $P(T > 4) = \frac{7}{12}$

3 a

s	1	2	3	6	10
$P(S=s)$	$\frac{1}{6}$	$\frac{2}{6}$	$\frac{1}{6}$	$\frac{1}{6}$	$\frac{1}{6}$

b $\frac{1}{2}$

4 a $c = \frac{1}{6}$ **b** $P(1 < X < 4) = \frac{1}{2}$

5 $c = \frac{1}{36}$

6 $k = \frac{1}{5}$ (k cannot be negative)

7 $k = \frac{27}{40}$

8 a $a = \frac{1}{8}$ and $b = \frac{5}{24}$

b $P(\text{Sum} > 7) = \frac{25}{144}$

9 a $P(C = 3) = P(A = 1 \text{ and } B = 2) + P(A = 2 \text{ and } B = 1)$

$= \frac{1}{3} \times \frac{2}{3} + \frac{1}{3} \times \frac{1}{6} = \frac{5}{18}$

b

c	2	3	4	5	6
P(C=c)	$\frac{1}{18}$	$\frac{5}{18}$	$\frac{6}{18}$	$\frac{5}{18}$	$\frac{1}{18}$

Exercise 14B

1 15.2 (3 s.f.)

2 $E(X) = \frac{16}{3}$ 3 $E(X) = 5\frac{2}{3}$

4 a $k = \frac{1}{25}$ b $E(X) = 5$

5 a $0.2 \le k \le 1$ b $E(x) = 1.6 + k$

6 a

$P(R = 1) = \frac{1}{5}$

$P(R = 2) = \frac{8}{45}$

$P(R = 3) = \frac{7}{45}$

$P(R = 4) = \frac{2}{15}$

$P(R = 5) = \frac{1}{9}$

$P(R = 6) = \frac{4}{45}$

$P(R = 7) = \frac{1}{15}$

$P(R = 8) = \frac{2}{45}$

b $E(R) = 3\frac{2}{3}$ c 1

7 a $P(R = 2) = P(\textit{blue then red})$

$= \frac{8}{10} \times \frac{2}{10} = \frac{4}{25}$

b $P(R > 3) = \frac{16}{125}$

c $P(R = n) = \left(\frac{8}{10}\right)^n \times \frac{2}{10}$

d 1

8 a $P(Z = 0) = 0.7489$

b $E(Z) = 1.7$

\$1.70 is the expected winnings on a ticket.

c Lose \$0.30 per ticket

Exercise 14C

1 a $P(X = 1) = \frac{1}{4}$

b $P(X < 1) = \frac{1}{16}$

c $P(X \le 1) = \frac{5}{16}$

d $P(X \ge 1) = \frac{15}{16}$

2 a $P(X = 2) = 0.329$

b $P(X < 2) = 0.351$

c $P(X \le 2) = 0.680$

d $P(X \ge 2) = 0.649$

3 a $P(X = 5) = 0.0389$

b $P(X < 5) = 0.952$

c $P(X > 5) = 0.00910$

d $P(X \ge 1) = 0.932$

Exercise 14D

1 Most likely outcome is 1 red face with probability 0.422.

2 a $P(X = 5) = 0.257$

b $P(X \le 3) = 0.260$

3 a $P(X = 0) = 0.851$

b If 13 are not faulty then 3 are faulty, $P(X = 3) = 0.000491$

c $P(X \ge 2) = 0.0109$

4 a $P(X = 5) = 0.0584$

b $P(13 \text{ not faulty}) = P(X \ge 3) = 0.474$

5 $P(X \le 3) = 0.913$

6 a $P(X > 1) = 0.224$

b $P(X = 1) = 0.399$

7 a i $P(X = 3) = 0.0307$

ii $P(X = 0) = 0.463$

iii $P(X \ge 2) = 0.171$

b i $[P(X = 0)]^2 = 0.215$

ii $[P(X \ge 2)]^2 = 0.0292$

iii $P(X = 0) \times P(X \ge 2) \times 2 = 0.158$

Exercise 14E

1 $n = 4$ 2 68 3 7 4 9 5 7

Exercise 14F

1 a $E(X) = 20$

b $E(X) = 6\frac{2}{3}$ c $E(X) = 10$

2 $n = 25$

3 a $X \sim B(15, 0.25)$

b $E(X) = 3.75$

c $P(X \ge 10) = 0.000795$

4 a $P(\text{girl}) = \frac{153}{300}$ b 38.2

Exercise 14G

1 $p = \frac{3}{4}$ and $n = 16$

2 a $X \sim B(20, 0.2)$

b $E(X) = 4$ $\text{Var}(X) = 3.2$

c $P(X \ge 10) = 0.00259$

3 $p = 0.8$ or 0.2

Exercise 14H

1 a $P(1 < Z < 2) + P(-2 < Z < -1) = 0.272$

b $P(0.5 < Z < 1.5) + P(-1.5 < Z < -0.5) = 0.483$

2 a $P(Z > 1) = 0.159$

b $P(Z > 2.4) = 0.00820$

c $P(-1 < Z < 0) = 0.3413$

d $P(-1.75 < Z < 0) = 0.4599$

3 a 0.742 b 0.2358

c 0.0359 d 0.397

4 a 0.3057 b 0.5952
c 0.2853

5 a 0.311 b 0.215

Exercise 14I

1 a 0.655 c 0.186
b 0.841 d 0.5

2 a 0.672 b 0.748
c 0.345

3 a 0.994 b 0.977
c 0.494

Exercise 14J

1 a $P(X < 130) = 0.933$
b $P(X > 90) = 0.692$
c $P(80 < X < 125) = 0.736$

2 477

3 a $P(X > 20) = 0.0668$
b $P(X < 10) = 15.87\%$

4 53.5%

5 a $P(X < 475) = 0.106$
b $[P(X < 475)]^3 = 0.001\,19$

Exercise 14K

1 a $a = 1.42$ b $a = 0.407$
c $a = 2.58$

2 a $a = 1.77$ b $a = -1.00$
c $a = -0.841$

3 a $a = 0.385$
b $a = 1.60$

4 a $P(Z < a) = 0.95$
$\therefore a = 1.64$
b $P(Z < a) = 0.8$
$\therefore a = 0.841$

Exercise 14L

1 $a = 5.64$

2 a $a = 413$ b $a = 433$

3 a $P(X < 500) = 0.106$
b 86.4%
c $a = 498.8$ $b = 505.1$

4 a $P(520 < X < 570) = 0.673$
b 582

5 a $d = 79.7$ b $f = 35.8$

Exercise 14M

1 $\sigma = 8.33$

2 $\mu = 15.4$

3 $\sigma = 4.23$ $\mu = 49.1$

4 $\sigma = 13.8$ $\mu = 71.3$

5 $\sigma = 7.66$ cm

6 $\mu = 546.5$ g

7 a $\sigma = 0.389$ kg
b 35%

8 $\mu = 54.3$ cm

9 $\sigma = 0.260$ m

10 a $\sigma = 33.7$ $\mu = 125.66$
b $P(X > 117) = 0.601 = 60.1\%$
Yes – this is consistent with the normal distribution.

11 $\sigma = 7.34$ $\mu = 507.1$

Chapter review

1 a 6 b $-\frac{7}{15}$

2 a $\frac{1}{35}$ b 3

3 $x = \frac{3}{8}, \frac{13}{64}$

4 a 2,4,6,8,12,16
b $\frac{1}{8}, \frac{2}{8}, \frac{1}{8}, \frac{2}{8}, \frac{1}{8}, \frac{1}{8}$
c 7.5 d £62.50

5 $\frac{40}{243}$

6 0.2

7 a 85 b 0.023

8 a $\frac{19}{27}$

x	−5	1
$P(X = x)$	$\frac{8}{27}$	$\frac{19}{27}$

b i $-\$\frac{7}{9}$ ii $-\$7$

9 a 0.254 b 0.448

10 0.0244

11 a i 0.0881 ii 0.006 37
b 2 c 14

12 0.935

13 a 8.69 b 0.755

14 38.9; 8.63

15 a 33.3 b 0.328 c 0.263

Exam-style questions

16 a 0,1,2 (1 mark)

b $P(X = 2) = \frac{10}{18} \times \frac{10}{18} = \frac{25}{81}$ (2 marks)

c

x	0	1	2
$P(X = x)$	$\frac{16}{81}$	$\frac{40}{81}$	$\frac{25}{81}$

(2 marks)

17 a $0.2 + k + 0.25 + k - 0.05 + 0.3 = 1 \Rightarrow k = 0.15$ (3 marks)

b $E(X) = 0 \times 0.2 + 1 \times 0.4 + 2 \times 0.1 + 3 \times 0.3 = 1.5.$ (2 marks)

18 a $0.05 + 0.22 + 0.27 + a + b = 1 \Rightarrow a + b = 0.46$ (2 marks)

b $E(X) = 2.46 \Rightarrow 0 \times 0.05 + 1 \times 0.22 + 2 \times 0.27 + 3a + 4b = 2.46$ (2 marks)

$3a + 4b = 1.7$

Solve simultaneously $a + b = 0.46$ and $3a + 4b = 1.7$ (1 mark)

$a = 0.14,\ b = 0.32$ (2 marks)

19 Let X be the number of defective batteries in a pack of 10 selected at random.

$X \sim B(10, 0.005)$

a $X \sim B(10, 0.005)$ (1 mark)

$P(X = 1) = 0.0478$ (3 s.f.) (1 mark)

b $P(X \le 1) = 0.999$ (3 s.f.) (2 marks)

c $P(X = 1 | X \le 1) = \dfrac{P(X = 1)}{P(X \le 1)}$

$= 0.0478$

(3 marks)

20 Let $X \sim B(n, p)$.

$np = 3$ and $npq = 1.2$ (2 marks)

Solve simultaneously (1 mark)

$q = 0.4 \Rightarrow p = 0.6$ (1 mark)

$n = 5$ (1 mark)

21 $X \sim N(50.1, 0.4^2)$

a $P(X < 49.5) = 0.0668$

(3 s.f.) (2 marks)

b $P(49.5 < X < 50.5) = 0.775$

(3 s.f.) (2 marks)

c $P(X > 49 | X < 49.5)$

$= \dfrac{P(49 < X < 49.5)}{P(X < 49.5)} = 0.955$

(3 marks)

22 $X \sim N(\mu, 5^2)$

a $P(X < 5) = 0.754 \Rightarrow$

$P\left(Z < \dfrac{5 - \mu}{3}\right) = 0.754$

(1 mark)

$\dfrac{5 - \mu}{3} = 0.6871... \Rightarrow \mu = 2.94$

(2 marks)

b $P(4 < X < 5) = 0.116$ (2 marks)

23 a i Let X be the number of correct answers in the 12 questions answered at random.

$X \sim B(12, 0.5)$ (1 mark)

$P(X = 2) = \begin{pmatrix} 12 \\ 2 \end{pmatrix} (0.5)^{12}$

$= 0.0161$ (2 marks)

ii $P(X = 12) = \begin{pmatrix} 12 \\ 12 \end{pmatrix} (0.5)^{12}$

$= 0.000244$

(1 mark)

b $E(X) = 12 \times 0.5 \times 0.5 = 3$

correct answers (1 mark)

3 correct random answers = 6 marks

9 incorrect random answers = −9 marks

8 answers known = 16 marks

If the student answers all the questions the expected number of marks is 13 marks which is 3 less than the total marks if he just answers the questions he knows the correct answer.

(1 mark)

24 a i $W \sim N(\mu, \sigma^2)$

$P(W < 65) = 0.27 \Rightarrow$

$P\left(Z < \dfrac{65 - \mu}{\sigma}\right) = 0.27$

(1 mark)

$P(W > 96) = 0.25 \Rightarrow$

$P\left(Z < \dfrac{96 - \mu}{\sigma}\right) = 0.75$

$\dfrac{65 - \mu}{\sigma} = -0.6128...,$

$\dfrac{96 - \mu}{\sigma} = 0.6744...$

(2 marks)

ii Solve simultaneously

$\dfrac{65 - \mu}{\sigma} = -0.6128...,$

$\dfrac{96 - \mu}{\sigma} = 0.6744...$

(1 mark)

$\mu = 79.8$ and $\sigma = 24.1$

(2 marks)

b $P(W > 100) = 0.20$

(2 marks)

c Let $Y \sim N(80.5, 10.1^2)$

$P(75 < Y < 85) = 0.379.$

(2 marks)

d $630P(Y > 85) = 207$

(2 marks)

e $\dfrac{630 \times 80.5 + 370m}{1000} = 79.8$

(1 mark)

$m = 78.5$ (1 mark)

25 a i $T \sim N(45, 9^2)$

$P(T \ge 55) = 0.133$

(2 marks)

ii $P(T \ge 65 | T > 55)$

$= \dfrac{P(T > 65)}{P(T \ge 55)}$

$= \dfrac{0.0131}{0.133}$

$= 0.0984$ (3 marks)

b $(0.133...)^3 = 0.00235$

(2 marks)

c $N \sim B(50, 0.133)$

i $E(N) = 50 \times 0.133 = 6.65$

(2 marks)

ii $P(N \ge 5) = 1 - P(N \le 4)$

$= 0.812$

(3 marks)

26 $P(X > 82) = 0.1 \Rightarrow P(X < 82)$

$= 0.9 \Rightarrow P\left(Z < \dfrac{82 - \mu}{\sigma}\right) = 0.9$

(2 marks)

$P(X < 40) = 0.2 \Rightarrow$

$P\left(Z < \frac{40-\mu}{\sigma}\right) = 0.2$ (1 mark)

$\frac{82-\mu}{\sigma} = 1.28...,$

$\frac{40-\mu}{\sigma} = -0.841...$ (1 mark)

Solve simultaneously (1 mark)

$\mu = 56.7$ and $\sigma = 19.8$

(2 marks)

Paper 1: Section A

1 **a** $4x^3$ (1 mark)

b $-2x^{-3}$ (1 mark)

c $\sin x + x\cos x$ (2 marks)

d $4e^{4x}$ (1 mark)

2 Wire length = 8
arc length = $r\theta$ (2 marks)

$8 = 4\theta$ $\theta = 2\text{rad}$ (2 marks)

3 **a**

$(f \circ g)(x) = 2(3x-2)+1 = 6x-3$

(2 marks)

b

$(g \circ f)(x) = 3(2x+1)-2 = 6x+1$

(2 marks)

c

$f : x \to y = 2x+1$ $y-1 = 2x$

$x = \frac{1}{2}y - \frac{1}{2}$

$f^{-1} : y \to \frac{1}{2}y - \frac{1}{2}$

$f^{-1}(x) = \frac{1}{2}x - \frac{1}{2}$ (3 marks)

4 **a** $\log x + \log y = p + q$

(2 marks)

b $\log x^2 - \log y = 2\log x - \log y$
$= 2p - q$

(2 marks)

c $\frac{1}{2}\log x = \frac{1}{2}p$ (2 marks)

d $\log 100 + \log y = 2 + q$

(2 marks)

5 $y = 2x + 11$ has gradient of 2.
Perpendicular gradient is $-\frac{1}{2}$

(2 marks)

$y = -\frac{1}{2}x + c$ through (4, 3)
$\Rightarrow 3 = -\frac{1}{2}\times 4 + c \Rightarrow c = 5$

(2 marks)

$y = -\frac{1}{2}x + 5$ (1 mark)

6 **a** This is an arithmetic progression. Let the first term be a. (1 mark)

$290 = \frac{10}{2}(2a + 9\times 2)$

(2 marks)

$290 = 10a + 90$ $a = 20$ cm

(1 mark)

a $u_{10} = 20 + 9\times 2 = 38$ cm

(2 marks)

7 Integral equals $\left[\frac{1}{2}\ln(2x+1)\right]_0^4$

(2 marks)

$= \frac{1}{2}\ln 9 - 0$ (2 marks)

$= \ln 3$ (1 mark)

Paper 1: Section B

8 **a** **i**

$\sum P(X = x) = 1 \Rightarrow a + 0.2 + 0.3 + 0.1$
$= 1 \Rightarrow a = 0.4$

(2 marks)

ii

$E(x) = \sum xP(X = x)$
$= 1\times 0.2 + 2\times 0.3 + 3\times 0.1$

$= 1.1$ (2 marks)

b $\sum P(X = x) = 1 \Rightarrow b + c + 0.3$
$= 1 \Rightarrow b + c = 0.7$

(2 marks)

Fair game

$\Rightarrow E(x) = 0$

$\Rightarrow -2b + c + 2\times 0.1 + 3\times 0.2 = 0$

$\Rightarrow -2b + c = -0.8$ (2 marks)

$\Rightarrow -3b = -1.5$ $b = 0.5$

$c = 0.2$ (3 marks)

9 **a** $D = 36 - 4(8+k) = 4 - 4k$

(2 marks)

b **i** Concave up quadratic, always positive, so no roots. (2 marks)

So $\{D \in \mathbb{R} | D < 0\}$

(1 mark)

ii $4 - 4k < 0 \Rightarrow k > 1$

$\{k \in \mathbb{R} | k > 1\}$ (2 marks)

c $f(x) = (x+3)^2 + (k-1)$

(2 marks)

d $(x+3)^2 + (k-1) > 0$

$\Rightarrow k - 1 > 0$

Since $(x+3)^2 \geq 0$, for all x

Hence $k > 1$ as before.

(2 marks)

e for $k = 4, f(x) = (x+3)^2 + 3$

So min point is $(-3, 3)$

(2 marks)

10 **a**

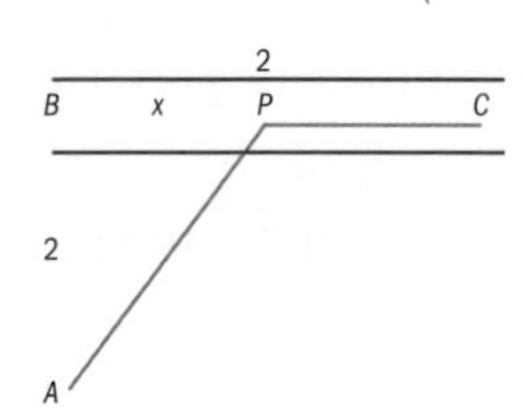

b $AP = \sqrt{4 + x^2}$ $PC = 2 - x$

(2 marks)

Time in field is $\frac{\sqrt{4+x^2}}{12}$.

Time on road is $\frac{2-x}{20}$

(2 marks)

So $T = \frac{\sqrt{4+x^2}}{12} + \frac{2-x}{20}$ (1 mark)

$\frac{dT}{dx} = \frac{\frac{1}{2}(4+x^2)^{-\frac{1}{2}}2x}{12} + \frac{-1}{20}$

(3 marks)

At a min $\frac{dT}{dx} = 0$ (1 mark)

$\frac{x}{12\sqrt{4+x^2}} - \frac{1}{20} = 0 \Rightarrow \frac{x}{12\sqrt{4+x^2}}$

$= \frac{1}{20} \Rightarrow 5x = 3\sqrt{4+x^2}$

(2 marks)

$\Rightarrow 25x^2 = 9\left(4+x^2\right) \Rightarrow 25x^2$

$= 36 + 9x^2 \Rightarrow 16x^2 = 36$

(2 marks)

$\Rightarrow x^2 = \frac{36}{16} \Rightarrow x = \frac{\pm 6}{4} \Rightarrow x = \frac{3}{2}$

as x cannot be negative.

(3 marks)

Paper 2: Section A

1 a i $(x+2)(x+7)$ (1 mark)

ii $x = -2$ or -7 (2 marks)

b i $x = \frac{-4 \pm \sqrt{16-8}}{2} = -2 \pm \sqrt{2}$

(2 marks)

ii -3.414 or -0.5858 (4s.f.)

(2 marks)

2 $P(X > 7) = 0.69146 \Rightarrow$

$P(X < 7) = 0.30854$ (1 mark)

Applying substitution

$Z = \frac{X-8}{\sigma}$ Z is $N(0, 1^2)$

(1 mark)

$P\left(Z < \frac{-1}{\sigma}\right) = 0.30854$

(1 mark)

Using inverse normal

$\frac{-1}{\sigma} = -0.49999...$ (2 marks)

So $\sigma = 2$ (1 mark)

3 a Binomial (10, 0.8)

(2 marks)

b $\mu = np = 10 \times 0.8 = 8$

(1 mark)

c $P(X = 7) = 0.201\,326...$

$= 0.201$ (3 s.f.)

(1 mark)

d

$P(X > 6) = 1 - P(X \le 6)$

$= 0.879126... = 0.879$ (3 s.f.)

(2 marks)

e

$P(X = 7 | X > 6) = \frac{P(X = 7 \cap X > 6)}{P(X > 6)}$

$= \frac{P(X = 7)}{P(X > 6)}$

(2 marks)

$= \frac{0.201\,326...}{0.879\,126...} = 0.229$ (3 s.f.)

(1 mark)

4 a

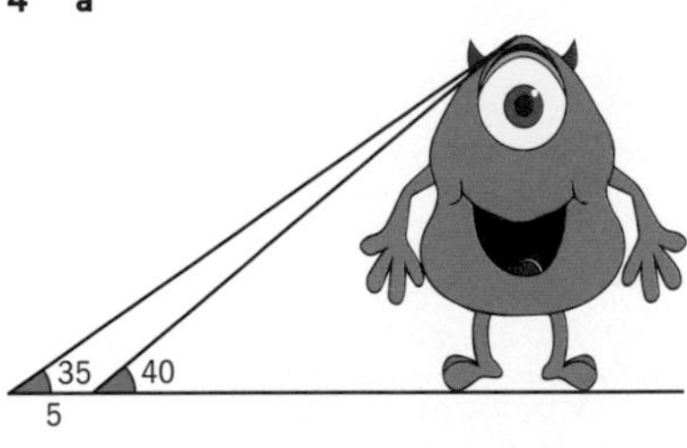

(1 mark)

b Let the monster's height be h and the original distance between Maria and the monster be x.

$\frac{h}{x} = \tan 40$

$\frac{h}{x+5} = \tan 35$ (2 marks)

$x \tan 40 = (x+5) \tan 35°$

$\frac{x+5}{x} = \frac{\tan 40°}{\tan 35°} = 1.19835...$

(2 marks)

$x + 5 = (1.19835...)x$

$5 = (0.19835...)x$

$x = 25.206...$ (1 mark)

$h = 21.2\,m$ (3 s.f.) (1 mark)

Note: can also be solved using the sine rule.

5 a $t = 0 \Rightarrow v = 0$ (1 mark)

b Total distance is

$\int_0^3 \left|5 \sin\left(t^2\right)\right| dt = 8.51\,m$ (3 s.f.)

(3 marks)

c Displacement is

$\int_0^3 5 \sin\left(t^2\right) dt = +3.87\,m$ (3 s.f.)

(3 marks)

6 General term is $C_r^6 x^{6-r} \left(\frac{2}{x}\right)^r$

require $6 - 2r = 0$ so $r = 3$

(2 marks)

$C_3^6 x^3 \left(\frac{2}{x}\right)^3 = 20 \times 8 = 160$

(2 marks)

Paper 2: Section B

1 a i 120.5 million GBP

ii 64.7 million GBP (3 s.f.)

(3 marks)

b i 67.7 (3 s.f.)

ii 19.2 (3 s.f.) (3 marks)

c i $r = 0.950$ (3 s.f.)

(2 marks)

ii Strong, positive

(2 marks)

iii $y = 0.282x + 33.7$ (3 s.f.)

(2 marks)

d $0.282 \times 80 + 33.7 = 56$

(2 marks)

e Not very valid, should use w on g line and extrapolation. (2 marks)

8 **a** **i** $x = 5$

ii $y = 3$ (2 marks)

b **i** $\left(0, \frac{-3}{10}\right)$

ii $\left(\frac{-1}{2}, 0\right)$ (2 marks)

c **i**

$$f'(x) = \frac{6(2x-10)-(6x+3)2}{(2x-10)^2}$$

$$= \frac{-66}{(2x-10)^2}$$ (2 marks)

ii $f'(x)$ is always negative so graph is always decreasing.

(2 marks)

d

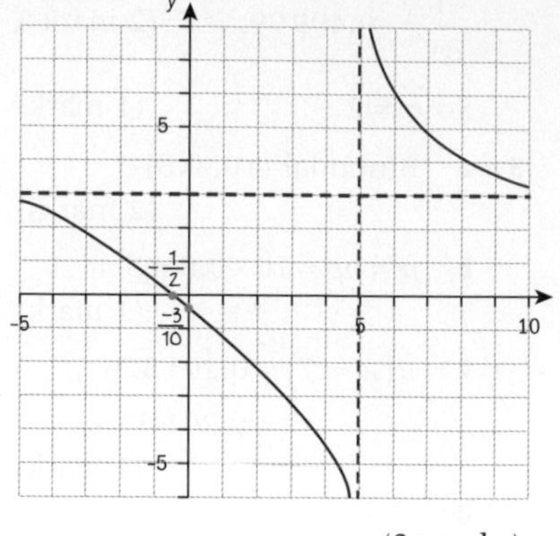

(3 marks)

e $y = -1.83x + 1.17$ (3 s.f.)

(2 marks)

9 **a** **i** $200(1.05)^3 = 231.53$

(2 marks)

ii 15 years (415.79)

(2 marks)

b **i** 3 years (290.7)

(2 marks)

ii Solving

$200(1.05)^t \geq 350(0.94)^t$

6 years $(268.02 > 241.45)$

(2 marks)

c Solving

$200(1.05)^2(1.02)^{t-2} \geq$
$350(0.94)^2(0.96)^{t-2}$

(2 marks)

8 years $(248.32 > 242.08)$

(1 mark)

Note: could also be solved with graphs or logs rather than the preferred "table" method.

Index